Carbon Monoxide in Drug Discovery

Wiley Series in Drug Discovery and Development
Binghe Wang, Series Editor

Computer Applications in Pharmaceutical Research and Development
Edited by Sean Ekins

Glycogen Synthase Kinase-3 (GSK-3) and Its Inhibitors: Drug Discovery and Development
Edited by Ana Martinez, Ana Castro, and Miguel Medina

Aminoglycoside Antibiotics: From Chemical Biology to Drug Discovery
Edited by Dev P. Arya

Drug Transporters: Molecular Characterization and Role in Drug Disposition
Edited by Guofeng You and Marilyn E. Morris

Drug–Drug Interactions in Pharmaceutical Development
Edited by Albert P. Li

Dopamine Transporters: Chemistry, Biology, and Pharmacology
Edited by Mark L. Trudell and Sari Izenwasser

Carbohydrate-Based Vaccines and Immunotherapies
Edited by Zhongwu Guo and Geert-Jan Boons

ABC Transporters and Multidrug Resistance
Edited by Ahcène Boumendjel, Jean Boutonnat, and Jacques Robert

Drug Design of Zinc-Enzyme Inhibitors: Functional, Structural, and Disease Applications
Edited by Claudiu T. Supuran and Jean-Yves Winum

Kinase Inhibitor Drugs
Edited by Rongshi Li and Jeffrey A. Stafford

Evaluation of Drug Candidates for Preclinical Development: Pharmacokinetics, Metabolism, Pharmaceutics, and Toxicology
Edited by Chao Han, Charles B. Davis, and Binghe Wang

HIV-1 Integrase: Mechanism and Inhibitor Design
Edited by Nouri Neamati

Carbohydrate Recognition: Biological Problems, Methods, and Applications
Edited by Binghe Wang and Geert-Jan Boons

Chemosensors: Principles, Strategies, and Applications
Edited by Binghe Wang and Eric V. Anslyn

Medicinal Chemistry of Nucleic Acids
Edited by Li He Zhang, Zhen Xi, and Jyoti Chattopadhyaya

Plant Bioactives and Drug Discovery: Principles, Practice, and Perspectives
Edited by Valdir Cechinel Filho

Dendrimer-Based Drug Delivery Systems: From Theory to Practice
Edited by Yiyun Cheng

Cyclic-Nucleotide Phosphodiesterases in the Central Nervous System: From Biology to Drug Discovery
Edited by Nicholas J. Brandon and Anthony R. West

Drug Transporters: Molecular Characterization and Role in Drug Disposition, Second Edition
Edited by Guofeng You and Marilyn E. Morris

Drug Delivery: Principles and Applications, Second Edition
Edited by Binghe Wang, Longqin Hu, and Teruna J. Siahaan

Drug Transporters: Molecular Characterization and Role in Drug Disposition, 3rd Edition
Edited by Guofeng You and Marilyn E. Morris

Carbon Monoxide in Drug Discovery
Edited by Binghe Wang and Leo E. Otterbein

Carbon Monoxide in Drug Discovery

Basics, Pharmacology, and Therapeutic Potential

Edited by

BINGHE WANG
Georgia State University,
Atlanta, GA, USA

LEO E. OTTERBEIN
Beth Israel Deaconess Medical Center,
Boston, MA, USA

This edition first published 2022
© 2022 John Wiley & Sons, Inc.

All rights reserved. No part of this publication may be reproduced, stored in a retrieval system, or transmitted, in any form or by any means, electronic, mechanical, photocopying, recording or otherwise, except as permitted by law. Advice on how to obtain permission to reuse material from this title is available at http://www.wiley.com/go/permissions.

The right of Binghe Wang and Leo E. Otterbein to be identified as the authors of the editorial material in this work has been asserted in accordance with law.

Registered Office
John Wiley & Sons, Inc., 111 River Street, Hoboken, NJ 07030, USA

Editorial Office
111 River Street, Hoboken, NJ 07030, USA

For details of our global editorial offices, customer services, and more information about Wiley products visit us at www.wiley.com.

Wiley also publishes its books in a variety of electronic formats and by print-on-demand. Some content that appears in standard print versions of this book may not be available in other formats.

Limit of Liability/Disclaimer of Warranty
While the publisher and authors have used their best efforts in preparing this work, they make no representations or warranties with respect to the accuracy or completeness of the contents of this work and specifically disclaim all warranties, including without limitation any implied warranties of merchantability or fitness for a particular purpose. No warranty may be created or extended by sales representatives, written sales materials or promotional statements for this work. The fact that an organization, website, or product is referred to in this work as a citation and/or potential source of further information does not mean that the publisher and authors endorse the information or services the organization, website, or product may provide or recommendations it may make. This work is sold with the understanding that the publisher is not engaged in rendering professional services. The advice and strategies contained herein may not be suitable for your situation. You should consult with a specialist where appropriate. Further, readers should be aware that websites listed in this work may have changed or disappeared between when this work was written and when it is read. Neither the publisher nor authors shall be liable for any loss of profit or any other commercial damages, including but not limited to special, incidental, consequential, or other damages.

Library of Congress Cataloging-in-Publication Data

Names: Wang, Binghe, 1962-author. | Otterbein, Leo E., author.
Title: Carbon monoxide in drug discovery : basics, pharmacology, and therapeutic potential /
edited by Binghe Wang, Georgia State University, Atlanta, GA, USA, Leo E. Otterbein,
Beth Israel Deaconess Medical Center, Boston, MA, USA.
Description: Hoboken, NJ: John Wiley & Sons, 2022. |
Series: Wiley series in drug discovery and development | Includes bibliographical references and index.
Identifiers: LCCN 2021053082 (print) | LCCN 2021053083 (ebook) | ISBN 9781119783404 (hardback) |
  ISBN 9781119783411 (pdf) | ISBN 9781119783428 (epub) | ISBN 9781119783435 (ebook)
Subjects: LCSH: Drugs--Research. | Carbon monoxide.
Classification: LCC RS122 .W36 2022 (print) | LCC RS122 (ebook) |
  DDC 615.1072--dc23/eng/20211201
LC record available at https://lccn.loc.gov/2021053082
LC ebook record available at https://lccn.loc.gov/2021053083

Cover image created with BioRender.com and courtesy of Grace E. Otterbein
Cover design by Wiley

Set in 10/12pt WarnockPro-Regular by Integra Software Services Pvt. Ltd, Pondicherry, India

10 9 8 7 6 5 4 3 2 1

Contents

List of Contributors *viii*

Preface: Carbon Monoxide: Promises and Challenges in Its Pharmaceutical Development *xii*

Section I   General Background and Physiological Actions *1*

1 **Endogenous CO Production in Sickness and in Health** *3*
   Ladie Kimberly De La Cruz and Binghe Wang

2 **Molecular Mechanisms of Actions for CO: An Overview** *27*
   Rodrigo W. Alves de Souza, Leo E. Otterbein, and Nils Schallner

3 **Pharmacokinetic Characteristics of Carbon Monoxide** *44*
   Xiaoxiao Yang, Mingjia Wang, Chalet Tan, Wen Lu, and Binghe Wang

4 **Carbon Monoxide and Energy Metabolism** *88*
   Daniela Dias-Pedroso, Nuno Soares, and Helena L.A. Vieira

5 **Role of CO in Circadian Clock** *97*
   Hiroaki Kitagishi and Ikuko Sagami

6 **Carbon Monoxide and Mitochondria** *108*
   Claude A. Piantadosi

7 **Carbon Monoxide, Oxygen, and Pseudohypoxia** *118*
   Grace E. Otterbein, Michael S. Tift, and Ghee Rye Lee

8 **Nitric Oxide in Human Physiology: Production, Regulation, and Interaction with Carbon Monoxide Signaling** *136*
   Maryam K. Mohammed and Brian S. Zuckerbraun

9 **When Carbon Monoxide Meets Hydrogen Sulfide** *160*
   Rui Wang

10 **Biliverdin and Bilirubin as Parallel Products of CO Formation: Not Just Bystanders** *175*
   Libor Vitek

Section II Delivery Forms 195

11 Delivery Systems and Noncarrier Formulations *197*
James Byrne, Christoph Steiger, Jakob Wollborn, and Giovanni Traverso

12 Metal-Based Carbon Monoxide-Releasing Molecules (CO-RMs) as Pharmacologically Active Therapeutics *203*
Roberta Foresti, Djamal Eddine Benrahla, Shruti Mohan, and Roberto Motterlini

13 Organic CO Donors that Rely on Photolysis for CO Release *223*
Yi Liao

14 Organic Carbon Monoxide Prodrugs that Release CO Under Physiological Conditions *232*
Zhengnan Yuan and Binghe Wang

15 Targeted Delivery of Carbon Monoxide *259*
Lisa M. Berreau

16 Anesthesia-Related Carbon Monoxide Exposure *286*
Richard J. Levy

17 Natural Products that Generate Carbon Monoxide: Chemistry and Nutritional Implications *302*
Ladie Kimberly De La Cruz and Binghe Wang

Section III Carbon Monoxide Sensing and Scavenging 319

18 Fluorescent Probes for Intracellular Carbon Monoxide Detection *321*
Ryan R. Walvoord, Morgan R. Schneider, and Brian W. Michel

Section IV Therapeutic Applications 345

19 CO in Solid Organ Transplantation *347*
Roberta Foresti, Roberto Motterlini, and Stephan Immenschuh

20 Carbon Monoxide in Lung Injury and Disease *360*
Stefan W. Ryter

21 Carbon Monoxide in Acute Brain Injury and Brain Protection *377*
Alexandra Mazur, Madison Fangman, Rani Ashouri, Hannah Pamplin, Shruti Patel, and Sylvain Doré

22 CO as a Protective Mediator of Liver Injury: The Role of PERK in HO-1/CO-Mediated Maintenance of Cellular Homeostasis in the Liver *385*
Yeonsoo Joe, Jeongmin Park, Mihyang Do, Stefan W. Ryter, Young-Joon Surh, Uh-Hyun Kim, and Hun Taeg Chung

23 CO and Cancer *401*
James N. Arnold and Joanne E. Anstee

24 CO and Diabetes *423*
Rebecca P. Chow and Hongjun Wang

25 **Carbon Monoxide and Acute Kidney Injury** *434*
    Mark de Caestecker

26 **CO as an Antiplatelet Agent: An Energy Metabolism Perspective** *453*
    Patrycja Kaczara, Kamil Przyborowski, Roberto Motterlini, and Stefan Chlopicki

27 **CO in Gastrointestinal Physiology and Protection** *466*
    Katarzyna Magierowska and Marcin Magierowski

28 **Carbon Monoxide and Sickle Cell Disease** *482*
    Edward Gomperts, John Belcher, Howard Levy, and Greg Vercellotti

29 **CO and Pain Management** *497*
    Olga Pol

30 **Clinical Trials of Low-Dose Carbon Monoxide** *511*
    Edward Gomperts, Andrew Gomperts, and Howard Levy

    Index *528*

List of Contributors

Joanne E. Anstee, Faculty of Life Sciences and Medicine, School of Cancer and Pharmaceutical Sciences, King's College London, Guy's Hospital, London SE1 1UL, UK

James N. Arnold, Faculty of Life Sciences and Medicine, School of Cancer and Pharmaceutical Sciences, King's College London, Guy's Hospital, London SE1 1UL, UK

Rani Ashouri, Department of Anesthesiology, Center for Translational Research in Neurodegenerative Disease and McKnight Brain Institute, University of Florida College of Medicine, 1275 Center Drive, Biomed Sci J493, Gainesville, FL 32610, USA

John Belcher, Division of Hematology, Oncology and Transplantation, Vascular Research Center, Department of Medicine, University of Minnesota, Minneapolis, MN 55408, USA

Djamal Eddine Benrahla, Mondor Institute for Biomedical Research (IMRB), Université Paris-Est Créteil, INSERM U955, F-94010 Créteil, France

Lisa M. Berreau, Department of Chemistry & Biochemistry, Utah State University, Logan, UT 84322-0300, USA

James Byrne, David H. Koch Institute for Integrative Cancer Research, Massachusetts Institute of Technology, Cambridge, MA 02142, USA; Harvard Radiation Oncology Program, Brigham and Women's Hospital, Harvard Medical School, Boston, MA 02114, USA; Division of Gastroenterology, Brigham and Women's Hospital, Harvard Medical School, Boston, MA 02115, USA

Stefan Chlopicki, Jagiellonian Centre for Experimental Therapeutics (JCET), Jagiellonian University, Krakow, Poland

Rebecca P. Chow, Department of Surgery, Medical University of South Carolina, 173 Ashley Avenue, Charleston, SC 29425, USA

Hun Taeg Chung, Department of Biological Sciences, University of Ulsan, Ulsan 44610, Republic of Korea; Mycos Therapeutics Inc., Ulsan 44610, Republic of Korea

Mark de Caestecker, Vanderbilt University, Nashville, TN, USA

Ladie Kimberly De La Cruz, Department of Chemistry and Center for Diagnostics and Therapeutics, Georgia State University, Atlanta, GA 30303, USA

Rodrigo Alves de Souza, Beth Israel Deaconess Medical Center in Boston, MA, USA

Daniela Dias-Pedroso, UCIBIO, Faculdade de Ciências e Tecnologia, Universidade Nova de Lisboa, Lisbon, Portugal; CEDOC, Faculdade de Ciência Médicas/NOVA Medical School, Universidade Nova de Lisboa, 1169-056 Lisbon, Portugal

Mihyang Do, Department of Biological Sciences, University of Ulsan, Ulsan 44610, Republic of Korea

Sylvain Doré, Department of Anesthesiology, Center for Translational Research in Neurodegenerative Disease and McKnight Brain Institute, University of Florida College of Medicine, 1275 Center Drive,

Carbon Monoxide in Drug Discovery: Basics, Pharmacology, and Therapeutic Potential, First Edition. Edited by Binghe Wang and Leo E. Otterbein.
© 2022 John Wiley & Sons, Inc. Published 2022 by John Wiley & Sons, Inc.

Biomed Sci J493, Gainesville, FL 32610, USA; Departments of Neurology, Psychiatry, Pharmaceutics, and Neuroscience, University of Florida College of Medicine, Gainesville, FL, USA

Madison Fangman, Department of Anesthesiology, Center for Translational Research in Neurodegenerative Disease and McKnight Brain Institute, University of Florida College of Medicine, 1275 Center Drive, Biomed Sci J493, Gainesville, FL 32610, USA

Roberta Foresti, Mondor Institute for Biomedical Research (IMRB), Université Paris-Est Créteil, INSERM U955, F-94010 Créteil, France

Andrew Gomperts, Hillhurst Biopharmaceuticals, Inc., 2029 Verdugo Blvd, Montrose, CA 91020, USA

Edward Gomperts, Children's Hospital Los Angeles, Los Angeles, CA 90027, USA; Hillhurst Biopharmaceuticals, Inc., Montrose, CA 91020, USA; Division of Hematology, Oncology and Transplantation, Vascular Research Center, Department of Medicine, University of Minnesota, Minneapolis, MN 55408, USA

Stephan Immenschuh, Institute of Transfusion Medicine and Transplant Engineering, Hannover Medical School, Hannover, Germany

Yeonsoo Joe, Department of Biological Sciences, University of Ulsan, Ulsan 44610, Republic of Korea; Mycos Therapeutics Inc., Ulsan 44610, Republic of Korea

Patrycja Kaczara, Jagiellonian Centre for Experimental Therapeutics (JCET), Jagiellonian University, Krakow, Poland

Uh-Hyun Kim, National Creative Research Laboratory for Ca^{2+} Signaling Network, Chonbuk National University Medical School, Jeonju 54907, Republic of Korea

Hiroaki Kitagishi, Department of Molecular Chemistry, Faculty of Science and Engineering, Doshisha University, Kyotanabe, Kyoto 610-0321, Japan

Ghee Rye Lee, Beth Israel Deaconess Medical Center, Center For Life Science, 3 Blackfan Circle, 617D, Boston, MA 02215, USA

Howard Levy, Hillhurst Biopharmaceuticals, Inc., 2029 Verdugo Blvd, Montrose, CA 91020, USA

Richard J. Levy, Department of Anesthesiology, Columbia University Medical Center, 622 W. 168th Street, New York, NY 10032, USA

Yi Liao, Department of Biomedical and Chemical Engineering and Sciences, Florida Institute of Technology, Melbourne, FL, USA

Wen Lu, Department of Chemistry and Center for Diagnostics and Therapeutics, Georgia State University, Atlanta, GA 30303, USA

Katarzyna Magierowska, Department of Physiology, Jagiellonian University Medical College, Krakow, Poland

Marcin Magierowski, Department of Physiology, Jagiellonian University Medical College, Krakow, Poland

Alexandra Mazur, Department of Anesthesiology, Center for Translational Research in Neurodegenerative Disease and McKnight Brain Institute, University of Florida College of Medicine, 1275 Center Drive, Biomed Sci J493, Gainesville, FL 32610, USA

Brian W. Michel, Department of Chemistry and Biochemistry, University of Denver, Denver, CO 80210, USA

Maryam K. Mohammed, Department of Surgery, University of Pittsburgh Medical Center, Pittsburgh, PA 15213, USA

Shruti Mohan, Mondor Institute for Biomedical Research (IMRB), Université Paris-Est Créteil, INSERM U955, F-94010 Créteil, France

Roberto Motterlini, Mondor Institute for Biomedical Research (IMRB), Université Paris-Est Créteil, INSERM U955, F-94010 Créteil, France

Grace E. Otterbein, University of Aberdeen School of Medical Sciences, Polwarth Building, Foresterhill, Aberdeen AB25 2ZD, UK

Leo E. Otterbein, Beth Israel Deaconess Medical Center in Boston, MA, USA

Hannah Pamplin, Department of Anesthesiology, Center for Translational Research in Neurodegenerative Disease and McKnight Brain Institute, University of Florida College of Medicine, 1275 Center Drive, Biomed Sci J493, Gainesville, FL 32610, USA

Jeongmin Park, Department of Biological Sciences, University of Ulsan, Ulsan 44610, Republic of Korea

Shruti Patel, Department of Anesthesiology, Center for Translational Research in Neurodegenerative Disease and McKnight Brain Institute, University of Florida College of Medicine, 1275 Center Drive, Biomed Sci J493, Gainesville, FL 32610, USA

Claude A. Piantadosi, Department of Medicine, Duke University School of Medicine, 200 Trent Drive, Durham, NC 27710, USA

Olga Pol, Grup de Neurofarmacologia Molecular, Institut d'Investigació Biomèdica Sant Pau, Hospital de la Santa Creu i Sant Pau, 08041 Barcelona, Spain; Grup de Neurofarmacologia Molecular, Institut de Neurociències, Universitat Autònoma de Barcelona, 08193 Barcelona, Spain

Kamil Przyborowski, Jagiellonian Centre for Experimental Therapeutics (JCET), Jagiellonian University, Krakow, Poland

Stefan W. Ryter, Joan and Sanford I. Weill Department of Medicine, and Division of Pulmonary and Critical Care Medicine, Weill Cornell Medical Center, New York, NY 10065, USA; Proterris, Inc., Boston, MA, USA

Ikuko Sagami, Graduate School of Life and Environmental Sciences, Kyoto Prefectural University, Sakyo-ku, Kyoto 606-8522, Japan

Nils Schallner, University of Freiburg, Freiburg, Germany

Morgan R. Schneider, Department of Chemistry and Biochemistry, University of Denver, Denver, CO 80210, USA

Nuno Soares, UCIBIO, Faculdade de Ciências e Tecnologia, Universidade Nova de Lisboa, Lisbon, Portugal; CEDOC, Faculdade de Ciência Médicas/NOVA Medical School, Universidade Nova de Lisboa, 1169-056 Lisbon, Portugal

Christoph Steiger, David H. Koch Institute for Integrative Cancer Research, Massachusetts Institute of Technology, Cambridge, MA 02142, USA

Young-Joon Surh, Tumor Microenvironment Global Core Research Center and Research Institute of Pharmaceutical Sciences, College of Pharmacy, Seoul National University, Seoul 08733, Republic of Korea

Chalet Tan, Departmental of Pharmaceutics and Drug Delivery, University of Mississippi School of Pharmacy, University, MS 38677, USA

Michael S. Tift, Department of Biology and Marine Biology, University of North Carolina, Wilmington, 601 S. College Road, Wilmington, NC 28403, USA

Giovanni Traverso, Division of Gastroenterology, Brigham and Women's Hospital, Harvard Medical School, Boston, MA 02115, USA; Department of Mechanical Engineering, Massachusetts Institute of Technology, Cambridge, MA 02139, USA

Greg Vercellotti, Division of Hematology, Oncology and Transplantation, Vascular Research Center, Department of Medicine, University of Minnesota, Minneapolis, MN 55408, USA

Helena L.A. Vieira, UCIBIO, Faculdade de Ciências e Tecnologia, Universidade Nova de Lisboa, Lisbon, Portugal; CEDOC, Faculdade de Ciência Médicas/NOVA Medical School, Universidade Nova de Lisboa, 1169-056 Lisbon, Portugal

Libor Vítek, 4th Department of Internal Medicine and Institute of Clinical Biochemistry and Laboratory Diagnostics, University General Hospital and 1st Faculty of Medicine, Charles University, Prague, Czech Republic

Ryan R. Walvoord, Department of Chemistry, Ursinus College, Collegeville, PA 19426, USA

Binghe Wang, Department of Chemistry and Center for Diagnostics and Therapeutics, Georgia State University, Atlanta, GA 30303, USA

Hongjun Wang, Department of Surgery, Medical University of South Carolina, 173 Ashley Avenue, Charleston, SC 29425, USA; Ralph Johnson Veteran Medical Center, Charleston, SC, USA

Minjia Wang, Departmental of Pharmaceutics and Drug Delivery, University of Mississippi School of Pharmacy, University, MS 38677, USA

Rui Wang, Department of Biology, York University, Toronto, Ontario M3J 1P3, Canada

Jakob Wollborn, Department of Anesthesiology, Perioperative and Pain Medicine, Brigham and Women's Hospital, Harvard Medical School, Boston, MA 02114, USA

Xiaoxiao Yang, Department of Chemistry and Center for Diagnostics and Therapeutics, Georgia State University, Atlanta, GA 30303, USA

Zhengnan Yuan, Department of Chemistry and Center for Diagnostics and Therapeutics, Georgia State University, Atlanta, GA 30303, USA

Brian S. Zuckerbraun, Department of Surgery, University of Pittsburgh Medical Center, Pittsburgh, PA 15213, USA

Preface

Carbon Monoxide: Promises and Challenges in Its Pharmaceutical Development

Binghe Wang and Leo E. Otterbein

Carbon monoxide (CO), one of the smallest organic natural molecules, is widely known for its toxicity. Formation of CO via incomplete combustion is a major contributing factor to accidental or intentional CO poisoning, leading to severe health consequences or death. In addition, CO is a by-product of tobacco smoking, and has been associated with some of the harmful effects of smoking. However, less known and probably far more important is the recognition of the essential physiological roles of CO as a signaling molecule in mammals. Against over more than a century of negative connotation, the last few decades have proven that CO possesses a multitude of physiological roles and therapeutic functions, including regulation of the immune response, cellular proliferation, and control of cell survival. This concept is supported by the discovery that CO is produced by all cells and more so under conditions of stress. This book comprehensively summarizes key aspects of CO's endogenous roles, therapeutic functions, and challenges that we face in its development as a therapeutic agent. We hope this preface will provide a thread for reading this book and a bird's-eye view of the landscape for understanding this field, and more importantly lay out the challenges ahead in understanding the detailed mechanisms of action of CO and in its development as a therapeutic agent. We have divided the book into four sections that provide a framework for the reader to follow the evolution of CO from an accepted poison to a bioactive molecule that may offer enormous clinical benefits.

Section I begins with "foundational" knowledge of the CO field, including a general background and known physiological mechanisms. Endogenous production is a prerequisite for a molecule to be an endogenous signaling molecule and CO is no exception. Therefore, the book commences with a detailed discussion of CO's endogenous production in all cells, the enzymes involved, and the detailed chemistry of the major pathway leading to endogenous CO production during the degradation of heme by the heme oxygenases (Chapter 1), a principal player in how CO is generated by all cells. Other products from heme degradation are also discussed so as to contextualize CO's functions (Chapter 10) as an endogenous signaling molecule. Section I also includes a comprehensive examination of the molecular targets of CO (Chapter 2), providing the molecular basis for later discussion on its physiological and therapeutic functions. The physiological, therapeutic, and/or toxicological functions of any molecule are only meaningful in the context of concentrations. Therefore, the examination of the pharmacokinetic and pharmacodynamic characteristics of CO is discussed (Chapter 3). This section also includes information on CO's role in energy metabolism (Chapter 4), regulation of the circadian clock (Chapter 5), and mitochondrial function (Chapter 6) because these three aspects impact a large number of the reported CO effects. One very important aspect of CO is its ability to signal and its relationship with other gaseous signaling molecules: nitric oxide (Chapter 8) and hydrogen sulfide (Chapter 9). Along a similar line, CO's functions overlap with that of oxygen in very intriguing ways, which go beyond simply competing for binding with hemoproteins (Chapter 7). Therefore, there are three chapters in this section devoted to the interplay among these four gas molecules.

Section II focuses on the development of various delivery forms of CO, which are critical to studying CO's mechanism(s) of actions, validating its pharmacological functions, and developing CO-based therapeutics. All such delivery forms focus on going beyond inhalational delivery of CO. Chapter 11 discusses noncarrier formulations, including CO in solution, CO donors encapsulated in various types of materials, and extracorporeal delivery of CO. Chapter 12 focuses on examples of metal-immobilized carbonyls, which are also commonly referred to as CO-releasing molecules (CORMs). For a period of three decades, there was a very high level of activity in this area. Chapter 12 is only able to describe a few select examples of immobilized carbonyls as CO donors. In recent years, there has been a rapid increase in the level of interest in developing metal-free CO donors for reasons of diversity and for avoiding metal-related issues. Chapter 13 discusses metal-free CO donors that rely on photolysis for CO release. A major new direction in the field of CO donors over the past 5 years has been the development of organic CO prodrugs with tunable release rates, triggered release, and the ability to deliver multiple payloads in a single prodrug. Chapter 14 comprehensively examines this area, including discussions of the unique chemistry employed and pharmacological validation studies. This chapter also has a very important section on the proper use of controls in studying CO donors, including both organic prodrugs and metal-based CORMs. Though using proper controls is normally considered routine practice in scientific research, the unique challenges of CO-independent effects observed with some metal-based CORMs indeed elevate this issue to a prominent position. In the field of CO delivery, there is always the question of whether there is the need for targeted CO delivery. Chapter 15 summarizes recent developments in this area. In terms of CO delivery, there is one area that may offer very unique opportunities for efficacy studies and toxicity assessment. Fluorinated general anesthetics such as sevoflurane, desflurane, and isoflurane are known to decompose under basic conditions, leading to the production of CO. Such basic conditions are needed in a ventilator to remove carbon dioxide. Therefore, there is the issue of anesthesia-related CO exposure, which is the focus of Chapter 16. Chapter 17 explores various aspects of natural product-based CO production, including the mechanism(s) of the chemical reactions involved and its potential implications in terms of nutritional and/or therapeutic values. These discussions also bring in a sense of effective concentrations needed for observed activities whenever possible. The last point is very important because discussions of pharmacologic activities outside of the context related to concentration and potency have very little meaning.

Section III has only one chapter, but it represents a very important aspect of the CO field. One unique challenge in studying a gaseous signaling molecule is difficulty in detection and concentration determination. Chapter 18 describes in detail available fluorescent probes for CO detection, including intracellular detection. The chapter is meticulously written with information on detection limits and signal to noise ratios for various probes. This chapter also includes discussions of reported "CO probes" that are only able to detect certain metal-based CORMs because of metal-mediated reactions, but not CO itself.

Section IV examines the pharmacologic effects and mechanistic understandings of CO in various cell culture and animal models, including organ transplantation (Chapter 19), lung injury (Chapter 20), brain injury (Chapter 21), liver injury (Chapter 22), cancer (Chapter 23), diabetes (Chapter 24), kidney injury (Chapter 25), platelet function (Chapter 26), gastrointestinal protection (Chapter 27), sickle cell disease (Chapter 28), and pain management (Chapter 29). The title for each chapter is sufficiently self-explanatory; collectively, these chapters show the breadth of clinical applications of CO. The last chapter (Chapter 30) summarizes all human clinical trials reported so far.

With its vast therapeutic potential, major challenges remain in understanding CO's molecular mechanism(s) of action and in its pharmaceutical development. Here, we would like to highlight some of these challenges to aid future studies. First, studying the dose–response relationship of a gaseous molecule (CO) is much more challenging than that of a traditional small molecule. The volatility of CO means that the concentration of delivery may or may not be directly related to the effective concentration of CO under a given set of conditions. Second, the mode in which it is administered, the route of delivery, the resulting tissue distribution, and the elimination are standard development matters, but fortunately CO is not metabolized to an appreciable degree, making standard ADME (absorption, distribution, metabolism, and excretion) studies unique. This delivery issue becomes more complex, however, when administering CO in a form other than a gas or a saturated liquid. For instance, the release kinetics of a CO donor or prodrug is known to affect the CO concentration and duration profiles, even in simple buffer solution. In animals, the effects may appear as the same as that observed with CO gas, but with added complexity, including the need to deconvolute the effects of CO from that of the CO donor molecule or its metabolic by-product. There have been a number of reports in recent years that attributed some of the widely reported pharmacological effects of certain CORMs to CO-independent effects. The absorption and tissue distribution may differ depending on physiology or pathophysiology, e.g., lung disease where inhaled CO diffusion will be different or liver disease where metabolism of CO donor molecules and therefore the release of CO may be altered. Other considerations include the fact that some metal-based CORMs have a wide range of chemical reactivities driven by the carrier molecule. The third challenge is the unique difficulty in studying the pharmacokinetic properties of CO. Current studies use carboxyhemoglobin (COHb) as a surrogate indicator of CO concentration. This may not be sufficient. The free concentration of CO in the blood is a very important factor to consider. There is a widely held perception that free CO concentration is always low. However, this is not correct. Though Chapter 3 addresses some of these issues, more work is needed to understand what it means to use COHb as an indicator of CO concentrations. A fourth challenge is the availability of a large number of hemoproteins in a cell, which are all potential targets for CO. How to deconvolute the effects of engaging such a large number of targets that are in constant flux is an important question, but has hardly been examined. Moreover, there are likely differences in hemoprotein distributions across different species. A fifth challenge is the issue of allometric scaling. CO's efficacy has been widely validated in various pharmacological models in different animal species such as mice and rats and large animals including pigs and dogs. However, the large number of clinical trials has yet to demonstrate efficacy in humans in a convincing fashion. In one kidney transplant study, the results seem to be very positive (Chapter 25); however, the trials were terminated prematurely without conclusive demonstration of efficacy in a statistically significant manner. The need to understand allometric scaling is a critical first step in translating success in animal models to humans.

We hope that this book will allow the readers to see the vast potential of CO and stimulate much needed research to assess its therapeutic potential. Collectively, we would like to thank all the authors for their contributions as experts in their respective fields related to CO as well as their diligence and patience in working with us. We would also like to express our sincere appreciation of Ms Andrea Mahone in the office of B. Wang for her assistance in coordinating all aspects of this book project as well as the Wiley team for their support.

May 2021

Section I.

General Background and Physiological Actions

1

Endogenous CO Production in Sickness and in Health

Ladie Kimberly De La Cruz and Binghe Wang

Department of Chemistry and Center for Diagnostics and Therapeutics, Georgia State University, Atlanta, GA, 30303, USA

Introduction

In 1949, Sjöstrand published results supporting carbon monoxide (CO) production in humans [1], ushering in the era of studying CO as a human metabolite, a biomarker for pathological events and metabolic processes [2–4], an endogenous signaling molecule [5,6], and a potential therapeutic agent [7–9]. Over the decades, a large amount of information has been gathered to allow for a deep understanding of its production and its physiological and pathological roles. Specific to this chapter, many possible endogenous sources of CO have been described, including heme degradation, lipid oxidation, degradation of various natural products, photochemical production, and microbial production. Among these, the photochemical process does not lead to *de novo* production because it is a process to dissociate CO from hemoglobin (Hb). Further, there are intriguing variations of CO production rates among individuals depending on gender, menstrual cycle, developmental stage (neonates and infants), physical exercise, and various pathological conditions. All such findings are consistent with the physiological and regulatory roles of CO and its potential therapeutic implications. At this time there is limited knowledge to fully analyze all the reasons for the observed variations in CO production, their physiological and pathological implications, and molecular links among different events. However, we would like to summarize these variations in one place so as to stimulate discussions and future studies of their biological implications.

CO Production Through Heme Degradation

Very impressive studies in the 1940s and 1950s identified heme degradation as the major source of CO production in cells [10–12]. Sjöstrand demonstrated the increased CO production accompanying erythrocyte destruction, laying the foundation for studying CO production under pathological conditions [3]. The relationship between hemolysis and CO production was substantiated by the work of Engstedt in 1957 [2] and Gydell in 1960 [4]. Later, Coburn measured the rate of CO production as 0.42 ± 0.07 ml/h (range: 0.35–0.57 ml/h) based on results of 10 men ranging from 18 to 70 years of age and from 125 to 170 lb in weight [13]. Such results were said to be somewhat consistent with calculated rates of production of 0.3 ml/h from heme catabolism based on the life span of erythrocytes being approximately 120 days. Endogenous CO production due to heme degradation as first suggested by Sjöstrand [10,11] and examined by Ludwig and coworkers showed that it comes from the α-methene of heme [12]. These earlier findings laid the foundation for what will be discussed next on the subject of endogenous CO production via heme degradation. The subsequent section discusses the chemistry of the enzymatic reactions and the enzymes as well as their roles and regulations.

CO is produced in a stoichiometric ratio through heme oxygenase (HO)-mediated degradation of the heme prosthetic group of hemoproteins such as Hb. The enzymatic function of HO is the removal of heme

Carbon Monoxide in Drug Discovery: Basics, Pharmacology, and Therapeutic Potential, First Edition. Edited by Binghe Wang and Leo E. Otterbein.
© 2022 John Wiley & Sons, Inc. Published 2022 by John Wiley & Sons, Inc.

by breaking it down to produce CO and biliverdin while also recovering free iron as end products. The cytoprotective function of HO is attributed not only to the removal of the cytotoxic free heme but also to the production of its by-products: CO and biliverdin (and then bilirubin). Two isozymes of HO are known to exist. HO-1, a member of heat shock protein 32 family, is the inducible form wherein exposure to stress stimuli can increase enzymatic activity by 100-fold [14]. In contrast, HO-2 is believed to be the constitutive form [14]. While these two enzymes catalyze the same reaction, they are products of two different genes and have different roles, regulation, and post-translational modifications [15]. A third isoform has been described but was found to be a pseudogene derived from HO-2 [16]. The biological implications of HO-1 are examined in detail in various chapters in this book and are not discussed here. There is also a chapter on the production of biliverdin and bilirubin and associated implications (Chapter 10). This section solely focuses on CO production by HO-mediated heme degradation and other pathways. However, it is important to note that CO can also stimulate HO-1 expression [17–19], forming a forward-feeding loop. Though there must be at least one "braking" mechanism in such a forward-feeding loop, it is not entirely clear yet where this "brake" is and why this mechanism is only active in certain situations. The ability for CO to bind to HO already suggests inhibitory effects [20], which may serve as one of the checkpoints. Further, there is the question as to whether the limited availability of heme could also serve as a way of "check and balance." There may be others that have not been identified.

The chemistry of the enzymatic reactions of HO to catabolize heme is very interesting and unique in many aspects. Exclusively encountered with the unique enzyme architecture of HO, the substrate heme also functions as a prosthetic group for oxygen activation [21,22]. Overall, heme degradation takes place in three successive oxygenation steps with seven electrons (Figure 1.1) that in mammalian systems can be supplied by NADPH-cytochrome P450 reductase. Upon formation of the heme–HO complex and in the presence of an electron donor, ferric heme is reduced to ferrous heme, which then binds oxygen to form a metastable ferrous oxygenated species. Acquisition of an electron followed by protonation leads to the formation of a ferric low-spin species consistent with a ferric hydroperoxide [23]. The ferric hydroperoxide intermediate may also be generated from the direct reaction between ferric heme and H_2O_2; however, the reaction with O_2 is faster (6.9×10^6 M^{-1} s^{-1} with O_2, while 1.3×10^3 M^{-1} s^{-1} with H_2O_2) [24,25]. This ferric hydroperoxide intermediate then acts as the activated oxygen species that leads to the self-hydroxylation of the α-meso carbon of the porphyrin ring of heme. The proposed mechanism by which the hydroxylation of the α-meso carbon occurs includes (i) a concerted hydroxylation mechanism [26], (ii) a stepwise homolytic O bond cleavage followed by addition of resulting hydroxy radical [27], and (iii) a stepwise heterolytic bond cleavage followed by addition of generated hydroxide [28]. Upon hydroxylation, an isoporphyrin π-cation intermediate [29] loses water to finally give ferric α-*meso*-hydroxyheme. The first few steps are common in HO and other heme-containing monooxygenases or peroxidases such as cytochrome P450. However, while HO promotes self-hydroxylation, P450-type monooxygenases go through an alternative pathway wherein the O–O bond cleaves to give a ferryl species without hydroxylation [24]. Restrictions imposed by the HO environment, specifically a distal helix directly above the heme plane where it physically shields the three other meso carbons, leads to the regiospecific hydroxylation of the α-meso carbon only [30].

The second step is the CO-producing phase wherein α-*meso*-hydroxyheme is converted to verdoheme in the presence of one equivalent of O_2. While the chemistry of the first step is fairly well studied, there are contentions regarding the need for exogenous electrons in the second step [31,32]. Furthermore, unlike the first step, oxygen activation by HO may not be required; rather ferric α-*meso*-hydroxyheme can undergo spontaneous autooxidation wherein O_2 adds to the ring carbon next to the α-*meso* carbon [24,31]. As a result, there are no regiospecificity restrictions in the second step [33]. The α-*meso*-hydroxyheme intermediate has various resonance structures involving keto–enol and radical forms. One pathway by which the second step occurs is through the reaction of the ferrous neutral radical form with O_2 in the presence of an electron followed by protonation to form the ferrous α-hydroperoxy oxophlorin [21]. This intermediate then attacks one of the pyrrole rings to form a four-membered ring, which then undergoes heterolytic rearrangement to extrude CO and form ferrous verdoheme. Another possible pathway begins with the addition of oxygen to the radical form of α-*meso*-hydroxyheme to generate the ferrous hydroperoxyl radical intermediate followed by an intramolecular transfer of electron from the iron center to the peroxyl radical to generate a ferric peroxide intermediate [34]. The peroxide moiety then forms a complex with the iron center to form a peroxo-bridged intermediate that subsequently undergoes heterolytic cleavage to convert the

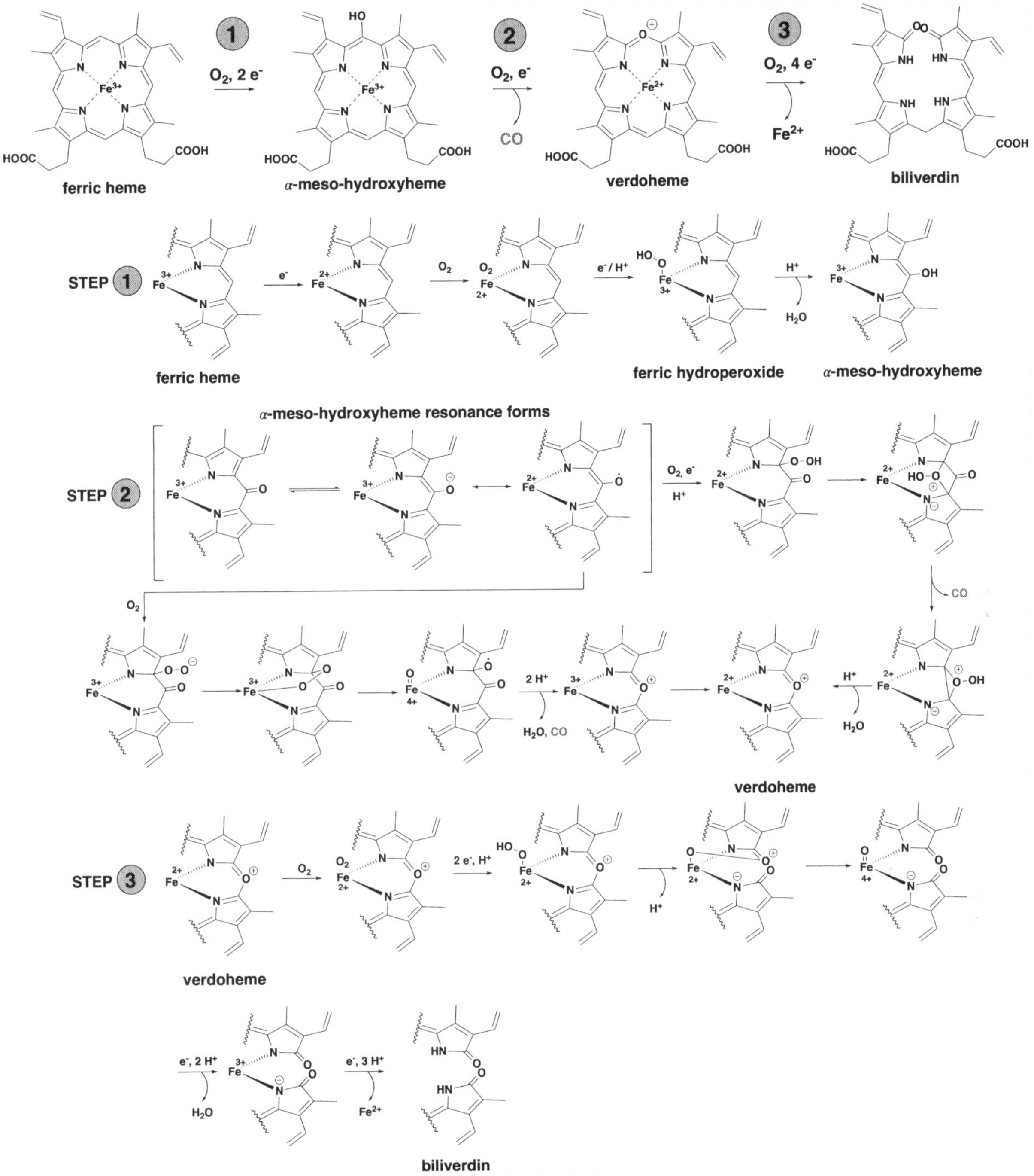

Figure 1.1 Three successive oxygenation steps of heme to CO, biliverdin, and the ferrous ion by HO action.

iron center to a ferryl oxo species and subsequently generating an alkoxy radical. This species undergoes extrusion of CO to form ferric verdoheme that in the presence of reducing equivalents is converted to ferrous verdoheme.

In the final step of heme degradation, four reducing equivalents and one equivalent of oxygen are required to achieve ring opening and demetallation of ferrous verdoheme to produce bilirubin and free iron. While the oxygen activation mechanism is similar to that of the first step, the binding of oxygen with ferrous verdoheme is less favorable. Furthermore, this step is also susceptible to inhibition by CO, a species that is produced in the prior step. In a study using 20% CO gas, CO–verdoheme complex forms indicating that high CO concentration does not completely inhibit the first two steps of HO-1 catalysis [35]. As opposed to other heme-containing globins such as myoglobin with a large ratio of affinities between CO and O_2 (41 for myoglobin), the HO enzyme has a low ratio of affinities between CO and O_2 (1.2–5.6) pointing to an inherent mechanism in HO to discriminate these two gases [20]. Binding of verdoheme to oxygen produces an intermediate similar to the ferrous hydroperoxide of the first step, which then proceeds to produce ring-opened intermediates that ultimately produce HO-bound ferric biliverdin. Consumption of one more electron leads to demetallation to produce free iron and biliverdin.

While HO is primarily known for its enzymatic degradation of heme, its nonenzymatic functions have also been reported. Despite the huge amount of literature attributing the cytoprotective effects of HO to its by-products, some argue that this is unlikely based on the limited supply of HO substrates in many tissues [15]. A mutant of the HO-1 gene of human monoblastic lymphoma cell U937, which lacks the heme-degrading activity of wild-type HO-1, exhibited protection against oxidative stress through upregulation of catalase and glutathione contents [36]. The nuclear translocation of HO-1 in response to exposure to stress conditions, although associated with reduced enzymatic activity, brings about changes to the binding of transcription factors involved in oxidative stress leading to protection against H_2O_2-mediated injury [37]. Furthermore, HO-1 was shown to alter gene expression and promote its transcriptional regulation via a feed-forward mechanism, which is postulated to be one way in which HO-1's basal levels are maintained under limiting substrate concentration [38]. All these also suggest that HO-1 has CO-independent activity.

In the subsequent sections, the manifestation of endogenous CO production in healthy adults, neonates, pregnant women, during exercise, and under different pathological conditions is discussed in detail. While the origin of endogenously produced CO arises primarily from heme catabolism, other sources such as induction of P450 will also be discussed.

CO Production in Healthy Adults

To understand the roles of CO in human physiology and pathology as well as its therapeutic potential, it is important to first look at CO production in somewhat quantitative terms. The first issue is to examine CO production in healthy individuals under "normal" conditions. In a landmark study, Coburn measured the rate of CO production (from COHb levels) as 0.42 ± 0.07 ml/h (range: 0.35–0.57 ml/h) and 6.1 µl/kg/h based on results of 10 men ranging from 18 to 70 years of age and from 125 to 170 lb in weight [13]. Though the dataset was not sufficiently large to conduct multivariable analysis, there did not seem to be a direct relationship between age and CO production among this group of adults. Later studies led to levels of 6.6 ± 1.9 µl/kg/h [39,40] which substantiated the work by Coburn. If one assumes the average Hb concentration, blood volume, and life span of red blood cells to be 7.5 mm (monomer), 4 l, and 120 days respectively, then each day there is about 0.25 mmol of Hb available for CO production (about 5.6 ml/day, 0.23 ml/h). In Coburn's report, the calculated rate of CO production from red blood cell turnover was 0.3 ml/h from heme catabolism, which was said to be in general agreement with the measured 0.42 ml/h. As an approximation, they are indeed in general agreement.

For a long time, CO production was thought to be solely from heme catabolism [41]. If so, one would expect a stoichiometric ratio between CO and bilirubin production (Figure 1.1). In a milestone study, Berk and colleagues examined 37 individuals and demonstrated that the ratio between CO production and bilirubin product was 1.14 ± 0.03, indicating the production of CO exceeds that of bilirubin by 14%. The results also suggest 12.3% of the CO produced being from nonheme source(s) [42]. The study population includes 12 normal individuals, 10 with Gilbert's syndrome, 9 with Gilbert's syndrome and hemolysis, and the rest with various conditions such as acute intermittent porphyria, refractory sideroblastic anemia, polycythemia vera, and nonspherocytic hemolytic anemia. Now it is well understood that there is CO production from other metabolic activities such as exercise, as will be discussed in a later section. Further, not all the heme molecules come from Hb either; i.e., not all heme molecules for CO production come from erythrocytes. In an extensive study, Berk and colleagues concluded that about 75% of the bilirubin production is derived from Hb and thus red blood cell turnover with the remaining 25% coming from other heme-containing proteins including myoglobin, neuroglobin, cytoglobin, and various heme-containing enzymes as possible candidates [43]. Knowing the stoichiometric CO production ratio from heme degradation, one can assume that 25% of the endogenously produced CO from heme degradation is from non-Hb heme sources.

In remarkable studies of the differences between males and females, it was found that the rate of CO production (V_{CO}) in young normal males averaged 0.42 ± 0.07 ml/h and 6.1 μl/kg/h [13]. Further, the production rate was found to be constant over a period of days or months. In contrast, V_{CO} was found to vary with the menstrual cycle in females with the progesterone phase producing CO at twice the rate as in the estrogen phase [44]. Specifically, in a study involving nine young and healthy women, it was found that the average rate of CO production was 5.24 ± 0.66 μl/kg/h during the estrogen phase and 10.2 ± 0.99 μl/kg/h in the progesterone phase. In an effort to understand the controlling factors, male subjects ($n = 7$) were also shown to increase their CO production rate from 5.85 ± 0.5 to 9.9 ± 0.9 μl/kg/h upon administration of 17α-hydroxy-6α-methylprogesterone acetate (14–25 μg/kg/day). At the time of this study, the steroid's ability to induce heme synthesis was already known [45]. Therefore, it was proposed that the increased CO production was likely due to progesterone's ability to induce HO-1 biosynthesis and therefore endogenous CO production. Exposure of human nonlaboring pregnant myometrial samples to progesterone but not estradiol induced HO-1 in a dose-dependent manner and increased CO production in the same manner as hemin [46]. These studies implicate the HO/CO pathway in the maintenance of uterine quiescence during pregnancy; however, a subsequent study directly contradicted this finding [47].

HO-1 expression in the uterus varies depending on the hormonal changes during the estrous cycle [48]. Because HO's presence is required in the different stages of the estrous cycle as well as pregnancy, the HO-1/CO axis is postulated to play a crucial role in reproduction, including proper placentation and fetal development [49]. Furthermore, the gestation of *Hmox-1* null mice that survive to adulthood is compromised as a result of improper placentation, intrauterine fetal growth restriction, and fetal lethality [50]. Exogenous administration of CO overcame the effects of *hmox1* deletion wherein a higher percentage of viable embryos were found in the CO treatment (50 ppm) group versus air control. Now it is well recognized that the implantation process for a fertilized egg involves controlled inflammation [51]. The central role CO plays in the implantation of fertilized eggs may be through regulation of immune responses and inflammatory processes. It is also important to note that progesterone levels remain high during pregnancy. Much more work is needed to firmly establish the mechanisms by which CO impacts pregnancy [52,53].

All these studies have helped in defining boundary conditions in understanding the relative quantities of CO production under normal conditions. Such results also suggest that CO from red blood cell turnover accounts for the major source in maintaining the basal level of CO and COHb. Such basal level of activities normally leads to a COHb level of ≤2% [54].

These discussions also suggest one thing: the availability of heme seems to be the rate-limiting factor in CO production. It is well known that stress often leads to the overexpression of HO-1. In a study where HO-1 is induced by surgical stress, both urinary bilirubin and aortic CO levels were elevated on the first day, which correlated with the decrease in plasma concentration of heme substrate and increase in tissue HO levels [55]. In experiments using ^{14}C-labeled heme, the production of ^{14}CO from aortic tissues on day 1 after surgery shot up to ~ 21 cpm × 10^{-7}/mg aortic tissue but gradually decreased to basal levels on day 5 (~4 cpm × 10^{-7}/mg aortic tissue) after surgery. However, when the animals were pretreated with zinc protoporphyrin IX (50 μmol/kg i.p.), an HO-1 inhibitor, ^{14}CO levels decreased to basal levels on day 1 indicating the HO-1 origin of released ^{14}CO.

CO Production in Neonates and Infants

In neonates, the production of CO is elevated compared to adults. This is attributed to the shorter life span of red blood cells in neonates (70–90 days versus 110–120 days in adults) [13,56,57]. Through extensive studies, CO production rate was determined to be 13.7 ± 3.6 μl/kg/h in normal infants [58]. Much of the increased CO production studies have been focused on its correlation with hyperbilirubinemia or other pathological conditions [59–62]. Because of the lack of the needed enzyme for bilirubin metabolism and thus excretion, the consequence of the increased RBC turnover and Hb degradation is the rapid accumulation of bilirubin in newborn infants, which can lead to neurotoxicity. This issue is exacerbated in newborns who are preterm or have mismatched maternal blood type (ABO incompatibility), infection, hemolytic diseases, or liver diseases [63]. Incidentally, neonates with Down syndrome seem to have elevated levels of COHb [64]. Elevated levels of COHb are also observed in preterm infants with late-onset sepsis [65]. At this time, it is not clear what the underlying molecular implications are with the elevated COHb levels.

It should be noted that there are some structural differences between adult and newborn Hb molecules. The dominant Hb form in newborns is termed as fetal hemoglobin (HbF). In terms of structure, while adult Hb is a tetramer of two α-globin peptides and two β-globin peptides, HbF is a tetramer of two α-globin peptides and two β-like γ-globin peptides [66]. This difference in structure gives rise to striking differences in their biophysical characteristics including varying binding affinities of ligands such as O_2 and CO wherein HbF has 20% lower preferential binding to CO over O_2 ($K = 174$) compared to adult Hb ($K = 217$) [67].

CO Production During Pregnancy

In nonsmoking pregnant women, maternal COHb level ranges between 0.4% and 2.6% [68], with the average being $1.1 \pm 0.2\%$ [69,70]. Fetal COHb levels tend to be higher than the maternal level at about $1.8 \pm 0.3\%$ [69], with a range of 0.4–2.8% [68]. However, some reports show higher COHb levels in mothers [71]. It is difficult to make meaningful comparisons with these data as there are many contributing factors to the discrepancy, such as variations in ambient CO levels and slow equilibration half-time for CO transfer between maternal and fetal circulation [67]. It is well known that fetal oxygen levels are lower than that of the mother. Could the elevated levels of CO provide protective effects much the same way as proposed in elephant seals, which can stay in hypoxic conditions for an extended period of time without the same kind of hypoxic/reperfusion damage that one would expect [72]? In another example, increased pulmonary CO production and HO-1 expression were demonstrated to be a protective adaptive response in newborn llamas against pulmonary hypertension commonly observed in mammals at high altitude [73]. In murine models, CO production has been shown to increase significantly during pregnancy [74] and CO has been shown to increase angiogenesis without impacting pregnancy-specific adaptations [75]. In humans, increased CO production during pregnancy (0.92 ml/h versus 0.62 ml/h in the progesterone phase and 0.32 ml/h in the estrogen phase) was attributed to fetal production [76–78]. Along a similar line, progesterone has been shown to increase HO-1 expression during pregnancy, leading to inhibition of human myometrial contractility via CO [46]. As discussed in the section entitled "CO Production in Healthy Adults," HO-1 and CO play critical roles in maintaining healthy pregnancy [79–81].

CO also seems to be related to pathological conditions unique to pregnancy. Preeclampsia is known to lead to elevated COHb in pregnant women. For example, in one study of 15 preeclamptic and 15 normal pregnant women, it was found that the average COHb level in preeclamptic women was 2.8% as compared to 0.7% in the normal control group [82]. Preeclamptic women also had a significantly lower (24.4 mmHg) P_{50} than normal pregnant women ($P_{50} = 30.1$ mmHg) [82] or normal women without pregnancy [83]. COHb is known to shift the oxygen binding curve to the left and known to inhibit the production of 2,3-diphosphoglyceric acid, which has a direct effect on the conformation of Hb, leading to a transition from the high-affinity state to the low-affinity state (see Chapter 3 on pharmacokinetics for details) [84]. Incidentally, CO has been reported as having beneficial effect in experimental preeclamptic mouse models [85,86]. One of the proposed mechanisms involves soluble *fms*-like tyrosine kinase-1 (sFlt-1) and its antiangiogenic activity wherein in adenovirus-expressed sFlt-1 preeclampsia-like mouse model CO (250 ppm) prevented the development of pregnancy hypertension, proteinuria, and Bowman's space. Another mechanism by which CO may afford protection from preeclampsia is through its anti-apoptotic effects. A CO-saturated culture medium was shown to confer 60% less apoptosis induced in villous explants from human placenta exposed to hypoxia/reoxygenation insults [87]. In another study, pregnant women with hypertension and preeclampsia were reported to exhibit decreased end-tidal breath CO levels [88]. In a large population-based cohort study of pregnant women exposed to increasing ambient levels of CO from 0.01 to 0.29 ppm, a corresponding decrease in rates of preeclampsia was also observed [89]. However, it should be noted there are some studies that report either no or even an increased risk for preeclampsia upon CO exposure, although the exposure doses to CO and other gaseous pollutants are higher [90,91]. So at this time, it seems that there are indications that CO production and activity are implicated in preeclampsia, but more studies are needed to disentangle the intricate relationships among various molecular players [75,92].

CO Production During Exercise

Exercise is known to induce HO-1 expression in humans [93]. However, this does not seem to automatically lead to increased CO levels in exhaled breath. For example, in one study in human athletes ($n = 9$) (as well as in equines and canines), decreased CO levels in breath collections were detected after maximal exercise [94]. However, this change in CO levels does not necessarily take into consideration the increased respiratory rates and exhalation volumes after exercise.

In another study, while subjects engaged in ramp bicycle exercise exhibited a gradual decrease in the fraction of CO in the exhaled air, the total amount of exhaled CO increased in a linear fashion relative to exercise intensity when the exhalation volume was taken into consideration [95]. This observed exercise-induced increase in endogenous CO production is corroborated in another study wherein blood COHb was used as an indicator of HO-1 activity. With 10 healthy nonsmoker volunteers who underwent cycling exercises with 15-min cycling/rest intervals ($n = 10$, 5 females and 5 males), the blood COHb levels doubled from $1.1 \pm 1.6\%$ to $2.1 \pm 1.6\%$ indicating a positive correlation between exercise and HO-1 activity and thus endogenous CO production [96]. These results present interesting questions as to the possible correlations among exercise, HO-1 expression, CO levels in exhaled air, COHb levels in blood, total CO production, and health consequences.

On one hand, exercise-induced inflammation is well recognized [97,98]. Does the increased COHb level indicate an effort by the host to attenuate inflammation? Along a different line, the upregulation of HO-1 after exercise has been linked to improved outcome in cancer chemotherapy [99–101]. One possible molecular link (this may not be the only link [102]) is the increased level of COHb. Although COHb levels were not determined in these studies, increased COHb levels have been correlated with exercise as described earlier [96,103]. Along a similar line, Otterbein and others have published data indicating that CO inhibits tumor growth in animal models [104], in part by sensitizing cancer cells to chemotherapy by stimulating mitochondria biogenesis, leading to metabolic exhaustion [105]. In contrast, CO has also been reported to offer cytoprotective effects against chemotherapy-induced cardiotoxicity [106,107]. Such findings suggest the need to further explore the roles that CO plays in cancer chemotherapy and how exercise improves chemotherapy outcome at the molecular level.

CO Production and Levels in Diseases

An Overview of Pathological Conditions that Lead to Changes in CO Production

Under pathological conditions, endogenous CO production may be drastically altered as heme substrates are made more available as a result of increased hemoprotein presence and hemoprotein damage. Regardless, HO-1 is increased as a result of stress.

Hematological diseases characterized by red blood cell destruction (hemolysis) such as anemia, hematoma, thalassemia, sickle cell disease, and Gilbert's syndrome lead to pronounced elevations of CO concentrations with COHb levels reaching as high as 10%, as summarized in an excellent review by Owens [78]. COHb levels have also been reported to show small, but obvious and significant elevations compared to that of normal subjects under certain airway illnesses, including chronic obstructive pulmonary disease (0.81% under stable conditions, 0.95% at severe exacerbation versus 055% of the controls) [108], inflammatory pulmonary diseases (0.93–1.13% versus 0.65% in controls) [109,110], and pneumonia (1.70% versus 1.40% in controls) with increased HO-1 expression in acute respiratory distress syndrome [111,112]. While the variations in COHb levels associated with various conditions are interesting and may point to certain diagnostic and/or signaling values, efforts to understand the true meaning of such variations will take much more work. Along the line of airway inflammation, it seems that infection by SARS-COV-2 also leads to elevated COHb levels, possibly due to increased hemolysis [113,114]. For example, COHb levels ranged from about 2.3 to 2.8 for nonsurvivors on day 30 as compared to a normal range of 0.5–1.5% [115,116]. However, the mechanistic implications of such increases are not clear.

Higher levels of exhaled CO have been reported in patients with other inflammatory airway disorders, including cystic fibrosis (6.7 ppm versus 2.4 ppm for healthy controls) [117] and asthma (5.8 ppm versus 2.9 ppm for healthy controls) [118]. In obstructive sleep apnea, a chronic respiratory disorder characterized by airway obstruction during sleep, venous blood CO levels are elevated post-sleep that can be returned to normal levels by continuous positive airway pressure therapy [119]. These studies have shown the potential of circulating CO levels as a biomarker for the assessment of disease burden. To this end, considerable effort has been dedicated to assess and use exhaled CO as a biomarker of diseases [120]. However, there are some studies reporting that there is no elevation in COHb or exhaled CO in patients with asthma and cystic fibrosis [121], and sepsis (in preterm neonates) [122]. Indeed, using COHb as a disease marker is a complicated one because the molecular implications of COHb level changes are not always clear. However, there have been many studies of CO's pharmacology as it relates to various pathological conditions. Results of such pharmacological assessments may lend some insights into the molecular underpinning of CO's role(s) in various pathological processes. Table 1.1 lists select examples of such studies.

Table 1.1 Association of levels of COHb, exhaled CO, and/or HO-1 with pathological conditions and pharmacological evidence in support of more than coincidental relationships.

Entry	Human conditions	COHb/exhaled CO/HO-1	Studies using pharmacological models	CO donors used	Comments
1	Smoking	↑/↑/— [123–125]			
	(a) Smoking and colitis: a possible inverse relationship [126]		CO's beneficial roles in treating chemically induced colitis	CO gas [127,128], CO prodrugs [129], CORM-A1 [130], CORM-2 [131,132], CORM-3 [133]	
	(b) Smoking and preeclampsia: a possible inverse relationship [134–137]		CO's ability to blunt placental ischemia-induced hypertension [138]; induction of HO-1 attenuates sFlt-1-induced hypertension in pregnant rats [139]	CORM-3 [138], CO gas [140], HO-1 induction [139]	
	(c) Smoking and Parkinson's disease: a possible inverse relationship [141]		None	None	
2	Systemic inflammation	↑/↑/—	LPS-induced systemic inflammation	CO gas [142,143], CO prodrug [144], CORM-2 [145]	
3	Sepsis	↑/↑/—[122,146,147]	Sepsis models (CLP)	CO gas [148], CORM-2 [8,149]	
4	Hemorrhagic shock		Attenuation of acute lung injury after hemorrhagic shock [150–152]; attenuation of kidney injury after hemorrhagic shock [153]	CO gas [150,151], CORM-3 [152]	
5	Ulcerative colitis		See entry #1a		Chapter 27
6	Cancer		Complex roles for HO-1, including nuclear localization [154–161]		Chapter 23
	(a) Colorectal cancer	↑/—/↑ [162]	HO inhibitor increased sensitivity to pirarubicin [162]		
	(b) Colorectal cancer		HO expression has anticancer effects [163]		
	(c) Colorectal cancer (C26 cells)			CORM-2 [164]	
	(d) Prostate cancer	HO-1 nuclear localization [105]	CO increased sensitivity to chemotherapy [105]	CO gas [105]	Chapter 23
	(e) Pancreatic cancer (CAPAN-2/PaTu-8902)		Antiproliferative effects of CO on pancreatic cancer [165]	CO gas and CORM [165]	
	(f) Breast cancer (4T1)		CO strengthens cooperative bioreductive antitumor therapy [166]	Encapsulated CORM (Fe-based) [167]	

(*Continued*)

Table 1.1 (Continued)

Entry	Human conditions	COHb/exhaled CO/HO-1	Studies using pharmacological models	CO donors used	Comments
	(g) Breast cancer (MCF-7)		CO expedited the metabolic exhaustion of cancer cells toward reversal of chemotherapy resistance [167]	Encapsulated CORM (Fe-based) [166]	
	(h) Lung cancer	↑/—/—	COHb, a prognostic indicator of lung cancer [168], an indicator of sensitivity to chemotherapy [169], and correlates with tumor size [170]	CORM-2 [171], CO gas [104]	
	(i) Lung cancer–smoking	Increase in COHb is associated with "inhalation" [125]	"Risk of lung cancer is less among heavy cigarette smokers who say that they inhale than it is among those who say that they do not inhale" [172]		This is intriguing, but uncorroborated by other studies
	(j) Chemotherapy-induced organ toxicity		CO attenuated chemotherapy-induced toxicity in the heart [107,173] and kidney [174–176]	CO gas [107], CORM-2 [173], CORM-3 [174,176]	
7	Liver injury	↑/—/— [177]	Warm IRI in swine [178], transplant-induced IRI injury (rats) [179,180], and chemically induced livery injury (mice) [181,182]	CO gas [178–181], CO prodrug (mitochondrion-targeted) [182]	Chapter 22
8	Sickle cell disease	↑/↑/— [183]	Inhaled CO reduced leukocytosis [184], vaso-occlusion, and inflammation [185,186]	CO gas [184], oral CO [185], MP4CO [186]	Chapter 28
9	AKI		See entries #4 and #6i; and reduced IRI injury [187]	CO gas [187]	Chapter 25
10	Pancreatitis		CO's protective effects on SAP in rats	CORM-2 [188–190]	
11	Respiratory diseases				Chapter 20
	(a) Acute respiratory distress syndrome	↑/—/↑ [111,191]	Phase I safety clinical trials for inhaled CO [192]; ongoing phase II clinical trial (NCT03799874)	Inhaled CO gas	
	(b) Chronic obstructive pulmonary disease	↑/↑/↑↓ [108,193–195]	Inhalation of CO led to reduction in sputum eosinophils and improved response to methacholine [196]; HO-1 overexpression prevented porcine pancreatic elastase-induced emphysema [197]	Inhaled CO gas, HO-1 induction	
	(c) Idiopathic pulmonary fibrosis (IPF)	↑/—/↑ [198,199]	HO-1 overexpression prevents IPF [200]; phase II clinical trial did not show efficacy [201]	Inhaled CO gas, HO-1 induction	
	(d) Cystic fibrosis	—/↑—/↑ [121,202,203]	HO-mediated protective effects against *Pseudomonas aeruginosa*-induced injury/apoptosis [203]	HO-1 induction	

Elevated levels of COHb have been reported to be associated with traumatic events [147,204,205]. For example, in cardiothoracic ICU patients who survived, the minimum level of COHb was reported to be significantly higher than those who did not survive [206,207]. Further systemic inflammation and sepsis are also associated with increased levels of CO [146]. Although the mechanistic links between CO and the various pathologies in all these cases are not clear, there have been reports of CO's therapeutic roles in animal models of systemic inflammation (entry 2, Table 1.1) and sepsis (entry 3) [142–144], hemorrhagic shock (entry 4) [150,151,208], acute lung injury [209], neuroprotection after cardiac arrest [210], extracorporeal resuscitation [210], and traumatic brain injury [211,212].

In cancer, many studies have documented the overexpression of HO-1 and its role in the initiation, progression, and metastasis [160,213–215]. For example, colorectal cancer patients have significantly higher COHb levels than noncancer patients [162]. Notably, nuclear expression of HO-1 is linked to malignancy and disease progression in certain cancers such as prostate [105], colorectal [162], and oral carcinoma [216]. However, antitumor effects of HO-1/CO have also been documented [163,217,218]. These reports suggest that the role HO-1 takes in cancer is dependent on several factors such as tumor type and subcellular localization [154,218].

In all these documented elevations from baseline COHb levels and exhaled CO levels in various diseases, the question arises regarding the pathophysiological significance of this observation. Is the elevation a result of the body's protective response against the stress stimuli, or is HO-1 overstimulation one of the contributing factors to the severity of the disease? While the answer to this question is not yet unambiguously identified, the relevance of HO/CO therapeutics is brought to the spotlight. As most of these studies have shown, elevated COHb levels are a result of HO-1 induction. Therefore, therapeutic strategies based on HO-1 induction may further put pressure on the body that is already under constant HO-1 stimulation. Exogenous CO administration by means of inhalation or through CO donors, on the other hand, provides CO without other by-products of heme degradation. Therefore, while HO/CO therapeutics are often discussed hand in hand, it is very important to distinguish these two therapeutic strategies.

Cigarette Smoking and CO Levels

While CO from cigarette smoking does not fall under "endogenous production," the discussion of CO levels in smokers does fall into the category of "diseases," as addiction to cigarette smoking is medically considered as a pathological condition leading to millions of deaths each year [219].

Cigarette smoking is one major source of exogenous CO through inhalation. CO from cigarette smoke arises from the incomplete combustion of tobacco and consists of more than 2% composition of tobacco smoke, translating into around 400 ppm inhaled into the lungs [220]. Smokers may commonly have elevated COHb levels of around 6% (heavy smokers up to 10–15%) as opposed to the <2% COHb steady levels in nonsmokers [221,124]. In extreme cases, there have been reports of CO poisoning from cigarette/cigar or hookah smoking with COHb levels averaging about 27% [222–224] and reaching as high as 39% [225–227]. Obviously, passing through water is not an efficient way of removing CO in tobacco smoke, which is understandable given the low solubility of CO in water. There has been much interest in both the toxicological and physiological implications of having such elevated COHb levels in smokers. In a study wherein the same levels of COHb (~6%) under CO inhalation and smoking were compared, CO from inhalation did not have any effect on heart rate, catecholamine release, and platelet activation observed with cigarette smoking. Such results indicate that the aforementioned cardiovascular effects of smoking cannot be attributed to CO exposure from smoking but rather to its other components [228]. In inflammatory bowel diseases, on the other hand, the effect of cigarette smoking is double-edged wherein the development and clinical outcome of Crohn's disease are exacerbated while the risk for ulcerative colitis is reduced in smokers [126,229]. Both nicotine and CO, among the >5000 components of cigarette smoke, have been identified as probable major players in the observed effects [230]. CO's protective effect against ulcerative colitis has been attributed to its anti-inflammatory effects in the gut by published studies [231,232]. In the case of Parkinson's disease, there have been reports of an inverse and very likely causal relationship with smoking as well [141,233,234]. As a postulation to study, could the anti-inflammatory effects of CO be a contributing factor in attenuating molecular events that facilitate progression of Parkinson's disease? Another epidemiological connection established is between preeclampsia and smoking, as discussed in the section entitled "CO Production During Pregnancy."

These published data raise a very important and cautionary point in drawing direct correlations between observed beneficial or harmful effects with any particular chemical entities in tobacco smoke.

Because of the complexity in the composition of tobacco smoke, COHb levels might just be a surrogate indicator of the level of exposure to tobacco smoke, not necessarily the causative agent for the observed effects. Studies at the molecular level are needed for establishing mechanistic links. Another issue to consider is the comparison of e-cigarettes with regular cigarettes. Will the lack of combustion in e-cigarettes mean less CO production [235,236]? If so, what will be the effect of this change in CO exposure on smoking-induced airway inflammation [237] considering CO's anti-inflammatory functions?

It should be emphasized that the overall harmful effects of smoking have been firmly established. Even if CO proves to be a beneficial component under certain conditions and at certain concentrations, it should never be interpreted as to suggest beneficial effects of smoking. These are two different matters. The data on the harmful effects of smoking are overwhelmingly strong and are widely accepted by experts in the relevant fields and government agencies [238–241]. Table 1.1 (entry 1) lists some of the observed inverse relationships between smoking and pathological conditions as well as animal model experiments supporting CO being one possible molecular link in such relationships.

CO Production from the Induction of P450

There are compounds that are known to induce P450 and other hepatic heme-containing enzymes. In an experiment to understand the correlation of such induction with CO production, Coburn and coworkers examined nine human subjects over a period of 7 days on phenobarbital or diphenylhydantoin, both of which are known to induce hepatic enzymes. It was found that the rate of CO production increased an average of 174% and 214% from control levels with these two compounds, leading to the catabolism of hepatic heme to account for over 50% of the total CO production [242]. At that time, CO production was fully attributed to heme catabolism either by the HO enzyme or through chemical degradation promoted by H_2O_2 [243]. However, it was shown that CO formation arises in part due to cytochrome P450 and lipid peroxidation [244]. Contrary to prior reports, NADPH- and iron-dependent CO generation from P450 in microsomes cannot be attributed to HO-mediated catabolism of heme derived from microsomes since HO inhibitors did not inhibit CO formation. Furthermore, CO formation was not accompanied by heme destruction. Instead, P450 inhibitors led to a dose-dependent decrease in CO formation. In addition, CO formation was found to be concomitant with malonyl dialdehyde formation, one of the final products of lipid peroxidation. The iron dependence of CO formation is also tied to lipid peroxidation as lipid peroxidation is also iron dependent. Is it possible that CO from lipid peroxidation also plays a regulatory role? The quantity certainly suggests this possibility, especially considering the effect of local enrichment through lipid peroxidation to the level that could engage other hemoprotein targets. Although more studies are needed before one can establish molecular links, the effects of the locally elevated levels of CO should be carefully examined.

Other Sources

Circulating CO in the human body may also arise from sources independent of host heme catabolism as indicated by around 12.3% deficit in comparing CO production and the stoichiometric formation of nonvolatile components from heme degradation reactions. The most obvious source of external CO is through inhalation of CO of outdoor and/or indoor origin mostly from incomplete combustion. Aside from breathing in air, the food humans eat may also be a minor source of CO through chemical or enzymatic pathways (thoroughly discussed in Chapter 17). One often overlooked source of CO inside mammals would include the intestinal microbiome where CO-producing catabolism of a wide array of substrates (i.e., heme, lipids, and natural products) can occur [245–248]. The microbiome can potentially act as a sink through utility of CO as an alternative energy source, or a source of CO either thorough bacterial HO activity or through other bacterial enzymatic processes that produce CO [249]. As mentioned earlier, membrane lipids have also been identified as sources of CO through peroxidative degradation of unsaturated fatty acids [250]. Another aspect of increased CO availability involves photochemical dissociation of CO from COHb [251–254], which could be involved in the external regulation of the circadian rhythm [255,256]. Chapter 5 of this book focuses on CO and the circadian clock.

Conclusions

Overall, from the results discussed, there are several important points to contemplate in giving an overview of where things stand. First, CO production predominantly comes from heme degradation as part of red blood cell turnover. The stable nature of red blood cell turnover means that under normal physiological conditions, there is steady production of CO, which is consistent with CO's "maintenance roles" in normal physiology. Results from knockout mice and human data have clearly established increased pathologies associated with a lack of HO-1 [257,258]. Second, stress and trauma are known to lead to increased expression of HO-1 as a result of increased availability of heme that leads to increased CO production. Importantly, unless there is severe systemic stress, the amount of CO produced may have profound local effects, even if not reflected in changes in COHb levels. Such responses are consistent with the "stress-response" and cyto-/organ-protective roles of HO-1/CO. Third, the CO production rate seems to be steady in men and varies in women depending on the menstrual cycle and pregnancy. Such results suggest regulatory functions for CO in the context of menstrual cycle and pregnancy. It is possible that the immunoregulatory effects of CO offer benefits in pregnancy initiation (implantation) and/or maintenance as well as healthy menstrual cycles. Fourth, there is a significant percentage of CO that is not derived from heme degradation. Such findings may suggest the possibility for signaling roles beyond the HO–CO axis and include events such as lipid peroxidation and the gut microbiome. Fifth, the possibility of CO being the molecular link for the inverse relationship between smoking and certain pathologies such as ulcerative colitis, Parkinson's disease, and pre-eclampsia is an area that is very intriguing and requires more work. Cigarette smoke has a large number of components; understanding the roles of key ingredients, especially possible beneficial ingredients in their pure form, may have public health implications. Sixth, the possibility for CO to serve as a molecular link to explain the improved cancer chemotherapy outcome after physical exercise is very interesting and deserves some attention. Further, there are animal model data to substantiate such a possible link. From the discussion of these points, one can see many interesting mechanistic connections at the molecular level, which are very likely, but have not been fully explored. These many well-recognized roles of CO in many cell culture and animal model studies are described in detail in the various chapters of this book in Section IV. Moreover, the quantitative COHb levels in blood and CO concentrations in exhaled breath under different scenarios may have varying degrees of accuracy and precision depending on sample size, method and instrument used, and study population. They should only serve as reference points. It would be completely understandable and reasonable to expect some variations from different studies. It is more important to look at these numbers in the context of what they might mean, instead of the precise numerical values.

Acknowledgements

We gratefully acknowledge the partial financial support by the National Institutes of Health (R01DK119202) for our CO-related work, the Georgia Research Alliance for an Eminent Scholar endowment, a Center for Diagnostics and Therapeutics graduate fellowship to LKDLC, and internal resources at Georgia State University.

References

1. Sjöstrand, T. (1949). Endogenous formation of carbon monoxide in man. *Nature* 164: 580.
2. Engstedt, L. (1957). Endogenous formation of carbon monoxide in hemolytic disease; with special regard to quantitative comparisons to other hemolytic indices. *Acta Med. Scand. Suppl.* 332: 1–63.
3. Sjöstrand, T. (1949). Endogenous formation of carbon monoxide in man under normal and pathological conditions. *Scan. J. Clin. Lab. Investig.* 1: 201–214.
4. Gydell, K. (1960). Transient effect of nicotinic acid on bilirubin metabolism and formation of carbon monoxide. *Acta Med. Scand.* 167: 431–441.
5. Marks, G.S., Brien, J.F., Nakatsu, K., and McLaughlin, B.E. (1991). Does carbon monoxide have a physiological function? *Trends Pharmacol. Sci.* 12: 185–188.
6. Wu, L. and Wang, R. (2005). Carbon monoxide: endogenous production, physiological functions, and pharmacological applications. *Pharmacol. Rev.* 57: 585–630.
7. Motterlini, R. and Otterbein, L.E. (2010). The therapeutic potential of carbon monoxide. *Nat. Rev. Drug Discov.* 9: 728–743.
8. Ji, X., Damera, K., Zheng, Y., Yu, B., Otterbein, L.E., and Wang, B. (2016). Toward carbon monoxide-based therapeutics: Critical drug delivery and developability issues. *J. Pharm. Sci.* 105: 406–416.
9. Yang, X., Lu, W., Hopper, C.P., Ke, B., and Wang, B. (2021). Nature's marvels endowed in gaseous molecules I: Carbon monoxide and its physiological and therapeutic roles. *Acta Pharm. Sinica B* 12. https://doi.org/10.1016/j.apsb.2020.10.010.
10. Sjöstrand, T. (1951). The in vitro formation and disposal of carbon monoxide in blood. *Nature* 168: 729–730.

11. Sjöstrand, T. (1951). A preliminary report on the in vitro formation of carbon monoxide in blood. *Acta. Physiol. Scand* 22: 142–143.
12. Ludwig, G.D., Blakemore, W.S., and Drabkin, D.L. (1957). Production of carbon monoxide by hemin oxidation. *J. Clin. Invest.* 36: 912.
13. Coburn, R.F., Blakemore, W.S., and Forster, R.E. (1963). Endogenous carbon monoxide production in man. *J. Clin. Invest.* 42: 1172–1178.
14. Maines, M.D. (1988). Heme oxygenase: function, multiplicity, regulatory mechanisms, and clinical applications. *FASEB J* 2: 2557–2568.
15. Dennery, P.A. (2014). Signaling function of heme oxygenase proteins. *Antioxid. Redox Signal.* 20: 1743–1753.
16. Hayashi, S., Omata, Y., Sakamoto, H., Higashimoto, Y., Hara, T., Sagara, Y., and Noguchi, M. (2004). Characterization of rat heme oxygenase-3 gene. Implication of processed pseudogenes derived from heme oxygenase-2 gene. *Gene* 336: 241–250.
17. Lee, B.-S., Heo, J., Kim, Y.-M., Shim, S.M., Pae, H.-O., Kim, Y.-M., and Chung, H.-T. (2006). Carbon monoxide mediates heme oxygenase 1 induction via Nrf2 activation in hepatoma cells. *Biochem. Biophys. Res. Commun.* 343: 965–972.
18. Kim, H.P. and Choi, A.M. (2007). A new road to induce heme oxygenase-1 expression by carbon monoxide. *Circ. Res.* 101: 862–864.
19. Kim, K.M., Pae, H.O., Zheng, M., Park, R., Kim, Y.M., and Chung, H.T. (2007). Carbon monoxide induces heme oxygenase-1 via activation of protein kinase R-like endoplasmic reticulum kinase and inhibits endothelial cell apoptosis triggered by endoplasmic reticulum stress. *Circ. Res.* 101: 919–927.
20. Sugishima, M., Moffat, K., and Noguchi, M. (2012). Discrimination between CO and O_2 in heme oxygenase: comparison of static structures and dynamic conformation changes following CO photolysis. *Biochemistry* 51: 8554–8562.
21. Yoshida, T. and Migita, C.T. (2000). Mechanism of heme degradation by heme oxygenase. *J. Inorg. Biochem.* 82: 33–41.
22. Matsui, T., Iwasaki, M., Sugiyama, R., Unno, M., and Ikeda-Saito, M. (2010). Dioxygen activation for the self-degradation of heme: Reaction mechanism and regulation of heme oxygenase. *Inorg. Chem.* 49: 3602–3609.
23. Davydov, R.M., Yoshida, T., Ikeda-Saito, M., and Hoffman, B.M. (1999). Hydroperoxy-heme oxygenase generated by cryoreduction catalyzes the formation of α-meso-hydroxyheme as detected by EPR and ENDOR. *J. Am. Chem. Soc.* 121: 10656–10657.
24. Matsui, T., Unno, M., and Ikeda-Saito, M. (2010). Heme oxygenase reveals its strategy for catalyzing three successive oxygenation reactions. *Acc. Chem. Res.* 43: 240–247.
25. Migita, C.T., Matera, K.M., Ikeda-Saito, M., Olson, J.S., Fujii, H., Yoshimura, T., Zhou, H., and Yoshida, T. (1998). The Oxygen and Carbon Monoxide Reactions of Heme Oxygenase. *J. Biol. Chem.* 273: 945–949.
26. Davydov, R., Matsui, T., Fujii, H., Ikeda-Saito, M., and Hoffman, B.M. (2003). Kinetic isotope effects on the rate-limiting step of heme oxygenase catalysis indicate concerted proton transfer/heme hydroxylation. *J. Am. Chem. Soc.* 125: 16208–16209.
27. Kumar, D., De Visser, S.P., and Shaik, S. (2005). Theory favors a stepwise mechanism of porphyrin degradation by a ferric hydroperoxide model of the active species of heme oxygenase. *J. Am. Chem. Soc.* 127: 8204–8213.
28. Kamachi, T. and Yoshizawa, K. (2005). Water-Assisted Oxo Mechanism for Heme Metabolism. *J. Am. Chem. Soc.* 127: 10686–10692.
29. Evans, J.P., Niemevz, F., Buldain, G., and De Montellano, P.O. (2008). Isoporphyrin intermediate in heme oxygenase catalysis: oxidation of α-meso-phenylheme. *J. Biol. Chem.* 283: 19530–19539.
30. Hirotsu, S., Chu, G.C., Unno, M., Lee, D.-S., Yoshida, T., Park, S.-Y., Shiro, Y., and Ikeda-Saito, M. (2004). The crystal structures of the ferric and ferrous forms of the heme complex of HmuO, a heme oxygenase of *Corynebacterium diphtheriae*. *J. Biol. Chem.* 279: 11937–11947.
31. Sakamoto, H., Omata, Y., Palmer, G., and Noguchi, M. (1999). Ferric α-hydroxyheme bound to heme oxygenase can be converted to verdoheme by dioxygen in the absence of added reducing equivalents. *J. Biol. Chem.* 274: 18196–18200.
32. Matera, K.M., Takahashi, S., Fujii, H., Zhou, H., Ishikawa, K., Yoshimura, T., Rousseau, D.L., Yoshida, T., and Ikeda-Saito, M. (1996). Oxygen and one reducing equivalent are both required for the conversion of α-hydroxyhemin to verdoheme in heme oxygenase. *J. Biol. Chem.* 271: 6618–6624.
33. Zhang, X., Fujii, H., Matera, K.M., Migita, C.T., Sun, D., Sato, M., Ikeda-Saito, M., and Yoshida, T. (2003). Stereoselectivity of each of the three steps of the heme oxygenase reaction: Hemin to meso-hydroxyhemin, meso-hydroxyhemin to verdoheme, and verdoheme to biliverdin. *Biochemistry* 42: 7418–7426.
34. Liu, Y., Moënne-Loccoz, P., Loehr, T.M., and De Montellano, P.R.O. (1997). Heme oxygenase-1, intermediates in verdoheme formation and the requirement for reduction equivalents. *J. Biol. Chem.* 272: 6909–6917.
35. Yoshida, T., Noguchi, M., and Kikuchi, G. (1980). A new intermediate of heme degradation catalyzed by the heme oxygenase system1. *J. Biochem.* 88: 557–563.

36. Hori, R., Kashiba, M., Toma, T., Yachie, A., Goda, N., Makino, N., Soejima, A., Nagasawa, T., Nakabayashi, K., and Suematsu, M. (2002). Gene transfection of H25A mutant heme oxygenase-1 protects cells against hydroperoxide-induced cytotoxicity. *J. Biol. Chem.* 277: 10712–10718.
37. Lin, Q., Weis, S., Yang, G., Weng, Y.H., Helston, R., Rish, K., Smith, A., Bordner, J., Polte, T., Gaunitz, F., and Dennery, P.A. (2007). Heme oxygenase-1 protein localizes to the nucleus and activates transcription factors important in oxidative stress. *J. Biol. Chem.* 282: 20621–20633.
38. Lin, Q.S., Weis, S., Yang, G., Zhuang, T., Abate, A., and Dennery, P.A. (2008). Catalytic inactive heme oxygenase-1 protein regulates its own expression in oxidative stress. *Free Radic. Biol. Med.* 44: 847–855.
39. Coltman, C.A. and Dudley, G.M. (1969). The relationship between endogenous carbon monoxide production and total heme mass in normal and abnormal subjects. *Am. J. Med. Sci.* 258: 374–385.
40. Lynch, S.R. and Moede, A.L. (1972). Variation in the rate of endogenous carbon monoxide production in normal human beings. *J. Lab. Clin. Med.* 79: 85–95.
41. Coburn, R.F., Williams, W.J., and Kahn, S.B. (1966). Endogenous carbon monoxide production in patients with hemolytic anemia. *J. Clin. Invest.* 45: 460–468.
42. Berk, P.D., Rodkey, F.L., Blaschke, T.F., Collison, H.A., and Waggoner, J.G. (1974). Comparison of plasma bilirubin turnover and carbon monoxide production in man. *J. Lab. Clin. Med.* 83: 29–37.
43. Berk, P.D., Blaschke, T.F., Scharschmidt, B.F., Waggoner, J.G., and Berlin, N.I. (1976). A new approach to quantitation of the various sources of bilirubin in man. *J. Lab. Clin. Med.* 87: 767–780.
44. Delivoria-Papadopoulos, M., Coburn, R.F., and Forster, R.E. (1974). Cyclic variation of rate of carbon monoxide production in normal women. *J. Appl. Physiol.* 36: 49–51.
45. Granick, S. and Kappas, A. (1967). Steroid induction of porphyrin synthesis in liver cell culture. I. Structural basis and possible physiological role in the control of heme formation. *J. Biol. Chem.* 242: 4587–4593.
46. Acevedo, C.H. and Ahmed, A. (1998). Hemeoxygenase-1 inhibits human myometrial contractility via carbon monoxide and is upregulated by progesterone during pregnancy. *J. Clin. Invest.* 101: 949–995.
47. Barber, A., Robson, S.C., and Lyall, F. (1999). Hemoxygenase and nitric oxide synthase do not maintain human uterine quiescence during pregnancy. *Am. J. Pathol.* 155: 831–840.
48. Zenclussen, M.L., Casalis, P.A., Jensen, F., Woidacki, K., and Zenclussen, A.C. (2014). Hormonal fluctuations during the estrous cycle modulate heme oxygenase-1 expression in the uterus. *Front. Endocrinol. (Lausanne)* 5: 32–32.
49. Zenclussen, M.L., Linzke, N., Schumacher, A., Fest, S., Meyer, N., Casalis, P.A., and Zenclussen, A.C. (2015). Heme oxygenase-1 is critically involved in placentation, spiral artery remodeling, and blood pressure regulation during murine pregnancy. *Front. Pharmacol.* 5: 291–291.
50. Zenclussen, M.L., Casalis, P.A., El-Mousleh, T., Rebelo, S., Langwisch, S., Linzke, N., Volk, H.D., Fest, S., Soares, M.P., and Zenclussen, A.C. (2011). Haem oxygenase-1 dictates intrauterine fetal survival in mice via carbon monoxide. *J. Pathol.* 225: 293–304.
51. Griffith, O.W., Chavan, A.R., Protopapas, S., Maziarz, J., Romero, R., and Wagner, G.P. (2017). Embryo implantation evolved from an ancestral inflammatory attachment reaction. *Proc. Natl. Acad. Sci. U.S.A* 114: E6566–e6575.
52. Rengarajan, A., Mauro, A.K., and Boeldt, D.S. (2020). Maternal disease and gasotransmitters. *Nitric Oxide* 96: 1–12.
53. Guerra, D.D. and Hurt, K.J. (2019). Gasotransmitters in pregnancy: from conception to uterine involution. *Biol. Reprod.* 101: 4–25.
54. Von Burg, R. (1999). Carbon monoxide. *J. Appl. Toxicol.* 19: 379–386.
55. Motterlini, R., Gonzales, A., Foresti, R., Clark, J.E., Green, C.J., and Winslow, R.M. (1998). Heme oxygenase-1-derived carbon monoxide contributes to the suppression of acute hypertensive responses in vivo. *Circ. Res.* 83: 568–577.
56. Fällström, S.P. (1968). Endogenous formation of carbon monoxide in newborn infants. IV. On the relation between the blood carboxyhaemoglobin concentration and the pulmonary elimination of carbon monoxide. *Acta Paediatr. Scand.* 57: 321–329.
57. Wranne, L. (1967). Studies on erythro-kinetics in infancy. VII. Quantitative estimation of the haemoglobin catabolism by carbon monoxide technique in young infants. *Acta Paediatr. Scand.* 56: 381–390.
58. Maisels, M.J., Pathak, A., Nelson, N.M., Nathan, D.G., and Smith, C.A. (1971). Endogenous production of carbon monoxide in normal and erythroblastotic newborn infants. *J. Clin. Invest.* 50: 1–8.
59. Stevenson, D.K., Wong, R.J., Ostrander, C.R., Maric, I., Vreman, H.J., and Cohen, R.S. (2020). Increased Carbon Monoxide Washout Rates in Newborn Infants. *Neonatology* 117: 118–122.
60. Smith, D.W., Hopper, A.O., Shahin, S.M., Cohen, R.S., Ostrander, C.R., Ariagno, R.L., and Stevenson, D.K. (1984). Neonatal bilirubin production estimated from "end-tidal" carbon monoxide concentration. *J. Pediatr. Gastroenterol. Nutr.* 3: 77–80.

61. Ostrander, C.R., Cohen, R.S., Hopper, A.O., Cowan, B.E., Stevens, G.B., and Stevenson, D.K. (1982). Paired determinations of blood carboxyhemoglobin concentration and carbon monoxide excretion rate in term and preterm infants. *J. Lab. Clin. Med.* 100: 745–755.
62. Cohen, R.S., Ostrander, C.R., Cowan, B.E., Stevens, G.B., Hopper, A.O., and Stevenson, D.K. (1982). Pulmonary excretion rates of carbon monoxide using a modified technique: differences between premature and full-term infants. *Biol. Neonate* 41: 289–293.
63. Huang, M.J., Kua, K.E., Teng, H.C., Tang, K.S., Weng, H.W., and Huang, C.S. (2004). Risk factors for severe hyperbilirubinemia in neonates. *Pediatr. Res.* 56: 682–689.
64. Kaplan, M., Vreman, H.J., Hammerman, C., and Stevenson, D.K. (1999). Neonatal bilirubin production, reflected by carboxyhaemoglobin concentrations, in Down's syndrome. *Arch. Dis. Child Fetal Neonatal Ed.* 81: F56–60.
65. Guney Varal, I. and Dogan, P. (2020). Serial carboxyhemoglobin levels and its relationship with late onset sepsis in preterm infants: An observational cohort study. *Fetal Pediatr. Pathol.* 39: 145–155.
66. Sankaran, V.G., Menne, T.F., Xu, J., Akie, T.E., Lettre, G., Van Handel, B., Mikkola, H.K.A., Hirschhorn, J.N., Cantor, A.B., and Orkin, S.H. (2008). Human fetal hemoglobin expression is regulated by the developmental stage-specific repressor BCL11A. *Science* 322: 1839–1842.
67. Engel, R.R., Rodkey, F.L., O'Neal, J.D., and Collison, H.A. (1969). Relative affinity of human fetal hemoglobin for carbon monoxide and oxygen. *Blood* 33: 37–45.
68. Longo, L.D. (1977). The biological effects of carbon monoxide on the pregnant woman, fetus, and newborn infant. *Am. J. Obstet. Gynecol.* 129: 69–103.
69. Longo, L.D. (1976). Carbon monoxide: effects on oxygenation of the fetus in utero. *Science* 194: 523–525.
70. Zavorsky, G.S., Blood, A.B., Power, G.G., Longo, L.D., Artal, R., and Vlastos, E.J. (2010). CO and NO pulmonary diffusing capacity during pregnancy: Safety and diagnostic potential. *Respir. Physiol. Neurobiol.* 170: 215–225.
71. Gemzell, C.A., Robbe, H., and Strøm, G. (1958). On the equilibration of carbon monoxide between human maternal and fetal circulation in vivo. *Scand. J. Clin. Lab. Invest.* 10: 372–378.
72. Tift, M.S., Ponganis, P.J., and Crocker, D.E. (2014). Elevated carboxyhemoglobin in a marine mammal, the northern elephant seal. *J. Exp. Biol.* 217: 1752–1757.
73. Herrera, E.A., Reyes, R.V., Giussani, D.A., Riquelme, R.A., Sanhueza, E.M., Ebensperger, G., Casanello, P., Méndez, N., Ebensperger, R., Sepúlveda-Kattan, E., Pulgar, V.M., Cabello, G., Blanco, C.E., Hanson, M.A., Parer, J.T., and Llanos, A.J. (2007). Carbon monoxide: a novel pulmonary artery vasodilator in neonatal llamas of the Andean altiplano. *Cardiovasc. Res.* 77: 197–201.
74. Zhao, H., Wong, R.J., Doyle, T.C., Nayak, N., Vreman, H.J., Contag, C.H., and Stevenson, D.K. (2008). Regulation of maternal and fetal hemodynamics by heme oxygenase in mice1. *Biol. Reprod.* 78: 744–751.
75. Dickson, M.A., Peterson, N., McRae, K.E., Pudwell, J., Tayade, C., and Smith, G.N. (2020). Carbon monoxide increases utero-placental angiogenesis without impacting pregnancy specific adaptations in mice. *Reprod. Biol. Endocrinol.* 18: 49.
76. Hill, E.P., Hill, J.R., Power, G.G., and Longo, L.D. (1977). Carbon monoxide exchanges between the human fetus and mother: a mathematical model. *Am. J. Physiol.* 232: H311–23.
77. Longo, L.D. (1970). Carbon monoxide in the pregnant mother and fetus and its exchange across the placenta. *Ann. N.Y. Acad. Sci.* 174: 312–341.
78. Owens, E.O. (2010). Endogenous carbon monoxide production in disease. *Clin. Biochem.* 43: 1183–1188.
79. Hendler, I., Baum, M., Kreiser, D., Schiff, E., Druzin, M., Stevenson, D.K., and Seidman, D.S. (2004). End-tidal breath carbon monoxide measurements are lower in pregnant women with uterine contractions. *J. Perinatol.* 24: 275–278.
80. Ahmed, A., Rahman, M., Zhang, X., Acevedo, C.H., Nijjar, S., Rushton, I., Bussolati, B., and St John, J. (2000). Induction of placental heme oxygenase-1 is protective against TNFalpha-induced cytotoxicity and promotes vessel relaxation. *Mol. Med.* 6: 391–409.
81. Sandrim, V.C., Caldeira-Dias, M., Bettiol, H., Barbieri, M.A., Cardoso, V.C., and Cavalli, R.C. (2018). Circulating heme oxygenase-1: Not a predictor of preeclampsia but highly expressed in pregnant women who subsequently develop severe preeclampsia. *Oxid. Med. Cell Longev.* 2018: 6035868.
82. Kambam, J.R., Entman, S., Mouton, S., and Smith, B.E. (1988). Effect of preeclampsia on carboxyhemoglobin levels: a mechanism for a decrease in P50. *Anesthesiology* 68: 433–434.
83. Kambam, J.R., Handte, R.E., Brown, W.U., and Smith, B.E. (1986). Effect of normal and preeclamptic pregnancies on the oxyhemoglobin dissociation curve. *Anesthesiology* 65: 426–427.
84. Roughton, F.J.W. and Darling, R.C. (1944). The effect of carbon monoxide on the oxyhemoglobin dissociation curve. *Am. J. Physiol.* 142: 17–20.

85. Ahmed, A., Rezai, H., and Broadway-Stringer, S. (2017). Evidence-based revised view of the pathophysiology of preeclampsia. *Adv. Exp. Med. Biol.* 956: 355–374.
86. Venditti, C.C., Casselman, R., Young, I., Karumanchi, S.A., and Smith, G.N. (2014). Carbon monoxide prevents hypertension and proteinuria in an adenovirus sFlt-1 preeclampsia-like mouse model. *PLoS One* 9: e106502.
87. Bainbridge, S.A., Belkacemi, L., Dickinson, M., Graham, C.H., and Smith, G.N. (2006). Carbon monoxide inhibits hypoxia/reoxygenation-induced apoptosis and secondary necrosis in syncytiotrophoblast. *Am. J. Pathol.* 169: 774–783.
88. Baum, M., Schiff, E., Kreiser, D., Dennery, P.A., Stevenson, D.K., Rosenthal, T., and Seidman, D.S. (2000). End-tidal carbon monoxide measurements in women with pregnancy-induced hypertension and preeclampsia. *Am. J. Obstet. Gynecol.* 183: 900–903.
89. Zhai, D., Guo, Y., Smith, G., Krewski, D., Walker, M., and Wen, S.W. (2012). Maternal exposure to moderate ambient carbon monoxide is associated with decreased risk of preeclampsia. *Am J Obstet Gynecol.* 207: 57.e1–9.
90. Rudra, C.B., Williams, M.A., Sheppard, L., Koenig, J.Q., and Schiff, M.A. (2011). Ambient carbon monoxide and fine particulate matter in relation to preeclampsia and preterm delivery in western Washington State. *Environ. Health Perspect.* 119: 886–892.
91. Nobles, C.J., Williams, A., Ouidir, M., Sherman, S., and Mendola, P. (2019). Differential effect of ambient air pollution exposure on risk of gestational hypertension and preeclampsia. *Hypertension* 74: 384–390.
92. Bakrania, B.A., Spradley, F.T., Satchell, S.C., Stec, D.E., Rimoldi, J.M., Gadepalli, R.S.V., and Granger, J.P. (2018). Heme oxygenase-1 is a potent inhibitor of placental ischemia-mediated endothelin-1 production in cultured human glomerular endothelial cells. *Am. J. Physiol. Regul. Integr. Comp. Physiol.* 314: R427–r432.
93. Islam, H., Bonafiglia, J.T., Turnbull, P.C., Simpson, C.A., Perry, C.G.R., and Gurd, B.J. (2020). The impact of acute and chronic exercise on Nrf2 expression in relation to markers of mitochondrial biogenesis in human skeletal muscle. *Eur. J. Appl. Physiol.* 120: 149–160.
94. Wyse, C., Cathcart, A., Sutherland, R., Ward, S., McMillan, L., Gibson, G., Padgett, M., and Skeldon, K. (2005). Effect of maximal dynamic exercise on exhaled ethane and carbon monoxide levels in human, equine, and canine athletes. *Comp. Biochem. Physiol. A Mol. Integr. Physiol.* 141: 239–246.
95. Yasuda, Y., Ito, T., Miyamura, M., and Niwayama, M. (2011). Effect of ramp bicycle exercise on exhaled carbon monoxide in humans. *J. Physiol. Sci.* 61: 279–286.
96. Ghio, A.J., Case, M.W., and Soukup, J.M. (2018). Heme oxygenase activity increases after exercise in healthy volunteers. *Free Radic. Res.* 52: 267–272.
97. Pyne, D.B. (1994). Exercise-induced muscle damage and inflammation: a review. *Aust. J. Sci. Med. Sport* 26: 49–58.
98. Aoi, W., Naito, Y., Takanami, Y., Kawai, Y., Sakuma, K., Ichikawa, H., Yoshida, N., and Yoshikawa, T. (2004). Oxidative stress and delayed-onset muscle damage after exercise. *Free Radic. Biol. Med.* 37: 480–487.
99. Ashcraft, K.A., Warner, A.B., Jones, L.W., and Dewhirst, M.W. (2019). Exercise as adjunct therapy in cancer. *Semin. Radiat. Oncol.* 29: 16–24. references cited therein.
100. Ashcraft, K.A., Peace, R.M., Betof, A.S., Dewhirst, M.W., and Jones, L.W. (2016). Efficacy and mechanisms of aerobic exercise on cancer initiation, progression, and metastasis: A critical systematic review of in vivo preclinical data. *Cancer Res.* 76: 4032–4050.
101. Idorn, M. and Thor Straten, P. (2017). Exercise and cancer: From "healthy" to "therapeutic"? *Cancer Immunol. Immunother.* 66: 667–671. and references cited therein.
102. Zelenka, J., Koncošová, M., and Ruml, T. (2018). Targeting of stress response pathways in the prevention and treatment of cancer. *Biotechnol. Adv.* 36: 583–602. and references cited therein.
103. Miyagi, M.Y., Seelaender, M., Castoldi, A., De Almeida, D.C., Bacurau, A.V., Andrade-Oliveira, V., Enjiu, L.M., Pisciottano, M., Hayashida, C.Y., Hiyane, M.I., Brum, P.C., Camara, N.O., and Amano, M.T. (2014). Long-term aerobic exercise protects against cisplatin-induced nephrotoxicity by modulating the expression of IL-6 and HO-1. *PLoS One* 9: e108543.
104. Nemeth, Z., Csizmadia, E., Vikstrom, L., Li, M., Bisht, K., Feizi, A., Otterbein, S., Zuckerbraun, B., Costa, D.B., Pandolfi, P.P., Fillinger, J., Döme, B., Otterbein, L.E., and Wegiel, B. (2016). Alterations of tumor microenvironment by carbon monoxide impedes lung cancer growth. *Oncotarget* 7: 23919–23932.
105. Wegiel, B., Gallo, D., Csizmadia, E., Harris, C., Belcher, J., Vercellotti, G.M., Penacho, N., Seth, P., Sukhatme, V., Ahmed, A., Pandolfi, P.P., Helczynski, L., Bjartell, A., Persson, J.L., and Otterbein, L.E. (2013). Carbon monoxide expedites metabolic exhaustion to inhibit tumor growth. *Cancer Res.* 73: 7009–7021.

106. Piantadosi, C.A., Carraway, M.S., Babiker, A., and Suliman, H.B. (2008). Heme oxygenase-1 regulates cardiac mitochondrial biogenesis via Nrf2-mediated transcriptional control of nuclear respiratory factor-1. *Circ. Res.* 103: 1232–1240.
107. Suliman, H.B., Carraway, M.S., Ali, A.S., Reynolds, C.M., Welty-Wolf, K.E., and Piantadosi, C.A. (2007). The CO/HO system reverses inhibition of mitochondrial biogenesis and prevents murine doxorubicin cardiomyopathy. *J. Clin. Invest.* 117: 3730–3741.
108. Yasuda, H., Yamaya, M., Nakayama, K., Ebihara, S., Sasaki, T., Okinaga, S., Inoue, D., Asada, M., Nemoto, M., and Sasaki, H. (2005). Increased arterial carboxyhemoglobin concentrations in chronic obstructive pulmonary disease. *Am. J. Respir. Crit. Care Med.* 171: 1246–1251.
109. Yasuda, H., Yamaya, M., Yanai, M., Ohrui, T., and Sasaki, H. (2002). Increased blood carboxyhaemoglobin concentrations in inflammatory pulmonary diseases. *Thorax* 57: 779–783.
110. Yasuda, H., Sasaki, T., Yamaya, M., Ebihara, S., Maruyama, M., Kanda, A., and Sasaki, H. (2004). Increased arteriovenous carboxyhemoglobin differences in patients with inflammatory pulmonary diseases. *Chest* 125: 2160–2168.
111. Mumby, S., Upton, R.L., Chen, Y., Stanford, S.J., Quinlan, G.J., Nicholson, A.G., Gutteridge, J.M.C., Lamb, N.J., and Evans, T.W. (2004). Lung heme oxygenase-1 is elevated in acute respiratory distress syndrome. *Crit. Care Med.* 32: 1130–1135.
112. Corbacioglu, S.K., Kilicaslan, I., Bildik, F., Guleryuz, A., Bekgoz, B., Ozel, A., Keles, A., and Demircan, A. (2013). Endogenous carboxyhemoglobin concentrations in the assessment of severity in patients with community-acquired pneumonia. *Am. J. Emerg. Med.* 31: 520–523.
113. Scholkmann, F., Restin, T., Ferrari, M., and Quaresima, V. (2021). The Role of Methemoglobin and Carboxyhemoglobin in COVID-19: A Review. *J. Clin. Med.* 10: 1–15.
114. Kwong, K.K. and Chan, S.T. (2020). The role of carbon monoxide and heme oxygenase-1 in COVID-19. *Toxicol. Rep.* 7: 1170–1171.
115. Paccaud, P., Castanares-Zapatero, D., Gerard, L., Montiel, V., Wittebole, X., Collienne, C., Laterre, P.-F., and Hantson, P. (2020). Arterial Carboxyhemoglobin Levels in Covid-19 Critically Ill Patients. Research Square.
116. Wilbur, S., Williams, M., Williams, R., Scinicariello, F., Klotzbach, J.M., Diamond, G.L., and Citra, M. (2012). Agency for Toxic Substances and Disease Registry (ATSDR) toxicological profiles. Toxicological Profile for Carbon Monoxide. Atlanta (GA): Agency for Toxic Substances and Disease Registry (US).
117. Paredi, P., Shah, P.L., Montuschi, P., Sullivan, P., Hodson, M.E., Kharitonov, S.A., and Barnes, P.J. (1999). Increased carbon monoxide in exhaled air of patients with cystic fibrosis. *Thorax* 54: 917–920.
118. Horváth, I., Donnelly, L.E., Kiss, A., Paredi, P., Kharitonov, S.A., and Barnes, P.J. (1998). Raised levels of exhaled carbon monoxide are associated with an increased expression of heme oxygenase-1 in airway macrophages in asthma: a new marker of oxidative stress. *Thorax* 53: 668–672.
119. Kobayashi, M., Miyazawa, N., Takeno, M., Murakami, S., Kirino, Y., Okouchi, A., Kaneko, T., and Ishigatsubo, Y. (2008). Circulating carbon monoxide level is elevated after sleep in patients with obstructive sleep apnea. *Chest* 134: 904–910.
120. Ryter, S.W. (2020). Chapter 6 – Exhaled carbon monoxide. In: Breathborne Biomarkers and the Human Volatilome, 2e (ed. J. Beauchamp, C. Davis and J. Pleil). Boston: Elsevier.
121. Zetterquist, W., Marteus, H., Johannesson, M., Nordval, S.L., Ihre, E., Lundberg, J.O., and Alving, K. (2002). Exhaled carbon monoxide is not elevated in patients with asthma or cystic fibrosis. *Eur. Respir. J.* 20: 92–99.
122. McArdle, A.J., Webbe, J., Sim, K., Parrish, G., Hoggart, C., Wang, Y., Kroll, J.S., Godambe, S., and Cunnington, A.J. (2016). Determinants of carboxyhemoglobin levels and relationship with sepsis in a retrospective cohort of preterm neonates. *PLoS One* 11: e0161784.
123. Baglole, C.J., Sime, P.J., and Phipps, R.P. (2008). Cigarette smoke-induced expression of heme oxygenase-1 in human lung fibroblasts is regulated by intracellular glutathione. *Am. J. Physiol. Lung Cell Mol. Physiol.* 295: L624–36.
124. Hart, C.L., Smith, G.D., Hole, D.J., and Hawthorne, V.M. (2006). Carboxyhaemoglobin concentration, smoking habit, and mortality in 25 years in the Renfrew/Paisley prospective cohort study. *Heart* 92: 321–324.
125. Wald, N., Idle, M., and Bailey, A. (1978). Carboxyhaemoglobin levels and inhaling habits in cigarette smokers. *Thorax* 33: 201–206.
126. Berkowitz, L., Schultz, B.M., Salazar, G.A., Pardo-Roa, C., Sebastián, V.P., Álvarez-Lobos, M.M., and Bueno, S.M. (2018). Impact of cigarette smoking on the gastrointestinal tract inflammation: Opposing effects in Crohn's disease and ulcerative colitis. *Front. Immunol.* 9: 74–74.
127. Takagi, T., Naito, Y., Mizushima, K., Akagiri, S., Suzuki, T., Hirata, I., Omatsu, T., Handa, O., Kokura, S., Ichikawa, H., and Yoshikawa, T. (2010). Inhalation of carbon monoxide ameliorates TNBS-induced colitis in mice through the inhibition of TNF-α expression. *Dig. Dis. Sci.* 55: 2797–2804.

128. Sheikh, S.Z., Hegazi, R.A., Kobayashi, T., Onyiah, J.C., Russo, S.M., Matsuoka, K., Sepulveda, A.R., Li, F., Otterbein, L.E., and Plevy, S.E. (2011). An anti-inflammatory role for carbon monoxide and heme oxygenase-1 in chronic Th2-mediated murine colitis. *J. Immunol.* 186: 5506–5513.

129. Ji, X., Zhou, C., Ji, K., Aghoghovbia, R.E., Pan, Z., Chittavong, V., Ke, B., and Wang, B. (2016). Click and release: A chemical strategy toward developing gasotransmitter prodrugs by using an intramolecular Diels-Alder reaction. *Angew. Chem. Int. Ed. Engl.* 55: 15846–15851.

130. Takagi, T., Naito, Y., Tanaka, M., Mizushima, K., Ushiroda, C., Toyokawa, Y., Uchiyama, K., Hamaguchi, M., Handa, O., and Itoh, Y. (2018). Carbon monoxide ameliorates murine T-cell-dependent colitis through the inhibition of Th17 differentiation. *Free Radic. Res.* 52: 1328–1335.

131. Takagi, T., Naito, Y., Uchiyama, K., Suzuki, T., Hirata, I., Mizushima, K., Tsuboi, H., Hayashi, N., Handa, O., Ishikawa, T., Yagi, N., Kokura, S., Ichikawa, H., and Yoshikawa, T. (2011). Carbon monoxide liberated from carbon monoxide-releasing molecule exerts an anti-inflammatory effect on dextran sulfate sodium-induced colitis in mice. *Dig. Dis. Sci.* 56: 1663–1671.

132. Yin, H., Fang, J., Liao, L., Nakamura, H., and Maeda, H. (2014). Styrene-maleic acid copolymer-encapsulated CORM2, a water-soluble carbon monoxide (CO) donor with a constant CO-releasing property, exhibits therapeutic potential for inflammatory bowel disease. *J. Control. Release* 187: 14–21.

133. Fukuda, W., Takagi, T., Katada, K., Mizushima, K., Okayama, T., Yoshida, N., Kamada, K., Uchiyama, K., Ishikawa, T., Handa, O., Konishi, H., Yagi, N., Ichikawa, H., Yoshikawa, T., Cepinskas, G., Naito, Y., and Itoh, Y. (2014). Anti-inflammatory effects of carbon monoxide-releasing molecule on trinitrobenzene sulfonic acid-induced colitis in mice. *Dig. Dis. Sci.* 59: 1142–1151.

134. Zhai, D., Guo, Y., Smith, G., Krewski, D., Walker, M., and Wen, S.W. (2012). Maternal exposure to moderate ambient carbon monoxide is associated with decreased risk of preeclampsia. *Am. J. Obstet. Gynecol.* 207: 57.e1–57.e9.

135. Alpoim, P.N., Godoi, L.C., Pinheiro, M.B., Freitas, L.G., Carvalho, M.D.G., and Dusse, L.M. (2016). The unexpected beneficial role of smoking in preeclampsia. *Clin. Chim. Acta* 459: 105–108.

136. Luque-Fernandez, M.A., Zoega, H., Valdimarsdottir, U., and Williams, M.A. (2016). Deconstructing the smoking-preeclampsia paradox through a counterfactual framework. *Eur. J. Epidemiol.* 31: 613–623.

137. England, L. and Zhang, J. (2007). Smoking and risk of preeclampsia: a systematic review. *Front. Biosci.* 12: 2471–2483.

138. George, E.M., Cockrell, K., Arany, M., Stec, D.E., Rimoldi, J.M., Gadepalli, R.S.V., and Granger, J.P. (2017). Carbon monoxide releasing molecules blunt placental ischemia-induced hypertension. *Am. J. Hypertens.* 30: 931–937.

139. George, E.M., Arany, M., Cockrell, K., Storm, M.V., Stec, D.E., and Granger, J.P. (2011). Induction of heme oxygenase-1 attenuates sFlt-1-induced hypertension in pregnant rats. *Am. J. Physiol. Regul. Integr. Comp. Physiol.* 301: R1495–500.

140. McRae, K.E., Pudwell, J., Peterson, N., and Smith, G.N. (2019). Inhaled carbon monoxide increases vasodilation in the microvascular circulation. *Microvasc. Res.* 123: 92–98.

141. Allam, M.F., Campbell, M.J., Hofman, A., Del Castillo, A.S., and Fernández-Crehuet Navajas, R. (2004). Smoking and Parkinson's disease: systematic review of prospective studies. *Mov. Disord.* 19: 614–621. and references cited therein.

142. Mazzola, S., Forni, M., Albertini, M., Bacci, M.L., Zannoni, A., Gentilini, F., Lavitrano, M., Bach, F.H., Otterbein, L.E., and Clement, M.G. (2005). Carbon monoxide pretreatment prevents respiratory derangement and ameliorates hyperacute endotoxic shock in pigs. *FASEB J.* 19: 2045–2047.

143. Sarady, J.K., Zuckerbraun, B.S., Bilban, M., Wagner, O., Usheva, A., Liu, F., Ifedigbo, E., Zamora, R., Choi, A.M., and Otterbein, L.E. (2004). Carbon monoxide protection against endotoxic shock involves reciprocal effects on iNOS in the lung and liver. *FASEB J.* 18: 854–856.

144. Ji, X., Pan, Z., Li, C., Kang, T., De La Cruz, L.K.C., Yang, L., Yuan, Z., Ke, B., and Wang, B. (2019). Esterase-sensitive and pH-controlled carbon monoxide prodrugs for treating systemic inflammation. *J. Med. Chem.* 62: 3163–3168.

145. Cepinskas, G., Katada, K., Bihari, A., and Potter, R.F. (2008). Carbon monoxide liberated from carbon monoxide-releasing molecule CORM-2 attenuates inflammation in the liver of septic mice. *Am. J. Physiol. Gastrointest. Liver Physiol.* 294: G184–91.

146. Morimatsu, H., Takahashi, T., Matsusaki, T., Hayashi, M., Matsumi, J., Shimizu, H., Matsumi, M., and Morita, K. (2010). An increase in exhaled CO concentration in systemic inflammation/sepsis. *J. Breath Res.* 4: 047103.

147. Moncure, M., Brathwaite, C.E., Samaha, E., Marburger, R., and Ross, S.E. (1999). Carboxyhemoglobin elevation in trauma victims. *J. Trauma* 46: 424–427.

148. Hoetzel, A., Dolinay, T., Schmidt, R., Choi, A.M., and Ryter, S.W. (2007). Carbon monoxide in sepsis. *Antioxid. Redox Signal.* 9: 2013–2026.
149. Qin, W., Zhang, J., Lv, W., Wang, X., and Sun, B. (2013). Effect of carbon monoxide-releasing molecules II-liberated CO on suppressing inflammatory response in sepsis by interfering with nuclear factor kappa B activation. *PLoS One* 8: e75840.
150. Kawanishi, S., Takahashi, T., Morimatsu, H., Shimizu, H., Omori, E., Sato, K., Matsumi, M., Maeda, S., Nakao, A., and Morita, K. (2013). Inhalation of carbon monoxide following resuscitation ameliorates hemorrhagic shock-induced lung injury. *Mol. Med. Rep.* 7: 3–10.
151. Kanagawa, F., Takahashi, T., Inoue, K., Shimizu, H., Omori, E., Morimatsu, H., Maeda, S., Katayama, H., Nakao, A., and Morita, K. (2010). Protective effect of carbon monoxide inhalation on lung injury after hemorrhagic shock/resuscitation in rats. *J. Trauma* 69: 185–194.
152. Kumada, Y., Takahashi, T., Shimizu, H., Nakamura, R., Omori, E., Inoue, K., and Morimatsu, H. (2019). Therapeutic effect of carbon monoxide-releasing molecule-3 on acute lung injury after hemorrhagic shock and resuscitation. *Exp. Ther. Med.* 17: 3429–3440.
153. Guerci, P., Ergin, B., Kapucu, A., Hilty, M.P., Jubin, R., Bakker, J., and Ince, C. (2019). Effect of polyethylene-glycolated carboxyhemoglobin on renal microcirculation in a rat model of hemorrhagic shock. *Anesthesiology* 131: 1110–1124.
154. Mascaró, M., Alonso, E.N., Alonso, E.G., Lacunza, E., Curino, A.C., and Facchinetti, M.M. (2021). Nuclear localization of heme oxygenase-1 in pathophysiological conditions: Does it explain the dual role in cancer? *Antioxidants (Basel)* 10: 87.
155. Chau, L.Y. (2015). Heme oxygenase-1: emerging target of cancer therapy. *J. Biomed. Sci.* 22: 22.
156. Salerno, L., Romeo, G., Modica, M.N., Amata, E., Sorrenti, V., Barbagallo, I., and Pittalà, V. (2017). Heme oxygenase-1: A new druggable target in the management of chronic and acute myeloid leukemia. *Eur. J. Med. Chem.* 142: 163–178.
157. Sorrenti, V., Pittalà, V., Romeo, G., Amata, E., Dichiara, M., Marrazzo, A., Turnaturi, R., Prezzavento, O., Barbagallo, I., Vanella, L., Rescifina, A., Floresta, G., Tibullo, D., Di Raimondo, F., Intagliata, S., and Salerno, L. (2018). Targeting heme oxygenase-1 with hybrid compounds to overcome Imatinib resistance in chronic myeloid leukemia cell lines. *Eur. J. Med. Chem.* 158: 937–950.
158. Tibullo, D., Barbagallo, I., Giallongo, C., La Cava, P., Parrinello, N., Vanella, L., Stagno, F., Palumbo, G.A., Li Volti, G., and Di Raimondo, F. (2013). Nuclear translocation of heme oxygenase-1 confers resistance to imatinib in chronic myeloid leukemia cells. *Curr. Pharm. Des.* 19: 2765–2770.
159. Loboda, A., Jozkowicz, A., and Dulak, J. (2015). HO-1/CO system in tumor growth, angiogenesis and metabolism – Targeting HO-1 as an anti-tumor therapy. *Vascul. Pharmacol.* 74: 11–22.
160. Was, H., Cichon, T., Smolarczyk, R., Rudnicka, D., Stopa, M., Chevalier, C., Leger, J.J., Lackowska, B., Grochot, A., Bojkowska, K., Ratajska, A., Kieda, C., Szala, S., Dulak, J., and Jozkowicz, A. (2006). Overexpression of heme oxygenase-1 in murine melanoma: increased proliferation and viability of tumor cells, decreased survival of mice. *Am. J. Pathol.* 169: 2181–2198.
161. Was, H., Dulak, J., and Jozkowicz, A. (2010). Heme oxygenase-1 in tumor biology and therapy. *Curr. Drug Targets* 11: 1551–1570.
162. Yin, H., Fang, J., Liao, L., Maeda, H., and Su, Q. (2014). Upregulation of heme oxygenase-1 in colorectal cancer patients with increased circulation carbon monoxide levels, potentially affects chemotherapeutic sensitivity. *BMC Cancer* 14: 436.
163. Andrés, N.C., Fermento, M.E., Gandini, N.A., Romero, A.L., Ferro, A., Donna, L.G., Curino, A.C., and Facchinetti, M.M. (2014). Heme oxygenase-1 has antitumoral effects in colorectal cancer: involvement of p53. *Exp. Mol. Pathol.* 97: 321–331.
164. Lv, C., Su, Q., Fang, J., and Yin, H. (2019). Styrene-maleic acid copolymer-encapsulated carbon monoxide releasing molecule-2 (SMA/CORM-2) suppresses proliferation, migration and invasion of colorectal cancer cells in vitro and in vivo. *Biochem. Biophys. Res. Commun.* 520: 320–326.
165. Vítek, L., Gbelcová, H., Muchová, L., Váňová, K., Zelenka, J., Koníčková, R., Suk, J., Zadinova, M., Knejzlík, Z., Ahmad, S., Fujisawa, T., Ahmed, A., and Ruml, T. (2014). Antiproliferative effects of carbon monoxide on pancreatic cancer. *Dig. Liver Dis.* 46: 369–375.
166. Li, Y., Dang, J., Liang, Q., and Yin, L. (2019). Carbon monoxide (CO)-Strengthened cooperative bioreductive anti-tumor therapy via mitochondrial exhaustion and hypoxia induction. *Biomaterials* 209: 138–151.
167. Li, Y., Dang, J., Liang, Q., and Yin, L. (2019). Thermal-responsive carbon monoxide (CO) delivery expedites metabolic exhaustion of cancer cells toward reversal of chemotherapy resistance. *ACS Cent. Sci.* 5: 1044–1058.
168. Yasuda, H., Nakayama, K., Ebihara, S., Asada, M., Sasaki, T., Suzuki, T., Inoue, D., Yoshida, M., Yamanda, S., and Yamaya, M. (2006). Arterial

carboxyhemoglobin concentrations as a prognostic predictor in elderly patients with advanced non-small-cell lung cancer. *J. Am. Geriatr. Soc.* 54: 712–713.
169. Yasuda, H., Yamaya, M., Nakayama, K., Ebihara, S., Asada, M., Sasaki, T., Inoue, D., Yoshida, M., Kubo, H., and Sasaki, H. (2006). Arterial carboxyhemoglobin concentrations as a predictor of chemosensitivity in elderly patients with advanced lung cancer. *J. Am. Geriatr. Soc.* 54: 373–375.
170. Yasuda, H., Yamaya, M., Ebihara, S., Sasaki, T., Inoue, D., Kubo, H., Sasaki, H., and Suzuki, S. (2004). Arterial carboxyhemoglobin concentrations in elderly patients with operable non-small cell lung cancer. *J. Am. Geriatr. Soc.* 52: 1592–1593.
171. Shao, L., Gu, Y.Y., Jiang, C.H., Liu, C.Y., Lv, L.P., Liu, J.N., and Zou, Y. (2018). Carbon monoxide releasing molecule-2 suppresses proliferation, migration, invasion, and promotes apoptosis in non-small cell lung cancer Calu-3 cells. *Eur. Rev. Med. Pharmacol. Sci.* 22: 1948–1957.
172. Wald, N.J., Idle, M., Boreham, J., and Bailey, A. (1983). Inhaling and lung cancer: an anomaly explained. *Br. Med. J. (Clin. Res. Ed.)* 287: 1273–1275.
173. Soni, H., Pandya, G., Patel, P., Acharya, A., Jain, M., and Mehta, A.A. (2011). Beneficial effects of carbon monoxide-releasing molecule-2 (CORM-2) on acute doxorubicin cardiotoxicity in mice: role of oxidative stress and apoptosis. *Toxicol. Appl. Pharmacol.* 253: 70–80.
174. Yoon, Y.E., Lee, K.S., Lee, Y.J., Lee, H.H., and Han, W.K. (2017). Renoprotective effects of carbon monoxide-releasing molecule 3 in ischemia-Reperfusion injury and cisplatin-induced toxicity. *Transplant. Proc.* 49: 1175–1182.
175. Yang, X., De Caestecker, M., Otterbein, L.E., and Wang, B. (2020). Carbon monoxide: An emerging therapy for acute kidney injury. *Med. Res. Rev.* 40: 1147–1177.
176. Tayem, Y., Johnson, T.R., Mann, B.E., Green, C.J., and Motterlini, R. (2006). Protection against cisplatin-induced nephrotoxicity by a carbon monoxide-releasing molecule. *Am. J. Physiol. Renal Physiol.* 290: F789–94.
177. Godai, K., Hasegawa-Moriyama, M., Kuniyoshi, T., Matsunaga, A., and Kanmura, Y. (2013). Increased carboxyhemoglobin level during liver resection with inflow occlusion. *J. Anesth.* 27: 306–308.
178. Murokawa, T., Sahara, H., Sekijima, M., Pomposelli, T., Iwanaga, T., Ichinari, Y., Shimizu, A., and Yamada, K. (2020). The protective effects of carbon monoxide against hepatic warm ischemia-reperfusion injury in MHC-inbred miniature swine. *J. Gastrointest. Surg.* 24: 974–982.
179. Kaizu, T., Ikeda, A., Nakao, A., Tsung, A., Toyokawa, H., Ueki, S., Geller, D.A., and Murase, N. (2008). Protection of transplant-induced hepatic ischemia/reperfusion injury with carbon monoxide via MEK/ERK1/2 pathway downregulation. *Am. J. Physiol. Gastrointest. Liver Physiol.* 294: G236–44.
180. Kaizu, T., Nakao, A., Tsung, A., Toyokawa, H., Sahai, R., Geller, D.A., and Murase, N. (2005). Carbon monoxide inhalation ameliorates cold ischemia/reperfusion injury after rat liver transplantation. *Surgery* 138: 229–235.
181. Chen, Y., Park, H.J., Park, J., Song, H.C., Ryter, S.W., Surh, Y.J., Kim, U.H., Joe, Y., and Chung, H.T. (2019). Carbon monoxide ameliorates acetaminophen-induced liver injury by increasing hepatic HO-1 and Parkin expression. *FASEB J.* 33: 13905–13919.
182. Zheng, Y., Ji, X., Yu, B., Ji, K., Gallo, D., Csizmadia, E., Zhu, M., Choudhury, M.R., De La Cruz, L.K.C., Chittavong, V., Pan, Z., Yuan, Z., Otterbein, L.E., and Wang, B. (2018). Enrichment-triggered prodrug activation demonstrated through mitochondria-targeted delivery of doxorubicin and carbon monoxide. *Nat. Chem.* 10: 787–794.
183. Caboot, J.B., Jawad, A.F., McDonough, J.M., Bowdre, C.Y., Arens, R., Marcus, C.L., Mason, T.B., Smith-Whitley, K., Ohene-Frempong, K., and Allen, J.L. (2012). Non-invasive measurements of carboxyhemoglobin and methemoglobin in children with sickle cell disease. *Pediatr. Pulmonol.* 47: 808–815. and references cited therein.
184. Beckman, J.D., Belcher, J.D., Vineyard, J.V., Chen, C., Nguyen, J., Nwaneri, M.O., O'Sullivan, M.G., Gulbahce, E., Hebbel, R.P., and Vercellotti, G.M. (2009). Inhaled carbon monoxide reduces leukocytosis in a murine model of sickle cell disease. *Am. J. Physiol. Heart Circ. Physiol.* 297: H1243–53.
185. Belcher, J.D., Gomperts, E., Nguyen, J., Chen, C., Abdulla, F., Kiser, Z.M., Gallo, D., Levy, H., Otterbein, L.E., and Vercellotti, G.M. (2018). Oral carbon monoxide therapy in murine sickle cell disease: Beneficial effects on vaso-occlusion, inflammation and anemia. *PLoS One* 13: e0205194.
186. Belcher, J.D., Young, M., Chen, C., Nguyen, J., Burhop, K., Tran, P., and Vercellotti, G.M. (2013). MP4CO, a pegylated hemoglobin saturated with carbon monoxide, is a modulator of HO-1, inflammation, and vaso-occlusion in transgenic sickle mice. *Blood* 122: 2757–2764.
187. Yoshida, J., Ozaki, K.S., Nalesnik, M.A., Ueki, S., Castillo-Rama, M., Faleo, G., Ezzelarab, M., Nakao, A., Ekser, B., Echeverri, G.J., Ross, M.A., Stolz, D.B., and Murase, N. (2010). Ex vivo application of carbon monoxide in UW solution prevents transplant-induced renal ischemia/reperfusion injury in pigs. *Am. J. Transplant.* 10: 763–772.

188. Chen, P., Sun, B., Chen, H., Wang, G., Pan, S., Kong, R., Bai, X., and Wang, S. (2010). Effects of carbon monoxide releasing molecule-liberated CO on severe acute pancreatitis in rats. *Cytokine* 49: 15–23.
189. Xue, J. and Habtezion, A. (2014). Carbon monoxide-based therapy ameliorates acute pancreatitis via TLR4 inhibition. *J. Clin. Invest.* 124: 437–447.
190. Wu, J., Zhang, R., Hu, G., Zhu, H.H., Gao, W.Q., and Xue, J. (2018). Carbon Monoxide Impairs CD11b(+) Ly-6C(hi) Monocyte Migration from the Blood to Inflamed Pancreas via Inhibition of the CCL2/CCR2 Axis. *J. Immunol.* 200: 2104–2114.
191. Hsia, W.J., Kunsoruska, M. and Jafari, B. (2018). Carboxyhemoglobin and survival outcome in patients with acute respiratory distress syndrome. *C45. Critical Care: Against the Wind – ARDS from Identification to Management to Outcomes, Am J Respir Crit Care Med.* 197: A5074–A5074.
192. Fredenburgh, L.E., Perrella, M.A., Barragan-Bradford, D., Hess, D.R., Peters, E., Welty-Wolf, K.E., Kraft, B.D., Harris, R.S., Maurer, R., Nakahira, K., Oromendia, C., Davies, J.D., Higuera, A., Schiffer, K.T., Englert, J.A., Dieffenbach, P.B., Berlin, D.A., Lagambina, S., Bouthot, M., Sullivan, A.I., Nuccio, P.F., Kone, M.T., Malik, M.J., Porras, M.A.P., Finkelsztein, E., Winkler, T., Hurwitz, S., Serhan, C.N., Piantadosi, C.A., Baron, R.M., Thompson, B.T., and Choi, A.M. (2018). A phase I trial of low-dose inhaled carbon monoxide in sepsis-induced ARDS. *JCI Insight* 3: e124039.
193. Montuschi, P., Kharitonov, S.A., and Barnes, P.J. (2001). Exhaled carbon monoxide and nitric oxide in COPD. *Chest* 120: 496–501.
194. Slebos, D.J., Kerstjens, H.A., Rutgers, S.R., Kauffman, H.F., Choi, A.M., and Postma, D.S. (2004). Haem oxygenase-1 expression is diminished in alveolar macrophages of patients with COPD. *Eur. Respir. J.* 23: 652–653. author reply 653.
195. Tsoumakidou, M., Tzanakis, N., Chrysofakis, G., and Siafakas, N.M. (2005). Nitrosative stress, heme oxygenase-1 expression and airway inflammation during severe exacerbations of COPD. *Chest* 127: 1911–1918.
196. Bathoorn, E., Slebos, D.-J., Postma, D.S., Koeter, G.H., Van Oosterhout, A.J.M., Van Der Toorn, M., Boezen, H.M., and Kerstjens, H.A.M. (2007). Anti-inflammatory effects of inhaled carbon monoxide in patients with COPD: a pilot study. *Eur. Respir. J.* 30: 1131–1137.
197. Shinohara, T., Kaneko, T., Nagashima, Y., Ueda, A., Tagawa, A., and Ishigatsubo, Y. (2005). Adenovirus-mediated transfer and overexpression of heme oxygenase 1 cDNA in lungs attenuates elastase-induced pulmonary emphysema in mice. *Hum. Gene Ther.* 16: 318–327.
198. Hara, Y., Shinkai, M., Kanoh, S., Fujikura, Y.K., Rubin, B., Kawana, A., and Kaneko, T. (2017). Arterial carboxyhemoglobin measurement is useful for evaluating pulmonary inflammation in subjects with interstitial lung disease. *Intern. Med.* 56: 621–626.
199. Lakari, E., Pylkäs, P., Pietarinen-Runtti, P., Pääkkö, P., Soini, Y., and Kinnula, V.L. (2001). Expression and regulation of hemeoxygenase 1 in healthy human lung and interstitial lung disorders. *Hum. Pathol.* 32: 1257–1263.
200. Tsuburai, T., Suzuki, M., Nagashima, Y., Suzuki, S., Inoue, S., Hasiba, T., Ueda, A., Ikehara, K., Matsuse, T., and Ishigatsubo, Y. (2002). Adenovirus-mediated transfer and overexpression of heme oxygenase 1 cDNA in lung prevents bleomycin-induced pulmonary fibrosis via a Fas-Fas ligand-independent pathway. *Hum. Gene Ther.* 13: 1945–1960.
201. Rosas, I.O., Goldberg, H.J., Collard, H.R., El-Chemaly, S., Flaherty, K., Hunninghake, G.M., Lasky, J.A., Lederer, D.J., Machado, R., Martinez, F.J., Maurer, R., Teller, D., Noth, I., Peters, E., Raghu, G., Garcia, J.G.N., and Choi, A.M.K. (2018). A phase II clinical trial of low-dose inhaled carbon monoxide in idiopathic pulmonary fibrosis. *Chest* 153: 94–104.
202. Antuni, J.D., Kharitonov, S.A., Hughes, D., Hodson, M.E., and Barnes, P.J. (2000). Increase in exhaled carbon monoxide during exacerbations of cystic fibrosis. *Thorax* 55: 138–142.
203. Zhou, H., Lu, F., Latham, C., Zander, D.S., and Visner, G.A. (2004). Heme oxygenase-1 expression in human lungs with cystic fibrosis and cytoprotective effects against Pseudomonas aeruginosa in vitro. *Am. J. Respir. Crit. Care Med.* 170: 633–640.
204. Yanagawa, Y. (2012). Significance of the carboxyhemoglobin level for out-of-hospital cardiopulmonary arrest. *J. Emerg. Trauma Shock* 5: 338–341.
205. Schober, P., Kalmanowicz, M., Schwarte, L.A., and Loer, S.A. (2009). Cardiopulmonary bypass increases endogenous carbon monoxide production. *J. Cardiothorac. Vasc. Anesth.* 23: 802–806.
206. Melley, D.D., Finney, S.J., Elia, A., Lagan, A.L., Quinlan, G.J., and Evans, T.W. (2007). Arterial carboxyhemoglobin level and outcome in critically ill patients. *Crit. Care Med.* 35: 1882–1887.
207. Fazekas, A.S., Wewalka, M., Zauner, C., and Funk, G.-C. (2012). Carboxyhemoglobin levels in medical intensive care patients: a retrospective, observational study. *Crit. Care* 16: R6–R6.
208. Gomez, H., Kautza, B., Escobar, D., Nassour, I., Luciano, J., Botero, A.M., Gordon, L., Martinez, S., Holder, A., Ogundele, O., Loughran, P., Rosengart, M.R., Pinsky, M., Shiva, S., and Zuckerbraun, B.S.

(2015). Inhaled carbon monoxide protects against the development of shock and mitochondrial injury following hemorrhage and resuscitation. *PLoS One* 10: e0135032.

209. Faller, S. and Hoetzel, A. (2012). Carbon monoxide in acute lung injury. *Curr. Pharm. Biotechnol.* 13: 777–786.

210. Wollborn, J., Steiger, C., Doostkam, S., Schallner, N., Schroeter, N., Kari, F.A., Meinel, L., Buerkle, H., Schick, M.A., and Goebel, U. (2020). Carbon Monoxide Exerts Functional Neuroprotection After Cardiac Arrest Using Extracorporeal Resuscitation in Pigs. *Crit. Care Med.* 48: e299–e307.

211. Choi, Y.K., Maki, T., Mandeville, E.T., Koh, S.-H., Hayakawa, K., Arai, K., Kim, Y.-M., Whalen, M.J., Xing, C., Wang, X., Kim, K.-W., and Lo, E.H. (2016). Dual effects of carbon monoxide on pericytes and neurogenesis in traumatic brain injury. *Nat. Med.* 22: 1335–1341.

212. Che, X., Fang, Y., Si, X., Wang, J., Hu, X., Reis, C., and Chen, S. (2018). The role of gaseous molecules in traumatic brain injury: An updated review. *Front. Neurosci.* 12: 392.

213. Maines, M.D. and Abrahamsson, P.A. (1996). Expression of heme oxygenase-1 (HSP32) in human prostate: normal, hyperplastic, and tumor tissue distribution. *Urology* 47: 727–733.

214. Deininger, M.H., Meyermann, R., Trautmann, K., Duffner, F., Grote, E.H., Wickboldt, J., and Schluesener, H.J. (2000). Heme oxygenase (HO)-1 expressing macrophages/microglial cells accumulate during oligodendroglioma progression. *Brain Res* 882: 1–8.

215. Tsai, J.-R., Wang, H.-M., Liu, P.-L., Chen, Y.-H., Yang, M.-C., Chou, S.-H., Cheng, Y.-J., Yin, W.-H., Hwang, -J.-J., and Chong, I.-W. (2012). High expression of heme oxygenase-1 is associated with tumor invasiveness and poor clinical outcome in non-small cell lung cancer patients. *Cell. Oncol.* 35: 461–471.

216. Gandini, N.A., Fermento, M.E., Salomón, D.G., Blasco, J., Patel, V., Gutkind, J.S., Molinolo, A.A., Facchinetti, M.M., and Curino, A.C. (2012). Nuclear localization of heme oxygenase-1 is associated with tumor progression of head and neck squamous cell carcinomas. *Exp. Mol. Pathol.* 93: 237–245.

217. Zou, C., Zou, C., Cheng, W., Li, Q., Han, Z., Wang, X., Jin, J., Zou, J., Liu, Z., Zhou, Z., Zhao, W., and Du, Z. (2016). Heme oxygenase-1 retards hepatocellular carcinoma progression through the microRNA pathway. *Oncol. Rep.* 36: 2715–2722.

218. Gueron, G., Giudice, J., Valacco, P., Paez, A., Elguero, B., Toscani, M., Jaworski, F., Leskow, F.C., Cotignola, J., Marti, M., Binaghi, M., Navone, N., and Vazquez, E. (2014). Heme-oxygenase-1 implications in cell morphology and the adhesive behavior of prostate cancer cells. *Oncotarget* 5: 4087–4102.

219. Hatsukami, D.K., Stead, L.F., and Gupta, P.C. (2008). Tobacco addiction. *Lancet* 371: 2027–2038.

220. Goldsmith, J.R. and Landaw, S.A. (1968). Carbon monoxide and human health. *Science* 162: 1352–1359.

221. Ernst, A. and Zibrak, J.D. (1998). Carbon monoxide poisoning. *N. Engl. J. Med.* 339: 1603–1608.

222. Eichhorn, L., Michaelis, D., Kemmerer, M., Jüttner, B., and Tetzlaff, K. (2018). Carbon monoxide poisoning from waterpipe smoking: a retrospective cohort study. *Clin. Toxicol.* 56: 264–272.

223. Von Rappard, J., Schönenberger, M., and Bärlocher, L. (2014). Carbon monoxide poisoning following use of a water pipe/hookah. *Dtsch. Arztebl. Int.* 111: 674–679.

224. La Fauci, G., Weiser, G., Steiner, I.P., and Shavit, I. (2012). Carbon monoxide poisoning in narghile (water pipe) tobacco smokers. *C.J.E.M.* 14: 57–59.

225. Dorey, A., Scheerlinck, P., Nguyen, H., and Albertson, T. (2020). Acute and chronic carbon monoxide toxicity from tobacco smoking. *Mil. Med.* 185: e61–e67.

226. Maalem, R., Alali, A., and Alqahtani, S. (2019). Tobacco hookah smoking-induced carbon monoxide poisoning: A case report of non-ambient exposure. *Clin. Case Rep* 7: 1178–1180.

227. Misek, R. and Patte, C. (2014). Carbon monoxide toxicity after lighting coals at a hookah bar. *J. Med. Toxicol.* 10: 295–298.

228. Zevin, S., Saunders, S., Gourlay, S.G., Jacob, P., and Benowitz, N.L. (2001). Cardiovascular effects of carbon monoxide and cigarette smoking. *J. Am. Coll. Cardiol.* 38: 1633–1638.

229. Verschuere, S., De Smet, R., Allais, L., and Cuvelier, C.A. (2012). The effect of smoking on intestinal inflammation: What can be learned from animal models? *J. Crohns Colitis* 6: 1–12.

230. Papoutsopoulou, S., Satsangi, J., Campbell, B.J., and Probert, C.S. (2020). Review article: impact of cigarette smoking on intestinal inflammation—direct and indirect mechanisms. *Aliment. Pharmacol. Ther.* 51: 1268–1285.

231. Naito, Y., Takagi, T., Uchiyama, K., Katada, K., and Yoshikawa, T. (2016). Multiple targets of carbon monoxide gas in the intestinal inflammation. *Arch. Biochem. Biophys.* 595: 147–152.

232. Sebastián, V.P., Salazar, G.A., Coronado-Arrázola, I., Schultz, B.M., Vallejos, O.P., Berkowitz, L., Álvarez-Lobos, M.M., Riedel, C.A., Kalergis, A.M., and Bueno, S.M. (2018). Heme oxygenase-1 as a

modulator of intestinal inflammation development and progression. *Front. Immunol.* 9: 1956.

233. Breckenridge, C.B., Berry, C., Chang, E.T., Sielken, R.L., Jr., and Mandel, J.S. (2016). Association between Parkinson's disease and cigarette smoking, rural living, well-water consumption, farming and pesticide use: Systematic review and meta-analysis. *PLoS One* 11: e0151841. and references cited therein.

234. Gallo, V., Vineis, P., Cancellieri, M., Chiodini, P., Barker, R.A., Brayne, C., Pearce, N., Vermeulen, R., Panico, S., Bueno-de-mesquita, B., Vanacore, N., Forsgren, L., Ramat, S., Ardanaz, E., Arriola, L., Peterson, J., Hansson, O., Gavrila, D., Sacerdote, C., Sieri, S., Kühn, T., Katzke, V.A., Van Der Schouw, Y.T., Kyrozis, A., Masala, G., Mattiello, A., Perneczky, R., Middleton, L., Saracci, R., and Riboli, E. (2019). Exploring causality of the association between smoking and Parkinson's disease. *Int. J. Epidemiol.* 48: 912–925. and references cited therein.

235. McRobbie, H., Phillips, A., Goniewicz, M.L., Smith, K.M., Knight-West, O., Przulj, D., and Hajek, P. (2015). Effects of switching to electronic cigarettes with and without concurrent smoking on exposure to nicotine, carbon monoxide, and acrolein. *Cancer Prev. Res. (Phila)* 8: 873–878.

236. Son, Y., Bhattarai, C., Samburova, V., and Khlystov, A. (2020). Carbonyls and carbon monoxide emissions from electronic cigarettes affected by device type and use patterns. *Int. J. Environ. Res. Public Health* 17: 2767.

237. Kou, Y.R., Kwong, K., and Lee, L.Y. (2011). Airway inflammation and hypersensitivity induced by chronic smoking. *Respir. Physiol. Neurobiol.* 178: 395–405.

238. West, R. (2017). Tobacco smoking: Health impact, prevalence, correlates and interventions. *Psychol. Health* 32: 1018–1036.

239. Warren, G.W., Alberg, A.J., Kraft, A.S., and Cummings, K.M. (2014). The 2014 Surgeon General's report: "The health consequences of smoking–50 years of progress": a paradigm shift in cancer care. *Cancer* 120: 1914–1916. and references cited therein.

240. Antman, E., Arnett, D., Jessup, M., and Sherwin, C. (2014). The 50th anniversary of the US surgeon general's report on tobacco: what we've accomplished and where we go from here. *J. Am. Heart Assoc* 3: e000740.

241. Lushniak, B.D. (2014). A historic moment: The 50th anniversary of the first Surgeon General's Report on Smoking and Health. *Public Health Rep* 129: 5–6.

242. Coburn, R.F. (1970). Endogenous carbon monoxide production. *N. Engl. J. Med.* 282: 207–209.

243. Schaefer, W.H., Harris, T.M., and Guengerich, F.P. (1985). Characterization of the enzymic and nonenzymic peroxidative degradation of iron porphyrins and cytochrome P-450 heme. *Biochemistry* 24: 3254–3263.

244. Archakov, A.I., Karuzina, I.I., Petushkova, N.A., Lisitsa, A.V., and Zgoda, V.G. (2002). Production of carbon monoxide by cytochrome P450 during iron-dependent lipid peroxidation. *Toxicol. In Vitro* 16: 1–10.

245. Adams, M. and Jia, Z. (2005). Structural and biochemical analysis reveal pirins to possess quercetinase activity. *J. Biol. Chem.* 280: 28675–28682.

246. Deshpande, A.R., Pochapsky, T.C., Petsko, G.A., and Ringe, D. (2017). Dual chemistry catalyzed by human acireductone dioxygenase. *Protein Eng. Des. Sel.* 30: 197–204.

247. Chatterjee, A., Hazra, A.B., Abdelwahed, S., Hilmey, D.G., and Begley, T.P. (2010). A "radical dance" in thiamin biosynthesis: mechanistic analysis of the bacterial hydroxymethylpyrimidine phosphate synthase. *Angew. Chem. Int. Ed. Engl.* 49: 8653–8656.

248. Voordouw, G. (2002). Carbon monoxide cycling by *Desulfovibrio vulgaris* Hildenborough. *J. Bacteriol.* 184: 5903–5911.

249. Hopper, C.P., De La Cruz, L.K., Lyles, K.V., Wareham, L.K., Gilbert, J.A., Eichenbaum, Z., Magierowski, M., Poole, R.K., Wollborn, J., and Wang, B. (2020). Role of carbon monoxide in host–gut microbiome communication. *Chem. Rev.* 120: 13273–13311.

250. Wolff, D.G. and Bidlack, W.R. (1976). The formation of carbon monoxide during peroxidation of microsomal lipids. *Biochem. Biophys. Res. Commun.* 73: 850–857.

251. Oren, D.A., Sit, D.K., Goudarzi, S.H., and Wisner, K.L. (2020). Carbon monoxide: a critical physiological regulator sensitive to light. *Translation. Psych.* 10: 87. and references cited therein.

252. Ermakov Iu, A., Pasechnik, V.I., and Tul'skiĭ, S.V. (1975). [Photodissociation of carboxyhemoglobin with different degrees of saturation]. *Biofizika* 20: 591–595.

253. Tanaka, T., Kashimura, T., Ise, M., Lohman, B.D., and Taira, Y. (2016). Light irradiation for treatment of acute carbon monoxide poisoning: an experimental study. *J. Intensive Care* 4: 58.

254. Sharma, V.S., Schmidt, M.R., and Ranney, H.M. (1976). Dissociation of CO from carboxyhemoglobin. *J. Biol. Chem.* 251: 4267–4272.

255. Oren, D.A. (1996). Humoral phototransduction: Blood is a messenger. *Neuroscientist* 2: 207–210.

256. Klemz, R., Reischl, S., Wallach, T., Witte, N., Jürchott, K., Klemz, S., Lang, V., Lorenzen, S., Knauer, M., Heidenreich, S., Xu, M., Ripperger, J.A., Schupp, M., Stanewsky, R., and Kramer, A. (2017). Reciprocal regulation of carbon monoxide metabolism and the circadian clock. *Nat. Struct. Mol. Biol.* 24: 15–22.

257. Yachie, A., Niida, Y., Wada, T., Igarashi, N., Kaneda, H., Toma, T., Ohta, K., Kasahara, Y., and Koizumi, S. (1999). Oxidative stress causes enhanced endothelial cell injury in human heme oxygenase-1 deficiency. *J. Clin. Invest.* 103: 129–135.

258. Kawashima, A., Oda, Y., Yachie, A., Koizumi, S., and Nakanishi, I. (2002). Heme oxygenase-1 deficiency: the first autopsy case. *Hum. Pathol.* 33: 125–130.

2

Molecular Mechanisms of Actions for CO: An Overview

Rodrigo W. Alves de Souza[1], Leo E. Otterbein[1], and Nils Schallner[2]

[1]Beth Israel Deaconess Medical Center in Boston, MA, USA
[2]University of Freiburg, Freiburg, Germany

Introduction and Therapeutic Perspectives

As we acknowledge extensive preclinical *in vivo* data demonstrating the therapeutic efficacy of exogenous carbon monoxide (CO) administration, the debate regarding the mechanisms of these beneficial actions remains ongoing. The effects of CO are heterogeneous and can be somewhat contradictory at first: CO regulates the immune response [1], cellular survival [2], tissue regeneration [3–5], and also proliferation [6,7]. CO can act as an anti- and proinflammatory [8–12], pro- and antiapoptotic [13–16], and pro- and antiproliferative molecule [17,18]. Its ambiguous biological activity can be explained with specificity toward cell and tissue types, as well as the injury model. A unifying theme that may begin to explain the diverse effects of the gas is its ability to avidly bind to various cellular proteins that have heme as a common structural component. Heme as iron- and porphyrin-containing coordination complex is abundantly present within the cell; hence, the effects of CO are diverse. The published protective effects afforded by CO and the heme oxygenase (HO) enzyme system, especially the inducible HO-1 isoform, which cleaves heme into biliverdin, iron, and CO, are well defined. However, the potential mechanisms of action for CO are just as heterogeneous. As a gas, CO diffuses freely across organ boundaries such as the blood–brain barrier and cell membranes. Therefore, its effect is independent of typical receptors and transmembrane transporters, and thus target organs and cell compartments are reached very quickly.

From a therapeutic paradigm perspective, the beneficial preclinical effects have to be weighed against potential toxicity where CO binding to heme impairs oxygen transport and cellular respiration, leading to internal asphyxiation at high, industrial concentrations. Pilot trials in humans have been completed reporting absence of toxicity at significant carboxyhemoglobin (COHb) levels of 10–15% [19,20]. Additionally, several clinical trials have been completed that while not yet showing beneficial effects at least echo safety and tolerability of CO, including the influence on aerobic performance [21], chronic lung disease [22,23], and acute respiratory distress syndrome [24]. The latter might also be relevant in COVID-associated lung failure ("CARDS" – COVID-ARDS), even though CARDS shows some distinct differences from classical ARDS: COVID patients with severe disease progression requiring intensive care unit admission with mechanical ventilation and/or extracorporeal membrane oxygenation show elevated endogenous CO levels (unpublished data). Additionally, high maximum and mean COHb levels during the initial phase of ICU treatment show a significant correlation with the patients' likelihood for fatal outcome, indicating that CO may be involved either in disease progression or perhaps in recovery. In this chapter, we will address the current state of the field with regard to potential mechanisms by which CO modulates cell survival, inflammation, proliferation, and metabolism.

Physiological Responses to CO

The major physiological source of CO in the body is through the breakdown of heme. Degradation of heme occurs in a tightly controlled manner involving the enzyme HO (EC 1.14.99.3), which cleaves the porphyrin IX ring in the presence of NADP and molecular oxygen, resulting in primary or first-order

Carbon Monoxide in Drug Discovery: Basics, Pharmacology, and Therapeutic Potential, First Edition. Edited by Binghe Wang and Leo E. Otterbein.
© 2022 John Wiley & Sons, Inc. Published 2022 by John Wiley & Sons, Inc.

heme degradation products, consisting of biliverdin IXα, with the concomitant liberation of CO and reduced heme iron. Heme metabolism concludes with the enzymatic reduction of biliverdin IXα by NAD(P)H:biliverdin reductase [25,26]. Because this represents primarily a book on the role of CO in intracellular signaling, we refer the reader to other sources for discussion of the possible roles of biliverdin/bilirubin and iron metabolism in reported HO-mediated protection.

CO was formerly considered a catabolic elimination product with no physiological significance [27]. Indeed, elevated CO concentrations cause tissue hypoxia by virtue of competitive binding with the oxygen binding sites of hemoglobin (Hb) [28,29]. CO is toxic to humans and animals at high concentrations, and lethality results from tissue hypoxia following Hb saturation. CO binds to Hb approximately 200 times more strongly than O_2, and the resulting COHb thus limits oxygen transport in the body. With 50% of human Hb occupied by CO, seizures and coma may result, sometimes with fatal consequences. Despite this toxicity, CO is increasingly studied as a diverse cell signaling molecule [30].

The CO messenger literature is enormous, involving nearly every organ and regulatory system. To understand how and where CO is functioning as part of a physiological process, a critical examination of the unusual features of CO that give this gas the ability to serve as a typical intracellular messenger is essential [31–33]. Most signaling molecules, such as nitric oxide (NO), are produced solely for a signaling purpose and are then rapidly catabolized (half-time of less than 2 s), and the intracellular background activity of the signaling molecule is maintained at a very low level. In contrast, there is minimal catabolism of CO, and endogenously produced CO is removed from the cell entirely via diffusion to the blood, where Hb avidly binds it. Homeostasis of blood CO is established by pulmonary excretion, a relatively slow process with a half-time of approximately 4 h. To provide a signal, CO has to be produced endogenously by HO activity or by exogenous application of CO at a rate that appreciably raises the CO concentration over the basal level, ruling out the possibility that trivial CO production could serve as a messenger. The lack of catabolism of CO, its removal from the cell by diffusion to the blood, and its excretion solely via the lungs make CO uniquely amenable to quantitative physiological modeling. The total body production of CO is easily assessed, via breath measurements and COHb concentrations in the blood. Moreover, the intracellular P_{CO} at baseline and following CO exposure can be determined since intracellular P_{CO} is in equilibrium with the blood P_{CO}. Most importantly, the diffusion of CO between the cell and blood can be quantitatively modeled, making it possible to estimate the intracellular production rate required to maintain a given cellular P_{CO} [34]. In addition to all these features that give CO a capacity to modulate several biological processes, it is essential to emphasize the effects of CO as a messenger or as a gasotransmitter, involving nearly every organ and regulatory system. We need to pay close attention to differences in CO concentration, timing, molecular specificity, and subcellular localization.

Interaction with NO and H$_2$S

The two other important gases involved in biological signaling as gasotransmitters are NO and hydrogen sulfide (H_2S), which are discussed in other chapters in this book. All three gases share some of the molecular targets described later, yet display notable differences in binding capabilities (Figure 2.1). Heme-based gas sensors within the target structures provide intrinsic gaseous selectivity due to its biochemical design and conformational changes taking place once gas binding has been initiated [35–39]. For example, this "sliding scale rule" provides insight into hemoprotein selectivity for NO over CO and O_2 and why gasotransmitter binding to soluble guanylate cyclase (sGC) and other hemoproteins is not solely dependent on intracellular gas concentration. It might also explain the profound effects of CO on intracellular O_2 sensing with concomitant downstream changes in cellular respiration and energy production.

Between all three gases, significant crosstalk exists. CO and NO not only compete for heme binding sites but simultaneously can reinforce activity and production of each another: CO has an activating effect on NO synthase (NOS) enzymes and mediates both vasoactive and organ-protective effects through increased NO production in the lung, liver, and vasculature [40–44]. On the contrary, protective anti-inflammatory effects are mediated via a NOS-induced activation of the HO-1/CO system [45]. These additive CO/NO functions explain the "on-demand" effect observed for CO: its influence on cellular targets is part of the tissue-specific response to a given physiological or pathophysiological stimulus, and it, therefore, serves the need of the tissue in a given situation.

Regarding H_2S, CO produced via the HO enzyme system or given exogenously can bind to the heme-containing prosthetic group in cystathionine β-synthase, one of the enzymatic sources of endogenous H_2S,

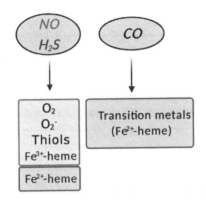

Figure 2.1 Chemical reactivity of NO, H_2S, and CO in biological systems. Despite NO and H_2S having multiple intracellular targets and CO being more specific to reacting uniquely with transition metals in a specific redox state (ferrous – Fe^{2+}), all three gases present a crosstalk.

leading to its inhibition and reduction in H_2S production [46]. On the other hand, H_2S can enhance the availability and release of NO [47], emphasizing the complex interplay between the three gases. The relationship of CO and O_2 and the concept of pseudohypoxia are covered in Chapter 7.

Molecular Effector Systems for CO: Heme Versus Nonheme Targets

The targets for CO are well described yet are extensively diverse within the set of cellular heme-containing structures. CO binds rapidly and with high affinity to heme-containing proteins such as Hb, myoglobin, enzymes of the mitochondrial respiratory chain, or enzymes involved in free radical generation. Due to the high affinity of CO to the presence of reduced iron (Fe^{2+}), typically heme iron, all heme-containing cellular structures must be considered as CO binding targets. In quiescent cells, any CO produced will target hemoproteins necessary for basal function such as sGC, oxidases, NO synthases, and the heme-containing transcription factors, including Bach1 and NPAS2 [48–50]. In the cardiovascular tissues, these heme-based proteins sensitive to gaseous molecules differ in prominence between cell types. For instance, guanylate cyclase is highly prevalent in smooth muscle cells while virtually absent in macrophages. Endothelial NOS (eNOS) is constitutively active in the endothelium while absent in most other cells. Mitochondria are perhaps present among all cell types but differ in terms of number per cell type. Cardiac myocytes contain many more mitochondria per cell than smooth muscle and endothelial cells. In each cell type, the function of the hemoproteins can be increased or blocked by binding CO. Moreover, many secondary downstream pathways are subsequently influenced, leading to functional changes in cellular energy production and ultimately inflammatory and apoptosis-modulating outputs that result from a common unifying upstream signature of a heme–CO interaction. Therefore, cellular targets and their cellular signaling detailed in the literature can be divided into primary (hemoproteins) and secondary (nonheme-containing) proteins. Even if the exact mechanisms of CO-mediated organ and cellular protection have not yet been clarified, one common distinguishing feature of the CO target structures is the capacity to influence heme and nonheme cellular signaling.

Heme Targets

In each cell type, the function of the hemoproteins can be increased or blocked by binding CO. Proteins such as sGC (EC 4.6.1.2), NOS, ion channels, and NADPH oxidase all bind CO that results in altered activity. Others are activated by reactive oxygen species (ROS) such as peroxisome proliferator-activated receptor-γ (PPAR-γ), whereas a third set of targets relates to the oxygen sensors that result in a pseudohypoxia resulting in stabilization of hypoxia-inducible factor-1α (HIF-1α) (Table 2.1). In hemoproteins, when CO binds sGC or eNOS, it activates the enzyme generating more cGMP or NO, respectively [18,41,43,44]. On the other hand, when CO binds to cytochrome c oxidase in mitochondria, CO inhibits their activity, resulting in increased superoxide ions that rapidly provoke signaling cascades as ROS leading to changes in gene regulation and ultimately influence cellular behavior, as described later and in more detail in Chapter 6 [51–53]. Regarding the enzyme sGC, it is a heterodimeric complex (α/β) that catalyzes GTP conversion to cGMP. cGMP is a potent intracellular messenger involved in vascular tone, gene regulation, neurotransmission, and many other cellular responses [54–57]. Each subunit possesses multiple domains: from the N to C termini, there is an H-NOX domain (heme–NO/oxygen binding domain), a PAS domain, a CC (coiled–coiled) domain, and a CAT (catalytic) domain [58]. The activity of sGC under reducing conditions is markedly enhanced greater than 100-fold, by the gaseous messenger NO, and this regulation accounts for the well-known vasodilatory influence of NO. CO has been reported to increase sGC activity, albeit much less effectively than NO, approximately three- to fourfold comparatively [59,60]. The postulated CO-mediated increase in sGC activity may contribute in part to its vasodilatory effects at low CO concentrations [61,62]. The structural and physicochemical bases of the interactions of NO and CO with sGC have been well studied [37]. The N-terminal region of the β-subunit of sGC, which presents an H-NOX domain, contains a heme whose propionate groups are surrounded by the conserved sequence motif Tyr-X-Ser-X-Arg, where X is any amino acid [63]. CO binding to sGC leads to a six-coordinate low-spin heme adduct, with an intact His–Fe bond. The atomic structural changes induced by binding of CO are subtle, where a very slight pivoting of the heme cofactor (<1 Å) has been suggested to account for the enhanced enzymatic activity [37]. There is evidence supporting CO effects through sGC. First, CO-induced vasorelaxation can be blocked by inhibiting sGC activity [64], consistent with a direct role for CO and sGC *in vivo*. Second, there is striking evidence that HO, the source of endogenous CO in nerve cells, colocalizes with sGC in cells with little or no NOS expression (Figure 2.2) [65,66]. Interestingly, CO and NO not only compete for heme binding sites but can also simultaneously reinforce activity and production of one another. CO activates NOS enzymes and mediates both vasoactive and organ-protective effects through increased NO production in the lung, liver, and vasculature [40–44]. On the contrary, protective anti-inflammatory effects are mediated via NOS-induced activation of the HO-1/CO system [45]. These additive CO/NO functions explain the "on-demand" effect observed for CO. The influence of CO on the cell is, in part, a tissue-specific response to a given physiological or pathophysiological stimulus and it therefore serves the need of the tissue in a given situation.

A compelling link has been shown to exist between the circadian autoregulatory feedback loop and the HO-1/CO enzyme system. Transcription factors *NPAS2*, *CLOCK*, and *RevErb-α* contain heme and thus, their turnover requires heme degradation via HO-1. *RevErb-α* regulates many physiological functions, including circadian rhythms and metabolic gene pathways [67]. Transcriptional activity of *NPAS2* [68] and *CLOCK* [69], two of transcription factors that regulate expression of circadian regulatory protein families *Per* (Period) and *Cry* (Cryptochrome), have been shown to be influenced by CO as heme-based gas sensors (Figure 2.2). The HO-1/CO system can also contribute to the restoration of circadian rhythmicity and neuronal outcome after hemorrhagic stroke [70]. In addition to HO-1, the enzyme HO-2 also generates CO in neurons [71,72]. Neuronal impulses dynamically regulate HO-2 activity through a kinase cascade in which protein kinase C activates casein kinase 2, which in turn phosphorylates and activates HO-2 [73]. HO-2 activity, which is constitutive, generates low micromolar concentrations of CO in the brain, which are sufficient to regulate the DNA binding activity of NPAS2. The production of CO fluctuates according to the circadian nature of heme metabolism [68]. Heme biosynthesis is also rate limited by δ-aminolevulinate synthase-1 (ALAS1) in the mitochondria and circadian [74,75]. Indeed, heme controls the activity of the BMAL1–NPAS2 transcriptional complex, in part by inhibiting DNA binding in response to CO. Heme also differentially modulates expression of the mammalian Period genes mPer1 and mPer2 through a mechanism involving NPAS2 and mPER2. mPER2 positively stimulates the activity of the BMAL1–NPAS2 transcription complex. In turn, NPAS2 transcriptionally regulates ALAS1. Vitamin B12 and heme compete for binding to NPAS2 and mPER2 and they exert opposite effects on mPer2 and mPer1 expression *in vivo* [75]. These data show that the circadian clock and heme biosynthesis

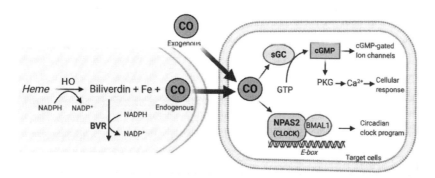

Figure 2.2 Activation of sGC by CO. CO is synthesized by HO. HO degrades heme (iron protoporphyrin IX), bringing about free iron, biliverdin, and gaseous CO. CO diffuses into the neighboring target cell, for example, smooth muscle cell in the cardiovascular system, and binds to sGC, which converts GTP into cGMP. cGMP acts as a second messenger of cells and activates a cascade of events. The produced cGMP activates a cGMP-dependent kinase, PKG, and triggers numerous cellular responses. NPAS2 and CLOCK form heterodimers with BMAL1 that bind to the circadian E-box elements to activate the expression of molecular circadian clock genes, such as Per and Cry. Illustration was created using BioRender (biorender.com).

are reciprocally regulated and suggest porphyrin-containing molecules as potential candidates for therapy of circadian disorders. Elevated heme levels are likely to inhibit heme synthesis by modulating circadian transcription factors through mPer2 or through HO-derived CO, which thus may constitute a cellular feedback loop to ensure heme homeostasis.

In addition to direct CO effects, there is increasing evidence that CO influences directly and indirectly intracellular signals by modulating ROS production by specific hemoproteins within the cell. The two principal sources of cellular ROS include (i) the membrane-bound NADPH oxidase and (ii) mitochondria.

CO and NADPH Oxidase

NADPH oxidase and mitochondrial proteins found in the respiratory chain [76–82] can bind CO and increase ROS production [83,84]. The ROS generated are likely converted to H_2O_2 and act as signaling molecules to alter cellular respiration, energy metabolism, biogenesis, and various protective signal transduction pathways. For example, an investigation showed that ROS modulation from NADPH oxidase and other sources from within the mitochondrial respiratory chain by CO occurs upstream of cyclin D1 and ERK1/2 MAPK in airway smooth muscle cells maintaining proliferative homeostasis [81].

CO and Mitochondria

The property of CO to interact with several metalloproteins implicates this property to mitochondrial cytochrome c oxidase. When bound to cytochrome c oxidase in mitochondria, CO inhibits their activity resulting in increased superoxide ions that rapidly provoke signaling cascades as ROS leading to changes in gene regulation and ultimately influencing cellular behavior [51–53]. In fact, CO-induced low levels of mitochondrial ROS act as signaling molecules, as a preconditioning-like effect, promoting different biological responses: cytoprotection, anti-inflammatory, modulation of cell metabolism, or cellular differentiation. In this section, the molecular mechanisms of mitochondrial ROS generation are discussed. It is worth noting that this topic will be explained in further detail in Chapter 6.

Mitochondrial Permeabilization and Cell Death Control

CO influences mitochondrial membrane permeabilization (MMP) and cell death regulation in part through modulation of Ca^{2+} regulation and membrane permeabilization of mitochondria. In nonsynaptic mitochondria isolated from rat cortex [85] and from liver [86], CO prevented Ca^{2+}-induced MMP by inhibiting mitochondrial swelling, mitochondrial depolarization, and mitochondrial inner membrane permeabilization. Furthermore, CO prevented astrocytic cell death by inhibiting MMP. Low amounts of CO promoted mitochondrial ROS generation, with an increase in levels of oxidized glutathione, which induces glutathionylation of adenine nucleotide translocator protein [85]. In summary, CO appears to increase mitochondrial capacity to take up Ca^{2+} and thus limit Ca^{2+}-induced MMP, which in turn inhibits cell death.

CO and Mitochondrial Biogenesis

CO-mediated mitochondrial H_2O_2 production oxidizes functional thiol groups on counterregulatory phosphatases, like PTEN and PTP1B, inactivating them [87–89]. This phosphatase inactivation permits the unopposed activity of the prosurvival kinase Akt—as observed after CO [90,91]. In the mitochondrial biogenesis program, Akt is regulatory, for example, in the phosphorylation of nuclear respiratory factor 1 (NRF-1), a key transcription factor for some 100 genes involved in oxidative phosphorylation, heme synthesis, and mitochondrial biogenesis [92]. The nuclear-encoded mitochondrial transcription factor A (Tfam) is essential for mitochondrial DNA (mtDNA) transcription, replication, and the maintenance of mtDNA copy number [93]. CO also stimulates the Nrf2 transcription factor as well as the coactivator PGC-1α, also required for the expression of mitochondrial transcriptome proteins needed in mitochondrial biogenesis [94]. Moreover, the CO-induced mitochondrial biogenesis was dependent on H_2O_2 generation and independent of NOS activation [90]. Still, in a mice sepsis model, CORM-3 treatment improved cardiac mitochondrial function via stimulation of mitochondrial biogenesis in a PGC-1α- and ROS-dependent manner [95]. Interestingly, HO-1 overexpression and CO production activated the mitochondrial biogenesis Nrf2 transcription factor, along with upregulation of the anti-inflammatory protein IL-10 and the IL-1 receptor antagonist [96]. Furthermore, activation of mitochondrial biogenesis by HO-1 rescued mice from lethal *Staphylococcus aureus* sepsis, via redox-regulated Nrf2 signaling [97]. Although a lot of work has been conducted regarding the Nrf2 regulation, its exact activation by CO remains unclear. It is not clear whether cGMP production is directly involved in Nrf2 activation by CO. Other findings have suggested the central role of ROS induced by low levels of CO in

CO-initiated preconditioning and protection by induction of antioxidant enzymes and protective signaling pathways [98]. Bing and collaborators showed that 18 h of CO exposure may create such an oxidative preconditioning to induce Nrf2 activation for modulation of cerebral ischemia [98]. Meanwhile, the ROS formation in response to low CO concentrations may have positive (signaling) and negative (damaging) effects depending on the amount of ROS formed and the cell type under investigation. It is important to point out that the redox-dependent Nrf2 system plays a central role in HO-1 induction in response to oxidative stress. Several critical regulatory domains are present in a 10-kb region of the 5′-flanking sequence of the HO-1 gene. Two of the most highly studied enhancer regions, E1 and E2, contain stress-response elements (StREs) and structurally resemble the antioxidant response element (ARE). The major positive transcriptional regulator of HO-1 gene acting on the StREs is represented by Nrf2. Nrf2 is ubiquitinated by forming a complex with the Kelch-like ECH-associated protein 1 (Keap1). Keap1 targets Nrf2 for ubiquitin-dependent degradation and hence represses Nrf2-dependent gene expression. Some HO-1 inducers or oxidative stress may prevent Keap1-dependent degradation of Nrf2, resulting in Nrf2 nuclear accumulation. Accumulated or stabilized Nrf2 may bind to the StRE regions to form complexes with the small Maf (*Musca domestica* antifungal) proteins leading to initiation of HO-1 gene transcription [99].

Finally, we recently published that the lack of a HO-1/CO system in skeletal muscle promotes decreased mitochondrial content and function, concomitantly decreasing oxidative metabolism. Moreover, exercise training enhanced skeletal muscle abnormalities in skeletal muscle-specific *Hmox1 knock-out* mice, such as muscle atrophy and dysregulation of the myogenic program (Figure 2.3) [100]. Nevertheless, little data are available in the literature demonstrating a direct effect of CO on skeletal muscle signaling in health and disease. Further studies correlating skeletal muscle signaling and the biological function of CO are urgently needed, given the therapeutical potential of CO to combat a wide range of different diseases (see Chapter 30).

CO and Mitochondrial Quality Control

The CO-mediated mitochondrial function also implicates modulation of mitochondrial quality control. Mitochondrial quality control can be defined as the balance between mitochondrial biogenesis and elimination of dysfunctional mitochondria via autophagy (mitophagy) [101,102]. In hepatocytes, CO activates protein kinase R-like endoplasmic reticulum kinase, which promotes nuclear translocation of transcription factor EB (TFEB) [102]. TFEB translocation is involved in lysosomal and mitochondrial biogenesis. In TFEB-deficient cells, CO failed to promote Parkin translocation into mitochondria and to induce mitophagy. Likewise, knockdown expression of TFEB also decreased expression of lysosomal genes (Lamp1, cathepsin B, and TPP1), along with a decrease in mtDNA, mitochondrial biogenesis markers (PGC-1α, Nrf1, and TFAM), and mitochondrial proteins, such

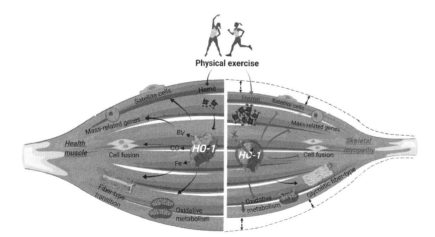

Figure 2.3 Proposed model for how skeletal muscle HO-1 activity regulates muscle performance. Skeletal muscle microtraumas that occur during aerobic exercise result in elevated heme levels inducing the expression of HO-1. Aerobic exercise training increases HO-1, which exerts a powerful contribution to improve muscle function by modulating satellite cell activation/proliferation, myoblast fusion, hypertrophy/atrophy-related genes, fiber-type transition, and mitochondrial function (left panel). Note that exercise training was unable to function appropriately in the absence of skeletal muscle HO-1 (right panel). Illustration was created using icons from BioRender (biorender.com) and Servier Medical Art (smart.servier.com).

as COX II, COX IV, and cytochrome *c* [102]. The effects of CO on autophagy are double-edged and strictly follow dose dependence: CO poisoning can be devastating with death occurring in a high percentage of patients and long-term sequalae in patients who survive especially due to apoptotic cell death in the central nervous system with concomitant long-term neurological deficits [103,104]. However, CO in physiological doses can serve as a potent inhibitor of autophagy cell death after an injurious stimulus. Protective effects via CO have been demonstrated in the lung [105–108], heart [78,109,110], liver [111,112], kidney [113,114], and central nervous system [15,70,115–120], which can explain in part the cyto- and organ-protective properties of the gas. Especially in solid organ transplantation, autophagy and apoptosis in the transplant induced by cold and warm ischemia after organ removal followed by reperfusion after transplantation limit transplant function and can be a decisive factor in long-term outcome.

The term mitochondrial quality control can also be applied for the maintenance of mitochondrial function via regulation of fusion/fission processes, which promote dynamic changes in mitochondrial morphology in response to environment, nutritional conditions, and potential stress. In fact, CO modulation of mitochondrial quality control also involves fusion and fission processes. Actually, increased expression of HO-1, caused by hemin exposure, protects astroglial C6 cell line against manganese (Mn^{2+})-induced oxidative stress and cell death via regulation of mitochondrial quality control [121]. Hemin treatment limited mitochondrial fragmentation and mitophagy and improved mitochondrial interconnectivity, along with a decrease in hydrogen peroxide and anion superoxide levels [121]. Nevertheless, it is not clear which are the downstream pathways related to mitochondrial fusion/fission and mitophagy, and which are the involved HO-1 products. In addition, Hull and colleagues have described how mice overexpressing HO-1 are protected against doxorubicin-induced cardiotoxicity via modulation of mitochondrial quality control [122]. Overexpression of HO-1 led to mitochondrial biogenesis activation assessed by increased expression of TFAM, PGC-1α, and Nrf1, and presence of mtDNA. Moreover, HO-1 overexpression also decreased fission and increased fusion levels, measured by morphology and expression of mitochondrial fission 1 and mitofusin 1 and 2. Finally, HO-1 improved basal levels of mitophagy immediately after doxorubicin treatment, while 14 days later HO-1 prevented mitophagy progression since it is much higher in wild-type animals [122].

In search for a unifying mechanistic signature in these diverse effector organs, it becomes clear that the antiapoptotic properties of CO are closely related to the above-discussed interaction with mitochondrial function. The intrinsic pathway of apoptosis induction depends on the loss of mitochondrial potential, mitochondrial membrane disruption, and release of proapoptotic factors such as cytochrome *c* into the cytoplasm. By stabilizing mitochondrial function, increasing cellular respiration, and via enhanced mitochondrial biogenesis and mitochondrial autophagy, CO directly interferes with intrinsic induction of apoptosis, providing potent cytoprotection. Thus, the exact role of how CO is involved in a tightly controlled crosstalk between all the mitochondrial processes is still unclear and more details of how CO acts as a key gasotransmitter in the control of several cellular functions of mitochondria are discussed in Chapter 6.

Nonheme Targets

In addition to these primary heme-containing target structures, CO influences a variety of indirect targets void of heme structures. A selection of the physiologically most relevant indirect nonheme target structures will be discussed herewith. These include nonheme proteins such as the p38 and extracellular signal-regulated kinase MAP kinases [123,124], PPAR-γ [90], Nrf2 [125], heat shock proteins [126], adenosine receptors [127], and HIF-1α (Figure 2.4, Table 2.1) [51,128]. By increasing ROS by CO, as discussed earlier, there is activation of PPAR-γ and HIF-1α that in turn regulates the gene expression toward a more tolerant anti-inflammatory phenotype that prevents TLR4 expression, MAP kinase activation, ion channel activation/inhibition, and NADPH oxidase complex formation. It has been reported that CORM-2 and HO-1 induction significantly increases PPAR-γ activation via ERK5 activation. Therefore, the induction of HO-1 and subsequent activation of ERK5 are critical in upregulating PPAR-γ transcriptional activation as well as subsequent inhibition on inflammation [129]. Moreover, the treatment with CO, before LPS stimulation, increases expression of PPAR-γ in the nuclei of macrophages [130], suggesting that the ability of CO to cause a relatively low-intensity oxidative burst from the mitochondria would lead to PPAR-γ activation.

Additionally, ROS, reactive nitrogen species (RNS), and electrophilic compounds produced in response to CO exposure increase the expression of immunomodulatory pathways, such as NF-κB, Activator

Table 2.1 Nonheme intracellular pathways altered by CO and/or HO-1.

Pathway	Upregulated (↑) or downregulated (↓)	Downstream effects
p38 MAPK	↑	Antioxidant and anti-inflammatory responses [30]; decreased proinflammatory cytokine release in rat and mouse lung [138,139]; decreased apoptosis in mouse lung [139,140], mouse lung fibroblasts [126], mouse lung endothelial cells (mLEC) [91,126], and rat primary pulmonary endothelial cells (PAEC)[140]; and increased stress response factors in mouse lung fibroblasts [126]
PI3K/Akt	↑	Prosurvival and antiapoptotic activities [91]
HIF-1	↑	Prosurvival and neovascularization activities [128]
NF-κB	↓	Cellular homeostasis [136]; deceased binding to iNOS gene and RNS in mouse lung graft [141]
PPAR-γ	↑	Mitochondrial biogenesis [90,96,97]; decreased release of proinflammatory cytokines by RAW 246.7 macrophages [130]; and decreased monocyte migration to the airway of mice [130,142]

protein-1 (AP-1), and Bach1 [77,131–135]. Under basal conditions, NF-κB is contained in the cytoplasm by inhibitor of NF-κB (IκB). In response to a wide range of signals, the regulatory NF-κB subunits p50 and p65 dissociate from IκB and subsequently translocate to the nucleus. Oxidative stress including ROS is produced in response to CO or depletion of reduced glutathione and subsequent increase in cytosolic oxidized form: GSSG in response to oxidative stress causes rapid ubiquitination and phosphorylation and thus subsequent degradation of IκB, which is a critical step for NF-κB activation. Given that HO-1/CO is induced by different stimuli, the regulation of NF-κB by HO-1 may be important for cellular homeostasis in many pathophysiological conditions, including inflammation. Interestingly, it appears that the NF-κB inhibitors such as pyrrolidine dithiocarbamate increase HO-1 mRNA and protein expression in human colon cancer HT29 cells [136]. Recently, it has been reported that (i) a functional κB element in the mouse HO-1 promoter and (ii) the NF-κB subunits p50 and p65 and the inducible NO synthase mediated HO-1 upregulation *in vivo* [137].

The AP-1 family of transcription factors consists of Jun family oncoproteins that homo- or heterodimerize with other members of the Jun or Fos protein families. Inducible gene expression via AP-1 has been shown to be involved in a diverse range of cellular responses, including immunological and antioxidant stress responses. Similar to NF-κB, AP-1 is upregulated by a wide variety of pro-oxidant and proinflammatory stimuli. Functional AP-1 sites, which mediated inducer-dependent gene expression of HO-1, have been identified in the promoter regions of the human HO-1 genes [143]. Age-related differences in the activation of AP-1 have been shown to contribute to the age-dependent increase in vascular smooth muscle cell (VSMC) proliferation and migration. The signaling networks that modulate all the critical changes in VSMC during the aging process such as AP-1 transcription factor are associated with the redox regulation of HO-1 [144].

It has been demonstrated that the transcription factor Bach1 (BTB and CNC homology 1, basic leucine zipper transcription factor 1), which inhibits oxidative stress-inducible genes, is a crucial negative regulator of oxidative stress-induced cellular senescence [145]. Bach1 forms a heterodimer with the small Maf oncoproteins and binds to the Maf recognition element (MARE) to inhibit target genes, including that encoding HO-1 [50]. The transcription repressor Bach1 has been recognized to be a key regulator of the inducible HO-1 gene expression. The DNA binding activity of Bach1 is negatively regulated by heme *in vitro*, and this may account for the substrate-dependent activation of HO-1 [146]. Bach1 has initially been shown to repress HO-1 gene expression in the presence of low levels of intracellular heme. When intracellular heme levels are elevated, Bach1 changes its conformation and dissociates from the HO-1 promoter, allowing Nrf2 to bind to AREs and to activate HO-1 gene expression [146]. In conclusion, inactivation of Bach1 by heme or oxidative stresses results in detachment of Bach1 from DNA and export into the cytoplasm, which allows Nrf2 to access MAREs to upregulate HO-1 (Figure 2.4) [147].

CO is able to regulate nonheme-containing ion channels, including calcium-activated K^+ [148],

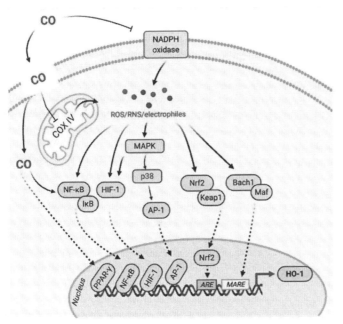

Figure 2.4 Main signaling pathways modulated by CO. Induction of antioxidative and HO-1 genes in adaptive responses that enhance the resistance of cells to environmental stresses mediated by oxidative stress (ROS and RNS) and electrophilic compounds. ROS, RNS, and endogenous and exogenous electrophiles/activators (see text for more details) can alter the AP-1, HIF-1, NF-κB/IκB, Nrf2 (NFE2L2)–Keap1, and Bach1 signaling complexes. Subsequent nuclear translocation exits, and the induction of ARE-driven genes results in upregulation of HO-1. Bach1 forms a heterodimer with Maf oncoproteins and binds to the MARE. Illustration was created using BioRender (biorender.com).

voltage-activated K^+ [149], and Ca^{2+}-channel (L-type) families [150], ligand-gated P2X receptors (P2X2 and P2X4) [151,152], tandem P domain K^+ channels (TREK1) [153], and the epithelial Na^+ channel [154]. The mechanisms by which CO regulates these ion channels are still unclear [155]. The evidence is consistent with the idea that phosphorylation of the potassium channels by PKG at selected cytoplasmic Ser residues (Ser855, Ser86968, and Ser107269 in human Slo1 AAB65837) increases the overall probability that the channel gate is open. Exactly how phosphorylation of these Ser residues in the cytoplasmic domain alters the energetics of the ion conduction gate in the transmembrane domain located several nanometers away is unclear [156]. Because CO stimulates sGC, albeit weakly, leading to activation of PKG, it is expected that CO promotes PKG-mediated phosphorylation of potassium channels and thus increases the channel activity. CO is also reported to stimulate another K^+ channel type (TREK1) through activation of the cGMP/PKG pathway [153]. Additionally, in porcine arteriolar smooth muscle cells, CO has been postulated to directly alter K^+ channel activity in a cGMP-independent activation fashion, resulting in membrane hyperpolarization [157]. The activation of large conductance KCa channels in smooth muscle cells accounts for a significant part of the relaxation induced by CO. Paradoxically, the KCa channel-mediated vasorelaxant effect of CO was significantly decreased to approximately 40% in the presence of sodium nitroprusside, a NO donor. These results suggest that NO might desensitize KCa channels to CO. The nature of the interaction between NO and CO on KCa channels is complex. These KCa channels are composed of two noncovalently linked subunits: the pore-forming α-subunit and the accessory β-subunit, which affect the electrophysiological and pharmacological properties of KCa channel organization [138]. The interactions of CO with α- and β-subunits of KCa channels may be the determining factor for the selective modulation of KCa channels and downstream effector molecules.

Conclusions

The biological effects of CO are diverse due to its fast pharmacokinetics as a gaseous molecule and due to its diverse pharmacodynamics. CO avidly binds to literally every heme-containing cellular structure and therefore will have versatile therapeutic effects depending on tissue and pathophysiological features of the specific injury. The unifying mechanistic approach to the therapeutic effects on a molecular level is best described as the gas's interaction with the mitochondria with all downstream implications related to changes in cellular respiration, metabolism, and cell signaling. Future research must focus on the dose and time kinetics in specific disease states and how these can best be modulated using CO as a therapeutic agent.

References

1. Wegiel, B., Larsen, R., Gallo, D., Chin, B.Y., Harris, C., Mannam, P., Kaczmarek, E., Lee, P.J., Zuckerbraun, B.S., Flavell, R., Soares, M.P., and Otterbein, L.E. (2014). Macrophages sense and kill bacteria through carbon monoxide-dependent inflammasome activation. *J. Clin. Invest.* 124: 4926–4940.
2. Otterbein, L.E., Foresti, R., and Motterlini, R. (2016). Heme oxygenase-1 and carbon monoxide in the heart: The balancing act between danger signaling and pro-survival. *Circ. Res.* 118: 1940–1959.
3. Lakkisto, P., Kyto, V., Forsten, H., Siren, J.M., Segersvard, H., Voipio-Pulkki, L.M., Laine, M., Pulkki, K., and Tikkanen, I. (2010). Heme oxygenase-1 and carbon monoxide promote neovascularization after myocardial infarction by modulating the expression of HIF-1alpha, SDF-1alpha and VEGF-B. *Eur. J. Pharmacol.* 635: 156–164.
4. Li, J., Song, L., Hou, M., Wang, P., Wei, L., and Song, H. (2018). Carbon monoxide releasing molecule3 promotes the osteogenic differentiation of rat bone marrow mesenchymal stem cells by releasing carbon monoxide. *Int. J. Mol. Med.* 41: 2297–2305.
5. Stifter, J., Ulbrich, F., Goebel, U., Bohringer, D., Lagreze, W.A., and Biermann, J. (2017). Neuroprotection and neuroregeneration of retinal ganglion cells after intravitreal carbon monoxide release. *PLoS One* 12: e0188444.
6. Choi, Y.K., Maki, T., Mandeville, E.T., Koh, S.H., Hayakawa, K., Arai, K., Kim, Y.M., Whalen, M.J., Xing, C., Wang, X., Kim, K.W., and Lo, E.H. (2016). Dual effects of carbon monoxide on pericytes and neurogenesis in traumatic brain injury. *Nat. Med.* 22: 1335–1341.
7. Wegiel, B., Gallo, D., Csizmadia, E., Harris, C., Belcher, J., Vercellotti, G.M., Penacho, N., Seth, P., Sukhatme, V., Ahmed, A., Pandolfi, P.P., Helczynski, L., Bjartell, A., Persson, J.L., and Otterbein, L.E. (2013). Carbon monoxide expedites metabolic exhaustion to inhibit tumor growth. *Cancer Res.* 73: 7009–7021.
8. Belcher, J.D., Gomperts, E., Nguyen, J., Chen, C., Abdulla, F., Kiser, Z.M., Gallo, D., Levy, H., Otterbein, L.E., and Vercellotti, G.M. (2018). Oral carbon monoxide therapy in murine sickle cell disease: Beneficial effects on vaso-occlusion, inflammation and anemia. *PLoS One* 13: e0205194.
9. Leake, A., Salem, K., Madigan, M.C., Lee, G.R., Shukla, A., Hong, G., Zuckerbraun, B.S., and Tzeng, E. (2020). Systemic vasoprotection by inhaled carbon monoxide is mediated through prolonged alterations in monocyte/macrophage function. *Nitric Oxide* 94: 36–47.
10. Lee, S.S., Gao, W., Mazzola, S., Thomas, M.N., Csizmadia, E., Otterbein, L.E., Bach, F.H., and Wang, H. (2007). Heme oxygenase-1, carbon monoxide, and bilirubin induce tolerance in recipients toward islet allografts by modulating T regulatory cells. *FASEB J.* 21: 3450–3457.
11. Lin, C.C., Hsiao, L.D., Cho, R.L., and Yang, C.M. (2019). Carbon monoxide releasing molecule-2-upregulated ROS-dependent heme oxygenase-1 axis suppresses lipopolysaccharide-induced airway inflammation. *Int. J. Mol. Sci.* 20: 3157.
12. Wang, J., Zhang, D., Fu, X., Yu, L., Lu, Z., Gao, Y., Liu, X., Man, J., Li, S., Li, N., Chen, X., Hong, M., Yang, Q., and Wang, J. (2018). Carbon monoxide-releasing molecule-3 protects against ischemic stroke by suppressing neuroinflammation and alleviating blood-brain barrier disruption. *J. Neuroinflammation* 15: 188.
13. Abe, T., Yazawa, K., Fujino, M., Imamura, R., Hatayama, N., Kakuta, Y., Tsutahara, K., Okumi, M., Ichimaru, N., Kaimori, J.Y., Isaka, Y., Seki, K., Takahara, S., Li, X.K., and Nonomura, N. (2017). High-pressure carbon monoxide preserves rat kidney grafts from apoptosis and inflammation. *Lab Invest.* 97: 468–477.
14. Almeida, A.S., Soares, N.L., Vieira, M., Gramsbergen, J.B., and Vieira, H.L. (2016). Carbon monoxide releasing molecule-A1 (CORM-A1) improves neurogenesis: Increase of neuronal differentiation yield by preventing cell death. *PLoS One* 11: e0154781.
15. Schallner, N., Pandit, R., LeBlanc, R., 3rd, Thomas, A.J., Ogilvy, C.S., Zuckerbraun, B.S., Gallo, D., Otterbein, L.E., and Hanafy, K.A. (2015). Microglia regulate blood clearance in subarachnoid hemorrhage by heme oxygenase-1. *J. Clin. Invest.* 125: 2609–2625.
16. Wu, M.S., Chien, C.C., Chang, J., and Chen, Y.C. (2019). Pro-apoptotic effect of haem oxygenase-1 in human colorectal carcinoma cells via endoplasmic reticular stress. *J. Cell. Mol. Med.* 23: 5692–5704.
17. Dreyer-Andersen, N., Almeida, A.S., Jensen, P., Kamand, M., Okarmus, J., Rosenberg, T., Friis, S.D., Martinez Serrano, A., Blaabjerg, M., Kristensen, B.W., Skrydstrup, T., Gramsbergen, J.B., Vieira, H.L.A., and Meyer, M. (2018). Intermittent, low dose carbon monoxide exposure enhances survival and dopaminergic differentiation of human neural stem cells. *PLoS One* 13: e0191207.
18. Otterbein, L.E., Zuckerbraun, B.S., Haga, M., Liu, F., Song, R., Usheva, A., Stachulak, C., Bodyak, N., Smith, R.N., Csizmadia, E., Tyagi, S., Akamatsu, Y., Flavell, R.J., Billiar, T.R., Tzeng, E., Bach, F.H., Choi, A.M., and Soares, M.P. (2003). Carbon monoxide suppresses arteriosclerotic lesions associated with chronic graft rejection and with balloon injury. *Nat. Med.* 9: 183–190.
19. Bathoorn, E., Slebos, D.J., Postma, D.S., Koeter, G.H., Van Oosterhout, A.J., Van Der Toorn, M., Boezen,

H.M., and Kerstjens, H.A. (2007). Anti-inflammatory effects of inhaled carbon monoxide in patients with COPD: a pilot study. *Eur. Respir. J.* 30: 1131–1137.

20. Mayr, F.B., Spiel, A., Leitner, J., Marsik, C., Germann, P., Ullrich, R., Wagner, O., and Jilma, B. (2005). Effects of carbon monoxide inhalation during experimental endotoxemia in humans. *Am. J. Respir. Crit. Care Med.* 171: 354–360.

21. Schmidt, W.F.J., Hoffmeister, T., Haupt, S., Schwenke, D., Wachsmuth, N.B., and Byrnes, W.C. (2020). Chronic exposure to low-dose carbon monoxide alters hemoglobin mass and V O2max. *Med. Sci. Sports Exerc.* 52: 1879–1887.

22. Casanova, N., Zhou, T., Gonzalez-Garay, M.L., Rosas, I.O., Goldberg, H.J., Ryter, S.W., Collard, H.R., El-Chemaly, S., Flaherty, K.R., Hunninghake, G.M., Lasky, J.A., Lederer, D.J., Machado, R.F., Martinez, F.J., Noth, I., Raghu, G., Choi, A.M.K., and Garcia, J.G.N. (2019). Low dose carbon monoxide exposure in idiopathic pulmonary fibrosis produces a CO signature comprised of oxidative phosphorylation genes. *Sci. Rep.* 9: 14802.

23. Rosas, I.O., Goldberg, H.J., Collard, H.R., El-Chemaly, S., Flaherty, K., Hunninghake, G.M., Lasky, J.A., Lederer, D.J., Machado, R., Martinez, F.J., Maurer, R., Teller, D., Noth, I., Peters, E., Raghu, G., Garcia, J.G.N., and Choi, A.M.K. (2018). A phase II clinical trial of low-dose inhaled carbon monoxide in idiopathic pulmonary fibrosis. *Chest* 153: 94–104.

24. Fredenburgh, L.E., Perrella, M.A., Barragan-Bradford, D., Hess, D.R., Peters, E., Welty-Wolf, K.E., Kraft, B.D., Harris, R.S., Maurer, R., Nakahira, K., Oromendia, C., Davies, J.D., Higuera, A., Schiffer, K.T., Englert, J.A., Dieffenbach, P.B., Berlin, D.A., Lagambina, S., Bouthot, M., Sullivan, A.I., Nuccio, P.F., Kone, M.T., Malik, M.J., Porras, M.A.P., Finkelsztein, E., Winkler, T., Hurwitz, S., Serhan, C.N., Piantadosi, C.A., Baron, R.M., Thompson, B.T., and Choi, A.M. (2018). A phase I trial of low-dose inhaled carbon monoxide in sepsis-induced ARDS. *JCI Insight* 3: e124039.

25. Tenhunen, R., Marver, H.S., and Schmid, R. (1968). The enzymatic conversion of heme to bilirubin by microsomal heme oxygenase. *Proc. Natl. Acad. Sci. U.S.A.* 61: 748–755.

26. Tenhunen, R., Ross, M.E., Marver, H.S., and Schmid, R. (1970). Reduced nicotinamide-adenine dinucleotide phosphate dependent biliverdin reductase: partial purification and characterization. *Biochemistry* 9: 298–303.

27. Barinaga, M. (1993). Carbon monoxide: killer to brain messenger in one step. *Science* 259: 309.

28. Coburn, R.F. (1979). Mechanisms of carbon monoxide toxicity. *Prev. Med.* 8: 310–322.

29. Von Burg, R. (1999). Carbon monoxide. *J. Appl. Toxicol.* 19: 379–386.

30. Kim, H.P., Ryter, S.W., and Choi, A.M. (2006). CO as a cellular signaling molecule. *Annu. Rev. Pharmacol. Toxicol.* 46: 411–449.

31. Gozzelino, R., Jeney, V., and Soares, M.P. (2010). Mechanisms of cell protection by heme oxygenase-1. *Annu. Rev. Pharmacol. Toxicol.* 50: 323–354.

32. Heinemann, S.H., Hoshi, T., Westerhausen, M., and Schiller, A. (2014). Carbon monoxide–physiology, detection and controlled release. *Chem. Commun. (Camb)* 50: 3644–3660.

33. Wu, L. and Wang, R. (2005). Carbon monoxide: endogenous production, physiological functions, and pharmacological applications. *Pharmacol. Rev.* 57: 585–630.

34. Levitt, D.G. and Levitt, M.D. (2015). Carbon monoxide: a critical quantitative analysis and review of the extent and limitations of its second messenger function. *Clin. Pharmacol.* 7: 37–56.

35. Martin, E., Berka, V., Sharina, I., and Tsai, A.L. (2012). Mechanism of binding of NO to soluble guanylyl cyclase: implication for the second NO binding to the heme proximal site. *Biochemistry* 51: 2737–2746.

36. Wu, G., Martin, E., Berka, V., Liu, W., Garcin, E.D., and Tsai, A.L. (2020). A new paradigm for gaseous ligand selectivity of hemoproteins highlighted by soluble guanylate cyclase. *J. Inorg. Biochem.* 214: 111267.

37. Ma, X., Sayed, N., Beuve, A., and Van Den Akker, F. (2007). CO differentially activate soluble guanylyl cyclase via a heme pivot-bend mechanism. *EMBO J.* 26: 578–588.

38. Stone, J.R. and Marletta, M.A. (1998). Synergistic activation of soluble guanylate cyclase by YC-1 and carbon monoxide: implications for the role of cleavage of the iron-histidine bond during activation by nitric oxide. *Chem. Biol.* 5: 255–261.

39. Zhong, F., Pan, J., Liu, X., Wang, H., Ying, T., Su, J., Huang, Z.X., and Tan, X. (2011). A novel insight into the heme and NO/CO binding mechanism of the alpha subunit of human soluble guanylate cyclase. *J. Biol. Inorg. Chem.* 16: 1227–1239.

40. Marazioti, A., Bucci, M., Coletta, C., Vellecco, V., Baskaran, P., Szabo, C., Cirino, G., Marques, A.R., Guerreiro, B., Goncalves, A.M., Seixas, J.D., Beuve, A., Romao, C.C., and Papapetropoulos, A. (2011). Inhibition of nitric oxide-stimulated vasorelaxation by carbon monoxide-releasing molecules. *Arterioscler. Thromb. Vasc. Biol.* 31: 2570–2576.

41. Sarady, J.K., Zuckerbraun, B.S., Bilban, M., Wagner, O., Usheva, A., Liu, F., Ifedigbo, E., Zamora, R., Choi, A.M., and Otterbein, L.E. (2004). Carbon monoxide protection against endotoxic shock involves

reciprocal effects on iNOS in the lung and liver. *FASEB J.* 18: 854–856.

42. Wegiel, B., Gallo, D.J., Raman, K.G., Karlsson, J.M., Ozanich, B., Chin, B.Y., Tzeng, E., Ahmad, S., Ahmed, A., Baty, C.J., and Otterbein, L.E. (2010). Nitric oxide-dependent bone marrow progenitor mobilization by carbon monoxide enhances endothelial repair after vascular injury. *Circulation* 121: 537–548.

43. Zuckerbraun, B.S., Billiar, T.R., Otterbein, S.L., Kim, P.K., Liu, F., Choi, A.M., Bach, F.H., and Otterbein, L.E. (2003). Carbon monoxide protects against liver failure through nitric oxide-induced heme oxygenase 1. *J. Exp. Med.* 198: 1707–1716.

44. Zuckerbraun, B.S., Chin, B.Y., Wegiel, B., Billiar, T.R., Czsimadia, E., Rao, J., Shimoda, L., Ifedigbo, E., Kanno, S., and Otterbein, L.E. (2006). Carbon monoxide reverses established pulmonary hypertension. *J. Exp. Med.* 203: 2109–2119.

45. Iwata, M., Inoue, T., Asai, Y., Hori, K., Fujiwara, M., Matsuo, S., Tsuchida, W., and Suzuki, S. (2020). The protective role of localized nitric oxide production during inflammation may be mediated by the heme oxygenase-1/carbon monoxide pathway. *Biochem. Biophys. Rep.* 23: 100790.

46. Omura, T. (2005). Heme-thiolate proteins. *Biochem. Biophys. Res. Commun.* 338: 404–409.

47. Ondrias, K., Stasko, A., Cacanyiova, S., Sulova, Z., Krizanova, O., Kristek, F., Malekova, L., Knezl, V., and Breier, A. (2008). H(2)S and HS(-) donor NaHS releases nitric oxide from nitrosothiols, metal nitrosyl complex, brain homogenate and murine L1210 leukaemia cells. *Pflugers Arch.* 457: 271–279.

48. Sun, J., Hoshino, H., Takaku, K., Nakajima, O., Muto, A., Suzuki, H., Tashiro, S., Takahashi, S., Shibahara, S., Alam, J., Taketo, M.M., Yamamoto, M., and Igarashi, K. (2002). Hemoprotein Bach1 regulates enhancer availability of heme oxygenase-1 gene. *EMBO J.* 21: 5216–5224.

49. Gilles-Gonzalez, M.A. and Gonzalez, G. (2005). Heme-based sensors: defining characteristics, recent developments, and regulatory hypotheses. *J. Inorg. Biochem.* 99: 1–22.

50. Igarashi, K. and Sun, J. (2006). The heme-Bach1 pathway in the regulation of oxidative stress response and erythroid differentiation. *Antioxid. Redox Signal.* 8: 107–118.

51. Chin, B.Y., Jiang, G., Wegiel, B., Wang, H.J., Macdonald, T., Zhang, X.C., Gallo, D., Cszimadia, E., Bach, F.H., Lee, P.J., and Otterbein, L.E. (2007). Hypoxia-inducible factor 1alpha stabilization by carbon monoxide results in cytoprotective preconditioning. *Proc. Natl. Acad. Sci. U.S.A.* 104: 5109–5114.

52. Lo Iacono, L., Boczkowski, J., Zini, R., Salouage, I., Berdeaux, A., Motterlini, R., and Morin, D. (2011). A carbon monoxide-releasing molecule (CORM-3) uncouples mitochondrial respiration and modulates the production of reactive oxygen species. *Free Radic. Biol. Med.* 50: 1556–1564.

53. Zuckerbraun, B.S., Chin, B.Y., Bilban, M., d'Avila, J.C., Rao, J., Billiar, T.R., and Otterbein, L.E. (2007). Carbon monoxide signals via inhibition of cytochrome c oxidase and generation of mitochondrial reactive oxygen species. *FASEB J.* 21: 1099–1106.

54. Arias-Diaz, J., Vara, E., Garcia, C., Villa, N., and Balibrea, J.L. (1995). Evidence for a cyclic guanosine monophosphate-dependent, carbon monoxide-mediated, signaling system in the regulation of TNF-alpha production by human pulmonary macrophages. *Arch. Surg.* 130: 1287–1293.

55. Brune, B., Schmidt, K.U., and Ullrich, V. (1990). Activation of soluble guanylate cyclase by carbon monoxide and inhibition by superoxide anion. *Eur. J. Biochem.* 192: 683–688.

56. Christodoulides, N., Durante, W., Kroll, M.H., and Schafer, A.I. (1995). Vascular smooth muscle cell heme oxygenases generate guanylyl cyclase-stimulatory carbon monoxide. *Circulation* 91: 2306–2309.

57. Durante, W., Kroll, M.H., Christodoulides, N., Peyton, K.J., and Schafer, A.I. (1997). Nitric oxide induces heme oxygenase-1 gene expression and carbon monoxide production in vascular smooth muscle cells. *Circ. Res.* 80: 557–564.

58. Derbyshire, E.R. and Marletta, M.A. (2012). Structure and regulation of soluble guanylate cyclase. *Annu. Rev. Biochem.* 81: 533–559.

59. Friebe, A., Schultz, G., and Koesling, D. (1996). Sensitizing soluble guanylyl cyclase to become a highly CO-sensitive enzyme. *EMBO J.* 15: 6863–6868.

60. Stone, J.R. and Marletta, M.A. (1994). Soluble guanylate cyclase from bovine lung: activation with nitric oxide and carbon monoxide and spectral characterization of the ferrous and ferric states. *Biochemistry* 33: 5636–5640.

61. Decaluwe, K., Pauwels, B., Verpoest, S., and Van De Voorde, J. (2012). Divergent mechanisms involved in CO and CORM-2 induced vasorelaxation. *Eur. J. Pharmacol.* 674: 370–377.

62. Dubuis, E., Potier, M., Wang, R., and Vandier, C. (2005). Continuous inhalation of carbon monoxide attenuates hypoxic pulmonary hypertension development presumably through activation of BKCa channels. *Cardiovasc. Res.* 65: 751–761.

63. Schmidt, P.M., Rothkegel, C., Wunder, F., Schroder, H., and Stasch, J.P. (2005). Residues stabilizing the heme moiety of the nitric oxide sensor soluble guanylate cyclase. *Eur. J. Pharmacol.* 513: 67–74.

64. Fujita, T., Toda, K., Karimova, A., Yan, S.F., Naka, Y., Yet, S.F., and Pinsky, D.J. (2001). Paradoxical rescue from ischemic lung injury by inhaled carbon monoxide driven by derepression of fibrinolysis. *Nat. Med.* 7: 598–604.
65. Verma, A., Hirsch, D.J., Glatt, C.E., Ronnett, G.V., and Snyder, S.H. (1993). Carbon monoxide: a putative neural messenger. *Science* 259: 381–384.
66. Vincent, S.R., Das, S., and Maines, M.D. (1994). Brain heme oxygenase isoenzymes and nitric oxide synthase are co-localized in select neurons. *Neuroscience* 63: 223–231.
67. Li, M., Gallo, D., Csizmadia, E., Otterbein, L.E., and Wegiel, B. (2014). Carbon monoxide induces chromatin remodelling to facilitate endothelial cell migration. *Thromb. Haemost.* 111: 951–959.
68. Dioum, E.M., Rutter, J., Tuckerman, J.R., Gonzalez, G., Gilles-Gonzalez, M.A., and McKnight, S.L. (2002). NPAS2: a gas-responsive transcription factor. *Science* 298: 2385–2387.
69. Lukat-Rodgers, G.S., Correia, C., Botuyan, M.V., Mer, G., and Rodgers, K.R. (2010). Heme-based sensing by the mammalian circadian protein CLOCK. *Inorg. Chem.* 49: 6349–6365.
70. Schallner, N., Lieberum, J.L., Gallo, D., LeBlanc, R.H., 3rd, Fuller, P.M., Hanafy, K.A., and Otterbein, L.E. (2017). Carbon monoxide preserves circadian rhythm to reduce the severity of subarachnoid hemorrhage in mice. *Stroke* 48: 2565–2573.
71. Artinian, L.R., Ding, J.M., and Gillette, M.U. (2001). Carbon monoxide and nitric oxide: interacting messengers in muscarinic signaling to the brain's circadian clock. *Exp. Neurol.* 171: 293–300.
72. Ewing, J.F. and Maines, M.D. (1991). Rapid induction of heme oxygenase 1 mRNA and protein by hyperthermia in rat brain: heme oxygenase 2 is not a heat shock protein. *Proc. Natl. Acad. Sci. U.S.A.* 88: 5364–5368.
73. Boehning, D., Moon, C., Sharma, S., Hurt, K.J., Hester, L.D., Ronnett, G.V., Shugar, D., and Snyder, S.H. (2003). Carbon monoxide neurotransmission activated by CK2 phosphorylation of heme oxygenase-2. *Neuron* 40: 129–137.
74. Ben-Shlomo, R., Akhtar, R.A., Collins, B.H., Judah, D.J., Davies, R., and Kyriacou, C.P. (2005). Light pulse-induced heme and iron-associated transcripts in mouse brain: a microarray analysis. *Chronobiol. Int.* 22: 455–471.
75. Kaasik, K. and Lee, C.C. (2004). Reciprocal regulation of haem biosynthesis and the circadian clock in mammals. *Nature* 430: 467–471.
76. Desmard, M., Boczkowski, J., Poderoso, J., and Motterlini, R. (2007). Mitochondrial and cellular heme-dependent proteins as targets for the bioactive function of the heme oxygenase/carbon monoxide system. *Antioxid. Redox Signal.* 9: 2139–2155.
77. Kaiser, S., Selzner, L., Weber, J., and Schallner, N. (2020). Carbon monoxide controls microglial erythrophagocytosis by regulating CD36 surface expression to reduce the severity of hemorrhagic injury. *Glia* 68: 2427–2445.
78. Lavitrano, M., Smolenski, R.T., Musumeci, A., Maccherini, M., Slominska, E., Di Florio, E., Bracco, A., Mancini, A., Stassi, G., Patti, M., Giovannoni, R., Froio, A., Simeone, F., Forni, M., Bacci, M.L., D'Alise, G., Cozzi, E., Otterbein, L.E., Yacoub, M.H., Bach, F.H., and Calise, F. (2004). Carbon monoxide improves cardiac energetics and safeguards the heart during reperfusion after cardiopulmonary bypass in pigs. *FASEB J.* 18: 1093–1095.
79. Piantadosi, C.A. (2008). Carbon monoxide, reactive oxygen signaling, and oxidative stress. *Free Radic. Biol. Med.* 45: 562–569.
80. Piantadosi, C.A., Carraway, M.S., and Suliman, H.B. (2006). Carbon monoxide, oxidative stress, and mitochondrial permeability pore transition. *Free Radic. Biol. Med.* 40: 1332–1339.
81. Taille, C., El-Benna, J., Lanone, S., Boczkowski, J., and Motterlini, R. (2005). Mitochondrial respiratory chain and NAD(P)H oxidase are targets for the antiproliferative effect of carbon monoxide in human airway smooth muscle. *J. Biol. Chem.* 280: 25350–25360.
82. Tsui, T.Y., Obed, A., Siu, Y.T., Yet, S.F., Prantl, L., Schlitt, H.J., and Fan, S.T. (2007). Carbon monoxide inhalation rescues mice from fulminant hepatitis through improving hepatic energy metabolism. *Shock* 27: 165–171.
83. Basuroy, S., Bhattacharya, S., Leffler, C.W., and Parfenova, H. (2009). Nox4 NADPH oxidase mediates oxidative stress and apoptosis caused by TNF-alpha in cerebral vascular endothelial cells. *Am. J. Physiol. Cell Physiol.* 296: C422–32.
84. Basuroy, S., Tcheranova, D., Bhattacharya, S., Leffler, C.W., and Parfenova, H. (2011). Nox4 NADPH oxidase-derived reactive oxygen species, via endogenous carbon monoxide, promote survival of brain endothelial cells during TNF-alpha-induced apoptosis. *Am. J. Physiol. Cell Physiol.* 300: C256–65.
85. Queiroga, C.S., Almeida, A.S., Martel, C., Brenner, C., Alves, P.M., and Vieira, H.L. (2010). Glutathionylation of adenine nucleotide translocase induced by carbon monoxide prevents mitochondrial membrane permeabilization and apoptosis. *J. Biol. Chem.* 285: 17077–17088.
86. Queiroga, C.S., Almeida, A.S., Alves, P.M., Brenner, C., and Vieira, H.L. (2011). Carbon monoxide prevents hepatic mitochondrial membrane permeabilization. *BMC Cell Biol.* 12: 10.

87. Lee, S.R., Yang, K.S., Kwon, J., Lee, C., Jeong, W., and Rhee, S.G. (2002). Reversible inactivation of the tumor suppressor PTEN by H2O2. *J. Biol. Chem.* 277: 20336–20342.

88. Chen, K., Thomas, S.R., Albano, A., Murphy, M.P., and Keaney, J.F., Jr. (2004). Mitochondrial function is required for hydrogen peroxide-induced growth factor receptor transactivation and downstream signaling. *J. Biol. Chem.* 279: 35079–35086.

89. Leslie, N.R., Bennett, D., Lindsay, Y.E., Stewart, H., Gray, A., and Downes, C.P. (2003). Redox regulation of PI 3-kinase signalling via inactivation of PTEN. *EMBO J.* 22: 5501–5510.

90. Suliman, H.B., Carraway, M.S., Tatro, L.G., and Piantadosi, C.A. (2007). A new activating role for CO in cardiac mitochondrial biogenesis. *J. Cell. Sci.* 120: 299–308.

91. Zhang, X., Shan, P., Alam, J., Fu, X.Y., and Lee, P.J. (2005). Carbon monoxide differentially modulates STAT1 and STAT3 and inhibits apoptosis via a phosphatidylinositol 3-kinase/Akt and p38 kinase-dependent STAT3 pathway during anoxia-reoxygenation injury. *J. Biol. Chem.* 280: 8714–8721.

92. Virbasius, J.V. and Scarpulla, R.C. (1994). Activation of the human mitochondrial transcription factor A gene by nuclear respiratory factors: a potential regulatory link between nuclear and mitochondrial gene expression in organelle biogenesis. *Proc. Natl. Acad. Sci. U.S.A.* 91: 1309–1313.

93. Piantadosi, C.A. and Suliman, H.B. (2006). Mitochondrial transcription factor A induction by redox activation of nuclear respiratory factor 1. *J. Biol. Chem.* 281: 324–333.

94. Wu, Z., Puigserver, P., Andersson, U., Zhang, C., Adelmant, G., Mootha, V., Troy, A., Cinti, S., Lowell, B., Scarpulla, R.C., and Spiegelman, B.M. (1999). Mechanisms controlling mitochondrial biogenesis and respiration through the thermogenic coactivator PGC-1. *Cell* 98: 115–124.

95. Lancel, S., Hassoun, S.M., Favory, R., Decoster, B., Motterlini, R., and Neviere, R. (2009). Carbon monoxide rescues mice from lethal sepsis by supporting mitochondrial energetic metabolism and activating mitochondrial biogenesis. *J. Pharmacol. Exp. Ther.* 329: 641–648.

96. Piantadosi, C.A., Withers, C.M., Bartz, R.R., MacGarvey, N.C., Fu, P., Sweeney, T.E., Welty-Wolf, K.E., and Suliman, H.B. (2011). Heme oxygenase-1 couples activation of mitochondrial biogenesis to anti-inflammatory cytokine expression. *J. Biol. Chem.* 286: 16374–16385.

97. MacGarvey, N.C., Suliman, H.B., Bartz, R.R., Fu, P., Withers, C.M., Welty-Wolf, K.E., and Piantadosi, C.A. (2012). Activation of mitochondrial biogenesis by heme oxygenase-1-mediated NF-E2-related factor-2 induction rescues mice from lethal Staphylococcus aureus sepsis. *Am. J. Respir. Crit. Care Med.* 185: 851–861.

98. Wang, B., Cao, W., Biswal, S., and Dore, S. (2011). Carbon monoxide-activated Nrf2 pathway leads to protection against permanent focal cerebral ischemia. *Stroke* 42: 2605–2610.

99. MacLeod, A.K., McMahon, M., Plummer, S.M., Higgins, L.G., Penning, T.M., Igarashi, K., and Hayes, J.D. (2009). Characterization of the cancer chemopreventive NRF2-dependent gene battery in human keratinocytes: demonstration that the KEAP1-NRF2 pathway, and not the BACH1-NRF2 pathway, controls cytoprotection against electrophiles as well as redox-cycling compounds. *Carcinogenesis* 30: 1571–1580.

100. Alves de Souza, R.W., Gallo, D., Lee, G.R., Katsuyama, E., Schaufler, A., Weber, J., Csizmadia, E., Tsokos, G.C., Koch, L.G., Britton, S.L., Wisløff, U., Brum, P.C., Otterbein, L.E. (2021). Skeletal muscle heme oxygenase-1 activity regulates aerobic capacity. *Cell Rep.* 35 (3): 109018.

101. Wang, Z., Figueiredo-Pereira, C., Oudot, C., Vieira, H.L., and Brenner, C. (2017). Mitochondrion: a common organelle for distinct cell deaths? *Int. Rev. Cell Mol. Biol.* 331: 245–287.

102. Kim, H.J., Joe, Y., Rah, S.Y., Kim, S.K., Park, S.U., Park, J., Kim, J., Ryu, J., Cho, G.J., Surh, Y.J., Ryter, S.W., Kim, U.H., and Chung, H.T. (2018). Carbon monoxide-induced TFEB nuclear translocation enhances mitophagy/mitochondrial biogenesis in hepatocytes and ameliorates inflammatory liver injury. *Cell Death Dis.* 9: 1060.

103. Rose, J.J., Wang, L., Xu, Q., McTiernan, C.F., Shiva, S., Tejero, J., and Gladwin, M.T. (2017). Carbon monoxide poisoning: Pathogenesis, management, and future directions of therapy. *Am. J. Respir. Crit. Care Med.* 195: 596–606.

104. Weaver, L.K. (2009). Clinical practice. Carbon monoxide poisoning. *N. Engl. J. Med.* 360: 1217–1225.

105. Clayton, C.E., Carraway, M.S., Suliman, H.B., Thalmann, E.D., Thalmann, K.N., Schmechel, D.E., and Piantadosi, C.A. (2001). Inhaled carbon monoxide and hyperoxic lung injury in rats. *Am. J. Physiol. Lung Cell. Mol. Physiol.* 281: L949–57.

106. Goebel, U., Siepe, M., Schwer, C.I., Schibilsky, D., Brehm, K., Priebe, H.J., Schlensak, C., and Loop, T. (2011). Postconditioning of the lungs with inhaled carbon monoxide after cardiopulmonary bypass in pigs. *Anesth. Analg.* 112: 282–291.

107. Otterbein, L.E., Mantell, L.L., and Choi, A.M. (1999). Carbon monoxide provides protection

against hyperoxic lung injury. *Am. J. Physiol.* 276: L688–94.
108. Song, R., Kubo, M., Morse, D., Zhou, Z., Zhang, X., Dauber, J.H., Fabisiak, J., Alber, S.M., Watkins, S.C., Zuckerbraun, B.S., Otterbein, L.E., Ning, W., Oury, T.D., Lee, P.J., McCurry, K.R., and Choi, A.M. (2003). Carbon monoxide induces cytoprotection in rat orthotopic lung transplantation via anti-inflammatory and anti-apoptotic effects. *Am. J. Pathol.* 163: 231–242.
109. Segersvard, H., Lakkisto, P., Hanninen, M., Forsten, H., Siren, J., Immonen, K., Kosonen, R., Sarparanta, M., Laine, M., and Tikkanen, I. (2018). Carbon monoxide releasing molecule improves structural and functional cardiac recovery after myocardial injury. *Eur. J. Pharmacol.* 818: 57–66.
110. Stein, A.B., Bolli, R., Dawn, B., Sanganalmath, S.K., Zhu, Y., Wang, O.L., Guo, Y., Motterlini, R., and Xuan, Y.T. (2012). Carbon monoxide induces a late preconditioning-mimetic cardioprotective and antiapoptotic milieu in the myocardium. *J. Mol. Cell. Cardiol.* 52: 228–236.
111. Chen, Y., Park, H.J., Park, J., Song, H.C., Ryter, S.W., Surh, Y.J., Kim, U.H., Joe, Y., and Chung, H.T. (2019). Carbon monoxide ameliorates acetaminophen-induced liver injury by increasing hepatic HO-1 and Parkin expression. *FASEB J.* 33: 13905–13919.
112. Kim, H.J., Joe, Y., Yu, J.K., Chen, Y., Jeong, S.O., Mani, N., Cho, G.J., Pae, H.O., Ryter, S.W., and Chung, H.T. (2015). Carbon monoxide protects against hepatic ischemia/reperfusion injury by modulating the miR-34a/SIRT1 pathway. *Biochim. Biophys. Acta* 1852: 1550–1559.
113. Ozaki, K.S., Yoshida, J., Ueki, S., Pettigrew, G.L., Ghonem, N., Sico, R.M., Lee, L.Y., Shapiro, R., Lakkis, F.G., Pacheco-Silva, A., and Murase, N. (2012). Carbon monoxide inhibits apoptosis during cold storage and protects kidney grafts donated after cardiac death. *Transpl. Int.* 25: 107–117.
114. Wang, P., Huang, J., Li, Y., Chang, R., Wu, H., Lin, J., and Huang, Z. (2015). Exogenous carbon monoxide decreases sepsis-induced acute kidney injury and inhibits NLRP3 inflammasome activation in rats. *Int. J. Mol. Sci.* 16: 20595–20608.
115. Cheng, Y., Mitchell-Flack, M.J., Wang, A., and Levy, R.J. (2015). Carbon monoxide modulates cytochrome oxidase activity and oxidative stress in the developing murine brain during isoflurane exposure. *Free Radic. Biol. Med.* 86: 191–199.
116. Kaiser, S., Frase, S., Selzner, L., Lieberum, J.L., Wollborn, J., Niesen, W.D., Foit, N.A., Heiland, D.H., and Schallner, N. (2019). Neuroprotection after hemorrhagic stroke depends on cerebral heme oxygenase-1. *Antioxidants (Basel)* 8: 496.
117. Mahan, V.L., Zurakowski, D., Otterbein, L.E., and Pigula, F.A. (2012). Inhaled carbon monoxide provides cerebral cytoprotection in pigs. *PLoS One* 7: e41982.
118. Wollborn, J., Steiger, C., Doostkam, S., Schallner, N., Schroeter, N., Kari, F.A., Meinel, L., Buerkle, H., Schick, M.A., and Goebel, U. (2020). Carbon monoxide exerts functional neuroprotection after cardiac arrest using extracorporeal resuscitation in pigs. *Crit. Care. Med.* 48: e299–e307.
119. Wu, J., Li, Y., Yang, P., Huang, Y., Lu, S., and Xu, F. (2019). Novel role of carbon monoxide in improving neurological outcome after cardiac arrest in aged rats: Involvement of inducing mitochondrial autophagy. *J. Am. Heart Assoc.* 8: e011851.
120. Zhang, L.M., Zhang, D.X., Fu, L., Li, Y., Wang, X.P., Qi, M.M., Li, C.C., Song, P.P., Wang, X.D., and Kong, X.J. (2019). Carbon monoxide-releasing molecule-3 protects against cortical pyroptosis induced by hemorrhagic shock and resuscitation via mitochondrial regulation. *Free Radic. Biol. Med.* 141: 299–309.
121. Gorojod, R.M., Alaimo, A., Porte Alcon, S., Martinez, J.H., Cortina, M.E., Vazquez, E.S., and Kotler, M.L. (2018). Heme Oxygenase-1 protects astroglia against manganese-induced oxidative injury by regulating mitochondrial quality control. *Toxicol. Lett.* 295: 357–368.
122. Hull, T.D., Boddu, R., Guo, L., Tisher, C.C., Traylor, A.M., Patel, B., Joseph, R., Prabhu, S.D., Suliman, H.B., Piantadosi, C.A., Agarwal, A., and George, J.F. (2016). Heme oxygenase-1 regulates mitochondrial quality control in the heart. *JCI Insight* 1: e85817.
123. Brouard, S., Otterbein, L.E., Anrather, J., Tobiasch, E., Bach, F.H., Choi, A.M., and Soares, M.P. (2000). Carbon monoxide generated by heme oxygenase 1 suppresses endothelial cell apoptosis. *J. Exp. Med.* 192: 1015–1026.
124. Peers, C. and Steele, D.S. (2012). Carbon monoxide: a vital signalling molecule and potent toxin in the myocardium. *J. Mol. Cell. Cardiol.* 52: 359–365.
125. Piantadosi, C.A., Carraway, M.S., Babiker, A., and Suliman, H.B. (2008). Heme oxygenase-1 regulates cardiac mitochondrial biogenesis via Nrf2-mediated transcriptional control of nuclear respiratory factor-1. *Circ. Res.* 103: 1232–1240.
126. Kim, H.P., Wang, X., Zhang, J., Suh, G.Y., Benjamin, I.J., Ryter, S.W., and Choi, A.M. (2005). Heat shock protein-70 mediates the cytoprotective effect of carbon monoxide: involvement of p38 beta MAPK and heat shock factor-1. *J. Immunol.* 175: 2622–2629.
127. Haschemi, A., Wagner, O., Marculescu, R., Wegiel, B., Robson, S.C., Gagliani, N., Gallo, D., Chen, J.F.,

Bach, F.H., and Otterbein, L.E. (2007). Cross-regulation of carbon monoxide and the adenosine A2a receptor in macrophages. *J. Immunol.* 178: 5921–5929.

128. Choi, Y.K., Kim, C.K., Lee, H., Jeoung, D., Ha, K.S., Kwon, Y.G., Kim, K.W., and Kim, Y.M. (2010). Carbon monoxide promotes VEGF expression by increasing HIF-1alpha protein level via two distinct mechanisms, translational activation and stabilization of HIF-1alpha protein. *J. Biol. Chem.* 285: 32116–32125.

129. Woo, C.H., Massett, M.P., Shishido, T., Itoh, S., Ding, B., McClain, C., Che, W., Vulapalli, S.R., Yan, C., and Abe, J. (2006). ERK5 activation inhibits inflammatory responses via peroxisome proliferator-activated receptor delta (PPARdelta) stimulation. *J. Biol. Chem.* 281: 32164–32174.

130. Bilban, M., Bach, F.H., Otterbein, S.L., Ifedigbo, E., d'Avila, J.C., Esterbauer, H., Chin, B.Y., Usheva, A., Robson, S.C., Wagner, O., and Otterbein, L.E. (2006). Carbon monoxide orchestrates a protective response through PPARgamma. *Immunity* 24: 601–610.

131. Choi, Y.K., Park, J.H., Yun, J.A., Cha, J.H., Kim, Y., Won, M.H., Kim, K.W., Ha, K.S., Kwon, Y.G., and Kim, Y.M. (2018). Heme oxygenase metabolites improve astrocytic mitochondrial function via a Ca^{2+}-dependent HIF-1alpha/ERRalpha circuit. *PLoS One* 13: e0202039.

132. Faleo, G., Neto, J.S., Kohmoto, J., Tomiyama, K., Shimizu, H., Takahashi, T., Wang, Y., Sugimoto, R., Choi, A.M., Stolz, D.B., Carrieri, G., McCurry, K.R., Murase, N., and Nakao, A. (2008). Carbon monoxide ameliorates renal cold ischemia-reperfusion injury with an upregulation of vascular endothelial growth factor by activation of hypoxia-inducible factor. *Transplantation* 85: 1833–1840.

133. Nizamutdinova, I.T., Kim, Y.M., Kim, H.J., Seo, H.G., Lee, J.H., and Chang, K.C. (2009). Carbon monoxide (from CORM-2) inhibits high glucose-induced ICAM-1 expression via AMP-activated protein kinase and PPAR-gamma activations in endothelial cells. *Atherosclerosis* 207: 405–411.

134. Park, D.W., Jiang, S., Tadie, J.M., Stigler, W.S., Gao, Y., Deshane, J., Abraham, E., and Zmijewski, J.W. (2013). Activation of AMPK enhances neutrophil chemotaxis and bacterial killing. *Mol. Med.* 19: 387–398.

135. Tsoyi, K., Ha, Y.M., Kim, Y.M., Lee, Y.S., Kim, H.J., Kim, H.J., Seo, H.G., Lee, J.H., and Chang, K.C. (2009). Activation of PPAR-gamma by carbon monoxide from CORM-2 leads to the inhibition of iNOS but not COX-2 expression in LPS-stimulated macrophages. *Inflammation* 32: 364–371.

136. Min, K.J., Lee, J.T., Joe, E.H., and Kwon, T.K. (2011). An IkappaBalpha phosphorylation inhibitor induces heme oxygenase-1(HO-1) expression through the activation of reactive oxygen species (ROS)-Nrf2-ARE signaling and ROS-PI3K/Akt signaling in an NF-kappaB-independent mechanism. *Cell. Signal.* 23: 1505–1513.

137. Li, Q., Guo, Y., Ou, Q., Cui, C., Wu, W.J., Tan, W., Zhu, X., Lanceta, L.B., Sanganalmath, S.K., Dawn, B., Shinmura, K., Rokosh, G.D., Wang, S., and Bolli, R. (2009). Gene transfer of inducible nitric oxide synthase affords cardioprotection by upregulating heme oxygenase-1 via a nuclear factor-{kappa}B-dependent pathway. *Circulation* 120: 1222–1230.

138. Wu, L., Cao, K., Lu, Y., and Wang, R. (2002). Different mechanisms underlying the stimulation of K(Ca) channels by nitric oxide and carbon monoxide. *J. Clin. Invest.* 110: 691–700.

139. Otterbein, L.E., Otterbein, S.L., Ifedigbo, E., Liu, F., Morse, D.E., Fearns, C., Ulevitch, R.J., Knickelbein, R., Flavell, R.A., and Choi, A.M. (2003). MKK3 mitogen-activated protein kinase pathway mediates carbon monoxide-induced protection against oxidant-induced lung injury. *Am. J. Pathol.* 163: 2555–2563.

140. Zhang, X., Shan, P., Otterbein, L.E., Alam, J., Flavell, R.A., Davis, R.J., Choi, A.M., and Lee, P.J. (2003). Carbon monoxide inhibition of apoptosis during ischemia-reperfusion lung injury is dependent on the p38 mitogen-activated protein kinase pathway and involves caspase 3. *J. Biol. Chem.* 278: 1248–1258.

141. Minamoto, K., Harada, H., Lama, V.N., Fedarau, M.A., and Pinsky, D.J. (2005). Reciprocal regulation of airway rejection by the inducible gas-forming enzymes heme oxygenase and nitric oxide synthase. *J. Exp. Med.* 202: 283–294.

142. Hoetzel, A., Dolinay, T., Vallbracht, S., Zhang, Y., Kim, H.P., Ifedigbo, E., Alber, S., Kaynar, A.M., Schmidt, R., Ryter, S.W., and Choi, A.M. (2008). Carbon monoxide protects against ventilator-induced lung injury via PPAR-gamma and inhibition of Egr-1. *Am. J. Respir. Crit. Care Med.* 177: 1223–1232.

143. Hock, T.D., Liby, K., Wright, M.M., McConnell, S., Schorpp-Kistner, M., Ryan, T.M., and Agarwal, A. (2007). JunB and JunD regulate human heme oxygenase-1 gene expression in renal epithelial cells. *J. Biol. Chem.* 282: 6875–6886.

144. Li, M. and Fukagawa, N.K. (2010). Age-related changes in redox signaling and VSMC function. *Antioxid. Redox Signal.* 12: 641–655.

145. Dohi, Y., Ikura, T., Hoshikawa, Y., Katoh, Y., Ota, K., Nakanome, A., Muto, A., Omura, S., Ohta, T., Ito,

A., Yoshida, M., Noda, T., and Igarashi, K. (2008). Bach1 inhibits oxidative stress-induced cellular senescence by impeding p53 function on chromatin. *Nat. Struct. Mol. Biol.* 15: 1246–1254.

146. Ogawa, K., Sun, J., Taketani, S., Nakajima, O., Nishitani, C., Sassa, S., Hayashi, N., Yamamoto, M., Shibahara, S., Fujita, H., and Igarashi, K. (2001). Heme mediates derepression of Maf recognition element through direct binding to transcription repressor Bach1. *EMBO J.* 20: 2835–2843.

147. Kaspar, J.W. and Jaiswal, A.K. (2010). Antioxidant-induced phosphorylation of tyrosine 486 leads to rapid nuclear export of Bach1 that allows Nrf2 to bind to the antioxidant response element and activate defensive gene expression. *J. Biol. Chem.* 285: 153–162.

148. Riddle, M.A. and Walker, B.R. (2012). Regulation of endothelial BK channels by heme oxygenase-derived carbon monoxide and caveolin-1. *Am. J. Physiol. Cell Physiol.* 303: C92–C101.

149. Dallas, M.L., Boyle, J.P., Milligan, C.J., Sayer, R., Kerrigan, T.L., McKinstry, C., Lu, P., Mankouri, J., Harris, M., Scragg, J.L., Pearson, H.A., and Peers, C. (2011). Carbon monoxide protects against oxidant-induced apoptosis via inhibition of Kv2.1. *FASEB J.* 25: 1519–1530.

150. Choi, Y.K., Kim, J.H., Lee, D.K., Lee, K.S., Won, M.H., Jeoung, D., Lee, H., Ha, K.S., Kwon, Y.G., and Kim, Y.M. (2017). Carbon monoxide potentiation of L-type Ca^{2+} channel activity increases HIF-1alpha-independent VEGF expression via an AMPKalpha/SIRT1-Mediated PGC-1alpha/ERRalpha Axis. *Antioxid. Redox Signal.* 27: 21–36.

151. Wilkinson, W.J., Gadeberg, H.C., Harrison, A.W., Allen, N.D., Riccardi, D., and Kemp, P.J. (2009). Carbon monoxide is a rapid modulator of recombinant and native P2X(2) ligand-gated ion channels. *Br. J. Pharmacol.* 158: 862–871.

152. Wilkinson, W.J. and Kemp, P.J. (2011). The carbon monoxide donor, CORM-2, is an antagonist of ATP-gated, human P2X4 receptors. *Purinergic Signal.* 7: 57–64.

153. Dallas, M.L., Scragg, J.L., and Peers, C. (2008). Modulation of hTREK-1 by carbon monoxide. *Neuroreport* 19: 345–348.

154. Althaus, M., Fronius, M., Buchackert, Y., Vadasz, I., Clauss, W.G., Seeger, W., Motterlini, R., and Morty, R.E. (2009). Carbon monoxide rapidly impairs alveolar fluid clearance by inhibiting epithelial sodium channels. *Am. J. Respir. Cell Mol. Biol.* 41: 639–650.

155. Wilkinson, W.J. and Kemp, P.J. (2011). Carbon monoxide: an emerging regulator of ion channels. *J. Physiol.* 589: 3055–3062.

156. Wu, Y., Yang, Y., Ye, S., and Jiang, Y. (2010). Structure of the gating ring from the human large-conductance Ca^{2+}-gated K^+ channel. *Nature* 466: 393–397.

157. Kanu, A. and Leffler, C.W. (2007). Carbon monoxide and Ca^{2+}-activated K^+ channels in cerebral arteriolar responses to glutamate and hypoxia in newborn pigs. *Am. J. Physiol. Heart Circ. Physiol.* 293: H3193–200.

3

Pharmacokinetic Characteristics of Carbon Monoxide

Xiaoxiao Yang, Minjia Wang, Chalet Tan, Wen Lu, and Binghe Wang

Department of Chemistry and Center for Diagnostics and Therapeutics, Georgia State University, Atlanta, GA 30303, USA
Departmental of Pharmaceutics and Drug Delivery, University of Mississippi School of Pharmacy, University, MS 38677, USA

Introduction

As discussed in the various chapters of this book, carbon monoxide (CO) is an endogenous signaling molecule and is toxic at high levels. In this context, CO is fundamentally not different from other signaling molecules, such as insulin, epinephrin, neurotransmitters, and many others, in the body if one examines the safety margins of all these molecules. We have discussed this safety margin issue previously [1]. Therefore, CO concentration is very important in determining its beneficial versus harmful effects. For this reason, understanding the pharmacokinetic (PK) profiles of CO, whether produced endogenously or administered exogenously, is very important for the development of CO-based therapeutics. Along this line, it is important to recognize at the outset that under normal physiological conditions, CO exists in the blood at high micromolar concentrations in a hemoglobin (Hb)-bound form, commonly referred to as carboxyhemoglobin (COHb). Although this nomenclature is technically incorrect and carbonyl-hemoglobin would be the appropriate name, it would cause confusion in the field to introduce a new and different term. Therefore, we shall stay with the name carboxyhemoglobin in this chapter and this book.

In a historical context, it is important to note that CO was first recognized as a poisonous gas long before the discovery of its physiological signaling roles. The toxicity of CO has been acknowledged by humans for thousands of years. The Greek philosopher Aristotle (384–322 BCE) recorded that toxic fumes could be generated through burning coals. It took almost two millennia and the efforts from generations of alchemists and chemists to characterize the chemical nature of this toxic gas as carbonic oxide [2]. After a series of experiments carried out between the 1840s and 1860s, the reason for CO poisoning was revealed by Claude Bernard as displacement of oxygen in the blood cell by CO [3]. It seemed that CO "fixed" in the blood could no longer be displaced by oxygen or any other gases, leading to the cessation of physiological functions [3]. The oxygen-carrying molecule in the blood, Hb, was discovered by Hünefeld in 1840. The affinity of CO to Hb relative to that of O_2 was determined by John Haldane and Lorrain Smith in 1897 [4]. The detailed COHb dissociation curve was determined later in 1912 by Haldane as well [5]. Before the endogenous source of CO was revealed in the late 1940s, toxicity was thought to be the primary biological activity known for CO. Until the paradigm-shifting discovery of CO's signaling functions in the 1990s, PK studies of CO in the early period were primarily focused on determining the relationship between systemic CO exposure and its toxicity, i.e., toxicokinetics. These pioneering studies laid out the critical issues in understanding Hb's binding affinity toward CO and helped understanding CO's toxicity at the molecular level [1]. Such studies also helped in the development of CO exposure guidelines and in providing suggested remedies to protect people from CO poisoning. However, the fact that the earliest studies of CO were almost entirely in the context of toxicokinetics also leaves a question open. Would reanalyses of the same earlier data lead to new interpretations, if such analyses take into consideration the discovery of the endogenous signaling and physiological roles of CO at an appropriate concentration?

CO at low levels (1–2 ppm) was first detected in the alveolar air of humans under normal and pathological conditions by Sjöstrand in 1949 [6], leading to the discovery of endogenous CO generation by *in vivo* degradation of Hb [7]. The discovery marked the beginning of the modern era of research into the physiological and pharmacological functions of CO [8]. In the following decades, PK models of CO were established to predict levels of COHb after systemic exposure to CO, elimination rate under different physiological and pathological conditions, and factors that influence the elimination rate such as respiratory rate and oxygen partial pressure. The distribution of CO in different tissues has also been assessed.

With the establishment of the physiological and therapeutic functions of CO, the toxicokinetic data can serve as part of the PK considerations in using CO as a potential therapeutic agent. As such, there is already some basic understanding of the absorption, distribution, metabolism, and toxicity of the active ingredient, CO, in various species, including humans, with data collected in labs around the world.

In the development of any drug, assessment of its PK profiles is as important as assessing its pharmacodynamic (PD) characteristics. The PK profile of CO is special when compared with nonvolatile small-molecule drugs. As a gaseous molecule, CO has unique PK characteristics, including high diffusivity, high binding affinity with Hb, metabolic stability, and excretion predominately through breathing. The pharmacological functions of CO are believed to be mostly through binding with heme-containing proteins. However, this property also contributes to its toxicity by binding to Hb and other hemoproteins. The formation of a high level of COHb through inhalation of high concentrations of CO gas decreases not only the oxygen-carrying capacity of the blood but also Hb's ability to release oxygen in the peripheral tissue by prohibiting the conformational changes needed for unloading the remaining oxygen. At a low COHb concentration achieved through inhaling low doses of CO gas, the remaining Hb is sufficient for the basic need for oxygen transport. Hb can act as a carrier of CO in the form of COHb, which can transfer the loaded CO to targets in the remote tissue, presumably in a binding affinity-dependent fashion. Specifically, the dissociation constant of CO with Hb is between 1.7 nM and 1.1 µM, corresponding to the relaxed (R) and taut/tense (T) states, respectively. Theoretically, such affinity values indicate that under equilibrium conditions without competition from O_2, the free CO concentration is between 1.7 nM and 1.1 µM when 50% Hb is bound with CO. If competition from O_2 is taken into consideration, the free CO concentration is expected to be higher, depending on the oxygen partial pressure. Therefore, if the binding affinity of a hemoprotein for CO in the remote tissue is higher than that of Hb, the equilibrium of $CO + Hb \rightleftharpoons COHb$ can be driven to the dissociation direction. The dissociated CO can diffuse into the tissue to bind with hemoproteins with high or comparable intrinsic CO binding affinities, such as myoglobin (Mb) (K_d = 28 nM for CO) [9] and neuroglobin (Nb) (K_d = 0.2 nM for CO) [10,11].

The distribution of CO between the blood and various tissues is an important consideration. It should be noted when CO is delivered through a non-inhalation form that allows for higher localized concentrations in the gastrointestinal (GI) tract or in the peritoneal cavity via oral administration or intraperitoneal injection, CO can diffuse into the circulation and bind with Hb. Hb then plays the role of a "scavenger" and carries CO to other parts of the body. In any case, CO finally eliminates through exhalation as the dominant route. Thus, in order to assess the relationship between CO dosage (by gas or by a donor) and CO exposure level (in the target tissue and systemic circulation), PK studies are essential for understanding the combined impact of convoluted factors, including formulation, route of administration, dosage, and CO release rates from various CO donors (Chapters 11–17).

Quantification of CO in Biological Samples

Before the discussion of CO PK studies, it is important to devote some space to discussing the analytical tools available because of the unique properties of CO and associated challenges in its detection with the needed sensitivity, selectivity, robustness, and reliability. Typically, biological samples in the liquid, solid, or gaseous form, such as venous/arterial blood and tissue homogenates, need to be collected and examined. Expired breath is also frequently analyzed because exhalation is the major route of CO excretion [12]. The *in vivo* CO level is of great interest in examining CO's pharmacological effects. However, due to the gaseous nature of CO, its *in vivo* measurements remain a long-standing challenge. Further, the often relatively small amount of exogenous CO compared with the variable endogenous CO body stores adds extra complexity in the measurement of CO amount in a biospecimen. Over the past decades, numerous efforts have been devoted to developing different quantification methods, which in general include radioisotope labeling, gas chromatography (GC), infrared absorption spectroscopy, spectrophotometry/CO-oximetry, immunochemical assay (ELISA), colorimetry [13], and electrochemical assays [14–20].

Due to the high affinity of CO to Hb, until recently CO had been believed to exist primarily in the form of COHb in the blood, with less than 2% of CO in the free unbound form under most circumstances [21,22]. Of course, this percentage depends on the CO exposure duration, the baseline endogenous CO levels, and the physiological state of the subject, which could affect blood pH, oxygen contents, and the affinity states of the various hemoprotein targets for CO. More recent studies have shown large variations in the percentage of CO in the free form [23,24], which are discussed in the section entitled "Pharmacokinetics of CO After Inhalation." Nevertheless, the COHb level in whole blood samples (most often venous blood) is widely considered as an important and convenient readout for CO exposure after CO administration [25]. In pharmaceutical and forensic analysis where blood samples are of interest, CO-oximetry and spectrophotometry are the most frequently used techniques for CO quantification *in vivo* due to their simplicity and specificity. However, GC analysis allows for higher accuracy at low COHb levels (≤2%) and high sensitivity (≤±0.1% COHb) [25]. For tissue samples, where CO exists in both hemoprotein-bound and dissolved forms, sample pretreatments such as protein denaturation and CO liberation are needed to convert all CO into the gas form. The liberated CO collected in the sealed headspace vial can be directly sampled and analyzed by headspace GC techniques. Exhaled breath samples are considered to reflect the partial pressure of CO in the alveoli, the blood circulation, or the average of all tissues, depending on breath holding time [24]. Similar to CO–air mixtures, the concentration of CO in exhaled breath samples can be directly measured by GC analysis or by infrared laser absorption spectroscopy [26,27]. Endogenously produced CO in exhaled breath samples has also been analyzed by isotope-labeling techniques using ^{14}C-heme precursors [17]. However, in modern days, this is only applicable in animal model studies. More recently, there is a rising interest in developing fluorescent probes for imaging CO in cell cultures. Fluorescent probes may also offer the advantage of high sensitivity, reproducibility, and broad applicability for *in vivo* and *in vitro* samples [28] (Chapter 18).

Spectrophotometry/CO-Oximetry

By applying the principles of the Lambert–Beer law, spectrophotometric methods can quantify distinct forms of Hb in blood, which have stable and distinguishable chromophores, allowing for direct and rapid measurements of COHb, reduced hemoglobin (HHb), oxyhemoglobin (O_2Hb), and methemoglobin (MetHb). Each of the four hemes in the tetrameric Hb contains an individual chromophore and has distinct spectrophotometric properties depending on the binding and redox status. In a conventional spectrophotometric method, blood samples are pretreated with a reducing agent (e.g., sodium dithionite) that converts O_2Hb and MetHb into HHb without altering the COHb level. The percentage of COHb is then calculated using absorbance values at two wavelengths with the following equation: percentage of COHb in the blood sample (%COHb) = [COHb]/([COHb] + [HHb]) [29,30]. The second type of spectrophotometric method employs an automated and commercially available spectrophotometer called CO-oximeter, which requires no sample pretreatment, and can quickly measure total Hb (tHb), COHb, and other bound forms of Hb simultaneously [18,19]. To ensure accuracy, close attention needs to be paid to temperature control, sample quality, and assay validation using standard controls. In most scenarios in studying the pharmacokinetics of COHb, the precision and accuracy of CO-oximeter (<±0.5%) are sufficient when standard protocols are followed [14].

Gas Chromatography

Due to its high sensitivity and selectivity, GC is widely considered the gold standard for CO analysis. Before GC analysis, blood or homogenized tissue samples are mixed with a cell lysis solution prior to acidification (e.g., saponin) or oxidation (e.g., ferricyanide) to liberate CO from the Hb/Mb-bound forms. CO released from the samples is directed to the headspace and then separated and quantified by GC. Headspace analysis is an important sampling technique for the GC measurement of volatile substances in biological samples in which a known volume of headspace gas in a sealed vial and in equilibrium with the liquid phase is directly injected into the GC [31]. Careful handling of gas samples is necessary to avoid any leak or contamination. The separated CO can be detected by a thermal conductivity detector [32,33], a flame ionization detector (FID) after catalytic reduction of CO to CH_4 (commonly referred to as methanizer/FID method) [34,35], a UV detector/reduction gas analyzer (RGA) of the released mercury vapor resulting from the combination of CO with mercuric oxide (commonly referred to as the RGA or mercury detection method) [36,37], or mass spectroscopy [38] for improved sensitivity. A calibration curve should be established using prepared standards from 0% to 100% of CO mixture in the carrier gas (e.g., nitrogen).

When whole blood samples are analyzed by GC, COHb% can be converted from the GC parameters using the following equation:

$$\mathrm{COHb}(\%) = 100 \times [\mathrm{CO}(\mathrm{mol}/\mathrm{ml}) \times M_r \mathrm{Hb}(\mathrm{g}/\mathrm{mol})] / [\mathrm{Hb}(\mathrm{g}/\mathrm{ml}) \times 4], \quad (3.1)$$

where M_r is 64 400 g/mol for Hb and Hb (g/ml) concentration can be determined by a hematology analyzer [35,39]. The whole blood sample is analyzed for CO concentration in mol/ml [35]. Others also reported methods of COHb level conversion from GC readings [36,38]. For example, Vreman et al. use Equation 3.2 in calculating COHb levels in their studies [36].

$$\mathrm{COHb}(\%) = \frac{\mathrm{VolCO}}{[\mathrm{Hb}] \times 1.34}, \quad (3.2)$$

where VolCO is the amount (ml) of CO contained in 100 ml of blood, [Hb] is the quantity (g) of all Hb species per 100 ml blood, and 1.34 is the Hüfner factor expressing the total CO binding capacity of 1 g of total Hb [40].

The reported sensitivities for most GC methods were <0.1% COHb and the lowest reported level was 0.005% using an RGA detector, compared with 0.5–1% sensitivity and precision of spectrophotometric methods [14]. The GC method can also directly measure the CO gas concentration in exhaled breath with approximately 0.1 ppm resolution and 7 ppm corresponding to 1% of whole blood COHb level [41].

Electrochemical Detection

Electrochemical sensors have long been widely used in CO detection for environmental air quality monitoring and indoor CO alarm in the commercial market [42–44]. Due to its low cost, ease of operation, and relatively high resolution, the electrochemical method is also frequently used in research and clinical bioanalysis, especially for detecting CO levels in end-tidal breath [45,46]. It can detect CO dissolved in solution and allow for direct and real-time determination of in vivo CO levels in organ tissues [47]. An electrochemical sensor is often called amperometric gas or micro-fuel cell sensor because its working principle resembles that of a fuel cell in which a small electric current is generated via electrochemical reaction in response to diffusion-limited exposure to CO [42,45]. Typically, an electrochemical sensor can produce an adequate signal/noise level (50:1), high resolution (1 ppm or 0.5%, or lower), short response time per measurement (mostly <30 s), adjustable flow rate, and high precision (<±2%) [45,46]. In order to obtain reproducible and reliable analytical results, issues such as nonspecific interference by H_2 and NO [48], fluctuations due to ambient CO levels [47], and specialized modification for different sample types and applications should be taken into consideration.

Fluorescent Probes

There are several excellent fluorescent probes that have been developed for the sensitive and intracellular detection of CO. Chapter 18 specifically focuses on this subject. Therefore, this topic is not duplicated in this chapter.

The selection of the CO quantification methodology should be made based on sample types and instrumentation availability. While GC has better sensitivity and is considered the gold standard for CO analysis in biological samples, CO-oximetry is commonly used for the analysis of CO from blood samples. The development of fluorescent CO probes will likely lead to new CO detection tools in the future, especially intracellularly.

Pharmacokinetics of CO After Inhalation

Because the absorption and distribution of CO after inhalation have been extensively reviewed and summarized by others in the context of toxicology [49–51], this chapter primarily focuses on the analysis of CO pharmacokinetics in the context of its therapeutic potential. The examination of CO's PK profiles has some unique challenges because of its gaseous nature and its partition between dissolved and Hb-bound forms. In drawing analogies with studying the PK profiles of traditional small-molecule drugs, one can view Hb as playing a similar role to albumin in its binding properties and in its ability to affect the availability of "free drugs" and associated PK profiles. For example, warfarin is about 97% plasma protein bound, with only 3% in the free form. As discussed earlier, CO has a strong binding affinity to Hb, which is dependent on the conformational state of Hb. Very importantly, the conformation of Hb is influenced significantly by many factors, including pH, cooperative binding, and the presence of other ligands such as oxygen and 2,3-diphosphoglycerate (2,3-DPG). As a result, the proportion of free CO in the blood varies within a fairly large range depending on CO's partial pressure, oxygen partial pressure, and physiological conditions. In a 2019 study, Oliverio and Varlet found by GC–MS studies that when healthy volunteers were given 57–105 ml of CO gas mixed in pure oxygen for 10 min, leading to a COHb value of about 10%, about 10–60% of the total blood CO (TBCO) was in the free form [23]. The results were drawn from the comparison of the differences between TBCO before and after flushing the blood sample with nitrogen. This significant amount of free CO was in disagreement with the common belief that CO mainly exists in the form of COHb in the blood. The study attributed the reported discrepancies between symptoms and measured COHb levels in CO poisoning studies to the level of free CO in the blood. Although the result could be biased due to the sample amount and experimental errors, it is understandable that certain amount of free CO exists in the blood due to the presence of varying amounts of O_2 and other factors that might affect CO binding. Again, like the competitive binding of small-molecule drug to plasma protein, oxygen binding can induce the release of CO from COHb and increase free CO in the blood. This is likely to happen when the oxygen partial pressure is high, for example, in the lung. Indeed, this competitive binding is the underlying principle for using hyperbaric oxygen (HBO) to treat CO poisoning.

Amid the debate on the quantitative proportions of free CO present in the blood at different CO exposure levels, there is emerging recognition of free CO in the blood in the context of toxicology and pharmacology [23,52,53]. This is not surprising. Much like the discussion of traditional small-molecule drugs, CO has to be in solution in order to establish a binding equilibrium with the intended target(s). As such, in solution means CO is a solute in a single molecule form, i.e., not a bound form. Therefore, there is a need to assess the solution concentration of CO for understanding its pharmacology. Direct assessment of free CO in the blood can be tricky due to the dynamic nature of Hb binding and differences in oxygen partial pressure at various locations, including artery and venous blood in different tissues. However, due to a lack of erythrocytes, the cerebrospinal fluid (CSF) could serve as a surrogate site for assessing an approximate basal level of free CO in the blood. This is with the assumption of free diffusion across the blood–brain barrier by CO the same way as other gas molecules such as oxygen and CO_2 [54]. Currently, relationships between free CO concentrations in human CSF and CO exposure levels have not been reported yet, but an initial analysis can be done using data generated in studying piglet cerebral vascular functions. It is known that the basal blood COHb level of the piglet is about 0% or below the instrument detection limit using an optical oximetry method [55]. The basal CO concentration in the CSF has been determined to be 20–80 nM using GC–MS detection [56–58]. Therefore, it is reasonable to conclude that free CO does exist in the blood and CSF, and the level of free CO should be a major factor to consider when studying pharmacological functions of CO derived from either endogenous production through HO-1 stimulation [56,58] or exogenous exposures.

There are many ways to deliver CO, including gas inhalation and using CO donors via intravenous infusion and GI administration. In animal studies, drug administration methods such as intraperitoneal, subcutaneous, and transdermal have also been reported. Among these administration routes, gas inhalation has been extensively studied in examining CO toxicology and in human clinical trials of CO safety and therapy (see Chapter 30). Therefore, the discussion starts with gas inhalation. In doing so, we use the small molecule–albumin binding model as a reference point.

PK Profiles

There was a very interesting experiment by Haldane in 1895 showing that the poisonous action of CO diminished if given in a mixture with pure oxygen [59]. In his experiment, mice were given high concentrations of CO (0.85–50%) in a combination of various gas mixtures, including air, nitrogen, hydrogen, and pure oxygen at normal and elevated pressures (1–1.9 bar). It was found that when CO was given with oxygen, CO toxicity diminished with increasing oxygen concentrations and pressure. Under atmospheric pressure, a CO concentration of 2000 ppm (0.2%) in an air mixture was sufficient to kill the mice within 30 min [60]. With pure oxygen pressure at 1.9 bar, about 25% of CO in the gas mixture did not show toxic effect as seen with the vitality of the mice. In comparison, when oxygen pressure decreased under otherwise the same conditions, the experimental mice died with characteristic cherry-red colored blood that was known to be the result of CO poisoning. However, determination of COHb concentrations was not made at that time. Haldane and others in later publications [61] attributed the reduced toxicity of CO to the compensatory effects of elevated oxygen pressures in mitigating CO binding to Hb and possibly other targets. One has to acknowledge that there have been no additional in-depth experimental studies to further elucidate the observed phenomenon for over 100 years. There are many questions. Is anoxia the only toxic effect of CO? Does the oxygen pressure affect COHb level? What is the interplay between oxygen and CO in the body (Chapter 7)? A detailed analysis of CO's PK profiles should help to answer these questions and define parameters important in assessing the correlation between COHb and various therapeutic indications.

Absorption

When delivered by gas inhalation, absorption of CO is mainly through the airway and subsequent gas exchange in the pulmonary alveoli. Besides the often inappreciable amount of CO absorbed directly in the upper respiratory tract [62], there are mainly three phases of absorption before CO enters the blood circulation.

The first phase is gas diffusion between the airway opening and the alveoli. The exchange rate in this phase is dependent on CO concentration in the inspired air and physical pulmonary conditions, such as tidal volume, inspiratory reserve volume, and respiratory rate. The breathing process under normal conditions is a combination of inflation and diffusion. As stated in Fick's first law in physics (Equation 3.3), the diffusion flux (J) is driven by the concentration gradient (∂c). Therefore, higher CO concentrations in the air should cause the alveoli CO concentration to increase more rapidly. The pulmonary function parameters such as ventilation rate and body position also determine the amount of CO that gets into the body within a given amount of time. For example, it was found that the CO diffusion rate is higher in a supine position at rest than in a sitting position; exercise can also significantly increase diffusion by enhancing ventilation rate [63].

$$J = -D \frac{\partial C}{\partial x}, \quad (3.3)$$

where J is the diffusion flux ($m^{-2} s^{-1}$). It measures the amount of substance that flows through a unit area during a unit time interval. D is the diffusion coefficient or diffusivity ($m^2 s^{-1}$) of the specific gas. It is proportional to the squared velocity of the diffusing particles. The velocity of the particles depends on the temperature, viscosity of the diffusing fluid, and the size of the particles. C (for ideal mixtures) is the concentration of the diffusing particles per given volume (m^{-3}), x is the position, and ∂x denotes the distance the particles travel in a given time.

The second phase is CO transfer in a liquid phase across the air–blood interface. During the transfer from the gas phase to the liquid phase, CO needs to be dissolved in the aqueous milieu of the alveoli membrane surface. This process can be determined by Henry's law. The quantity of an ideal gas that dissolves in a definite volume of liquid is directly proportional to the partial pressure of the gas. Therefore, a higher CO concentration in the inspired air translates into a higher partial pressure of CO, thus higher solubility in the aqueous phase. The rate of CO to diffuse and dissolve in an aqueous medium is also proportional to the concentration difference stated by Fick's law and the molecular mass of the gas as stated by Graham's law. For a gas molecule diffusing across a permeable membrane, Fick's law can be derived into Equation 3.4:

$$V'_{gas} = D \cdot A \cdot \frac{\Delta P}{T}, \quad (3.4)$$

where V'_{gas} is the rate of gas diffusion across the permeable membrane, D is the diffusion coefficient of the gas for the particular membrane, A is the surface area of the membrane, ΔP is the difference in the partial pressure of the gas across the membrane, and T is the thickness of the membrane.

For a gas mixture such as CO in the air, the rate for a gas molecule to effuse through a permeable membrane is inversely proportional to the square root of the molecular weight (Equation 3.5). Therefore, CO has a higher effusion rate than oxygen due to its lower molecular weight.

$$\frac{r_1}{r_2} = \sqrt{\frac{M_2}{M_1}}, \quad (3.5)$$

where r_1 and r_2 are the rates of effusion of gas 1 and gas 2, respectively, and M_1 and M_2 are the molar masses of gas 1 and gas 2, respectively.

As the circulating blood continuously dilutes the dissolved CO, the alveolar CO continuously gets transferred from the alveolar space across layers of membrane and diffuses into the blood. Therefore, factors that may affect the CO absorption rate include cardiac output, the total surface area of the alveolar, and the physiological conditions of the alveolar microenvironment and structure.

The solubility of a gas in a given medium is denoted by Henry's law (Equation 3.6):

$$C = k \cdot P, \quad (3.6)$$

where C is the concentration of a dissolved gas, k is the Henry's law constant referring to the particular gas and the medium [64], and P is the partial pressure of the gas.

The solubility of pure oxygen at 1 bar in water at 25 °C is 42 mg/l ($k = 1.3 \times 10^{-8}$ M/Pa) [64], and the solubility of CO in water under the same conditions is 27.6 mg/l ($k = 9.7 \times 10^{-9}$ M/Pa) [64]. In an air mixture containing 78% N_2, 21% O_2, and 100 ppm CO, the solubility of CO in water should be 2.75 µg/l (about 100 nM) according to Henry's law, which is much less than the solubility of oxygen (8.8 mg/l, 275 µM) under the same condition. This solubility ratio (~2750) exceeds that of the average difference in Hb's binding affinity for CO and oxygen (246) [5]. This has implications on the achievable equilibrium COHb levels. In a theoretical calculation, these numbers mean a maximal COHb level of 8.9% without considering CO's effects on human physiology, which could in turn affect Hb's affinity for CO and O_2 and shift the equilibrium point. In experimental studies, Sheps et al. found that it took an average of about 80 min of exposure to 100 ppm of CO for COHb levels to reach 6% [65]. In a different study, exposure of healthy human subjects to 100 ppm of CO for 8 h led to an average of 12.5% of COHb [66,67]. Assuming the average breathing rate of 6 l/min, 7.5 mM of Hb concentration, and 4 l of blood volume, it would take less than 2 h to provide a sufficient amount of CO to reach 10% COHb. These experimental and theoretical numbers not only give some important reference points for consideration in experimental designs, but also indicate kinetic barriers to reaching equilibrium points. Further supporting this point is the results of a human clinical trial (NCT00122694); inhalation exposure to 100–125 ppm of CO for 2 h/day for four consecutive days led to a median COHb peak level of 2.6–3.1% [68]. These points are further discussed later using examples and computational modeling work.

The third phase is the diffusion of dissolved CO into the erythrocyte and binding with Hb. Though the water solubility of oxygen and CO is low, the actual solubility (equivalent) of oxygen in the arterial blood is about 300 mg/l (~9.4 mM) under normal atmospheric pressure and physiological conditions as a result of Hb binding [69] and the high Hb concentrations in human blood, at about 1.9–2.8 mM for the Hb tetramers in adults [70,71].

Binding with Hb is the additional driving force that helps CO to diffuse into the bloodstream. If one considers the CO partial pressure according to the total CO concentration in the blood as a whole and compares it with the CO partial pressure in the inspired air, CO seems to diffuse against the concentration gradient. Because of Hb's high binding affinity to CO by forming COHb, it helps to "remove" or "sequester" CO in the red blood cell and in the plasma, thus maintaining a high pressure differential between air and blood, and consequently helps CO diffusion from air into the blood.

When inhaled CO concentration is constant and enough time is allowed for an establishment of a steady state (the section entitled "The CFK Model" discusses the time needed to reach steady state, which can be as long as 250 min in humans), the ratio of COHb and oxyhemoglobin (O_2Hb) is equal to the ratio of their respective partial pressures multiplied by the affinity constant M (also known as the Haldane coefficient). This relationship was described by Haldane in 1898 (Equation 3.7):

$$\frac{COHb}{O_2Hb} = M \cdot \frac{P_{CO}}{P_{O_2}}. \quad (3.7)$$

The affinity constant M is temperature dependent and was determined to be about 246 at body temperature [5]. However, it needs to be noted that the apparent affinity constant M in this equation is different from the ratio of actual binding constant for CO and oxygen to Hb, and does not take into consideration possible variations of other physiological conditions such as pH, partial pressure of CO_2, and the presence of cofactor 2,3-DPG.

Distribution

As stated earlier, CO is highly diffusible in the free state and can bind with Hb in the circulation system. Carried in the blood in both dissolved free and COHb-bound forms, CO is distributed to tissues in the body and binds with hemoproteins in the cell. The dynamic equilibrium among different states of Hb is almost certainly that allows CO to be transferred from Hb to other targets with a much lower affinity (higher K_d) than the lowest K_d of Hb to CO (1.3 nM). Before CO reaches a hemoprotein in the intracellular milieu, it has to be in the free form to diffuse out of the blood and diffuse across several biological membranes, including the vascular wall, cellular membrane, and certain organelle membrane. For CO in the form of COHb, it has to dissociate from Hb into free form before engaging other targets. This process can be driven by the Bohr effect or competitive binding with oxygen. In the latter case, the competition would depend on oxygen partial pressure and would involve binding to both Hb and other hemoproteins, unless the target hemoprotein has a much higher binding affinity with CO than oxygen.

CO Binding with Hb and Mb

As stated at the beginning of the section entitled "Quantification of CO in Biological Samples," CO and O_2 competitively bind with hemoproteins. As two most abundant hemoproteins in the body, Hb and Mb have major roles in regulating CO's distributions.

As a tetrameric protein, there are four heme molecules in one Hb, which gives a heme concentration of about 7.5–11.3 mM in the blood. The overall apparent dissociation constant for CO to Hb is approximately 1.3 nM, and is 234 times lower than that of O_2 [72]. This value, however, is not a constant but is dependent on the level of oxygen saturation, the Hb conformational state, pH, and presence of other molecules such as 2,3-DPG [73]. It is well understood that the four heme molecules work cooperatively in binding with oxygen (and CO). It is generally believed that Hb exists in two conformations, a low-affinity T form and a high-affinity R form, the interconversion of which involves some complex processes with the detailed mechanism still subject to debate and interpretations [74–78]. Briefly, deoxyHb adopts the T state at low oxygen concentrations due to the allosteric effect from increased CO_2 concentrations, increased acidity in remote tissues (Bohr effect), and a stabilization effect by 2,3-DPG. In the well-oxygenated pulmonary capillaries, the elevated pH and reduced CO_2 concentration destabilize the T state. Binding of the first oxygen molecule to the T state of Hb has a K_d of 422 µM [79]. However, this binding changes the ferrous iron to a six-coordination state and an octahedral conformation, leading to allosteric changes of the Hb tetramer to the high-affinity R state [80], with a K_d of 0.3–1 µM for O_2. As a result of the conformational changes, binding affinity increases by about 300-fold for oxygen [81]. This cooperative binding phenomenon gives a sigmoidal shape O_2Hb dissociation curve (Figure 3.1A) and empowers Hb to bind fast and tightly with oxygen in the lung where the oxygen partial pressure is high. The same cooperativity also readily allows unloading oxygen to the peripheral tissue where the oxygen partial pressure is low. CO binds with Hb in a similar manner as oxygen. Binding of the first CO molecule to the T form of deoxyHb transforms it into the R form with increased affinity for the successive CO or O_2 molecules. This is part of CO's toxic effect. At a low CO concentration, when O_2 is more capable of competing against CO binding, the consequence of CO binding is the formation of partially CO-bound Hb, including $Hb_4(O_2)_2(CO)_2$ and $Hb_4(O_2)_3CO$ [82]. Such binding results in not only a decreased oxygen binding capacity but also a shift of the oxyHb dissociation curve to the left with a decreased ability for Hb to unload oxygen at peripheral tissue where the oxygen partial pressure is low (Figure 3.1B) [83,84]. At high CO concentrations, CO can occupy the entire tetrameric Hb to form $Hb_4(CO)_4$ [82], leading to a complete loss of oxygen-carrying ability of Hb. The K_d for CO with the T state of Hb was determined to be 1.8 µM for the α subunit and 4.5 µM for the β subunit. In contrast, the K_d of CO binding to Hb in the R state was determined to be 1.7 nM for the α subunit and 0.7 nM for the β subunit [81,85]. The difference in affinity between the R and T states is 1000–6400-fold, which is far greater than that of oxygen. This means that the difference in affinity between CO and O_2 with Hb is higher (about 600-fold) in the R state, compared to the difference in the T state (about 230-fold). The higher relative affinity for CO in the R state (lung) than in the T state (peripheral tissue) has been used as one way to explain why inhaled CO is more prone to causing toxicity than CO administered in other ways, such as infusion of COHb in experimental animal models [86,87].

The mutual influence of CO and O_2 on each other's binding affinity to Hb not only is part of CO's toxicity but also affects the pharmacokinetics of CO. The apparent oxygen binding/dissociation characteristics are illustrated by the Bohr effect, and this phenomenon is the reason for the efficient delivery of O_2 to remote tissues. The Bohr effect for oxygen is expressed by the Bohr factor: $\Delta\log(P_{50})/\Delta pH$, in which $\Delta\log(P_{50})$ is the change of log value of the partial pressure of O_2 to saturate 50% of Hb (P_{50}) and ΔpH is the change of the pH value. Hlastala et al. conducted a series of experiments on the influence of CO on the Bohr effect of human Hb [84]. It was found that besides the lower O_2 saturation level and P_{50}, increased COHb level could significantly increase the Bohr factor. Especially at high COHb levels (>25%), a small pH decrease can lead to a significant decrease of oxygen's binding affinity to Hb, thus facilitating the unloading of remaining oxygen. It seems that this effect offers

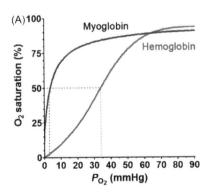
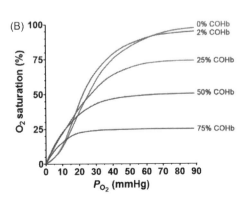

Figure 3.1 Schematic plots of oxygen dissociation curves of Hb and Mb (A) and O_2 saturation curve at different COHb levels (B). Source: Modified from [84].

protection under CO intoxication conditions [84]. It also indicates that the partially CO-bound Hb tetramer could result in a "pseudo-oxygenated" state of Hb, which resembles the conformation of a completely oxygenated Hb and its high propensity to release the remaining oxygen as a result of the Bohr effect. It should be noted that the Bohr effect also applies to CO's binding to Hb [80]. Sawicki and Gibson used a flow-laser flash experiment to demonstrate that CO's cooperativity is also pH dependent, though less pronounced than that of O_2 [88]. At pH 7 in phosphate buffer, Hb is completely switched to the R state after binding three CO molecules. In contrast, at pH 9, significant conformational changes occur after binding with one CO molecule [88]. However, the physiological relevance of pH 9 is not clear. At a lower physiological pH, upon CO binding, mammalian Hb still exists as a mixture of T and R states, due to the high binding affinity of CO and cooperativity of mammalian Hb. On the contrary, fish Hb appears to have an exaggerated Bohr effect under acidic pH. For example, carp Hb is "frozen" in the T state even upon ligand binding (Root effect). This phenomenon helps fish to use oxygen more efficiently and maintain gas in the swim bladder. As a research tool, it is easier to study the CO's binding property at a lower pH and the T state. Saffran and Gibson used menhaden Hb and flash photolysis to assess the affinity of Hb in the T state for CO at acidic pH [89]. The K_d for CO at pH 6.0 was determined to be 0.58 µM for the high-affinity chains and 11 µM for the low-affinity chains. Although the values from fish do not directly apply to mammalian Hb, they are still within the similar range of the T state CO affinity of human Hb.

Because the K_d of Hb for CO can be as low as 0.7 nM, if oxygen as the binding competitor is not present in the air, the water solubility of CO under 100 ppm is sufficient to drive to full saturation of Hb. However, as oxygen is present in the air and the partial pressure is over 2000-fold higher than that of CO at 100 ppm, steady-state COHb level is not able to reach such full CO saturation. In a phase I clinical trial (NCT00122694) of inhalation of 100–125 ppm CO gas for 2 h/day for four consecutive days, a median COHb peak level of 2.6–3.1% was observed [68]. This is far below saturation and clearly shows that other factors such as oxygen partial pressure and elimination through exhalation, among others, contribute to the lower COHb levels. Details will be discussed in the PK modeling section.

In 1980, Agostoni et al. utilized computer modeling to simulate MbCO saturation levels at different COHb concentrations and different oxygen partial pressures [90]. Steady state was assumed with constant CO exposure and constant O_2 consumption, without considering kinetic effects. A three-compartment model was adopted, including arterial blood compartment, venous capillary blood compartment, and tissue Mb compartment (Figure 3.2). Among these compartments, due to oxygen consumption, P_{O_2} was different and was set as an independent parameter in a range of physiologically relevant values. CO was entered as COHb saturation level at the arterial level and capillary P_{CO} was computed assuming that no COHb saturation changes occurred at the capillary level. Since a steady state was assumed, the total amount of CO in the blood, which is the COHb form plus dissolved CO, was considered to be

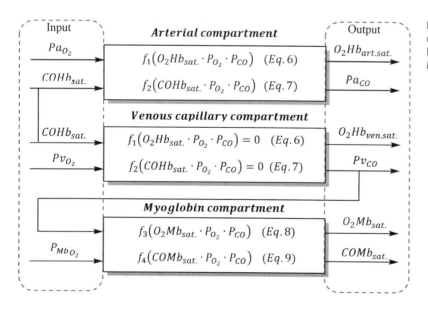

Figure 3.2 A three-compartment model for the CO/O_2 distribution between blood and Mb. Source: Modified from [90].

a constant. Because of the high binding affinity of CO with Hb, most of the CO was expected to be in the COHb form. COHb in the blood was considered to be constant. The COHb and O$_2$Hb levels can be expressed as Equations 3.8 and 3.9 based on the Adair equation to reflect the cooperative binding between four heme units in the Hb:

$$O_2Hb_{sat.} = \frac{x}{x+y} \cdot \frac{A_1(x+y) + 2A_2(x+y)^2 + 3A_3(x+y)^3 + 4A_4(x+y)^4}{4[1 + A_1(x+y) + A_2(x+y)^2 + A_3(x+y)^3 + A_4(x+y)^4]}, \quad (3.8)$$

$$COHb_{sat.} = \frac{y}{x+y} \cdot \frac{A_1(x+y) + 2A_2(x+y)^2 + 3A_3(x+y)^3 + 4A_4(x+y)^4}{4[1 + A_1(x+y) + A_2(x+y)^2 + A_3(x+y)^3 + A_4(x+y)^4]}, \quad (3.9)$$

where A_1, A_2, A_3, and A_4 are the Adair's constants ($A_1 = 0.0218$, $A_2 = 0.00912$, $A_3 = 3.75 \times 10^{-6}$, and $A_4 = 2.47 \times 10^{-6}$) [91], $x = P_{O_2}$, and $y = M_{Hb} \cdot P_{CO}$ (M is the Haldane's coefficient, $M_{Hb} = 245$; see Equation 3.7).

Due to the lack of cooperativity in binding, the binding of O$_2$ and CO with Mb can be expressed as Equations 3.10 and 3.11:

$$MbO_{2sat.} = \frac{P_{O_2}}{P_{O_2} + M_{Mb} \cdot P_{CO} + P_{50}}, \quad (3.10)$$

$$MbCO_{sat.} = \frac{M_{Mb} \cdot P_{CO}}{P_{O_2} + M_{Mb} \cdot P_{CO} + P_{50}}, \quad (3.11)$$

where P_{50} is the P_{O_2} giving 50% MbO$_2$ saturation, and the value was given by Antonini and Brunori in their book [92], despite that there was some uncertainty about this constant [93]. Here, the values are assigned to be $M_{Mb} = 39$ and $P_{50_{(Mb)}} = 2.7$ mmHg [90,92].

According to these equations, modeling work is conducted here using modern technical computing software (Wolfram Mathematica 12) to provide various simulations. The results are in agreement with the original work done by Agostoni et al. [90], with additional insights and details due to the ability of the modern software to empower a broader and more comprehensive view of the various scenarios. Plotting Equation 3.9 gives the relationship between P_{CO} and P_{O_2} under different COHb levels (Figure 3.3). Plotting Equation 3.8 together with Equation 3.10 gives the distribution of CO to Mb at different COHb levels (Figures 3.4 and 3.5). It should be noted that Equation 3.9 is regression based and should be considered as an approximation (so is Figure 3.3). The equation itself is not exact [94,95]; there are variations of the parameters such as the Adair constant and the Haldane coefficient. The Adair constant of CO is slightly different from that of O$_2$ [91]. However, these parameters are generated through rigorous regression analyses and represent a good approximation that can be used to analyze the dynamic distribution of CO in the body [95].

In Figure 3.3, a notable inflection point can be seen at around or below 20 mmHg of oxygen partial pressure on the contour of different levels of COHb. Toward the higher P_{O_2} end from that point, a higher P_{CO} is needed to achieve a higher COHb level. In other words, at the same COHb level, the higher P_{O_2} (arterial blood) means a higher P_{CO} in the blood as

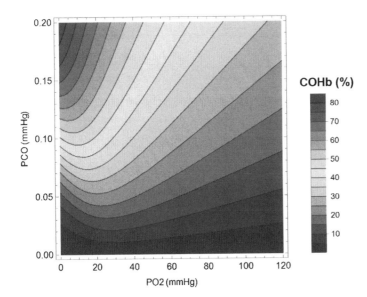

Figure 3.3 P_{CO} in arterial and venous compartments as a function of P_{O_2} at different COHb levels according to Equation 3.9.

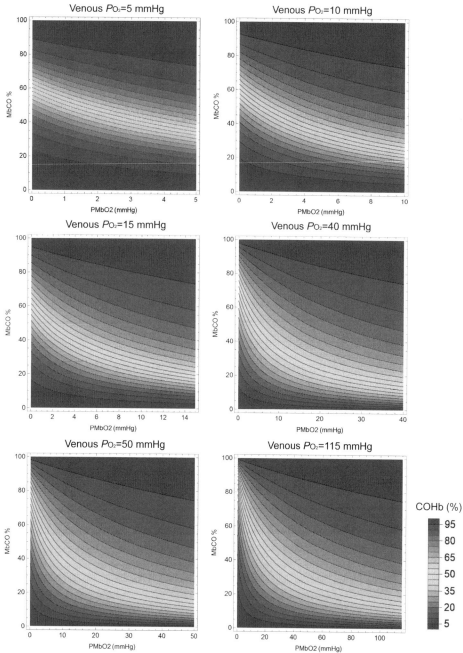

Figure 3.4 MbCO saturation as a function of at various predefined COHb levels with the blood level between 5 and 115 mmHg. (The plots are based on Equations 3.9 and 3.11 with the parameters and constants detailed in the text).

well. Toward the lower P_{O_2} end (venous blood) from that inflection point, the congested curves indicate that it is easier to increase P_{CO} to achieve a higher COHb level, and lowering P_{O_2} can increase CO partial pressure with the same COHb level. From this plot, one can also see that in the high-P_{O_2} region, increasing P_{CO} does not result in much increase in the COHb level. In other words, if P_{CO} remains the same, the higher the P_{O_2}, the lower the COHb level. This phenomenon is attributed to the competition between O_2 and CO in binding with Hb [90]. The inflection point reflects the minimum P_{O_2} to dominate Hb binding effectively.

There is no doubt that blood COHb levels are the most important factor to consider in evaluating the PK properties of CO. However, in assessing tissue distributions, CO binding to Mb is also a critical factor to consider because of Mb's high abundance, second only to Hb. While results from several earlier CO-rebreathing studies [96–98] estimated about 10–15% of the CO being in the extravascular space under normal physiological conditions, Coburn

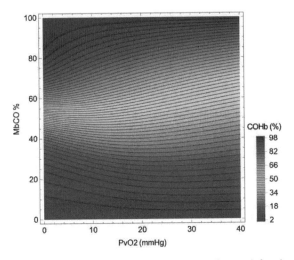

Figure 3.5 MbCO saturation as a function of at predefined COHb levels and of 1 mmHg (mitochondria vicinity). (The plots are based on Equations 3.9 and 3.11 with the parameters and constants detailed in the text)

studied physiological variables affecting CO binding to Mb and its subsequent distribution to the extravascular compartment [17]. From such studies, it is clear that the picture is more convoluted than what a constant ratio seems to indicate, the percentage of CO in the extravascular compartment can change depending on the physiological states and experimental conditions, and many factors need to be considered. Further, binding kinetics may also be a factor, which could affect the experimental outcome and often is not considered in equilibrium binding studies. In the subsequent sections, we look at CO binding to Mb in detail in the context of its effects on the PK properties of CO.

Due to the high binding affinity of Mb for CO (K_d = 29 nM), it binds to CO as CO is carried to muscle tissues by blood circulation. Mb, as the second most abundant hemoprotein in the body, has a concentration of 0.9–2.2 g per 100 g dry muscle, depending on the anatomical position [99]. On average, a man with a body weight of 70 kg has about 120–150 g of Mb, corresponding to about 6.6–8.4 mmol. The total heme from Mb is calculated to be about one-fifth of that from the Hb pool in the body. The binding affinity of oxygen to Mb is higher than that of Hb. Therefore, Mb acts as a reservoir to store oxygen for cellular metabolism. Its binding affinity to CO is less than that of Hb at the highest affinity R state, with the K_d of human Mb being 0.8 μM for O_2 and 29 nM for CO, leading to a 28-fold difference [100,101]. The Haldane's coefficient of Mb, M_{Mb}, is 39 [90,92]. As a result, oxygen has a higher potential to compete with CO in binding with Mb than in the case of Hb. However, in the tissue, a much lower O_2 partial pressure is presented, diminishing this potential.

Theoretically, according to the three-compartment model illustrated in Figure 3.2 and the corresponding equation (Equations 3.9 and 3.11), a correlation between tissue compartment (Mb) oxygen partial pressure (P_{MbO_2}) and MbCO saturation level at different COHb levels can be generated based on the assumption of the P_{MbO_2} being equal to or less than the capillary oxygen partial pressure (P_{O_2}). The capillary can be either venous or arterial, and the value for P_{O_2} should be fixed to decrease the number of variables to three, allowing for the generation of a 3D contour plot (Figure 3.4).

From the plots in Figure 3.4, it is clear that Mb can get substantially saturated by CO, especially at low blood oxygen partial pressures. For example, at venous P_{O_2} = 5 mmHg, the MbCO level is significantly higher than the COHb level throughout the range of P_{MbO_2}. Due to the diminished competition from oxygen, it is more favorable for CO to "transfer" from COHb to Mb at low capillary oxygen levels. As the blood oxygen partial pressure increases, it becomes harder for Mb to be saturated by CO.

This model can potentially offer explanations to many experimental findings in CO's distributions that are hard to explain without such an in-depth analysis. For example, Coburn found that the canine myocardium has a higher COMb level than the skeletal muscle under normal ambient air conditions [102]. The ratio (COMb/COHb) was determined to be about 3 for myocardium and 1 for resting skeletal muscle. On the oxygen partial pressure side, Whalen's study offered evidence that may help to explain this difference. P_{O_2} in resting skeletal tissue is generally higher than that in the cardiac tissue. For cats, the mean P_{O_2} is about 6.9 mmHg in the cardiac (ventricle) muscle, and 19.7 mmHg in the resting cat skeletal (soleus) muscle, which gives a ratio close to 1:3 [103]. The lower oxygen partial pressure in the cardiac muscle may be due to the higher metabolic consumption rate of oxygen. Therefore, according to the model presented in Figure 3.4, under such P_{O_2} conditions, cardiac muscle with a lower oxygen partial pressure should have a higher COMb level than the skeletal muscle, which is indeed the case based on experimental findings. In the same experiment, Coburn also found that after the dog was kept in a closed system and oxygen was removed to create hypoxic conditions (arterial P_{O_2} < 40 mmHg), skeletal and myocardium COMb levels increased by three- and twofold, respectively. At the same time, the blood COHb decreased from 0.8% to about 0.5%. CO seemed to shift from the blood compartment to the tissue (Mb) compartment. Such experimental findings further support the model described earlier: COMb levels should increase when oxygen partial pressure drops. Hence, the model has a good correlation with the experimental data and can be used to predict CO distribution in the Mb compartment at different COHb levels, at least under low-P_{O_2} conditions such as muscle tissues.

Similarly, a model can be derived to reflect the correlation between COMb saturation levels and the capillary venous P_{O_2} by fixing the oxygen partial pressure in the intracellular space (P_{MbO_2}), where Mb resides. It has been reported that the oxygen partial pressure near mitochondria is about 1 mmHg [104], which is presumed to represent the partial pressure in the intracellular space. Therefore, the modeling work is done with $P_{MbO_2} = 1\,\text{mmHg}$ (Figure 3.5). The model indicates that cellular Mb is partially saturated by CO to a high degree, and COMb level is higher than COHb when the COHb level is less than approximately 50%, which is in accordance with the aforementioned experimental result in cats [102]. Beyond this level, the inverse dependence of COMb levels on P_{VO_2} seems to be reversed. Although the implications of this high level of COMb are unclear, it does indicate that the intracellular Mb is hard to get fully saturated at high COHb levels. As explained by the modeling work, the background intracellular CO in the form of COMb is nonnegligible, and a calculation can be made: Assuming a man with a COHb level of 1%, which is at a physiological level sustained by heme turnover, the human heme concentration in the blood derived from Hb is about 8 mM based on the presence of four heme molecules per tetrameric Hb. Therefore, there is about 36 mmol heme in the blood in the form of Hb, assuming a blood volume of 4.5 l. One percent COHb corresponds to 0.36 mmol of bound CO. According to the model, at this COHb level and muscle P_{VO_2} of 20–40 mmHg [105,106], the calculated COMb level is about 2%. Total Mb is in the range of about 120–150 g for a 70 kg man [99], corresponding to 6.6–8.4 mmol. Taking 7 mmol as a reference point, 2% COMb is about 0.14 mmol of Mb-bound CO. Therefore, at this COHb level, the amount of CO stored in the blood is about 2.5-fold compared with that stored in Mb. In other words, as the first and second most abundant hemoproteins, under normoxic conditions, about 75% of hemoprotein-bound CO is in the COHb form, while most of the remaining 25% is in the COMb form. Under lower P_{VO_2} hypoxic conditions, the COMb storage can be even higher, which may render the amount of CO stored in the COMb form being the same as that in the COHb form. The results are generally in line with the experimental and calculation results reported by Coburn in that the COMb/COHb ratio is about 1 for resting skeletal muscle (higher PO$_2$) and 3 for myocardium (lower PO$_2$) under normoxic conditions. Under hypoxic conditions, the COMb/COHb concentration ratio could be much higher at about 3.2 for resting skeletal muscle and 6.5 for myocardium muscle [17]. The assumption that there is a considerable portion of body CO storage in the form of COMb is also supported by the tissue CO concentration measurements as discussed in the last part of this section (Table 3.1) [107]. Normal human muscle CO concentrations are about 1/12 of the blood CO concentration (in pmol/mg). Assuming average muscle weight is about 50% of the body weight [108], and the blood volume is about 7% of body weight [109], CO stored in muscle could be as much as about two-thirds of that stored in the blood. However, it should be noted that these modeling works made certain assumptions as to body muscle weight percentage, blood volume index, Mb content, and oxygen partial pressure. Certainly, these numbers may change depending on individual variability and the physical states of a specific experimental subject. Nevertheless, such modeling work provides some boundary parameters to consider in designing CO-related experiments.

These theoretical and experimental findings indicate that systemic delivery of CO at a low COHb level would lead to a high percentage of the CO being in the form of COMb. Such findings may indicate the important "maintenance roles" of CO in muscle biology. This phenomenon was also used to explain the cardiac toxicity of CO in coronary patients caused by smoking [93]. It should be noted that Mb content is different among various species. COMb concentration in skeletal muscle (in fresh tissue) is about 0.08–1.52 mg/g for rats [110], 3.67–5.47 mg/g for canines [111], and 2–10 mg/g for humans [112]. Such findings may indicate different CO distribution profiles in different species. Therefore, specific attention is needed in analyzing preclinical PK data of CO generated in different animal models. Another issue to be mentioned is that there is a huge difference in Mb concentrations among different tissues, including muscle tissues at different anatomical locations. Studies even led to the discovery that Mb is absent in some smooth muscle tissues such as the uterus and taenia coli muscle [112]. The difference in Mb distribution is expected to result in differences of CO distributions in the tissue, which again requires detailed attention when evaluating CO for various therapeutic indications.

CO Distribution in the Tissue with Neuroglobin

CO's distributions in the blood and muscle tissues are relatively easy to define due to the abundance of heme proteins in both. As proposed by Luomanmaki and Coburn, CO equilibrates rapidly with extravascular tissues (the section entitled "The CFK Model" discusses the time needed to reach a steady state, which can be as long as 250 min in humans) [113]. Therefore, under a steady state, the difference in CO distribution in tissues should be dependent on the concentration of CO-binding proteins in the tissue and their respective affinities. Hemoproteins bind with CO with various affinities. Among them, one of the newly identified globin family members, neuroglobin (Nb), was found to possess the highest known binding affinity to CO. Nb has a K_d of 0.21 nM for CO and 3.2 nM for oxygen [10,11]. Its binding affinity to NO is also high, with a K_d of 1 nM [114]. As the name implies, Nb is predominantly expressed in the neuron cell [115], as well as other tissues such as retina and endocrine glands, including adenohypophysis, adrenal gland, testes, and pancreatic islets [116–119]. It is proposed that Nb serves a similar role to Mb in the muscle tissue, storing and transporting oxygen in neuron cells. It may also have other functions, including nitrite reductase activity [120] and NO scavenging ability [121], among other neuroprotective effects [122–124]. Under physiological conditions, the average Nb concentration in a mouse brain was estimated to be about 1 μM [116], and was said to be upregulated under hypoxia and traumatic conditions [125]. It is worth noting that a high concentration of Nb was found in mouse retinal tissue with a concentration of about 100 μM, which is almost equivalent to the Mb content in muscle cells [126]. Nb is found in the plexiform cell layer and in the inner segment of the photoreceptors, which coincide with regions of high metabolic activity and O_2 consumption rate. Such findings suggest important roles of Nb in regulating cellular metabolism [126]. This tissue-specific concentration coupled with the high binding affinity toward CO indicates that Nb may contribute to the distribution of CO in the nervous system and specific tissues such as the retina and endocrine glands.

Even though quantitative studies of direct binding competitions between CO and O_2 have not been reported yet, Nb is similar to Mb and does not have cooperative binding with CO or O_2 because there is only one heme binding site. Therefore, modeling work can also be done the same way as for Mb. Here, oxygen P_{50} of Nb is reported to be 1 mmHg [116]. The Haldane constant M_{Nb} can be estimated to be about 16 according to the relevant binding affinities. By plotting the simulation results from Equations 3.9 and 3.12, correlations between oxygen partial pressures and NbCO saturation levels under different COHb% levels are shown in Figure 3.6.

$$NbCO_{sat.} = \frac{M_{Nb} \cdot P_{CO}}{P_{O_2} + M_{Nb} \cdot P_{CO} + P_{50}}. \quad (3.12)$$

From Figure 3.6, one can see some similarities between the patterns of the plots of CO binding to Mb (Figure 3.4) and Nb. However, it is "harder" for Nb to get near full saturation when compared with Mb. This is because of the higher binding affinity to oxygen for Nb and a smaller difference in binding affinity between CO and oxygen compared with that of Mb. It may indicate a protective mechanism that decreases the chance for CO's toxic effect to the nervous system and/or a modulation role in maintaining homeostasis of CO as the signaling molecule. Another layer of this modulation role comes from the homeostasis of the vascular oxygen partial pressure in the nervous system. Oxygen delivery to the brain is mainly controlled by oxygen consumption and blood CO_2 levels. Since cerebral oxygen consumption rate is steady in normal adult humans, cerebral vasculature can respond to decreases in oxygen availability by increasing cerebral blood flow in order to maintain cerebral oxygen delivery. Many model studies in dog [127], rabbit [128], cat [129], and sheep [130] suggest that this regulatory mechanism applies to not only hypoxic hypoxia conditions but also CO-induced hypoxia [127,128]. For example, Traystman and Fitzgerald examined the cerebral blood flow responses to CO-induced hypoxia in anesthetized dogs, particularly when COHb is less than 20% [127]. A COHb level as low as 2.5% resulted in a small but significant increase in cerebral blood flow to 102% of the control. With reductions in the oxygen-carrying capacity of 5%, 10%, 20%, and 30% (5%, 10%, 20%, and 30% COHb), cerebral blood flow increased to approximately 105%, 110%, 120%, and 130% of the control, respectively. At each of these levels, cerebral oxygen consumption remained unchanged [100]. At COHb levels above 30%, increases in cerebral blood flow in response to CO-induced hypoxia are no longer proportional to the COHb levels, leading to decreased cerebral oxygen consumption. Studies also showed that cortical tissue oxygen partial pressure ranges from 30 to 48 mmHg [105], higher than that in muscle tissues (28.9 ± 3.4 mmHg). Therefore, theoretically, it is reasonable to see Mb having a higher CO saturation level than Nb at the same COHb level.

A particular case of Nb distribution is in the retina cells. As stated earlier, Nb concentration in the retina cells can be as high as 100 μM. Combining with a relatively low oxygen partial pressure (22 mmHg) [105], CO should be able to affect Nb functions in the retina cells. Various clinical case reports indicate that CO exposure by poisoning and smoking can induce visual dysfunction and optic neuropathy such as blurred

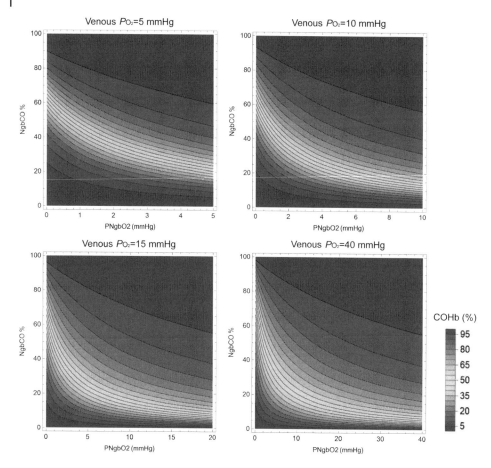

Figure 3.6 NbCO saturation as a function of P_{NbO_2} at predefined COHb and P_{VO_2} levels. (The plots are based on Equations 3.9 and 3.12 with the parameters and constants detailed in the text).

vision, photophobia, diplopia, and disturbances of the optic nerve, including optic neuritis and atrophy [131–136]. In a controlled study of dark adaptation time and light sensitivity of dark-adapted eyes in five young and healthy smokers and nonsmokers [132], subjects were allowed to breathe in 70 and 100 ppm CO in the inspired air after a priming dose of 5000 ppm for 5 or 8 min. The observed iCOHb levels increased to 19.1 ± 1.0% in smokers and 17.5 ± 1.9% in nonsmokers. Dark adaptation time was longer, and the light sensitivity of dark-adapted eyes was reduced in smokers as compared to nonsmokers at comparable levels of both inspired CO and COHb levels. These findings indicate that CO can potentially perturb retinal functions even at low CO exposure levels. Although all reported cases attributed the retinal dysfunction to CO-induced hypoxia, disruption of Nb by CO could be another reason that needs further investigation.

CO Distribution in Tissues with Cytochrome c Oxidase

Following Hb, Mb, and Nb is cytochrome *c* oxidase (COX) as a major hemoprotein that binds to CO. COX has a moderate binding affinity toward CO with a K_d of ~0.3 μM. The numerical ratio in the relative binding affinity (M, in the same context of Haldane's constant) between CO and oxygen toward COX commonly ranges between 0.5 and 2.5 [95,137,138]. However, much higher numbers (>10^3) have also been reported when the ratio is calculated from the reported apparent K_d for oxygen [139,140]. This variation in experimentally observed affinity is probably intrinsic to the nature of COX and the functional dependence of COX on other factors. COX is an enzyme that consumes the substrate oxygen by reducing it to water via a series of redox reactions mediated by metal centers (Fe and Cu). The initial reversible binding of oxygen with COX is driven by subsequent irreversible reactions. COX is the last step of the electron transportation chain in the mitochondrial respiration mechanism. As such, the Michaelis constant of COX is also dependent on the energy backpressure (proton motive force) and the redox potential difference across the respiratory chain [141]. Therefore, the affinity of COX to oxygen also changes as reflected by its variable Michaelis constant (1–10 μM) reported in the literature [142,143]. Factors such as assay temperature [144], oxygen saturation level [139], protein purity, and composition [145], among others [146], could complicate the results and cause variations. A relative binding affinity of about 0.5–2.5 is probably a reasonable estimation for COX based on direct competition experiments. In one example, when HEK 293 cells were treated with 20 μM CO, cellular respiration was inhibited by 40% when O_2 concentration was between 10 and 30 μM [147], indicating a comparable apparent binding affinity of CO and O_2 toward COX. Therefore, in the following modeling work, 1 and 2.5 were assigned as representative M_{COX} values for the generation of two sets of plots. As COX is located in the mitochondria, the oxygen partial pressure is fairly low, as stated earlier [104]. By assuming P_{50} = 0.5 mmHg for CO [95,148], a correlation between COHb% and COX-CO saturation under various venous oxygen partial pressure can be modeled (Figure 3.7).

Under normoxic conditions, the plot indicates that COX is unlikely to be saturated to a high level by CO even under high COHb saturation levels (Figure 3.7A and C). This phenomenon can be attributed to the highly competitive binding between CO and O_2 to COX due to their comparable binding affinities. Under hypoxic conditions, the oxygen partial pressure may become as low as the P_{50} (0.5 mmHg) [149], while mitochondrial respiration still functions. Under such hypoxic conditions, COX-CO levels may increase substantially (Figure 3.7B and C). For instance, judging from the plots (Figure 3.7), under normoxic conditions, 20% COHb can result in a COX-CO saturation of about 2% (M_{COX} = 1, P_{VO_2} = 30 mmHg) or 4.5% (M_{COX} = 2.5, P_{VO_2} = 30 mmHg); under hypoxic conditions, the same 20% COHb can lead to 7% (M_{COX} = 1, P_{VO_2} = 10 mmHg) or 10% (M_{COX} = 2.5, P_{VO_2} = 10 mmHg) COX-CO saturation. Under this condition, the function of the respiratory chain could be altered substantially. This analysis agrees with previous reports that COX inhibition by CO is significant only under hypoxic conditions [137,138,147,150]. Because target engagement does not seem to be significant under normoxic conditions, COX should not significantly affect the distribution of CO under physiological conditions, even though the presence of COX is ubiquitous in cells. However, under hypoxic conditions, one would need to consider the impact of significant COX inhibition and COX-dependent redistribution of CO, whether endogenously generated or exogenously delivered [147].

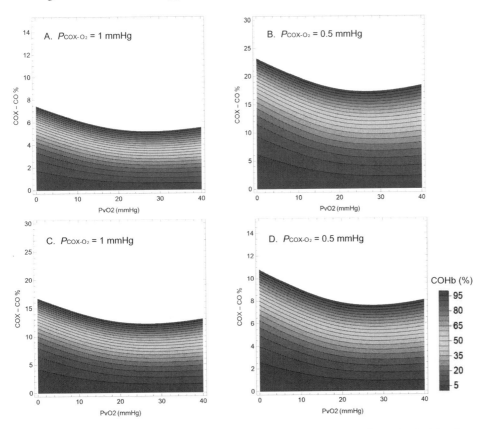

Figure 3.7 COX-CO saturation as a function of P_{VO_2} at predefined COHb and P_{COX-O_2} levels. (The plots are based on Equations 3.9 and 3.12 with MCOX = 1 for A and B, and MCOX = 2.5 for C and D. Other parameters and constants are detailed in the text).

Direct Assessment of CO Distribution in Tissues

Sensitive instrumental methods have been developed to detect trace amounts of CO, such as GC using a methanizer-coupled FID [152,153] or a mercury reduction gas detector (RGD) [36,107,154,155]. According to the manufacturers' specifications, the detection limits are 0.2 ppm and 50 ppb for the methanizer FID and RGD, respectively. Studies in CO inhalation animal models and CO-poisoned human victims have led to rich information of tissue distributions of CO under various COHb levels (Table 3.1). Along this line, the meticulous work by Vreman and coworkers is especially commendable.

From the results by Vreman et al. in human postmortem samples, only blood CO concentration showed definitive correlations with the COHb level (R = 0.94). Generally speaking, tissue CO concentration increases at higher COHb levels. However, the numerical correlation is moderate at best (R = 0.48–0.76), indicating complicated individual variations among samples. Many factors could have contributed to the large experimental variations, including physiological conditions, postmortem time, blood contents (tissue unperfused), instrument variations [107], and environmental factors such as temperature and light. In animal models, the conditions are expected to be better controlled. Therefore, variations were indeed much smaller than in the postmortem human samples. In the heme arginate administration models mimicking endogenously generated CO, CO concentration only increased proportionally with COHb in the blood, heart, spleen, and lung [122]. In the case of exogenous CO administration, CO concentrations were observed to increase in all tissues. However, the numerical ratio did not increase proportionally to that of COHb. Among all the tested tissues, lung showed the highest increase ratio, which was even more than the blood. Muscle exhibited the lowest increase in ratio, which was only 1/42 of the increase ratio of blood CO concentration. Even though blood perfusion was considered a factor that could cause inaccurate determination of the "true" tissue concentration of CO, concentrations of CO-binding proteins and oxygenation level, among others, are considered important factors that could render different tissue distribution patterns. In the end, one has to consider the complexity of the various convoluting factors that could affect CO distribution in analyzing CO's PK issues, and much more work is needed to allow for a detailed understanding of the issues described earlier.

Elimination

In studying CO as a therapeutic agent, pharmacokinetics/elimination issues are mostly discussed in the context of exogenously administered CO in various forms. As such, discussions also need to consider two basic boundary conditions: one is elimination after reaching a steady state and the other after acute exposure of relatively high doses of CO [156]. Typical cases of post-steady-state elimination include endogenous CO generation and prolonged administration of CO for therapeutic and research purposes. On the other hand, acute exposure scenarios may include unintentional exposure (including CO poisoning) and deliberate administration of CO in research settings. When the steady state of COHb is not reached under acute CO exposure, the equilibrium between COHb and tissue CO level is not reached. Then, the distribution of CO to the peripheral tissue, such as muscle, contributes to the elimination kinetics [156].

$$Y = \underbrace{A \cdot e^{-BX}}_{\text{distribution}} + \underbrace{C \cdot e^{-DX}}_{\text{elimination}}. \quad (3.13)$$

There have been many studies that provided a wealth of information on CO elimination, including mathematical modeling. For example, studies in sheep have established a COHb elimination model in both short- and long-term exposures [157]. Studies also demonstrated that the two-compartment model, which is commonly used in the PK analysis of conventional small-molecule drugs, can also be used to analyze the elimination of CO in the body. It combines the distribution of CO from the central compartment to the peripheral compartment and elimination from the central compartment (Figure 3.8 and Equation 3.13). In one study using sheep, it was found that inhalation of 500 ppm CO in air for 5 h was sufficient to lead to a maximum COHb level of about 35%. The animal was ventilated in normal air after reaching this maximal COHb level and the COHb level was recorded and fitted into Equation 3.12. Such studies yielded the following constants: A = 7.8, B = 0.043, C = 27.2, and D = 0.0065. Due to the small value of B, the model is nearly linear with a distribution half-life of 21.5 min and an elimination half-life of 117.8 min (Figure 3.9).

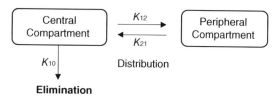

Figure 3.8 A two-compartment model of CO.

Table 3.1 Tissue CO concentrations from animal models and postmortem human subjects.

Author and references	Method	Subject	COHb[a] (%)	CO concentration (pmol/mg)									
				Blood	Spleen	Heart	Kidney	Muscle	Liver	Brain	Lung	Intestine	Testes
Vreman et al. [36,151]	GC	Rat	0.5	47 ± 10	11 ± 3	6 ± 3	5 ± 2	4 ± 4	4 ± 1	2 ± 1	2 ± 1	2 ± 1	1 ± 1
		Mouse	0.5	45 ± 5	6 ± 1	6 ± 1	7 ± 2	10 ± 1	5 ± 1	2 ± 0	3 ± 1	4 ± 2	2 ± 1
		Mouse (30 μM heme)	0.9	88 ± 10	11 ± 1	14 ± 3	7 ± 2	7 ± 1	5 ± 1	2 ± 0	8 ± 3	3 ± 1	2 ± 0
		Mouse (500 ppm CO, 30 min)	28	2648 ± 400	229 ± 55	100 ± 18	120 ± 12	14 ± 1	115 ± 31	18 ± 4	250 ± 2	9 ± 7	6 ± 3
Vreman et al. [107]	GC	Human (normal)	1.4 ± 1.2	165 ± 143	79 ± 75	31 ± 23	23 ± 18	15 ± 9	NA	3 ± 3	57 ± 59	NA	NA
		Human (death from fire + CO)	40.7 ± 28.8	3623 ± 1975	2290 ± 1409	128 ± 63	721 ± 427	168 ± 172	NA	17 ± 14	1097 ± 697	NA	NA
		Human (death from CO suicide)	56.4 ± 28.9	5196 ± 2625	3455 ± 1347	527 ± 249	885 ± 271	265 ± 157	NA	72 ± 38	2694 ± 1730	NA	NA

[a] For mouse and rat experiments, mixed venous and arterial blood was used; for human studies, the study did not indicate how blood was collected.

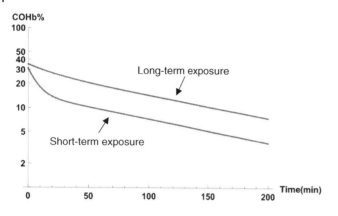

Figure 3.9 Elimination curves of CO after short- and long-term exposures in sheep. (Plotted based on the model with fitted constants).

In the case of studying acute CO exposure, sheep inhaled a mixture of 12.6% O_2, 5.2% CO_2, 80.2% N_2, and 2.0% CO for 3 and 1.5 min. Since the animal was anesthetized and the total amount of CO intake was equalized by controlling the tidal volume of the ventilator, the average peak COHb level was consistently at about 30%. The fitted constants for the acute exposure are $A = 16.7$, $B = 0.13$, $C = 14.4$, and $D = 0.0069$. Here, B is substantially higher than the steady-state case, and the curve obviously followed a biphasic pattern. A rapid decrease was found within a short period in the initial phase. The half-lives were calculated to be 5.65 min for the distribution phase and 102.8 min for the elimination phase. The elimination-phase kinetics in both cases are almost the same. Therefore, it can be assumed that in this case the elimination rate is independent of the CO exposure mode and the peak level of COHb.

As a highly diffusible and generally stable gas, CO is mainly eliminated unchanged through exhalation, with only a small percentage being metabolized. An early experiment using radioactive ^{11}CO showed less than 0.1% of the inhaled CO being oxidized to CO_2 in the human body [158]. Luomanmaki and Coburn reassessed the oxidation of CO using a more sensitive method by breathing in radioactive ^{14}CO in a rebreathing system [113]. It was found that 0.30 ± 0.18% and 0.16 ± 0.03% of the total administered ^{14}CO were oxidized to $^{14}CO_2$ every hour in dogs and humans, respectively, in a closed system. By increasing CO storage in the dog through addition of nonradioactive ^{12}CO to a COHb level of about 35%, the average rates of oxidation of ^{14}CO did not change, indicating the rates of oxidation of ^{12}CO to $^{12}CO_2$ were proportional to COHb level and body CO stores. It suggested that the oxidation of CO in the body was likely to be a first-order reaction. At a normal basal condition, the average metabolic rate was calculated to be 0.02 ml/h in dogs and 0.015 ml/h in humans. CO oxidation rate appeared to be insignificant when compared to the endogenous CO generation rate. Under normal basal conditions, the rates of CO oxidation were found to be about 10% and 3% of the rate of endogenous CO production in dogs and humans, respectively [113].

Factors that influence the elimination rate involve respiration functions [159], blood pH [61], oxygen concentration and pressure [61,160,161], and cardiovascular function, among others [162]. Increasing alveolar ventilation decreases CO elimination time [162]. Bruce and Bruce found that the half-life of COHb has a higher correlation ($r = 0.714$) with Vb/VAwo (blood volume/ventilation during washout) than simple ventilation alone [163]. By monitoring COHb in the femoral artery blood, Stadie and Martin [61] reported in 1925 that in anesthetized dogs on an artificial ventilator, hyperventilation slightly increased COHb elimination rate. Experimentally, lowering blood pH from 7.4 to 7.0 by inhaling 10% CO_2 or to pH 6.8 by infusion of a dilute HCl solution could significantly increase the rate of COHb elimination. It was found that CO_2 could enhance CO elimination not only by decreasing blood pH, but also by significantly stimulating respiration rate. Although at that time the study did not further explain the mechanism of why lowering pH could increase COHb elimination, it is reasonable to assume that by allosteric modulation of the Hb binding affinity, the lower pH and higher CO_2 concentration can lower the binding affinity of CO to Hb and therefore facilitate its elimination through ventilation. However, it is obvious that the decreased binding affinity of oxygen to Hb at low pH may further exacerbate anoxemic conditions. It is worth investigating whether amelioration of CO poisoning could be achieved by modulating blood pH alone or in combination with HBO therapy to supply more oxygen in the dissolved form while lowering the affinity of CO to Hb. However, it should be noted that this elimination is only in reference to the COHb levels and not tissue CO elimination. While decreasing the COHb level is important in treating CO poisoning, it is probably more important to quickly remove CO in tissues. Much of this needs further studies.

PK Models

After endogenous CO generation was confirmed and used as a biomarker for hemolysis, there have been great interests in developing mathematical models to predict the pharmacokinetics of CO in the body so that it can be used to calculate the severity of hemolysis. As mentioned in the earlier sections, CO's absorption and elimination also very much depend on respiratory functions. Therefore, a model can also be used to evaluate respiratory functions by simply breathing low concentrations of CO and monitoring COHb levels. It is a common diagnostic method known as $D_L CO$ (diffusion capacity of the lung for CO) used in hospitals to determine lung functions. Even though the therapeutic potential of CO had not been acknowledged yet in the 1960s, several models had incorporated a term to account for endogenous CO generation in modeling the PK profiles of exogenously administered CO via different delivery forms.

The CFK Model

One of the most widely accepted and sophisticated models for PK studies of COHb levels is the CFK (Coburn–Forster–Kane) model (Figure 3.10 and Equation 3.14) [164]. This model is based on four dynamic processes, including exogenous CO (e.g., inhalation), endogenous CO (e.g., heme degradation), elimination through respiration, and total blood/tissue storage. It covers the major absorption/distribution/elimination pathways and can give satisfactory predictability in general. Since the model takes endogenously generated CO into consideration, it is possible to calculate COHb levels after external administration of CO as well. One of the limitations of CFK model is that it does not distinguish COHb levels between the arterial blood and venous blood, and it does not take into consideration the free CO content in the blood.

$$[COHb]_t = \frac{\{A[COHb]_0 - (B\dot{V}_E CO + P_I CO)\} \cdot e^{-tA/v_b B} + B\dot{V}_E CO + P_I CO}{A},$$
(3.14)

where

$$A = \frac{P_C O_2}{M[O_2 Hb]} \text{ and } B = \frac{1}{D_L CO} + \frac{P_L}{V_A}.$$

$[COHb]_t$ is the concentration (ml/ml) of CO-bound Hb in the blood at the given time. It should be noted that the relative percentage value of COHb% should be converted to the volumetric concentration by using $[COHb]=COHb\% * 0.2$ (0.2 is a normal value for saturated volumetric CO concentration). $P_C O_2$ is the average partial pressure of oxygen in lung capillaries in mmHg, and is 110 at sea level with water vapor pressure being 49 mmHg at 37 °C. M is the ratio of the affinity of Hb for CO to that for O_2, which is approximately 246 for humans. $[O_2 Hb]$ is the concentration (ml/ml) of oxygen-bound Hb in whole blood. $D_L CO$ is the diffusivity of the lung for CO (ml/min/mmHg). It can be experimentally determined. In humans, it is typically referred to as $D_L CO = 35 V_{O_2} \cdot e^{0.33}$, where $V_{O_2} = (RMV/22.4) - 0.0309$ and RMV is the respiratory volume (ml) per minute. $P_L = P_B - P_{H_2 O}$, where P_B is barometric pressure in mmHg and $P_{H_2 O}$ is water vapor pressure at body temperature in mmHg (typically it is 49 mmHg). $\dot{V}_E CO$ is the rate of endogenous CO production in ml/min, which is approximately 0.007 ml/min for a healthy human. It was suggested the basal level of COHb in rat is 0.7%, which

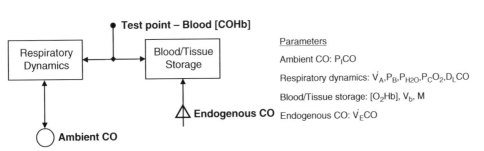

Figure 3.10 The CFK model.

corresponds to the initial amount of CO in the blood of 0.117 mg/kg body weight, and the rate of endogenous production (V_{CO}) is 35 μg CO/h/kg body weight. P_ICO is the partial pressure of CO in inhaled air in mmHg. $\dot{V}_A$ is alveolar ventilation rate in ml/min, and is equal to $0.933 V_E - 132f$, where V_E is the ventilation volume in ml/min and f is the ventilation frequency. $[COHb]_0$ is the basal COHb level at the beginning of the exposure (for a human nonsmoker, it is approximately 0.8%, or 0.0016 ml/ml). e is the base of the natural logarithm, and t is the duration of exposure. V_b is blood volume (approximately 74 ml/kg body weight).

Under steady-state conditions after inhaling a steady concentration of CO for an infinite period of time to reach a plateau of COHb concentration, which results in $e^{-tA/V_b B}$ to become zero, the model can be simplified as

$$[COHb]_t = \frac{A[COHb]_0 + BV_{CO} + P_ICO}{A}.$$

It consists of two parts: $[COHb_{end}]$ (COHb from endogenous CO) and $[COHb_{ex}]$ (COHb from exogenous CO).

$$[COHb] = [COHb]_0 + \frac{B \cdot \dot{V}_E CO + P_ICO}{A}.$$

We can define endogenously generated COHb as $[COHb_{end}] = (B \cdot \dot{V}_E CO)/A$, and exogenous CO generated COHb as $[COHb_{ex}] = P_ICO/A$.

The CFK model showed good predictive power in experimental CO inhalation studies in humans as well as in small animals. As reported in a study in human volunteers by Peterson and Stewart, CO exposure at a concentration between 1 and 1000 ppm led to several important findings regarding CO absorption through inhalation (Figure 3.11, CFK model, parameters for the CFK model used are listed in Table 3.2) [165]. First, COHb levels are proportional to the inhaled CO concentrations and exposure time in a nonlinear fashion. Second, the CFK model has better predicting power of the COHb–time relationship under normal oxygen partial pressure than under hypoxic conditions. Third, according to the regression formula, the steady-state COHb level is proportional to the exposed CO concentration. Fourth, the time to reach a steady state is similar (around 250 min) regardless of differences in exposure CO concentrations. Such results indicate that the half-life for COHb to eliminate after CO exposure is independent of the peak COHb level, exposure duration, and exposure CO concentration, but dependent on the individual difference of respiratory function, physiological conditions, and atmosphere O_2 pressure. It should be noted the CFK model equation in the original paper had an error that an extra variable M was mistakenly put to the power of e [165].

A study of long-term exposures to 50 ppm CO showed excellent agreement with CFK predicted data (Figure 3.11) [165]. It also correlated well under other CO concentrations. Though plots for the experimental data of 100 and 200 ppm inhalation were included in the original paper, the lack of the original data means that we are unable to generate the graphical results for comparison under these two levels. Another main advantage of this model is that it

Table 3.2 Parameters used for calculations based on the CFK model

M	$[O_2Hb]$	P_CO_2	$\dot{V}_E CO$	$D_L CO$	P_L	V_b	$\dot{V}_A$
218	0.2	100	0.007	30	713	5500	6000 (awake)
							4000 (sleeping)

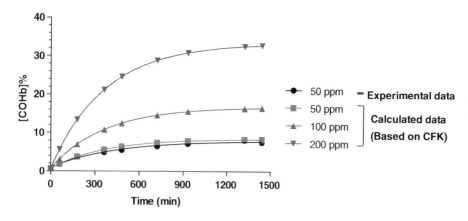

Figure 3.11 Comparison of experimental data and calculated data using the CFK model for predicting COHb–time profiles upon inhalation exposure to CO at different concentrations. Source: Adapted from [165].

takes into consideration endogenously generated CO, which makes it possible to calculate COHb levels after external administration of CO via various delivery methods, by assuming CO administered through other route being part of endogenously generated CO.

Other Models
Though the CFK model is powerful in predicting COHb in long-term CO inhalation studies, it also has several limitations. Because the CFK equation provides a first-order approximation to the slow dynamics of CO transport and storage, it is unlikely to predict a rapid increase in COHb levels accurately. The CFK model does not differentiate arterial or venous COHb levels but mostly predicts a value between the two [167]. In the case of assessing COHb levels after short CO administration, "ad hoc" modifications of the CFK equations often are necessary to improve the fit of the model. This means the necessity of trial experiments before applying the model. Another limitation of the CFK model is that it does not include extravascular storage sites for CO, such as the other substantial hemoprotein pool – Mb. Therefore, in predicting the washout phase of CO, the CFK model typically generates a shorter half-life because it does not include the elimination of CO from the Mb compartment. To address these limitations, a more comprehensive model was generated by Bruce and Bruce [163,167], which showed improved prediction and applicability over the CFK model (Figure 3.12).

As one can see in Figure 3.12, the model comprises multiple compartments, including lung, arterial, venous, nonmuscle tissue, and muscle tissue. Therefore, the parameters include the volume of the compartment (V), partial pressure of oxygen and CO (P), and oxygen or CO capacity of Hb and Mb (O_2Hb, COHb, MbCO) in the compartments and capillary vessels, and blood flow in the vessel (Q). The model was tested to be fairly accurate in predicting the oxyhemoglobin dissociation curve under various COHb levels (0–60%). It is capable of predicting COHb levels in the muscle, arterial, and mixed venous flows as well as MbCO levels. The prediction results have been shown to correlate well with experimental values from human samples. In prolonged CO inhalation studies, this model generated similar results as the CFK model. This multicomponent model is also capable of predicting COHb and MbCO levels with repetitive administrations of high concentrations of CO in a short-pulse manner, which is hard to do with the CFK model. Aside from all the advantages of this model in predicting COHb with various inhalation protocols, it also takes the endogenous CO generation into consideration by assuming all of it goes into the mixed venous compartment at a consistent rate.

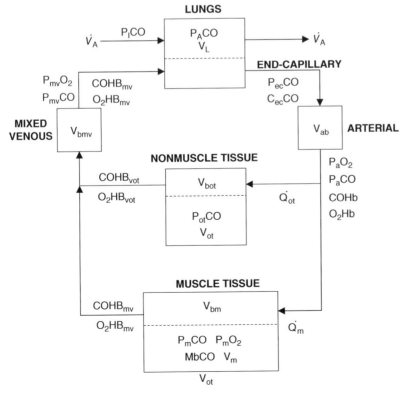

Figure 3.12 A multicompartment model by Bruce and Bruce. Source: Adapted from [156].

However, it does not have a dedicated operator to define the kinetics of such endogenous CO generation. Therefore, additional research is needed to address this issue to build a more inclusive model.

Another simple model for predicting steady-state COHb levels was derived by Coburn et al. [164]. It was based on the single compartment of alveolar/pulmonary capillary (Equation 3.15):

$$(COHb)_{ss} = \frac{M_{Hb} \cdot O_2Hb \cdot P_ICO}{A_{O_2}}, \quad (3.15)$$

where P_ICO is the partial pressure of CO in inhaled air in mmHg, M_{Hb} is the Haldane constant of Hb with a typical value of 218, and A_{O_2} is the average pulmonary capillary P_{O_2}, with a typical value of 100 mmHg.

Similarly, time-dependent changes in blood COHb in response to inhaled CO can be approximated by a simple exponential regression, based on the-compartment model (Equation 3.16). However, pretesting to get the steady-state COHb level is a prerequisite for using this model.

$$COHb = (COHb)_{ss}\left(1 - e^{-t/T}\right). \quad (3.16)$$

In this equation, t is the postexposure time, and T is the time constant, depending on the species-related physiological conditions such as diffusivity of the lung for CO, blood volume, and alveolar ventilation rate. For humans, $T = 360$ min can give good regression results that agree with experimental data [165]. The limitation of this model is that it does not take into consideration endogenous CO generation and it can only predict mixed venous COHb with a steady CO inhalation rate.

PK Properties of CO from Endogenous Production and Externally Administered CO Donors

PK Characteristics of Endogenously Generated CO

As a part of normal biological and physiological processes, CO is continuously produced at low levels in the body and is primarily exhaled from the lung. Endogenous generation of CO is mostly attributed to the turnover of heme-containing proteins catalyzed by heme oxygenases (HOs), leading to stoichiometric production of CO [166,168,169]. By employing ^{14}C-labeled Hb and ^{14}C-glycine (as a precursor of heme biosynthesis), extensive work conducted decades ago collectively demonstrated that the degradation of Hb during the destruction of senescent erythrocytes is responsible for approximately 75% of endogenous CO production in the body [17,170–174]. The turnover of other intracellular hemoproteins, such as Mb and cytochromes, is another significant source of CO production. In 70-kg healthy males, the average rate of endogenous CO production is 19 μmol/h, giving the equivalent of 6.1 μl of CO gas per hour per kilogram [170]. In females, the rate is dependent on the menstrual cycle [175]. In newborns, the rate averages 16.3 μl of CO gas per hour per kilogram [176]. Chapter 1 discusses these aspects in detail.

As an relative inert, poorly water-soluble molecule with a high binding affinity for heme, CO is present in the body in three forms: (i) gaseous CO, for which the level is denoted by its CO partial pressure (P_{CO}) or ppm concentration; (ii) free CO dissolved in the blood and other tissue fluids, which is the biologically active form; and (iii) Hb-bound (COHb) and other hemoprotein-bound CO [17]. Tissue distribution and accumulation of CO are driven by passive diffusion of CO across cellular membrane and partition of CO among these interchangeable forms. The distribution of endogenous CO has been discussed in the section entitled "Distribution." A very small fraction of CO in the body is metabolized to CO_2 [177]. CO is predominately eliminated via pulmonary excretion at rates that are dependent on alveolar ventilation, CO's diffusivity in the capillary blood and alveolar gas, pulmonary capillary oxygen tension, and other variables. Under normal conditions in humans, an average COHb level of 0.8% is commonly achieved due to endogenous CO production [178]. Therefore, the P_{CO} in the venous blood was estimated to be 0.0018 mmHg according to Haldane's law (Equation 3.7) [178]. These are important parameters and reference points in understanding the PK profiles of endogenously produced and exogenously administered CO.

From Endogenous CO to CO Gas and CO Donors: Target Engagements and Therapeutic Implications

In a classical sense, CO is no different from other small-molecule drugs; it has to reach a sufficiently high concentration that is comparable to the affinity for a given target (K_d) in order to achieve target engagement. Most of the hemoproteins found up to date have a higher P_{50} than that of CO partial pressure in the blood pool (0.0018 mmHg, 0.8% COHb) generated by endogenous CO production under normal conditions [95]. As a result, endogenous CO should not significantly perturb the biological functions of those hemoproteins with a lower CO binding affinity than Hb. Thus far, the lowest effective concentration reported for *in vivo* treatments using CO gas was in a piglet pial vessel dilatation model, which yielded sensitive responses to the treatment by 0.1 µM CO gas dissolved in an artificial cerebrospinal fluid (CSF) [56]. This CO concentration is comparable to that used in cell culture experiments when exposed to 100 ppm of CO gas, as calculated by Henry's law. However, it is higher than the basal CO concentration in the CSF, which is about 20–80 nM as found in the same study as well as studies by others [56,179–181]. Assuming the dissolved CO gas is in equilibrium with the free CO in the blood due to diffusion, the basal CO level is only slightly lower than the effective CO concentration that showed vessel dilation activity. Thus, the physiological implications for such a sensitive response are an interesting question for further exploration.

For those targets with the need for a higher *in vitro* effective concentration, the level of COHb needed to achieve the same activity as seen in the *in vitro* experiments would be life-threatening. Does this mean CO is not able to function as a therapeutic agent? The short answer is no. The pharmacological activities are not entirely dictated by the K_d of hemoprotein to CO and systemic COHb concentrations for reasons of enriched local CO concentrations, stability, and presence of other modulators. The difference in the K_d of different hemoproteins allows for the selectivity of CO at different concentrations – from the lower endogenously generated CO concentrations to exogenously administered CO gas or CO donors. Our understanding of the intricate relationship between pharmacology and pharmacokinetics will be critical to the successful development of CO-based therapeutics. As a matter of fact, the difference in target engagement is the foundation of using CO as a therapeutic agent rather than a promiscuous "reagent." The desired therapeutic activity of CO is associated with the inherent binding affinity of the target protein, blood supply conditions, the oxygen level, local pH, the route of administration, CO release rate from a donor, and probably binding kinetics, among other factors. It is also important to mention that the biological activity of CO is not simply an "on" and "off" scenario. CO can modulate the signaling pathway by tuning the activity of the CO-binding protein in a competitive manner. As a signaling molecule, the effects of CO are expected to reflect certain physiological or pathological situations. In a study in healthy volunteers, exercise was reported to induce HO-1 expression [182,183] and increase COHb levels by onefold from the basal level to about 1.1% [184]. Under pathological conditions, HO-1 expression can be induced leading to increased CO production. Wang et al. showed in a rat bile duct ligation model that both mRNA and protein expressions of HO-1 were increased compared with the control sham group. As a result, COHb level increased by twofold. Stimulation of HO-1 expression by cobalt protoporphyrin further increased COHb to about 1%, which was threefold of the background level of the control group [185]. Chapter 1 of this book discusses variations of COHb levels and HO-1 expression under various pathological conditions.

Exogenously delivered CO may bring in fundamental changes to mitochondrial functions. Here, COX is a good example to show the difference between endogenous and exogenous CO in terms of CO-mediated signaling. COX is crucial to the oxidative respiratory process to sustain energy supply and mitochondrial homeostasis of the cell. Binding affinity (K_d) of CO to COX was reported to be 0.3 µM (0.027 mmHg) [186], which is about 15 times higher than the background CO concentration. As shown in Figure 3.7, with competition from oxygen under normoxic conditions, the maximum CO saturation level of COX should be no more than 20%, and typically should be less than 10% when COHb is <20%. Although these numbers seem small, CO can initiate cell signaling cascades under these conditions by inducing reactive oxygen species (ROS) formation and causing secondary ROS-related functions [187]. Furthermore, a moderate elevation of ROS formation may be involved in the anti-inflammation mechanism of CO. The effect of 250 ppm of CO in macrophages could be abrogated by the addition of *N*-acetylcysteine (an antioxidant) and inhibition of ROS generation by complex III inhibitor, antimycin A [188].

There is one thing that needs to be considered in designing and conducting cell culture-based experiments and in extrapolating such results to animal models. In cell culture, there is no competition from Hb, and the oxygen level is extremely low at the site of COX [95]. As a result, CO saturation can reach a higher level than that in an *in vivo* environment. For cell culture conditions, Levitt and Levitt proposed an equation to calculate COX-CO saturation levels based on the partial pressures of CO and oxygen (Equation 3.17) [95]. Here, we plot this equation for intuitive interpretations based on the given parameters (Figure 3.13). A calculation based on Equation 3.17 shows that 10 ppm (0.0076 mmHg) of CO can induce 2.7% of COX inhibition according to the conditions used in the experiment by Thom et al. [189]. Zuckerbraun et al. showed in their study that cell exposed to 250 ppm CO

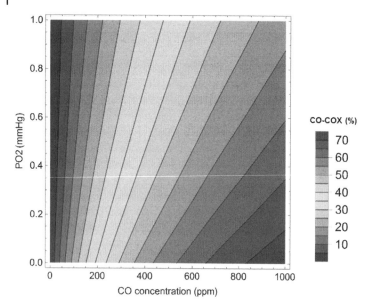

Figure 3.13 Correlations among CO-COX saturation level, CO concentration, and intracellular oxygen partial pressure in cell culture experiments. P_{VO_2} is intracellular oxygen partial pressure in the proximity of mitochondria where COX is located (0–1 mmHg); CO concentration in the cell culture atmosphere: 0–1000 ppm; and model is based on $M_{COX} = 2.5$ and $K_i = 0.27$ mmHg.

for 1 h induced a significant amount of ROS formation, which was attributed to the COX inhibition [188]. In such experiments, according to the calculation, 50% COX could be inhibited by 250 ppm (0.19 mmHg) CO [95]. In a human body, inhalation of 10 ppm CO (0.0074 mmHg) is about four times of the normal background CO level, and can be easily achieved in smokers [190], and of course by exogenous CO administration. However, with the competition from oxygen and Hb pool, a higher CO dosage would be needed to achieve the same COX saturation level as seen in the *in vitro* studies [189]. There has not been a reported study on the *in vivo* COX activity with CO in therapeutic settings. However, most of the research on COX inhibition by CO was based on CO's toxicology, from which the indication of therapeutic pharmacokinetics could still be inferred. Mice exposed to 1000 ppm of CO for 3 h increased COHb levels to 61 ± 4% compared to 1 ± 1% of the air control group. This high COHb level did not induce myocardial tissue hypoxia in mice; however, myocardial COX activity decreased by about 25% [191]. The results indicate that the degree of COX inhibition by CO falls within the general framework of the model proposed in Figure 3.7. The results clearly show that COX activity can be modulated by CO generated endogenously or administered exogenously.

$$\text{CO-COX}(\%) = \frac{P_{CO}}{P_{CO} + K_i + (P_{O_2} / M_{COX})}. \quad (3.17)$$

Owing to CO's stability and its high diffusivity, it is reasonable to envision that CO generated from heme degradation by the induction of HO-1 could function as a remote signaling molecule. Therefore, other parts of the body could partially reflect the conditioning effects of oxidative stress at a distal location, which provides prophylactic conditioning in advance of an injury or inflammation event. CO administered exogenously may also be able to mimic this preconditioning benefit. In an *in vitro* cellular study, a 1-h exposure to 10 ppm CO was found to lead to a threefold increase in the superoxide dismutase level, a defense mechanism of the cell to ameliorate oxidative stress [189]. A study of CO intoxication in humans showed that the CO-poisoned group with a median COHb level of 28% had a significant elevation of uncoupled mitochondria respiration in their peripheral blood mononuclear cells (PBMCs). After treating patients with HBO, the CO-poisoned group showed a significant increase in H_2O_2 production in PBMCs when compared to the unpoisoned control group with prior HBO treatment [192]. Along the same line is the finding that by directly delivering CO to the mitochondria through an innovative enrichment-triggered release mechanism, a much lower dose of the CO donor (5 μM) is needed to show a strong suppressive effect of TNF-α production in RAW 264.7 cells when compared to the higher concentrations (25–100 μM) needed to show similar activities for other nontargeted CO donors [193]. In a liver protection experiment in mice, a dosage of 4 mg/kg i.v. of the mitochondria-targeted CO prodrug resulted in a sustained level of 4–4.3% COHb (basal level: 2%) for about 60 min from the 20-min point. At this dosage, it showed a significant protective effect against acetaminophen-induced acute liver injury.

To this end, it is understandable that the high level of COHb achievable with exogenous administration may significantly shift CO's signaling functions, which are not commonly seen with endogenous CO levels. One has to be very careful in extrapolating results from cell culture studies to animal models: the lack of Hb competition in cell culture leads to very different requirements for the concentration of CO needed for engaging the same target. Detailed analyses are discussed in the following sections.

PK Properties of CO from Various Donors

In developing therapeutics, pharmacokinetics is a critical consideration. The correlation between pharmacokinetics and pharmacodynamics forms an important foundation for the design of dosing schemes. In studying traditional small-molecule-based therapeutics, often there is a clear understanding of the concentration profiles needed for a particular pharmacological effect or at least there are ways to arrive at such a point. However, for CO-based therapeutics, there has not been a consensus as to what an efficacious COHb level is or even whether COHb is the most relevant surrogate indicator for available CO concentrations. As discussed in earlier sections, many factors could affect the pharmacological effect of CO at a given level of COHb, including delivery form (gas versus CO donor), delivery route, targeted delivery versus nontargeted delivery, rate of delivery, pathophysiological states of the subject, pH, and the interplay among CO and other possible ligands for hemoproteins such as NO, H_2S, O_2, and 2,3-DPG, among others. Additionally, the combinatorial use of total dosage in mg/kg and COHb levels allows for descriptions of both total CO exposure and the body's response to such exposures as reflected in COHb concentrations. Depending on the therapeutic indications, special attention may also be needed to consider CO distributions into peripheral tissues and organs. We discuss these issues in detail later.

The average COHb level in healthy humans is around 1%, which is similar to that observed in various animal models [16,194]. FDA has set an upper limit of COHb level at 12–14% for certain human clinical studies [169,195,196]. Incidentally, this is also the level of COHb that can be achieved in heavy cigarette smokers [197]. In studies using CO donors in animal models, pharmacological effects have been observed at a much lower level of COHb (4–6%) [198–203]. Such examples illustrate the difficulty in establishing a universally agreed therapeutic threshold of the COHb level. As a matter of fact, there is not much discussion of a need to establish or examine the validity of the concept for a "therapeutic threshold of COHb levels" for various pharmacological indications. This is partly because of the lack of understanding of PK behaviors of CO in the context of its therapeutic relevancy. There are many issues to consider. First, most of the pharmacological studies on CO suggest that the duration of CO exposure should be in the range of minutes to hours to have any noticeable therapeutic effects [198,204–207]. Quick (seconds to minutes) CO release is often applied in transient CO exposure observations [208,209], such as ion-channel kinetic studies [208,209]. However, there has not been a controlled study that helps to establish a duration "standard." Second, there have been a large number of pharmacological studies of CO in different animal species. However, the issue of allometric scaling has not been thoroughly examined, though this is clearly an important issue for various reasons. Third, various other factors could impact the relationship between the COHb level and pharmacological effects, including therapeutic indications, local versus systemic CO delivery, targeted versus nontargeted delivery, CO gas versus CO donor versus CO solution, other pathological conditions, and the subject's physiological state. For example, the dose needed in one study for treating chemically induced gastric damage through a locally delivered CO prodrug [153] is much lower than that needed to treat systemic inflammation in another study [210]. Several studies using dextran sulfate sodium- or 2,4,6-trinitrobenzene sulfonic acid (TNBS)-induced murine colitis models [211,212] demonstrated the therapeutic COHb levels for inhaled CO gas at 14–18% of COHb. Wang et al. developed a series of organic prodrugs as CO donor molecules and showed marked anti-inflammatory efficacy in a similar TNBS-induced mouse colitis model via i.p. administration with a much lower therapeutic COHb level (2–3%), though the COHb levels were determined separately from the efficacy experiments [198,200]. Additionally, detailed PK analyses further showed that >3 µM of prodrug and the inactive metabolite were detected in the plasma, consistent with up to 33% of the prodrug being absorbed and being systemically available after i.p. administration. Such results indicate that the prodrug distributed in tissue and subsequently locally released CO along with the systemic dissolved CO may contribute to the observed pharmacological activity.

For a certain blood COHb level caused by exogenous CO, the equilibrium local CO concentrations can be vastly different in various tissues/organs. In addition, the effective local CO concentrations for different cellular responses and biological effects are rarely the same as demonstrated in various disease models [178]. A study conducted by Steiger et al. demonstrated a marked anti-inflammatory effect in a similar TNBS-induced mouse colitis model using an encapsulated oral carbon monoxide release system (OCOR) [200,213]. The OCOR was designed to release CO while keeping the active CO donor (e.g., CORM-2) from being absorbed before CO release [214]. The orally administered OCOR only led to a slight elevation in the COHb level (<0.8 %) in mice suggesting that the locally generated high concentrations of CO within the GI system rather than in the circulation accounted for most of the observed anti-inflammatory effects.

CO Drinks/Homogeneous Formulations
A straightforward way of delivering CO, other than gas inhalation, is to dissolve CO in a solution and to administer it through the GI system. According to Henry's constant of CO in water ($k = 9.7 \times 10^{-9}$ M/Pa) [64], the solubility of CO in water under 1 atm is about 1 mM. However, this relatively high concentration in aqueous solution is hard to achieve for medicinal applications. Since CO in the air is mostly around 1 ppm, the partial pressure is only about 0.1 Pa, and therefore, the equilibrium concentration is about 1 nM. Though a pressurized container can be used to keep the concentration during storage, once opened, the CO concentration is expected to drop instantly, much the same way as a carbonated beverage. Special formulations can be developed to increase the solubility of CO in the solution. For example, CO has a higher solubility in alcoholic solution ($k = 1.3 \times 10^{-7}$ M/Pa) [215]. There were several studies in preparing and testing CO solutions for *in vivo* activities. Takagi et al. [216] developed a CO-saturated solution by bubbling 50% CO in the air into saline, leading to a solution concentration of approximately 500 μM. This CO solution was tested for effectiveness in treating gastric mucosal wound healing. Oral administration of a CO solution (500 μM CO, about 4 μmol/kg) at a dosage of 0.2 ml per day for 4 days accelerated healing of acetic acid-induced gastric ulcer in mice, and significantly promoted reepithelialization of the wound. However, the COHb level achieved was not reported. According to the dose, it is reasonable to expect the increase of COHb level to be no more than 1%. Therefore, the observed activity is unlikely to be attributed to an increase in COHb level or systemic CO exposure. Therefore, direct contact with the lesion site seemed to be effective for therapeutic purposes.

Along this line, Hillhurst Biopharmaceuticals has developed "CO drinks" for clinical applications [201]. One of the candidates, HBI-002, is a CO solution containing components that are defined as "generally recognized as safe (GRAS)," which are kept confidential. HBI-002 has been studied in models of sickle cell disease (SCD) [201] and kidney ischemia–reperfusion injury [217]. In the study of HBI-002 in a mouse SCD model, a single gavage at a dosage of 10 mg/kg led to the COHb level to peak 0–10 min after administration and to completely clear in approximately 3 h [201]. The peak COHb level varied depending on the mouse breed. In NY1DD transgenic mice (SCD), COHb levels increased from 1.6 ± 0.2% to 5.4 ± 0.8%; in Townes-SS transgenic mice (SCD), COHb levels increased from 1.8 ± 0.2% to 4.7 ± 0.4%; and in Townes-AS (control) mice, COHb levels increased from 0.6 ± 0.2% to 3.0 ± 0.6%. The basal COHb level of non-sickle Townes-AS control mice was reported to be significantly lower than that of the sickle mice, presumably because of a markedly reduced hemolytic rate than the sickle mice. The result showed that CO delivered by this homogeneous solution was readily absorbed from the gut into the systemic circulation, leading to measurable COHb increases within minutes [201]. Even though the actual CO concentration in this formulation was not disclosed in the study, according to our own experience in CO delivery, one would expect the CO concentration to be in the millimolar range to achieve the reported PK profiles. Indeed, Hillhurst Biopharmaceuticals disclosed the CO concentration in their patent to be in the range of 2.6–157 mM [218].

Dihalomethane Exposures
In 1972, Stewart et al. found an elevated COHb level following human exposure to a commonly used organic solvent, methylene chloride (DCM) [219]. Upon uptake, DCM can be metabolized by cytochrome P450 into formyl chloride, leading to CO production as the major metabolite [220]. Besides DCM, other dihalomethanes (DHMs) such as CH_2F_2, CH_2FCl, and CH_2BrCl can also undergo similar conversions to generate CO *in vivo* [220]. Despite the toxicity caused by the electrophilic nature of its metabolites, DHMs can be considered as metabolically activated CO donors. Many PK studies have been done to address their toxicological profiles thereafter. As a result, it is feasible to extract key pieces of information to guide PK studies of pharmaceutical CO donors.

Because DCM has been widely used in household paint removers, human exposure through inhalation is a real issue. Peterson studied the PK profiles of DCM in human volunteers and came up with a model by nonlinear regression analysis (Equation 3.18) [221].

$$\Delta\%\text{COHb} = \frac{0.0842(D \cdot C_i)^{0.72}}{(T+220)^{0.775}}, \quad (3.18)$$

where D is the DCM exposure duration in min, C_i is the inhaled DCM concentration in ppm, and T is the postexposure time. Figure 3.14 shows plots of the relationship between ΔCOHb% and post-DCM exposure time. The model was fully generated by a regression method, and the regression coefficient was 0.794. As an approximation, the model only showed the elimination phase of COHb levels. Results from this model provide close resemblance to the elimination phase of a physiologically based pharmacokinetic (PBPK) model obtained by Andersen et al. [222], which is discussed next.

Unlike inhaled CO gas, CO from DCM is metabolically generated at a rate depending on P450 activities. A competitive non-CO generating pathway is the glutathione transferase (GST)-mediated pathway that generates CO_2 as the final product. It was found that depending on the dosage and sustained time of DCM exposure, the P450 metabolic pathway can be saturated, leading to a constant rate of CO generation. Andersen et al. found that rats exposed to 200 or 1014 ppm of DCM for 4 h resulted in the same COHb plateau level of about 7.5% [222]. However, exposure to 1014 ppm of DCM showed about 1.5 h longer plateau region than 200 ppm exposure. In the latter case, COHb started to drop immediately after 4 h DCM exposure. With an even higher DCM dosage (5199 ppm) and a shorter exposure time (0.5 h), a plateau at about 7.5% was also reached 2 h after exposure. In human volunteers, Andersen et al. combined the DHM metabolism model and the CFK model developed for endogenously generated CO, with the metabolic conversion from DHM to CO being a correlative factor (Figure 3.15) [222].

In this model, CO is produced from DHM oxidation in the liver and is transferred to the blood. CO also enters blood by endogenous production (REN_{CO}) and eliminates by diffusion across alveolar surfaces in response to a partial pressure difference between the alveolar (PA_{CO}) and capillary (PC_{CO}) CO tension. Pulmonary transmembrane CO flux is given by the concentration gradient ($PA_{CO} - PC_{CO}$) times the diffusing capacity of the lung for CO (D_L). The inhaled partial pressure of CO (P_ICO) and the ventilation rate (Q_P) determine the total amount of inhaled CO. Blood CO apportions between free CO and COHb. Therefore, the generalized mass balance differential equation of the rate of CO generation in the blood (AM_{CO}) is expressed as Equation 3.19:

$$\frac{dAM_{CO}}{dt} = REN_{CO} + METAB \times P_1 + UPTAKE, \quad (3.19)$$

where REN_{CO} is the rate of endogenous CO production from heme degradation, METAB is the metabolic oxidation rate of DCM in the body (depending on P450 activity), and P_1 is the yield of the CO formed by formyl chloride decomposition. As a matter of fact, the METAB × P_1 entry can be considered as part of the endogenous CO production in the original differential equation form of the CFK model. By solving the integrated form of this equation with the substitution of corresponding parameters, Andersen et al. developed a PK model to predict the COHb level after inhalation of DCM. The predicted COHb–time curve fitted well with the documented experimental data acquired in human subjects exposed to 50–500 ppm of DCM, and in rats exposed to 200–5159 ppm of DCM [222]. It should be noted that a higher maximal COHb level was found in humans

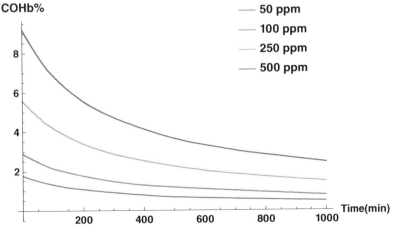

Figure 3.14 Simulated time-dependent COHb% changes after DCM exposure. (Modeled based on Equation 3.18 after 450 min inhalation with different DCM concentrations in humans).

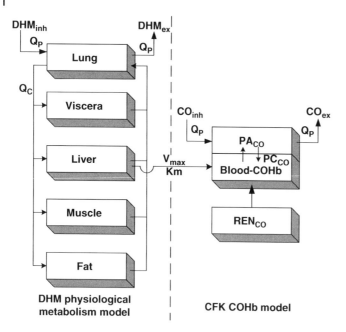

Figure 3.15 PK description of DHMs and CO. Source: Adapted from [222].

than in rats when given DCM, presumably due to the higher level of P450 capacity and lower GST activity in humans than in rats. Instead of having a plateau at around 7.5% as found in human volunteers exposed to DCM, COHb levels of up to 30–40% have been reported after accidental exposures to DCM [223,224]. The resulting PBPK model has been widely used and adapted to estimate COHb levels. In several studies, the parameters were rigorously optimized, leading to a good fit with the experimental data [225–227].

CO-Releasing Molecules
Metal-based CO-releasing molecules (CORMs) occupy a special place in the development of CO donors. Chapter 12 specifically focuses on this subject. In early studies of one such CORM (CORM-3), Motterlini and coworkers found no increase in COHb levels after intravenous infusion of CORM-3 at a dosage of 3.54 mg/kg over 1 h (Figure 3.16). *In vitro* mixing of CORM-3 with human blood at a concentration of 20–400 μM showed a dose-dependent increase of COHb levels, ranging between 0% and 2.2%, with the COHb level being 2.2% at a CORM-3 concentration of 400 μM [228]. However, if one assumes 1 equiv. of CO being released from CORM-3 and being completely bound to Hb subsequently, 400 μM would give rise to about 5.6% COHb. Therefore, the yield of CO transferred from CORM-3 to Hb *in vitro* is calculated to be about 40%. Further, the blood CORM-3 concentration at a dosage of 3.54 mg/kg in mice was estimated to be 20 μM by the study, which was unlikely to significantly increase the COHb level. Consequently, the study attributed the activity of CORM-3 at the low dosage to its putative ability to deliver CO to tissues at a dosage that does not increase blood COHb [228]. Chapter 12 provides detailed discussions of CORM-3.

Another CORM with boranocarbonate chemistry, CORM-A1, showed stoichiometric release of CO *in vitro* using a caboxyMb formation assay [229]. In an *in vitro* COHb formation test, it was found 18 mM CORM-A1 increased COHb levels to about 70% after 120 min of incubation. The millimolar concentration of CORM-A1 used was much higher than those used in the CORM-3 experiments. Assuming blood Hb concentration of about 14 g/dl, 70% COHb is equal to about 6 mM Therefore, the COHb formation yield is the equivalent of about 33%. However, no further *in vivo* PK work was conducted in the study [229]. In another *in vivo* study of the renal blood flow in mice, CORM-A1 dosed by bolus i.v. infusion at 0.96 μmol (0.1 mg) per mouse increased COHb level by 17 ± 7% within 15 min. Then, COHb levels returned to baseline within 1 h [230]. A comparison between CORM-3 and CORM-A1 can be made by analyzing the results from the work by De Backer et al. [231]. Intraperitoneal administration of CORM-3 and CORM-A1 at 40 and 15 mg/kg gave peak COHb levels of 2.3% and 8.4% in mice, respectively. For CORM-A1, the time for COHb to peak was similar to the other studies discussed earlier [230]. CORM-3 showed a shorter time to peak and CORM-A1 demonstrated a much higher level of CO availability than CORM-3 (Figure 3.17).

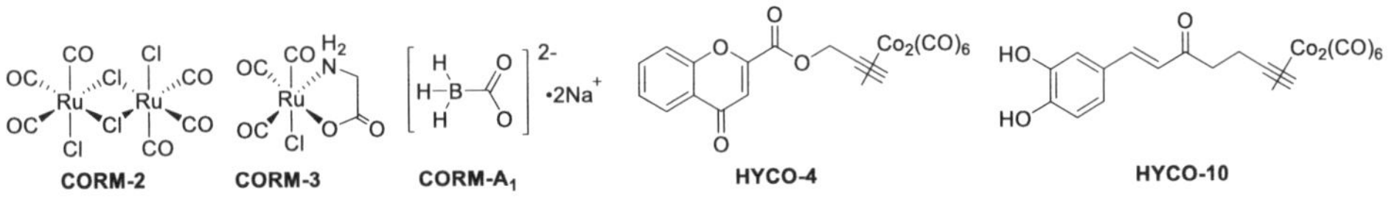

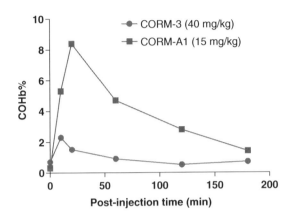

Figure 3.16 Structures of CORMs involved in the discussion.

Figure 3.17 Blood COHb levels after intraperitoneal administration of CORM-3 and CORM-A1. Source: Plot based on reported data [231].

Competitive binding to Hb between CO released from CORM-2 and oxygen was shown by Lang et al. [232]. In their *in vitro* erythrocyte CO incubation study at a hematocrit of 0.4% (normal range in adults: 38–54%), the basal COHb level was 1%. After incubation with 10 μM of CORM-2 for 24 h, the COHb level increased to about 4.2%. However, when the 10 μM CORM-2 was incubated in 100% oxygen environment, the COHb level only reached to about 1.5%.

Magierowski's group generated a good amount of data on COHb levels and CO concentrations in the gastric tissue in their studies of CO's healing effects on gastric damages (Table 3.3) [233]. Efficacy was observed in GI protection and the increase in COHb level was small. Such results again suggest the possibility of CO's effect through delivery to the site of action instead of systemic availability.

Motterlini's group developed a series of cobalt-based hybrid molecules that combine an Nrf2 inducer with a CO releaser (HYCOs) [235]. Among them, HYCO-10 showed the ability to induce HO-1 expression *in vivo*. HYCO-4 and HYCO-10 administered via oral gavage at a dosage of 3 mg/mouse (about 100 mg/kg) increased the COHb level from 0.5% to about 3.6 ± 1.0% and 1.8 ± 0.2%, respectively. However, no further PK result was provided in the study.

Despite the large amount of research work on the biological evaluation of CORMs, studies of their pharmacokinetics are lagging far behind. Especially worth noting is the fact that there has not been a peer-reviewed systematic PK study of the CO released, the donor, and the post-release product(s) among all the metal-based CORMs. Clearly, more PK work is needed in the further development of such CORMs into clinically useful CO therapies.

Organic CO Prodrugs

There have been ongoing efforts in designing organic CO donors. In an attempt to develop oral delivery platforms as "CO in a pill," Wang et al. have developed a series of organic CO prodrugs by taking advantage of a cheletropic extrusion reaction (Figure 3.18) (Chapter 14) [195,200,210,236]. Those prodrugs were designed to release CO with tunable rates under physiological conditions and are responsive to various triggers such as temperature, chemical reagents, esterase, ROS, and pH.

Recent studies [198,199,210] have examined the pharmacological efficacy and PK properties of select CO prodrugs in various preclinical models (Chapter 14). Specifically relevant to the theme of this chapter, Wang et al. reported a detailed PK study on five organic CO prodrugs in mice. With slight chemical modifications, the *in vitro* CO release half-lives of the prodrugs ranged from 20 min to 5 h in organic/aqueous cosolvents and 1–10 h when they are formulated in micelles. Because the aqueous medium is known to promote the key reaction needed for CO release from such prodrugs [237], CO release can occur during the absorption, distribution, and metabolism phases of CO prodrugs. As such, the PK

Table 3.3 COHb data of oral administration of CORMs.

Administration route	CORMs (dose)	COHb%	CO concentration[ml (CO)/g (tissue)]	Reference
Oral (intragastric)	CORM-2 (5 mg/kg)	1.55 ± 0.12	1.90 ± 0.35	[233]
	CORM-2 (50 mg/kg)	2.20 ± 0.11	3.68 ± 0.61	
	Vehicle (basal)	0.78 ± 0.03	1.04 ± 0.11	
Oral (intragastric)	CORM-2 (1 mg/kg)	0.95	1.3 ± 0.2	[234]
	Vehicle (basal)	0.65	1.0 ± 0.2	
Oral (intragastric)	CORM-2 (5 mg/kg)	2.4 ± 0.1	3.8 ± 0.6% (sat.%)	[152]
	Vehicle (basal)	1.8 ± 0.1	1.8 ± 0.3% (sat.%)	

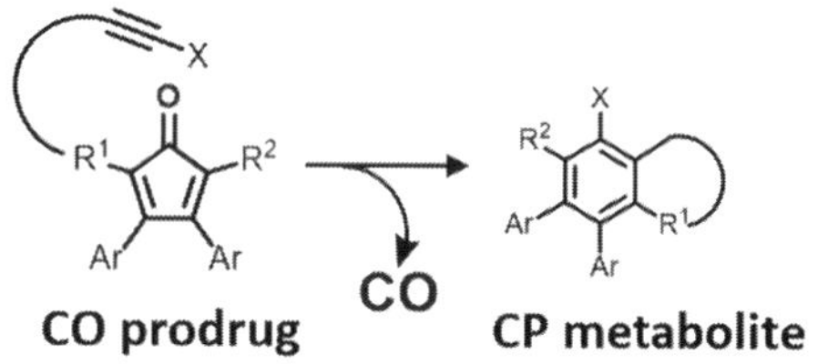

Figure 3.18 The general structures of the prodrugs, metabolites, and the schematic illustration of CO release.

properties of CO are the combined results of the interplay among the CO release kinetics of the prodrugs at the administration site and in the systemic circulation, CO diffusion/Hb binding/elimination, and the biodisposition of the CO prodrug. The *in vivo* studies showed that all prodrugs elevated the blood COHb level by 1.5–3% and sustained the elevation above the baseline for 1.5 h. It is also interesting to note that the CO prodrugs tested have different *in vitro* CO release rates, but similar peak COHb levels and t_{max} of 1–2 h. However, those with a slower release rate tend to have lower COHb area under the curve. A brief GI transit time might limit CO absorption and delivery efficiency, especially for prodrugs with slower CO release rates. Such hypothesis is supported by the fecal analysis and murine GI transit studies that indicate approximately 70% of the compound leaves the small intestine 6 h after oral application and 60% of the prodrug/metabolite was recovered in the fecal collection 10 h after oral administration [198,238].

These prodrugs were able to increase the COHb levels regardless of the route of administration: intravenous, intraperitoneal, and oral routes. Additionally, detailed plasma analysis showed micromolar concentrations of both the prodrugs and inactive metabolites in the circulation with quick elimination half-lives of 1–2 h. The demonstrated quick elimination, low accumulation of the prodrugs, and the inactive metabolites in the circulation help to minimize the number of safety concerns. These findings provide critical evidence supporting therapeutic applications of such prodrugs via oral delivery and toward the eventual goal of "CO in a pill" [236].

Relevant to the demonstration of adequate PK profiles in support of efficacy, organic prodrug CO-103 showed marked anti-inflammatory efficacy in murine colitis models at 15 mg/kg via i.p. administration [200]. Additionally, i.v. administration of 4 mg/kg of an organic CO prodrug had a peak COHb level of 4.5% in mice and showed protection against chemically induced liver injury [199]. An esterase-sensitive and pH-controlled CO prodrug exhibited efficacy at 10 mg/kg dose in mouse models of systemic inflammation and inflammation-induced liver injury [210]. Organic prodrug CO-111 showed strong protective effects in chemically induced gastric injury models in rats at 0.1 mg/kg via p.o. administration [153]. At this dosage level, it is not expected that the COHb level would change. Together, these findings demonstrate that the therapeutically relevant COHb levels of organic prodrugs are around 2–4% in various murine models. Further, CO prodrugs delivered locally to the injury site are shown to offer protection without the need to elevate systemic COHb levels. More work is needed to establish the PK–PD relationship for these organic CO prodrugs.

CO-Releasing Materials

CO-releasing materials differ from the aforementioned approaches that dose CO either in the form of gas or as a small molecule that can release CO under physiological conditions; they encapsulate CO donor in a vessel by either covalent tethers or adsorption that can change the CO release profile according to the features of the vessel and sometimes help to retain the toxic by-product after CO release. These approaches include micelles [202,239], polymer conjugates [240], iron- and zirconium-based metal–organic frameworks (also known as porous coordination polymers) [241,242], mesoporous silica [243], poly(L-lactide-co-D/L-lactide) nanofiber nonwovens [244], SiO_2 nanoparticles [245], phenylboronic acid-containing framboidal nanoparticles [246], graphene oxide nanoparticles [247], detonation nanodiamond [248], and peptide dendrimer nanogel [249].

Yin et al. developed styrene–maleic acid (SMA) copolymer micelles that encapsulate CORM-2 (SMA/CORM-2) [202]. Due to the hydrophobic core and hydrophilic surface structure, the formed micelles are water-soluble and thus allow for resolution of the solubility problem of CORM-2. *In vitro* tests in fetal bovine serum showed the release half-life of the encapsulated SMA/CORM-2 being about 24 h, which was slower than the free CORM-2 capable of instant CO release. In *in vivo* PK studies of SMA/CORM-2, the GC method was used to determine the CO concentration in the blood and colitis tissue. An i.v. injection of 20 mg/kg SMA/CORM-2 (equivalent to 2 mg/kg CORM-2) increased blood CO concentration from 10 to about 36 nM. Peak time was found to be about 24 h. The higher level of CO availability was accompanied by the longer retention of SMA/CORM-2 than that of the free CORM-2. However, it should be noted that the CO concentration value of 10 nM only corresponds to about 0.000125% COHb, which is much lower (about 1000-fold) than the reported basal CO level discussed in the former sections. Considering 0.5% COHb as a basal level, the total CO concentration is estimated to be about 10 μM. It is not clear where the discrepancy comes from, but it might be due to the use of NO to liberate CO as described in the method or other unknown factors such as a simple unit error. The same issue could also be found in the tissue concentration, where the study showed around 5 pmol/g, while most studies in the literature showed about 5 pmol/mg. Nevertheless, for blood CO concentrations, SMA/CORM-2 showed a longer time to peak (24 h) than free CORM-2 (1 h). Compared with CORM-2, SMA/CORM-2 also showed higher sustained blood CO concentrations at above threefold of the basal level from 12 to 48 h. It is also interesting to find that oral gavage of the same dosage (20 mg/kg) gave a similar blood CO PK profile, indicating high oral bioavailability of CO generated from the SMA/CORM-2 system. It is intriguing to see that even though i.v. injection of CORM-2 increased blood CO concentration to two- to threefold of the basal level between 1 and 24 h, it did not lead to an increase in CO concentration in the colon tissue. On the other hand, SMA/CORM-2 significantly increased colonic tissue CO concentrations to about ninefold after 4 h of i.v. injection, which persisted for about 24 h. The authors attributed the superior CO PK profiles to the enhanced permeability and retention (EPR) effect of the micelle, which helped to keep the encapsulated drug in circulation for a longer period of time. The CO peak time of the blood and tissue was different, indicating that SMA/COMR-2 may have unique distribution profiles as well. CO concentration in colon tissues after oral gavage was about twofold lower than that of the i.v. administration using the same CO equivalent dose. The study also found that the SMA/CORM-2 treatment group had a higher CO concentration in inflamed colon tissues than in normal colon tissues of the control group. The author also attributed this distribution difference to the EPR effect.

Aside from the example discussed earlier, most of the reported approaches stated the biocompatibility of the materials developed. However, most studies only proved their feasibility to release CO and efficacy in CO-dependent bioactivity *in vitro*. To explore their therapeutic potential, the PK profiles of the CO released from these encapsulated CORMs need to be carefully studied.

Summary and Perspectives

CO's PK characteristics have been extensively studied since the 1950s, with the goal of understanding its toxicology and establishing safety criteria for accidental and occupational exposures. With the recognition of CO's therapeutic potential, it has become obvious that the PK properties of CO in the context of therapeutic effects also need to be studied. Studies with CO gas inhalation have clearly established boundary conditions for safety and tolerance. However, studying the PK properties of CO from various delivery forms clearly has its unique challenges. Compared with studying the PK properties of traditional small molecules, additional challenges in studying CO include its volatility, binding to various hemoproteins, changing affinity for various hemoproteins depending on physiological conditions, the need to analyze "carrier" molecules from various donor forms, and interference of CO's binding by a large number of other molecules such as O_2, H_2S and other thiols, NO, CO_2, and 2,3-DPG. At this time, there is no clearly established dose–response relationship for a pharmacological indication of CO, no consensus on a therapeutic COHb threshold level, no clear understanding of a specific sustained duration of elevated COHb levels needed for efficacy, a lack of understanding of the adequacy of COHb as a CO surrogate for PK studies, and a lack of quantitative correlation of COHb levels with PD outcome/efficacy. Therefore, much more work is needed in this area.

Acknowledgments

CO-related work in the Wang lab was partially supported by a grant from the National Institutes of Health (R01DK119202). The authors also acknowledge the support of the Georgia Research Alliance for an Eminent Scholar endowment and internal financial sources at Georgia State University. The authors would also like to express their appreciation of the following colleagues for reviewing this chapter and their suggestions for revisions: Leo Otterbein, Lisa Berreau, Stefan Ryter, and Albert Donnay.

References

1. Yang, X., Lu, W., Hopper, C.P., Ke, B., and Wang, B. (2020). Nature's marvels endowed in gaseous molecules I: carbon monoxide and its physiological and therapeutic roles. *Acta Pharma. Sin. B* 10.1016/j.apsb.2020.10.010.
2. Cruickshank, W. (1801). Some observations on different hydrocarbons and combinations of carbon with oxygen, etc. in reply to some of Dr. Priestley's late objections to the new system of chemistry. *J. Nat. Phil. Chem. Arts* 5: 1–9.
3. Bernard, C. (1865). Introduction a L'etude De La Medecine Experimentale. Paris: J.B. Bailliere et Fils.
4. Haldane, J.S. and Smith, L. (1897). The absorption of oxygen by the lungs. *J. Physiol.* 22: 231.
5. Douglas, C.G., Haldane, J.S., and Haldane, J.B. (1912). The laws of combination of haemoglobin with carbon monoxide and oxygen. *J. Physiol.* 44: 275–304.
6. Sjöstrand, T. (1949). Endogenous formation of carbon monoxide in man under normal and pathological conditions. *Scand. J. Clin. Lab. Invest.* 1: 201–214.
7. Sjöstrand, T. (1952). The formation of carbon monoxide by the decomposition of haemoglobin in vivo. *Acta Physiol. Scand.* 26: 338–344.
8. Yang, X.-X., Ke, B.-W., Lu, W., and Wang, B.-H. (2020). CO as a therapeutic agent: discovery and delivery forms. *Chin. J. Nat. Med.* 18: 284–295.
9. Springer, B.A., Sligar, S.G., Olson, J.S., and Phillips, G.N., Jr. (1994). Mechanisms of ligand recognition in myoglobin. *Chem. Rev.* 94: 699–714.
10. Dewilde, S., Kiger, L., Burmester, T., Hankeln, T., Baudin-Creuza, V., Aerts, T., Marden, M.C., Caubergs, R., and Moens, L. (2001). Biochemical characterization and ligand binding properties of neuroglobin, a novel member of the globin family. *J. Biol. Chem.* 276: 38949–38955.
11. Azarov, I., Wang, L., Rose, J.J., Xu, Q., Huang, X.N., Belanger, A., Wang, Y., Guo, L., Liu, C., Ucer, K.B., McTiernan, C.F., O'Donnell, C.P., Shiva, S., Tejero, J., Kim-Shapiro, D.B., and Gladwin, M.T. (2016). Five-coordinate H64Q neuroglobin as a ligand-trap antidote for carbon monoxide poisoning. *Sci. Transl. Med.* 8: 368ra173.
12. Zavorsky, G.S., Tesler, J., Rucker, J., Fedorko, L., Duffin, J., and Fisher, J.A. (2014). Rates of carbon monoxide elimination in males and females. *Physiol. Rep.* 2: e12237.
13. Contostavlos, D.L. and Lichtenwalner, M. (2003). A simple field test to detect elevated concentrations of carboxyhemoglobin in autopsy blood. *J. Clin. Forensic Med.* 10: 77–80.
14. Boumba, V.A. and Vougiouklakis, T. (2005). Evaluation of the methods used for carboxyhemoglobin analysis in postmortem blood. *Int. J. Toxicol.* 24: 275–281.
15. Vreman, H.J., Wong, R.J., and Stevenson, D.K. (2001). Sources, sinks, and measurements of carbon monoxide. In: Carbon Monoxide and Cardiovascular Functions (ed. R. Wang), 273–307. CRC press.
16. Coburn, R., Forster, R., and Kane, P. (1965). Considerations of the physiological variables that determine the blood carboxyhemoglobin concentration in man. *J. Clin. Invest.* 44: 1899–1910.
17. Coburn, R.F. (1970). The carbon monoxide body stores. *Ann. N. Y. Acad. Sci.* 174: 11–22.
18. Brunelle, J.A., Degtiarov, A.M., Moran, R.F., and Race, L.A. (1996). Simultaneous measurement of total hemoglobin and its derivatives in blood using CO-oximeters: analytical principles; their application in selecting analytical wavelengths and reference methods; a comparison of the results of the choices made. *Scand. J. Clin. Lab. Invest.* 56: 47–69.
19. Brown, L.J. (1980). A new instrument for the simultaneous measurement of total hemoglobin, % oxyhemoglobin, % carboxyhemoglobin, % methemoglobin, and oxygen content in whole blood. *IEEE Trans. Biomed. Eng.* 27: 132–138.
20. Brehmer, C. and Iten, P.X. (2003). Rapid determination of carboxyhemoglobin in blood by Oximeter. *Forensic. Sci. Int.* 133: 179–181.
21. Hardy, K.R. and Thom, S.R. (1994). Pathophysiology and treatment of carbon monoxide poisoning. *J. Toxicol. Clin. Toxicol.* 32: 613–629.
22. Suematsu, M., Goda, N., Sano, T., Kashiwagi, S., Egawa, T., Shinoda, Y., and Ishimura, Y. (1995). Carbon monoxide: an endogenous modulator of

sinusoidal tone in the perfused rat liver. *J. Clin. Invest.* 96: 2431–2437.

23. Oliverio, S. and Varlet, V. (2019). Total blood carbon monoxide: Alternative to carboxyhemoglobin as biological marker for carbon monoxide poisoning determination. *J. Anal. Toxicol.* 43: 79–87.

24. Coburn, R.F. (2020). Coronary and cerebral metabolism-blood flow coupling and pulmonary alveolar ventilation-blood flow coupling may be disabled during acute carbon monoxide poisoning. *J. Appl. Physiol.* 129: 1039–1050.

25. Mahoney, J., Vreman, H., Stevenson, D., and Van Kessel, A. (1993). Measurement of carboxyhemoglobin and total hemoglobin by five specialized spectrophotometers (CO-oximeters) in comparison with reference methods. *Clin. Chem.* 39: 1693–1700.

26. Stepanov, E.V., Zyrianov, P.V., Khusnutdinov, A.N., Kouznetsov, A.I., and Ponurovskii, Y.Y. (1996). In *Multicomponent gas analyzers based on tunable diode lasers*, Application of Tunable Diode and Other Infrared Sources for Atmospheric Studies and Industrial Process Monitoring, International Society for Optics and Photonics. 270–280.

27. Kouznetsov, A.I., Stepanov, E.V., Shulagin, Y.A., and Skrupskii, V.A. (1996). In *Endogenous CO dynamics monitoring in breath by tunable diode laser*, Laser Diodes and Applications II, International Society for Optics and Photonics. 247–256.

28. Mukhopadhyay, S., Sarkar, A., Chattopadhyay, P., and Dhara, K. (2020). Recent advances in fluorescence light-up endogenous and exogenous carbon monoxide detection in biology. *Chem. Asian J.* 15: 3162–3179.

29. Katsumata, Y., Aoki, M., Sato, K., Suzuki, O., Oya, M., and Yada, S. (1982). A simple spectrophotometry for determination of carboxyhemoglobin in blood. *J. Forensi. Sci.* 27: 928–934.

30. Brown, S. (1986). Clarke's isolation and identification of drugs. *J. Clin. Pathol.* 39: 1368.

31. Sakata, M. and Haga, M. (1980). Determination of carbon monoxide in blood by head space analysis. *J. Toxicol. Sci.* 5: 35–43.

32. Van Dam, J. and Daenens, P. (1994). Microanalysis of carbon monoxide in blood by head-space capillary gas chromatography. *J. Forensic Sci.* 39: 473–478.

33. Goldbaum, L., Chace, D., and Lappas, N. (1986). Determination of carbon monoxide in blood by gas chromatography using a thermal conductivity detector. *J. Forensic Sci.* 31: 133–142.

34. Griffin, B.R. (1979). A sensitive method for the routine determination of carbon monoxide in blood using flame ionization gas chromatograpy. *J. Anal. Toxicol.* 3: 102–104.

35. Sundin, A.-M. and Larsson, J.E. (2002). Rapid and sensitive method for the analysis of carbon monoxide in blood using gas chromatography with flame ionisation detection. *J. Chromatogr. B* 766: 115–121.

36. Vreman, H.J., Wong, R.J., Kadotani, T., and Stevenson, D.K. (2005). Determination of carbon monoxide (CO) in rodent tissue: effect of heme administration and environmental CO exposure. *Anal. Biochem.* 341: 280–289.

37. Vreman, H.J., Kwong, L.K., and Stevenson, D.K. (1984). Carbon monoxide in blood: an improved microliter blood-sample collection system, with rapid analysis by gas chromatography. *Clin. Chem.* 30: 1382–1386.

38. Oritani, S., Zhu, B.-L., Ishida, K., Shimotouge, K., Quan, L., Fujita, M.Q., and Maeda, H. (2000). Automated determination of carboxyhemoglobin contents in autopsy materials using head-space gas chromatography/mass spectrometry. *Forensic Sci. Int.* 113: 375–379.

39. Whitehead, R.D., Jr., Mei, Z., Mapango, C., and Jefferds, M.E.D. (2019). Methods and analyzers for hemoglobin measurement in clinical laboratories and field settings. *Ann N Y Acad Sci* 1450: 147–171.

40. Ostrander, C.R., Cohen, R.S., Hopper, A.O., Cowan, B.E., Stevens, G.B., and Stevenson, D.K. (1982). Paired determinations of blood carboxyhemoglobin concentration and carbon monoxide excretion rate in term and preterm infants. *J. Lab. Clin. Med.* 100: 745–755.

41. Peterson, J.E. (1970). Postexposure relationship of carbon monoxide in blood and expired air. *Arch. Environ. Health* 21: 172–173.

42. Penney, D.G. (2000). Carbon Monoxide Toxicity. CRC Press.

43. Afshar-Mohajer, N., Zuidema, C., Sousan, S., Hallett, L., Tatum, M., Rule, A.M., Thomas, G., Peters, T.M., and Koehler, K. (2018). Evaluation of low-cost electro-chemical sensors for environmental monitoring of ozone, nitrogen dioxide, and carbon monoxide. *J. Occup. Environ. Hyg.* 15: 87–98.

44. Wei, P., Ning, Z., Ye, S., Sun, L., Yang, F., Wong, K.C., Westerdahl, D., and Louie, P.K.K. (2018). Impact analysis of temperature and humidity conditions on electrochemical sensor response in ambient air quality monitoring. *Sensors (Basel)* 18.

45. Cao, Z., Buttner, W.J., and Stetter, J.R. (1992). The properties and applications of amperometric gas sensors. *Electroanalysis* 4: 253–266.

46. Vreman, H.J., Stevenson, D.K., Oh, W., Fanaroff, A.A., Wright, L.L., Lemons, J.A., Wright, E., Shankaran, S., Tyson, J.E., and Korones, S.B. (1994). Semiportable electrochemical instrument for determining carbon monoxide in breath. *Clin. Chem.* 40: 1927–1933.

47. Vreman, H.J., Mahoney, J.J., and Stevenson, D.K. (1993). Electrochemical measurement of carbon monoxide in breath: interference by hydrogen. *Atoms Environ.* 27: 2193–2198.
48. Park, S.S., Kim, J., and Lee, Y. (2012). Improved electrochemical microsensor for the real-time simultaneous analysis of endogenous nitric oxide and carbon monoxide generation. *Anal. Chem.* 84: 1792–1796.
49. Penney, D.G. (1990). Acute carbon monoxide poisoning: animal models: a review. *Toxicology* 62: 123–160.
50. Wu, L. and Wang, R. (2005). Carbon monoxide: endogenous production, physiological functions, and pharmacological applications. *Pharmacol. Rev.* 57: 585–630.
51. Gorman, D., Drewry, A., Huang, Y.L., and Sames, C. (2003). The clinical toxicology of carbon monoxide. *Toxicology* 187: 25–38.
52. Rose, J.J., Wang, L., Xu, Q., McTiernan, C.F., Shiva, S., Tejero, J., and Gladwin, M.T. (2017). Carbon monoxide poisoning: Pathogenesis, management, and future directions of therapy. *Am. J. Respir. Crit. Care Med.* 195: 596–606.
53. Oliverio, S. and Varlet, V. (2020). What are the limitations of methods to measure carbon monoxide in biological samples? *Forensic Toxicol.* 38: 1–14.
54. Kazemi, H., Klein, R.C., Turner, F.N., and Strieder, D.J. (1968). Dynamics of oxygen transfer in the cerebrospinal fluid. *Respir. Physiol.* 4: 24–31.
55. Morris, G.L., Curtis, S.E., and Simon, J. (1985). Perinatal piglets under sublethal concentrations of atmospheric carbon monoxide. *J. Anim. Sci.* 61: 1070–1079.
56. Knecht, K.R., Milam, S., Wilkinson, D.A., Fedinec, A.L., and Leffler, C.W. (2010). Time-dependent action of carbon monoxide on the newborn cerebrovascular circulation. *Am. J. Physiol. Heart Circ. Physiol.* 299: H70–5.
57. Kanu, A. and Leffler, C.W. (2007). Carbon monoxide and Ca2+-activated K+ channels in cerebral arteriolar responses to glutamate and hypoxia in newborn pigs. *Am. J. Physiol. Heart Circ. Physiol.* 293: H3193–200.
58. Carratu, P., Pourcyrous, M., Fedinec, A., Leffler, C.W., and Parfenova, H. (2003). Endogenous heme oxygenase prevents impairment of cerebral vascular functions caused by seizures. *Am. J. Physiol. Heart Circ. Physiol.* 285: H1148–57.
59. Haldane, J. (1895). The relation of the action of carbonic oxide to oxygen tension. *J. Physiol.* 18: 201–217.
60. Zazzeron, L., Liu, C., Franco, W., Nakagawa, A., Farinelli, W.A., Bloch, D.B., Anderson, R.R., and Zapol, W.M. (2015). Pulmonary phototherapy for treating carbon monoxide poisoning. *Am. J. Respir. Crit. Care Med.* 192: 1191–1199.
61. Stadie, W.C. and Martin, K.A. (1925). The elimination of carbon monoxide from the blood: a theoretical and experimental study. *J. Clin. Invest.* 2: 77–91.
62. Guyatt, A.R., Holmes, M.A., and Cumming, G. (1981). Can carbon monoxide be absorbed from the upper respiratory tract in man? *Eur. J. Respir. Dis.* 62: 383–390.
63. Tikuisis, P., Kane, D.M., McLellan, T.M., Buick, F., and Fairburn, S.M. (1992). Rate of formation of carboxyhemoglobin in exercising humans exposed to carbon monoxide. *J. Appl. Physiol.* 72: 1311–1319.
64. Sander, R. (2015). Compilation of Henry's law constants (version 4.0) for water as solvent. *Atmos. Chem. Phys.* 15: 4399–4981.
65. Sheps, D.S., Herbst, M.C., Hinderliter, A.L., Adams, K.F., Ekelund, L.G., O'Neil, J.J., Goldstein, G.M., Bromberg, P.A., Ballenger, M., Davis, S.M. et al. (1991). Effects of 4 percent and 6 percent carboxyhemoglobin on arrhythmia production in patients with coronary artery disease. *Res. Rep. Health Eff. Inst.* 1–46. discussion 47–58.
66. Stewart, R.D., Peterson, J.E., Baretta, E.D., Bachand, R.T., Hosko, M.J., and Herrmann, A.A. (1970). Experimental human exposure to carbon monoxide. *Arch. Environ. Health* 21: 154–164.
67. National Research Council (US) Committee on Acute Exposure Guideline Levels. (2010). Acute Exposure Guideline Levels for Selected Airborne Chemicals, Vol. 8. Washington (DC): National Academies Press (US).
68. Bathoorn, E., Slebos, D.J., Postma, D.S., Koeter, G.H., Van Oosterhout, A.J., Van Der Toorn, M., Boezen, H.M., and Kerstjens, H.A. (2007). Anti-inflammatory effects of inhaled carbon monoxide in patients with COPD: a pilot study. *Eur. Respir. J.* 30: 1131–1137.
69. Zander, R. (1981). Oxygen solubility in normal human blood. In: Oxygen Transport to Tissue (ed. A.G.B. Kovách, E. Dóra, M. Kessler and I.A. Silver), 331–332. Pergamon.
70. Beutler, E. and Waalen, J. (2006). The definition of anemia: what is the lower limit of normal of the blood hemoglobin concentration? *Blood* 107: 1747–1750.
71. Lodemann, P., Schorer, G., and Frey, B.M. (2010). Wrong molar hemoglobin reference values-a longstanding error that should be corrected. *Ann. Hematol.* 89: 209.
72. Chakraborty, S., Balakotaiah, V., and Bidani, A. (2004). Diffusing capacity reexamined: relative roles of diffusion and chemical reaction in red cell uptake of O2, CO, CO2, and NO. *J. Appl. Physiol.* 97: 2284–2302.
73. Zock, J.P. (1990). Carbon monoxide binding in a model of hemoglobin differs between the T and the R conformation. *Adv. Exp. Med. Biol.* 277: 199–207.

74. Mihailescu, M.R. and Russu, I.M. (2001). A signature of the T —> R transition in human hemoglobin. *Proc. Natl. Acad. Sci. U.S.A.* 98: 3773–3777.
75. Fan, J.S., Zheng, Y., Choy, W.Y., Simplaceanu, V., Ho, N.T., Ho, C., and Yang, D. (2013). Solution structure and dynamics of human hemoglobin in the carbonmonoxy form. *Biochemistry* 52: 5809–5820.
76. Cho, H.S., Schotte, F., Stadnytskyi, V., DiChiara, A., Henning, R., and Anfinrud, P. (2018). Dynamics of quaternary structure transitions in R-state carbonmonoxyhemoglobin unveiled in time-resolved X-ray scattering patterns following a temperature jump. *J. Phys. Chem. B* 122: 11488–11496.
77. Yonetani, T., Park, S.I., Tsuneshige, A., Imai, K., and Kanaori, K. (2002). Global allostery model of hemoglobin. Modulation of O(2) affinity, cooperativity, and Bohr effect by heterotropic allosteric effectors. *J. Biol. Chem.* 277: 34508–34520.
78. Yuan, Y., Tam, M.F., Simplaceanu, V., and Ho, C. (2015). New look at hemoglobin allostery. *Chem. Rev.* 115: 1702–1724.
79. Sharma, V.S., Geibel, J.F., and Ranney, H.M. (1978). "Tension" on heme by the proximal base and ligand reactivity: conclusions drawn from model compounds for the reaction of hemoglobin. *Proc. Natl. Acad. Sci. U.S.A.* 75: 3747–3750.
80. Brunori, M., Bonaventura, J., Bonaventura, C., Antonini, E., and Wyman, J. (1972). Carbon monoxide binding by hemoglobin and myoglobin under photodissociating conditions. *Proc. Natl. Acad. Sci. U.S.A.* 69: 868–871.
81. Unzai, S., Eich, R., Shibayama, N., Olson, J.S., and Morimoto, H. (1998). Rate constants for O2 and CO binding to the alpha and beta subunits within the R and T states of human hemoglobin. *J. Biol. Chem.* 273: 23150–23159.
82. Sharma, V.S., Schmidt, M.R., and Ranney, H.M. (1976). Dissociation of CO from carboxyhemoglobin. *J. Biol. Chem.* 251: 4267–4272.
83. Andersen, C.C. and Stark, M.J. (2012). Haemoglobin transfusion threshold in very preterm newborns: a theoretical framework derived from prevailing oxygen physiology. *Med. Hypotheses* 78: 71–74.
84. Hlastala, M.P., McKenna, H.P., Franada, R.L., and Detter, J.C. (1976). Influence of carbon monoxide on hemoglobin-oxygen binding. *J. Appl. Physiol.* 41: 893–899.
85. Vandegriff, K.D., Le Tellier, Y.C., Winslow, R.M., Rohlfs, R.J., and Olson, J.S. (1991). Determination of the rate and equilibrium constants for oxygen and carbon monoxide binding to R-state human hemoglobin cross-linked between the alpha subunits at lysine 99. *J. Biol. Chem.* 266: 17049–17059.
86. Romao, C.C., Blattler, W.A., Seixas, J.D., and Bernardes, G.J. (2012). Developing drug molecules for therapy with carbon monoxide. *Chem. Soc. Rev.* 41: 3571–3583.
87. Goldbaum, L.R., Ramirez, R.G., and Absalon, K.B. (1975). What is the mechanism of carbon monoxide toxicity? *Aviat. Space. Environ. Med.* 46: 1289–1291.
88. Sawicki, C.A. and Gibson, Q.H. (1978). The relation between carbon monoxide binding and the conformational change of hemoglobin. *Biophys. J.* 24: 21–33.
89. Saffran, W.A. and Gibson, Q.H. (1978). The effect of pH on carbon monoxide binding to Menhaden hemoglobin. Allosteric transitions in a root effect hemoglobin. *J. Biol. Chem.* 253: 3171–3179.
90. Agostoni, A., Stabilini, R., Viggiano, G., Luzzana, M., and Samaja, M. (1980). Influence of capillary and tissue PO2 on carbon monoxide binding to myoglobin: a theoretical evaluation. *Microvasc. Res.* 20: 81–87.
91. Winslow, R.M., Swenberg, M.L., Berger, R.L., Shrager, R.I., Luzzana, M., Samaja, M., and Rossi-Bernardi, L. (1977). Oxygen equilibrium curve of normal human blood and its evaluation by Adair's equation. *J. Biol. Chem.* 252: 2331–2337.
92. Antonini, E. and Brunori, M. (1971). Hemoglobin and Myoglobin in Their Reactions with Ligands, 436. Amsterdam: North-Holland.
93. Agostoni, A., Stabilini, R., Viggiano, G., Luzzana, M., and Samaja, M. (1980). Influence of capillary and tissue PO2 on carbon monoxide binding to myoglobin: a theoretical evaluation. *Microvasc. Res.* 20: 81–87.
94. Roughton, F.J. (1970). The equilibrium of carbon monoxide with human hemoglobin in whole blood. *Ann. N. Y. Acad. Sci.* 174: 177–188.
95. Levitt, D.G. and Levitt, M.D. (2015). Carbon monoxide: a critical quantitative analysis and review of the extent and limitations of its second messenger function. *Clin. Pharmacol.* 7: 37–56.
96. Root, W.S., Roughton, F.J.W., and Gregersen, M.I. (1946). Simultaneous determinations of blood volume by co and dye (T-1824) under various conditions. *Am. J. Physiol.* 146: 739–755.
97. Root, W.S., Allen, T.H., and Gregersen, M.I. (1953). Simultaneous determinations in splenectomized dogs of cell volume with CO and P32 and plasma volume with T-1824. *Am. J. Physiol.* 175: 233–235.
98. Nomof, N., Hopper, J., Jr., Brown, E., Scott, K., and Wennesland, R. (1954). Simultaneous determinations of the total volume of red blood cells by use of carbon monoxide and chromium in healthy and diseased human subjects. *J. Clin. Invest.* 33: 1382–1387.

99. Akeson, A., Biorck, G., and Simon, R. (1968). On the content of myoglobin in human muscles. *Acta Med. Scand.* 183: 307–316.
100. Moffet, D.A., Case, M.A., House, J.C., Vogel, K., Williams, R.D., Spiro, T.G., McLendon, G.L., and Hecht, M.H. (2001). Carbon monoxide binding by de novo heme proteins derived from designed combinatorial libraries. *J. Am. Chem. Soc.* 123: 2109–2115.
101. Gibson, Q.H., Olson, J.S., McKinnie, R.E., and Rohlfs, R.J. (1986). A kinetic description of ligand binding to sperm whale myoglobin. *J Biol Chem* 261: 10228–10239.
102. Coburn, R. (1970). Biological effects of carbon monoxide. *Ann. N. Y. Acad. Sci.* 174: 11–22.
103. Whalen, W.J. (1971). Intracellular PO2 in heart and skeletal muscle. *Physiologist* 14: 69–82.
104. Coburn, R.F., Ploegmakers, F., Gondrie, P., and Abboud, R. (1973). Myocardial myoglobin oxygen tension. *Am. J. Physiol.* 224: 870–876.
105. Ortiz-Prado, E., Dunn, J.F., Vasconez, J., Castillo, D., and Viscor, G. (2019). Partial pressure of oxygen in the human body: a general review. *Am. J. Blood Res.* 9: 1–14.
106. Boekstegers, P., Riessen, R., and Seyde, W. (1990). Oxygen partial pressure distribution within skeletal muscle: indicator of whole body oxygen delivery in patients? *Adv. Exp. Med. Biol.* 277: 507–514.
107. Vreman, H.J., Wong, R.J., Stevenson, D.K., Smialek, J.E., Fowler, D.R., Li, L., Vigorito, R.D., and Zielke, H.R. (2006). Concentration of carbon monoxide (CO) in postmortem human tissues: effect of environmental CO exposure. *J. Forensic Sci.* 51: 1182–1190.
108. Rynkiewicz, M. and Rynkiewicz, T. (2010). Bioelectrical impedance analysis of body composition and muscle mass distribution in advanced Kayakers. *Hum. Mov.* 11: 11–16.
109. Cameron, J.R., Skofornick, J.G., and Grant, R.M. (1999). Physics of the Body, 2e. Medical Physics Pub Corp.
110. Ohno, T. and Kuroshima, A. (1986). Muscle myoglobin as determined by electrophoresis in thermally acclimated rat. *Jpn. J. Physiol.* 36: 733–744.
111. Gimenez, M., Sanderson, R.J., Reiss, O.K., and Banchero, N. (1977). Effects of altitude on myoglobin and mitochondrial protein in canine skeletal muscle. *Respiration* 34: 171–176.
112. Fasold, H., Riedl, G., and Jaisle, F. (1970). Evidence for an absence of myoglobin from human smooth muscle. *Eur J Biochem* 15: 122–126.
113. Luomanmaki, K. and Coburn, R.F. (1969). Effects of metabolism and distribution of carbon monoxide on blood and body stores. *Am. J. Physiol.* 217: 354–363.
114. Trashin, S., De Jong, M., Luyckx, E., Dewilde, S., and De Wael, K. (2016). Electrochemical evidence for neuroglobin activity on NO at physiological concentrations. *J. Biol. Chem.* 291: 18959–18966.
115. Dewilde, S., Mees, K., Kiger, L., Lechauve, C., Marden, M.C., Pesce, A., Bolognesi, M., and Moens, L. (2008). Chapter nineteen - expression, purification, and crystallization of neuro- and cytoglobin. In: Methods in Enzymology (ed. R.K. Poole), Vol. 436, 341–357. Academic Press.
116. Burmester, T., Weich, B., Reinhardt, S., and Hankeln, T. (2000). A vertebrate globin expressed in the brain. *Nature* 407: 520–523.
117. Pesce, A., Bolognesi, M., Bocedi, A., Ascenzi, P., Dewilde, S., Moens, L., Hankeln, T., and Burmester, T. (2002). Neuroglobin and cytoglobin. Fresh blood for the vertebrate globin family. *EMBO Rep.* 3: 1146–1151.
118. Reuss, S., Saaler-Reinhardt, S., Weich, B., Wystub, S., Reuss, M.H., Burmester, T., and Hankeln, T. (2002). Expression analysis of neuroglobin mRNA in rodent tissues. *Neuroscience* 115: 645–656.
119. Geuens, E., Brouns, I., Flamez, D., Dewilde, S., Timmermans, J.P., and Moens, L. (2003). A globin in the nucleus! *J. Biol. Chem.* 278: 30417–30420.
120. Petersen, M.G., Dewilde, S., and Fago, A. (2008). Reactions of ferrous neuroglobin and cytoglobin with nitrite under anaerobic conditions. *J Inorg Biochem* 102: 1777–1782.
121. Jin, K., Mao, X.O., Xie, L., Khan, A.A., and Greenberg, D.A. (2008). Neuroglobin protects against nitric oxide toxicity. *Neurosci. Lett.* 430: 135–137.
122. Wakasugi, K., Nakano, T., and Morishima, I. (2003). Oxidized human neuroglobin acts as a heterotrimeric Galpha protein guanine nucleotide dissociation inhibitor. *J Biol Chem* 278: 36505–36512.
123. Tiwari, P.B., Astudillo, L., Pham, K., Wang, X., He, J., Bernad, S., Derrien, V., Sebban, P., Miksovska, J., and Darici, Y. (2015). Characterization of molecular mechanism of neuroglobin binding to cytochrome c: a surface plasmon resonance and isothermal titration calorimetry study. *Inorg. Chem. Commun.* 62: 37–41.
124. Fago, A., Mathews, A.J., Moens, L., Dewilde, S., and Brittain, T. (2006). The reaction of neuroglobin with potential redox protein partners cytochrome b 5 and cytochrome c. *FEBS Lett.* 580: 4884–4888.
125. Shang, A., Feng, X., Wang, H., Wang, J., Hang, X., Yang, Y., Wang, Z., and Zhou, D. (2012). Neuroglobin upregulation offers neuroprotection in traumatic brain injury. *Neurol. Res.* 34: 588–594.
126. Hankeln, T., Ebner, B., Fuchs, C., Gerlach, F., Haberkamp, M., Laufs, T.L., Roesner, A., Schmidt,

M., Weich, B., Wystub, S., Saaler-Reinhardt, S., Reuss, S., Bolognesi, M., De Sanctis, D., Marden, M.C., Kiger, L., Moens, L., Dewilde, S., Nevo, E., Avivi, A., Weber, R.E., Fago, A., and Burmester, T. (2005). Neuroglobin and cytoglobin in search of their role in the vertebrate globin family. *J. Inorg. Biochem.* 99: 110–119.

127. Traystman, R.J. and Fitzgerald, R.S. (1977). Cerebral circulatory responses to hypoxic hypoxia and carbon monoxide hypoxia in carotid baroreceptor and chemoreceptor denervated dogs. *Acta Neurol. Scand. Suppl.* 64: 294–295.

128. Ludbrook, G.L., Helps, S.C., Gorman, D.F., Reilly, P.L., North, J.B., and Grant, C. (1992). The relative effects of hypoxic hypoxia and carbon monoxide on brain function in rabbits. *Toxicology* 75: 71–80.

129. Okeda, R., Matsuo, T., Kuroiwa, T., Nakai, M., Tajima, T., and Takahashi, H. (1987). Regional cerebral blood flow of acute carbon monoxide poisoning in cats. *Acta Neuropathol.* 72: 389–393.

130. Langston, P., Gorman, D., Runciman, W., and Upton, R. (1996). The effect of carbon monoxide on oxygen metabolism in the brains of awake sheep. *Toxicology* 114: 223–232.

131. Simmons, I.G. and Good, P.A. (1998). Carbon monoxide poisoning causes optic neuropathy. *Eye (Lond)* 12 (Pt 5): 809–814.

132. Von Restorff, W. and Hebisch, S. (1988). Dark adaptation of the eye during carbon monoxide exposure in smokers and nonsmokers. *Aviat. Space Environ. Med.* 59: 928–931.

133. Ferguson, L.S., Burke, M.J., and Choromokos, E.A. (1985). Carbon monoxide retinopathy. *Arch. Ophthalmol.* 103: 66–67.

134. Wilmer, W.H. (1921). Effects of carbon monoxid upon the eye. *Am. J. Ophthalmol.* 4: 73–90.

135. Fink, A.I. (1951). Carbon-monoxide asphyxia with visual sequelae★: with report of a case. *Am. J. Ophthalmol.* 34: 1024–1027.

136. Bi, W.-K., Wang, J.-L., Zhou, X.-D., Li, Z.-K., Jiang, -W.-W., Zhang, S.-B., Zou, Y., Bi, M.-J., and Li, Q. (2020). Clinical characteristics of visual dysfunction in carbon monoxide poisoning patients. *J. Ophthalmol.* 2020: 9537360.

137. Haab, P. (1990). The effect of carbon monoxide on respiration. *Experientia* 46: 1202–1206.

138. Queiroga, C.S.F., Almeida, A.S., and Vieira, H.L.A. (2012). Carbon monoxide targeting mitochondria. *Biochem. Res. Int.* 2012: 749845.

139. Hill, B.C. and Greenwood, C. (1984). The reaction of fully reduced cytochrome c oxidase with oxygen studied by flow-flash spectrophotometry at room temperature. Evidence for new pathways of electron transfer. *Biochem. J.* 218: 913–921.

140. Gibson, Q.H. and Greenwood, C. (1963). Reactions of cytochrome oxidase with oxygen and carbon monoxide. *Biochem. J.* 86: 541–554.

141. Krab, K., Kempe, H., and Wikstrom, M. (2011). Explaining the enigmatic K(M) for oxygen in cytochrome c oxidase: a kinetic model. *Biochim. Biophys. Acta* 1807: 348–358.

142. Petersen, L.C., Nicholls, P., and Degn, H. (1974). The effect of energization on the apparent Michaelis-Menten constant for oxygen in mitochondrial respiration. *Biochem. J.* 142: 247–252.

143. Bienfait, H.F., Jacobs, J.M., and Slater, E.C. (1975). Mitochondrial oxygen affinity as a function of redox and phosphate potentials. *Biochim. Biophys. Acta* 376: 446–457.

144. Chance, B., Saronio, C., Leigh, J.S., Jr., Ingledew, W.J., and King, T.E. (1978). Low-temperature kinetics of the reaction of oxygen and solubilized cytochrome oxidase. *Biochem. J.* 171: 787–798.

145. Gibson, Q.H., Palmer, G., and Wharton, D.C. (1965). The binding of carbon monoxide by cytochrome C oxidase and the ratio of the cytochromes a and A3. *J. Biol. Chem.* 240: 915–920.

146. Queiroga, C.S., Almeida, A.S., and Vieira, H.L. (2012). Carbon monoxide targeting mitochondria. *Biochem. Res. Int.* 2012: 749845.

147. D'Amico, G., Lam, F., Hagen, T., and Moncada, S. (2006). Inhibition of cellular respiration by endogenously produced carbon monoxide. *J. Cell Sci.* 119: 2291–2298.

148. Petersen, L.C. (1977). The effect of inhibitors on the oxygen kinetics of cytochrome c oxidase. *Biochim. Biophys. Acta* 460: 299–307.

149. Scandurra, F.M. and Gnaiger, E. (2010). Cell respiration under hypoxia: facts and artefacts in mitochondrial oxygen kinetics. *Adv. Exp. Med. Biol.* 662: 7–25.

150. Cooper, C.E. and Brown, G.C. (2008). The inhibition of mitochondrial cytochrome oxidase by the gases carbon monoxide, nitric oxide, hydrogen cyanide and hydrogen sulfide: chemical mechanism and physiological significance. *J. Bioenerg. Biomembr.* 40: 533–539.

151. Kadotani, T., Vreman, H.J., Wong, R.J., and Stevenson, D.K. (1999). Concentration of Carbon Monoxide (CO) in tissue. *Pediatr. Res.* 45: 67–67.

152. Magierowska, K., Korbut, E., Hubalewska-Mazgaj, M., Surmiak, M., Chmura, A., Bakalarz, D., Buszewicz, G., Wojcik, D., Sliwowski, Z., Ginter, G., Gromowski, T., Kwiecien, S., Brzozowski, T., and Magierowski, M. (2019). Oxidative gastric mucosal damage induced by ischemia/reperfusion and the mechanisms of its prevention by carbon monoxide-releasing tricarbonyldichlororuthenium (II) dimer. *Free Radic. Biol. Med.* 145: 198–208.

153. Bakalarz, D., Surmiak, M., Yang, X., Wójcik, D., Korbut, E., Śliwowski, Z., Ginter, G., Buszewicz, G., Brzozowski, T., Cieszkowski, J., Głowacka, U., Magierowska, K., Pan, Z., Wang, B., and Magierowski, M. (2021). Organic carbon monoxide prodrug, BW-CO-111, in protection against chemically-induced gastric mucosal damage. *Acta Pharma. Sin. B* 11: 456–475.
154. Cronje, F.J., Carraway, M.S., Freiberger, J.J., Suliman, H.B., and Piantadosi, C.A. (2004). Carbon monoxide actuates O(2)-limited heme degradation in the rat brain. *Free Radic. Biol. Med.* 37: 1802–1812.
155. Vreman, H.J., Stevenson, D.K., Henton, D., and Rosenthal, P. (1988). Correlation of carbon monoxide and bilirubin production by tissue homogenates. *J. Chromatogr.* 427: 315–319.
156. Wagner, J.A., Horvath, S.M., and Dahms, T.E. (1975). Carbon monoxide elimination. *Respir. Physiol.* 23: 41–47.
157. Shimazu, T., Ikeuchi, H., Sugimoto, H., Goodwin, C.W., Mason, A.D., Jr., and Pruitt, B.A., Jr. (2000). Half-life of blood carboxyhemoglobin after short-term and long-term exposure to carbon monoxide. *J. Trauma* 49: 126–131.
158. Tobias, C.A., Lawrence, J.H. et al. (1945). The elimination of carbon monoxide from the human body with reference to the possible conversion of CO to CO2. *Am. J. Physiol.* 145: 253–263.
159. Kreck, T.C., Shade, E.D., Lamm, W.J., McKinney, S.E., and Hlastala, M.P. (2001). Isocapnic hyperventilation increases carbon monoxide elimination and oxygen delivery. *Am. J. Respir. Crit. Care. Med.* 163: 458–462.
160. Weaver, L.K., Howe, S., Hopkins, R., and Chan, K.J. (2000). Carboxyhemoglobin half-life in carbon monoxide-poisoned patients treated with 100% oxygen at atmospheric pressure. *Chest* 117: 801–808.
161. Ernst, A. and Zibrak, J.D. (1998). Carbon monoxide poisoning. *N. Engl. J. Med.* 339: 1603–1608.
162. Pan, K.T., Leonardi, G.S., and Croxford, B. (2020). Factors contributing to CO uptake and elimination in the body: A critical review. *Int. J. Environ. Res. Public Health* 17: 528–541.
163. Bruce, M.C. and Bruce, E.N. (2006). Analysis of factors that influence rates of carbon monoxide uptake, distribution, and washout from blood and extravascular tissues using a multicompartment model. *J. Appl. Physiol.* 100: 1171–1180.
164. Coburn, R.F., Forster, R.E., and Kane, P.B. (1965). Considerations of the physiological variables that determine the blood carboxyhemoglobin concentration in man. *J. Clin. Invest.* 44: 1899–1910.
165. Peterson, J.E. and Stewart, R.D. (1970). Absorption and elimination of carbon monoxide by inactive young men. *Arch. Environ. Health* 21: 165–171.
166. Ryter, S.W., Alam, J., and Choi, A.M. (2006). Heme oxygenase-1/carbon monoxide: from basic science to therapeutic applications. *Physiol. Rev.* 86: 583–650.
167. Bruce, E.N. and Bruce, M.C. (2003). A multicompartment model of carboxyhemoglobin and carboxymyoglobin responses to inhalation of carbon monoxide. *J. Appl. Physiol.* 95: 1235–1247.
168. Wu, L. and Wang, R. (2005). Carbon monoxide: endogenous production, physiological functions, and pharmacological applications. *Pharmacol. Rev.* 57: 585–630.
169. Rochette, L., Cottin, Y., Zeller, M., and Vergely, C. (2013). Carbon monoxide: mechanisms of action and potential clinical implications. *Pharmacol Ther* 137: 133–152.
170. Coburn, R., Williams, W., White, P., and Kahn, S. (1967). The production of carbon monoxide from hemoglobin in vivo. *J. Clin. Invest.* 46: 346–356.
171. Landaw, S.A. and Winchell, H.S. (1970). Endogenous production of 14CO: a method for calculation of RBC life-span in vivo. *Blood* 36: 642–656.
172. Coburn, R.F. (1970). Endogenous carbon monoxide production. *N. Engl. J. Med.* 282: 207–209.
173. White, P. (1970). Carbon monoxide production and heme catabolism. *Ann. N. Y. Acad. Sci.* 174: 23–31.
174. Landaw, S.A. (1970). Kinetic aspects of endogenous carbon monoxide production in experimental animals. *Ann. N. Y. Acad. Sci.* 174: 32.
175. Lynch, S.R. and Moede, A.L. (1972). Variation in the rate of endogenous carbon monoxide production in normal human beings. *J. Lab. Clin. Med.* 79: 85–95.
176. Maisels, M.J., Pathak, A., Nelson, N.M., Nathan, D.G., and Smith, C.A. (1971). Endogenous production of carbon monoxide in normal and erythroblastotic newborn infants. *J. Clin. Invest.* 50: 1–8.
177. Luomanmaki, K. and Coburn, R. (1969). Effects of metabolism and distribution of carbon monoxide on blood and body stores. *Am. J. Physiol.* 217: 354–363.
178. Levitt, D.G. and Levitt, M.D. (2015). Carbon monoxide: a critical quantitative analysis and review of the extent and limitations of its second messenger function. *Clin. Pharmacol.: Adv. Appl.* 7: 37.
179. Carratu, P., Pourcyrous, M., Fedinec, A., Leffler, C.W., and Parfenova, H. (2003). Endogenous heme oxygenase prevents impairment of cerebral vascular functions caused by seizures. *Am. J. Physiol. Heart Circ. Physiol.* 285: H1148–H1157.

180. Leffler, C.W., Parfenova, H., Fedinec, A.L., Basuroy, S., and Tcheranova, D. (2006). Contributions of astrocytes and CO to pial arteriolar dilation to glutamate in newborn pigs. *Am. J. Physiol. Heart Circ. Physiol.* 291: H2897–H2904.

181. Kanu, A. and Leffler, C.W. (2007). Carbon monoxide and Ca2+-activated K+ channels in cerebral arteriolar responses to glutamate and hypoxia in newborn pigs. *Am. J. Physiol. Heart Circ. Physiol.* 293: H3193–H3200.

182. Atamaniuk, J., Stuhlmeier, K.M., Vidotto, C., Tschan, H., Dossenbach-Glaninger, A., and Mueller, M.M. (2008). Effects of ultra-marathon on circulating DNA and mRNA expression of pro- and anti-apoptotic genes in mononuclear cells. *Eur. J. Appl. Physiol.* 104: 711–717.

183. Thompson, D., Basu-Modak, S., Gordon, M., Poore, S., Markovitch, D., and Tyrrell, R.M. (2005). Exercise-induced expression of heme oxygenase-1 in human lymphocytes. *Free Radic. Res.* 39: 63–69.

184. Ghio, A.J., Case, M.W., and Soukup, J.M. (2018). Heme oxygenase activity increases after exercise in healthy volunteers. *Free Radic. Res.* 52: 267–272.

185. Wang, Q.M., Du, J.L., Duan, Z.J., Guo, S.B., Sun, X.Y., and Liu, Z. (2013). Inhibiting heme oxygenase-1 attenuates rat liver fibrosis by removing iron accumulation. *World J. Gastroenterol.* 19: 2921–2934.

186. Yoshikawa, S., Choc, M.G., O'Toole, M.C., and Caughey, W.S. (1977). An infrared study of CO binding to heart cytochrome c oxidase and hemoglobin A. Implications re O2 reactions. *J. Biol. Chem.* 252: 5498–5508.

187. Almeida, A.S., Figueiredo-Pereira, C., and Vieira, H.L. (2015). Carbon monoxide and mitochondria-modulation of cell metabolism, redox response and cell death. *Front. Physiol.* 6: 33.

188. Zuckerbraun, B.S., Chin, B.Y., Bilban, M., d'Avila, J.C., Rao, J., Billiar, T.R., and Otterbein, L.E. (2007). Carbon monoxide signals via inhibition of cytochrome c oxidase and generation of mitochondrial reactive oxygen species. *FASEB J.* 21: 1099–1106.

189. Thom, S.R., Fisher, D., Xu, Y.A., Notarfrancesco, K., and Ischiropoulos, H. (2000). Adaptive responses and apoptosis in endothelial cells exposed to carbon monoxide. *Proc. Natl. Acad. Sci. U.S.A.* 97: 1305–1310.

190. Middleton, E.T. and Morice, A.H. (2000). Breath carbon monoxide as an indication of smoking habit. *Chest* 117: 758–763.

191. Iheagwara, K.N., Thom, S.R., Deutschman, C.S., and Levy, R.J. (2007). Myocardial cytochrome oxidase activity is decreased following carbon monoxide exposure. *Biochim. Biophys. Acta* 1772: 1112–1116.

192. Jang, D.H., Khatri, U.G., Shortal, B.P., Kelly, M., Hardy, K., Lambert, D.S., and Eckmann, D.M. (2018). Alterations in mitochondrial respiration and reactive oxygen species in patients poisoned with carbon monoxide treated with hyperbaric oxygen. *Intensive Care Med. Exp.* 6: 4.

193. Zheng, Y., Ji, X., Yu, B., Ji, K., Gallo, D., Csizmadia, E., Zhu, M., Choudhury, M.R., De La Cruz, L.K.C., Chittavong, V., Pan, Z., Yuan, Z., Otterbein, L.E., and Wang, B. (2018). Enrichment-triggered prodrug activation demonstrated through mitochondria-targeted delivery of doxorubicin and carbon monoxide. *Nat. Chem.* 10: 787–794.

194. Tyuma, I., Ueda, Y., Imaizumi, K., and Kosaka, H. (1981). Prediction of the carbonmonoxyhemoglobin levels during and after carbon monoxide exposures in various animal species. *Jpn. J. Physiol.* 31: 131–143.

195. Ji, X., Damera, K., Zheng, Y., Yu, B., Otterbein, L.E., and Wang, B. (2016). Toward carbon monoxide-based therapeutics: Critical drug delivery and developability issues. *J. Pharm. Sci.* 105: 406–416.

196. Motterlini, R. and Otterbein, L.E. (2010). The therapeutic potential of carbon monoxide. *Nat. Rev. Drug Discov.* 9: 728–743.

197. Nordenberg, D., Yip, R., and Binkin, N.J. (1990). The effect of cigarette smoking on hemoglobin levels and anemia screening. *JAMA* 264: 1556–1559.

198. Wang, M., Yang, X., Pan, Z., Wang, Y., De La Cruz, L.K., Wang, B., and Tan, C. (2020). Towards "CO in a pill": pharmacokinetic studies of carbon monoxide prodrugs in mice. *J. Control. Release* 327: 174–185.

199. Zheng, Y., Ji, X., Yu, B., Ji, K., Gallo, D., Csizmadia, E., Zhu, M., Choudhury, M.R., De La Cruz, L.K.C., Chittavong, V., Pan, Z., Yuan, Z., Otterbein, L.E., and Wang, B. (2018). Enrichment-triggered prodrug activation demonstrated through mitochondria-targeted delivery of doxorubicin and carbon monoxide. *Nature Chem.* 10: 787–794.

200. Steiger, C., Uchiyama, K., Takagi, T., Mizushima, K., Higashimura, Y., Gutmann, M., Hermann, C., Botov, S., Schmalz, H.-G., and Naito, Y. (2016). Prevention of colitis by controlled oral drug delivery of carbon monoxide. *J. Control. Release* 239: 128–136.

201. Belcher, J.D., Gomperts, E., Nguyen, J., Chen, C., Abdulla, F., Kiser, Z.M., Gallo, D., Levy, H., Otterbein, L.E., and Vercellotti, G.M. (2018). Oral carbon monoxide therapy in murine sickle cell disease: beneficial effects on vaso-occlusion, inflammation and anemia. *PLoS One* 13: e0205194.

202. Yin, H., Fang, J., Liao, L., Nakamura, H., and Maeda, H. (2014). Styrene-maleic acid

copolymer-encapsulated CORM2, a water-soluble carbon monoxide (CO) donor with a constant CO-releasing property, exhibits therapeutic potential for inflammatory bowel disease. *J. Control. Release* 187: 14–21.

203. Zhang, J., Cao, S., Kwansa, H., Crafa, D., Kibler, K.K., and Koehler, R.C. (2012). Transfusion of hemoglobin-based oxygen carriers in the carboxy state is beneficial during transient focal cerebral ischemia. *J. Appl. Physiol.* 113: 1709–1717.

204. Ji, X., Zhou, C., Ji, K., Aghoghovbia, R.E., Pan, Z., Chittavong, V., Ke, B., and Wang, B. (2016). Click and release: A chemical strategy toward developing gasotransmitter prodrugs by using an intramolecular Diels-Alder reaction. *Angew. Chem. Int. Ed. Engl.* 55: 15846–15851.

205. Tinajero-Trejo, M., Rana, N., Nagel, C., Jesse, H.E., Smith, T.W., Wareham, L.K., Hippler, M., Schatzschneider, U., and Poole, R.K. (2016). Antimicrobial activity of the manganese photoactivated carbon monoxide-releasing molecule [Mn(CO)$_3$(tpa-kappa(3)N)]$^+$ against a pathogenic Escherichia coli that causes urinary infections. *Antioxid. Redox Signal.* 24: 765–780.

206. Zhang, W.Q., Atkin, A.J., Thatcher, R.J., Whitwood, A.C., Fairlamb, I.J., and Lynam, J.M. (2009). Diversity and design of metal-based carbon monoxide-releasing molecules (CO-RMs) in aqueous systems: revealing the essential trends. *Dalton Trans.* 22: 4351–4358.

207. Zobi, F., Degonda, A., Schaub, M.C., and Bogdanova, A.Y. (2010). CO releasing properties and cytoprotective effect of cis-trans-[Re(II)(CO)$_2$Br$_2$L$_2$]n complexes. *Inorg. Chem.* 49: 7313–7322.

208. Dallas, M.L., Scragg, J.L., and Peers, C. (2008). Modulation of hTREK-1 by carbon monoxide. *Neuroreport* 19: 345–348.

209. Ling, K., Men, F., Wang, W.C., Zhou, Y.Q., Zhang, H.W., and Ye, D.W. (2018). Carbon monoxide and its controlled release: Therapeutic application, detection, and development of carbon monoxide releasing molecules (CORMs). *J. Med. Chem.* 61: 2611–2635.

210. Ji, X., Pan, Z., Li, C., Kang, T., De La Cruz, L.K.C., Yang, L., Yuan, Z., Ke, B., and Wang, B. (2019). Esterase-sensitive and pH-controlled carbon monoxide prodrugs for treating systemic inflammation. *J. Med. Chem.* 62: 3163–3168.

211. Hegazi, R.A., Rao, K.N., Mayle, A., Sepulveda, A.R., Otterbein, L.E., and Plevy, S.E. (2005). Carbon monoxide ameliorates chronic murine colitis through a heme oxygenase 1-dependent pathway. *J. Exp. Med.* 202: 1703–1713.

212. Sheikh, S.Z., Hegazi, R.A., Kobayashi, T., Onyiah, J.C., Russo, S.M., Matsuoka, K., Sepulveda, A.R., Li, F., Otterbein, L.E., and Plevy, S.E. (2011). An anti-inflammatory role for carbon monoxide and heme oxygenase-1 in chronic Th2-mediated murine colitis. *J. Immunol.* 186: 5506–5513.

213. Steiger, C., Hermann, C., and Meinel, L. (2017). Localized delivery of carbon monoxide. *Eur. J. Pharm. Biopharm.* 118: 3–12.

214. Steiger, C., Lühmann, T., and Meinel, L. (2014). Oral drug delivery of therapeutic gases—Carbon monoxide release for gastrointestinal diseases. *J. Control. Release* 189: 46–53.

215. Tonner, S.P., Wainwright, M.S., Trimm, D.L., and Cant, N.W. (1983). Solubility of carbon monoxide in alcohols. *J. Chem. Eng. Data* 28: 59–61.

216. Takagi, T., Naito, Y., Uchiyama, K., Mizuhima, K., Suzuki, T., Horie, R., Hirata, I., Tsuboi, H., and Yoshikawa, T. (2016). Carbon monoxide promotes gastric wound healing in mice via the protein kinase C pathway. *Free Radic. Res.* 50: 1098–1105.

217. Correa-Costa, M., Gallo, D., Csizmadia, E., Gomperts, E., Lieberum, J.L., Hauser, C.J., Ji, X., Wang, B., Camara, N.O.S., Robson, S.C., and Otterbein, L.E. (2018). Carbon monoxide protects the kidney through the central circadian clock and CD39. *Proc. Natl. Acad. Sci. U.S.A.* 115: E2302–E2310.

218. Gomperts, E.D. and Forman, H.J. (2018). Solution of carbon monoxide for the treatment of disease, including sickle cell disease, US9980981B2.

219. Stewart, R.D., Fisher, T.N,, Hosko, M.J., Peterson, J.E. Baretta, E.D., Dodd, H.C. Carboxyhemoglobin elevation after exposure to dichloromethan. *Science.* 176:295–296.

220. Gargas, M.L., Clewell, H.J., 3rd, and Andersen, M.E. (1986). Metabolism of inhaled dihalomethanes in vivo: differentiation of kinetic constants for two independent pathways. *Toxicol. Appl. Pharmacol.* 82: 211–223.

221. Peterson, J.E. (1978). Modeling the uptake, metabolism and excretion of dichloromethane by man. *Am. Ind. Hyg. Assoc. J.* 39: 41–47.

222. Andersen, M.E., Clewell, H.J., 3rd, Gargas, M.L., MacNaughton, M.G., Reitz, R.H., Nolan, R.J., and McKenna, M.J. (1991). Physiologically based pharmacokinetic modeling with dichloromethane, its metabolite, carbon monoxide, and blood carboxyhemoglobin in rats and humans. *Toxicol. Appl. Pharmacol.* 108: 14–27.

223. Langehennig, P.L., Seeler, R.A., and Berman, E. (1976). Paint removers and carboxyhemoglobin. *N. Engl. J. Med.* 295: 1137.

224. Manno, M., Chirillo, R., Daniotti, G., Cocheo, V., and Albrizio, F. (1989). Carboxyhaemoglobin and fatal methylene chloride poisoning. *Lancet* 2: 274.
225. Jonsson, F., Bois, F., and Johanson, G. (2001). Physiologically based pharmacokinetic modeling of inhalation exposure of humans to dichloromethane during moderate to heavy exercise. *Toxicol. Sci.* 59: 209–218.
226. Bos, P.M., Zeilmaker, M.J., and Van Eijkeren, J.C. (2006). Application of physiologically based pharmacokinetic modeling in setting acute exposure guideline levels for methylene chloride. *Toxicol. Sci.* 91: 576–585.
227. Amsel, J., Soden, K.J., Sielken, R.L., Jr., and Valdez-Flora, C. (2001). Observed versus predicted carboxyhemoglobin levels in cellulose triacetate workers exposed to methylene chloride. *Am. J. Ind. Med.* 40: 180–191.
228. Guo, Y., Stein, A.B., Wu, W.J., Tan, W., Zhu, X., Li, Q.H., Dawn, B., Motterlini, R., and Bolli, R. (2004). Administration of a CO-releasing molecule at the time of reperfusion reduces infarct size in vivo. *Am. J. Physiol. Heart Circ. Physiol.* 286: H1649–53.
229. Motterlini, R., Sawle, P., Hammad, J., Bains, S., Alberto, R., Foresti, R., and Green, C.J. (2005). CORM-A1: a new pharmacologically active carbon monoxide-releasing molecule. *FASEB J.* 19: 284–286.
230. Ryan, M.J., Jernigan, N.L., Drummond, H.A., McLemore, G.R., Jr., Rimoldi, J.M., Poreddy, S.R., Gadepalli, R.S., and Stec, D.E. (2006). Renal vascular responses to CORM-A1 in the mouse. *Pharmacol. Res.* 54: 24–29.
231. De Backer, O., Elinck, E., Blanckaert, B., Leybaert, L., Motterlini, R., and Lefebvre, R.A. (2009). Water-soluble CO-releasing molecules reduce the development of postoperative ileus via modulation of MAPK/HO-1 signalling and reduction of oxidative stress. *Gut.* 58: 347–356.
232. Lang, E., Qadri, S.M., Jilani, K., Zelenak, C., Lupescu, A., Schleicher, E., and Lang, F. (2012). Carbon monoxide-sensitive apoptotic death of erythrocytes. *Basic Clin. Pharmacol. Toxicol.* 111: 348–355.
233. Magierowska, K., Magierowski, M., Hubalewska-Mazgaj, M., Adamski, J., Surmiak, M., Sliwowski, Z., Kwiecien, S., and Brzozowski, T. (2015). Carbon Monoxide (CO) released from tricarbonyldichlororuthenium (II) Dimer (CORM-2) in gastroprotection against experimental ethanol-induced gastric damage. *PLoS One* 10: e0140493.
234. Magierowska, K., Magierowski, M., Surmiak, M., Adamski, J., Mazur-Bialy, A.I., Pajdo, R., Sliwowski, Z., Kwiecien, S., and Brzozowski, T. (2016). The protective role of carbon monoxide (CO) produced by heme oxygenases and derived from the CO-releasing molecule CORM-2 in the pathogenesis of stress-induced gastric lesions: Evidence for non-involvement of Nitric Oxide (NO). *Int. J. Mol. Sci.* 17: 442.
235. Nikam, A., Ollivier, A., Rivard, M., Wilson, J.L., Mebarki, K., Martens, T., Dubois-Rande, J.L., Motterlini, R., and Foresti, R. (2016). Diverse Nrf2 activators coordinated to cobalt carbonyls induce heme oxygenase-1 and release carbon monoxide in vitro and in vivo. *J. Med. Chem.* 59: 756–762.
236. Pan, Z., Chittavong, V., Li, W., Zhang, J., Ji, K., Zhu, M., Ji, X., and Wang, B. (2017). Organic CO prodrugs: Structure-CO-release rate relationship studies. *Chemistry* 23: 9838–9845.
237. Blokzijl, W., Blandamer, M.J., and Engberts, J.B.F.N. (1991). Diels-Alder reactions in aqueous solutions. Enforced hydrophobic interactions between diene and dienophile. *J. Am. Chem. Soc.* 113: 4241–4246.
238. Peng, P., Wang, C., Shi, Z., Johns, V.K., Ma, L., Oyer, J., Copik, A., Igarashi, R., and Liao, Y. (2013). Visible-light activatable organic CO-releasing molecules (PhotoCORMs) that simultaneously generate fluorophores. *Org. Biomol. Chem.* 11: 6671–6674.
239. Hasegawa, U., Van Der Vlies, A.J., Simeoni, E., Wandrey, C., and Hubbell, J.A. (2010). Carbon monoxide-releasing micelles for immunotherapy. *J Am Chem Soc* 132: 18273–18280.
240. Brückmann, N.E., Wahl, M., Reiß, G.J., Kohns, M., Wätjen, W., and Kunz, P.C. (2011). Polymer conjugates of photoinducible CO-releasing molecules. *Eur. J. Inorg. Chem.* 2011: 4571–4577.
241. Ma, M., Noei, H., Mienert, B., Niesel, J., Bill, E., Muhler, M., Fischer, R.A., Wang, Y., Schatzschneider, U., and Metzler-Nolte, N. (2013). Iron metal-organic frameworks MIL-88B and NH2-MIL-88B for the loading and delivery of the gasotransmitter carbon monoxide. *Chemistry* 19: 6785–6790.
242. Diring, S., Carne-Sanchez, A., Zhang, J., Ikemura, S., Kim, C., Inaba, H., Kitagawa, S., and Furukawa, S. (2017). Light responsive metal-organic frameworks as controllable CO-releasing cell culture substrates. *Chem. Sci.* 8: 2381–2386.
243. Carmona, F.J., Rojas, S., Sánchez, P., Jeremias, H., Marques, A.R., Romão, C.C., Choquesillo-Lazarte, D., Navarro, J.A.R., Maldonado, C.R., and Barea, E. (2016). Cation exchange strategy for the encapsulation of a photoactive CO-releasing organometallic molecule into anionic porous frameworks. *Inorg. Chem.* 55: 6525–6531.

244. Bohlender, C., Glaser, S., Klein, M., Weisser, J., Thein, S., Neugebauer, U., Popp, J., Wyrwa, R., and Schiller, A. (2014). Light-triggered CO release from nanoporous non-wovens. *J. Mater. Chem. B* 2: 1454–1463.
245. Dordelmann, G., Pfeiffer, H., Birkner, A., and Schatzschneider, U. (2011). Silicium dioxide nanoparticles as carriers for photoactivatable CO-releasing molecules (PhotoCORMs). *Inorg. Chem.* 50: 4362–4367.
246. Van Der Vlies, A.J., Inubushi, R., Uyama, H., and Hasegawa, U. (2016). Polymeric framboidal nanoparticles loaded with a carbon monoxide donor via phenylboronic acid-catechol complexation. *Bioconjug. Chem.* 27: 1500–1508.
247. He, Q., Kiesewetter, D.O., Qu, Y., Fu, X., Fan, J., Huang, P., Liu, Y., Zhu, G., Liu, Y., Qian, Z., and Chen, X. (2015). NIR-responsive on-demand release of CO from metal carbonyl-caged graphene oxide nanomedicine. *Adv. Mater.* 27: 6741–6746.
248. Dordelmann, G., Meinhardt, T., Sowik, T., Krueger, A., and Schatzschneider, U. (2012). CuAAC click functionalization of azide-modified nanodiamond with a photoactivatable CO-releasing molecule (PhotoCORM) based on [Mn(CO)$_3$(tpm)]$^+$. *Chem. Commun.* 48: 11528–11530.
249. Yang, G., Fan, M., Zhu, J., Ling, C., Wu, L., Zhang, X., Zhang, M., Li, J., Yao, Q., Gu, Z., and Cai, X. (2020). A multifunctional anti-inflammatory drug that can specifically target activated macrophages, massively deplete intracellular H2O2, and produce large amounts CO for a highly efficient treatment of osteoarthritis. *Biomaterials* 255: 120155.

Appendix

Mathematica code for generating the plots in the chapter (code was executed with Wolfram Mathematica 12).

Fig. *3*

```
C1 = ContourPlot[(100*245 y)/(
x + 245 y)*(0.0218*(x + 245 y) + 2*0.000912*(x +
245 y)^2 +
3*(0.0000038)*(x + 245 y)^3 +
4*(0.00000247)*(x + 245 y)^4)/(4*(1 + 0.0218*(x +
245 y) +
0.000912*(x + 245 y)^2 + (0.0000038)*(x +
245 y)^3 + (0.00000247)*(x + 245 y)^4)), {x, 0,
120}, {y,
0, 0.2}, ColorFunction -> "ThermometerColors",
Contours -> 19,
PlotLegends ->
Placed[BarLegend[Automatic, LegendMarkerSize
-> 150,
LabelStyle -> {FontSize -> 12, Black},
LegendLabel -> "COHb(%)"], {After, Top}],
PlotRange -> {0, 100}];
Show[C1, FrameLabel -> {{HoldForm[PCO
"(mmHg)"],
None}, {HoldForm[PO2 "(mmHg)"], None}},
LabelStyle -> {FontSize -> 12, Black}]
```

Fig. *4-7*

Note: Different hemoproteins can be plotted by changing the value of M, P50, and PO2.

```
M = 2.5;
P50 = 0.5;
PO2 = 1;
c = P50 + PO2;
cmb = ContourPlot[(100*245 (c*0.01 y)/(M*(1
- 0.01 y)))/(
x + 245 (c*0.01 y)/(
M*(1 - 0.01 y)))*((0.0218*(x +
245 (c*0.01 y)/(M*(1 - 0.01 y))) +
2*0.000912*(x + 245 (c*0.01 y)/(M*(1 - 0.01 y)))^2
+
3*(0.0000038)*(x + 245 (c*0.01 y)/(M*(1 - 0.01
y)))^3 +
4*(0.00000247)*(x +
245 (c*0.01 y)/(M*(1 - 0.01 y)))^4)/(4*(1 +
0.0218*(x + 245 (c*0.01 y)/(M*(1 - 0.01 y))) +
0.000912*(x +
245 (c*0.01 y)/(M*(1 - 0.01 y)))^2 + (0.0000038)*(x
+
245 (c*0.01 y)/(M*(1 - 0.01 y)))^3 +
(0.00000247)*(x +
245 (c*0.01 y)/(39*(1 - 0.01 y)))^4))), {x, 0, 40}, {y,
0, 20}, ColorFunction -> "ThermometerColors",
Contours -> 19,
PlotLegends ->
Placed[BarLegend[Automatic, LegendMarkerSize
-> 150,
LabelStyle -> {FontSize -> 12, Black},
LegendLabel -> "COHb(%)"], {After, Top}],
PlotRange -> {0, 100}];
Show[cmb,
FrameLabel -> {{HoldForm[COX - CO (%)],
None}, {HoldForm[PvO2 "(mmHg)"], None}},
LabelStyle -> {FontSize -> 12, Black}]
```

Fig. *9*

```
slow = Plot[7.8*Exp[-0.043*x] + 27.2*Exp[-
0.0065*x], {x, 0, 200},
PlotRange -> {1, 100}, ScalingFunctions -> "Log10",
```

Ticks -> {Automatic, {1, 2, 5, 10, 20, 30, 40, 50, 100}}];
acute = Plot[16.7*Exp[-0.13*x] + 14.4*Exp[-0.0069*x], {x, 0, 200},
PlotRange -> {1, 100}, ScalingFunctions -> "Log10"];
Show[slow, acute, AxesLabel -> {"Time(min)", "COHb%"},
AxesStyle -> Directive[Black, Bold],
TicksStyle -> Directive[Black, Bold]]

Fig. 13

Mcox = 2.5;
Ki = 0.27
cmb = ContourPlot[
100*((x*10^-6)*760)/(((x*10^-6)*760) + Ki + y/Mcox), {x, 0,
1000}, {y, 0, 1}, ColorFunction -> "ThermometerColors",
Contours -> 19,
PlotLegends ->
Placed[BarLegend[Automatic, LegendMarkerSize -> 150,

LabelStyle -> {FontSize -> 12, Black},
LegendLabel -> "CoxCO(%)"], {After, Top}],
PlotRange -> {0, 100}];
Show[cmb,
FrameLabel -> {{HoldForm[PO2 "(mmHg)"],
None}, {HoldForm["CO concentration (ppm)"],
None}},
LabelStyle -> {FontSize -> 12, Black}]

Fig. 14

Plot[{(0.0842*(450*50)^0.72)/(x + 220)^0.775, (0.0842*(450*100)^0.72)/(x + 220)^0.775, (0.0842*(450*250)^0.72)/(x + 220)^0.775, (0.0842*(450*500)^0.72)/(x + 220)^0.775}, {x, 1,
1000}, PlotLegends -> {"50", "100", "250", "500"},
AxesLabel -> {"Time(min)", "COHb%"},
AxesStyle -> Directive[Black, Bold],
TicksStyle -> Directive[Black, Bold],
Ticks -> {Automatic, Automatic}]

4

Carbon Monoxide and Energy Metabolism

Daniela Dias-Pedroso[1,2], Nuno Soares[1,2], and Helena L.A. Vieira[1,2,3]

[1] UCIBIO, Applied Molecular Biosciences Unit, Department of Chemistry, NOVA School of Science and Technology, Universidade Nova de Lisboa, Caparica, Portugal
[2] CEDOC, NOVA Medical School, Universidade Nova de Lisboa, 1169-056 Lisbon Portugal
[3] Associate Laboratory i4HB - Institute for Health and Bioeconomy, NOVA School of Science and Technology, Universidade Nova de Lisboa, Caparica, Portugal

Introduction to Cell Energy Metabolism and ATP

In living organisms, free energy is needed for three main processes: (i) mechanical work (muscle contraction, cell movements); (ii) active transport of molecules and ions through membranes; and (iii) generation of biomolecules (DNA, fat, glucose) from precursors. The universal energy molecule in all life forms is adenosine triphosphate (ATP). ATP is a highly accessible molecule that links energy-releasing and energy-requiring metabolic pathways. In mammals, ATP is generated from the oxidation of food (carbohydrates and fat) in catabolic pathways and is consumed in anabolic pathways for the biosynthesis of complex molecules. Oxidation of carbon fuel generates ATP directly (glycolysis followed by anaerobic generation of lactate) or by promoting ion gradients (mitochondrial aerobic processes via tricarboxylic acid [TCA] cycle and oxidative phosphorylation [OXPHOS]) [1]. Therefore, ATP quantification is an indirect way to assess cellular energy metabolism status and a first step to disclose energy metabolic response.

ATP and Carbon Monoxide

About two decades ago, endogenous or low concentrations of exogenous carbon monoxide (CO) were first described as an anti-inflammatory and antiapoptotic molecule [2,3]. Since then, mitochondria have been extensively implicated in CO-induced cytoprotective pathways [4–8]. Although mitochondria are key organelles in the control of cell metabolism and bioenergy, not much attention has been given to the potential role of CO in the regulation of cell metabolism. CO modulation of cellular ATP levels was first mentioned by Lavitrano and colleagues in 2004 [9]. Pigs pretreated with CO gas at 250 ppm, followed by cardiopulmonary bypass with cardioplegic arrest, presented higher heart levels of ATP, as well as phosphocreatine, which is a high-energy phosphate cellular reserve [9]. These data indicate that pretreatment of CO improved cardiac metabolic status after ischemia and reperfusion injury [9]. Nevertheless, CO-induced higher ATP levels can also be a consequence of increased cell viability rather than an improvement of specific cellular metabolism. Therefore, CO enhancement of ATP levels must be assessed under physiological conditions, without stimulation of any kind of biological injury. In fact, Tsui and colleagues measured ATP following CO treatment in liver models [10,11]. In culture of hepatocytes, stimulation of heme oxygenase-1 (HO-1) activity or exogenous CO administration increased ATP production, which in turn activated p38 MAPK (mitogen-activated protein kinase) signaling [10]. Likewise, mouse livers exposed to CO gas presented higher levels of ATP. These higher concentrations of ATP were associated with increased mouse survival in response to fulminant hepatitis [11]. Moreover, the CO-induced improvement of energy status was dependent on soluble guanylyl cyclase [11]. In primary culture of mouse astrocytes, CO also increased cellular levels of ATP under basal conditions [12].

In 1982, before the great boom of CO as a cytoprotective molecule with pharmacological potential, an interesting study about energy status in rats following exposure to CO was published. Rats exposed to 1.3% of CO gas for 4 min presented reduced levels of ATP and phosphocreatine in the brain, which was accompanied by ultrastructural alterations of mitochondria [13]. These data are in accordance with the cytotoxic properties of CO and point to the fact that CO binds to and inhibits heme proteins, such as cytochrome *c* oxidase, which in turn limits mitochondrial production of ATP. Nevertheless, when rats were chronically exposed to 0.13–0.15% of CO gas for 12 h, there was an increase of cerebral ATP and phosphocreatine levels [13]. Despite not exploring any mechanism, this effect is in agreement with data described 30 years later, suggesting an improvement of cell energy status by endogenous or low concentrations of exogenous CO.

CO Targets Mitochondrial Metabolism: Focus on CO Modulation of OXPHOS

In mammalian cells, the primary source of ATP is mitochondria. Glucose is the main energy fuel, which is converted by glycolysis into pyruvate for entering into the TCA cycle. Although glucose is the main carbon source, amino acids and fatty acids can also be converted into pyruvate. In the TCA cycle, under aerobic conditions, the reducing species NADH and $FADH_2$ are generated in the mitochondrial matrix. Then, NADH and $FADH_2$ electrons are transferred to O_2 during OXPHOS, through the electron-transport chain complexes in the inner mitochondrial membrane, generating mitochondrial membrane potential. ATP production occurs when protons flow back to the matrix through a channel in an ATP-synthesizing complex, called F_0F_1ATPase. Thus, under aerobic conditions, cellular ATP is mainly synthetized in mitochondria [1].

In the last two decades, it has been extensively demonstrated that physiological CO levels or exogenous cytoprotective low concentrations of CO target mitochondria and modulate several different processes: (i) mitochondrial biogenesis; (ii) mitochondrial reactive oxygen species (ROS) generation for cell signalling; (iii) mild uncoupling of mitochondria; and (iv) OXPHOS [14–18] (see Chapter 6 on CO and mitochondria). Much evidence in the literature has demonstrated that CO targets cytochrome *c* oxidase (complex IV), and by partially inhibiting it, promotes generation of small amounts of ROS [4–6]. Nevertheless, only later it was shown that CO binding to cytochrome *c* oxidase also modulates mitochondrial metabolism, specifically OXPHOS. In fact, in isolated hepatic mitochondria, low concentrations of exogenous CO present a two-step response in the control of cytochrome *c* oxidase activity: in the first 5–10 min following treatment, CO inhibits cytochrome *c* oxidase, while after 30 min, there is an increase in this respiratory complex specific activity [8]. Likewise in nonsynaptic isolated mitochondria or in astrocytic primary cultures, CO also presents this double effect on cytochrome *c* oxidase activity [12]. Moreover, this later increase in cytochrome *c* oxidase activity is associated with higher levels of ATP production and oxygen consumption, along with a decrease in lactate generation [12]. Thus, CO appears to reinforce metabolism toward OXPHOS rather than glycolysis with anaerobic lactate formation. This CO effect is accompanied by an increase in mitochondrial population and prevention against oxidative stress-induced apoptosis in astrocytes [12]. Still, CO-driven ATP production not only is cytoprotective due to cellular energetic status but can also modulate cell signaling. In fact, CO-enhanced ATP generation in astrocytes promotes neuroprotection via purinergic astrocyte-to-neuron signaling [19].

Cancer cells present greater glycolytic metabolism (with lactate production for ATP generation), even in the presence of oxygen. This phenomenon is described as the Warburg effect. In prostate cancer cells, Wegiel and colleagues have described that overexpression of HO-1 or exogenous CO treatment reinforces OXPHOS, decreasing glycolytic metabolism [20]. In fact, the CO-induced metabolic shift toward mitochondrial metabolism acts as an anti-Warburg effect, which facilitates cancer cell death during chemotherapy by mitotic catastrophe and cellular metabolic exhaustion [20]. Moreover, this CO-induced metabolic effect seems to be ROS dependent; whenever ROS are scavenged by pegylated catalase and superoxide dismutase, there is a reversion on the CO-promoted anti-Warburg effect [20]. Accordingly, in human prostate cancer cells, HO-1 is mainly located in the nucleus with low levels of activity, and this is correlated with worse clinical outcomes [20].

Inflammatory cellular response is also regulated by cell metabolism. There is a reprograming of cell metabolism shifting away from OXPHOS during inflammatory activation in macrophages, dendritic cells, and T cells [21]. In fact, immune cells appear to be mainly glycolytic, since under aerobic conditions glycolysis rate increases for providing ATP and intermediates for biosynthesis of immune response proteins [21]. The same effect was observed in microglia, the immune resident cells in the brain. Inflammatory activation of microglia toward M1 phenotype, with production and release of proinflammatory factors, is accompanied by a metabolic shift favoring glycolysis over OXPHOS [22]. Likewise, CO promotes an anti-neuroinflammatory response in microglia by acting on cell metabolism. Administration of low doses (10 μM) of CORM-401 (a manganese-containing CORM – CO-releasing molecule) in microglia challenged with lipopolysaccharides (LPSs) maintains mitochondrial oxygen consumption and ATP levels, but decreases glycolysis [23]. Nevertheless, when CORM-401 was applied at higher doses (50 µM) or during longer periods, there was a partial inhibition of mitochondrial respiration and ATP production, suggesting a dual response for CO regulation of cell metabolism and inflammation [23]. Accordingly, in cell line models of fibroblasts, carcinoma, or breast cancer, long-term exposure to high concentrations of CORM-401 (50 µM) also inhibits mitochondrial respiration, reduces ATP production, and promotes electron leakage [24]. Therefore, CO modulation of OXPHOS is highly dependent on its concentration, which is not surprising considering CO's ability to bind to heme proteins.

During neuronal differentiation from pluripotent cells to mature neurons, there is a great cellular metabolic shift from glycolysis to OXPHOS. In fact, similar to cancer cells, proliferative undifferentiated cells present higher levels of glycolysis and lactate production for the generation of ATP, whereas mature neurons are mostly oxidative with greater mitochondrial population [25–27]. Likewise, ROS and oxygen levels are important signaling factors during neuronal differentiation processes via modulation of oxidative metabolism and mitochondrial function [28]. Using a neuronal differentiation SH-SY5Y cell model, CORM-A1 administration increased final yield of neuronal production in a manner that was dependent on mitochondrial ROS production [29]. Furthermore, in another model for neurogenesis (NT2 cell line), CORM-A1 also improved neuronal differentiation by reinforcing OXPHOS and mitochondrial metabolism [30]. This effect was assessed by the higher expression of pyruvate and lactate dehydrogenase, by the increased ^{13}C incorporation from ^{13}C-glucose into the TCA cycle metabolites, and by the greater mitochondrial population [30]. Accordingly, in NT2 cell culture under hypoxia (5% of oxygen), there is a reversion of CO-induced neuronal differentiation, indicating the importance of oxidative metabolism [30]. Finally, functional cardiomyocyte differentiation from stem cells also requires the presence of HO-1/CO axis and is dependent on mitochondrial biogenesis [31].

In conclusion, one may speculate that mitochondrial ROS are generated by CO due to partial and transient inhibition of cytochrome c oxidase, which in turn act as signaling molecules and improve mitochondrial function and metabolism [7,8,12]. Although the hypothesis about CO regulation of mitochondrial metabolism through ROS generation and signaling is deeply discussed in several publications cited therein [15,32,33], the molecular mechanisms involved in this dual CO effect on cytochrome c oxidase activity (inhibition and stimulation) are not yet understood.

Finally, and in accordance with the experimental data, a clinical trial using inhaled CO in patients with idiopathic pulmonary fibrosis has demonstrated an upregulation of 23 genes associated with OXPHOS in peripheral blood mononuclear cells [34].

In summary, CO reinforces oxidative mitochondrial metabolism in diverse models (heart, brain, kidney, cancer, immune cells) for the improvement of several different processes: neuronal differentiation, anti-inflammation, cytoprotection, and anticancer effect (Figure 4.1).

Nonoxidative Metabolism, Glycolysis, and CO: The Other Side of the Same Coin

In accordance with CO reinforcement of oxidative metabolism for ATP generation and bioenergy production, CO might reduce glycolysis, which has proved to be true in four different models.

In a pig model of ischemic myocardium, pre-exposure of animals to CO gas (up to reaching 5% of carboxyhemoglobin [COHb]) improved heart tissue metabolism [35]. During ischemia, lactate production is higher due to glycolytic metabolism for the maintenance of energy supply. In this work, ischemia-induced lactate production in heart tissues was about 50% lower in CO-treated animals. Likewise, pyruvate levels increased in the presence of CO and there were similar levels of ATP in CO-exposed animals and controls. Thus, CO promotes a decrease in glycolysis during ischemia, but maintains energy reserve [35].

The interaction between breast cancer cells and lung endothelial cells is decisive for the development of metastasis. A coculture system with breast cancer cells and lung endothelial cells was used to follow the CO effect on cell adhesion and transmigration, with a particular focus on cell metabolism. In fact, CORM-401 exposure limited transmigration by distinct regulation of breast cancer cells' metabolism [36]. While in lung endothelial cells CORM-401 decreased glycolysis in favor of mitochondrial respiration, in breast cancer cells CORM-401 inhibited both metabolic pathways: glycolytic and oxidative [36]. Because cell adhesion and transmigration are highly bioenergy-demanding cellular processes, cell metabolism is a promising target for inhibition of metastasis.

Likewise, CORM-401 also altered endothelial cell (EA.hy926 cell line) metabolism, shifting from glycolysis toward OXPHOS, along with a mild

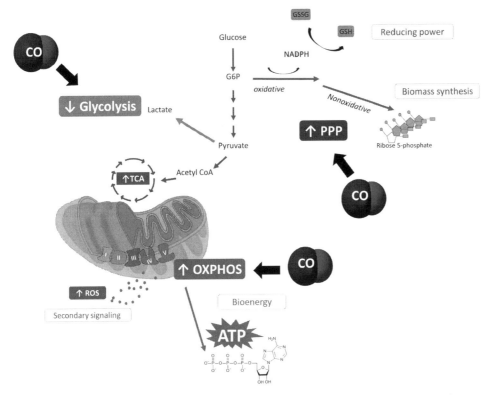

Figure 4.1 CO modulation of cell metabolism: OXPHOS, glycolysis and pentose phosphate pathway.

mitochondrial depolarization [37]. CORM-401 increased the levels of mitochondrial Ca^{2+} and accelerated the activity of mitochondrial respiratory complexes I and II [37].

Finally, in macrophages CO limits inflammatory activity in response to LPS and ATP, as assessed by a decrease in caspase-1 activation and the release of interleukin-1β and interleukin-18, by modulation of NLRP3 inflammasome [38,39]. The underlying mechanism for CO inhibition of NLRP3 inflammasome activation is the reduction of glycolysis, which was assessed by acidification of culture medium [39].

In contrast, CO via oral administration of CORM-401 appears to promote glycolysis in adipocytes of obese mice fed with high-fat diet [40]. The CO-induced glycolysis reinforcement led to body weight reduction by acceleration of glucose metabolism [40]. In conclusion, CO emerges as a key metabolic modulator decreasing glycolysis in cardiomyocytes [35], endothelial cells [36,37], and macrophages [39] (Figure 4.1). Nevertheless, depending on the context and cell type, CO also improved glycolysis, such as described in adipocytes [40]. Furthermore, one may speculate that CO modulates cell metabolism for promoting cellular and tissue homeostasis.

CO Control of the Pentose Phosphate Pathway and Antioxidant Defense

Glutathione (GSH) is the major antioxidant system in the cell. Redox reactions convert reduced GSH into oxidized GSH (GSSG), which promotes an antioxidant effect by reducing other molecules. Hence, GSSG must then be reduced again; this GSSG–GSH recycling depends on the availability of electrons, which are carried by NADPH. NADPH is mostly generated by the oxidative phase of the pentose phosphate pathway (PPP). PPP is an important pathway for glucose oxidation and is linked to glycolysis at the level of glucose 6-phosphate. Moreover, the nonoxidative phase of PPP converts pentose phosphates into phosphorylated aldoses and ketones and also produces ribose 5-phosphates, which are precursors for nucleotide synthesis [41,42]. It is well established that CO targets mitochondria and generates small amounts of ROS for signaling [14,15,16,17,18] (see Chapter 6 on CO and mitochondria). Therefore, CO might directly or indirectly modulate PPP, in order to maintain cellular redox balance, via reinforcing GSH recycling (Figure 4.1).

In 2014, CO modulation of PPP was shown for the first time. CO promoted increased levels of reduced GSH in red blood cells by three distinct mechanisms [43]. CO directly stimulates PPP, and whenever PPP is inhibited by 2-deoxyglucose, there is a reversion of CO-induced elevation of GSH levels. Second, CO leads to deglutathionylation of hemoglobin at Cys93 and Cys112, which generates great amounts of reduced GSH. Finally, glutathione reductase activity is promoted by CO, increasing the concentration of reduced GSH [43]. Still in the same year, but using a different model, Yamamoto and colleagues demonstrated that the HO-1/CO axis induces PPP against glycolysis in leukemia cells [44]. In fact, CO stimulation of PPP generates NADPH, leading to production of reduced GSH, which in turn protects these cancer cells against oxidative stress. CO or HO-1 induction reduces methylation of PFKFB3, suppressing the activity of fructose 2,6-bisphosphate (F-2,6-BP). Because F-2,6-BP is an allosteric activator of the rate-limiting enzyme of glycolysis, phosphofructokinase-1, there is a shift from glycolysis toward PPP [44].

In the human endothelial cell line EA.hy926, CORM-401 also stimulates PPP for NADPH production, which is needed for nitric oxide (NO) synthase activity [45]. CO promotes Ca^{2+} signaling via NO production that triggers Ca^{2+} release from endoplasmic reticulum. Likewise, whenever PPP is inhibited by 6-aminonicotinamide, CO-induced NO generation is prevented [45].

A metabolic switch from glycolytic to oxidative mitochondrial metabolism occurs during the neuronal differentiation process [28,46]. Therefore, neurons are more oxidative than their precursors or pluripotent cells. Consequently, one can speculate that neurons must present an effective antioxidant cellular defense. Because GSH is the major nonenzymatic antioxidant player in the cell, it is not surprising that the neuronal differentiation process may also affect and stimulate PPP for supplying NADPH for GSH reduction. In the human neuroblastoma SH-SY5Y cell model for neuronal differentiation, CORM-A1 treatment enhanced the final yield of neurons by stimulation of PPP and improving the GSH/GSSG ratio [47]. PPP was assessed by the analysis of its metabolites in GC–MS and by the increased expression of the rate-limiting enzyme glucose 6-phosphate dehydrogenase (G6PD). Likewise, CO-induced neuronal yield improvement did not occur whenever G6PD expression was knocked down [47]. Of note, in this neuronal differentiation cell model, no metabolic shift between glycolysis and OXPHOS was observed. As neuronal precursors, neuroblastoma SH-SY5Y cells are already fully committed to their neuronal fate, when compared to cells with higher differentiation potency. Thus, one can speculate that their metabolism is already mainly oxidative, the final metabolic adaptation being related to antioxidant cellular defense.

Release of free heme during hemolysis causes tissue damage by oxidative stress, inflammation, and vascular dysfunction. Macrophages are key cells for clearance of free heme, by the action of HO that uses reduced NADPH for converting heme into iron, biliverdin, and CO. Moreover, macrophage heme clearance occurs concomitantly with cellular metabolic adaptation. Loading macrophages with heme promotes a shift from mitochondrial OXPHOS toward glucose consumption, in particular toward PPP [48]. HO-derived CO or exogenous administration of CO (CORM-3) increases the expression of three key genes for PPP activity: glucose transporter Glut1 and the PPP regulating enzymes G6PD and 6-phosphogluconate dehydrogenase [48]. Furthermore, CORM-3 administration improves *in vivo* heme clearance in a mouse model of sickle cell disease via stimulation of PPP [48]. In another very recent publication using various cancer and fibroblasts cell lines, CORM-401 exposure for short periods generates ROS and shifts glycolysis into PPP along with NADPH production [24].

In contrast to the great amount of research concerning CO's biology affecting mitochondria, the control of PPP by CO is in its initial steps of development, and further work is needed for a clear understanding of the underlying molecular mechanisms.

CO and Lipids

The role of CO in lipid metabolism is much less explored. Some of the studies were designed to uncover the molecular mechanisms of CO-triggered toxicity, using higher concentrations than those used for studying endogenous CO role in cytoprotection.

There are some studies indicating that CO modulates cholesterol synthesis and fatty acid β-oxidation. In a 40-year-old study, it was hypothesized that smokers present atherosclerosis due to high concentrations of CO [49]. For testing it, rat livers were perfused with 30% of COHb, corresponding to high levels of CO inhalation. COHb-perfused livers presented lower secretion levels of lipoprotein triacylglycerol (very low density lipoproteins) and increased ketogenesis, despite no effect being observed in the concentration of free fatty acids [49]. Likewise, in the macrophage cell line J774A.1, few minutes of exposure to CO at high concentrations (1000 ppm), which is at the borderline for toxicity, modulates lipid metabolism. CO significantly increased triglyceride levels and reduced the high-density lipoprotein-mediated cholesterol efflux [50]. Thus, one may speculate that CO at high concentrations may have a proatherogenic effect on macrophages [50]. Furthermore, in primary cultures of astrocytes, human HO-1 overexpression and CORM-3 treatment both increased cholesterol biosynthesis and efflux [51]. Much research is still needed to clarify the role of CO in cholesterol synthesis, lipoprotein formation and secretion, and its effect on atherosclerosis or cholesterol-derived hormone signaling. Concerning β-oxidation, in 1978 it was published that CO altered mitochondrial β-oxidation (measured by oxidation of palmitoyl-CoA), but did not change peroxisome β-oxidation in brown adipocyte tissue [52].

It is well established that low concentrations of CO stimulate PGC-1α (peroxisome proliferator-activated receptor gamma coactivator 1 alpha) expression, increasing mitochondrial population. In addition, enhanced mitochondrial mass and function improve fatty acid β-oxidation [53]. Thus, one can speculate that the CO modulation of lipid metabolism could be indirectly mediated by the enhanced mitochondrial function. Likewise, PGC-1α also stimulates peroxisome activity, which is important for the oxidation of very long and long-chain fatty acids for further metabolization in the mitochondria [54]. Therefore, the role of CO in peroxisome is a biological subject yet to be explored and can be associated with CO control of lipid metabolism.

CO in Metabolic Diseases: Diabetes and Obesity

CO modulation of metabolism goes much beyond the cell. In fact, CO plays a key role in diabetes and obesity etiology and control. Although diabetes is the topic of Chapter 24, herein the focus is the role of CO in cell metabolism affecting obesity and diabetes.

In intact mouse islet, exogenous CO gas facilitates glycogen hydrolysis, which increases glucose reserve, by stimulating the activities of the lysosomal/vacuolar enzymes: acid glucan-1,4-α-glucosidase and acid α-glucosidase (acid α-glucoside hydrolases) [55]. In the same model, CO was also shown to improve insulin secretion in a cGMP-dependent manner [55].

In a high-fat-diet mouse model of obesity, intraperitoneal administration of CORM-A1 reverted obesity and decreased fasted glucose and insulin in a food intake-independent manner. Following CORM-A1 treatment, adipocytes were smaller in size and presented higher levels of mitochondrial markers: the mitochondrial biogenesis transcription factors PGC-1α and NRF-1, and uncoupling protein-1 [56,57]. Accordingly, CORM-A1-treated mice present higher levels of oxygen consumption [57]. Likewise, in a similar model with oral administration of CORM-401, CO also decreased mouse obesity, and improved glucose tolerance and insulin sensitivity. These effects are accompanied by a decrease of white adipocyte size, and reduction of macrophage infiltration and the levels of proinflammatory factors [40]. In contrast to the previous work, CO modulation of cell metabolism is based on reinforcement of glycolysis [40]. CO stimulates a mild mitochondrial uncoupling effect, which in turn decreases mitochondrial ATP production and accelerates glycolysis and cytosolic ATP generation in adipocytes [40]. In summary, Hosick's and Braud's teams have described similar CO effects concerning obesity, insulin resistance, and adipocyte size. Nevertheless, the underlying molecular mechanisms concerning adipocyte metabolic shift seem to diverge.

Likewise, CO gas inhalation also improved insulin sensitivity, lowered blood glucose levels, and reduced body weight of the high-fat-diet mouse model for diabetes by acting on the liver in a fibroblast growth factor 21 (FGF-21)-dependent manner [58]. CO favors liver expression of FGF-21 by mitochondrial ROS signaling. In hepatocytes derived from CO gas-treated mice, there was an enhancement of oxygen consumption and mitochondrial population, which was assessed by the expression of mitochondrial proteins: complexes III and IV, as well as Nrf1 (nuclear respiratory factor 1), TFAM (mitochondrial transcription factor A), and PGC-1α, proteins linked to mitochondrial biogenesis control [58]. Furthermore, the same authors have found that white adipocytes are converted into beige adipocytes via a CO-induced metabolic shift toward OXPHOS and increased mitochondrial population [58]. In conclusion, CO protects against obesity and diabetes type 2 by improving mitochondrial metabolism in hepatocytes and adipocytes [58]. In another work performed by Upadhyay and colleagues, CORM-A1 improves hepatic functions in mice with a high-fat and high-glucose diet, which is a model for nonalcoholic steatohepatitis [59]. CORM-A1 promotes translocation of Nrf2 (nuclear factor erythroid 2-related factor 2) into the nucleus and stimulation of antioxidant responsive genes, which in turn increases mitochondrial biogenesis, mitochondrial oxidative metabolism, and ATP production. The

CORM-A1-induced improvement of oxidative metabolism protects hepatocytes against cell death [59].

Final Remarks

In summary, CO does regulate cell metabolism and bioenergy production. In the vast majority of tissues and cell models, CO is shown to improve ATP production by facilitating mitochondrial metabolism – OXPHOS – and by reducing glycolysis. Still, PPP, which is associated with ROS signaling and nucleotide generation, is also modulated by CO. The potential CO control of lipid metabolism, such as cholesterol synthesis or fatty acid β-oxidation, is far from being understood and is still a fledgling field of research studies. In conclusion, a tight control of energy metabolism is crucial for cellular functions, affecting its role in the tissue and organism. Thus, the use of CO for modulation of bioenergetic cellular pathways can be a promising pharmacological strategy for several disorders.

References

1. Berg, J.M., Tymoczko, J.L., and Stryer, L. (2012). *Biochemistry*. Kate Ahr Parker.
2. Otterbein, L.E. et al. (2000). Carbon monoxide has anti-inflammatory effects involving the mitogen-activated protein kinase pathway. *Nat. Med.* 6: 422–428.
3. Brouard, S. et al. (2000). Carbon monoxide generated by heme oxygenase 1 suppresses endothelial cell apoptosis. *J. Exp. Med.* 192: 1015–1026.
4. D'Amico, G., Lam, F., Hagen, T., and Moncada, S. (2006). Inhibition of cellular respiration by endogenously produced carbon monoxide. *J. Cell Sci.* 119: 2291–2298.
5. Zuckerbraun, B.S. et al. (2007). Carbon monoxide signals via inhibition of cytochrome c oxidase and generation of mitochondrial reactive oxygen species. *FASEB J.* 21: 1099–1106.
6. Taillé, C., El-Benna, J., Lanone, S., Boczkowski, J., and Motterlini, R. (2005). Mitochondrial respiratory chain and NAD(P)H oxidase are targets for the antiproliferative effect of carbon monoxide in human airway smooth muscle. *J. Biol. Chem.* 280: 25350–25360.
7. Queiroga, C.S.F. et al. (2010). Glutathionylation of Adenine nucleotide translocase induced by carbon monoxide prevents mitochondrial membrane permeabilization and apoptosis. *J. Biol. Chem.* 285: 17077–17088.
8. Queiroga, C.S.F., Almeida, A.S., Alves, P.M., Brenner, C., and Vieira, H.L.A. (2011). Carbon monoxide prevents hepatic mitochondrial membrane permeabilization. *BMC Cell Biol.* 12.
9. Lavitrano, M. et al. (2004). Carbon monoxide improves cardiac energetics and safeguards the heart during reperfusion after cardiopulmonary bypass in pigs. *FASEB J.* 18: 1093–1095.
10. Tsui, T.Y., Siu, Y.T., Schlitt, H.J., and Fan, S.T. (2005). Heme oxygenase-1-derived carbon monoxide stimulates adenosine triphosphate generation in human hepatocyte. *Biochem. Biophys. Res. Commun.* 336: 898–902.
11. Tsui, T.Y. et al. (2007). Carbon monoxide inhalation rescues mice from fulminant hepatitis through improving hepatic energy metabolism. *Shock* 27: 165–171.
12. Almeida, A.S., Queiroga, C.S.F., Sousa, M.F.Q., Alves, P.M., and Vieira, H.L.A. (2012). Carbon monoxide modulates apoptosis by reinforcing oxidative metabolism in astrocytes: role of Bcl-2. *J. Biol. Chem.* 287: 10761–10770.
13. Sokal, J.A., Opacka, J., Górny, R., and Kolakowski, J. (1982). Effect of different conditions of acute exposure to carbon monoxide on the cerebral high-energy phosphates and ultrastructure of brain mitochondria in rats. *Toxicol. Lett.* 11: 213–219.
14. Wegiel, B., Nemeth, Z., Correa-Costa, M., Bulmer, A.C., and Otterbein, L.E. (2014). Heme oxygenase-1: a metabolic nike. *Antioxid. Redox Signal.* 20: 1709–1722.
15. Almeida, A.S., Figueiredo-Pereira, C., and Vieira, H.L.A. (2015). Carbon monoxide and mitochondria-modulation of cell metabolism, redox response and cell death. *Front. Physiol.* 6.
16. Figueiredo-Pereira, C., Dias-Pedroso, D., Soares, N.L., and Vieira, H.L.A. (2020). CO-mediated cytoprotection is dependent on cell metabolism modulation. *Redox Biol.* 32.
17. Motterlini, R. and Foresti, R. (2017). Biological signaling by carbon monoxide and carbon monoxide-releasing molecules. *Am. J. Physiol. Physiol.* 312: C302–C313.
18. Bilban, M. *et al.* (2008). Heme oxygenase and carbon monoxide initiate homeostatic signaling. *J. Mol. Med.* 86: 267–279.
19. Queiroga, C.S.F., Alves, R.M.A., Conde, S.V., Alves, P.M.P., and Vieira, H.L.A. (2016). Paracrine effect of carbon monoxide: astrocytes promote neuroprotection via purinergic signaling. *J. Cell Sci.* 129: 3178–3188.
20. Wegiel, B. et al. (2013). Carbon monoxide expedites metabolic exhaustion to inhibit tumor growth. *Cancer Res.* 73: 7009–7021.

21. Palsson-Mcdermott, E.M. and O'Neill, L.A.J. (2013). The Warburg effect then and now: from cancer to inflammatory diseases. *BioEssays*. 10.1002/bies.201300084.
22. Orihuela, R., McPherson, C.A., and Harry, G.J. (2016). Microglial M1/M2 polarization and metabolic states. *Br. J. Pharmacol.* 173: 649–665.
23. Wilson, J.L. et al. (2017). Carbon monoxide reverses the metabolic adaptation of microglia cells to an inflammatory stimulus. *Free Radic. Biol. Med.* 104: 311–323.
24. Stucki, D. *et al.* (2020). Endogenous carbon monoxide signaling modulates mitochondrial function and intracellular glucose utilization: Impact of the heme oxygenase substrate hemin. *Antioxidants* 9: 652.
25. Kasahara, A. and Scorrano, L. (2014). Mitochondria: from cell death executioners to regulators of cell differentiation. *Trends Cell Biol.* 24: 761–770.
26. Almeida, A.S. and Vieira, H.L.A. (2017). Role of cell metabolism and mitochondrial function during adult neurogenesis. *Neurochem. Res.* 42: 1787–1794.
27. Agostini, M. *et al.* (2016). Metabolic reprogramming during neuronal differentiation. *Cell Death Differ.* 10.1038/cdd.2016.36.
28. Vieira, H.L.A., Alves, P.M., and Vercelli, A. (2011). Modulation of neuronal stem cell differentiation by hypoxia and reactive oxygen species. *Prog. Neurobiol.* 93: 444–455.
29. Almeida, A.S., Soares, N.L., Vieira, M., Gramsbergen, J.B., and Vieira, H.L.A. (2016). Carbon monoxide releasing molecule-A1 (CORM-A1) improves neurogenesis: Increase of neuronal differentiation yield by preventing cell death. *PLoS One* 11: e0154781.
30. Almeida, A.S.A.S., Sonnewald, U., Alves, P.M., and Vieira, H.L.A. (2016). Carbon monoxide improves neuronal differentiation and yield by increasing the functioning and number of mitochondria. *J. Neurochem.* 138: 423–435.
31. Suliman, H.B., Zobi, F., and Piantadosi, C.A. (2016). Heme oxygenase-1/carbon monoxide system and embryonic stem cell differentiation and maturation into cardiomyocytes. *Antioxid. Redox Signal.* 24: 345–360.
32. Queiroga, C.S.F., Almeida, A.S., and Vieira, H.L.A. (2012). Carbon monoxide targeting mitochondria. *Biochem. Res. Int.* 10.1155/2012/749845.
33. Oliveira, S.R. et al. (2016). Mitochondria and carbon monoxide: cytoprotection and control of cell metabolism – a role for Ca2+? *J. Physiol.* 594: 4131–4138.
34. Casanova, N. *et al.* (2019). Low dose carbon monoxide exposure in idiopathic pulmonary fibrosis produces a CO signature comprised of oxidative phosphorylation genes. *Sci. Rep.* 9: 14802.
35. Ahlstrom, K. et al. (2009). Metabolic responses in ischemic myocardium after inhalation of carbon monoxide. *Acta Anaesthesiol. Scand.* 53: 1036–1042.
36. Stojak, M., Kaczara, P., Motterlini, R., and Chlopicki, S. (2018). Modulation of cellular bioenergetics by CO-releasing molecules and NO-donors inhibits the interaction of cancer cells with human lung microvascular endothelial cells. *Pharmacol. Res.* 136: 160–171.
37. Kaczara, P. et al. (2016). Carbon monoxide shifts energetic metabolism from glycolysis to oxidative phosphorylation in endothelial cells. *FEBS Lett.* 590: 3469–3480.
38. Jung, S.S. et al. (2015). Carbon monoxide negatively regulates NLRP3 inflammasome activation in macrophages. *Am. J. Physiol. - Lung Cell. Mol. Physiol.* 308.
39. Lee, D.W. et al. (2017). Carbon monoxide regulates glycolysis-dependent NLRP3 inflammasome activation in macrophages. *Biochem. Biophys. Res. Commun.* 493: 957–963.
40. Braud, L. et al. (2018). Carbon monoxide–induced metabolic switch in adipocytes improves insulin resistance in obese mice. *JCI Insight* 3.
41. Dringen, R., Hoepken, H.H., Minich, T., and Ruedig, C. (2007). Handbook of neurochemistry and molecular neurobiology: Brain energetics. In: Integration of Molecular and Cellular Processes (ed. A. Lajtha, G.E. Gibson and G.A. Dienel), 41–62. Springer US. 10.1007/978-0-387-30411-3_3.
42. Nelson, D.L. and Cox, M.M. (2005). Lehninger Principles of Biochemistry 4e, Publisher W.H. Freeman. ISBN-10: 0716743396.
43. Metere, A. *et al.* (2014). Carbon monoxide signaling in human red blood cells: Evidence for pentose phosphate pathway activation and protein deglutathionylation. *Antioxid. Redox Signal.* 20: 403–416.
44. Yamamoto, T. et al. (2014). Reduced methylation of PFKFB3 in cancer cells shunts glucose towards the pentose phosphate pathway. *Nat. Commun.* 5: 3480.
45. Kaczara, P. et al. (2018). CORM-401 induces calcium signalling, NO increase and activation of pentose phosphate pathway in endothelial cells. *FEBS J.* 10.1111/febs.14411.
46. Varum, S. et al. (2011). Energy metabolism in human pluripotent stem cells and their differentiated counterparts. *PLoS One* 6: e20914.
47. Almeida, A.S. *et al.* (2018). Improvement of neuronal differentiation by carbon monoxide: role of pentose phosphate pathway. *Redox Biol.* 17: 338–347.
48. Bories, G.F.P. *et al.* (2020). Macrophage metabolic adaptation to heme detoxification involves CO-dependent activation of the pentose phosphate pathway. *Blood* 136: 1535–1548.

49. Gardner, R.S., Topping, D.L., and Mayes, P.A. (1978). Immediate effects of carbon monoxide on the metabolism of chylomicron remnants by perfused rat liver. *Biochem. Biophys. Res. Commun.* 82: 526–531.
50. Petrick, L., Rosenblat, M., and Aviram, M. (2016). In vitro effects of exogenous carbon monoxide on oxidative stress and lipid metabolism in macrophages. *Toxicol. Ind. Health* 32: 1318–1323.
51. Hascalovici, J.R. et al. (2009). Impact of heme oxygenase-1 on cholesterol synthesis, cholesterol efflux and oxysterol formation in cultured astroglia. *J. Neurochem.* 108: 72–81.
52. Kramar, R., Hüttinger, M., Gmeiner, B., and Goldenberg, H. (1978). β-oxidation in peroxisomes of brown adipose tissue. *Biochim. Biophys. Acta - Lipids Lipid Metab.* 531: 353–356.
53. Cheng, C.-F., Ku, H.-C., and Lin, H. (2018). PGC-1α as a pivotal factor in lipid and metabolic regulation. *Int. J. Mol. Sci.* 19: 3447.
54. Huang, T.-Y. et al. (2017). Overexpression of PGC-1α increases peroxisomal activity and mitochondrial fatty acid oxidation in human primary myotubes. *Am. J. Physiol. Metab.* 312: E253–E263.
55. Mosen, H. (2006). Nitric oxide inhibits, and carbon monoxide activates, islet acid -glucoside hydrolase activities in parallel with glucose-stimulated insulin secretion. *J. Endocrinol.* 190: 681–693.
56. Hosick, P.A. et al. (2014). Chronic carbon monoxide treatment attenuates development of obesity and remodels adipocytes in mice fed a high-fat diet. *Int. J. Obes.* 38: 132–139.
57. Hosick, P.A., AlAmodi, A.A., Hankins, M.W., and Stec, D.E. (2016). Chronic treatment with a carbon monoxide releasing molecule reverses dietary induced obesity in mice. *Adipocyte* 5: 1–10.
58. Joe, Y. et al. (2018). FGF21 induced by carbon monoxide mediates metabolic homeostasis via the PERK/ATF4 pathway. *FASEB J.* 32: 2630–2643.
59. Upadhyay, K.K. et al. (2020). Carbon monoxide releasing molecule-A1 improves nonalcoholic steatohepatitis via Nrf2 activation mediated improvement in oxidative stress and mitochondrial function. *Redox Biol.* 28: 101314.

5

Role of CO in Circadian Clock

Hiroaki Kitagishi[1] and Ikuko Sagami[2]

[1]*Department of Molecular Chemistry, Faculty of Science and Engineering, Doshisha University, Kyotanabe, Kyoto 610-0321, Japan*
[2]*Graduate School of Life and Environmental Sciences, Kyoto Prefectural University, Sakyo-ku, Kyoto 606-8522, Japan*

Introduction

The biological clock is present in almost all living organisms from unicellular cyanobacteria to all higher organisms such as fungi, plants, and mammals. It regulates diverse behavioral and physiological functions with a period of approximately 24 h [1]. In mammals, the master clock is located in the suprachiasmatic nucleus (SCN) of the brain, and the rhythm is synchronized according to the optical information received by the retina. Through direct and indirect signals, the master clock hierarchically entrains other biological clocks of peripheral tissues. As a result, circadian rhythms allow appropriate temporal regulation of major vital activities, including sleep, feeding, temperature changes, hormone secretion, and metabolism.

At the molecular level, the basis of the biological clock is the transcription–translation feedback loop (TTFL) regulated by a group of clock genes and clock proteins in a cycle of approximately 24 h [2]. The Nobel Prize in Physiology or Medicine 2017 was awarded to three researchers for the discovery of clock genes and their mechanisms that regulate the internal clock. As shown in Figure 5.1, the core factors, CLOCK and NPAS2, form a heterodimer with BMAL1, which binds to DNA E-box (CACGTG) present in the upstream region of target clock genes such as *Pers* and *Crys* to activate their transcription. When the translation products PER and CRY proteins accumulate over time, they migrate from the cytoplasm to the nucleus, and the PER/CRY heterodimer inhibits transcription of their own genes in a feedback manner. PER and CRY proteins further undergo proteasomal degradation; consequent decrease of PERs and CRYs restarts a new cycle of transcription. CLOCK, NPAS2, and BMAL1 also regulate gene expression of the nuclear receptors ROR and REV-ERB, whereas expression of CLOCK, NPAS2, and BMAL1 is promoted by ROR, and is suppressed by REV-ERB (Figure 5.1). In addition to the oscillations in gene expression generated by TTFL, several post-translational modifications such as phosphorylation are involved in the regulation of activity, stability, and localization of these clock proteins [1]. Heme binding to some of these clock proteins affects their functions [3,4]. Furthermore, the heme iron of the protein functions as the binding site for gaseous signaling molecules such as carbon monoxide (CO) and nitric oxide (NO) [3,4]. Therefore, the biological clock controlled by heme and CO has gained considerable attention.

Association of Heme and CO with Circadian Regulatory Factors

Heme performs various functions *in vivo* as a prosthetic group of heme proteins such as hemoglobin, myoglobin, and cytochromes participating in O_2 transport, O_2 storage, and biochemical reactions mediated by electron transfer. Heme also plays regulatory role in gene expression, mRNA stabilization, splicing, and protein synthesis. In contrast to the positive functions, excess heme causes serious oxidative stress due to reactive oxygen species production that leads to DNA or protein damage. Thus, the biosynthesis and degradation of heme *in vivo* must be strictly regulated (Figure 5.2).

The biosynthesis of heme starts in the mitochondrial matrix with the condensation of succinyl-CoA and glycine to form aminolevulinic acid (ALA). This is the rate-determining step catalyzed by the enzymes ALA

Carbon Monoxide in Drug Discovery: Basics, Pharmacology, and Therapeutic Potential, First Edition. Edited by Binghe Wang and Leo E. Otterbein.
© 2022 John Wiley & Sons, Inc. Published 2022 by John Wiley & Sons, Inc.

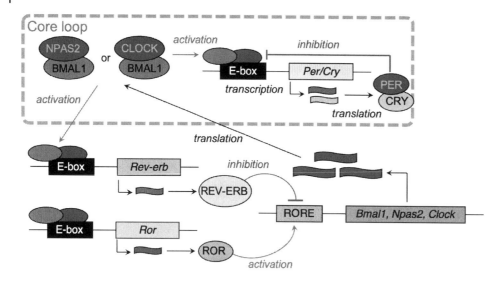

Figure 5.1 Core TTFL in the biological clock. CLOCK, NPAS2, BMAL1, and ROR function as activators, whereas PER, CRY, and REV-ERB function as suppressors. Among these control factors, CLOCK, NPAS2, PER2, and REV-ERBs are recognized as heme-binding proteins.

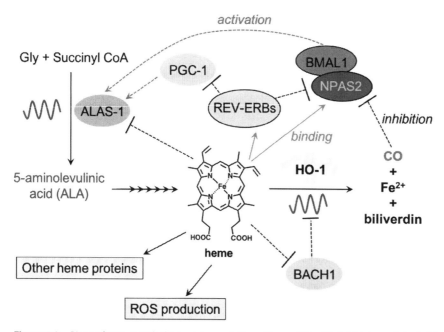

Figure 5.2 Biosynthesis, metabolism, and association of heme/CO with the biological clock. Biosynthesis and degradation of heme are controlled by the biological circadian clock. Heme and CO also control the functions of clock regulatory proteins in a feedback manner. HO-2, a constitutive isoform of HO, is expressed as the erythroid-specific enzyme and not suppressed by heme. Among the HO isoforms, HO-1 is experimentally confirmed the relations to the circadian system. This figure is adapted from Ref. [10].

synthases (ALAS-1 and ALAS-2). ALAS-1 is expressed ubiquitously, whereas ALAS-2 is an erythroid-specific isozyme. The expression of *Alas-1* gene is regulated by NPAS2/BMAL1, PER1, and PER2, resulting in its daily oscillations [5]. Furthermore, transcription of *Npas2* and *Bmal1* genes is suppressed by REV-ERB, and the stability of REV-ERB is altered by heme binding (see later) [6]. The expression of PGC-1, an activator of ALAS-1, is also suppressed by REV-ERB, and thus, it is regulated by heme.

Excess heme in cells is continuously decomposed by heme oxygenase enzymes (HO-1 and HO-2), producing CO, biliverdin, and Fe^{2+} as by-products (Figure 5.2). HO-1 is highly inducible by a variety of stimuli, including oxidative stress, heavy metals, cytokines, and its substrate heme, whereas HO-2 is constitutively expressed and participates in the basal level regulation of heme. HO activity oscillates daily, and it is higher at night than during the day [7]. Expression of *Ho-1* gene is suppressed by the transcription factor BACH1. Binding of heme to BACH1 inhibits its DNA binding activity, promoting proteasome-dependent degradation of BACH1, resulting in activation of *Ho-1* gene expression [8]. Therefore, BACH1 indirectly regulates the biological clock through the endogenous CO production. In the HO-1 knockout fibroblasts (*Ho-1*$^{-/-}$), expression of CLOCK(NPAS2)/BMAL1 target genes, such as *Per2* and *Rev-erbα*, increases and that of REV-ERB target genes, such as *Bmal1*, decreases [9]. Interestingly, these effects are alleviated by CO. Furthermore, when the expression of both *Ho-1* and *Ho-2* genes is functionally inhibited to suppress CO production, the genes involved in gluconeogenesis and glucose metabolism, which are the targets of clock transcription factors, are also affected: The mRNA levels of *Pck1* and *G6pc*, both are important players in gluconeogenesis, were upregulated in *Ho-1*$^{-/-}$ and *Ho-2*^{KD} hepatocytes and thus caused significant increase in the glucose production [9]. Many genes involved in gluconeogenesis and glucose metabolism are rhythmically expressed under the control of clock genes such as CLOCK, NPAS2, BMAL1, and REV-ERBs. These results indicate that while the biological clock controls the biosynthesis and degradation of heme, heme and CO, a degradation product of heme, both affect the biological clock and regulate glucose metabolism.

Binding of Heme and CO to the Regulatory Factors

CLOCK and NPAS2

Among the clock controlling factors, NPAS2, a paralogue of CLOCK, is the first transcription factor recognized as heme-binding protein [11]. NPAS2 is present in the SCN and almost all cells, and it forms a heterodimer with BMAL1, similar to CLOCK. The NPAS2/BMAL1 heterodimer binds to E-box to regulate the biological clock. Since the CLOCK knockout mice (*Clock*$^{-/-}$) maintains the circadian rhythm with a slightly shorter biological cycle than usual, NPAS2 is considered to be a complementary factor of CLOCK in SCN and peripheral cells such as hepatocytes [12,13]. On the other hand, *Npas2*-knockout mice, not *Clock*-knockout mice, demonstrated complicated sleep patterns at night, and they were unable to adapt to irregular feeding schedules [14]. Additionally, NPAS2 knockdown mice of *Clock*$^{-/-}$ cannot maintain the day and night cycle. These observations indicate that both CLOCK and NPAS2 play a key role in the biological rhythm related to light and food availability. However, the underlying mechanisms explaining the difference between CLOCK and NPAS2 have not been elucidated yet.

CLOCK and NPAS2 as well as BMAL1 are the transcriptional factors belonging to the basic-helix–loop–helix–PER–ARNT–SIM (bHLH–PAS) family (Figure 5.3). Recently, the 3D structures of heme-unbound apo-CLOCK homodimer and heterodimer with BMAL1 were reported without any 3D structures of the heme-bound holo-CLOCK (Figure 5.4) [15,16]. So far, there is no information of the actual structure for either apo- or holo-NPAS2. The bHLH domain, composed of approximately 50 amino acids, forms a DNA-binding motif containing two α-helix structures linked by a loop composed of four to six basic amino acids (Figure 5.4A). The PAS domain commonly forms a cage structure composed of five β-sheets and one α-helix (Figure 5.4B), which is essential for interacting with another protein to form homo- or heterodimers. CLOCK and NPAS2 possess two PAS domains (PASA and PASB) with high sequence homology (see Figure 5.3). Each of CLOCK and NPAS2 homodimers also binds to the DNA E-box. However, the transcriptional activation occurs only when these proteins bind to E-box as a heterodimer with BMAL1 (Figure 5.4C). Interestingly, the heterodimer formation and subsequent binding to DNA E-box are activated by NAD(P)H that functions as a reducing agent involved in many biosynthetic and antioxidant reactions in cells [17,18].

NPAS2 binds two heme molecules in its PASA and PASB domains (see Figure 5.4B). The binding of CO to the ferrous (Fe^{2+}) heme in PAS domains inhibits the interaction between NPAS2/BMAL1 heterodimer and E-box sequence [19]. The CO binding affinity of heme bound to PASA is 10 times higher than that of PASB-bound heme. Therefore, heme in the PASA domain predominantly functions as a CO-sensing motif. Based on the UV–Vis and resonance Raman spectroscopic analyses of the bHLH–PASA domain, reduced (Fe^{2+}) or oxidized (Fe^{3+}) heme in the PAS domain is bound to two histidine residues (His119/His171), whereas the heme in the PASA domain without the bHLH unit is bound to His119/Cys170 [20,21]. These suggest that the coordination structure of heme in the PASA domain can be transformed by an interdomain interaction between the bHLH and PASA domains. Notably, DNA binding ability of NPAS2/BMAL1 heterodimer is not affected by the heme redox state (Fe^{2+} or Fe^{3+}; Figure 5.5) [22]. In contrast to NO, CO can bind to only reduced (Fe^{2+}) but not to oxidized (Fe^{3+}) heme in the protein. During CO binding, one ligand (His119 or His171) in the PASA domain should be replaced with CO. More flexible His171 can be preferably replaced with CO, causing a

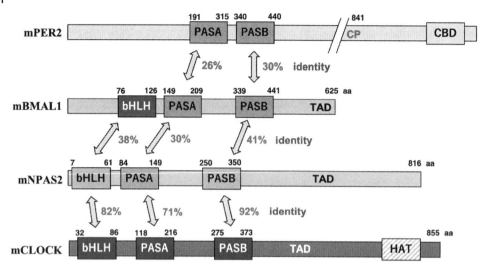

Figure 5.3 Domain structures of murine (m) core circadian clock factors. CP: Cys–Pro heme-binding motif; CBD: CRY-binding domain; TAD: transactivation domain; and HAT: histone acetyltransferase. A part of this figure is adapted from Ref. [10].

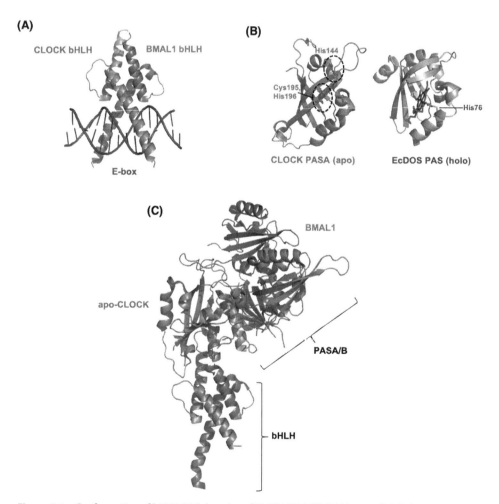

Figure 5.4 Conformation of bHLH–PAS domains of BMAL1/CLOCK. (A) Human CLOCK bHLH/BMAL1 bHLH with DNA E-box (PDB: 4H10). (B) Comparison between apo-CLOCK PASA (PDB: 4F3L) and heme-bound EcDOS PAS domain (PDB: 1V9Y). His144, Cys195, and His196 residues of CLOCK correspond to His119, Cys170, and His171 of NPAS2, respectively. (C) The crystal structure of murine CLOCK/BMAL1 (PDB: 4F3L). A part of this figure is adapted from Ref. [10].

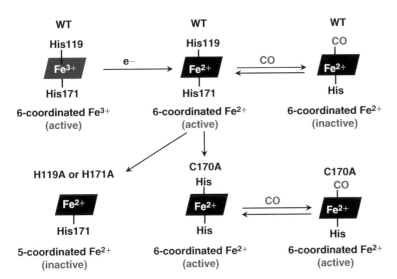

Figure 5.5 Reactivity of the NPAS2 PASA heme sensor domain in the wild-type and its mutants.

structural change in the bHLH domain. The replacement of His119 and His171 to Ala, i.e., H119A and H171A mutants, inactivated the DNA binding. Although C170A mutant was still active, its activity was not affected by CO, and thus C170A lost its CO-responsive DNA binding function (Figure 5.5). These results indicate that the amino acid residues near the CO binding site, such as C170, are essential for exerting the CO-sensing ability of NPAS2. Structural perturbation of the PASA domain upon binding of CO to heme must be transmitted to the bHLH domain, which alters the whole protein structure, thereby regulating the binding affinity to DNA E-box.

Concerning CLOCK protein, the ferric (Fe^{3+}) heme binds to the CLOCK bHLH–PASA domain to form a six-coordinated Fe^{3+} complex [23]. Similar to His119 in NPAS2, His144 in CLOCK has been identified as an axial ligand to heme (see Figure 5.4B), whereas another ligand has not been identified [16]. In contrast to NPAS2, the DNA binding ability of CLOCK bHLH–PASA homodimer is interrupted by heme binding. Additionally, the function of CLOCK as a CO-responsive transcriptional factor has not been identified, probably due to a much lower CO binding affinity of CLOCK (K_d = 0.1 mM at 22 °C) [23] than that of NPAS2 (K_d = 1–2 μM at 25 °C) [19]. The overexpressed NPAS2 in cells colocalizes with heme in the nuclei, whereas the colocalization of CLOCK with heme is unclear [24]. This observation suggests that NPAS2 regulates the biological clock along with the heme cofactor in the nuclei. However, the involvement of heme and CO in the CLOCK functions needs to be investigated further.

PER

PERs (PER1 and PER2) are negative components of the core clock machinery and each protein contains two PAS domains (Figure 5.3) that can bind to heme, suggesting to be heme-responsive sensors [25,26]. Due to a difference in the binding affinities, heme can be transferred from NPAS2 to PER2 when apo-PER2 is mixed with holo-NPAS2. Thus, the binding of heme to NPAS2 can be regulated by PER2 in the nuclei. In addition to the PAS domain, the Cys–Pro (CP) motif functions as another heme binding site (Figure 5.3) [27]. The binding of oxidized (Fe^{3+}) heme to the CP motif of PER2 inhibits heterodimer formation with CRY, and it induces PER2 protein degradation. Therefore, the CP site in PER2 is regarded as a redox sensor that regulates the life span of proteins, thereby affecting the circadian clock system. However, another report suggested that binding of heme to PER2 is nonspecific [28]. Further studies are needed to elucidate the heme binding to PER2 and the effects of CO on PER2 function.

REV-ERB

Nuclear receptor REV-ERBα and REV-ERBβ proteins, whose expressions are activated by CLOCK(NPAS2)/BMAL1, function as transcriptional repressors of target genes such as *Bmal1*, *Npas2*, and *Clock* (Figure 5.1). REV-ERBα protein localizes with heme in the nucleus, suggesting actual heme binding *in vivo* [24]. In the reconstituted system, heme associates with the ligand-binding domain (LBD) of the REV-ERB proteins with a 1:1 stoichiometry (Figure 5.6) and enhances the stability

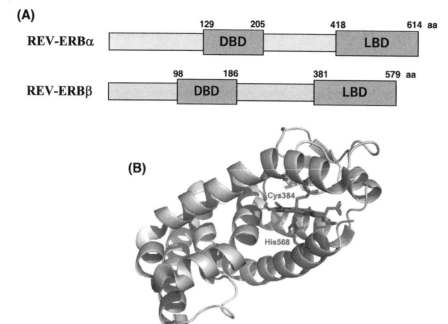

Figure 5.6 Structure of REV-ERBα/β. (A) DBD: DNA-binding domain (Zn finger); LBD: ligand-binding domain. (B) Crystal structure of the holo-LBD domain in REV-ERBβ (PDB: 3CQV). A part of this figure is adapted from Ref. [10].

of the proteins. In human REV-ERBβ, His568 and Cys384 residues coordinate as axial ligands of oxidized (Fe^{3+}) heme (Figure 5.7) [29–32]. The heme binding to REV-ERBs causes the recruitment of the corepressor NCoR, modulating the expression of the REV-ERB target genes such as *Bmal1* and *Npas2* [29,33]. In the reduced (Fe^{2+}) form, Cys384 of human REV-ERBβ dissociates from heme, and thus, gaseous ligands such as CO and NO can bind to the reduced (Fe^{2+}) heme (Figure 5.7). The binding of CO or NO to heme in REV-ERB inhibits its repressor activity in the transcription of *Bmal1* and *Npas2*. The effect of CO is approximately six times lower than that of NO. These results suggest that REV-ERB functions not only as a heme sensor but also as a sensor for gaseous molecules (NO and/or CO).

Effects of CO on Biological Rhythms

CO in the Circadian System

As discussed earlier, heme binds to several circadian factors through their heme-binding domains. Gaseous ligands such as CO and NO can bind to the heme cofactor in these proteins, possibly regulating their transcriptional activities. The effect of endogenous or exogenous CO on regulation of the yeast metabolic cycle has been examined previously [34]. Pathological studies have revealed that application of exogenous CO can adjust the disrupted circadian rhythms in injured cells [35,36]. The role of endogenous CO in the circadian system has been revealed using HO-knockout/knockdown systems [9], suggesting that a rhythmic heme degradation generating endogenous CO is required for maintaining the E-box-controlled circadian rhythms. However, heme itself, whose internal level is controlled by the activities of HO-1 and ALAS-1 (see Figure 5.2), influences the regulation of circadian rhythms [5,37]. Additionally, heme degradation by HO consumes NAD(P)H, and it generates not only CO, but also biliverdin and iron. Intracellular NAD(P)H levels are also an influencing factor in the regulation of circadian rhythms [18,38]. In addition to CO depletion, genetic or pharmacological inhibition of HO has a potential to affect various biological events. Thus, further research is required to reveal the role of CO in the circadian clock.

Selective Removal of Endogenous CO by HemoCD1

To investigate the biological role of endogenous CO, hemoCD1, a highly selective CO scavenger *in vivo* (Figure 5.8), was used. HemoCD1 is a synthetic heme protein model compound consisting of Fe^{2+} porphyrin and cyclodextrin dimer, whose CO binding affinity in aqueous solution is the highest among the reported heme proteins [39–45]. HemoCD1 injected into animals (mice and rats) quantitatively removes

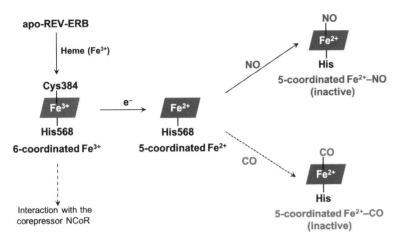

Figure 5.7 Reactivity of the human REV-ERBβ heme sensor domain.

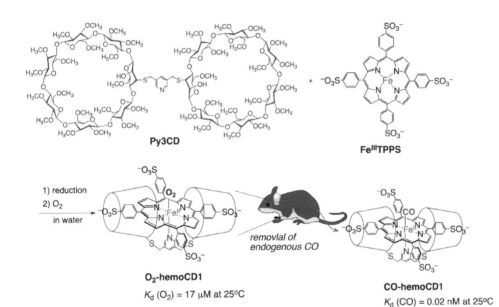

Figure 5.8 A CO-scavenging agent, hemoCD1, 1:1 supramolecular inclusion complex of an Fe^{2+} porphyrin (FeTPPS) with per-O-methylated cyclodextrin dimer (Py3CD). HemoCD1 is used for the selective removal of endogenous CO in mice via the ligand exchange reaction of O_2 with CO. The binding affinity for CO is much higher than that for O_2. This figure is adapted from Ref. [43].

endogenous CO, and it is excreted in the urine in the CO-bound form [39,42]. Therefore, it is possible to easily produce a pseudo-knockdown state of endogenous CO *in vivo* using hemoCD1 [42–44]. The removal of endogenous CO by hemoCD1 causes a feedback response to compensate endogenous CO *in vivo* by inducing *Ho-1*. Recently, hemoCD1 was used to study the effects of CO on the circadian clock system [43]. Intraperitoneal administration of hemoCD1 to mice immediately reduced endogenous CO. The removal of CO promoted the binding of NPAS2 and CLOCK to DNA E-box in the murine liver, resulting in upregulation of the E-box-controlled clock genes (*Per1*, *Per2*, *Cry1*, *Cry2*, and *Rev-erb*) (first phase, Figure 5.9). Within 3 h of administration, most hemoCD1 in mice was excreted in the urine, and *Ho-1* expression was significantly induced in the liver. An increased endogenous CO production due to an overexpression of HO-1 caused the dissociation of NPAS2 and CLOCK from E-box, which in turn induced downregulation of the clock genes (second phase, Figure 5.9). The downregulation continued for over 12 h even after recovery of the internal CO level to normal. The delayed downregulation was attributed to an increase in TNF-α expression induced by free heme [43], which accumulated due to removal of CO from COHb in the blood (third phase, Figure 5.9) [42]. The temporal decrease in endogenous CO disrupted the clock gene expression for over 19 h, and subsequently, it returned to normal (fourth phase, Figure 5.9). The proposed mechanism for disruption of clock gene expression caused by the temporal decrease of endogenous CO is summarized in Figure 5.10. The CO pseudo-knockdown study provided clear evidence that endogenous CO contributes to the regulation of the mammalian circadian clock.

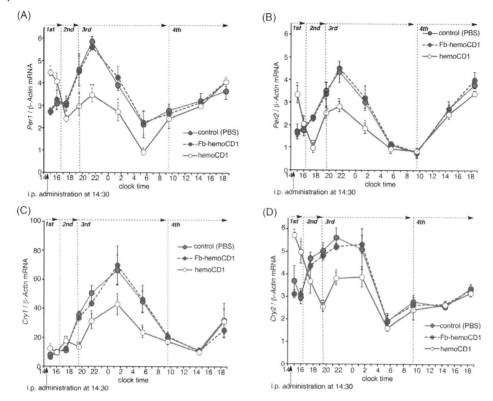

Figure 5.9 Changes in the mRNA levels of clock genes (Per1, A; Per2, B; Cry1, C; Cry2, D) in the murine liver after an intraperitoneal administration of hemoCD1. Phosphate-buffered saline (PBS) and iron-free hemoCD1 (Fb-hemoCD1) were administered in a similar manner to control samples. Each bar represents the mean ± SE (n = 3 mice per group). The mice were housed for two weeks under a 12 h light/dark cycle (lights on at 7:00 and light off at 19:00) until the day before the experiments. The mice were then housed under constant dark conditions while recording observations. Asterisk denotes statistical significance in the mRNA levels of clock genes ($^*P < 0.05$, $^{**}P < 0.01$) as compared to those of the controls. Note that the values on the vertical axes cannot be directly compared between the different panels (A–D) because the amount of cDNA used for real-time PCR varies with the gene of interest. This figure is adapted from Ref. [43].

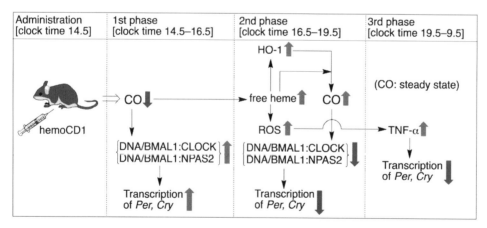

Figure 5.10 Mechanism of circadian clock disruption caused by hemoCD1-mediated endogenous CO depletion in mice. This figure is adapted from Ref. [43].

Summary

In the biological system, production and degradation of heme are significantly involved in the circadian clock system. Therefore, the *in vivo* concentration of heme and CO, a by-product of heme degradation, oscillates daily. Furthermore, heme and CO interact with the proteins of core circadian regulatory factors such as NPAS2, CLOCK, PER2, and REV-ERB, regulating their transcriptional activities in a feedback manner. As overviewed in the first part of this chapter, investigations of the structural and functional relationships of core clock components provided important insights into the feedback regulation by heme and CO.

The role of endogenous CO in the *in vivo* regulation of circadian clock was clarified using a highly selective CO scavenger, hemoCD1. Temporal reduction of endogenous CO by hemoCD1 in mice significantly affected the circadian rhythms of the E-box-controlled clock genes. The CO-dependent transcriptional activity changes of NPAS2 and CLOCK and subsequent inflammatory responses to produce TNF-α were both responsible for the CO removal-induced circadian rhythm disruption. In principle, selective depletion of small biomolecules such as CO by genetic or pharmacological methods is not possible without any side effects. As demonstrated in the work with hemoCD1, the pseudo-knockdown approach based on the highly selective CO binding will clarify the roles and functions of such small molecules not only in the circadian clock but also in the entire biological system.

References

1. Koike, N., Yoo, S.-H., Huang, H.-C., Kumar, V., Lee, C., Kim, T.-K., and Takahashi, J.S. (2012). Transcriptional architecture and chromatin landscape of the core circadian clock in mammals. *Science* 338: 349–354.
2. Gustafson, C.L. and Partch, C.L. (2015). Emerging models for the molecular basis of mammalian circadian timing. *Biochemistry* 54: 134–149.
3. Shimizu, T., Huang, D., Yan, F., Stranava, M., Bartosova, M., Fojtíková, V., and Martínková, M. (2015). Gaseous O_2, NO, and CO in signal transduction: structure and function relationships of heme-based gas sensors and heme-redox sensors. *Chem. Rev.* 115: 6491–6533.
4. Shimizu, T., Lengalova, A., Martínek, V., and Martínková, M. (2019). Heme: emergent roles of heme in signal transduction, functional regulation and as catalytic centres. *Chem. Soc. Rev.* 48: 5624–5657.
5. Kaasik, K. and Lee, C.C. (2004). Reciprocal regulation of haem biosynthesis and the circadian clock in mammals. *Nature* 430: 467–471.
6. Zhao, X., Cho, H., Yu, R.T., Atkins, A.R., Downes, M., and Evans, R.M. (2014). Nuclear receptors rock around the clock. *EMBO Rep.* 15: 518–528.
7. Rubio, M.F., Agostino, P.V., Ferreyra, G.A., and Golombek, D.A. (2003). Circadian heme oxygenase activity in the hamster suprachiasmatic nuclei. *Neurosci. Lett.* 353: 9–12.
8. Igarashi, K. and Watanabe-Matsui, M. (2014). Wearing red for signaling: the heme-bach axis in heme metabolism, oxidative stress response and iron immunology. *Tohoku J. Exp. Med.* 232: 229–253.
9. Klemz, R., Reischl, S., Wallach, T., Witte, N., Jürchott, K., Klemz, S., Lang, V., Lorenzen, S., Knauer, M., Heidenreich, S., Xu, M., Ripperger, J.A., Schupp, M., Stanewsky, R., and Kramer, A. (2017). Reciprocal regulation of carbon monoxide metabolism and the circadian clock. *Nat. Struct. Mol. Biol.* 24: 15–22.
10. Sagami, I. (2020). Regulation of mammalian circadian rhythm by heme, CO and NO signaling. *KAGAKU TO SEIBUTU* 58: 172–180 (written in Japanese).
11. Reick, M., Garcia, J.A., Dudley, C., and McKnight, S.L. (2001). NPAS2: An analog of clock operative in the mammalian forebrain. *Science* 293: 506–509.
12. DeBruyne, J.P., Weaver, D.R., and Reppert, S.M. (2007). CLOCK and NPAS2 have overlapping roles in the suprachiasmatic circadian clock. *Nat. Neurosci.* 10: 543–545.
13. Landgraf, D., Wang, L.L., Diemer, T., and Welsh, D.K. (2016). NPAS2 compensates for loss of CLOCK in peripheral circadian oscillators. *PLOS Genet.* 12: e1005882.
14. Dudley, C.A., Erbel-Sieler, C., Estill, S.J., Reick, M., Franken, P., Pitts, S., and McKnight, S.L. (2003). Altered patterns of sleep and behavioral adaptability in NPAS2-deficient mice. *Science* 301: 379–383.
15. Wang, Z., Wu, Y., Li, L., and Su, X.-D. (2013). Intermolecular recognition revealed by the complex structure of human CLOCK-BMAL1 basic helix-loop-helix domains with E-box DNA. *Cell Res.* 23: 213–224.
16. Freeman, S.L., Kwon, H., Portolano, N., Parkin, G., Girija, U.V., Basran, J., Fielding, A.J., Fairall, L., Svistunenko, D.A., Moody, P.C.E., Schwabe, J.W.R., Kyriacou, C.P., and Raven, E.L. (2019). Heme binding to human CLOCK affects interactions with the E-box. *Proc. Nat. Acad. Sci. U. S. A.* 116: 19911–19916.
17. Rutter, J., Reick, M., Wu, L.C., and McKnight, S.L. (2001). Regulation of Clock and NPAS2 DNA binding by the redox state of NAD cofactors. *Science* 293: 510–514.
18. Yoshii, K., Tajima, F., Ishijima, S., and Sagami, I. (2015). Changes in pH and NADPH regulate the DNA binding activity of neuronal PAS Domain protein 2, a mammalian circadian transcription factor. *Biochemistry* 54: 250–259.
19. Dioum, E.M., Rutter, J., Tuckerman, J.R., Gonzalez, G., Gilles-Gonzalez, M.A., and McKnight, S.L. (2002). NPAS2: a gas-responsive transcription factor. *Science* 298: 2385–2387.

20. Uchida, T., Sato, E., Sato, A., Sagami, I., Shimizu, T., and Kitagawa, T. (2005). CO-dependent activity-controlling mechanism of heme-containing CO-sensor protein, neuronal PAS domain protein 2. *J. Biol. Chem.* 280: 21358–21368.

21. Uchida, T., Sagami, I., Shimizu, T., Ishimori, K., and Kitagawa, T. (2012). Effects of the bHLH domain on axial coordination of heme in the PAS-A domain of neuronal PAS domain protein 2 (NPAS2): conversion from His119/Cys170 coordination to His119/His171 coordination. *J. Inorg. Biochem.* 108: 188–195.

22. Ishida, M., Ueha, T., and Sagami, I. (2008). Effects of mutations in the heme domain on the transcriptional activity and DNA-binding activity of NPAS2. *Biochem. Biophys. Res. Commun.* 368: 292–297.

23. Lukat-Rodgers, G.S., Correia, C., Botuyan, M.V., Mer, G., and Rodgers, K.R. (2010). Heme-based sensing by the mammalian circadian protein CLOCK. *Inorg. Chem.* 49: 6349–6365.

24. Itoh, R., Fujita, K.-I., Mu, A., Kim, D.H.T., Tai, T.T., Sagami, I., and Taketani, S. (2013). Imaging of heme/hemeproteins in nucleus of the living cells expressing heme-binding nuclear receptors. *FEBS Lett.* 587: 2131–2136.

25. Kitanishi, K., Igarashi, J., Hayasaka, K., Hikage, N., Saiful, I., Yamauchi, S., Uchida, T., Ishimori, K., and Shimizu, T. (2008). Heme-binding characteristics of the isolated PAS-A domain of mouse Per2, a transcriptional regulatory factor associated with circadian rhythms. *Biochemistry* 47: 6157–6168.

26. Hayasaka, K., Kitanishi, K., Igarashi, J., and Shimizu, T. (2011). Heme-binding characteristics of the isolated PAS-B domain of mouse Per2, a transcriptional regulatory factor associated with circadian rhythms. *Biochim. Biophys. Acta* 1814: 326–333.

27. Yang, J., Kim, K.D., Lucas, A., Drahos, K.E., Santos, C.S., Mury, S.P., Capelluto, D.G.S., and Finkielstein, C.V. (2008). A novel heme-regulatory motif mediates heme-dependent degradation of the circadian factor period 2. *Mol. Cell. Biol.* 28: 4697–4711.

28. Airola, M.V., Du, J., Dawson, J.H., and Crane, B.R. (2010). Heme binding to the mammalian circadian clock protein Period 2 is nonspecific. *Biochemistry* 49: 4327–4338.

29. Raghuram, S., Stayrook, K.R., Huang, P., Rogers, P.M., Nosie, A.K., McClure, D.B., Burris, L.L., Khorasanizadeh, S., Burris, T.P., and Rastinejad, F. (2007). Identification of heme as the ligand for the orphan nuclear receptors REV-ERBα and REV-ERBβ. *Nat. Struct. Mol. Biol.* 14: 1207–1213.

30. Woo, E.-J., Jeong, D.G., Lim, M.-Y., Kim, S.J., Kim, K.-J., Yoon, S.-M., Park, B.-C., and Ryu, S.E. (2007). Structural insight into the constitutive repression function of the nuclear receptor Rev-erbβ. *J. Mol. Biol.* 373: 735–744.

31. Pardee, K.I., Xu, X., Reinking, J., Schuetz, A., Dong, A., Liu, S., Zhang, R., Tiefenbach, J., Lajoie, G., Plotnikov, A.N., Botchkarev, A., Krause, H.M., and Edwards, A. (2009). The structural basis of gas-responsive transcription by the human nuclear hormone receptor REV-ERBβ. *PLOS Biol.* 7: e1000043.

32. Matta-Camacho, E., Banerjee, S., Hughes, T.S., Solt, L.A., Wang, Y., Burris, T.P., and Kojetin, D.J. (2014). Structure of REV-ERBβ ligand-binding domain bound to a porphyrin antagonist. *J. Biol. Chem.* 289: 20054–20066.

33. Yin, L., Wu, N., Curtin, J.C., Qatanani, M., Szwergold, N.R., Reid, R.A., Waitt, G.M., Parks, D.J., Pearce, K.H., Wisely, G.B., and Lazar, M.A. (2007). Rev-erbα, a heme sensor that coordinates metabolic and circadian pathways. *Science* 318: 1786–1789.

34. Tu, B.P. and McKnight, S.L. (2009). Evidence of carbon monoxide-mediated phase advancement of the yeast metabolic cycle. *Proc. Nat. Acad. Sci. U. S. A.* 106: 14293–14296.

35. Schallner, N., Lieberum, J.L., Gallo, D., LeBlanc, R.H., III, Fuller, P.M., Hanafy, K.A., and Otterbein, L.E. (2017). Carbon monoxide preserves circadian rhythm to reduce the severity of subarachnoid hemorrhage in mice. *Stoke* 48: 2565–2573.

36. Correa-Costa, M., Gallo, D., Csizmadia, E., Gomperts, E., Lieberum, J.L., Hauser, C.J., Ji, X., Wang, B., Câmara, N.O.S., Robson, S.C., and Otterbein, L.E. (2018). Carbon monoxide protects the kidney through the central circadian clock and CD39. *Proc. Natl. Acad. Sci. U. S. A.* 115: E2302–E2310.

37. Damulewicz, M., Loboda, A., Jozkowicz, A., Dulak, J., and Pyza, E. (2017). Interactions between the circadian clock and heme oxygenase in the retina of drosophila melanogaster. *Mol. Neurobiol.* 54: 4953–4962.

38. Rey, G., Valekunja, U.K., Feeney, K.A., Wulund, L., Milev, N.B., Stangherlin, A., Ansel-Bollepalli, L., Velagapudi, V., O'Neill, J.S., and Reddy, A.B. (2016) The pentose phosphate pathway regulates the circadian clock. *Cell. Metab.* 24: 462–473.

39. Kitagishi, H., Negi, S., Kiriyama, A., Honbo, A., Sugiura, Y., Kawaguchi, A.T., and Kano, K. (2010). A diatomic molecule receptor that removes CO in a living organism. *Angew. Chem. Int. Ed.* 49: 1312–1315.

40. Kitagishi, H. and Minegishi, S. (2017). Iron(II) porphyrin–cyclodextrin supramolecular complex as a carbon monoxide-depleting agent in living organisms. *Chem. Pharm. Bull.* 65: 336–340.

41. Kano, K., Kitagishi, H., Dagallier, C., Kodera, M., Matsuo, T., Hayashi, T., Hisaeda, Y., and Hirota, S. (2006). Iron porphyrin-cyclodextrin supramolecular

complex as a functional model of myoglobin in aqueous solution. *Inorg. Chem.* 45: 4448–4460.

42. Kitagishi, H., Minegishi, S., Yumura, A., Negi, S., Taketani, S., Amagase, Y., Mizukawa, Y., Urushidani, T., Sugiura, Y., and Kano, K. (2016). Feedback response to selective depletion of endogenous carbon monoxide in the blood. *J. Am. Chem. Soc.* 138: 5417–5425.

43. Minegishi, S., Sagami, I., Negi, S., Kano, K., and Kitagishi, H. (2018). Circadian clock disruption by selective removal of endogenous carbon monoxide. *Sci. Rep.* 8: 11996.

44. Minegishi, S., Yumura, A., Miyoshi, H., Negi, S., Taketani, S., Motterlini, R., Foresti, R., Kano, K., and Kitagishi, H. (2017). Detection and removal of endogenous carbon monoxide by selective and cell-permeable hemoprotein-model complexes. *J. Am. Chem. Soc.* 139: 5984–5991.

45. Mao, Q., Kawaguchi, A.T., Mizobata, S., Motterlini, R., Foresti, R., and Kitagishi, H. (2021). Highly sensitive quantification of carbon monoxide (CO) in vivo reveals a protective role of circulating hemoglobin in CO intoxication. *Commun. Biol.* 4: 425.

6

Carbon Monoxide and Mitochondria

Claude A. Piantadosi

Department of Medicine, Duke University School of Medicine, 200 Trent Drive, Durham, NC 27710, USA

Introduction

Carbon monoxide (CO) is not only an asphyxiant and metabolic poison, but also a dependable research tool for studying the biochemical reactions of a major class of proteins, the heme proteins (Hp). Typically, the CO research focus has been on its reactions with hemoglobin, myoglobin, P450-type cytochromes, and, in the mitochondrion, cytochrome *c* oxidase [1–3].

The Heme Protein Database contained 294 nonredundant Hp as of 2021. This list discloses some of the most notable hemoproteins in biology (Table 6.1). Each is synthesized within the mitochondria and, moreover, many of them involve the function of the mitochondrion itself [4]. Structurally, heme is a planar organic ring molecule that forms a tetrapyrrole, which grasps a metal ion to form a metal porphyrin. In heme, this metal is molecular iron, and the addition of an apoprotein creates the hemoprotein. These pigments have varied roles in biology, but their biochemical functions are mostly related to oxidation–reduction (redox) reactions involving the transport, storage, or activation of molecular oxygen (O_2).

The value of CO to the scientist comes from its propensity to interact with iron in the heme center to alter the absorption spectrum of the protein, which is easily measured. CO binds iron only in its reduced state (Fe^{2+}), which produces characteristic shifts in the Hp optical peaks that have proven useful in their identification. Moreover, this Hp effect of CO accounts for many of the mechanisms of actions of CO in biology because it interferes with protein function by inhibiting the binding of O_2 or by preventing redox reactions involving the transfer of electrons through the heme center to molecular O_2. Several well-known examples are found in the cytochromes. The group of cytochrome P450s is actually named according to the CO absorption peak at 450 nm [5]. CO is also capable of binding to some proteins containing other transition metals at their active sites, for instance, cobalt, nickel, and copper, thereby interfering with their functions. Here, I provide a summary of the biochemistry of CO as it pertains to Hp in mitochondria and their relationships to changes in mitochondrial and cellular function.

History of CO and the Mitochondrial Electron Transport Chain

The cytochrome system was discovered by an eminent parasitologist, David Keilin, of the Molteno Institute at the University of Cambridge in 1925. Keilin devised the name "cytochrome," which sprung from his observations using a microspectroscope to examine cells under various conditions, including CO exposure [6]. This instrument is a simple, but ingenious direct-vison spectroscope that replaces the ocular of a monocular microscope.

In the 1920s, Keilin had been studying the evolutionary adaptations of dipterous larvae, and noticed hypoxia-related changes in the reddish color of the wings of new adults. Soon thereafter, he turned to a mycelial suspension of yeast, which after a vigorous shaking in water was placed immediately under the microspectroscope. Only a few small faint absorption bands in the visible region were detectable. Surprisingly, however, within a few seconds, four

Carbon Monoxide in Drug Discovery: Basics, Pharmacology, and Therapeutic Potential, First Edition. Edited by Binghe Wang and Leo E. Otterbein.
© 2022 John Wiley & Sons, Inc. Published 2022 by John Wiley & Sons, Inc.

Table 6.1 Important Hp.

Protein	Location	Function
Hemoglobin	Blood	O_2 transport
Myoglobin	Muscle cells	O_2 storage and transport
Cytoglobin	Cytoplasm	Intracellular O_2 storage or transfer
Neuroglobin	Brain mitochondria	Brain O_2 transport
Cytochrome c oxidase	Mitochondria	O_2 reduction to H_2O in mitochondria
Cytochrome bc_1 complex	Mitochondria	Mitochondrial electron transport
Cytochrome c	Mitochondria	Mitochondrial electron transport
Succinate dehydrogenase	Mitochondria	Mitochondrial electron transport
P450 cytochromes	Liver cells; other ER	Catalyze hydroxylation of C–H bonds
Nitric oxide synthases	Vascular endothelium, nervous system, immune cells	Generate nitric oxide from L-arginine
Guanylate cyclases	Cytoplasm	Generate cGMP
Myeloperoxidases	Neutrophil lysosome	Generate hypohalous acids
Catalases	Peroxisomes	Detoxify hydrogen peroxide

sharp bands appeared, representing the pigments that he named cytochromes. He could repeatedly demonstrate this reversibility on the same yeast sample, as well as on samples of mammalian tissue, such as the heart muscle.

Keilin interpreted his findings as follows: the yeast contained pigments (cytochromes) that became oxidized when shaking the yeast preparation in air, and which had minimal absorption bands. When the preparation was allowed to sit, respiration by the yeast consumed the dissolved oxygen in the cells, and the cytochromes became reduced, resulting in a sharp, four-band spectrum. This meant that one or more cytochrome-containing proteins were serving as a linking protein for oxygen to the carbon substrates in the living cell.

Careful studies of the respiratory poisons, cyanide, azide, and CO supported this idea, for as the rate of cellular oxygen consumption fell, the spectroscopic changes appeared that accompany the normal redox cycle of the cytochromes. In other words, the slower the rate of electron transport, the more reduced the cytochromes became. He deduced that the terminal acceptor cytochrome had an alpha band at 603–605 nm, and was the site of action of cyanide and CO (Figure 6.1).

Keilin could not understand why biologists had overlooked such a conspicuous system, and by perusing older literature, he discovered the work of physician C.A. MacMunn (1852–1911), who had observed the four spectral peaks in 1887. MacMunn had attributed these to tissue pigments and named them myohematin and histohematin. His pigments had the same bands as Keilin's cytochromes, and the same increase in intensity with hypoxia as Keilin's did. However, the influential physiologists of his time, particularly Hoppe-Selye, claimed that he was simply measuring the breakdown products of hemoglobin in the tissues, called hemochromogens, even though the absorption bands were dissimilar and still visible in insects lacking hemoglobin. In this way, MacMunn's work was relegated to the dustbin of science, until Keilin rediscovered the existence and isolated the functions of the cytochromes almost 15 years after MacMunn's death.

Keilin's work was also questioned, notably by renowned German physiologist Otto Warburg, who received the Nobel Prize in 1931 for his work on tissue respiration. In 1928, Warburg concluded, by measuring the oxygen consumption of living cells and studying the proteins that reacted, that the respiration enzyme, which he called Atsmungsferment, was an iron compound related to hemoglobin. Aware of Keilin's work, he did not connect the two and determine that cytochrome c oxidase was the enzyme he sought. Meanwhile, Keilin had published in 1925, "On cytochrome, a respiratory pigment common to animals, yeast, and higher plants," where he had drawn, based on his work with cyanide and CO, the critical diagram for oxygen consumption, R ← cytochrome ← O_2, where R is the carbon substrate [7].

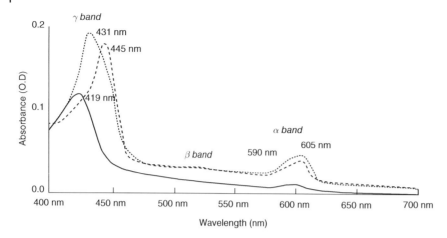

Figure 6.1 The reduced minus oxidized optical spectrum of the mitochondrial cytochrome c oxidase. The absorption bands in two visible regions and one UV region of the spectrum are shown and labeled α, β, and γ, respectively. The α-band is the visible band of cytochrome c oxidase at 603–605 nm, which is modified by CO binding (dashed line). X-axis is wavelength in nanometers and y-axis is absorption in units of optical density (OD). Not shown are cytochrome c at 550 nm and the bc1 complex at 564 nm.

CO, Cytochrome c Oxidase, and Mitochondrial Oxygen Availability

When physiologist Claude Bernard brought CO to scientific attention in 1857, he had established that the gas produces asphyxia by combining with hemoglobin [8]. By 1895, J.S. Haldane had demonstrated that CO binding to hemoglobin was antagonized by high partial pressures of O_2 and that animals survived fatal CO poisoning by rapidly eliminating CO if massive quantities of O_2 were dissolved in their plasma [1]. This work preceded Keilin's discovery of cytochrome oxidase and the effects of CO on it by 30 years, meaning that no mechanism was apparent for direct CO toxicity in tissue.

Later, it was realized that carboxyhemoglobin (COHb) shifted the oxyhemoglobin dissociation curve to the left by imparting a greater affinity for O_2 to the unoccupied hemoglobin heme after CO was bound to one. This meant that CO made it more difficult for unoccupied hemoglobin to unload oxygen into the tissue. This was interpreted to mean that CO toxicity was caused by the falling tissue P_{O_2} in the tissues (CO hypoxia) [9].

Although these ideas formed a basis for CO biology and the O_2 therapy of CO poisoning for a long time, they did not explain the observed differences between the effects of hypoxic hypoxia and those of CO hypoxia on living tissues, because the latter involves simultaneous inhibition of respiration [10]. For both events, O_2 therapy is effective because CO binds competitively to both Hp, hemoglobin and cytochrome oxidase, and competes with O_2 [10].

In the cell, cytochrome c oxidase (cytochromes a and a_3), the terminal enzyme of the mitochondrial electron transport system, reduces molecular O_2 to water in a four-electron reaction. Only the fully reduced enzyme (a_3 heme and Cu reduced) binds CO [11]. This competitive relationship is described by the Warburg partition coefficient:

$$K = (n/1-n)(CO/O_2),$$

where n, the fraction of a compound bound to CO, is equal to 0.5. Hence, K is the ratio of CO/O_2 necessary to load one-half of the binding sites with CO. Warburg coefficients *in vitro* are ~0.4 for myoglobin, 1.0 for cytochrome P450, and 5–15 for cytochrome c oxidase [12]. However, the *in vitro* coefficients do not translate well to the cell because the CO/O_2 ratio is challenging to measure due to the difficulty of measuring P_{CO} and P_{O_2} inside cells. However, cellular P_{O_2} is equal to or less than venous P_{O_2}; thus, it is quite low in active organs such as brain, heart, and liver (5–30 mmHg) [13]. As tissue P_{O_2} falls, less CO is needed to produce an effect, and when all the O_2 has been consumed, CO inhibits cytochrome c oxidase noncompetitively. In other words, the CO effect on O_2 uptake is not seen until O_2 is reintroduced to the system.

The cytochrome oxidase reaction accounts for more than 90% of the oxygen utilized by tissues. The enzyme has a Michaelis–Menten constant (K_m) for O_2 of less than 1 mmHg *in vitro* [14]. Because intracellular P_{O_2} is normally greater than 1 mmHg, cytochrome oxidase should remain oxidized until serious hypoxia is present. This state has been observed under low ADP conditions (state 4), where the enzyme remains oxidized until O_2 concentration falls to 10^{-6} M, but ADP addition (state 3) increases substrate turnover, and the reduction level of the oxidase increases significantly [15].

In summary, CO affects respiratory function in cells because the O_2 gradients are steep; this avoids high mitochondrial P_{O_2}, which causes untoward protein, lipid, and nucleic acid oxidation. In this context, it is important to realize that the oxidase is an O_2 sink that irreversibly reduces O_2 to water, thereby limiting oxidant stress at the mitochondrion [15,16]. Accordingly, in uncoupled mitochondria, CO/O_2 ratios as low as 0.2 markedly delayed recovery of cytochrome redox state from anoxia and normoxia [17]. This implies that a mitochondrial P_{O_2} of about 0.1 mmHg would allow one-half of the cytochrome c oxidase to bind CO at COHb levels equivalent to that of a heavy smoker (10%).

In respiring tissues, CO/O_2 ratios 5–10 times higher than those mentioned earlier were required to reduce O_2 uptake by one-half [15]. This means CO must inhibit more than half of the oxidase to inhibit O_2 consumption by half because unblocked cytochrome oxidase may accept electrons from multiple cytochrome c molecules. In other words, inhibition of some of the oxidase increases the reduction state of the electron transport system, but O_2 consumption falls more slowly than for a strictly linear respiratory chain. This effect is called cushioning, and disguises the effect of CO on the electron transport system.

Nonetheless, CO binding to cytochrome c oxidase is found *in vivo* because tissues with high O_2 requirements tend to have the steepest P_{O_2} gradients, allowing a small amount of the oxidase to always be reduced. The degree of reduced enzyme increases as hypoxia increases [16], eventually allowing CO binding to the oxidase regardless of the P_{O_2}. The a_3–CO complex is detected in the brain cortex of blood-circulated animals exposed to CO [18], and the CO-dependent increase in the reduction levels in the upstream electron carriers generates oxidant stress [19].

CO is oxidized to CO_2 by mitochondria, but much more slowly than the rate of endogenous CO production and in proportion to the tissue CO stores [20]. Oxidation of CO to CO_2 by reduced cytochrome oxidase was first reported in the 1960s, and later, oxidized cytochrome oxidase too was found to facilitate CO oxidation [21]. A high CO/O_2 ratio favors the ferrous carbonyl, while intermediate CO/O_2 ratios favor the CO-oxygenating species, and low CO/O_2 ratios permit the oxidized state. In any case, after CO binding, that oxidase molecule cannot transfer electrons to O_2 until CO oxygenation (oxidation) occurs. Although the rates of CO oxygenation are low physiologically, the ease by which cytochrome c oxidase oxidizes CO to CO_2 supports a molecular role for CO oxygenation in maintaining mitochondrial health.

CO and Mitochondrial Oxidant Production

The CO levels in the cell appear to be in the picomolar range [22] and this would allow CO binding at the physiological P_{O_2} of metabolically active tissues such as the brain and the heart. CO binding causes electrons to "back up" measurably into the cytochrome bc_1 region of the chain, which enhances superoxide anion production mainly at complex III [8]. This leads to hydrogen peroxide (H_2O_2) leakage into the matrix and outside the mitochondrion. The amount of H_2O_2 generated is sufficient to deplete mitochondrial glutathione stores independently of hypoxia [23]. This proclivity of CO to enhance mitochondrial oxidant production has been shown for endogenous CO in intact hepatic blood vessels, resulting in increased susceptibility to further oxidant stress [24].

Mitochondrial oxidant production has been implicated in diverse cell regulatory processes such as stabilization of hypoxia-inducible factor-1α, cell proliferation, angiogenesis factor expression, superoxide dismutase 2 (SOD2) and other antioxidant enzyme production, metalloproteinase expression in tumor cells, mitochondrial biogenesis, and mitochondrial quality control (MQC) (reviewed in [23]).

The rate of superoxide production, and hence H_2O_2 release at complex III, is enhanced by the slow rate of electron transfer to O_2 when CO is bound to cytochrome c oxidase because the Q cycle electron pool fills up, and the rate of electron transfer from semireduced Q to molecular O_2 generates superoxide in proportion to the product of the concentrations of semireduced Q and the concentration of O_2 [25]. Moreover, SOD2 induction (by H_2O_2) [26] serves not only to protect mitochondria from oxidant damage [27], e.g., superoxide-driven hydroxyl radical formation in the presence of iron, but also as a redox signal generator.

CO and Mitochondrial Quality Control

To illustrate CO's capacity to regulate mitochondrial health, it is perhaps best to show an example of CO-dependent genetic pathways, mainly the genetic induction of genes involved in MQC, and leave the nonmitochondrial pathways to other authors in this book. MQC refers to the combined processes of mitochondrial biogenesis, mitochondrial autophagy (mitophagy), and, to some extent, mitochondrial dynamics, i.e., mitochondrial movement, fusion, and fission. Since mammalian cells cannot make new mitochondria *de novo*, they depend on mitochondrial replenishment and elimination of damaged mitochondria through biogenesis and mitophagy, respectively [28,29]. Each of these processes requires the coordinated expression of nuclear and mitochondrial genes. The roles of the genes for mitochondrial biogenesis are reasonably well understood, while those for mitophagy are currently under intense investigation.

Let us consider what happens if energy-transducing pathways become compromised. Then, energy sensor proteins are activated that play crucial roles in regulating cellular energy metabolism. The archetype kinase is the 5′-AMP-activated protein kinase (AMPK), an energy-sensitive kinase that is activated by low energy status sensed by increases in the AMP:ATP ratio [30] and by low $NAD^+/NADH$ ratio [31]. AMPK phosphorylates the peroxisome proliferator-activated receptor gamma coactivator 1-alpha (PGC-1α) [32] to promote mitochondrial biogenesis, along with the NAD-dependent protein deacetylase SIRT1 responsible for deacetylation of PGC-1α [30]. It also inhibits energy-consuming processes, including protein, carbohydrate, and lipid biosynthesis, as well as cell growth and proliferation by direct phosphorylation of metabolic enzymes, and by longer term effects on the phosphorylation of transcriptional regulators. In other words, this pathway is activated by impending energy failure in the cell. There is also experimental support for an AMPK-dependent switch between PGC-1α-dependent and PGC-1α-independent mitochondrial biogenesis pathways, for example, those activated by SIRT1 [33] or, in our example, by CO.

CO is principally involved in a different mechanism, the redox pathway for mitochondrial biogenesis, which is thought to precede energy compromise. This mechanism involves retrograde activation from the mitochondrion to the nucleus of a transcriptional pathway that regulates expression levels of mitochondrial biogenesis genes as well as other genes necessary for MQC [34]. CO-generated mitochondrial H_2O_2 oxidizes reduced functional thiol groups on phosphatases that inactivate specific cell progrowth pathways. Two of these phosphatases, PTEN (phosphatidylinositol 3,4,5-trisphosphate 3-phosphatase and dual-specificity protein phosphatase) and PTP1B (tyrosine-protein phosphatase non-receptor type 1), are sensitive to inactivation by oxidants, including mitochondrial H_2O_2 [35,36]. After inactivation by mitochondrial oxidants, the loss of phosphatase activity allows the prosurvival serine/threonine-protein kinase Akt1 (RAC-alpha serine/threonine-protein kinase or protein kinase B) activity to increase, as found after CO exposure [37–39].

Akt1 supports at least two important regulatory steps in the redox pathway for mitochondrial biogenesis. One is the phosphorylation of nuclear respiratory factor-1 (NRF-1), a strategic transcription factor for the expression of a group of genes involved in oxidative phosphorylation and mitochondrial biogenesis [40]. The phosphorylation step is necessary for the nuclear accumulation of NRF-1. The other step is the inhibition of GSK3B (glycogen synthase kinase B), a ubiquitous serine/threonine kinase involved in the regulation of many cell processes. Akt1 (PKB) phosphorylates both known GSK3B, regulatory sites to downregulate GSK-3 activity through the PI3K/PKB pathway [41]. This relieves the GSK3B blockade to the nuclear accumulation of the Nfe2l2 transcription factor [42].

The above process is illustrated in Figure 6.2, where a set of gene readouts, such as the nuclear-encoded mitochondrial transcription factor A (Tfam), is used to illustrate the pathway. Tfam is essential for mitochondrial DNA (mtDNA) transcription, replication, and maintenance of mtDNA copy number. Of course, other nuclear mitochondrial and nonmitochondrial genes are regulated by these transcription factors.

Accompanying NRF-1 phosphorylation is the upregulation of the transcriptional coactivator PGC-1α, which is involved in the expression of steroid and nuclear receptors, as well as mitochondrial genes that contribute to adaptive thermogenesis. PGC-1α is required for full expression of mitochondrial transcriptome proteins, without which fully functional organelles cannot be replicated.

Mitochondrial pro-oxidants lead to the rapid induction and nuclear accumulation of NFE2L2, the major driver of antioxidant and anti-stress enzyme expression in the cell [43]. The NFE2L2 protein is docked to its Keap1 stress-sensor protein, which contains oxidant-sensitive thiols, which while reduced force the complex to be retained in the cytoplasm until it is ubiquitinated and degraded [44]. Oxidation of these thiols causes Keap1 to release NFE2L2, which permits it to translocate to the nucleus where it activates the expression of cytoprotective genes through a cis-acting element called the antioxidant/electrophile responsive element (ARE/EpRE). AREs are present in the promoter regions of many of stress-related genes, including phase 2 detoxifying enzymes. Subsequently, the expression of these genes and their proteins allows neutralization of reactive electrophiles.

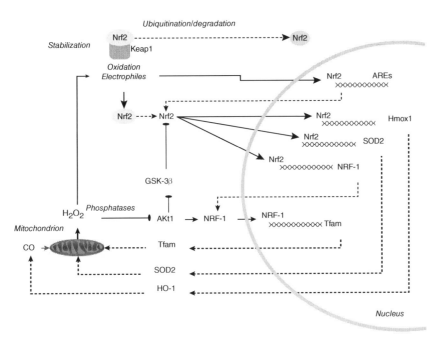

Figure 6.2 Diagram of the basic redox pathway for mitochondrial biogenesis. The CO binding to cytochrome c oxidase in the mitochondrion (lower left) increases mitochondrial H_2O_2 release. H_2O_2 activates Akt1 by thiol inhibition of inhibitory phosphatases (see text for description). This releases the activity of Akt1, leading to phosphorylation of the NRF-1 transcription factor. In the nucleus, NRF-1 increases the expression of Tfam gene, a gene for a protein involved in mitochondrial genome transcription and mtDNA replication. Akt1 also inhibits GSK-1b (glycogen synthase-1b), which inhibits the nuclear translocation of the NFE2L2 transcription factor. NFE2L2 (sometimes called Nrf2) is a rapid response transcription factor docked constitutively in the cytoplasm to the Keap1 electrophile sensor. Normally, NFE2L2 is degraded after ubiquitination by an E3 ubiquitin ligase. Oxidation of Keap1 releases NFE2L2, which enters the nucleus where it activates multiple stress response genes, including heme oxygenase-1 (HO-1), SOD2 and NRF-1. NFE2L2 protein is also able to activate its own gene.

More than 100 known gene promoters contain functional ARE sites [26], including NRF-1, which has several conserved ARE sites in its proximal promotor region. This allows NFE2L2 to trigger the mitochondrial stress response by supporting MQC gene activation partly by upregulation of the NRF-1 gene.

Another important factor in the redox pathway for the mitochondrial biogenesis program is the NFE2L2-driven induction of HO-1 [45], a major cytoprotective gene that utilizes O_2 and NADPH to metabolize heme and limit free-heme-induced cellular damage [46]. This includes heme released from mitochondria, the site of heme synthesis [47]. Free heme provides the substrate for the ongoing production of CO, thereby stabilizing the CO signal until heme levels fall [22]. In this case, endogenous CO, like nitric oxide, may activate soluble guanylate cyclase, which allows vasodilation, but which also has an important gene regulation role in mitochondrial biogenesis [48–50]. Although the importance of mitochondrial biogenesis to the protection of cellular energy homeostasis is clear, the activation of the process by the above redox pathway enables mitochondria to recover from early oxidative or nitrosative damage prior to complete energy failure, for example, during sepsis [51,52] as well as from inflammatory damage that interferes with energy-requiring cell repair processes [53,54].

Other Mitochondrial Effects of CO

Other mitochondrial effects of CO have been reported that are not as well delineated biologically as those mentioned earlier. Until now, the principle of heme targeting by CO has been sacrosanct. There are, however, reports of mitochondrial potassium channels being involved in the mitochondrial responses to CO by affecting coupling of oxidative phosphorylation, based on measurements loosely made in the presence of a respiratory inhibitor, CO [55]. These observations have not been confirmed by all investigators [56], and the physiological relevance of uncoupling measurements in the presence of terminal oxidase inhibitors at micromolar concentrations (usually as CO-releasing molecules), a respiratory poison, is moot.

The reason for some excitement here is that mitochondrial potassium channels are important for cell protection [57]. The inner mitochondrial membrane (IMM) has minimal permeability to charged molecules, thereby allowing the maintenance of an electrochemical proton gradient that drives ATP synthesis. However, the exchange of simple ions (and a few more complex molecules) between matrix and cytosol is vital for proper mitochondrial function. As many as eight mitochondrial potassium channels in a variety of cell types are known that affect inner membrane integrity, supporting and regulating ion exchange and leading to changes in oxidative phosphorylation or oxidant production [57].

Work in plasma membranes has shown CO activation of KATP channels, and heme binding to a specific motif on the SUR2A (ATP-binding cassette subfamily C member 9) receptor, which is required for a CO-dependent increase in channel activity [58]. In the IMM, the channels of interest to CO biology include the ATP-regulated potassium (mitoKATP) channel, the calcium-activated potassium (mitoBKCa) channel, the voltage-gated Kv1.3 potassium channel, and the two-pore domain TASK-3 potassium channel [59–61]. Micromolar concentrations of CO that reportedly produce "mild" mitochondrial uncoupling are attributed to CO-dependent activation of the calcium-activated K^+ (BKCa) channels or perhaps other IMM proteins, including the mitochondrial phosphate carrier, the adenine nucleotide translocase (ADT1). None of those proteins, however, supports heme or non-iron transition metals required for direct CO binding.

The mitoBKCa channel is normally activated by membrane depolarization or an increase in cytosolic Ca^{2+} that mediates K^+ export [57]. There is, however, a requirement for heme binding to the mitoBKCa channel for channel modulation (inhibitory), which is relieved by molecular CO, suggesting that CORMs have complex heme-dependent effects on mitoBKCa channels [62]. In other words, molecular heme inhibits this channel, while CO binding to that heme relieves the block and increases channel activation.

In the plasma membrane BKCa channel, CO may be interacting with the hemes involved with this protein, giving it a role of redox regulation of the channel. This high-level mechanism allows the cell to detect changes in redox state via the BKCa channel along with heme oxygenase, heme, and CO. Similar mechanisms are known to regulate the mitoBKCa channel. Thus, a redox state-sensitive ion channel (e.g., the mitoBKCa channel) could play an important role in the respiratory function of mitochondria.

The redox regulation of the mitoKATP channel in cardiomyocytes has also been observed and supports a relationship between the mitoBKCa channel and the PTP. Potassium channel activation may reduce calcium uptake through the Ca^{2+} uniporter. Mitochondrial substrate accumulation may reduce channel activity and support PTP activation, leading to apoptosis. Moreover, if CO is uncoupling respiration, it should be accompanied by loss of membrane potential, which acts as a potent inducer of mitophagy [63].

Summary and Conclusions

The biochemistry of CO is relatively well understood when it is simplified on the basis of the limited chemical reactivity of the gas, for instance, with reduced transition metals, especially iron in Hp. This feature leads to its distinctive binding to the family of Hp to block or alter their functions. Specifically, CO activity in mitochondria depends on the concentrations of CO and the reduced transition metal, e.g., Fe^{2+}, in relation to the concentration of molecular O_2. This principle holds for both endogenous and exogenous CO. The CO/O_2 ratio and O_2 concentration-dependent changes in the redox state of cytochrome oxidase are critical in mitochondria, as is the case for Hp in other cellular compartments that are known to be impacted by CO.

CO influences the reaction of mitochondrial cytochrome *c* oxidase, the terminal electron acceptor, as opposed to the other cytochromes, which are generally sheltered from its effects, just as they are from O_2. This leads to a backup of electrons preceding the failure of ATP production, and at high CO concentrations, failure of oxidative phosphorylation. Thus, at low CO concentrations, the initial CO-mediated redox changes in the electron transport chain are sensed as oxidant stress, generating a mitochondrial stress response that includes mitochondrial biogenesis and other elements of MQC. Other reported mitochondrial effects of CO on mitochondria, such as loose uncoupling, should be corroborated by the identification of a molecular Hp target, or a receptor that interacts with a heme. These are not unique in biology and have been demonstrated for several conditions.

References

1. Haldane, J. (1895). The relation of the action of carbonic oxide to oxygen tension. *J. Physiol.* 18: 201–217.
2. Keilin, D. and Hartree, E.F. (1947). Activity of the cytochrome system in heart muscle preparations. *Biochem. J.* 41: 500–502.

3. Keilin, D. and Hartree, E.F. (1939). Cytochrome and cytochrome oxidase. *Proc. Royal Soc. (London) Ser. B* 127: 167–191.
4. Piel, R.B., 3rd, Dailey, H.A., Jr., and Medlock, A.E. (2019). The mitochondrial heme metabolon: insights into the complex(ity) of heme synthesis and distribution. *Mol. Genet. Metab.* 128: 198–203.
5. Estabrook, R.W., Franklin, M.R., and Hildebrandt, A.G. (1970). Factors influencing the inhibitory effect of carbon monoxide on cytochrome P-450-catalyzed mixed function oxidation reactions. *Ann. N.Y. Acad. Sci.* 174: 218–232.
6. Keilin, D. (1966). The History of Cell Respiration and Cytochrome. U.K.: Cambridge Press Cambridge.
7. Hartree, E.F. (1973). The discovery of cytochrome. *Biochem. Ed.* 1: 69–70.
8. Piantadosi, C.A. (2002). Biological chemistry of carbon monoxide. *Antioxid. Redox Signal.* 4: 259–270.
9. Stewart, R.D. (1975). The effect of carbon monoxide on humans. *Annu. Rev. Pharmacol.* 15: 409–423.
10. Brown, S.D. and Piantadosi, C.A. (1992). Recovery of energy metabolism in rat brain after carbon monoxide hypoxia. *J. Clin. Investig.* 89: 666–672.
11. Wohlrab, H. and Ogunmola, G.B. (1971). Carbon monoxide binding studies of cytochrome a3 hemes in intact rat liver mitochondria. *Biochem* 10: 1103–1106.
12. Coburn, R.F. and Forman, H.J. (1987). Carbon monoxide toxicity. *Handbook Physiol.* 4: 439–456.
13. Tenney, S.M. (1977). A theoretical analysis of the relationships between venous blood and mean tissue oxygen pressure. *Resp. Physiol.* 20: 293–296.
14. Chance, B.A.W. G.R. (1956). The respiratory chain and oxidative phosphorylation. *Adv. Enzymol.* 17: 65.
15. Oshino, N., Sugano, T., Oshino, R., and Chance, B. (1974). Mitochondrial function under hypoxic conditions: the steady states of cytochrome alpha+alpha3 and their relation to mitochondrial energy states. *Biochim. Biophys. Acta* 368: 298–310.
16. Kreisman, N.R., Sick, T.J., LaManna, J.C., and Rosenthal, M. (1981). Local tissue oxygen tension-cytochrome a,a3 redox relationships in rat cerebral cortex in vivo. *Brain Res.* 218: 161–174.
17. Oshino, N., Sugano, T., Oshino, R., and Chance, B. (1974). Mitochondrial function under hypoxic conditions: the steady states of cytochrome alpha+alpha3 and their relation to mitochondrial energy states. *Biochim. Biophys. Acta* 368: 298–310.
18. Brown, S.D. and Piantadosi, C.A. (1990). In vivo binding of carbon monoxide to cytochrome c oxidase in rat brain. *J. Appl. Physiol.* 68: 604–610.
19. Piantadosi, C.A., Zhang, J., and Demchenko, I.T. (1997). Production of hydroxyl radical in the hippocampus after CO hypoxia or hypoxic hypoxia in the rat. *Free Radic. Biol. Med.* 22: 725–732.
20. Luomanmaki, K. and Coburn, R.F. (1969). Effects of metabolism and distribution of carbon monoxide on blood and body stores. *Am. J. Physiol.* 217: 354–363.
21. Young, L.J. and Caughey, W.S. (1986). Mitochondrial oxygenation of carbon monoxide. *Biochem. J.* 239: 225–227.
22. Cronje, F.J., Carraway, M.S., Freiberger, J.J., Suliman, H.B., and Piantadosi, C.A. (2004). Carbon monoxide actuates O(2)-limited heme degradation in the rat brain. *Free Radic. Biol. Med.* 37: 1802–1812.
23. Piantadosi, C.A. (2008). Carbon monoxide, reactive oxygen signaling, and oxidative stress. *Free Radic. Biol. Med.* 45: 562–569.
24. Suematsu, M., Kashiwagi, S., Sano, T., Goda, N., Shinoda, Y., and Ishimura, Y. (1994). Carbon monoxide as an endogenous modulator of hepatic vascular perfusion. *Biochem. Biophys. Res. Commun.* 205: 1333–1337.
25. Trumpower, B.L. (1990). The protonmotive Q cycle. Energy transduction by coupling of proton translocation to electron transfer by the cytochrome bc1 complex. *J. Biol. Chem.* 265: 11409–11412.
26. Rushmore, T.H., Morton, M.R., and Pickett, C.B. (1991). The antioxidant responsive element. Activation by oxidative stress and identification of the DNA consensus sequence required for functional activity. *J. Biol. Chem.* 266: 11632–11639.
27. Kokoszka, J.E., Coskun, P., Esposito, L.A., and Wallace, D.C. (2001). Increased mitochondrial oxidative stress in the Sod2 (±) mouse results in the age-related decline of mitochondrial function culminating in increased apoptosis. *Proc. Natl. Acad. Sci. U.S.A.* 98: 2278–2283.
28. Scarpulla, R.C. (2008). Transcriptional paradigms in Mammalian mitochondrial biogenesis and function. *Physiol. Rev.* 88: 611–638.
29. Lemasters, J.J. (2005). Selective mitochondrial autophagy, or mitophagy, as a targeted defense against oxidative stress, mitochondrial dysfunction, and aging. *Rejuvenation Res.* 8: 3–5.
30. Canto, C., Gerhart-Hines, Z., Feige, J.N., Lagouge, M., Noriega, L., Milne, J.C., Elliott, P.J., Puigserver, P., and Auwerx, J. (2009). AMPK regulates energy expenditure by modulating NAD+ metabolism and SIRT1 activity. *Nature* 458: 1056–1060.
31. Rafaeloff-Phail, R., Ding, L., Conner, L., Yeh, W.K., McClure, D., Guo, H., Emerson, K., and Brooks, H. (2004). Biochemical regulation of mammalian AMP-activated protein kinase activity by NAD and NADH. *J. Biol. Chem.* 279: 52934–52939.
32. Jager, S., Handschin, C., St-Pierre, J., and Spiegelman, B.M. (2007). AMP-activated protein kinase (AMPK)

action in skeletal muscle via direct phosphorylation of PGC-1alpha. *Proc. Natl. Acad. Sci. U.S.A.* 104: 12017–12022.
33. Valero, T. (2014). Mitochondrial biogenesis: pharmacological approaches. *Curr. Pharm. Des.* 20: 5507–5509.
34. Suliman, H.B., Keenan, J.E., and Piantadosi, C.A. (2017). Mitochondrial quality-control dysregulation in conditional HO-1-/- mice. *JCI Insight* 2: e89676.
35. Leslie, N.R., Bennett, D., Lindsay, Y.E., Stewart, H., Gray, A., and Downes, C.P. (2003). Redox regulation of PI 3-kinase signalling via inactivation of PTEN. *Embo. J.* 22: 5501–5510.
36. Lee, S.R., Yang, K.S., Kwon, J., Lee, C., Jeong, W., and Rhee, S.G. (2002). Reversible inactivation of the tumor suppressor PTEN by H2O2. *J. Biol. Chem.* 277: 20336–20342.
37. Fujimoto, H., Ohno, M., Ayabe, S., Kobayashi, H., Ishizaka, N., Kimura, H., Yoshida, K., and Nagai, R. (2004). Carbon monoxide protects against cardiac ischemia–reperfusion injury in vivo via MAPK and Akt–eNOS pathways. *Arterioscler. Thromb. Vasc. Biol.* 24: 1848–1853.
38. Zhang, X., Shan, P., Alam, J., Fu, X.Y., and Lee, P.J. (2005). Carbon monoxide differentially modulates STAT1 and STAT3 and inhibits apoptosis via a phosphatidylinositol 3-kinase/Akt and p38 kinase-dependent STAT3 pathway during anoxia-reoxygenation injury. *J. Biol. Chem.* 280: 8714–8721.
39. Suliman, H.B., Carraway, M.S., Tatro, L.G., and Piantadosi, C.A. (2007). A new activating role for CO in cardiac mitochondrial biogenesis. *J. Cell Sci.* 120 (Pt 2): 299–308.
40. Virbasius, J.V. and Scarpulla, R.C. (1994). Activation of the human mitochondrial transcription factor A gene by nuclear respiratory factors: a potential regulatory link between nuclear and mitochondrial gene expression in organelle biogenesis. *Proc. Natl. Acad. Sci. U.S.A.* 91: 1309–1313.
41. Fang, X., Yu, S.X., Lu, Y., Bast, R.C., Jr., Woodgett, J.R., and Mills, G.B. (2000). Phosphorylation and inactivation of glycogen synthase kinase 3 by protein kinase A. *Proc. Natl. Acad. Sci. U.S.A.* 97: 11960–11965.
42. Piantadosi, C.A., Carraway, M.S., Babiker, A., and Suliman, H.B. (2008). Heme oxygenase-1 regulates cardiac mitochondrial biogenesis via Nrf2-mediated transcriptional control of nuclear respiratory factor-1. *Circ. Res.* 103: 1232–1240.
43. Itoh, K., Chiba, T., Takahashi, S., Ishii, T., Igarashi, K., Katoh, Y., Oyake, T., Hayashi, N., Satoh, K., Hatayama, I., Yamamoto, M., and Nabeshima, Y. (1997). An Nrf2/small Maf heterodimer mediates the induction of phase II detoxifying enzyme genes through antioxidant response elements. *Biochem. Biophy. Res. Commun.* 236: 313–322.
44. Itoh, K., Wakabayashi, N., Katoh, Y., Ishii, T., Igarashi, K., Engel, J.D., and Yamamoto, M. (1999). Keap1 represses nuclear activation of antioxidant responsive elements by Nrf2 through binding to the amino-terminal Neh2 domain. *Genes Dev.* 13: 76–86.
45. Alam, J., Stewart, D., Touchard, C., Boinapally, S., Choi, A.M., and Cook, J.L. (1999). Nrf2, a Cap'n'Collar transcription factor, regulates induction of the heme oxygenase-1 gene. *J. Biol. Chem.* 274: 26071–26078.
46. Otterbein, L.E., Lee, P.J., Chin, B.Y., Petrache, I., Camhi, S.L., Alam, J., and Choi, A.M. (1999). Protective effects of heme oxygenase-1 in acute lung injury. *Chest* 116 (1 Suppl): 61S–63S.
47. Desmard, M., Boczkowski, J., Poderoso, J., and Motterlini, R. (2007). Mitochondrial and cellular heme-dependent proteins as targets for the bioactive function of the heme oxygenase/carbon monoxide system. *Antioxid. Redox Signal.* 9: 2139–2155.
48. Nisoli, E., Clementi, E., Paolucci, C., Cozzi, V., Tonello, C., Sciorati, C., Bracale, R., Valerio, A., Francolini, M., Moncada, S., and Carruba, M.O. (2003). Mitochondrial biogenesis in mammals: the role of endogenous nitric oxide. *Science* 299: 896–899.
49. Nisoli, E., Clementi, E., Tonello, C., Moncada, S., and Carruba, M.O. (2004). Can endogenous gaseous messengers control mitochondrial biogenesis in mammalian cells? *Prostaglandins Other Lipid Mediat.* 73: 9–27.
50. Nisoli, E., Falcone, S., Tonello, C., Cozzi, V., Palomba, L., Fiorani, M., Pisconti, A., Brunelli, S., Cardile, A., Francolini, M., Cantoni, O., Carruba, M.O., Moncada, S., and Clementi, E. (2004). Mitochondrial biogenesis by NO yields functionally active mitochondria in mammals. *Proc. Natl. Acad. Sci. U.S.A.* 101: 16507–16512.
51. Haden, D.W., Suliman, H.B., Carraway, M.S., Welty-Wolf, K.E., Ali, A.S., Shitara, H., Yonekawa, H., and Piantadosi, C.A. (2007). Mitochondrial biogenesis restores oxidative metabolism during Staphylococcus aureus sepsis. *Am. J. Respir. Crit. Care Med.* 176: 768–777.
52. Suliman, H.B., Welty-Wolf, K.E., Carraway, M., Tatro, L., and Piantadosi, C.A. (2004). Lipopolysaccharide induces oxidative cardiac mitochondrial damage and biogenesis. *Cardiovasc. Res.* 64: 279–288.
53. Piantadosi, C.A., Withers, C.M., Bartz, R.R., Macgarvey, N.C., Fu, P., Sweeney, T.E., Welty-Wolf, K.E., and Suliman, H.B. (2011). Heme oxygenase-1 couples activation of mitochondrial biogenesis to anti-inflammatory cytokine expression. *J. Biol. Chem.* 286: 16374–16385.

54. Suliman, H.B., Carraway, M.S., Ali, A.S., Reynolds, C.M., Welty-Wolf, K.E., and Piantadosi, C.A. (2007). The CO/HO system reverses inhibition of mitochondrial biogenesis and prevents murine doxorubicin cardiomyopathy. *J. Clin. Invest.* 117: 3730–3741.
55. Lanza, I.R. and Nair, K.S. (2009). Functional assessment of isolated mitochondria in vitro. *Methods Enzymol.* 457: 349–372.
56. Gasier, H.G., Dohl, J., Suliman, H.B., Piantadosi, C.A., and Yu, T. (2020). Skeletal muscle mitochondrial fragmentation and impaired bioenergetics from nutrient overload are prevented by carbon monoxide. *Am. J. Physiol. Cell. Physiol.* 319: C746–C756.
57. Szabo, I. and Zoratti, M. (2014). Mitochondrial channels: ion fluxes and more. *Physiol. Rev.* 94: 519–608.
58. Kapetanaki, S.M., Burton, M.J., Basran, J., Uragami, C., Moody, P.C.E., Mitcheson, J.S., Schmid, R., Davies, N.W., Dorlet, P., Vos, M.H., Storey, N.M., and Raven, E. (2018). A mechanism for CO regulation of ion channels. *Nat. Commun.* 9: 907.
59. Lo Iacono, L., Boczkowski, J., Zini, R., Salouage, I., Berdeaux, A., Motterlini, R., and Morin, D. (2011). A carbon monoxide-releasing molecule (CORM-3) uncouples mitochondrial respiration and modulates the production of reactive oxygen species. *Free Radic. Biol. Med.* 50: 1556–1564.
60. Long, R., Salouage, I., Berdeaux, A., Motterlini, R., and Morin, D. (2014). CORM-3, a water soluble CO-releasing molecule, uncouples mitochondrial respiration via interaction with the phosphate carrier. *Biochim. Biophys. Acta* 1837: 201–209.
61. Kaczara, P., Motterlini, R., Rosen, G.M., Augustynek, B., Bednarczyk, P., Szewczyk, A., Foresti, R., and Chlopicki, S. (2015). Carbon monoxide released by CORM-401 uncouples mitochondrial respiration and inhibits glycolysis in endothelial cells: a role for mitoBKCa channels. *Biochim. Biophys. Acta* 1847: 1297–1309.
62. Rotko, D., Bednarczyk, P., Koprowski, P., Kunz, W.S., Szewczyk, A., and Kulawiak, B. (2020). Heme is required for carbon monoxide activation of mitochondrial BKCa channel. *Eur. J. Pharmacol.* 881: 173191.
63. Lemasters, J.J. (2014). Variants of mitochondrial autophagy: types 1 and 2 mitophagy and micromitophagy (Type 3). *Redox Biol.* 2: 749–754.

7

Carbon Monoxide, Oxygen, and Pseudohypoxia

Grace E. Otterbein[1], Michael S. Tift[2], and Ghee Rye Lee[3]

[1]*University of Aberdeen School of Medical Sciences, Polwarth Building, Foresterhill, Aberdeen AB25 2ZD, UK*
[2]*Department of Biology and Marine Biology, University of North Carolina, Wilmington, 601 S. College Road, Wilmington, NC 28403, USA*
[3]*Center for Life Science, Blackfan Circle, 617D, Boston, MA 02215, USA*

Introduction

Carbon monoxide (CO) and oxygen (O_2) as small gaseous molecules play a wide range of physiological roles in organisms and have been discussed in detail throughout this book. CO as a by-product to heme degradation is involved in modulating immune responses, vasomotor tone, cell proliferation, and cell survival and death. Amid these ascribed functions, often overlooked is the impact on O_2 delivery, availability, and storage. O_2 is needed for organisms to generate sufficient energy in the form of mitochondria-derived ATP. Our understanding of the evolutionary role of these two gases has expanded due to discoveries made in animals and humans living under extreme conditions, such as deep-diving marine mammals and mammals living at high altitudes. While oxygen is clearly required for life and used every day by medical centers to ensure acceptable blood oxygen saturation in patients, the sophisticated roles that CO plays are also emerging. The potential of CO for therapeutic benefit is being actively investigated based on numerous reports showing its biological benefits, as described in the various chapters of this book. While oxygen use seems logical, there are indeed toxicities associated with its overuse given that its levels in most tissues would be considered hypoxic relative to concentrations in inspired air. Moreover, one of the hallmarks of inflammation is hypoxia and a rapid induction of the heme oxygenases (HOs) resulting in increased generation of CO locally, which raises the question as to why generate CO in an environment dependent on the availability of O_2? Since CO binds to many of the same sites in hemoproteins as O_2, such as mitochondrial oxidases, it may create a scenario whereby the cell responds as if under oxygen deficit. We define this concept as pseudohypoxia. Numerous studies have shed light on the intricate biological roles that CO and O_2 play under various physiological and pathophysiological conditions, both independently and interdependently. Numerous scientific discoveries in animals and humans have shifted and expanded our views of these two gases. Depending on the context, conditions, and the duration and the amount of exposure, O_2 and CO can exert a variety of effects that are contradictory to their stereotypes. In fact, they may require one another. This chapter discusses the interplay between CO and O_2 from a comparative physiological standpoint and discusses how their signaling pathways impact one another to modulate specific cellular and physiological responses.

Physiological Roles of the Two Gases: Oxygen and Carbon Monoxide

Oxygen

In order for cells to continuously support their function and maintain tissue health, and ultimately survival of the organism, they require a constant supply of energy, which is provided primarily through aerobic cellular respiration that occurs in mitochondria. This process requires O_2 as the last electron acceptor of the mitochondrial electron transport chain (ETC), generating water and ATP. Conditions that result in a large or chronic reduction in O_2 supply to tissues could therefore result in inadequate ATP production, and possibly tissue injury or death. On the other hand, too much O_2 supplied to tissues can overwhelm mitochondria, resulting in oxidative stress and generation of reactive oxygen species

Carbon Monoxide in Drug Discovery: Basics, Pharmacology, and Therapeutic Potential, First Edition. Edited by Binghe Wang and Leo E. Otterbein.
© 2022 John Wiley & Sons, Inc. Published 2022 by John Wiley & Sons, Inc.

(ROS), including superoxide anions, hydrogen peroxide, and hydroxyl radicals that can alter lipids, proteins, and DNA resulting in organ failure and disease. Fortunately, cells are equipped with a battery of antioxidant enzymes that neutralize excess radicals. Small amounts of free radicals are excellent signaling molecules. While organisms depend heavily on aerobic cellular respiration for energy supply, they can also utilize other metabolic processes in the absence of O_2 known as anaerobic cellular respiration, sometimes exclusively in cases where O_2 is not available (e.g., hypoxia or anoxia). However, aerobic cellular respiration is much more efficient than anaerobic respiration, yielding a greater number of ATP. Complete reliance on anaerobic respiration is not sustainable and will ultimately deprive organisms of sufficient energy to ensure survival.

The "Oxygen Paradox"

Under normal conditions, the amount of oxygen entering the lung during inspiration is 20.9%, but once it enters the bloodstream, this level quickly drops. Most cells in the human body exist at 5% O_2 [1]. Table 7.1 lists O_2 tensions in various compartments in the body. Importantly, O_2 levels become even more extreme when considering the mitochondria and the extracellular environment where the gradient is considerably lower at 1.3% as a result of the consumption of O_2 by cytochrome c oxidase (COX) [1]. These levels make one posit whether there is something unique about our physiology that requires an inspired O_2 concentration of 21% (159 mmHg at sea level). For example, how does the overall health of the human population respond to fluctuations in O_2 concentrations in our environment? Why is it that the inclination is to increase O_2 delivery, particularly in hospital settings in situations where tissue and mitochondrial levels may be sufficient to maintain cellular function? Does the dramatic fall in O_2 across tissue environments require 20.9% inspired oxygen or would inspiring 15% suffice?

In cases of inflammation and disease, the human body most often creates a hypoxic environment localized around the site of tissue injury as a result of the immune response [2,3]. This also occurs around a tumor and some refer to tumors as unresolved inflammation. There is also a concomitant increase in temperature and cellular infiltrates, all of which are hallmarks of an innate response. This can be explained by a positive correlation between an increase in cellular activity (inflammation) and O_2 availability or energy supply. However, it has been observed in many studies that inspiration of *reduced* O_2 (hypoxia) results in improved resistance to inflammation in a variety of diseases, as well as a reduction in

Table 7.1 Normal values of pO_2 in various human tissues expressed in mmHg and in percentage of pO_2 in the microenvironment [8].

Environment or tissues	pO_2 (mmHg)	pO_2 (%)
Air	160	20.9
Inspired air in the trachea	150	19.7
Air in the alveoli	110	14.5
Arterial blood	100	13.2
Venous blood	40	5.3
Cell	9.9–19	1.3–2.5
Mitochondria	<9.9	<1.3
Brain	33.8 ± 2.6	4.4 ± 0.3
Lung	42.8	5.6
Skin (subpapillary plexus)	35.2 ± 8	4.6 ± 1.1
Skin (dermal papillae)	24 ± 6.4	3.2 ± 0.8
Skin (superficial region)	8 ± 3.2	1.1 ± 0.4
Intestinal tissue	57.6 ± 2.3	7.6 ± 0.3
Liver	40.6 ± 5.4	5.4 ± 0.7
Kidney	72 ± 20	9.5 ± 2.6
Muscle	29.2 ± 1.8	3.8 ± 0.2
Bone marrow	48.9 ± 4.5	6.4 ± 0.6

proinflammatory mediators and lymphocytes [4–7]. Indeed, this negates the well-accepted dogma and teachings that patients experiencing lower than normal O_2 levels should always be placed on a higher amount of inspired O_2. Perhaps, the response and purposeful intention by the body to create a hypoxic environment in scenarios of injury is the means by which humans and other animals (see later) have evolved to deal with stress and inflammation, particularly in the acute setting. Of course, poor lung function or tissue perfusion caused by pathologies such as stenotic vessels or chronic lung disease requires treatment strategies to deliver supplemental oxygen. It is well accepted, however, that elevated levels of oxygen, while life sustaining, promote ROS generation and therein lies the paradox and it is the unavoidable consequence of living in an aerobic environment.

Hyperoxia and the Delicate Balance Between Cell Survival and Death

The upper limit dose of O_2 that human lungs can tolerate without injury is not well known, yet hyperoxic gases (defined as oxygen levels greater than 20.9% O_2) are routinely delivered to patients as part of supportive care and regardless of need. This is of particular importance in instances such as bacterial or viral pneumonia where the lungs become filled with fluid leading to poor oxygen and CO_2 exchange. Of note, COVID-19 patients suffer from pulmonary vascular hyperinflammatory syndrome caused by a cytokine storm where the only treatment that is constant is to increase O_2. Hyperoxia is defined as any instance where cells, tissues, or organs are exposed to a partial pressure of O_2 (pO_2) that is higher than normal atmospheric pO_2 at 40 mmHg (in tissues) or 100 mmHg (in blood; Table 7.1) [9]. Animal studies have clearly shown that exposure to higher than normal oxygen concentrations results in acute lung injury within 3–4 days with significant morbidity and mortality at >95% O_2 [10,11]. High O_2 concentrations injure the epithelial lining of the respiratory tract leading to sterile inflammation, leukocyte infiltration, edema, and ultimately ventilation/perfusion mismatch [12,13]. Persistent exposure to hyperoxia causes irreversible tissue damage [9]. Importantly, other remote organs such as the central nervous system can be affected by the elevated O_2 levels in the blood [14]. Exposure to 50–75% O_2, while not lethal, does result in chronic lung injury and fibrosis [15]. There have been no clinical studies assessing the effects of elevated O_2 exposure to healthy volunteers nor studies showing that O_2 exposure in compromised patients is detrimental to the lung or other tissues. The large preclinical literature, however, showing increased ROS production and tissue damage with O_2 exposure alone supports a critical balance between providing necessary O_2 bioavailability for survival and the risks of increasing tissue damage.

In certain instances, hyperbaric oxygen therapy is used to treat a number of conditions such as anemia, arterial gas embolism, decompression sickness, burns, gangrene, nonhealing wounds, and CO poisoning [16]. This typically involves inhalation of 100% O_2 under an increased environmental pressure (~2–3 times ambient pressure) in order to more rapidly increase oxygen delivery into the blood and delivery to tissues [17]. With regard to CO poisoning, the goal is to rapidly eliminate CO from occupying the O_2 binding sites on hemoglobin. Blood levels of CO are measured as carboxyhemoglobin (COHb) and it is the most commonly used readout for determining CO levels in the body. One must ask, however, by delivering more O_2, does this result in such a rapid shift in partial pressure that CO is forced into tissues and that once the O_2 therapy is stopped the COHb levels reemerge from the tissues and COHb becomes elevated again until eliminated through the lung. Increasing barometric pressure, in addition to elevating tissue O_2 levels, can also result in toxicity in the central nervous system with symptoms of nausea, dizziness, muscle twitching, visual disturbances, and disorientation before progressing to convulsions. Such an effect is also experienced by deep-sea divers [18].

Hypoxia

Low O_2 availability, or hypoxia, can be detrimental to cells and tissues and its level must be constantly monitored by the chemosensory systems so that the host can respond appropriately to changes in O_2 availability and optimize the supply of O_2 to tissues. Airway neuroepithelial bodies sense the O_2 level in the inspired air and carotid bodies located at the bifurcation of the common carotid artery monitor the O_2 level in the circulation [19]. In response to low O_2 level, pulmonary blood vessels immediately constrict to divert blood from the poorly oxygenated regions in the lung to better oxygenated ones in order to match ventilation to perfusion. In contrast, peripheral blood vessels dilate in order to deliver oxygenated blood to O_2-deprived tissues [20]. This also signals to increase rate and systemic blood pressure. To optimize the energy use and production, metabolism shifts from oxidative phosphorylation to glycolysis and fatty acid synthesis and consumption [21]. Cells also undergo changes at the molecular level, mainly through the action of hypoxia-inducible factor-1α (HIF-1α), a transcription factor that regulates a myriad of signaling pathways and gene expression patterns that act to increase the chance of survival of the cells and the host under hypoxia [22–28]. Increased HIF-1α promotes erythropoiesis by upregulating erythropoietin in the kidney, angiogenesis by increasing the production of VEGF, and glycolytic enzyme activity and production. The details of HIF-1α are described later. Hypoxia also activates NF-κB through HIF-1α-dependent and HIF-1α-independent pathways, which then activates acute inflammatory responses [29].

Hypoxia is implicated in or contributes to the pathogenesis of various human diseases [28]. For example, obstructive sleep apnea is a disorder characterized by recurrent cessation of breathing, which leads to global hypoxemia, systemic inflammation, and oxidative stress [30]. In pulmonary fibrosis, extensive tissue scarring arising as damaged pulmonary epithelium is repaired can lead to chronic inflammatory responses, accumulation of scar tissue, and decreased ability of the lung to deliver O_2 [31]. Hypoxia or ischemia (or anoxia) can also cause significant tissue injuries as cells become starved of O_2. Ischemia/reperfusion injury (IRI) is a significant medical challenge that occurs as a result of organ transplantation [32]. It is important here that IRI is considered a two-hit injury primarily due to sudden changes in O_2. The lack of oxygen with ischemia followed by reperfusion and oxygenation results in increased ROS and subsequent tissue injury [33,34]. Similar pathology results after a blocked coronary artery is suddenly opened with a balloon catheter. The reoxygenation is thought to be as injurious as the ischemia [35]. Perhaps in these settings, a slow transition to normal oxygen levels would be more effective. Ischemia/reperfusion increases HO-1 and thus CO, which begs the following question: How does CO contribute to the injury? Indeed animals or tissues without HO-1 show significantly greater IRI [36]. Additionally, it has been suggested that hypoxia contributes to atherosclerosis [37]. The accumulation of macrophages with increased oxygen consumption, or foam cells, and inflammatory responses in the atherosclerotic lesion create a hypoxic zone and such chronic hypoxia is thought to contribute to atherosclerosis [38].

Benefits of Intermittent Hypoxia

While often thought of as being associated with pathologies, such as obstructive sleep apnea, pulmonary fibrosis, and atherosclerosis, reduced concentrations of O_2 delivered to tissue can be beneficial in certain circumstances. Intermittent hypoxia (IH) describes the periodic alternation between exposure to hypoxic conditions (below 21% O_2) and normoxia (21% O_2) or hyperoxia (above 21% O_2) [39]. Along with the anti-inflammatory, antiapoptotic, and strengthened innate immune responses, IH, described as ranging from 2% to 16% inspired O_2, is associated with reduced arterial hypertension [40], increased aerobic capacity [41], increased bone mineral density [42], and improved recovery of contractile function in postischemic myocardial tissue [43,44].

Normobaric hypoxia treatment is a form of IH treatment where the amount of inspired O_2 is decreased to levels below 21% for a period of time at normal atmospheric pressure, effectively creating a pseudo-altitude response. Hypertensive patients can be intermittently exposed to normobaric hypoxia (4–10 times 3-min cycles/day, at 10% inspired O_2) and maintain normotensive pressures [44,45]. This is most likely a result of the increased nitric oxide (NO) production by vascular endothelium, which acts as a potent vasodilator [46]. There is a similar response seen in high-altitude human populations that thrive above 4200 m and sojourners acclimatizing to altitudes above 2500 m [47,48]. IH increases production of erythropoietin, which stimulates erythropoiesis, increasing O_2-carrying capacity and, in turn, aerobic capacity due to increased erythrocyte numbers [49,50]. Thus, IH is routinely used as a method to train and improve athletic performance [51]. Vascular endothelial growth factor (VEGF) is regulated by hypoxia and known to promote angiogenesis in situations where there is poor tissue oxygenation leading to generation of new vessels. An increase in blood vessel development to improve blood flow to muscles during exercise, for example, is another potential mode of action and benefit of IH. Repetitive IH has been shown to impart salutary effects in patients suffering respiratory complications as a result of spinal cord injury (SCI) [52]. This treatment leads to recovery of synaptic pathways to phrenic motor neurons below the SCI location. IH is also neuroprotective, able to improve function and recovery of damaged neurons and increasing neuroplasticity [53–55]. Similarly, following an ischemic event in the brain, IH enhances cognitive recovery in mice through prolonged activation of HIF-1α and its neuroprotective effects through attenuating neuroinflammation, in part, by reducing the expression of one or more of the NO synthases (NOS) [55,56].

Indeed, the many benefits of IH are evident across numerous systems and depend on the level of hypoxia during IH, the duration of hypoxia, the number of treatment cycles, recovery period, and condition, for instance, recovering from IH in normoxia or hyperoxia, and with set cycle pattern such as consecutive or alternating days [44].

Oxygen and HIF-1α

HIF-1α is considered a master transcription factor that regulates hypoxic responses in the host [23]. Instances where cells signal that there is insufficient O_2 lead to stimulation of alternative metabolic pathways including AMP-activated protein kinase and glycolysis to enable sufficient ATP for survival. This acute phase response includes the stabilization of HIF-1α, which has a complex regulation [1]. Under normoxic conditions, the two prolyl subunits that comprise HIF-1α are hydroxylated by prolyl hydroxylases (PHDs) that sense O_2 and degrade the HIF subunits by proteasome activity [28]. In contrast, when the cell experiences hypoxic conditions, hydroxylase activity is inhibited, resulting in stabilization and accumulation of HIF-1α in the cell. HIF-1α is then able to translocate to the nucleus, combine with HIF-1β to form a complex, and regulate gene expression. This DNA-binding complex binds to hypoxia-responsive elements (HREs) of target genes [22,23], activating a wide array of genes involved in various adaptive processes, including but not limited to angiogenesis, vasomotor tone, O_2 transport, glycolysis, red blood cell production, heme metabolism, apoptosis, cell growth, and cell differentiation (Figure 7.1) [22,24]. In order to ensure survival of the cell under hypoxic conditions, HIF-1α orchestrates the expression of numerous genes that compensate and protect when O_2 availability is low such as 0–0.5% [57,58]. Once oxygen levels are restored and in sufficient amounts to activate PHDs, HIF-1α is rapidly degraded and HIF-1α-dependent gene transcription is downregulated [58]. It is important to note that the cell is highly sensitive to changes in O_2 tension such as the PHDs and mitochondrial oxidases, yet cellular pO_2 is already relatively low, and even more so at the mitochondria. This suggests enormous regulation and sensitivity to even subtle changes in O_2 availability.

Understanding the Roles of CO in Nature

This book provides an excellent update of CO chemistry, biology, and medicine. A brief overview is presented here for context. CO is odorless and colorless and known for its ability to bind hemoglobin, forming COHb in the body with greater than 200 times affinity when compared to O_2 [59]. Binding of CO to Hb shifts the oxygen–hemoglobin dissociation curve to the left, diminishing oxygen release to tissues. While many know the gas as being found in sources of incomplete fuel combustion (e.g., car exhaust), CO is

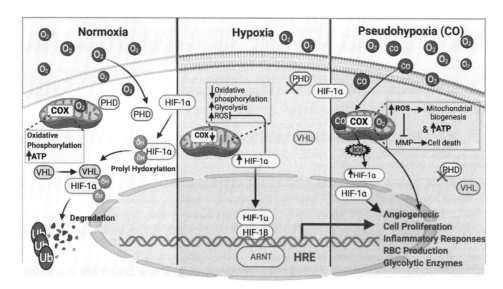

Figure 7.1 Cellular physiology under conditions of normoxia, hypoxia, and pseudohypoxia. Depicted are the cellular response to conditions of normoxia, hypoxia, and pseudohypoxia. Under normoxic conditions (left), the mitochondria utilize COX to generate adequate ATP via oxidative phosphorylation and the O_2-sensing PHDs degrade the Von Hippel–Lindau and HIF-1α complex. Under hypoxia (middle), there is decreased COX activity causing cellular metabolism to shift to glycolysis, and increasing ROS production. PHDs do not sense sufficient levels of O_2, leading to an increase in active HIF-1α, which enters the nucleus complexes with HIF-1β, which together bind the aryl hydrocarbon receptor nuclear translocator (ARNT) protein. This binding triggers HREs to begin transcription of glycolytic enzymes. In pseudohypoxia involving CO (right), COX activity is increased and ROS levels rise, leading to mitochondrial biogenesis. The ROS is a signaling amount that among other signaling can inhibit mitochondrial membrane permeability (MMP) and therefore prevent cell death. CO stabilizes HIF-1α increases leading to regulation of angiogenesis, cell proliferation, modulation of inflammatory responses and metabolism, and erythropoiesis.

generated endogenously by HO enzyme action on heme [60]. Not only is HO-1 responsible for the catabolism of heme, but induction of HO-1 or treatment with CO also confers potent cytoprotective effects under various pathophysiological conditions [61]. HO-1 can be induced by a plethora of stressors such as heme, pathogens, inflammatory cytokines, oxidants, and both hypoxia and hyperoxia [61]. The other products of heme catabolism are also bioactive and the reader is referred to Chapter 10 on bilirubin that has enormous antioxidant capabilities.

The rate of endogenous CO production is limited by the availability of heme and the induction of HO-1 [62,63]. Mice that have the inducible HO enzyme (HO-1; *Hmox1*) knocked out have a 5% fecundity, and those that are born exhibit striking pathophysiology with premature death [64]. The phenotype includes elevations in oxidative stress, growth retardation, anemia, and increased vulnerability to injury with a shortened life span [64,65]. In humans, HO-1 deficiency is associated with a similar phenotype of slowed growth and various blood disorders, including anemia, abnormal coagulation, and endothelial damage [66]. The knockout mouse and human provide insight into the critical role of HO-1 in mammalian tissue viability [65]. Moreover, challenging HO-1-deficient animals with a stressor such as bacteria or acute tissue injury results in enhanced damage and administration of CO can rescue and effectively substitute for the lack of HO-1 [67,68]. HO-2-deficient mice show no overt phenotype, but when challenged with stress exhibit increased free radical-induced lipid peroxidation [69]. Further, HO-2-null mice show decreased gastric motility and susceptibility to diabetes [70,71]. Individuals that adapt well at high altitude and therefore hypoxia show a unique adaptive response expressing higher levels of HO-2 [72,73]. Collectively, the HOs are intimately involved with hypoxia and O_2-dependent processes.

Elevated Endogenous Carbon Monoxide: Examples from Nature

Marine Mammals

There are a few specific examples from natural systems (animal and human) that have evolved to maintain chronically elevated endogenous CO levels in the body, which appear to confer some selective advantage(s) for the individuals. In general, these groups have certain similarities in that several are chronically exposed to hypoxia. The most extreme and well-known example is deep-diving seals. The first reported measure of endogenous CO in deep-diving seals occurred by accident in 1959, when Lewis 'Griffith' Pugh measured the levels of the gas in the blood of deep-diving Weddell seals (*Leptonychotes weddellii*) in Antarctica [74]. The purpose of these measurements was for Pugh to test out his CO measurement device using opportunistic blood samples from the seals, which were being slaughtered at the time to provide food for the sled dogs that accompanied the men living and working at Scott Base, Antarctica. At the time, there was concern over CO poisoning in Antarctic explorers from using primus stoves as a heat source, to melt snow, or to cook food inside tents [75]. Pugh's findings revealed that the blood from Weddell seals had six times the content of CO compared to nonsmokers and levels that were similar to the heavy smokers living at Scott Base. Similarly, Pugh found levels of bilirubin, another by-product of heme degradation, in the plasma of Weddell seals that were twice those found in the plasma of healthy men. Sjöstrand had recently published his findings of endogenous CO production from heme degradation in 1949 [76], and therefore, it was postulated by Pugh that the high CO levels in the blood of the seals were attributable to the fact that the animals had much higher heme stores in the blood (hemoglobin) and muscle (myoglobin), compared to men. However, Pugh also suggested the high CO content in the blood of the seals could be due to infrequent removal of CO through the lungs due to the natural long-duration breath holds of the animals, or even due to the animals breathing air pockets trapped under the sea ice where they forage, which might contain high concentrations of CO that was possibly produced by microorganisms in the seabed.

Since Pugh's study in 1959, the investigation of endogenous CO in wildlife animals was largely ignored until 2014 when Tift and colleagues measured COHb levels in another deep-diving seal, the northern elephant seal (*Mirounga angustirostris*) [77]. Like the Weddell seals, elephant seals have extremely high concentrations of hemoglobin and myoglobin, along with one of the highest reported mass-specific blood volumes reported in nature (216 ml/kg) [78]. Interestingly, these blood and muscle heme stores are known to increase with the age of the animals, making them an ideal model system to understand the impact of on-board heme stores in relation to endogenous CO production. Tift and colleagues showed that as hemoglobin stores increased with the age of animals, so did the percent of COHb. In adult animals with average hemoglobin concentration of ~25 g/dl, COHb was reported to be as high as 11%. In young pups with hemoglobin concentrations between 15 and 22 g/dl, COHb was still elevated compared to healthy humans, with an average of 7%. To date, the COHb levels in adult elephant seals are the highest ever reported. The only values of COHb due to endogenous CO production that come close are human patients with hemolytic anemia (mean = 5.9%, maximum = 9.7%) [79].

The leading hypothesis for the primary source of the high endogenous CO in these deep-diving animals is their increased heme content in blood (hemoglobin) and muscle (myoglobin). As stated previously, it is well established that the turnover of red blood cells, and subsequent degradation of heme from hemoglobin by HO enzymes, results in approximately 85% of endogenous CO production in healthy humans at sea level. This natural turnover of red blood cells and the heme from hemoglobin results in COHb of less than 1% in healthy humans. Therefore, turnover of the much larger of pool of red blood cells and hemoglobin in deep-diving seals could very well be the source of CO that results in higher COHb levels. To test this, red blood cell life span and hemoglobin turnover would need to be measured. To date, this has never been reported in the literature for any marine mammal species.

Another example of the HO/CO pathway playing a role in cytoprotection during chronic hypoxia exposure can be seen in the Camelidae family. Members of the Camelidae family include four wild species (*Camelus dromedarius*, *Camelus bactrianus*, *Lama guanicoe*, and *Vicugna vicugna*), and two domesticated species: *Lama glama* (llama) and *Lama pacos* (alpaca). The two domesticated species are believed to be hybrids of the guanaco and vicuña, which both evolved to live at high altitudes in South America [80,81]. Some adaptations these animals have that allow them to thrive at high altitude include large hearts and hemoglobin with a high oxygen affinity [82]. Currently, the two domesticated species are used as pack animals for several human populations living at high altitude.

Hypoxic pulmonary hypertension is prevalent in most mammals exposed to chronic hypoxia at altitude [83]. However, camelids living with chronic hypoxia appear to not be susceptible to hypoxic pulmonary hypertension. In 2008, Herrera and colleagues aimed to investigate whether two endogenous gases (CO and NO) produced by the lungs, both known to be vasodilators, could explain how certain camelid species avoid hypoxic pulmonary hypertension [84]. They found that neonatal sheep that underwent gestation at altitude (3600 m, 480 mmHg barometric pressure) developed pulmonary hypertension, while neonatal llamas that underwent gestation at the same altitude did not. Their data showed that neonatal llamas gestated at altitude upregulated pulmonary CO production and HO-1 expression, but did not increase pulmonary NO production. However, the neonatal sheep that underwent gestation at altitude upregulated pulmonary NO production, but significantly reduced pulmonary CO production and HO-1 expression in pulmonary tissues. They concluded that the upregulation of pulmonary NO production in the neonatal sheep at altitude was insufficient to prevent hypoxic pulmonary hypertension. Collectively, the data in deep-diving animals and those that live at altitude suggest that (i) CO may be beneficial in preventing hypertension and likely right heart failure and (ii) these animals also exist under conditions of hypoxia, which is particularly impressive in the diving animals where O_2 is unavailable. How does CO influence such an O_2 tolerance? One distinct possibility is that the ability to generate CO conveys a selective advantage that must involve either increased availability of O_2 from the elevated Hb and red cell numbers, a unique metabolic switch that permits more efficient use of O_2, or an alternative energy source.

Effects of Altitude on Endogenous CO Production

Rats taken from an altitude of 330 to 10 000 or 15 000 ft for six weeks show an increased COHb level from 0.7% to 1.2% or 1.7%, respectively [85]. However, it appears that the same spike in COHb can occur within only 24 h of laboratory rats being exposed to a simulated altitude of 17 000 ft, where the animals also experience a rapid increase in pulmonary HO-1 protein expression and activity [86]. For the next 20 days at that altitude, HO-1 protein expression and activity in pulmonary tissues remained blunted; COHb levels dropped significantly after 72 h at altitude, and then slowly began to rise at day 14, with COHb levels at day 21 resembling levels similar to those first observed during the initial 24-h period. This highlights that the HO/CO pathway can be quickly upregulated in response to acute hypoxia, but can be blunted during chronic hypoxia. Similarly, humans who move from an elevation of 330 to 11 540 ft increase their COHb level from 0.8% to 1.0% after 20 h at altitude [87]. Yet, native Andeans living at altitude for their entire lives experience elevations in their resting COHb levels [72]. As noted earlier, the elevated COHb is thought to occur from increased HO-2 expression in those who adapt to altitude, suggesting a complex regulatory process where CO is increased as O_2 levels drop, which seems inherently contradictory, but may be explained by the ability of CO to drive more effective use of available O_2.

Why Produce CO if Constantly Exposed to Hypoxia?

If heme degradation via HO activity is indeed the leading source of CO production in these groups that are consistently exposed to hypoxia such as those at depth and those at altitude, one intriguing question is why would they sacrifice three molecules of O_2 to produce one molecule of CO? One obvious reason could be linked to the body avoiding the cytotoxicity of free heme [88]. In groups with large heme pools (e.g., marine mammals and certain high-altitude human populations), the turnover of a large pool of heme proteins has the potential to release free heme into circulation. Free heme will aggregate proteins and degrade them into small peptides, and can induce oxidative stress through the formation of reactive lipid peroxides. Similarly, HO-1 is a stress response gene and upregulated in response to inflammation and hypoxia, so why would nature have designed a biological system whereby one is produced at the expense of the other. In other words, CO levels produced at the site of inflammation would presumably further tissue hypoxia. Alternatively, is it possible that CO exerts other signaling functions independent of modulating O_2 availability? If the affinity of CO for heme targets such as PHDs is greater than that for O_2, might CO displace O_2, thus making it more available to support aerobic metabolism, even if transiently?

Physiological Similarities Between CO and Hypoxia

Both CO and O_2, when delivered at appropriate concentrations and at the right time, can protect various organs from different types of insults or injury that are characterized by oxidative stress, inflammation, or mitochondrial dysfunction [61,89]. For example, hypoxia is protective in animal models of IRI [32,90], sepsis [5], infection [91,92], myocardial infarction [93,94], and hypertension [95]. Likewise, CO administration, when given at safe low doses, is also protective against IRI [96], sepsis [97], hemorrhagic shock [98], arteriosclerosis [99], and infection (bacterial and viral) [100–103]. These two gases activate different signaling pathways to elicit unique biological responses, or do they? They also elicit similar protective effects because their signaling pathways involve similar hemoproteins [104].

Interplay Between HIF-1α and HO-1

HIF-1α and HO-1 are closely related as one signaling pathway regulates the other. In vascular and pulmonary smooth muscle cells (SMCs) from rats, exposure to 0% oxygen for up to 48 h resulted in increased HO-1 activity with higher bilirubin and CO production [105]. HIF-1α activation in microvascular endothelial cells also increases HO-1 expression [26]. However, the effect of hypoxia and subsequent HIF activation on HO-1 is tissue dependent because HO-1 mRNA levels are suppressed in human umbilical vein endothelial cells treated with hypoxia [106]. In contrast, HO-1 mRNA expression is increased during hypoxia in vascular SMCs of the lung, liver, heart, and aorta of rats [107]. Further, hypoxia-induced HO-1 was not observed in liver cells that lacked HIF-1 DNA binding activity [107]. Czibik et al. demonstrated that the cardioprotective property of HIF-1α requires HO-1 activity because inhibition in HO-1 activity using the selective pharmacological inhibitor, zinc deuteroporphyrin 2,4-bis-glycol, abrogated HIF-1α-mediated protection of cardiomyocytes from hydrogen peroxide-induced cell death *in vitro* [25].

Chin et al. reported that CO can rapidly stabilize HIF-1α in macrophages within minutes and does so, in part by transiently increasing mitochondria-derived ROS production that contributes to the stabilization of HIF-1α [108]. CO-induced HIF-1α expression increases production of TGF-β in macrophages, cardiomyocytes, and transplanted kidneys [109,110]. These findings show that CO requires HIF-1α to mediate its protective effects as studies in animals show that the benefits of CO are lost in *HIF-1α$^{-/-}$* macrophages. In a model of kidney IRI, CO-induced VEGF expression promotes recovery from IRI, and without HIF-1α, CO fails to increase VEGF production, abrogating the protective effects of CO against kidney IRI [110]. It is important to note that a low dose of CO exposure, at a concentration of 250 ppm, does not create tissue hypoxia [111,112]. Therefore, CO-induced HIF-1α expression is unlikely to be due to diminished O_2 availability [110]. In other words, CO effectively activates HIF even when O_2 is available at adequate physiological concentrations. We refer to this as pseudohypoxia. Such pseudohypoxic preconditioning allows tissues to develop tolerance and protection from injury such as organ transplantation, hemorrhagic shock, and ischemic stroke [113,114]. Reperfusion injury is believed to underlie tissue injury after a hypoxic/ischemic event due to the sudden availability of O_2. One mechanism by which CO protects against reperfusion injury may be its ability to limit ROS generation displacing O_2 within ischemic cells and tissues, thus making it available for aerobic metabolism.

CO in Pulmonary Hypertension

Pulmonary arterial hypertension (PAH) is a disease characterized by vascular remodeling and proliferation of the pulmonary arteries that ultimately causes an increase in vascular resistance, progressive heart failure, and death [115]. HO-1 shows increased activity in rodents that have PAH [116], suggesting that its role is important in the response to stress injury. This is also indicative of the increased production of CO that must occur within the lungs, an interesting contrast to the dogma of optimizing O_2 uptake. Overexpression of HO-1 in epithelial cells of the lungs of transgenic mice is cytoprotective against pulmonary inflammation, vessel hypertrophy, and hypertension caused by hypoxia [117]. The use of CO as a prophylactic molecule for vascular disorders such as PAH is well described [118]. Unlike NO, CO is a relatively poor vasodilator and thus its activity in models of vascular disease such as PAH involves an alternative mode of action, which interestingly involves NO [118].

Zuckerbraun et al. found that 1-h daily exposure to exogenous CO at nontoxic concentrations in rodent models of PAH can substitute for increased expression of HO-1 and in fact reverse established hypoxia-induced PAH pathology, in addition to its prophylactic effects. The mechanism of action of exogenous CO on PAH was found to be dependent on the generation of NO. NO is not only a powerful vasodilator that brings transient relief to those suffering from PAH, but also a potent inducer of HO-1 by endothelial NOS (NOS3/eNOS) [118]. Impaired expression of eNOS as a result of endothelial dysfunction results in vascular remodeling and impaired pulmonary vascular tone and right heart hypertrophy [119,120]. The vascular retro-remodeling of hypertrophic vessels and right heart observed with administration of CO that involved yet another gas, NO, underscores the complex relationships among the gasotransmitters that have now been observed in multiple disease models [121–123]. It is important to emphasize that each gas molecule has bioactive activity that allows them to immediately influence one another. CO, NO, O_2, hydrogen sulfide (H_2S), and to some extent CO_2 all have been shown to modulate one another vis-à-vis the enzymes and gases that generate them and ultimately mediate their generation and/or consumption or reactivity (Figure 7.2). The HOs, NOS, cystathionine β-synthase (CBS), and cystathionine γ-lyase (CSE) all contain heme and are thus targets for CO, NO, and O_2. Each of the gases can regulate their enzyme activity. Mitochondria also contain heme and thus can influence O_2 consumption and are exquisitely sensitive to changes in O_2 levels, which is required for HO-1 activity.

CO and Oxygen in the Vasculature: Their Interplay in Controlling Vasomotor Tone

The relationship between CO and hypoxia is quite complex and can vary by cell type and tissue. One of the shared tissues heavily influenced by CO and O_2 is the vasculature. The blood vessels are sensitive to the concentration of various biological gases in the blood and can adjust their tone accordingly to ensure tissue perfusion and sufficient O_2 delivery. Indeed, the oxygen sensors are located in the carotid body where glomus cells sense O_2 and CO levels and relay signals

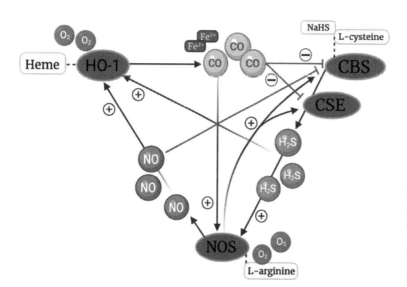

Figure 7.2 Bioactivity and interconnected enzymatic modulation of the gaseous triumvirate: CO, NO, O_2, and H_2S. Diagram depicts the complex regulation of the gases and the enzyme systems that they modulate. Note the complex and overlapping interrelationships. In the presence of O_2, HO-1 efficiently catabolizes heme to generate CO, along with free iron and biliverdin. CO can inhibit CBS and CSE, two enzymes responsible for the production of H_2S [124]. H_2S is involved in the generation of NO through activation of NOS enzymes in the presence of O_2. CO stimulates NOS, which modulate CBS and CSE proteins. Simultaneously, NO production is capable of inhibiting CBS proteins [125]. Both NO and H_2S induce HO-1 activity and generation of CO [126,127].

to the respiratory sensors in the brain through the cranial nerves. Both CO and O_2 can effectively modulate vasomotor tone, which may result from direct effects on the carotid body. When the O_2 level in the bloodstream and in cells deviates from basal physiological levels, vasculature tone may change, either by constricting or by dilating, so as to maintain sufficient O_2 availability and delivery to cells. In hyperoxic conditions, a supraphysiological amount of O_2 leads to increased systemic vascular resistance and vasoconstriction to limit O_2 availability that is toxic to cells [128]. The production of CO by HO-2 in the carotid body may influence glomus cell O_2 sensing and thus drive respiration [129]. In contrast, lack of O_2 under hypoxic conditions would induce systemic vasodilation to increase blood flow and O_2 delivery to cells, except in pulmonary arteries. Pulmonary arterial SMCs constrict in response to alveolar hypoxia. This results in distributing blood flow to better oxygenated parts of the lung and maximize ventilation/perfusion matching and O_2 loading. CO, by binding to the same hemoprotein sensor in the carotid body, may activate or inhibit the sensors in the carotid body depending on the metabolic need. This may explain anemic hypoxia, where the vasculature predominantly becomes sensitive to lack of O_2 availability, but CO, at a low concentration, could influence vasomotor tones in a hypoxia-independent (pseudohypoxia) manner. In general, CO has weak effects on vasodilation of the great vessels, having very little effects on central arterial pressure unlike NO [130]. In contrast, CO is effective as a vasodilator of the microvasculature, dilating small arterioles and capillaries [97]. This difference may reflect competition with other gases, differences in heme sensors, such as guanylate cyclase, or a variation in endothelial cell and vascular smooth muscle phenotype. In contrast to central vasomotor tone, the brain responds to changes in CO and O_2 in a different manner to maximize O_2 delivery. Controlling vascular tone in the brain requires intricate interplay between CO and O_2. In the brain, CO is largely produced by HO-2, which accounts for approximately 80% of total brain HO activity [131,132].

Similar to CO and O_2, H_2S gas can also influence vasomotor tone. It elicits vasodilation in both peripheral and cerebral circulation. Under physiological condition, CO tonically blocks the activity of CBS, synthesis of H_2S, and therefore H_2S-mediated vasodilation. Morikawa et al. showed that hypoxia induces vasodilation in the brain by inhibiting HO-2-mediated production of CO, which then promotes the activity of CBS and therefore vasodilation induced by H_2S [132]. The authors show that vasodilatation caused by hypoxia requires HO-2, as hypoxia-induced cerebral vasodilatory effect is lost in HO-2-null mice [132]. It is important to note that even within the brain, different vascular beds may respond differently to hypoxia, which may involve differential HO-2 expression [133]. For example, cerebral arterioles that are embedded in the parenchymal region of the brain continue to exhibit hypoxia-induced vasodilation, despite the lack of HO-2 expression [132]. In contrast, HO-2-expressing arterioles located on the surface lose the vasodilatory effect caused by hypoxia [132,133]. Interestingly, there are other studies that describe the vasodilatory effects of CO on cerebral vasculature. For example, compared to hypoxic hypoxia, CO-induced hypoxia resulted in greater cerebral blood flow across different species, including humans, dogs, sheep, goats, rabbits, and rodents [134]. Increased cerebral blood flow during CO-induced hypoxia is reported to be due to vasodilation [135]. Perhaps, the varying effects of CO on vasomotor tone depend on various factors, including how CO is generated (endogenously or exogenously), how much CO is present in the bloodstream (anemic hypoxia or pseudohypoxia), and from which isoform of HO it is generated (HO-1 or HO-2). These factors all influence how cells respond to O_2 and how CO modulates HIF-1α to regulate adaptive processes of the host.

CO and O_2 in Mitochondria: Their Interplay in Modulating Mitochondrial Respiration

Oxidative phosphorylation requires O_2 to generate energy because COX in the ETC transfers electrons from cytochrome c to the terminal electron acceptor O_2 and generates water [136,137]. Therefore, COX activity and the overall oxidative phosphorylation process depend on and are sensitive to availability of O_2. Unsurprisingly, acute hypoxia reduces the activity of COX but paradoxically increases ROS production (Figure 7.1) [138]. Under chronic hypoxic conditions, mitochondria adapt to prolonged lack of O_2 availability by replacing specific proteins in the ETC to ensure that mitochondrial composition and function are maintained [27]. Activation of HIF-1α switches cellular metabolism from oxidative phosphorylation to glycolysis and suppresses the activity of the tricarboxylic acid cycle [139]. HIF-1α knockout cells continue to generate ROS during hypoxia, resulting in cell death [140]. CO modulates cellular metabolism by also targeting mitochondrial COX (Figure 7.1). Early studies have shown that CO binds to COX and inhibits the ETC activity, leading to reduced ATP production and ultimately cell death [141–145].

Importantly, this observation and conclusion are made from studies that utilized a high concentration of CO and in many cases isolated mitochondrial preparations [141–144]. In contrast, low concentrations of CO induce biphasic responses that not only are nontoxic, but also preserve cellular function and viability. During the first 30 min of CO exposure, CO elicits a modest increase in mitochondrial ROS production that likely act as signaling molecules versus detrimental radical production [146]. Chronic CO exposure over the next 0.5–24 h leads to increased mitochondrial biogenesis, improving mitochondrial metabolism in a manner that maintains ATP production and preventing mitochondrial membrane permeabilization and thus cell death [108,145–147]. The degree to which CO inhibits COX activity will depend on ambient O_2 concentrations [148]. D'Amico et al. showed that the extent to which CO exerts changes in mitochondrial functions depends on the O_2 level [148]. CO exposure decreased macrophage activation in part by blocking COX activity and mitochondrial respiration, an effect that becomes much more pronounced under hypoxia. One explanation is that ambient O_2 concentration will determine the extent to which mitochondrial ETC is in a reduced state. The lower the O_2 availability, the more the reduced state of the ETC and CO will bind to COX under reducing conditions. Hypoxia significantly favors binding of CO to COX and therefore affects COX activity [148]. This study demonstrates that the inhibitory effect of CO on COX activity is dependent on local O_2 levels. These findings are another example of the interplay between CO and O_2 in regulating mitochondrial function and the impact on cellular metabolism and function [148].

Independent Effects of CO and Oxygen

While the interplay between CO and O_2 has been well described under various physiological settings and in animal models of injury, both hypoxia and CO can elicit cardioprotective effects [65,93–95,99,149,150]. IH conditioning is cardioprotective in several animal models through antihypertensive effects as described in spontaneously hypertensive rats [95] that results in reduced myocardial infarct size [93,94], mortality rates, and the risk of arrhythmia [149,150]. Pretreatment with CO by inhalation at a low dose (250 ppm) reduced intimal hyperplasia formation in mice, rats [99], and pigs [151] due to its anti-inflammatory properties and antiproliferative effects on SMCs. It is possible that in such studies, the protective effects of CO and hypoxia are mediated through activation of other signaling pathways. Leake et al. have shown that inhaled CO increases the release of VEGF in macrophages, a pro-endothelial factor, and high mobility group box 1 (HMGB1), which stimulates endothelial cell proliferation and angiogenesis. Thus, inhaled CO may drive systemic hypoxia, but hypoxia-treated macrophages did not show increased HMGB1 production. Additionally, lipopolysaccharide (LPS)-treated macrophages treated with CO released NO, but hypoxia-treated macrophages produce significantly less NO upon LPS stimulation [152]. Collectively, these data show that the effects of inhaled CO are likely independent of hypoxia [152]. Early studies by Otterbein et al. showed that CO blocked LPS-induced proinflammatory cytokine TNF in macrophages and mice compared to air-treated controls [153]. Treatment with hypoxia did not confer the same anti-inflammatory effects and thus the effects of CO were concluded to be independent of hypoxia and related to specific activation of p38 MAP kinase. Finally, in a bacterial model of sepsis the ability of CO to enhance the innate immune response was also not mediated through hypoxia [100].

Conclusions

CO and O_2 are necessary for life and contribute to both physiological and pathophysiological effects in the organism. Numerous examples including those documented in marine mammals living in deep sea environments, humans living at high altitudes, and various animal models of organ injury support a complex interrelationship between these two diatomic gases. Cells and tissues have become exquisitely sensitive to the levels of these gases and therefore can adapt to changes in their concentrations in a manner that maximizes energy production and optimizes survival. CO and hypoxia are thought to act independently of each other through different signaling pathways, activating HO-1 and HIF-1, respectively. However, studies have shown that not only do they share the same signaling pathways, but they also depend on each other and the activation of their associated enzymatic production to induce desired physiological effects in the host. Also, one should appreciate that cells and tissues would respond profoundly differently to these gases, depending on the activation state of the cells and the environment they find themselves in. More studies are needed in order to understand the biology of CO and O_2 as well as their sister gases NO and H_2S, and their interplay with one another so that their clinical therapeutic values can be properly assessed.

References

1. Solaini, G., Baracca, A., Lenaz, G., and Sgarbi, G. (2010). Hypoxia and mitochondrial oxidative metabolism. *Biochim. Biophys. Acta* 1797: 1171–1177.
2. Colgan, S.P. and Taylor, C.T. (2010). Hypoxia: an alarm signal during intestinal inflammation. *Nat. Rev. Gastroenterol. Hepatol.* 7: 281–287.
3. Hong, W.X., Hu, M.S., Esquivel, M., Liang, G.Y., Rennert, R.C., McArdle, A., Paik, K.J., Duscher, D., Gurtner, G.C., Lorenz, H.P., and Longaker, M.T. (2014). The role of hypoxia-inducible factor in wound healing. *Adv. Wound Care (New Rochelle)* 3: 390–399.
4. Ohta, A. (2018). Oxygen-dependent regulation of immune checkpoint mechanisms. *Int. Immunol.* 30: 335–343.
5. Thompson, A.A.R., Dickinson, R.S., Murphy, F., Thomson, J.P., Marriott, H.M., Tavares, A., Willson, J., Williams, L., Lewis, A., Mirchandani, A., Dos Santos Coelho, P., Doherty, C., Ryan, E., Watts, E., Morton, N.M., Forbes, S., Stimson, R.H., Hameed, A.G., Arnold, N., Preston, J.A., Lawrie, A., Finisguerra, V., Mazzone, M., Sadiku, P., Goveia, J., Taverna, F., Carmeliet, P., Foster, S.J., Chilvers, E.R., Cowburn, A.S., Dockrell, D.H., Johnson, R.S., Meehan, R.R., Whyte, M.K.B., and Walmsley, S.R. (2017). Hypoxia determines survival outcomes of bacterial infection through HIF-1alpha dependent re-programming of leukocyte metabolism. *Sci. Immunol.* 2.
6. Eltzschig, H.K., Köhler, D., Eckle, T., Kong, T., Robson, S.C., and Colgan, S.P. (2009). Central role of Sp1-regulated CD39 in hypoxia/ischemia protection. *Blood* 113: 224–232.
7. Rosenberger, P., Schwab, J.M., Mirakaj, V., Masekowsky, E., Mager, A., Morote-Garcia, J.C., Unertl, K., and Eltzschig, H.K. (2009). Hypoxia-inducible factor-dependent induction of netrin-1 dampens inflammation caused by hypoxia. *Nat. Immunol.* 10: 195–202.
8. Carreau, A., El Hafny-Rahbi, B., Matejuk, A., Grillon, C., and Kieda, C. (2011). Why is the partial oxygen pressure of human tissues a crucial parameter? Small molecules and hypoxia. *J. Cell. Mol. Med.* 15: 1239–1253.
9. Horncastle, E. and Lumb, A.B. (2019). Hyperoxia in anaesthesia and intensive care. *BJA Educ.* 19: 176–182.
10. Smith, J.L. (1899). The pathological effects due to increase of oxygen tension in the air breathed. *J. Physiol.* 24: 19–35.
11. Paine, J.R., Lynn, D., and Keys, A. (1941). Observations on the effects of the prolonged administration of high oxygen concentration to dogs. *J. Thorac. Cardiovasc. Surg.* 11: 151–168.
12. Sato, T., Paquet-Fifield, S., Harris, N.C., Roufail, S., Turner, D.J., Yuan, Y., Zhang, Y.-F., Fox, S.B., Hibbs, M.L., Wilkinson-Berka, J.L., Williams, R.A., Stacker, S.A., Sly, P.D., and Achen, M.G. (2016). VEGF-D promotes pulmonary oedema in hyperoxic acute lung injury. *J. Pathol.* 239: 152–161.
13. Fox, R.B., Hoidal, J.R., Brown, D.M., and Repine, J.E. (1981). Pulmonary inflammation due to oxygen toxicity: involvement of chemotactic factors and polymorphonuclear leukocytes. *Am. Rev. Respir. Dis.* 123: 521–523.
14. Dean, J.B., Mulkey, D.K., Garcia, A.J., 3rd, Putnam, R.W., and Henderson, R.A., 3rd. (2003). Neuronal sensitivity to hyperoxia, hypercapnia, and inert gases at hyperbaric pressures. *J. Appl. Physiol. (1985)* 95: 883–909.
15. Waheed, S., D'Angio, C.T., Wagner, C.L., Madtes, D.K., Finkelstein, J.N., Paxhia, A., and Ryan, R.M. (2002). Transforming growth factor alpha (TGF(alpha)) is increased during hyperoxia and fibrosis. *Exp. Lung Res.* 28: 361–372.
16. Jones, M.W., Brett, K., Han, N., and Wyatt, H.A. (2020). Hyperbaric Physics. Treasure Island (FL): StatPearls Publishing.
17. Dejmek, J., Kohoutová, M., Kripnerová, M., Čedíková, M., Tůma, Z., Babuška, V., Bolek, L., and Kuncová, J. (2018). Repeated exposure to hyperbaric hyperoxia affects mitochondrial functions of the lung fibroblasts. *Physiol. Res.* 67: S633–S643.
18. Cooper, J.S., Phuyal, P., and Shah, N. (2020). Oxygen toxicity. In: StatPearls. Treasure Island (FL): StatPearls Publishing.
19. Peers, C. and Kemp, P.J. (2001). Acute oxygen sensing: diverse but convergent mechanisms in airway and arterial chemoreceptors. *Respir. Res.* 2: 145–149.
20. Waypa, G.B. and Schumacker, P.T. (2010). Hypoxia-induced changes in pulmonary and systemic vascular resistance: where is the O2 sensor? *Respir. Physiol. Neurobiol.* 174: 201–211.
21. Fuhrmann, D.C., Olesch, C., Kurrle, N., Schnütgen, F., Zukunft, S., Fleming, I., and Brüne, B. (2019). Chronic hypoxia enhances β-oxidation-dependent electron transport via electron transferring flavoproteins. *Cells* 8.
22. Brahimi-Horn, M.C. and Pouysségur, J. (2007). Harnessing the hypoxia-inducible factor in cancer and ischemic disease. *Biochem. Pharmacol.* 73: 450–457.
23. Lee, J.-W., Bae, S.-H., Jeong, J.-W., Kim, S.-H., and Kim, K.-W. (2004). Hypoxia-inducible factor (HIF-1) alpha: its protein stability and biological functions. *Exp. Mol. Med.* 36: 1–12.
24. Bernhardt, W.M., Warnecke, C., Willam, C., Tanaka, T., Wiesener, M.S., and Eckardt, K.-U. (2007). Organ

protection by hypoxia and hypoxia-inducible factors. *Methods Enzymol.* 435: 221–245.

25. Czibik, G., Sagave, J., Martinov, V., Ishaq, B., Sohl, M., Sefland, I., Carlsen, H., Farnebo, F., Blomhoff, R., and Valen, G. (2009). Cardioprotection by hypoxia-inducible factor 1 alpha transfection in skeletal muscle is dependent on haem oxygenase activity in mice. *Cardiovasc. Res.* 82: 107–114.

26. Ockaili, R., Natarajan, R., Salloum, F., Fisher, B.J., Jones, D., Fowler, A.A., 3rd, and Kukreja, R.C. (2005). HIF-1 activation attenuates postischemic myocardial injury: role for heme oxygenase-1 in modulating microvascular chemokine generation. *Am. J. Physiol. Heart Circ. Physiol.* 289: H542–548.

27. Fukuda, R., Zhang, H., Kim, J.-W., Shimoda, L., Dang, C.V., and Semenza, G.L. (2007). HIF-1 regulates cytochrome oxidase subunits to optimize efficiency of respiration in hypoxic cells. *Cell* 129: 111–122.

28. Lee, J.W., Ko, J., Ju, C., and Eltzschig, H.K. (2019). Hypoxia signaling in human diseases and therapeutic targets. *Exp. Mol. Med.* 51: 1–13.

29. Culver, C., Sundqvist, A., Mudie, S., Melvin, A., Xirodimas, D., and Rocha, S. (2010). Mechanism of hypoxia-induced NF-kappaB. *Mol. Cell. Biol.* 30: 4901–4921.

30. Gabryelska, A., Łukasik, Z.M., Makowska, J.S., and Białasiewicz, P. (2018). Obstructive sleep apnea: From intermittent hypoxia to cardiovascular complications via blood platelets. *Front. Neurol.* 9: 635.

31. Aquino-Gálvez, A., González-Ávila, G., Jiménez-Sánchez, L.L., Maldonado-Martínez, H.A., Cisneros, J., Toscano-Marquez, F., Castillejos-López, M., Torres-Espíndola, L.M., Velázquez-Cruz, R., Rodríguez, V.H.O., Flores-Soto, E., Solís-Chagoyán, H., Cabello, C., Zúñiga, J., and Romero, Y. (2019). Dysregulated expression of hypoxia-inducible factors augments myofibroblasts differentiation in idiopathic pulmonary fibrosis. *Respir. Res.* 20: 130.

32. Guo, H.-C., Zhang, Z., Zhang, L.-N., Xiong, C., Feng, C., Liu, Q., Liu, X., Shi, X.-L., and Wang, Y.-L. (2009). Chronic intermittent hypobaric hypoxia protects the heart against ischemia/reperfusion injury through upregulation of antioxidant enzymes in adult guinea pigs. *Acta Pharmacol. Sin.* 30: 947–955.

33. Guarnieri, C., Flamigni, F., and Caldarera, C.M. (1980). Role of oxygen in the cellular damage induced by re-oxygenation of hypoxic heart. *J. Mol. Cell Cardiol.* 12: 797–808.

34. Granger, D.N., Rutili, G., and McCord, J.M. (1981). Superoxide radicals in feline intestinal ischemia. *Gastroenterology* 81: 22–29.

35. Turer, A.T. and Hill, J.A. (2010). Pathogenesis of myocardial ischemia-reperfusion injury and rationale for therapy. *Am. J. Cardiol.* 106: 360–368.

36. Liu, X., Wei, J., Peng, D.H., Layne, M.D., and Yet, S.F. (2005). Absence of heme oxygenase-1 exacerbates myocardial ischemia/reperfusion injury in diabetic mice. *Diabetes* 54: 778–784.

37. Hultén, L.M. and Levin, M. (2009). The role of hypoxia in atherosclerosis. *Curr. Opin. Lipidol.* 20: 409–414.

38. Savransky, V., Nanayakkara, A., Li, J., Bevans, S., Smith, P.L., Rodriguez, A., and Polotsky, V.Y. (2007). Chronic intermittent hypoxia induces atherosclerosis. *Am. J. Respir. Crit. Care Med.* 175: 1290–1297.

39. Gangwar, A., Paul, S., Ahmad, Y., and Bhargava, K. (2020). Intermittent hypoxia modulates redox homeostasis, lipid metabolism associated inflammatory processes and redox post-translational modifications: benefits at high altitude. *Sci. Rep.* 10: 7899.

40. Serebrovskaya, T.V., Manukhina, E.B., Smith, M.L., Downey, H.F., and Mallet, R.T. (2008). Intermittent hypoxia: cause of or therapy for systemic hypertension? *Exp. Biol. Med. (Maywood)* 233: 627–650.

41. Urdampilleta, A., González-Muniesa, P., Portillo, M.P., and Martínez, J.A. (2012). Usefulness of combining intermittent hypoxia and physical exercise in the treatment of obesity. *J. Physiol. Biochem.* 68: 289–304.

42. Guner, I., Uzun, D.D., Yaman, M.O., Genc, H., Gelisgen, R., Korkmaz, G.G., Hallac, M., Yelmen, N., Sahin, G., Karter, Y., and Simsek, G. (2013). The effect of chronic long-term intermittent hypobaric hypoxia on bone mineral density in rats: role of nitric oxide. *Biol. Trace Elem. Res.* 154: 262–267.

43. Wang, Z.-H., Chen, Y.-X., Zhang, C.-M., Wu, L., Yu, Z., Cai, X.-L., Guan, Y., Zhou, Z.-N., and Yang, H.-T. (2011). Intermittent hypobaric hypoxia improves postischemic recovery of myocardial contractile function via redox signaling during early reperfusion. *Am. J. Physiol. Heart Circ. Physiol.* 301: H1695–1705.

44. Navarrete-Opazo, A. and Mitchell, G.S. (2014). Therapeutic potential of intermittent hypoxia: a matter of dose. *Am. J. Physiol. Regul. Integr. Comp. Physiol.* 307: R1181–1197.

45. Lyamina, N.P., Lyamina, S.V., Senchiknin, V.N., Mallet, R.T., Downey, H.F., and Manukhina, E.B. (2011). Normobaric hypoxia conditioning reduces blood pressure and normalizes nitric oxide synthesis in patients with arterial hypertension. *J. Hypertens.* 29: 2265–2272.

46. Moncada, S. and Higgs, E.A. (2006). The discovery of nitric oxide and its role in vascular biology. *Br. J. Pharmacol.* 147 (Suppl 1): S193–201.

47. Beall, C.M., Laskowski, D., and Erzurum, S.C. (2012). Nitric oxide in adaptation to altitude. *Free Radic. Biol. Med.* 52: 1123–1134.

48. Beall, C.M., Laskowski, D., Strohl, K.P., Soria, R., Villena, M., Vargas, E., Alarcon, A.M., Gonzales, C., and Erzurum, S.C. (2001). Pulmonary nitric oxide in mountain dwellers. *Nature* 414: 411–412.

49. Brugniaux, J.V., Pialoux, V., Foster, G.E., Duggan, C.T., Eliasziw, M., Hanly, P.J., and Poulin, M.J. (2011). Effects of intermittent hypoxia on erythropoietin, soluble erythropoietin receptor and ventilation in humans. *Eur. Respir. J.* 37: 880–887.

50. Mancini, D.M., Katz, S.D., Lang, C.C., LaManca, J., Hudaihed, A., and Androne, A.S. (2003). Effect of erythropoietin on exercise capacity in patients with moderate to severe chronic heart failure. *Circulation* 107: 294–299.

51. Dale, E.A., Ben Mabrouk, F., and Mitchell, G.S. (2014). Unexpected benefits of intermittent hypoxia: enhanced respiratory and nonrespiratory motor function. *Physiology (Bethesda)* 29: 39–48.

52. Lovett-Barr, M.R., Satriotomo, I., Muir, G.D., Wilkerson, J.E.R., Hoffman, M.S., Vinit, S., and Mitchell, G.S. (2012). Repetitive intermittent hypoxia induces respiratory and somatic motor recovery after chronic cervical spinal injury. *J. Neurosci.* 32: 3591–3600.

53. Streeter, K.A., Sunshine, M.D., Patel, S., Gonzalez-Rothi, E.J., Reier, P.J., Baekey, D.M., and Fuller, D.D. (2017). Intermittent hypoxia enhances functional connectivity of midcervical spinal interneurons. *J. Neurosci.* 37: 8349–8362.

54. Satriotomo, I., Nichols, N.L., Dale, E.A., Emery, A.T., Dahlberg, J.M., and Mitchell, G.S. (2016). Repetitive acute intermittent hypoxia increases growth/neurotrophic factor expression in non-respiratory motor neurons. *Neuroscience* 322: 479–488.

55. Qiao, Y., Liu, Z., Yan, X., and Luo, C. (2015). Effect of intermittent hypoxia on neuro-functional recovery post brain ischemia in mice. *J. Mol. Neurosci.* 55: 923–930.

56. Li, S., Hafeez, A., Noorulla, F., Geng, X., Shao, G., Ren, C., Lu, G., Zhao, H., Ding, Y., and Ji, X. (2017). Preconditioning in neuroprotection: from hypoxia to ischemia. *Prog. Neurobiol.* 157: 79–91.

57. Jiang, B.H., Semenza, G.L., Bauer, C., and Marti, H.H. (1996). Hypoxia-inducible factor 1 levels vary exponentially over a physiologically relevant range of O2 tension. *Am. J. Physiol.* 271: C1172–80.

58. Yu, A.Y., Frid, M.G., Shimoda, L.A., Wiener, C.M., Stenmark, K., and Semenza, G.L. (1998). Temporal, spatial, and oxygen-regulated expression of hypoxia-inducible factor-1 in the lung. *Am. J. Physiol.* 275: L818–26.

59. Zock, J.P. (1990). Carbon monoxide binding in a model of hemoglobin differs between the T and the R conformation. *Adv. Exp. Med. Biol.* 277: 199–207.

60. Coburn, R.F., Blakemore, W.S., and Forster, R.E. (1963). Endogenous carbon monoxide production in man. *J. Clin. Invest.* 42: 1172–1178.

61. Motterlini, R. and Otterbein, L.E. (2010). The therapeutic potential of carbon monoxide. *Nat. Rev. Drug Discov.* 9: 728–743.

62. Johnson, R.A., Kozma, F., and Colombari, E. (1999). Carbon monoxide: from toxin to endogenous modulator of cardiovascular functions. *Braz. J. Med. Biol. Res.* 32: 1–14.

63. Ryter, S.W. and Otterbein, L.E. (2004). Carbon monoxide in biology and medicine. *Bioessays* 26: 270–280.

64. Poss, K.D. and Tonegawa, S. (1997). Reduced stress defense in heme oxygenase 1-deficient cells. *Proc. Natl. Acad. Sci. U. S. A.* 94: 10925–10930.

65. Otterbein, L.E., Foresti, R., and Motterlini, R. (2016). Heme oxygenase-1 and carbon monoxide in the heart: The balancing act between danger signaling and pro-survival. *Circ. Res.* 118: 1940–1959.

66. Kawashima, A., Oda, Y., Yachie, A., Koizumi, S., and Nakanishi, I. (2002). Heme oxygenase-1 deficiency: the first autopsy case. *Hum. Pathol.* 33: 125–130.

67. True, A.L., Olive, M., Boehm, M., San, H., Westrick, R.J., Raghavachari, N., Xu, X., Lynn, E.G., Sack, M.N., Munson, P.J., Gladwin, M.T., and Nabel, E.G. (2007). Heme oxygenase-1 deficiency accelerates formation of arterial thrombosis through oxidative damage to the endothelium, which is rescued by inhaled carbon monoxide. *Circ. Res.* 101: 893–901.

68. Chung, S.W., Liu, X., Macias, A.A., Baron, R.M., and Perrella, M.A. (2008). Heme oxygenase-1-derived carbon monoxide enhances the host defense response to microbial sepsis in mice. *J. Clin. Invest.* 118: 239–247.

69. Chang, E.F., Wong, R.J., Vreman, H.J., Igarashi, T., Galo, E., Sharp, F.R., Stevenson, D.K., and Noble-Haeusslein, L.J. (2003). Heme oxygenase-2 protects against lipid peroxidation-mediated cell loss and impaired motor recovery after traumatic brain injury. *J. Neurosci.* 23: 3689–3696.

70. Goodman, A.I., Chander, P.N., Rezzani, R., Schwartzman, M.L., Regan, R.F., Rodella, L., Turksaven, S., Lianos, E.A., Dennery, P.A., and Abraham, N.G. (2006). Heme oxygenase-2 deficiency contributes to diabetes-mediated increase in superoxide anion and renal dysfunction. *J. Am. Soc. Nephrol.* 17: 1073–1081.

71. Zakhary, R., Poss, K.D., Jaffrey, S.R., Ferris, C.D., Tonegawa, S., and Snyder, S.H. (1997). Targeted gene deletion of heme oxygenase 2 reveals neural role for carbon monoxide. *Proc. Natl. Acad. Sci. U. S. A.* 94: 14848–14853.

72. Tift, M.S., Alves De Souza, R.W., Weber, J., Heinrich, E.C., Villafuerte, F.C., Malhotra, A., Otterbein, L.E.,

and Simonson, T.S. (2020). Adaptive potential of the heme oxygenase/carbon monoxide pathway during hypoxia. *Front. Physiol.* 11: 886.

73. Wuren, T., Simonson, T.S., Qin, G., Xing, J., Huff, C.D., Witherspoon, D.J., Jorde, L.B., and Ge, R.L. (2014). Shared and unique signals of high-altitude adaptation in geographically distinct Tibetan populations. *PLoS One* 9: e88252.

74. Pugh, L.G. (1959). Carbon monoxide content of the blood and other observations on Weddell seals. *Nature* 183: 74–76.

75. Pugh, L.G. (1959). Carbon monoxide hazard in Antarctica. *Br. Med. J.* 1: 192–196.

76. Sjöstrand, T. (1949). Endogenous formation of carbon monoxide in man. *Nature* 164: 580.

77. Tift, M.S., Ponganis, P.J., and Crocker, D.E. (2014). Elevated carboxyhemoglobin in a marine mammal, the northern elephant seal. *J. Exp. Biol.* 217: 1752–1757.

78. Hassrick, J.L., Crocker, D.E., Teutschel, N.M., McDonald, B.I., Robinson, P.W., Simmons, S.E., and Costa, D.P. (2010). Condition and mass impact oxygen stores and dive duration in adult female northern elephant seals. *J. Exp. Biol.* 213: 585–592.

79. Hampson, N.B. (2007). Carboxyhemoglobin elevation due to hemolytic anemia. *J. Emerg. Med.* 33: 17–19.

80. Wheeler, J.C. (1995). Evolution and present situation of the South American Camelidae. *Biol. J. Linn. Soc.* 54: 271–295.

81. Stanley, H.F., Kadwell, M., and Wheeler, J.C. (1994). Molecular evolution of the family Camelidae: a mitochondrial DNA study. *Proc. Biol. Sci.* 256: 1–6.

82. Jürgens, K.D., Pietschmann, M., Yamaguchi, K., and Kleinschmidt, T. (1988). Oxygen binding properties, capillary densities and heart weights in high altitude camelids. *J. Comp. Physiol. B* 158: 469–477.

83. Penaloza, D. and Arias-Stella, J. (2007). The heart and pulmonary circulation at high altitudes: healthy highlanders and chronic mountain sickness. *Circulation* 115: 1132–1146.

84. McGrath, J.J. (1992). Effects of altitude on endogenous carboxyhemoglobin levels. *J. Toxicol. Environ. Health* 35: 127–133.

85. Carraway, M.S., Ghio, A.J., Carter, J.D., and Piantadosi, C.A. (2000). Expression of heme oxygenase-1 in the lung in chronic hypoxia. *Am. J. Physiol. Lung Cell. Mol. Physiol.* 278: L806–812.

86. McGrath, J.J., Schreck, R.M., and Lee, P.S. (1993). Carboxyhemoglobin levels in humans: Effects of altitude. *Inhal. Toxicol.* 5: 241–249.

87. Kumar, S. and Bandyopadhyay, U. (2005). Free heme toxicity and its detoxification systems in human. *Toxicol. Lett.* 157: 175–188.

88. Verges, S., Chacaroun, S., Godin-Ribuot, D., and Baillieul, S. (2015). Hypoxic conditioning as a new therapeutic modality. *Front. Pediatr.* 3: 58.

89. Chen, L., Lu, X.-Y., Li, J., Fu, J.-D., Zhou, Z.-N., and Yang, H.-T. (2006). Intermittent hypoxia protects cardiomyocytes against ischemia-reperfusion injury-induced alterations in Ca2+ homeostasis and contraction via the sarcoplasmic reticulum and Na+/Ca2+ exchange mechanisms. *Am. J. Physiol. Cell Physiol.* 290: C1221–1229.

90. Schaible, B., McClean, S., Selfridge, A., Broquet, A., Asehnoune, K., Taylor, C.T., and Schaffer, K. (2013). Hypoxia modulates infection of epithelial cells by Pseudomonas aeruginosa. *PLoS One* 8: e56491.

91. Schaible, B., Taylor, C.T., and Schaffer, K. (2012). Hypoxia increases antibiotic resistance in Pseudomonas aeruginosa through altering the composition of multidrug efflux pumps. *Antimicrob. Agents Chemother.* 56: 2114–2118.

92. Neckár, J., Ostádal, B., and Kolár, F. (2004). Myocardial infarct size-limiting effect of chronic hypoxia persists for five weeks of normoxic recovery. *Physiol. Res.* 53: 621–628.

93. Zong, P., Setty, S., Sun, W., Martinez, R., Tune, J.D., Ehrenburg, I.V., Tkatchouk, E.N., Mallet, R.T., and Downey, H.F. (2004). Intermittent hypoxic training protects canine myocardium from infarction. *Exp. Biol. Med. (Maywood)* 229: 806–812.

94. Manukhina, E.B., Jasti, D., Vanin, A.F., and Downey, H.F. (2011). Intermittent hypoxia conditioning prevents endothelial dysfunction and improves nitric oxide storage in spontaneously hypertensive rats. *Exp. Biol. Med. (Maywood)* 236: 867–873.

95. Ozaki, K.S., Kimura, S., and Murase, N. (2012). Use of carbon monoxide in minimizing ischemia/reperfusion injury in transplantation. *Transplant. Rev. (Orlando)* 26: 125–139.

96. Shen, W.-C., Wang, X., Qin, W.-T., Qiu, X.-F., and Sun, B.-W. (2014). Exogenous carbon monoxide suppresses Escherichia coli vitality and improves survival in an Escherichia coli-induced murine sepsis model. *Acta Pharmacol. Sin.* 35: 1566–1576.

97. Nassour, I., Kautza, B., Rubin, M., Escobar, D., Luciano, J., Loughran, P., Gomez, H., Scott, J., Gallo, D., Brumfield, J., Otterbein, L.E., and Zuckerbraun, B.S. (2015). Carbon monoxide protects against hemorrhagic shock and resuscitation-induced microcirculatory injury and tissue injury. *Shock* 43: 166–171.

98. Otterbein, L.E., Zuckerbraun, B.S., Haga, M., Liu, F., Song, R., Usheva, A., Stachulak, C., Bodyak, N., Smith, R.N., Csizmadia, E., Tyagi, S., Akamatsu, Y., Flavell, R.J., Billiar, T.R., Tzeng, E., Bach, F.H., Choi, A.M., and Soares, M.P. (2003). Carbon monoxide

suppresses arteriosclerotic lesions associated with chronic graft rejection and with balloon injury. *Nat. Med.* 9: 183–190.

99. Wegiel, B., Larsen, R., Gallo, D., Chin, B.Y., Harris, C., Mannam, P., Kaczmarek, E., Lee, P.J., Zuckerbraun, B.S., Flavell, R., Soares, M.P., and Otterbein, L.E. (2014). Macrophages sense and kill bacteria through carbon monoxide-dependent inflammasome activation. *J. Clin. Invest.* 124: 4926–4940.

100. Nobre, L.S., Seixas, J.D., Romão, C.C., and Saraiva, L.M. (2007). Antimicrobial action of carbon monoxide-releasing compounds. *Antimicrob. Agents Chemother.* 51: 4303–4307.

101. Ma, Z., Pu, F., Zhang, X., Yan, Y., Zhao, L., Zhang, A., Li, N., Zhou, E.-M., and Xiao, S. (2017). Carbon monoxide and biliverdin suppress bovine viral diarrhoea virus replication. *J. Gen. Virol.* 98: 2982–2992.

102. Zhang, A., Zhao, L., Li, N., Duan, H., Liu, H., Pu, F., Zhang, G., Zhou, E.-M., and Xiao, S. (2017). Carbon monoxide inhibits porcine reproductive and respiratory syndrome virus replication by the cyclic GMP/Protein kinase G and NF-κB signaling pathway. *J. Virol.* 91.

103. Motterlini, R. and Foresti, R. (2017). Biological signaling by carbon monoxide and carbon monoxide-releasing molecules. *Am. J. Physiol. Cell Physiol.* 312: C302–C313.

104. Morita, T., Perrella, M.A., Lee, M.E., and Kourembanas, S. (1995). Smooth muscle cell-derived carbon monoxide is a regulator of vascular cGMP. *Proc. Natl. Acad. Sci. U. S. A.* 92: 1475–1479.

105. Nakayama, M., Takahashi, K., Kitamuro, T., Yasumoto, K., Katayose, D., Shirato, K., Fujii-Kuriyama, Y., and Shibahara, S. (2000). Repression of heme oxygenase-1 by hypoxia in vascular endothelial cells. *Biochem. Biophys. Res. Commun.* 271: 665–671.

106. Lee, P.J., Jiang, B.H., Chin, B.Y., Iyer, N.V., Alam, J., Semenza, G.L., and Choi, A.M. (1997). Hypoxia-inducible factor-1 mediates transcriptional activation of the heme oxygenase-1 gene in response to hypoxia. *J. Biol. Chem.* 272: 5375–5381.

107. Chin, B.Y., Jiang, G., Wegiel, B., Wang, H.J., Macdonald, T., Zhang, X.C., Gallo, D., Cszimadia, E., Bach, F.H., Lee, P.J., and Otterbein, L.E. (2007). Hypoxia-inducible factor 1alpha stabilization by carbon monoxide results in cytoprotective preconditioning. *Proc. Natl. Acad. Sci. U. S. A.* 104: 5109–5114.

108. Lakkisto, P., Kytö, V., Forsten, H., Siren, J.-M., Segersvärd, H., Voipio-Pulkki, L.-M., Laine, M., Pulkki, K., and Tikkanen, I. (2010). Heme oxygenase-1 and carbon monoxide promote neovascularization after myocardial infarction by modulating the expression of HIF-1alpha, SDF-1alpha and VEGF-B. *Eur. J. Pharmacol.* 635: 156–164.

109. Faleo, G., Neto, J.S., Kohmoto, J., Tomiyama, K., Shimizu, H., Takahashi, T., Wang, Y., Sugimoto, R., Choi, A.M.K., Stolz, D.B., Carrieri, G., McCurry, K.R., Murase, N., and Nakao, A. (2008). Carbon monoxide ameliorates renal cold ischemia-reperfusion injury with an upregulation of vascular endothelial growth factor by activation of hypoxia-inducible factor. *Transplantation* 85: 1833–1840.

110. Meilin, S., Rogatsky, G.G., Thom, S.R., Zarchin, N., Guggenheimer-Furman, E., and Mayevsky, A. (1996). Effects of carbon monoxide on the brain may be mediated by nitric oxide. *J. Appl. Physiol. (1985)* 81: 1078–1083.

111. Resch, H., Zawinka, C., Weigert, G., Schmetterer, L., and Garhöfer, G. (2005). Inhaled carbon monoxide increases retinal and choroidal blood flow in healthy humans. *Invest. Ophthalmol. Vis. Sci.* 46: 4275–4280.

112. Lu, G.-W. and Shao, G. (2014). Hypoxic preconditioning: effect, mechanism and clinical implication (Part 1). *Zhongguo Ying Yong Sheng Li Xue Za Zhi* 30: 489–501.

113. Francis, K.R. and Wei, L. (2010). Human embryonic stem cell neural differentiation and enhanced cell survival promoted by hypoxic preconditioning. *Cell Death Dis.* 1: e22.

114. Mandegar, M., Fung, Y.-C.B., Huang, W., Remillard, C.V., Rubin, L.J., and Yuan, J.X.J. (2004). Cellular and molecular mechanisms of pulmonary vascular remodeling: role in the development of pulmonary hypertension. *Microvasc. Res.* 68: 75–103.

115. Voelkel, N.F., Bogaard, H.J., Al Husseini, A., Farkas, L., Gomez-Arroyo, J., and Natarajan, R. (2013). Antioxidants for the treatment of patients with severe angioproliferative pulmonary hypertension? *Antioxid. Redox Signal.* 18: 1810–1817.

116. Minamino, T., Christou, H., Hsieh, C.M., Liu, Y., Dhawan, V., Abraham, N.G., Perrella, M.A., Mitsialis, S.A., and Kourembanas, S. (2001). Targeted expression of heme oxygenase-1 prevents the pulmonary inflammatory and vascular responses to hypoxia. *Proc. Natl. Acad. Sci. U. S. A.* 98: 8798–8803.

117. Zuckerbraun, B.S., Chin, B.Y., Wegiel, B., Billiar, T.R., Czsimadia, E., Rao, J., Shimoda, L., Ifedigbo, E., Kanno, S., and Otterbein, L.E. (2006). Carbon monoxide reverses established pulmonary hypertension. *J. Exp. Med.* 203: 2109–2119.

118. Giaid, A. and Saleh, D. (1995). Reduced expression of endothelial nitric oxide synthase in the lungs of

patients with pulmonary hypertension. *N. Engl. J. Med.* 333: 214–221.

119. Jeffery, T.K. and Morrell, N.W. (2002). Molecular and cellular basis of pulmonary vascular remodeling in pulmonary hypertension. *Prog. Cardiovasc. Dis.* 45: 173–202.

120. Zuckerbraun, B.S., Billiar, T.R., Otterbein, S.L., Kim, P.K.M., Liu, F., Choi, A.M.K., Bach, F.H., and Otterbein, L.E. (2003). Carbon monoxide protects against liver failure through nitric oxide-induced heme oxygenase 1. *J. Exp. Med.* 198: 1707–1716.

121. Ushiyama, M., Morita, T., and Katayama, S. (2002). Carbon monoxide regulates blood pressure cooperatively with nitric oxide in hypertensive rats. *Heart Vessels* 16: 189–195.

122. Hervera, A., Gou, G., Leánez, S., and Pol, O. (2013). Effects of treatment with a carbon monoxide-releasing molecule and a heme oxygenase 1 inducer in the antinociceptive effects of morphine in different models of acute and chronic pain in mice. *Psychopharmacology (Berl)* 228: 463–477.

123. Yuan, G., Vasavda, C., Peng, Y.J., Makarenko, V.V., Raghuraman, G., Nanduri, J., Gadalla, M.M., Semenza, G.L., Kumar, G.K., Snyder, S.H., and Prabhakar, N.R. (2015). Protein kinase G-regulated production of H2S governs oxygen sensing. *Sci. Signal.* 8: ra37.

124. Taoka, S. and Banerjee, R. (2001). Characterization of NO binding to human cystathionine beta-synthase: possible implications of the effects of CO and NO binding to the human enzyme. *J. Inorg. Biochem.* 87: 245–251.

125. Farrugia, G. and Szurszewski, J.H. (2014). Carbon monoxide, hydrogen sulfide, and nitric oxide as signaling molecules in the gastrointestinal tract. *Gastroenterology* 147: 303–313.

126. Liu, X.M., Peyton, K.J., Ensenat, D., Wang, H., Hannink, M., Alam, J., and Durante, W. (2007). Nitric oxide stimulates heme oxygenase-1 gene transcription via the Nrf2/ARE complex to promote vascular smooth muscle cell survival. *Cardiovasc. Res.* 75: 381–389.

127. Smit, B., Smulders, Y.M., Eringa, E.C., Oudemans-van Straaten, H.M., Girbes, A.R.J., Wever, K.E., Hooijmans, C.R., and Spoelstra-de Man, A.M.E. (2018). Effects of hyperoxia on vascular tone in animal models: systematic review and meta-analysis. *Crit. Care* 22: 189.

128. Prabhakar, N.R., Dinerman, J.L., Agani, F.H., and Snyder, S.H. (1995). Carbon monoxide: a role in carotid body chemoreception. *Proc. Natl. Acad. Sci. U. S. A.* 92: 1994–1997.

129. Evgenov, O.V., Ichinose, F., Evgenov, N.V., Gnoth, M.J., Falkowski, G.E., Chang, Y., Bloch, K.D., and Zapol, W.M. (2004). Soluble guanylate cyclase activator reverses acute pulmonary hypertension and augments the pulmonary vasodilator response to inhaled nitric oxide in awake lambs. *Circulation* 110: 2253–2259.

130. Ishikawa, M., Kajimura, M., Adachi, T., Maruyama, K., Makino, N., Goda, N., Yamaguchi, T., Sekizuka, E., and Suematsu, M. (2005). Carbon monoxide from heme oxygenase-2 Is a tonic regulator against NO-dependent vasodilatation in the adult rat cerebral microcirculation. *Circ. Res.* 97: e104–114.

131. Morikawa, T., Kajimura, M., Nakamura, T., Hishiki, T., Nakanishi, T., Yukutake, Y., Nagahata, Y., Ishikawa, M., Hattori, K., Takenouchi, T., Takahashi, T., Ishii, I., Matsubara, K., Kabe, Y., Uchiyama, S., Nagata, E., Gadalla, M.M., Snyder, S.H., and Suematsu, M. (2012). Hypoxic regulation of the cerebral microcirculation is mediated by a carbon monoxide-sensitive hydrogen sulfide pathway. *Proc. Natl. Acad. Sci. U. S. A.* 109: 1293–1298.

132. Hanafy, K.A., Oh, J., and Otterbein, L.E. (2013). Carbon Monoxide and the brain: time to rethink the dogma. *Curr. Pharm. Des.* 19: 2771–2775.

133. Koehler, R.C. and Traystman, R.J. (2002). Cerebrovascular effects of carbon monoxide. *Antioxid. Redox Signal.* 4: 279–290.

134. Jaggar, J.H., Leffler, C.W., Cheranov, S.Y., Tcheranova, D.E.S., and Cheng, X. (2002). Carbon monoxide dilates cerebral arterioles by enhancing the coupling of Ca^{2+} sparks to Ca^{2+}-activated K^+ channels. *Circ. Res.* 91: 610–617.

135. Lee, G.R., Shaefi, S., and Otterbein, L.E. (2019). HO-1 and CD39: It takes two to protect the realm. *Front. Immunol.* 10: 1765.

136. Bock, F.J. and Tait, S.W.G. (2020). Mitochondria as multifaceted regulators of cell death. *Nat. Rev. Mol. Cell Biol.* 21: 85–100.

137. Bell, E.L., Klimova, T.A., Eisenbart, J., Moraes, C.T., Murphy, M.P., Budinger, G.R.S., and Chandel, N.S. (2007). The Qo site of the mitochondrial complex III is required for the transduction of hypoxic signaling via reactive oxygen species production. *J. Cell Biol.* 177: 1029–1036.

138. Fuhrmann, D.C. and Brüne, B. (2017). Mitochondrial composition and function under the control of hypoxia. *Redox Biol.* 12: 208–215.

139. Kim, J.-W., Tchernyshyov, I., Semenza, G.L., and Dang, C.V. (2006). HIF-1-mediated expression of pyruvate dehydrogenase kinase: a metabolic switch required for cellular adaptation to hypoxia. *Cell Metab.* 3: 177–185.

140. Miró, O., Casademont, J., Barrientos, A., Urbano-Márquez, A., and Cardellach, F. (1998). Mitochondrial cytochrome c oxidase inhibition

during acute carbon monoxide poisoning. *Pharmacol. Toxicol.* 82: 199–202.

141. Brown, S.D. and Piantadosi, C.A. (1992). Recovery of energy metabolism in rat brain after carbon monoxide hypoxia. *J. Clin. Invest.* 89: 666–672.

142. Iheagwara, K.N., Thom, S.R., Deutschman, C.S., and Levy, R.J. (2007). Myocardial cytochrome oxidase activity is decreased following carbon monoxide exposure. *Biochim. Biophys. Acta* 1772: 1112–1116.

143. Alonso, J.-R., Cardellach, F., López, S., Casademont, J., and Miró, O. (2003). Carbon monoxide specifically inhibits cytochrome c oxidase of human mitochondrial respiratory chain. *Pharmacol. Toxicol.* 93: 142–146.

144. Almeida, A.S., Figueiredo-Pereira, C., and Vieira, H.L.A. (2015). Carbon monoxide and mitochondria-modulation of cell metabolism, redox response and cell death. *Front. Physiol.* 6: 33.

145. Queiroga, C.S.F., Almeida, A.S., Alves, P.M., Brenner, C., and Vieira, H.L.A. (2011). Carbon monoxide prevents hepatic mitochondrial membrane permeabilization. *BMC Cell Biol.* 12: 10.

146. Almeida, A.S., Queiroga, C.S.F., Sousa, M.F.Q., Alves, P.M., and Vieira, H.L.A. (2012). Carbon monoxide modulates apoptosis by reinforcing oxidative metabolism in astrocytes: role of Bcl-2. *J. Biol. Chem.* 287: 10761–10770.

147. D'Amico, G., Lam, F., Hagen, T., and Moncada, S. (2006). Inhibition of cellular respiration by endogenously produced carbon monoxide. *J. Cell Sci.* 119: 2291–2298.

148. Meerson, F.Z., Gomzakov, O.A., and Shimkovich, M.V. (1973). Adaptation to high altitude hypoxia as a factor preventing development of myocardial ischemic necrosis. *Am. J. Cardiol.* 31: 30–34.

149. Meerson, F.Z., Ustinova, E.E., and Manukhina, E.B. (1989). Prevention of cardiac arrhythmias by adaptation to hypoxia: regulatory mechanisms and cardiotropic effect. *Biomed. Biochim. Acta* 48: S83–88.

150. Raman, K.G., Barbato, J.E., Ifedigbo, E., Ozanich, B.A., Zenati, M.S., Otterbein, L.E., and Tzeng, E. (2006). Inhaled carbon monoxide inhibits intimal hyperplasia and provides added benefit with nitric oxide. *J. Vasc. Surg.* 44: 151–158.

151. Leake, A., Salem, K., Madigan, M.C., Lee, G.R., Shukla, A., Hong, G., Zuckerbraun, B.S., and Tzeng, E. (2020). Systemic vasoprotection by inhaled carbon monoxide is mediated through prolonged alterations in monocyte/macrophage function. *Nitric Oxide* 94: 36–47.

152. Otterbein, L.E., Bach, F.H., Alam, J., Soares, M., Tao Lu, H., Wysk, M., Davis, R.J., Flavell, R.A., and Choi, A.M. (2000). Carbon monoxide has anti-inflammatory effects involving the mitogen-activated protein kinase pathway. *Nat. Med.* 6: 422–428.

153. Herrera, E.A., R.V. Reyes, R.V., Giussani, D.A., Riquelme, R.A., Sanhueza, E.M., Ebensperger, G., Casanello, P., Méndez, N., Ebensperger, R., Sepúlveda-Kattan, E., Pulgar, V.M., Cabello, G., Blanco, C.E., Hanson, M.A., Parer, J.T., Llanos, A.J. (2008). Carbon monoxide: a novel pulmonary artery vasodilator in neonatal llamas of the Andean altiplano. *Cardiovasc Res.* 77(1): 197–201.

8

Nitric Oxide in Human Physiology: Production, Regulation, and Interaction with Carbon Monoxide Signaling

Maryam K. Mohammed and Brian S. Zuckerbraun

Department of Surgery, University of Pittsburgh Medical Center, Pittsburgh, PA 15213, USA

Introduction

Carbon monoxide (CO) and nitric oxide, alternatively known by its systemic name, nitrogen monoxide (NO), are diatomic species existing in the environment as odorless, colorless gases. They have been recognized for most of history almost solely for their toxic and pollutant qualities [1]. CO, the most abundant pollutant in the lower atmosphere, accumulates at toxic levels in the environment from incomplete combustion of hydrocarbons and in the human body from xenobiotic metabolism of inhaled methylene chloride. Its toxicity is mediated by its ability to bind hemoglobin and increase its affinity for O_2, thereby precluding dissociation of O_2 and causing tissue hypoxia [2]. NO is primarily present in nature as a pollutant from automobile exhaust and as an industrial by-product of nitric acid (HNO_3) production via the Ostwald process. The uncatalyzed combustion of oxygen (O_2) and nitrogen (N_2) at temperatures >2000 °C during lighting strikes and photodissociation of nitrogen dioxide (NO_2) by sunlight are lesser, secondary sources of NO.

The historical depiction of CO and NO exclusively as toxins has proven to be an incomplete characterization. Identification of endogenously generated NO as endothelium-derived relaxing factor led to not only its designation of "Molecule of the Year" in 1992 and the 1998 Nobel Prize in Medicine, but also interest in the biological importance of CO and NO [3]. Since then, an extensive list of the physiological roles of CO and NO has been generated across multiple organ systems, with many areas of overlap in signaling pathways and physiological effects. They are arguably the most biologically relevant, and certainly the best characterized, of the known gaseous signaling molecules other than oxygen. Perturbations in CO and NO signaling have subsequently been identified in the pathophysiology underlying many diseases, leading to interest in the potential therapeutic role of CO and NO and further emphasizing the importance of these molecules in biology and medicine.

Chemical Properties and Chemical Biology of CO and NO

Chemical Properties

CO and NO share several similarities in chemical and physical properties. They exist as odorless, colorless gases in the environment at atmospheric pressure but are soluble at physiologically relevant concentrations. Both are uncharged species with similar molecular mass (CO: 28.01; NO: 30.01) and water solubility (CO: 2.6 ml/100 ml; NO: 4.7 ml/100 ml) [4,5]. Like all gaseous signaling molecules, they are freely permeable through the hydrophobic cell membrane on account of their small size and uncharged state, obviating the need for a cognate membrane receptor or channel for intracellular transport [6].

CO and NO have a similar molecular structure and bond characteristics. Both are linear, heteronuclear, diatomic species containing elements from the second row of the periodic table. NO has a bond length of 1.154 Å, in between that of a double and a triple bond [7]. CO has a bond length of 1.128 Å, consistent with a partial triple bond [8].

One notable difference in structure is that NO is a free radical. Its unpaired electron resides in a high-energy π^* (antibonding) molecular orbital [7]. Thus, while NO is a stable molecule, it reacts rapidly in the presence of other free radical species and with

Carbon Monoxide in Drug Discovery: Basics, Pharmacology, and Therapeutic Potential, First Edition. Edited by Binghe Wang and Leo E. Otterbein.
© 2022 John Wiley & Sons, Inc. Published 2022 by John Wiley & Sons, Inc.

a broader range of substrates than does the comparatively inert CO [1,9].

Chemical Biology of CO and NO

The biological reactivity of CO is limited to covalent bonds with transition metals in a specific redox state, most commonly ferrous heme. The reactivity of NO is diverse by comparison and includes covalent bonding to transition metals, reduction–oxidation (redox) chemistry with metal complexes or metallo-oxo complexes, and free radical chemistry [10]. The chemistry of these reactions and common examples of each are described below.

Covalent Bonding of CO and NO with Transition Metals

Both CO and NO form coordinate covalent bonds with transition metals, resulting in compounds known as metal carbonyls and metal nitrosyls, respectively. Metal carbonyl and nitrosyl compounds are used as catalysts and reagents in synthetic and organometallic chemistry, but are also commonly found in human and mammalian biology from the reaction of CO and NO with endogenous metalloproteins containing transition metal moieties, most often heme proteins [11].

NO bonds with transition metals via the nitrogen atom, while CO does through the carbon atom [12,13]. In the case of NO, while the π^b (bonding) orbitals are polarized toward the more electronegative oxygen atom, which contributes the majority of their electron density, the π^* orbital containing the unpaired electron is polarized toward nitrogen. Given the reactivity of the unpaired electron, bonding interactions with transition metals occur via the nitrogen atom [14]. CO bonds with transition metals via the carbon atom. Despite the greater electronegativity of oxygen relative to carbon, the C–O bond is characterized by a small dipole moment along its bond with the net negative end associated with the carbon atom [8,15].

The shared affinity of NO and CO for transition metals can be explained by molecular orbital theory [7]. Both NO and CO donate electron density to the metal and form a σ covalent bond that is stabilized by back-donation of electron density from the metal to NO or CO. The back-donation is possible because the lowest unoccupied molecular orbital of both CO and NO is a π^* molecular orbital, which is similar in energy and shares symmetry with the partially filled dxy, dxz, and dyz orbitals of transition metals. The donation of electron density from the metal d orbitals to the π^*, known as "π-backbonding," creates a shared orbital that is lower in energy and stabilizes the σ bond [7,16].

Given that NO is more electronegative than CO, it is a better acceptor of electron density from the metal via π-backbonding and thus demonstrates a greater binding affinity with metalloproteins [7]. The affinity of NO for hemoglobin is approximately 1500 times greater than that of CO [17]. NO is able to react with more than one oxidation state of iron. Both NO and CO have been shown to bind ferrous (Fe^{2+}) heme, while only NO has been shown to bind ferric (Fe^{3+}) heme, albeit with lower affinity than it does to ferrous heme [11].

Redox Chemistry of NO with Metal Complexes

The reactivity of NO with biological metals extends beyond covalent interactions with transition metals to include redox chemistry with dioxygen–metal complexes and high-valent metallo-oxo complexes. A common example of the former is the deoxygenation reaction, in which NO reduces oxyhemoglobin, containing ferrous heme, to methemoglobin, containing ferric heme. Nitrate (NO_3^-) is produced from the oxidation of NO. The rate of the reaction is rapid, limited only by diffusion of NO into the heme pocket of hemoglobin [9,18].

NO also reacts rapidly with high-valent metallo-oxo complexes, many of which are produced from oxidation of metals or metal–oxygen complexes by hydrogen peroxide. High-valent metallo-oxo complexes are powerful oxidants and mediators of peroxide-associated tissue damage. Reaction of these compounds with NO produces a lower-valent state metallo-oxo compound and nitrite (NO_2^-) as a by-product. This is thought to be the primary mechanism by which NO is able to reduce the deleterious effects of peroxide [9,19,20].

Reactions of NO with Free Radicals

The reactions of NO with other free radical species include both termination and propagation reactions, leading to seemingly contradictory antioxidant and pro-oxidant behavior. A notable example is the dual roles of NO in oxidative stress. Lipid peroxyl radicals are produced by free radical or enzyme-mediated lipid oxidation, such as by lipoxygenase. At lower concentrations, NO impedes lipid oxidation via termination reactions with the atherogenic lipid peroxyl radicals and by inhibiting the nonheme iron-containing enzyme, lipoxygenase. NO can inhibit lipoxygenase by reducing the ferric iron to the inactive ferrous state or by directly competing at ligand binding sites [21–23]. However, at higher concentrations, NO reacts with O_2 and superoxide to produce several radical pro-oxidant intermediate species with deleterious effects. Oxidation of NO generates the nitrosonium cation, while reduction generates the nitroxyl anion, a short-lived species in solution that readily decomposes via dimerization and dehydration to yield nitrous oxide [4,5]. Reaction with superoxide yields peroxynitrite (ONOO$^-$) [24]. These intermediates are part of a group collectively termed reactive nitrogen oxide species (RNOS). Like reactive oxygen species (ROS), RNOS participate in reactions that predominantly result in local tissue injury, including the initiation of lipid peroxidation, nitrosation, nitration, and oxidation [9,25]. In addition to being concentration dependent, these reactions are spatially and temporally dependent, in that tissues distant from the source of NO production may be relatively spared due to diffusion and consumption of NO over distance and time [9].

The net effect of NO as an antioxidant or pro-oxidant is dependent on the balance of competing factors, including the local concentration of NO. At higher concentrations, pro-oxidant activity of NO is more common.

Endogenous Production and Regulation of NO

Pathways for endogenous CO and NO production exhibit several similarities and have a similar basal rate of production (NO: 850 mmol/day; CO: 500 mmol/day) [17]. Both systems use inducible and constitutively expressed isoforms of heme proteins that activate O_2 to form a redox-active center at heme with NADPH as a cofactor [4,26]. CO is primarily produced by heme oxygenases (HOs) (Chapter 1) and NO is produced by nitric oxide synthases (NOSs). Differences include unique substrates, the use of (6R)-5,6,7,8-tetrahydro-L-biopterin (BH$_4$) as an additional cofactor by NOS, the presence of a cytochrome P450 reductase domain within the same polypeptide chain as the oxygenase domain in NOS, production of superoxide by NOS, and the role of calcium/calmodulin (Ca^{2+}/CaM) signaling in regulating NOS activity [4,27].

Activity and Regulation of Nitric Oxide Synthases

NO is produced endogenously by NOS, a family of homodimeric, heme-containing flavoproteins [28]. They catalyze the production of endogenous NO from the oxidation of L-arginine to L-citrulline. Three isoforms are expressed in mammals: endothelial (eNOS) and neuronal (nNOS), which are constitutively expressed with Ca^{2+}/CaM-dependent activity, and a third isoform with inducible expression (iNOS) and Ca^{2+}-independent activity, primarily involved in immune response [29]. Compared to eNOS and nNOS, iNOS produces large quantities of NO for sustained periods of time; iNOS is estimated to produce 100–1000-fold more NO than eNOS over hours to days until it is proteolytically degraded [30,31]. While cytoprotective effects include antimicrobial and antithrombotic activity, excess NO is cytotoxic, causing injury to host tissues and playing a role in the pathophysiology of many disease states, including septic shock, inflammatory arthritis, and cerebral ischemia [32].

NOS Expression and Regulation

Each isoform is transcribed from a unique gene, with approximately 51–57% homology across the three isoforms in humans [33]. The genes encoding NOS are NOS1 (nNOS), NOS2 (iNOS), and NOS3 (eNOS). Expression of NOS1 and NOS3 is generally constitutive, although there is accumulating evidence of transcriptional regulation under some conditions [34]. NOS2 expression is induced in response to stress stimuli.

NOS1

Expression of NOS1 predominantly occurs in neuronal cells of the central and peripheral nervous systems, but has also been identified in nonneuronal cell types. Upregulation of NOS1 mRNA has been identified in the response to noxious stimuli such as heat, electrical stimulation, and allergic substances [35–37]. Downregulation of NOS1 has been demonstrated in response to *ex vivo* lipopolysaccharide (LPS) in rat brain, spleen, stomach, and rectum, with a corresponding increase in NOS2 expression, suggesting a shift in isoforms as a response to stress [38]. Of note, there are conflicting reports, with an earlier *in vivo* study reporting increased NOS1 expression in rat hypothalamic tissue after systemic LPS [39]. Increased NOS1 and NOS3 expression has been seen with hypoxia in rats *in vivo*. The presence of putative hypoxia-inducible factor 1 (HIF-1) binding sites in the NOS1 transcript has raised the possibility of this being from direct activation of NOS1 transcription rather than a generalized stress response, but this remains to be investigated further [34,40,41]. Expression of NOS1 is increased in rat osmoresponsive neurons with hypernatremia and is decreased in rat kidney with salt restriction [42–44].

NOS2

NOS2 plays a key role in mediating the response to tissue damage and infection. Although first discovered in macrophages, its expression has since found to be inducible in nearly all cell types [45]. Outside of rare examples of constitutive expression in rat kidney and rat ovarian follicular cells, NOS2 expression is induced by inflammatory cytokines and microbial products such as type I interferons (IFN-α/β), interferon-γ (IFN-γ), interleukin-1β, interleukin-6, LPS, and tumor necrosis factor [32,46]. Expression is inhibited by glucocorticoids [47].

The primary method of controlling NOS2 expression has proven to be transcriptional modulation, although some post-transcriptional modifications have been reported. Cytokines increase NOS2 expression by activating kinase signaling pathways, including cyclic adenosine monophosphate/protein kinase A (cAMP/PKA), extracellular regulated kinase 1/extracellular regulated kinase 2, mitogen-activated protein kinases (MAPKs), and Janus kinase–signal transducer and activator of transcription pathways, which converge on activation of specific transcription factors [48]. Negative regulators of NOS2 expression include suppressor of cytokine signaling proteins. It is worth noting that the effects of some signaling pathways on NOS2 expression are cell type specific [48].

The promoter region of human NOS2, estimated at 16 kb in length, is a complex region with areas of homology to binding sites for multiple transcription factors [49,50]. Due to the complexity of the NOS2 promoter and binding by multiple transcription factors, induction of NOS2 expression in cultured human cells requires a synergistic combination of cytokines; individual cytokines produce minimal to no increase in expression [51]. Transcription factors that have been shown to be necessary for optimal expression of NOS2 include nuclear factor-κB (NF-κB) and signal inducer activation of transcription 1 (STAT1). Several other transcription factors, including octamer factor, interferon regulatory factor 1, and HIF-1 have also been shown to induce NOS2 expression but are not as critical [48].

A recently identified mechanism for synergistic induction of NOS2 expression by LPS and IFN-α/β in mouse macrophages explains the need for NF-κB and STAT-1/2 for optimal induction of NOS2. LPS activates NF-κB through membrane-bound receptor binding and induces production of IFN-α/β through cytosolic receptor binding. NF-κB binds the NOS2 promoter and recruits the transcription factor IIH (TFIIH) and cyclin-dependent kinase 7 (CDK7). In parallel, IFN signaling leads to formation of interferon-stimulated gamma complex 3 (ISGF3), consisting of STAT1, STAT2, and interferon regulating factor 9. ISGF3 binds the NOS2 promoter sequence and recruits RNA polymerase II, which is then phosphorylated at its C-terminal domain (CTD) by the TFIIH–CDK7 complex [52]. CTD phosphorylation has been shown to be a necessary step for RNA polymerase II to associate with proteins required for promoter clearance, mRNA capping, and transcription elongation [53,54].

NOS2 expression and iNOS activity are decreased with activation of the transcription factors peroxisome proliferator-activated receptors, tumor suppressor p53 (p53), and Smad [48]. Glucocorticoids decrease NOS2 expression, likely by inhibiting binding of NF-κB to the NOS2 promoter [55]. Autoinhibition of NOS2 expression by NO has been described in the literature and is thought to occur through multiple mechanisms including enhancement of p53 activity, reduction of HIF-1 activity, inhibition of NF-κB activity by interfering with degradation of IκB-α, and tyrosine nitration of STAT-1α [48,56–59].

While regulation of transcription has long been considered the primary means of iNOS regulation, there is increasing evidence of post-transcriptional regulation of NOS2. Marked differences between promoter activity and levels of NOS2 mRNA in humans suggest that regulation of mRNA stability may play a role in NOS2 regulation [48]. AU-rich elements (AREs), sequence motifs associated with mRNA destabilization, have been identified in the 3′ untranslated region of human NOS2 mRNA [60,61]. There is evidence of both stabilizing and destabilizing interactions with NOS2 ARE. The evolutionarily conserved RNA-binding protein T-cell intracellular antigen 1-related protein has been shown to stabilize NOS2 mRNA transcripts in human colon carcinoma cells and enhance cytokine-induced NOS2 expression [62]. Transforming growth factor-β has been shown to reduce murine macrophage iNOS mRNA without decreasing expression, suggesting that it may reduce translation or stability of NOS2 mRNA. This may have evolved as a cytoprotective mechanism in a cell line exposed to high concentrations of NO [63].

NOS3

NOS3 is primarily expressed in endothelial cells, where it was first discovered, but its expression has since been noted in nonendothelial tissues, including neuronal tissue, hepatocytes, and gastrointestinal mucosa [34]. NOS3 expression is increased with exercise training and shear stress. While a putative shear stress-response element has been proposed in the NOS3 promoter sequence, it functionality remains to be proven [34,64,65]. NOS3 expression in pulmonary endothelial cells is downregulated across multiple species, including humans [66–68]. Results are not consistent in extrapulmonary endothelial cells [34].

NOS Structure

All isoforms share a carboxyl-terminal (C-terminal) reductase domain homologous to the cytochrome P450 reductase and an amino-terminal (N-terminal) oxygenase domain containing a heme group. The heme group is linked to a CaM-binding domain in the middle of the protein. Binding of CaM functionally links the reductase and oxygenase domains, enabling electron flow from flavin groups in the reductase domain to heme in the oxygenase domain, where they facilitate the conversion of O_2 and L-arginine to NO and L-citrulline. The oxygenase domain of each NOS isoform contains a BH_4 prosthetic group used as a reducing equivalent [29].

Steps of NO Production

NO is produced by NOS from five-electron oxidation of a guanidino nitrogen on L-arginine. The oxidation of the guanidino nitrogen on L-arginine is the result of two separate monooxygenation steps, with N-hydroxy-L-arginine (NOHA) produced as an intermediate. NOS utilizes 1.5 molecules of NADPH and 2 molecules of O_2 as cosubstrates per molecule of L-arginine. Additional cofactors include flavin adenine dinucleotide (FAD), flavin mononucleotide (FMN), and BH_4 [29].

The reaction begins with two-electron reduction of FAD at the C-terminal reductase domain by NADPH, followed by electron transfer to FMN, producing $FMNH_2$. $FMNH_2$ transfers electrons to the N-terminal oxygenase domain heme moiety on the opposite monomer, reducing ferric (Fe^{3+}) heme to ferrous (Fe^{2+}) heme. O_2 binds heme, forming a ferrous–dioxygen species that is reduced by BH_4 of the N-terminal oxygenase domain to the oxidative ferric-peroxo and ferryl (Fe^{4+})-oxo heme species [29,69–72]. While it was long held that the active oxidant in NOS is a ferryl heme species as it is in cytochrome P450, newer evidence suggests that it may in fact be the ferric-peroxo complex [72]. In the first monooxygenation step, L-arginine is oxidized at a guanidino nitrogen to produce NOHA, which contains a N^G-OH-guanidine group. In the second step, oxygenation of NOHA yields NO from the N^G-OH-guanidine group and L-citrulline [72].

NOSs are uniquely catalytically self-sufficient. In between monooxygenation steps, BH_4 is regenerated from the BH_3 radical using an electron supplied by a flavin moiety, resetting the catalytic activity of NOS. This is in contrast to other BH_4-utilizing enzymes, which require extrinsic dihydrobiopterin reductase for BH_4 regeneration [29,73].

Post-translational Regulation of NOS

Once thought to be exclusively controlled by Ca^{2+}/CaM signaling, the understanding of NOS regulation has evolved into a model with multiple levels of regulation, including protein–protein interactions and post-translational modifications. Key examples of these are described later [74].

Ca^{2+}/CaM Signaling

Intracellular Ca^{2+} concentration is arguably the most significant, and certainly best studied, regulator of eNOS and nNOS catalytic activity. Activity of iNOS is independent of Ca^{2+} signaling.

CaM (also known as calcium-modulating protein) is a small (16.7 kDa in vertebrates), highly conserved eukaryotic intracellular protein that is responsible for the protein modulatory effects of Ca^{2+}. CaM is often described as a "calcium sensor" and is largely responsible for Ca^{2+} signal transduction [75]. The structure of CaM includes N-terminal and C-terminal globular domains, each containing a pair of Ca^{2+}-binding helix–loop–helix domains known as EF hands and methionine-rich hydrophobic pockets that bind complementary hydrophobic domains of a variety of target proteins [76]. The N- and C-terminal domains are connected by a flexible linker region [77]. In the inactive state, CaM exists in a compact orientation, with hydrophobic binding sites inaccessible to ligand [75]. Binding of Ca^{2+} induces conformational changes that expose these hydrophobic pockets, allowing them to bind target proteins [78].

In NOS, the CaM-binding domain is present in the intervening domain between the C-terminal reductase and the N-terminal heme. Binding of CaM is thought to induce conformational changes that facilitate electron transfer from flavin moieties to heme, the rate-limiting step in NO production. Without CaM bound, NOS is in an "input state" where the

FMN domain is locked to the NADPH/FAD domain to facilitate electron transfer between FMN and FAD, but inaccessible to heme. Binding of CaM shifts conformational equilibrium toward a state in which the FMN(H$_2$) subdomain is exposed and accessible to bind heme while also limiting the conformational space available to the FMN domain, further facilitating FMN–heme interaction [69,79–81].

The differential responses of eNOS/nNOS and iNOS to Ca^{2+} signaling are attributed to differences in CaM binding affinity. The constitutively expressed isoforms bind only to activated CaM, while iNOS binds both activated and inactive CaM. Differences in primary sequence underly this difference in CaM affinity. The reductase FMN subdomain in eNOS and nNOS contains a 40–50 amino acid insert, known as an autoinhibitory (AI) element, that destabilizes CaM binding. CaM thus binds transiently to eNOS/nNOS only when activated by high intracellular concentrations of Ca^{2+}, producing nanomolar amounts of NO over short periods of time before CaM dissociates [33]. INOS, which lacks an AI element, has a higher affinity for CaM. CaM is thus tightly bound to iNOS regardless of intracellular Ca^{2+} concentration, and the catalytic activity of iNOS is instead limited by its own production and degradation. Micromolar amounts of NO are produced over longer periods of time until iNOS is proteolytically degraded or inactivated [31,51,82].

Heat Shock Protein 90

Heat shock protein 90 (Hsp90) is a highly conserved molecular chaperone, so name for its molecular mass of approximately 90 kDa, that binds target proteins to facilitate protein folding, ligand binding, or assembly of multiprotein complexes. Although notably identified first by its elevated expression as part of the heat shock response, it has since been identified as a key regulator of proteostasis in both physiological and stress conditions for several hundred target proteins. It is well studied in cancer and neurodegenerative disorders, where mutations of Hsp90 leading to protein misfolding have been implicated in pathogenesis of disease [83].

Regulation of NOS activity by Hsp90 was initially demonstrated in 1998 after it was identified as a coprecipitate in a complex with eNOS. Hsp90 was subsequently found to associate with and increase the catalytic activity of eNOS in response to NOS agonists such as vascular endothelial growth factor, histamine, and fluid shear stress [84]. It has since been found to associate with and increase activity of nNOS and iNOS as well [85,86]. A putative mechanism was identified recently using a combination of electron transfer kinetics and molecular modeling to demonstrate that Hsp90 increases interdomain electron transfer (IET) between the FMN and heme domains by binding to heme and narrowing the conformational space available to FMN. The effect is thought to augment similar effects by CaM, described previously [81].

Hsp90 has also been shown to decrease NOS susceptibility to proteolytic degradation when associated in a heterocomplex, although this effect is more pronounced with nNOS than with eNOS [87].

Serine/Threonine Phosphorylation

Modulation of NOS catalytic activity by phosphorylation of key serine and threonine residues has been described for eNOS and nNOS. Although phosphorylation sites have been described in iNOS, the significance remains unknown [61].

Five phosphorylation sites have been identified in eNOS [88]. Phosphorylation of Ser^{1177}, located in the C-terminus, and Ser^{633}, in the CaM AI segment, has been associated with increased eNOS activity. Phosphorylation of Thr^{495}, located in the CaM-binding domain, is inhibitory. Phosphorylation of Ser^{615} and Ser^{114} is of unclear significance [88]. Phosphorylation of Ser^{1177} is the best characterized of these modifications and was originally identified as a substrate of adenosine monophosphate kinase during myocardial ischemia [89]. Since then, phosphorylation of Ser^{1177} has been associated with many stimuli, including mechanical shear stress, bradykinin, insulin, and statin drugs [90–93]. Phosphorylation of Ser^{1177} is hypothesized to increase eNOS activity by inducing a conformational change that removes the C-terminus of eNOS from a position between the two monomers of eNOS, where it is thought to normally inhibit electron transfer between the monomers [94].

Phosphorylation of nNOS has been described at Ser^{847}, Ser^{1412}, and Ser^{741} [95]. CaM kinase IIα-mediated phosphorylation of Ser^{847}, located in the AI element, decreases CaM binding and inhibits nNOS activity [96,97]. Phosphorylation at Ser^{1412} has been associated with increased nNOS activity [98]. Although not yet described in *in vivo*, phosphorylation of Ser^{741} in *Escherichia coli* and SR9 cells transfected with rat nNOS has been associated with decreased nNOS activity [99].

Regulation of Dimerization

Dimerization of NOS is required for enzyme activity and is largely a result of interactions between the oxygenase domains. Incorporation of heme is necessary for dimerization, while binding of BH_4 and L-arginine is thought to stabilize the dimer [100–102]. Zinc has also been shown to optimize dimerization in all three isoforms by maintaining hydrogen bonds that stabilize the dimer interface and the structure of the BH_4 binding site [103–105]. Crystal structure analysis suggests that binding of zinc plays a greater role than BH_4 in dimer stabilization [104].

Dimerization is required for optimal CaM-mediated electron transfer between the reductase and oxygenase domains [106,107]. While eNOS and nNOS are largely constitutively expressed as inactive monomers, iNOS is primarily present as an active homodimer after induction. Screening of murine macrophage cDNA libraries revealed a 110-kDa protein, known as NOS-associated protein 110, that interacts with the iNOS N-terminal region to prevent dimerization and inhibit iNOS activity [108]. The CNS protein kalirin-7 has also been shown to decrease iNOS activity by binding to iNOS monomers and preventing dimerization [109]. These proteins may play cytoprotective roles during inflammatory response.

NO-Induced Autoinactivation

The affinity of NO for heme, particularly ferrous heme, alludes to a mechanism for autoinactivation of NOS. NO bound to ferric heme at the end of the catalytic cycle can be reduced to ferrous heme by the reductase domain if it does not dissociate. The NO–ferrous heme complex can only be removed by reacting with O_2 in a futile cycle that produces nitrate but regenerates the catalytically active ferric heme. Of note, this reaction is an inherent part of the NOS catalytic cycle. The relevance of autoinactivation varies by isoform. In eNOS, slow IET precludes reduction of the NO–ferric heme complex and formation of the NO–ferrous heme complex [27]. The NO–ferrous heme complex forms in both iNOS and nNOS, but only accumulates to a significant degree in nNOS given a proximal tryptophan residue that lowers redox potential and thus reduces oxidation of the NO–ferrous heme complex [110,111].

Another method of NO-mediated NOS inactivation, independent of heme, has been demonstrated in iNOS. NO was shown to inactive ZnS_4 cluster by S-nitrosation, causing loss of zinc and irreversible dissociation of the iNOS dimer [112].

Inhibition by CO

CO exerts largely inhibitory effects on NOS. At low concentrations in rat vascular endothelium, CO led to increased NO release, although this is partly attributable to release of heme-bound NO that is maintained at a steady state by constitutive eNOS production. At higher concentrations (above 1 μM), CO was shown to inhibit eNOS activity [113]. CO likely inhibits NOS activity by multiple approaches. Indirectly, it may decrease availability of heme for the NOS active site, as it does with cytochrome P450, or consume the shared cofactor NADPH [4,114]. Increased availability of free iron, a known suppressor of NOS expression, may also play a role [115]. Direct binding of CO to NOS has also been reported to occur and decrease NOS activity [116].

Superoxide Production by NOS

Production of superoxide was first noted in nNOS and has since been identified with all three isoforms of NOS [117–120]. The underlying phenomenon is functional "uncoupling" of the NOS from its substrate or BH_4 cofactor, but the mechanism varies slightly between isoforms [121].

Uncoupling of both nNOS and eNOS occurs at the heme center [122]. Uncoupling in nNOS is primarily secondary to depletion of L-arginine but can also occur with BH_4 depletion. In eNOS, uncoupling occurs only with depletion of BH_4 [121]. In both isoforms, O_2 is able to bind ferrous heme and form a ferrous dioxygen heme complex, but dissociates into heme and superoxide when it is unable to be further reduced to an active redox center by BH_4 or react with a substrate [121,123]. As expected, superoxide production is Ca^{2+}/CaM dependent in both eNOS and nNOS. Absence of bound CaM precludes electron transfer to oxygenase domain heme, the site of uncoupling [120,122,124]. In nNOS, uncoupling is ameliorated with addition of L-arginine or BH_4 [125]. Uncoupling of eNOS is not ameliorated with a high concentration of L-arginine [121,122].

Uncoupling of iNOS occurs at the flavin moieties of the reductase domain, which produce substantial quantities of superoxide even in the presence of L-arginine [124]. There is also evidence to suggest that iNOS produces superoxide when not uncoupled. Superoxide and NO are produced simultaneously and combine to form peroxynitrite [126]. This is in contrast to nNOS, which although has been shown to produce peroxynitrite does so using NO and superoxide derived from nNOS molecules in different coupling states or using NO from coupled nNOS and superoxide from a different pathway, such as NADPH oxidase [124]. This evidence reclassifies iNOS as a peroxynitrite-producing enzyme and has implications for its physiological relevance as a producer of RNS/ROS [121].

Regulation of CO and Heme Oxygenase by NO

HOs are a unique family of monomeric proteins involved in heme degradation that, although not heme proteins per se, use heme as both a substrate and a cofactor (Chapter 1) [127,128]. Two isoforms, inducible (HO-1) and constitutively expressed (HO-2), have been identified in humans [129]. HO catalyzes the rate-limiting step in the oxidative cleavage of heme to ferrous iron, CO, and biliverdin [130]. HO is the primary producer of endogenous CO and is considered an antioxidant both for its metabolism of heme, a powerful pro-oxidant, and for its production of biliverdin, a precursor to the antioxidant bilirubin [114,131]. As other chapters have described the structure, activity, and regulation of HO-1 in detail, we will limit our discussion to the overlap between NO/NOS and CO/HO signaling, specifically the role of NO in regulation of HO.

Transcriptional Regulation of HO-1 by NO

NO has been shown to induce HO-1 expression in multiple cell/tissue types [132–134]. Changes in the cellular redox state, including oxidative or nitrosative stress, are well-known inducers of HO-1 expression. In early studies of vascular smooth muscle cells (VSMCs), NO-induced HO-1 expression was reversed with the addition of the antioxidant *N*-acetyl-L-cysteine, unchanged with addition of peroxynitrite, and not subject to autocrine downregulation by CO [134]. Since then, the nuclear factor erythroid-related factor 2/antioxidant response element (Nrf2/ARE) pathway has been identified as the signaling cascade responsible for HO-1 induction in response to oxidative and nitrosative stress [135–137]. Nrf2 is a basic leucine zipper transcription factor that is constitutively inactivated by the cytoskeleton-associated protein Kelch-like ECH-associated protein 1 (Keap1), which binds Nrf2 and targets it for ubiquitination [138]. Keap1 has been proposed to be a component of a novel E3 ubiquitin ligase complex. The association of Nrf2 with Keap1 is mediated by key cysteine residues [139]. NO has been proposed to activate Nrf2 through two mechanisms, including *S*-nitrosylation of these key cysteine residues in Keap1, thereby facilitating dissociation of Nrf2, and protein kinase C-mediated phosphorylation of Nrf2 [140]. When released and activated, Nrf2 undergoes nuclear translocation, where it binds consensus sequences known as AREs in the promoters of several antioxidant genes, including HMOX1 [141,142].

The upregulation of HO-1 by NO has been proposed as a cytoprotective mechanism both as an adaptive response to the nitrosative stress generated by iNOS expression and at lower, nontoxic concentrations [143].

Post-translational Regulation of HO by NO

While it has been shown to induce expression of HO, NO has also been shown to bind and inhibit catalytic activity of heme-bound HO. This has been demonstrated in HO-1, where NO binds the catalytic heme in both the ferric and ferrous states to reversibly reduce catalytic activity [144]. In contrast, the interaction of NO with HO-2 has been shown to be irreversible and likely does not occur through the catalytic heme [144,145]. The differences in interaction may be secondary to C-terminal structural differences between HO-1 and HO-2.

Unlike HO-1, HO-2 contains two C-terminal heme regulatory motifs (HRMs), conserved sequences with a cysteine–proline dipeptide at the center that binds ferric heme through the Cys residue and a downstream hydrophobic region [146]. The HRMs of HO-2, centered at Cys^{265} and Cys^{282}, constitute a thiol/disulfide redox switch given their proximity. They bind heme in reducing conditions but form a disulfide bond in oxidizing conditions. Of note, the heme bound to HRMs is not a catalytic heme, but has been proposed to function in a regulatory capacity by modifying activity of the constitutively expressed HO-2 in a substrate-dependent manner. As the concentration of intracellular heme increases and the heme-binding capacity of HRMs is exceeded, the remaining heme is able to bind the catalytic site of HO-2, which exists at a conserved histidine residue [147]. HRM motifs have also been proposed to function as a binding site or reservoir for gaseous signaling molecules such as NO [145,148].

While the inhibitory interactions of NO and HO have been demonstrated *in vitro*, their physiological significance remains unclear.

CO and NO Signal Transduction

Both CO and NO have demonstrated roles as neurotransmitters. Transduction of CO and NO signaling is primarily mediated by a cytosolic ferrous heme protein, soluble guanylate cyclase (sGC) [149]. CO and NO bind and activate sGC, catalyzing the conversion of guanosine 5′-triphosphate (GTP) to the second messenger cyclic guanosine 3′,5′-monophosphate (cGMP) [149]. Effectors of cGMP include protein kinase G I/II (PKG, also known as cGMP-dependent protein kinase), cyclic nucleotide-regulated ion channels, and phosphodiesterases. The latter regulate the concentrations of cyclic nucleotides by hydrolyzing cGMP and cAMP [150]. Activated PKG and cyclic nucleotide-regulated ion channels phosphorylate target proteins, including transcription factors, and modulate cell membrane potential, respectively, to facilitate the physiological effects of CO and NO. At high concentrations, cGMP also exhibits cross-reactivity with PKA, an analogue of PKG that is principally activated by cAMP [151].

Activation of sGC by CO and NO

Activation of sGC occurs when the ligand, CO or NO, binds to the distal side of ferrous heme to initially form an unstable six-coordinate adduct in which the heme iron is displaced distally, into the heme plane, by its bonding with the electronegative ligand [152]. The proximal Fe–His bond then dissociates, yielding a five-coordinate adduct. The associated conformational change, purported to position the C-terminal catalytic domain in closer proximity to the N-terminal heme, activates sGC [153,154].

Binding and activation of sGC by NO is efficient, requiring only nanomolar concentrations of NO to create a 100–400-fold increase in sGC activity within milliseconds [153]. Desensitization of sGC is similarly rapid, leading to a decrease in maximal activation with repeat exposure to NO. In contrast, dissociation of NO from sGC is slow and is influenced differently by allosteric modulators such as GTP, ATP, and 3-(5′-hydroxymethyl-2′-furyl)-1-benzylindazole [155]. These differences have been used to suggest that a binary model of sGC activation and deactivation is incomplete and that an intermediate species exists between fully activated and deactivated states or that a nonheme NO binding site exists on sGC and allosterically modulates Fe–His dissociation [156]. These sites may correspond to cysteine residues, as thiol oxidation results in a loss of sGC activity [156].

CO binds to sGC with significantly lower affinity than NO, leading to only a fourfold increase in activity following binding of CO to sGC. Mechanistic studies suggest that the difference in activation is secondary to decreased efficiency of Fe–His bond cleavage [157]. Allosteric regulators that promote dissociation of the Fe–His bond lead to a rate of cGMP production by sGC–CO comparable to that of sGC–NO [158]. Kinetic parameters for the binding of CO and NO with GC and heme are summarized in Table 8.1 [159].

Deactivation of sGC occurs with dissociation of CO or NO. However, given the relative excess of sGC compared to intracellular NO and the affinity of NO and heme, it has been proposed that, once activated, all intracellular sGC molecules remain bound to one NO molecule at heme while the nonheme NO dissociates. This occurs rapidly (Table 8.1) following removal of excess intracellular NO. The residual activity of sGC with heme-bound NO and no allosteric modulation by the nonheme NO remains ~15% of its maximal activity [152].

Interplay of NO and CO in Pathophysiology

While CO and NO were originally discovered as mediators of vascular tone and have been well characterized in cardiovascular physiology, their importance has since been demonstrated in all organ

Table 8.1 Rate constants for association (k_{on}) and dissociation (k_{off}) of NO and CO with sGC and heme. Source: Adapted from [159].

Source and ligand		k_{on} (M^{-1} s^{-1})	k_{off} (s^{-1})
GC	NO	1.4×10^8	8×10^{-4}
	CO	1.2×10^5	28
Heme	NO[a]	—	3×10^{-2}
	CO	2×10^{8}[b]	1×10^{-2}[b]
		6×10^{8}[c]	1.2×10^{-1}[c]

[a] Rate constants for heme–NO interaction obtained with excess imidazole.
[b] In cetyltrimethylammonium bromide.
[c] In benzene.

systems. The physiological roles of CO are discussed in detail in separate chapters. Select examples of the physiological roles of NO with corresponding roles in pathology and therapeutic potential will be reviewed later, focusing on examples with interplay between NO and CO.

Interactions in Cardiovascular Pathophysiology

CO and NO modulate many aspects of cardiovascular physiology with the sum of their effects largely cardioprotective. They have been found to induce vasodilation and reduce intravascular pressure, reduce risk of thrombosis by inhibiting platelet aggregation, and decrease migration and proliferation of VSMCs, a hallmark of atherosclerosis. Accordingly, the therapeutic application of NO and CO in cardiovascular disease is an area of great interest and active exploration [160].

Vascular Tone

CO and NO are able to regulate arterial tone by inducing VSMC relaxation [161]. The effect has been demonstrated across a variety of arterial sizes. There is also *in vivo* evidence of NO contributed to resting and regulation of venous tone in humans [162]. However, the roles of CO and NO in mediating venodilation remain to be completely understood, as there is also evidence to suggest decreased eNOS expression in venous endothelium relative to arterial endothelium and increased sensitivity of venous endothelium to organic nitrates.

Vascular dilation by CO and NO is a cGMP-dependent process. Activation of sGC signaling and production of cGMP lead to binding and activation of cGMP-dependent PKG. PKG reduces intracellular Ca^{2+} by several mechanisms, including (i) inhibiting voltage-dependent Ca^{2+} channels and reducing Ca^{2+} influx, (ii) activating plasma membrane ATPases to increase ATP-dependent Ca^{2+} efflux, (iii) inhibiting inositol triphosphate receptors to reduce Ca^{2+} release from the sarcoplasmic reticulum, and (iv) activating sarcoplasmic calcium ATPases. Reduction of intracellular Ca^{2+} reduces CaM-mediated activation of myosin light chain kinase and increases the activity of myosin light chain phosphatase, which mediates VSMC relaxation [163]. NO may also induce vasodilation through cGMP-independent mechanisms, such as activation of large conductance Ca^{2+}-activated K^+ (BK_{Ca}) channels, which mediate smooth muscle relaxation when activated. There is evidence to show both that BK_{Ca} activation by NO is a cGMP-dependent process and that it is a result of direct binding of NO to the BK_{Ca} [164].

While both CO and NO have been shown to induce vasodilation, NO is thought to play a more significant role in modulating vascular tone, as the sensitivity of VSMCs to CO is less consistent [165]. In addition, NO is a more potent vasorelaxant than CO given its higher affinity for sGC and ability to generate a greater increase in cGMP concentration [161]. Decreased NO bioavailability secondary to endothelial dysfunction has frequently been demonstrated in patients with a history of essential hypertension [166,167]. Decreased expression of eNOS has also been found in patients with primary pulmonary hypertension and is suspected to be the mechanism underlying glucocorticoid-mediated hypertension [55,168]. Supplementation with L-arginine improved blood pressure and anginal symptoms in patients with essential hypertension and normal coronary arteries [169].

While upregulation of NO/sGC/cGMP signaling has become a target for development of new antihypertensives, it is also becoming increasingly apparent that several existing antihypertensive drugs mediate their effects by this pathway [170].

Platelet Aggregation and Thrombosis

Cardioprotective effects of NO and CO include inhibition of platelet aggregation, albeit through different mechanisms. NO inhibits platelet aggregation in a cGMP-dependent pathway [171–173]. NO produced by eNOS or by platelets induced cGMP/PKG, which inhibits platelet activation by decreasing intracellular Ca^{2+} flux, phosphorylating and inhibiting TxA_2 receptor, and inhibiting the activity of PI3K, which activated GP Ib–IIIa fibrinogen receptors [174]. NO also decreases platelet activity in a cGMP-independent manner by S-nitrosylation of N-ethylmaleimide-sensitive factor, which decreases platelet degranulation [174].

Although the antithrombotic effect of CO has been known for decades, the mechanism was only recently identified. CO-mediated inhibition of platelet activity was found to be secondary to inhibition of mitochondrial cytochrome *c* oxidase and glycolysis, resulting in depletion of cytosolic NAD^+ [175]. This, in turn, reduces activity of platelet NADH/NADPH oxidase, which is required for production of ROS to enhance the availability of released ADP for platelet activation [176].

Loss of HO-1 has been shown to markedly increase the rate of vascular graft thrombosis in a murine model, while repletion of CO reverses this prothrombotic potential [177]. A NO-releasing aspirin derivative, NCX4215, demonstrated approximately sevenfold higher antithrombotic activity against thrombin-mediated platelet aggregated relative to aspirin in a rat model [178]. NO may also play a role in feedback loops inhibiting overstimulation of platelet aggregation; porcine platelets have been found to express iNOS on activation and aggregation [179].

VSMC Proliferation and Migration

In addition to their role in maintaining vascular tone, VSMCs facilitate vessel renewal and repair. In response to environmental cues such as chemoattractants released from injured endothelium, they transiently dedifferentiate from the mature, contractile state into migratory–proliferative and synthetic states that lose contractility, become migratory, and produce large quantities of extracellular matrix. Under normal physiology, a low steady state of flux exists between quiescent and proliferative VSMCs [180]. Loss of control in phenotype transition of VSMCs leads to increased intimal presence of proliferative VSMCs, a hallmark of initiation in many vascular pathologies, including atherosclerosis, in-stent stenosis, and vascular graft remodeling [181,182].

Endogenous and exogenous CO and NO have demonstrated cGMP-dependent inhibition of mitogen-stimulated VSMC proliferation and migration in a cGMP-dependent mechanism *in vitro* [183–185]. Activation of Ca^{2+}/CaM-dependent protein kinase II (CaM kinase II) is a necessary step for migration of activated VSMCs, and blocking CaM kinase II impairs VSMC migration [186]. Thus, cGMP may modulate VSMC activation and migration by modulating intracellular Ca^{2+} concentrations.

The therapeutic relevance of CO and NO antiproliferative and antimigratory activity in VSMCs is promising. Platelet-derived growth factor, a known activator of VSMCs, has been shown to induce HO-1/CO expression in rat aortic VSMCs [187]. Endovascular balloon injury in a rat carotid injury model induced HO-1 expression and inhibited VSMC proliferation through endogenous CO [188]. Upregulation of HMOX1 expression in patients undergoing balloon angioplasty reduced the postangioplasty inflammatory response and reduced the incidence of restenosis [189]. Taken together, these findings suggest that HO-1/CO induction may exist as a cytoprotective mechanism in vascular injury and can be exploited as a therapeutic target in vascular disease. NO signaling has shown similar results. Animal models have consistently demonstrated decreased intimal hyperplasia in balloon-mediated vascular injury following pretreatment with L-arginine or NO donors or catheter-directed delivery of NO directly to the site of vascular injury [190–192].

The effects of L-arginine in a rabbit thoracic aortic injury model were reduced with coadministration of the NOS antagonist, L-N^G-nitro arginine methyl ester [190]. NO-releasing coatings and liposomal delivery have been explored as potential modalities for incorporating NO into vascular grafts to reduce incidence of graft failure [193,194].

VSMC Apoptosis

As reviewed earlier, both CO and NO have several salutary effects on VSMC physiology, largely mediated by sGC/cGMP-dependent pathways [195,196]. However, excess NO, secondary to induction of iNOS in VSMCs after vascular injury, leads to cytotoxicity. Oxidative stress and nitrosative stress have been shown to induce VSMC apoptosis, a key process that increases the vulnerability of atherosclerotic plaques to rupture [197,198]. Induction of HO-1 by NO, reviewed in earlier sections, has been demonstrated in VSMCs and has been shown to be an important negative feedback mechanism that reduces apoptosis mediated by cytotoxic concentrations of NO [136]. This has led to interest in HO-1 induction as a therapeutic modality for prevention of atherosclerotic plaque rupture. Of note, in addition to decreased incidence of plaque rupture, HO-1 induction in a rabbit model of vulnerable atherosclerotic plaques was shown to increase eNOS expression while reducing expression of iNOS [199]. HO-1 expression has also been shown to be more prevalent in carotid plaques obtained from asymptomatic patients [200]. Proposed mechanisms for the protective effects of HO-1 include inhibition of iNOS activity by CO binding to the catalytic heme and consumption of heme by HO-1, reducing availability for incorporation in iNOS [201].

Roles in Innate Immunity and Septic Shock

NO is produced in large quantities by macrophages, and to a lesser degree by neutrophils, from the iNOS, as part of the innate immune response to microbial invasion [202,203]. Transcriptional induction of iNOS, reviewed earlier as well, occurs secondary to stimulation by IFN-γ, proinflammatory cytokines, or bacterial products such as LPS. Induction is synergistic, with exposure to single cytokines insufficient to induce NOS2 expression due to the complexity of the NOS2 promoter and the need for sequential, cooperative binding by multiple transcription factors to induce transcription [203].

Once produced, NOS remains in primary and tertiary granules, in the mitochondria, or in association with the actin cytoskeleton until released

[203]. Under conditions of normoxia, NO is rapidly oxidized to its stable products, nitrite and nitrate, or combines with superoxide from NADPH oxidase or from the oxidase domain of iNOS (reviewed in earlier sections) to produce peroxynitrite and other RNS. The antimicrobial activity of NO is thought to be mediated by nonspecific reactions with structural elements or components of replication machinery, including DNA damage by oxidation or nitrosation and protein damage by direct binding to metalloenzymes or *S*-nitrosylation of thiol residues [204]. These effects can be directly cytotoxic or can cause sufficient impairment that the host immune response can resolve the infection [205].

While part of the host defense, excessive NO produced in the innate immune response has been implicated in the hypotension and vasoplegia characteristic of septic shock [206]. Circulating metabolites of NO rise progressively in animal models of septic shock [207]. NOS inhibitors have been shown to reduce hypotension and increase systemic vascular resistance in shock [208]. However, in animal models and clinical trials, NOS inhibitors have been associated with increased end-organ damage, decreased cardiac output, and increased mortality resulting in premature termination of a phase III clinical trial [209,210]. This is possibly due to inhibition of both iNOS and eNOS, resulting in excessive vasoconstriction leading to adverse effects.

As in many cells subject to prolonged exposure to NO, NO-dependent induction of HO-1 expression has been demonstrated in macrophages. In one study, NO-induced HO-1 expression was noted in LPS-treated rat Kupffer cells, the largest population of tissue macrophages but not in rat liver parenchymal cells [211]. Expression of HO-1 in macrophages has been shown to dampen inflammation by reducing STAT-1, MAPK, and NF-κB signaling and promoting the anti-inflammatory M2 phenotype [212,213]. Expression of HO-1 in tissue-resident macrophages in animal models has been shown to exert cytoprotective effects and reduce ischemia–reperfusion injury in both hepatic and renal parenchymal cells [214,215]. These findings have led to interest in augmenting CO/HO-1 signaling as a therapeutic intervention to mediate end-organ damage in endotoxemia-induced shock. In fact, in a study of mechanically ventilated patients with severe sepsis, endogenous CO production was higher in those who survived [216]. Low-dose CO administered by inhalation or by chemical CO-donor compounds improved survival in septic rats, mice, swine, and nonhuman primates [217]. CO was also shown in animal models to reduce the severity of LPS-mediated coagulopathy, renal and hepatic injury, and glycemic disorders [218,219]. Low-dose inhaled CO as a therapeutic in sepsis has been an area of interest and is being explored through phase I clinical trials [217].

Insulin Secretion and Pathogenesis of Diabetes

NO likely plays a dual role in the physiology of pancreatic islet β-cells. It has been shown to induce β-cell death and initiate the pathogenesis of diabetes as the primary effector of inflammatory cytokines, which induce iNOS expression [220]. However, at low concentrations, it has been shown to selectively promote β-cell survival by attenuating the DNA damage response; this protective response was not seen in other cell lines, including macrophages, hepatocytes, and fibroblasts [221].

NO has also been proposed to play a role in insulin secretion from pancreatic β-cells in concert with CO. Constitutive expression of nNOS has previously been demonstrated in β-cells [222]. Recently, it was shown that glucagon-like peptide 1 increased NO synthesis by phosphorylating and activating nNOS via PKA, which further increased insulin secretion by activation of HO-2. CO produced by HO-2 was shown to activate sGC, with downstream effects, including increased cytosolic Ca^{2+} concentration inducing insulin secretion. In the same study, administration of NO or CO donors to diabetic mice increased glucose tolerance, although concurrent administration did not. The latter observation was proposed to be likely due to negative feedback of CO on nNOS activity [223].

Conclusions

CO and NO are highly conserved signaling molecules with similarities in chemical structure leading to shared reactivity with metalloproteins. While initially found to be vasoactive substances, over the past three decades, our understanding of their physiological importance has expanded to include roles in every organ system. Their reactivity with common targets has led to complex relationships and regulatory feedback loops, including synergistic, counterregulatory, and compensatory interactions. This is particularly the case in inflammatory disorders, where NO has demonstrated both pathogenic roles and HO-1-mediated cytoprotective activity. As knowledge of these relationships continues to expand, we anticipate that the therapeutic potential of NO and CO signaling will be increasingly explored in the coming decades.

References

1. Lancaster, J.R. (2015). Nitric oxide: A brief overview of chemical and physical properties relevant to therapeutic applications. *Future Sci. OA* 1 (1): fso.15.59. https://doi.org/10.4155/fso.15.59.
2. Ernst, A. and Zibrak, J.D. (1998). Carbon monoxide poisoning. *N. Engl. J. Med.* 339 (22): 1603–1608. https://doi.org/10.1056/NEJM199811263392206.
3. Ignarro, L.J., Byrns, R.E., Buga, G.M., and Wood, K.S. (1987). Endothelium-derived relaxing factor from pulmonary artery and vein possesses pharmacologic and chemical properties identical to those of nitric oxide radical. *Circ. Res.* 61 (6): 866–879. https://doi.org/10.1161/01.RES.61.6.866.
4. Hartsfield, C.L. (2002). Cross talk between carbon monoxide and nitric oxide. *Antioxid. Redox Signal.* 4 (2): 301–307. https://doi.org/10.1089/152308602753666352.
5. Hughes, M.N. (1999). Relationships between nitric oxide, nitroxyl ion, nitrosonium cation and peroxynitrite. *Biochim. Biophys. Acta BBA – Bioenerg.* 1411 (2–3): 263–272. https://doi.org/10.1016/S0005-2728(99)00019-5.
6. Wang, R. (2002). Two's company, Three's a crowd: Can H2S be the third endogenous gaseous transmitter? *FASEB J.* 16 (13): 1792–1798. https://doi.org/10.1096/fj.02-0211hyp.
7. McCleverty, J.A. (2004). Chemistry of nitric oxide relevant to biology. *Chem. Rev.* 104 (2): 403–418. https://doi.org/10.1021/cr020623q.
8. Kaczorowski, D. and Zuckerbraun, B. (2007). Carbon monoxide: Medicinal chemistry and biological effects. *Curr. Med. Chem.* 14 (25): 2720–2725. https://doi.org/10.2174/092986707782023181.
9. Wink, D.A. and Mitchell, J.B. (1998). Chemical biology of nitric oxide: Insights into regulatory, cytotoxic, and cytoprotective mechanisms of nitric oxide. *Free Radic. Biol. Med.* 25 (4–5): 434–456. https://doi.org/10.1016/S0891-5849(98)00092-6.
10. Motterlini, R. and Foresti, R. (2017). Biological signaling by carbon monoxide and carbon monoxide-releasing molecules. *Am. J. Physiol.-Cell Physiol.* 312 (3): C302–C313. https://doi.org/10.1152/ajpcell.00360.2016.
11. Cooper, C.E. (1999). Nitric oxide and iron proteins. *Biochim. Biophys. Acta BBA – Bioenerg.* 1411 (2–3): 290–309. https://doi.org/10.1016/S0005-2728(99)00021-3.
12. Enemark, J.H. and Feltham, R.D. (1974). Principles of structure, bonding, and reactivity for metal nitrosyl complexes. *Coord. Chem. Rev.* 13 (4): 339–406. https://doi.org/10.1016/S0010-8545(00)80259-3.
13. Diefenbach, A., Bickelhaupt, F.M., and Frenking, G. (2000). The nature of the transition metal–carbonyl bond and the question about the valence orbitals of transition metals. A bond-energy decomposition analysis of $TM(CO)_6^q$ ($TM^q = Hf^{2-}$, Ta^-, W, Re^+, Os^{2+}, Ir^{3+})†. *J. Am. Chem. Soc.* 122 (27): 6449–6458. https://doi.org/10.1021/ja000663g.
14. Ford, P.C. and Lorkovic, I.M. (2002). Mechanistic aspects of the reactions of nitric oxide with transition-metal complexes. *Chem. Rev.* 102 (4): 993–1018. https://doi.org/10.1021/cr0000271.
15. Frenking, G., Loschen, C., Krapp, A., Fau, S., and Strauss, S.H. (2007). Electronic structure of CO—An exercise in modern chemical bonding theory. *J. Comput. Chem.* 28 (1): 117–126. https://doi.org/10.1002/jcc.20477.
16. Toledo, J.C. and Augusto, O. (2012). Connecting the chemical and biological properties of nitric oxide. *Chem. Res. Toxicol.* 25 (5): 975–989. https://doi.org/10.1021/tx300042g.
17. Foresti, R. and Motterlini, R. (1999). The heme oxygenase pathway and its interaction with nitric oxide in the control of cellular homeostasis. *Free Radic. Res.* 31 (6): 459–475. https://doi.org/10.1080/10715769900301031.
18. Eich, R.F., Li, T., Lemon, D.D., Doherty, D.H., Curry, S.R., Aitken, J.F., Mathews, A.J., Johnson, K.A., Smith, R.D., Phillips, G.N., and Olson, J.S. (1996). Mechanism of NO-Induced oxidation of myoglobin and hemoglobin. *Biochemistry* 35 (22): 6976–6983. https://doi.org/10.1021/bi960442g.
19. Wink, D.A., Hanbauer, I., Laval, F., Cook, J.A., Krishna, M.C., and Mitchell, J.B. (1994). Nitric oxide protects against the cytotoxic effects of reactive oxygen species. *Ann. N. Y. Acad. Sci.* 738: 265–278. https://doi.org/10.1111/j.1749-6632.1994.tb21812.x.
20. Kanner, J., Harel, S., and Granit, R. (1991). Nitric oxide as an antioxidant. *Arch. Biochem. Biophys.* 289 (1): 130–136. https://doi.org/10.1016/0003-9861(91)90452-o.
21. Epstein, F.H., Steinberg, D., Parthasarathy, S., Carew, T.E., Khoo, J.C., and Witztum, J.L. (1989). Beyond cholesterol. *N. Engl. J. Med.* 320 (14): 915–924. https://doi.org/10.1056/NEJM198904063201407.
22. Kanner, J., Harel, S., and Granit, R. (1992). Nitric oxide, an inhibitor of lipid oxidation by lipoxygenase, cyclooxygenase and hemoglobin. *Lipids* 27 (1): 46–49. https://doi.org/10.1007/BF02537058.
23. Wood, I., Trostchansky, A., and Rubbo, H. (2020). Structural considerations on lipoxygenase function, inhibition and crosstalk with nitric oxide pathways. *Biochimie* 178: 170–180. https://doi.org/10.1016/j.biochi.2020.09.021.

24. Beckman, J.S. and Koppenol, W.H. (1996). Nitric oxide, superoxide, and peroxynitrite: The good, the bad, and ugly. *Am. J. Physiol.-Cell Physiol.* 271 (5): C1424–C1437. https://doi.org/10.1152/ajpcell.1996.271.5.C1424.
25. Hogg, N. and Kalyanaraman, B. (1999). Nitric oxide and lipid peroxidation. *Biochim. Biophys. Acta BBA – Bioenerg.* 1411 (2–3): 378–384. https://doi.org/10.1016/S0005-2728(99)00027-4.
26. Denisov, I.G., Ikeda-Saito, M., Yoshida, T., and Sligar, S.G. (2002). Cryogenic absorption spectra of hydroperoxo-ferric heme oxygenase, the active intermediate of enzymatic heme oxygenation. *FEBS Lett.* 532 (1–2): 203–206. https://doi.org/10.1016/S0014-5793(02)03674-8.
27. Gorren, A.C.F. and Mayer, B. (2007). Nitric-Oxide synthase: a cytochrome P450 family foster child. *Biochim. Biophys. Acta BBA – Gen. Subj.* 1770 (3): 432–445. https://doi.org/10.1016/j.bbagen.2006.08.019.
28. Ortiz De Montellano, P.R., Nishida, C., Rodriguez-Crespo, I., and Gerber, N. (1998). Nitric oxide synthase structure and electron transfer. *Drug Metab. Dispos. Biol. Fate Chem.* 26 (12): 1185–1189.
29. Stuehr, D.J. (2004). Enzymes of the L-Arginine to nitric oxide pathway. *J. Nutr.* 134 (10): 2748S–2751S. https://doi.org/10.1093/jn/134.10.2748S.
30. Morris, S.M. and Billiar, T.R. (1994). New insights into the regulation of inducible nitric oxide synthesis. *Am. J. Physiol.-Endocrinol. Metab.* 266 (6): E829–E839. https://doi.org/10.1152/ajpendo.1994.266.6.E829.
31. Vodovotz, Y., Kwon, N.S., Pospischil, M., Manning, J., Paik, J., and Nathan, C. (1994). Inactivation of nitric oxide synthase after prolonged incubation of mouse macrophages with IFN-Gamma and bacterial lipopolysaccharide. *J. Immunol. Baltim. Md 1950* 152 (8): 4110–4118.
32. MacMicking, J., Xie, Q., and Nathan, C. (1997). Nitric oxide and macrophage function. *Annu. Rev. Immunol.* 15 (1): 323–350. https://doi.org/10.1146/annurev.immunol.15.1.323.
33. Alderton, W.K., Cooper, C.E., and Knowles, R.G. (2001). Nitric oxide synthases: Structure, function and inhibition. *Biochem. J.* 357 (3): 593–615. https://doi.org/10.1042/bj3570593.
34. Förstermann, U., Boissel, J., and Kleinert, H. (1998). expressional control of the 'constitutive' isoforms of nitric oxide synthase (nos i and nos iii). *FASEB J.* 12 (10): 773–790. https://doi.org/10.1096/fasebj.12.10.773.
35. Sharma, H.S., Westman, J., Alm, P., Sjöquist, P.O., Cervós-Navarro, J., and Nyberg, F. (1997). Involvement of nitric oxide in the pathophysiology of acute heat stress in the rat. Influence of a new antioxidant compound H-290/51. *Ann. N. Y. Acad. Sci.* 813: 581–590. https://doi.org/10.1111/j.1749-6632.1997.tb51749.x.
36. Reiser, P.J., Kline, W.O., and Vaghy, P.L. (1997). Induction of neuronal type nitric oxide synthase in skeletal muscle by chronic electrical stimulation in vivo. *J. Appl. Physiol.* 82 (4): 1250–1255. https://doi.org/10.1152/jappl.1997.82.4.1250.
37. Calza, L., Giardino, L., Pozza, M., Micera, A., and Aloe, L. (1997). Time-course changes of nerve growth factor, corticotropin-releasing hormone, and nitric oxide synthase isoforms and their possible role in the development of inflammatory response in experimental allergic encephalomyelitis. *Proc. Natl. Acad. Sci.* 94 (7): 3368–3373. https://doi.org/10.1073/pnas.94.7.3368.
38. Bandyopadhyay, A., Chakder, S., and Rattan, S. (1997). Regulation of inducible and neuronal nitric oxide synthase gene expression by interferon-gamma and VIP. *Am. J. Physiol.-Cell Physiol.* 272 (6): C1790–C1797. https://doi.org/10.1152/ajpcell.1997.272.6.C1790.
39. Lee, S., Barbanel, G., and Rivier, C. (1995). Systemic endotoxin increases steady-state gene expression of hypothalamic nitric oxide synthase: Comparison with corticotropin-releasing factor and vasopressin gene transcripts. *Brain Res.* 705 (1–2): 136–148. https://doi.org/10.1016/0006-8993(95)01142-0.
40. Kvietikova, I., Wenger, R.H., Marti, H.H., and Gassmann, M. (1995). The transcription factors ATF-1 and CREB-1 bind constitutively to the hypoxia-inducible factor-1 (HIF-1)DNA recognition site. *Nucleic Acids Res.* 23 (22): 4542–4550. https://doi.org/10.1093/nar/23.22.4542.
41. Shaul, P.W., North, A.J., Brannon, T.S., Ujiie, K., Wells, L.B., Nisen, P.A., Lowenstein, C.J., Snyder, S.H., and Star, R.A. (1995). Prolonged in vivo hypoxia enhances nitric oxide synthase type I and type III gene expression in adult rat lung. *Am. J. Respir. Cell Mol. Biol.* 13 (2): 167–174. https://doi.org/10.1165/ajrcmb.13.2.7542896.
42. Kadowaki, K., Kishimoto, J., Leng, G., and Emson, P.C. (1994). Up-regulation of nitric oxide synthase (NOS) gene expression together with NOS activity in the rat hypothalamo-hypophysial system after chronic salt loading: Evidence of a neuromodulatory role of nitric oxide in arginine vasopressin and oxytocin secretion. *Endocrinology* 134 (3): 1011–1017. https://doi.org/10.1210/endo.134.3.7509733.
43. O'Shea, R.D. and Gundlach, A.L. (1996). Food or water deprivation modulate nitric oxide synthase (NOS) activity and gene expression in rat hypothalamic neurones: Correlation with neurosecretory activity? *J.*

44. Singh, I., Grams, M., Wang, W.H., Yang, T., Killen, P., Smart, A., Schnermann, J., and Briggs, J.P. (1996). Coordinate regulation of renal expression of nitric oxide synthase, renin, and angiotensinogen MRNA by dietary salt. *Am. J. Physiol.* 270 (6 Pt 2): F1027–1037. https://doi.org/10.1152/ajprenal.1996.270.6.F1027.

45. Galea, E. and Feinstein, D.L. (1999). Regulation of the expression of the inflammatory nitric oxide synthase (NOS2) by cyclic AMP. *FASEB J.* 13 (15): 2125–2137. https://doi.org/10.1096/fasebj.13.15.2125.

46. Nussler, A.K., Di Silvio, M., Billiar, T.R., Hoffman, R.A., Geller, D.A., Selby, R., Madariaga, J., and Simmons, R.L. (1992). Stimulation of the nitric oxide synthase pathway in human hepatocytes by cytokines and endotoxin. *J. Exp. Med.* 176 (1): 261–264. https://doi.org/10.1084/jem.176.1.261.

47. Geller, D.A., Nussler, A.K., Di Silvio, M., Lowenstein, C.J., Shapiro, R.A., Wang, S.C., Simmons, R.L., and Billiar, T.R. (1993). Cytokines, endotoxin, and glucocorticoids regulate the expression of inducible nitric oxide synthase in hepatocytes. *Proc. Natl. Acad. Sci.* 90 (2): 522–526. https://doi.org/10.1073/pnas.90.2.522.

48. Kleinert, H., Schwarz, P.M., and Förstermann, U. (2003). Regulation of the expression of inducible nitric oxide synthase. *Biol. Chem.* 384: 10–11. https://doi.org/10.1515/BC.2003.152.

49. De Vera, M.E., Shapiro, R.A., Nussler, A.K., Mudgett, J.S., Simmons, R.L., Morris, S.M., Billiar, T.R., and Geller, D.A. (1996). Transcriptional regulation of human inducible nitric oxide synthase (NOS2) gene by cytokines: initial analysis of the human NOS2 promoter. *Proc. Natl. Acad. Sci.* 93 (3): 1054–1059. https://doi.org/10.1073/pnas.93.3.1054.

50. Taylor, B.S., De Vera, M.E., Ganster, R.W., Wang, Q., Shapiro, R.A., Morris, S.M., Billiar, T.R., and Geller, D.A. (1998). Multiple NF-ΚB enhancer elements regulate cytokine induction of the human inducible nitric oxide synthase gene. *J. Biol. Chem.* 273 (24): 15148–15156. https://doi.org/10.1074/jbc.273.24.15148.

51. Geller, D.A. and Billiar, T.R. (1998). Molecular biology of nitric oxide synthases. *Cancer Metastasis Rev.* 17 (1): 7–23. https://doi.org/10.1023/A:1005940202801.

52. Farlik, M., Reutterer, B., Schindler, C., Greten, F., Vogl, C., Müller, M., and Decker, T. (2010). Nonconventional initiation complex assembly by STAT and NF-KappaB transcription factors regulates nitric oxide synthase expression. *Immunity* 33 (1): 25–34. https://doi.org/10.1016/j.immuni.2010.07.001.

53. Chapman, R.D., Heidemann, M., Hintermair, C., and Eick, D. (2008). Molecular evolution of the RNA polymerase II CTD. *Trends Genet. TIG* 24 (6): 289–296. https://doi.org/10.1016/j.tig.2008.03.010.

54. Hirose, Y. and Ohkuma, Y. (2007). Phosphorylation of the C-terminal domain of RNA polymerase II plays central roles in the integrated events of eucaryotic gene expression. *J. Biochem. (Tokyo)* 141 (5): 601–608. https://doi.org/10.1093/jb/mvm090.

55. Kleinert, H., Euchenhofer, C., Ihrig-Biedert, I., and Förstermann, U. (1996). Glucocorticoids inhibit the induction of nitric oxide synthase II by down-regulating cytokine-induced activity of transcription factor nuclear factor-kappa B. *Mol. Pharmacol.* 49 (1): 15–21.

56. Forrester, K., Ambs, S., Lupold, S.E., Kapust, R.B., Spillare, E.A., Weinberg, W.C., Felley-Bosco, E., Wang, X.W., Geller, D.A., Tzeng, E., Billiar, T.R., and Harris, C.C. (1996). Nitric oxide-induced P53 accumulation and regulation of inducible nitric oxide synthase expression by wild-type P53. *Proc. Natl. Acad. Sci.* 93 (6): 2442–2447. https://doi.org/10.1073/pnas.93.6.2442.

57. Yin, J.H., Yang, D.I., Ku, G., and Hsu, C.Y. (2000). INOS expression inhibits hypoxia-inducible factor-1 activity. *Biochem. Biophys. Res. Commun.* 279 (1): 30–34. https://doi.org/10.1006/bbrc.2000.3896.

58. Katsuyama, K., Shichiri, M., Marumo, F., and Hirata, Y. (1998). NO inhibits cytokine-induced INOS expression and NF-KappaB activation by interfering with phosphorylation and degradation of IkappaB-Alpha. *Arterioscler. Thromb. Vasc. Biol.* 18 (11): 1796–1802. https://doi.org/10.1161/01.atv.18.11.1796.

59. Llovera, M., Pearson, J.D., Moreno, C., and Riveros-Moreno, V. (2001). Impaired response to interferon-γ in activated macrophages due to tyrosine nitration of STAT1 by endogenous nitric oxide: Tyrosine nitration impairs macrophage STAT1 signalling. *Br. J. Pharmacol.* 132 (2): 419–426. https://doi.org/10.1038/sj.bjp.0703838.

60. Caput, D., Beutler, B., Hartog, K., Thayer, R., Brown-Shimer, S., and Cerami, A. (1986). Identification of a common nucleotide sequence in the 3′-untranslated region of MRNA molecules specifying inflammatory mediators. *Proc. Natl. Acad. Sci.* 83 (6): 1670–1674. https://doi.org/10.1073/pnas.83.6.1670.

61. Geller, D.A., Lowenstein, C.J., Shapiro, R.A., Nussler, A.K., Di Silvio, M., Wang, S.C., Nakayama, D.K., Simmons, R.L., Snyder, S.H., and Billiar, T.R. (1993). Molecular cloning and expression of inducible nitric oxide synthase from human hepatocytes. *Proc. Natl. Acad. Sci.* 90 (8): 3491–3495. https://doi.org/10.1073/pnas.90.8.3491.

62. Fechir, M., Linker, K., Pautz, A., Hubrich, T., and Kleinert, H. (2005). The RNA binding protein TIAR is involved in the regulation of human INOS expression. *Cell. Mol. Biol. Noisy–Gd. Fr.* 51 (3): 299–305.

63. Vodovotz, Y., Bogdan, C., Paik, J., Xie, Q.W., and Nathan, C. (1993). Mechanisms of suppression of macrophage nitric oxide release by transforming growth factor beta. *J. Exp. Med.* 178 (2): 605–613. https://doi.org/10.1084/jem.178.2.605.

64. Xiao, Z., Zhang, Z., Ranjan, V., and Diamond, S.L. (1997). Shear stress induction of the endothelial nitric oxide synthase gene is calcium-dependent but not calcium-activated. *J. Cell. Physiol.* 171 (2): 205–211. https://doi.org/10.1002/(SICI)1097-4652(199705)171:2<205::AID-JCP11>3.0.CO;2-C.

65. Marsden, P.A., Heng, H.H., Scherer, S.W., Stewart, R.J., Hall, A.V., Shi, X.M., Tsui, L.C., and Schappert, K.T. (1993). Structure and chromosomal localization of the human constitutive endothelial nitric oxide synthase gene. *J. Biol. Chem.* 268 (23): 17478–17488.

66. Ziesche, R., Petkov, V., Williams, J., Zakeri, S.M., Mosgöller, W., Knöfler, M., and Block, L.H. (1996). Lipopolysaccharide and interleukin 1 augment the effects of hypoxia and inflammation in human pulmonary arterial tissue. *Proc. Natl. Acad. Sci. U. S. A.* 93 (22): 12478–12483. https://doi.org/10.1073/pnas.93.22.12478.

67. Dai, A., Zhang, Z., and Niu, R. (1995). [The effects of hypoxia on nitric oxide synthase activity and mRNA expression of pulmonary artery endothelial cells in pigs]. *Zhonghua Jie He He Hu Xi Za Zhi Zhonghua Jiehe He Huxi Zazhi Chin. J. Tuberc. Respir. Dis.* 18 (3): 164–166, 191.

68. Liao, J.K., Zulueta, J.J., Yu, F.S., Peng, H.B., Cote, C.G., and Hassoun, P.M. (1995). Regulation of Bovine endothelial constitutive nitric oxide synthase by oxygen. *J. Clin. Invest.* 96 (6): 2661–2666. https://doi.org/10.1172/JCI118332.

69. Stuehr, D.J. and Haque, M.M. (2019). Nitric oxide synthase enzymology in the 20 years after the nobel prize. *Br. J. Pharmacol.* 176 (2): 177–188. https://doi.org/10.1111/bph.14533.

70. Garcin, E.D., Bruns, C.M., Lloyd, S.J., Hosfield, D.J., Tiso, M., Gachhui, R., Stuehr, D.J., Tainer, J.A., and Getzoff, E.D. (2004). Structural basis for isozyme-specific regulation of electron transfer in nitric-oxide synthase. *J. Biol. Chem.* 279 (36): 37918–37927. https://doi.org/10.1074/jbc.M406204200.

71. Smith, B.C., Underbakke, E.S., Kulp, D.W., Schief, W.R., and Marletta, M.A. (2013). Nitric oxide synthase domain interfaces regulate electron transfer and calmodulin activation. *Proc. Natl. Acad. Sci.* 110 (38): E3577–E3586. https://doi.org/10.1073/pnas.1313331110.

72. Zhu, Y. and Silverman, R.B. (2008). Revisiting heme mechanisms. A perspective on the mechanisms of nitric oxide synthase (NOS), heme oxygenase (HO), and cytochrome P450s (CYP450s). *Biochemistry* 47 (8): 2231–2243. https://doi.org/10.1021/bi7023817.

73. Tejero, J. and Stuehr, D. (2013). Tetrahydrobiopterin in nitric oxide synthase. *IUBMB Life* 65 (4): 358–365. https://doi.org/10.1002/iub.1136.

74. Kavya, R., Saluja, R., Singh, S., and Dikshit, M. (2006). Nitric oxide synthase regulation and diversity: Implications in Parkinson's disease. *Nitric Oxide* 15 (4): 280–294. https://doi.org/10.1016/j.niox.2006.07.003.

75. Chin, D. and Means, A.R. (2000). Calmodulin: A prototypical calcium sensor. *Trends Cell Biol.* 10 (8): 322–328. https://doi.org/10.1016/S0962-8924(00)01800-6.

76. Kawasaki, H., Soma, N., and Kretsinger, R.H. (2019). Molecular dynamics study of the changes in conformation of calmodulin with calcium binding and/or target recognition. *Sci. Rep.* 9 (1): 10688. https://doi.org/10.1038/s41598-019-47063-1.

77. Mehler, E.L., Pascual-Ahuir, J.-L., and Weinstein, H. (1991). Structural dynamics of calmodulin and troponin C. *Protein Eng. Des. Sel.* 4 (6): 625–637. https://doi.org/10.1093/protein/4.6.625.

78. Kretsinger, R.H., Rudnick, S.E., and Weissman, L.J. (1986). Crystal structure of calmodulin. *J. Inorg. Biochem.* 28 (2–3): 289–302. https://doi.org/10.1016/0162-0134(86)80093-9.

79. Iyanagi, T., Xia, C., and Kim, -J.-J.P. (2012). NADPH–Cytochrome P450 oxidoreductase: Prototypic member of the diflavin reductase family. *Arch. Biochem. Biophys.* 528 (1): 72–89. https://doi.org/10.1016/j.abb.2012.09.002.

80. Salerno, J.C., Ray, K., Poulos, T., Li, H., and Ghosh, D.K. (2013). Calmodulin activates neuronal nitric oxide synthase by enabling transitions between conformational states. *FEBS Lett.* 587 (1): 44–47. https://doi.org/10.1016/j.febslet.2012.10.039.

81. Zheng, H., Li, J., and Feng, C. (2020). Heat shock protein 90 enhances the electron transfer between the FMN and heme cofactors in neuronal nitric oxide synthase. *FEBS Lett.* 594 (17): 2904–2913. https://doi.org/10.1002/1873-3468.13870.

82. Alderton, W.K., Cooper, C.E., and Knowles, R.G. (2001). Nitric oxide synthases: structure, function and inhibition. *Biochem. J.* 357 (Pt 3): 593–615. https://doi.org/10.1042/0264-6021:3570593.

83. Schopf, F.H., Biebl, M.M., and Buchner, J. (2017). The HSP90 chaperone machinery. *Nat. Rev. Mol. Cell Biol.* 18 (6): 345–360. https://doi.org/10.1038/nrm.2017.20.

84. García-Cardeña, G., Fan, R., Shah, V., Sorrentino, R., Cirino, G., Papapetropoulos, A., and Sessa, W.C. (1998). Dynamic activation of endothelial nitric oxide synthase by Hsp90. *Nature* 392 (6678): 821–824. https://doi.org/10.1038/33934.

85. Bender, A.T., Silverstein, A.M., Demady, D.R., Kanelakis, K.C., Noguchi, S., Pratt, W.B., and Osawa, Y. (1999). Neuronal nitric-oxide synthase is regulated by the Hsp90-Based chaperone system in vivo. *J. Biol. Chem.* 274 (3): 1472–1478. https://doi.org/10.1074/jbc.274.3.1472.
86. Yoshida, M. and Xia, Y. (2003). Heat shock protein 90 as an endogenous protein enhancer of inducible nitric-oxide synthase. *J. Biol. Chem.* 278 (38): 36953–36958. https://doi.org/10.1074/jbc.M305214200.
87. Averna, M., Stifanese, R., De Tullio, R., Salamino, F., Pontremoli, S., and Melloni, E. (2008). In vivo degradation of nitric oxide synthase (NOS) and heat shock protein 90 (HSP90) by calpain is modulated by the formation of a NOS-HSP90 heterocomplex: In vivo degradation of NOS and HSP90 by calpain. *FEBS J.* 275 (10): 2501–2511. https://doi.org/10.1111/j.1742-4658.2008.06394.x.
88. Mount, P.F., Kemp, B.E., and Power, D.A. (2007). Regulation of endothelial and myocardial NO synthesis by multi-site ENOS phosphorylation. *J. Mol. Cell. Cardiol.* 42 (2): 271–279. https://doi.org/10.1016/j.yjmcc.2006.05.023.
89. Chen, Z.-P., Mitchelhill, K.I., Michell, B.J., Stapleton, D., Rodriguez-Crespo, I., Witters, L.A., Power, D.A., Ortiz De Montellano, P.R., and Kemp, B.E. (1999). AMP-Activated protein kinase phosphorylation of endothelial NO synthase. *FEBS Lett.* 443 (3): 285–289. https://doi.org/10.1016/S0014-5793(98)01705-0.
90. Boo, Y.C., Sorescu, G., Boyd, N., Shiojima, I., Walsh, K., Du, J., and Jo, H. (2002). Shear stress stimulates phosphorylation of endothelial nitric-oxide synthase at Ser1179 by Akt-independent mechanisms. *J. Biol. Chem.* 277 (5): 3388–3396. https://doi.org/10.1074/jbc.M108789200.
91. Harris, M.B., Ju, H., Venema, V.J., Liang, H., Zou, R., Michell, B.J., Chen, Z.-P., Kemp, B.E., and Venema, R.C. (2001). Reciprocal phosphorylation and regulation of endothelial nitric-oxide synthase in response to bradykinin stimulation. *J. Biol. Chem.* 276 (19): 16587–16591. https://doi.org/10.1074/jbc.M100229200.
92. Montagnani, M., Chen, H., Barr, V.A., and Quon, M.J. (2001). Insulin-stimulated activation of ENOS is independent of Ca^{2+} but requires phosphorylation by Akt at Ser^{1179}. *J. Biol. Chem.* 276 (32): 30392–30398. https://doi.org/10.1074/jbc.M103702200.
93. Harris, M.B., Blackstone, M.A., Sood, S.G., Li, C., Goolsby, J.M., Venema, V.J., Kemp, B.E., and Venema, R.C. (2004). Acute activation and phosphorylation of endothelial nitric oxide synthase by HMG-CoA reductase inhibitors. *Am. J. Physiol. Heart Circ. Physiol.* 287 (2): H560–566. https://doi.org/10.1152/ajpheart.00214.2004.
94. Lane, P. and Gross, S.S. (2002). Disabling a C-terminal autoinhibitory control element in endothelial nitric-oxide synthase by phosphorylation provides a molecular explanation for activation of vascular NO synthesis by diverse physiological stimuli. *J. Biol. Chem.* 277 (21): 19087–19094. https://doi.org/10.1074/jbc.M200258200.
95. Zhou, L. and Zhu, D.-Y. (2009). Neuronal nitric oxide synthase: Structure, subcellular localization, regulation, and clinical implications. *Nitric Oxide* 20 (4): 223–230. https://doi.org/10.1016/j.niox.2009.03.001.
96. Hayashi, Y., Nishio, M., Naito, Y., Yokokura, H., Nimura, Y., Hidaka, H., and Watanabe, Y. (1999). Regulation of neuronal nitric-oxide synthase by Calmodulin Kinases. *J. Biol. Chem.* 274 (29): 20597–20602. https://doi.org/10.1074/jbc.274.29.20597.
97. Komeima, K., Hayashi, Y., Naito, Y., and Watanabe, Y. (2000). Inhibition of neuronal nitric-oxide synthase by calcium/ calmodulin-dependent protein kinase IIα through Ser847 phosphorylation in NG108-15 neuronal cells. *J. Biol. Chem.* 275 (36): 28139–28143. https://doi.org/10.1074/jbc.M003198200.
98. Adak, S., Santolini, J., Tikunova, S., Wang, Q., Johnson, J.D., and Stuehr, D.J. (2001). Neuronal nitric-oxide synthase mutant (Ser-1412 → Asp) demonstrates surprising connections between heme reduction, NO complex formation, and catalysis. *J. Biol. Chem.* 276 (2): 1244–1252. https://doi.org/10.1074/jbc.M006857200.
99. Song, T., Hatano, N., Horii, M., Tokumitsu, H., Yamaguchi, F., Tokuda, M., and Watanabe, Y. (2004). Calcium/calmodulin-dependent protein kinase I inhibits neuronal nitric-oxide synthase activity through serine 741 phosphorylation. *FEBS Lett.* 570 (1–3): 133–137. https://doi.org/10.1016/j.febslet.2004.05.083.
100. Klatt, P., Pfeiffer, S., List, B.M., Lehner, D., Glatter, O., Bächinger, H.P., Werner, E.R., Schmidt, K., and Mayer, B. (1996). Characterization of heme-deficient neuronal nitric-oxide synthase reveals a role for heme in subunit dimerization and binding of the amino acid substrate and tetrahydrobiopterin. *J. Biol. Chem.* 271 (13): 7336–7342. https://doi.org/10.1074/jbc.271.13.7336.
101. Klatt, P., Schmidt, K., Lehner, D., Glatter, O., Bächinger, H.P., and Mayer, B. (1995). Structural analysis of porcine brain nitric oxide synthase reveals a role for tetrahydrobiopterin and L-arginine in the formation of an SDS-resistant dimer. *EMBO J.* 14 (15): 3687–3695.
102. Baek, K.J., Thiel, B.A., Lucas, S., and Stuehr, D.J. (1993). Macrophage nitric oxide synthase subunits. Purification, characterization, and role of prosthetic

groups and substrate in regulating their association into a dimeric enzyme. *J. Biol. Chem.* 268 (28): 21120–21129.

103. Hemmens, B., Goessler, W., Schmidt, K., and Mayer, B. (2000). Role of bound zinc in dimer stabilization but not enzyme activity of neuronal nitric-oxide synthase. *J. Biol. Chem.* 275 (46): 35786–35791. https://doi.org/10.1074/jbc.M005976200.

104. Li, H., Raman, C.S., Glaser, C.B., Blasko, E., Young, T.A., Parkinson, J.F., Whitlow, M., and Poulos, T.L. (1999). Crystal structures of zinc-free and -bound heme domain of human inducible nitric-oxide synthase. *J. Biol. Chem.* 274 (30): 21276–21284. https://doi.org/10.1074/jbc.274.30.21276.

105. Zou, M.-H., Shi, C., and Cohen, R.A. (2002). Oxidation of the zinc-thiolate complex and uncoupling of endothelial nitric oxide synthase by peroxynitrite. *J. Clin. Invest.* 109 (6): 817–826. https://doi.org/10.1172/JCI0214442.

106. Panda, K., Ghosh, S., and Stuehr, D.J. (2001). Calmodulin activates intersubunit electron transfer in the neuronal nitric-oxide synthase dimer. *J. Biol. Chem.* 276 (26): 23349–23356. https://doi.org/10.1074/jbc.M100687200.

107. Siddhanta, U., Presta, A., Fan, B., Wolan, D., Rousseau, D.L., and Stuehr, D.J. (1998). Domain swapping in inducible nitric-oxide synthase. Electron transfer occurs between flavin and heme groups located on adjacent subunits in the dimer. *J. Biol. Chem.* 273 (30): 18950–18958. https://doi.org/10.1074/jbc.273.30.18950.

108. Ratovitski, E.A., Bao, C., Quick, R.A., McMillan, A., Kozlovsky, C., and Lowenstein, C.J. (1999). An inducible nitric-oxide synthase (NOS)-Associated protein inhibits NOS dimerization and activity. *J. Biol. Chem.* 274 (42): 30250–30257. https://doi.org/10.1074/jbc.274.42.30250.

109. Youn, H., Ji, I., Ji, H.P., Markesbery, W.R., and Ji, T.H. (2007). Under-expression of Kalirin-7 increases INOS activity in cultured cells and correlates to elevated INOS activity in alzheimer's disease hippocampus1. *J. Alzheimers Dis.* 12 (3): 271–281. https://doi.org/10.3233/JAD-2007-12309.

110. Fernández, M.L., Martí, M.A., Crespo, A., and Estrin, D.A. (2005). Proximal effects in the modulation of nitric oxide synthase reactivity: a QM-MM study. *JBIC J. Biol. Inorg. Chem.* 10 (6): 595–604. https://doi.org/10.1007/s00775-005-0004-6.

111. Tejero, J., Hunt, A.P., Santolini, J., Lehnert, N., and Stuehr, D.J. (2019). Mechanism and regulation of ferrous heme-nitric oxide (NO) oxidation in NO synthases. *J. Biol. Chem.* 294 (19): 7904–7916. https://doi.org/10.1074/jbc.RA119.007810.

112. Smith, B.C., Fernhoff, N.B., and Marletta, M.A. (2012). Mechanism and kinetics of inducible nitric oxide synthase Auto- *S* -Nitrosation and inactivation. *Biochemistry* 51 (5): 1028–1040. https://doi.org/10.1021/bi201818c.

113. Thorup, C., Jones, C.L., Gross, S.S., Moore, L.C., and Goligorsky, M.S. (1999). Carbon monoxide induces vasodilation and nitric oxide release but suppresses endothelial NOS. *Am. J. Physiol.-Ren. Physiol.* 277 (6): F882–F889. https://doi.org/10.1152/ajprenal.1999.277.6.F882.

114. Maines, M.D. (1997). The heme oxygenase system: A regulator of second messenger gases. *Annu. Rev. Pharmacol. Toxicol.* 37 (1): 517–554. https://doi.org/10.1146/annurev.pharmtox.37.1.517.

115. Weiss, G., Werner-Felmayer, G., Werner, E.R., Grünewald, K., Wachter, H., and Hentze, M.W. (1994). Iron regulates nitric oxide synthase activity by controlling nuclear transcription. *J. Exp. Med.* 180 (3): 969–976. https://doi.org/10.1084/jem.180.3.969.

116. White, K.A. and Marletta, M.A. (1992). Nitric oxide synthase is a cytochrome P-450 type hemoprotein. *Biochemistry* 31 (29): 6627–6631. https://doi.org/10.1021/bi00144a001.

117. Stroes, E., Hijmering, M., Van Zandvoort, M., Wever, R., Rabelink, T.J., and Van Faassen, E.E. (1998). Origin of superoxide production by endothelial nitric oxide synthase. *FEBS Lett.* 438 (3): 161–164. https://doi.org/10.1016/S0014-5793(98)01292-7.

118. Xia, Y., Roman, L.J., Masters, B.S.S., and Zweier, J.L. (1998). Inducible nitric-oxide synthase generates superoxide from the reductase domain. *J. Biol. Chem.* 273 (35): 22635–22639. https://doi.org/10.1074/jbc.273.35.22635.

119. Vásquez-Vivar, J., Kalyanaraman, B., Martásek, P., Hogg, N., Masters, B.S.S., Karoui, H., Tordo, P., and Pritchard, K.A. (1998). Superoxide generation by endothelial nitric oxide synthase: The influence of cofactors. *Proc. Natl. Acad. Sci.* 95 (16): 9220–9225. https://doi.org/10.1073/pnas.95.16.9220.

120. Heinzel, B., John, M., Klatt, P., Böhme, E., and Mayer, B. (1992). Ca2+/Calmodulin-dependent formation of hydrogen peroxide by brain nitric oxide synthase. *Biochem. J.* 281 (3): 627–630. https://doi.org/10.1042/bj2810627.

121. Luo, S., Lei, H., Qin, H., and Xia, Y. (2014). Molecular mechanisms of endothelial NO synthase uncoupling. *Curr. Pharm. Des.* 20 (22): 3548–3553. https://doi.org/10.2174/13816128113196660746.

122. Xia, Y., Tsai, A.-L., Berka, V., and Zweier, J.L. (1998). Superoxide generation from endothelial nitric-oxide synthase. *J. Biol. Chem.* 273 (40): 25804–25808. https://doi.org/10.1074/jbc.273.40.25804.

123. Gebhart, V., Reiß, K., Kollau, A., Mayer, B., and Gorren, A.C.F. (2019). Site and mechanism of uncoupling of nitric-oxide synthase: Uncoupling by monomerization and other misconceptions. *Nitric Oxide* 89: 14–21. https://doi.org/10.1016/j.niox.2019.04.007.

124. Xia, Y., Dawson, V.L., Dawson, T.M., Snyder, S.H., and Zweier, J.L. (1996). Nitric oxide synthase generates superoxide and nitric oxide in arginine-depleted cells leading to peroxynitrite-mediated cellular injury. *Proc. Natl. Acad. Sci.* 93 (13): 6770–6774. https://doi.org/10.1073/pnas.93.13.6770.

125. Vásquez-Vivar, J., Hogg, N., Martásek, P., Karoui, H., Pritchard, K.A., and Kalyanaraman, B. (1999). Tetrahydrobiopterin-dependent inhibition of superoxide generation from neuronal nitric oxide synthase. *J. Biol. Chem.* 274 (38): 26736–26742. https://doi.org/10.1074/jbc.274.38.26736.

126. Xia, Y. and Zweier, J.L. (1997). Superoxide and peroxynitrite generation from inducible nitric oxide synthase in macrophages. *Proc. Natl. Acad. Sci.* 94 (13): 6954–6958. https://doi.org/10.1073/pnas.94.13.6954.

127. Ortiz De Montellano, P. (2000). The mechanism of heme oxygenase. *Curr. Opin. Chem. Biol.* 4 (2): 221–227. https://doi.org/10.1016/S1367-5931%2899%2900079-4.

128. Unno, M., Matsui, T., and Ikeda-Saito, M. (2007). Structure and catalytic mechanism of heme oxygenase. *Nat. Prod. Rep.* 24 (3): 553. https://doi.org/10.1039/b604180a.

129. Wilks, A. (2002). Heme oxygenase: Evolution, structure, and mechanism. *Antioxid. Redox Signal.* 4 (4): 603–614. https://doi.org/10.1089/15230860260220102.

130. Zuckerbraun, B. (2003). Heme oxygenase-1: A cellular hercules. *Hepatology* 37 (4): 742–744. https://doi.org/10.1053/jhep.2003.50139.

131. Jeney, V., Balla, J., Yachie, A., Varga, Z., Vercellotti, G.M., Eaton, J.W., and Balla, G. (2002). Pro-oxidant and cytotoxic effects of circulating heme. *Blood* 100 (3): 879–887. https://doi.org/10.1182/blood.V100.3.879.

132. Min, K.-S., Hwang, Y.-H., Ju, H.-J., Chang, H.-S., Kang, K.-H., Pi, S.-H., Lee, S.-K., Lee, S.-K., and Kim, E.-C. (2006). Heme oxygenase-1 mediates cytoprotection against nitric oxide-induced cytotoxicity via the CGMP pathway in human pulp cells. *Oral Surg. Oral Med. Oral Pathol. Oral Radiol. Endod.* 102 (6): 803–808. https://doi.org/10.1016/j.tripleo.2005.11.036.

133. Yee, E.L., Pitt, B.R., Billiar, T.R., and Kim, Y.M. (1996). Effect of nitric oxide on heme metabolism in pulmonary artery endothelial cells. *Am. J. Physiol.* 271 (4 Pt 1): L512–518. https://doi.org/10.1152/ajplung.1996.271.4.L512.

134. Hartsfield, C.L., Alam, J., Cook, J.L., and Choi, A.M.K. (1997). Regulation of heme oxygenase-1 gene expression in vascular smooth muscle cells by nitric oxide. *Am. J. Physiol.-Lung Cell. Mol. Physiol.* 273 (5): L980–L988. https://doi.org/10.1152/ajplung.1997.273.5.L980.

135. Buckley, B.J., Marshall, Z.M., and Whorton, A.R. (2003). Nitric oxide stimulates Nrf2 nuclear translocation in vascular endothelium. *Biochem. Biophys. Res. Commun.* 307 (4): 973–979. https://doi.org/10.1016/S0006-291X%2803%2901308-1.

136. Liu, X., Peyton, K., Ensenat, D., Wang, H., Hannink, M., Alam, J., and Durante, W. (2007). Nitric oxide stimulates heme oxygenase-1 gene transcription via the Nrf2/ARE complex to promote vascular smooth muscle cell survival. *Cardiovasc. Res.* 75 (2): 381–389. https://doi.org/10.1016/j.cardiores.2007.03.004.

137. Li, C.-Q., Kim, M.Y., Godoy, L.C., Thiantanawat, A., Trudel, L.J., and Wogan, G.N. (2009). Nitric oxide activation of keap1/Nrf2 signaling in human colon carcinoma cells. *Proc. Natl. Acad. Sci.* 106 (34): 14547–14551. https://doi.org/10.1073/pnas.0907539106.

138. Kang, M.-I., Kobayashi, A., Wakabayashi, N., Kim, S.-G., and Yamamoto, M. (2004). Scaffolding of keap1 to the actin cytoskeleton controls the function of Nrf2 as key regulator of cytoprotective phase 2 genes. *Proc. Natl. Acad. Sci. U. S. A.* 101 (7): 2046–2051. https://doi.org/10.1073/pnas.0308347100.

139. Zhang, D.D. and Hannink, M. (2003). Distinct cysteine residues in keap1 are required for keap1-dependent ubiquitination of Nrf2 and for stabilization of Nrf2 by chemopreventive agents and oxidative stress. *Mol. Cell. Biol.* 23 (22): 8137–8151. https://doi.org/10.1128/MCB.23.22.8137-8151.2003.

140. Um, H.-C., Jang, J.-H., Kim, D.-H., Lee, C., and Surh, Y.-J. (2011). Nitric oxide activates Nrf2 through S-nitrosylation of keap1 in PC12 cells. *Nitric Oxide* 25 (2): 161–168. https://doi.org/10.1016/j.niox.2011.06.001.

141. Alam, J., Stewart, D., Touchard, C., Boinapally, S., Choi, A.M., and Cook, J.L. (1999). Nrf2, a Cap'n'Collar transcription factor, regulates induction of the heme oxygenase-1 gene. *J. Biol. Chem.* 274 (37): 26071–26078. https://doi.org/10.1074/jbc.274.37.26071.

142. Sasaki, H., Sato, H., Kuriyama-Matsumura, K., Sato, K., Maebara, K., Wang, H., Tamba, M., Itoh, K., Yamamoto, M., and Bannai, S. (2002). Electrophile

143. Polte, T., Abate, A., Dennery, P.A., and Schröder, H. (2000). Heme oxygenase-1 is a CGMP-inducible endothelial protein and mediates the cytoprotective action of nitric oxide. *Arterioscler. Thromb. Vasc. Biol.* 20 (5): 1209–1215. https://doi.org/10.1161/01.ATV.20.5.1209.

response element-mediated induction of the cystine/glutamate exchange transporter gene expression. *J. Biol. Chem.* 277 (47): 44765–44771. https://doi.org/10.1074/jbc.M208704200.

144. Wang, J., Lu, S., Moënne-Loccoz, P., and Ortiz De Montellano, P.R. (2003). Interaction of nitric oxide with human heme oxygenase-1. *J. Biol. Chem.* 278 (4): 2341–2347. https://doi.org/10.1074/jbc.M211131200.

145. Ding, Y., McCoubrey, W.K., and Maines, M.D. (1999). Interaction of heme oxygenase-2 with nitric oxide donors. is the oxygenase an intracellular "sink" for NO? *Eur. J. Biochem.* 264 (3): 854–861. https://doi.org/10.1046/j.1432-1327.1999.00677.x.

146. Zhang, L. and Guarente, L. (1995). Heme binds to a short sequence that serves a regulatory function in diverse proteins. *EMBO J.* 14 (2): 313–320. https://doi.org/10.1002/j.1460-2075.1995.tb07005.x.

147. Liu, L., Dumbrepatil, A.B., Fleischhacker, A.S., Marsh, E.N.G., and Ragsdale, S.W. (2020). Heme oxygenase-2 is post-translationally regulated by heme occupancy in the catalytic site. *J. Biol. Chem.* 295 (50): 17227–17240. https://doi.org/10.1074/jbc.RA120.014919.

148. Huang, T.J., McCoubrey, W.K., and Maines, M.D. (2001). Heme oxygenase-2 interaction with metalloporphyrins: Function of heme regulatory motifs. *Antioxid. Redox Signal.* 3 (4): 685–696. https://doi.org/10.1089/15230860152543023.

149. Denninger, J.W. and Marletta, M.A. (1999). Guanylate cyclase and the · NO/CGMP signaling pathway. *Biochim. Biophys. Acta BBA – Bioenerg.* 1411 (2–3): 334–350. https://doi.org/10.1016/S0005-2728(99)00024-9.

150. Pilz, R.B. and Casteel, D.E. (2003). Regulation of gene expression by cyclic GMP. *Circ. Res.* 93 (11): 1034–1046. https://doi.org/10.1161/01.RES.0000103311.52853.48.

151. Cornwell, T.L., Arnold, E., Boerth, N.J., and Lincoln, T.M. (1994). Inhibition of smooth muscle cell growth by nitric oxide and activation of CAMP-dependent protein kinase by CGMP. *Am. J. Physiol.-Cell Physiol.* 267 (5): C1405–C1413. https://doi.org/10.1152/ajpcell.1994.267.5.C1405.

152. Horst, B.G. and Marletta, M.A. (2018). Physiological activation and deactivation of soluble guanylate cyclase. *Nitric Oxide* 77: 65–74. https://doi.org/10.1016/j.niox.2018.04.011.

153. Stone, J.R. and Marletta, M.A. (1996). Spectral and kinetic studies on the activation of soluble guanylate cyclase by nitric oxide †. *Biochemistry* 35 (4): 1093–1099. https://doi.org/10.1021/bi9519718.

154. Kharitonov, V.G., Sharma, V.S., Pilz, R.B., Magde, D., and Koesling, D. (1995). Basis of guanylate cyclase activation by carbon monoxide. *Proc. Natl. Acad. Sci.* 92 (7): 2568–2571. https://doi.org/10.1073/pnas.92.7.2568.

155. Cary, S.P.L., Winger, J.A., and Marletta, M.A. (2005). Tonic and acute nitric oxide signaling through soluble guanylate cyclase is mediated by nonheme nitric oxide, ATP, and GTP. *Proc. Natl. Acad. Sci.* 102 (37): 13064–13069. https://doi.org/10.1073/pnas.0506289102.

156. Fernhoff, N.B., Derbyshire, E.R., and Marletta, M.A. (2009). A nitric oxide/cysteine interaction mediates the activation of soluble guanylate cyclase. *Proc. Natl. Acad. Sci.* 106 (51): 21602–21607. https://doi.org/10.1073/pnas.0911083106.

157. Friebe, A., Schultz, G., and Koesling, D. (1996). Sensitizing soluble guanylyl cyclase to become a highly CO-sensitive enzyme. *EMBO J.* 15 (24): 6863–6868.

158. Makino, R., Obata, Y., Tsubaki, M., Iizuka, T., Hamajima, Y., Kato-Yamada, Y., Mashima, K., and Shiro, Y. (2018). Mechanistic insights into the activation of soluble guanylate cyclase by carbon monoxide: A multistep mechanism proposed for the BAY 41-2272 induced formation of 5-Coordinate CO–Heme. *Biochemistry* 57 (10): 1620–1631. https://doi.org/10.1021/acs.biochem.7b01240.

159. Sharma, V.S. and Magde, D. (1999). Activation of soluble guanylate cyclase by carbon monoxide and nitric oxide: A mechanistic model. *Methods San. Diego. Calif.* 19 (4): 494–505. https://doi.org/10.1006/meth.1999.0892.

160. Kim, -H.-H. and Choi, S. (2018). Therapeutic aspects of carbon monoxide in cardiovascular disease. *Int. J. Mol. Sci.* 19 (8): 2381. https://doi.org/10.3390/ijms19082381.

161. Furchgott, R.F. and Jothianandan, D. (1991). Endothelium-dependent and -independent vasodilation involving cyclic GMP: Relaxation induced by nitric oxide, carbon monoxide and light. *Blood Vessels* 28 (1–3): 52–61. https://doi.org/10.1159/000158843.

162. Blackman, D.J., Morris-Thurgood, J.A., Atherton, J.J., Ellis, G.R., Anderson, R.A., Cockcroft, J.R., and Frenneaux, M.P. (2000). Endothelium-derived nitric oxide contributes to the regulation of venous tone in humans. *Circulation* 101 (2): 165–170. https://doi.org/10.1161/01.CIR.101.2.165.

163. Khalaf, D., Krüger, M., Wehland, M., Infanger, M., and Grimm, D. (2019). The Effects Of Oral L-Arginine and l-Citrulline supplementation on blood pressure. *Nutrients* 11 (7): 1679. https://doi.org/10.3390/nu11071679.

164. Mistry, D.K. and Garland, C.J. (1998). Nitric Oxide (NO)-Induced activation of large conductance Ca^{2+}-Dependent K^+ Channels (BK_{Ca}) in smooth muscle cells isolated from the rat mesenteric artery: NO activates BK_{Ca} channels in rat mesenteric artery. *Br. J. Pharmacol.* 124 (6): 1131–1140. https://doi.org/10.1038/sj.bjp.0701940.

165. Villamor, E., Pérez-Vizcaíno, F., Cogolludo, A.L., Conde-Oviedo, J., Zaragozá-Arnáez, F., López-López, J.G., and Tamargo, J. (2000). Relaxant effects of carbon monoxide compared with nitric oxide in pulmonary and systemic vessels of newborn piglets. *Pediatr. Res.* 48 (4): 546–553. https://doi.org/10.1203/00006450-200010000-00021.

166. Puddu, P., Puddu, G.M., Zaca, F., and Muscari, A. (2000). Endothelial dysfunction in hypertension. *Acta Cardiol.* 55 (4): 221–232. https://doi.org/10.2143/AC.55.4.2005744.

167. Node, K., Kitakaze, M., Yoshikawa, H., Kosaka, H., and Hori, M. (1997). Reduced plasma concentrations of nitrogen oxide in individuals with essential hypertension. *Hypertens. Dallas Tex 1979* 30 (3 Pt 1): 405–408. https://doi.org/10.1161/01.hyp.30.3.405.

168. Giaid, A. and Saleh, D. (1995). Reduced expression of endothelial nitric oxide synthase in the lungs of patients with pulmonary hypertension. *N. Engl. J. Med.* 333 (4): 214–221. https://doi.org/10.1056/NEJM199507273330403.

169. Palloshi, A., Fragasso, G., Piatti, P., Monti, L.D., Setola, E., Valsecchi, G., Galluccio, E., Chierchia, S.L., and Margonato, A. (2004). Effect of oral L-Arginine on blood pressure and symptoms and endothelial function in patients with systemic hypertension, positive exercise tests, and normal coronary arteries. *Am. J. Cardiol.* 93 (7): 933–935. https://doi.org/10.1016/j.amjcard.2003.12.040.

170. Kalinowski, L., Dobrucki, L.W., Szczepanska-Konkel, M., Jankowski, M., Martyniec, L., Angielski, S., and Malinski, T. (2003). Third-generation β-Blockers stimulate nitric oxide release from endothelial cells through ATP Efflux: A novel mechanism for antihypertensive action. *Circulation* 107 (21): 2747–2752. https://doi.org/10.1161/01.CIR.0000066912.58385.DE.

171. Brüne, B. and Ullrich, V. (1987). Inhibition of platelet aggregation by carbon monoxide is mediated by activation of guanylate cyclase. *Mol. Pharmacol.* 32 (4): 497–504.

172. Radomski, M.W., Palmer, R.M.J., and Moncada, S. (1987). Endogenous nitric oxide inhibits human platelet adhesion to vascular endothelium. *The Lancet* 330 (8567): 1057–1058. https://doi.org/10.1016/S0140-6736%2887%2991481-4.

173. Radomski, M.W., Palmer, R.M., and Moncada, S. (1990). An L-Arginine/nitric oxide pathway present in human platelets regulates aggregation. *Proc. Natl. Acad. Sci. U. S. A.* 87 (13): 5193–5197. https://doi.org/10.1073/pnas.87.13.5193.

174. Gkaliagkousi, E., Ritter, J., and Ferro, A. (2007). Platelet-derived nitric oxide signaling and regulation. *Circ. Res.* 101 (7): 654–662. https://doi.org/10.1161/CIRCRESAHA.107.158410.

175. Kaczara, P., Sitek, B., Przyborowski, K., Kurpinska, A., Kus, K., Stojak, M., and Chlopicki, S. (2020). Antiplatelet effect of carbon monoxide is mediated by NAD^+ and ATP depletion. *Arterioscler. Thromb. Vasc. Biol.* 40 (10): 2376–2390. https://doi.org/10.1161/ATVBAHA.120.314284.

176. Krötz, F., Sohn, H.Y., Gloe, T., Zahler, S., Riexinger, T., Schiele, T.M., Becker, B.F., Theisen, K., Klauss, V., and Pohl, U. (2002). NAD(P)H oxidase-dependent platelet superoxide anion release increases platelet recruitment. *Blood* 100 (3): 917–924. https://doi.org/10.1182/blood.v100.3.917.

177. Chen, B., Guo, L., Fan, C., Bolisetty, S., Joseph, R., Wright, M.M., Agarwal, A., and George, J.F. (2009). Carbon monoxide rescues heme oxygenase-1-deficient mice from arterial thrombosis in allogeneic aortic transplantation. *Am. J. Pathol.* 175 (1): 422–429. https://doi.org/10.2353/ajpath.2009.081033.

178. Wallace, J.L., McKnight, W., Del Soldato, P., Baydoun, A.R., and Cirino, G. (1995). Anti-thrombotic effects of a nitric oxide-releasing, gastric-sparing aspirin derivative. *J. Clin. Invest.* 96 (6): 2711–2718. https://doi.org/10.1172/JCI118338.

179. Berkels, R., Bertsch, A., Zuther, T., Dhein, S., Stockklauser, K., Rösen, P., and Rösen, R. (1997). Evidence for a NO synthase in porcine platelets which is stimulated during activation/aggregation. *Eur. J. Haematol.* 58 (5): 307–313. https://doi.org/10.1111/j.1600-0609.1997.tb01676.x.

180. Owens, G.K. (1995). Regulation of differentiation of vascular smooth muscle cells. *Physiol. Rev.* 75 (3): 487–517. https://doi.org/10.1152/physrev.1995.75.3.487.

181. Owens, G.K., Kumar, M.S., and Wamhoff, B.R. (2004). Molecular regulation of vascular smooth muscle cell differentiation in development and disease. *Physiol. Rev.* 84 (3): 767–801. https://doi.org/10.1152/physrev.00041.2003.

182. Ross, R. (1986). The pathogenesis of atherosclerosis — An update. *N. Engl. J. Med.* 314 (8): 488–500. https://doi.org/10.1056/NEJM198602203140806.
183. Morita, T., Perrella, M.A., Lee, M.E., and Kourembanas, S. (1995). Smooth muscle cell-derived carbon monoxide is a regulator of vascular CGMP. *Proc. Natl. Acad. Sci.* 92 (5): 1475–1479. https://doi.org/10.1073/pnas.92.5.1475.
184. Garg, U.C. and Hassid, A. (1989). Nitric oxide-generating vasodilators and 8-bromo-cyclic guanosine monophosphate inhibit mitogenesis and proliferation of cultured rat vascular smooth muscle cells. *J. Clin. Invest.* 83 (5): 1774–1777. https://doi.org/10.1172/JCI114081.
185. Bouallegue, A., Daou, G.B., and Srivastava, A.K. (2007). Nitric oxide attenuates endothelin-1-induced activation of ERK1/2, PKB, and Pyk2 in vascular smooth muscle cells by a CGMP-dependent pathway. *Am. J. Physiol.-Heart Circ. Physiol.* 293 (4): H2072–H2079. https://doi.org/10.1152/ajpheart.01097.2006.
186. Pauly, R.R., Bilato, C., Sollott, S.J., Monticone, R., Kelly, P.T., Lakatta, E.G., and Crow, M.T. (1995). Role of calcium/calmodulin-dependent protein kinase II in the regulation of vascular smooth muscle cell migration. *Circulation* 91 (4): 1107–1115. https://doi.org/10.1161/01.CIR.91.4.1107.
187. Durante, W., Peyton, K.J., and Schafer, A.I. (1999). Platelet-derived growth factor stimulates heme oxygenase-1 gene expression and carbon monoxide production in vascular smooth muscle cells. *Arterioscler. Thromb. Vasc. Biol.* 19 (11): 2666–2672. https://doi.org/10.1161/01.ATV.19.11.2666.
188. Togane, Y., Morita, T., Suematsu, M., Ishimura, Y., Yamazaki, J.-I., and Katayama, S. (2000). Protective roles of endogenous carbon monoxide in neointimal development elicited by arterial injury. *Am. J. Physiol.-Heart Circ. Physiol.* 278 (2): H623–H632. https://doi.org/10.1152/ajpheart.2000.278.2.H623.
189. Schillinger, M., Exner, M., Mlekusch, W., Ahmadi, R., Rumpold, H., Mannhalter, C., Wagner, O., and Minar, E. (2002). Heme oxygenase-1 genotype is a vascular anti-inflammatory factor following balloon angioplasty. *J. Endovasc. Ther.* 9 (4): 385–394. https://doi.org/10.1177/152660280200900401.
190. Mcnamara, D.B., Bedi, B., Aurora, H., Tena, L., Ignarro, L.J., Kadowitz, P.J., and Akers, D.L. (1993). L-arginine inhibits balloon catheter-induced intimal hyperplasia. *Biochem. Biophys. Res. Commun.* 193 (1): 291–296. https://doi.org/10.1006/bbrc.1993.1622.
191. Marks, D.S., Vita, J.A., Folts, J.D., Keaney, J.F., Welch, G.N., and Loscalzo, J. (1995). Inhibition of neointimal proliferation in rabbits after vascular injury by a single treatment with a protein adduct of nitric oxide. *J. Clin. Invest.* 96 (6): 2630–2638. https://doi.org/10.1172/JCI118328.
192. Groves, P.H., Banning, A.P., Penny, W.J., Newby, A.C., Cheadle, H.A., and Lewis, M.J. (1995). The effects of exogenous nitric oxide on smooth muscle cell proliferation following porcine carotid angioplasty. *Cardiovasc. Res.* 30 (1): 87–96.
193. Huang, S.-L., Kee, P.H., Kim, H., Moody, M.R., Chrzanowski, S.M., MacDonald, R.C., and McPherson, D.D. (2009). Nitric oxide-loaded echogenic liposomes for nitric oxide delivery and inhibition of intimal hyperplasia. *J. Am. Coll. Cardiol.* 54 (7): 652–659. https://doi.org/10.1016/j.jacc.2009.04.039.
194. Sugimoto, M., Yamanouchi, D., and Komori, K. (2009). Therapeutic approach against intimal hyperplasia of vein grafts through endothelial nitric oxide synthase/nitric oxide (ENOS/NO) and the Rho/Rho-Kinase pathway. *Surg. Today* 39 (6): 459–465. https://doi.org/10.1007/s00595-008-3912-6.
195. Frismantiene, A., Philippova, M., Erne, P., and Resink, T.J. (2018). Smooth muscle cell-driven vascular diseases and molecular mechanisms of VSMC plasticity. *Cell. Signal.* 52: 48–64. https://doi.org/10.1016/j.cellsig.2018.08.019.
196. Majesky, M.W., Dong, X.R., Regan, J.N., and Hoglund, V.J. (2011). Vascular smooth muscle progenitor cells: Building and repairing blood vessels. *Circ. Res.* 108 (3): 365–377. https://doi.org/10.1161/CIRCRESAHA.110.223800.
197. Pollman, M.J., Yamada, T., Horiuchi, M., and Gibbons, G.H. (1996). Vasoactive substances regulate vascular smooth muscle cell apoptosis: countervailing influences of nitric oxide and angiotensin II. *Circ. Res.* 79 (4): 748–756. https://doi.org/10.1161/01.RES.79.4.748.
198. Clarke, M.C.H., Figg, N., Maguire, J.J., Davenport, A.P., Goddard, M., Littlewood, T.D., and Bennett, M.R. (2006). Apoptosis of vascular smooth muscle cells induces features of plaque vulnerability in atherosclerosis. *Nat. Med.* 12 (9): 1075–1080. https://doi.org/10.1038/nm1459.
199. Li, T., Tian, H., Zhao, Y., An, F., Zhang, L., Zhang, J., Peng, J., Zhang, Y., and Guo, Y. (2011). Heme oxygenase-1 inhibits progression and destabilization of vulnerable plaques in a rabbit model of atherosclerosis. *Eur. J. Pharmacol.* 672 (1–3): 143–152. https://doi.org/10.1016/j.ejphar.2011.09.188.
200. Ameriso, S.F., Villamil, A.R., Zedda, C., Parodi, J.C., Garrido, S., Sarchi, M.I., Schultz, M., Boczkowski, J., and Sevlever, G.E. (2005). Heme oxygenase-1 is

expressed in carotid atherosclerotic plaques infected by *Helicobacter Pylori* and is more prevalent in asymptomatic subjects. *Stroke* 36 (9): 1896–1900. https://doi.org/10.1161/01.STR.0000177494.43587.9e.

201. Durante, W., Kroll, M.H., Christodoulides, N., Peyton, K.J., and Schafer, A.I. (1997). Nitric oxide induces heme oxygenase-1 gene expression and carbon monoxide production in vascular smooth muscle cells. *Circ. Res.* 80 (4): 557–564. https://doi.org/10.1161/01.RES.80.4.557.

202. Wheeler, M.A., Smith, S.D., García-Cardeña, G., Nathan, C.F., Weiss, R.M., and Sessa, W.C. (1997). Bacterial infection induces nitric oxide synthase in human neutrophils. *J. Clin. Invest.* 99 (1): 110–116. https://doi.org/10.1172/JCI119121.

203. Bogdan, C. (2015). Nitric oxide synthase in innate and adaptive immunity: An update. *Trends Immunol.* 36 (3): 161–178. https://doi.org/10.1016/j.it.2015.01.003.

204. Fang, F.C. (1997). Perspectives series: Host/Pathogen interactions. Mechanisms of nitric oxide-related antimicrobial activity. *J. Clin. Invest.* 99 (12): 2818–2825. https://doi.org/10.1172/JCI119473.

205. Müller, A.J., Aeschlimann, S., Olekhnovitch, R., Dacher, M., Späth, G.F., and Bousso, P. (2013). Photoconvertible pathogen labeling reveals nitric oxide control of leishmania major infection in vivo via dampening of parasite metabolism. *Cell Host Microbe* 14 (4): 460–467. https://doi.org/10.1016/j.chom.2013.09.008.

206. Thiemermann, C. (1997). Nitric oxide and septic shock. *Gen. Pharmacol. Vasc. Syst.* 29 (2): 159–166. https://doi.org/10.1016/S0306-3623(96)00410-7.

207. Feihl, F., Waeber, B., and Liaudet, L. (2001). Is nitric oxide overproduction the target of choice for the management of septic shock? *Pharmacol. Ther.* 91 (3): 179–213. https://doi.org/10.1016/s0163-7258(01)00155-3.

208. Petros, A., Bennett, D., and Vallance, P. (1991). Effect of nitric oxide synthase inhibitors on hypotension in patients with septic shock. *Lancet Lond. Engl.* 338 (8782–8783): 1557–1558. https://doi.org/10.1016/0140-6736(91)92376-d.

209. López, A., Lorente, J.A., Steingrub, J., Bakker, J., McLuckie, A., Willatts, S., Brockway, M., Anzueto, A., Holzapfel, L., Breen, D., Silverman, M.S., Takala, J., Donaldson, J., Arneson, C., Grove, G., Grossman, S., and Grover, R. (2004). Multiple-center, randomized, placebo-controlled, double-blind study of the nitric oxide synthase inhibitor 546C88: Effect on survival in patients with septic shock. *Crit. Care Med.* 32 (1): 21–30. https://doi.org/10.1097/01.CCM.0000105581.01815.C6.

210. Klabunde, R.E. and Ritger, R.C. (1991). NG-Monomethyl-l-Arginine (NMA) restores arterial blood pressure but reduces cardiac output in a canine model of endotoxic shock. *Biochem. Biophys. Res. Commun.* 178 (3): 1135–1140. https://doi.org/10.1016/0006-291x(91)91010-a.

211. Immenschuh, S., Tan, M., and Ramadori, G. (1999). Nitric oxide mediates the lipopolysaccharide dependent upregulation of the heme oxygenase-1 gene expression in cultured Rat Kupffer cells. *J. Hepatol.* 30 (1): 61–69. https://doi.org/10.1016/S0168-8278(99)80008-7.

212. Jadhav, A., Tiwari, S., Lee, P., and Ndisang, J.F. (2013). The heme oxygenase system selectively enhances the anti-inflammatory macrophage-M2 phenotype, reduces pericardial adiposity, and ameliorated cardiac injury in diabetic cardiomyopathy in Zucker diabetic fatty rats. *J. Pharmacol. Exp. Ther.* 345 (2): 239–249. https://doi.org/10.1124/jpet.112.200808.

213. Otterbein, L.E., Bach, F.H., Alam, J., Soares, M., Tao Lu, H., Wysk, M., Davis, R.J., Flavell, R.A., and Choi, A.M. (2000). Carbon monoxide has anti-inflammatory effects involving the mitogen-activated protein kinase pathway. *Nat. Med.* 6 (4): 422–428. https://doi.org/10.1038/74680.

214. Devey, L., Ferenbach, D., Mohr, E., Sangster, K., Bellamy, C.O., Hughes, J., and Wigmore, S.J. (2009). Tissue-resident macrophages protect the liver from ischemia reperfusion injury via a heme oxygenase-1-dependent mechanism. *Mol. Ther.* 17 (1): 65–72. https://doi.org/10.1038/mt.2008.237.

215. Ferenbach, D.A., Nkejabega, N.C.J., McKay, J., Choudhary, A.K., Vernon, M.A., Beesley, M.F., Clay, S., Conway, B.C., Marson, L.P., Kluth, D.C., and Hughes, J. (2011). The induction of macrophage hemeoxygenase-1 is protective during acute kidney injury in aging Mice. *Kidney Int.* 79 (9): 966–976. https://doi.org/10.1038/ki.2010.535.

216. Zegdi, R., Perrin, D., Burdin, M., Boiteau, R., and Tenaillon, A. (2002). Increased endogenous carbon monoxide production in severe Sepsis. *Intensive Care Med.* 28 (6): 793–796. https://doi.org/10.1007/s00134-002-1269-7.

217. Nakahira, K. and Choi, A.M.K. (2015). Carbon monoxide in the treatment of sepsis. *Am. J. Physiol.-Lung Cell. Mol. Physiol.* 309 (12): L1387–L1393. https://doi.org/10.1152/ajplung.00311.2015.

218. Kyokane, T., Norimizu, S., Taniai, H., Yamaguchi, T., Takeoka, S., Tsuchida, E., Naito, M., Nimura, Y., Ishimura, Y., and Suematsu, M. (2001). Carbon monoxide from heme catabolism protects against hepatobiliary dysfunction in endotoxin-treated rat

liver. *Gastroenterology* 120 (5): 1227–1240. https://doi.org/10.1053/gast.2001.23249.

219. Sarady, J.K., Zuckerbraun, B.S., Bilban, M., Wagner, O., Usheva, A., Liu, F., Ifedigbo, E., Zamora, R., Choi, A.M.K., and Otterbein, L.E. (2004). Carbon monoxide protection against endotoxic shock involves reciprocal effects on INOS in the lung and liver. *FASEB J.* 18 (7): 854–856. https://doi.org/10.1096/fj.03-0643fje.

220. Lukic, M.L., Stosic-Grujicic, S., Ostojic, N., Chan, W.L., and Liew, F.Y. (1991). Inhibition of nitric oxide generation affects the induction of diabetes by streptozocin in mice. *Biochem. Biophys. Res. Commun.* 178 (3): 913–920. https://doi.org/10.1016/0006-291X(91)90978-G.

221. Oleson, B.J., Broniowska, K.A., Naatz, A., Hogg, N., Tarakanova, V.L., and Corbett, J.A. (2016). Nitric oxide suppresses β-cell apoptosis by inhibiting the DNA damage response. *Mol. Cell. Biol.* 36 (15): 2067–2077. https://doi.org/10.1128/MCB.00262-16.

222. Lajoix, A.-D., Reggio, H., Chardès, T., Péraldi-Roux, S., Tribillac, F., Roye, M., Dietz, S., Broca, C., Manteghetti, M., Ribes, G., Wollheim, C.B., and Gross, R. (2001). A neuronal isoform of nitric oxide synthase expressed in pancreatic β-cells controls insulin secretion. *Diabetes* 50 (6): 1311–1323. https://doi.org/10.2337/diabetes.50.6.1311.

223. Rahman, F.U., Park, D.-R., Joe, Y., Jang, K.Y., Chung, H.T., and Kim, U.-H. (2019). Critical roles of carbon monoxide and nitric oxide in Ca^{2+} Signaling for insulin secretion in pancreatic islets. *Antioxid. Redox Signal.* 30 (4): 560–576. https://doi.org/10.1089/ars.2017.7380.

9

When Carbon Monoxide Meets Hydrogen Sulfide

Rui Wang

Department of Biology, York University, Toronto, Ontario M3J 1P3, Canada

Introduction

The arrival of gasotransmitters to the scientific world is no accident. Decades of research discoveries on cellular signaling mechanisms have built up experimental evidence and theoretical expectations that many gas molecules in our body function in similar but not redundant ways. Six criteria have been established to characterize these gas molecules as gasotransmitters to reveal their common intrinsic attributes and differentiate them from other signaling molecules, such as neurotransmitters. *First, gasotransmitters are small molecules that exist in the gaseous state at physiological temperature and under atmospheric pressure. Second, they are freely permeable across biological membrane. As such, their intracellular and intercellular movements do not exclusively rely on cognate membrane receptors or other transportation machineries. Third, they are endogenously generated in mammalian cells with specific substrates and enzymes; more than the products of metabolism, their production is regulated to fulfill signaling messenger functions. Fourth, they have well-defined specific functions at physiologically relevant concentrations. Fifth, their functions can be mimicked by their exogenously applied counterparts. Sixth, they are involved in signal transduction and have specific cellular and molecular targets* [1,2].

Carbon monoxide (CO) is an odorless gas and hydrogen sulfide (H_2S) has a stench odor, contributing to the bad smell of flatus and rotten eggs. Both are known for their toxicities as respiratory poisons. Natural events, such as volcano eruptions or natural fermentation, and human activities in agriculture or industry generate both CO and H_2S. H_2S has been assigned a critical role in early life on Earth [3]. A similar role may have also been undertaken by CO [4]. In biological systems, CO is involved in protoporphyrin metabolism and H_2S is a critical component of transsulfuration or reverse transsulfuration in microbiota or mammals. CO and H_2S are two well-characterized gasotransmitters. They possess all aforementioned six characteristics. Yet, they differ from each other in their chemical natures, metabolism pathways, functional targets, and molecular mechanisms for signaling processes, to name a few. Moreover, these two gasotransmitters do not live in totally separate worlds. They exist in the same biological system, in terms of the whole body or a single cell, and interact with each other to jointly impact the destiny or the net outcome of the system. Without considering the interactions among gasotransmitters, our understanding of the biology of CO would be incomplete and inaccurate. From this lens, this chapter offers an integrated view on what happens when CO meets H_2S and why this close encounter beyond the fifth kind is important.

Intertwined Productions of CO and H_2S

Production of CO and H_2S in Mammalian Cells

CO production in mammalian cells is mostly catalyzed by heme oxygenase (HO). HO is a monooxygenase that catalyzes the production of CO and biliverdin IXα, using heme and molecular oxygen (O_2) as substrates. HO-1 is inducible under various stress conditions, while HO-2 is a constitutive enzyme. CO stimulates cGMP production and opens calcium-dependent K channels in the cardiovascular system, leading to vasodilation [5,6]. The production and metabolism of CO as well as its widespread impact on numerous physiological and pathophysiological processes in our body have been extensively covered in this comprehensive book and readers are referred to the relevant chapters.

H_2S was traditionally viewed as a toxic gas detected in contaminated environment [7]. Over the last decade, physiological importance of endogenously produced H_2S has been realized [8,9]. Two key enzymes in the reverse transsulfuration pathway, cystathionine β-synthase (CBS) and cystathionine γ-lyase (CSE), produce H_2S, pyruvate, and ammonium using homocysteine and/or L-cysteine as substrates [1,10,11]. Both CBS and CSE use pyridoxal L-phosphate as cofactor. Human CBS is a tetramer and is allosterically stimulated by S-adenosyl methionine, which binds to a conserved domain in the C-terminal end of the protein. CBS is also a heme-containing protein, which reacts with CO. CSE is a homotetramer. It can catalyze the production of H_2S from cysteine and homocysteine alone or in combination.

The third enzyme 3-mercaptopyruvate sulfurtransferase (MST) for enzymatic H_2S production has also been identified. MST takes two steps to produce H_2S. First, the reaction of L-cysteine and α-ketoglutarate generates 3-mercaptopyruvate in the presence of cysteine aminotransferase. MST causes the desulfuration of 3-mercaptopyruvate and transfers the sulfur to a nucleophilic cysteine in the active site to form a bound persulfide. Under reducing conditions or in the presence of acceptors like thioredoxin, H_2S would be released (Figure 9.1).

CBS is the primary enzyme producing H_2S in the central nervous system. It is expressed in neurons and astrocytes, more in the hippocampus and cerebellum than in cerebral cortex and brain stem. In the cardiovascular system, respiratory system, testes, spleen, and the pancreas, CBS expression is rare or absent [1,7], but that of CSE is abundant [1,12]. CSE is also expressed in different regions of the brain, such as cortex, striatum, cerebellum, brain stem, hippocampus, and hypothalamus [13]. In the cardiovascular system, MST has been detected in the endothelium and other types of cells [14], but not in vascular smooth muscle cells or cardiomyocytes. MST is expressed in the liver and kidney. In the central nervous system, MST is localized in hippocampal pyramidal neurons, cerebellar Purkinje cells, and mitral cells in the olfactory bulb of the brain [14].

Endogenous level of H_2S in human plasma is around 3 μM, determined by measuring the headspace H_2S gas [15]. Using the monobromobimane method coupled with reverse-phase high-performance liquid chromatography, the same authors recorded free H_2S levels of 0.2–0.8 μM and acid-labile sulfur levels of 1.8–3.8 μM in mouse or human blood [15]. With a modified gas chromatography/mass spectrometry technique, H_2S levels of 0.5–2.5 μM were detected in swine and mouse blood [16]. Thus, as a conservative estimate based on the results from different measuring methods, plasma H_2S level in healthy humans or animals is in the lower micromolar to higher nanomolar

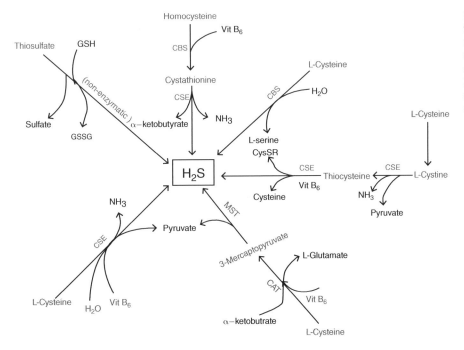

Figure 9.1 Biosynthesis and transformation of H2S in mammalian cells [7]. CSE, cystathionine γ-lyase; CBS, cystathionine β-synthase; MST, 3-mercaptopyruvate sulfurtransferase; CAT, cysteine aminotransferase.

range [15,17–20]. The physiological concentration of H_2S in tissues where it is produced, however, remains to be determined, but it is expected to be 10 times greater than that in the blood [21]. It is worth noting that vascular tissues, such as aortae, may contain free H_2S at the level ~20–100-fold higher than that of other tissues [22].

Lack of endogenous H_2S production is known to lead to the development of hypertension [12]. We showed that H_2S suppressed early development of atherosclerosis [23] and inhibited the proliferation of vascular smooth muscle cells [7,24], but promoted angiogenesis and endothelial proliferation [25,26]. H_2S relaxes vascular tissues by opening K_{ATP} channels in vascular smooth muscle cells [8,27,28], or by functioning as an endothelium-derived hyperpolarizing factor to cause endothelium-dependent vasorelaxation [29,30]. H_2S protects the heart from ischemia/reperfusion damage [31–34]. It also offers anti-inflammation and antioxidant protections [7,8,12]. H_2S can form a persulfide with the cysteine residue(s) of its target proteins in a reaction named S-sulfhydration [25,35]. This post-translational modification usually increases protein activities. Increased endogenous production of H_2S beyond the physiological range may also be detrimental. For example, high level of endogenous H_2S was shown to enhance hypoxic pulmonary vasoconstriction in rats [36]. Increased endogenous production of H_2S from pancreatic beta cells also contributes to the pathogenesis of diabetes [37].

The gasotransmitter roles of CO and H_2S are related to their interactions with metalloproteins, which contain a metal cofactor essential for function. Heme proteins, metalloenzymes, and copper proteins are examples of metalloproteins. Some metalloproteins containing iron–sulfur clusters play important role in the redox reactions of electron transport in mitochondria, such as complex I, complex II, and coenzyme Q–cytochrome c reductase (cytochrome $bc1$ complex). H_2S interacts with metals in electron transfer reactions, which occur between the sulfide species and the metal. In addition, binding of H_2S to a metal is often important for the formation of a stable metal complex. Many protective effects of H_2S against heavy metal toxicity (such as mercury toxicity) result from its interactions with heavy metals in mammals, plants, and bacteria.

Reciprocal Effects of CO and H_2S on Their Own Production

CO binds to the heme-containing CBS and inhibits its activity. This inhibition results in decreased endogenous production of H_2S and accumulation of homocysteine in mammalian cells [38]. Conversely, H_2S affects endogenous production of CO in two ways. The first one is to inhibit the synthesis of heme, the substrate of HO. This action of H_2S will be discussed in more detail later in the following section. The second mechanism for H_2S-impacted CO production is the activation of nuclear factor-erythroid-2 p45-related factor 2 (Nrf2).

Nrf2 is a constitutive transcription factor and a master regulator of the antioxidant response. Kelch-like ECH-associated protein-1 (Keap1) is a redox-sensitive ubiquitin ligase substrate adaptor, which causes ubiquitination and degradation of Nrf2. Our previous study showed that mouse embryonic fibroblasts (MEFs) isolated from CSE knockout (KO) mice displayed increased oxidative stress and accelerated cellular senescence in comparison with MEFs from wild-type (WT) mice. The protein levels of p53 and p21 were significantly increased in MEFs from CSE-KO mice, and knockdown of p53 or p21 reversed CSE deficiency-induced senescence. Incubation of the cells with NaHS, an H_2S salt, significantly increased glutathione level and rescued CSE-KO MEFs from cell senescence. NaHS also induced S-sulfhydration of Keap1 at cysteine-151, facilitating Nrf2 dissociation from Keap1 so that Nrf2 could translocate to the nucleus where it stimulates mRNA expression of Nrf2-targeted downstream genes [39]. Among Nrf2-targeted antioxidant and phase II detoxifying enzymes are NAD(P)H:quinone oxidoreductase-1, glutamate cysteine ligase catalytic subunit, thioredoxin reductase-1, and HO-1. In short, H_2S seems to have a double-edged effect on CO production. The net outcome of the effects of H_2S on decreased heme synthesis and upregulated HO-1 consequent to activation of Nrf2 would depend on the types and age of the targeted cells.

Heme and Iron Metabolism

Heme, or known as reduced hematin, is both the substrate of HO for CO production and an inducer of HO expression [40]. Upregulation of HO and increased endogenous CO production cause vasorelaxation and lower blood pressure in experimental hypertensive rats [5,41,42]. Chronic hemin treatment also upregulates the expression of HO and inhibits the proliferation of vascular smooth muscle cells *in vitro* [43], or increases intracavernosal pressure response in hypertensive rats [44]. A balance between heme and CO plays important roles in regulating cellular survival and functions.

The Structural and Functional Roles of Heme

Heme is a prosthetic group that consists of a complex of ferrous iron (Fe^{2+}) and protoporphyrin IX (PP-IX) [40]. The Fe^{3+} oxidation product of heme is termed hemin. Hemin (Fe^{3+}) is reduced to heme (Fe^{2+}) *in vivo* prior to oxidation by HO (Figure 9.2). The physiological level of free heme in normal cells is below 1 µM [45]. At physiological concentration range, free heme downregulates δ-aminolevulinate synthase and reduces the expression of Bach1. The latter would lift or reverse the inhibited gene expression of HO-1 [45]. Due to its lipophilic nature, heme readily moves around among different organelles [46], and interacts with many cellular membranes and organelles, including lipid bilayers, mitochondria, cytoskeleton, nuclei, and several intracellular enzymes [47]. Heme also reversibly binds selective gas molecules. The heme in hemoglobin binds NO, CO, and H_2S, and the heme in soluble guanylate cyclase binds NO and CO with different affinities.

The xenobiotic detoxification function of the liver relies on the incorporation of heme into cytochrome P450 proteins. Oxidative phosphorylation is also maintained by hemoproteins, such as cytochrome *c* oxidase. The elevated circulatory heme level results either from excessive filtration of heme proteins as would occur in rhabdomyolysis or from the destabilization of intracellular heme proteins (e.g., cytochromes) in ischemia–reperfusion and nephrotoxin-induced renal injury [48–50]. Under oxidative stress, some hemoproteins can release their heme prosthetic groups [51].

The nonprotein-bound (free) heme is highly cytotoxic, most probably due to the iron atom contained within its PP-IX ring, which can act as a Fenton's reagent to catalyze in an unfettered manner the production of free radicals. It catalyzes the oxidation and aggregation of protein and DNA, and induces the formation of cytotoxic lipid peroxide via lipid peroxidation, triggering site-directed oxidative damage [52]. Due to its lipophilic properties, free heme impairs lipid bilayers in organelles such as mitochondria and nuclei. Heme-responsive genes remain repressed when heme at low concentrations binds to Bach1 with MARE sequences [53]. However, at higher concentrations heme inactivates the binding of Bach1, allowing access of transcription factors such as Nrf2 to interact with the MARE sequences [53]. This in turn activates the heme-responsive genes. In this regard, intact HO activity is crucial for the removal of the pro-oxidant heme [54].

The availability of heme also influences the metabolism of hemoglobin, myoglobin, cytochromes, prostaglandin endoperoxide synthase, CBS, NO synthase, catalase, peroxidases, respiratory burst oxidase,

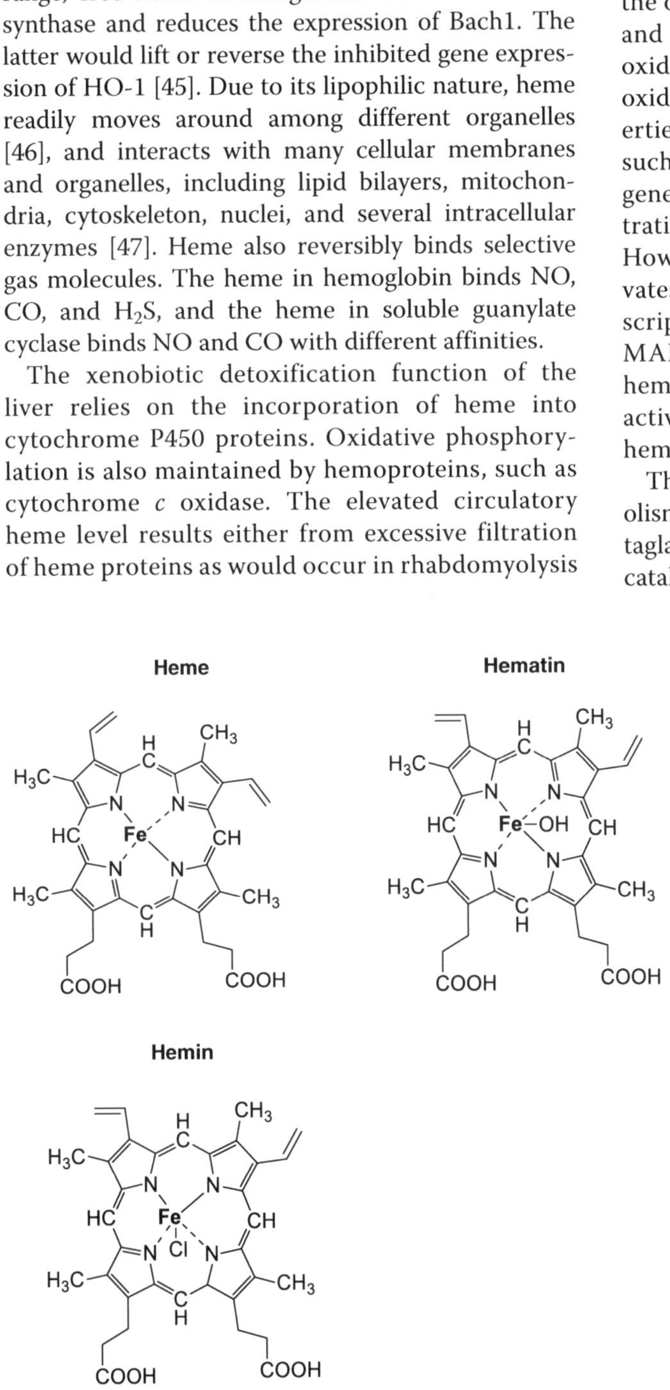

Figure 9.2 Chemical structures of heme, hematin, and hemin.

guanylyl cyclase, tryptophan dioxygenase, pyrrolases, cytochrome, and many others [40]. These heme proteins play important roles in regulating cellular functions, from oxygen delivery and mitochondrial respiration to signal transduction [55].

Increased biosynthesis of heme as well as abnormal release of heme from heme-containing proteins presents pro-oxidant threats. The physiological importance of heme is far more than its structural role in protein complex. It is also a dynamic signaling molecule due to its cytotoxicity and hydrophobicity [56]. Free heme is an efficient trigger of low-density lipoprotein oxidation and it greatly amplifies oxidant-mediated endothelial damage [57]. Heme is also chemotactic for neutrophils and has been shown to induce inflammation in mice [58]. These heme-triggered events contribute to the etiology of diverse diseases, including atherosclerosis, kidney malfunction, sickle cell disease, and malaria [59]. The cell copes with these threats by upregulating HO-1 expression to degrade heme into CO and biliverdin/bilirubin.

Decreased biosynthesis of heme, on the other hand, presents a different hazard to negatively affect numerous physiological processes at cellular and whole body levels. One such health hazard is porphyria. Porphyrias are characterized by decreased heme synthesis and accumulation of porphyrins or porphyrin precursors. Deficiency in heme synthesis seriously suppresses the synthesis of numerous heme-containing proteins and damages many critical cellular events, such as the mitochondrial function due to decreased level of cytochromes. The accumulation of porphyrins and their precursors causes liver damage, skin photosensitivity (cutaneous porphyria), and neurological disorders. Porphyria attack can be triggered by hormones, such as estrogen, infection, alcohol, and some drugs. Moreover, heme causes cell aging and shortening of telomeres. The stress-induced premature senescence might be induced by heme accumulation.

To regulate the physiological process of heme biosynthesis is, therefore, important for the homeostasis of numerous cellular and organ functions. With decreased heme synthesis, the construction of important hemoproteins would be jeopardized but the resulting low level of free heme would reduce the risk of oxidative stress. Thus, an optimal physiological level of heme synthesis and free heme must be maintained at the optimal "setting point." How is this "setting point" set and what regulates it in different organs and tissues? Our understanding on the regulation of heme synthesis is yet limited to the genetic alterations of the key heme-synthesizing enzymes.

Heme Synthesis

Heme is synthesized in all nucleated cells using glycine and succinyl-CoA from the citric acid cycle as the precursors. Heme biosynthesis involves eight enzymes, four of which are cytoplasmic enzymes, including ALA dehydratase, porphobilinogen deaminase, uroporphyrinogen III synthase, and uroporphyrinogen III decarboxylase. The final four enzymes are localized in the mitochondrion, including coproporphyrinogen oxidase (CPOX), ALA synthase, protoporphyrin III oxidase, and ferrochelatase. Iron eventually combines with PP-IX, catalyzed by ferrochelatase, to complete heme formation in the mitochondrion. CPOX is involved in the aerobic oxidative decarboxylation of propionate groups of rings A and B of coproporphyrinogen III to yield the vinyl groups in protoporphyrinogen IX.

Heme is mostly synthesized in erythroid cells and the liver. In erythroid cells (hematopoietic stem cells, erythroblasts, and reticulocytes), heme is required for hemoglobin synthesis. Heme is required in the liver for the synthesis of a variety of hemoproteins, particularly cytochrome P450 [60]. About 85% of organismal heme is synthesized in erythroid cells whose total number is considerably lower than that of hepatocytes. Both iron metabolism and the regulation of heme synthesis are different in hemoglobin-synthesizing erythroid cells from those in hepatocytes.

The Effects of H_2S on Heme Synthesis

Free heme level in the liver is critical for heme protein synthesis, redox balance, and mitochondrial functions. The epigenetic regulation heme synthesis is unknown. Tenhunen et al. showed that H_2S inhibited ferrochelatase activity in rat liver mitochondrial fraction with an apparent K_i of 3.4 mM [61]. Although the H_2S exposure level reported in this study was clearly far higher than endogenous H_2S level, the inhibitory effect of H_2S on heme biosynthesis was nevertheless indicated. H_2S-induced inhibition of heme synthesis may be one of the protective mechanisms for limiting free heme level and mitigating the development of oxidative damages. This notion matches well with many published observations that the mice with CSE expression deficiency (CSE-KO mice) have abnormal metabolism of hemoglobin and other heme proteins.

Compared to WT mice, the expression levels of hemoglobin, erythropoietin (EPO), and CBS are lower in CSE-KO mice after a 72-h period of hypoxia (11% O_2). This downregulation of heme proteins was reversed by exogenous H_2S supplementation. The anemic behavior was also observed in patients who suffered from chronic kidney disease (CKD) with associated comorbidities, including anemia. Anemic CKD patients who required exogenous EPO exhibited lower urinary thiosulfate levels compared to nonanemic CKD patients of similar CKD classification. Relative deficiency in renal EPO production is thought to be a primary cause of anemia. Interestingly, CKD patients display low renal levels of H_2S. Previous *in vitro* experiments have revealed that H_2S-deficient renal cell lines produce less EPO than WT renal cell lines during hypoxia [62]. Using an *in vivo* murine model of whole body hypoxia and in clinical samples obtained from CKD patients, Leigh et al. further demonstrated that H_2S might be a primary mediator of EPO synthesis during hypoxia [62]. These results confirm an interplay between the actions of H_2S during hypoxia and EPO production.

Increased erythropoiesis has also been observed in CSE-KO mice via the upregulation of one of the heme biosynthetic enzymes, CPOX, and consequent stimulation of heme biosynthesis [63]. CSE-KO mice exhibit elevated red blood cell counts and red blood cell mean corpuscular volumes compared to WT mice. These changes are associated with elevated plasma and liver heme levels as well as the induction of CPOX (the sixth enzyme involved in heme biosynthesis) in CSE-KO mice. H_2S directly inhibits the promoter activation of CPOX. Furthermore, a functional interaction between the CSE/H_2S system and the CPOX system in terms of mitochondrial bioenergetics has been suggested. CSE-KO hepatocytes and the isolated mitochondria exhibited increased oxidative phosphorylation and CPOX silencing partially diminished this increase. Although heme is essential for the biosynthesis of mitochondrial electron chain complexes, and CPOX is required for heme biosynthesis, the observed functional mitochondrial alterations were not associated with detectable changes in mitochondrial electron transport chain protein expression. To reiterate, physiological levels of H_2S prevent heme accumulation in the mitochondrion and related oxidative stress and damage to liver functions. On the other hand, overproduction of H_2S leads to hepatic coproporphyria. These effects are mediated by H_2S-induced downregulation of CPOX expression and *S*-sulfhydration of the same, which leads to decreased heme synthesis [63]. The elucidation of H_2S–heme interaction would unveil one of the first identified mechanisms by which a gasotransmitter regulates heme biosynthesis, hemoprotein synthesis, and cellular redox balance.

Oxygen Sensing

Our body senses changes in oxygen levels. The carotid bodies are an oxygen-sensing organ that detects the changes in oxygenation of the arterial blood and triggers chemoreflex to maintain homeostasis. Oxygen sensing and reaction are critical for hypoxia-induced increase in blood flow in vital organs and homeostatic regulation of breathing. Recent studies reveal that the carotid bodies also sense changes in CO and H_2S levels as part of the key elements of the oxygen-sensing reflex arc. Elevated CO and decreased H_2S render the carotid bodies insensitive to hypoxia, attenuating ventilatory adaptations to high-altitude hypoxia. On the other hand, hypoxia decreases HO-2-dependent CO production and increases H_2S production [64]. The consequent depolarization of type I cells leads to sensory excitation of the carotid bodies [65].

Intermittent hypoxia stimulates the formation of reactive oxygen species and increased intracellular Ca^{2+} levels in the carotid bodies. Reactive oxygen species inhibit HO-2 by oxidizing the same. This leads to decreased production of CO [66]. CBS possesses a prosthetic heme that allows CO binding to inhibit the enzyme activity and to regulate H_2S generation. As such, CBS is a CO-responsive protein [67]. Decreased CO level during hypoxia in fact would result in increased production of H_2S that mediates the arterial vasodilation [64]. The levels of CO and H_2S in the carotid bodies are inversely correlated. A higher CO level is companied with a lower H_2S level, and vice versa [68]. Under physiological conditions, the carotid bodies of Brown–Norway (BN) rats have higher CO levels due to higher heme affinity to HO-2 and lower H_2S levels than Sprague–Dawley (SD) rats. BN rats cannot sense oxygen changes at the carotid bodies appropriately during hypobaric hypoxia, developing pulmonary edema. On the other hand, in comparison with SD rats, a lower CO level due to lower heme affinity to HO-2 and a higher H_2S level are observed in the carotid bodies of spontaneously hypertensive rats (SHRs). Upon hypoxic challenge, SHRs react with hypersensitivity of the carotid bodies, developing hypertension. Altering CO levels in BN rats or SHRs leads to opposite changes in H_2S levels [68].

Hypoxia can also affect CBS-generated H_2S level independent of the cellular CO level. At the normal oxygen level, CBS in hepatocytes is degraded inside mitochondria by Lon protease. Once hypoxic attack occurs, Lon protease cannot recognize the deoxygenated heme group in CBS, and as such it fails to degrade CBS. The accumulation of CBS proteins in the mitochondrion leads to increased mitochondrial production of H_2S [69]. Morikawa et al. examined vascular disposition in cerebellar slices and in intact mouse brains using two-photon intravital laser scanning microscopy [64]. They found that coordinated actions of H_2S formed by CBS and CO generated by HO-2 are responsible for hypoxia-induced cerebral vasodilation. Mice with targeted deletion of HO-2 or CBS display impaired vascular responses to hypoxia. Thus, in intact adult brain cerebral cortex of HO-2-null mice, imaging mass spectrometry reveals an impaired ability to maintain ATP levels on hypoxia.

Changes in the oxygenation status of cells affect endogenous H_2S levels in different ways. Under hypoxic conditions, oxidation of H_2S in mitochondria is decreased, leading to a net elevation of H_2S levels. In vascular smooth muscle cells, hypoxia or calcium overloading also leads to the translocation of CSE from the cytosol to mitochondria, where CSE uses approximately threefold higher concentrations of L-cysteine to produce H_2S [70]. Consequently, ATP production in the mitochondrion is increased under hypoxic conditions.

Different tissues handle changes in the partial pressure of oxygen differently by altering their H_2S production and oxidation, as well as by changing their responses to H_2S [19]. For example, H_2S dilates systemic blood vessels and/or promotes angiogenesis to increase blood supply to hypoxic tissues [26]. In contrast, in hypoxic rat lungs, H_2S constricts vascular smooth muscle in order to achieve regional ventilation–perfusion matching through a mechanism known as hypoxic pulmonary vasoconstriction [36,71]. In this way, H_2S helps the diversion of blood from oxygen-deprived areas to oxygen-rich areas [72]. The ability of cells to utilize H_2S to drive mitochondrial production of ATP appears to be an important rescue mechanism during hypoxia or anoxia. At the same time, H_2S dilates airway smooth muscle to increase lung ventilation in humans [73], and to reduce airway resistance in mice [74]. These compensatory changes induced by H_2S in response to hypoxia help improve the efficiency of pulmonary gas exchange. Moreover, H_2S mediates the response of carotid body chemoreceptors to hypoxia by modulating BK_{Ca} channels [75]. As far as bioenergy production is concerned, under hypoxic conditions the upregulated mitochondrial production of H_2S, which can act as an electron donor in the mitochondrial respiratory chain, contributes to the generation of bioenergy [76]. Thus, in facing a hypoxic challenge, mitochondrial production of ATP can be maintained by hypoxia-increased H_2S levels. It should be noted that a decrease in the partial pressure of oxygen in pulmonary circulation does not always lead to an increase in H_2S levels, and H_2S does not always cause the constriction of pulmonary arteries. For instance, in hypoxic pulmonary hypertension, hypoxia and a reduction of H_2S levels in the pulmonary circulation occur in parallel [77]. In another example, in contrast to the assumed vasoconstrictive effect of H_2S on pulmonary arteries, one recent study showed that, under normoxic conditions, H_2S actually dilated preconstricted human lobar pulmonary artery rings and reduced pulmonary artery pressure [78]. It is not yet clear whether the vasoactive effects of H_2S on pulmonary circulation, and the interaction of H_2S with hypoxia, are species specific.

In contrast to oxygen sensing by the carotid body during hypoxia, which is accomplished by opposite changes in CO and H_2S levels, preeclampsia is characterized with the same pattern of decreased endogenous levels of CO and H_2S. Preeclamptic women have both decreased exhaled air CO level and lower plasma H_2S level [79]. Chorionic villous sampling from women at 11-week gestation shows that HO-1 mRNA expression is decreased in women who go on to develop preeclampsia. There are ample data to support the notion that the pathogenesis of preeclampsia is largely due to decreased endogenous CO production and HO expression. CSE expression is downregulated at mRNA and protein levels in human placenta with preeclampsia. Inhibition of CSE activity by DL-propargylglycine (PPG) in pregnant mice induced hypertension and liver damage, and promoted abnormal labyrinth vascularization in the placenta and decreased fetal growth. With the first trimester (8–12 weeks of gestation) human placental explants, PPG had decreased the production of placenta growth factor.

The mechanisms for the anti-preeclampsia effects of CO and H_2S are related to their proangiogenic ability. Soluble Flt-1 (sFlt-1) and soluble endoglin (sEng) are two key antiangiogenic factors responsible for the clinical signs of preeclampsia. Animals exposed to high circulating levels of sFlt-1 and sEng elicit severe preeclampsia-like symptoms. Upregulation of HO-1 and increased production of CO inhibit sFlt-1 and sEng release. The CSE–H_2S axis exerts the same inhibition of sFlt-1 and sEng release. Endothelial CSE knockdown by small interfering RNA (siRNA) transfection increased the endogenous release of sFlt-1 and sEng from human umbilical vein endothelial cells, while adenoviral-mediated CSE overexpression inhibited their release. Pharmacological intervention with GYY4137, an H_2S slow-releasing agent, inhibited circulating sFlt-1 and sEng levels and restored fetal growth. All these observations point to sFlt-1 and sEng as the common node of CO and H$_2$S actions individually [79]. To date, we have little knowledge on how CO and H$_2$S inhibit sFlt-1 and sEng release and which steps are comprised, being their transcription, translation, assembly and packing, or releasing. We also do not know whether the effects of CO and H$_2$S rely on the presence of each other. A better understanding of the interaction between the HO-1/CO and CSE/H$_2$S systems in pregnancy would help devise a more integrated approach for the treatment of preeclampsia.

Microbiota, CO, and H$_2$S

Microbiota are the collection of bacteria, viruses, and fungi in a given environment. Human beings are inhabited by microbiota from the skin and mouth to the gastrointestinal tract under physiological conditions. The coexistence of microbiota with their host is important for the homeostasis of human health. The metabolism of microbiota produces various metabolites for supporting and regulating host metabolism. The appropriate population and colonization of microbiota also protect host from the invasion and proliferation of foreign microorganisms, i.e., pathogens. Microbiota also serve as signaling sources to inform the host of the metabolism and functional changes. Immune priming is another critical role of microbiota that assists the development of the host immune system. Disruptions of the homeostasis of microbiota directly impact human health. Well-known examples include obesity, diabetes, inflammatory bowel diseases, asthma, and a wide spectrum of viral infections.

Bacterial and viral infections are often accompanied by changes in the redox physiology of both the host and the pathogen. Gasotransmitters can be produced by both microbiota and the host mammalian cells.

The Production of CO and H$_2$S in Host Cells and Their Effects on Microbiota

Microbiota produce gasotransmitters for utilization by themselves and for informing their hosts. The human host does the same and communicates with microbiota reciprocally. Host-produced CO and H$_2$S offer antioxidant and anti-inflammation protection in most mammalian cells.

Bactericidal Effects of CO and H$_2$S

Macrophages produce multiple antimicrobial molecules, including nitric oxide (NO), hydrogen peroxide (H$_2$O$_2$), and acid (H$^+$). These molecules in macrophages have the mission and capability to kill engulfed bacteria. Macrophages also produce endogenously H$_2$S. The activation of macrophages by lipopolysaccharide (LPS) upregulates the expression of CSE mRNA and protein and increases the production of proinflammatory cytokines and NO [80]. After applying siRNA to specifically decrease CSE mRNA and proteins in LPS-activated macrophages, the levels of proinflammatory cytokines were significantly lower than those in untreated cells. However, the production levels of NO by the transfected cells were higher, suggesting that CSE activity has an inhibitory effect on NO production. It appears that CSE and its product H$_2$S inhibit the activation of macrophages, decrease NO production, and increase proinflammatory cytokines.

Macrophage production of CO via the activity of HO-1 can be induced by the enteric microbiota, leading to increased bactericidal activities of macrophages [81]. CO has been used as an antibacterial agent [82]. CORMs, a family of metal-based CO donors, have stronger bactericidal effect than CO gas, perhaps due to the ability of CORMs to deliver CO selectively to intracellular targets or due to the thiol binding of metal ions released from CORMs [83]. Recently reported transcriptomic consequences of CORM treatment of *Escherichia coli* revealed a myriad of unexpected targets for CO and potential CO sensors [82]. CO per se protects against intestinal inflammation in experimental models of colitis. Onyiah et al. investigated the interactions among CO, HO-1, and the enteric microbiota. By promoting bacterial clearance, CO and HO-1 mitigated intestinal inflammation in mice [84]. The colitis was induced in WT mice housed in specific pathogen-free (SPF) conditions by infection of mice with *Salmonella typhimurium*. Different strains of mice were intraperitoneal injected with cobalt(III) protoporphyrin IX chloride (CoPP) to upregulate HO-1 or the CO-releasing molecule Alfama-186. In colons of germ-free wild-type mice, colonization with SPF microbiota transfer induced production of HO-1 via activation of Nrf2-, interleukin-10 (IL-10)-, and Toll-like receptor-dependent pathways. Administration of CoPP to IL-10$^{-/-}$ mice before transition from germ-free to SPF conditions reduced their development of colitis. In mice with *S. typhimurium*-induced enterocolitis, CoPP reduced the numbers of live *S. typhimurium* recovered from the lamina propria, mesenteric lymph nodes, spleen, and liver. Knockdown of HO-1 in mouse macrophages impaired their bactericidal activity against *E. coli*, *Enterococcus faecalis*, and *S. typhimurium*, whereas exposure to CO or overexpression of HO-1 increased their bactericidal activity. HO-1 induction and CO increased acidification of phagolysosomes. This study and others show that colonic HO-1 and CO prevent acute and chronic colonic inflammation in mice.

CO not only kills planktonic bacteria but also prevents biofilm maturation and kills bacteria within the established biofilm. Biofilm is layer of mucilage, consisting of bacterial species like *E. coli* and *Streptococcus aureus* adhering to the surface of the cells, those lining intestine, bladder, or skin. Multidrug-resistant bacteria, such as extended-spectrum β-lactamase (ESBL)-producing Enterobacteriaceae, are involved in biofilm-related infections. It has been shown that a CO-releasing molecule (CORM-2) has antibacterial effects on ESBL-producing uropathogenic *E. coli* (UPEC) in the biofilm mode of growth and following colonization of host bladder epithelial cells. The studied bacteria grew within a biofilm formed for 24 h on plastic surface. CORM-2 (500 µM) exposure for 24 h significantly suppressed the ESBL-producing and non-ESBL-producing UPEC isolates grown in biofilms. The antibacterial effect of CORM-2 on planktonic bacteria was reduced and delayed in the stationary growth phase compared to the exponential growth phase. In human bladder epithelial cell colonization experiments, CORM-2 exposure for 4 h significantly reduced the bacterial counts of an ESBL-producing UPEC isolate [85]. A previous study also demonstrated that CORM-2 inhibited biofilm formation and planktonic growth of the majority of clinical *Pseudomonas aeruginosa* isolates tested, for both mucoid and nonmucoid strains [86].

Biofilm containing *S. aureus*, a leading pathogen for skin and skin structure infections, is resistant to conventional antibiotics. A photoresponsive CO-releasing molecule [CORM-1, $Mn_2(CO)_{10}$] produced 70% bactericidal efficacy against the biofilm-embedded *S. aureus* after photostimulation. The underlying mechanisms for the antimicrobial effect of the released CO include the inhibition of the electron transport chain and increased concentration of reactive oxygen species in the biofilms [87]. Moreover, a water-triggered CO-releasing compound EBOR-CORM-1 ([NEt_4][$MnBr_2(CO)_4$]) was shown to clear planktonic and biofilm cells of *P. aeruginosa* strain PAO1 in a concentration-dependent manner. EBOR-CORM-1 also decreased the growth of cystic fibrosis isolate coculture populations harboring intraspecific variation [88].

Biofilms serve as a barrier of the host cells from allergens or pathogens and as a microenvironment for microbiota to live and thrive. Microbiota dysbiosis and impaired barrier function are among the most prominent features of inflammatory bowel disease. In the gastrointestinal tract, H_2S participates in the regulation of the integrity of gut biofilms and mucosal homeostasis. Motta et al. examined colonic microbiota biofilm formation in rodents, using fluorescent *in situ* hybridization, and quantified colonic mucus granules with periodic acid–Alcian Blue staining [89]. They found that intestinal microbiota formed linear biofilms in the colon of healthy rodents. Colonic microbiota biofilms were fragmented and mucus granule production was decreased in dinitrobenzene sulfonic acid (DNBS)-induced colitis in C57Bl/6 mice and Wistar rats. Colonic tissue from CSE-KO mice produced significantly less H_2S and had mild granulocyte infiltration in the colon. Endogenous H_2S contributed to mucus layer thickness in the colon, and the inhibition of CSE with β-cyanoalanine in mice aggravated inflammation and altered colonic microbiota biofilm. Treatment with an H_2S donor (diallyl disulfide, DADS) of DNBS-induced colitis in rats decreased neutrophil infiltration, restored microbiota biofilm, and increased the production of mucus granules.

The effects of H_2S on human microbiota biofilms and planktonic bacteria were also examined to determine changes in their growth, viability, and biomass. In these *ex vivo* studies, biofilms ($n > 40$ per group) were exposed anaerobically to H_2S donors for 24 h, including NaHS or DADS. Biofilm metabolic activity (2,3-bis(2-methoxy-4-nitro-5-sulfophenyl)-2H-tetrazolium-5-carboxanilide inner salt assay) and biofilm biomass (crystal violet assay) were determined. It was found that H_2S promoted human colonic multispecies microbiota biofilms from colon biopsies from healthy volunteers. H_2S has antimicrobial activity against human planktonic bacteria. These observations show that H_2S supports the correction of microbiota biofilm dysbiosis and mucus layer reconstitution [89].

Bacterial Protection Effects of CO and H_2S

The protective effects of host-derived H_2S on microbiota have been reported in recent years, most noticeably on *Mycobacterium tuberculosis* (Mtb) that causes tuberculosis, a devastating human infectious disease causing about 2 million deaths annually. Saini et al. reported that H_2S stimulates Mtb respiration, growth, and pathogenesis. Mtb-infected CBS-deficient mice survived longer with reduced organ burden, and pharmacological inhibition of CBS reduced Mtb bacillary load in mice. High-resolution respirometry, transcriptomics, and mass spectrometry established that H_2S stimulated Mtb respiration and bioenergetics [90].

Cytochrome *bd* oxidase promotes sulfide-resistant bacterial respiration and growth, including those of Mtb [91]. The same terminal oxidase is also required for bacterial tolerance to CO [92]. CSE-generated H_2S also deteriorates the development of Mtb pathogenesis. CSE-KO mice survived longer than WT mice after Mtb infection, having decreased tuberculosis pathology and lower bacterial burdens in the lung, spleen, and liver [93]. CSE-deficient macrophages had lower colony forming units after *in vitro* Mtb infection than WT macrophages. The application of an H_2S slow-releasing agent GYY3147 and a CSE inhibitor PPG to Mtb-infected cells generated opposite changes in Mtb survival in macrophages, stimulation and inhibition, respectively. CSE and H_2S triggered excessive innate immune responses and inhibited the adaptive immune response, decreasing circulating IL-1β, IL-6, TNF-α, and IFN-γ levels in response to Mtb infection. Moreover, H_2S in the infected macrophages inhibited flux through glycolysis and the pentose phosphate pathway. These animal studies and *in vitro* cell culture studies were reconciled with the spatial distribution of H_2S-producing enzymes in human necrotic, nonnecrotic, and cavitary pulmonary tuberculosis lesions. These recent reports demonstrate that Mtb exploits host-derived H_2S, resulting from CBS or CSE activations, to exacerbate tuberculosis pathogenesis by altering immunometabolism.

Host-derived CO also promotes the development of tuberculosis but via a different mechanism from that of H_2S. Mtb upregulates host HO-1 expression to produce more CO and adapts its transcriptome during CO exposure. After screening an Mtb transposon library for CO-susceptible mutants, Zacharia et al. found that disruption of Rv1829 (CO resistance, Cor) led to marked CO sensitivity. Cor is essential for Mtb pathogenesis [94]. Heterologous expression of Cor in *E. coli* rescued it from CO toxicity. In a mouse model of tuberculosis, the virulence of the *cor* mutant was attenuated. CO resistance is essential for mycobacterial survival *in vivo* and this resistance is realized by Cor that shields bacteria from host-derived CO [94]. It remains to be determined whether Mtb resistance to CO is an evolutionary adaptation of mycobacteria for survival within macrophages. Future studies are merited to detect the expression of CO-resistant genes in other species of microbiota.

The Production of CO and H_2S in Microbiota for Their Self-protection

Microbial CO formation occurs under the catalysis of an array of enzymes [95]. The most noticeable CO-generating enzymes in bacteria are canonical heme-degrading enzymes, i.e., HOs. In terms of CO production and iron acquisition, HO performs the same way in prokaryotes as in eukaryotes [96]. Other CO-generating enzymes include various dioxygenases, methane monooxygenase, nitrogenase, nitrile hydratase, CO-methylating acetyl-CoA synthase, 5-methyltetrahydrofolate:corrinoid/iron–sulfur protein co-methyltransferase, etc. Bacterial production of CO spreads widely among numerous types of prokaryotes. Readers are referred to a recent comprehensive review on this topic by Hopper et al. [95]. Interestingly, CO production by pathogenic bacteria appears to be rare, which may be an evolutionary self-protection mechanism for these pathogens.

Microbiota-derived CO and H_2S constitute an intertwined defense system against antibiotics and oxidative stress, serving a protective role for themselves and a detrimental role for the host. Many species of microbiota can survive and thrive under culture headspaces sometimes exceeding 1 atm of CO. The presence of CO dehydrogenase, a Mo-containing CO oxidoreductase, in many strains of bacteria allows them to use molecular oxygen as the electron acceptor for aerobic oxidation. Numerous species of microbiota, both aerobic and anaerobic, utilize CO as a substrate, producing organic compounds such as acetate, ethanol, 2,3-butanediol, and butyrate [82]. Phylogenetically diverse members of the bacteria and Archaea anaerobically oxidize CO using CooS enzymes that contain Ni/Fe catalytic centers. This CO oxidation process is coupled to numerous respiratory processes, such as desulfurication, hydrogenogenesis, acetogenesis, and methanogenesis [97]. It appears that carboxydotrophs are adapted to provide a metabolic "currency exchange" system in microbial communities in which CO is converted to CO_2 and H_2 that feed major metabolic pathways for energy conservation or carbon fixation [98].

Some or all of the homologues of H_2S-producing enzymes (CSE, CBS, MST) participate in bacterial production of H_2S. Bacteria also use sulfur amino acids (L-cysteine and methionine) as substrates for H_2S production. For most bacteria, L-cysteine is produced enzymatically via CysE, CysK, and CysM from L-serine [99]. Key bacteria associated with the metabolism of methionine are *Clostridium* spp., *Peptostreptococcus* spp., *Eubacterium*, *Salmonella enterica*, *E. coli*, *Enterobacter aerogenes*, *Klebsiella pneumoniae*, and *Desulfovibrio* spp. They mostly reside in human large intestine. Nonenzymatic pathways also contribute to endogenous H_2S levels in bacteria. These include degradation of cysteine and other sulfur-containing amino acids/peptides, and dissimilatory reduction of inorganic sulfur compounds by sulfate-reducing bacteria (SRBs). Being one of the oldest species of microbiota on Earth [3], SRBs represent a major class of the normal gut microbiota. Human gut-inhabited SRBs include *Desulfovibrio, Desulfobacter, Desulfolobus,* and *Desulfotomaculum*. The nonenzymatically generated H_2S by gut SRB also contributes to the circulatory H_2S level in human host.

Among the germ-killing mechanisms of bactericidal antibiotics, such as spectinomycin, gentamycin, amikacin, and ampicillin, are the stimulation of respiration and the consequential increase in toxic hydroxyl radicals via Fe^{2+}-catalyzed Fenton reaction and oxidative damage to bacterial DNA. The antibiotic resistance and tolerance to oxidative stress of Gram-negative and Gram-positive bacteria, including *Bacillus anthracis, P. aeruginosa, S. aureus,* and *E. coli,* have been linked to the ability of these bacteria to produce H_2S. Multidrug-resistant strains of *E. coli* isolated from patients suffering from urinary tract infection have enhanced endogenous H_2S production. In comparison with WT bacterial strains, the strains with MST, CBS, or CSE deficiency exhibited decreased endogenous H_2S production and higher susceptibility to structurally and functionally different classes of antibiotics. Overexpression of MST equipped these bacteria with self-protection against spectinomycin. On the other hand, pharmacologically inhibiting the activities of MST, CBS, and CSE rendered them sensitive to a range of antibiotics. The application of exogenous H_2S salt, NaHS, at low concentration makes these pathogens resistant to antibiotics [100]. H_2S-enhanced bacterial resistance to antibiotics and oxidative stress may be explained by the antioxidant effects of H_2S against metal ions, such as direct scavenge of Fe^{2+} and suppression of the related antibiotic-induced oxidative damage. H_2S can upregulate the antioxidant enzymes such as catalase and superoxide dismutase, leading to accelerated degradation of H_2O_2. This effect may also contribute to the self-protection of bacteria against damages by reactive oxygen species.

The Involvement of CO and H_2S in Host–Virus Interaction

Being composed of only genetic materials in RNA or DNA, viruses do not have machineries or milieus to produce gasotransmitters. On the other hand, virus infection of the host cells constitutes a critical stress challenge. In response to the challenge, induction of HO-1 and increased CO production by the host cells may occur. When the viral infection targets at the upper and lower respiratory tract, the consequentially increased host production of CO may be detected in the exhaled air [101]. To date, we have no knowledge of whether viral infection affects host production of H_2S.

The oxidative state of host cells provides an environment permissive for the replication of both RNA viruses (including influenza, paramyxovirus, and COVID-19) and DNA viruses (including poxviruses, parvoviruses, and hepadnaviruses). In this context, the antioxidant properties of CO and H_2S may be exploited to suppress oxidative stress inside host to curtail viral infection and long-term viral replication.

The antiviral activities of CO in infected host cells have been shown. HO-1 and CO inhibited the replication of porcine reproductive and respiratory syndrome virus (PRRSV) in both a PRRSV permissive cell line, MARC-145, and the predominant cell type targeted during *in vivo* PRRSV infection, porcine alveolar macrophages. Intercellular spread of PRRSV, but not PRRSV entry into host cells, was suppressed by CO. This inhibitory effect of CO on PRRSV replication is mediated by the activation of the cyclic GMP/protein kinase G signaling pathway [102]. PRRSV replication required the activation of transcription factor NF-κB, which is suppressed by CO. Moreover, CO significantly lowered the mRNA levels of PRRSV-induced proinflammatory cytokine.

Similar to the effect of CO, H_2S inhibited NF-κB activation and the replication of paramyxovirus respiratory syncytial virus (RSV). RSV infection resulted in downregulation of expression and impaired activity of CSE in host cells and decreased endogenous levels of H_2S [103]. Consistent with these findings, exogenous administration of H_2S reduces the secretion of virus-induced chemokines and cytokine through inhibition of NF-κB-mediated activation of genes encoding proinflammatory cytokines. Treatment with GYY4137, an H_2S slow-releasing agent, significantly blocked RSV replication *in vitro* and *in vivo* by targeting viral assembly, release, spread, and replication [103].

Interestingly, a previous report showed that sulforaphane had the ability to block HIV-1 infection in primary macrophages [104]. Sulforaphane is an isothiocyanate that can induce Nrf2. This sulfur-rich compound can be found in cruciferous vegetables like broccoli, bok choy, kale, cabbage, cabbage, cauliflower, brussels sprouts, collards, mustard greens, and watercress [105]. Just like diallyl sulfide, DADS, diallyl trisulfide in garlic, sulforaphane is a natural H_2S donor [106]. When it was added into cell culture medium or mixed with mouse liver homogenates, respectively, sulforaphane released significant amounts of H_2S. Both sulforaphane and NaHS decreased the viability of PC-3 cells (a human prostate cancer cell line) in a dose-dependent manner, and supplementing methemoglobin or oxidized glutathione, two H_2S scavengers, reversed sulforaphane-suppressed cell viability. Sulforaphane activated p38 mitogen-activated protein kinases (MAPK) and c-Jun N-terminal kinase. Pretreatment of PC-3 cells with methemoglobin decreased sulforaphane-stimulated MAPK activities [106]. Sulforaphane can also enter cerebral circulation, as a brain-permeable compound, to increase cerebral blood flow via the activation of K_{ATP} and B_K channels in arteriolar smooth muscle [107].

Bazhanov et al. examined the broad-range antiviral activity of H_2S on pathogenic RNA viruses, enveloped RNA viruses from Ortho-, Filo-, Flavi-, and Bunyavirus families [108]. H_2S slow-releasing agent GYY4137 significantly reduced replication of all tested viruses. In a model of influenza infection, GYY4137 treatment resulted in decreased expression of viral proteins and mRNA, suggesting inhibition of an early step of replication. The antiviral activity coincided with the decrease of virus-induced proinflammatory mediators and nuclear translocation of transcription factors, such as NF-κB and members of interferon regulatory factor families.

COVID-19 is a pleomorphic enveloped virus containing single-stranded (positive-sense) RNA associated with a nucleoprotein within a capsid comprised of matrix protein. The envelope bears club-shaped glycoprotein projections. Although H_2S may possess antiviral capability against a wide spectrum of enveloped RNA viruses, the effects of gasotransmitters on COVID-19 virus are unknown. This fledging but important area merits for future exploration, especially in the wake of COVID-19 pandemic.

Conclusions

CO and H_2S do not and cannot exist or act alone in nature or in our body. Being antagonistic or agonistic, CO and H_2S interact as part of a complicated and coordinated signaling network. The investigation of this interplay began decades ago and has achieved some initial success. We still have a long way to go. An end of the beginning marks a new beginning.

Acknowledgments

This study has been supported by research grants from the Canadian Institutes of Health Research, Natural Science and Engineering Research Council of Canada, and the Heart and Stroke Foundation of Canada.

References

1. Wang, R. (2002). *FASEB J.* 16 (13): 1792–1798.
2. Wang, R. (2014). *Trends Biochem. Sci.* 39 (5): 227–232.
3. Wang, R. (2010). *Sci. American* 302 (3): 66–71.
4. Miyakawa, S., Yamanashi, H., Kobayashi, K., Cleaves, H.J., and Miller, S.L. (2002). *Proc. Natl. Acad. Sci. USA.* 99: 14628–14631.
5. Ndisang, J.F., Wu, L., Zhao, W., and Wang, R. (2003). *Blood* 101: 3893–3900.
6. Shamloul, R. and Wang, R. (2005). *Cell. Mol. Biol.* 51: 507–512.
7. Wang, R. (2012). *Physiol. Rev.* 92 (2): 791–896.
8. Zhao, W., Zhang, J., Lu, Y., and Wang, R. (2001). *EMBO J.* 20: 6008–6016.
9. Zhao, W. and Wang, R. (2002). *Am. J. Physiol. Heart Circ. Physiol.* 283 (2): H474–H480.
10. Yang, G., Cao, K., Wu, L., and Wang, R. (2004). *J. Biol. Chem.* 279: 49199–49205.
11. Yang, G., Sun, X., and Wang, R. (2004). *FASEB J.* 18 (14): 1782–1784.
12. Yang, G., Wu, L., Jiang, B., Yang, W., Qi, J., Cao, K., Meng, Q., Mustafa, A.K., Mu, W., Zhang, S., Snyder, S.H., and Wang, R. (2008). *Science* 322: 587–590.
13. Paul, B.D., Sbodio, J.I., Xu, R., Vandiver, M.S., Cha, J.Y., Snowman, A.M., and Snyder, S.H. (2014). *Nature* 509 (7498): 96–100.
14. Shibuya, N., Mikami, Y., Kimura, Y., Nagahara, N., and Kimura, H. (2009). *J. Biochem.* 146 (5): 623–626.
15. Shen, X., Peter, E.A., Bir, S., Wang, R., and Kevil, C.G. (2012). *Free Radic. Biol. Med.* 52 (11–12): 2276–2283.
16. McCook, O., Radermacher, P., Volani, C., Asfar, P., Ignatius, A., Kemmler, J., Möller, P., Szabó, C., Whiteman, M., Wood, M.E., Wang, R., Georgieff, M., and Wachter, U. (2014). *Nitric Oxide* 41: 48–61.
17. Peter, E.A., Shen, X., Shah, S.H., Pardue, S., Glawe, J.D., Zhang, W.W., Reddy, P., Akkus, N.I., Varma, J., and Kevil, C.G. (2013). *J. Am. Heart Assoc* 2 (5): e000387.
18. Wang, K., Ahmad, S., Cai, M., Rennie, J., Fujisawa, T., Crispi, F., Baily, J., Miller, M.R., Cudmore, M., Hadoke, P.W., Wang, R., Gratacós, E., Buhimschi, I.A., Buhimschi, C.S., and Ahmed, A. (2013). *Circulation* 127 (25): 2514–2522.
19. Wallace, J.L. and Wang, R. (2015). *Nat. Rev. Drug Discov.* 14 (5): 329–345.
20. Wang, R., Szabo, C., Ichinose, F., Ichinose, F., Ahmed, A., Whiteman, M., and Papapetropoulos, A. (2015). *Trends Pharmacol. Sci.* 36 (9): 568–578.

21. Beauchamp, R.O., Bus, J.S., Popp, J.A., Boreiko, C.J., and Andjelkovich, D.A. (1984). *CRC Crit. Rev. Toxicol.* 13: 25–97.
22. Levitt, M.D., Abdel-Rehim, M.S., and Furne, J. (2011). *Antioxid. Redox Signal.* 15: 373–378.
23. Mani, S., Li, H., Untereiner, A., Wu, L., Yang, G.D., Austin, R.C., Dickhout, J.G., Lhotak, S., Meng, Q.H., and Wang, R. (2013). *Circulation* 127 (25): 2523–2534.
24. Yang, G., Wu, L., Bryan, S., Khaper, N., Mani, S., and Wang, R. (2010). *Cardiovasc. Res.* 86 (3): 487–495.
25. Altaany, Z., Ju, Y., Yang, G., and Wang, R. (2014). *Sci. Signal.* 7 (342): ra87.
26. Papapetropoulos, A., Pyriochou, A., Altaany, Z., Yang, G., Marazioti, A., Zhou, Z., Jeschke, M.G., Branski, L.K., Herndon, D.N., Wang, R., and Szabó, C. (2009). *Proc. Natl. Acad. Sci. USA.* 106 (51): 21972–21977.
27. Jiang, B., Tang, G., Cao, K., Wu, L., and Wang, R. (2010). *Antioxid. Redox Signal.* 12 (10): 1167–1178.
28. Tang, G., Wu, L., Liang, W., and Wang, R. (2005). *Mol. Pharmacol.* 68 (6): 1757–1764.
29. Tang, G., Yang, G., Jiang, B., Ju, Y., Wu, L., and Wang, R. (2013). *Antioxid. Redox Signal.* 19 (14): 1634–1646.
30. Mustafa, A., Sikka, G., Gazi, S.K., Steppan, J., Jung, S.M., Bhunia, A.K., Barodka, V.M., Gazi, F.K., Barrow, R.K., Wang, R., Amzel, L.M., Berkowitz, D.E., and Snyder, S.H. (2011). *Circ. Res.* 109 (11): 1259–1268.
31. Elrod, J.W., Calvert, J.W., Morrison, J., Doeller, J.E., Kraus, D.W., Tao, L., Jiao, X., Scalia, R., Kiss, L., Szabo, C., Kimura, H., Chow, C.W., and Lefer, D.J. (2007). Hydrogen sulfide attenuates myocardial ischemia-reperfusion injury by preservation of mitochondrial function. *Proc. Natl. Acad. Sci. USA.* 104 (39): 15560–15565.
32. Johansen, D., Ytrehus, K., Baxter, G.F., Johansen, D., Ytrehus, K., and Baxter, G.F. (2006). *Basic Res. Cardiol.* 101 (1): 53–60.
33. Kondo, K., Bhushan, S., King, A.L., Prabhu, S.D., Hamid, T., Koenig, S., Murohara, T., Predmore, B.L., Gojon, G., Sr., Gojon, G., Jr., Wang, R., Karusula, N., Nicholson, C.K., Calvert, J.W., and Lefer, D.J. (2013). *Circulation* 127 (10): 1116–1127.
34. Zhu, Y.Z., Wang, Z.J., Ho, P., Loke, Y.Y., Zhu, Y.C., Huang, S.H., Tan, C.S., Whiteman, M., Lu, J., and Moore, P.K. (2007). *J. Appl. Physiol.* 102 (1): 261–268.
35. Mustafa, A.K., Gadalla, M.M., Sen, N., Kim, S., Mu, W., Gazi, S.K., Barrow, R.K., Yang, G., Wang, R., and Snyder, S.H. (2009). *Sci. Signal.* 2 (96): ra72.
36. Madden, J.A., Ahlf, S.B., Dantuma, M.W., Olson, K.R., and Roerig, D.L. (2012). *J. Appl. Physiol.* 112 (3): 411–418.
37. Wu, L., Yang, W., Jia, X., Yang, G., Duridanova, D., Cao, K., and Wang, R. (2009). *Lab. Invest.* 89 (1): 59–67.
38. Kabil, O., Yadav, V., and Banerjee, R. (2016). *J. Biol. Chem.* 291: 16418–16423.
39. Yang, G., Zhao, K., Ju, Y., Mani, S., Cao, Q., Puukila, S., Khaper, N., Wu, L., and Wang, R. (2013). *Antioxid. Redox Signal.* 18: 1906–1919.
40. Wu, L. and Wang, R. (2005). *Pharmacol. Rev.* 57: 585–630.
41. Ndisang, J.F., Zhao, W., and Wang, R. (2002). *Hypertension* 40: 315–321.
42. Wang, R., Shamloul, R., Wang, X., Meng, Q., and Wu, L. (2006). *Hypertension* 48: 685–692.
43. Chang, T., Wu, L., and Wang, R. (2008). *Am. J. Physiol. Heart Circ. Physiol.* 295 (3): H999–H1007.
44. Shamloul, R. and Wang, R. (2006). *J. Sex Med.* 3: 619–627.
45. Sassa, S. (2004). *Antioxid. Redox Signal.* 6: 819–824.
46. Ingi, T., Cheng, J., and Ronnett, G.V. (1996). *Neuron* 16: 835–842.
47. Ryter, S.W. and Tyrrell, R.M. (2000). *Free Radic. Biol. Med.* 28: 289–309.
48. Agarwal, A., Balla, J., Alam, J., Croatt, A.J., and Nath, K.A. (1995). *Kidney Int.* 48 (4): 1298–1307.
49. Balla, J., Jacob, H.S., Balla, G., Nath, K., Eaton, J.W., and Vercellotti, G.M. (1993). *Proc. Natl. Acad. Sci. USA.* 90 (20): 9285–9289.
50. Shimizu, H., Takahashi, T., Suzuki, T., Yamasaki, A., Fujiwara, T., Odaka, Y., Hirakawa, M., Fujita, H., and Akagi, R. (2000). *Crit. Care Med.* 28 (3): 809–817.
51. Kumar, S. and Bandyopadhyay, U. (2005). *Toxicol. Lett.* 157 (3): 175–188.
52. Atamna, H. (2004). *Ageing Res. Rev.* 3 (3): 303–318.
53. Ogawa, K., Sun, J., Taketani, S., Nakajima, O., Nishitani, C., Sassa, S., Hayashi, N., Yamamoto, M., Shibahara, S., Fujita, H., and Igarashi, K. (2001). *EMBO J.* 20 (11): 2835–2843.
54. Jeney, V., Balla, J., Yachie, A., Varga, Z., Vercellotti, G.M., Eaton, J.W., and Balla, G. (2002). *Blood* 100 (3): 879–887.
55. Platt, J.L. and Nath, K.A. (1998). *Nat. Med.* 4 (12): 1364–1365.
56. Hanna, D.A., Martinez-Guzman, O., and Reddi, A.R. (2017). *Biochemistry* 56 (13): 1815–1823.
57. Julius, U. and Pietzsch, J. (2005). *Antioxid. Redox Signal.* 7: 1507–1512.
58. Silva, G., Jeney, V., Chora, A., Larsen, R., Balla, J., and Soares, M.P. (2009). *J. Biol. Chem.* 284: 29582–29595.
59. Balla, J., Vercellotti, G.M., Jeney, V., Yachie, A., Varga, Z., Jacob, H.S., Eaton, J.W., and Balla, G. (2007). *Antioxid. Redox Signal.* 9: 2119–2137.
60. Marks, G.S. (1985). *Crit. Rev. Toxicol.* 15 (2): 151–179.
61. Tenhunen, R., Savolainen, H., and Jäppinen, P. (1983). *Clin. Sci. (Lond)* 64 (2): 187–191.
62. Leigh, J., Juriasingani, S., Akbari, M., Shao, P., Saha, M.N., Lobb, I., Bachtler, M., Fernandez, B., Qian, Z.,

62. Van Goor, H., Pasch, A., Feelisch, M., Wang, R., and Sener, A. (2018). *Can. Urol. Assoc. J.* 13 (7): E210–E219.
63. Módis, K., Ramanujam, S., Govar, A., Lopez, E., Anderson, K.E., Wang, R., and Szabo, C. (2019). *Biochem. Pharmacol.* 169: 113604.
64. Morikawa, T., Kajimura, M., Nakamura, T., Hishiki, T., Nakanishi, T., Yukutake, Y., Nagahata, Y., Ishikawa, M., Hattori, K., Takenouchi, T., Takahashi, T., Ishii, I., Matsubaram, K., Kabem, Y., Uchiyama, S., Nagata, E., Gadalla, M.M., Snyder, S.H., and Suematsu, M. (2012). *Proc. Natl. Acad. Sci. USA.* 109 (4): 1293–1298.
65. Prabhakhar, N.R. and Joyner, M.J. (2015). *Front. Physiol.* 5: 524.
66. Semenza, G.L. and Prabhakar, N.R. (2018). *J. Physiol.* 596 (15): 2977–2983.
67. Kabe, Y., Yamamoto, T., Kajimura, M., Sugiura, Y., Koike, I., Ohmura, M., Nakamura, T., Tokumoto, Y., Tsugawa, H., Handa, H., Kobayashi, T., and Suematsu, M. (2016). *Free Radic. Biol. Med.* 99: 333–344.
68. Peng, Y.J., Makarenko, V.V., Nanduri, J., Vasavda, C., Raghuraman, G., Yuan, G., Gadalla, M.M., Kumar, G.K., Snyder, S.H., and Prabhakar, N.R. (2014). *Proc. Natl. Acad. Sci. U S A.* 111 (3): 1174–1179.
69. Teng, H., Wu, B., Zhao, K., Yang, G., Wu, L., and Wang, R. (2013). *Proc. Natl Acad. Sci. USA* 110: 12679–12684.
70. Fu, M., Zhang, W., Wu, L., Yang, G., Li, H., and Wang, R. (2012). *Proc. Natl. Acad. Sci. USA* 109: 2943–2948.
71. Skovgaard, N. and Olson, K.R. (2012). *Am. J. Physiol. Regul. Integr. Comp. Physiol.* 303: R487–R494.
72. Goubern, M., Andriamihaja, M., Nübel, T., Blachier, F., and Bouillaud, F. (2007). *FASEB J.* 21: 1699–1706.
73. Fitzgerald, R., De Santiago, B., Lee, Y.D., Yang, G., Kim, J.K., Foster, D.B., Chan-Li, Y., Horton, M.R., Panettieri, R.A., Wang, R., and An, S.S. (2014). *Biochem. Biophys. Res. Commun.* 446: 393–398.
74. Zhang, G., Wang, P., Yang, G., Cao, Q., and Wang, R. (2013). *Am. J. Pathol.* 182: 1188–1195.
75. Li, Q., Sun, B., Wang, X., Jin, Z., Zhou, Y., Dong, L., Jiang, L.H., and Rong, W. (2010). *Antioxid. Redox Signal.* 12: 1179–1189.
76. Módis, K., Asimakopoulou, A., Coletta, C., Papapetropoulos, A., and Szabo, C. (2013). *Biochem. Biophys. Res. Commun.* 433: 401–407.
77. Chunyu, Z., Junbao, D., Dingfang, B., Hui, Y., Xiuying, T., and Chaoshu, T. (2003). *Biochem. Biophys. Res. Commun.* 302: 810–816.
78. Ariyaratnam, P., Loubani, M., and Morice, A.H. (2013). *Microvasc. Res.* 90: 135–137.
79. Ahmed, A. (2014). *Pregnancy Hypertens.* 4 (3): 243–244.
80. Badiei, A., Rivers-Auty, J., Ang, A.D., and Bhatia, M. (2013). *Appl. Microbiol. Biotechnol.* 97 (17): 7845–7852.
81. Onyiah, J.C., Sheikh, S.Z., Maharshak, N., Otterbein, L.E., and Plevy, S.E. (2014). *Gut Microbes* 5 (2): 220–224.
82. Wilson, J.L., Jesse, H.E., Poole, R.K., and Davidge, K.S. (2012). *Curr. Pharm. Biotechnol.* 13 (6): 760–768.
83. Southam, H.H., Smith, T.W., Lyon, R.L., Liao, C., Trevitt, C.R., Middlemiss, L.A., Cox, F.L., Chapman, J.A., El-Khamisy, S.F., Hippler, M.,., Williamson, M.P., Henderson, P.J.F., and Poole, R.K. (2018). *Redox Biol.* 18: 114–123.
84. Onyiah, J.C., Sheikh, S.Z., Maharshak, N., Steinbach, E.C., Russo, S.M., Kobayashi, T., Mackey, L.C., Hansen, J.J., Moeser, A.J., Rawls, J.F., Borst, L.B., Otterbein, L.E., and Plevy, S.E. (2013). *Gastroenterology* 144 (4): 789–798.
85. Bang, S.C., Kruse, R., Johansson, K., and Persson, K. (2016). *BMC Microbiol.* 16: 64.
86. Murray, T.S., Okegbe, C., Gao, Y., Kazmierczak, B.I., Motterlini, R., Dietrich, L.E., and Bruscia, E.M. (2012). *PLoS One* 7 (4): e35499.
87. Klinger-Strobel, M., Gläser, S., Makarewicz, O., Wyrwa, R., Weisser, J., Pletz, M.W., and Schiller, A. (2016). *Antimicrob. Agents Chemother.* 60 (7): 4037–4046.
88. Flanagan, L., Steen, R.R., Saxby, K., Klatter, M., Aucott, B.J., Winstanley, C., Fairlamb, I.J.S., Lynam, J.M., Parkin, A., and Friman, V.P. (2018). *Front. Microbiol.* 9: 195.
89. Motta, J.P., Flannigan, K.L., Agbor, T.A., Beatty, J.K., Blackler, R.W., Workentine, M.L., Da Silva, G.J., Wang, R., Buret, A.G., and Wallace, J.L. (2015). *Inflamm. Bowel Dis.* 21 (5): 1006–1017.
90. Saini, V., Chinta, K.C., Reddy, V.P., Glasgow, J.N., Stein, A., Lamprecht, D.A., Rahman, M.A., Mackenzie, J.S., Truebody, B.E., Adamson, J.H., Kunota, T.T.R., Bailey, S.M., Moellering, D.R., Lancaster, J.R., Jr., and Steyn, A.J.C. (2020). *Nat. Commun.* 11 (1): 557.
91. Forte, E., Borisov, V.B., Falabella, M., Colaco, H.G., Tinajero-Trejo, M., Poole, R.K., Vicente, J.B., Sarti, P., and Giuffrè, A. (2016). *Sci. Rep.* 6: 23788.
92. Jesse, H.E., Nye, T.L., McLean, S., Green, J., Mann, B.E., and Poole, R.K. (2013). *Biochim. Biophys. Acta* 1834: 1693–1703.
93. Rahman, M.A., Cumming, B.M., Addicott, K.W., Pacl, H.T., Russell, S.L., Nargan, K., Naidoo, T., Ramdial, P.K., Adamson, J.H., Wang, R., and Steyn, A.J.C. (2020). *Proc. Natl. Acad. Sci. U S A.* 117 (12): 6663–6674.
94. Zacharia, V.M., Manzanillo, P.S., Nair, V.R., Marciano, D.K., Kinch, L.N., Grishin, N.V., Cox, J.S., and Shiloh, M.U. (2013). *mBio.* 4 (6): e00721–13.

95. Hopper, C.P., De La Cruz, L.K., Lyles, K.V., Wareham, L.K., Gilbert, J.A., Eichenbaum, Z., Magierowski, M., Poole, R.K., Wollborn, J., and Wang, B. (2020). *Chem. Rev.* 10.1021/acs.chemrev.0c00586. Online ahead of print.
96. Frankenberg-Dinkel, N. (2004). *Antioxid. Redox Signal.* 6: 825–834.
97. Oelgeschläger, E. and Rother, M. (2008). *Arch. Microbiol.* 90 (3): 257–269.
98. Robb, F.T. and Techtmann, S.M. (2018). *F1000Res* 7: F1000 Faculty Rev-1981.
99. Sawa, T., Ono, K., Tsutsuki, H., Zhang, T., Ida, T., Nishida, M., and Akaike, T. (2018). *Adv. Microb. Physiol.* 72: 1–28.
100. Shatalin, K., Shatalina, E., Mironov, A., and Nudler, E. (2011). *Science* 334 (6058): 986–990.
101. Yamaya, M., Sekizawa, K., Ishizuka, S., Monma, M., Mizuta, K., and Sasaki, H. (1998). *Am. J. Respir. Crit. Care Med.* 158 (1): 311–314.
102. Zhang, A., Zhao, L., Li, N., Duan, H., Liu, H., Pu, F., Zhang, G., Zhou, E.-M., and Xiao, S. (2017). *J. Virol.* 91 (1): e01866–16.
103. Li, H., Ma, Y., Escaffre, O., Ivanciuc, T., Komaravelli, N., Kelley, J.P., Coletta, C., Szabo, C., Rockx, B., Garofalo, R.P., and Casola, A. (2015). *J. Virol.* 89 (10): 5557–5568.
104. Furuya, A.K., Sharifi, H.J., Jellinger, R.M., Cristofano, P., Shi, B., and De Noronha, C.M. (2016). *PLoS Pathog* 12: e1005581.
105. Wu, L., Noyan Ashraf, M.H., Facci, M., Wang, R., Paterson, P.G., Ferrie, A., and Juurlink, B.H. (2004). *Proc. Natl. Acad. Sci. U S A.* 101 (18): 7094–7099.
106. Pei, Y., Wu, B., Cao, Q., Wu, L., and Yang, G. (2011). *Toxicol. Appl. Pharmacol.* 257 (3): 420–428.
107. Parfenova, H., Liu, J., Hoover, D.T., and Fedinec, A.L. (2020). *J. Cereb. Blood Flow Metab.* 40 (10): 1987–1996.
108. Bazhanov, N., Escaffre, O., Freiberg, A.N., Garofalo, R.P., and Casola, A. (2017). *Sci. Rep.* 7: 41029.

10

Biliverdin and Bilirubin as Parallel Products of CO Formation: Not Just Bystanders

Libor Vítek

4th Department of Internal Medicine and Institute of Clinical Biochemistry and Laboratory Diagnostics, General University Hospital in Prague and 1st Faculty of Medicine, Charles University, Prague, Czech Republic

Evolutionary Aspects of Biliverdin and Bilirubin

Heme (iron–protoporphyrin IX) belongs to the phylogenetically old superfamily of tetrapyrrolic compounds, which are among the most conserved structures in nature. It has even been postulated that tetrapyrroles might have been formed prebiotically during the period of the formation of life on Earth using the presumed primeval atmospheric prebiotic pathways [1]. Nevertheless, in humans heme serves as a prosthetic group for ubiquitously occurring hemoproteins, with hemoglobin and myoglobin being the most abundant [2]. Besides the key role of heme in biology across the whole range of living things in nature, its degradation products (bilirubin and biliverdin) as well as other related linear tetrapyrroles and their derivatives play an enormous role in the regulation of various biological processes, from the simplest organisms to human beings (Table 10.1) [3]. These include

i. the bilins such as phytochromobilin or phycocyanobilin, which serve as accessory light-harvesting pigments in higher plants, cyanobacteria, and blue-green algae [4,5], but they are also involved in oxidative stress defense and the regulation of other biological processes [6] with possible implications for human health [7,8];

ii. chlorophyll degradation products (such as phylloleucobilin or phylloxanthobilin, among others), which possibly contribute to defense mechanisms from both biotic and abiotic stress stimuli in higher plants, algae, and cyanobacteria by acting as radical scavengers and preventing toxicity associated with increased oxidative stress (although these processes and the detailed mechanisms are still only poorly understood) [9,10];

iii. biliverdin, bilirubin, and their reductive derivatives such as urobilinoids, exerting a number of biological functions in the human body (see later), that have evolved through evolution's long history from simpler organisms. For instance, it is well known that biliverdin and its precursor (protoporphyrin IX) are responsible for the pigmentation of birds' eggs and feathers. Although originally thought to have only a camouflage function, it is more apparent now that these pigments in egg shells provide protection against various harmful entities, including infectious agents [11]. In fact, light-dependent antimicrobial activity against Gram-positive bacteria by the tetrapyrrolic pigments present in egg shells has been well documented [12]. It is quite interesting that hen's eggshell color, due to deposits of porphyrins/biliverdin, is related to the feared coronaviral poultry infections caused by the infectious bronchitis virus [13]. It has also been reported in Japanese quails that biliverdin concentration in the eggshells is influenced by postnatal stress with biliverdin possibly serving as a potential adaptive strategy to cope with developmental stress [14]. Interestingly, biliverdin pigmentation of the eggs has even recently been discovered in the oviraptorid dinosaurs of the Late Cretaceous period [15,16]. It is also possible that biliverdin and bilirubin may play an important role in ontogenesis as suggested by studies on sea lampreys [17]. Although detailed data are lacking, it is also possible that biliverdin may play a developmental role in the human fetus, since it is a predominant bile pigment in fetal life [18]. If true, this seems to resemble the role of biliverdin in oogenesis and embryonic dorsal axis development described in *Xenopus laevis*, in which biliverdin binds to the

Carbon Monoxide in Drug Discovery: Basics, Pharmacology, and Therapeutic Potential, First Edition. Edited by Binghe Wang and Leo E. Otterbein.
© 2022 John Wiley & Sons, Inc. Published 2022 by John Wiley & Sons, Inc.

Table 10.1 Functional activities of linear tetrapyrroles and their oxidative derivatives.

Linear tetrapyrroles and related pigments	Examples	Functions
"Classical" bile pigments	Bilirubin Biliverdin Urobilinoids	• Defense against oxidative stress • Anti-inflammatory effects • Antimicrobial effects • Cell apoptosis-modulating effects • Gene transcription-modulating effects • Effects on protein post-translational modifications • Endocrine effects due to their binding to various nuclear receptors • Cell signaling functions • Possible roles in ontogenesis • Visual functions in some vertebrates
Bilins	Phytochromobilin Phycocyanobilin	• Light harvesting • Defense against oxidative stress
Chlorophyll degradation products	Phylloleukobilin Phylloxanthobilin	• Defense against biotic/abiotic stress
Bilirubin oxidation products	Bilirubin photoisomers Biopyrrins Propentdyopents Monopyrrolic bilirubin oxidation products (BOXes)	• Roles in oxidative stress defense/propagation • Roles in inflammatory processes • Vasoactive activities

highly conserved yolk proteins vitellogenin and lipovitellin that are important for embryogenesis [19–21]. Another example of (thus far only vaguely explored) biological role of biliverdin is its binding to the serpin family of proteins in green frogs [22] or its presence in the blue coral *Heliopora coerulea* [23]. It is also noteworthy to mention the recently discovered bilirubin-fluorescent proteins present in marine organisms [24]. These evolutionarily old bilirubin-binding proteins are acknowledged to have visual functions, and also to protect the muscle tissue of migrating eels from increased oxidative stress [25]. Interestingly, bilirubin, putatively believed to only be a human bile pigment, has also been identified in the seeds of the flower *Strelitzia reginae* [26,27], with possible biological consequences. Its seed extracts are already used in skin revitalizing lotions, and further applications can soon be expected. Although we are only at the beginning of our understanding of the physiological functions of these unique pyrrolic compounds, all of these data suggest their broad complexity and versatility.

It is important to note that also carbon monoxide (CO), a parallel product of heme degradation, is likely to contribute to the developmental processes as well, including conception, placentation/fetoplacental angiogenesis and gametogenesis [28] as well as retinogenesis [29], or myeloid cell maturation [30].

Biliverdin and Bilirubin Metabolism

Bilirubin and its precursor biliverdin are derived entirely from the degradation of heme, catabolized by the unique and ubiquitous microsomal enzyme heme oxygenase (HMOX). Up to 80% of bilirubin derives from hemoglobin released during destruction of senescent red blood cells in the splenic sinusoids, while the remainder is produced by myoglobin and the other hemoproteins that mainly originate in the liver. The daily production of bilirubin is quite high, reaching almost 5 mg of bilirubin per kg of body weight [31]. This represents approximately 300 mg of bilirubin in a 70 kg human (500 μmol), with physiological blood concentrations between 5 and 17 μmol/l [32]. Interestingly, no biliverdin is present in vascular bed due to its rapid conversion by biliverdin reductase (BLVR), its catabolic enzyme [33–35], which is in contrast to some lower animals such as frogs [22], lizards [36], or fish [37]. Such a high amount of bilirubin generated in a human body each day can hardly be considered to only be a waste product. In fact, data from recent decades have clearly demonstrated the versatile biological functions of this pigment across the entire range of human body organs and compartments [38–41]. The key enzyme responsible for initiation of heme catabolism is HMOX, having two isoforms (HMOX1, OMIM 141250; and HMOX2, OMIM 141251, respectively) [42,43].

HMOX1, also known as heat shock protein 32, is considered the most inducible enzyme in the human body [44,45]. In fact, it is upregulated by a wide array of endogenous stimuli as well as a variety of xenobiotics and exogenous noxious substances, mostly associated with increased oxidative stress, inflammation, proliferation, and apoptosis [42–45]. It is interesting to note that not only tetrapyrroles, heme, biliverdin, and bilirubin, but also HMOX is a very old enzyme phylogenetically, with a high degree of structural conservation even in cyanobacteria and higher plants [43,46,47]. The HMOX pathway, through its activation role in gene transcriptions, is closely associated with another evolutionarily conserved cell system, which is the multifunctional nuclear factor erythroid 2-related factor, considered not only as a regulator of expression of gene coding for antioxidant, anti-inflammatory, and detoxifying proteins, but also as a powerful modulator of species longevity [48]. All these data further expand the importance of the whole heme catabolic pathway. While HMOX1 is almost ubiquitously expressed in the human body, with predominance in the reticuloendothelial system, HMOX2 is an isoenzyme constitutively expressed in the central nervous system, vascular endothelium, liver, and intestine, having its highest demonstrated activity in the testes [42]. Although generally considered as a non-inducible isoform, HMOX2 is in fact upregulated by corticosteroids [44,49].

HMOX1 is implicated in the pathogenesis of a wide array of human diseases, including cardiovascular, cancer, infectious, and autoimmune conditions, whereas HMOX2 exhibits especially neuroprotective properties and a modulating activity in male reproduction [50,51]. The brain is a typical organ to demonstrate the differential roles of both isoforms as to their biological functions. While HMOX1 is exclusively expressed in specific brain regions such as the dentate gyrus, certain parts of the hypothalamus, or brain stem nuclei, HMOX2 is abundantly expressed in the forebrain, hippocampus, midbrain, basal ganglia, thalamic regions, cerebellum, and brain stem [51]. Functionally, HMOX1 is also believed to protect astrocytes, but not neurons, from increased oxidative stress [52], while HMOX2 in particular protects neuronal cells [53–55]. Due to these facts, both enzymes (as well as the whole heme catabolic pathway) have become important therapeutic targets [50,56,57]. Apart from the products of the whole heme catabolic pathway (whose biological functions can be exemplified in the central nervous system) such as biliverdin and bilirubin [58,59] plus obviously CO [60], the enzymes per se have additional biological regulatory functions. Within the cells, HMOX is localized in the endoplasmic reticulum (functioning primarily as the heme detoxifying enzyme), but it is also present in the nucleus (activating transcription factors and yet still also retaining its enzymatic activity), mitochondria (regulating mitochondrial heme content, but also playing a role in apoptosis), and the caveolae of the plasma membrane (involved in redox and other signaling pathways) [61,62]. HMOX also undergoes various post-translational modifications, which are important for its cellular compartmentalization as well as proper functioning [61] and have important pathophysiological implications as suggested by oxidative stress-induced post-translational modifications of cerebral HMOX1 in patients with Alzheimer's disease [63].

The primary function of HMOX enzymes is to break the cyclic tetrapyrrolic structure of heme into linear tetrapyrrole biliverdin, releasing ferrous iron and CO as additional products. Biliverdin is in turn converted to bilirubin by reduction of the central methylene bridge by BLVR, occurring in two isoforms: BLVR A and B (BLVRA/B, OMIM Nos. 109750 and 600941, respectively). Again, BLVR is an evolutionarily old enzyme found ubiquitously within the human body, thus accounting for the absence of biliverdin in systemic concentrations [64]. It is a special enzyme with a unique dual pH- and NADH/NADPH cofactor-dependent activity [65], possesses an autophosphorylation feedback mechanism to augment its reductase activity [66], and exerts the rare multispecific (serine/threonine/tyrosine) kinase activity, thus accounting for its cell signaling functions, particularly in the insulin signaling pathway [67] with beneficiary metabolic consequences [68]. In a sophisticated manner, BLVR is compartmentalized within the cell, being present in the cytosol (exerting its enzymatic activity), translocating into the nucleus (acting as a transcription factor in a variety of signaling pathways) [69], and being localized at the cell membrane as demonstrated in macrophages (acting as a negative adapter of toll-like receptor 4 [TLR4], inhibiting pro-inflammatory cytokine production) [70]. It has also been postulated that biliverdin is an important constituent of the bilirubin–biliverdin redox amplifying antioxidant potential of bilirubin [71].

Unconjugated bilirubin as a nonpolar molecule must be solubilized in the systemic circulation; this is achieved through binding to albumin under normal physiological conditions [72]. However, in patients with analbuminemia, a congenital disorder of albumin synthesis, bilirubin-solubilizing activities are substituted by other agents such as high-density lipoprotein [73]. This observation points to additional bilirubin-binding molecules within the human body, which seem to have potentially wide pathophysiological implications. Such bilirubin-binding activities have been detected for plasma proteins such as apolipoprotein D [74] and α1-fetoprotein (ontogenetically a fetal albumin) [75], as well as intracellular proteins including hepatic ligandin (protein Y, an alpha isoform of glutathione S-transferase B) [76], fatty acid binding protein-1 [77], and other members of the lipocalin protein superfamily (for a review, see [41]). Bilirubin, after being taken up by sinusoidal transport proteins of the hepatocytes, is conjugated within the liver cell with glucuronic acid by bilirubin UDP-glucuronosyltransferase (UGT1A1), a phylogenetically old biotransforming enzyme. Its congenital deficiency, which is mostly due to promoter gene variation in the so-called TATA box (at least in Caucasians), leads to the manifestation of mild unconjugated hyperbilirubinemia (Gilbert syndrome) [78]. Based on recent research, it has become clear that *UGT1A1* expression is substantially under the control of multifunctional nuclear receptors, including constitutive androstane receptor (CAR), pregnane X receptor (PXR), glucocorticoid receptor, aryl hydrocarbon receptor (AhR), and hepatocyte nuclear receptor 1α [79], which regulate *UGT1A1* transcription via the phenobarbital-responsive enhancer module [79]. It is also important to note that bilirubin per se serves as a ligand for most of these receptors, as well as other nuclear receptors involved not only in biotransformation but also in metabolic pathways, thus forming a complex regulatory network of cell metabolism. In particular, these additional metabolic nuclear receptors include peroxisome proliferator-activated receptors (PPAR) α and γ, being involved in various energy homeostasis pathways and the pathogenesis of various metabolic diseases such as obesity, diabetes, or metabolic syndrome [41] (see later).

Bilirubin conjugated with glucuronic acid, after being actively secreted into bile, enters the lumen of the intestinal tract where it is further converted into a series of reduction products known as urobilinoids [80].

Biological Functions of Bilirubin and Biliverdin

Since the first sporadic and anecdotal reports of the anti-inflammatory and antioxidant effects of bilirubin were published in the 1930s to 1950s [81–84], an explosive increase in interest in this specific field started after confirmation of its antioxidant activities by Stocker and colleagues in 1987 [85]. The role of bilirubin in protection against increased oxidative stress has now been generally accepted (for a review, see [86]), but both bilirubin and biliverdin exert a number of other biological activities with important clinical consequences. These include cell signaling functions [41,87,88], resulting in substantial atheroprotective effects [40], intermediary metabolism-modulating [89,90], antiproliferative [91], antimutagenic [92], antigenotoxic [93], immunomodulatory (mostly anti-inflammatory and immunosuppressive) [94], antineurodegenerative [59], and antiaging activities [95,96]. Such broad and versatile activities are accounted for by the fact that bilirubin has potent and widespread inhibitory activities against protein phosphorylation [97], serves as a ligand of multiple biologically active molecules within the human body, including various nuclear as well as cytoplasmic receptors [41], modulates transcription of a battery of genes [98], and exerts mitochondrial toxicity in high concentrations [99], while having beneficiary effects on mitochondrial metabolism when only mildly elevated [95,100,101]. Biliverdin, although not so well investigated, likely also possesses potent metabolic activities, partially due to being a precursor of bilirubin [102] and a member of the bilirubin–biliverdin redox cycle [71], as well as due to its direct effects [103].

It must be stressed again that many of the effects mentioned earlier resemble those reported for CO, including the beneficial activities on atherogenesis [104], immunometabolism [105], inflammation [106], angiogenesis [107], or central nervous system pathologies [108–110], which further underlines the biological importance and complex action of the whole heme catabolic pathway [104,111–114].

Bilirubin and Immune System

As mentioned earlier, products of the heme catabolic pathway strongly affect the functions of both the innate and adaptive immunity [94,115]. Both bilirubin and biliverdin exert anticomplement activities [116,117]. Bilirubin has also been reported to stimulate macrophage phagocytic activity [118] and antigen-presenting functions [119,120], and to modulate the generation of acute-phase proteins [121], regulating M1/M2 macrophage polarization [122–124], interfering with the TLR4 signaling pathway [125,126], and inducing apoptosis and necrosis in immune cells by depleting cellular glutathione [127]. The complement and macrophage functions as targets of the anti-inflammatory action of biliverdin, BLVRA, and bilirubin have recently been reported [70,116–118,123,128].

Bilirubin is also a potent modulator of adaptive immunity as demonstrated by decreased antibody production in babies with severe neonatal jaundice [129]. In addition, bilirubin was reported to inhibit T-cell natural killer activity, antibody-dependent cellular cytotoxicity [130], and cytotoxic T lymphocyte activity [131]. Bilirubin also suppresses T-cell-mediated formation of proinflammatory cytokines and prevents NF-κB transcription activity, thus effectively inhibiting the Th1 immune response [94]. Bilirubin also reduces Th2 cell responsiveness as demonstrated in an experimental model of allergic asthma [132]. Additionally, bilirubin suppresses the proliferation of alloreactive T cells (the same effect was also observed for biliverdin treatment, which surprisingly was more potent than that of immunosuppressant cyclosporin A) [133]. Bilirubin was demonstrated to increase Foxp3$^+$ T regulatory (Treg) cells in an experimental model of pancreatic islet transplantation [134,135], while at the same time suppressing Th17 immunity; this is due to its effect on CD39 ectonucleotidase [136], as demonstrated in an experimental model of colitis [137]. Differentiation of these T cells is dependent on AhR [138]; thus, the agonistic activity of bilirubin on this nuclear receptor [139], together with its other widespread immunomodulatory functions, may account for its beneficial effects against various inflammatory, autoimmune, and neurodegenerative diseases (for a review, see [38,59,140]), thus being suggestive of its potential in transplantation medicine [120,135].

The anti-inflammatory effects of bile pigments are further potentiated by their direct antiviral activity against herpetic viruses, hepatitis C virus, enterovirus, and HIV, as has been shown for both biliverdin [141–144] and bilirubin [145], although for antibacterial activity the data for bilirubin are not as conclusive [146–148].

As mentioned earlier for other biological effects, modulation of the immune system effectors seems to be regulated by bilirubin and CO in a similar and synergistic way. The anti-inflammatory effects of CO include modulation of oxidative stress, leukocyte adhesion, cytokine production, NF-κB activity, or direct bactericidal activities (for reviews, see [106,149–152]). Apart from the effect of CO on innate immunity, it also affects adaptive immunity as demonstrated experimentally by induction of Foxp3$^+$ of Treg cells and shifting the Th1/Th17/M1 balance toward a Th2/M2 response in diabetic mice and murine lymph node-derived T cells [153–155]. These anti-inflammatory effects of CO are believed to be mediated, inter alia, via the modulation of mitogen-activated protein and extracellular signal-regulated kinases (MAPK/ERK) [153], cellular targets inhibited also by bilirubin [91,156].

Bilirubin and Metabolic Functions

The role of bilirubin in the pathogenesis of atherosclerotic and different cancer diseases has been explored in detail during the last two decades, and it has been the subject of recent comprehensive reviews [38,40,140]. However, studies in recent years have revealed much wider biological activities of bilirubin, with deep impacts on both metabolic and signaling pathways. The first review suggesting a protective effect of bilirubin on diabetes and associated diseases was published in 2012 [157]. That comprehensive review discussed the negative associations between systemic bilirubin concentrations and arterial hypertension [158], diabetes [159,160], obesity (in fact, body mass index) [161], and metabolic syndrome [162], and was based on observational studies reported up to that time (but the possible cellular mechanisms of bilirubin remained only speculative). Since then, additional clinical studies have confirmed and further expanded these early observations as described in a number of important original studies and extensive reviews [39,40,88,90,163–166]. Bilirubin was shown to be strongly negatively associated with both nonalcoholic steatohepatitis (NASH, commonly a part of metabolic syndrome) [167] and unfavorable lipid profiles [168]. Although protective effects of bilirubin were reported in subjects with constitutively elevated systemic bilirubin concentrations (i.e., in Gilbert syndrome) [159,169], it must be emphasized that also CO produced during heme degradation by the activity of HMOX has variable antidiabetic effects as demonstrated in various experimental studies [153,170,171]. Thus, it seems that both bilirubin and CO, as well as the other components of the whole heme catabolic pathway (such as HMOX1 and BLVRA), act protectively in a synergistic way.

Not surprisingly, recent epidemiological studies have also confirmed the relationship between serum bilirubin concentrations and disease-specific as well as overall mortality as summarized in several reviews

[140,165,172–174]. Based on such data, bilirubin was even suggested to have an antiaging effect [175]. Interestingly, the perception of looking older was also negatively associated with concentrations of serum bilirubin [176]. Also, the experimental data are consistent with this generally accepted idea. We ourselves have recently described about the antiaging effects of elevated serum bilirubin concentrations in hyperbilirubinemic Gunn rats [95]. These were widespread and involved an attenuation of oxidative stress, decreased inflammatory status, increased glucose tolerance, and enhanced mitochondrial function, with fewer signs of cellular senescence that was due in part to modulation of mTOR (mammalian target of rapamycin) signaling. Similarly, improved mitochondrial metabolism, with potential metabolic benefits, has also been described by both Gordon et al. [100] and our group [101]. Apart from targeting mTOR, an evolutionarily conserved nutrient-sensing protein kinase regulating metabolism, aging processes, and overall life span [177]; moreover, several other cellular targets might be responsible for the vast biological activities of bilirubin. These include PPAR α and γ [88], so important for the homeostasis of energy sources in the human body [178], as well as AMP-activated protein kinase (AMK) [179,180], a similarly important cellular energy sensor [181]. It is important to note that these pathways are interlinked [182], thus accounting for the effects of bilirubin on both of them. Bilirubin thus affects insulin signaling [183–186] (an immediately acting fed- and fasting-state hormone) as well as fibroblast growth hormone 21 [88,187], considered a late-acting hormone [188]. All of these bilirubin-modulated pathways are interlinked as demonstrated by the role of the CD39 ectonucleotidase/adenosine pathway in immunity and inflammation [136,189], also having important effects on glucose metabolism and insulin signaling [190], the AMP-activated protein kinase pathway [191], and pathogenesis of liver diseases, including NASH [192,193] – indicating the complexity of bilirubin-related modulation of cell signaling. In addition to the immune system-modulating effects of CO, it has also been demonstrated recently that CO, in parallel to the same effect of bilirubin, exhibited its anti-inflammatory effects partially through its impact on CD39 further underlining mutual interconnection of the whole catabolic pathway [194].

The broad metabolic activity has also been corroborated in recent observations by Gordon and colleagues, who reported almost 400 genes modulated by the biliverdin–bilirubin signaling axis in hepatic HepG2 cells, most of them in a PPARα-dependent fashion [98]. In fact, after being discovered to be a specific and potent ligand of PPARα structurally resembling fenofibrate, a pharmaceutically used PPARα agonist (inhibiting, inter alia, hepatic lipid accumulation [195], as well as remodeling white adipose tissue in a murine animal model [100]), bilirubin has been suggested to be a selective PPAR modulator [88], fulfilling the definition of selective nuclear receptor modulators despite its additional broad metabolic activities mediated via other cell targets [196]. Nevertheless, since also being active toward the PPARγ nuclear receptor [183], bilirubin might be considered a PPARα/γ dual agonist [197] (Table 10.2).

Additionally, bilirubin is a ligand of other nuclear receptors (see earlier), contributing to glucose and lipid metabolism, with important links to the pathogenesis of atherosclerosis and related diseases such as AhR [139], PXR, and CAR [198], with important metabolic cross-talking with PPARs [199]. Thus, bilirubin seems to act as a multifunctional modulator at multiple cellular metabolic checkpoints, which are often mutually interrelated. This is exemplified by bilirubin-induced suppression of p38 MAPK (but also other MAPKs) [156], known to be crucial for insulin signaling as well as atherogenesis [200].

It is also necessary to note that BLVRA, apart from its enzymatic role in biliverdin reduction, also exerts pleiotropic metabolic functions particularly as a novel modulator of the insulin/insulin-like growth factor 1/phosphatidylinositol 3-kinase/MAPK signaling cascade, with important implications for glucose homeostasis [201,202].

Biological Functions of Bilirubin Metabolites and Derivatives

Not only biliverdin and bilirubin, but also other bilirubin derivatives and tetrapyrrolic molecules formed in the human body or occurring in the world of microbes, plants, and animals across the totality of the living world have remarkable biological activities that might be of clinical importance for humans. This is true for bilirubin-10-sulfonate, occurring physiologically in frogs [203], but possibly also in the gut lumen due to microbial conversion of bile pigments [204], which exhibits apparent anti-inflammatory activities [205]. Another example is bilirubin ditaurate occurring physiologically in the bile of some fish [206] exerting protective activities against platelet mitochondrial reactive oxygen species production [207], low-density lipoprotein oxidation [208], genotoxicity [209], myeloperoxidase-produced hypochlorite [210], increasing antioxidant capacity [211], or suppressing complement action [117]. The same

Table 10.2 Cell targets of bilirubin.

Cell target	Function
NADPH oxidase	Antioxidant effects [260]
C1q complement binding	Anticomplement action [116]
Epigenetic regulation	Inhibition of protein phosphorylation [97]
	Inhibition of histone acetylation [261]
Nuclear receptor activation	
PPARα/γ	Broad metabolic effects [41]
AhR, CAR, PXR, LXR	Various immune system-modulating and metabolic effects [41]
Mitochondria	When mildly elevated, improvement of metabolic functions [100]
ApoD binding	Antiproliferative and metabolic effects [41]
FABP1	Metabolic and immune system signaling [41]
MRGPRX4	Sensory functions and immune effects [41]
Specific metabolic targets	
AMK [179,180]	Energy homeostasis [181]
Insulin signaling [183–186]	Glucose homeostasis
FGF21 [88,187]	Energy homeostasis [181]
CD39 [136]	Glucose homeostasis [190] as well as immune system-modulating effects [136,189]
MAPK [156]	Insulin signaling and lipid homeostasis [200]
mTOR [95]	Nutrient sensing, antiaging effects [177]

AMK, AMP-activated protein kinase; FABP, fatty acid binding protein; FGF21, fibroblast growth hormone 21; MAPK, mitogen-activated protein kinase; mTOR, mammalian target of rapamycin; MRGPRX4, Mas-related G protein-coupled receptor X4.

anticomplement effects have also been reported for bilirubin monoglucuronoside [212].

Additionally, a potent cytoprotective activity against ventricular myocytes was also reported for delta-bilirubin (bilirubin covalently bound to albumin) [213]. Substantial antioxidant [211,214], antiviral [144], and antimutagenic [209] effects were also observed for urobilinoids, which are reduction products of bilirubin formed in the gut lumen by intestinal microbiota.

Even more interestingly, the potent biological activities of bilirubin photoproducts, generated during phototherapy of neonatal jaundice (in particular lumirubin), were recently reported by us [101,215]. This is also true for other bilirubin oxidation products, such as tripyrrolic biopyrrins, clinically relevant markers of increased oxidative stress [216–218]. Similarly, the biological activities of dipyrrolic propentdyopents and monopyrrolic bilirubin oxidation products (BOX A–D) with potential clinical implications have also been reported [219–223]. Recent data point to the biological potential of molecules that possess a tetrapyrrolic structure, and it may account for the clinical observations from the last few decades related to the protective actions and effects of bilirubin.

Therapeutic Potential of the Modulation of Bilirubin Metabolism

Based on data reported during in the last few decades, it seems tempting pharmacologically, nutraceutically, or by some other means to modulate the heme catabolic pathway, especially to mildly increase intracellular and systemic concentrations of bilirubin (apart from other effector products of the heme catabolic pathway, such as CO). This therapeutic approach was first comprehensively reviewed by McCarty in 2007 [224] and also recently by us [38,140,225]. In this respect, several possible ways to modulate bilirubin concentrations have been suggested and investigated. Direct administration of bilirubin and bile pigments (such as biliverdin) has been demonstrated in animal models to elevate systemic bilirubin concentrations in numerous experimental studies with clear and mostly beneficial biological effects; such bilirubin administration has also been shown to be applicable for human subjects [226]. However, these approaches are not plausible in clinical settings.

As an alternative approach, it is possible to use xenobiotics known to inhibit the expression of or to compete with the activities of UGT1A1. Among the many candidate drugs [225], the most well known is the antiviral agent atazanavir for treatment of HIV infection. Treatment with this drug is often associated with unconjugated hyperbilirubinemia [227], mostly due to a decrease of UGT1A1 activity [228], and also partially due to the inhibition of OATP1B1-mediated bilirubin transport from the hepatic sinusoids into the liver cell [229]. Atazanavir treatment of HIV-infected patients accompanied with the elevation of systemic bilirubin concentrations has been shown to reduce carotid atherosclerosis [230]; a similar effect was also observed in hyperbilirubinemic subjects with Gilbert syndrome [231]. Simultaneously, atazanavir-induced hyperbilirubinemia in HIV-infected patients has been associated with decreased cardiovascular morbidity as well as overall mortality [232]. The same treatment, when given to diabetic patients, significantly improved plasma antioxidant capacity and endothelium-dependent vasodilation; additionally, it decreased plasmatic concentrations of von Willebrand factor [233]. A similar effect (i.e., partial inhibition of UGT1A1 with consequent mild elevation of bilirubin concentrations) to that of atazanavir was demonstrated in our own study with silymarin flavonolignans [234].

Another approach on how to modulate the heme catabolic pathway in a clinically beneficiary way is with HMOX1 induction. Due to the enormous induction potential of HMOX1 (*HMOX1* being considered a vitagene conferring protection against increased oxidative stress [235]), a variety of routinely used therapeutics have been shown to activate HMOX1, including nonsteroidal anti-inflammatory drugs or hypolipidemics as summarized in a review [225]. The same effect has been reported for many plant-originated polyphenolic compounds (such as resveratrol or curcumin), often used as nutraceuticals [225].

Not only bilirubin per se, but also other tetrapyrroles used by humans and present in our environment might have therapeutic potentials. This is certainly true for pulverized bovine gallstones (Calculus Bovis, Niu Huang in Chinese) used for centuries in China for a variety of clinical conditions [236,237]. Further, other tetrapyrrolic compounds that commonly occur in nature, such as the chlorophylls of green plants [238] or phycocyanobilin of blue-green algae [8,239], have obvious therapeutic potentials.

In the last five years, therapeutic approaches for bilirubin targeting at pathologically altered tissues and organs have been investigated. Bilirubin has been incorporated into various forms of nanoparticles, and the biological effectiveness of these particles has been experimentally verified in numerous recent studies focused on the treatment of pulmonary [240], colonic [124,240], lung [132], skin [241], and pancreatic tissue inflammation [242]. Additionally, the treatment of conditions associated with increased oxidative stress [243,244], cancer [245–249], liver fibrosis [250], and ischemia–reperfusion injury [251] has also been reported. These effects were seen not only in systemic administrations, but also when used topically. For example, in an experimental psoriasis model, bilirubin nanoparticles alleviated psoriatic skin changes by reducing oxidative stress and inflammation [241]. This is consistent with the clinical observation of the positive association of serum bilirubin concentrations with symptom alleviation of psoriatic patients [252]. It is also noteworthy to mention that nanoparticles often are used to combine bilirubin with other therapeutics such as folate [249], losartan [250], or anticancer agents [253]. Progress in this rapidly evolving field was the subject of several recent comprehensive reviews [253–256]. In addition to amelioration of clinical symptomatology with this bilirubin nanomedicine approach, it was also demonstrated that parenteral administration of bilirubin nanoparticles elevated systemic bilirubin concentrations to levels comparable to those seen in subjects with Gilbert syndrome [132]. Such observations may lend insights into the mechanism(s) of action for this therapeutic approach.

Interestingly, bilirubin was also used in an experimental study by Bae et al. to coat everolimus-containing vascular stents to prevent arterial neointimal hyperplasia and inflammation, with potential uses in human cardiovascular medicine [257]. A similar nanomedicine approach was also used for biliverdin. Apart from possible metabolic effects, biliverdin encapsulated into nanoparticles was also used as a photothermal agent with a high selectivity for tumor tissues, with the aim to visualize tumors by either photoacoustic or magnetic resonance imaging [258]. Also interesting is a study in mice by Gibbs et al. who demonstrated signaling activities with consequent beneficiary metabolic effects of a human BLVR-based peptide [259].

Conclusions

Bilirubin, the principal mammalian bile pigment, is a molecule with multiple biological functions in both animals and plants. Although its high serum concentrations may cause severe neurotoxicity in humans, mildly elevated levels of bilirubin exert a protective action, predominantly via the scavenging of ROS.

Additionally, an ever-increasing body of evidence suggests that the products of the heme catabolic pathway affect multiple biochemical and molecular targets with important pathophysiological consequences, although further robust, well-designed clinical studies are needed to confirm currently known, often experimental and proof-of-principle data.

Acknowledgment

This work was supported by grants NV18-07-00342 and MH CZ-DRO-VFN64165 from the Czech Ministry of Health.

References

1. Aylward, N. and Bofinger, N. (2005). Possible origin for porphin derivatives in prebiotic chemistry-a computational study. *Orig. Life Evol. Biosph.* 35 (4): 345–368.
2. Vitek, L. and Ostrow, J.D. (2009). Bilirubin chemistry and metabolism; harmful and protective aspects. *Curr. Pharm. Des.* 15 (25): 2869–2883.
3. Smith, A.G. and Witty, M. (2002). Laboratory methods for the study of tetrapyrroles. In: Heme, Chlorophyll, and Bilins. Methods and Protocols (ed. A.G. Smith and M. Witty). Humana Press, Inc.
4. Rockwell, N.C. and Lagarias, J.C. (2017). Ferredoxin-dependent bilin reductases in eukaryotic algae: Ubiquity and diversity. *J. Plant Physiol.* 217: 57–67.
5. Beale, S.I. (1994). Biosynthesis of open-chain tetrapyrroles in plants, algae, and cyanobacteria. In: The Biosynthesis of Tetrapyrrolic Pigments (ed. D.J. Chadwick and K. Ackrill). Chichester, England: John Willey & Sons Ltd.
6. Fujita, Y., Tsujimoto, R., and Aoki, R. (2015). Evolutionary aspects and regulation of tetrapyrrole biosynthesis in Cyanobacteria under aerobic and anaerobic environments. *Life (Basel)* 5 (2): 1172–1203.
7. Strasky, Z., Zemankova, L., Nemeckova, I., Rathouska, J., Wong, R.J., Muchova, L., Subhanova, I., Vanikova, J., Vanova, K., Vitek, L., and Nachtigal, P. (2013). Spirulina platensis and phycocyanobilin activate atheroprotective heme oxygenase-1: a possible implication for atherogenesis. *Food Funct.* 4 (11): 1586–1594.
8. Konickova, R., Vankova, K., Vanikova, J., Vanova, K., Muchova, L., Subhanova, I., Zadinova, M., Zelenka, J., Dvorak, A., Kolar, M., Strnad, H., Rimpelova, S., Ruml, T., Wong, R.J., and Vitek, L. (2014). Anti-cancer effects of blue-green alga Spirulina platensis, a natural source of bilirubin-like tetrapyrrolic compounds. *Ann. Hepatol.* 13 (2): 273–283.
9. Hortensteiner, S. and Krautler, B. (2011). Chlorophyll breakdown in higher plants. *Biochim. Biophys. Acta* 1807 (8): 977–988.
10. Moser, S. and Krautler, B. (2019). In search of bioactivity – phyllobilins, an unexplored class of abundant heterocyclic plant metabolites from breakdown of chlorophyll. *Isr. J. Chem.* 59: 420–431.
11. Kilner, R.M. (2006). The evolution of egg colour and patterning in birds. *Biol. Rev. Camb. Philos. Soc.* 81 (3): 383–406.
12. Ishikawa, S., Suzuki, K., Fukuda, E., Arihara, K., Yamamoto, Y., Mukai, T., and Itoh, M. (2010). Photodynamic antimicrobial activity of avian eggshell pigments. *FEBS Lett.* 584 (4): 770–774.
13. Samiullah, S., Roberts, J.R., and Chousalkar, K. (2015). Eggshell color in brown-egg laying hens - a review. *Poult. Sci.* 94 (10): 2566–2575.
14. Duval, C., Zimmer, C., Miksik, I., Cassey, P., and Spencer, K.A. (2014). Early life stress shapes female reproductive strategy through eggshell pigmentation in Japanese quail. *Gen. Comp. Endocrinol.* 208: 146–153.
15. Wiemann, J., Yang, T.R., Sander, P.N., Schneider, M., Engeser, M., Kath-Schorr, S., Muller, C.E., and Sander, P.M. (2017). Dinosaur origin of egg color: oviraptors laid blue-green eggs. *PeerJ.* 5: e3706.
16. Wiemann, J., Yang, T.R., and Norell, M.A. (2018). Dinosaur egg colour had a single evolutionary origin. *Nature* 563 (7732): 555–558.
17. Youson, J.H. (1993). Biliary atresia in lampreys. *Adv. Vet. Sci. Comp. Med.* 37: 197–255.
18. Krasner, J., Juchau, M.R., Niswander, K.R., and Yaffe, S.J. (1971). Theoretical considerations regarding the presence of biliverdin in amniotic fluid. *Am. J. Obstet. Gynecol.* 109: 159–165.
19. Montorzi, M., Dziedzic, T.S., and Falchuk, K.H. (2002). Biliverdin during Xenopus laevis oogenesis and early embryogenesis. *Biochemistry* 41 (31): 10115–10122.
20. Falchuk, K.H., Contin, J.M., Dziedzic, T.S., Feng, Z., French, T.C., Heffron, G.J., and Montorzi, M. (2002). A role for biliverdin IXalpha in dorsal axis development of Xenopus laevis embryos. *Proc. Natl. Acad. Sci. U. S. A.* 99 (1): 251–256.
21. Redshaw, M.R., Follett, B.K., and Lawes, G.J. (1970). Biliverdin - Component of yolk proteins in Xenopus laevis. *Int. J. Biochem.* 2 (7): 80–84.
22. Taboada, C., Brunetti, A.E., Lyra, M.L., Fitak, R.R., Faigon Soverna, A., Ron, S.R., Lagorio, M.G., Haddad, C.F.B., Lopes, N.P., Johnsen, S., Faivovich, J., Chemes, L.B., and Bari, S.E. (2020). Multiple origins of green coloration in frogs mediated by a novel

biliverdin-binding serpin. *Proc. Natl. Acad. Sci. U. S. A.* 117 (31): 18574–18581.
23. Hongo, Y., Yasuda, N., and Naga, I.S. (2017). Identification of genes for synthesis of the blue pigment, biliverdin IXalpha, in the blue coral Heliopora coerulea. *Biol. Bull.* 232 (2): 71–81.
24. Gruber, D.F., Gaffney, J.P., Mehr, S., DeSalle, R., Sparks, J.S., Platisa, J., and Pieribone, V.A. (2015). Adaptive evolution of eel fluorescent proteins from fatty acid binding proteins produces bright fluorescence in the marine environment. *PLoS One* 10 (11): e0140972.
25. Funahashi, A., Komatsu, M., Furukawa, T., Yoshizono, Y., Yoshizono, H., Orikawa, Y., Takumi, S., Shiozaki, K., Hayashi, S., Kaminishi, Y., and Itakura, T. (2016). Eel green fluorescent protein is associated with resistance to oxidative stress. *Comp. Biochem. Physiol. C Toxicol. Pharmacol.* 181–182: 35–39.
26. Pirone, C., Quirke, J.M., Priestap, H.A., and Lee, D.W. (2009). Animal pigment bilirubin discovered in plants. *J. Am. Chem. Soc.* 131 (8): 2830.
27. Pirone, C., Johnson, J.V., Quirke, J.M.E., Priestap, H.A., and Lee, D. (2010). The animal pigment bilirubin identified in strelitzia reginae, the bird of paradise flower. *Hortscience* 45 (9): 1411–1415.
28. Guerra, D.D. and Hurt, K.J. (2019). Gasotransmitters in pregnancy: from conception to uterine involution. *Biol. Reprod.* 101 (1): 4–25.
29. Kalinichenko, S.G., Matveeva, N.Y., and Pushchin, I.I. (2019). Gaseous transmitters in human retinogenesis. *Acta Histochem.* 121 (5): 604–610.
30. Wegiel, B., Hedblom, A., Li, M., Gallo, D., Csizmadia, E., Harris, C., Nemeth, Z., Zuckerbraun, B.S., Soares, M., Persson, J.L., and Otterbein, L.E. (2014). Heme oxygenase-1 derived carbon monoxide permits maturation of myeloid cells. *Cell Death Dis.* 5: e1139.
31. Berk, P.D., Rodkey, F.L., Blaschke, T.F., Collison, H.A., and Waggoner, J.G. (1974). Comparison of plasma bilirubin turnover and carbon monoxide production in man. *J. Lab. Clin. Med.* 83 (1): 29–37.
32. Vitek, L. (2019). Bilirubin as a predictor of diseases of civilization. Is it time to establish decision limits for serum bilirubin concentrations? *Arch. Biochem. Biophys.* 672: 108062.
33. McDonagh, A.F. (2010). Green jaundice revisited. *Am. J. Med.* 123 (9): e23.
34. Greenberg, A.J., Bossenmaier, I., and Schwartz, S. (1971). Green jaundice. A study of serum biliverdin, mesobiliverdin and other green pigments. *Am. J. Dig. Dis.* 16 (10): 873–880.
35. Larson, E.A., Evans, G.T., and Watson, C.J. (1947). A study of the serum biliverdin concentration in various types of jaundice. *J. Lab. Clin. Med.* 32 (5): 481–488.
36. Rodriguez, Z.B., Perkins, S.L., and Austin, C.C. (2018). Multiple origins of green blood in New Guinea lizards. *Sci. Adv.* 4 (5): eaao5017.
37. Fang, L.S. and Bada, J.L. (1990). The blue-green blood plasma of marine fish. *Comp. Biochem. Physiol. B* 97 (1): 37–45.
38. Gazzin, S., Vitek, L., Watchko, J., Shapiro, S.M., and Tiribelli, C. (2016). A novel perspective on the biology of bilirubin in health and disease. *Trends Mol. Med.* 22 (9): 758–768.
39. Gazzin, S., Masutti, F., Vitek, L., and Tiribelli, C. (2017). The molecular basis of jaundice: An old symptom revisited. *Liver Int.* 37 (8): 1094–1102.
40. Vitek, L. (2017). Bilirubin and atherosclerotic diseases. *Physiol. Res.* 66 (Suppl. 1): S11–S20.
41. Vitek, L. (2020). Bilirubin as a signaling molecule. *Med. Res. Rev.* 40 (4): 1335–1351.
42. Ryter, S.W., Alam, J., and Choi, A.M. (2006). Heme oxygenase-1/carbon monoxide: from basic science to therapeutic applications. *Physiol. Rev.* 86 (2): 583–650.
43. Wilks, A. (2002). Heme oxygenase: evolution, structure, and mechanism. *Antioxid. Redox Signal.* 4 (4): 603–614.
44. Maines, M.D. (1997). The heme oxygenase system: a regulator of second messenger gases. *Annu. Rev. Pharmacol. Toxicol.* 37: 517–554.
45. Morse, D. and Choi, A.M. (2002). Heme oxygenase-1: the "emerging molecule" has arrived. *Am. J. Respir. Cell Mol. Biol.* 27 (1): 8–16.
46. Cornejo, J., Willows, R.D., and Beale, S.I. (1998). Phytobilin biosynthesis: cloning and expression of a gene encoding soluble ferredoxin-dependent heme oxygenase from *Synechocystis* sp. *PCC 6803. Plant J.* 15 (1): 99–107.
47. Davis, S.J., Kurepa, J., and Vierstra, R.D. (1999). The Arabidopsis thaliana HY1 locus, required for phytochrome-chromophore biosynthesis, encodes a protein related to heme oxygenases. *Proc. Natl. Acad. Sci. U. S. A.* 96 (11): 6541–6546.
48. Loboda, A., Damulewicz, M., Pyza, E., Jozkowicz, A., and Dulak, J. (2016). Role of Nrf2/HO-1 system in development, oxidative stress response and diseases: an evolutionarily conserved mechanism. *Cell. Mol. Life Sci.* 73 (17): 3221–3247.
49. Raju, V.S., McCoubrey, W.K., Jr., and Maines, M.D. (1997). Regulation of heme oxygenase-2 by glucocorticoids in neonatal rat brain: characterization of a functional glucocorticoid response element. *Biochim. Biophys. Acta* 1351: 89–104.
50. Intagliata, S., Salerno, L., Ciaffaglione, V., Leonardi, C., Fallica, A.N., Carota, G., Amata, E., Marrazzo, A., Pittala, V., and Romeo, G. (2019). Heme oxygenase-2

(HO-2) as a therapeutic target: Activators and inhibitors. *Eur. J. Med. Chem.* 183: 111703.

51. Munoz-Sanchez, J. and Chanez-Cardenas, M.E. (2014). A review on hemeoxygenase-2: Focus on cellular protection and oxygen response. *Oxid. Med. Cell. Longev.* 2014: 604981.

52. Dwyer, B.E., Nishimura, R.N., and Lu, S.Y. (1995). Differential expression of heme oxygenase-1 in cultured cortical neurons and astrocytes determined by the aid of a new heme oxygenase antibody. Response to oxidative stress. *Brain Res. Mol. Brain Res.* 30: 37–47.

53. Parfenova, H. and Leffler, C.W. (2008). Cerebroprotective functions of HO-2. *Curr. Pharm. Des.* 14 (5): 443–453.

54. Dore, S., Takahashi, M., Ferris, C.D., Zakhary, R., Hester, L.D., Guastella, D., and Snyder, S.H. (1999). Bilirubin, formed by activation of heme oxygenase-2, protects neurons against oxidative stress injury. *Proc. Natl. Acad. Sci. U. S. A.* 96 (5): 2445–2450.

55. Dore, S., Sampei, K., Goto, S., Alkayed, N.J., Guastella, D., Blackshaw, S., Gallagher, M., Traystman, R.J., Hurn, P.D., Koehler, R.C., and Snyder, S.H. (1999). Heme oxygenase-2 is neuroprotective in cerebral ischemia. *Mol. Med.* 5 (10): 656–663.

56. Salerno, L., Floresta, G., Ciaffaglione, V., Gentile, D., Margani, F., Turnaturi, R., Rescifina, A., and Pittala, V. (2019). Progress in the development of selective heme oxygenase-1 inhibitors and their potential therapeutic application. *Eur. J. Med. Chem.* 167: 439–453.

57. Zhang, X., Yu, Y., Lei, H., Cai, Y., Shen, J., Zhu, P., He, Q., and Zhao, M. (2020). The Nrf-2/HO-1 signaling axis: a ray of hope in cardiovascular diseases. *Cardiol. Res. Pract.* 2020: 5695723.

58. Mancuso, C. (2017). Bilirubin and brain: a pharmacological approach. *Neuropharmacology* 118: 113–123.

59. Jayanti, S., Vítek, L., Tiribelli, C., and Gazzin, S. (2020). The role of bilirubin and other "yellow players" in neurodegenerative diseases. *Antioxidants* 9 (9): 900.

60. Mahan, V.L. (2019). Neurointegrity and neurophysiology: astrocyte, glutamate, and carbon monoxide interactions. *Med. Gas Res.* 9 (1): 24–45.

61. Dunn, L.L., Midwinter, R.G., Ni, J., Hamid, H.A., Parish, C.R., and Stocker, R. (2014). New insights into intracellular locations and functions of heme oxygenase-1. *Antioxid. Redox Signal.* 20 (11): 1723–1742.

62. Patel, H.H. and Insel, P.A. (2009). Lipid rafts and caveolae and their role in compartmentation of redox signaling. *Antioxid. Redox Signal.* 11 (6): 1357–1372.

63. Barone, E. and Butterfield, D.A. (2015). Insulin resistance in Alzheimer disease: is heme oxygenase-1 an Achille's heel? *Neurobiol. Dis.* 84: 69–77.

64. Maines, M.D. (1984). New developments in the regulation of heme metabolism and their implications. *CRC Crit. Rev. Toxicol.* 12 (3): 241–314.

65. Kutty, R.K. and Maines, M.D. (1981). Purification and characterization of biliverdin reductase from rat liver. *J. Biol. Chem.* 256: 3956–3962.

66. Salim, M., Brown-Kipphut, B.A., and Maines, M.D. (2001). Human biliverdin reductase is autophosphorylated, and phosphorylation is required for bilirubin formation. *J. Biol. Chem.* 276 (14): 10929–10934.

67. Lerner-Marmarosh, N., Shen, J., Torno, M.D., Kravets, A., Hu, Z., and Maines, M.D. (2005). Human biliverdin reductase: a member of the insulin receptor substrate family with serine/threonine/tyrosine kinase activity. *Proc. Natl. Acad. Sci. U. S. A.* 102 (20): 7109–7114.

68. O'Brien, L., Hosick, P.A., John, K., Stec, D.E., and Hinds, T.D., Jr. (2015). Biliverdin reductase isozymes in metabolism. *Trends Endocrinol. Metab.* 26 (4): 212–220.

69. Lerner-Marmarosh, N., Miralem, T., Gibbs, P.E., and Maines, M.D. (2008). Human biliverdin reductase is an ERK activator; hBVR is an ERK nuclear transporter and is required for MAPK signaling. *Proc. Natl. Acad. Sci. U. S. A.* 105 (19): 6870–6875.

70. Wegiel, B., Baty, C.J., Gallo, D., Csizmadia, E., Scott, J.R., Akhavan, A., Chin, B.Y., Kaczmarek, E., Alam, J., Bach, F.H., Zuckerbraun, B.S., and Otterbein, L.E. (2009). Cell surface biliverdin reductase mediates biliverdin-induced anti-inflammatory effects via phosphatidylinositol 3-kinase and Akt. *J. Biol. Chem.* 284 (32): 21369–21378.

71. Sedlak, T.W. and Snyder, S.H. (2004). Bilirubin benefits: cellular protection by a biliverdin reductase antioxidant cycle. *Pediatrics* 113 (6): 1776–1782.

72. Brodersen, R. (1980). Binding of bilirubin to albumin. *CRC Crit. Rev. Clin. Lab. Sci.* 11: 305–399.

73. Berger, G.M., Stephen, C.R., Finestone, A., and Beatty, D.W. (1985). Analbuminaemia. Clinical and laboratory features in a South African patient. *S. Afr. Med. J.* 67 (11): 418–422.

74. Goessling, W. and Zucker, S.D. (2000). Role of apolipoprotein D in the transport of bilirubin in plasma. *Am. J. Physiol. Gastrointest. Liver Physiol.* 279 (2): G356–G365.

75. Terentiev, A.A. and Moldogazieva, N.T. (2013). Alpha-fetoprotein: a renaissance. *Tumour Biol.* 34 (4): 2075–2091.

76. Vander Jagt, D.L., Dean, V.L., Wilson, S.P., and Royer, R.E. (1983). Regulation of the glutathione S-transferase activity of bilirubin transport protein (ligandin) from human liver. Enzymic memory

involving protein-protein interactions. *J. Biol. Chem.* 258: 5689–5694.

77. Theilmann, L., Stollman, Y.R., Arias, I.M., and Wolkoff, A.W. (1984). Does Z-protein have a role in transport of bilirubin and bromosulfophthalein by isolated perfused rat liver? *Hepatology* 4 (5): 923–926.

78. Bosma, P.J., Chowdhury, J.R., Bakker, C., Gantla, S., De Boer, A., Oostra, B.A., Lindhout, D., Tytgat, G.N., Jansen, P.L., and Oude Elferink, R.P. (1995). The genetic basis of the reduced expression of bilirubin UDP-glucuronosyltransferase 1 in Gilbert's syndrome. *N. Engl. J. Med.* 333: 1171–1175.

79. Sugatani, J., Mizushima, K., Osabe, M., Yamakawa, K., Kakizaki, S., Takagi, H., Mori, M., Ikari, A., and Miwa, M. (2008). Transcriptional regulation of human UGT1A1 gene expression through distal and proximal promoter motifs: implication of defects in the UGT1A1 gene promoter. *Naunyn Schmiedebergs Arch. Pharmacol.* 377 (4–6): 597–605.

80. Vitek, L., Kotal, P., Jirsa, M., Malina, J., Cerna, M., Chmelar, D., and Fevery, J. (2000). Intestinal colonization leading to fecal urobilinoid excretion may play a role in the pathogenesis of neonatal jaundice. *J. Pediatr. Gastroenterol. Nutr.* 30 (3): 294–298.

81. Hench, P.S. (1938). Effect of jaundice on rheumatoid arthritis. *Br. Med. J.* 2 (4050): 394–398.

82. Gorin, N. (1949). Temporary relief of asthma by jaundice; report of three cases. *J. Am. Med. Assoc.* 141 (1): 24.

83. Beer, H. and Bernard, K. (1959). Einfluss von bilirubin und vitamin E auf die oxidation ungesättigter Fettsäuren durch UV-bestrahlung. *Chimia* 13: 291–292.

84. Bernard, K., Ritzel, G., and Steiner, K.U. (1954). Über eine biologische bedeutung der gallenfarbstoffe: bilirubin und biliverdin als antioxydantien für das vitamin A und die essentiellen Fettsäuren. *Helv. Chim. Acta* 37: 306–313.

85. Stocker, R., Yamamoto, Y., McDonagh, A.F., Glazer, A.N., and Ames, B.N. (1987). Bilirubin is an antioxidant of possible physiological importance. *Science* 235 (4792): 1043–1046.

86. Tell, G. and Gustincich, S. (2009). Redox state, oxidative stress, and molecular mechanisms of protective and toxic effects of bilirubin on cells. *Curr. Pharm. Des.* 15 (25): 2908–2914.

87. Hamoud, A.R., Weaver, L., Stec, D.E., and Hinds, T.D., Jr. (2018). Bilirubin in the liver-gut signaling axis. *Trends Endocrinol. Metab.* 29 (3): 140–150.

88. Hinds, T.D., Jr. and Stec, D.E. (2018). Bilirubin, a cardiometabolic signaling molecule. *Hypertension* 72 (4): 788–795.

89. Novak, P., Jackson, A.O., Zhao, G.J., and Yin, K. (2020). Bilirubin in metabolic syndrome and associated inflammatory diseases: new perspectives. *Life Sci.* 257: 118032.

90. Hinds, T.D., Jr. and Stec, D.E. (2019). Bilirubin safeguards cardiorenal and metabolic diseases: a protective role in health. *Curr. Hypertens. Rep.* 21 (11): 87.

91. Ollinger, R., Kogler, P., Troppmair, J., Hermann, M., Wurm, M., Drasche, A., Konigsrainer, I., Amberger, A., Weiss, H., Ofner, D., Bach, F.H., and Margreiter, R. (2007). Bilirubin inhibits tumor cell growth via activation of ERK. *Cell Cycle* 6 (24): 3078–3085.

92. Bulmer, A.C., Ried, K., Blanchfield, J.T., and Wagner, K.H. (2008). The anti-mutagenic properties of bile pigments. *Mutat. Res.* 658 (1–2): 28–41.

93. Wallner, M., Antl, N., Rittmannsberger, B., Schreidl, S., Najafi, K., Mullner, E., Molzer, C., Ferk, F., Knasmuller, S., Marculescu, R., Doberer, D., Poulsen, H.E., Vitek, L., Bulmer, A.C., and Wagner, K.H. (2013). Anti-genotoxic potential of bilirubin in vivo: damage to DNA in hyperbilirubinemic human and animal models. *Cancer Prev. Res. (Phila)* 6 (10): 1056–1063.

94. Jangi, S., Otterbein, L., and Robson, S. (2013). The molecular basis for the immunomodulatory activities of unconjugated bilirubin. *Int. J. Biochem. Cell Biol.* 45 (12): 2843–2851.

95. Zelenka, J., Dvorak, A., Alan, L., Zadinova, M., Haluzik, M., and Vitek, L. (2016). Hyperbilirubinemia protects against aging-associated inflammation and metabolic deterioration. *Oxid. Med. Cell. Longev.* 2016: 6190609.

96. Kim, S.Y. and Park, S.C. (2012). Physiological antioxidative network of the bilirubin system in aging and age-related diseases. *Front. Pharmacol.* 3: 45.

97. Hansen, T.W., Mathiesen, S.B., and Walaas, S.I. (1996). Bilirubin has widespread inhibitory effects on protein phosphorylation. *Pediatr. Res.* 39: 1072–1077.

98. Gordon, D.M., Blomquist, T.M., Miruzzi, S.A., McCullumsmith, R., Stec, D.E., and Hinds, T.D., Jr. (2019). RNA sequencing in human HepG2 hepatocytes reveals PPAR-alpha mediates transcriptome responsiveness of bilirubin. *Physiol. Genomics* 51 (6): 234–240.

99. Watchko, J.F. (2006). Kernicterus and the molecular mechanisms of bilirubin-induced CNS injury in newborns. *Neuromolecular Med.* 8 (4): 513–529.

100. Gordon, D.M., Neifer, K.L., Hamoud, A.A., Hawk, C.F., Nestor-Kalinoski, A.L., Miruzzi, S.A., Morran, M.P., Adeosun, S.O., Sarver, J.G., Erhardt, P.W., McCullumsmith, R.E., Stec, D.E., and Hinds, T.D., Jr. (2020). Bilirubin remodels murine white adipose

100. tissue by reshaping mitochondrial activity and the coregulator profile of peroxisome proliferator-activated receptor alpha. *J. Biol. Chem.* 295 (29): 9804–9822.
101. Dvorak, A., Pospisilova, K., Zizalova, K., Capkova, N., Muchova, L., Vecka, M., Vrzackova, N., Krizova, J., Zelenka, J., and Vitek, L. (2021). The effects of bilirubin and lumirubin on metabolic and oxidative stress markers. *Front. Pharmacol.* 12: 567001.
102. Bisht, K., Wegiel, B., Tampe, J., Neubauer, O., Wagner, K.H., Otterbein, L.E., and Bulmer, A.C. (2014). Biliverdin modulates the expression of C5aR in response to endotoxin in part via mTOR signaling. *Biochem. Biophys. Res. Commun.* 449 (1): 94–99.
103. Zheng, J., Nagda, D.A., Lajud, S.A., Kumar, S., Mouchli, A., Bezpalko, O., O'Malley, B.W., Jr., and Li, D. (2014). Biliverdin's regulation of reactive oxygen species signalling leads to potent inhibition of proliferative and angiogenic pathways in head and neck cancer. *Br. J. Cancer* 110 (8): 2116–2122.
104. Siow, R.C., Sato, H., and Mann, G.E. (1999). Heme oxygenase-carbon monoxide signalling pathway in atherosclerosis: anti-atherogenic actions of bilirubin and carbon monoxide? *Cardiovasc. Res.* 41 (2): 385–394.
105. Park, J., Joe, Y., Ryter, S.W., Surh, Y.J., and Chung, H.T. (2019). Similarities and distinctions in the effects of metformin and carbon monoxide in immunometabolism. *Mol. Cells* 42 (4): 292–300.
106. Motterlini, R. (2007). Carbon monoxide-releasing molecules (CO-RMs): Vasodilatory, anti-ischaemic and anti-inflammatory activities. *Biochem. Soc. Trans.* 35 (Pt 5): 1142–1146.
107. Dulak, J., Deshane, J., Jozkowicz, A., and Agarwal, A. (2008). Heme oxygenase-1 and carbon monoxide in vascular pathobiology: Focus on angiogenesis. *Circulation* 117 (2): 231–241.
108. Qiao, L., Zhang, N., Huang, J.L., and Yang, X.Q. (2017). Carbon monoxide as a promising molecule to promote nerve regeneration after traumatic brain injury. *Med. Gas Res.* 7 (1): 45–47.
109. Lu, K., Wu, W.J., Zhang, C., Zhu, Y.L., Zhong, J.Q., and Li, J. (2020). CORM-3 Regulates microglia activity, prevents neuronal injury, and improves memory function during radiation-induced brain injury. *Curr. Neurovasc. Res.* 17 (4): 464–470.
110. Li, Y., Zhang, L.M., Zhang, D.X., Zheng, W.C., Bai, Y., Bai, J., Fu, L., and Wang, X.P. (2020). CORM-3 ameliorates neurodegeneration in the amygdala and improves depression- and anxiety-like behavior in a rat model of combined traumatic brain injury and hemorrhagic shock. *Neurochem. Int.* 140: 104842.
111. Durante, W. (2011). Protective role of heme oxygenase-1 against inflammation in atherosclerosis. *Front. Biosci. (Landmark Ed)* 16: 2372–2388.
112. Dulak, J., Loboda, A., and Jozkowicz, A. (2008). Effect of heme oxygenase-1 on vascular function and disease. *Curr. Opin. Lipidol.* 19 (5): 505–512.
113. Was, H., Dulak, J., and Jozkowicz, A. (2010). Heme oxygenase-1 in tumor biology and therapy. *Curr. Drug Targets* 11 (12): 1551–1570.
114. Wang, C.Y. and Chau, L.Y. (2010). Heme oxygenase-1 in cardiovascular diseases: molecular mechanisms and clinical perspectives. *Chang Gung Med. J.* 33 (1): 13–24.
115. Canesin, G., Hejazi, S.M., Swanson, K.D., and Wegiel, B. (2020). Heme-derived metabolic signals dictate immune responses. *Front. Immunol.* 11: 66.
116. Basiglio, C.L., Arriaga, S.M., Pelusa, F., Almara, A.M., Kapitulnik, J., and Mottino, A.D. (2010). Complement activation and disease: protective effects of hyperbilirubinaemia. *Clin. Sci.* 118 (2): 99–113.
117. Nakagami, T., Toyomura, K., Kinoshita, T., and Morisawa, S. (1993). A beneficial role of bile pigments as an endogenous tissue protector: anti-complement effects of biliverdin and conjugated bilirubin. *Biochim. Biophys. Acta* 1158: 189–193.
118. Bilej, M., Vetvicka, V., and Sima, P. (1989). The stimulatory effect of bilirubin on phagocytic activity. *Folia Microbiol. (Praha)* 34: 136–140.
119. Vetvicka, V., Miler, I., Sima, P., Taborsky, L., and Fornusek, L. (1985). The effect of bilirubin on the Fc receptor expression and phagocytic activity of mouse peritoneal macrophages. *Folia Microbiol. (Praha)* 30: 373–380.
120. Sundararaghavan, V.L., Binepal, S., Stec, D.E., Sindhwani, P., and Hinds, T.D., Jr. (2018). Bilirubin, a new therapeutic for kidney transplant? *Transplant. Rev. (Orlando)* 32 (4): 234–240.
121. Deetman, P.E., Bakker, S.J., and Dullaart, R.P. (2013). High sensitive C-reactive protein and serum amyloid A are inversely related to serum bilirubin: effect-modification by metabolic syndrome. *Cardiovasc. Diabetol.* 12 (1): 166.
122. Takei, R., Inoue, T., Sonoda, N., Kohjima, M., Okamoto, M., Sakamoto, R., Inoguchi, T., and Ogawa, Y. (2019). Bilirubin reduces visceral obesity and insulin resistance by suppression of inflammatory cytokines. *PLoS One* 14 (10): e0223302.
123. Bisht, K., Canesin, G., Cheytan, T., Li, M., Nemeth, Z., Csizmadia, E., Woodruff, T.M., Stec, D.E., Bulmer, A.C., Otterbein, L.E., and Wegiel, B. (2019).

123. Deletion of biliverdin reductase A in myeloid cells promotes chemokine expression and chemotaxis in part via a complement C5a-C5aR1 pathway. *J. Immunol.* 202 (10): 2982–2990.

124. Lee, Y., Sugihara, K., Gillilland, M.G., 3rd, Jon, S., Kamada, N., and Moon, J.J. (2020). Hyaluronic acid-bilirubin nanomedicine for targeted modulation of dysregulated intestinal barrier, microbiome and immune responses in colitis. *Nat. Mater.* 19 (1): 118–126.

125. Adin, C.A., VanGundy, Z.C., Papenfuss, T.L., Xu, F., Ghanem, M., Lakey, J., and Hadley, G.A. (2017). Physiologic doses of bilirubin contribute to tolerance of islet transplants by suppressing the innate immune response. *Cell Transplant.* 26 (1): 11–21.

126. Wegiel, B., Gallo, D., Csizmadia, E., Roger, T., Kaczmarek, E., Harris, C., Zuckerbraun, B.S., and Otterbein, L.E. (2011). Biliverdin inhibits Toll-like receptor-4 (TLR4) expression through nitric oxide-dependent nuclear translocation of biliverdin reductase. *Proc. Natl. Acad. Sci. U. S. A.* 108 (46): 18849–18854.

127. Khan, N.M. and Poduval, T.B. (2011). Immunomodulatory and immunotoxic effects of bilirubin: molecular mechanisms. *J. Leukoc. Biol.* 90 (5): 997–1015.

128. Bisht, K., Tampe, J., Shing, C., Bakrania, B., Winearls, J., Fraser, J., Wagner, K.H., and Bulmer, A.C. (2014). Endogenous tetrapyrroles influence leukocyte responses to lipopolysaccharide in human blood: pre-clinical evidence demonstrating the anti-inflammatory potential of biliverdin. *J. Clin. Cell. Immunol.* 5 (3): 218.

129. Nejedla, Z. (1970). The development of immunological factors in infants with hyperbilirubinemia. *Pediatrics* 45 (1): 102–104.

130. Haga, Y., Tempero, M.A., and Zetterman, R.K. (1996). Unconjugated bilirubin inhibits in vitro major histocompatibility complex-unrestricted cytotoxicity of human lymphocytes. *Biochim. Biophys. Acta* 1316: 29–34.

131. Haga, Y., Tempero, M.A., and Zetterman, R.K. (1996). Unconjugated bilirubin inhibits in vitro cytotoxic T lymphocyte activity of human lymphocytes. *Biochim. Biophys. Acta* 1317: 65–70.

132. Kim, D.E., Lee, Y., Kim, M., Lee, S., Jon, S., and Lee, S.H. (2017). Bilirubin nanoparticles ameliorate allergic lung inflammation in a mouse model of asthma. *Biomaterials* 140: 37–44.

133. Yamashita, K., McDaid, J., Ollinger, R., Tsui, T.Y., Berberat, P.O., Usheva, A., Csizmadia, E., Smith, R.N., Soares, M.P., and Bach, F.H. (2004). Biliverdin, a natural product of heme catabolism, induces tolerance to cardiac allografts. *FASEB J.* 18 (6): 765–767.

134. Rocuts, F., Zhang, X., Yan, J., Yue, Y., Thomas, M., Bach, F.H., Czismadia, E., and Wang, H. (2010). Bilirubin promotes de novo generation of T regulatory cells. *Cell Transplant.* 19 (4): 443–451.

135. Lee, S.S., Gao, W., Mazzola, S., Thomas, M.N., Csizmadia, E., Otterbein, L.E., Bach, F.H., and Wang, H. (2007). Heme oxygenase-1, carbon monoxide, and bilirubin induce tolerance in recipients toward islet allografts by modulating T regulatory cells. *FASEB J.* 21 (13): 3450–3457.

136. Lee, G.R., Shaefi, S., and Otterbein, L.E. (2019). HO-1 and CD39: it takes two to protect the realm. *Front. Immunol.* 10: 1765.

137. Longhi, M.S., Vuerich, M., Kalbasi, A., Kenison, J.E., Yeste, A., Csizmadia, E., Vaughn, B., Feldbrugge, L., Mitsuhashi, S., Wegiel, B., Otterbein, L., Moss, A., Quintana, F.J., and Robson, S.C. (2017). Bilirubin suppresses Th17 immunity in colitis by upregulating CD39. *JCI Insight* 2: 9.

138. Stevens, E.A. and Bradfield, C.A. (2008). Immunology: t cells hang in the balance. *Nature* 453 (7191): 46–47.

139. Bock, K.W. and Kohle, C. (2010). Contributions of the Ah receptor to bilirubin homeostasis and its antioxidative and atheroprotective functions. *Biol. Chem.* 391 (6): 645–653.

140. Wagner, K.H., Wallner, M., Molzer, C., Gazzin, S., Bulmer, A.C., Tiribelli, C., and Vitek, L. (2015). Looking to the horizon: the role of bilirubin in the development and prevention of age-related chronic diseases. *Clin. Sci. (Lond)* 129 (1): 1–25.

141. Lehmann, E., El Tantawy, W.H., Ocker, M., Bartenschlager, R., Lohmann, V., Hashemolhosseini, S., Tiegs, G., and Sass, G. (2010). The heme oxygenase 1 product biliverdin interferes with hepatitis C virus replication by increasing antiviral interferon response. *Hepatology* 51 (2): 398–404.

142. Zhu, Z., Wilson, A.T., Luxon, B.A., Brown, K.E., Mathahs, M.M., Bandyopadhyay, S., McCaffrey, A.P., and Schmidt, W.N. (2010). Biliverdin inhibits hepatitis C virus nonstructural 3/4A protease activity: mechanism for the antiviral effects of heme oxygenase? *Hepatology* 52 (6): 1897–1905.

143. Mori, H., Otake, T., Morimoto, M., Ueba, N., Kunita, N., Nakagami, T., Yamasaki, N., and Taji, S. (1991). In vitro anti-human immunodeficiency virus type 1 activity of biliverdin, a bile pigment. *Jpn. J. Cancer Res.* 82: 755–757.

144. Nakagami, T., Taji, S., Takahashi, M., and Yamanishi, K. (1992). Antiviral activity of a bile pigment, biliverdin, against human herpesvirus 6 (HHV-6) in vitro. *Microbiol. Immunol.* 36: 381–390.

145. Santangelo, R., Mancuso, C., Marchetti, S., Di, S.E., Pani, G., and Fadda, G. (2012). Bilirubin: an endogenous molecule with antiviral activity in vitro. *Front. Pharmacol.* 3: 36.
146. Ferrante, A. and Thong, Y.H. (1982). Inhibition of serum bactericidal activity by bilirubin. *Immunol. Lett.* 4 (2): 103–105.
147. Hansen, R., Gibson, S., De Paiva Alves, E., Goddard, M., MacLaren, A., Karcher, A.M., Berry, S., Collie-Duguid, E.S.R., El-Omar, E., Munro, M., and Hold, G.L. (2018). Adaptive response of neonatal sepsis-derived Group B Streptococcus to bilirubin. *Sci. Rep.* 8 (1): 6470.
148. Nobles, C.L., Green, S.I., and Maresso, A.W. (2013). A product of heme catabolism modulates bacterial function and survival. *PLoS Pathog.* 9 (7): e1003507.
149. Motterlini, R., Haas, B., and Foresti, R. (2012). Emerging concepts on the anti-inflammatory actions of carbon monoxide-releasing molecules (CO-RMs). *Med. Gas Res.* 2 (1): 28.
150. Babu, D., Motterlini, R., and Lefebvre, R.A. (2015). CO and CO-releasing molecules (CO-RMs) in acute gastrointestinal inflammation. *Br. J. Pharmacol.* 172 (6): 1557–1573.
151. Onyiah, J.C., Sheikh, S.Z., Maharshak, N., Otterbein, L.E., and Plevy, S.E. (2014). Heme oxygenase-1 and carbon monoxide regulate intestinal homeostasis and mucosal immune responses to the enteric microbiota. *Gut Microbes* 5 (2): 220–224.
152. Chin, B.Y. and Otterbein, L.E. (2009). Carbon monoxide is a poison. to microbes! CO as a bactericidal molecule. *Curr. Opin. Pharmacol.* 9 (4): 490–500.
153. Nikolic, I., Saksida, T., Vujicic, M., Stojanovic, I., and Stosic-Grujicic, S. (2015). Anti-diabetic actions of carbon monoxide-releasing molecule (CORM)-A1: immunomodulation and regeneration of islet beta cells. *Immunol. Lett.* 165 (1): 39–46.
154. Nikolic, I., Saksida, T., Mangano, K., Vujicic, M., Stojanovic, I., Nicoletti, F., and Stosic-Grujicic, S. (2014). Pharmacological application of carbon monoxide ameliorates islet-directed autoimmunity in mice via anti-inflammatory and anti-apoptotic effects. *Diabetologia* 57 (5): 980–990.
155. Nikolic, I., Vujicic, M., Stojanovic, I., Stosic-Grujicic, S., and Saksida, T. (2014). Carbon monoxide-releasing molecule-A1 inhibits Th1/Th17 and stimulates Th2 differentiation in vitro. *Scand. J. Immunol.* 80 (2): 95–100.
156. Bosch, F., Thomas, M., Kogler, P., Oberhuber, R., Sucher, R., Aigner, F., Semsroth, S., Wiedemann, D., Yamashita, K., Troppmair, J., Kotsch, K., Pratschke, J., and Ollinger, R. (2014). Bilirubin rinse of the graft ameliorates ischemia reperfusion injury in heart transplantation. *Transpl. Int.* 27 (5): 504–513.
157. Vitek, L. (2012). The role of bilirubin in diabetes, metabolic syndrome, and cardiovascular diseases. *Front. Pharmacol.* 3: 55.
158. Papadakis, J.A., Ganotakis, E.S., Jagroop, I.A., Mikhailidis, D.P., and Winder, A.F. (1999). Effect of hypertension and its treatment on lipid, lipoprotein(a), fibrinogen, and bilirubin levels in patients referred for dyslipidemia. *Am. J. Hypertens.* 12 (7): 673–681.
159. Inoguchi, T., Sasaki, S., Kobayashi, K., Takayanagi, R., and Yamada, T. (2007). Relationship between Gilbert syndrome and prevalence of vascular complications in patients with diabetes. *JAMA* 298 (12): 1398–1400.
160. Cheriyath, P., Gorrepati, V.S., Peters, I., Nookala, V., Murphy, M.E., Srouji, N., and Fischman, D. (2010). High total bilirubin as a protective factor for diabetes mellitus: an analysis of NHANES data from 1999–2006. *J. Clin. Med. Res.* 2 (5): 201–206.
161. Andersson, C., Weeke, P., Fosbol, E.L., Brendorp, B., Kober, L., Coutinho, W., Sharma, A.M., Van Gaal, L., Finer, N., James, W.P., Caterson, I.D., Rode, R.A., and Torp-Pedersen, C. (2009). Acute effect of weight loss on levels of total bilirubin in obese, cardiovascular high-risk patients: An analysis from the lead-in period of the Sibutramine cardiovascular outcome trial. *Metabolism* 58 (8): 1109–1115.
162. Choi, S.H., Yun, K.E., and Choi, H.J. (2013). Relationships between serum total bilirubin levels and metabolic syndrome in Korean adults. *Nutr. Metab. Cardiovasc. Dis.* 23 (1): 31–37.
163. Seyed Khoei, N., Grindel, A., Wallner, M., Molzer, C., Doberer, D., Marculescu, R., Bulmer, A., and Wagner, K.H. (2018). Mild hyperbilirubinaemia as an endogenous mitigator of overweight and obesity: implications for improved metabolic health. *Atherosclerosis* 269: 306–311.
164. Bulmer, A.C., Bakrania, B., Du Toit, E.F., Boon, A.C., Clark, P.J., Powell, L.W., Wagner, K.H., and Headrick, J.P. (2018). Bilirubin acts as a multipotent guardian of cardiovascular integrity: more than just a radical idea. *Am. J. Physiol. Heart Circ. Physiol.* 315 (3): H429–47.
165. McCallum, L., Panniyammakal, J., Hastie, C.E., Hewitt, J., Patel, R., Jones, G.C., Muir, S., Walters, M., Sattar, N., Dominiczak, A.F., and Padmanabhan, S. (2015). Longitudinal blood pressure control, long-term mortality, and predictive utility of serum liver enzymes and bilirubin in hypertensive patients. *Hypertension* 66 (1): 37–43.
166. Weaver, L., Hamoud, A.R., Stec, D.E., and Hinds, T.D., Jr. (2018). Biliverdin reductase and bilirubin in

hepatic disease. *Am. J. Physiol. Gastrointest. Liver Physiol.* 314 (6): G668–76.
167. Hinds, T.D., Jr., Adeosun, S.O., Alamodi, A.A., and Stec, D.E. (2016). Does bilirubin prevent hepatic steatosis through activation of the PPARalpha nuclear receptor? *Med. Hypotheses* 95: 54–57.
168. Bulmer, A.C., Verkade, H.J., and Wagner, K.H. (2013). Bilirubin and beyond: a review of lipid status in Gilbert's syndrome and its relevance to cardiovascular disease protection. *Prog. Lipid Res.* 52 (2): 193–205.
169. Abbasi, A., Deetman, P.E., Corpeleijn, E., Gansevoort, R.T., Gans, R.O., Hillege, H.L., Van Der Harst, P., Stolk, R.P., Navis, G., Alizadeh, B.Z., and Bakker, S.J. (2015). Bilirubin as a potential causal factor in type 2 diabetes risk: A Mendelian randomization study. *Diabetes* 64 (4): 1459–1469.
170. Rodella, L.F., Vanella, L., Peterson, S.J., Drummond, G., Rezzani, R., Falck, J.R., and Abraham, N.G. (2008). Heme oxygenase-derived carbon monoxide restores vascular function in type 1 diabetes. *Drug Metab. Lett.* 2 (4): 290–300.
171. Braud, L., Pini, M., Muchova, L., Manin, S., Kitagishi, H., Sawaki, D., Czibik, G., Ternacle, J., Derumeaux, G., Foresti, R., and Motterlini, R. (2018). Carbon monoxide-induced metabolic switch in adipocytes improves insulin resistance in obese mice. *JCI Insight* 3: 22.
172. Vitek, L., Hubacek, J.A., Pajak, A., Dorynska, A., Kozela, M., Eremiasova, L., Danzig, V., Stefler, D., and Bobak, M. (2019). Association between plasma bilirubin and mortality. *Ann. Hepatol.* 18 (2): 379–385.
173. Lind, L., Zanetti, D., Hogman, M., Sundman, L., and Ingelsson, E. (2020). Commonly used clinical chemistry tests as mortality predictors: results from two large cohort studies. *PLoS One* 15 (11): e0241558.
174. Suh, S., Cho, Y.R., Park, M.K., Kim, D.K., Cho, N.H., and Lee, M.K. (2018). Relationship between serum bilirubin levels and cardiovascular disease. *PLoS One* 13 (2): e0193041.
175. Adeosun, S.O. and Stec, D.E. (2017). Bilirubin protects the ageing heart. *Acta Physiol. (Oxford)* 220 (4): 402–403.
176. Bulpitt, C.J., Markowe, H.L.J., and Shipley, M.J. (2001). Why do some people look older than they should? *Postgrad. Med. J.* 77: 578–581.
177. Stallone, G., Infante, B., Prisciandaro, C., and Grandaliano, G. (2019). mTOR and aging: an old fashioned dress. *Int. J. Mol. Sci.* 20 (11): 2774.
178. Bragt, M.C. and Popeijus, H.E. (2008). Peroxisome proliferator-activated receptors and the metabolic syndrome. *Physiol. Behav.* 94 (2): 187–197.
179. Hinds, T.D., Jr., Hosick, P.A., Chen, S., Tukey, R.H., Hankins, M.W., Nestor-Kalinoski, A., and Stec, D.E. (2017). Mice with hyperbilirubinemia due to Gilbert's syndrome polymorphism are resistant to hepatic steatosis by decreased serine 73 phosphorylation of PPARalpha. *Am. J. Physiol. Endocrinol. Metab.* 312 (4): E244–52.
180. Molzer, C., Wallner, M., Kern, C., Tosevska, A., Schwarz, U., Zadnikar, R., Doberer, D., Marculescu, R., and Wagner, K.H. (2016). Features of an altered AMPK metabolic pathway in Gilbert's Syndrome, and its role in metabolic health. *Sci. Rep.* 6: 30051.
181. Lin, S.C. and Hardie, D.G. (2018). AMPK: Sensing glucose as well as cellular energy status. *Cell Metab.* 27 (2): 299–313.
182. Gonzalez, A., Hall, M.N., Lin, S.C., and Hardie, D.G. (2020). AMPK and TOR: The Yin and Yang of cellular nutrient sensing and growth control. *Cell Metab.* 31 (3): 472–492.
183. Liu, J., Dong, H., Zhang, Y., Cao, M., Song, L., Pan, Q., Bulmer, A., Adams, D.B., Dong, X., and Wang, H. (2015). Bilirubin increases insulin sensitivity by regulating cholesterol metabolism, adipokines and PPARgamma levels. *Sci. Rep.* 5: 9886.
184. Dong, H., Huang, H., Yun, X., Kim, D.S., Yue, Y., Wu, H., Sutter, A., Chavin, K.D., Otterbein, L.E., Adams, D.B., Kim, Y.B., and Wang, H. (2014). Bilirubin increases insulin sensitivity in leptin-receptor deficient and diet-induced obese mice through suppression of ER stress and chronic inflammation. *Endocrinology* 155 (3): 818–828.
185. Zhang, F., Guan, W., Fu, Z., Zhou, L., Guo, W., Ma, Y., Gong, Y., Jiang, W., Liang, H., and Zhou, H. (2020). Relationship between serum indirect bilirubin level and insulin sensitivity: Results from two independent cohorts of obese patients with impaired glucose regulation and type 2 diabetes mellitus in China. *Int. J. Endocrinol.* 2020: 1–10.
186. Lin, L.Y., Kuo, H.K., Hwang, J.J., Lai, L.P., Chiang, F.T., Tseng, C.D., and Lin, J.L. (2009). Serum bilirubin is inversely associated with insulin resistance and metabolic syndrome among children and adolescents. *Atherosclerosis* 203 (2): 563–568.
187. McCarty, M.F. (2015). Practical prospects for boosting hepatic production of the "pro-longevity" hormone FGF21. *Horm. Mol. Biol. Clin. Investig.* 30: 2.
188. Potthoff, M.J., Kliewer, S.A., and Mangelsdorf, D.J. (2012). Endocrine fibroblast growth factors 15/19 and 21: From feast to famine. *Genes Dev.* 26 (4): 312–324.
189. Antonioli, L., Pacher, P., Vizi, E.S., and Hasko, G. (2013). CD39 and CD73 in immunity and inflammation. *Trends Mol. Med.* 19 (6): 355–367.

190. Enjyoji, K., Kotani, K., Thukral, C., Blumel, B., Sun, X., Wu, Y., Imai, M., Friedman, D., Csizmadia, E., Bleibel, W., Kahn, B.B., and Robson, S.C. (2008). Deletion of cd39/entpd1 results in hepatic insulin resistance. *Diabetes* 57 (9): 2311–2320.

191. Da Silva, C.G., Jarzyna, R., Specht, A., and Kaczmarek, E. (2006). Extracellular nucleotides and adenosine independently activate AMP-activated protein kinase in endothelial cells: Involvement of P2 receptors and adenosine transporters. *Circ. Res.* 98 (5): e39–47.

192. Wang, P., Jia, J., and Zhang, D. (2020). Purinergic signalling in liver diseases: Pathological functions and therapeutic opportunities. *JHEP Rep.* 2 (6): 100165.

193. Wang, S., Gao, S., Zhou, D., Qian, X., Luan, J., and Lv, X. (2021). The role of the CD39-CD73-adenosine pathway in liver disease. *J. Cell. Physiol.* 236 (2): 851–862.

194. Correa-Costa, M., Gallo, D., Csizmadia, E., Gomperts, E., Lieberum, J.L., Hauser, C.J., Ji, X., Wang, B., Camara, N.O.S., Robson, S.C., and Otterbein, L.E. (2018). Carbon monoxide protects the kidney through the central circadian clock and CD39. *Proc. Natl. Acad. Sci. U. S. A.* 115 (10): E2302–E2310.

195. Stec, D.E., John, K., Trabbic, C.J., Luniwal, A., Hankins, M.W., Baum, J., and Hinds, T.D., Jr. (2016). Bilirubin binding to PPARalpha inhibits lipid accumulation. *PLoS One* 11 (4): e0153427.

196. Balint, B.L. and Nagy, L. (2006). Selective modulators of PPAR activity as new therapeutic tools in metabolic diseases. *Endocr. Metab. Immune. Disord. Drug Targets* 6 (1): 33–43.

197. Ahmed, I., Furlong, K., Flood, J., Treat, V.P., and Goldstein, B.J. (2007). Dual PPAR alpha/gamma agonists: promises and pitfalls in type 2 diabetes. *Am. J. Ther.* 14 (1): 49–62.

198. Xiao, L., Zhang, Z., and Luo, X. (2014). Roles of xenobiotic receptors in vascular pathophysiology. *Circ. J.* 78 (7): 1520–1530.

199. Xu, P., Zhai, Y., and Wang, J. (2018). The role of PPAR and its cross-talk with CAR and LXR in obesity and atherosclerosis. *Int. J. Mol. Sci.* 19 (4).

200. Liu, Z. and Cao, W. (2009). p38 mitogen-activated protein kinase: A critical node linking insulin resistance and cardiovascular diseases in type 2 diabetes mellitus. *Endocr. Metab. Immune. Disord. Drug Targets* 9 (1): 38–46.

201. Gibbs, P.E., Lerner-Marmarosh, N., Poulin, A., Farah, E., and Maines, M.D. (2014). Human biliverdin reductase-based peptides activate and inhibit glucose uptake through direct interaction with the kinase domain of insulin receptor. *FASEB J.* 28 (6): 2478–2491.

202. Miralem, T., Lerner-Marmarosh, N., Gibbs, P.E., Jenkins, J.L., Heimiller, C., and Maines, M.D. (2016). Interaction of human biliverdin reductase with Akt/protein kinase B and phosphatidylinositol-dependent kinase 1 regulates glycogen synthase kinase 3 activity: A novel mechanism of Akt activation. *FASEB J.* 30 (8): 2926–2944.

203. Tiribelli, C. and Ostrow, J.D. (1997). New concepts in bilirubin and jaundice. *Hepatology* 24: 1296–1311.

204. Shiels, R.G., Vidimce, J., Pearson, A.G., Matthews, B., Wagner, K.H., Battle, A.R., Sakellaris, H., and Bulmer, A.C. (2019). Unprecedented microbial conversion of biliverdin into bilirubin-10-sulfonate. *Sci. Rep.* 9 (1): 2988.

205. Shiels, R.G., Hewage, W., Pennell, E.N., Vidimce, J., Grant, G., Pearson, A.G., Wagner, K.H., Morgan, M., and Bulmer, A.C. (2020). Biliverdin and bilirubin sulfonate inhibit monosodium urate induced sterile inflammation in the rat. *Eur. J. Pharm. Sci.* 155: 105546.

206. Sakai, T., Watanabe, K., and Kawatsu, H. (1987). Occurrence of ditaurobilirubin, bilirubin conjugated with two moles of taurine, in the gallbladder bile of yellowtail, Seriola quinqueradiata. *J. Biochem.* 102 (4): 793–796.

207. Pennell, E.N., Wagner, K.H., Mosawy, S., and Bulmer, A.C. (2019). Acute bilirubin ditaurate exposure attenuates ex vivo platelet reactive oxygen species production, granule exocytosis and activation. *Redox Biol.* 26: 101250.

208. Wu, T.W., Fung, K.P., Wu, J., Yang, C.C., and Weisel, R.D. (1996). Antioxidation of human low density lipoprotein by unconjugated and conjugated bilirubins. *Biochem. Pharmacol.* 51: 859–862.

209. Molzer, C., Huber, H., Steyrer, A., Ziesel, G.V., Wallner, M., Hong, H.T., Blanchfield, J.T., Bulmer, A.C., and Wagner, K.H. (2013). Bilirubin and related tetrapyrroles inhibit food-borne mutagenesis: a mechanism for antigenotoxic action against a model epoxide. *J. Nat. Prod.* 76 (10): 1958–1965.

210. Stocker, R. and Peterhans, E. (1989). Antioxidant properties of conjugated bilirubin and biliverdin: biologically relevant scavenging of hypochlorous acid. *Free Radic. Res. Commun.* 6 (1): 57–66.

211. Molzer, C., Huber, H., Steyrer, A., Ziesel, G., Ertl, A., Plavotic, A., Wallner, M., Bulmer, A.C., and Wagner, K.H. (2012). In vitro antioxidant capacity and antigenotoxic properties of protoporphyrin and structurally related tetrapyrroles. *Free Radic. Res.* 46 (11): 1369–1377.

212. Arriaga, S.M., Mottino, A.D., and Almara, A.M. (1999). Inhibitory effect of bilirubin on complement-mediated hemolysis. *Biochim. Biophys. Acta* 1473 (2–3): 329–336.

213. Wu, T.W., Wu, J., Li, R.K., Mickle, D., and Carey, D. (1991). Albumin-bound bilirubins protect human ventricular myocytes against oxyradical damage. *Biochem. Cell Biol.* 69 (10–11): 683–688.

214. Nakamura, T., Sato, K., Akiba, M., and Ohnishi, M. (2006). Urobilinogen, as a bile pigment metabolite, has an antioxidant function. *J. Oleo Sci.* 55 (4): 191–197.

215. Jasprova, J., Dal Ben, M., Vianello, E., Goncharova, I., Urbanova, M., Vyroubalova, K., Gazzin, S., Tiribelli, C., Sticha, M., Cerna, M., and Vitek, L. (2016). The biological effects of bilirubin photoisomers. *PLoS One* 11 (2): e0148126.

216. Yamaguchi, T., Shioji, I., Sugimoto, A., Komoda, Y., and Nakajima, H. (1994). Chemical structure of a new family of bile pigments from human urine. *J. Biochem. Tokyo* 116: 298–303.

217. Vitek, L., Kraslova, I., Muchova, L., Novotny, L., and Yamaguchi, T. (2007). Urinary excretion of oxidative metabolites of bilirubin in subjects with Gilbert syndrome. *J. Gastroenterol. Hepatol.* 22 (6): 841–845.

218. Kunii, H., Ishikawa, K., Yamaguchi, T., Komatsu, N., Ichihara, T., and Maruyama, Y. (2009). Bilirubin and its oxidative metabolite biopyrrins in patients with acute myocardial infarction. *Fukushima J. Med. Sci.* 55 (2): 39–51.

219. Kranc, K.R., Pyne, G.J., Tao, L., Claridge, T.D., Harris, D.A., Cadoux-Hudson, T.A., Turnbull, J.J., Schofield, C.J., and Clark, J.F. (2000). Oxidative degradation of bilirubin produces vasoactive compounds. *Eur. J. Biochem.* 267 (24): 7094–7101.

220. Clark, J.F., Loftspring, M., Wurster, W.L., and Pyne-Geithman, G.J. (2008). Chemical and biochemical oxidations in spinal fluid after subarachnoid hemorrhage. *Front. Biosci.* 13: 1806–1812.

221. Koehler, R.C. (2019). Propentdyopents. A new culprit in delayed vasospasm after subarachnoid hemorrhage. *Circ. Res.* 124 (12): 1686–1688.

222. Ritter, M., Neupane, S., Seidel, R.A., Steinbeck, C., and Pohnert, G. (2018). In vivo and in vitro identification of Z-BOX C - a new bilirubin oxidation end product. *Org. Biomol. Chem.* 16 (19): 3553–3555.

223. Madea, D., Mahvidi, S., Chalupa, D., Mujawar, T., Dvořák, A., Muchová, L., Janoš, J., Slavíček, P., Švenda, J., Vítek, L., and Klán, P. (2020). Wavelength-dependent photochemistry and biological relevance of a bilirubin dipyrrinone subunit. *J Org Chem.* 85(20): 13015–13028. doi: 10.1021/acs.joc.0c01673. Epub 2020 Oct 1.

224. McCarty, M.F. (2007). "Iatrogenic Gilbert syndrome"- a strategy for reducing vascular and cancer risk by increasing plasma unconjugated bilirubin. *Med. Hypotheses* 69 (5): 974–994.

225. Vitek, L., Bellarosa, C., and Tiribelli, C. (2019). Induction of mild hyperbilirubinemia: Hype or real therapeutic opportunity? *Clin. Pharmacol. Ther.* 106 (3): 568–575.

226. Dekker, D., Dorresteijn, M.J., Welzen, M.E.B., Timman, S., Pickkers, P., Burger, D.M., Smits, P., Wagener, F., and Russel, F.G.M. (2018). Parenteral bilirubin in healthy volunteers: A reintroduction in translational research. *Br. J. Clin. Pharmacol.* 84 (2): 268–279.

227. Du, P., Wang, A., Ma, Y., and Li, X. (2019). Association between the UGT1A1*28 allele and hyperbilirubinemia in HIV-positive patients receiving atazanavir: A meta-analysis. *Biosci. Rep.* 39 (5).

228. Rodriguez-Novoa, S., Martin-Carbonero, L., Barreiro, P., Gonzalez-Pardo, G., Jimenez-Nacher, I., Gonzalez-Lahoz, J., and Soriano, V. (2007). Genetic factors influencing atazanavir plasma concentrations and the risk of severe hyperbilirubinemia. *AIDS* 21 (1): 41–46.

229. Chiou, W.J., De Morais, S.M., Kikuchi, R., Voorman, R.L., Li, X., and Bow, D.A. (2014). In vitro OATP1B1 and OATP1B3 inhibition is associated with observations of benign clinical unconjugated hyperbilirubinemia. *Xenobiotica* 44 (3): 276–282.

230. Chow, D., Kohorn, L., Souza, S., Ndhlovu, L., Ando, A., Kallianpur, K.J., Byron, M.M., Baumer, Y., Keating, S., and Shikuma, C. (2016). Atazanavir use and carotid intima media thickness progression in HIV: Potential influence of bilirubin. *AIDS* 30 (4): 672–674.

231. Vitek, L., Novotny, L., Sperl, M., Holaj, R., and Spacil, J. (2006). The inverse association of elevated serum bilirubin levels with subclinical carotid atherosclerosis. *Cerebrovasc. Dis.* 21 (5–6): 408–414.

232. Li, M., Chan, W.W., and Zucker, S.D. (2020). Association between atazanavir-induced hyperbilirubinemia and cardiovascular disease in patients infected with HIV. *J. Am. Heart Assoc.* 9 (19): e016310.

233. Dekker, D., Dorresteijn, M.J., Pijnenburg, M., Heemskerk, S., Rasing-Hoogveld, A., Burger, D.M., Wagener, F.A., and Smits, P. (2011). The bilirubin-increasing drug atazanavir improves endothelial function in patients with type 2 diabetes mellitus. *Arterioscler. Thromb. Vasc. Biol.* 31 (2): 458–463.

234. Suk, J., Jasprova, J., Biedermann, D., Petraskova, L., Valentova, K., Kren, V., Muchova, L., and Vitek, L. (2019). Isolated silymarin flavonoids increase systemic and hepatic bilirubin concentrations and

lower lipoperoxidation in mice. *Oxid. Med. Cell. Longev.* 2019: 6026902.

235. Calabrese, V., Cornelius, C., Mancuso, C., Barone, E., Calafato, S., Bates, T., Rizzarelli, E., and Kostova, A.T. (2009). Vitagenes, dietary antioxidants and neuroprotection in neurodegenerative diseases. *Front. Biosci.* 14: 376–397.

236. Qin, X.F. (2008). Bilirubin would be the indispensable component for some of the most important therapeutic effects of Calculus Bovis (Niuhuang). *Chin. Med. J. (Engl)* 121 (5): 480.

237. Yu, Z.J., Xu, Y., Peng, W., Liu, Y.J., Zhang, J.M., Li, J.S., Sun, T., and Wang, P. (2020). Calculus bovis: a review of the traditional usages, origin, chemistry, pharmacological activities and toxicology. *J. Ethnopharmacol.* 254: 112649.

238. Vankova, K., Markova, I., Jasprova, J., Dvorak, A., Subhanova, I., Zelenka, J., Novosadova, I., Rasl, J., Vomastek, T., Sobotka, R., Muchova, L., and Vitek, L. (2018). Chlorophyll-mediated changes in the redox status of pancreatic cancer cells are associated with its anticancer effects. *Oxid. Med. Cell. Longev.* 2018: 4069167.

239. Markova, I., Konickova, R., Vankova, K., Lenicek, M., Kolar, M., Strnad, H., Hradilova, M., Sachova, J., Rasl, J., Klimova, Z., Vomastek, T., Nemeckova, I., Nachtigal, P., and Vitek, L. (2020). Anti-angiogenic effects of the blue-green alga Arthrospira platensis on pancreatic cancer. *J. Cell. Mol. Med.* 24 (4): 2402–2415.

240. Lee, Y., Kim, H., Kang, S., Lee, J., Park, J., and Jon, S. (2016). Bilirubin nanoparticles as a nanomedicine for anti-inflammation therapy. *Angew. Chem. Int. Ed. Engl.* 55 (26): 7460–7463.

241. Keum, H., Kim, T.W., Kim, Y., Seo, C., Son, Y., Kim, J., Kim, D., Jung, W., Whang, C.H., and Jon, S. (2020). Bilirubin nanomedicine alleviates psoriatic skin inflammation by reducing oxidative stress and suppressing pathogenic signaling. *J. Control. Release* 325: 359–369.

242. Yao, Q., Jiang, X., Zhai, Y.Y., Luo, L.Z., Xu, H.L., Xiao, J., Kou, L., and Zhao, Y.Z. (2020). Protective effects and mechanisms of bilirubin nanomedicine against acute pancreatitis. *J. Control. Release* 322: 312–325.

243. Kim, M.J., Lee, Y., Jon, S., and Lee, D.Y. (2017). PEGylated bilirubin nanoparticle as an anti-oxidative and anti-inflammatory demulcent in pancreatic islet xenotransplantation. *Biomaterials* 133: 242–252.

244. Fullagar, B., Rao, W., Gilor, C., Xu, F., He, X., and Adin, C.A. (2017). Nano-encapsulation of bilirubin in pluronic F127-chitosan improves uptake in beta cells and increases islet viability and function after hypoxic stress. *Cell Transplant.* 26 (10): 1703–1715.

245. Lee, Y., Lee, S., and Jon, S. (2018). Biotinylated bilirubin nanoparticles as a tumor microenvironment-responsive drug delivery system for targeted cancer therapy. *Adv. Sci.* 1–8: 1800017.

246. Lee, Y., Lee, S., Lee, D.Y., Yu, B., Miao, W., and Jon, S. (2016). Multistimuli-responsive bilirubin nanoparticles for anticancer therapy. *Angew. Chem. Int. Ed. Engl.* 55 (36): 10676–10680.

247. Lee, S., Lee, Y., Kim, H., Lee, D.Y., and Jon, S. (2018). Bilirubin nanoparticle-assisted delivery of a small molecule-drug conjugate for targeted cancer therapy. *Biomacromolecules* 19 (6): 2270–2277.

248. Lee, D.Y., Kim, J.Y., Lee, Y., Lee, S., Miao, W., Kim, H.S., Min, J.J., and Jon, S. (2017). Black pigment gallstone-inspired platinum-chelated bilirubin nanoparticles for combined photoacoustic imaging and photothermal therapy of cancers. *Angew. Chem. Int. Ed. Engl.* 56 (44): 13684–13688.

249. Rathinaraj, P., Muthusamy, G., Prasad, N.R., Gunaseelan, S., Kim, B., and Zhu, S. (2020). Folate-gold-bilirubin nanoconjugate induces apoptotic death in multidrug-resistant oral carcinoma cells. *Eur. J. Drug Metab. Pharmacokinet.* 45 (2): 285–296.

250. Surendran, S.P., Thomas, R.G., Moon, M.J., Park, R., Lee, J.H., and Jeong, Y.Y. (2020). A bilirubin-conjugated chitosan nanotheranostics system as a platform for reactive oxygen species stimuli-responsive hepatic fibrosis therapy. *Acta Biomater.* 116: 356–367.

251. Kim, J.Y., Lee, D.Y., Kang, S., Miao, W., Kim, H., Lee, Y., and Jon, S. (2017). Bilirubin nanoparticle preconditioning protects against hepatic ischemia-reperfusion injury. *Biomaterials* 133: 1–10.

252. Zhou, Z.X., Chen, J.K., Hong, Y.Y., Zhou, R., Zhou, D.M., Sun, L.Y., Qin, W.L., and Wang, T.C. (2016). Relationship between the serum total bilirubin and inflammation in patients with psoriasis vulgaris. *J. Clin. Lab. Anal.* 30 (5): 768–775.

253. Chen, Z., Vong, C.T., Gao, C., Chen, S., Wu, X., Wang, S., and Wang, Y. (2020). Bilirubin nanomedicines for the treatment of reactive oxygen species (ROS)-mediated diseases. *Mol. Pharm.* 17 (7): 2260–2274.

254. Yao, Q., Chen, R., Ganapathy, V., and Kou, L. (2020). Therapeutic application and construction of bilirubin incorporated nanoparticles. *J. Control. Release* 328: 407–424.

255. Vitek, L. and Tiribelli, C.T. (2020). Bilirubin, intestinal integrity, the microbiome, and inflammation. *New Eng. J. Med.* 383 (7): 684–686.

256. Yao, Q., Jiang, X., Kou, L., Samuriwo, A.T., Xu, H.L., and Zhao, Y.Z. (2019). Pharmacological actions and therapeutic potentials of bilirubin in islet

transplantation for the treatment of diabetes. *Pharmacol. Res.* 145: 104256.

257. Bae, I.H., Park, D.S., Lee, S.Y., Jang, E.J., Shim, J.W., Lim, K.S., Park, J.K., Kim, J.H., Sim, D.S., and Jeong, M.H. (2018). Bilirubin coating attenuates the inflammatory response to everolimus-coated stents. *J. Biomed. Mater. Res. B Appl. Biomater.* 106 (4): 1486–1495.

258. Xing, R., Zou, Q., Yuan, C., Zhao, L., Chang, R., and Yan, X. (2019). Self-assembling endogenous biliverdin as a versatile near-infrared photothermal nanoagent for cancer theranostics. *Adv. Mater.* 31 (16): e1900822.

259. Gibbs, P.E., Miralem, T., Lerner-Marmarosh, N., and Maines, M.D. (2016). Nanoparticle delivered human biliverdin reductase-based peptide increases glucose uptake by activating IRK/Akt/GSK3 axis: The peptide is effective in the cell and wild-type and diabetic ob/ob mice. *J. Diabetes Res.* 2016: 4712053.

260. Lanone, S., Bloc, S., Foresti, R., Almolki, A., Taille, C., Callebert, J., Conti, M., Goven, D., Aubier, M., Dureuil, B., El Benna, J., Motterlini, R., and Boczkowski, J. (2005). Bilirubin decreases nos2 expression via inhibition of NAD(P)H oxidase: implications for protection against endotoxic shock in rats. *FASEB J.* 19 (13): 1890–1892.

261. Vianello, E., Zampieri, S., Marcuzzo, T., Tordini, F., Bottin, C., Dardis, A., Zanconati, F., Tiribelli, C., and Gazzin, S. (2018). Histone acetylation as a new mechanism for bilirubin-induced encephalopathy in the Gunn rat. *Sci. Rep.* 8 (1): 13690.

Section II.

Delivery Forms

11

Delivery Systems and Noncarrier Formulations

James Byrne[1,2,3,4,5], Christoph Steiger[3], Jakob Wollborn[6], and Giovanni Traverso[5,7]

[1]*Department of Biomedical Engineering, University of Iowa, Iowa City, IA 52240, USA*
[2]*Department of Radiation Oncology, University of Iowa, Iowa City, IA 52240, USA*
[3]*David H. Koch Institute for Integrative Cancer Research, Massachusetts Institute of Technology, Cambridge, MA 02142, USA*
[4]*Harvard Radiation Oncology Program, Brigham and Women's Hospital, Harvard Medical School, Boston, MA 02114, USA*
[5]*Division of Gastroenterology, Brigham and Women's Hospital, Harvard Medical School, Boston, MA 02115, USA*
[6]*Department of Anesthesiology, Perioperative and Pain Medicine, Brigham and Women's Hospital, Harvard Medical School, Boston, MA 02114, USA*
[7]*Department of Mechanical Engineering, Massachusetts Institute of Technology, Cambridge, MA 02139, USA*

Introduction

Delivery systems and noncarrier formulations of carbon monoxide (CO) are unique alternatives to current standard methods of CO delivery. These systems enable CO exposure without or with limited interference from the carrier molecule. In line with other CO-based technologies, these systems have been designed to mimic the local feedback loop of the heme oxygenase-1–carbon monoxide pathway [1].

Key advantages to a number of delivery systems and noncarrier formulations include the lack of use or limited release of transition metals systemically [2]. Transition metals are the major dose-limiting aspect to a number of CO-releasing molecules (CORMs) [3]. A number of novel technologies have been developed that take advantage of unique material properties and facilitate delivery through multiple avenues, including intravenous, intraperitoneal (IP), and gastrointestinal (GI) routes [4–6]. Herein, we describe these technologies and expand on the potential areas of work.

Oral CO Release Systems

Oral CO release systems (OCORS) were developed as oral formulations of CO. At the time of inception of the OCORS, inhaled CO gas, IP administration of a CORM or CO-rich solution, and bowel insufflation of CO gas were the primary methods of CO delivery for GI disorders. Due to these limited options, there was great interest in developing local delivery systems that could administer CO at the site of greatest need [1].

The OCORS were engineered as a tableted form of CO-releasing molecule-2 (CORM-2) that contained sodium sulfite as a trigger for release (Figure 11.1) [4]. The OCORS tablets were initially comprised of a semipermeable cellulose acetate shell, Eudragit® E PO-coated sodium sulfite crystals, and CORM-2.

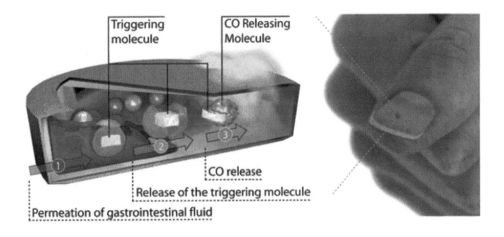

Figure 11.1 Oral CO-releasing system (depicted as a microscale system) was designed for controlled and oral delivery of CO within the GI tract. Controlled permeation of GI fluids through the system's coating triggers a cascade of reactions that ultimately result in CO release [7]. Source: Used with permission from Elsevier.

Carbon Monoxide in Drug Discovery: Basics, Pharmacology, and Therapeutic Potential, First Edition. Edited by Binghe Wang and Leo E. Otterbein.
© 2022 John Wiley & Sons, Inc. Published 2022 by John Wiley & Sons, Inc.

Controlled permeation of biorelevant media through the cellulose acetate shell enabled dissolution of the sodium sulfite crystals, which facilitated contact with CORM-2 and release of CO. The thickness of the cellulose acetate shell determined release rates of CO, as well as shelf life of the system. Citric acid was included in certain formulations to reduce the potential impact of the pH shift in the GI tract and enabled greater overall CO release.

Therapeutic levels of CO (up to 500 ppm) were released from the CORM-2 used in the OCORS. CO release from the OCORS was completed within 10 h, which is within the window for normal to delayed GI transit times [8]. Furthermore, it was determined that 90% of ruthenium, the primary functional transition metal in CORM-2, was released from the OCORS tablet within 13 h [4].

To evaluate the utility of the OCORS in a chemically induced colitis mouse model, a microscale OCORS (M-OCORS) was created [7]. The M-OCORS consisted of a similar composition to the original OCORS but was scaled down to a 1-mm tablet core. Different coatings around the tablet and sodium sulfite crystals were evaluated to slow the release of CO from the M-OCORS. Transitioning to cellulose acetate butyrate for the tablet and sodium sulfite coating prolonged near-maximal release from ~26 min to up ~17 h. Pharmacokinetic evaluation of the slowest release formulation of M-OCORS demonstrated peak carboxyhemoglobin (COHb) levels of 0.69% at 3 h and subsequent decline to basal levels of 0.27% after that. The formulation with the most prolonged CO release was further tested in a 2,4,6-trinitrobenzenesulfonic acid (TNBS)-induced colitis mouse model. Two M-OCORS tablets were administered 3 days prior and 2 days after TNBS exposure. Inactive M-OCORS tablets were used as controls. It was found that the active M-CORS tablets reduced TNF-alpha expression and myeloperoxidase (MPO) activity levels in colonic tissue [7].

The M-OCORS was further evaluated in a postoperative ileus mouse model. In 1 h prior to an operation, mice were administered twenty 0.5-mm M-OCORS tablets through oral gavage or 48 nmol of nitrite intravenously injected into the inferior vena cava. Postoperative ileus was caused by surgical manipulation of the small intestine using sterile moist cotton swabs. Intestinal transit was studied by using a liquid fluorescein-labeled dextran meal 1.5 h prior to euthanasia (24 h post-induction of ileus). The majority of the fluorescein-labeled dextran meal had progressed further into the small bowel compared to control animals that were not administered M-OCORS tablets. Tissues were also stained for MPO activity and IL-6 protein levels. There was no significant difference in MPO activity and IL-6 protein levels in animal treated with or without M-OCORS [9].

In an effort to replace ruthenium, an esterase-triggered OCORS was developed using iron-based rac-1. The release profiles were found to be comparable to the ruthenium-based OCORS [7].

Therapeutic Gas Releasing Systems

A major challenge in the field of transplant medicine has been the preservation of the organ to be transplanted. Many have investigated the use of gasotransmitters for organ preservation, including CO. However, all CO systems have involved CORMs. To overcome this potential limitation, therapeutic gas delivery systems (TGRS) have been developed for creation and release of therapeutic gases into solution for transplant perfusion [10].

The TGRS consisted of a stainless reaction chamber with one or two 0.04-in. polydimethylsiloxane membranes within the reaction chamber to enable delivery of therapeutic gases and restrict transport of nongaseous components into the perfusion solution (Figure 11.2). The gas releasing

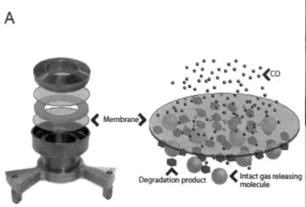

A

B

Figure 11.2 Therapeutic gas release system was designed to exclusively deliver CO gas by leveraging gas-permeable membranes (details in the text) [10]. Source: Used with permission from Elsevier.

molecule and trigger were contained on the contralateral side of the membrane of the perfusion fluid, and upon exposure of the gas releasing molecule and trigger to water, there was release of the gas into solution. The thickness of the membranes directly impacted the rate of gas release into solution. Release of CO and that of hydrogen sulfide were both evaluated using the TGRS. To evaluate the effect of the TGRS in a relevant liver reperfusion injury model, the hepatic artery and portal vein of rats were cannulated and the livers harvested. The livers that were perfused and stored in the TGRS-treated CO perfusion fluid (Custodiol) had lower ischemic injury, per cytoplasmic HMGB1, compared to CO-free fluid.

For translation of the TGRS, multiple all-in-one cylindrical designs were created for different healthcare settings, including single use in ambulances, small hospitals, and larger centers with existing transplant perfusion systems, as shown in Figure 11.3. Integration of CO detectors in the TGRS system was shown to provide excellent control of CO spikes in perfusion fluid [10].

Extracorporeal Delivery of CO

Extracorporeal circulation is an expanding concept for critical medical appliances (e.g., hemodialysis, extracorporeal membrane oxygenation, and cardiopulmonary bypass). By providing a circulating blood interface outside the body, gases can be directly applied to a circuit. For example, an oxygenator integrated into an extracorporeal circulation can facilitate oxygenation and decarboxylation of blood in the context of cardiopulmonary bypass. Using a similar technique as the TGRS, specific semipermeable membranes may be used to apply gases like CO. This concept is of particular interest when critical conditions are treated with extracorporeal systems (like acute respiratory distress syndrome), and the therapeutic aspects of a gas can be used to further treat the underlying pathology.

Herein, an extracorporeal delivery system was conceptualized – integrating a CO-permeable membrane into a perfusion circuit, which was capable of retaining the components of the releasing molecule (extracorporeal CO-releasing system, ECCORS) [11]. CO was released from the CORM and permeated through the membrane into the blood (see Figure 11.3). By using feedback loop algorithms, targeted CO administration to the body was feasible in a safe and controllable manner while retaining potentially toxic component of the CORM. To apply the concept of ECCORS into clinical context, a porcine model of ischemia–reperfusion and extracorporeal resuscitation strategy was used. CO was delivered using the CORM Beck-1. Targeting a systemic COHb concentration of 8–12%, beneficial effects were observed for myocardial performance [12], acute renal injury [13], and neurological damage after cardiac arrest [14] – emphasizing the potential for CO in the extracorporeal treatment of ischemia–reperfusion.

Electrospun Formulations of CO

The use of CORMs within biocompatible materials, including sutures and tissue engineering scaffolds, provides the potential to harness the benefits of CO

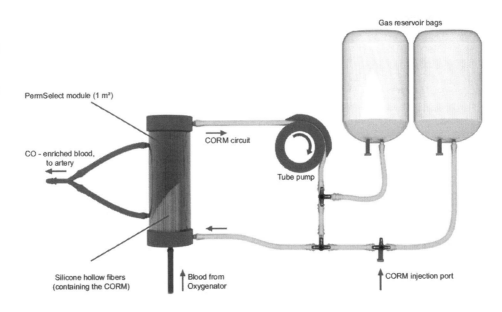

Figure 11.3 A membrane-based approach was also used for extracorporeal CO delivery. CO was generated from CORMs in a sealed circuit and delivered to a porcine model via gas-permeable tube membranes [11]. Source: Used with permission from Elsevier.

to improve wound healing and enhance cellular regeneration and proliferation [15]. The majority of materials for these applications are hydrophobic, and incorporation of a hydrophobic CORM would enable the retention of the residual metal fragments in the matrix after release of the CO. CORM-1, a highly hydrophobic CO-loaded dimanganese decacarbonyl compound, proved to be the ideal CORM for incorporation in biocompatible materials. Furthermore, CORM-1 releases CO upon exposure to light, and thus, light can be a trigger for release.

A CORM-1-containing polylactic acid (PLA) fiber was generated through electrospinning of a polymer solution onto a counter electrode (Figure 11.4). Fibers were created with prefiber polymer solutions containing up to 20 wt% of CORM-1. Inductively coupled plasma-mass spectrometry analysis determined that the highest CORM-1 concentration within these fibers was 14.8 wt%. Scanning electron microscopy (SEM) revealed that fiber size was independent of the CORM-1 wt%; however, pore size within the fiber was directly proportional to CORM-1 wt%. Additional materials analysis through SEM–energy-dispersive X-ray spectroscopy demonstrated the homogeneous distribution of CORM-1 within the material. Laser scanning microscopy enabled the evaluation of the CO release through gas bubble formation. Excellent control over CO release was demonstrated through discrete light pulses directed toward the fiber. Lastly, there was minimal leaching of CORM-1 into aqueous media over a period of 3 days.

Cytotoxicity experiments demonstrated that the CORM-1-entrapped PLA fibers were well tolerated by 3T3 mouse fibroblast cells after incubation up to 4 days under nonilluminated conditions. If the cells and fibers underwent illumination with a 365 nm laser, there was cytotoxicity noted. The pure CO gas was determined to be the cytotoxic component through a paired well setup where the cells and fibers were kept in different connected wells [15]. Repeat cytotoxicity in 3T3 cells demonstrated an increased number of dead cells up to 22% of cells in subsequent studies [16].

The electrospun formulation was subsequently tested for antimicrobial activity against *Staphylococcus aureus* [16]. The initial polymer solution contained 20 wt% CORM-1, and it was determined that 80 mm^2 pieces of the fiber released the entire CO content upon irradiation at 405 nm for 5 min in dry conditions, whereas under wet conditions, the release of the entire CO content took up to 60 min. The released CO accounted for 71.8–127.4 ppm in a 570-ml desiccator. Methicillin-resistant *S. aureus* was grown on top of these fibers and subsequently exposed to 405 nm radiation for 5 min. Twenty-four hours after irradiation, there was an approximate 80% reduction in viable bacteria, as confirmed by growth on agar plates. This effect was similar to exposure to daptomycin at 100 mg/l [16].

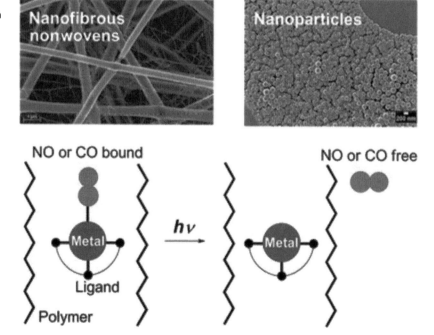

Figure 11.4 Light was used as a trigger to release CO from electrospun materials. While CO permeates through the material, degradation products of the CORM remain in the delivery system [15]. Source: Used with permission from The Royal Society of Chemistry.

CO-Enriched Solutions

Other noncarrier formulations that have been investigated rely on simple encapsulation strategies, such as CO-enriched solutions. The utility of CO-enriched solutions has been evaluated for use in postoperative ileus and protection against ischemia–reperfusion injury.

CO was bubbled into lactate ringers (LRs) for 5 min at room temperature and preserved in a liquid tight tube without a gas layer [5]. Up to 1200 μM CO was found to be contained within the solution. Intraperitoneal injection of 1.5 ml resulted in COHb levels up to 8% within 5 min after injection, which subsequently dropped down to baseline levels by 120 min. The CO-containing LR was subsequently tested in a postoperative ileus mouse model. An IP dose of the CO-containing LR was administered directly after intestinal manipulation. Intestinal transit, as determined by a fluorescent tracer, 24 h after manipulation was found to be similar between mice given the IP CO–LR and the sham-operated mice. In addition, reduced leukocytic infiltration and proinflammatory mediators (IL-1β, COX-2, iNOS, ICAM-1, and TLR-4) were seen in the muscularis propria of the small intestine of mice treated with the CO–LR compared to LR. It was believed that the postoperative ileus was prevented through a reduction in inflammation.

Preservation of orthotopic left rat lung for transplantation was evaluated in CO-containing organ preservation solutions. To generate the CO-containing solutions, CO gas was bubbled into University of Wisconsin (UW) solution at 4 °C for 5 min before use. CO concentrations of up to 1025 μM were achieved using 100% CO bubbled into solution. In a rat transplant model, the excised left lung was flushed with the CO-containing UW solution and stored in the CO-containing UW solution for 6 h at 4 °C. The organ was transplanted in recipient rats, and the function of the lung graft was assessed. Grafts preserved in at least 5% CO-containing UW solution resulted in improved lung function upon assessing P_{O_2} levels. Furthermore, inflammatory cell infiltration and levels of inflammatory mediators were found to be reduced in lung grafts that were exposed to 5% and 100% CO-containing UW solutions [17]. In addition, a 1% CO-containing UW solution was evaluated for preservation of a heart transplant in rats [18]. Five of six hearts that were preserved in the CO-containing solution for 24 h after removal remained functional upon transplantation, whereas all six hearts of untreated rats failed to function. The cytoprotective effects of CO were evident upon histological examination of the hearts, which showed extensive apoptosis in untreated hearts compared to the CO-treated hearts [18].

Furthermore, Hillhurst Biopharmaceuticals is developing CO-saturated solutions. There are multiple formulations in development, including oral (HBI-002) and intravenously administered (HBI-137) formulations. As compared to aqueous formulations, these lipid-based formulations can dissolve more CO and this can render them more relevant for clinical applications. For example, this could improve their shelf life, handling properties, and required application volume. HBI-002 has been evaluated in sickle cell disease and kidney ischemia–reperfusion injury models [6,19]. HBI-002 dosed at 10 mg/kg in NY1DD and Townes-SS transgenic mouse models of sickle cell disease model was found to have COHb levels of 5.4% and 4.7% within 5 min of administration. Townes-SS mice treated once daily for 10 doses were found to have an increase in hemoglobin from 5.3 to 6.3 g/dl and a reduction in white blood cell counts from 29.1×10^3 to $20.3 \times 10^3/\mu l$. In a kidney ischemia–reperfusion injury model, HBI-002 was dosed at 0.2 mg/kg prior to clamping of the renal pedicles. This pretreatment with HBI-002 significantly reduced the degree of kidney damage, as evaluated by blood urea nitrogen and creatinine 24 h after clamping.

Conclusions

There is a great opportunity to explore numerous noncarrier formulations that could provide benefit to patients. Less complex formulations may benefit patients by reducing the potential for adverse effects from carrier molecules, including drug–drug interaction [20].

Furthermore, opportunities to expand clinically relevant CO-releasing materials beyond injectable and inhaled dosage forms may increase the viability and applicability of CO therapeutics. The benefit from the noncarrier formulations described herein is a fraction of what is possible.

References

1. Steiger, C., Hermann, C., and Meinel, L. (2017). Localized delivery of carbon monoxide. *Eur. J. Pharm. Biopharm.* 118: 3–12.
2. Hopper, C.P., Meinel, L., Steiger, C., and Otterbein, L.E. (2018). Where is the clinical breakthrough of heme oxygenase-1/carbon monoxide therapeutics? *Curr. Pharm. Des.* 24 (20): 2264–2282.

3. Ismailova, A., Kuter, D., Bohle, D.S., and Butler, I.S. (2018). An overview of the potential therapeutic applications of CO-releasing molecules. *Bioinorg. Chem. Appl.* 2018: 8547364.
4. Steiger, C., Luhmann, T., and Meinel, L. (2014). Oral drug delivery of therapeutic gases - carbon monoxide release for gastrointestinal diseases. *J. Control. Release* 189: 46–53.
5. Nakao, A., Schmidt, J., Harada, T., Tsung, A., Stoffels, B., Cruz, R.J., Jr., Kohmoto, J., Peng, X., Tomiyama, K., Murase, N., Bauer, A.J., and Fink, M.P. (2006). A single intraperitoneal dose of carbon monoxide-saturated ringer's lactate solution ameliorates postoperative ileus in mice. *J. Pharmacol. Exp. Ther.* 319 (3): 1265–1275.
6. Correa-Costa, M., Gallo, D., Csizmadia, E., Gomperts, E., Lieberum, J.L., Hauser, C.J., Ji, X., Wang, B., Camara, N.O.S., Robson, S.C., and Otterbein, L.E. (2018). Carbon monoxide protects the kidney through the central circadian clock and CD39. *Proc. Natl. Acad. Sci. U. S. A.* 115 (10): E2302–E2310.
7. Steiger, C., Uchiyama, K., Takagi, T., Mizushima, K., Higashimura, Y., Gutmann, M., Hermann, C., Botov, S., Schmalz, H.G., Naito, Y., and Meinel, L. (2016). Prevention of colitis by controlled oral drug delivery of carbon monoxide. *J. Control. Release* 239: 128–136.
8. Degen, L.P. and Phillips, S.F. (1996). Variability of gastrointestinal transit in healthy women and men. *Gut.* 39 (2): 299–305.
9. Van Dingenen, J., Steiger, C., Zehe, M., Meinel, L., and Lefebvre, R.A. (2018). Investigation of orally delivered carbon monoxide for postoperative ileus. *Eur. J. Pharm. Biopharm.* 130: 306–313.
10. Steiger, C., Wollborn, J., Gutmann, M., Zehe, M., Wunder, C., and Meinel, L. (2015). Controlled therapeutic gas delivery systems for quality-improved transplants. *Eur. J. Pharm. Biopharm.* 97 (Pt A): 96–106.
11. Wollborn, J., Hermann, C., Goebel, U., Merget, B., Wunder, C., Maier, S., Schafer, T., Heuler, D., Muller-Buschbaum, K., Buerkle, H., Meinel, L., Schick, M.A., and Steiger, C. (2018). Overcoming safety challenges in CO therapy - Extracorporeal CO delivery under precise feedback control of systemic carboxyhemoglobin levels. *J. Control. Release* 279: 336–344.
12. Wollborn, J., Steiger, C., Ruetten, E., Benk, C., Kari, F.A., Wunder, C., Meinel, L., Buerkle, H., Schick, M.A., and Goebel, U. (2020). Carbon monoxide improves haemodynamics during extracorporeal resuscitation in pigs. *Cardiovasc. Res.* 116 (1): 158–170.
13. Wollborn, J., Schlueter, B., Steiger, C., Hermann, C., Wunder, C., Schmidt, J., Diel, P., Meinel, L., Buerkle, H., Goebel, U., and Schick, M.A. (2019). Extracorporeal resuscitation with carbon monoxide improves renal function by targeting inflammatory pathways in cardiac arrest in pigs. *Am. J. Physiol. Renal Physiol.* 317 (6): F1572–F1581.
14. Wollborn, J., Steiger, C., Doostkam, S., Schallner, N., Schroeter, N., Kari, F.A., Meinel, L., Buerkle, H., Schick, M.A., and Goebel, U. (2020). Carbon monoxide exerts functional neuroprotection after cardiac arrest using extracorporeal resuscitation in pigs. *Crit. Care Med.* 48 (4): e299–e307.
15. Bohlender, C., Glaser, S., Klein, M., Weisser, J., Thein, S., Neugebauer, U., Popp, J., Wyrwa, R., and Schiller, A. (2014). Light-triggered CO release from nanoporous non-wovens. *J. Mater. Chem. B* 2 (11): 1454–1463.
16. Klinger-Strobel, M., Glaser, S., Makarewicz, O., Wyrwa, R., Weisser, J., Pletz, M.W., and Schiller, A. (2016). Bactericidal effect of a photoresponsive carbon monoxide-releasing nonwoven against staphylococcus aureus biofilms. *Antimicrob. Agents Chemother.* 60 (7): 4037–4046.
17. Kohmoto, J., Nakao, A., Sugimoto, R., Wang, Y., Zhan, J., Ueda, H., and McCurry, K.R. (2008). Carbon monoxide-saturated preservation solution protects lung grafts from ischemia-reperfusion injury. *J. Thorac. Cardiovasc. Surg.* 136 (4): 1067–1075.
18. Akamatsu, Y., Haga, M., Tyagi, S., Yamashita, K., Graca-Souza, A.V., Ollinger, R., Czismadia, E., May, G.A., Ifedigbo, E., Otterbein, L.E., Bach, F.H., and Soares, M.P. (2004). Heme oxygenase-1-derived carbon monoxide protects hearts from transplant associated ischemia reperfusion injury. *FASEB J.* 18 (6): 771–772.
19. Belcher, J.D., Gomperts, E., Nguyen, J., Chen, C., Abdulla, F., Kiser, Z.M., Gallo, D., Levy, H., Otterbein, L.E., and Vercellotti, G.M. (2018). Oral carbon monoxide therapy in murine sickle cell disease: beneficial effects on vaso-occlusion, inflammation and anemia. *PLoS One* 13 (10): e0205194.
20. Reker, D., Shi, Y., Kirtane, A.R., Hess, K., Zhong, G.J., Crane, E., Lin, C.H., Langer, R., and Traverso, G. (2020). Machine learning uncovers food- and excipient-drug interactions. *Cell Rep.* 30 (11): 3710–3716 e4.

12

Metal-Based Carbon Monoxide-Releasing Molecules (CO-RMs) as Pharmacologically Active Therapeutics

Roberta Foresti, Djamal Eddine Benrahla, Shruti Mohan, and Roberto Motterlini

Mondor Institute for Biomedical Research (IMRB), Université Paris-Est Créteil, INSERM U955 F-94010 Créteil France

12.1 Introduction

In 2002, the first paper on the identification and characterization of carbon monoxide-releasing molecules (CO-RMs) was published by our group [1]. This study was the result of 4 years of work that started with the idea that a "solid form" of CO, easy to manipulate in the laboratory, would be very useful for exploring the physiological actions of CO in biological systems. The rationale was based on the emerging findings that CO produced by cells during the degradation of heme by heme oxygenase had important signaling and protective functions [2–4]. Our search for chemicals that could fulfill the role of CO-RMs led us to consider "transition metal carbonyls" as good candidates, also inspired by an article showing that photoexcitation of dimanganese decacarbonyl [$Mn_2(CO)_{10}$] in an alkane solution can result in dissociative loss of CO (see Scheme 12.1) [5,6].

$$Mn_2(CO)_{10} \xrightarrow{h\nu} Mn_2(CO)_9 + CO$$

Scheme 12.1 Release of CO from dimanganese decacarbonyl [$Mn_2(CO)_{10}$] following photoexcitation
Source: From [5].

The subsequent demonstration in our laboratory that [$Mn_2(CO)_{10}$] irradiated by light could also release CO to biological systems and exert *in vitro* and *in vivo* pharmacological effects typical of endogenously produced CO was the first step leading us to propose metal carbonyl complexes as a potential novel class of pharmaceuticals for the therapeutic delivery of CO [1]. We subsequently named this first CO releaser "CORM-1" [7]. This chapter summarizes our almost 20 years of research on CO-RMs, emphasizing the most salient aspects in the gradual characterization, development, and progression toward metal carbonyls that could be used to treat a variety of pathological conditions. Since the discovery of CO-RMs, approximately 600 articles have been published; due to space restriction, in this chapter we will highlight primarily studies using CO-RMs *in vivo* or *ex vivo* with less emphasis on publications in which data were obtained only in test tube experiments, using cells or with chemistry approaches.

12.2 Discovery of Transition Metal Carbonyls as Pharmacologically Active CORMs

Once we verified that CORM-1 was able to release CO upon light irradiation, we identified other metal carbonyls that possessed similar properties and developed specific assays for assessing their chemical reactivity in biological systems. Among them, a method measuring the conversion of reduced myoglobin to carbonmonoxy myoglobin (COMb) in solution in the presence of CO-RMs was optimized (Mb assay) alongside the use of an amperometric CO electrode for the detection of spontaneous CO release by metal carbonyls. With these methods, we then characterized tricarbonyldichlororuthenium(II) dimer ([$Ru(CO)_3Cl_2$]$_2$) as CORM-2 [1], tricarbonylchloro(glycinato)ruthenium(II) ([$Ru(CO)_3Cl$(glycinate)]) as CORM-3 [8], and sodium boranocarbonate ([$Na_2H_3BCO_2$]) as CORM-A1 [9]. Although CORM-A1 does not contain a transition metal, the characterization of its biological effects was instrumental in our studies on CO and therefore will be included in this chapter. The chemical structures of these four CO-RMs are reported in Figure 12.1 and their effects *in vitro* and in animal models of disease are summarized in the following sections.

Carbon Monoxide in Drug Discovery: Basics, Pharmacology, and Therapeutic Potential, First Edition. Edited by Binghe Wang and Leo E. Otterbein.
© 2022 John Wiley & Sons, Inc. Published 2022 by John Wiley & Sons, Inc.

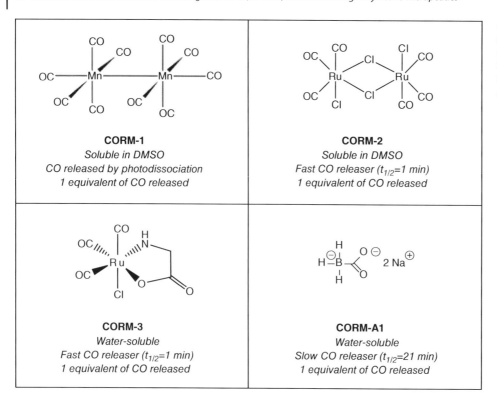

Figure 12.1 Chemical structures and properties of CORM-1, CORM-2, CORM-3, and CORM-A1.
Source: Modified from [7].

12.2.1 CORM-1 and CORM-2

In our first experiments, we demonstrated the capacity of CORM-1 to liberate CO in a concentration-dependent manner once stimulated by light [1,10]. This light stimulation step was instrumental to prove the pharmacological effects of CO released by the compound in isolated perfused rat hearts. Hearts were first infused with a nitric oxide synthase inhibitor (L-NAME) to increase coronary perfusion pressure and then CORM-1 was added to the perfusion buffer. It was found that perfusion of hearts in the presence of CORM-1 significantly reduced L-NAME-mediated vasoconstriction but only when illuminated by light, whereas no effect was observed when experiments were performed in the dark [1]. The vasodilatory properties of CORM-1 were then confirmed by others in isolated pressurized cerebral arterioles of newborn pigs [11] and in rat internal anal sphincter smooth muscle strips [12]. Thus, CO liberated by CORM-1 has the ability to exert relaxation in different tissue types. Despite these promising results, the fact that light is required to trigger the release of CO from CO-RMs represents a limitation in pharmacological studies (although it can also be seen as an advantage for certain applications).

We subsequently identified metal carbonyl complexes such as CORM-2 that could liberate CO to myoglobin in solution without the need for photoexcitation and that nevertheless exerted vasodilation and antihypertensive effects *ex vivo* and *in vivo* [1]. The absence of an activation step (light) and the "spontaneous" release of CO by CORM-2 clearly facilitated the use of this ruthenium carbonyl complex for *in vivo* studies. Accordingly, CORM-2 has been the mostly used compound in all articles published so far on metal-based CO-RMs. From these studies, it emerged that CORM-2 possesses two important properties: it protects against tissue damage and attenuates the inflammatory response in a variety of disease models. For instance, administration of CORM-2 to mice subjected to ischemia-induced acute renal failure significantly reduced plasma creatinine, a marker of renal injury [13]. In another study using mice lacking the CO-producing enzyme heme oxygenase-1 (HO-1), treatment with CORM-2 reduced arterial thrombosis and increased survival after abdominal aortic transplantation [14]. This was an important finding as it demonstrated that exogenous CO can compensate for the lack of endogenous CO production in protecting against aortic graft rejection. In a model of pancreatic tumor, mice treated with CORM-2 or exposed to CO gas exhibited decreased tumor proliferation and survived for longer periods of time [15]. Among other interesting studies, CORM-2 was shown to inhibit bacterial infection *in vivo* by prolonging the survival rate of mice infected with *Pseudomonas aeruginosa* [16]. The bactericidal action of CORM-2 appeared to involve inhibition of bacterial respiration by CO. In another article on microbial infection, injection of CORM-2 into wild-type mice increased phagocytosis and rescued HO-1-deficient mice from sepsis-induced lethality [17].

Concerning the effects of CORM-2 in inflammation, the first reports indicating an anti-inflammatory action of CO liberated by CORM-2 were described *in vitro* [18,19] but were soon followed by interesting *in vivo* studies. Sun and colleagues demonstrated in a model of full-thickness thermal injury, which causes a strong systemic inflammatory response, that this compound markedly diminished leukocytes' infiltration and the production of inflammatory markers in the lung and intestine [20,21]. Interestingly, CORM-2 was also effective in accelerating the healing of pre-existing gastric ulcers in rats in association with reducing the expression of inducible nitric oxide synthase, interleukin-1β, and tumor necrosis factor-α (TNF-α) [22]. In a mouse model of acute pancreatitis, treatment with CORM-2 reduced mortality, pancreatic damage, and lung injury. In addition, CORM-2 decreased systemic inflammatory cytokines, suppressed systemic and pancreatic macrophage TNF-α secretion, and inhibited macrophage TLR4 receptor complex expression [23]. Finally, it has also been shown that CORM-2 protects the skin against the suppression of the immune response in mice exposed to UVA irradiation [24]. The interesting aspect of this study is that CORM-2 was applied topically using a skin lotion, demonstrating the localized action of CO in addition to systemic effects and indicating that metal carbonyls can be adapted to different formulations for therapeutic delivery. Examples of this possibility are recent works showing that CORM-2 incorporated in nanoemulsions functionalized with folic acid to target lymphomas in mice exerted better antitumor effects and increased survival rates compared to free CORM-2 [25]; similarly, water-soluble styrene–maleic acid copolymers encapsulated with CORM-2 elicited a higher tissue CO content and prolonged half-life in circulation compared to free CORM-2, significantly improving symptoms of colitis and the inflammatory profile after *in vivo* challenge with dextran sulfate sodium in mice [26]. Because CORM-2 fully dissolves in dimethyl sulfoxide (DMSO) but is poorly soluble in water, the findings highlighted earlier suggest that increasing the solubility of metal carbonyl complexes in aqueous solutions may enhance their compatibility and efficacy in *in vivo* models. Indeed, this aspect was the main reason for the design and synthesis of CORM-3, the first water-soluble metal carbonyl developed by our group. Another important issue was that water solubility could decrease the potential toxicity of metal carbonyls in biological systems and, although we will not discuss this in this chapter, we proved over the years that water-soluble CO-RMs are in general less toxic in cells than compounds soluble in DMSO [27].

12.2.2 CORM-3

The synthesis of CORM-3 ([$Ru(CO)_3Cl(glycinate)$]), conducted in collaboration with Brian Mann (University to Sheffield, UK), consisted of coordinating a glycine residue to the ruthenium metal, and our first set of experiments to characterize its biochemical and biological properties revealed the following: (i) CORM-3 is stable in water and starts to liberate CO once dissolved in buffer solutions (i.e., phosphate buffer at pH 7.4) or cell culture media; (ii) it liberates one equivalent of CO very rapidly to myoglobin in solution (half-life ≈ 1 min); (iii) it decreases cellular damage in both cardiomyocytes and isolated hearts subjected to hypoxia or ischemia–reperfusion injury; and (iv) when administered *in vivo*, CORM-3 significantly prolonged heart graft survival in a model of cardiac allograft rejection, whereas an inactive CORM-3 (iCORM-3) depleted of CO was without effect [8,28]. Subsequent studies from our group and others confirmed that the effects mediated by CORM-3 and its mechanism(s) of action in biological preparations were due to CO. For instance, relaxation of isolated aortas in the presence of CORM-3 is abolished when reduced myoglobin, a known scavenger of CO, is added to the buffer prior to addition of CORM-3 [29,30]. Moreover, CORM-3-mediated vasodilation involved several mechanisms, including activation of soluble guanylate cyclase to produce the second messenger cGMP [29], activation of large conductance potassium channels present in the smooth muscle [30–32], and inhibition of sodium channels in a model of alveolar fluid clearance in the lung [33]. Thus, these initial investigations emphasize that the actions mediated by CO encompass different intracellular pathways that can be simultaneously targeted.

CORM-3 has been subsequently studied by different laboratories revealing its pleiotropic actions and the ability to counteract various pathological conditions. These include, among others, protection against organ damage induced by ischemic insults [34–36], bactericidal activity [37], and attenuation of the inflammatory response in the septic heart [38], the ischemic kidney [35], the intestine following postoperative ileus [39], the arthritic joints [40], and the brain after hemorrhagic stroke [41]. Of note, an interesting study revealed that administration of 30 mg/kg CORM-3 given intraperitoneally to mice daily for 24 days after myocardial infarction alleviated left ventricular remodeling, improved systolic function, and reduced ventricular hypertrophy [42]. In this study, it was also found that treatment with CORM-3 resulted in reduced cardiac tissue apoptosis by activating Bcl-2, a key antiapoptotic protein that regulates cell death. Another report showed that treatment with CORM-3 ameliorated the progression of neurological deficits in a mouse model of traumatic brain injury [43]. The authors found that CO exerted its effects by reducing pericyte death and stimulating an increase in neural stem cells in the injured mice leading to the recovery of a range of behavioral and motor tasks. A cross-talk between pericytes and neural stem cells elicited by CO, potentially stimulating neurogenesis, was the mechanism proposed from *in vitro* studies. Altogether, these positive results were very encouraging and substantiated the usefulness of metal carbonyls for the therapeutic delivery of CO. Our group also continued to explore other avenues to identify and diversify our portfolio of CO-RMs to understand their chemical reactivity in biology and improve their efficacy *in vivo* [44–46].

12.2.3 CORM-A1

CORM-A1 ($[Na_2H_3BCO_2]$) was the next in the series of compounds we characterized, being the first CORM of organic origin since it consists of a carboxylic acid containing the nonmetal boron [9]. As in the case of metal carbonyls, we came across this compound in the literature, from an article published by Roger Alberto (University of Zurich, Switzerland) who utilized $[Na_2H_3BCO_2]$ for a pure chemical application [47]. In his study, Alberto needed to introduce CO into technetium-99m (i.e., carbonylation reaction) for the synthesis of radiopharmaceuticals, and although this reaction can be achieved using CO gas, the challenge was to find a solid, air-stable, and more practical source of CO. Alberto's group successfully demonstrated that indeed $[Na_2H_3BCO_2]$ carbonylates technetium-99m, suggesting to us that this compound could be a novel CO-RM to be investigated in biological models. Indeed, we showed that CORM-A1 spontaneously liberates CO in physiological buffers (pH 7.4) with a half-life of approximately 20 min at 37 °C and this kinetic was reflected by gradual vasodilatory and hypotensive effects over time when the compound was applied *ex vivo* and *in vivo* [9]. In addition, we found that one equivalent of CO is released from CORM-A1 and that the rate is pH dependent, being faster at acidic pH. Importantly, also this compound elicited a variety of beneficial effects *in vivo*. In a mouse model of multiple sclerosis, CORM-A1 improved the clinical and histopathological signs of the disease in association with reduced inflammatory infiltrations of the spinal cords [48]. CORM-A1 also inhibited platelet aggregation and attenuated thrombosis in rats [49], and prevented hepatic damage and steatosis in mice fed a high fat-high fructose diet by improving lipid homeostasis and reducing oxidative stress [50]. Still in relation to metabolism, an important study investigated whether the effects of chronic, low-level treatment with CORM-A1 can reverse established obesity in mice fed a high-fat diet (HFD). It was found that CORM-A1 lowered fasting blood glucose and ameliorated insulin sensitivity while decreasing hepatic steatosis [51]. These results also highlighted that the effects observed are due to CO liberated by CORM-A1 since its inactive counterpart (iCORM-A1) did not have any effect. Notably, this was the first study in which CORM-A1 was administered intraperitoneally and daily for 30 weeks, indicating that the compound is well tolerated by animals during prolonged treatments. In the following sections, we will describe a new metal carbonyl that can be administered repeatedly to animals by oral route to reduce metabolic dysfunction caused by obesity and exerts effects similar to CORM-A1.

12.3 CORM-401, a Manganese-Based Carbonyl for the Therapeutic Delivery of CO with High Efficiency In Vitro and In Vivo

The initial metal carbonyls we studied (CORM-2 and CORM-3) worked very well in terms of both CO release and biological activities mediated by CO. However, they both contain Ru, a metal not found naturally in mammalian systems. This does not necessarily mean that a Ru-based CORM cannot be utilized for therapeutic purposes, especially for short periods of time, since some Ru complexes have been investigated for their anticancer properties and entered clinical trials [52]. Nevertheless, we considered Fe and Mn as alternative metals because there are a large number of mammalian proteins containing these two transition metals, which are also essential dietary elements for mammals [53,54]. Our attempts to work with Fe carbonyls were only partially successful: the compounds proved difficult to solubilize in aqueous buffers and, despite showing CO-dependent pharmacological effects when dissolved in organic solvents, most of them were unstable and precipitated out of solutions [27,55]. Instead, our efforts to develop Mn-based CO-RMs were more promising: not only we synthesized a set of compounds very stable in physiological solutions, but one of them liberated between three and four equivalents of CO per mole of compound, unlike all the metal carbonyls tested previously (>300) that released no more than 1 CO per mole. We named this molecule CORM-401 [56,57].

12.3.1 CORM-401: Mechanism(s) and Efficiency of CO Delivery

CORM-401 ($[Mn(CO)_4\{S_2CNMe(CH_2CO_2H)\}]$) is soluble and stable in Dulbecco phosphate buffer solutions (up to 5 mg/ml) and can be stored for long periods of time (several months) at −20 °C without any decomposition. As shown in the chemical structure (Figure 12.2A), CORM-401 contains four carbonyl groups coordinated to the metal. The mechanism of CO loss of the first carbonyl is dissociative and reversible. This reversible binding can be prevented by the presence of a CO acceptor (i.e., Mb or hemoglobin [Hb]) or a ligand leading to a rapid loss of CO [56]. The residual carbonyls can be released if an excess of the acceptor is present [57]. Moreover, in the presence of oxidants such as hydrogen peroxide (H_2O_2), the Mn is rapidly oxidized, thus favoring the loss of most of the CO groups from CORM-401, and leading us to use this approach to generate a type of inactive CORM-401 to be tested alongside the active form. The multiple release of carbonyls from CORM-401 observed in the cuvette using the Mb assay is reflected in a higher CO accumulation in cultured cells when compared to CORM-3 and CORM-A1, as detected by the CO-sensitive fluorescence probe COP-1 (see Figure 12.2B) [57].

As for other CO-RMs, we still do not know whether the CO is released from the metal complex prior to entering the cell, whether the metal complex is taken up by cells and CO is then released intracellularly, or whether both mechanisms occur at the same time. Nevertheless, the properties of CORM-401 indicate that less amounts of the metal carbonyl complex can

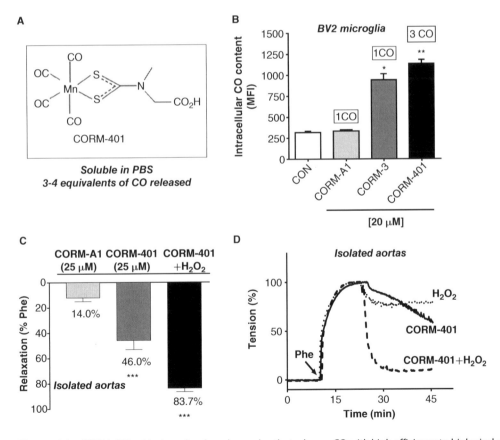

Figure 12.2 CORM-401, a Mn-based carbonyl complex that releases CO with high efficiency to biological systems. (A) Chemical structure of CORM-401. (B) CO accumulation in BV2 microglia cells incubated with 20 μM CORM-A1, CORM-3, or CORM-401 assessed by the CO-sensitive fluorescent probe COP-1. (C) Extent of vasodilation in isolated aortic rings treated with 25 μM CORM-A1, CORM-401, or a combination of CORM-401 + H_2O_2. (D) Representative traces of vascular tension in isolated aortas precontracted with phenylephrine (Phe) and then treated with 25 μM H_2O_2, CORM-401, or a combination of CORM-401 + H_2O_2. Note the synergistic vasodilatory effect of CORM-401 when CO release from this compound is accelerated by the presence of the oxidant. *Source:* Adapted from [57].

be used to achieve doses of CO (i.e., the active principle) that elicit important pharmacological effects, an important aspect in drug development. This concept is exemplified by our data in endothelial cells showing that, compared to CORM-A1, much less CORM-401 is required to promote migration and stimulation of proangiogenic factors. Similarly, vasorelaxation of precontracted rat isolated aortas observed in the presence of 25 µM CORM-401 is three times more pronounced than that exerted by 25 µM CORM-A1 (Figure 12.2C). In line with the observation that H_2O_2 stimulates the loss of CO, the extent and kinetics of vasorelaxation over time are markedly amplified when CORM-401 is applied simultaneously with H_2O_2 (Figure 12.2D). These data highlight our progression in the optimization of compounds for therapeutic delivery of CO; our next step was to assess the effects of CORM-401 in vivo.

12.3.2 CORM-401 in Models of Metabolic Dysfunction and Inflammation

Our laboratory has been investigating a role of CO in the modulation of energetic metabolism ever since our first observation in isolated renal mitochondria that different CO-RMs increase O_2 consumption by an uncoupling mechanism [58]. We confirmed this phenomenon in cells in culture showing that low micromolar concentrations of CORM-401 stimulate mitochondrial respiration in endothelial cells, microglia, and adipocytes (Figure 12.3A) [59–61]. The uncoupling phenomenon is mostly due to a leak of protons across the mitochondrial inner membrane, which leads to dissipation of energy by mitochondria to produce heat at the cost of ATP production. Decreasing mitochondrial efficiency by enhancing proton leak with uncoupling agents is an effective strategy to promote weight loss in humans, with 2,4-dinitrophenol being one of the most common uncouplers used despite its reported side effects [62]. Therefore, we reasoned that the uncoupling properties of CORM-401 could be exploited to reduce weight gain during obesity. In 2014, the group of David Stec had already reported the beneficial effects of CORM-A1 in a model of HFD-induced obesity in mice showing that daily intraperitoneal administration of this compound for 24 weeks led to a decrease in body weight gain while lowering plasma glucose and insulin levels [63].

Using a similar model, we treated mice with CORM-401 by oral gavage three times a week for 14 weeks during an HFD regime and found that (i) the levels of blood carboxyhemoglobin (COHb) and CO accumulation in adipose tissue markedly

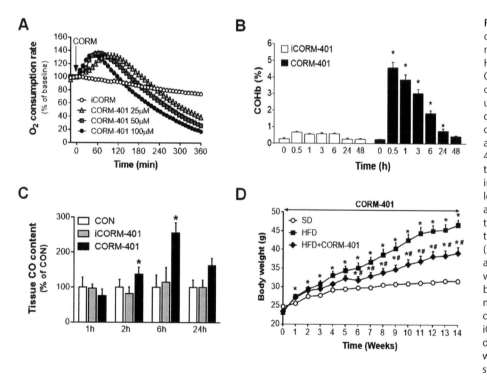

Figure 12.3 CORM-401 delivers CO and ameliorates metabolic dysfunction in HFD-induced obesity. (A) CORM-401 increases O_2 consumption rate via an uncoupling mechanism in a dose-dependent manner in cultured adipocytes. Oral administration of CORM-401 (30 mg/kg) results in a transient and marked increase in blood COHb levels (B) and CO accumulation in adipose tissue (C) in mice. (D) Mice treated with CORM-401 (30 mg/kg) orally administered three times a week for 14 weeks reduced body weight gain in a model of HFD-induced obesity. Note that iCORM-401, which is depleted of CO, was without effect. Mice fed a standard diet (SD) represented the control group.
Source: Adapted from [61].

increased with peaks at 1 and 6 h, respectively, demonstrating that CO can be delivered *in vivo* through CO-RMs also by oral administration (Figure 12.3B and C); (ii) CORM-401 significantly reduced body weight gain and improved glucose homeostasis (Figure 12.3D); (iii) CORM-401, but not its inactive counterpart (iCORM-401), increases adipocyte insulin sensitivity *in vitro* and *in vivo*; and (iv) the CO-mediated effect was accompanied by increased Akt phosphorylation in adipose tissue and a concomitant switch toward glycolysis [61]. Confirming the anti-inflammatory effects of CO, we also found that CORM-401 reduced HFD-induced macrophage infiltration in adipose tissue as well as circulating levels of the proinflammatory molecules IL-6 and IL-1β [61].

This study and the previous work by Stec and colleagues not only demonstrate that CORM-A1 and CORM-401 are very effective in counteracting the metabolic dysfunction caused by obesity but also reveal for the first time that these two compounds can be administered to mice for long periods of time without any major side effects.

The pharmacological properties of CORM-401 have been confirmed in other models of disease. For instance, a reduced production of proinflammatory markers by CORM-401 has been reported in different organs, including heart, liver, lung, and brain, of mice challenged with lipopolysaccharide (LPS) [64]. More recently, by using a peptide dendrimer nanogel (PDN) as a carrier, CORM-401 was integrated into a multifunctional anti-inflammatory drug to specifically target macrophages for the treatment of osteoarthritis [65]. The rationale behind this novel technology is that the cell membrane of activated macrophages overexpresses the folate receptor beta and hyaluronic acid receptor and, by simultaneously binding their corresponding ligands onto the surface of PDN, this technology can be used as a multifunctional delivery system that specifically targets activated macrophages. CORM-401 was finally encapsulated into the functionalized PDN that was injected into the joints of osteoarthritic mice to evaluate its efficacy. The findings showed that CORM-401 treatment greatly depletes the production of reactive oxygen species in the joints and effectively suppresses the degradation of articular cartilage and their extracellular matrix while remarkably reducing the release of the proinflammatory mediators TNF-α, IL-16, and IL-1β. Altogether, these studies reveal that the water-soluble CORM-401 delivers CO with high efficiency *in vivo* and can be used alone or encapsulated into delivery systems to counteract metabolic and inflammatory disorders.

12.4 COHb, a Natural Metal Carbonyl that Delivers CO?

Although metal carbonyls have been studied for over 140 years in the field of organometallic chemistry and industrial catalysis [66], we would like to highlight at this point that the best example of a naturally occurring metal carbonyl is found in biology. In fact, the heme (Fe^{2+})–CO complex formed when CO binds to Hb, known as COHb, is essentially an iron carbonyl [67]. This notion has extremely important implications not only to better understand the true nature of CO toxicity but also to appreciate the impact of using metal-based CO-RMs for the safe and efficacious delivery of CO. It is well known that the affinity of Hb for CO is much higher than the affinity for O_2, explaining why the formation of COHb can compromise O_2 delivery to tissues [68]. The level of COHb in blood is indeed an established parameter to assess CO poisoning in human subjects who have been exposed to high levels of CO gas. In general, while the basal levels of COHb in healthy individuals are less than 1% and no symptoms are usually manifested in subjects with values of 10–15% COHb, higher levels are often associated with increased nausea, dizziness, and difficulties in breathing and can be lethal when reaching 50–60% [69]. This has been documented in humans and confirmed in dogs, showing that when CO gas was administered by inhalation in increasing amounts, these animals died rapidly when COHb reached 60% and above. However, if dogs were transfused with blood previously saturated with CO gas, they survived indefinitely despite exhibiting 60% COHb levels [70]. These seemingly paradoxical findings indicate that (i) the percentage of COHb in blood is not the best index to determine CO toxicity; (ii) CO administered *in vivo* as "free gas" is far more toxic than when it is "bound to a transition metal" as in COHb; and (iii) the toxicity of free CO gas is most likely linked to CO accumulation in cells and tissues and to inhibition of mitochondrial respiration. Thus, these results open a key question on the physiological role of Hb as a potential sink of CO. In this regard, a recent article submitted by the group of Kitagishi in collaboration with our laboratory demonstrates that CO accumulation in tissues in rats exposed to high levels of CO is prevented by circulating Hb, revealing a protective role of Hb in CO intoxication (Q. Mao et al., paper under revision). This detailed investigation was possible thanks to the use of hemoCD1, a cyclodextrin–porphyrin complex with extremely high affinity for CO that can be used as a CO removal agent and to quantify reliably CO in tissues *in vivo* [71].

Whether COHb can efficiently deliver CO to tissues to exert CO-mediated pharmacological effects is also an interesting possibility. Hb-based O_2 carriers have been investigated for several years as blood substitutes for the treatment of traumatic emergencies or surgical interventions [72]. More recently, COHb formulations have been introduced in this class of therapeutics to take advantage of the vasodilatory and anti-inflammatory properties of CO. This strategy ultimately renders the COHb complex more stable since Hb outside red blood cells is highly susceptible to autoxidation and binding to CO strongly prevents this effect [73]. Few examples from the literature are worth mentioning. CO-MP4, a polyethylene glycol-conjugated Hb derivative carrying CO, was shown to be very stable over 30 days of storage at 37 °C, whereas oxy-MP4 rapidly oxidized to met-Hb at a rate of 29% per day. In addition, CO-MP4 transferred CO to Hb within red blood cells indicating the possibility that this complex may be able to deliver CO into the circulation. CO-MP4 also significantly reduced infarct size in a model of myocardial infarction in rats [74]. Using another pegylated COHb, Nugent and colleagues reported beneficial effects in preserving microvascular and systemic hemodynamic parameters after hemorrhagic shock as well as in a model of hypoxia-induced vaso-occlusion typical of sickle cell disease [75,76]. Finally, COHb has been encapsulated in phospholipid vesicles (CO-HbV) to further improve the stability and mimic the red blood cell structure. This formulation rapidly liberated CO in rats as the levels of CO in exhaled breath immediately increased after injection, and similarly to the study described earlier, protected against inflammation and oxidative damage of different organs in rat models of acute pancreatitis and hemorrhagic shock [77,78]. However, from the data presented in all these reports, it is difficult to conclude whether the protective effects of COHb formulations are due to CO liberated by the compounds or essentially due to their increased stability as O_2 carriers once the CO has been released. Ad hoc experiments to assess whether CO is accumulated in different tissues following treatment with the different COHb carriers described would help to better understand the mechanism(s) of actions of these compounds.

12.5 Chemical Reactivity and Biological Activity of Other Metal-Based CO-RMs

In addition to the CO-RMs described earlier, several other carbonyl compounds containing iron, manganese, ruthenium, molybdenum, rhenium, and tungsten have been reported in the literature; however, to the best of our knowledge, the majority of these studies have been only conducted in solutions or in cultured cells *in vitro*. Readers should refer to comprehensive reviews to learn more about the synthesis of such metal carbonyls, their chemical reactivity in solutions, and results obtained *in vitro* [79–82]. Only a few published articles have reported on the potential therapeutic properties of these additional CO-RMs in *in vivo* models. Studies conducted in mice infected with a malaria parasite and treated with ALF-492, a Ru-containing carbonyl complex, revealed that the compound delivered CO to tissues, reduced parasite accumulation and neuroinflammation in the brain, and prolonged survival [83]. It has also been reported that ALF-794, a molybdenum-based CO-RM, protects liver against acetaminophen-induced injury in mice resulting in a dose-dependent decrease in the levels of serum alanine aminotransferase, a standard marker of hepatic damage [84]. Another molybdenum carbonyl complex (ALF-186) was intravitreally applied into the eyes of rats directly after retinal ischemia–reperfusion injury and the effect was compared to its inactive counterpart. While the ischemic event led to a significant loss in retinal ganglion cells, eyes treated with ALF-186 were markedly protected from ganglion cell damage, apoptosis, and inflammation [85]. Parallel experiments *in vitro* indicated that ALF-186 may stimulate axon regeneration although this effect needs to be further investigated. Although not studied *in vivo*, metal-based "photoCO-RMs" have also attracted interest since they can be stimulated by light to release CO [86]. This is considered by some as a useful property to control the delivery of CO for therapeutic applications [79]. In this regard, we would like to point out that metal carbonyls in general are all sensitive to light and therefore all metal-based CO-RMs are essentially "photoCO-RMs." Second, the need of photoactivation may limit the application of these compounds and certainly more studies are needed to validate the feasibility and advantage of photoCO-RMs compared to CO-RMs that spontaneously liberate CO *in vivo* after interaction with biological components.

In summary, irrespective of the transition metal present in the numerous carbonyl complexes tested as CO-RMs, the majority of them show the ability to liberate CO at least in the cuvette using the Mb assay. In addition, for metal carbonyls that have been investigated *in vivo*, the delivery of CO is associated with salutary actions in a variety of pathological conditions even when the compounds are administered to animals repeatedly for several months as in the case of CORM-401 [61]. One issue still debated is whether the transition metal contained in CO-RMs exerts side effects or has potential harmful consequences; we will touch on this aspect in the following section.

12.6 Metal-Based CORMs: Concepts and Misconceptions

It is undisputable that transition metals are ideal carriers of CO, as also emphasized earlier with the example of COHb, a naturally occurring iron carbonyl in mammalian systems. Although this property has been exploited to deliver CO using metal-based CO-RMs in a variety of preclinical experimental models, the potential toxicity of the metal and the effectiveness of CO release by these compounds have been questioned concerning their applicability as therapeutics in the clinic.

12.6.1 Transition Metals: Perceived or Real Toxicity?

From a mere scientific perspective, we should not necessarily assume that the presence of a transition metal within a CO-RM is synonymous of "toxic effects" until the appropriate toxicological and pharmacokinetic studies (absorption, distribution, metabolism, and excretion [ADME]) have been carried out. As for all other substances, many factors, including dose and time of exposure to transition metal complexes, will determine their toxicity in tissues and organs once administered to mammalian organisms. To give an example, we consider the case of Mn, an essential element found in tissues of mammals [53]. Mn is involved in many biochemical functions such as nutrients' metabolism, immune function, growth, and development. Mn is also a crucial cofactor for several proteins, including the antioxidant enzyme, Mn superoxide dismutase, and pyruvate carboxylase, transferases, hydrolases, and kinases. If inadequate Mn intake is associated with adverse health effects such as diabetes and metabolic syndrome, excessive exposure to this element causes manganism, a condition characterized by neurological symptoms resembling those of Parkinson's disease [87]. If we now consider the use of a Mn-based CO-RM for clinical applications, we are not aware of any ADME study that has been performed so far; therefore, we are unable at present to determine whether a pharmacologically relevant dose of CO-RM with a specific administration protocol would or would not be associated with a toxic effect. Nevertheless, and to the best of our knowledge, two interesting reports are available on the toxicity of Ru, molybdenum, chromium, and tungsten carbonyl complexes [84,88]. In these studies, it was found that following acute oral administration in mice, LD_{50} (the dose causing a 50% mortality) was 800–1000 mg/kg for CORM-3, 300–500 mg/kg for a Mo-CORM, 150–200 mg/kg for a chromium-CORM, and 75–125 mg/kg for a tungsten-CORM, indicating that CORM-3 was the less toxic among the compounds examined [88]. In addition, the LD_{50} of a series of molybdenum-based CO-RMs (ALF-782, ALF-785, and ALF-795) was between 100 and 450 mg/kg [84]. We note that CORM-3 and molybdenum-based CO-RMs have been shown to exert pharmacological effects *in vivo* at doses ranging from 3 to 60 mg/kg. Thus, although more robust studies are required to fully assess the toxicity profile of metal-based CO-RMs, these preliminary observations suggest that the therapeutic window for this class of compounds is significantly below the doses causing toxic effects. We should also consider that the toxicity caused by metal carbonyls at high doses reported above may derive from the large quantities of CO liberated *in vivo*. This possibility has never been investigated but may prove to be an important aspect for assessing the toxicity of metal carbonyls in mammals.

12.6.2 The Case of CORM-3: To Release or Not Release CO!

The ability of CORM-3 (and CORM-2 as well) to release CO has been questioned due to our data obtained in solution using the Mb assay. We initially reported that addition of CORM-2 or CORM-3 to a solution containing deoxy-Mb resulted in the formation of COMb indicating liberation of CO by the metal carbonyls [1,8]. We must point out that commercially available Mb used in our assay is mostly in the oxidized form (met-Mb, Fe^{3+}) and thus incapable of binding CO. However, by addition of sodium dithionite in solution, met-Mb can be promptly converted to deoxy-Mb (Fe^{2+}), which strongly binds CO. Indeed, in the Mb assay we originally developed, sodium dithionite was added to the met-Mb solution prior to addition of CO-RMs [1]. Following our publications, Santos-Silva and colleagues showed that no CO can be detected by gas chromatography in the headspace of a closed vial containing a 10 mM solution of CORM-3 in different aqueous media, including water, phosphate buffer, and fetal bovine serum [89]. In addition, McLean and coworkers reported that CO is not liberated from CORM-2 or CORM-3 at an appreciable rate in the presence of deoxy-Mb alone in solution, but it is the presence of sodium dithionite that facilitates release of CO from these CO-RMs [90]. In particular, these last authors concluded that "it is now necessary to redefine CO-RMs in terms of the CO release rate" since we had defined CORM-2 and CORM-3 as "fast releasers" but their data showed that "in the absence of dithionite, these CO-RMs are actually very slow CO releasers." We argue that these studies have little relevance for the behaviors and effects of CO-RMs in biological systems as they were conducted only in solutions using the Mb or similar assays. In fact, these data diverge from important *ex vivo* and *in vivo* experimental evidence that typifies the function of CO and that we took into account, together with the Mb assay and detection of CO release using a CO electrode, to define whether CO-RMs were fast or slow CO releasers. Those authors completely neglected data from our group and others demonstrating that CORM-2 and CORM-3 caused a rapid relaxation of precontracted aortas that is prevented by the presence of the CO scavenger deoxy-Mb [1,8,30]. The rapid effect caused by CORM-2 and CORM-3 contrasts with the "slow" relaxing action mediated by CORM-A1, and it is perfectly in line with the rate of CO release from these compounds determined using the Mb assay [9]. Second, when injected in rats, CORM-3 exerted a rapid and transient decrease in mean arterial pressure, while a gradual hypotensive effect over time was observed in the case of CORM-A1 [9]; notably, these effects were absent when using the inactive counterparts (iCORMs). Third, treatment of mice with CORM-3 (40 mg/kg) injected intraperitoneally resulted in a transient but significant increase in blood COHb levels [42]. Finally, a considerable number of different CO-sensitive fluorescent probes have been recently reported and in all published studies the authors used either CORM-3 or CORM-2 to assess the effectiveness of their probes to detect CO, corroborating that these two metal carbonyls indeed deliver CO to living cells and, most importantly, *in vivo* [91–95]. Notably, in all these studies dithionite was absent and, therefore, the data strongly support that Ru-based CO-RMs have the intrinsic capacity to release CO once interacting with biological components.

To unequivocally reconcile the divergent interpretation of data on CO detection liberated by CORM-3 using the deoxy-Mb assay in the presence of dithionite, we have recently revisited a different method for measuring CO *in vitro* knowing that when CO is added to a solution containing oxy-Hb (O_2Hb, Fe^{2+}) in the absence of dithionite, CO can still efficiently displace O_2 and quickly bind to Hb to form COHb [96,97]. We have tested this experimentally and the data on the changes in the spectra of mouse O_2Hb in phosphate buffer after bubbling the solution with CO gas are presented later (R. Motterlini and R. Foresti, unpublished observations). As shown in Figure 12.4A, bubbling with CO for few seconds results in a rapid shift of the Soret band from 414 nm (O_2Hb) to 419 nm, which is typical of a COHb spectrum. Similarly, addition of CORM-3 to an O_2Hb solution clearly reveals the transition from O_2Hb to COHb over time demonstrating that CO is liberated by CORM-3 in phosphate buffer and trapped by Hb (Figure 12.4B).

This confirms our original results suggesting that CORM-3, although stable in pure water, rapidly loses CO in phosphate buffer solutions, culture media, and human plasma with different kinetics also indicating that dissociation of CO from this metal carbonyl is greatly affected by the various ligands present in the biological environment [28]. Most importantly, when CORM-3 was administered to mice by oral gavage, blood COHb significantly increased after 1 h and returned to basal levels by 3 h confirming the ability of this compound to deliver CO *in vivo* (Figure 12.4C). CORM-A1 also produced a transient increase in COHb, whereas in the presence of CORM-401 blood COHb takes much longer to return to basal levels. These results support once again the differential kinetics and amounts of CO released by these three CO-RMs as previously reported by our group [57].

Thus, we believe that over the years we and others have produced substantial convincing evidence that CORM-3 is indeed capable of rapidly liberating CO in cells and *in vivo* independently of the results obtained with the Mb assay in the presence of sodium dithionite. It is unfortunate that many researchers in the field of CO-RMs have not fully appreciated the importance of the chemical reactivity of transition metal carbonyls in biological systems evident in several publications [1,8,28,42,57,91,94] and keep making and publishing strong statements on the characteristics of CO-RMs based only on few experiments conducted either in solutions contained in a test tube or at the most *in vitro*.

12.6.3 Inactive CORMs: A Useful Tool to Assess the Biological Effect of CO from CORMs

It was clear from the beginning of our studies that a negative control was necessary to differentiate the effect of CO released by the metal-based CO-RMs from any secondary action caused by the rest of the molecule. This is a standard practice in pharmacological studies. We must emphasize that the metabolism of metal carbonyls in biological systems is virtually unknown and thus it is difficult to predict the (multiple) products of metal carbonyl decomposition once CO is liberated. Considering these limitations, we

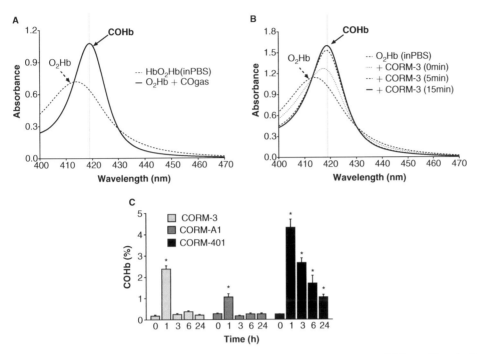

Figure 12.4 CORM-3, in the absence of dithionite, liberates CO to O_2Hb in vitro and delivers CO to the blood circulation in vivo. *(A) Spectra of mouse O_2Hb (5.5 µM) in phosphate buffer solution (PBS) without dithionite and after bubbling the solution with pure CO gas for 10 s. Note the shift of the Soret peak from 414 nm (O_2Hb) to 419 nm, which is typical of a COHb. (B) Changes in the spectra of O_2Hb (8.5 µM) after addition of CORM-3 (12.5 µM) showing the transition to COHb over time. (C) Oral administration of CORM-3 (30 mg/kg) results in a rapid and transient increase in blood COHb levels. The effect of CORM-A1 and CORM-401 administered to mice at the same dose is shown for comparison (n = 3 independent experiments).*
Source: R. Motterlini and R. Foresti (unpublished observations).

took a pragmatic approach and focused on generating compounds with a chemical structure as similar as possible to the original CO-RM but lacking the carbonyl groups. These compounds, which we named iCORMs, were obtained as follows. In the case of CORM-2 ([Ru(CO)$_3$Cl$_2$]$_2$), we synthesized an iCORM-2 in which the carbonyl groups were replaced by DMSO ([Ru(DMSO)$_3$Cl$_2$]$_2$) [1]. Concerning CORM-3 ([Ru(CO)$_3$Cl(glycinate)]), we took advantage of its ability to lose CO once solubilized in physiological buffers and prepared iCORM-3 by incubating CORM-3 overnight in phosphate buffer and then bubbling the solution with nitrogen to remove the residual CO gas [1]. To evaluate specifically the effect of the transition metal present in CORM-2 and CORM-3, in some experiments we and others used ruthenium chloride as an alternative iCORM [1,98]. For CORM-A1 ([Na$_2$H$_3$BCO$_2$]), which is not a metal carbonyl, we depleted CO by dissolving the compounds in weak acidic solutions after discovering that CO liberation is stimulated by acidic pH [9]. Finally, in the case of CORM-401 ([Mn(CO)$_4${S$_2$CNMe(CH$_2$CO$_2$H)}]), which is a Mn tetracarbonyl coordinated to a sarcosine moiety, iCORM-401 either consisted of a mixture of MnSO$_4$ and sarcosine [57,60] or was obtained after reacting CORM-401 with equimolar concentrations of H$_2$O$_2$, having found that oxidation of Mn triggers the loss of CO [57,61].

Thus, the production of an appropriate iCORM does not follow a general rule but needs to take into account the properties of the corresponding active CO-RM. The inability to release CO from all these iCORMs was confirmed with the Mb assay and, indeed, in most cases these iCORMs have been shown to lack the numerous biological and pharmacological effects exerted by the active CO-RMs. This is particularly true for the many studies conducted *in vivo* in which the ability of CORM-2, CORM-3, or CORM-401 to mitigate or abolish a given pathological condition in animals has not been recapitulated by their inactive counterparts. Nevertheless, in few examples limited to studies conducted in cells in culture or in test tubes, iCORMs or the transition metal per se appeared to exert some effects. For instance, Winburn and collaborators reported that both CORM-2 and iCORM-2 caused cell toxicity in two kidney cell lines as evidenced by increased apoptosis and necrosis, leading the authors to suggest that the ruthenium-based by-product, iCORM-2, is cytotoxic [99]. Similarly, ruthenium (100 µM) but not CORM-2 has been shown to inhibit procoagulant activities in human plasma induced by snake venoms from different species in reactions conducted in a test tube [100]. We would like to point out that, in both reports described earlier, measurements of CO liberation from CORM-2 (or iCORM-2) in the *in vitro* systems utilized by the authors have not been performed. Additional studies revealed that the active antimicrobial agent against *Escherichia coli* exposed to CORM-3 *in vitro* is a thiol-reactive Ru(II) ion, and not the CO donated by the compound [101]. On the other hand, Nobre and colleagues demonstrated using a CO-sensitive fluorescence probe (COP-1) that both CORM-3 and CORM-2, but not iCORMs, delivered CO inside *E. coli* and had an improved ability to kill these bacteria when compared to CO-RMs that are unable to promote CO accumulation in cells [102]. Another work has reported that CORM-2 activates nonselective cation current in human endothelial cells independently of CO release from the compound [103].

Altogether, these varied results emphasize the importance of carefully assessing the comparative effects of CORMs with their inactive counterparts to verify and confirm both the release and the contribution of CO on a specific effect in a given experimental system. Thus, despite the experimental limitations of iCORMs, we believe that the use of the transition metal per se and CO-RMs depleted of CO can still provide very important information. We encourage others to include them in experiments evaluating the different classes of CO-RMs, whether metal-based, organic CO donors, CO prodrugs, or photoCORMs.

12.7 Hybrid CO-Releasing Molecules: Amplifying the Therapeutic Potential of the Nrf2/HO-1/CO Axis

The rationale of studying CO as a therapeutic is based on the well-known cytoprotective actions of HO-1, the inducible isoform of heme oxygenase that degrades heme to CO and biliverdin and is recognized as an essential defensive system in animals and humans [104,105]. It is most likely that not only CO but all products of heme degradation contribute to this protective effect and thus upregulation of HO-1 expression may also be an attractive strategy for therapeutic intervention [45]. Knowing that Nrf2 is a key transcription factor in regulating the expression of HO-1 and other defensive genes [106,107], our group started to work on designing molecules possessing the ability to rapidly release CO and simultaneously stimulate the Nrf2–HO-1 axis. Thus, we synthesized hybrid compounds consisting of CO-RMs conjugated to known Nrf2 activators to provide a synergistic effect compared to CORMs or Nrf2-activating agents alone. The chemical features and biological functions of these "CO-RM derivatives," termed HYCOs (hybrid CO-releasing molecules), are briefly discussed next.

12.7.1 HYCOs as Dual Activity Molecules that Simultaneously Induce HO-1 and Release CO: Proof of Concept

The premise on HYCOs is that a molecule with a dual mode of action will provide greater tissue protection by first limiting damage through CO delivery and subsequently promoting the endogenous upregulation of Nrf2-dependent defensive genes and proteins, a process that takes several hours due to transcription and translation processes. Among the numerous Nrf2 activators reported in the literature, we focused on fumaric esters since dimethyl fumarate has been approved as a drug to treat multiple sclerosis, an autoimmune disease characterized by chronic inflammation [108,109]. We first showed that it was chemically feasible to combine a cobalt-based CO-RM with a fumaric ester derivative and that, although these initial compounds potently induced Nrf2 and HO-1 in different cell types, they were capable of releasing some CO to Mb but less efficient in donating CO to cells [110]. We also showed that cobalt-based HYCOs were more potent than dimethyl fumarate alone in reducing the production of nitrite, an inflammatory marker, in macrophages stimulated with LPS. Next, a series of HYCOs in which cobalt carbonyls were conjugated with different Nrf2/HO-1 inducers were prepared and tested *in vitro* and *in vivo*. From this group of compounds, two HYCOs (HYCO-4 and HYCO-10) were identified with considerable capacity to increase Nrf2/HO-1 expression in cells as well as in different organs *in vivo*. Most importantly, these two HYCOs were able to deliver CO to the blood circulation when orally administered to mice [111].

Thus, we provided the proof of concept that the synthesis of dual activity molecules capable of releasing CO and inducing HO-1 is feasible. We note that the choice of cobalt carbonyl complexes was dictated mainly by the easiness of synthesis and handling, although we were aware that cobalt is not a preferential metal for biological applications. Consequently, our next step was to prepare new HYCOs by replacing the less pharmacologically attractive cobalt carbonyls with more biocompatible Mn or Ru carbonyl complexes. This was a logical progress in our strategy due to the extensive preclinical results already obtained with CORM-3 (Ru) and, to a lesser extent, CORM-401 (Mn).

12.7.2 Mn- and Ru-Based HYCOs: Chemical Reactivity, Biological Actions, and Therapeutic Effects

The strategy summarizing the design of Mn- and Ru-based HYCOs is reported in Figure 12.5. As shown, HYCOs were synthesized by conjugating an ethyl fumaric moiety either to Mn- or Ru-CORMs using different linkers. Specifically, a Mn carbonyl similar to CORM-401 was used to prepare HYCO-3, HYCO-7, and HYCO-13, whereas a Ru-CORM served to obtain HYCO-6 and HYCO-11. The linkers containing an increasing number of nitrogen atoms were utilized in an attempt to increase water solubility. However, even in the presence of the highest number of nitrogen atoms (HYCO-13), the compounds were still poorly soluble in water and therefore they were all dissolved in DMSO [112].

Most of these HYCOs retained the ability to induce Nrf2/HO-1 in cells and tissues. In addition, replacement of cobalt with Mn-CORMs greatly improved the ability of these compounds to deliver CO to cells and consistently increase blood COHb when orally administered in vivo [64,113]. When tested in animals, HYCO-3 was shown to counteract LPS-mediated inflammation in several organs in mice. Interestingly, reduction of some inflammatory markers was dependent on the CO-releasing part of HYCO-3, while others relied on Nrf2 activation as demonstrated using Nrf2-deficient mice [64]. We note that the effect of HYCO-3 was more pronounced compared to that of dimethyl fumarate (Nrf2/HO-1 activator) or CORM-401 alone, especially concerning the activation of the anti-inflammatory response. Finally, selected HYCOs were evaluated in several models of disease characterized by inflammation. These studies revealed that HYCO-6 accelerated skin wound closure, HYCO-6 and HYCO-13 reduced psoriasis symptoms equaling or surpassing the effect of dimethyl fumarate, and HYCO-3 and HYCO-13 significantly ameliorated scores of motor dysfunction in a mouse model that mimics human multiple sclerosis (Figure 12.6).

These results also indicate that prolonged daily administration of HYCOs (up to 40 days) is well tolerated in animals. Our data clearly confirm that HYCOs possess a dual mode of action highlighting the notion that simultaneous Nrf2 targeting and CO delivery could be a clinically relevant application to combat inflammation. Recently, another group followed the same idea and synthesized hybrids bearing a Fe–carbonyl moiety and a fumaric acid derivative. The compounds exhibited potent anti-inflammatory actions in murine dendritic cells but were not tested *in vivo* [114].

Conclusions and Future Perspectives

The initial proposition in 2002 that metal-based CO-RMs could be used as effective CO carriers to mimic the physiological effects and exploit the therapeutic actions of CO has been corroborated by several preclinical studies using different models of disease. The progression toward biologically active CO-RMs has been gradual, moving from Ru-based

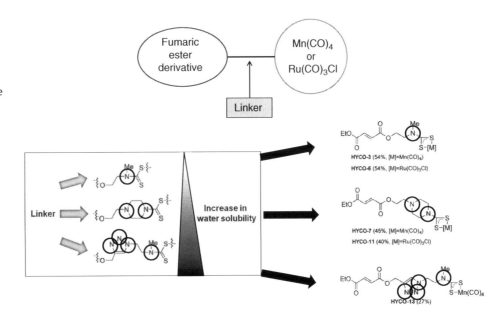

Figure 12.5 Varying the CO-releasing moiety and the linker to modulate the properties of HYCOs. A fumaric ester derivative was conjugated either to Mn-CORMs to prepare HYCO-3, HYCO-7, and HYCO-13 or to Ru-CORMs to obtain HYCO-6 and HYCO-11. Different linkers containing an increasing number of nitrogen atoms were utilized in an attempt to increase water solubility.
Source: Based on [112].

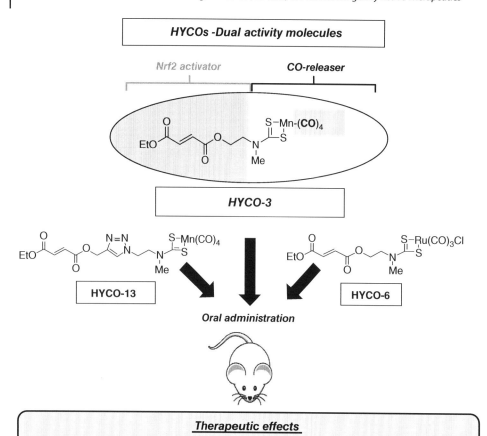

Figure 12.6 Therapeutic effects of HYCOs in vivo. HYCOs, hybrid molecules consisting of Nrf2 activators conjugated to a CO-RM moiety, possess a "dual activity action" by inducing Nrf2/HO-1 expression and simultaneously delivering CO in vitro and in vivo. Oral administration of HYCOs has recently been shown to exert therapeutic effects in models of systemic inflammation, psoriasis, skin wound, and multiple.
Source: Adapted from [64] and [113].

compounds poorly soluble in biological fluids to carbonyl complexes containing transition metals compatible with tissues and having improved solubility in aqueous solutions, increased stability, and enhanced efficiency in terms of both CO release and bioactivity. We and other laboratories have implemented and optimized new methodologies and learned important aspects of the chemical reactivity of CO-RMs within the biological systems, which are fundamental for continuing on the development of this class of molecules as pharmaceuticals. The results accumulated are very promising and clinical trials on CO-RMs may be envisioned soon knowing that emerging formulations proposed by different laboratories (nanoencapsulation, coordination to other drugs, and targeted delivery) may offer the right solutions for a safe, efficacious, and therapeutic delivery of CO in humans.

Acknowledgments

We wish to thank all our biology and chemistry colleagues who have collaborated with us over the years in the study of CO-RMs and HYCOs. The work conducted in our laboratory was supported by grants from the Agence National de la Recherche (ANR CO-HEAL, CARMMA, SWEET-CO), la Sociétés d'Accélération du Transfert de Technologies (SATT IDF Innov), INSERM, and Université Paris-Est Créteil.

References

1. Motterlini, R., Clark, J.E., Foresti, R., Sarathchandra, P., Mann, B.E., and Green, C.J. (2002). Carbon monoxide-releasing molecules: characterization of biochemical and vascular activities. *Circ. Res.* 90: e17–e24.

2. Verma, A., Hirsch, D.J., Glatt, C.E., Ronnett, G.V., and Snyder, S.H. (1993). Carbon monoxide: a putative neural messenger. *Science* 259: 381–384.

3. Suematsu, M., Goda, N., Sano, T., Kashiwagi, S., Egawa, T., Shinoda, Y., and Ishimura, Y. (1995). Carbon monoxide: an endogenous modulator of sinusoidal tone in the perfused rat liver. *J. Clin. Invest.* 96: 2431–2437.

4. Motterlini, R., Gonzales, A., Foresti, R., Clark, J.E., Green, C.J., and Winslow, R.M. (1998). Heme

oxygenase-1-derived carbon monoxide contributes to the suppression of acute hypertensive responses in vivo. *Circ. Res.* 83: 568–577.

5. Hepp, A.F. and Wrighton, M.S. (1983). Relative importance of metal-metal bond scission and loss of carbon monoxide from photoexcited dimanganese decacarbonyl: spectroscopic detection of a coordinatively unsaturated, CO-bridged dinuclear species in low-temperature alkane matrices. *J. Am. Chem. Soc.* 105: 5934–5935.

6. Herrick, R.S. and Brown, T.L. (1984). Flash photolytic investigation of photoinduced carbon monoxide dissociation from dinuclear manganese carbonyl compounds. *Inorg. Chem.* 23: 4550–4553.

7. Motterlini, R., Mann, B.E., and Foresti, R. (2005). Therapeutic applications of carbon monoxide-releasing molecules (CO-RMs). *Expert Opin. Investig. Drugs* 14: 1305–1318.

8. Clark, J.E., Naughton, P., Shurey, S., Green, C.J., Johnson, T.R., Mann, B.E., Foresti, R., and Motterlini, R. (2003). Cardioprotective actions by a water-soluble carbon monoxide-releasing molecule. *Circ. Res.* 93: e2–e8.

9. Motterlini, R., Sawle, P., Bains, S., Hammad, J., Alberto, R., Foresti, R., and Green, C.J. (2005). CORM-A1: a new pharmacologically active carbon monoxide-releasing molecule. *FASEB J.* 19: 284–286.

10. (2002). Motterlini, R., Foresti, R., and Green, C.J. Studies on the development of carbon monoxide-releasing molecules: potential applications for the treatment of cardiovascular dysfunction. In: *Carbon Monoxide and Cardiovascular Functions* (ed. R. Wang), 249–271. Boca Raton, Florida, USA: CRC Press.

11. Fiumana, E., Parfenova, H., Jagger, J.H., and Leffler, C.W. (2003). Carbon monoxide mediates vasodilator effects of glutamate in isolated pressurized cerebral arterioles of newborn pigs. *Am. J. Physiol. Heart Circ. Physiol.* 284: H1073–H1079.

12. Rattan, S., Haj, R.A., and De Godoy, M.A. (2004). Mechanism of internal anal sphincter relaxation by CORM-1, authentic CO, and NANC nerve stimulation. *Am. J. Physiol. Gastrointest. Liver Physiol.* 287: G605–G611.

13. Vera, T., Henegar, J.R., Drummond, H.A., Rimoldi, J.M., and Stec, D.E. (2005). Protective effect of carbon monoxide-releasing compounds in ischemia-induced acute renal failure. *J. Am. Soc. Nephrol.* 16: 950–958.

14. Chen, B., Guo, L., Fan, C., Bolisetty, S., Joseph, R., Wright, M.M., Agarwal, A., and George, J.F. (2009). Carbon monoxide rescues heme oxygenase-1-deficient mice from arterial thrombosis in allogeneic aortic transplantation. *Am. J. Pathol.* 175: 422–429.

15. Vitek, L., Gbelcova, H., Muchova, L., Vanova, K., Zelenka, J., Konickova, R., Suk, J., Zadinova, M., Knejzlik, Z., Ahmad, S., Fujisawa, T., Ahmed, A., and Ruml, T. (2014). Antiproliferative effects of carbon monoxide on pancreatic cancer. *Dig. Liver Dis.* 46: 369–375.

16. Desmard, M., Foresti, R., Morin, D., Dagoussat, M., Berdeaux, A., Denamur, E., Crook, S.H., Mann, B.E., Scapens, D., Montravers, P., Boczkowski, J., and Motterlini, R. (2012). Differential antibacterial activity against Pseudomonas aeruginosa by carbon monoxide-releasing molecules. *Antioxid. Redox Signal.* 16: 153–163.

17. Chung, S.W., Liu, X., Macias, A.A., Baron, R.M., and Perrella, M.A. (2008). Heme oxygenase-1-derived carbon monoxide enhances the host defense response to microbial sepsis in mice. *J. Clin. Invest.* 118: 239–247.

18. Sawle, P., Foresti, R., Mann, B.E., Johnson, T.R., Green, C.J., and Motterlini, R. (2005). Carbon monoxide-releasing molecules (CO-RMs) attenuate the inflammatory response elicited by lipopolysaccharide in RAW264.7 murine macrophages. *Br. J. Pharmacol.* 145: 800–810.

19. Megias, J., Busserolles, J., and Alcaraz, M.J. (2007). The carbon monoxide-releasing molecule CORM-2 inhibits the inflammatory response induced by cytokines in Caco-2 cells. *Br. J. Pharmacol.* 150: 977–986.

20. Sun, B., Sun, H., Liu, C., Shen, J., Chen, Z., and Chen, X. (2007). Role of CO-releasing molecules liberated CO in attenuating leukocytes sequestration and inflammatory responses in the lung of thermally injured mice. *J. Surg. Res.* 139: 128–135.

21. Sun, B.W., Jin, Q., Sun, Y., Sun, Z.W., Chen, X., Chen, Z.Y., and Cepinskas, G. (2007). Carbon liberated from CO-releasing molecules attenuates leukocyte infiltration in the small intestine of thermally injured mice. *World J. Gastroenterol.* 13: 6183–6190.

22. Magierowski, M., Magierowska, K., Hubalewska-Mazgaj, M., Sliwowski, Z., Ginter, G., Pajdo, R., Chmura, A., Kwiecien, S., and Brzozowski, T. (2017). Carbon monoxide released from its pharmacological donor, tricarbonyldichlororuthenium (II) dimer, accelerates the healing of pre-existing gastric ulcers. *Br. J. Pharmacol.* 174: 3654–3668.

23. Xue, J. and Habtezion, A. (2014). Carbon monoxide-based therapy ameliorates acute pancreatitis via TLR4 inhibition. *J. Clin. Invest* 124: 437–447.

24. Allanson, M. and Reeve, V.E. (2005). Ultraviolet A (320–400 nm) Modulation of ultraviolet B (290–320 nm)-Induced immune suppression is mediated by carbon monoxide. *J. Invest. Dermatol.* 124: 644–650.

25. Loureiro, A., Bernardes, G.J., Shimanovich, U., Sarria, M.P., Nogueira, E., Preto, A., Gomes, A.C., and Cavaco-Paulo, A. (2015). Folic acid-tagged protein nanoemulsions loaded with CORM-2 enhance the survival of mice bearing subcutaneous A20 lymphoma tumors. *Nanomedicine* 11: 1077–1083.

26. Yin, H., Fang, J., Liao, L., Nakamura, H., and Maeda, H. (2014). Styrene-maleic acid copolymer-encapsulated CORM2, a water-soluble carbon monoxide (CO) donor with a constant CO-releasing property, exhibits therapeutic potential for inflammatory bowel disease. *J. Control. Release* 187: 14–21.

27. Motterlini, R., Sawle, P., Hammad, J., Mann, B.E., Johnson, T.R., Green, C.J., and Foresti, R. (2013). Vasorelaxing effects and inhibition of nitric oxide in macrophages by new iron-containing carbon monoxide-releasing molecules (CO-RMs). *Pharmacol. Res.* 68: 108–117.

28. Motterlini, R., Mann, B.E., Johnson, T.R., Clark, J.E., Foresti, R., and Green, C.J. (2003). Bioactivity and pharmacological actions of carbon monoxide-releasing molecules. *Curr. Pharmacol. Design* 9: 2525–2539.

29. Foresti, R., Hammad, J., Clark, J.E., Johnson, R.A., Mann, B.E., Friebe, A., Green, C.J., and Motterlini, R. (2004). Vasoactive properties of CORM-3, a novel water-soluble carbon monoxide-releasing molecule. *Br. J. Pharmacol.* 142: 453–460.

30. Alshehri, A., Bourguignon, M.P., Clavreul, N., Badier-Commander, C., Gosgnach, W., Simonet, S., Vayssettes-Courchay, C., Cordi, A., Fabiani, J.N., Verbeuren, T.J., and Feletou, M. (2013). Mechanisms of the vasorelaxing effects of CORM-3, a water-soluble carbon monoxide-releasing molecule: interactions with eNOS. *Naunyn Schmiedebergs Arch. Pharmacol.* 386: 185–196.

31. Bolognesi, M., Sacerdoti, D., Piva, A., Di Pascoli, M., Zampieri, F., Quarta, S., Motterlini, R., Angeli, P., Merkel, C., and Gatta, A. (2007). Carbon monoxide-mediated activation of large-conductance calcium-activated potassium channels contributes to mesenteric vasodilatation in cirrhotic rats. *J. Pharmacol. Exp. Ther.* 321: 187–194.

32. Abid, S., Houssaini, A., Mouraret, N., Marcos, E., Amsellem, V., Wan, F., Dubois-Rande, J.L., Derumeaux, G., Boczkowski, J., Motterlini, R., and Adnot, S. (2014). p21-Dependent protective effects of a carbon monoxide-releasing molecule-3 in pulmonary hypertension. *Arterioscler. Thromb. Vasc. Biol.* 34: 304–312.

33. Althaus, M., Fronius, M., Buchackert, Y., Vadasz, I., Clauss, W.G., Seeger, W., Motterlini, R., and Morty, R.E. (2009). Carbon monoxide rapidly impairs alveolar fluid clearance by inhibiting epithelial sodium channels. *Am. J. Respir. Cell Mol. Biol.* 41: 639–650.

34. Guo, Y., Stein, A.B., Wu, W.J., Tan, W., Zhu, X., Li, Q.H., Dawn, B., Motterlini, R., and Bolli, R. (2004). Administration of a CO-releasing molecule at the time of reperfusion reduces infarct size in vivo. *Am. J. Physiol. Heart Circ. Physiol.* 286: H1649–H1653.

35. Bagul, A., Hosgood, S.A., Kaushik, M., and Nicholson, M.L. (2008). Carbon monoxide protects against ischemia-reperfusion injury in an experimental model of controlled nonheartbeating donor kidney. *Transplantation* 85: 576–581.

36. George, E.M., Cockrell, K., Arany, M., Stec, D.E., Rimoldi, J.M., Gadepalli, R.S.V., and Granger, J.P. (2017). Carbon monoxide releasing molecules blunt placental ischemia-induced hypertension. *Am. J. Hypertens.* 30: 931–937.

37. Desmard, M., Davidge, K.S., Bouvet, O., Morin, D., Roux, D., Foresti, R., Ricard, J.D., Denamur, E., Poole, R.K., Montravers, P., Motterlini, R., and Boczkowski, J. (2009). A carbon monoxide-releasing molecule (CORM-3) exerts bactericidal activity against Pseudomonas aeruginosa and improves survival in an animal model of bacteraemia. *FASEB J.* 23: 1023–1031.

38. Lancel, S., Hassoun, S.M., Favory, R., Decoster, B., Motterlini, R., and Neviere, R. (2009). Carbon monoxide rescues mice from lethal sepsis by supporting mitochondrial energetic metabolism and activating mitochondrial biogenesis. *J. Pharmacol. Exp. Ther.* 1329: 641–648.

39. De Backer, O., Elinck, E., Blanckaert, B., Leybaert, L., Motterlini, R., and Lefebvre, R.A. (2009). Water-soluble CO-releasing molecules (CO-RMs) reduce the development of postoperative ileus via modulation of MAPK/HO-1 signaling and reduction of oxidative stress. *Gut.* 58: 347–356.

40. Ferrandiz, M.L., Maicas, N., Garcia-Arnandis, I., Terencio, M.C., Motterlini, R., Devesa, I., Joosten, L.A., Van Den Berg, W.B., and Alcaraz, M.J. (2008). Treatment with a CO-releasing molecule (CORM-3) reduces joint inflammation and erosion in murine collagen-induced arthritis. *Ann. Rheum. Dis.* 67: 1211–1217.

41. Yabluchanskiy, A., Sawle, P., Homer-Vanniasinkam, S., Green, C.J., Foresti, R., and Motterlini, R. (2012). CORM-3, a carbon monoxide-releasing molecule, alters the inflammatory response and reduces brain damage in a rat model of hemorrhagic stroke. *Crit. Care Med.* 40: 544–552.

42. Wang, G., Hamid, T., Keith, R.J., Zhou, G., Partridge, C.R., Xiang, X., Kingery, J.R., Lewis, R.K., Li, Q., Rokosh, D.G., Ford, R., Spinale, F.G., Riggs, D.W., Srivastava, S., Bhatnagar, A., Bolli, R., and Prabhu,

S.D. (2010). Cardioprotective and antiapoptotic effects of heme oxygenase-1 in the failing heart. *Circulation* 121: 1912–1925.

43. Choi, Y.K., Maki, T., Mandeville, E.T., Koh, S.H., Hayakawa, K., Arai, K., Kim, Y.M., Whalen, M.J., Xing, C., Wang, X., Kim, K.W., and Lo, E.H. (2016). Dual effects of carbon monoxide on pericytes and neurogenesis in traumatic brain injury. *Nat. Med.* 22: 1335–1341.

44. Motterlini, R. and Otterbein, L.E. (2010). The therapeutic potential of carbon monoxide. *Nat. Rev. Drug Discov.* 9: 728–743.

45. Motterlini, R. and Foresti, R. (2014). Heme oxygenase-1 as a target for drug discovery. *Antioxid. Redox Signal.* 20: 1810–1826.

46. Motterlini, R. and Foresti, R. (2017). Biological signaling by carbon monoxide and carbon monoxide-releasing molecules (CO-RMs). *Am. J. Physiol. Cell Physiol.* 312: C302–C313.

47. Alberto, R., Ortner, K., Wheatley, N., Schibli, R., and Schubiger, A.P. (2001). Synthesis and properties of boranocarbonate: a convenient in situ CO source for the aqueous preparation of [(99m)Tc(OH$_2$)$_3$(CO)$_3$]$^+$. *J. Am. Chem. Soc.* 123: 3135–3136.

48. Fagone, P., Mangano, K., Quattrocchi, C., Motterlini, R., Di Marco, R., Magro, G., Penacho, N., Romao, C.C., and Nicoletti, F. (2011). Prevention of clinical and histological signs of proteolipid protein (PLP)-induced experimental allergic encephalomyelitis (EAE) in mice by the water-soluble carbon monoxide-releasing molecule (CORM)-A1. *Clin. Exp. Immunol.* 163: 368–374.

49. Kramkowski, K., Leszczynska, A., Mogielnicki, A., Chlopicki, S., Fedorowicz, A., Grochal, E., Mann, B., Brzoska, T., Urano, T., Motterlini, R., and Buczko, W. (2012). Antithrombotic properties of water-soluble carbon monoxide-releasing molecules. *Arterioscler. Thromb. Vasc. Biol.* 32: 2149–2157.

50. Upadhyay, K.K., Jadeja, R.N., Vyas, H.S., Pandya, B., Joshi, A., Vohra, A., Thounaojam, M.C., Martin, P.M., Bartoli, M., and Devkar, R.V. (2020). Carbon monoxide releasing molecule-A1 improves nonalcoholic steatohepatitis via Nrf2 activation mediated improvement in oxidative stress and mitochondrial function. *Redox. Biol.* 28: 101314.

51. Hosick, P.A., Alamodi, A.A., Hankins, M.W., and Stec, D.E. (2016). Chronic treatment with a carbon monoxide releasing molecule reverses dietary induced obesity in mice. *Adipocyte* 5: 1–10.

52. Hartinger, C.G., Jakupec, M.A., Zorbas-Seifried, S., Groessl, M., Egger, A., Berger, W., Zorbas, H., Dyson, P.J., and Keppler, B.K. (2008). KP1019, a new redox-active anticancer agent—preclinical development and results of a clinical phase I study in tumor patients. *Chem. Biodivers.* 5: 2140–2155.

53. Freeland-Graves, J.H., Mousa, T.Y., and Kim, S. (2016). International variability in diet and requirements of manganese: causes and consequences. *J. Trace Elem. Med. Biol.* 38: 24–32.

54. Freeland-Graves, J.H., Sachdev, P.K., Binderberger, A.Z., and Sosanya, M.E. (2020). Global diversity of dietary intakes and standards for zinc, iron, and copper. *J. Trace Elem. Med. Biol.* 61: 126515.

55. Hewison, L., Johnson, T.R., Mann, B.E., Meijer, A.J., Sawle, P., and Motterlini, R. (2011). A re-investigation of [Fe(l-cysteinate)(2)(CO)(2)](2-): an example of non-heme CO coordination of possible relevance to CO binding to ion channel receptors. *Dalton Trans.* 40: 8328–8334.

56. Crook, S.H., Mann, B.E., Meijer, J.A.H.M., Adams, H., Sawle, P., Scapens, D., and Motterlini, R. (2011). [Mn(CO)4{S2CNMe(CH2CO2H)}], a new water-soluble CO-releasing molecule. *Dalton Trans.* 40: 4230–4235.

57. Fayad-Kobeissi, S., Ratovonantenaina, J., Dabire, H., Wilson, J.L., Rodriguez, A.M., Berdeaux, A., Dubois-Rande, J.L., Mann, B.E., Motterlini, R., and Foresti, R. (2016). Vascular and angiogenic activities of CORM-401, an oxidant-sensitive CO-releasing molecule. *Biochem. Pharmacol.* 102: 64–77.

58. Sandouka, A., Balogun, E., Foresti, R., Mann, B.E., Johnson, T.R., Tayem, Y., Green, C.J., Fuller, B., and Motterlini, R. (2005). Carbon monoxide-releasing molecules (CO-RMs) modulate respiration in isolated mitochondria. *Cell. Mol. Biol.* 51: 425–432.

59. Kaczara, P., Motterlini, R., Rosen, G.M., Augustynek, B., Bednarczyk, P., Szewczyk, A., Foresti, R., and Chlopicki, S. (2015). Carbon monoxide released by CORM-401 uncouples mitochondrial respiration and inhibits glycolysis in endothelial cells: a role for mitoBK channels. *Biochim. Biophys. Acta* 1847: 1297–1309.

60. Wilson, J.L., Bouillaud, F., Almeida, A.S., Vieira, H.L., Ouidja, M.O., Dubois-Rande, J.L., Foresti, R., and Motterlini, R. (2017). Carbon monoxide reverses the metabolic adaptation of microglia cells to an inflammatory stimulus. *Free Rad. Biol. Med.* 104: 311–323.

61. Braud, L., Pini, M., Muchova, L., Manin, S., Kitagishi, H., Sawaki, D., Czibik, G., Ternacle, J., Derumeaux, G., Foresti, R., and Motterlini, R. (2018). Carbon monoxide-induced metabolic switch in adipocytes improves insulin resistance in obese mice. *JCI Insight* 3: e123485.

62. Harper, J.A., Dickinson, K., and Brand, M.D. (2001). Mitochondrial uncoupling as a target for drug development for the treatment of obesity. *Obes. Rev.* 2: 255–265.

63. Hosick, P.A., Alamodi, A.A., Storm, M.V., Gousset, M.U., Pruett, B.E., Gray, W., III, Stout, J., and Stec, D.E. (2014). Chronic carbon monoxide treatment attenuates development of obesity and remodels adipocytes in mice fed a high-fat diet. *Int. J. Obes. (Lond)* 38: 132–139.
64. Motterlini, R., Nikam, A., Manin, S., Ollivier, A., Wilson, J.L., Djouadi, S., Muchova, L., Martens, T., Rivard, M., and Foresti, R. (2019). HYCO-3, a dual CO-releaser/Nrf2 activator, reduces tissue inflammation in mice challenged with lipopolysaccharide. *Redox. Biol.* 20: 334–348.
65. Yang, G., Fan, M., Zhu, J., Ling, C., Wu, L., Zhang, X., Zhang, M., Li, J., Yao, Q., Gu, Z., and Cai, X. (2020). A multifunctional anti-inflammatory drug that can specifically target activated macrophages, massively deplete intracellular H2O2, and produce large amounts CO for a highly efficient treatment of osteoarthritis. *Biomaterials* 255: 120155.
66. Herrmann, W.A. (1990). 100 Years of metal carbonyls. A serendipitous chemical discovery of major scientific and industrial impact. *J. Organomet. Chem.* 383: 21–44.
67. Foresti, R. and Motterlini, R. (2010). Interaction of carbon monoxide with transition metals: evolutionary insights into drug target discovery. *Curr. Drug Targets* 11: 1595–1604.
68. Piantadosi, C.A. (2002). Biological chemistry of carbon monoxide. *Antioxid. Redox Signal.* 4: 259–270.
69. Foresti, R., Bani-Hani, M.G., and Motterlini, R. (2008). Use of carbon monoxide as a therapeutic agent: promises and challenges. *Intensive Care Med.* 34: 649–658.
70. Goldbaum, L.R., Ramirez, R.G., and Absalon, K.B. (1975). What is the mechanism of carbon monoxide toxicity? *Aviat. Space Environ. Med.* 46: 1289–1291.
71. Minegishi, S., Yumura, A., Miyoshi, H., Negi, S., Taketani, S., Motterlini, R., Foresti, R., Kano, K., and Kitagishi, H. (2017). Detection and removal of endogenous carbon monoxide by selective and cell permeable hemoprotein-model complexes. *J. Am. Chem. Soc.* 139: 5984–5991.
72. Sen Gupta, A. (2019). Hemoglobin-based oxygen carriers: Current state-of-the-art and novel molecules. *Shock* 52: 70–83.
73. Bisse, E., Schaeffer-Reiss, C., Van Dorsselaer, A., Alayi, T.D., Epting, T., Winkler, K., Benitez Cardenas, A.S., Soman, J., Birukou, I., Samuel, P.P., and Olson, J.S. (2017). Hemoglobin Kirklareli (alpha H58L), a new variant associated with Fe deficiency and increased CO binding. *J. Biol. Chem.* 292: 2542–2555.
74. Vandegriff, K.D., Young, M.A., Lohman, J., Bellelli, A., Samaja, M., and Malavalli, A. (2008). Winslow, R. M. CO-MP4, a polyethylene glycol-conjugated haemoglobin derivative and carbon monoxide carrier that reduces myocardial infarct size in rats. *Br. J. Pharmacol.* 154: 1649–1661.
75. Nugent, W.H., Sheppard, F.R., Dubick, M.A., Cestero, R.F., Darlington, D.N., Jubin, R., Abuchowski, A., and Song, B.K. (2020). Microvascular and systemic impact of resuscitation with PEGylated carboxyhemoglobin-based oxygen carrier or hetastarch in a rat model of transient hemorrhagic shock. *Shock* 53: 493–502.
76. Nugent, W.H., Jubin, R., Buontempo, P.J., Kazo, F., and Song, B.K. (2019). Microvascular and systemic responses to novel PEGylated carboxyhaemoglobin-based oxygen carrier in a rat model of vaso-occlusive crisis. *Artif. Cells Nanomed. Biotechnol.* 47: 95–103.
77. Nagao, S., Taguchi, K., Sakai, H., Yamasaki, K., Watanabe, H., Otagiri, M., and Maruyama, T. (2016). Carbon monoxide-bound hemoglobin vesicles ameliorate multiorgan injuries induced by severe acute pancreatitis in mice by their anti-inflammatory and antioxidant properties. *Int. J. Nanomedicine.* 11: 5611–5620.
78. Sakai, H., Horinouchi, H., Tsuchida, E., and Kobayashi, K. (2009). Hemoglobin vesicles and red blood cells as carriers of carbon monoxide prior to oxygen for resuscitation after hemorrhagic shock in a rat model. *Shock* 31: 507–514.
79. Gonzales, M.A. and Mascharak, P.K. (2014). Photoactive metal carbonyl complexes as potential agents for targeted CO delivery. *J. Inorg. Biochem.* 133: 127–135.
80. Ling, K., Men, F., Wang, W.C., Zhou, Y.Q., Zhang, H.W., and Ye, D.W. (2018). Carbon monoxide and its controlled release: Therapeutic application, detection, and development of carbon monoxide releasing molecules (CORMs). *J. Med. Chem.* 61: 2611–2635.
81. Ismailova, A., Kuter, D., Bohle, D.S., and Butler, I.S. (2018). An overview of the potential therapeutic applications of CO-releasing molecules. *Bioinorg. Chem. Appl.* 2018: 8547364.
82. Faizan, M., Muhammad, N., Niazi, K.U.K., Hu, Y., Wang, Y., Wu, Y., Sun, H., Liu, R., Dong, W., Zhang, W., and Gao, Z. (2019). CO-Releasing materials: An emphasis on therapeutic implications, as release and subsequent cytotoxicity are the part of therapy. *Materials (Basel)* 12.
83. Pena, A.C., Penacho, N., Mancio-Silva, L., Neres, R., Seixas, J.D., Fernandes, A.C., Romao, C.C., Mota, M.M., Bernardes, G.J., and Pamplona, A. (2012). A novel carbon monoxide-releasing molecule fully protects mice from severe malaria. *Antimicrob. Agents Chemother.* 56: 1281–1290.
84. Marques, A.R., Kromer, L., Gallo, D.J., Penacho, N., Rodrigues, S.S., Seixas, J.D., Bernardes, G.J., Reis, P.M., Otterbein, S.L., Ruggieri, R.A., Goncalves,

A.S.G., Goncalves, A.M.L., De Matos, M.N., Bento, I., Otterbein, L.E., Blattler, W.A., and Romao, C.C. (2012). Generation of carbon monoxide releasing molecules (CO-RMs) as drug candidates for the treatment of acute liver injury: Targeting of CO-RMs to the liver. *Organometallics* 31: 5810–5822.

85. Stifter, J., Ulbrich, F., Goebel, U., Bohringer, D., Lagreze, W.A., and Biermann, J. (2017). Neuroprotection and neuroregeneration of retinal ganglion cells after intravitreal carbon monoxide release. *PLoS One* 12: e0188444.

86. Wright, M.A. and Wright, J.A. (2016). PhotoCORMs: CO release moves into the visible. *Dalton Trans.* 45: 6801–6811.

87. Martins, A.C., Krum, B.N., Queiros, L., Tinkov, A.A., Skalny, A.V., Bowman, A.B., and Aschner, M. (2020). Manganese in the diet: Bioaccessibility, adequate intake, and neurotoxicological effects. *J. Agric. Food Chem.* 68: 12893–12903.

88. Wang, P., Liu, H., Zhao, Q., Chen, Y., Liu, B., Zhang, B., and Zheng, Q. (2014). Syntheses and evaluation of drug-like properties of CO-releasing molecules containing ruthenium and group 6 metal. *Eur. J. Med. Chem.* 74: 199–215.

89. Santos-Silva, T., Mukhopadhyay, A., Seixas, J.D., Bernardes, G.J., Romao, C.C., and Romao, M.J. (2011). CORM-3 Reactivity toward proteins: The crystal structure of a Ru(II) dicarbonyl-lysozyme complex. *J. Am. Chem. Soc.* 133: 1192–1195.

90. McLean, S., Mann, B.E., and Poole, R.K. (2012). Sulfite species enhance carbon monoxide release from CO-releasing molecules: implications for the deoxymyoglobin assay of activity. *Anal. Biochem.* 427: 36–40.

91. Gong, S., Hong, J., Zhou, E., and Feng, G. (2019). A near-infrared fluorescent probe for imaging endogenous carbon monoxide in living systems with a large stokes shift. *Talanta* 201: 40–45.

92. Zhang, C., Xie, H., Zhan, T., Zhang, J., Chen, B., Qian, Z., Zhang, G., Zhang, W., and Zhou, J. (2019). A new mitochondrion targetable fluorescent probe for carbon monoxide-specific detection and live cell imaging. *Chem. Commun. (Camb.)* 55: 9444–9447.

93. Liu, K., Kong, X., Ma, Y., and Lin, W. (2018). Preparation of a Nile Red-Pd-based fluorescent CO probe and its imaging applications in vitro and in vivo. *Nat. Protoc.* 13: 1020–1033.

94. Tian, X., Liu, X., Wang, A., Lau, C., and Lu, J. (2018). Bioluminescence imaging of carbon monoxide in living cells and nude mice based on Pd(0)-Mediated Tsuji-Trost reaction. *Anal. Chem.* 90: 5951–5958.

95. Wang, J., Li, C., Chen, Q., Li, H., Zhou, L., Jiang, X., Shi, M., Zhang, P., Jiang, G., and Tang, B.Z. (2019). An easily available ratiometric reaction-based AIE probe for carbon monoxide light-up imaging. *Anal. Chem.* 91: 9388–9392.

96. Small, K.A., Radford, E.P., Frazier, J.M., Rodkey, F.L., and Collison, H.A. (1971). A rapid method for simultaneous measurement of carboxy- and methemoglobin in blood. *J. Appl. Physiol.* 31: 154–160.

97. Rodkey, F.L., Hill, T.A., Pitts, L.L., and Robertson, R.F. (1979). Spectrophotometric measurement of carboxyhemoglobin and methemoglobin in blood. *Clin. Chem.* 25: 1388–1393.

98. Magierowska, K., Magierowski, M., Hubalewska-Mazgaj, M., Adamski, J., Surmiak, M., Sliwowski, Z., Kwiecien, S., and Brzozowski, T. (2015). Carbon Monoxide (CO) released from Tricarbonyldichlororuthenium (II) Dimer (CORM-2) in gastroprotection against experimental ethanol-induced gastric damage. *PLoS One* 10: e0140493.

99. Winburn, I.C., Gunatunga, K., McKernan, R.D., Walker, R.J., Sammut, I.A., and Harrison, J.C. (2012). Cell damage following carbon monoxide releasing molecule exposure: implications for therapeutic applications. *Basic Clin. Pharmacol. Toxicol.* 111: 31–41.

100. Nielsen, V.G. (2020). Ruthenium, not carbon monoxide, inhibits the procoagulant activity of atheris, echis, and pseudonaja venoms. *Int. J. Mol. Sci.* 21: 2970.

101. Southam, H.M., Smith, T.W., Lyon, R.L., Liao, C., Trevitt, C.R., Middlemiss, L.A., Cox, F.L., Chapman, J.A., El-Khamisy, S.F., Hippler, M., Williamson, M.P., Henderson, P.J.F., and Poole, R.K. (2018). A thiol-reactive Ru(II) ion, not CO release, underlies the potent antimicrobial and cytotoxic properties of CO-releasing molecule-3. *Redox. Biol.* 18: 114–123.

102. Nobre, L.S., Jeremias, H., Romao, C.C., and Saraiva, L.M. (2016). Examining the antimicrobial activity and toxicity to animal cells of different types of CO-releasing molecules. *Dalton Trans.* 45: 1455–1466.

103. Dong, D.L., Chen, C., Huang, W., Chen, Y., Zhang, X.L., Li, Z., Li, Y., and Yang, B.F. (2008). Tricarbonyldichlororuthenium (II) dimer (CORM2) activates non-selective cation current in human endothelial cells independently of carbon monoxide releasing. *Eur. J. Pharmacol.* 590: 99–104.

104. Poss, K.D. and Tonegawa, S. (1997). Reduced stress defense in heme oxygenase 1-deficient cells. *Proc. Natl. Acad. Sci. USA* 94: 10925–10930.

105. Yachie, A., Niida, Y., Wada, T., Igarashi, N., Kaneda, H., Toma, T., Ohta, K., Kasahara, Y., and Koizumi, S. (1999). Oxidative stress causes enhanced endothelial cell injury in human heme oxygenase-1 deficiency. *J. Clin. Invest.* 103: 129–135.

106. Niture, S.K., Kaspar, J.W., Shen, J., and Jaiswal, A.K. (2010). Nrf2 signaling and cell survival. *Toxicol. Appl. Pharmacol.* 244: 37–42.
107. Hayashi, R., Himori, N., Taguchi, K., Ishikawa, Y., Uesugi, K., Ito, M., Duncan, T., Tsujikawa, M., Nakazawa, T., Yamamoto, M., and Nishida, K. (2013). The role of the Nrf2-mediated defense system in corneal epithelial wound healing. *Free Radic. Biol. Med.* 61: 333–342.
108. Kappos, L., Gold, R., Miller, D.H., Macmanus, D.G., Havrdova, E., Limmroth, V., Polman, C.H., Schmierer, K., Yousry, T.A., Yang, M., Eraksoy, M., Meluzinova, E., Rektor, I., Dawson, K.T., Sandrock, A.W., and O'Neill, G.N. (2008). Efficacy and safety of oral fumarate in patients with relapsing-remitting multiple sclerosis: a multicentre, randomised, double-blind, placebo-controlled phase IIb study. *Lancet* 372: 1463–1472.
109. Burness, C.B. and Deeks, E.D. (2014). Dimethyl fumarate: a review of its use in patients with relapsing-remitting multiple sclerosis. *CNS Drugs* 28: 373–387.
110. Wilson, J.L., Fayad-Kobeissi, S., Oudir, S., Haas, B., Michel, B.W., Dubois-Rande, J.L., Ollivier, A., Martens, T., Rivard, M., Motterlini, R., and Foresti, R. (2014). Design and synthesis of novel hybrid molecules that activate the transcription factor Nrf2 and simultaneously release carbon monoxide. *Chemistry* 20: 14698–14704.
111. Nikam, A., Ollivier, A., Rivard, M., Wilson, J.L., Mebarki, K., Martens, T., Dubois-Rande, J.L., Motterlini, R., and Foresti, R. (2016). Diverse Nrf2 activators coordinated to cobalt carbonyls induce heme oxygenase-1 and release carbon monoxide in vitro and in vivo. *J. Med. Chem.* 59: 756–762.
112. Ollivier, A., Foresti, R., El Ali, Z., Martens, T., Kitagishi, H., Motterlini, R., and Rivard, M. (2019). Design and biological evaluation of manganese- and ruthenium-based hybrid CO-releasing molecules (HYCOs). *ChemMedChem* 14: 1684–1691.
113. El Ali, Z., Ollivier, A., Manin, S., Rivard, M., Motterlini, R., and Foresti, R. (2020). Therapeutic effects of CO-releaser/Nrf2 activator hybrids (HYCOs) in the treatment of skin wound, psoriasis and multiple sclerosis. *Redox. Biol.* 34: 101521.
114. Bauer, B., Goderz, A.L., Braumuller, H., Neudorfl, J.M., Rocken, M., Wieder, T., and Schmalz, H.G. (2017). Methyl fumarate-derived iron carbonyl complexes (FumET-CORMs) as powerful anti-inflammatory agents. *ChemMedChem* 12: 1927–1930.

13

Organic CO Donors that Rely on Photolysis for CO Release

Yi Liao

Department of Biomedical and Chemical Engineering and Sciences, Florida Institute of Technology, Melbourne, FL, USA

Introduction

As described in Chapters 12 and 14, CO-releasing molecules (CORMs) and organic CO prodrugs have been developed as CO donors. CORMs are often categorized based on either their chemical structures or the methods of release. The majority of CORMs studied contain transition metals, including both essential trace elements (manganese, iron, and cobalt) and nonphysiological metals (ruthenium, tungsten, and rhenium). Since this chapter is about organic CO donors that rely on photolysis (organic photoCORMs), metallic photoCORMs are briefly reviewed here.

The term "photoCORM" was introduced by Ford and coworkers. The group studied the photoreaction of $W(CO)_5(TPPTS)^{3-}$ (**1**, Figure 13.1), which releases CO upon and after photoirradiation [1]. Motterlini and coworkers used photosensitive $Fe(CO)_5$ and $Mn_2(CO)_{10}$ (**2, 3**, Figure 13.1) to generate CO for biological studies [2]. Different ligands have been introduced to metal carbonyls to increase their water solubility and lower toxicity. Complexes of *fac*-$\{Mn(CO)_3\}^+$ and a variety of ligands have been extensively studied as photoCORMs. For example, Schatzschneider and coworkers used tris(pyrazolyl)methane as a ligand to prepare a *fac*-$\{Mn(CO)_3\}^+$-based photoCORM (**4**, Figure 13.1) [3]. While early metallic photoCORMs are activated by UV light, visible-light-responsive metallic CORMs have been reported in recent years. For example, Westerhausen and coworkers reported a dicarbonyl bis(cysteamine) iron(II) complex (CORM-S1, **5**, Figure 13.1), which releases CO under blue light [4]. Mascharak and coworkers developed a series of visible-light-responsive CORMs with highly conjugated ligands on *fac*-$\{Mn(CO)_3\}^+$ (e.g., **6**, Figure 13.1) [5]. Although Fe and Mn carbonyls are the most studied, Ru and

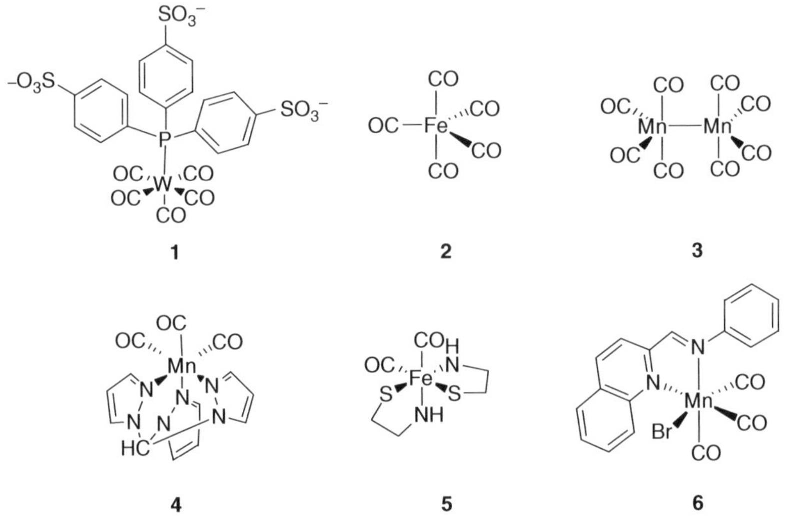

Figure 13.1 Structures of some metallic photoCORMs.

Carbon Monoxide in Drug Discovery: Basics, Pharmacology, and Therapeutic Potential, First Edition. Edited by Binghe Wang and Leo E. Otterbein.
© 2022 John Wiley & Sons, Inc. Published 2022 by John Wiley & Sons, Inc.

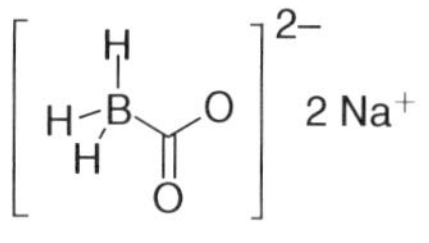

Figure 13.2 Structures of some boron-containing CORMs.

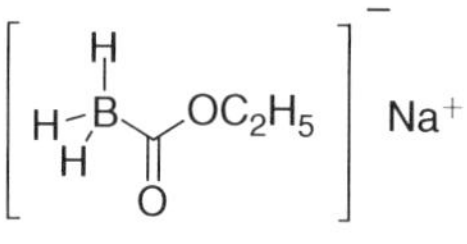

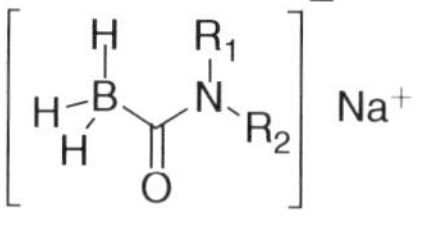

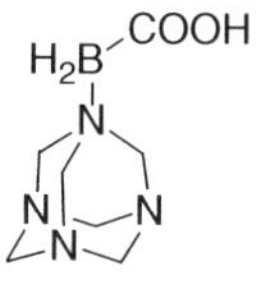

7, **8**, **9**, **10**

Re photoCORMs have also been reported [6,7]. The typical mechanism of CO release from these photoCORMs involves electron transfer from low-valent metal center to the π^* orbital of an ancillary ligands. The metal-to-ligand charge transfer weakens the bond between CO and the metal, which leads to the release of CO. This is by no means a thorough review of metallic photoCORMs. Interested readers may refer to recent reviews of this area [8].

Metallic CORMs contain heavy metals, which present a potential long-term health concern even though the CORMs themselves have shown relatively low cytotoxicity *in vitro*. Nonmetallic CORMs have been developed to address the challenges associated with metallic CORMs. Methylene chloride (MC) is among the earliest studied CO donors [9]. MC is metabolized by the liver and biotransformed to CO. MC administration prevented apoptosis and extends liver allograft survival. Studies also showed that MC administration reduced graft immunogenicity and chronic rejection in combination with HO-1 gene transfer. Several classes of novel organic CO prodrugs based on the ability for norbornadiene-7-ones to release CO via a cheletropic reaction were developed by Wang's group (Chapter 14) [10]. The norbornadiene-7-one-based CO donors can be formed via either a click reaction between a cyclopentadienone and an alkyne or an elimination reaction from a norborn-2-en-7-one moiety. Further, these prodrugs are capable of triggered release (pH-, reactive oxygen species-, esterase-, and enrichment-sensitive triggers), mitochondrion targeting, delivering two payloads with one prodrug, and employing dual-triggering mechanisms. A detailed description of this type of organic CO donors is the subject of Chapter 14. The current state of organic photoCORM is described in this chapter and the future direction of this field is also discussed.

In addition to organic CORMs, nonmetallic CORMs also include some boron compounds. Among them, CORM-A1 with a formula of $Na_2[H_3BCO_2]$ (**7**, Figure 13.2) is one of the most studied CORMs [11]. CORM-A1 has a high CO content with 27 wt% of CO. It decomposes in water and releases one equivalent of CO. Therapeutic applications of CORM-A1 on diseases such as vascular diseases, gastrointestinal diseases, obesity, diabetes, and seizures have been studied. CORM-A1 derivatives including a boranocarbonate ester and boranocarbamates (**8** and **9**, Figure 13.2) have been developed to tune the rate of CO release and improve its therapeutic potential [12]. Amine carboxyboranes have been studied by Dingra's group as novel CORMs [13]. For example, hexamethylenetetramine carboxyborane (HMTA-CB, **10**, Figure 13.2) slowly decomposes under physiological conditions, leading to the formation of CO, boric acid, and HMTA, which (in its salt form) is a drug for treatment of urinary tract infection. The study showed that HMTA-CB prevents the cytokine production in a dose-dependent manner. *meso*-BODIPY carboxylic acid is an organoboron photoCORM, which is described later.

Development of Organic PhotoCORMs

Many organic CORMs rely on photolysis to release CO. For comparison, the most extensively studied nonorganic CORMs, including CORM2, CORM3, CORM-A1, CORM-401, and their derivatives, release CO by reacting with water or other abundant biomolecules. There is no reported organic compound that can generate CO via hydrolysis. It is indeed desirable to develop organic CORMs with tunable and pharmaceutically meaningful hydrolysis rate. However, such molecules would be moisture sensitive and their purification may need special technique. Easy synthesis or commercial availability is always highly desirable.

Organic photoCORMs promise spatially and temporally controlled delivery of CO, and thus are promising prodrugs and powerful research tools for understanding the biological effects of CO. For the development of a photoCORM, a key question is whether it is possible to generate a biologically effective concentration of CO under *in vivo* or *in vitro* conditions using a readily available light source, which does minimal harm to the biological system. Answering this question will allow us to understand the criteria for a useful photoCORM. Herein, we use a simple model to explain these criteria.

First, it is necessary to define a biological effective concentration of CO. The normal concentration of CO in physiological condition is about 2 nM in the presence of ~1% COHb (carbonyl hemoglobin) [14]. Gases with CO concentration of several hundred ppm are commonly used in both *in vivo* and *in vitro* studies. For example, Otterbein and coworkers reported that 250 ppm CO significantly inhibited the expression of lipopolysaccharide-induced proinflammatory cytokine tumor necrosis factor-α and increased the accumulation of the anti-inflammatory cytokine interleukin-10 in *in vitro* and *in vivo* studies [15]. Using a Henry constant of 1100 L atm/mol, the concentration of CO in an aqueous solution equilibrated with 250 ppm CO gas is 2.3×10^{-7} M, i.e., 230 nM. The effective CO level is more than 100 times of the normal level, which indicates that localized delivery of CO is important. While this estimate is valid for an *in vitro* study, the case of *in vivo* study is more complicated. A thorough discussion of the CO level in *in vivo* studies is beyond the scope of this chapter. Interested readers may refer to an article by Levitt and Levitt [14]. Chapter 3 on CO pharmacokinetics also covers this subject. Herein, we use 230 nM as the effective concentration of CO in the model.

We assume a cubic sample with $1 \times 1 \times 1$ cm³ dimension filled with a photoCORM solution. Irradiation is applied straight to one side (1×1 cm²) of the cube. The intensity (I) of irradiation required to generate 230 nM CO in the sample is determined by the wavelength of the light (λ), fraction of light absorbed (f), quantum yield of photoreaction (ϕ), number of CO (n) generated from each photoCORM, and the time (t) needed to generate the concentration. Considering all these factors leads to Equation 13.1:

$$I = \frac{2.3 \times 10^{-10} \times N_A \times h \times c}{\lambda \times f \times \phi \times n \times t} (W/cm^2), \quad (13.1)$$

where N_A is the Avogadro constant, h is the Planck constant, and c is the speed of light.

For *in vivo* studies, light must be able to penetrate the tissue and thus the wavelength must be in the penetration window between 600 and 1100 nm [16]. Shorter wavelength can be used in *in vitro* studies. However, the wavelength still needs to be in the visible range to avoid harmful effects on the cells. The fraction of absorption (f) depends on the molar absorptivity (ε) of the photoCORM, the concentration of the photoCORM (C), and the external light loss. The external light loss can be minimized in an *in vitro* study. For *in vivo* studies or photodynamic therapy (PDT), the light source is often 1–2 cm from the targeted tissue. If near-infrared light is used, the light loss will be 75–95% based on a half-loss depth of 5 mm [17]. In other words, about 5–25% of light can arrive the sample. Since the concentration of the photoCORM is often tens of μM or lower, it is difficult for the sample to completely absorb the light that arrives. The percentage of the light absorbed by the photoCORMs can be calculated using the Beer–Lambert law, i.e., $1-10^{-\varepsilon l C}$.

From Equation 13.1, it is apparent that a high quantum yield can lower the required power of light and is desirable. A photoCORM may release more than one CO after absorbing one photon, which increases overall efficiency of the photorelease. The capability of releasing more than one CO also implies a higher CO content of the photoCORM and less side product from the photoreaction. In both *in vivo* and *in vitro* studies, the system is open, which means CO generated from the photoCORMs can diffuse out of the irradiated sample. Therefore, CO must be quickly generated in certain time so that it can accumulate to the effective concentration needed, despite diffusion.

To estimate the required I in a typical experiment, we assume that 800 nm light is used and each photoCORM releases one CO. We also assume that percent of the light that arrives the sample, is absorbed by the sample, and induces the photoreaction (ϕ) are all 10%, and thus $f \times \phi = 0.001$. Given the small size of the sample, the time that is required to generate 230 nM CO is assumed to be 5 s. Plugging in these values to Equation 13.1 results in an I of 7 mW/cm². For comparison, the upper limit of the power used for PDT is about 200 mW/cm², and the power of LED commonly used for *in vitro* studies is tens of mW/cm². Even if we assume ϕ is 1%, I will be 70 mW/cm², which is still in the allowable range for both *in vivo* and *in vitro* studies.

Besides the photochemical requirements, the toxicity of a photoCORM must be low. From the earlier discussion, the IC_{50} must be much larger than 230 nM. The photoreaction generates not only CO but also a side product. The effect of this side product is even a bigger concern than that of the photoCORM since it stays in the system for a longer period. The biological effects of a synthetic compound are difficult to predict. Therefore, it is desirable that the toxicity and, ideally, other bioeffects of the side product have been well studied before. Another approach is endowing the side product a useful function. For example, the increased or decreased fluorescence of the side product compared to that of the photoCORM has been utilized to track photoreaction. This has become a common feature of recently developed photoCORMs.

Figure 13.3 Structures of organic photoCORMs.

Aromatic diketones
11 R=H,
12 R=OC₈H₁₇
13 R= t-Bu
14

xanthene-9-carboxylic acid
15

meso-BODIPY carboxylic acids
16 R= Me
17 R= —⫽—⟨phenyl⟩—O(CH₂CH₂O)₃CH₃

3-hydroxyflavones
19 X=O, R=R'=H
20 X=O, R=NEt₂, R'=H
21 X=S, R=R'=H
22 X=S, R=NEt₂, R'=H
23 X=O, R=H, R'=SO₃⁻
18

3-hydroxyquinoline
24

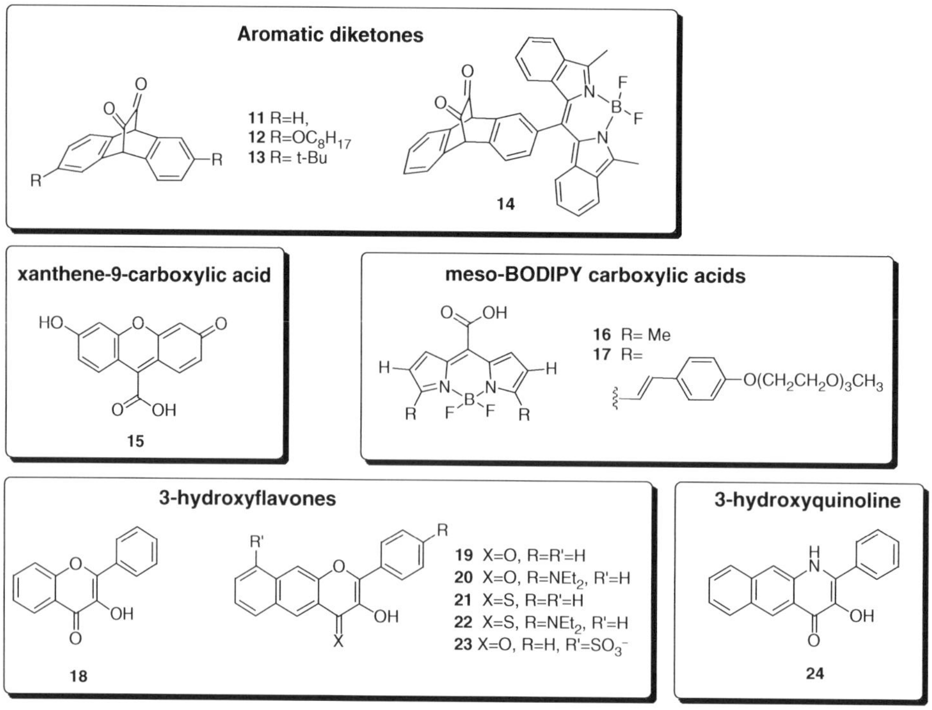

In summary, an ideal photoCORM can be easily synthesized and has a high photoactivity resulting from a high quantum yield and a high molar absorptivity. For *in vivo* studies, the activating wavelength must be in the penetration window for biological samples. For *in vitro* studies, the wavelength can be in the visible range. The toxicity of both the photoCORM and the side product(s) generated from the photoreaction must be low. A useful function of the side product is desirable. Capability of releasing multiple CO increases the photorelease efficiency and decreases the amount of side product. Targeting the photoCORM to the target is highly desirable for generating an effective and localized concentration. In addition, solubility in water is often a concern. Solubility of hydrophobic photoCORMs can be improved by encapsulating them in a water-soluble nanocarrier.

Current Organic PhotoCORMs

As discussed earlier, there are many requirements for a useful photoCORM. Fortunately, most of these requirements were considered since the early development of organic photoCORMs. Today, there are three types of organic compounds that have been studied as photoCORMs, including aromatic carboxylic acids, hydroxyflavones, and aromatic diketones. The aromatic carboxyl photoCORMs, including derivatives of xanthene carboxylic acid and *meso*-carboxy BODIPY, were developed and studied by Klan and coworkers [18,19]. The hydroxyflavones were extensively studied by Berreau and coworkers [20–23]. Such studies also include hydroxyquinolone derivatives [24]. Aromatic diketones were mostly studied by Liao and coworkers [25–27]. The structures of these photoCORMs are listed in Figure 13.3. These organic photoCORMs have been reviewed in several recent articles [28,29]. This chapter intends to give an overview of the area rather than detailed description of each individual organic photoCORM. Therefore, these photoCORMs are compared with each other in different aspects.

Discovery of the Organic PhotoCORMs

The three types of organic photoCORMs were independently developed and have different mechanisms of photorelease of CO. It is interesting to see how these CORMs were discovered. Around 2010, Liao's group was conducting a systematic study of photo-retro-Diels–Alder reaction. During this study, they noticed the photoreaction of compound **11** (Figure 13.3), which is an aromatic cyclic α-diketone. Its photoreaction had been previously studied by Neckers and coworkers [30]. Unlike some nonaromatic α-diketones, **11** quantitatively generates two equivalents of CO and anthracene with a moderate quantum yield (2%). Liao's group realized that it could be used as an organic photoCORM. At that time, no organic photoCORM had been reported although metallic photoCORMs were known. However, it was found that the photoactivity of **11** was completely diminished in aqueous condition due to hydration. They then encapsulated **11** in pluronic micelles, which protected **11** from hydration and increased the water solubility. The encapsulated **11** as well as its derivatives effectively released CO under visible light in aqueous conditions [25]. As can be seen here, this type of photoCORM was developed based on a photoreaction studied before. The hydration problem, which prevented the diketones to be used as photoCORMs, was solved using a materials chemistry approach. The resulting photoCORMs are actually functional nanomaterials rather than molecules.

Klan's group has been working on photoinduced drug release for years. According to Šebej, one of the major contributors of the photoCORM work, they observed "serendipitously" formed xanthene-9-carboxylic acid (**15**, Figure 13.3) as a photoproduct of a drug-releasing compound developed by the group. This fluorescein analogue was found to release one equivalent of CO under 500 nm light in aqueous buffer and was recognized as the first organic visible-light-triggered water-soluble CORM [18]. However, the synthesis of **15** is cumbersome, which prevented further studies of its biological effects. The group then designed meso-BODIPY carboxylic acid (**16**, Figure 13.3), which is isoelectronic to **15** [19]. Compound **16** is much easier to synthesize and allows photoactivation with a longer wavelength of light. Different from the diketones, this type of photoCORM was developed based on a new photoreaction discovered in a different study. Although the first one of this type was not ideal, further development led to a useful photoCORM.

It was well known that 3-hydroxyflavones generate one equivalent of CO upon catalysis by dioxgenases of fungal and bacterial origins [31]. Many synthetic model complexes have been developed to mimic enzyme/substrate adducts in dioxygenases. Berreau and coworkers studied a series of synthetic complexes with a 3-hydroxyflavonolato ligand and found that they underwent a clean photoreaction in the presence of O_2 and produced one equivalent of CO [32,33]. Actually, 3-hydroxyflavone itself undergoes similar oxygenation reaction and produces CO under irradiation [34,35]. However, either UV light or a photosensitizer must be used for this reaction, which makes it unsuitable for biological applications. Berreau's group designed a flavonol derivative (**18**, Figure 13.3) by substituting one of the phenyl moiety with a naphthalenyl group [20]. This modification extended the conjugation of the molecule and allowed the photoreaction to be induced by visible light. Hydroxyquinoline **24** with a similar structure was developed later to further improve the biocompatibility [24]. The development of this type of photoCORMs was based on a known photoreaction of a biomolecule. Modification of the structure solved the problem of activating wavelength and resulted in a useful organic photoCORM.

As described earlier, the criteria for photoCORM, including activating wavelength, water solubility, easy accessibility, etc., were considered from the beginning of the development, even though different approaches were used for these organic photoCORMs. The toxicity, which is hard to predict based on the molecular structure, was also somewhat considered. Flavonols are found in fruits and vegetables and have known beneficial effects on health. Their derivatives are expected to have relatively low toxicity. The aromatic carboxylic photoCORMs are analogues of fluorescein and BODIPY, which are commonly used fluorophores in biological studies. The diketones were not used in biological study before they were tested as photoCORMs. However, anthracene, the side product of the photoreaction, has been reported to have very low toxicity in animal tests.

Photoreactions and Photochemical Properties

The photoreactions of the three types of photoCORMs are shown in Figure 13.4. Irradiating an aqueous solution of xanthene-9-carboxylic acid **15** with a 500-nm LED resulted in formation of CO and 3,6-dihydroxy-9H-xanthen-9-one (Figure 13.4A) [18]. Based on the results from isotopic labeling experiments, it was proposed that an intermediate α-lactone formed from the photoreaction, which rapidly decomposed to release one equivalent of CO. The meso-carboxy BODIPY shared a similar mechanism [19]. This type of photoCORM generally has long maximum absorption wavelength (λ_{max}) near or above 500 nm and relatively high molar absorptivity of around 10^4 M^{-1} cm^{-1}. However, the quantum yields were only in the range of 10^{-5} to 10^{-4}. The combination of the low quantum yield and high molar absorptivity yields a reasonable cross section. Especially, **17** with extended conjugated structure has a $\lambda_{max} = 652$ nm and substantial absorption tailing up to ~750 nm. While the quantum yield is only 1.2×10^{-5}, the unique spectral properties enable it to release CO under 730 nm light, which is in the tissue penetration window.

Figure 13.4 Photoreactions of the three types of organic photoCORMs.

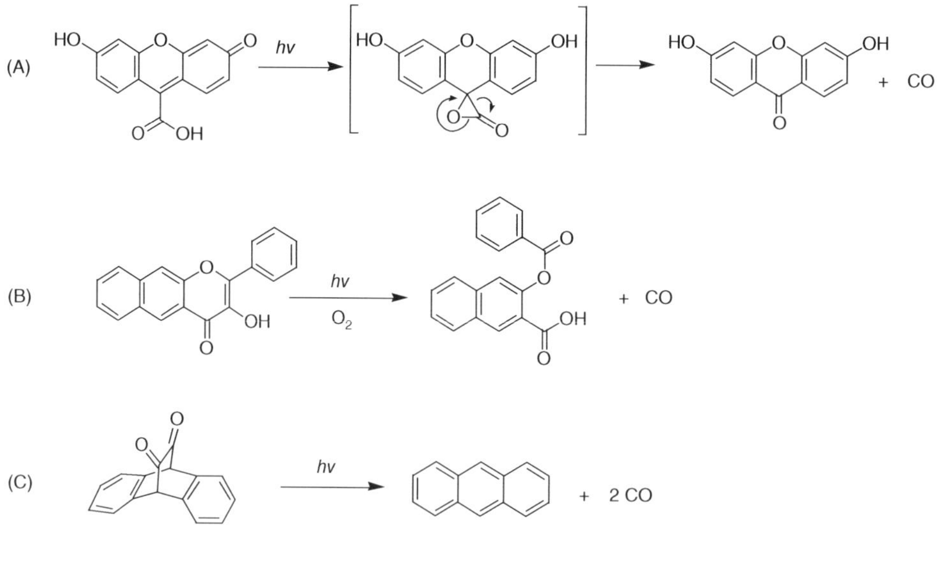

The CO release reaction of 3-hydroxyflavones is a photoinduced bimolecular reaction between the flavonol and oxygen. This reaction has been studied by different groups, but the mechanism is still not fully understood. Both singlet and triplet oxygen species have been proposed as the reactant, and the intermediate may be a 2 + 3 or 2 + 2 adduct [29,31,34,35]. However, the products of the photoreaction, CO and a O-benzoyl salicylic acid derivative, are well defined. It is known that 3-hydroxyflavone also undergoes photoinduced rearrangement in the absence of O_2 [36]. In some cases, the photoproducts decomposed to release CO. As expected, the photoreactions of **19** are similar to those of the 3-hydroxyflavones (Figure 13.4B). Under visible light with a wavelength near its λ_{max} (409 nm), **19** undergoes dioxygenase-type reaction and releases one equivalent of CO. In the presence of oxygen, the yield is essentially quantitative. The quantum yield is 0.006–0.01 depending on the solvent. Notably, **19** has been recently reported to be activated by two-photon excitation at 800 nm to trigger CO release [37]. The sulfur substituted **21** showed a much improved quantum yield of 0.426. The 3-hydroxybenzoquinolone **24** undergoes similar photoreaction with a quantum yield of 0.0045 [38].

The photoreaction of the diketone **11** yields two CO and an anthracene derivative with a quantum yield of 0.02 (Figure 13.4C) [25,30]. The reaction is likely to have a concerted mechanism, which explains the quantitative production of anthracene. The CO content of **11** is as high as 24 wt% due to the capability of releasing more than one CO and the relatively small molecular size. The photoreaction can be induced by visible light even though **11** does not have an extended conjugated structure. This is because photoexcitation induces n–π* instead of π–π* transition. The n–π* transition does not rely on large conjugated structure and allows small molecules to absorb long-wavelength light. However, the molar absorptivity for the n–π* transition is low (<1000 M^{-1} cm^{-1}) and it is difficult to further shift the activating wavelength to the tissue penetration window. Yamada's group introduced a BODIPY moiety on the diketone (**14** in Figure 13.3), which resulted in a molecule with an absorption band up to ~650 nm [38]. Upon excitation, the BODIPY moiety transfers an electron to the diketone moiety and triggers CO release. However, a metal halide lamp with a 390-nm filter was used as the light source for the CO release studies. It is not clear whether the molecule can be activated by light above 600 nm.

The photoreactions of the three types of organic photoCORMs can be monitored utilizing the difference of fluorescence between the photoCORMs and the corresponding photoproducts. The flavonols and benzoquinolonols are fluorescent, which allows monitoring of their distribution before the release of CO. The side product from the CO photoreleasing reaction is nonfluorescent and thus acts as a turn-off reporter. The diketones themselves are only weakly fluorescent, while the anthracene side products from the photoreaction are highly fluorescent and act as turn-on reporters. The xanthene and BODIPY carboxylic acids and the side products of their photoreactions are both fluorescent. However, the absorption and emission of the photoproducts are significantly blueshifted, which allows differentiation of the photoCORMs and the photoproducts and monitoring of the photoreaction.

Biological Studies

Low toxicity is required for a CORM. The flavonol **19** showed low toxicity to A549 (IC_{50} = 41 µM), human umbilical vein endothelial cells (HUVECs) (IC_{50} = 82 µM), and RAW 264.7 cells (nontoxic up to 100 µM) [20]. The benzoquinolonol **24** as well as its side product from the photoreaction was tested on HUVECs using an MTT assay. Cell viability of greater than 50% was observed for each compound up to 100 µM [24]. The cytotoxicity of xanthene-9-carboxylic acid **15** has never been tested likely due to the limited availability of this compound. The BODIPY carboxylic acids showed no toxicity on HepG2 and neuroblastoma SH-SY5Y cell lines at up to 100 µM [19]. Diketone **12** (Figure 13.4) encapsulated in pluronic micelles and the corresponding anthracene photoproduct showed no observable cytotoxicity on KG-1 cells at up to 40 µM [25]. Negligible toxicity of poly(butyl cyanoacrylate) nanoparticle containing the diketone **13** (Figure 13.4) on endothelial cells was observed up to 50 µg/ml [27]. In summary, all three types of organic photoCORMs have fairly low cytotoxicity at tens of µM at least in *in vitro* studies.

The benzoquinolonol **24** strongly binds to bovine serum albumin (BSA). Berreau, Benninghoff, and coworkers studied the anticancer effects of the CORM/BSA adduct on A549 cells [24]. A low IC_{50} value of 24 µM was observed after the CORM was activated by visible light. In addition, the adduct also produced significant anti-inflammatory effects on RAW 264.7 cells. Recently, the same group compared the effects of extracellular and intracellular CO release using the flavonol **19** and its sulfonated derivative **23** [23]. The former is lipophilic and can diffuse into the cells, while the latter has low lipophilicity and stays out of the cell. The results showed that extracellular CO release is less toxic and is sufficient to produce an anti-inflammatory effect similar to that of intracellularly released CO at nanomolar concentrations. This is an excellent example of using photoCORMs as research tools to understand the effects of CO. Works on targeted CO delivery are covered in Chapter 15.

As described earlier, BODIPY carboxylic acid **17** absorbs light in the tissue penetration window. This is the only organic photoCORM that has been used in *in vivo* studies [19]. The work by Klan's group showed that SKH1 mice treated with **17** followed by irradiation doubled the level of COHb in the blood, liver, and kidneys. Bashur, Liao, and coworkers incorporated diketone **12** in electrospun polycaprolactone scaffolds for cardiovascular tissue applications [26]. Results showed that the mesh allowed a time frame, in which vascular cells can be seeded prior to activation. This permitted the conduit to be used as a tissue engineered vascular graft for a cardiovascular application.

Summary and Outlook

As described earlier, different types of organic photoCORMs have been developed in the past few years. All these molecules and materials meet the basic requirements for a useful photoCORM at least for *in vitro* studies. In fact, they have been applied to both *in vitro* and *in vivo* studies and started to show utilities as research tools.

It is expected that organic photoCORMs and related innovations will be improved in different aspects and more fruitful results will appear in near future. Well-designed *in vivo* studies on the treatment of certain diseases will generate high impact in this area. It is desirable that light can be applied using a well-accepted technique, for example, laser setup in PDT. Tissue penetration is a common problem for photomedicine. The activating wavelength must be in the tissue penetration window and the targeted disease shall occur near skin. Although photoirradiation allows CO to be released in the targeted location, it does not guarantee a high local concentration. Nanotechnology for accumulating drugs in targeted location may be utilized to solve this problem. Development of CORMs with high CO contents will also help. CORMs generate both CO and side products, which is one of the major concerns for using them as therapeutics. It is desirable to develop organic CORMs that generate food or drug molecules as the side products. In this case, the function of the drug needs to contribute to that of CO.

Organic photoCORMs are likely to be used as research tools in the near future. They may be used to study the function of CO in cells. In this case, the organic photoCORMs may be modified to target different organelles or biomolecules [22]. The photoCORMs may also be applied to complex cell or tissue models, which will allow better understanding of the biological function of CO as a gasotransmitter. The future of organic photoCORM largely relies on the progress in CO-based medicine. After all, photorelease is a drug delivery method. It is important only if the drug is important.

References

1. Rimmer, R.D., Richter, H., and Ford, P.C. (2009). A photochemical precursor for carbon monoxide release in aerated aqueous media. *Inorg. Chem.* 49: 1180–1185.
2. Motterlini, R., Clark, J.E., Foresti, R., Sarathchandra, P., Mann, B.E., and Green, C.J. (2002). Carbon monoxide-releasing molecules characterization of biochemical and vascular activities. *Circ. Res.* 90: e17–e24.
3. Niesel, J. and Pinto, A. (2008). NDongo, H.W.P.; Merz, K.; Ott, I.; Gust, R.; Schatzschneider, U. Photoinduced CO release, cellular uptake and cytotoxicity of a tris(pyrazolyl)methane (tpm) manganese tricarbonyl complex. *Chem. Commun.* 1798–1800.
4. Kretschmer, R., Gessner, G., Görls, H., Heinemann, S.H., and Westerhausen, M. (2011). Dicarbonyl-Bis(Cysteamine) Iron (II): A light induced carbon monoxide releasing molecule based on iron (CORM-S1). *J. Inorg. Biochem.* 105: 6–9.
5. Gonzalez, M.A., Carrington, S.J., Fry, N.L., Martinez, J.L., and Mascharak, P.K. (2012). Syntheses, structures, and properties of new manganese carbonyls as photoactive CO-releasing molecules: Design strategies that lead to CO photolability in the visible region. *Inorg. Chem.* 51: 11930–11940.
6. Yang, S., Chen, M., Zhou, L., Zhang, G., Gao, Z., and Zhang, W. (2016). Photo-activated co-releasing molecules (PhotoCORMs) of robust Sawhorse Scaffolds [μ^2-OOCR^1, η^1-NH$_2$CHR^2(C=O)OCH$_3$, Ru(I)$_2$CO$_4$. *Dalton Trans.* 45: 3727–3733.
7. Carrington, S.J., Chakraborty, I., Bernard, J.M., and Mascharak, P.K. (2016). A theranostic two-tone luminescent photoCORM derived from Re (I) and (2-Pyridyl)-Benzothiazole: Trackable CO delivery to malignant cells. *Inorg. Chem.* 55: 7852–7858.
8. For example, Kottelat, E. and Zobi, F. (2017). Visible light-activated photoCORMs. *Inorganics* 5: 24.
9. Chauveau, C., Bouchet, D., Roussel, J.-C., Mathieu, P., Braudeau, C., Renaudin, K., Tesson, L., Soulillou, J.-P., Iyer, S., Buelow, R., and Anegon, I. (2002). Gene transfer of heme oxygenase-1 and carbon monoxide delivery inhibit chronic rejection. *Am. J. Transplant.* 2: 581–592.
10. Ji, X. and Wang, B. (2018). Strategies toward organic carbon monoxide prodrugs. *Acc. Chem. Res.* 51: 1377–1385. Hopper, C.P.; De La Cruz, L. K.; Lyles, K.V.; Wareham, L. K.; Gilbert, J. A.; Eichenbaum, Z.; Magierowski, M.; Poole, R. K.; Wollborn, J.; Wang, B. Role of Carbon Monoxide in Host-Gut Microbiome Communication. *Chem Rev.* **2020**, doi:10.1021/acs.chemrev.0c00586.
11. Motterlini, R., Sawle, P., Hammad, J., Bains, S., Alberto, R., Foresti, R., and Green, C.J. (2005). CORM-A1: A new pharmacologically active carbon monoxide-releasing molecule. *FASEB J.* 19: 284–286.
12. Pitchumony, T.S., Spingler, B., Motterlini, R., and Alberto, R. (2010). Syntheses, structural characterization and CO releasing properties of boranocarbonate [H$_3$BCO$_2$H]$^-$ derivatives. *Org. Biomol. Chem.* 8: 4849–4854.
13. Ayudhya, T.I., Raymond, C.C., and Dingra, N.N. (2017). Hexamethylenetetramine carboxyborane: Synthesis, structural characterization and CO releasing properties. *Dalton Trans.* 46: 882–889.
14. Levitt, D.G. and Levitt, M.D. (2015). Carbon monoxide: a critical quantitative analysis and review of the extent and limitations of its second messenger function. *Clin. Pharmacol.* 7: 37–56.
15. Otterbein, L.E., Bach, F.H., Alam, J., Soares, M., Lu, T., Wysk, M., Davis, R.J., Flavell, R.A., and Choi, A.M. (2000). Carbon monoxide has anti-inflammatory effects involving the mitogen-activated protein kinase pathway. *Nat. Med.* 6: 422–428.
16. Hamblin, M.R. and Huang, -Y.-Y. (2014). Handbook of Photomedicine, 1e. Boca Raton: CRC Press.
17. Ruggiero, E., Alonso-de Castro, S., Habtemariam, A., and Salassa, L. (2016). Upconverting nanoparticles for the near infrared photoactivation of transition metal complexes: new opportunities and challenges in medicinal inorganic photochemistry. *Dalton Trans.* 45: 13012–13020.
18. Sebej, P., Wintner, J., Müller, P., Slanina, T., Anshori, J.A., Antony, L.A., Klan, P., and Wirz, J. (2013). Fluorescein analogues as photoremovable protecting groups, absorbing at 520 nm. *J. Org. Chem.* 78: 1833.
19. Palao, E., Slanina, T., Muchova, L., Solomek, T., Vitek, L., and Klan, P. (2016). Transition-metal-free CO-releasing BODIPY derivatives activatable by visible to NIR light as promising bioactive molecules. *J. Am. Chem. Soc.* 138: 126–133.
20. Anderson, S.N., Richards, J.M., Esquer, H.J., Benninghoff, A.D., Arif, A.M., and Berreau, L.M. (2015). A structurally-tunable 3-Hydroxyflavone motif for visible light-induced carbon monoxide-releasing molecules (CORMs). *ChemistryOpen* 4: 590–594.
21. Soboleva, T. and Berreau, L.M. (2019). Tracking CO release in cells via the luminescence of donor molecules and/or their by-products. *Isr. J. Chem.* 59: 339–350.
22. Lazarus, L.S., Benninghoff, A.D., and Berreau, L.M. (2020). Development of triggerable, trackable, and targetable carbon monoxide releasing molecules. *Acc. Chem. Res.* 53: 2273–2285.

23. Lazarus, L.S., Simons, C.R., Arcidiacono, A., Benninghoff, A.D., and Berreau, L.M. (2019). Extracellular vs intracellular delivery of CO: Does it matter for a stable, diffusible gasotransmitter? *J. Med. Chem.* 62: 9990–9995.
24. Popova, M., Lazarus, L.S., Ayad, S., Benninghoff, A.D., and Berreau, L.M. (2018). Visible-light-activated quinolone carbon-monoxide-releasing molecule: Prodrug and albumin-assisted delivery enables anticancer and potent anti-inflammatory effects. *J. Am. Chem. Soc.* 140: 9721–9729.
25. Peng, P., Wang, C., Shi, Z., Johns, V.K., Ma, L., Oyer, J., Copik, A., Igarashi, R., and Liao, Y. (2013). Visible-light activatable organic CO-releasing molecules (PhotoCORMs) that simultaneously generate fluorophores. *Org. Biomol. Chem.* 11: 6671–6674.
26. Michael, E., Abeyrathna, N., Patel, A.V., Liao, Y., and Bashur, C.A. (2016). Incorporation of photo-carbon monoxide releasing materials into electrospun scaffolds for vascular tissue engineering. *Biomed. Mater.* 11: 25009.
27. Elgattar, A., Washington, K.S., Talebzadeh, S., Alwagdani, A., Khalil, T., Alghazwat, O., Alshammri, S., Pal, H., Bashur, C., and Liao, Y. (2019). Poly(butyl cyanoacrylate) nanoparticle containing an organic photoCORM. *Photochem. Photobiol. Sci.* 18: 2666–2672.
28. Abeyrathna, N., Washington, K., Bashur, C., and Liao, Y. (2017). Nonmetallic carbon monoxide releasing molecules (CORMs). *Org. Bio. Mol. Chem.* 15: 8692–8699.
29. Slanina, T. and Šebej, P. (2018). Visible-light-activated photoCORMs: rational design of CO-releasing organic molecules absorbing in the tissue-transparent window. *Photochem. Photobiol. Sci.* 17: 692–710.
30. Mondal, R., Okhrimenko, A.N., Shah, B.K., and Neckers, D.C. (2008). Photodecarbonylation of R-diketones: A mechanistic study of reactions leading to acenes. *J. Phys. Chem. B* 112: 11–15.
31. Soboleva, T. and Berreau, L.M. (2019). 3-Hydroxyflavones and 3-Hydroxy-4-oxoquinolines as carbon monoxide-releasing molecules. *Molecules* 24: 1252.
32. Grubel, K., Laughlin, B.J., Maltais, T.R., Smith, R.C., Arif, A.M., and Berreau, L.M. (2011). Photochemically-induced dioxygenase-type CO-release reactivity of group 12 metal flavonolate complexes. *Chem. Commun.* 47: 10431–10433.
33. Grubel, K., Marts, A.R., Greer, S.M., Tierney, D.L., Allpress, C.J., Anderson, S.N., Laughlin, B.J., Smith, R.C., Arif, A.M., and Berreau, L.M. (2012). Photoinitiated dioxygenase-type reactivity of open-shell 3d divalent metal flavonolato complexes. *Eur. J. Inorg. Chem.* 4750–4757.
34. Matsuura, T., Matsushima, H., and Nakashima, R. (1970). Photoinduced reactions—XXXVI: Photosensitized oxygenation of 3-hydroxyflavones as a nonenzymatic model for quercetinase. *Tetrahedron* 26: 435–443.
35. Studer, S.L., Brewer, W.E., Martinez, M.L., and Chou, P.T. (1989). Time-resolved study of the photooxygenation of 3-hydroxyflavone. *J. Am. Chem. Soc.* 111: 7643–7644.
36. Matsuura, T., Takemoto, T., and Nakashima, R. (1973). Photoinduced reactions—LXXI: Photorearrangement of 3-hydroxyflavones to 3-aryl-3-hydroxy-1,2-indandiones. *Tetrahedron* 29: 3337–3340.
37. Li, Y., Shu, Y., Liang, M., Xie, X., Jiao, X., Wang, X., and Tang, B. (2018). A two-photon H2O2-activated CO photoreleaser. *Angew. Chem. Int. Ed. Engl.* 57: 12415–12419.
38. Aotake, T., Suzuki, M., Tahara, K., Kuzuhara, D., Aratani, N., Tamai, N., and Yamada, H. (2015). An optically and thermally switchable electronic structure based on an anthracene-BODIPY conjugate. *Chem. - A Eur. J.* 21: 4966–4974.

14

Organic Carbon Monoxide Prodrugs that Release CO Under Physiological Conditions

Zhengnan Yuan and Binghe Wang

Department of Chemistry and Center for Diagnostics and Therapeutics, Georgia State University, Atlanta, GA 30303, USA

Introduction

As discussed in various chapters in this book, endogenously produced carbon monoxide (CO) has been shown to interact with various molecular targets and possesses cellular signaling roles in mammals [1–4]. Beyond that, Chapters 19–30 cover extensively the therapeutic applications of CO. For clinical applications of CO-based therapeutics, the development of pharmaceutically acceptable delivery methods is a critical issue because of the gaseous nature of CO. Depending on the therapeutic indications, the need for delivery features may vary. Therefore, the availability of a range of delivery methods is a critical strength in the field for various applications. Section II of this book covers the various CO delivery methods, including inhalation and CO solution (Chapter 11), metal-based CO-releasing molecules (CORMs) (Chapter 12), organic CO donors that rely on photolysis for CO release (Chapter 13), targeted delivery of CO (Chapter 15), anesthesia-derived CO (Chapter 16), and natural product-based CO sources (Chapter 17). In this chapter, we focus on the design, synthesis, and pharmacological validation of organic CO prodrugs that release CO under physiological conditions with tunable release rates, enrichment-triggered release, the ability to deliver two payloads with a single prodrug, and triggered release by pH, enzyme, and reactive oxygen species (ROS). Further, a new generation of prodrugs using FDA-approved artificial sweeteners as the "carrier" molecules is also included.

At this moment, it is important to define the issue of "prodrug," which is a very well-established pharmaceutical approach to delivering bioactive ingredients with a long tradition [5]. One important reason for us to use this term, prodrug, is to emphasize the need for considering pharmaceutical issues in developing and studying CO donors, which should go beyond the chemical property of being able to release a CO molecule. There is much more to a successful donor than such a chemical property. This developability issue [6,7] becomes prominent at the beginning of the "Validation of the Pharmacological Efficacy of the CO Prodrugs in Various Disease Models" section, in discussing the various factors to consider in designing biological experiments as validations of the pharmacological efficacy of various CO donors. Because of the vast experience of the pharmaceutical industry in developing small molecule drugs and prodrugs [6,7] and the well-established approaches of using prodrugs to deliver an active ingredient [5], we feel that these organic CO prodrugs will be very useful as research tools and potential therapeutic agents. Further, these prodrugs are structurally diverse and amendable to structural optimization, tethering with targeting molecules, and various formulation manipulations. These are all important features for pharmaceutical development. It should be noted that much of the science has been published and extensively reviewed [8–10]. This chapter will try to minimize overlaps with published reviews and will instead focus on our thought processes in designing these prodrugs, the chemistry principles, problems to overcome, pharmacological validations, and remaining issues. Below, we aim to provide an overview of these organic CO prodrugs by dividing this chapter into four subsequent sections: (i) the general chemistry principles of all reported organic CO prodrugs that fall into two types, A and B; (ii) descriptions of individual prodrugs of type A; (iii) description of individual prodrugs of type B; and (iv) pharmacological validations in cell cultures and animal models. This organization is intended to allow for easy

---

Carbon Monoxide in Drug Discovery: Basics, Pharmacology, and Therapeutic Potential, First Edition. Edited by Binghe Wang and Leo E. Otterbein.
© 2022 John Wiley & Sons, Inc. Published 2022 by John Wiley & Sons, Inc.

navigation of this chapter by readers with expertise in different subject areas. It is also important to note that the "Validation of the Pharmacological Efficacy of the CO Prodrugs in Various Disease Models" section starts with discussions of issues to pay attention to in designing experiments involving CO prodrugs, especially in relevance to proper controls needed for meaningful results.

Chemistry Concepts and Design Principles Used in Developing Organic CO Prodrugs

Prodrug development in general has had a long and successful history [5]. In designing traditional prodrugs, a functional group "handle" is needed for the drug molecule for bioreversible derivatization. In the case of CO, however, this "handle" does not exist. In a typical organic molecule, one would not expect to see a "CO" group in the right oxidation state for CO release unless prodrug activation involves breaking a carbon–carbon bond, which is hard under near-physiological conditions. Therefore, some unconventional approaches are needed. There are two types of CO prodrugs reported so far. The first one is based on cheletropic reaction of a norbornadienone structural unit [4,8] (Scheme 14.1A, type A) and the second on decarboxylation–decarbonylation chemistry of an "α-ketoacid" moiety (Scheme 14.1B, type B).

For the first type of CO prodrugs, the key concept is to take advantage of an extrusion reaction for CO release (Scheme 14.1A). In this case, it is the cheletropic reaction of a norbornadienone structural unit. As background information, the basic idea of the design was rooted in a seminar that Binghe Wang gave as a graduate student in spring 1985 at the Department of Chemistry, University of British Columbia. The seminar was entitled "Extrusion Reactions" (or something similar), which covered a range of such reactions, including those that lead to CO release; obviously, this unique class of reactions made quite an impression on a first-year graduate student. Who would have thought that approximately 30 years later, such extrusion reactions discussed in a graduate seminar would form the foundation of a major class of organic CO prodrugs?

Briefly, cheletropic reactions are pericyclic reactions that are bioorthogonal to reactions needed in normal biochemical transformations in living systems and thus are ideal for this type of applications in designing prodrugs. This specific type of cheletropic reactions involving norbornadienones has been known for a long time [11]. However, a key question is how to adapt a chemical transformation that is normally carried out under nonphysiological conditions for application *in vivo* or at least under near-physiological conditions.

The first chemistry issue to overcome is the spontaneous nature for norbornadienones to undergo extrusion reactions to release CO at near-physiological temperature. Such chemical reactivities mean that norbornadienones themselves are not suitable structural features as prodrugs. This also means that the design has to start with precursors to a norbornadienone intermediate (**1**) as the true prodrug (Scheme 14.1A). One can envision *in situ* "synthesis" of norbornadienones as a way to deliver CO. In doing so, there are three general approaches, all of which involve *in situ* activation through (A-i) a bimolecular Diels–Alder reaction (DAR), (A-ii) a unimolecular/intramolecular DAR, or (A-iii) an elimination reaction by involving an electron-withdrawing group (EWG) and a leaving group (LG) to yield the unstable norbornadienone **11** for CO generation (Scheme 14.1A). There are pros and cons as well as unique challenges for each approach, which are discussed in detail in the following sections. It should also be noted that with proper design of the R_2 and R_3 groups, the products after CO release (**6**, **9**, and **12**) can be made fluorescent. This allows for the formation of a fluorescent reporter for easy monitoring of the CO release reactions.

For the second type of CO prodrugs based on decarbonylation chemistry of an "α-ketoacid" moiety, the design takes advantage of a decarboxylation–decarbonylation sequence of reactions. A key driver for us to design the second type of CO prodrugs is our desire to use a "carrier" with minimal toxicity and/or known safety profiles. For this purpose, a chemical strategy should allow the installation of a benign moiety as the CO carrier. With proper LGs, 1,2-dicarbonyl compounds are well known to undergo hydrolysis, decarboxylation, and decarbonylation to release CO, CO_2, and the corresponding LGs in aqueous environments (Scheme 14.1B). By selecting a suitable and benign candidate as the LG, such simple and reliable chemistry can be adapted for developing CO prodrugs by using a "carrier" molecule with minimal and/or known safety profiles. Saccharin and acesulfame are both FDA-approved sweeteners, which are widely used in the food industry [12]. In applying the second strategy to the development of CO prodrugs, saccharin and acesulfame were installed as LGs. The release mechanism, kinetics, and biological applications of these prodrugs are discussed in the following sections as well [10].

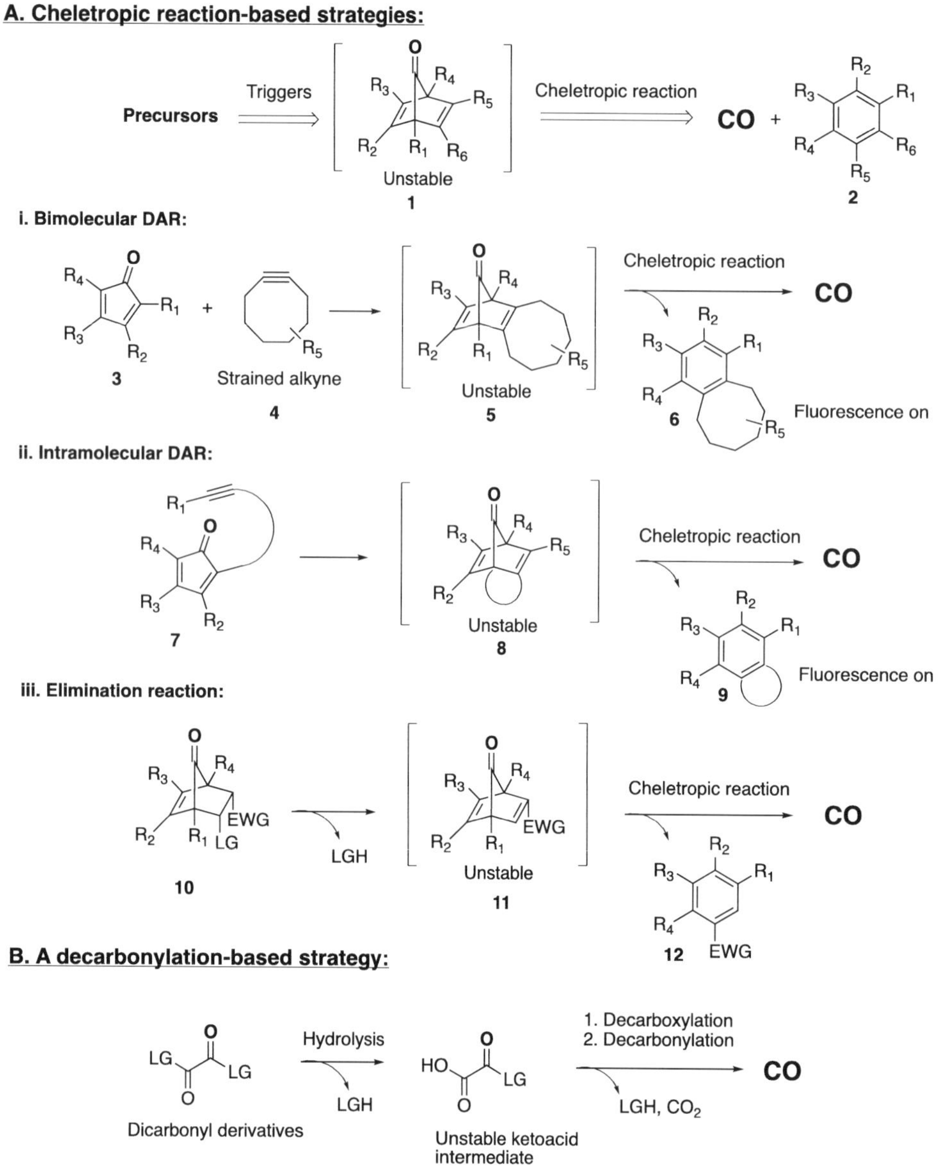

A. Cheletropic reaction-based strategies:

Scheme 14.1 Chemical strategies for delivering CO by (A) cheletropic and (B) decarbonylation reactions.

i. Bimolecular DAR:

ii. Intramolecular DAR:

iii. Elimination reaction:

B. A decarbonylation-based strategy:

CO Prodrugs Based on the Cheletropic Extrusion of CO from Norbornadienones

In this section, we discuss eight different classes of CO prodrugs based on cheletropic extrusion for CO release.

In Situ Generation of Norbornadienone Through a Bimolecular DAR

Conceivably, an inverse-electron demand DAR between an appropriately substituted cyclopentadienone and an alkyne should yield the desired norbornadienone (Scheme 14.1A). Precedents for such a reaction are not rare. However, they are often carried out at high temperature [13], which is far from being compatible with physiological conditions. In the field of click chemistry for various applications [14,15], the idea of an activated alkyne is often used to lower the activation energy through narrowing the HOMO–LUMO gap between the dienophile and diene [16]. Thus, we thought of testing various strained alkynes to achieve tunable reaction rates under near-physiological conditions. When tetraphenylcyclopentadienone (TPCPD, **BW-CO-13**, Scheme 14.2) was used as a model diene, it is interesting to find that bicyclo[6.1.0]nonyne (BCN, **BW-CO-14**) was the only one that gave appreciable reaction under near-physiological conditions among all the strained alkynes with comparable HOMO energy levels (Scheme 14.2)

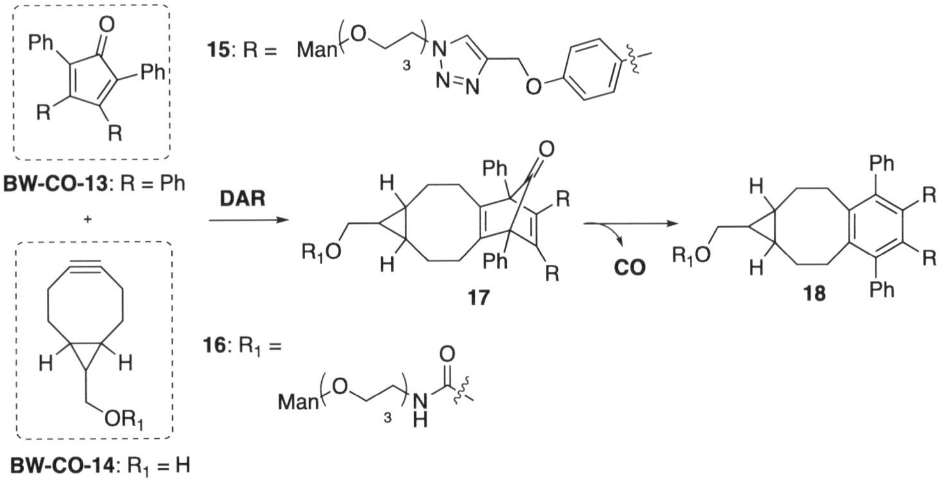

Scheme 14.2 DAR$_{inv}$ reactions between TPCPD and BCN to release CO.

[17]. This approach was shown to deliver a sufficient amount of CO at micromolar concentrations to achieve anti-inflammatory effects in cell culture models using RAW 264.7 cells after stimulation with lipopolysaccharides (LPSs) [18]. While the feasibility studies were considered successful in achieving CO delivery under physiological conditions, it was not entirely clear why BCN was so special because there are others that are considered just as "activated" [17], but did not lead to the same facile reaction. This approach offers the advantage of using two relatively inert reagents to allow for *in situ* generation of the needed norbornadienones. In a detailed study, the reaction rate constants between two sugar-modified water-soluble analogues of cyclopentadienone (**BW-CO-13**) and BCN (**BW-CO-14**) were determined to be 0.61 M^{-1} s^{-1} at 37 °C, which gives a reaction half-life of approximately 38 min when the cyclopentadienone concentration was 25 μM and BCN (**BW-CO-14**) was present in excess at 500 μM [18]. This is an important consideration since CO release half-life will be important for effective delivery of sufficient levels of CO. The concentrations used in this study are within the range that is feasible for dosing in humans, in the context of the peak plasma concentrations of examples of approved drugs: Tylenol, >100 μM; naproxen, >300 μM; Zantac, 1.5 μM; and 5-fluorouracil, >300 μM. Along this line, it should be noted that for a gaseous molecule such as CO, the dosage/concentration used in *in vitro* studies cannot be directly translated into animal studies for reasons of its volatility and rapid equilibrium/evaporation during cell culture experiments, which is different from the relatively enclosed system in animal models. In experiments using prodrugs for another gaseous molecule, H$_2$S, we have shown that the donor concentration does not equate H$_2$S concentration, understandably, and the maximal concentration and duration of sustained availability of H$_2$S depend on both the prodrug concentration and the release half-life [19]. Such characteristics are certainly different from nonvolatile small organic molecules. We attribute these observations to the volatility of such a gaseous molecule, which leads to its easy escape in an open system. The detailed dosages for various prodrugs are described in the pharmacological validation section.

On a different note, using an *in situ* bimolecular reaction to deliver CO has its own potential issues too. First, because the reaction rate of a bimolecular reaction is concentration dependent, the release rate is dependent on the concentration of the two reaction components. This means that dosage and release half-life cannot be controlled independent of each other. Second, using two components to deliver a single active principal (CO) has a more delicate requirement for the "synchronization" of the pharmacokinetic profiles of both, which may be hard to achieve. These are undesirable features of this delivery approach. However, the concentration-dependent nature of the bimolecular approach does offer an advantage of allowing for the development of an enrichment-triggered delivery approach, which would be hard to achieve in any other way as described below.

Enrichment-Triggered Release of CO for Targeting Mitochondria

In targeted delivery of drug molecules through a prodrug approach, one of the challenges is the activation of the prodrug and release of the drug molecule at the desired site after enrichment. A bimolecular approach would actually be ideally suited for the development of enrichment-triggered release because of the concentration-dependent nature of the activation reaction. Along this line, we developed an enrichment-triggered prodrug system for delivering CO to mitochondria [20].

It is well understood that the mitochondrion is the site of many of the known molecular targets for CO (Chapters 2 and 6) [21]. Therefore, it is conceivable that targeted delivery of CO to the mitochondrion would offer enhanced potency and possibly minimized side effects. Because of the electric potential created by the oxidative phosphorylation reactions across the inner membrane of the mitochondrion, positively charged species tends to enrich inside the mitochondrion. This is especially true for molecules with shielded positive charges. As such, triphenylphosphonium (TPP) is a commonly used targeting moiety for enrichment in the mitochondrion [22,23], though other positively charged species are also used for this purpose [24]. Experiments with TPP have shown enrichments of 100–500-fold at 37 °C [22,23,25]. Therefore, we decided to examine the possibility of CO delivery to the mitochondrion through TPP-mediated enrichment. The basic idea is to create a reactant pair, which would keep the DAR rate at the minimum at the low circulating concentration after administration (single-digit micromolar), while allow for a rapid reaction to release CO when the reactants are enriched by hundreds of times in the mitochondrion. This would allow for maximal delivery efficiency and minimal side effects resulting from engaging possible targets outside of the mitochondrion. For this idea to work, a key piece of chemistry is to tune the bimolecular reaction rate constant to the range that allows little reaction at single-digit micromolar concentrations and allow for rapid reaction with half-life less than a few hours once enriched in the mitochondrion. The detailed chemistry optimization and computational work is described in the original publication [26]. Below, we describe the basic ideas of the study.

As shown in Scheme 14.3, both the cyclopentadienone **BW-CO-21** and BCN **BW-CO-22** are attached to TPP through a linker. First, the CO release kinetics were studied *in vitro* by monitoring the fluorescence intensity from the released product (**BW-CO-23**). The second-order rate constants for **BW-CO-19/20** and **BW-CO-21/22** were found to be in the range of 0.14–0.20 $M^{-1} s^{-1}$, suggesting minimal effects on the reaction kinetics by the installation of a TPP moiety. CO release from this prodrug system was then evaluated in cell culture. Specifically, RAW 264.7 cells were incubated with 1 and 5 µM of nontargeted prodrugs (**BW-CO-19**, **BW-CO-20**) or targeted prodrugs (**BW-CO-21**, **BW-CO-22**), and the fluorescence signals from the release product were recorded (Figure 14.1). Without the TPP moiety, no fluorescence was observed within 3 h upon the treatment with 1 or 5 µM of the prodrug pair (**BW-CO-19**, **BW-CO-20**) (Figure 14.1A and B). Such observations are consistent with the reaction kinetics, with the anticipated first half-life of 5 µM of **BW-CO-19** and **BW-CO-20** each to be around 397 h. In contrast, a significant fluorescence increase was observed after incubation with RAW 264.7 cells with 1 or 5 µM of the TPP-conjugated prodrug pair (**BW-CO-21**, **BW-CO-22**) for 4 h (Figure 14.1C and D). The results from colocalization experiments using MT Deep Red, a mitochondrial tracker, also indicate CO release in the mitochondria (Figure 14.1E–H). Such results indicate that enrichment of the prodrug reaction pair in mitochondria was the reason for the significantly accelerated release of CO, further affirming the feasibility of the design principles. Pharmacological validation of this approach was conducted (see the "CO Prodrugs Based on the Cheletropic Extrusion of CO from Norbornadienones" section).

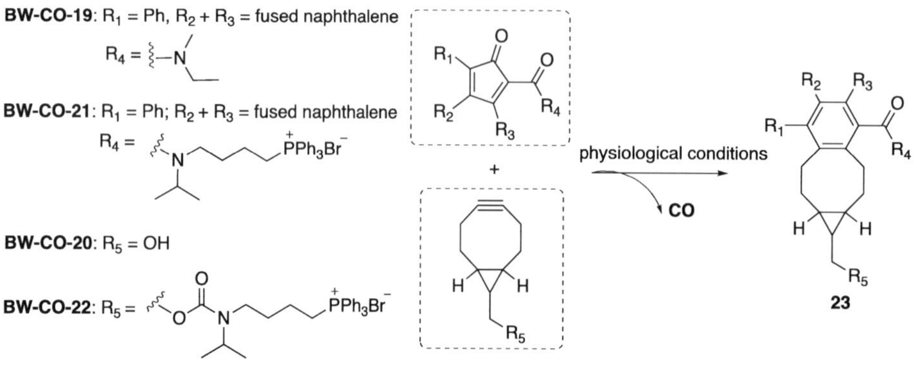

Scheme 14.3 An enrichment-triggered strategy for delivering CO.

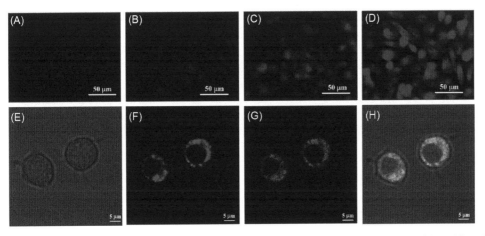

Figure 14.1 Fluorescence imaging studies of CO prodrug pairs (BW-CO-19, BW-CO-20) and (BW-CO-21, BW-CO-22) in RAW 264.7 cells: (A) RAW 264.7 cells treated with BW-CO-19 (1 μM) and BW-CO-20 (1 μM); (B) RAW 264.7 cells treated with BW-CO-19 (5 μM) and BW-CO-20 (5 μM); (C) RAW 264.7 cells treated with BW-CO-21 (1 μM) and BW-CO-22 (1 μM); and (D) RAW 264.7 cells treated with BW-CO-21 (5 μM) and BW-CO-22 (5 μM). Confocal images of RAW 264.7 cells treated with compounds BW-CO-21 (5 μM) and BW-CO-22 (2.5 μM) and MT Deep Red (50 nM): (E) bright field; (F) red channel; (G) DAPI channel; and (H) merged images of (E)–(G). Adapted with permission from Ref. [20]. Copyright (2018) Springer Nature.

CO Prodrugs that Take Advantage of an Intramolecular DAR

As discussed in the *"In Situ* Generation of Norbornadienone Through a Bimolecular DAR" section, the use of a bimolecular reaction for CO delivery presents some significant practical issues. Therefore, the development of a unimolecular prodrug system is desirable. This means to incorporate the diene and dienophile of the DAR into a single molecule. However, this approach carries a major difficulty, i.e., how to keep the diene and dienophile from reacting with each other during synthesis and storage and yet allow them to undergo the desired DAR in physiological media. At the beginning of the project, we thought of building prodrugs that would have augmented hydrophobicity for the dienophile portion to achieve hydrophobicity-facilitated DAR in aqueous solution while maintaining stability in organic solvents during preparation and purification. Upon testing a few cases, it proved to work well. Then, we realized that more than three decades earlier, Breslow and others had reported that increasing solvent hydrophilicity tremendously accelerates the rate of DARs, mostly driven by hydrophobicity [27–31]. For example, the reaction rate between cyclopentadiene and butenone showed a more than 700-fold increase in water compared with that in isooctane. Later, more studies suggested that this acceleration might come from both hydrophobic interactions and water-driven hydrogen-bonding interactions. Such reports point to ways to further improve our design by maximizing the difference in reaction rates between organic solvents and aqueous solution. Below, we describe some key technical aspects of our prodrug design.

Specifically, we built upon our success with the bimolecular CO prodrugs and designed various unimolecular CO prodrugs with tunable release rates and triggered release [32]. In doing so, the alkyne moiety was conjugated to the cyclopentadienone part through an ester or amide linker (Scheme 14.1A-ii). Under physiological conditions, an intramolecular DAR would allow the formation of the norbornadienone intermediate to undergo a cheletropic reaction for releasing CO. At this point, it is important to note that by using an intramolecular DAR, it was no longer necessary to use an activated alkyne as the dienophile. Regular linear alkynes are sufficiently reactive as long as the entropic factors provide the needed drive for the reaction.

The kind of cheletropic reactions in Schemes 14.1–14.3 are normally spontaneous, which leave the DAR to be the rate-determining step. In designing the system to tune the CO release rates, several factors were considered to modulate the DAR step. These include (i) controlling reactivity factors by using either a terminal or internal alkyne and (ii) controlling entropic factors through (a) variation of the ring size (A, Scheme 14.4), (b) using either an ester or an amide linker, (c) using *N*-alkylation of the amide group, and (d) imposing additional restrictions on the freedom of rotation of the linker. These designs are based on the several considerations (Scheme 14.4). First, internal alkynes are known to be generally less reactive than terminal alkynes [33,34]. Second, ring size is well known to provide strong entropic controls of reactions leading to such ring formation [35,36]. This is part of the reason why a 1,6-relationship between a nucleophile and an LG favors substitutions, and in the case of lactone and lactam formation, a 1,5-relationship is more favorable than a 1,6-relationship [36,37]. Third, the restricted rotation in an amide bond offers entropic advantages over an ester linker when ring formation is involved. Fourth, N-alkylation offers further restrictions on the freedom of rotation of the amide bond and thus provides added entropic advantages [38,39]. Fifth, the *gem*-dimethyl group on a linear alkyl chain is known to impose very significant restrictions on the freedom of rotation and thus offer additional entropic advantages [40,41]. Thus, three scaffolds of unimolecular DAR-based CO prodrugs were prepared to study the relationship between structures and release kinetics (Scheme 14.5).

For each prodrug, the CO release kinetics were characterized in DMSO/PBS (pH 7.4, 4:1) at 37 °C by monitoring the formation of the release product using fluorescence or ultraviolet light.

To tune the CO release rate by modulating the reaction (DAR) entropy, different structural modifications were implemented on the linker between the alkyne and dienone moieties (Scheme 14.6). By incorporating a methyl group on the alkyne moiety, the $t_{1/2}$ (6.2 h) of **BW-CO-104** was found to be much longer than that from **BW-CO-103** (1.2 h), which is a terminal alkyne [32]. However, for **BW-CO-101** and **BW-CO-102**, the $t_{1/2}$ values for both were short at 0.03 h, which was really at the limit of accurate determination of reaction rates when the high-performance liquid chromatography (HPLC) method was used by measuring product formation. In this case, we felt that the entropic factors imposed by the formation of a five-membered lactam

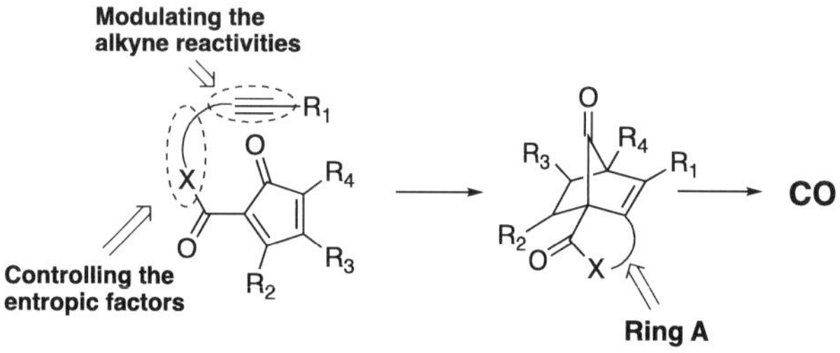

Scheme 14.4 Approaches for tuning the release rate of unimolecular CO prodrugs.

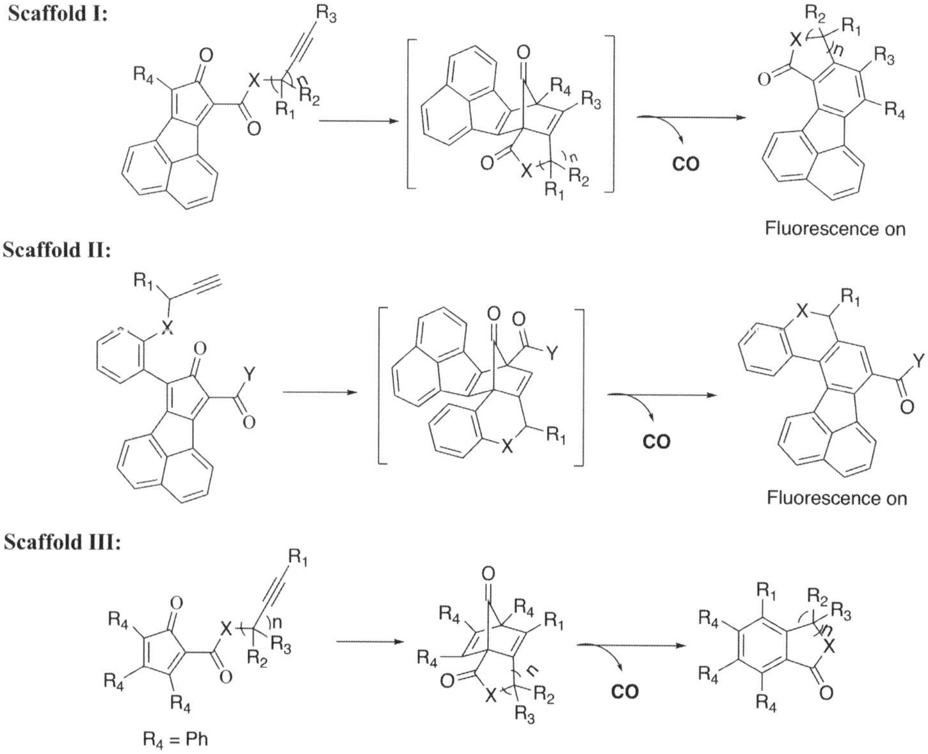

Scheme 14.5 Structural frameworks and release mechanisms of three scaffolds of unimolecular DAR-based CO prodrugs.

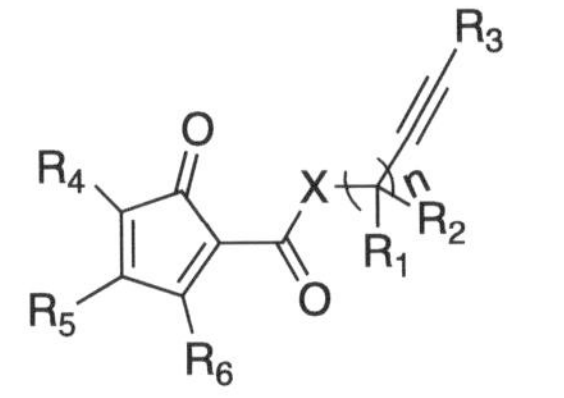

Scheme 14.6 Representative structures of CO prodrugs in scaffold I.

BW-CO-101: X=NMe, $R_1=R_2=R_3$=H, R_4=Ph, n=1
BW-CO-102: X=NMe, $R_1=R_2$=H, R_3=TBDPS, R_4=Ph, n=1
BW-CO-103: X=NMe, $R_1=R_2=R_3$=H, R_4=Ph, n=2
BW-CO-104: X=N-iso-Pr, $R_1=R_2$=H, R_3=Me, R_4=Ph, n=2
BW-CO-105: X=O, $R_1=R_2=R_3$=H, R_4=Ph, n=1
BW-CO-106: X=NH, $R_1=R_2=R_3$=H, R_4=Ph, n=1
BW-CO-108: X=O, R_1=Me, R_2 = H, R_3=H, R_4=Ph, n=1
BW-CO-109: X=O, R_1=H, R_2 = Me, R_3=H, R_4=Ph, n=1
BW-CO-110: X=NH, R_1=Me, R_2 = H, R_3=H, R_4=Ph, n=1
BW-CO-111: X=NH, R_1=Me, R_2 = Me, R_3=H, R_4=Ph, n=1
BW-CO-112: X=NH, R_1=Me, R_2 = Me, R_3=H, R_4=Methythio, n=1
BW-CO-113: X=O, R_1=Me, R_2 = Me, R_3=H, R_4=Methythio, n=1

$R_5 + R_6$ = fused naphthalene

Scheme 14.7 Representative structures of CO prodrugs in scaffold III.

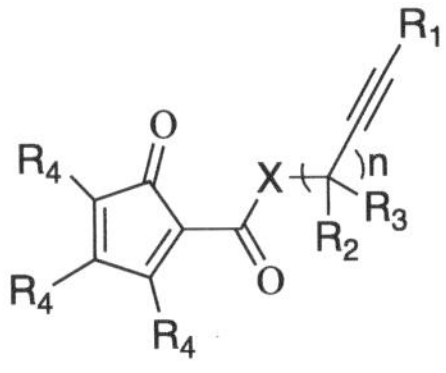

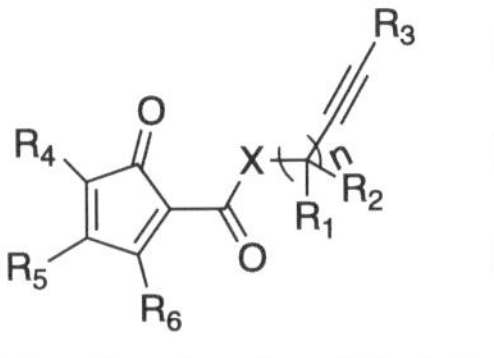

BW-CO-119: X=O, R_1=H, R_2 = H, R_3=H, n=1
BW-CO-120: X=O, R_1=Me, R_2 = H, R_3=H, n=1

R_4 = Ph

Scheme 14.8 Structures of CO-107 and CO-114.

BW-CO-107: X= 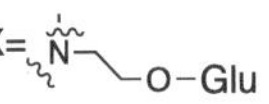 N-CH₂-O-Glu
$R_1=R_2=R_3$=H, R_4=Ph, n=2

BW-CO-114: X=O, R_1=H, R_2 = H, R_3=H, R_4= (PEG-phenyl group), n=1

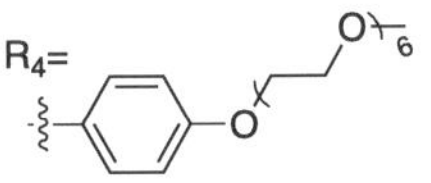

$R_5 + R_6$ = fused naphthalene

ring and the *N*-alkylation were so strong that even the introduction of a terminal bulky TBDMS (*t*-butyldimethylsilyl) group on the alkyne was not sufficient to increase the $t_{1/2}$.

Removal of the methyl group on the amide moiety tremendously decreased the CO release rate when comparing the half-lives of **BW-CO-101** ($t_{1/2}$ 0.03 h) and **BW-CO-106** ($t_{1/2}$ > 1 week). This is consistent with the lessened entropic drive for the DAR due to decreased conformational constraints. Prodrugs with an amide linker were found to have faster release rates as compared to those with an ester linker. For example, prodrugs (amide linkers) **BW-CO-110** and **BW-CO-111** exhibited $t_{1/2}$ of 12 and 0.2 h, respectively. In comparison, the $t_{1/2}$ values for the corresponding prodrugs with an ester linker, **BW-CO-108** and **BW-CO-109**, were determined to be 55 and 0.55 h, respectively.

The "*gem*-dialkyl" effect was also found to accelerate CO release rates. For example, the $t_{1/2}$ of **BW-CO-111** with a *gem*-dimethyl group is 0.20 h, which is much shorter than that of **BW-CO-106** ($t_{1/2}$ ~ 10 days). In scaffold III (Scheme 14.7), upon installment of a methyl group next to the ester group, the $t_{1/2}$ decreased from 0.12 h (**BW-CO-119**) to 0.07 h (**BW-CO-120**). The comparison of the $t_{1/2}$ of **BW-CO-101** and **BW-CO-102** (~0.03 h) with that of **BW-CO-103** (1.2 h) clearly shows that the formation of a five-membered ring favors the DAR. Interestingly, the electron density on the dienone ring did not show much of an effect on the release rate. Replacement of the phenyl group by a methyl thiol group on the dienone moiety (**BW-CO-113**) did not change the $t_{1/2}$ of the prodrugs as compared to that of **BW-CO-109**. This result is consistent with entropic factors being the dominating factor in tuning the release rate in the unimolecular CO prodrug systems studied.

One of the issues with these organic CO prodrugs is their hydrophobicity, which affects water solubility and thus application in biological studies. Various formulations have been developed to address this issue as described in the pharmacological validation section. To improve the intrinsic water solubility of these prodrugs, one can also tether a solubilizing structural moiety. As examples, a sugar moiety was conjugated on the amide group to afford **BW-CO-107**, and a PEG linker was attached on the dienone ring to afford **BW-CO-114** (Scheme 14.8). In PBS (1% DMSO), **BW-CO-107** exhibit decreased $t_{1/2}$ (0.18 h) compared to that (2.1 h) in DMSO/PBS (5:1) solution, suggesting accelerated intramolecular DAR in an aqueous solution.

In conclusion, DARs were successfully applied in developing a series of organic CO prodrugs. The release kinetics and structure–activity relationship of the prodrugs have been extensively studied, allowing for reaction rate tuning to suit various applications [8,32,34]. Most of these prodrugs can be stored at room temperature for weeks without stability problem [32]. Again, it also should be noted that most of the cyclized products after CO release are fluorescent. Such quality allows for tracking of CO release in cell culture and in animal model studies (e.g., Figure 14.1).

All the factors described above allow us to tune the release rate almost at will from a half-life of a few minutes to days, although compounds with very long half-lives would have little practical values unless one can control metabolism and excretion of the prodrug at the same time. Further, with all the prodrugs designed, there is still the question of what happens to the side product after CO release. Will they have their own pharmacological activity or toxicity? These are important questions, but can be adequately controlled in experiments. How to consider the issue of proper negative controls is also discussed at the beginning of the "CO Prodrugs Based on the Cheletropic Extrusion of CO from Norbornadienones" section on pharmacological validation. One thing that also presents a pharmaceutical issue is the lipophilic nature for most of the prodrugs. However, these can be addressed through formulation as indicated by the results from animal model studies in the "CO Prodrugs Based on the Cheletropic Extrusion of CO from Norbornadienones" section or through conjugation with structures that helps to enhance water solubility as demonstrated with the tethering of a sugar moiety [18]. Another practical aspect for these prodrugs is that they would start releasing CO upon dissolution in an aqueous solution especially with those with a half-life in minutes, which could pose challenges to experimental preparations. If the injection solution contains water, CO release during preparation could introduce variability. To address this issue, we have also developed prodrugs that have other triggers built in for prodrug activation, as described in the following sections.

High-Content Prodrugs

With the basic chemistry of using a DAR to construct a norbornadiene structure for CO release through extrusion, we have also designed prodrug approaches to allow for delivery of two payloads from a single prodrug. As a therapeutic agent, CO has been reported to synergize with other drug molecules such as doxorubicin [42,43] and metronidazole [44]. Therefore, we are interested in developing methods to deliver CO with doxorubicin and metronidazole, respectively, in a single prodrug.

As a first example, we designed a prodrug that uses a cascade of reactions to release CO and metronidazole with the formation of a fluorescent product as a reporter [45]. As shown in Scheme 14.9, under physiological conditions, the intramolecular DAR leads to the unstable norbornadienone intermediate **25**, which undergoes a subsequent cheletropic reaction to release CO. Then, the product from the first two reactions has a nucleophilic hydroxyl group in a 1,5-relationship with the ester group carrying metronidazole. This arrangement sets up a favorable and spontaneous lactonization reaction [46] to release metronidazole and to form a trackable product **27** with blue fluorescence to monitor metronidazole release from the prodrug **BW-CO-24**. When 25 μM of the prodrug was incubated in PBS buffer (pH 7) at 37 °C, 85% of metronidazole was released within 6 h, and no other side products were found, indicating that the prodrug was activated through the designated mechanism. CO release was verified in RAW 264.7 cells by incubating 20 or 40 μM prodrug **BW-CO-24** with COP-1 (a fluorescent CO

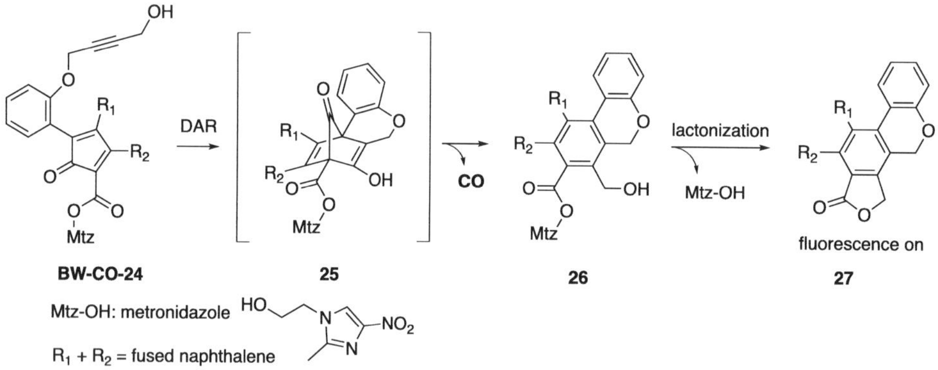

Scheme 14.9 A prodrug system for delivering CO and metronidazole.

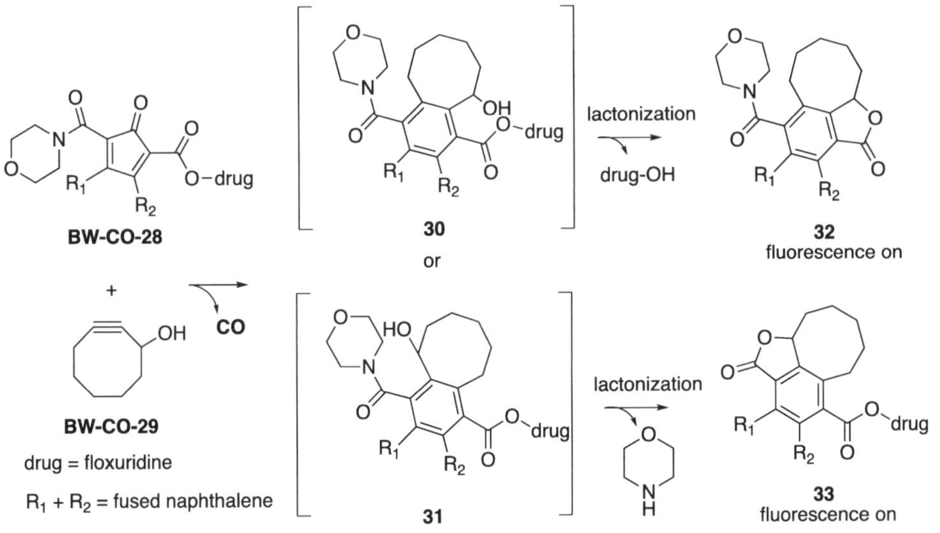

Scheme 14.10 A bimolecular prodrug system for delivering CO and floxuridine.

probe, Chapter 18) for 6 h. Increased green fluorescence from COP-1 was observed, suggesting CO release in cell culture. Additionally, the blue fluorescence from the release product was also detected in a dose-dependent manner. As an initial assessment of the synergistic effects between CO and metronidazole, we also observed enhanced antimicrobial effects by metronidazole when CO was released at the same time with one strain of *Helicobacter pylori*, but not all the strains. More mechanistic studies are needed in order to comprehensively examine whether there are true synergistic relationships between these two [44,45].

As a second example, we also designed a bimolecular prodrug system to deliver CO, floxuridine (an anticancer drug), and a fluorescent reporter through a cascade of bioorthogonal reactions (Scheme 14.10) [47]. As shown in Scheme 14.10, a drug payload is attached to cyclopentadienone **BW-CO-28** through an ester bond, and a strained alkyne **BW-CO-29** is modified by introducing a hydroxyl group at the propargyl position. Upon undergoing a DAR between the cyclopentadienone (**BW-CO-28**) and the strained alkyne (**BW-CO-29**) to release CO, the intermediate **30** underwent lactonization to release floxuridine and a product with blue fluorescence. The second-order reaction constant was determined to be 0.17 $M^{-1} s^{-1}$ by monitoring the fluorescence from reaction products (**32** and **33**). When 5 mM of the alkyne was incubated with 100 μM of the cyclopentadienone, the release was completed within around 4 h in PBS buffer (30% DMSO) with the yield of floxuridine being about 50%, which was due to the regioselectivity in the first cyclization step, leading to the trapping of drug molecule in the final product **33**. The prodrug and strained alkyne pair (**BW-CO-28**/**BW-CO-29**) were also incubated with MB-231 cells and were found to lead to the same release.

In this part, we have presented two approaches that allow for codelivery of CO and a drug payload with concomitant formation of a fluorescent product as a reporter compound for reaction monitoring, much the same as shown in Figure 14.1. Such prodrug systems provide a versatile set of tools for studying the combinatorial therapeutic effects of CO with other drug molecules.

Enzyme-Triggered CO Release from Prodrugs Based on Intramolecular DARs

In an effort to design a class of prodrugs that do not rely on exposure to water to trigger CO release, we were interested in employing a biological trigger. In doing so, we thought about using conformational constraints as a way to control CO release. In such a design, the conformational constraints prevent the intramolecular DAR and yet the constraints can be removed by an enzyme-catalyzed reaction (Scheme 14.11).

In the first set of examples, we used an ester group to impose the needed conformational constraints and to offer esterase sensitivity. Specifically, the dienophile (alkyne) is constrained in a cyclic structure to prevent its reaction with the diene (cyclopentadienone). As for the choice of ring structure, we selected a seven-membered ring for two reasons. First, it is small and rigid enough to offer a sufficient level of

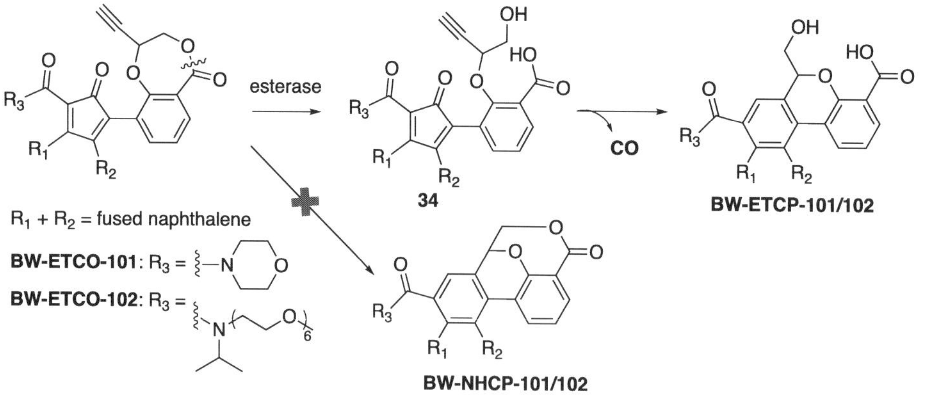

Scheme 14.11 Esterase-sensitive CO prodrugs.

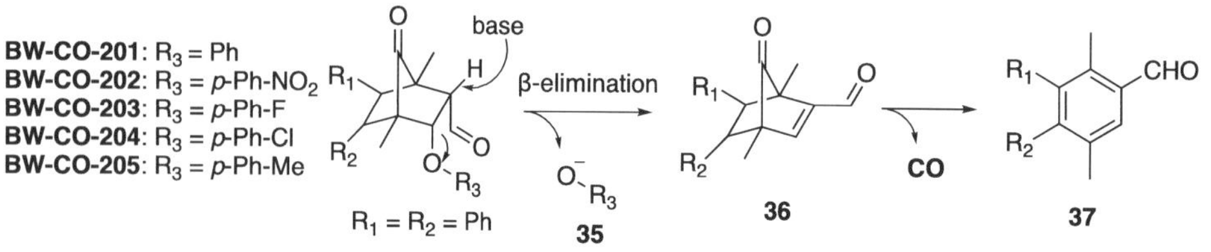

Scheme 14.12 β-Elimination-triggered CO prodrugs.

conformational constraints and yet it is easily cleavable. If we chose a five- or six-membered ring, then entropic factors would favor ring formation and make it hard for an esterase to achieve cleavage. Second, a seven-membered ring is fairly easy to construct synthetically. As such, compounds **BW-ETCO-101** and **BW-ETCO-102** were prepared with incorporation of a morpholine or a PEG moiety to increase water solubility (Scheme 14.11) [48]. In the presence of porcine liver esterase (PLE) (10 units/ml), the degradation half-lives of **ETCO-101** and **ETCO-102** were found to be 1 and 4 h in PBS (5% DMSO), respectively. Further, **BW-ETCP-101** and **BW-ETCP-102** were characterized as the only cyclized product by HPLC, and no **BW-NHCP-101** or **BW-NHCP-102** was detected. Such results suggest that the seven-membered ring structure effectively restrains the alkyne from reacting with the dienone part. In contrast, without an esterase, the hydrolyses of **BW-ETCO-101** and **BW-ETCO-102** were found to be much slower with half-lives of 24 and 17 h, respectively. Overall, the results support the esterase-catalyzed release being the mechanism for CO release.

A β-Elimination Strategy Starting from Norbornene-7-ones: Base-Catalyzed Elimination

As shown in Scheme 14.1A-iii, taking advantage of an elimination reaction to generate the norbornadienone intermediate allows for the subsequent cheletropic reaction for CO release [49]. Generally speaking, such elimination requires an EWG for easy deprotonation of the α-position and an appropriate LG. In this design, a key factor to consider is the ease of the elimination reaction for CO release under physiological conditions with little other assistance. As such, the requirements for the pair of EWG and LG are stringent. In one design, we used an aldehyde group as the EWG and substituted phenols as the LG (Scheme 14.12). The choice of substituted phenols serves multiple purposes. First, phenolates are stable anions and thus fairly good LGs. Second, substituted phenols allow for easy studies of electronic factors in the stabilization of the phenolic anion to tune the ease of the rate-limiting elimination step. Third, phenols are easily detectable in HPLC studies.

We synthesized five analogues (Scheme 14.12): **BW-CO-201** to **BW-CO-205**. These compounds were assessed in PBS (30% DMSO) and were found to undergo ready β-elimination to release CO with concomitant production of the phenol species and a new substituted aromatic compound (**37**). With various LGs, the CO release half-lives of **BW-CO-201** to **BW-CO-205** were in a range from 20 min to 1.2 h in PBS (pH 7.4, 30% DMSO) buffer. Additionally, a linear relationship was also observed between CO release rate constant and the Hammett constant of the LG. The pH value of the testing buffer was also found to affect the release rate, as expected. For example, the half-life of **BW-CO-203** in PBS buffer (pH 7.4) is around 0.65 h. However, at pH 3,

Scheme 14.13 Elimination-based water-soluble CO prodrugs.

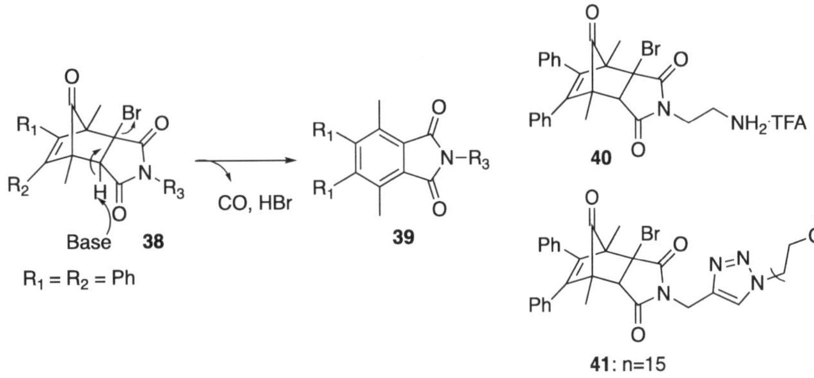

Scheme 14.14 ROS-sensitive CO prodrugs.

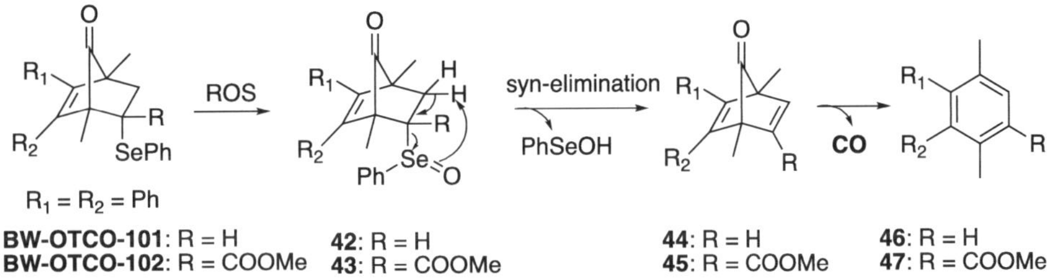

BW-CO-203 was found to be more stable with a half-life of 9 h. These results also further affirm the elimination being the rate-determining step in triggering CO release from the prodrug. In a simulated gastric fluid system (without pepsin, pH ~ 1), 80% of the prodrug **BW-CO-203** remains intact after 8 h of incubation at 37 °C, indicating the suitableness of this type of prodrugs for CO delivery to the lower gastrointestinal (GI) system, by taking advantage of the pH difference between the stomach and the lower GI system.

Along a similar line, Larsen and coworkers independently developed CO prodrugs based on this β-elimination for the formation of norbornene-7-ones and the subsequent CO release [50,51]. In Larsen's design, a halide (Br) was used as an LG and an amide group was used as the EWG (Scheme 14.13). For improving the water solubility of the prodrugs, a terminal amino group and a PEG side chain were installed to afford prodrugs **40** and **41**, respectively. In Tris–sucrose buffer (pH 7.4), the half-lives of **40** and **41** were measured to be 19 and 320 min, respectively. Nuclear magnetic resonance and dynamic light scattering characterizations indicated that prodrug **41** could form micelles of 7–8 nm in size at high concentrations (>2.5 mM) in aqueous solution. Micelle formation also offers protection against base-catalyzed elimination reaction and thus decreases the release rate. It should also be noted that, compared to **40** (IC_{50} 139 μM), prodrug **41** exhibits a lower level of cytotoxicity (IC_{50} > 1 mM) on AC16 (human cardiomyocytes) upon incubation for 1 h.

A β-Elimination Strategy Starting from Norbornene-7-ones: ROS-Triggered Elimination

One of the key therapeutic effects for CO is modulation of inflammation, which is commonly associated with elevated levels of ROS. Therefore, ROS-sensitive CO prodrugs offer the chance for enhanced potency at the site of inflammation and reduced off-target effects. Along this line, we developed a class of ROS-sensitive prodrugs by taking advantage of the sensitivity of selenium to ROS-mediated oxidation and the propensity for selenium oxide to undergo spontaneous elimination [52]. Specifically, two compounds were prepared: one with an EWG and one without. As shown in Scheme 14.14, in the presence of ROS, oxidation led to formation of the selenoxide intermediate (**42** and **43**), which underwent *syn*-elimination (selenoxide elimination) to release the phenylselenenic acid and to generate the double bond on the norbornadienone intermediate (**44** and **45**). Subsequent cheletropic reaction releases CO.

To test whether the prodrug is sensitive toward various ROS species, **BW-OTCO-102** was incubated with hypochlorite (NaClO), singlet oxygen (prepared by reaction of NaClO with H_2O_2), superoxide (KO_2), and other ROS at 37 °C [52]. CO release from prodrug **OTCO-102** was effectively triggered by hypochlorite, singlet oxygen, and superoxide. For example, 20 μM of **BW-OTCO-102** was totally consumed and converted to **47** within 30 min in the presence of hypochlorite (40 μM). In the control group without hypochlorite, no cyclized product was detectable after 48 h incubation at 37 °C. Other ROS species including hydrogen peroxide, hydroxyl radical (prepared by reaction of $FeSO_4$ with H_2O_2), *tert*-butyl hydroperoxide, and *tert*-butoxy radical (prepared by reaction of $FeSO_4$ with H_2O_2) only led to minimal formation (less than 5%) of the cyclized byproduct (**47**), indicating the requirement of certain select ROS to activate the prodrug for CO release. To further profile the release kinetics of **46** and **47** in the presence of hypochlorite, the consumption of the selenoxide intermediate and the formation of the product after CO release were monitored by HPLC at different time points. For prodrug **BW-OTCO-101**, the oxidation step was completed within 30 s upon incubation with 40 μM hypochlorite. However, the subsequent selenoxide elimination is much slower with a half-life of 4 h, as determined by detecting the formation of OTCP. For **BW-OTCO-102**, the situation was found to be different. The oxidized intermediate was not detected by HPLC, indicating a quick subsequent elimination, and CO release half-life was determined to be 30 s in the presence of 40 μM hypochlorite. Such results indicate that the release rates of these ROS-sensitive CO prodrugs are tunable through installation of different R groups. This tunability allows for tailoring the prodrug for specific therapeutic requirements.

Cancerous cells and LPS-stimulated macrophages are known to have elevated endogenous ROS levels as compared to normal cells. As such, the ROS-sensitive prodrug **BW-OTCO-102** was tested on two cell lines: HeLa, an ovarian cancer cell line, and H9c2, a rat cardiomyoblast cell line. CO release was monitored by COP-1 (a fluorescent CO probe) [53]. Cell imaging studies showed much stronger green fluorescence from COP-1 in HeLa cells as compared to that in H9c2 cells, indicating the degree of CO release being much higher in cancerous cells, presumably due to an increased ROS level. Similarly, fluorescence from COP-1 also significantly increased in LPS-treated RAW 267.4 cells. LPS has been widely used to induce inflammatory response in macrophages associated with increased production of ROS. Taking together, these results suggest the sensitivity of **BW-OTCO-102** toward endogenous ROS species and support the feasibility of using **BW-OTCO-102** to deliver CO to sites with elevated ROS levels. However, the possibility for these prodrugs, the products of these prodrugs, or ROS to induce HO-1 cannot be ruled out. If induction of HO-1 happens, it may also contribute to increased CO availability. It should also be noted that these ROS-sensitive prodrugs do contain a selenium, which has its own cytotoxicity issues [54]. Therefore, future efforts may focus on non-selenium-based approaches in order to minimize selenium-related issues.

A β-Elimination Strategy Starting from Norbornene-7-ones: An Esterase-Triggered Release System

Because of the facile elimination of the norbornene-7-one structural moiety when there is an LG attached to the β-position of an EWG, we designed a prodrug with enhanced stability for storage and sample preparation by building an additional trigger into the system. Specifically, we designed esterase-sensitive prodrugs that rely on the norbornene-7-one elimination chemistry (Scheme 14.15) [55]. In this case, the electron-withdrawing aldehyde group is masked as an acetal moiety. Upon incubation with PLE, hydrolysis would lead to unmasking of the aldehyde group and initiate the β-elimination for subsequent CO release. Four esterase-sensitive prodrugs (**BW-CO-48** to **BW-CO-51**) were synthesized bearing various structural features at the acetal moiety to modulate the hydrolysis rate. Studies of the reaction kinetic were carried out in PBS buffer with 3 U/ml of PLE. The hydrolysis rates were found to be correlated with the steric hindrance of the R group, suggesting the esterase-catalyzed hydrolysis being the rate-limiting step. The prodrug (**BW-CO-48**) with the least sterically hindered group was totally converted to the intermediate within 2 min. Increasing bulkiness of the R group led to a decrease in the rate of PLE-mediated hydrolysis and CO release. For the sterically bulky ester prodrug **BW-CO-51**, no hydrolysis product was observed after 36 h of incubation with PLE. In the absence of PLE, no hydrolysis product was formed in either PBS or stimulated gastric fluid after incubation for up to 6 h. To study the hydrolysis in biological media, prodrug **BW-CO-48** was incubated in the serum of rabbit, rat, mouse, or human, respectively. In rabbit, rat, and mouse sera, the half-lives of the esterase-triggered hydrolysis ranged from 3 to 30 min. However, the CO release rate of **BW-CO-48** in human serum was found to be slower with a half-life of 14 h. CO release from these prodrugs was also confirmed in cell culture studies by using a selective CO fluorescent probe, COP-1 (Chapter 18)

Scheme 14.15 Esterase-sensitive CO prodrugs.

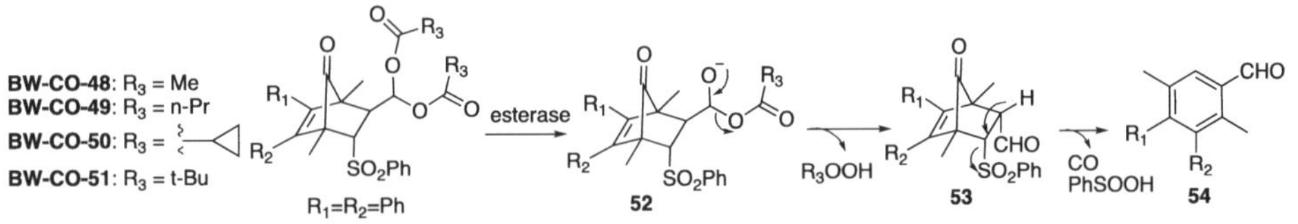

BW-CO-48: R₃ = Me
BW-CO-49: R₃ = n-Pr
BW-CO-50: R₃ = cyclopropyl
BW-CO-51: R₃ = t-Bu

R₁=R₂=Ph

Scheme 14.16 CO prodrugs based on decarboxylation–decarbonylation chemistry.

BW-CO-306: LG:
BW-CO-307: LG:

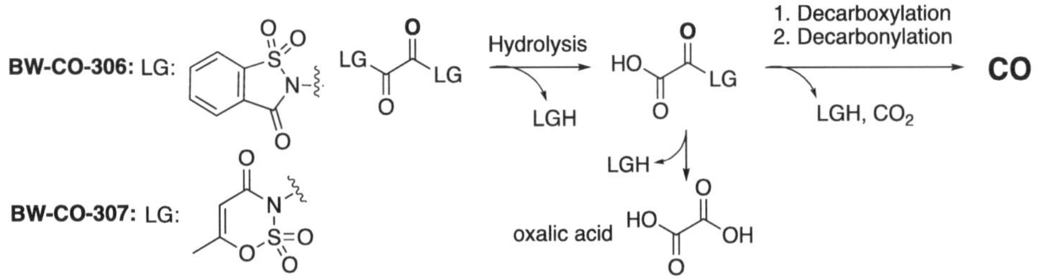

[53]. Upon incubation of RAW 264.7 cells with 50 μM of prodrug **BW-CO-48**, the fluorescence from COP-1 significantly increased, indicating intracellular CO release. Additionally, no cytotoxicity was observed at up to 200 μM of the prodrugs. Such results demonstrate the feasibility of using these esterase-sensitive prodrugs to deliver CO intracellularly.

A Decarboxylation–Decarbonylation Strategy for CO Prodrug Development

To consider using decarbonylation chemistry in CO prodrug design, one of the key aspects is to search for a suitable decarbonylation reaction, which can be initiated under near-physiological conditions. With a proper LG, 1,2-dicarbonyl compounds are known to undergo hydrolysis and decarboxylation to trigger the subsequential decarbonylation for CO release in aqueous solutions (Scheme 14.16). In our studies, we extensively examined the pK_a requirements for the LG and decided on FDA-approved sweeteners, saccharin and acesulfame [10]. First, the pK_a values of saccharin and acesulfame are within the range of 1.6 and 3, which is preferable for the decarbonylation pathway, based on the examination of CO release from various 1,2-dicarbonyl compounds. Second, after CO release, the by-products include only CO_2 and saccharin or acesulfame. This ensures a "clean" and safe delivery of CO to biological systems. As such, **BW-CO-306** and **BW-CO-307** were prepared in one step by using saccharin or acesulfame as an LG (Scheme 14.16) [10]. The solid forms of **BW-CO-306** and **BW-CO-307** are stable for over 24 days (the duration of the study) of exposure to ambient light at room temperature. In an acetonitrile and water solution (4:1), the CO release yields of **BW-CO-306** and **BW-CO-307** were measured to be around 76% and 92%, respectively, using a gas chromatography system. As the initial step, hydrolysis can be affected by the solution pH. Under lower pH, enhanced rate of the hydrolysis might lead to the generation of oxalic acid without involving the decarbonylation step (Scheme 14.16). In the case of **BW-CO-106**, CO release yield was decreased from 60% to 30% after adjusting the pH from 7.4 to 1. However, for **BW-CO-307**, the CO release yield was consistently over 90% within the pH range of 1–7.4. Further kinetic studies suggest the release half-lives of **BW-CO-306** and **BW-CO-307** being about 1.28 and 9.5 min in 60% PBS buffer, respectively. CO release from **BW-CO-306** and **BW-CO-307** was also validated *in vivo* through pharmacokinetic studies. As compared to the control group, oral administration of 50 and 100 mg/kg of **BW-CO-306** significantly elevated the carboxyhemoglobin (COHb) level to around 5% after 10 min (t_{max}) in a mouse model (Figure 14.2A). Considering the short release half-life (1.2 min) of **BW-CO-306**, a CO prodrug (**BW-CO-103**) with a longer release half-life (~72 min) was also studied as a comparison. Results from **BW-CO-103** clearly showed a more sustained increase of COHb over a period of 60 min, which is consistent with the *in vitro* release kinetics of these two CO prodrugs (Figure 14.2B). Taken together, these results firmly established the feasibility of using **BW-CO-306** for CO delivery *in vivo* for pharmacological applications.

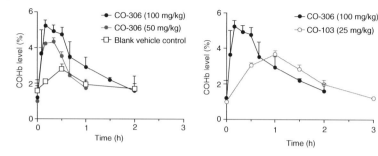

Figure 14.2 COHb elevation in mice upon administration of BW-CO-306 and BW-CO-103. Reproduced from Ref. [10].

Validation of the Pharmacological Efficacy of the CO Prodrugs in Various Disease Models

In this part, we briefly discuss the pharmacological efficacy validation of several organic CO prodrugs described above. As compared to inhaled CO gas, organic CO prodrugs allow for precisely controlled dosage and administration via oral, intraperitoneal, or intravascular (i.v.) administration. Specifically, selective CO prodrugs have been examined in models of various pathologies through collaborations with different labs described in the "Acknowledgments" section and the respective publications. Such studies include LPS-induced systemic inflammation [55], acetaminophen (APAP)-induced liver injury [20], 2,4,6-trinitrobenzenesulfonic acid (TNBS)-induced colitis [32], kidney reperfusion injury [56], chemically induced gastric injury [57], kidney injuries induced chemically or by rhabdomyolysis, and sensitizing *Helicobacter pylori* toward metronidazole [45] (Figure 14.3). Additionally, the vasorelaxation and combinatory therapeutic effects of the prodrugs developed by the Larsen lab were also studied in *ex vivo* models [50,51]. One of the difficult issues with CO in animal models is the lack of clarity of the issue of dose dependence because of the gaseous nature of this molecule. As a result, the amount of CO available to various molecular targets is proportional to the amount of CO administered, but drawing quantitative relationship is harder than with traditional small molecules. Some of these issues are examined in the pharmacokinetics chapter of this book (Chapter 3).

Because of the recent discovery of chemical reactivity and possibly CO-independent biological effects with some Ru-based CORMs revealed in solution-phase and *in vitro* studies [58–77], we would like to comment on the implications and limitations of such results. On one hand, chemical properties and reactivities (including binding) form the molecular underpinnings of all biological events with no exception. Thus, mechanistic studies at the molecular level are critical. On the other hand, the higher level of complexity at the whole organism level may mean that the relative importance of various molecular events may need to be analyzed in a different context. Therefore, studies in solution, in cell culture, at the organ level, and *in vivo* should augment each other in our studies to achieve a more perfect understanding of the biological consequence of a given molecule and its properties in different context. Chapter 12 also has a section on this subject. For the purpose of this chapter, it is important to comment on control experiments when using CO donors (metal-based CORMs or CO prodrugs). First of all, it is important to recognize that control experiments in a prodrug project rarely, if ever, perfectly duplicate all the factors other than the active ingredient (i.e., serving as a perfect negative control). This is true no matter whether it is a metal-based CORM, an organic prodrug, or just a prodrug in general. Therefore, it is not surprising that out of over 600 studies, sporadic CO-independent effects have been reported with some metal-based CORMs. However, not being able to provide "perfect" controls does not mean that this is a "black box" either. The nature of scientific control experiments is such that it requires us to pay close attention to what have not been controlled adequately in an experiment. At the minimum, we need to examine interactions at the molecular level, which form the mechanistic underpinnings of biological consequences at cellular and whole organism levels. Therefore, we raise a few issues for consideration, especially those that may pose additional challenges in designing proper negative controls and performing data analysis. First, chemical reactivity is a major factor to consider, especially in the context of its relevance in a biological system. Ideally, one would hope to use molecular entities that are chemically inert for drug discovery, unless the chemical reaction is part of the intended functions. However, chemical inertia in the absolute sense is hard to achieve. Technically, almost everything our body encounters is chemically reactive, depending on the circumstance, including the essential nutrient, glucose, the chemical reactivity of which contributes to protein glycation and the formation of glycated hemoglobin and albumin [78–80]. Therefore, this issue is most relevant only when the reactivity of the CO donor is high and/or the reaction timescale is shorter than or about the same as the intended biological response (and the biological half-life when the study is in animal models). Depending on what the chemical reactivity is, the impact could be broad and multifactorial. For example, some ruthenium complexes are known to be very chemically reactive in redox reactions and toward nucleophiles such as a thiol group on an amino acid or proteins. For organic CO prodrugs, the presence of a Michael acceptor in some structures requires attention for general cytotoxicity. Michael acceptors are also able to react with thiols with varying reactivity depending on the specific structure. It should be noted that there are clinically approved drugs that contain the Michael acceptor moiety, including ibrutinib, Tamiflu, drospirenone, budesonide, dimethyl fumarate, and dustasteride. Thus, the presence of a Michael acceptor by itself is not a "showstopper." However, such structural moieties and their possible interference need careful attention. All the chemical reactivity issues need to be considered in the context of concentration and timescale because bimolecular reactions are concentration dependent. This leads to the second point, CO release half-life. If the half-life is long relative to its reactivity (reaction rate) or the biological half-life (in cell culture and animal model experiments), then one has to also consider the effects of the prodrug or CORM as a distinct entity, whether in binding to a biological target or in chemical reactions. In a way, off-target effects are common in drug discovery in general and they are not "showstoppers." Some are even explored for drug repurposing applications [81]. However, it is important that we understand such off-target effects at the earliest possible time for both the prodrug as a distinct entity and its product(s) after drug release. Third, if the CO release reaction leads to multiple components, then one needs to be very careful in designing control experiments, as the level of complexity increases significantly. Such are the cases with some metal-based CORMs [75] and the organic prodrugs in the categories of pH-, ROS-, and enrichment-sensitive prodrugs. Fourth, if the CO release reaction leads to multiple components, for which the exact chemical identity and ratio are hard to establish or if the composition shifts with time [75], this could lead to chemical intractability problems in the drug discovery and developmental process. With all this said, most of the factors can be adequately controlled, minimized, or even eliminated starting with a good understanding of the chemistry in terms of both the reaction(s) and its mechanism(s). When unexpected issues are discovered, they are to be avoided or at least taken into consideration in result interpretation.

In the next section, we describe the results from various pharmacological validation experiments in cell culture models and animal models, which are summarily described in Figure 14.3. As is generally true, increasing attention is required for unexpected or idiosyncratic effects from such prodrugs as more sophisticated *in vivo* experiments are designed and conducted. In the case of CO prodrugs, increasing attention is also needed for possible non-CO-related effects from each prodrug at every stage. Specific to the CO prodrugs described in this section, the respective side products after CO release were used as the negative controls in all biological assessments. These controls so far have proven to be sufficient.

LPS-Stimulated Systemic Inflammation

LPS is known to induce acute inflammatory responses in macrophages, resulting in elevation of proinflammatory cytokine productions. Endogenous administration of CO gas has been reported to inhibit TNF-α production from macrophages [82,83]. In this case, organic CO prodrugs were also assessed in LPS-induced inflammation models.

BW-CO-103 (**CO-103**) is one of the unimolecular click reaction-based CO prodrugs with a half-life of around 1.2 h [32]. In using a CO prodrug for evaluating its biological effects (from CO), it is necessary to have a control compound to understand the effects of CO. As such, the reaction product after CO release, namely **BW-CP-103** (**CP-103**), was used as the negative control in the studies. The cytotoxicity profile showed that both **CO-103** and **CP-103** did not affect the viability of RAW 264.7 cells (a macrophage cell line) at up to 100 μM. For such studies, RAW 264.7 cells were preincubated with 12.5–100 μM **CO-103** or **CP-103** for 5 h for completion of CO release from the prodrug. Subsequently, the cells were challenged with 1 μg/ml LPS for 1 h, and TNF-α secretion was detected by the ELISA assay. Compared to the control, 12.5–100 μM of **CO-103** significantly inhibited the production of TNF-α in a dose-dependent manner (Figure 14.4). The fluorescence from **CP-103** was also monitored after incubating cells with the

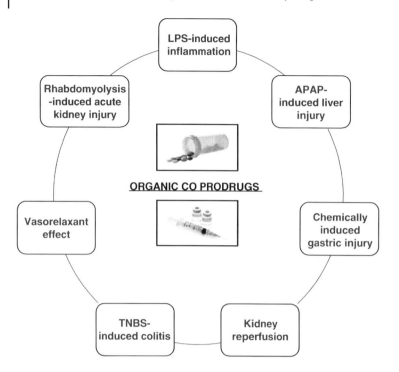

Figure 14.3 Applications of organic CO prodrugs in disease models.

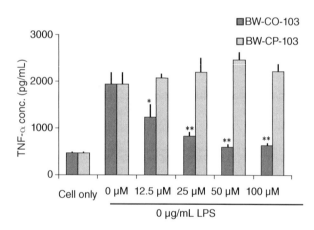

Figure 14.4 Anti-inflammatory effects of BW-CO-103 on RAW 264.7 cells. Adapted with permission from Ref. [32]. Copyright (2016) John Wiley & Sons.

prodrug, and the fluorescence intensity was found to be dose and time dependent. Such results further indicate that the observed inhibitory effects from **CO-103** on TNF-α production were due to intracellular CO release.

The esterase-sensitive CO prodrug (**BW-CO-48**) was studied in mouse models of LPS-induced systemic inflammation to examine its therapeutic efficacy (Scheme 14.15) [55]. Specifically, prior to injection of LPS (10 mg/kg), C57 mice were treated with 5 or 10 mg/kg of CO prodrug **BW-CO-48**. Mice were sacrificed after 6 h, and biomarkers, including TNF-α, alanine transaminase (ALT), and aspartate transaminase (AST), were analyzed. It was found that 5 and 10 mg/kg prodrug decreased the TNF-α level by 16% and 33%, respectively, as compared to the control groups, indicating a dose-dependent anti-inflammation effect. ALT and AST are important biomarkers for liver injury. The CO prodrug-treated group also showed dose-dependent decreases in ALT and AST concentrations, indicating liver protective effects from **BW-CO-48** (Figure 14.5A–C). In contrast, the product after releasing CO (**54**) did not exhibit such effect (Figure 14.5A–C). Further, liver hematoxylin and eosin (H&E) staining results also confirmed that the esterase-sensitive CO prodrug protected liver tissue from inflammatory damage (Figure 14.5D–F).

Elimination-based organic CO prodrugs, **BW-CO-203 (CO-203)** and **BW-CO-205 (CO-205)**, were also tested for their effects in LPS-induced inflammation model in cell culture. Specifically, prior to LPS

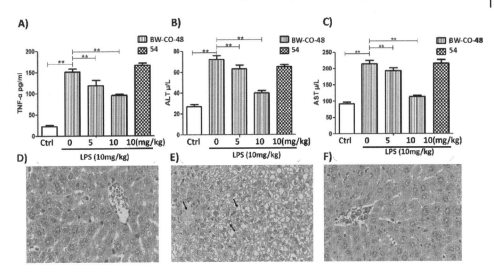

Figure 14.5 Protective effects of prodrug BW-CO-48 against LPS-induced liver injuries as indicated by the suppression of TNF-α (A), ALT (B), and AST levels (C). H&E staining of liver tissue in the vehicle control (D), LPS (E), and BW-CO-48 (10 mg/kg) treatment group (F). $^{**}P < 0.01$. Adapted with permission from Ref. [55]. Copyright (2019) American Chemical Society.

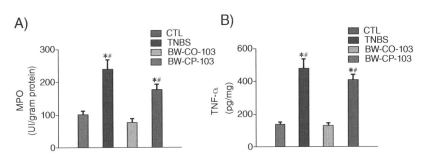

Figure 14.6 The effects of BW-CO-103 on (A) MPO activity and (B) TNF-α expression level in colon. $^{*}P < 0.05$ versus the CTL (sham control) group; $^{\#}P < 0.05$ versus the BW-CO-103 group. n = 10 mice for each group. Adapted with permission from Ref. [32]. Copyright (2016) John Wiley & Sons.

treatment, CO prodrugs were incubated in RAW 264.7 for 4 h. Both **CO-203** and **CO-205** at 50 μM effectively inhibited TNF-α production by about 80% and 75%, respectively. In contrast, the control groups did not exhibit effects, indicating the role of the released CO in inhibiting TNF-α production.

CO is known to have molecular targets such as cytochrome c oxidase (COX) and the complex of cytochrome c and cardiolipin located in mitochondria [84–86]. Therefore, targeted delivery of CO to mitochondria is expected to result in enhanced biological responses. As such, mitochondria-targeted CO prodrugs were studied for their activity *in vitro* and *in vivo*. In RAW 264.7 cells, pretreatment with mitochondria-targeted CO prodrugs (**BW-CO-21** and **BW-CO-22,** Scheme 14.3) significantly inhibited TNF-α production as a result of stimulation by LPS with an IC_{50} of about 5 μM [20], which is the lowest among other organic CO prodrugs (IC_{50} range: 25–100 μM), suggesting the potential importance of targeted delivery of CO to mitochondria. Chapter 15 of this book specifically addresses targeted delivery of CO.

Taking together, the therapeutic effects of DAR and elimination-based CO prodrugs have been validated on LPS-induced inflammation models. Such results indicate the potential of using these prodrugs to treat inflammatory diseases.

TNBS-Induced Colitis

Ulcerative colitis is a common inflammatory bowel disease that can increase the risk of colon cancer [87,88]. CO has been reported to attenuate inflammation in colitis models [89]. As such, the therapeutic effects of **CO-103** were also examined in a TNBS-induced colitis mouse model [32]. Specifically, healthy mice were administrated with **CO-103** (15 mg/kg), **CP-103** (15 mg/kg), or vehicle (30/70 Solutol/saline) through intraperitoneal injection 1 h before applying TNBS. After that, the mice were treated again with **CO-103** or control compound at 24, 48, and 72 h. **CO-103** treatment increased the survival rate from 48% to 75%. When examined using other parameters such as colon length, colon thickness, and clinical scoring, as well as the inflammatory biomarkers myeloperoxidase (MPO) and TNF-α, the effect of **CO-103** treatment was also very pronounced (Figure 14.6). **CO-103** greatly decreased MPO and TNF-α levels by around 60% and 75%, respectively, when compared with those animals treated with **CP-103**.

APAP-Induced Liver Injury

Inhaled CO gas has been reported to alleviate APAP-induced acute liver injury [2,90]. To pharmacologically validate the efficacy of our organic CO prodrugs, we developed an enrichment-triggered CO prodrug system for mitochondria targeting and CO release. Such a prodrug was tested in an APAP-induced acute liver injury model in mice [20].

There are two phases to APAP-induced liver injury [91,92]. The first phase involves rapid, free radical-induced cell injury leading to the release of inflammatory mediators. The second phase involves massive cell death caused by the subsequent cytokine storm, which can eventually lead to organ failure. Specifically, mice were treated with APAP for 4 h prior to administration of CO prodrug **BW-CO-19** or **BW-CO-21** (dosage: 4 or 0.4 mg/kg, i.v.; Scheme 14.3), followed by the administration of the corresponding prodrug partner, **BW-CO-20** or **BW-CO-22**, with a 3-min delay [20]. Twenty-four hours later, serum and liver tissue were collected for ALT analysis as a measure of liver damage and immunostaining for cell death. Treatment with the prodrug led to dose-dependent reduction in ALT concentration (Figure 14.7). Even at a low dose of 0.4 mg/kg, the prodrug (**BW-CO-21** and **BW-CO-22**) was able to reduce ALT levels by more than 50% (Figure 14.7A). An increase in COHb levels occurred between 20 and 60 min after prodrug administration (Figure 14.7B). CO prodrug treatment also resulted in decreased cell death as reflected by histological staining [20]. These beneficial effects of CO are consistent with earlier studies using inhaled CO gas, and provide pharmacological validation of the efficacy of CO prodrugs in this model. Further, these data highlight the effects of targeting a CO prodrug to mitochondria.

Creating Synergy in Delivering Two Payloads from a Single Prodrug: Antibacterial Studies

CO was reported to have sensitization effects on some bacteria toward existing antibiotics [44]. Therefore, we were interested in examining the ability to deliver CO together with metronidazole to evaluate its inhibition against *H. pylori* [45]. In cell culture studies, 0.31 μg/ml of **BW-CO-24** (Scheme 14.9) achieved more than 90% inhibition of *H. pylori* growth. In contrast, the MIC_{90} of metronidazole alone was determined to be 2.5 μg/ml under the same conditions. Considering that the molecular weight of metronidazole is only one-third of that of **BW-CO-24**, the increased potency of the prodrug was around 26-fold in the context of molar concentrations. Additionally, the cyclized product **27** did not exhibit an inhibitory effect at 0.31 μg/ml. Consistent with previous reports, these results indicate that CO can act to sensitize metronidazole toward inhibition of bacterial growth. Such high-content prodrug system could also be used to codeliver CO and other antibiotics of interest. However, it is important to note that the results were from a single strain of *H. pylori*. More extensive studies are needed to truly establish synergy.

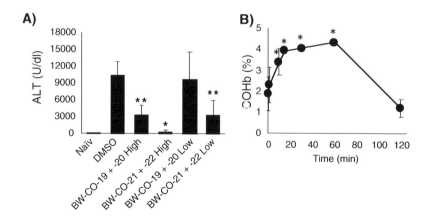

Figure 14.7 CO prodrugs protect against acute liver failure. (A) Effect of CO prodrugs on APAP-induced liver injury. Mice were treated with APAP (300 mg/kg, intraperitoneally) followed by administration of compounds BW-CO-19 + BW-CO-20 or BW-CO-21 + BW-CO-22 (4 mg/kg [high] or 0.4 mg/kg [low], i.v.) or DMSO. Three minutes separated the first and second injections. Serum ALT levels were determined 24 h after APAP. Results are presented as mean ± SD. n = 6–8 animals per group. $^{**}P < 0.03$ and $^{*}P < 0.001$ versus DMSO. (B) COHb levels over time after administration of BW-CO-21 + BW-CO-22 (4 mg/kg, i.v.). Results are presented as mean ± SD of three mice per time point. $^{*}P < 0.005$ versus baseline. Adapted with permission from Ref. [20]. Copyright (2018) Springer Nature.

Along a similar line, we also designed a prodrug capable of delivering both CO and floxuridine and tested it in the human breast cancer cell line MB-231 [47]. Specifically, 100 μM of the strained alkyne **BW-CO-29** was incubated with 0.5–10 μM of **BW-CO-28** (Scheme 14.10) with MB-231 cells for 72 h, followed by assessment of the cell viability. The control group was treated with the prodrug release product or floxuridine only. At 5 or 10 μM, the prodrug showed comparable cytotoxicity as compared to that in the floxuridine-treated group. It should be noted that only 50% of floxuridine can be released from the prodrug pairs (Scheme 14.10). Without the strained alkyne part (**BW-CO-29**), prodrug **BW-CO-28** also exhibited toxicity (weaker than the combinatory treatment with **28** and **29** and floxuridine). Since the caged floxuridine (**BW-CO-28**) was found to be quite stable for 24 h in DMSO/PBS at 37 °C, it is reasonable to assume that the observed toxicity comes from the prodrug **BW-CO-28** itself. It is also possible that **BW-CO-28** might interact with various enzymes in cells, resulting in release of the caged drug. As such, detailed degradation profiles of the prodrug need to be studied in the future for answering this question.

Vasorelaxation Effects

CO has been reported to modulate the activity of ion channels, and consequently exhibit vasorelaxation effects [93,94]. As such, elimination-based CO prodrugs by the Larsen lab were studied for their effects on the relaxation responses from isolated rat aortic rings [50,51]. It was found that both **40** and **41** (Scheme 14.13) exhibited dose-dependent vasorelaxation effects, and the EC_{50} values were measured to be around 1.6 and 33 μM, respectively. However, in the presence of a soluble guanylate cyclase inhibitor (1H-(1,2,4)oxadiazolo[4,3-a]quinoxalin-1-one), such activity was abolished, indicating that the cGMP pathway might be involved in modulating the vasodilation effect of such CO prodrugs.

Kidney Ischemia–Reperfusion Injury

During organ transplantation, the sudden lack of blood flow followed by subsequent reperfusion with blood leads to a cascade of damaging responses, including rapid generation of ROS and leukocyte infiltration. Such events are known as ischemia–reperfusion injury (IRI) and can cause severe organ damage [95]. Several chapters in this book deal with models of ischemia and reperfusion in detail and are suggested readings on this topic. In rat and pig models, inhaled CO was found to possess protective effects against IRI [96–98]. One organic CO prodrug, **BW-CO-101** (**CO-101**), was also evaluated in a kidney IRI model in mice [56]. Prior to bilateral kidney ischemia, mice were treated with 100 mg/kg **CO-101**, and injury biomarkers (serum creatinine and blood urea nitrogen [BUN]) levels were determined 24 h post-reperfusion. As shown in Figure 14.8, administration of **CO-101** effectively reduced both serum creatinine and BUN concentrations by about 50% as compared to controls. Such results are also comparable to those of inhaled CO gas and liquid CO (HBI-002), validating the efficacy of **CO-101** as a CO donor in therapeutic applications. However, **CO-101** has a short half-life (~2 min). Thus, its handling is difficult. A delay of a few minutes between dissolution and prodrug administration could lead to a substantial loss of CO, and thus potency. Therefore, we suggest the use of prodrugs with a half-life in the range of 30 min to 2 h for future work and development.

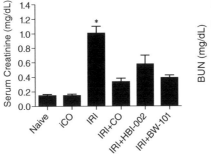
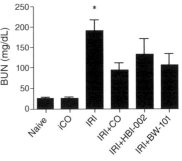

Figure 14.8 Mice were treated with inhaled CO (iCO), oral CO (HBI-002), or a CO-prodrug (CO-101) 1 h before a 45-min bilateral kidney ischemia. (A) Serum creatinine and (B) BUN were measured 24 h after reperfusion. Results represent mean ± SD of 5–10 mice per group. *P < 0.001 versus iCO, P < 0.05 versus HBI-002, and P < 0.01 versus BW-CO-101. Adapted with permission from Ref. [56]. Copyright (2018) PNAS.

Chemically Induced Gastric Injury

Inhaled CO and metal-based CORMs have been reported to possess gastroprotective effects (Chapters 11 and 12) [89,99–103]. Along a similar line, we were interested in validating the efficacy of an organic CO prodrug, **BW-CO-111 (CO-111)**, in ethanol- and aspirin-induced gastric damage in rats [57].

As shown in Figure 14.9A, 0.1 and 0.5 mg/kg **CO-111** were found to significantly decrease the area of lesion in ethanol-induced gastric damage, while 0.02 and 5 mg/kg **CO-111** did not show a benefit, thus indicating some dose dependence and biphasic effects. As such, understanding the therapeutic window is important when evaluating these molecules for therapeutic use. The cyclized product (**BW-CP-111**) after CO release did not show observable effects either. Based on the results from **CO-111**, it is reasonable to conclude that the observed protective activity came from CO release. Importantly, the studies also led to an initial understanding of the effective therapeutic index for future prodrugs. In an aspirin-induced gastric injury model, 0.1 mg/kg **CO-111** also alleviated lesions in the stomach by around 85% (Figure 14.9B). Studies were also carried out to probe the pathway by which **CO-111** protects the gastric damage in ethanol- and aspirin-induced injury models. In ethanol-induced gastric damage model, the protective effects from **CO-111** were found to result from the modulation of the COX2/PGE$_2$ pathway.

Rhabdomyolysis-Induced Acute Kidney Injury

Induction of HO-1 to accelerate heme degradation has been validated as an efficient therapeutic strategy for rhabdomyolysis-induced acute kidney injury (AKI) in animal models [104]. As one of the products from heme catabolism, CO is known to be an HO-1 inducer and to offer cytoprotective effects of its own [4,105]. As such, delivery of CO by organic prodrugs might also provide a new way to treat rhabdomyolysis-induced AKI. For this purpose, **BW-CO-306** was formulated in activated charcoal for CO administration in a glycerol-induced rhabdomyolysis mouse model. To induce AKI, 5.4 ml/kg of 50% glycerol in water was intramuscularly injected into the hindleg muscle of mice [106]. A CO prodrug or control, **BW-CO-306** (50 mg/kg) or **BW-CP-306** (saccharin), was orally administered to mice 24 h prior to glycerol injection and then daily. When compared to the **BW-CP-306**-treated controls, **BW-CO-306** treatment was observed to decrease the level of BUN and increase the survival rate of mice from AKI (Figure 14.10A and B). The tubular injury scores from the **BW-CO-306**-treated group were also at a lower degree based on histological studies (Figure 14.10C). Furthermore, the marker of renal tubular injury, namely kidney injury molecule 1 (Kim-1), was monitored by using qRT-PCR (mRNA) and immunofluorescence staining. Results from both assays suggest attenuated expression of Kim-1 in the **BW-CO-306** treatment group, but not the **BW-CP-306** treatment control group (Figure 14.10D and E). Such findings strongly indicate the therapeutic effects of CO in rhabdomyolysis-induced AKI. Additionally, the side product from the CO prodrug, **BW-CO-306**, is an FDA-approved food additive, saccharin.

This new class of CO prodrugs has the advantages of easy synthesis, using benign "carrier" molecules, being orally active and highly stable, and high conversion yield.

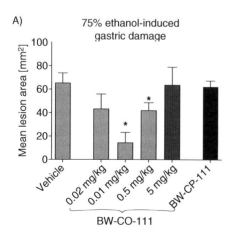

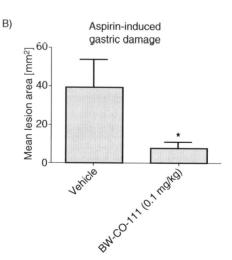

Figure 14.9 (A) Ethanol- and (B) aspirin-induced gastric lesion areas in rats pretreated i.g. with vehicle, CO prodrug BW-CO-111 (0.02–5 mg/kg), or BW-CP-111 (0.1 mg/kg). Results are mean ± SEM of five rats per group. *$P < 0.05$ compared with the vehicle control group. Adapted with permission from Ref. [57]. Copyright (2020) Elsevier.

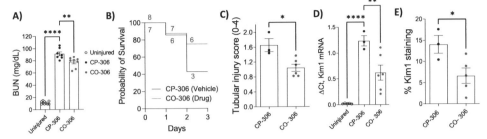

Figure 14.10 BW-CO-306 protects against rhabdomyolysis-induced AKI. (A) BUN 24 h after glycerol injection. (B) Survival curves with numbers of mice at each time point indicated. (C) Semiquantitative analysis of renal tubular injury (tubular injury score 0–4). (D) qRT-PCR for Kim-1 mRNA in the kidney. (E) Quantification of the % surface area immunofluorescence staining for Kim-1. Mouse numbers as indicated, with individual data points, mean ± SEM indicated. *$P < 0.05$; **$P < 0.01$; ****$P < 0.0001$. Reproduced from Ref. [10].

Conclusions

One of the often overlooked issues in drug discovery and development is pharmaceutical developability, which often becomes the "showstopper" issue despite promising biological activity [7]. The CO field has advanced to a stage that this issue needs to be carefully considered. Furthermore, for different therapeutic indications, the desirable pharmaceutical properties may vary. Therefore, the structural diversity of "drug" candidates is an important issue. Along this line, organic CO prodrugs represent a new direction in the field of CO donors. There are two general types of organic CO prodrugs: (i) those based on cheletropic chemistry and (ii) those based on decarboxylation–decarbonylation chemistry. Our lab and that of Larsen independently advanced various designs, which rely on the cheletropic reaction of an organic molecule containing a ketone moiety as the latent CO molecule. These CO prodrugs offer the possibility of tunable release rates, controlled release, targeted delivery, structural optimizations for pharmaceutical properties and easy conjugation with other molecules for various purposes such as high-content prodrug preparation, targeting and improved water solubility. Additionally, our lab also developed new CO prodrugs using decarboxylation–decarbonylation chemistry. One key advantage of these new CO prodrugs is benign "carrier" molecules such as artificial sweeteners saccharin and acesulfame. The structural characterizations and confirmation of these organic CO prodrugs and their side products after CO release are straightforward. This chemical tractability is an important aspect in the further development of a pharmaceutical agent [5,7]. Furthermore, the pharmaceutical industry has a vast amount of experience in developing small molecules as therapeutic agents, including formulation, pharmacokinetic, and metabolism studies. These are important traits for these organic CO prodrugs. Very importantly, the efficacy of these organic CO prodrugs has been pharmacologically validated in various animal models. There have also been extensive studies of the pharmacokinetic properties for some of the CO prodrugs (Chapter 3) [107]. All these give us hope that the newly developed organic CO prodrugs will contribute significantly to the development of CO-based therapeutics for various applications.

Acknowledgments

We gratefully acknowledge the partial financial support by the National Institutes of Health (R01DK119202) for our CO-related work, the Georgia Research Alliance through an Eminent Scholar endowment, the Brains & Behaviors program through a graduate fellowship to Z.Y., and internal resources at Georgia State University. We are grateful to our collaborators on various projects over the last decade for their contributions to studying CO prodrugs in various disease models described in the respective original research publications, including Leo Otterbein of Harvard Medical School, Bowen Ke of Huaxi Medical College, Marcin Magierowski of Jagiellonian University Medical College, Mark De Caestecker of Vanderbilt University, and Robert Maier of the University of Georgia.

References

1. Wu, L. and Wang, R. (2005). Carbon monoxide: endogenous production, physiological functions, and pharmacological applications. *Pharmacol. Rev.* 57: 585–630.
2. Motterlini, R. and Otterbein, L.E. (2010). The therapeutic potential of carbon monoxide. *Nat. Rev. Drug Discov.* 9: 728–743.

3. Romão, C.C., Blättler, W.A., Seixas, J.D., and Bernardes, G.J. (2012). Developing drug molecules for therapy with carbon monoxide. *Chem. Soc. Rev.* 41: 3571–3583.

4. Yang, X., Lu, W., Hopper, C.P., Ke, B., and Wang, B. (2021). Nature's marvels endowed in gaseous molecules I: Carbon monoxide and its physiological and therapeutic roles. *Acta Pharm. Sin. B* 11: 1434–1445.

5. Wang, B., Hu, L., and Siahaan, T. (2016). Drug Delivery: Principles and Applications, 2e. Hoboken, New Jersey: John Wiley and Sons.

6. Han, C., David, C., and Wang, B. (2010). Evaluation of Drug Candidates for Preclinical Development: Pharmacokinetics, Metabolism, Pharmaceutics, and Toxicology. New York: John Wiley and Sons.

7. Han, C. and Wang, B. (2016). Factors that impact the developability of drug candidates-an overview. In: Drug Delivery: Principles and Applications, 2e (ed. B. Wang, L. Hu and T. Siahaan), 1–15. Hoboken, Jew Jersey: John Wiley and Sons.

8. Ji, X. and Wang, B. (2018). Strategies toward organic carbon monoxide prodrugs. *Acc. Chem. Res.* 51: 1377–1385.

9. Yang, X.X., Ke, B.W., Lu, W., and Wang, B.H. (2020). CO as a therapeutic agent: discovery and delivery forms. *Chin. J. Nat. Med.* 18: 284–295.

10. Cruz, L.K.D.L., Yang, X., Menshikh, A., Brewer, M., Lu, W., Wang, M., Wang, S., Ji, X., Cachuela, A., Yang, H., Gallo, D., Tan, C., Otterbein, L., Caestecker, M.D., and Wang, B. (2021). Adapting decarbonylation chemistry for the development of prodrugs capable of in-vivo delivery of carbon monoxide utilizing sweeteners as carrier molecules. *Chem. Sci.* 12: 10649–10654.

11. Birney, D.M., Wiberg, K.B., and Berson, J.A. (1988). Geometry of the transition state of the decarbonylation of bicyclo[2.2.1]hepta-2,5-dien-7-one. Experimental and ab initio theoretical studies. *J. Am. Chem. Soc.* 110: 6631–6642.

12. Grenby, T.H. (1991). Intense sweeteners for the food industry: an overview. *Trends Food Sci. Tech.* 2: 2–6.

13. Pearson, A.J. and Zhou, Y. (2009). Diels-Alder reactions of cyclopentadienones with aryl alkynes to form biaryl compounds. *J. Org. Chem.* 74: 4242–4245.

14. Kolb, H.C., Finn, M.G., and Sharpless, K.B. (2001). Click chemistry: Diverse chemical function from a few good reactions. *Angew. Chem. Int. Ed. Engl.* 40: 2004–2021.

15. Ji, X., Pan, Z., Yu, B., De La Cruz, L.K., Zheng, Y., Ke, B., and Wang, B. (2019). Click and release: bioorthogonal approaches to "on-demand" activation of prodrugs. *Chem. Soc. Rev.* 48: 1077–1094.

16. Schaub, T.A., Margraf, J.T., Zakharov, L., Reuter, K., and Jasti, R. (2018). Strain-promoted reactivity of alkyne-containing cycloparaphenylenes. *Angew. Chem. Int. Ed. Engl.* 57: 16348–16353.

17. Chen, W., Wang, D., Dai, C., Hamelberg, D., and Wang, B. (2012). Clicking 1,2,4,5-tetrazine and cyclooctynes with tunable reaction rates. *Chem. Commun.* 48: 1736–1738.

18. Wang, D., Viennois, E., Ji, K., Damera, K., Draganov, A., Zheng, Y., Dai, C., Merlin, D., and Wang, B. (2014). A click-and-release approach to CO prodrugs. *Chem. Commun.* 50: 15890–15893.

19. Zheng, Y., Yu, B., Ji, K., Pan, Z., Chittavong, V., and Wang, B. (2016). Esterase-sensitive prodrugs with tunable release rates and direct generation of hydrogen sulfide. *Angew. Chem. Int. Ed. Engl.* 55: 4514–4518.

20. Zheng, Y., Ji, X., Yu, B., Ji, K., Gallo, D., Csizmadia, E., Zhu, M., Choudhury, M.R., De La Cruz, L.K.C., Chittavong, V., Pan, Z., Yuan, Z., Otterbein, L.E., and Wang, B. (2018). Enrichment-triggered prodrug activation demonstrated through mitochondria-targeted delivery of doxorubicin and carbon monoxide. *Nat. Chem.* 10: 787–794.

21. Schallner, N. and Otterbein, L.E. (2015). Friend or foe? Carbon monoxide and the mitochondria. *Front. Physiol.* 6: 17.

22. Smith, R.A.J., Porteous, C.M., Gane, A.M., and Murphy, M.P. (2003). Delivery of bioactive molecules to mitochondria in vivo. *Proc. Natl. Acad. Sci. U.S.A.* 100: 5407–5412.

23. Murphy, M.P. (2008). Targeting lipophilic cations to mitochondria. *Biochim. Biophys. Acta* 1777: 1028–1031.

24. Battogtokh, G., Choi, Y.S., Kang, D.S., Park, S.J., Shim, M.S., Huh, K.M., Cho, Y.Y., Lee, J.Y., Lee, H.S., and Kang, H.C. (2018). Mitochondria-targeting drug conjugates for cytotoxic, anti-oxidizing and sensing purposes: current strategies and future perspectives. *Acta Pharm. Sin. B* 8: 862–880.

25. Murphy, M.P. and Smith, R.A.J. (2007). Targeting antioxidants to mitochondria by conjugation to lipophilic cations. *Ann. Rev. Pharmacol. Toxicol.* 47: 629–656.

26. Zheng, Y., Ji, X., Yu, B., Ji, K., Gallo, D., Csizmadia, E., Zhu, M., De La Cruz, L.K.C., Choudhary, M.R., Chittavong, V., Pan, Z., Yuan, Z., Otterbein, L.E., and Wang, B. (2018). Enrichment-triggered prodrug activation demonstrated through mitochondria-targeted delivery of doxorubicin and carbon monoxide. *Nat. Chem.* 10: 787–794.

27. Breslow, R., Maitra, U., and Rideout, D. (1983). Selective Diels–Alder reactions in aqueous solutions and suspensions. *Tetrahedron Lett.* 24: 1901–1904.

28. Rideout, D.C. and Breslow, R. (1980). Hydrophobic acceleration of Diels-Alder reactions. *J. Am. Chem. Soc.* 102: 7816–7817.
29. Breslow, R. (1991). Hydrophobic effects on simple organic reactions in water. *Acc. Chem. Res.* 24: 159–164.
30. Engberts, J.B.F.N. (1995). Diels-Alder reactions in water: Enforced hydrophobic interaction and hydrogen bonding. *Pure Appl. Chem.* 67: 823–828.
31. Otto, S. and Engberts, J.B.F.N. (2000). Diels–Alder reactions in water. *Pure Appl. Chem.* 72: 1365–1372.
32. Ji, X., Zhou, C., Ji, K., Aghoghovbia, R.E., Pan, Z., Chittavong, V., Ke, B., and Wang, B. (2016). Click and release: A chemical strategy toward developing gasotransmitter prodrugs by using an intramolecular Diels–Alder reaction. *Angew. Chem. Int. Ed.* 55: 15846–15851.
33. Patterson, D.M., Nazarova, L.A., and Prescher, J.A. (2014). Finding the right (bioorthogonal) chemistry. *ACS Chem. Biol.* 9: 592–605.
34. Pan, Z., Chittavong, V., Li, W., Zhang, J., Ji, K., Zhu, M., Ji, X., and Wang, B. (2017). Organic CO prodrugs: Structure–CO-release rate relationship studies. *Chem. Eur. J.* 23: 9838–9845.
35. Illuminati, G. and Mandolini, L. (1981). Ring closure reactions of bifunctional chain molecules. *Acc. Chem. Res.* 14: 95–102.
36. Di Martino, A., Galli, C., Gargano, P., and Mandolini, L. (1985). Ring-closure reactions. Part 23. Kinetics of formation of three- to seven-membered-ring N-tosylazacycloalkanes. The role of ring strain in small- and common-sized-ring formation. *J. Chem. Soc., Perkin Trans.* 2: 1345–1349.
37. Galli, C., Illuminati, G., Mandolini, L., and Tamborra, P. (1977). Ring-closure reactions. 7. Kinetics and activation parameters of lactone formation in the range of 3- to 23-membered rings. *J. Am. Chem. Soc.* 99: 2591–2597.
38. Page, M.I. and Jencks, W.P. (1971). Entropic contributions to rate accelerations in enzymic and intramolecular reactions and the chelate effect. *Proc. Natl. Acad. Sci. U.S.A.* 68: 1678.
39. Bunnett, J.F. and Hauser, C.F. (1965). Steric acceleration of the lactonization of 2-(Hydroxymethyl)benzoic Acids1,2. *J. Am. Chem. Soc.* 87: 2214–2220.
40. Milstien, S. and Cohen, L.A. (1972). Stereopopulation control. I. Rate enhancement in the lactonizations of o-hydroxyhydrocinnamic acids. *J. Am. Chem. Soc.* 94: 9158–9165.
41. Wang, B., Nicolaou, M.G., Liu, S., and Borchardt, R.T. (1996). Structural analysis of a facile lactonization system facilitated by a "Trimethyl Lock". *Bioorg. Chem.* 24: 39–49.
42. Suliman, H.B., Carraway, M.S., Ali, A.S., Reynolds, C.M., Welty-Wolf, K.E., and Piantadosi, C.A. (2007). The CO/HO system reverses inhibition of mitochondrial biogenesis and prevents murine doxorubicin cardiomyopathy. *J. Clin. Invest.* 117: 3730–3741.
43. Otterbein, L.E., Hedblom, A., Harris, C., Csizmadia, E., Gallo, D., and Wegiel, B. (2011). Heme oxygenase-1 and carbon monoxide modulate DNA repair through ataxia-telangiectasia mutated (ATM) protein. *Proc. Natl. Acad. Sci. U. S. A.* 108: 14491–14496.
44. Tavares, A.F., Parente, M.R., Justino, M.C., Oleastro, M., Nobre, L.S., and Saraiva, L.M. (2013). The bactericidal activity of carbon monoxide–releasing molecules against helicobacter pylori. *PLOS ONE* 8: e83157.
45. De La Cruz, L.K.C., Benoit, S.L., Pan, Z., Yu, B., Maier, R.J., Ji, X., and Wang, B. (2018). Click, release, and fluoresce: A chemical strategy for a cascade prodrug system for codelivery of carbon monoxide, a drug payload, and a fluorescent reporter. *Org. Lett.* 20: 897–900.
46. Shan, D., Nicolaou, M.G., Borchardt, R.T., and Wang, B. (1997). Prodrug strategies based on intramolecular cyclization reactions. *J. Pharm. Sci.* 86: 765–767.
47. Ji, X., Aghoghovbia, R.E., De La Cruz, L.K.C., Pan, Z., Yang, X., Yu, B., and Wang, B. (2019). Click and release: A high-content bioorthogonal prodrug with multiple outputs. *Org. Lett.* 21: 3649–3652.
48. Ji, X., Ji, K., Chittavong, V., Yu, B., Pan, Z., and Wang, B. (2017). An esterase-activated click and release approach to metal-free CO-prodrugs. *Chem. Commun.* 53: 8296–8299.
49. Ji, X., De La Cruz, L.K.C., Pan, Z., Chittavong, V., and Wang, B. (2017). pH-Sensitive metal-free carbon monoxide prodrugs with tunable and predictable release rates. *Chem. Commun.* 53: 9628–9631.
50. Kueh, J.T.B., Stanley, N.J., Hewitt, R.J., Woods, L.M., Larsen, L., Harrison, J.C., Rennison, D., Brimble, M.A., Sammut, I.A., and Larsen, D.S. (2017). Norborn-2-en-7-ones as physiologically-triggered carbon monoxide-releasing prodrugs. *Chem. Sci.* 8: 5454–5459.
51. Thiang Brian Kueh, J., Seifert-Simpson, J.M., Thwaite, S.H., Rodgers, G.D., Harrison, J.C., Sammut, I.A., and Larsen, D.S. (2020). Studies towards non-toxic, water soluble, vasoactive norbornene organic carbon monoxide releasing molecules. *Asian J. Org. Chem.* 9: 2127–2135.
52. Pan, Z., Zhang, J., Ji, K., Chittavong, V., Ji, X., and Wang, B. (2018). Organic CO prodrugs activated by endogenous ROS. *Org. Lett.* 20: 8–11.

53. Michel, B.W., Lippert, A.R., and Chang, C.J. (2012). A reaction-based fluorescent probe for selective imaging of carbon monoxide in living cells using a palladium-mediated carbonylation. *J. Am. Chem. Soc.* 134: 15668–15671.
54. Wallenberg, M., Misra, S., and Björnstedt, M. (2014). Selenium cytotoxicity in cancer. *Basic Clin. Pharmacol. Toxicol.* 114: 377–386.
55. Ji, X., Pan, Z., Li, C., Kang, T., De La Cruz, L.K.C., Yang, L., Yuan, Z., Ke, B., and Wang, B. (2019). Esterase-sensitive and pH-controlled carbon monoxide prodrugs for treating systemic inflammation. *J. Med. Chem.* 62: 3163–3168.
56. Correa-Costa, M., Gallo, D., Csizmadia, E., Gomperts, E., Lieberum, J.-L., Hauser, C.J., Ji, X., Wang, B., Câmara, N.O.S., Robson, S.C., and Otterbein, L.E. (2018). Carbon monoxide protects the kidney through the central circadian clock and CD39. *Proc. Natl. Acad. Sci. U.S.A.* 115: E2302–E2310.
57. Bakalarz, D., Surmiak, M., Yang, X., Wójcik, D., Korbut, E., Śliwowski, Z., Ginter, G., Buszewicz, G., Brzozowski, T., Cieszkowski, J., Głowacka, U., Magierowska, K., Pan, Z., Wang, B., and Magierowski, M. (2021). Organic carbon monoxide prodrug, BW-CO-111, in protection against chemical-induced gastric mucosal damage. *Acta Pharm. Sin. B.* 11: 456–475.
58. Santos-Silva, T., Mukhopadhyay, A., Seixas, J.D., Bernardes, G.J., Romão, C.C., and Romão, M.J. (2011). Towards improved therapeutic CORMs: understanding the reactivity of CORM-3 with proteins. *Curr. Med. Chem.* 18: 3361–3366.
59. Southam, H.M., Smith, T.W., Lyon, R.L., Liao, C., Trevitt, C.R., Middlemiss, L.A., Cox, F.L., Chapman, J.A., El-Khamisy, S.F., Hippler, M., Williamson, M.P., Henderson, P.J.F., and Poole, R.K. (2018). A thiol-reactive Ru(II) ion, not CO release, underlies the potent antimicrobial and cytotoxic properties of CO-releasing molecule-3. *Redox Biol.* 18: 114–123.
60. Juszczak, M., Kluska, M., Wysokiński, D., and Woźniak, K. (2020). DNA damage and antioxidant properties of CORM-2 in normal and cancer cells. *Sci. Rep.* 10: 12200.
61. Nielsen, V.G. (2020). The anticoagulant effect of *Apis mellifera* phospholipase A(2) is inhibited by CORM-2 via a carbon monoxide-independent mechanism. *J. Thromb. Thrombolysis* 49: 100–107.
62. Nielsen, V.G., Wagner, M.T., and Frank, N. (2020). Mechanisms responsible for the anticoagulant properties of neurotoxic *Dendroaspis* venoms: A viscoelastic analysis. *Int. J. Mol. Sci.* 21: 2082.
63. Nielsen, V.G. (2020). Ruthenium, not carbon monoxide, inhibits the procoagulant activity of *Atheris*, *Echis*, and *Pseudonaja* venoms. *Int. J. Mol. Sci.* 21: 2970.
64. Stucki, D., Krahl, H., Walter, M., Steinhausen, J., Hommel, K., Brenneisen, P., and Stahl, W. (2020). Effects of frequently applied carbon monoxide releasing molecules (CORMs) in typical CO-sensitive model systems - A comparative in vitro study. *Arch. Biochem. Biophys.* 687: 108383.
65. Rossier, J., Delasoie, J., Haeni, L., Hauser, D., Rothen-Rutishauser, B., and Zobi, F. (2020). Cytotoxicity of Mn-based photoCORMs of ethynyl-α-diimine ligands against different cancer cell lines: the key role of CO-depleted metal fragments. *J. Inorg. Biochem.* 209: 111122.
66. Santos-Silva, T., Mukhopadhyay, A., Seixas, J.D., Bernardes, G.J., Romão, C.C., and Romão, M.J. (2011). CORM-3 reactivity toward proteins: The crystal structure of a Ru(II) dicarbonyl-lysozyme complex. *J. Am. Chem. Soc.* 133: 1192–1195.
67. Nobre, L.S., Jeremias, H., Romao, C.C., and Saraiva, L.M. (2016). Examining the antimicrobial activity and toxicity to animal cells of different types of CO-releasing molecules. *Dalton Trans.* 45: 1455–1466.
68. Yuan, Z., Yang, X., De La Cruz, L.K.C., and Wang, B. (2020). Nitro reduction-based fluorescent probes for carbon monoxide require reactivity involving a ruthenium carbonyl moiety. *Chem. Commun.* 56: 2190–2193.
69. Wareham, L.K., Poole, R.K., and Tinajero-Trejo, M. (2015). CO-releasing metal carbonyl compounds as antimicrobial agents in the post-antibiotic era. *J. Biol. Chem.* 290: 18999–19007.
70. Gessner, G., Sahoo, N., Swain, S.M., Hirth, G., Schönherr, R., Mede, R., Westerhausen, M., Brewitz, H.H., Heimer, P., Imhof, D., Hoshi, T., and Heinemann, S.H. (2017). CO-independent modification of K(+) channels by tricarbonyldichlororuthenium(II) dimer (CORM-2). *Eur. J. Pharmacol.* 815: 33–41.
71. Dong, D.L., Chen, C., Huang, W., Chen, Y., Zhang, X.L., Li, Z., Li, Y., and Yang, B.F. (2008). Tricarbonyldichlororuthenium (II) dimer (CORM2) activates non-selective cation current in human endothelial cells independently of carbon monoxide releasing. *Eur. J. Pharmacol.* 590: 99–104.
72. Nielsen, V.G. and Garza, J.I. (2014). Comparison of the effects of CORM-2, CORM-3 and CORM-A1 on coagulation in human plasma. *Blood Coagul. Fibrinolysis* 25: 801–805.
73. Tavares, A.F., Teixeira, M., Romão, C.C., Seixas, J.D., Nobre, L.S., and Saraiva, L.M. (2011). Reactive oxygen

species mediate bactericidal killing elicited by carbon monoxide-releasing molecules. *J. Biol. Chem.* 286: 26708–26717.

74. Tavares, A.F., Nobre, L.S., and Saraiva, L.M. (2012). A role for reactive oxygen species in the antibacterial properties of carbon monoxide-releasing molecules. *FEMS Microbiol. Lett.* 336: 1–10.

75. Seixas, J.D., Santos, M.F., Mukhopadhyay, A., Coelho, A.C., Reis, P.M., Veiros, L.F., Marques, A.R., Penacho, N., Gonçalves, A.M., Romão, M.J., Bernardes, G.J., Santos-Silva, T., and Romão, C.C. (2015). A contribution to the rational design of Ru(CO)$_3$Cl$_2$L complexes for in vivo delivery of CO. *Dalton Trans.* 44: 5058–5075.

76. Yuan, Z., Yang, X., Ye, Y., Tripathi, R., and Wang, B. (2021). Chemical reactivities of two widely used ruthenium-based CO-releasing molecules with a range of biologically important reagents and molecules. *Anal. Chem.* 93: 5317–5326.

77. Southam, H.M., Williamson, M.P., Chapman, J.A., Lyon, R.L., Trevitt, C.R., Henderson, P.J.F., and Poole, R.K. (2021). 'Carbon-Monoxide-Releasing Molecule-2 (CORM-2)' is a misnomer: ruthenium toxicity, not CO release, accounts for its antimicrobial effects. *Antioxidants* 10: 915.

78. Leslie, R.D. and Cohen, R.M. (2009). Biologic variability in plasma glucose, hemoglobin A1c, and advanced glycation end products associated with diabetes complications. *J. Diabetes Sci. Technol.* 3: 635–643.

79. Hay-Lombardie, A., Kamel, S., and Bigot-Corbel, E. (2019). Insights on glycated albumin. *Ann. Biol. Clin. (Paris)* 77: 407–414.

80. Anguizola, J., Matsuda, R., Barnaby, O.S., Hoy, K.S., Wa, C., DeBolt, E., Koke, M., and Hage, D.S. (2013). Review: Glycation of human serum albumin. *Clin. Chim. Acta* 425: 64–76.

81. Pushpakom, S., Iorio, F., Eyers, P.A., Escott, K.J., Hopper, S., Wells, A., Doig, A., Guilliams, T., Latimer, J., McNamee, C., Norris, A., Sanseau, P., Cavalla, D., and Pirmohamed, M. (2019). Drug repurposing: progress, challenges and recommendations. *Nat. Rev. Drug Discov.* 18: 41–58.

82. Otterbein, L.E., Bach, F.H., Alam, J., Soares, M., Tao Lu, H., Wysk, M., Davis, R.J., Flavell, R.A., and Choi, A.M.K. (2000). Carbon monoxide has anti-inflammatory effects involving the mitogen-activated protein kinase pathway. *Nat. Med.* 6: 422–428.

83. Zuckerbraun, B.S., Chin, B.Y., Bilban, M., d'Avila, J.D.C., Rao, J., Billiar, T.R., and Otterbein, L.E. (2007). Carbon monoxide signals via inhibition of cytochrome c oxidase and generation of mitochondrial reactive oxygen species. *FASEB J.* 21: 1099–1106.

84. Alonso, J.-R., Cardellach, F., López, S., Casademont, J., and Miró, Ò. (2003). Carbon monoxide specifically inhibits cytochrome C oxidase of human mitochondrial respiratory chain. *Pharmacol. Toxicol.* 93: 142–146.

85. Brown, S.D. and Piantadosi, C.A. (1990). In vivo binding of carbon monoxide to cytochrome c oxidase in rat brain. *J. Appl. Physiol.* 68: 604–610.

86. Kapetanaki, S.M., Silkstone, G., Husu, I., Liebl, U., Wilson, M.T., and Vos, M.H. (2009). Interaction of carbon monoxide with the apoptosis-inducing cytochrome c–cardiolipin complex. *Biochemistry* 48: 1613–1619.

87. Eaden, J.A., Abrams, K.R., and Mayberry, J.F. (2001). The risk of colorectal cancer in ulcerative colitis: a meta-analysis. *Gut* 48: 526.

88. Ekbom, A., Helmick, C., Zack, M., and Adami, H.-O. (1990). Ulcerative Colitis and Colorectal Cancer. *N. Engl. J. Med.* 323: 1228–1233.

89. Steiger, C., Uchiyama, K., Takagi, T., Mizushima, K., Higashimura, Y., Gutmann, M., Hermann, C., Botov, S., Schmalz, H.-G., Naito, Y., and Meinel, L. (2016). Prevention of colitis by controlled oral drug delivery of carbon monoxide. *J. Contr. Release* 239: 128–136.

90. Zuckerbraun, B.S., Billiar, T.R., Otterbein, S.L., Kim, P.K.M., Liu, F., Choi, A.M.K., Bach, F.H., and Otterbein, L.E. (2003). Carbon monoxide protects against liver failure through nitric oxide–induced heme oxygenase 1. *J. Exp. Med.* 198: 1707–1716.

91. Robinson, M.W., Harmon, C., and O'Farrelly, C. (2016). Liver immunology and its role in inflammation and homeostasis. *Cell. Mol. Immunol.* 13: 267–276.

92. Bradley, M.J., Vicente, D.A., Bograd, B.A., Sanders, E.M., Leonhardt, C.L., Elster, E.A., and Davis, T.A. (2017). Host responses to concurrent combined injuries in non-human primates. *J. Inflamm.* 14: 23.

93. Wang, R., Wang, Z., and Wu, L. (1997). Carbon monoxide-induced vasorelaxation and the underlying mechanisms. *Br. J. Pharmacol.* 121: 927–934.

94. Wilkinson, W.J. and Kemp, P.J. (2011). Carbon monoxide: an emerging regulator of ion channels. *J. Physiol.* 589: 3055–3062.

95. Kerrigan, C.L. and Stotland, M.A. (1993). Ischemia reperfusion injury: a review. *Microsurgery* 14: 165–175.

96. Hou, J., Cai, S., Kitajima, Y., Fujino, M., Ito, H., Takahashi, K., Abe, F., Tanaka, T., Ding, Q., and Li, X.-K. (2013). 5-Aminolevulinic acid combined with ferrous iron induces carbon monoxide generation in mouse kidneys and protects from renal

ischemia-reperfusion injury. *Am. J. Physiol. Renal Physiol.* 305: F1149–F1157.

97. Hanto, D.W., Maki, T., Yoon, M.H., Csizmadia, E., Chin, B.Y., Gallo, D., Konduru, B., Kuramitsu, K., Smith, N.R., Berssenbrugge, A., Attanasio, C., Thomas, M., Wegiel, B., and Otterbein, L.E. (2010). Intraoperative administration of inhaled carbon monoxide reduces delayed graft function in kidney allografts in swine. *Am. J. Transplant.* 10: 2421–2430.

98. Caumartin, Y., Stephen, J., Deng, J.P., Lian, D., Lan, Z., Liu, W., Garcia, B., Jevnikar, A.M., Wang, H., Cepinskas, G., and Luke, P.P.W. (2011). Carbon monoxide-releasing molecules protect against ischemia–reperfusion injury during kidney transplantation. *Kidney Int.* 79: 1080–1089.

99. Magierowska, K., Brzozowski, T., and Magierowski, M. (2018). Emerging role of carbon monoxide in regulation of cellular pathways and in the maintenance of gastric mucosal integrity. *Pharmacol. Res.* 129: 56–64.

100. Sheikh, S.Z., Hegazi, R.A., Kobayashi, T., Onyiah, J.C., Russo, S.M., Matsuoka, K., Sepulveda, A.R., Li, F., Otterbein, L.E., and Plevy, S.E. (2011). An anti-inflammatory role for carbon monoxide and heme oxygenase-1 in chronic Th2-mediated murine colitis. *J. Immunol.* 186: 5506.

101. Takagi, T., Naito, Y., Uchiyama, K., Suzuki, T., Hirata, I., Mizushima, K., Tsuboi, H., Hayashi, N., Handa, O., Ishikawa, T., Yagi, N., Kokura, S., Ichikawa, H., and Yoshikawa, T. (2011). Carbon monoxide liberated from carbon monoxide-releasing molecule exerts an anti-inflammatory effect on dextran sulfate sodium-induced colitis in mice. *Dig. Dis. Sci.* 56: 1663–1671.

102. Magierowski, M., Magierowska, K., Hubalewska-Mazgaj, M., Sliwowski, Z., Ginter, G., Pajdo, R., Chmura, A., Kwiecien, S., and Brzozowski, T. (2017). Carbon monoxide released from its pharmacological donor, tricarbonyldichlororuthenium (II) dimer, accelerates the healing of pre-existing gastric ulcers. *Bri. J. Pharmacol.* 174: 3654–3668.

103. Hegazi, R.A.F., Rao, K.N., Mayle, A., Sepulveda, A.R., Otterbein, L.E., and Plevy, S.E. (2005). Carbon monoxide ameliorates chronic murine colitis through a heme oxygenase 1–dependent pathway. *J. Exp. Med.* 202: 1703–1713.

104. Nath, K.A. (2006). Heme oxygenase-1: a provenance for cytoprotective pathways in the kidney and other tissues. *Kidney Int.* 70: 432–443.

105. Kim, K.M., Pae, H.-O., Zheng, M., Park, R., Kim, Y.-M., and Chung, H.-T. (2007). Carbon monoxide induces heme oxygenase-1 via activation of protein kinase R–like endoplasmic reticulum kinase and inhibits endothelial cell apoptosis triggered by endoplasmic reticulum stress. *Circ. Res.* 101: 919–927.

106. Menshikh, A., Scarfe, L., Delgado, R., Finney, C., Zhu, Y., Yang, H., and De Caestecker, M.P. (2019). Capillary rarefaction is more closely associated with CKD progression after cisplatin, rhabdomyolysis, and ischemia-reperfusion-induced AKI than renal fibrosis. *Am. J. Physiol. Renal Physiol.* 317: F1383–F1397.

107. Wang, M., Yang, X., Pan, Z., Wang, Y., De La Cruz, L.K., Wang, B., and Tan, C. (2020). Towards "CO in a pill": pharmacokinetic studies of carbon monoxide prodrugs in mice. *J. Contr. Release* 327: 174–185.

15

Targeted Delivery of Carbon Monoxide

Lisa M. Berreau

Department of Chemistry & Biochemistry, Utah State University, Logan, UT 84322-0300, USA

Introduction

Carbon monoxide (CO) is widely known for its strong binding affinity to heme proteins and for its potential toxicity [1]. Importantly, CO is also a signaling molecule produced endogenously in humans, primarily via the O_2-dependent breakdown of heme in a reaction catalyzed by heme oxygenase (HO) enzymes [2,3]. Prior research has shown that administration of low concentrations of CO gas produces beneficial health effects, including anti-inflammatory, anti-apoptotic, antihypertensive, vasodilation, and cytoprotective outcomes [4]. On this basis, CO is of significant current interest as a potential therapeutic for treating several diseases, including for possibly addressing the acute respiratory distress syndrome associated with COVID-19 [5,6]. Administration of targeted, controlled amounts of CO is especially of interest for antitumor applications [7,8].

In contrast to the reactive properties of other small signaling molecules in humans, such as the radical character of NO and the acid/base chemistry of H_2S, CO is a stable diatomic molecule [9]. It is diffusible in biological environments, readily permeating cell membranes. When introduced as a gas, systemic distribution of CO increases the concentration of carboxyhemoglobin (COHb) and the amount delivered exogenously is not the amount delivered in cells [10]. CO-releasing molecules (CORMs) can replicate the biological effects of CO gas with minimal effects on COHb levels and offer better control over CO delivery [11,12]. Spatiotemporal control of CO release using molecular donors offers the possibility of using lower concentrations to produce desired biological effects.

This chapter examines the current landscape of targeted CO delivery using CO donors, including metal-based CORMs, organic CO prodrugs, and others that are sensitive to various stimuli. As a framework to consider targeted CO delivery, this contribution starts with a brief summary of the endogenous production of CO by HO enzymes, focusing on the subcellular and tissue localization of these enzymes. Subsequently described are the structural motifs associated with well-known molecular CO donors that have been used extensively in biological applications as well as emerging structural motifs that offer enhanced control for targeted CO delivery. Included in the latter are CO donors that are triggered by endogenous or exogenous stimuli, or a combination of these stimuli. Elaboration of triggered molecular CO donors into structures capable of subcellular localization prior to CO release is then described. These types of CO donors enable experiments to probe the intracellular levels of CO needed to produce desired biological effects. Incorporation of CO donors into biomolecules or nanomaterials to enhance targeting is then examined. Such constructs offer the possibility of both tissue and subcellular localization based on passive and active targeting effects. Overall, research directed at targeted CO delivery is in its early stages of development and is an important area for further investigation to help bring forward enhanced opportunities for use of CO in medicine.

Endogenous CO Production: Reaction and Localization

CO is produced endogenously primarily through the oxidative ring opening of heme in a dioxygenase-type reaction catalyzed by HO enzymes (Figure 15.1) [13,14]. HO-1 is a stress-induced isoform that is

---

Carbon Monoxide in Drug Discovery: Basics, Pharmacology, and Therapeutic Potential, First Edition. Edited by Binghe Wang and Leo E. Otterbein.
© 2022 John Wiley & Sons, Inc. Published 2022 by John Wiley & Sons, Inc.

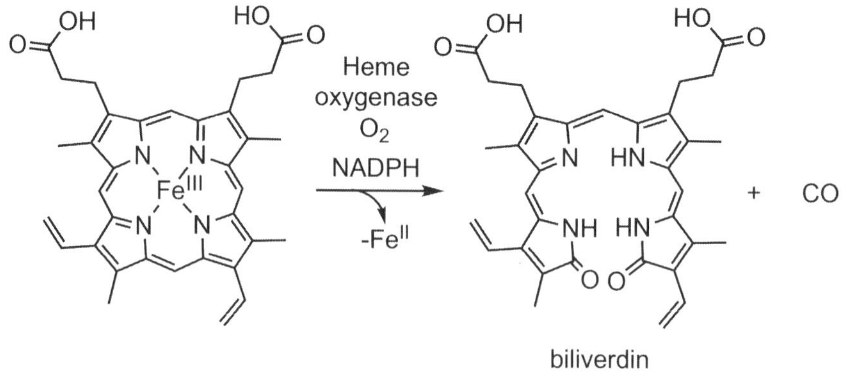

Figure 15.1 HO-catalyzed release of CO from heme.

upregulated in all tissues in response to stimuli associated with pathological states. The activity of HO-1 protects organs and tissues after injury [15]. HO-1 is an endoplasmic reticulum-associated protein under normal conditions, but translocates to mitochondria, caveolae, and nucleus under stress conditions [16]. HO-2 is a constitutively expressed form of the enzyme found in specific tissues (brain, testes, and endothelium) [17]. Further information on HO enzymes is available in Chapter 1.

CO Delivery Using Spontaneously Releasing Metal Carbonyl and Metal-Free CO Donors

Spontaneous CO-releasing metal carbonyl complexes such as $[RuCl_2(CO)_3]_2$ (CORM-2) [18], $[Ru(CO)_3Cl(glycinate)]$ (CORM-3) [19], and $[Mn(CO)_4(S_2CNMe(CH_2CO_2H)]$ (CORM-401) [20] (Figure 15.2A) have been used to reproduce CO gas-induced biological responses [21–23]. These complexes and structurally related analogues have been used extensively to probe the biological effects of CO via nonlocalized delivery (see Chapter 12). Metal carbonyl CORMs can be described as having a "CORM sphere" and a "drug sphere" with the latter corresponding to the components of the supporting organic ligands [24]. Notably, through studies of ligand modification, a spontaneously releasing Mo(0) CORM, $Mo(CO)_3(CNCMe_2CO_2H)_3$ (ALF794), was identified that exhibits enhanced localization to the liver. This compound enabled targeted CO delivery with therapeutic efficacy in an animal model of acetaminophen-induced acute liver injury [25].

While acknowledging that significant advances have been made using metal carbonyl-based CORMs as CO delivery tools [1], it is important to note that such complexes will exhibit chemistry inherent to the presence of the metal center. Reactivity such as CO leakage and protein binding occurs due to metal-centered ligand exchange, and redox activity involving the metal can occur either prior to or following CO release [26–29]. These factors complicate the interpretation of the biological effects produced by these CORMs [29].

As an approach toward enhancing the drug-like properties of CO delivery molecules, metal-free CO donors have been recently developed based on cycloadducts that exhibit spontaneous CO release (Figure 15.2B and C; see Chapter 14) [30–32]. Derivatives of the norborn-2-ene-7-one framework (Figure 15.2B) have been constructed with R_2 and R_3 being a naphthalene group that enables fluorescence tracking of the organic product following CO release [33].

Controlling CO Release: Triggered CO Donors

For targeted CO delivery, a key prerequisite is a stable CO donor that releases CO only upon triggering via an internal or external stimulus [34,35]. Notably, internal and external stimuli can also be combined to produce highly controlled CO-releasing motifs that can be defined as following a molecular logic gate approach [36]. Described below are representative examples of triggered CO donors with the acknowledgment that those discussed are not inclusive of all compounds that have been previously reported within each category. Prioritized in this discussion are triggered CO donors that contain a fluorophore that enables tracking in biological environments prior to and/or following CO release [37].

Internal Stimuli-Sensitive CO Donors

Internal stimuli include physiological triggers such as pH, oxidants, and thiol- or enzyme-mediated reactivity that result in CO release.

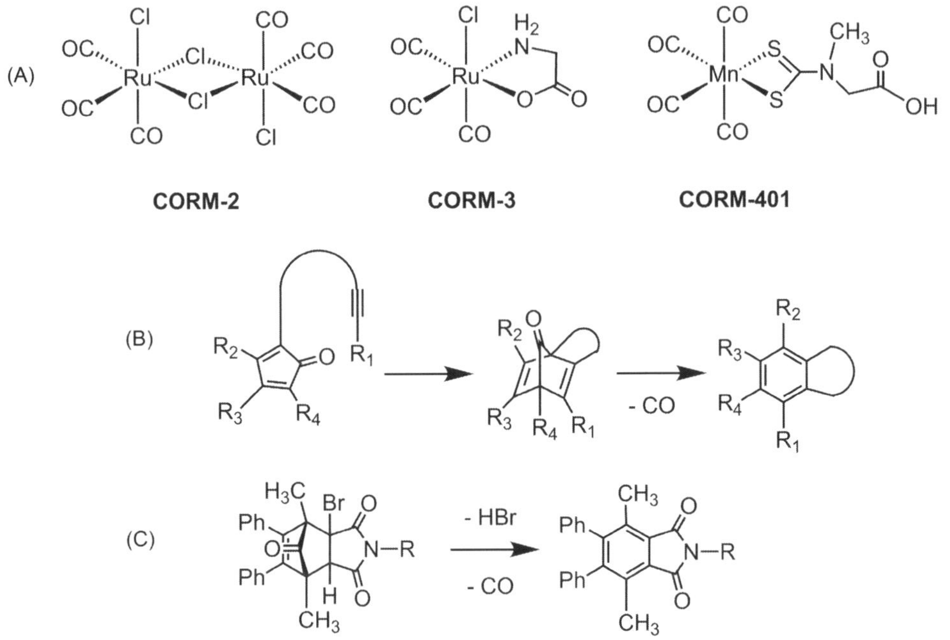

Figure 15.2 Examples of spontaneously releasing CO donors.

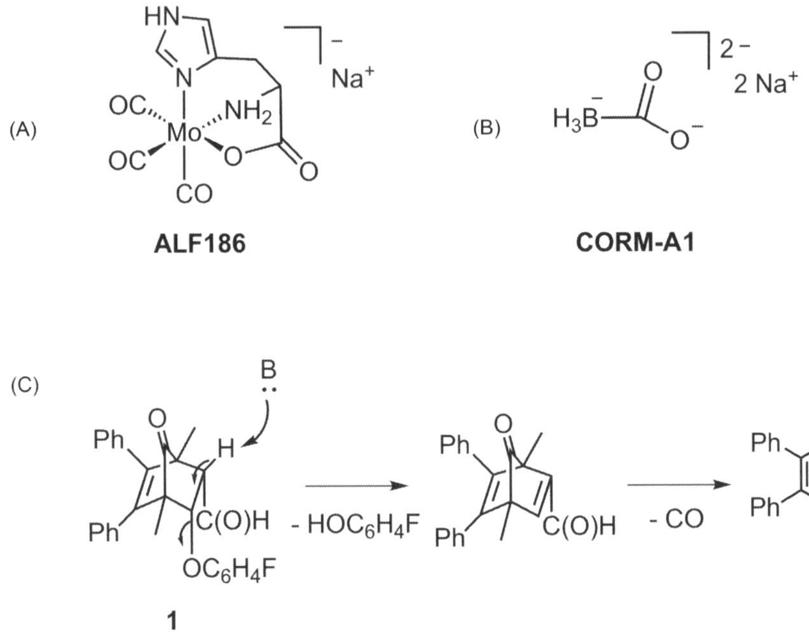

Figure 15.3 Examples of pH-sensitive CO donors.

pH

Spontaneous CO donors that exhibit changes in CO release rate as a function of pH are known for both metal carbonyl and metal-free CO donors. ALF186 (Figure 15.3A) exhibits faster O_2-induced CO release at higher pH due to the presence of the amine in the primary coordination sphere [38]. CORM-A1, a boronocarbonate sodium salt (Figure 15.3B), exhibits faster CO release at lower pH (pH 7.4, $t_{1/2}$ = 27.06 min; pH = 5.5, $t_{1/2}$ = 2.01 min; 37 °C) [39]. A pH-modulated organic CO donor with a p-FC$_6$H$_4$O leaving group (**1**, Figure 15.3C) is stable in simulated gastric fluid (pH 1) for >9 h but exhibits CO release at pH 7.4 ($t_{1/2}$ = 0.65 h, PBS buffer) [40]. Bromo-functionalized norbornene CORMs (Figure 15.2C) do not exhibit CO release at pH < 7 [31].

Oxidation

Oxidation-sensitive CO donors release CO following exposure to a molecular oxidant (e.g., O_2, H_2O_2, or other reactive oxygen species [ROS]). ALF186 (Figure 15.3A) is an example of a metal carbonyl CORM that is triggered via oxidation involving O_2 [38]. When dissolved in aqueous media at 37 °C under anaerobic

Figure 15.4 Examples of oxidation-triggered CORMs.

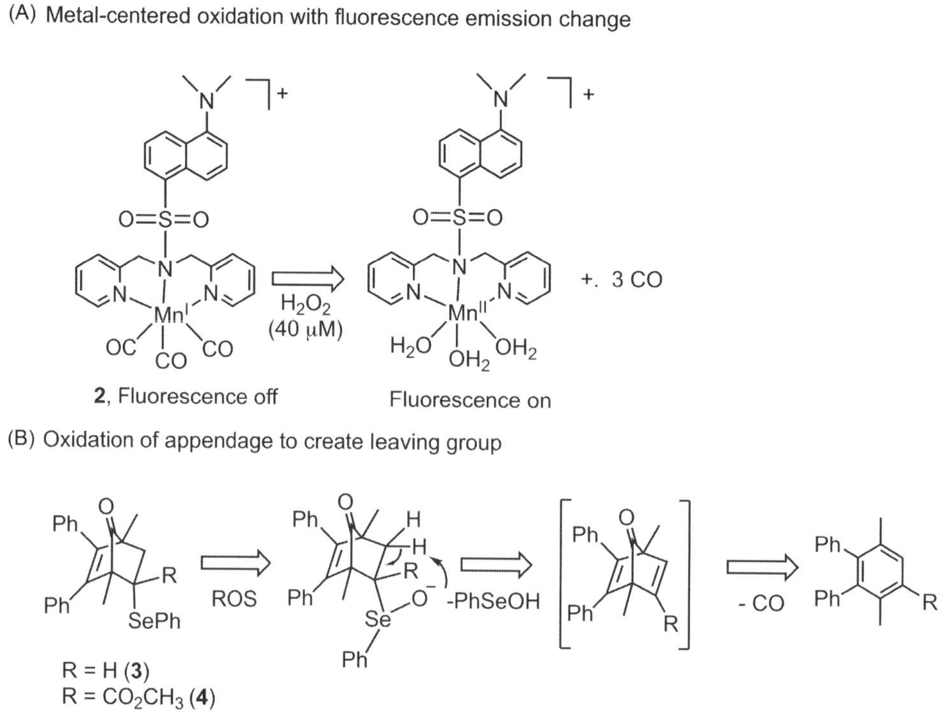

(A) Metal-centered oxidation with fluorescence emission change

2, Fluorescence off → Fluorescence on +· 3 CO

(B) Oxidation of appendage to create leaving group

R = H (**3**)
R = CO_2CH_3 (**4**)

conditions, <0.1 equiv. CO is released. However, under similar conditions in an aerobic environment, ~2.6 equiv. CO is released after 4 h. ALF186 exhibits low toxicity and produces a decrease in arterial blood pressure following administration *in vivo*. The oxidized molybdenum ion from this complex becomes part of a polyoxomolybdate cluster $[PMo_{12}O_{40}]^{3-}$. The O_2-induced reactivity of ALF186 is also associated with the formation of hydroxyl radical, an ROS. O_2 reactivity and hydroxyl radical generation are also known to occur with other metal carbonyl CORMs, including CORM-2 [41,42].

Several disease states are associated with elevated levels of ROS [43–45]. In an approach to leverage the cellular differences in ROS in normal versus cancer cells, the ROS-triggered CO release reactivity of manganese carbonyl CORMs in the presence of H_2O_2 has been examined. Incorporation of $Mn_2(CO)_{10}$ in a mesoporous silica nanoparticle enables intratumoral H_2O_2-triggered release of CO resulting in Fenton-type reactivity and formation of hydroxyl radicals [46]. No cytotoxicity for $Mn_2(CO)_{10}$ was observed in normal cells presumably due to their lower H_2O_2 content. H_2O_2-induced CO release also occurs for the $[LMn(I)(CO)_3]^+$ complex (L = 5-(dimethylamino)-*N,N*-bis(pyridine-2-ylmethyl)naphthalene-1-sulfonamide; **1**, Figure 15.4A) in aqueous buffer [47]. EC_{50} values (10–30 μM) for this complex in five cancer cell lines are below that found in normal cells (EC_{50} ~ 80 μM), which was attributed to lower

amounts of H_2O_2 in normal cells. The changes in the coordination environment involving CO release from **2** (Figure 15.4A) produce a fluorescent $[LMn(II)(OH_2)_3]^+$ by-product.

Incorporation of an ROS-sensitive group in an organic CO donor offers a possible approach to trigger CO release in environments with enhanced ROS production. With this in mind, Wang and coworkers developed norborn-2-ene-7-one precursors containing an oxidizable phenylselenium group (**3** and **4**, Figure 15.4B) [48]. In the presence of ROS (hypochlorite, superoxide, or singlet oxygen [$^1O_2$]), oxidation of the phenylselenium moiety leads to the formation of selenium oxide species that undergoes *syn*-elimination to form a double bond between the C5 and C6 positions triggering spontaneous CO release. Delivery of CO from **3** sensitizes cancer cells to doxorubicin (DOX) treatment.

Reactivity with Glutathione

Thiols such as glutathione (GSH) can be used as triggers for drug release as the concentration of GSH is higher in cancer cells than in normal cells [7,49,50]. In this regard, a water-soluble iron carbonyl complex, $[Fe_2\{\mu\text{-}SCH_2CH(OH)CH_2(OH)\}_2(CO)_6]$ (**5**, Figure 15.5), undergoes reaction with GSH to release CO in a concentration-dependent manner [51]. Comparison of CO release from **5** versus CORM-3 in two cancer cell lines (BEL-7402 and HeLa) using a fluorescent CO probe (COP-1) [52] showed similar levels of CO

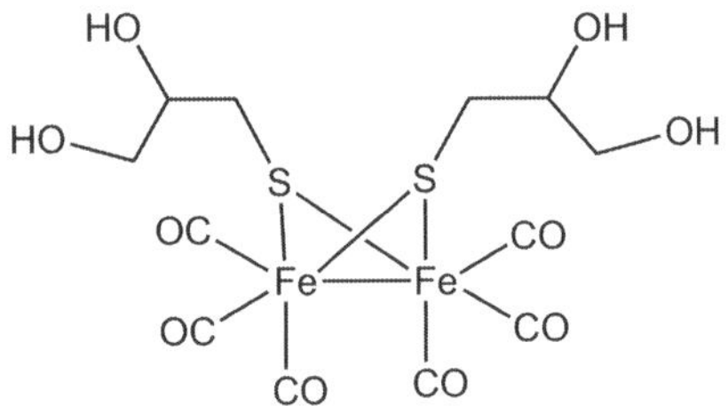

Figure 15.5 GSH-triggered CORM.

generation after 10 min. While no further change in fluorescence intensity was found with extended time for the CORM-3-treated cells, those treated with **5** exhibited continuing enhancement of fluorescence intensity for the COP-1 probe over 60 min, indicating continued CO release. Comparison of the fluorescence intensity of COP-1-mediated detection of CO derived from **5** in normal cells (HL-7702) versus the signal in cancer cells provided evidence of minimal CO release in the latter, indicating cancer cell-induced enhanced release.

Enzyme-Triggered CORMs

The strategy of enzyme-triggered CORMs focuses on engaging cell-specific enzymes to produce controlled CO release in specific tissues. Pursuing this approach, Schmalz and coworkers synthesized a series of air-stable acyloxydiene–Fe(CO)$_3$ complexes (**6–10**, Figure 15.6A) that vary in the substitution pattern on the π-ligand and in the nature of the acyloxy substituent [53,54]. Members of this family of complexes undergo enzyme-catalyzed hydrolysis in the presence of porcine liver esterase (PLE) and/or a lipase from *Candida rugosa* followed by oxidation to release CO as detected using the myoglobin assay (Figure 15.6B). In cell-based experiments, consistent with CO release, each acetyl ester compound was found to inhibit NO production by inducible nitric oxide synthase in LPS-stimulated RAW 267.4 cells, with the diacetate compound being the most active. Members of this family of compounds and analogues were also examined for their ability to offer CO-induced protection against hypothermic preservation damage as well as inhibition of TNF-α-mediated VCAM-1 expression in human umbilical vein endothelial cells (HUVECs) and renal proximal tubular epithelial cells [55]. Overall, these studies indicated that structural features of the acyloxydiene–Fe(CO)$_3$ complexes significantly affect their biological activity, with the compounds also exhibiting different effects in various cell lines. The variation in biological effects may be due to structural influences on cellular uptake or on the ability of cellular esterases to hydrolyze the compounds. A goal in further studies of enzyme-triggered CORMs is the identification of molecules with restricted cell specificity. A limitation in this regard is that the ester appendages may be susceptible to multiple esterases, including extracellular enzymes.

As an approach toward using enzyme-triggered CORMs to both deliver CO and activate the transcription factor Nrf2, acyloxydiene–Fe(CO)$_3$ complexes containing a methyl fumarate ester were prepared and investigated (**11** and **12**, Figure 15.6C) [56]. This work builds on prior reports of hybrid alkyne-ligated [Co$_2$(CO)$_6$] compounds termed "HYCOs" that were designed to target and maximize the anti-inflammatory activity of the HO-1 pathway for cellular protection [57].

Esterase-triggered metal-free click-and-release CO donors (**13** and **14**, Figure 15.6D) have been developed based on the hydrolysis of a seven-membered lactone followed by spontaneous CO release [58]. These compounds exhibit different CO release rates depending on the R$_1$ substituent (half-lives of 1 and 4 h, respectively, in 5% DMSO [pH 7.4] at 37 °C), with the ester cleavage mediated by PLE being the rate-determining step. Both **13** and **14** inhibit the LPS-induced TNF-α production in RAW 264.7 cells, providing evidence for an anti-inflammatory effect resulting from CO delivery.

Combining Two Internal Stimuli to Enhance Control of CO Release

Wang and coworkers combined esterase cleavage with pH effects to develop a compound (**15**, Figure 15.6E) wherein CO release is mediated by two endogenous triggers [59]. As shown in Figure 15.6E, PLE-mediated cleavage of the ester appendage of **15** followed by expulsion of PhSOOH leads to the formation of a norborn-2-en-7-one compound that undergoes spontaneous CO release. *In vitro* (RAW 264.7 cells) and *in vivo* studies in mice showed that CO release from **15** could produce significant anti-inflammatory effects.

External Stimuli

Examples of external stimuli used to induce CO release from CORMs are light, magnetic heating, and ultrasound.

PhotoCORMs

Light-sensitive CORMs (photoCORMs) are of considerable current interest for targeted CO delivery as they enable spatiotemporal control of CO release

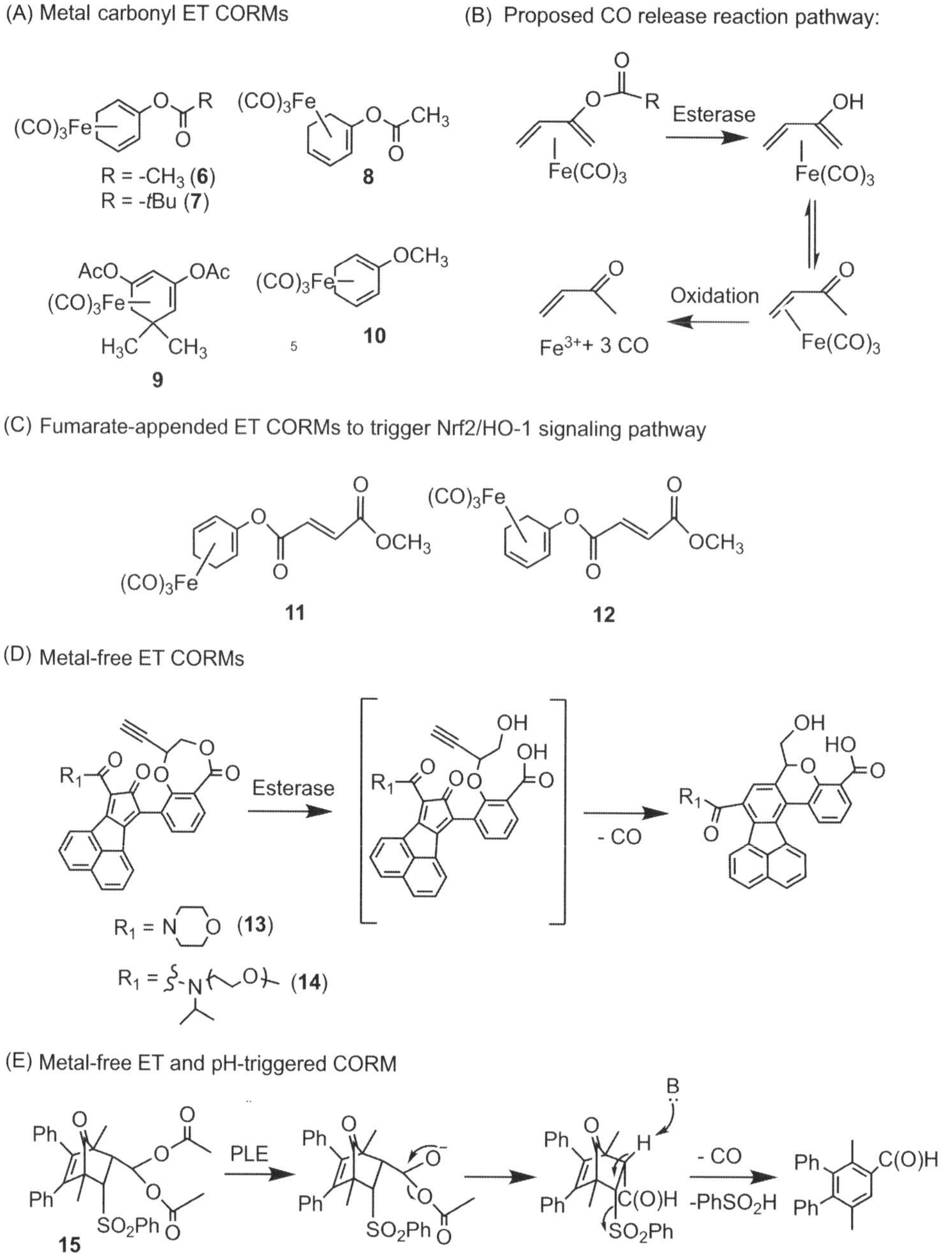

Figure 15.6 Enzyme-triggered CORMs.

(A) Metal carbonyl ET CORMs

(B) Proposed CO release reaction pathway:

(C) Fumarate-appended ET CORMs to trigger Nrf2/HO-1 signaling pathway

(D) Metal-free ET CORMs

(E) Metal-free ET and pH-triggered CORM

using a noninvasive approach that can be tuned for specific applications. More information on photoCORMs can be found in Chapter 13. Early photoCORMs for biological applications (e.g., $Fe(CO)_5$, $Mn_2(CO)_{10}$ [CORM-1]) [18] were based on the known metal carbonyl chemistry of first-row transition metals and involved ultraviolet (UV) light-triggered CO release. Due to the toxicity of these simple metal carbonyl compounds, and limitations associated with the use of UV light, subsequent studies focused on the development of transition metal CORMs supported by multidentate chelate ligands and triggered by visible light (Figure 15.7) [60]. A common structural type in this class of compounds is manganese(I) tricarbonyl species supported by nitrogen donor chelate ligands [60–64]. Modulation of the supporting ligand in such complexes enables redshifting of the absorption features and triggering with low-energy visible light (Figure 15.7D) [65,66]. A terpyridine-ligated Mn(I) photoCORM (**23**, Figure 15.7D) can be triggered for CO release using two-photon excitation [67]. Incorporation of a Mn(I) photoCORM into an upconverting nanoparticle encapsulated by an amphiphilic

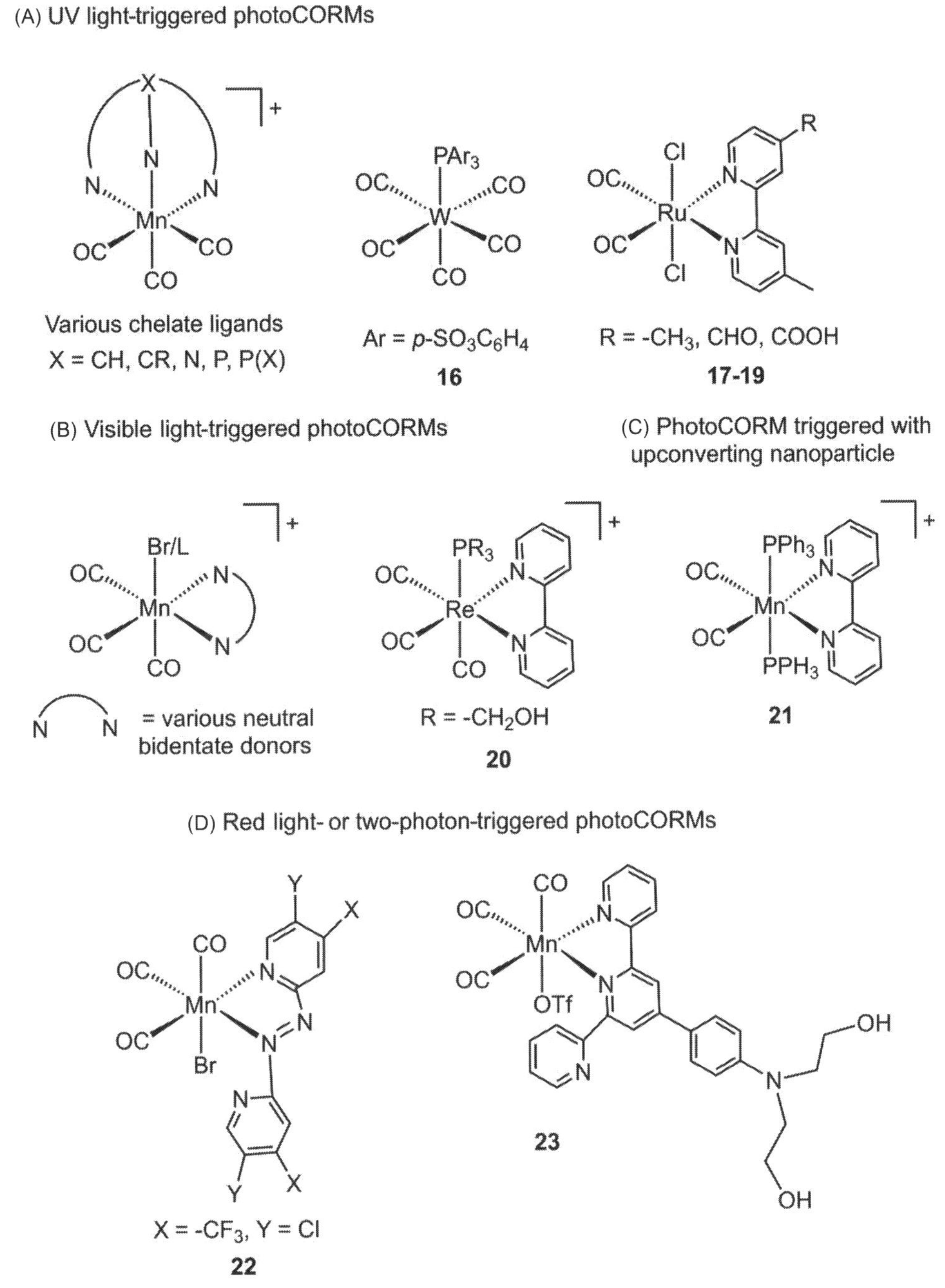

Figure 15.7 Metal carbonyl-based photoCORMs.

(A) UV light-triggered photoCORMs

Various chelate ligands
X = CH, CR, N, P, P(X)

Ar = p-SO$_3$C$_6$H$_4$
16

R = -CH$_3$, CHO, COOH
17-19

(B) Visible light-triggered photoCORMs

N⌒N = various neutral bidentate donors

R = -CH$_2$OH
20

(C) PhotoCORM triggered with upconverting nanoparticle

21

(D) Red light- or two-photon-triggered photoCORMs

X = -CF$_3$, Y = Cl
22

23

polymer promotes water solubility and enables triggered CO release using near-infrared (NIR) light [68]. These advances are important toward triggering the compounds with light in the tissue penetrating window [69]. Polymer composites with encapsulated [Mn(I)(CO)$_3$]-containing complexes have been combined with fiber-optic technology for localized CO release, including to cancer cells [70,71].

An important component of targeted CO delivery using a photoCORM is the ability to track the location and extent of CO release *in vivo* [37]. Metal carbonyl photoCORMs of four types relevant to tracking have been reported: (i) luminescent Re(I) and Mn(I) CORMs that exhibit no change upon CO release [72–75]; (ii) turn-off Re(I) photoCORMs [76,77]; (iii) turn-on Mn(I)-based photoCORMs [47,77,78]; and (iv) emissive metal carbonyl photoCORMs with unique fluorescence signals pre- and post-CO release [79,80]. Overall, the luminescent photoCORMs reported to date all have a similar formulation, [MX(CO)$_3$(α-diimine)] with M = Re(I) or Mn(I). The range of emission wavelengths spanned by these compounds (or their photoinduced released ligand) is from 400 to 605 nm.

Despite the advances enabled by metal carbonyl photoCORMs in terms of controlled and trackable CO delivery, challenges remain toward advancing their use for targeted CO delivery. These include the leakiness (spontaneous release) of the metal carbonyl unit due to ligand exchange reactivity, the need for some complexes of UV light to trigger CO release, and the possible redox reactivity of the metal-containing CO release by-products. These limitations have led to significant efforts toward incorporation of such motifs in materials to enhance control of CO release, increase payload of the CORM, improve fluorescence tracking, and sequester CO release by-products [35].

In terms of metal-free photoCORMs, cyclopropenones [81], 1,3-cyclobutanediones [82], and 1,2-dioxolane-3,5-diones [83] are known to release CO upon exposure to UV or near-UV (<420 nm) illumination. Several examples of organic photoCORMs triggered by visible light have also been recently developed. These include molecules based on fluorescein [84], 9,10-dihydro-9,10-ethanoanthracene-11,12-dione [85], BODIPY [86], 3-hydroxyflavone [87–89], 3-hydroxyquinolone [90], and cyanine–flavonol [88] frameworks (Figure 15.8; see Chapter 13). Many of these molecules can be triggered for CO release using low-energy visible (>500 nm) to NIR light, or via two-photon excitation in solution, and in cellular and animal model environments [86,91,92]. A subset of the visible light-triggered organic photoCORMs (30–34; Figure 15.8) are trackable via fluorescence prior to CO release [86–90]. The coumarin-appended 3-hydroxyflavone 33 (Figure 15.8) offers ratiometric fluorescence tracking pre- and post-CO release in solution, cells, and mice [89]. Product tracking is also available with the α–diketone photoCORMs (25–27, Figure 15.8) as CO release results in the formation of a fluorescent anthracene by-product.

A distinct advantage of visible light-triggered metal-free photoCORMs is the typically well-defined release of a known amount of CO (1–2 equiv. depending on structure) without leakage prior to illumination. Additionally, in most cases, a well-defined CO

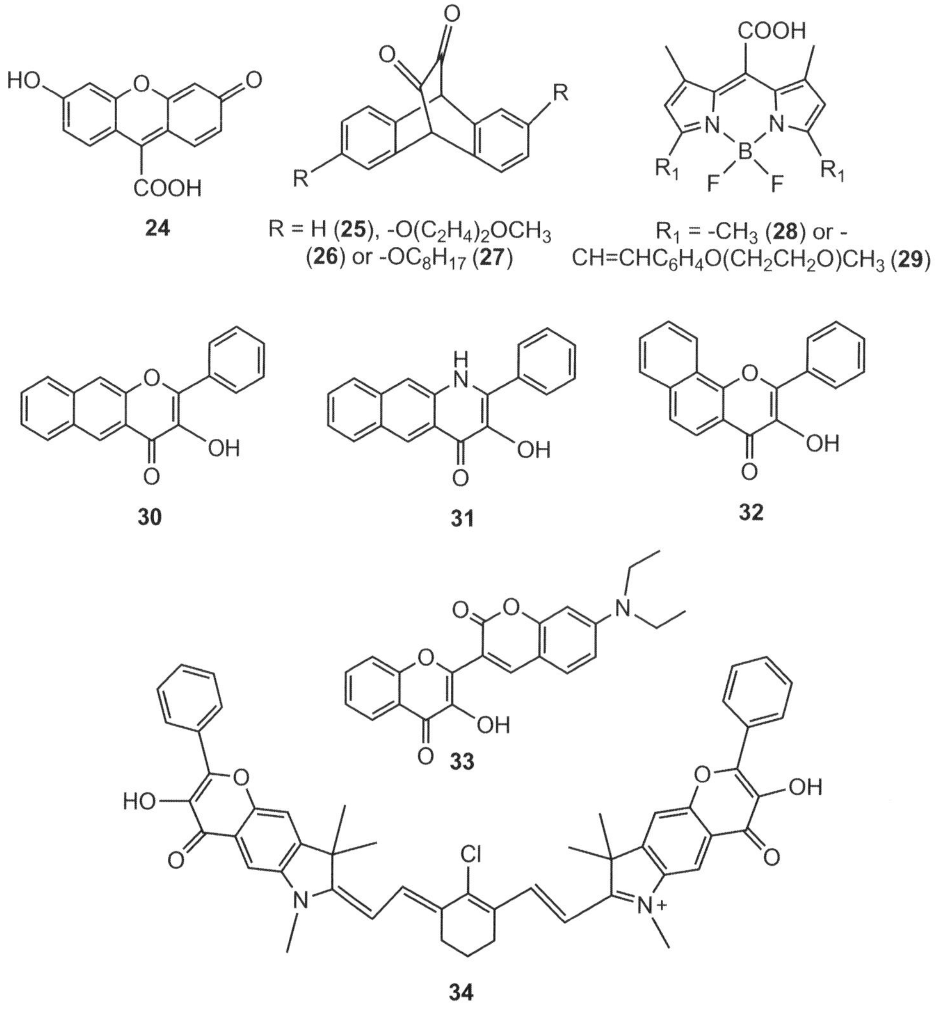

Figure 15.8 Metal-free photoCORMs.

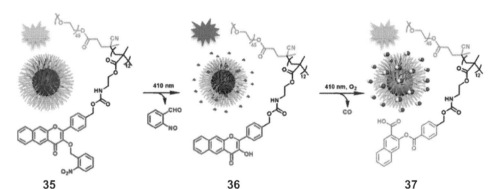

Figure 15.9 Visible light-triggered tandem deprotection and CO release from flavonol-containing micelles. Adapted with permission from Ref. [96]. Copyright (2020) The Royal Society of Chemistry.

release by-product (iCORM) is produced. For metal carbonyl photoCORMs, the exact amount of CO delivered can be influenced by conditions, and the iCORM in many cases is actually a mixture and is not easily identified. The excited-state reaction pathways of some organic photoCORMs have been examined using complementary spectroscopic and computational approaches. For example, mechanistic studies of the O_2-dependent CO-releasing reactions of the 3-hydroxyflavone (**30**) and cyanine–flavonol derivatives (**34**) have revealed that the CO release process primarily involves reaction between a triplet excited state and $^3O_2$ [91,93,94].

Nonmetal flavonol-based photoCORMs that also contain a light-sensitive protecting group on the CO release unit have been used to construct polymeric micelles. These constructs can be used to enable CO-induced anti-inflammatory effects and targeted CO delivery for wound healing (Figure 15.9), as well as for NO-sensing/CO-releasing "breathing micelles" via tandem light-induced reactions [95–97]. These approaches demonstrate that one wavelength of light can be used to trigger two consecutive chemical reactions (**35** → **36** and **36** → **37**; Figure 15.9), with the stepwise sequence being trackable by fluorescence emission associated with the flavonol. A red emission associated with the polymeric extended flavonol **36** is attributed to the formation of flavonol aggregates within the micelle core.

Combining Internal and External Stimuli in Highly Controlled, Environment-Sensing CORMs

PhotoCORMs that enable sensing of the environment prior to light-triggered CO release offer a secondary level of control for targeted CO delivery. We have termed such systems sense-of-logic (SL-photoCORMs) as they can best be described using a molecular logic gate model [98]. The systems described below can be viewed as operating by an AND logic gate sequence [99,100] with distinct input parameters and outputs for the sense-and-release sequences.

ROS Sensing

As H_2O_2 is elevated in cancer cells [7], molecular structures that exhibit H_2O_2-dependent reactivity coupled with triggered CO release are of interest for targeted CO delivery. Tang et al. developed a flavonol–boronate conjugate (**38**, Figure 15.10) that detects H_2O_2 (DL = 66 nM), undergoing reaction to produce the extended flavonol **30** that can then be triggered using two-photon illumination (800 nm) for CO release [92]. This stepwise reactivity is evident in the fluorescent emission properties of the compounds, with **38** exhibiting emission in the blue region and **30** emissive in the green channel. Loss of the fluorescence signal from **30** provides evidence for the CO release reaction and the formation of the non-emissive depside product (**41**). Compound **38** was used to examine the effect of CO release on angiotensin II-induced H_2O_2 fluctuation and vasoconstriction in zebrafish. Deprotection of **38** by H_2O_2 followed by two-photon-induced (800 nm) CO release from **30** reversed the observed vasoconstriction.

Thiol Sensing

Increased levels of GSH, cysteine, and other intracellular thiols in the environment of cancer cells have been used as triggers to induce release of drug molecules, including CO [101]. To enable thiol sensing prior to visible light-triggered CO release, the extended flavonol **30** was functionalized with a Michael acceptor to give **39** (Figure 15.10). Solution and cell-based studies demonstrate logic gate-type deprotection reactivity with cysteine, with distinct fluorescence signals evident for **39** and **30**, which enabled sensing of endogenous cellular thiols prior to triggered CO release. Notably, while the latter step involves dioxygenase-type CO release reactivity from **30**, the reaction proceeds under hypoxic (1% O_2) conditions [100].

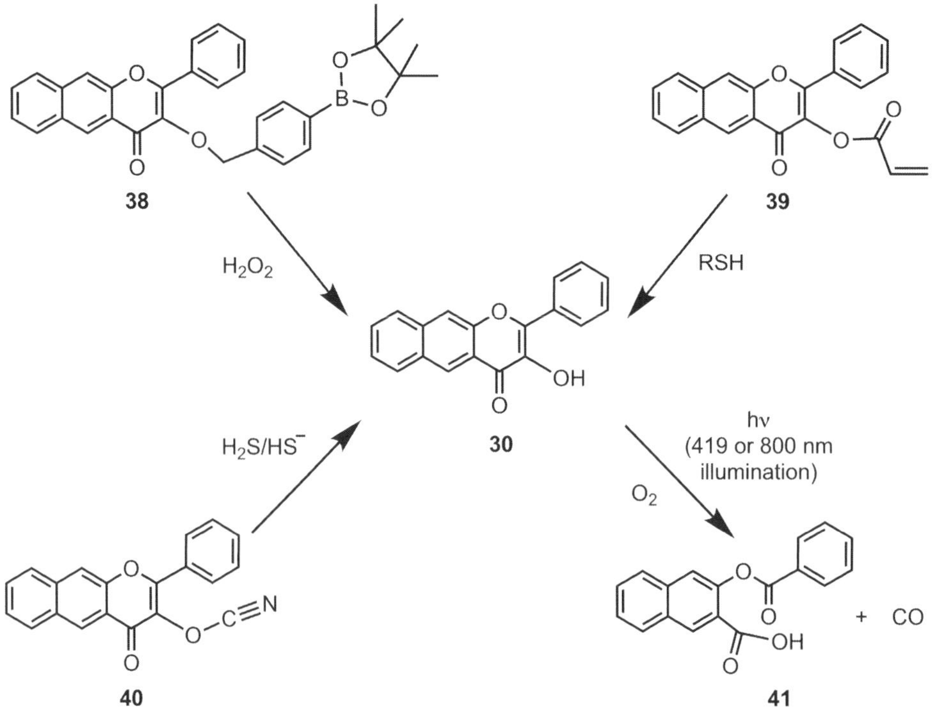

Figure 15.10 Logic gate reactivity of appended flavonols. The blue coloration of 38–40 and the green coloration of 30 represent the trackable emission features during triggered deprotection and light-induced CO release of the compounds.

H₂S Sensing

Interplay between CO and H$_2$S is involved in maintaining cell homeostasis and in various disease states, including sleep apnea [102]. H$_2$S is also involved in colitis, an inflammatory disorder of the colon, and the H$_2$S-producing enzyme CSE is upregulated in colon cancer cells [103]. Development of molecular tools to investigate H$_2$S/CO interplay, and to trigger CO release via H$_2$S sensing, remains largely unexplored. To address this area, incorporation of an H$_2$S-sensing motif to the framework of **30** produced **40** (Figure 15.10), which enabled AND-type logic gate H$_2$S-sensing reactivity. This reaction is characterized by the same photochromatic changes as noted for the thiol-sensing system above [104]. Notably, while the H$_2$S sensor **40** (Figure 15.10) is unstable with respect to hydrolysis in solution, when taken up in cells it is stable and exhibits the expected sense-and-release reactivity. CO release from **30** (derived from **40** via H$_2$S reactivity) was detected in HUVECs using a Nile Red sensor [105]. It is noted that a pyrene–flavonol hybrid containing a 2,4-dinitrobenzene group for ratiometric H$_2$S sensing was subsequently reported [106].

Enzyme-Triggered PhotoCORMs for Targeted CO Delivery

Westerhausen and coworkers targeted intracellular delivery of CO via uptake of lipophilic acetoxymethyl-appended Mn(I)(CO)$_3$ derivatives [107]. Complexes such as **42** (Figure 15.11A) are designed to undergo intracellular enzyme-catalyzed cleavage leaving a charged hydrophilic complex **43** within the cell, which is then triggered for CO release using UV light (Figure 15.11A). Fourier transform IR (FTIR) microspectroscopy was used to monitor uptake of **42** in LX-2 cells, with the N-FINDR algorithm being used to separate cell and CORM spectra to visualize intracellular compound distribution via the ν_{CO} vibrations. Uptake of the compound was identified after washing of the cells, providing evidence for the proposed trapping of the esterase-deprotected form. The cellular distribution of the complex was found to be nearly uniform. Illumination of cells resulted in loss of the CO vibrations, consistent with intracellular CO release.

Luan and coworkers developed an extended flavonol motif (Figure 15.11B) to target the use of bacterial lipases to deprotect an ester-modified flavonol photoCORM that can subsequently be triggered with UV light to release CO [88]. The core flavonol moiety of this compound (**32**) is an isomer of **30** and exhibits similar excited-state intramolecular proton transfer fluorescence properties, thus enabling fluorescence tracking of the enzyme-mediated hydrolysis of the acetyl moiety, followed by light-induced CO release. The reactivity occurred only in the presence of lipases versus other proteins (e.g., alkaline phosphatase and fibrinogen) or PBS solution, which were used as controls. The CO released from **44** via esterase activity showed significant bactericidal abilities on

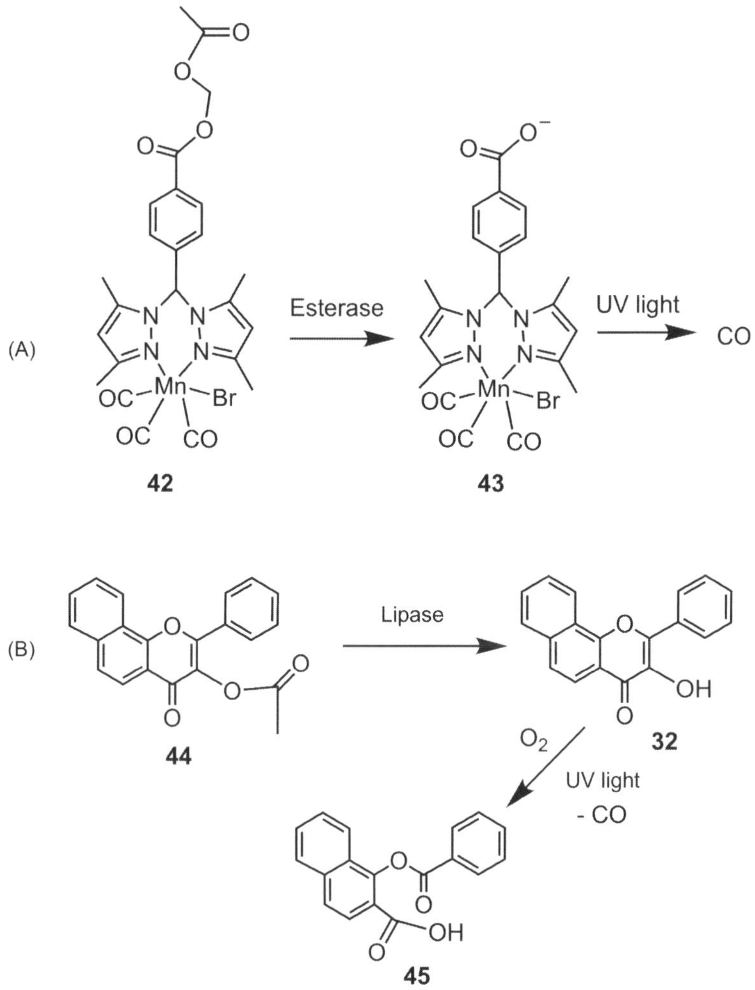

Figure 15.11 Enzyme-triggered photoCORMs for localized CO delivery.

lipase-secreting Gram-positive *Staphylococcus aureus* and methicillin-resistant *S. aureus*, as well as eradication of bacterial cells within biofilms. Incorporation of the compound on gauze was found to eradicate wound infections in a mouse model.

Tumor Microenvironment and NIR-Mediated CO Release Using a Cu(II) Coordination Polymer

Building on the known GSH-mediated reduction of Cu(II) to Cu(I) in cancer cells and subsequent Cu(I)-mediated Fenton reactivity with endogenous H_2O_2 [108], a coordination polymer was developed for environment and NIR-mediated formation of OH• and CO in tumor cells [109]. Specifically, the NCu-FleCP coordination polymer (Figure 15.12) was constructed using a carboxy-appended extended flavonol. FTIR analysis indicates Cu(II) coordination for the flavonol 3-OH/4-keto chelate motif and the 4′-carboxylate moiety within the coordination polymer. While stable in PBS buffer and Dulbecco's modified Eagle's medium for at least 3 days, in the presence of GSH in PBS, the coordination polymer undergoes cleavage to give free flavonol and Cu(I) along with the disulfide of GSH. Introduction of H_2O_2 to the mixture of NCu-FleCP and GSH in PBS buffer at pH 7.4 results in the formation OH•, as evidenced by a Methylene Blue assay. CO release from the free carboxyflavonol is induced by illumination of the NCu-FleCP/GSH mixture using an 808 nm laser. CO release is negligible directly from the coordination polymer, and from the combination of NCu-FleCP and GSH without illumination, thus indicating sequential reactivity (NCu-FleCP + GLU → Cu(I) + Fle) wherein only the free flavonol undergoes NIR-triggered CO release. NCu-FleCP is cytotoxic to multiple cancer cells lines (4T1, Hepa1-6, and A549) with IC_{50} values of 13–36 μg/ml. The toxicity is enhanced (IC_{50} 7–12 μg/ml) upon illumination of the cells at 808 nm to induce CO release. Comparison to a normal cell line (NIH-373) showed that the coordination polymer, including under illumination, has minimal effect on cell viability (>80% viable at

Figure 15.12 Targeted synergistic OH• and CO generation in cancer cells using a Cu(II)–flavonol coordination polymer. (A) Synthesis of the flavonol and coordination polymer. (B) Proposed intracellular reactivity of NCu-FleCP upon reaction with GSH and illumination with 808 nm light. Reprinted with permission from Ref. [108]. Copyright (2020) John Wiley & Sons.

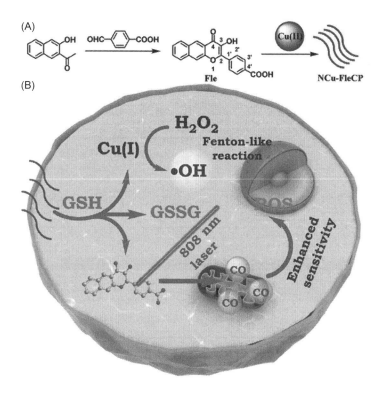

20 h with 20 μg/ml NCu-FleCP). The differentiation between cancer cells and normal cells is attributed to the higher concentration of GSH in the former, as it is only under conditions wherein Cu(II) is reduced to Cu(I) that the free flavonol becomes available to undergo light-induced CO release. Studies of the tumor-reducing capability of synergistic OH• and CO delivery from NCu-FleCP upon illumination with 808 nm light in a 4T1 tumor-bearing mouse model showed that tumor size reduction exceeded that produced only by NCu-FleCP, indicating a synergistic antitumor effect resulting from OH• generation and CO release.

Magnetic Heating-Promoted CO Release

With limitations in terms of the depth of penetration of light in tissue [110], other physical methods are being investigated to trigger CO release from CORMs. Localized magnetic field-to-heat conversion offers an approach for targeted CO delivery, particularly for targeted treatment of tumors [111]. In this regard, conjugation of a CORM-3 analogue to an iron oxide (Fe_2O_3) nanoparticle produces CORM@IONP (**46**, Figure 15.13), which was then encased in a polymer shell (dextran) to promote water solubility. Compartmentalization in an alginate followed by application of an alternating magnetic field results in heat-driven CO release from alginate@dextran@

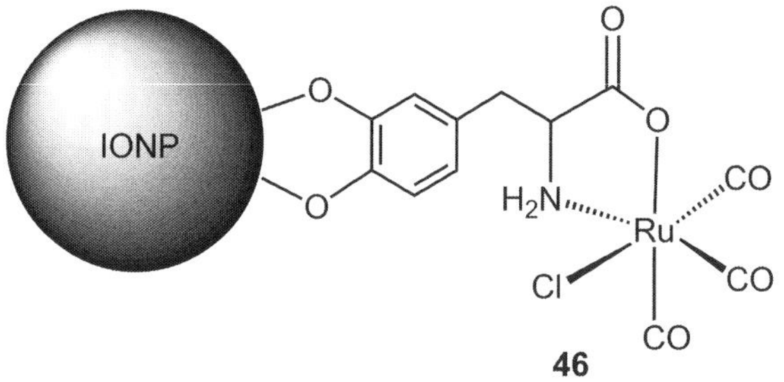

Figure 15.13 CORM@IONP.

CORM@IONP without heating the surrounding solution [112–114].

Ultrasound-Mediated CO Release

Ultrasound can be used as a trigger for drug delivery from micelles for cancer therapy [115]. Encapsulation of CORM-2 in Pluronic F-127 micelles provides a more stable complex with respect to CO release [116]. In the presence of cysteine, which is known to induce CO release via ligand exchange in Ru(II)-based CORMs [117], the micelle-encapsulated CORM-2 releases only a minimal amount of CO. However, application of low-intensity ultrasound to the same sample increases the amount of CO release by fourfold, nearly to that found for CORM-2 in the absence of the micelle. Treatment of prostate cancer cells

(PC-3) with the CORM-2-loaded micelles and application of ultrasound reduced cellular proliferation.

Mitochondria-Targeted CO Delivery Using Small Molecule CO Donors and PhotoCORMs

Cytochrome c oxidase (COX) in the mitochondrial electron transport chain is major heme protein site related to CO signaling and toxicity effects [9]. As such, efforts have been initiated to target CO delivery to mitochondria. Initial studies involved the use of CORM-2, CORM-3, and CORM-401 as spontaneous CORMs for mitochondrial bioenergetics studies [118–121]. As these CORMs release CO spontaneously via ligand exchange and cannot be tracked in terms of the site of CO delivery, the effective local CO concentration needed to modulate mitochondrial function remains to be fully defined. While it is known that CO binding to COX is toxic, less investigated is how low concentrations promote cytoprotection by inducing electron accumulation at complex III, resulting in the generation of ROS. As an approach toward targeting CO release to mitochondria, Wang and coworkers pursued enrichment-triggered CO release via click-and-release reactivity [122]. Triphenylphosphonium-tailed Diels–Alder reaction precursors **47** and **48** were constructed with inclusion of a naphthalene moiety to enable fluorescence turn-on when the precursors click together and CO is released (Figure 15.14). After bimolecular reactivity resulting in CO release was confirmed, the compounds were introduced to RAW 264.7 cells at concentrations of 1 or 5 μM and imaged using the DAPI channel. Appropriate controls showed that only in the presence of both reaction partners was blue emission produced indicating the formation of the product (**49**) and release of CO. This emission colocalizes with that of MT Deep Red mitochondrial tracker indicating mitochondrial localization. In a chemically induced liver injury model in mice, the **47**–**48** pair showed potent activity in offering organ protection with EC_{50} being ~0.4 mg/kg, which is the lowest that has ever been observed for an organic CO prodrug in offering organ protection among different models, including chemically induced colitis, ischemia–reperfusion injury of the kidney, chemically induced gastric damage, and LPS-induced systemic inflammation (Chapter 14). Such results are consistent with enhanced potency by targeting mitochondria.

In order to examine in more detail the effects of CO release on mitochondrial function, an extended flavonol compound **50** (Figure 15.15) was prepared containing a triphenylphosphonium tail for mitochondrial targeting [123]. Compound **50** retains the CO release and photophysical properties of the parent flavonol **30** (Figure 15.8) and exhibits good cellular uptake in A549 and HUVECs. Localization to mitochondria was confirmed via confocal

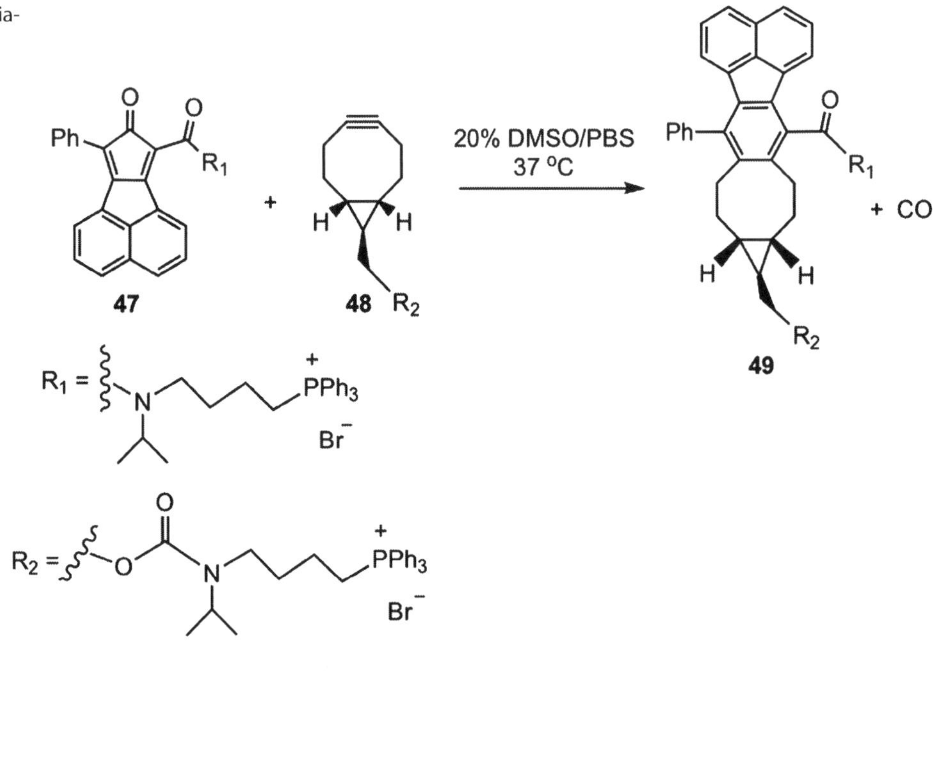

Figure 15.14 Mitochondria-targeted CO release using click-and-release reactivity.

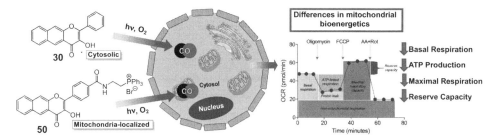

Figure 15.15 Mitochondria-targeted CO release versus cytosolic CO release evaluated by mitochondrial bioenergetics. At a concentration of 10 µM, both compounds produce similar decreases in basal respiration, ATP production, maximal respiration and reserve capacity. Adapted with permission from Ref. [123]. Copyright (2020) American Chemical Society.

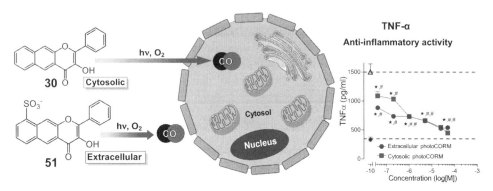

Figure 15.16 Evaluation of cytosolic versus extracellular CO release using visible light-triggered photoCORMs. Adapted with permission from Ref. [124]. Copyright (2020) American Chemical Society.

microscopy and colocalization studies with MitoTracker Red (Pearson's colocalization coefficient = 0.707 ± 0.14 for 29 cells evaluated). Localization of **50** to the mitochondria significantly increased its toxicity (A549) relative to the parent flavonol **30** (IC$_{50}$ values: **50**:14 µM; **30**:81 µM). Visible light-induced CO release further increased toxicity only for **50** (IC$_{50}$ ~ 5 µM; **30**:76 µM).

Mitochondrial bioenergetics studies were performed to examine the effect of cytosolic CO release from **30** versus mitochondria-localized CO from **50** at concentrations of 0–10 µM using an extracellular flux analyzer (Agilent Seahorse XF). These studies showed that localization at a concentration of 10 µM did not produce notable differences as CO release from both **30** and **50** yielded similar effects in terms of decreasing basal respiration, ATP production, maximal respiration, and reserve capacity. Notably, the concentrations used in this study were at least twofold below those needed to produce the same effects using nonlocalized CORM-2 and CORM-3. Overall, these results indicate that intracellular CO release produces more potent biological effects than traditionally used metal carbonyl CORMs such as CORM-3, which may be due to the latter being primarily an extracellular CO release compound [118].

Targeted Intracellular Versus Extracellular CO Delivery Using PhotoCORMs

An essential question in the targeted delivery of CO is how intracellular versus extracellular delivery of CO impacts the observed biological effects. As shown in Figure 15.16, the extended flavonol **30** and a sulfonated analogue were used to directly examine the impact of extracellular versus intracellular CO delivery on cytotoxicity and anti-inflammatory effects [124]. Sulfonation of **30** produces a water-soluble extended flavonol **51** (Figure 15.16) that exhibits similar absorption/emission and CO release properties to the unsubstituted parent compound. Compound **51** exhibits minimal cellular uptake as determined by fluorescence microscopy. Delivery of CO to cells using **51** was confirmed using a Nile Red CO sensor [105]. Comparison of the toxicity produced upon CO release from intracellular **30** and extracellular **51** in

RAW 264.7 cells showed that the former exhibited significant toxicity (IC$_{50}$ 22.39 ± 2.98 μM), whereas the latter produced no toxicity up to 100 μM. Notably, the location of CO release did not impact the anti-inflammatory properties of the compounds, with both suppressing LPS-induced TNF-α levels to a similar extent, and a 50% reduction in TNF-α was found for **30** and **51** starting at 1 and 0.04 mM, respectively. The difference in these values relates to an effect of the delivery scaffold, with the sulfonated flavonol providing enhanced anti-inflammatory effects in the absence of CO release.

Targeted CO Delivery Using CORMs or PhotoCORMs Associated with Biomolecules

Peptide Conjugation to PhotoCORMs

Peptide-appended photoCORMs wherein the amino acid sequence is designed to transport CORMs via recognition for targeted delivery were developed by Schatzschneider and coworkers [125–127]. Sonagashira and "click" coupling were used to couple an alkyne-appended photoCORM, [(tpm)Mn(CO)$_3$]$^+$ (tpm = tris(pyrazolyl)methane), to the five-amino acid sequence Thr–Phe–Ser–Asp–Leu [125]. Subsequently, a [Mo(CO)$_4$(bpy)] photoCORM containing an aldehyde on the periphery of the bpy ligand was coupled to a transforming growth factor (TGF) β-recognizing sequence via N-terminus aminoxyacetic acid using oxime ligation [126]. Similarly, Mn(I) tricarbonyl complexes [Mn(bpea^{CH_2C_6H_4R})(CO)$_3$]PF$_6$ (bpea = 2,2-bis(pyrazolyl)ethylamine functionalized with R = –I, –CCH, or –CHO) were used to explore various coupling methods to the TGF-β sequence [127]. The obtained photoCORM–peptide conjugates exhibit UV- or visible light-triggered CO release similar to the free photoCORMs.

Folic Acid-Tagged Protein Nanoemulsions Containing CORM-2

Folic acid (FA) binds with high affinity to folate receptor (FR)-expressing cells, with FR-α being overexpressed in cancer cells [128,129]. FA is an optimal targeting ligand with high affinity for the FR (K_d = 10^{-10} M) even after conjugation [130]. FA-functionalized bovine serum albumin (BSA) nanoemulsions containing CORM-2 were developed that demonstrated targeted delivery to FR-positive cancer cells producing anticancer activity and enhanced survival of mice having A20 lymphoma tumors [131]. The induced anticancer effects were suggested to result from the anti-Warburg effect of CO wherein excess O$_2$ consumption leads to growth inhibition, cellular exhaustion, and death [132].

Albumin-Based Delivery of CORM to Cancer Cells and Tumors

Albumin is the most abundant protein in plasma and is of significant interest as a targeted drug carrier system for tumor cells [133]. In this regard, a BSA–Ru(CO)$_2$ conjugate derived from CORM-3 was prepared and found to exhibit CO release in solution and in cells [134]. Anti-inflammatory effects were found in HeLa and Caco-2 cells in terms of the downregulation of TNF-α, IL-6, and IL-10 in both cell lines. BSA can target proteins overexpressed in cancer cells, stromal cells, and tumor vessels [135] With this in mind, tumor delivery of CO from the BSA–Ru(CO)$_2$ conjugate was evaluated in BALB/c athymic nude mice having CT-26 colon carcinoma tumors. Four hours after injection of the BSA–Ru(CO)$_2$ conjugate, CO levels at the tumor were threefold higher as compared to the blood level. CO accumulation at the tumor was also 7-fold higher than at the kidney or liver, and 17-fold higher than other tissues. These results suggest that BSA-mediated delivery of CO offers targeted delivery to tumors.

BSA-mediated delivery of a 3-hydroxyquinolone-based, visible light-triggered photoCORM (**31**, Figure 15.8) produces potent anticancer effects in A549 cells (IC$_{50}$ 24 μM) [90]. This albumin:photoCORM complex also produces significant anti-inflammatory effects in RAW 264.7 cells when introduced at nanomolar concentrations.

Notably, the BSA–Ru(CO)$_2$ and the BSA–3-hydroxyquinolone-based CO delivery systems present two different approaches toward combining a CORM and BSA protein. Each BSA–Ru(CO)$_2$ conjugate interaction involves an Ru–N$_{His}$ bond resulting from ligand exchange involving CORM-3 in solution, with about seven binding sites per protein. The BSA–3-hydroxyquinolone conjugate involves noncovalent binding of approximately one organic photoCORM per protein, with the strongest interactions being at sites I and II, which are the known warfarin and ibuprofen binding sites, respectively [136]. The binding constant for the BSA–3-hydroxyquinolone interaction (2.9 × 10^6) is in the range of the strongest binding naturally occurring flavonoids (10^5–10^7 M^{-1}), including quercetin [137–139]. The BSA binding affinity of **31** is ~900-fold higher than that of the extended flavonol **30** (Figure 15.8; 3.2 × 10^3 M^{-1}) [140], demonstrating that modest structural changes in the organic photoCORM can be used to tune protein transport of these molecules.

Intracellular CO Delivery Using a Vitamin B_{12}-Appended Manganese PhotoCORM

Vitamin B_{12} conjugates have been investigated for a variety of therapeutics [141]. As an approach toward improving cellular uptake of a photoCORM, a macrocycle-ligated Mn(I)(CO)$_3$ photoCORM was conjugated to the 5′-OH ribose group of vitamin B_{12} [142]. This approach was pursued as the 5′-OH moiety is the only functional group exposed to the exterior aqueous environment when vitamin B_{12} is docked into the human transcobalamin II transporter that mediates cellular B_{12} uptake [143]. The B_{12}-conjugated photoCORM is actively transported and internalized in 3T3 fibroblasts. This process was tracked using synchrotron IR spectromicroscopy in living cells using the ν_{CO} region of spectra to elucidate localization information. The B_{12}-conjugated photoCORM distributes to both the nuclei and cytosol. Visible light illumination (LED, λ_{max} 470 nm) of the cells produced CO release resulting in cytoprotective effects under conditions of hypoxia and metabolic depletion.

Antibody–PhotoCORM Conjugates for CO Delivery to Cancer Cells

Antibody–drug conjugates are designed to target chemotherapeutics to selective targets, such as cancer cells [144]. Recently, a biotinylated photoCORM (**52**, Figure 15.17) was linked to streptavidin-conjugated mouse monoclonal immunoglobulin G antibodies to prepare a family of antibody–photoCORM conjugates (Ab–photoCORMs) for CO delivery to ovarian cancer cells [145]. The biotin-appended compound is stable in dark for 48 h and in human serum for 24 h, but undergoes visible light-induced CO release reactivity as evidenced by the myoglobin assay. CO release from the biotin-appended compound induces a reduction in cell viability in two ovarian cancer cell lines OVCAR-5 and SKOV-3 (ED$_{50}$ = 48 and 25 μM, respectively) assayed 24 h post-treatment. The family of antibody conjugates was synthesized using commercial antibodies with one to four streptavidin molecules, with each streptavidin binding four biotin-appended photoCORMs. These constructs were used to examine delivery of CO to ovarian cancer cells, using a nonspecific Ab–photoCORM conjugate as a control. The Ab–photoCORM-mediated delivery of CO significantly decreased cell viability, whereas the nonspecific Ab–photoCORM, as well as the free CORM biotinylated CORM, did not produce similar toxic effects. Notably, the effects produced by the Ab–photoCORMs occur at concentrations of hundreds of picomoles versus higher concentrations (10–50 μM) needed to produce similar effects using nontargeted CORMs. Overall, while the observed effects on cell viability could be due to other mechanisms of action that affect signaling and inflammatory responses, the ability of Ab–photoCORMs to accumulate in cancer cells does enhance the targeted delivery of cytotoxic amounts of CO.

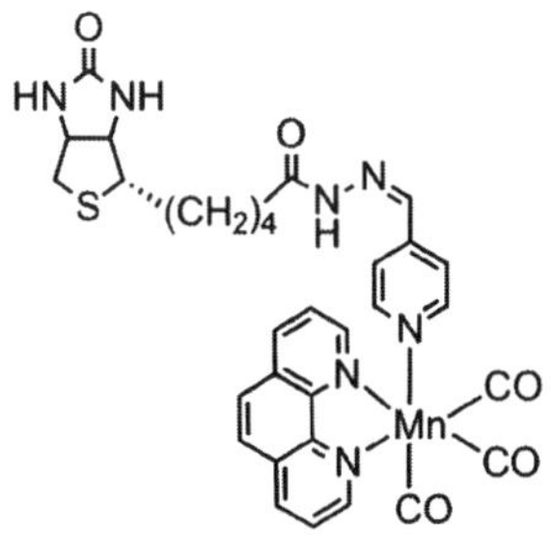

Figure 15.17 Biotin-appended photoCORM used for development of Ab–photoCORM conjugates.

Nanomaterials and Targeted CO Delivery

Nanomaterial-based approaches for the delivery of gasotransmitters, including CO, are a very active field of research, with several recent reviews highlighting advancements in this area [8,146–149]. Incorporation of CORMs into nanomaterials can be used to (1) enhance the solution stability of the CORM; (2) increase CO payload; (3) improve cellular uptake and biocompatibility; (4) minimize toxicity, including for CO release by-products; (5) take advantage of properties of the materials to enhance or tune CO release; and (6) enable targeted delivery for biomedical applications. Nanomaterials offer the possibility of targeted CO delivery to tumor tissue both through the enhanced permeation and retention (EPR) effect and/or via active targeting. The examples below are reported systems involving active targeting to cancer cells or sites of inflammation.

CO Nanogenerator for Tumor Therapy and Anti-inflammatory Applications

Partially oxidized tin disulfide (SnS$_2$) nanosheets (POS NSs) coated with a tumor-targeting polymer (PEG-cRGD) and loaded with the chemotherapy medication DOX were used to construct a nanogenerator PPOSD (polymer@POS@DOX) for applications in tumor therapy combined with anti-inflammatory effects (Figure 15.18) [150]. The targeting ability of PPOSD was evaluated using two cell lines, αvβ3-positive

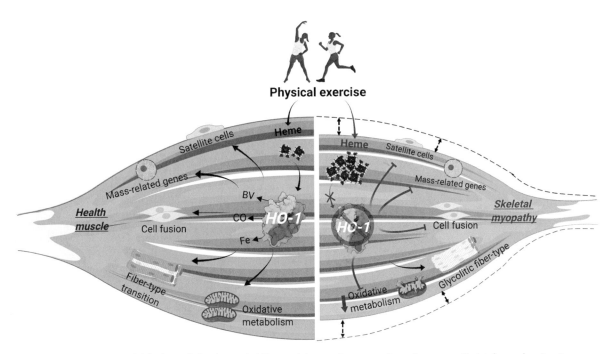

Figure 2.3 Proposed model for how skeletal muscle HO-1 activity regulates muscle performance. Skeletal muscle microtraumas that occur during aerobic exercise result in elevated heme levels inducing the expression of HO-1. Aerobic exercise training increases HO-1, which exerts a powerful contribution to improve muscle function by modulating satellite cell activation/proliferation, myoblast fusion, hypertrophy/atrophy-related genes, fiber-type transition, and mitochondrial function (left panel). Note that exercise training was unable to function appropriately in the absence of skeletal muscle HO-1 (right panel). Illustration was created using icons from BioRender (biorender.com) and Servier Medical Art (smart.servier.com).

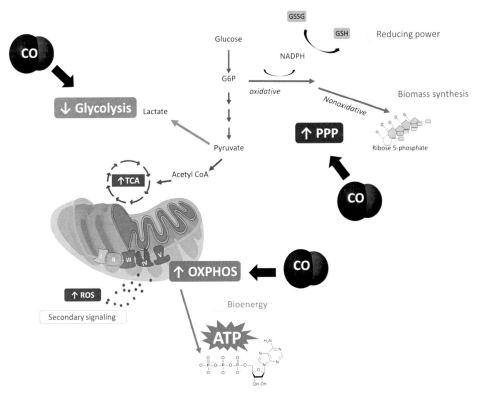

Figure 4.1 CO modulation of cell metabolism: OXPHOS, glycolysis and pentose phosphate pathway.

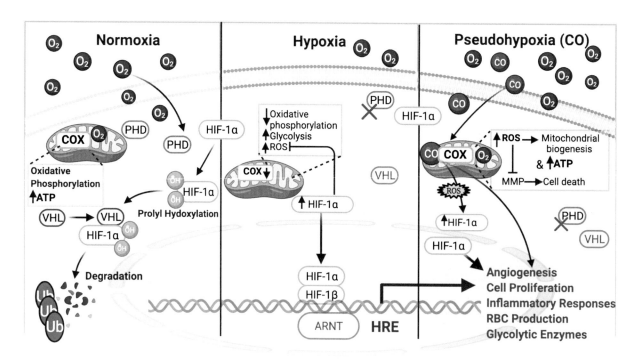

Figure 7.1 Cellular physiology under conditions of normoxia, hypoxia, and pseudohypoxia. Depicted are the cellular response to conditions of normoxia, hypoxia, and pseudohypoxia. Under normoxic conditions (left), the mitochondria utilize COX to generate adequate ATP via oxidative phosphorylation and the O_2-sensing PHDs degrade the Von Hippel–Lindau and HIF-1α complex. Under hypoxia (middle), there is decreased COX activity causing cellular metabolism to shift to glycolysis, and increasing ROS production. PHDs do not sense sufficient levels of O_2, leading to an increase in active HIF-1α, which enters the nucleus complexes with HIF-1β, which together bind the aryl hydrocarbon receptor nuclear translocator (ARNT) protein. This binding triggers HREs to begin transcription of glycolytic enzymes. In pseudohypoxia involving CO (right), COX activity is increased and ROS levels rise, leading to mitochondrial biogenesis. The ROS is a signaling amount that among other signaling can inhibit mitochondrial membrane permeability (MMP) and therefore prevent cell death. CO stabilizes HIF-1α increases leading to regulation of angiogenesis, cell proliferation, modulation of inflammatory responses and metabolism, and erythropoiesis.

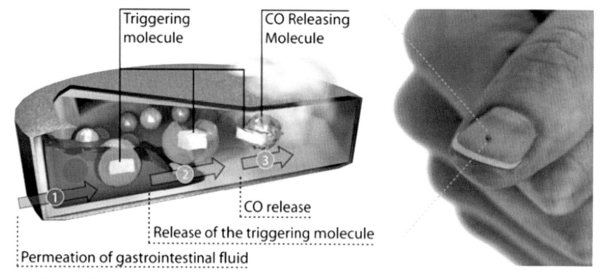

Figure 11.1 Oral CO-releasing system (depicted as a microscale system) was designed for controlled and oral delivery of CO within the GI tract. Controlled permeation of GI fluids through the system's coating triggers a cascade of reactions that ultimately result in CO release [7]. Source: Used with permission from Elsevier.

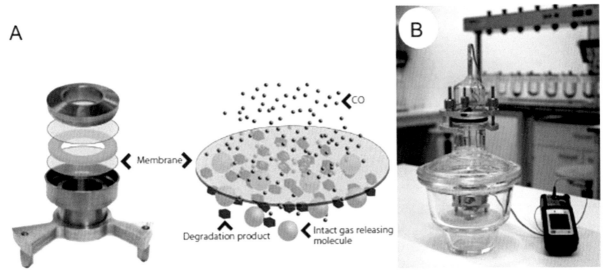

Figure 11.2 Therapeutic gas release system was designed to exclusively deliver CO gas by leveraging gas-permeable membranes (details in the text) [10]. Source: Used with permission from Elsevier.

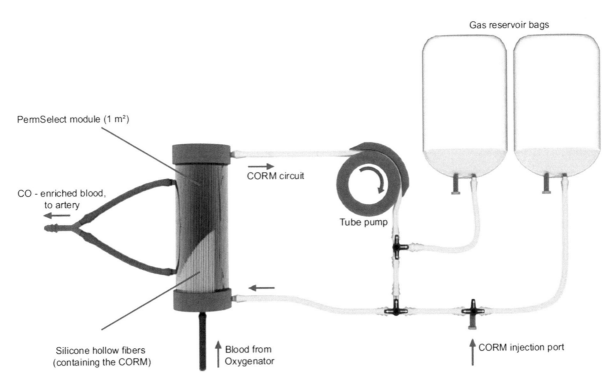

Figure 11.3 A membrane-based approach was also used for extracorporeal CO delivery. CO was generated from CORMs in a sealed circuit and delivered to a porcine model via gaspermeable tube membranes [11]. Source: Used with permission from Elsevier.

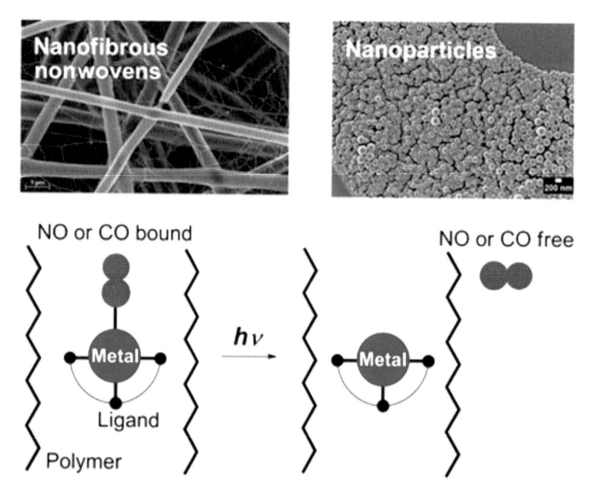

Figure 11.4 Light was used as a trigger to release CO from electrospun materials. While CO permeates through the material, degradation products of the CORM remain in the delivery system [15]. Source: Used with permission from The Royal Society of Chemistry.

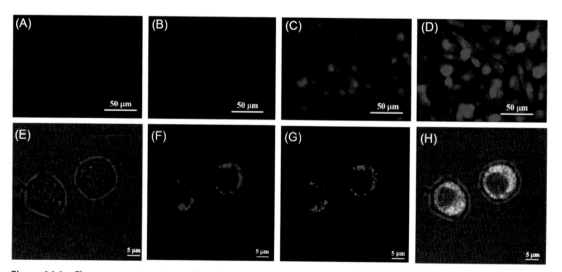

Figure 14.1 Fluorescence imaging studies of CO prodrug pairs (BW-CO-19, BW-CO-20) and (BW-CO-21, BW-CO-22) in RAW 264.7 cells: (A) RAW 264.7 cells treated with BW-CO-19 (1 μM) and BW-CO-20 (1 μM); (B) RAW 264.7 cells treated with BW-CO-19 (5 μM) and BW-CO-20 (5 μM); (C) RAW 264.7 cells treated with BW-CO-21 (1 μM) and BW-CO-22 (1 μM); and (D) RAW 264.7 cells treated with BW-CO-21 (5 μM) and BW-CO-22 (5 μM). Confocal images of RAW 264.7 cells treated with compounds BW-CO-21 (5 μM) and BW-CO-22 (2.5 μM) and MT Deep Red (50 nM): (E) bright field; (F) red channel; (G) DAPI channel; and (H) merged images of (E)–(G). Adapted with permission from Ref. [20]. Copyright (2018) Springer Nature.

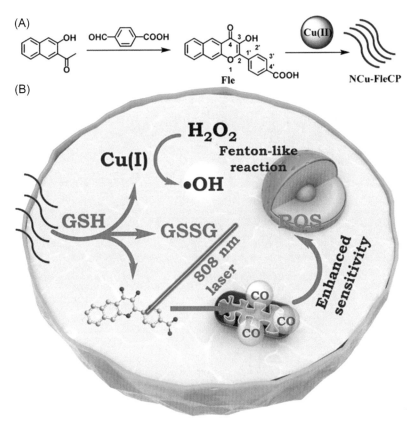

Figure 15.12 Targeted synergistic OH• and CO generation in cancer cells using a Cu(II)–flavonol coordination polymer. (A) Synthesis of the flavonol and coordination polymer. (B) Proposed intracellular reactivity of NCu-FleCP upon reaction with GSH and illumination with 808 nm light. Reprinted with permission from Ref. [108]. Copyright (2020) John Wiley & Sons.

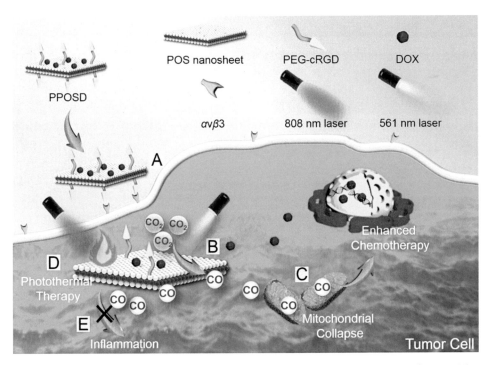

Figure 15.18 Components of CO nanogenerator. (A) cRGD-mediated tumor targeting of PPOSD. (B) CO generation via CO_2 photoreduction. (C) CO-induced mitochondrial damage and chemosensitization. (D) PTT of tumor by the POS in PPOSD via illumination. (E) Inhibition of PTT-induced inflammatory reaction by generated CO. Adapted with permission from Ref. [150]. Copyright (2020) American Chemical Society.

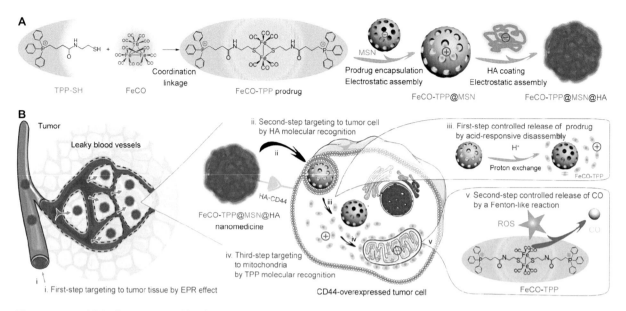

Figure 15.19 (A) Multistage assembly of FeCO–TPP@MSN@HA. (B) Mechanism of tissue, cell, and mitochondrial targeting followed by multistage disassembly with ROS generation and CO release. Reproduced under Creative Commons CC-BYNC license from Ref. [151]. Copyright (2020) American Association for the Advancement of Science.

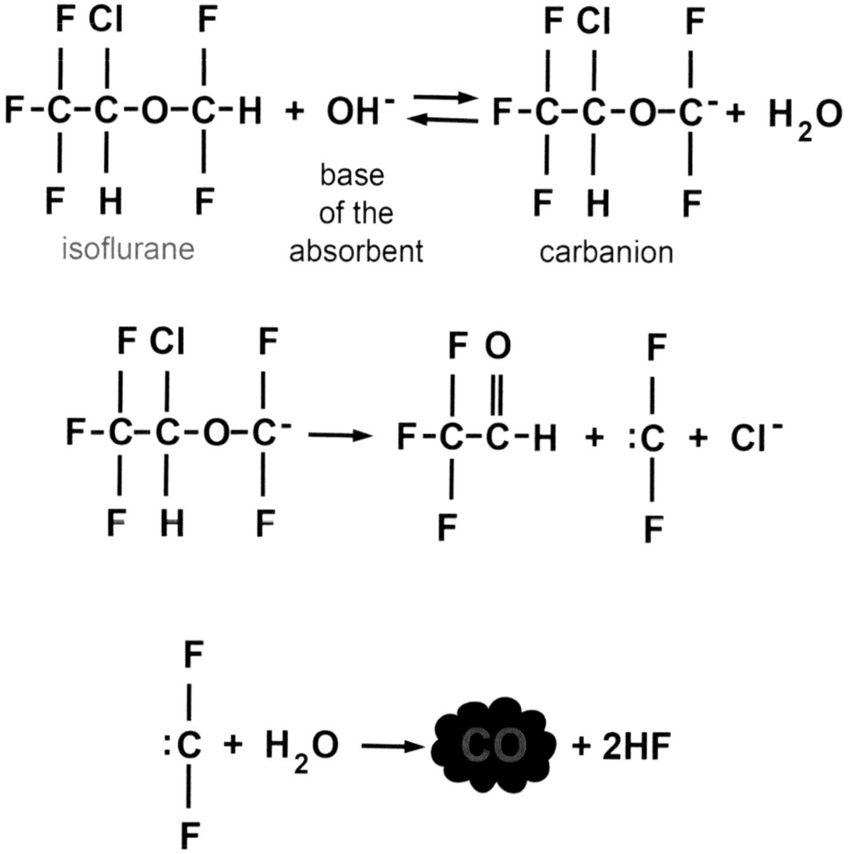

Figure 16.2 Mechanism of CO generation from inhaled anesthetic agents. Proton (H^+) abstraction is catalyzed by the base (OH^-) of the carbon dioxide absorbent. With inadequately hydrated absorbent, the carbanion decomposes and subsequently interacts with hydroxide or residual water (H_2O) within the absorbent, generating CO. Adapted from Ref. [6] with permission.

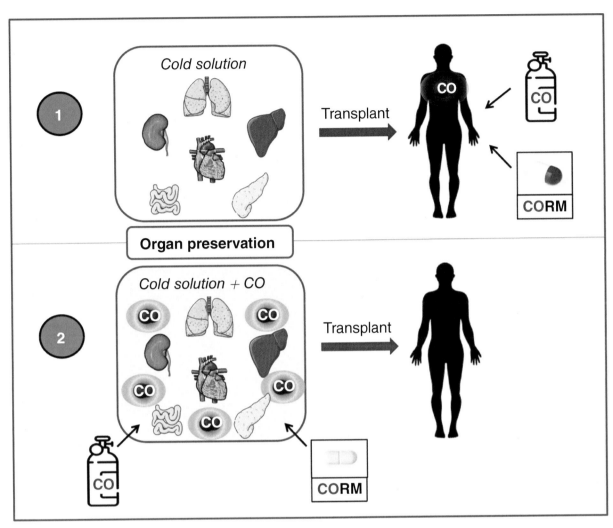

Figure 19.1 Strategies to improve the outcome of organ transplantation by CO. (1) Recipients are treated with either CO gas or CORM prior to transplantation to increase their resistance to graft rejection. (2) Organs are stored in cold preservation solutions saturated with CO gas or supplemented with CORMs before grafts are transplanted into the recipients.

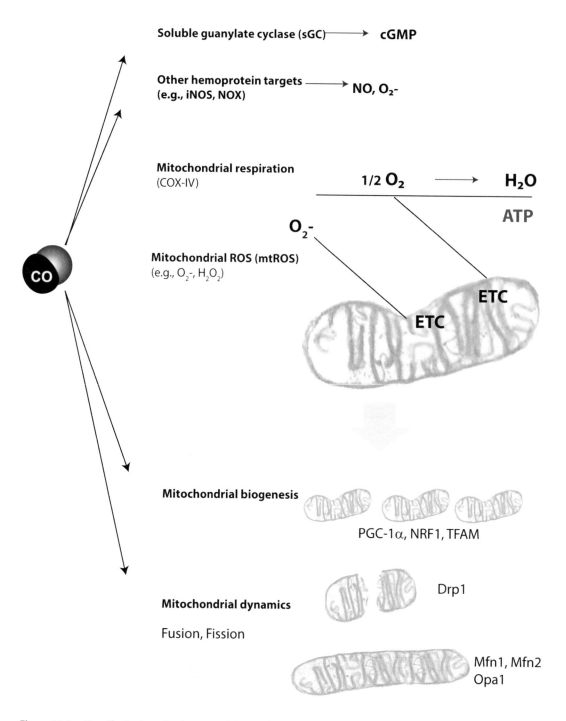

Figure 20.1 Hypothetical proximal targets of CO as relevant to pulmonary cells. CO is known to interact with several biochemical targets in the various cell types of the lung. The classical target is sGC that, upon gaseous ligand binding, can increase the production of cGMP. CO can potentially target other cellular hemoproteins, including iNOS and NOX isoforms, potentially leading to modulation of NO and superoxide anion (O_2^-) production, respectively. Mitochondria have emerged as a primary target for CO action, based on numerous in vitro studies. CO can inhibit cellular respiration at high concentration and modulate mtROS production via interactions with cytochrome c oxidase and other potential sites of the electron transport chain. CO can also impact the mitochondrial biogenesis program through modulation of its various regulatory factors, including PGC-1α, NRF1, and TFAM. Modulation of mitochondrial dynamics (fusion and fission) is an area of potential novel investigations.

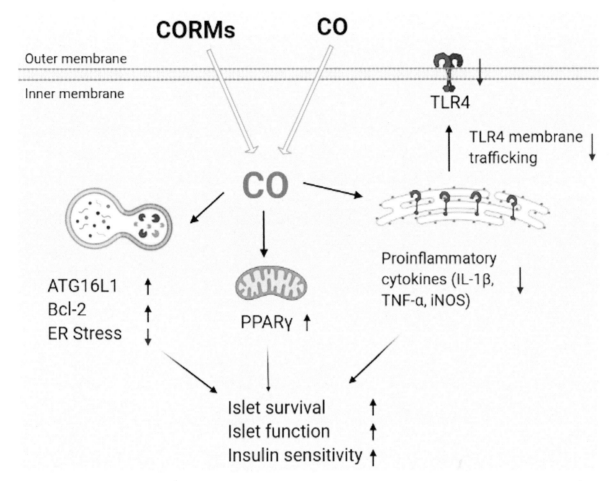

Figure 24.1 Nonheme-dependent mechanisms involved in CO-mediated islet function. Gaseous CO or CORMs promote islet function via multiple pathways. These include (i) ATG16L1 and blc-2 expression and reduction of ER stress; (ii) the activation of PPARγ; and (iii) the inhibition of membrane TLR4 expression. CO modulates these pathways in islet cells, contributing to reduced inflammation and thus prolonged islet survival.

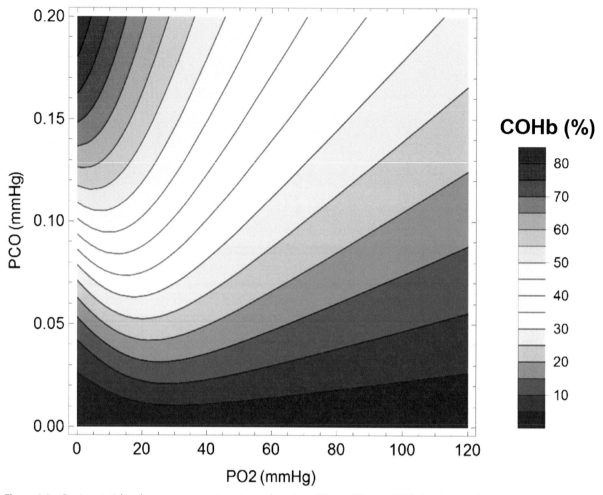

Figure 3.3 P_{CO} in arterial and venous compartments as a function of P_{O_2} at different COHb levels according to Equation 3.9.

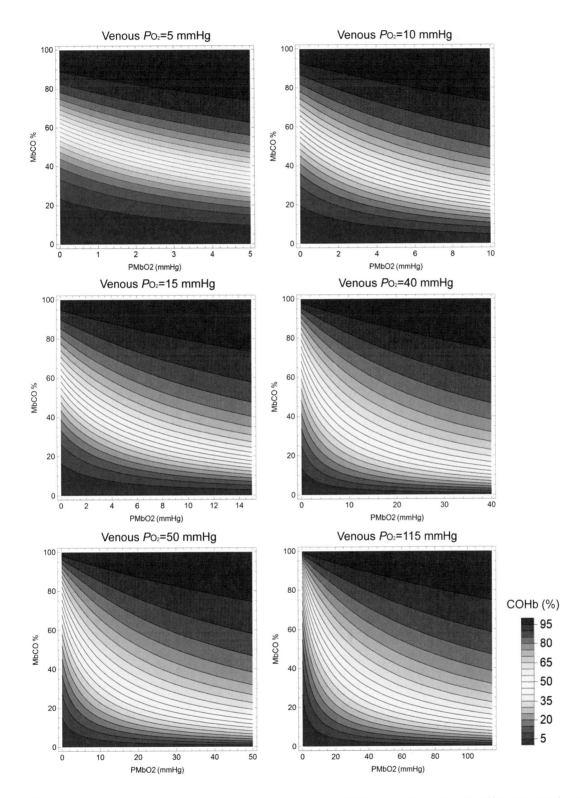

Figure 3.4 MbCO saturation as a function of at various predefined COHb levels with the blood level between 5 and 115 mmHg. (The plots are based on Equations 3.9 and 3.11 with the parameters and constants detailed in the text).

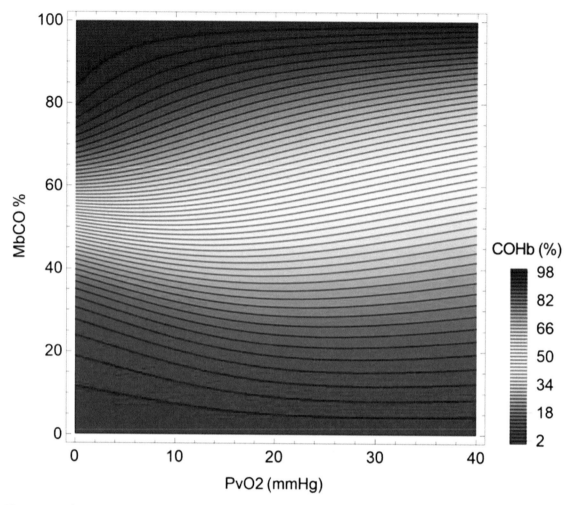

Figure 3.5 MbCO saturation as a function of at predefined COHb levels and of 1 mmHg (mitochondria vicinity). (The plots are based on Equations 3.9 and 3.11 with the parameters and constants detailed in the text).

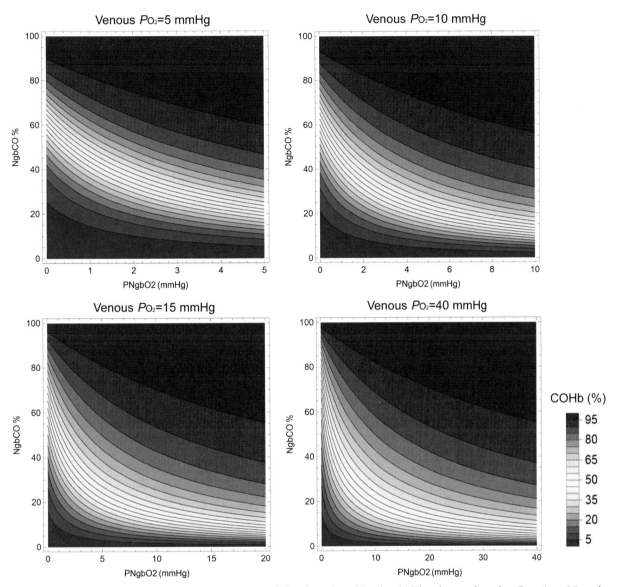

Figure 3.6 NbCO saturation as a function of P_{NbO_2} at predefined COHb and P_{VO_2} levels. (The plots are based on Equations 3.9. and 3.12 with the parameters and constants detailed in the text).

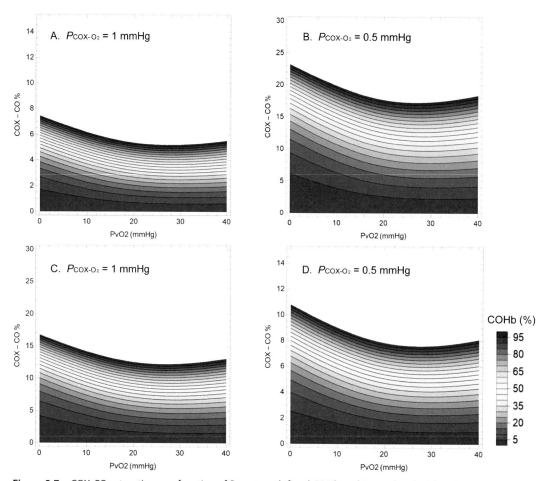

Figure 3.7 COX-CO saturation as a function of P_{VO_2} at predefined COHb and P_{COX-O_2} levels. (The plots are based on Equations 3.9 and 3.12 with MCOX = 1 for A and B, and MCOX = 2.5 for C and D. Other parameters and constants are detailed in the text).

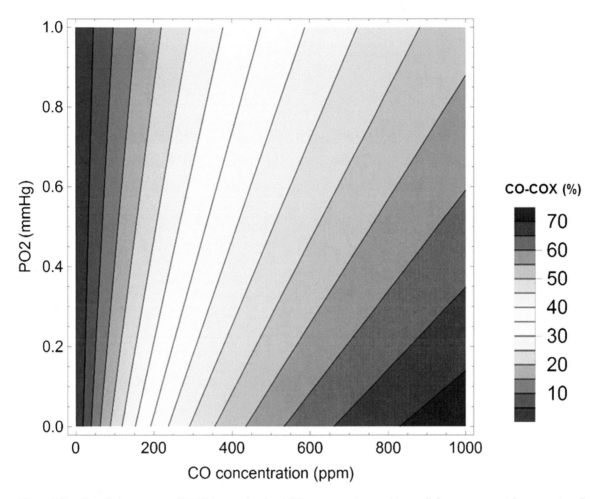

Figure 3.13 Correlations among CO-COX saturation level, CO concentration, and intracellular oxygen partial pressure in cell culture experiments. P_{VO_2} is intracellular oxygen partial pressure in the proximity of mitochondria where COX is located (0–1 mmHg); CO concentration in the cell culture atmosphere: 0–1000 ppm; and model is based on M_{COX} = 2.5 and Ki = 0.27 mmHg.

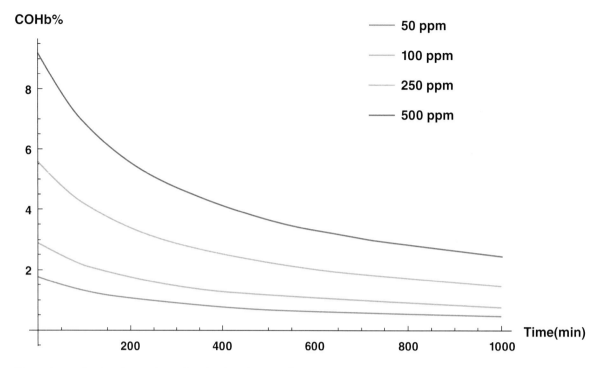

Figure 3.14 Simulated time-dependent COHb% changes after DCM exposure. (Modeled based on Equation 3.18 after 450 min inhalation with different DCM concentrations in humans).

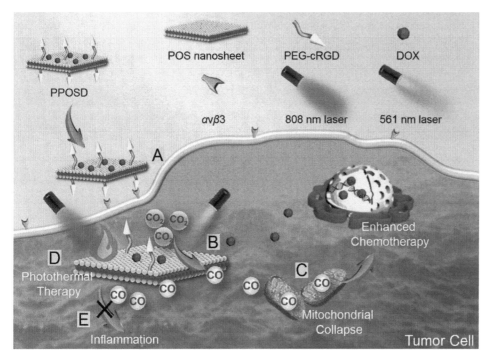

Figure 15.18 Components of CO nanogenerator. (A) cRGD-mediated tumor targeting of PPOSD. (B) CO generation via CO_2 photoreduction. (C) CO-induced mitochondrial damage and chemosensitization. (D) PTT of tumor by the POS in PPOSD via illumination. (E) Inhibition of PTT-induced inflammatory reaction by generated CO. Adapted with permission from Ref. [150]. Copyright (2020) American Chemical Society.

murine mammary carcinoma (4T1) cells and αvβ3-negative human breast adenocarcinoma (MCF-7) cells, with the former showing better uptake, indicating the specific tumor targeting capability of PPOSD. With intravenous injection in mice, the nanosheets target and localize in tumors via cRGD-mediated recognition. Upon illumination at 561 nm, the SnS_2 portion mediates the catalytic conversion of CO_2 to CO. The CO produced sensitizes the chemotherapeutic effect of DOX via CO-induced effects on mitochondrial function, and mitigates the inflammatory response produced upon photothermal therapy via illumination of the POS at 808 nm. These combined activities were found to occur with high biosafety as determined by examination of organs via hematoxylin–eosin staining and a blood test.

Multistage Strategy Assembly/Disassembly Strategy for Tumor-Targeted CO Delivery

Mesoporous silica nanoparticles that offer tissue targeting via the EPR effect, tumor cell targeting, mitochondrial targeting of a CORM, and multitriggered environment-induced release of CO have been developed as an approach for targeted CO delivery for cancer applications (Figure 15.19) [151]. The CO-releasing motif employed in this construct, FeCO (Figure 15.19), is a positively charged diiron carbonyl complex containing triphenylphosphonium-tailed bridging thiolate ligands for mitochondrial targeting. Incorporation of this motif into negatively charged silica nanoparticles, followed by coating of the particle with hyaluronic acid (HA), provides mesoporous silica nanoparticles (MSN; FeCO–TPP@MSN@HA) for passive targeting via the EPR effect and active targeting to CD44-overexpressed tumor cells [152]. The construct exhibits high stability in aqueous solution. Uptake of a fluorescently labeled version was found to be better in cancer cells overexpressing CD44 (HeLa and 4T1) than in normal cells (MCF-10A and HEK293T). In HeLa, endocytosis results in FeCO–TPP@MSN@HA in lysosome. The acidic environment of the lysosome causes release of FeCO–TPP. MitoTracker Green was used to assess mitochondrial colocalization with FeCO–TPP (labeled with blue-QL) and MSN (labeled with red-RITC). The FeCO–TPP prodrug gradually accumulates in mitochondria after lysosomal-triggered release. Release of CO, which is proposed to occur via ROS-triggered oxidative reactivity, was confirmed using the fluorescent CO sensor COP-1 [52]. Mitochondrial function in HeLa cells was impacted, but protected in noncancerous cells to maintain normal energy

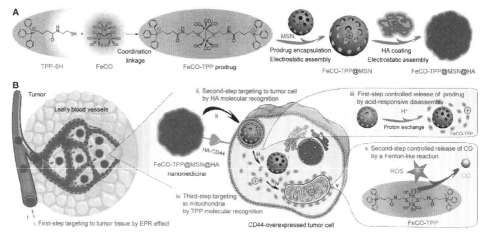

Figure 15.19 (A) Multistage assembly of FeCO–TPP@MSN@HA. (B) Mechanism of tissue, cell, and mitochondrial targeting followed by multistage disassembly with ROS generation and CO release.
Reproduced under Creative Commons CC-BY-NC license from Ref. [151]. Copyright (2020) American Association for the Advancement of Science.

metabolism. *In vivo* studies provided evidence that FeCO–TPP@MSN@HA inhibited tumor growth and metastasis. Overall, the combined results provide evidence for a new multistage approach toward targeting CO delivery and resulting effects to cancer cells.

Targeted Synergistic Anti-inflammatory Effects Involving ROS Alleviation and CO Release

As an approach toward targeting inflammatory sites with controlled CO release, mesoporous silica nanoparticles coated with MnO_2 followed by silica etching were loaded with the extended flavonol **30** (Figure 15.8) and then coated with an inflammation-targeting neutrophil membrane (CD11b/c, neutrophil-specific surface proteins) to give Neu–MnO_2/Fla (Figure 15.20) [153]. The resulting particles exhibited minimal flavonol loss in a variety of solvents and undergo the expected visible light-induced CO release, with a 58% yield of CO generated from the encapsulated flavonol. The Neu–MnO_2/Fla particles exhibit low toxicity in PC12 cells. The uptake of Neu–MnO_2/Fla in LPS-induced PC12 cells is higher than that found for structurally similar red blood cell membrane-coated nanoparticles (RBC–MnO_2/Fla) and is attributed to targeting by the neutrophil membrane coating. Intracellular CO release from the Fla portion of Neu–MnO_2/Fla was tracked in cells via loss of the flavonol fluorescent emission. Under hypoxic conditions, light-induced CO release from the extended flavonol **30** is enhanced in the presence of H_2O_2 through O_2-producing catalase activity mediated by the MnO_2 coating. Neu–MnO_2/Fla with illumination (visible light or two-photon) exhibits anti-inflammatory effects in LPS-stimulated PC12 cells and a paw inflammation model, as evidenced by decreases in the amount of ROS present and decreased levels of inflammatory cytokines TNF-α and IL-1β.

Mitochondria-Targeted Micelles for Combined CO Delivery with an Iron Prodrug for Ferroptosis

A mitochondria-targeted, light-responsive polymer (PNBE) that self-assembles into micelles encapsulating the organic photoCORM diphenylcyclopropenone and H_2O_2-responsive aminoferrocene prodrugs (FeCO@Mito-PNBE) was prepared and characterized [154]. In HeLa, colocalization studies using MitoTracker Red and micelles incorporating fluorescein showed accumulation of FeCO@Mito-PNBE at mitochondria. UV light illumination triggers CO release from the CORM resulting in increased mitochondrial ROS production. Reactivity of the aminoferrocene complex with H_2O_2 results in release of Fe^{2+}, which produces Fenton-type reactivity (OH• generation and ferroptosis [155]) and mitochondrial damage. The effects of FeCO@Mito-PNBE were evaluated in two cancer cell lines (HeLa and 4T1). ROS generation in 4T1 cancer cells was monitored using the fluorescent dye 2′,7′-dichlorofluorescein diacetate (DCFH-DA). An increase in the DCFH-DA fluorescence signal was observed in the presence of FeCO@Mito-PNBE with UV light illumination, indicating an increase in the amount of intracellular ROS. A decrease in mitochondrial membrane potential was also identified, indicating that the combined, targeted effects of FeCO@Mito-PNBE produced mitochondrial damage. Overall, this multifunctional targeting system is designed to damage mitochondria via combined CO delivery and ferroptosis as an approach toward enhanced cancer therapy.

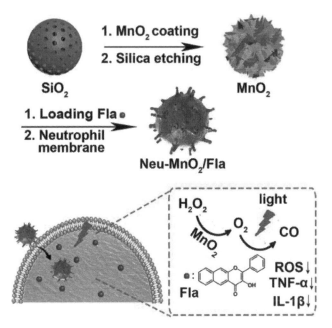

Figure 15.20
Representation of synergistic anti-inflammation. (Top) Assembly of Neu–MnO$_2$/Fla. (Bottom) Neutrophil membrane-targeted CO delivery with catalytic O$_2$ generation by MnO$_2$ nanozymes for targeted anti-inflammatory applications. Reproduced under Creative Commons CC-BY license of Ref. [153]. Copyright (2020) Cell Press.

NIR-Triggered Nanoparticles for Targeted CO Delivery

Upconverting nanoparticles offer the possibility of using NIR illumination to trigger CO release in metal carbonyl complexes [68]. Coating of an upconverting nanoparticle (NaYF$_4$:Yb, Tm with NaYF4:Nd coating) with a FA-functionalized lipid bilayer for targeting cancer cells that also contains a hydrophobically tailed Mn(I)(CO)$_3$ moiety enables an approach for targeted CO delivery [156]. Illumination at 808 or 980 nm is converted to 360 nm illumination to trigger CO release. Evaluation of the coated nanoparticles in HCT116 cells showed that UV light-triggered CO release induces dose-dependent cytotoxicity, ROS generation, and apoptosis of cancer cells.

Summary and Perspectives

Over the past ~20 years, interest in the therapeutic potential of CO for addressing a variety of health conditions has grown rapidly. Challenges with using CO gas, including a lack of target selectivity and poor accumulation of CO in target tissue, led to the evaluation of metal carbonyl compounds as CORMs. A limitation of several early CORMs (e.g., CORM-3) is their rapid CO release in solution, which precludes targeted CO delivery. The development of metal carbonyl CORMs that are triggered for CO release by either internal or external stimuli has in part addressed this issue. Most notable in terms of controlled metal carbonyl CORMs for targeted CO delivery are photoCORMs. Through supporting ligand design strategies, these molecules can be structurally tuned to modulate the wavelength of light needed to induce CO release, fluorescence properties, and linking groups needed for incorporation into biomolecules or nanomaterials for targeting applications. Recent advances in targeted CO delivery using metal carbonyl photoCORMs have included their incorporation into nanomaterials for passive targeting via EPR effects. Examples of active targeting through combining a metal carbonyl photoCORM with biological or synthetic materials containing tissue or intracellular targeting motifs are emerging. Potent cytotoxicity effects are produced, particularly for mitochondria-targeted or antibody-delivered motifs.

The development of metal-free CO donors or pro-drugs has changed the landscape in terms of CO delivery by enabling core CO-releasing motifs to be easily modified using straightforward synthetic organic chemistry approaches. The simplicity of some of these systems allows for the introduction of a variety of features, including components for fluorescence trackability and triggers for the CO process, and provides well-defined CO release by-products. In this regard, Wang and coworkers have demonstrated that spontaneous CO-releasing norborn-2-en-7-one compounds can be structurally modified to introduce environment-based triggers (e.g., oxidation, pH, esterase) to enhance control of the CO release process and thereby potentially enhance localized CO release.

Organic photoCORMs represent the most controlled small molecules reported to date for CO delivery. Lacking the leakage reactivity of metal carbonyl CORMs, such molecules offer many advantages to significantly advance targeted CO delivery. Most notably, this includes tunability of structural features to enhance biological targetability. In this regard, functionalized extended flavonols have been developed that enable targeted CO delivery to mitochondria, and direct examination of intracellular versus extracellular CO delivery. These studies allowed examination of how the site of CO release relates to the amount needed to induce biological effects, such as changes in mitochondrial bioenergetics or CO-induced anti-inflammatory effects. Many opportunities remain in terms of further defining CO release/biological response relationships. For example, molecular photoCORMs exhibiting intracellular localization at endoplasmic reticulum similar to HO-1 remain rare [157]. Such molecules would be useful to further probe how intracellular localization of CO release relates to observed biological effects. These types of studies may provide further insight into how to use targeted delivery of CO to reduce the dosage needed to produce desired effects.

Environment-sensing flavonol-based photoCORMs are prototypes for additional advancements in targeted CO delivery. Only through sensing of the biological environment in terms of ROS, thiols, or other signaling agents (e.g., H_2S) is the CO-releasing unit revealed for visible light-induced CO release. As flavonol-based photoCORMs offer multiple sites for functionalization, such motifs can be incorporated into polymeric micelles, nanomaterials, or other conjugate systems for targeted delivery that includes environmental sensing.

Development of nanomaterial-based systems for delivery of CORMs offers the possibility for retention and active targeting. Recent examples demonstrate that the incorporation of CORMs in nanomaterials enables tissue and/or intracellular targeting with enhanced biological effects for potential therapeutic applications. Overall, pursuit of continued enhancement of CORM properties coupled with advances in biological- and materials-based delivery approaches offers a rich environment for further investigation of targeted CO delivery.

Acknowledgment

Research on CO donor molecule development in the Berreau lab is supported by the National Institutes of Health (R15GM124596).

References

1. Motterlini, R. and Otterbein, L.E. (2010). The therapeutic potential of carbon monoxide. *Nat. Rev. Drug Discov.* 9: 728–743.
2. Tenhunen, R., Marver, H.S., and Schmid, R. (1968). The enzymatic conversion of heme to bilirubin by microsomal heme oxygenase. *Proc. Natl. Acad. Sci. U.S.A.* 61: 748–755.
3. Araujo, J.A., Zhang, M., and Yin, F. (2012). Heme oxygenase-1, oxidation, inflammation, and atherosclerosis. *Front. Pharmacol.* 3: 119.
4. Mann, B.E. and Motterlini, R. (2007). CO and NO in medicine. *Chem. Commun.* 4197–4208.
5. Ryter, S.W. (2020). Therapeutic potential of heme oxygenase-1 and carbon monoxide in acute organ injury, critical illness, and inflammatory disorders. *Antioxidants* 9, 1153.
6. Gonzalez, H., Horie, S., and Laffey, J.G. (2021). Emerging cellular and pharmacologic therapies for acute respiratory distress syndrome. *Curr. Opin. Crit. Care* 27: 20–28.
7. Jin, Q., Deng, Y., Jia, F., Tang, Z., and Ji, J. (2018). Gas therapy: An emerging "green" strategy for anticancer therapeutics. *Adv. Therapeutics* 1: 1800084.
8. Zhou, Y., Yu, W., Cao, J., and Gao, H. (2020). Harnessing carbon monoxide-releasing platforms for cancer therapy. *Biomaterials* 255: 120193.
9. Motterlini, R. and Foresti, R. (2017). Biological signaling by carbon monoxide and carbon monoxide-releasing molecules. *Am. J. Physiol. Cell. Physiol.* 312: C302–C313.
10. Foresti, R., Bani-Hani, M.G., and Motterlini, R. (2008). Use of carbon monoxide as a therapeutic agent: Promises and challenges. *Intensive Care Med.* 34: 649–658.
11. Hu, H.J., Sun, Q., Ye, Z.H., and Sun, X.J. (2016). Characteristics of exogenous carbon monoxide deliveries. *Med. Gas Res.* 6: 96–101.
12. Motterlini, R. (2007). Carbon monoxide-releasing molecules (CO-RMs): Vasodilatory, anti-ischaemic and anti-inflammatory activities. *Biochem. Soc. Trans.* 35: 1142–1146.
13. Tenhunen, R., Marver, H.S., and Schmid, R. (1969). Microsomal heme oxygenase. characterization of the enzyme. *J. Biol. Chem.* 244: 6388–6394.
14. Wegiel, B. and Otterbein, L.E. (2012). Go green: The anti-inflammatory effects of biliverdin reductase. *Front. Pharmacol.* 3: 47.
15. Ayer, A., Zarjou, A., Agarwal, A., and Stocker, R. (2016). Heme oxygenases in cardiovascular health and disease. *Physiol. Rev.* 96: 1449–1508.
16. Ryter, S.W., Alam, J., and Choi, A.M.K. (2006). Heme oxygenase-1/carbon monoxide: From basic science to therapeutic applications. *Physiol. Rev.* 86: 583–650.
17. Maines, M.D. (1997). The heme oxygenase system: A regulator of second messenger gases. *Annu. Rev. Pharmacol. Toxicol.* 37: 517–554.
18. Motterlini, R., Clark, J.E., Foresti, R., Sarathchandra, P., Mann, B.E., and Green, C.J. (2002). Carbon

monoxide-releasing molecules: Characterization of biochemical and vascular activities. *Circ. Res.* 90: E17–24.
19. Foresti, R., Hammad, J., Clark, J.E., Johnson, T.R., Mann, B.E., Friebe, A., Green, C.J., and Motterlini, R. (2004). Vasoactive properties of CORM-3, a novel water-soluble carbon monoxide-releasing molecule. *Br. J. Pharmacol.* 142: 453–460.
20. Crook, S.H., Mann, B.E., Meijer, A.J.H.M., Adams, H., Sawle, P., Scapens, D., and Motterlini, R. (2011). [Mn(CO){S_2CNMe(CHCO$_2$H)}], a new water-soluble CO-releasing molecule. *Dalton Trans.* 40: 4230–4235.
21. Motterlini, R., Haas, B., and Foresti, R. (2012). Emerging concepts on the anti-inflammatory actions of carbon monoxide-releasing molecules (CO-RMs). *Med. Gas Res.* 2: 28.
22. Motterlini, R., Mann, B.E., and Foresti, R. (2005). Therapeutic applications of carbon monoxide-releasing molecules. *Expert Opin. Investig. Drugs* 14: 1305–1318.
23. Motterlini, R., Mann, B.E., Johnson, T.R., Clark, J.E., Foresti, R., and Green, C.J. (2003). Bioactivity and pharmacological actions of carbon monoxide-releasing molecules. *Curr. Pharm. Des.* 9: 2525–2539.
24. Romao, C.C., Blattler, W.A., Seixas, J.D., and Bernardes, G.J.L. (2012). Developing drug molecules for therapy with carbon monoxide. *Chem Soc. Rev.* 41: 3571–3583.
25. Marques, A.R., Kromer, L., Gallo, D., Penacho, N., Rodrigues, S.S., Seixas, J.D., Bernardes, G.J., Reis, P.M., Otterbein, L.E., Ruggieri, R.A., Goncalves, A.S.G., Goncalves, A.M.L., De Matos, M.N., Bento, I., Otterbein, L.E., Blattler, W.A., and Romao, C.C. (2012). Generation of carbon monoxide releasing molecules (CO-RMs) as drug candidates for the treatment of acute liver injury: Targeting of CO-RMs to the liver. *Organometallics* 16: 5810–5822.
26. Santos-Silva, T., Mukhopadhyay, A., Seixas, J.D., Bernardes, G.J.L., Romao, C.C., and Romao, M.J. (2011). CORM-3 reactivity toward proteins: The crystal structure of a Ru(II) dicarbonyl-lysozyme complex. *J. Am. Chem. Soc.* 133: 1192–1195.
27. Yuan, Z., Yang, X., De La Cruz, L.K., and Wang, B. (2020). Nitro reduction-based fluorescent probes for carbon monoxide require reactivity involving a ruthenium carbonyl moiety. *Chem. Commun.* 56: 2190–2193.
28. Southam, H.M., Smith, T.W., Lyon, R.L., Liao, C., Trevitt, C.R., Middlemiss, L.A., Cox, F.L., Chapman, J.A., El-Khamisy, S.F., Hippler, M., Williamson, M.P., Henderson, P.J.F., and Poole, R.K. (2018). A thiol-reactive Ru(II) ion, not CO release, underlies the potent antimicrobial and cytotoxic properties of Co-releasing molecule-3. *Redox Biol.* 18: 114–123.
29. Stucki, D., Krahl, H., Walter, M., Steinhausen, J., Hommel, K., Brenneisen, P., and Stahl, W. (2020). Effects of frequently applied carbon monoxide releasing molecules (CORMs) in typical CO-sensitive model systems - a comparative in vitro study. *Arch. Biochem. Biophys.* 687: 108383.
30. Ji, X. and Wang, B. (2018). Strategies toward organic carbon monoxide prodrugs. *Acc. Chem. Res.* 51: 1377–1385.
31. Kueh, J.T.B., Stanley, N.J., Hewitt, R.J., Woods, L.M., Larsen, L., Harrison, J.C., Rennison, D., Brimble, M.A., Sammut, I.A., and Larsen, D.S. (2017). Norborn-2-en-7-ones as physiologically-triggered carbon monoxide-releasing prodrugs. *Chem. Sci.* 8: 5454–5459.
32. Kueh, J.T.B., Seifert-Simpson, J.M., Thwaite, S.H., Rodgers, G.D., Harrison, J.C., Sammut, I.A., and Larsen, D.S. (2020). Studies towards non-toxic, water soluble, vasoactive norbornene organic carbon monoxide releasing molecules. *Asian J. Chem.* 9: 2127–2135.
33. Ji, X., Zhou, C., Ji, K., Aghoghovbia, R.E., Pan, Z., Chittavong, V., Ke, B., and Wang, B. (2016). Click and release: A chemical strategy toward developing gasotransmitter prodrugs by using an intramolecular Diels-Alder reaction. *Angew. Chem. Int. Ed. Engl.* 55: 15846–15851.
34. Ling, K., Men, F., Wang, W.C., Zhou, Y.Q., Zhang, H.W., and Ye, D.W. (2018). Carbon monoxide and its controlled release: Therapeutic application, detection, and development of carbon monoxide releasing molecules (CORMs). *J. Med. Chem.* 61: 2611–2635.
35. Yan, H., Du, J., Zhu, S., Nie, G., Zhang, H., Gu, Z., and Zhao, Y. (2019). Emerging delivery strategies of carbon monoxide for therapeutic applications: From CO gas to CO releasing nanomaterials. *Small* 15: e1904382.
36. Erbas-Cakmak, S., Kolemen, S., Sedgwick, A.C., Gunnlaugsson, T., James, T.D., Yoon, J., and Akkaya, E.U. (2018). Molecular logic gates: The past, present and future. *Chem. Soc. Rev.* 47: 2228–2248.
37. Soboleva, T. and Berreau, L.M. (2019). Tracking CO release in cells via the luminescence of donor molecules and/or their by-products. *Isr. J. Chem.* 59: 339–350.
38. Seixas, J.D., Mukhopadhyay, A., Santos-Silva, T., Otterbein, L.E., Gallo, D.J., Rodrigues, S.S., Guerreiro, B.H., Goncalves, A.M., Penacho, N., Marques, A.R., Coelho, A.C., Reis, P.M., Romao, M.J., and Romao, C.C. (2013). Characterization of a versatile organometallic pro-drug (CORM) for experimental CO based therapeutics. *Dalton Trans.* 42: 5985–5998.
39. Motterlini, R., Sawle, P., Hammad, J., Bains, S., Alberto, R., Foresti, R., and Green, C.J. (2005).

CORM-A1: a new pharmacologically active carbon monoxide-releasing molecule. *FASEB J.* 19: 284–286.

40. Ji, X., De La Cruz, L.K.C., Pan, Z., Chittavong, V., and Wang, B. (2017). Ph-sensitive metal-free carbon monoxide prodrugs with tunable and predictable release rates. *Chem. Commun.* 53: 9628–9631.

41. Marazioti, A., Bucci, M., Coletta, C., Vellecco, V., Baskaran, P., Szabo, C., Cirino, G., Marques, A.R., Guerreiro, B., Goncalves, A.M.L., Seixas, J.D., Beuve, A., Romao, C.C., and Papapetropoulos, A. (2011). Inhibition of nitric oxide-stimulated vasorelaxation by carbon monoxide-releasing molecules. *Arterioscler. Thromb. Vasc. Biol.* 31: 2570–2576.

42. Tavares, A.F.N., Teixeira, M., Romao, C.C., Seixas, J.D., Nobre, L.S., and Saraiva, L.M. (2011). Reactive oxygen species mediate bactericidal killing elicited by carbon monoxide-releasing molecules. *J. Biol. Chem.* 286: 26708–26717.

43. Habtemariam, S. (2019). Modulation of reactive oxygen species in health and disease. *Antioxidants* 8, 513.

44. Alfadda, A.A. and Sallam, R.M. (2012). Reactive oxygen species in health and disease. *J. Biomed. Biotechnol.* 2012: 936486.

45. Bardaweel, S.K., Gul, M., Alzweiri, M., Ishaqat, A., ALSalamat, H.A.-, and Bashatwah, R.M. (2018). Reactive oxygen species: The dual role in physiological and pathological conditions of the human body. *Eurasian J. Med.* 50: 193–201.

46. Jin, Z., Wen, Y., Xiong, L., Yang, T., Zhao, P., Tan, L., Wang, T., Qian, Z., Su, B.L., and He, Q. (2017). Intratumoral H_2O_2-triggered release of CO from a metal carbonyl-based nanomedicine for efficient CO therapy. *Chem. Commun.* 53: 5557–5560.

47. G, U.R., Axthelm, J., Hoffmann, P., Taye, N., Glaser, S., Gorls, H., Hopkins, S.L., Plass, W., Neugebauer, U., Bonnet, S., and Schiller, A. (2017). Co-registered molecular logic gate with a CO-releasing molecule triggered by light and peroxide. *J. Am. Chem. Soc.* 139: 4991–4994.

48. Pan, Z., Zhang, J., Ji, K., Chittavong, V., Ji, X., and Wang, B. (2018). Organic CO prodrugs activated by endogenous ROS. *Org. Lett.* 20: 8–11.

49. Hassan, S.S.M. and Rechnitz, G.A. (1982). Determination of glutathione and glutathione reductase with a silver sulfide membrane electrode. *Anal. Chem.* 54: 1972–1976.

50. Kuppusamy, P., Li, H., Ilangovan, G., Cardounel, A.J., Zweier, J.L., Yamada, K., Krishna, M.C., and Mitchell, J.B. (2002). Noninvasive imaging of tumor redox status and its modification by tissue glutathione levels. *Cancer Res.* 62: 307–312.

51. Gao, C., Liang, X., Guo, Z., Jiang, B.P., Liu, X., and Shen, X.C. (2018). Diiron hexacarbonyl complex induces site-specific release of carbon monoxide in cancer cells triggered by endogenous glutathione. *ACS Omega* 3: 2683–2689.

52. Michel, B.W., Lippert, A.R., and Chang, C.J. (2012). A reaction-based fluorescent probe for selective imaging of carbon monoxide in living cells using a palladium-mediated carbonylation. *J. Am. Chem. Soc.* 134: 15668–15671.

53. Romanski, S., Kraus, B., Schatzschneider, U., Neudorfl, J.M., Amslinger, S., and Schmalz, H.G. (2011). Acyloxybutadiene iron tricarbonyl complexes as enzyme-triggered CO-releasing molecules (ET-CORMs). *Angew. Chem. Int. Ed. Engl.* 50: 2392–2396.

54. Romanski, S., Kraus, B., Guttentag, M., Schlundt, W., Rucker, H., Adler, A., Neudorfl, J.M., Alberto, R., Amslinger, S., and Schmalz, H.G. (2012). Acyloxybutadiene tricarbonyl iron complexes as enzyme-triggered CO-releasing molecules (ET-CORMs): A structure-activity relationship study. *Dalton Trans.* 41: 13862–13875.

55. Romanski, S., Stamellou, E., Jaraba, J.T., Storz, D., Kramer, B.K., Hafner, M., Amslinger, S., Schmalz, H.G., and Yard, B.A. (2013). Enzyme-triggered CO-releasing molecules (ET-CORMs): Evaluation of biological activity in relation to their structure. *Free Radic. Biol. Med.* 65: 78–88.

56. Bauer, B., Goderz, A.L., Braumuller, H., Neudorfl, J.M., Rocken, M., Wieder, T., and Schmalz, H.G. (2017). Methyl fumarate-derived iron carbonyl complexes (FumET-CORMs) as powerful anti-inflammatory agents. *ChemMedChem* 12: 1927–1930.

57. Wilson, J.L., Fayad Kobeissi, S., Oudir, S., Haas, B., Michel, B., Dubois Rande, J.L., Ollivier, A., Martens, T., Rivard, M., Motterlini, R., and Foresti, R. (2014). Design and synthesis of new hybrid molecules that activate the transcription factor Nrf2 and simultaneously release carbon monoxide. *Chemistry* 20: 14698–14704.

58. Ji, X., Ji, K., Chittavong, V., Yu, B., Pan, Z., and Wang, B. (2017). An esterase-activated click and release approach to metal-free CO-prodrugs. *Chem Commun.* 53: 8296–8299.

59. Ji, X., Pan, Z., Li, C., Kang, T., De La Cruz, L.K.C., Yang, L., Yuan, Z., Ke, B., and Wang, B. (2019). Esterase-sensitive and pH-controlled carbon monoxide prodrugs for treating systemic inflammation. *J. Med. Chem.* 62: 3163–3168.

60. Schatzschneider, U. (2011). PhotoCORMs: Light-triggered release of carbon monoxide from the coordination sphere of transition metal complexes for biological applications. *Inorg. Chim. Acta* 374: 19–23.

61. Niesel, J., Pinto, A., Peindy N'Dongo, H.W., Merz, K., Ott, I., Gust, R., and Schatzschneider, U. (2008). Photoinduced CO release, cellular uptake and

cytotoxicity of a tris(pyrazolyl)methane (Tpm) manganese tricarbonyl complex. *Chem. Commun.* 1798–1800.

62. Kunz, P.C., Huber, W., Rojas, A., Schatzschnedier, U., and Spingler, B. (2009). Tricarbonylmanganese(I) and –rhenium(I) complexes of imidazol-based phosphane ligands: Influence of the substitution pattern on the CO release properties. *Eur. J. Inorg. Chem.* 2009: 5358–5366.

63. Gonzalez, M.A., Carrington, S.J., Fry, N.L., Martinez, J.L., and Mascharak, P.K. (2012). Syntheses, structures, and properties of new manganese carbonyls as photoactive CO-releasing molecules: Design strategies that lead to CO photolability in the visible region. *Inorg. Chem.* 51: 11930–11940.

64. Gonzalez, M.A., Yim, M.A., Cheng, S., Moyes, A., Hobbs, A.J., and Mascharak, P.K. (2012). Manganese carbonyls bearing tripodal polypyridine ligands as photoactive carbon monoxide-releasing molecules. *Inorg. Chem.* 51: 601–608.

65. Chakraborty, I., Carrington, S.J., and Mascharak, P.K. (2014). Design strategies to improve the sensitivity of photoactive metal carbonyl complexes (photoCORMs) to visible light and their potential as CO-donors to biological targets. *Acc. Chem. Res.* 47: 2603–2611.

66. Gonzales, M.A. and Mascharak, P.K. (2014). Photoactive metal carbonyl complexes as potential agents for targeted CO delivery. *J. Inorg. Biochem.* 133: 127–135.

67. Jiang, Q., Xia, Y., Barrett, J., Mikhailovsky, A., Wu, G., Wang, D., Shi, P., and Ford, P.C. (2019). Near-infrared and visible photoactivation to uncage carbon monoxide from an aqueous-soluble photoCORM. *Inorg. Chem.* 58: 11066–11075.

68. Pierri, A.E., Huang, P.J., Garcia, J.V., Stanfill, J.G., Chui, M., Wu, G., Zheng, N., and Ford, P.C. (2015). A photoCORM nanocarrier for CO release using NIR light. *Chem. Commun.* 51: 2072–2075.

69. Smith, A.M., Mancini, M.C., and Nie, S. (2009). Bioimaging: Second window for *in vivo* imaging. *Nat. Nanotechnol.* 4: 710–711.

70. Pinto, M.N., Chakraborty, I., Sandoval, C., and Mascharak, P.K. (2017). Eradication of HT-29 colorectal adenocarcinoma cells by controlled photorelease of CO from a CO-releasing polymer (photoCORP-1) triggered by visible light through an optical fiber-based device. *J. Control. Release* 264: 192–202.

71. Glaser, S., Mede, R., Gorls, H., Seupel, S., Bohlender, C., Wyrwa, R., Schirmer, S., Dochow, S., Reddy, G.U., Popp, J., Westerhausen, M., and Schiller, A. (2016). Remote-controlled delivery of CO via photoactive CO-releasing materials on a fiber optical device. *Dalton Trans.* 45: 13222–13233.

72. Chakraborty, I., Jimenez, J., Sameera, W.M.C., Kato, M., and Mascharak, P.K. (2017). Luminescent Re(I) carbonyl complexes as trackable photoCORMs for CO delivery to cellular targets. *Inorg. Chem.* 56: 2863–2873.

73. Chakraborty, I., Jimenez, J., and Mascharak, P.K. (2017). CO-induced apoptotic death of colorectal cancer cells by a luminescent photoCORM grafted on biocompatible carboxymethyl chitosan. *Chem. Commun.* 53: 5519–5522.

74. Jimenez, J., Pinto, M.N., Martinez-Gonzalez, J., and Mascharak, P.K. (2019). Photo-induced eradication of human colorectal adenocarcinoma HT-29 cells by carbon monoxide (CO) delivery from a Mn-based green luminescent photoCORM. *Inorg. Chim. Acta* 485: 112–117.

75. Jimenez, J., Chakraborty, I., Dominguez, A., Martinez-Gonzalez, J., Sameera, W.M.C., and Mascharak, P.K. (2018). A luminescent manganese photoCORM for CO delivery to cellular targets under the control of visible light. *Inorg. Chem.* 57: 1766–1773.

76. Chakraborty, I., Carrington, S.J., Hauser, J., Oliver, S.R., and Mascharak, P.K. (2015). Rapid eradication of human breast cancer cells through trackable light-triggered CO delivery by mesoporous silica nanoparticles packed with a designed photoCORM. *Chem. Mater.* 27: 8387–8397.

77. Chakraborty, I., Carrington, S.J., Roseman, G., and Mascharak, P.K. (2017). Synthesis, structures, and CO release capacity of a family of water-soluble photoCORMs: Assessment of the biocompatibility and their phototoxicity toward human breast cancer cells. *Inorg. Chem.* 56: 1534–1545.

78. Carrington, S.J., Chakraborty, I., Bernard, J.M., and Mascharak, P.K. (2014). Synthesis and characterization of a "turn-on" photoCORM for trackable CO delivery to biological targets. *ACS Med. Chem. Lett.* 5: 1324–1328.

79. Pierri, A.E., Pallaoro, A., Wu, G., and Ford, P.C. (2012). A luminescent and biocompatible photoCORM. *J. Am. Chem. Soc.* 134: 18197–18200.

80. Carrington, S.J., Chakraborty, I., Bernard, J.M., and Mascharak, P.K. (2016). A theranostic two-tone luminescent photoCORM derived from Re(I) and (2-pyridyl)-benzothiazole: Trackable CO delivery to malignant cells. *Inorg. Chem.* 55: 7852–7858.

81. Kuzmanich, G., Gard, M.N., and Garcia-Garibay, M.A. (2009). Photonic amplification by a singlet-state quantum chain reaction in the photodecarbonylation of crystalline diarylcyclopropenones. *J. Am. Chem. Soc.* 131: 11606–11614.

82. Kuzmanich, G. and Garcia-Garibay, M.A. (2011). Ring strain release as a strategy to enable the

singlet state photodecarbonylation of crystalline 1, 4-cyclobutanediones. *J. Phys. Org. Chem.* 24: 883–888.

83. Chapman, O.L., Wojtkowski, P.W., Adam, W., Rodriguez, O., and Rucktaeschel, R. (1972). Photochemical transformations. XIIV. Cyclic peroxides. Synthesis and chemistry of a-lactones. *J. Am. Chem. Soc.* 94: 1365–1367.

84. Antony, L.A.P., Slanina, T., Sebej, P., Solomek, T., and Klan, P. (2013). Fluorescein analogue xanthene-9-carboxylic acid: A transition-metal-free CO releasing molecule activated by green light. *Org. Lett.* 15: 4552–4555.

85. Peng, P., Wang, C., Shi, Z., Johns, V.K., Ma, L., Oyer, J., Copik, A., Igarashi, R., and Liao, Y. (2013). Visible-light activatable organic CO-releasing molecules (photoCORMs) that simultaneously generate fluorophores. *Org. Biomol. Chem.* 11: 6671–6674.

86. Palao, E., Slanina, T., Muchova, L., Solomek, T., Vitek, L., and Klan, P. (2016). Transition-metal-free CO-releasing bodipy derivatives activatable by visible to NIR light as promising bioactive molecules. *J. Am. Chem. Soc.* 138: 126–133.

87. Anderson, S.N., Richards, J.M., Esquer, H.J., Benninghoff, A.D., Arif, A.M., and Berreau, L.M. (2015). A structurally-tunable 3-hydroxyflavone motif for visible light-induced carbon monoxide-releasing molecules (CORMs). *ChemistryOpen* 4: 590–594.

88. Wang, X., Chen, X., Song, L., Zhou, R., and Luan, S. (2020). An enzyme-responsive and photoactivatable carbon-monoxide releasing molecule for bacterial infection theranostics. *J. Mater. Chem. B* 8: 9325–9334.

89. Feng, W., Feng, S., and Feng, G. (2019). CO release with ratiometric fluorescence changes: A promising visible-light-triggered metal-free CO-releasing molecule. *Chem. Commun.* 55: 8987–8990.

90. Popova, M., Lazarus, L.S., Ayad, S., Benninghoff, A.D., and Berreau, L.M. (2018). Visible-light-activated quinolone carbon-monoxide-releasing molecule: Prodrug and albumin-assisted delivery enables anticancer and potent anti-inflammatory effects. *J. Am. Chem. Soc.* 140: 9721–9729.

91. Stackova, L., Russo, M., Muchova, L., Orel, V., Vitek, L., Stacko, P., and Klan, P. (2020). Cyanine-flavonol hybrids for near-infrared light-activated delivery of carbon monoxide. *Chemistry* 26: 13184–13190.

92. Li, Y., Shu, Y., Liang, M., Xie, X., Jiao, X., Wang, X., and Tang, B. (2018). A two-photon H_2O_2-activated CO photoreleaser. *Angew. Chem. Int. Ed. Engl.* 57: 12415–12419.

93. Russo, M., Stacko, P., Nachtigallova, D., and Klan, P. (2020). Mechanisms of orthogonal photodecarbonylation reactions of 3-hydroxyflavone-based acid-base forms. *J. Org. Chem.* 85: 3527–3537.

94. Szakacs, Z., Bojtar, M., Drahos, L., Hessz, D., Kallay, M., Vidoczy, T., Bitter, I., and Kubinyi, M. (2016). The kinetics and mechanism of photooxygenation of 4'-diethylamino-3-hydroxyflavone. *Photochem. Photobiol. Sci.* 15: 219–227.

95. Tao, S., Cheng, J., Su, G., Li, D., Shen, Z., Tao, F., You, T., and Hu, J. (2020). Breathing micelles for combinatorial treatment of rheumatoid arthritis. *Angew. Chem. Int. Ed. Engl.* 59: 21864–21869.

96. Cheng, J., Zheng, B., Cheng, S., Zhang, G., and Hu, J. (2020). Metal-free carbon monoxide-releasing micelles undergo tandem photochemical reactions for cutaneous wound healing. *Chem. Sci.* 11: 4499–4507.

97. Zhang, M., Cheng, J., Huang, X., Zhang, G., Ding, S., Hu, J., and Qiao, R. (2020). Photo-degradable micelles capable of releasing of carbon monoxide under visible light irradiation. *Macromol. Rapid Commun.* 41: e2000323.

98. Andreasson, J. and Pischel, U. (2015). Molecules with a sense of logic: A progress report. *Chem. Soc. Rev.* 44: 1053–1069.

99. Lazarus, L.S., Benninghoff, A.D., and Berreau, L.M. (2020). Development of triggerable, trackable, and targetable carbon monoxide releasing molecules. *Acc. Chem. Res.* 53: 2273–2285.

100. Lazarus, L.S., Esquer, H.J., Benninghoff, A.D., and Berreau, L.M. (2017). Sense and release: A thiol-responsive flavonol-based photonically driven carbon monoxide-releasing molecule that operates via a multiple-input AND logic gate. *J. Am. Chem. Soc.* 139: 9435–9438.

101. Dalzoppo, D., Di Paolo, V., Calderan, L., Pasut, G., Rosato, A., Caccuri, A.M., and Quintieri, L. (2017). Thiol-activated anticancer agents: The state of the art. *Anticancer Agents Med. Chem.* 17: 4–20.

102. Peng, Y.J., Zhang, X., Gridina, A., Chupikova, I., McCormick, D.L., Thomas, R.J., Scammell, T.E., Kim, G., Vasavda, C., Nanduri, J., Kumar, G.K., Semenza, G.L., Snyder, S.H., and Prabhakar, N.R. (2017). Complementary roles of gasotransmitters CO and H_2S in sleep apnea. *Proc. Natl. Acad. Sci. U. S. A.* 114: 1413–1418.

103. Guo, F.F., Yu, T.C., Hong, J., and Fang, J.Y. (2016). Emerging roles of hydrogen sulfide in inflammatory and neoplastic colonic diseases. *Front. Physiol.* 7: 156.

104. Soboleva, T., Benninghoff, A.D., and Berreau, L.M. (2017). An H_2S-sensing/CO-releasing flavonol that operates via logic gates. *Chempluschem* 82: 1408–1412.

105. Liu, K., Kong, X., Ma, Y., and Lin, W. (2018). Preparation of a Nile Red-Pd-based fluorescent CO

probe and its imaging applications *in vitro* and *in vivo*. *Nat. Protoc.* 13: 1020–1033.

106. Li, Y., Shu, Y., Wang, X., Jiao, X., Xie, X., Zhang, J., and Tang, B. (2019). An H_2S-activated ratiometric CO Photoreleaser enabled by excimer/monomer conversion. *Chem. Commun.* 55: 6301–6304.

107. Mede, R., Hoffmann, P., Neumann, C., Gorls, H., Schmitt, M., Popp, J., Neugebauer, U., and Westerhausen, M. (2018). Acetoxymethyl concept for intracellular administration of carbon monoxide with $Mn(CO)_3$-based photoCORMs. *Chemistry* 24: 3321–3329.

108. Ma, B., Wang, S., Liu, F., Zhang, S., Duan, J., Li, Z., Kong, Y., Sang, Y., Liu, H., Bu, W., and Li, L. (2019). Self-assembled copper-amino acid nanoparticles for in situ glutathione "and" H_2O_2 sequentially triggered chemodynamic therapy. *J. Am. Chem. Soc.* 141: 849–857.

109. Sun, P., Jia, L., Hai, J., Lu, S., Chen, F., Liang, K., Sun, S., Liu, H., Fu, X., Zhu, Y., and Wang, B. (2021). Tumor microenvironment-"and" near-infrared light-activated coordination polymer nanoprodrug for on-demand CO-sensitized synergistic cancer therapy. *Adv. Healthc. Mater.* 10: e2001728.

110. Pinto, M.N. and Mascharak, P.K. (2020). Light-assisted and remote delivery of carbon monoxide to malignant cells and tissues: Photochemotherapy in the spotlight. *J. Photochem. Photobiol. C* 42: 100341.

111. Price, P.M., Mahmoud, W.E., Al-Ghamdi, A.A., and Bronstein, L.M. (2018). Magnetic drug delivery: Where the field is going. *Front. Chem.* 6: 619.

112. Kunz, P.C., Meyer, H., Barthel, J., Sollazzo, S., Schmidt, A.M., and Janiak, C. (2013). Metal carbonyls supported on iron oxide nanoparticles to trigger the CO-gasotransmitter release by magnetic heating. *Chem. Commun.* 49: 4896–4898.

113. Meyer, H., Winkler, F., Kunz, P., Schmidt, A.M., Hamacher, A., Kassack, M.U., and Janiak, C. (2015). Stabilizing alginate confinement and polymer coating of CO-releasing molecules supported on iron oxide nanoparticles to trigger the CO release by magnetic heating. *Inorg. Chem.* 54: 11236–11246.

114. Meyer, H., Brenner, M., Hofert, S.P., Knedel, T.O., Kunz, P.C., Schmidt, A.M., Hamacher, A., Kassack, M.U., and Janiak, C. (2016). Synthesis of oxime-based CO-releasing molecules, CORMs and their immobilization on maghemite nanoparticles for magnetic-field induced CO release. *Dalton Trans.* 45: 7605–7615.

115. Al Sawaftah, N.M. and Husseini, G.A. (2020). Ultrasound-mediated drug delivery in cancer therapy: A review. *J. Nanosci. Nanotechnol.* 20: 7211–7230.

116. Alghazwat, O., Talebzadeh, S., Oyer, J., Copik, A., and Liao, Y. (2021). Ultrasound responsive carbon monoxide releasing micelle. *Ultrason. Sonochem.* 72: 105427.

117. van der Vlies, A.J., Inubushi, R., Uyama, H., and Hasegawa, U. (2016). Polymeric framboidal nanoparticles loaded with a carbon monoxide donor via phenylboronic acid-catechol complexation. *Bioconjug. Chem.* 27: 1500–1508.

118. Lo Iacono, L., Boczkowski, J., Zini, R., Salouage, I., Berdeaux, A., Motterlini, R., and Morin, D. (2011). A carbon monoxide-releasing molecule (CORM-3) uncouples mitochondrial respiration and modulates the production of reactive oxygen species. *Free Radic. Biol. Med.* 50: 1556–1564.

119. Wilson, J.L., Bouillaud, F., Almeida, A.S., Vieira, H.L., Ouidja, M.O., Dubois-Rande, J.L., Foresti, R., and Motterlini, R. (2017). Carbon monoxide reverses the metabolic adaptation of microglia cells to an inflammatory stimulus. *Free Radic. Biol. Med.* 104: 311–323.

120. Kaczara, P., Motterlini, R., Rosen, G.M., Augustynek, B., Bednarczyk, P., Szewczyk, A., Foresti, R., and Chlopicki, S. (2015). Carbon monoxide released by CORM-401 uncouples mitochondrial respiration and inhibits glycolysis in endothelial cells: A role for mitoBKCa channels. *Biochim. Biophys. Acta* 1847: 1297–1309.

121. Reiter, C.E. and Alayash, A.I. (2012). Effects of carbon monoxide (CO) delivery by a CO donor or hemoglobin on vascular hypoxia inducible factor 1a and mitochondrial respiration. *FEBS Open Bio.* 2: 113–118.

122. Zheng, Y., Ji, X., Yu, B., Ji, K., Gallo, D., Csizmadia, E., Zhu, M., Choudhury, M.R., De La Cruz, L.K.C., Chittavong, V., Pan, Z., Yuan, Z., Otterbein, L.E., and Wang, B. (2018). Enrichment-triggered prodrug activation demonstrated through mitochondria-targeted delivery of doxorubicin and carbon monoxide. *Nat. Chem.* 10: 787–794.

123. Lazarus, L.S., Esquer, H.J., Anderson, S.N., Berreau, L.M., and Benninghoff, A.D. (2018). Mitochondrial-localized versus cytosolic intracellular CO-releasing organic photoCORMs: Evaluation of CO effects using bioenergetics. *ACS Chem. Biol.* 13: 2220–2228.

124. Lazarus, L.S., Simons, C.R., Arcidiacono, A., Benninghoff, A.D., and Berreau, L.M. (2019). Extracellular vs intracellular delivery of CO: Does it matter for a stable, diffusible gasotransmitter?. *J. Med. Chem.* 62: 9990–9995.

125. Pfeiffer, H., Rojas, A., Niesel, J., and Schatzschneider, U. (2009). Sonogashira and "click" reactions for the *N*-terminal and side-chain functionalization of

126. Pfeiffer, H., Sowik, T., and Schatzschnedier, U. (2013). Bioorthogonal oxime ligation of a Mo(CO)$_4$(N-N) CO-releasing molecule (CORM) to a Tgf beta-binding peptide. *J. Organomet. Chem.* 734: 17–24.

peptides with [Mn(CO)$_3$(Tpm)]$^+$-based CO releasing molecules (Tpm = tris(pyrazolyl)methane). *Dalton Trans.* 4292–4298.

127. Pai, S., Radacki, K., and Schatzschnedier, U. (2014). Sonogashira, CuAAC, and oxime ligations for the synthesis of Mn^I tricarbonyl photoCORM peptide conjugates. *Eur. J. Inorg. Chem.* 2014: 2886–2895.

128. Low, P.S. and Kularatne, S.A. (2009). Folate-targeted therapeutic and imaging agents for cancer. *Curr. Opin. Chem. Biol.* 13: 256–262.

129. Elnakat, H. and Ratnam, M. (2004). Distribution, functionality and gene regulation of folate receptor isoforms: Implications in targeted therapy. *Adv. Drug Deliv. Rev.* 56: 1067–1084.

130. Low, P.S., Henne, W.A., and Doorneweerd, D.D. (2008). Discovery and development of folic-acid-based receptor targeting for imaging and therapy of cancer and inflammatory diseases. *Acc. Chem. Res.* 41: 120–129.

131. Loureiro, A., Bernardes, G.J.L., Shimanovich, U., Sarria, M.P., Nogueira, E., Preto, A., Gomes, A.C., and Cavaco-Paulo, A. (2015). Folic acid-tagged protein nanoemulsions loaded with CORM-2 enhance the survival of mice bearing subcutaneous A20 lymphoma tumors. *Nanomedicine* 11: 1077–1083.

132. Wegiel, B., Gallo, D., Csizmadia, E., Harris, C., Belcher, J., Vercellotti, G.M., Penacho, N., Seth, P., Sukhatme, V., Ahmed, A., Pandolfi, P.P., Helczynski, L., Bjartell, A., Persson, J.L., and Otterbein, L.E. (2013). Carbon monoxide expedites metabolic exhaustion to inhibit tumor growth. *Cancer Res.* 73: 7009–7021.

133. Lamichhane, S. and Lee, S. (2020). Albumin nanoscience: Homing nanotechnology enabling targeted drug delivery and therapy. *Arch. Pharm. Res.* 43: 118–133.

134. Chaves-Ferreira, M., Albuquerque, I.S., Matak-Vinkovic, D., Coelho, A.C., Carvalho, S.M., Saraiva, L.M., Romao, C.C., and Bernardes, G.J.L. (2015). Spontaneous CO release from Ru(II)(CO)$_2$-protein complexes in aqueous solution, cells, and mice. *Angew. Chem. Int. Ed. Engl.* 54: 1172–1175.

135. Hoogenboezem, E.N. and Duvall, C.L. (2018). Harnessing albumin as a carrier for cancer therapies. *Adv. Drug. Deliv. Rev.* 130: 73–89.

136. Pal, S., Saha, C., Hossain, M., Dey, S.K., and Kumar, G.S. (2012). Influence of galloyl moiety in interaction of epicatechin with bovine serum albumin: A spectroscopic and thermodynamic characterization. *PLoS One* 7: e43321.

137. Dufour, C. and Dangles, O. (2005). Flavonoid-serum albumin complexation: Determination of binding constants and binding sites by fluorescence spectroscopy. *Biochim. Biophys. Acta* 1721: 164–173.

138. Xiao, J., Suzuki, M., Jiang, X., Chen, X., Yamamoto, K., Ren, F., and Xu, M. (2008). Influence of B-ring hydroxylation on interactions of flavonols with bovine serum albumin. *J. Agric. Food Chem.* 56: 2350–2356.

139. Liu, E.H., Qi, L.W., and Li, P. (2010). Structural relationship and binding mechanisms of five flavonoids with bovine serum albumin. *Molecules* 15: 9092–9103.

140. Popova, M., Soboleva, T., Arif, A.M., and Berreau, L.M. (2017). Properties of a flavonol-based photoCORM in aqueous buffered solutions: influence of metal ions, surfactants and proteins on visible-light induced release. *RSC Adv.* 7: 21997–22007.

141. Clardy, S.M., Allis, D.G., Fairchild, T.J., and Doyle, R.P. (2011). Vitamin B12 in drug delivery: Breaking through the barriers to a B$_{12}$ bioconjugate pharmaceutical. *Expert Opin. Drug Deliv.* 8: 127–140.

142. Zobi, F., Quaroni, L., Santoro, G., Zlateva, T., Blacque, O., Sarafimov, B., Schaub, M.C., and Bogdanova, A.Y. (2013). Live-fibroblast IR imaging of a cytoprotective photoCORM activated with visible light. *J. Med. Chem.* 56: 6719–6731.

143. Wuerges, J., Garau, G., Geremia, S., Fedosov, S.N., Petersen, T.E., and Randaccio, L. (2006). Structural basis for mammalian vitamin B$_{12}$ transport by transcobalamin. *Proc. Natl. Acad. Sci. U. S. A.* 103: 4386–4391.

144. Theocharopoulos, C., Lialios, P.P., Gogas, H., and Ziogas, D.C. (2020). An overview of antibody-drug conjugates in oncological practice. *Ther. Adv. Med. Oncol.* 12: 1758835920962997.

145. Kawahara, B., Gao, L., Cohn, W., Whitelegge, J.P., Sen, S., Janzen, C., and Mascharak, P.K. (2020). Diminished viability of human ovarian cancer cells by antigen-specific delivery of carbon monoxide with a family of photoactivatable antibody-photoCORM conjugates. *Chem. Sci.* 11: 467–473.

146. He, Q. (2017). Precision gas therapy using intelligent nanomedicine. *Biomater. Sci.* 5: 2226–2230.

147. Manoharan, D., Li, W.-P., and Yeh, C.-S. (2019). Advances in controlled gas-releasing nanomaterials for therapeutic applications. *Nanoscale Horiz.* 4: 557–578.

148. Nguyen, D. and Boyer, C. (2015). Macromolecular and inorganic nanomaterials scaffolds for carbon monoxide delivery: Recent developments and future trends. *ACS Biomater. Sci. Eng.* 1: 895–913.
149. Kautz, A.C., Kunz, P.C., and Janiak, C. (2016). CO-releasing molecule (CORM) conjugate systems. *Dalton Trans.* 45: 18045–18063.
150. Wang, S.B., Zhang, C., Chen, Z.X., Ye, J.J., Peng, S.Y., Rong, L., Liu, C.J., and Zhang, X.Z. (2019). A versatile carbon monoxide nanogenerator for enhanced tumor therapy and anti-inflammation. *ACS Nano* 13: 5523–5532.
151. Meng, J., Jin, Z., Zhao, P., Zhao, B., Fan, M., and He, Q. (2020). A multistage assembly/disassembly strategy for tumor-targeted CO delivery. *Sci. Adv.* 6: eaba1362.
152. Yoon, H.Y., Koo, H., Choi, K.Y., Lee, S.J., Kim, K., Kwon, I.C., Leary, J.F., Park, K., Yuk, S.H., Park, J.H., and Choi, K. (2012). Tumor-targeting hyaluronic acid nanoparticles for photodynamic imaging and therapy. *Biomaterials* 33: 3980–3989.
153. Liu, C., Du, Z., Ma, M., Sun, Y., Ren, J., and Qu, X. (2020). Carbon monoxide controllable targeted gas therapy for synergistic anti-inflammation. *iScience* 23: 101483.
154. Gao, F., Wang, F., Nie, X., Zhang, Z., Chen, G., Xia, L., Wang, L.-H., Wang, C.-H., Hao, Z.-Y., Zhang, W.-J., Hong, C.-Y., and You, Y.-Z. (2020). Mitochondria-targeted delivery and light controlled release of iron prodrug and CO to enhance cancer therapy by ferroptosis. *New J. Chem.* 44: 3478–3486.
155. Li, J., Cao, F., Yin, H.L., Huang, Z.J., Lin, Z.T., Mao, N., Sun, B., and Wang, G. (2020). Ferroptosis: past, present and future. *Cell Death Dis.* 11: 88.
156. Opoku-Damoah, Y., Zhang, R., Ta, H.T., Amilan Jose, D., Sakla, R., and Ping Xu, Z. (2021). Lipid-encapsulated upconversion nanoparticle for near-infrared light-mediated carbon monoxide release for cancer gas therapy. *Eur. J. Pharm. Biopharm.* 158: 211–221.
157. Geri, S., Krunclova, T., Janouskova, O., Panek, J., Hruby, M., Hernandez-Valdes, D., Probst, B., Alberto, R.A., Mamat, C., Kubeil, M., and Stephan, H. (2020). Light-activated carbon monoxide prodrugs based on bipyridyl dicarbonyl ruthenium(II) complexes. *Chemistry* 26: 10992–11006.

16

Anesthesia-Related Carbon Monoxide Exposure

Richard J. Levy

Department of Anesthesiology, Columbia University Medical Center, 622 W. 168th Street, New York, NY 10032, USA

Introduction

Surgery would be impossible without the advent of inhalational anesthesia to render patients unconscious and insensitive to the noxious and painful stimulation of surgical trauma. Centuries ago, it was recognized that compounds, such as diethyl ether, induce a reversible, sleep-like state in animals [1]. However, it was not until the mid-1800s that ether was used to induce anesthesia in humans [1]. In 1842, Crawford Williamson Long successfully administered an inhaled anesthetic to James Venable using an ether-soaked towel to facilitate the surgical excision of two small tumors from his neck [1]. William Thomas Green Morton then began experimenting with liquid ether, recognizing the potential for the drug to be used as an inhaled anesthetic given its volatility and safety profile [1]. Following his use of ether to enable minor dental procedures, Morton successfully anesthetized Edward Gilbert Abbott with an ether-soaked sponge contained within a glass bulb delivery device to permit Dr. John C. Warren to surgically excise Abbott's jaw tumor [1]. The public demonstration of this event in the Bullfinch Amphitheater of the Massachusetts General Hospital on October 16, 1846, is recognized as the moment when inhaled anesthesia was invented and represents one of the most important medical advances in history [1].

Over the next century, researchers and pharmacologists sought to develop the "ideal inhaled anesthetic" [1]. Several different agents were tested, including chloroform, ethylene gas, propylene, cyclopropane, and divinyl ether [1]. However, many of these agents were found to be flammable and explosive [1]. Thus, in the 1930s, the field honed in on organic fluorinated compounds, recognizing that fluorine substitution reduced the boiling point and increased stability [2]. Trifluoroethyl vinyl ether was the first fluorinated inhaled anesthetic developed in the late 1940s and synthesis of several other agents, such as halothane, methoxyflurane, enflurane, isoflurane, desflurane, and sevoflurane, soon followed [1]. Although fluorination generally enhanced safety, several fluorinated anesthetics were subsequently phased out of practice because of toxicity concerns. For example, halothane induced hepatitis, methoxyflurane and enflurane caused fluoride-induced nephrotoxicity, and enflurane caused seizures [1]. Today, sevoflurane, desflurane, and isoflurane are the most commonly used inhaled anesthetic agents; however, toxicity associated with their use remains a concern.

Numerous devices have been developed throughout history to optimize inhaled delivery of anesthetics. Adaptation of these devices culminated in the creation of a continuous flow apparatus in the early 1900s, establishing the basis of the contemporary system used today: the anesthesia machine (Figure 16.1) [3]. The anesthesia machine is the most important piece of equipment in anesthesia practice and is designed to deliver oxygen, precisely mix and supply anesthetic gases and vapors, allow patient ventilation, and minimize risk while maximizing safety [3]. The machine is a continuous flow apparatus that receives a pressurized supply of gases (such as oxygen, air, and nitrous oxide) and allows flow control through anesthesia variable bypass vaporizers to deliver specific gas mixtures to the patient via the breathing circuit [3].

The breathing circuit is a semi-closed circle system, designed to permit unidirectional gas flow in a circular pathway [4]. The circle system can function as a completely closed circuit, avoiding exposure to ambient atmosphere to limit volatile anesthetic usage and conserve heat [4]. Conceptually, the completely closed anesthesia circuit relies on the fact that oxygen

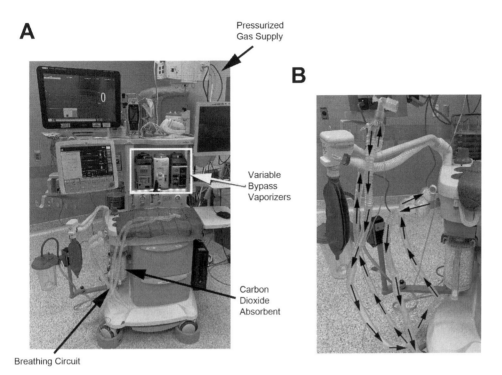

Figure 16.1 The anesthesia machine. (A) A modern-day anesthesia machine is shown. Pressurized gas supply lines (oxygen [green], air [yellow], and nitrous oxide [blue]), the anesthesia breathing circuit, and carbon dioxide absorbent are indicated with arrows. Location of the anesthesia variable bypass vaporizers is highlighted with a white box. (B) The semi-closed circle system breathing circuit is shown. Arrows indicate the unidirectional circular gas flow within the circuit.

is consumed by the patient and is the only gas removed from the breathing system [4]. Thus, inhaled anesthetics can be administered safely in a closed system if oxygen is replaced within the circuit at the same rate that it is consumed [4].

However, exhaled carbon dioxide must be actively removed from within the circuit to avoid the toxic threat of carbon dioxide buildup [4]. This is achieved by chemical absorption. Experiments involving carbon dioxide absorption began in the 1700s using lime water to sustain the life of confined bees [5]. Long-lasting carbon dioxide absorbent for human use was established during World War I using soda lime to remove a soldier's exhaled carbon dioxide in military gas masks [5]. Chemical carbon dioxide absorption was first applied to the closed-circuit anesthesia breathing system in 1923 [5]. Since then, carbon dioxide absorbents have been a mainstay of anesthesia breathing systems and remain a cost-effective measure to efficiently remove exhaled carbon dioxide while conserving volatile anesthetic agents, especially during low-flow anesthesia (LFA).

Soda lime and other absorbents convert exhaled carbon dioxide to carbonate and water based on the principle that bases neutralize acids (see formula) [4]. In the reaction, carbon dioxide first reacts with water to form carbonic acid. Carbonic acid then interacts with hydroxide (usually in the form of potassium or sodium hydroxide) to form soluble salts and water in an exothermic manner. Finally, sodium or potassium carbonate react with calcium hydroxide to form calcium carbonate.

$$CO_2 + H_2O \rightarrow H_2CO_3$$

$$H_2CO_3 + 2KOH \rightarrow K_2CO_3 + 2H_2O$$

$$H_2CO_3 + 2NaOH \rightarrow Na_2CO_3 + 2H_2O$$

$$K_2CO_3 \text{ (or } Na_2CO_3) + Ca(OH)_2 \rightarrow 2KOH \text{ (or } 2NaOH) + CaCO_3$$

There are several different carbon dioxide absorbents in use today. They can be classified as high alkali, low alkali, or alkali free and vary based on their hydroxide chemical composition. The exact chemical composition is critically important because carbon dioxide absorbents can degrade inhaled anesthetics and generate carbon monoxide (CO) [6]. The amount of CO generated during a clinical anesthetic depends on the water content and strong alkali hydroxide composition of the absorbent as well as the volatile anesthetic agent used [6]. As will be discussed, anesthetic degradation by conventional carbon dioxide absorbents can be a significant source of exogenous CO within the anesthesia breathing circuit [6].

The ability to control the rate of fresh gas flow (FGF) is a unique characteristic of the anesthesia breathing machine. Anesthesiologists usually adjust FGF rate based on patient needs and practice paradigms. Flow rates that exceed rates of patient minute ventilation (the product of tidal volume and respiratory rate) result in venting and scavenging of excess gas from the breathing circuit [4]. Administration of an inhaled anesthetic in this manner is termed high-flow anesthesia, but is usually avoided to limit the wastage of anesthetic gases. LFA, on the other hand, is a commonly used, economical approach to permit rebreathing and conserve inhaled anesthetics [7]. LFA has many definitions; however, it generally refers to techniques that set FGF to rates that are less than patient alveolar ventilation rates [4]. Some authors have attempted to define LFA more specifically. For example, Baum and Aitkenhead defined LFA as a paradigm when at least 50% of exhaled gas is rebreathed [7]. Baker subdivided the LFA technique into flow categories: metabolic flow (~250 ml/min), minimal flow (250–500 ml/min), low flow (500–1000 ml/min), and medium flow (1–2 l/min) [8]. Most anesthesiologists consider LFA as setting FGF to less than 2 l/min in adults or below the rate of minute ventilation in children [4,9,10]. FGF is a major determinant of CO formation within the anesthesia breathing circuit and CO exposure occurs commonly during low-flow general endotracheal anesthesia [6]. In this chapter, we will review the concept of CO exposure in the setting of an anesthetic, identify the sources of production, and discuss the potential therapeutic role for CO as a part of routine anesthetic management.

CO Exposure During General Endotracheal Anesthesia

CO was first detected within a closed-circuit anesthesia system in 1965 [11]. The investigators who reported the finding documented up to 810 ppm CO within the anesthesia breathing circuit [11]. At the time, they concluded that the phenomenon of CO "buildup" resulted from "complete closure of the anesthetic system" [11]. Although the report was published over a half-century ago, the authors insightfully suspected the patient's own exhaled breath as a potential source of CO [11].

In the early to mid-1990s, the concept of CO generation within the anesthesia breathing circuit and the potential for anesthesia-related CO exposure became more widely recognized [12]. During that period, a number of reports emerged, describing high CO concentrations within the anesthesia breathing circuit, evidence of CO toxicity in patients, and elevated carboxyhemoglobin (COHb) levels [13–15]. The first report of CO toxicity described a rise in COHb in a 76-year-old, nonsmoking elderly woman who underwent general endotracheal anesthesia for surgical resection of her thyroid [14]. Blood aliquots sampled 25 min after the initiation of anesthesia and then 1 h later demonstrated COHb levels of 9.1% and 28%, respectively [14]. The patient complained of a headache postoperatively, which resolved with hyperbaric oxygen therapy [14]. Initially, the anesthesiologists suspected fresh gas contamination with CO, but this was ruled out by thorough testing of the institutional gas supply lines [14].

Several weeks later, another case of anesthesia-related CO exposure was identified at the same hospital [14]. In this second case, a COHb level of 24.7% was detected in a patient undergoing total hip replacement under general anesthesia [14]. Although investigators initially failed to identify the source of exogenous CO, a comprehensive assessment detected CO concentrations in excess of 500 ppm in the gas mixture exiting the carbon dioxide absorbent [14]. In 1990, eight cases of anesthesia-related CO exposure were subsequently reported and a total of 31 cases were published [16]. Alarmingly, CO concentrations within the breathing circuit were found to exceed 1000 ppm in certain instances and some patient COHb levels reached 30% or greater [17].

These cases of anesthesia-related CO exposure were presented and reviewed at the American Society of Anesthesiologists annual meeting in 1990. Analysis revealed that the majority of these cases shared a common characteristic: each occurred on a Monday morning, as the first surgical case of the day, or resulted from the use of an anesthesia machine that had been previously idle for at least 2 days [14]. This observation raised concern for a potential chemical reaction between the carbon dioxide absorbent and the inhaled anesthetic [14]. Subsequent research investigation conclusively demonstrated that CO is, in fact, produced when volatile anesthetics are degraded by conventional carbon dioxide absorbents [15,18–20]. Importantly, researchers found that the amount of CO formed is inversely proportional to the water content of the absorbent [12].

In 2003, Abbott Laboratories along with the US Food and Drug Administration published a "Dear Health Care Professional" letter to alert practitioners of the risk of adverse events related to a sevoflurane–carbon dioxide absorbent interaction [21]. The letter focused mostly on the fire risk that manifests from the exothermic reaction, but also alerted anesthesiologists of the risk of CO production. In 2005, the Anesthesia Patient Safety Foundation (APSF) called for a meeting of experts to address the fire risk and the potential for CO exposure (Carbon Dioxide Absorbent Desiccation Safety Conference Convened by APSF, accessed September 23, 2020). The discussion focused mostly on the formation of toxic by-products via the chemical degradation of volatile anesthetics (Carbon Dioxide Absorbent Desiccation Safety Conference Convened by APSF, accessed September 23, 2020). As a result, the APSF made two specific recommendations to anesthesiologists: use carbon dioxide absorbents that lack strong alkali hydroxides and do not degrade inhalational anesthetics or adopt strict protocols to prevent conventional absorbent desiccation (Carbon Dioxide Absorbent Desiccation Safety Conference Convened by APSF, accessed September 23, 2020) [22].

Exogenous Sources

CO is generated exogenously within the anesthesia breathing circuit when inhaled anesthetic agents are degraded and broken down by conventional carbon dioxide absorbents [17,23]. CO is formed in this context by proton abstraction of the halogenated anesthetic difluoromethyl ether moiety [23]. This is facilitated by the absorbent to yield a carbanion intermediate (Figure 16.2) [23]. In the absence of water, the carbanion decomposes and subsequently interacts with hydroxide or residual water, generating CO [23]. Agents that lack a difluoromethyl ether group (such as sevoflurane and halothane) can also generate CO, suggesting that other mechanisms may be involved [24,25]. Importantly, it has been shown that CO produced by degradation can occur with all of the volatile anesthetic agents in use today [20,24–26].

As mentioned, water content is a key variable that determines the magnitude of anesthetic breakdown and exogenous CO formation within the anesthesia breathing circuit [13,20]. Specifically, the amount of CO generated is inversely related to the amount of water in the absorbent [20]. Completely dried absorbent produces the greatest amount of CO compared to absorbent that is partially hydrated [13,20]. For example, partially hydrated barium hydroxide lime that contains 8–10% water (13% is fully hydrated) generates only minimal amounts of CO, while desiccated barium hydroxide lime with 5% water or less produces significant concentrations of CO [26]. This property explains why toxic CO exposures were observed on Monday mornings: anesthesia machines were likely left idle over the weekend with fresh gases flowing [26]. Two days of continuous gas flow permitted the carbon dioxide absorbent within the anesthesia machine to lose water and dry out, setting the stage for inhaled anesthetic degradation and CO production.

Tests of Baralyme® absorbent (Chemetron Medical Division, Allied Healthcare Products, St Louis, MO) suggested that, when desiccated, it could generate up to 20 000 ppm CO during a desflurane anesthetic, while experiments with partially hydrated Baralyme®

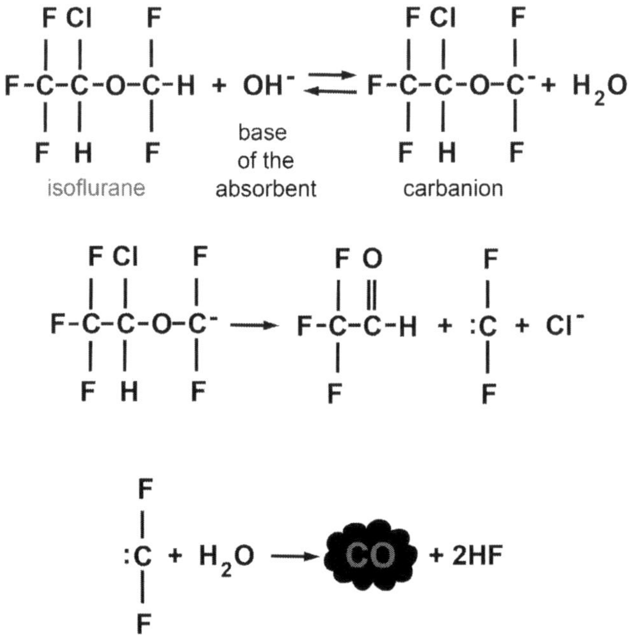

Figure 16.2 Mechanism of CO generation from inhaled anesthetic agents. Proton (H^+) abstraction is catalyzed by the base (OH^-) of the carbon dioxide absorbent. With inadequately hydrated absorbent, the carbanion decomposes and subsequently interacts with hydroxide or residual water (H_2O) within the absorbent, generating CO. Adapted from Ref. [6] with permission.

(containing 1.6% or 3.2% water) demonstrated the potential to produce lower amounts of CO (peak concentrations of 15 000 ppm or less) [20]. With soda lime, fully dried out absorbents generate ~2500 ppm CO from desflurane and ~500 ppm CO from isoflurane, while partially hydrated absorbents produce ~60 and ~100 ppm CO with each agent, respectively [20]. Rehydrating desiccated barium hydroxide lime prevented CO formation from desflurane, which suggests that fully hydrated conventional absorbents could be safer [13,20]. However, *in vitro* experimentation demonstrated that fully hydrated soda lime does, indeed, break down desflurane during a 2-h exposure and generates up to 23 ppm CO [27]. Although the amount of CO produced from hydrated absorbent is markedly less than that generated by desiccated absorbent, CO is still generated and FGF inversely correlates with the amount of CO formed [27]. One explanation is that fresh gas dilutes CO that is generated during anesthetic degradation and lower FGFs may limit the dilutional effect, resulting in higher concentrations within the breathing circuit [27].

In addition to the water content of the carbon dioxide absorbent, other important variables that affect the degree of CO production include type of inhaled anesthetic used, anesthetic concentration, absorbent temperature, patient carbon dioxide production, and the chemical composition of the specific carbon dioxide absorbent [20,27]. The amount of CO produced for the different volatile anesthetics is generally described as desflurane ≥ enflurane > isoflurane ≥ halothane = sevoflurane, with higher agent doses producing higher peak concentrations of CO for each anesthetic [17,20,23,27]. However, this relationship is not absolute. For example, CO can be produced from sevoflurane in an exponential manner, reaching levels that exceed 11 000 ppm when the temperature of desiccated Baralyme® rises to 80 °C [28]. These concentrations exceed those produced by equipotent enflurane and isoflurane at 45 °C when using the same absorbent [20]. Thus, comparisons made between anesthetic agents must be interpreted cautiously with regard to the potential for CO generation.

Degradation of anesthetics and CO production are accelerated by heat, resulting in greater CO production [20]. This characteristic is particularly important because inhaled anesthetic degradation is an exothermic process [28,29]. Thus, heat generated from anesthetic breakdown raises the temperature of the absorbent, leading to further anesthetic degradation and greater CO formation [20,28,29]. Sevoflurane undergoes 13% degradation over 1 h at 40 °C versus 56% degradation at 60 °C [30]. This effect is greater when desiccated or partially dried out absorbent is used [20].

Carbon dioxide absorption is also exothermic and reduces the water content within the upstream absorbent over time [31]. The combined effects of heating and drying indirectly increase the potential for CO to be formed during absorption of high carbon dioxide concentrations within the circuit [27,31]. This may explain why patients who excrete more carbon dioxide may encounter greater CO exposures during an anesthetic [27,31]. Although absorbent temperature is an important variable, the lack of a rise in soda lime temperature does not predict the amount of CO that is generated [24,25].

Chemical composition of the carbon dioxide absorbent is another critical factor that determines the degree of inhaled anesthetic breakdown and CO formation. Specifically, strong alkali hydroxides, such as potassium hydroxide and sodium hydroxide, are the key components within conventional carbon dioxide absorbents that are responsible for anesthetic degradation [32]. Since base-catalyzed proton abstraction occurs to a greater degree with potassium versus sodium hydroxide, dried barium hydroxide lime produces more CO than desiccated soda lime [17,23]. This is because Baralyme® contains 4.6% potassium hydroxide while conventional soda lime is composed of 2.6% potassium hydroxide and ~1.5% sodium hydroxide [23]. Of note, Baralyme® was removed from the commercial market in 2004 due to "concerns regarding [its] use ... in conjunction with ... newer inhalation anesthetics when Baralyme® [was] ... allowed, contrary to recommended practice, to become desiccated" (US SEC Allied Healthcare Products, Inc., accessed September 23, 2020).

More modern carbon dioxide absorbents lack potassium and sodium hydroxide [24,25]. These newer absorbents contain calcium or lithium hydroxide and generate approximately 10% as much CO as conventional absorbents [17]. However, CO production can still occur and reports suggest significant CO formation with some of the calcium-containing absorbents when they become desiccated [24,25]. On the other hand, lithium hydroxide-based absorbents appear to lack CO-generating capability, even when completely dried out [24,25].

Endogenous Sources

CO can also arise within the anesthesia breathing circuit from endogenous patient sources [26,33]. CO is generated naturally within the liver, spleen, kidney, and in tissues within the central nervous system and reticuloendothelial system as a result of heme catabolism by heme oxygenase [34]. Chapter 1 discusses this subject in detail. CO formed in this manner diffuses into the circulation, binds to hemoglobin to form COHb, and is excreted by the lungs in exhaled breath [34].

During LFA, exhaled gases (other than carbon dioxide) are neither scavenged nor removed from the breathing circuit. Thus, in this setting patients rebreathe their own exhaled CO [13]. Children have been shown to rebreathe up to 20 ppm CO during general endotracheal anesthesia when FGF is set to rates lower than minute ventilation [9,10]. Such exposures are associated with a concomitant rise in COHb [9,10]. CO rebreathing also occurs in adults during LFA and, in smokers, concentrations within the breathing circuit reach levels as high as 145 ppm [35]. CO exposure has been shown to inversely correlate with FGF rate such that higher CO levels result from lower FGFs [9,35]. Importantly, little or no CO rebreathing occurs when FGFs exceed the rate of minute ventilation and CO will decrease within the circuit by ~5.9 ppm for each liter per minute increase in gas flow [9,35].

There are several pathological processes that result in increased endogenous CO production and elevated COHb levels. These include high heme turnover states such as autoimmune-mediated hemolysis and sickle cell disease and conditions that increase heme oxygenase activity and heme breakdown, such as trauma, sepsis, and shock [33,36–38]. Furthermore, exposure to environmental sources of CO, as occurs with inhalation of smoke or air pollution, also increases COHb [39]. In addition, transfusion of blood with an elevated COHb content results in a rise in recipient COHb [40]. In these scenarios, elevated COHb levels can lead to greater CO concentrations in exhaled breath, resulting in an even more profound CO exposure when rebreathing is permitted during LFA.

Biological Effects of CO

Policy

The United States Environmental Protection Agency set the National Ambient Air Quality Standard for CO in 1971 that established the limit for a time-weighted average environmental CO exposure at 9 ppm for 8 h and 35 ppm for 1 h (US Environmental Protection Agency. Carbon monoxide, accessed September 24, 2020). This regulation was based on research that found a correlation between CO exposure and the time to the onset of angina in exercising adults with coronary artery disease (US Environmental Protection Agency. Carbon monoxide, accessed September 24, 2020) [41–48]. Limits for anesthesia-related CO exposure have never been established because of a paucity of research and understanding of the phenomenon. However, one prior study linked preoperative smoking with cardiac risk in patients undergoing general endotracheal anesthesia [49]. In this work, smoking tobacco prior to surgery led to a mean exhaled CO concentration of 52.4 ppm (versus 9.4 ppm CO in nonsmokers) and was a significant predictor of intraoperative myocardial ischemia when considered simultaneously with rate pressure product [49]. This indicated that CO exposure during a general endotracheal anesthetic has the potential to disrupt myocardial tissue oxygenation [49].

There are currently no policies regulating levels of indoor CO exposure. However, several US agencies provide guidelines regarding upper limits of exposure. For example, the Occupational Safety and Health Administration suggests limiting exposure to 50 ppm CO for 8 h, the National Institute for Occupational Safety and Health recommends a time-weighted limit of 35 ppm CO over 8 h with an upper ceiling of 200 ppm CO, and the American Conference of Governmental Industrial Hygienists recommends limiting exposure to 25 ppm CO for 8 h (US Environmental Protection Agency. Carbon monoxide, accessed September 24, 2020).

It should be noted that CO concentrations generated by completely desiccated and partially dehydrated carbon dioxide absorbent may exceed these recommended concentration limits [20,24,25]. However, with strict adherence to the APSF guidelines, the risk of anesthesia-related overt CO poisoning can be reduced. In the context of LFA, levels of anesthesia-related CO exposure can surpass these time-weighted limits depending on the concentration of CO in the circuit and the duration of exposure [9,10,27,35]. Unfortunately, anesthesia-related CO exposure has been poorly studied, so the consequences of such exposures remain generally unknown.

Overt Toxicity Versus Low-Dose Exposure

The mechanisms of CO's actions, including toxicity in different contexts, are discussed in various chapters and should be reviewed. Organs with the highest aerobic activity, such as the brain and the heart, are most vulnerable to CO toxicity [50]. Cardiovascular manifestations of CO poisoning include hypotension, vasodilation, arrhythmias, ischemia, infarction, and cardiac arrest, while neurological effects include headache, dizziness, impaired judgment, confusion, altered mental status, seizures, syncope, stroke, and coma [50,51]. CO toxicity can manifest with nausea, vomiting, and lethargy in children [52,53]. Interestingly, many of these symptoms mimic common side effects seen with the majority of anesthetic agents during emergence from general anesthesia. Thus, without monitoring for CO levels within the anesthesia breathing circuit or measuring COHb levels, toxic CO exposures during an anesthetic can easily be missed. It should be noted that anesthesia-related CO exposures reported in the 1990s commonly resulted in COHb levels that were in excess of 10–20% and were associated with clinical signs of overt CO toxicity [12–14,16]. In at least one of these cases, hyperbaric oxygen therapy was used as a therapy [14].

The concentration of CO and the duration of exposure determine the degree of toxicity [51,54]. Environmental exposure to less than 120 ppm CO for up to 4 h, for example, is not life threatening, does not elicit symptoms, and does not usually result in tissue hypoxia [51,54,55]. Lack of signs and symptoms following such an exposure defines it as subclinical and subtoxic [51]. Alternatively, symptoms of toxicity begin to manifest following exposure to concentrations that exceed 200 ppm CO and exposure to concentrations greater than 800 ppm can be rapidly lethal [51].

With regard to anesthesia-related CO exposure, the amount of COHb formed has been previously modeled using a validated mathematical equation, accounting for clinically relevant conditions during a general isoflurane or desflurane anesthetic in the context of desiccated barium hydroxide lime [26,56]. Postexposure COHb levels were calculated to be greatest in smaller patients with anemia who were inspiring lower oxygen concentrations [26,56]. For example, the model found that, with 7.5% desflurane and partially desiccated barium hydroxide lime, COHb levels reached 60% within 20 min of exposure in a 25-kg patient compared to 30% COHb in a 100-kg patient over the same duration [26,56].

Because the phenomenon of anesthesia-related CO exposure was not widely known in the era of routine Baralyme* use, the role of toxic CO exposures in causing perioperative morbidity and mortality is unknown. Importantly, removal of Baralyme* from the market and institutional adherence to APSF guidelines has likely reduced the risk of overt anesthesia-related CO toxicity. However, conventional carbon dioxide absorbents are still in use today and complete and partial desiccation can still occur. Therefore, exposure to high concentrations of CO during an anesthetic remains a plausible risk. Nevertheless, the rate of toxic anesthesia-related CO exposure is unknown because anesthesiologists do not currently monitor for CO exposure during an anesthetic as part of the standard of care.

With regard to subtoxic CO, exposure to low concentrations of CO commonly occurs during low-flow endotracheal anesthesia due to exogenous CO production, rebreathing of endogenously generated exhaled CO, or both. The impact of such a low-concentration CO exposure during an anesthetic is not known. Unlike toxic levels of CO, however, low-dose CO is a signaling molecule and can exert a range of diverse and complex cytoprotective effects [57]. These mechanisms have been discussed in multiple chapters throughout this textbook and should be reviewed. As such, low-dose CO is currently being developed and investigated as a novel therapy to treat a variety of disease processes. Because surgical patients often inspire low concentrations of CO during LFA, the perioperative beneficial effects of CO may be clinically relevant and could carry therapeutic potential. However, future research is necessary to determine whether CO is a common anesthesia-related pollutant with deleterious effects or whether low-dose exposures during an anesthetic have the potential to affect and improve patient outcomes.

Biological Effect of LFA

Despite the lack of rigorous investigation into the effects of low-dose CO exposure during LFA, several publications suggest a potential effect of LFA, itself. For example, when compared with high-flow anesthesia, LFA with desflurane and nitrous oxide preserved pulmonary function and mucociliary clearance in an adult population undergoing tympanomastoidectomy [58]. These effects of LFA were consistent with known cytoprotective effects of low-dose CO on lung inflammation and oxidative injury [59]. In other work, LFA with 100% oxygen was associated with a significant decrease in postoperative hypoxic events, hospital mortality, surgical site infections, and postoperative nausea and vomiting compared with high-flow anesthesia using 70% nitrous oxide and 30% oxygen in a large surgical cohort [60]. Even though there were obvious confounders and a lack of appropriate controls in this study, the hint of a potential benefit of LFA is interesting. In another observational study, LFA with sevoflurane was associated with lower plasma viscosity and prolonged activated partial thromboplastin times compared with a high-flow technique [61]. Other research assessed for an association between LFA and the immune response.

This work found that LFA with sevoflurane had minimal effect on neutrophils and T cells in adults undergoing abdominal surgery, while LFA with desflurane significantly affected number of lymphocytes and neutrophils, percentage of T helper lymphocytes (CD4) and cytotoxic T lymphocytes (CD8), and CD4/CD8 ratio [62]. While these results may have been due to a direct desflurane effect, it is possible that anesthesia-related CO exposure may have contributed given that low-dose CO is known to impact the immune system [63–65]. In other work, LFA with desflurane preserved and significantly increased plasma nitric oxide (NO) levels in adults undergoing thyroidectomy, 24 h after surgery, compared with high-flow anesthesia [66]. Although CO can cause a rapid release of bound NO from hemoproteins, the increase in circulating NO seen following LFA could also have resulted from CO-mediated activation of NO synthase [67–72]. Of further interest, recent investigation found an effect of anesthetic technique on survival in cancer patients following surgery [73]. In this study, inhaled anesthesia was associated with greater mortality versus total intravenous anesthesia [73]. Although the authors did not comment on the type of flow paradigm used, the findings raise questions about of the role of LFA and anesthesia-related CO exposure in promoting survival of circulating tumor cells and depression of cell-mediated immunity.

Potential Therapeutic Role of Perioperative CO Exposure

Preclinical studies using various animal models of injury and disease indicate that inhaled CO confers cytoprotection in many different tissues and organ systems [74–76]. The clinical scenarios that have been experimentally modeled represent human disease states that anesthesiologists and intensivists commonly encounter in the perioperative setting. Thus, the perioperative therapeutic potential of CO may prove to be clinically relevant in the near future. In a rat hyperoxia-induced lung injury model, 250 ppm CO attenuated neutrophil accumulation and apoptosis in the lungs after exposure and improved survival. In a rat model of ventilator-induced lung injury, concentrations of up to 250 ppm CO decreased tumor necrosis factor-α (TNF-α) and total cell count in bronchoalveolar lavage fluid while increasing levels of interleukin (IL)-10 [77]. Mechanisms of CO-mediated protection from ventilator-induced lung injury appear to be mediated via p38 MAPK signaling, differential regulation of anti- and proinflammatory transcription factors such as the proliferator-activated receptor-gamma and early growth response (Egr)-1, and increased caveolin-1 expression [77–79]. In mechanically ventilated baboons with pneumococcal pneumonia, inhaled CO (100–300 ppm) given in conjunction with antibiotics accelerated acute lung injury resolution [80]. In a murine model of aspiration, 500 ppm CO inhaled for 6 h significantly decreased the number of neutrophils in the lavage fluid and reduced the degree of lung injury [81]. Furthermore, in an aeroallergen model of asthma, exposure to 250 ppm CO substantially reduced the amount of eosinophils and IL-5 in the bronchoalveolar lavage fluid and reduced airway reactivity via a cGMP-dependent mechanism [82,83]. Inhaled low-dose CO has also been shown to protect the lungs following ischemia and reperfusion via attenuation of the inflammatory response and by preventing apoptosis [84–89]. Such cytoprotective CO effects have been shown in clinically relevant animal models of lung transplantation, cardiopulmonary bypass (CPB), and secondary lung injury after remote ischemia/reperfusion [86,89–91]. The prosurvival mechanisms involve the p38 MAPK pathway, the phosphatidylinositol 3-kinase/Akt pathway, and Egr-1 expression [84–87].

In human patients with chronic obstructive pulmonary disease, the anti-inflammatory effects of inspired CO have been tested preliminarily [92]. Exposure to 100–125 ppm CO for 2 h daily for 4 consecutive days resulted in a downward trend in eosinophil number within sputum and improved airway responsiveness to methacholine [92]. In other work, a phase I trial recently assessed the safety of inhaled CO (100–200 ppm) administered to intubated ICU patients with sepsis-induced acute respiratory distress syndrome [93]. There were no CO-associated adverse events in CO treated patients and COHb never exceeded 10% [93]. Importantly, investigators noted a decrease in circulating mtDNA levels in the exposed cohort [93].

The cardiovascular system has also been a successful therapeutic experimental target of low-dose CO. In a rodent left anterior descending coronary artery occlusion model, pre-occlusion CO exposure limited infarct size, suppressed macrophage and monocyte migration, and decreased the expression of TNF-α [94]. CO-mediated protection was related to endothelial NOS and cGMP and activation of the Akt and p38 MAPK pathways within myocardium [94]. In a pig CPB model that included 2 h of cardiac arrest, exposure to 250 ppm CO for 2 h prior to application of the aortic cross clamp along with administration of CO-saturated cardioplegia solution enhanced myocardial bioenergetics, decreased tissue edema and cell death, and led to fewer instances of electrical cardioversions required post-reperfusion [95]. Thus, inhaled CO has potential to protect the heart in the context of ischemia and reperfusion.

As with the lungs, low-dose CO also protected the heart in the setting of experimental organ transplantation. In mouse-to-rat cardiac transplantation model, exposure of the donor for 2 days before transplantation and the recipient for 14–16 days post-transplantation to 250–400 ppm CO limited rejection and improved long-term graft survival by inhibiting platelet aggregation, thrombosis, myocardial infarction, and cardiomyocyte apoptosis [96]. In related work, pre-explant exposure of rat donors to 400 ppm CO and placing the heart in storage solution containing 1000 ppm CO protected the graft from reperfusion injury via an antiapoptotic mechanism [97]. Furthermore, exposing rat heart transplant recipients to 20 ppm CO continuously, for up to 100 days after transplantation, markedly improved allograft survival [98]. In this model, CO treatment reduced the amount of vascular inflammation, fibrosis, and cellular infiltration in the transplanted heart, and inhibited the expression of proinflammatory cytokines and mediators [98]. Thus, CO has the potential to be developed as a novel therapeutic agent to be used in conjunction with standard myocardial protective strategies in the setting of cardiac surgery and cardiac transplantation.

Pulmonary arterial hypertension (PAH), a pathological state that is caused by a wide range of disease processes, carries significant perioperative risk [99]. In a rat PAH model, chronic and continuous exposure to 50 ppm CO for 21 days was shown to attenuate hypoxia-induced PAH by stimulating calcium-activated potassium channels within pulmonary artery smooth myocytes [100]. In other investigation, daily exposure to 250 ppm CO for 1 h over a 3-week period resolved rodent PAH and right ventricle hypertrophy and restored pulmonary vascular architecture [101]. CO-mediated reversal of PAH in this model was mediated by endothelial NOS and resulted in an increase in smooth myocyte programmed cell death and a decrease in cellular proliferation [101]. With regard to clinical studies, two trials attempted to evaluate CO as a potential therapy for PAH (ClinicalTrials.gov, accessed September 24, 2020). The first, a phase I/phase II trial in adults with severe PAH, proposed to evaluate the safety and efficacy of inspiring 150 ppm CO for 3 h at variable frequencies over a 16-week study period (ClinicalTrials.gov, accessed September 24, 2020). Unfortunately, this study was aborted due to lack of funding and enrolled subject acuity (ClinicalTrials.gov, accessed September 24, 2020). The second study, a phase I trial in neonates with PAH, was designed to evaluate for adverse events associated with inspiring CO (ClinicalTrials.gov, accessed September 24, 2020). Success of such a trial, if completed, could lead to the further development of inhaled CO as a therapeutic agent for PAH.

The brain is another important organ system that could be therapeutically targeted with low-concentration CO. In preclinical work, exposure to 250 ppm CO for 1 h successfully preconditioned mouse neurons cultured *in vitro*, protecting cells from experimentally induced cell death via an antiapoptotic manner [102]. The prosurvival mechanism in this study was dependent on CO-mediated generation of reactive oxygen species and involved activation of guanylate cyclase and NOS [102]. *In vivo* preconditioning with low-dose CO has also been shown to be neuroprotective in a variety of animal models. For example, in a piglet model of CPB, exposure to 250 ppm CO for 3 h, 1 day prior to surgery, completely prevented neocortical and hippocampal cell death after deep hypothermic circulatory arrest [103]. In other work, preconditioning newborn mice over a 3-day period with 250 ppm CO for 1 h prevented hypoxia- and ischemia-induced cell death in the hippocampus [104]. Postconditioning with CO has also been shown to provide some degree of neuroprotection in brain ischemia models. For example, mice exposed to 250 ppm CO for 18 h demonstrated a 30% reduction in infarct size after complete occlusion of the middle cerebral artery [105]. Therefore, low-concentration CO has future potential to be employed as a therapy to protect the brain from injury.

In contrast to the beneficial effects of CO, however, are the pathological disruptive effects of low concentrations of CO on proliferation, differentiation, myelination, and physiological cell death in the brain and neuronal tissue during development [106–110]. For example, continuous perinatal exposure to 75, 150, or 300 ppm CO, from conception until postnatal day 10, caused a dose-dependent decrease in cerebellar weight and the number of γ-aminobutyric acid (GABA)-ergic neurons in exposed rat pups [109,110]. Such CO exposures also increased DNA content and glial proliferation in the neostriatum in response to loss of neurons [109,110]. Continuous prenatal exposure to 75 ppm CO from conception until day 20 of gestation interrupted differentiation by reducing the number of glutamic acid decarboxylase/GABA positive neuronal bodies and axon terminals in the rat cerebellar cortex [106]. With regard to myelination, rats continuously exposed to either 75 or 150 ppm CO during their gestation (conception to day 20) demonstrated a decrease in sciatic nerve myelin sheath thickness later in life [107]. Although CO has been shown to impair these critical developmental processes, the exact mechanisms remain unknown. Because COHb levels in these CO exposure models were below the threshold necessary to impair oxygen delivery, CO-mediated disruption was not the result of tissue hypoxia.

With regard to natural apoptosis in the developing murine brain, postnatal exposure to 5 or 100 ppm CO for 3 h impaired cytochrome c release from forebrain mitochondria, decreased caspase-3 activity, reduced the number of activated caspase-3-positive cells in the neocortex and hippocampus, and decreased programmed cell death in a dose-dependent manner [108]. CO-mediated inhibition of neuronal apoptosis in this work resulted from a concentration-dependent inhibition of cytochrome c peroxidase [108]. The postnatal phase of programmed cell death is physiological and important for selective elimination of aberrant and excess neurons [111]. Consequently, CO-mediated inhibition of apoptosis in the neonatal brain impaired murine neuron elimination as evidenced by an increased number of neurons and megalencephaly 1 week postexposure [108].

Experimental research has demonstrated that such perinatal low-dose CO exposure impairs memory, learning, and behavior later in life [112–114]. For example, continuous maternal exposure to 75 or 150 ppm CO from conception to day 20 of gestation led to abnormal habituation and working memory in prenatally exposed rats while sparing motor activity [114]. In addition, continuous gestational exposure to 150 ppm CO during the same period permanently impaired acquisition of a two-way active avoidance task in rats tested 3 months postexposure [112]. Postnatal exposure to low-concentration CO has also been shown to disrupt rodent neurodevelopment [108]. For example, 3-h exposure to 5 or 100 ppm CO in the newborn period impaired murine reference memory, memory retention, and spatial working memory in a dose-dependent manner weeks after exposure and caused avoidance activity and abnormal socialization [108]. Thus, the experimental evidence suggests that perinatal exposure to low concentrations of CO can interfere with normal brain development. The findings raise concern about the safety of such exposures in infants and children. Today, levels of CO encountered during a general anesthetic fall largely within the subtoxic, low-concentration range. However, due to a paucity of research, the consequences of such anesthesia-related CO exposures in the pediatric population are unknown.

Despite the vulnerability of the developing nervous system to low-concentration CO, the potential exists, however, for CO to paradoxically provide neuroprotection in certain clinical settings or when combined with neurotoxins. For example, the prosurvival, antiapoptotic effects of low-dose CO could feasibly limit and offset proapoptotic and harmful effects of anesthetics in the developing brain. The majority of commonly used inhaled anesthetic agents have been shown to induce widespread neuronal cell death in the developing mammalian brain, resulting in cognitive defects and behavioral deficits later in life [115–119]. This anesthesia-induced toxic effect has been demonstrated in a number of different newborn animal models, including nonhuman primates [115,120,121]. Although the exact mechanisms of anesthesia-induced neurotoxicity have not been fully elucidated, downstream, the process appears to involve activation of the oxidative stress-associated mitochondrial apoptosis pathway [120,122–124]. Thus, therapeutically targeting this pathway is a viable strategy that could result in the development of novel therapeutic agents to prevent anesthetic toxicity in the immature brain. In a neonatal murine model of anesthesia-induced neurotoxicity, postnatal exposure to 100 ppm CO for 1 h was shown to limit and prevent isoflurane-induced neurotoxicity by an antiapoptotic mechanism and modulated lipid peroxidation within forebrain mitochondria [125,126]. Thus, the prosurvival effects of low-concentration CO have differential effects in the immature brain depending on the context of exposure. Further work is certainly necessary to determine the safety of CO exposure during states of health and disease before low-concentration CO can be adopted as a therapeutic agent for infants and children.

Conclusions

Patients can be exposed to CO during a general endotracheal anesthetic. Although the risk of overt CO toxicity is reduced by strict adherence to APSF guidelines, generation of high concentrations of CO within the circuit can still occur during an anesthetic with the use of certain carbon dioxide absorbents when they become desiccated. Anesthesiologists should be aware of the type of carbon dioxide absorbent they are using and understand institutional practice and policies with regard to changing the absorbent and preventing desiccation. Use of CO monitoring and CO detection technology within the anesthesia breathing circuit should be considered in the future to prevent toxic exposures.

Low-flow general endotracheal anesthesia results in low-concentration CO exposures. Although the effects of such low-dose exposures have not been well studied, preclinical investigation indicates that they might be cytoprotective, but may also have pathological consequences depending on the context of exposure. Given this gap in our knowledge, further investigation is necessary. Preclinical evidence demonstrating the prosurvival benefits of low-dose CO, however, is certainly intriguing. If developed as a novel perioperative therapeutic approach, anesthesia-related CO exposure could seamlessly be incorporated into the armamentarium of every anesthesiologist. CO exposure, whether achieved by the LFA paradigm or by exogenous CO administration, could be titrated and guided by real-time monitoring to permit precise dosing and desired time-weighted exposures. With further investigation, we will better understand whether perioperative CO exposure prevents perioperative-related injury, improves patient recovery, and enhances outcome and survival.

References

1. Whalen, F.X., Bacon, D.R., and Smith, H.M. (2005). Inhaled anesthetics: An historical overview. *Best Pract. Res. Clin. Anaesthesiol.* 19 (3): 323–330.
2. Calverly, R.K. and Fluorinated Anesthetics, I. (1986). The early years 1932–1946. *Survey Anesthesiol.* 30 (3): 170–172.
3. Gurudatt, C. (2013). The basic anaesthesia machine. *Indian J. Anaesth.* 57 (5): 438–445.
4. Parthasarathy, S. (2013). The closed circuit and the low flow systems. *Indian J. Anaesth.* 57 (5): 516–524.
5. Morris, L.E. (1994). Closed carbon dioxide filtration revisited. *Anaesth. Intensive Care* 22 (4): 345–358.
6. Levy, R.J. (2016). Anesthesia-related carbon monoxide exposure: Toxicity and potential therapy. *Anesth. Analg.* 123 (3): 670–681.
7. Baum, J.A. and Aitkenhead, A.R. (1995). Low-flow anaesthesia. *Anaesthesia* 50 (Suppl): 37–44.
8. Baker, A.B. (1994). Low flow and closed circuits. *Anaesth. Intensive Care* 22 (4): 341–342.
9. Levy, R.J., Nasr, V.G., Rivera, O., Roberts, R., Slack, M., Kanter, J.P., Ratnayaka, K., Kaplan, R.F., and McGowan, F.X., Jr. (2010). Detection of carbon monoxide during routine anesthetics in infants and children. *Anesth. Analg.* 110 (3): 747–753.
10. Nasr, V., Emmanuel, J., Deutsch, N., Slack, M., Kanter, J., Ratnayaka, K., and Levy, R. (2010). Carbon monoxide re-breathing during low-flow anaesthesia in infants and children. *Br. J. Anaesth.* 105 (6): 836–841.
11. Middleton, V., Van Poznak, A., Artusio, J.F., Jr., and Smith, S.M. (1965). Carbon monoxide accumulation in closed circle anesthesia systems. *Anesthesiology* 26 (6): 715–719.
12. Baxter, P.J. and Kharasch, E.D. (1997). Rehydration of desiccated Baralyme prevents carbon monoxide formation from desflurane in an anesthesia machine. *Anesthesiology* 86 (5): 1061–1065.
13. Lentz, R.E. (1995). Carbon monoxide poisoning during anesthesia poses puzzles. *J. Clin. Monit.* 11 (1): 66–67.
14. Moon, R.E. (1995). Cause of CO poisoning, relation to halogenated agents still not clear. *J. Clin. Monit.* 11 (1): 67–71.
15. Woehick, H.J., Dunning, M., 3rd, Nithipatikom, K., Kulier, A.H., and Henry, D.W. (1996). Mass spectrometry provides warning of carbon monoxide exposure via trifluoromethane. *Anesthesiology* 84 (6): 1489–1493.
16. Moon, R.E., Ingram, C., Brunner, E.A., and Meyer, A.F. (1991). In *Spontaneous generation of carbon monoxide within anesthetic circuits*, The Journal of the American Society of Anesthesiologists, The American Society of Anesthesiologists, A873-A873.
17. Coppens, M.J., Versichelen, L.F., Rolly, G., Mortier, E.P., and Struys, M.M. (2006). The mechanisms of carbon monoxide production by inhalational agents. *Anaesthesia* 61 (5): 462–468.
18. Frink, E.J., Jr., Nogami, W.M., Morgan, S.E., and Salmon, R.C. (1997). High carboxyhemoglobin concentrations occur in swine during desflurane anesthesia in the presence of partially dried carbon dioxide absorbents. *Anesthesiology* 87 (2): 308–316.
19. Woehlck, H.J., Dunning, M., 3rd, Gandhi, S., Chang, D., and Milosavljevic, D. (1995). Indirect detection of intraoperative carbon monoxide exposure by mass spectrometry during isoflurane anesthesia. *Anesthesiology* 83 (1): 213–217.
20. Fang, Z.X., Eger, E.I., 2nd, Laster, M.J., Chortkoff, B.S., Kandel, L., and Ionescu, P. (1995). Carbon monoxide production from degradation of desflurane, enflurane, isoflurane, halothane, and sevoflurane by soda lime and Baralyme. *Anesth. Analg.* 80 (6): 1187–1193.
21. Laster, M., Roth, P., and Eger, E.I., 2nd (2004). Fires from the interaction of anesthetics with desiccated absorbent. *Anesth. Analg.* 99 (3): 769–774. table of contents.
22. Murray, J.M., Renfrew, C.W., Bedi, A., McCrystal, C.B., Jones, D.S., and Fee, J.P. (1999). Amsorb: a new carbon dioxide absorbent for use in anesthetic breathing systems. *Anesthesiology* 91 (5): 1342–1348.
23. Baxter, P.J., Garton, K., and Kharasch, E.D. (1998). Mechanistic aspects of carbon monoxide formation from volatile anesthetics. *Anesthesiology* 89 (4): 929–941.
24. Keijzer, C., Perez, R.S., and De Lange, J.J. (2005). Carbon monoxide production from desflurane and

six types of carbon dioxide absorbents in a patient model. *Acta Anaesthesiol. Scand.* 49 (6): 815–818.

25. Keijzer, C., Perez, R.S., and De Lange, J.J. (2005). Carbon monoxide production from five volatile anesthetics in dry sodalime in a patient model: halothane and sevoflurane do produce carbon monoxide; temperature is a poor predictor of carbon monoxide production. *BMC Anesthesiol.* 5 (1): 6.

26. Woehlck, H.J. (2001). Carbon monoxide rebreathing during low flow anesthesia. *Anesth. Analg.* 93 (2): 516–517.

27. Fan, S.Z., Lin, Y.W., Chang, W.S., and Tang, C.S. (2008). An evaluation of the contributions by fresh gas flow rate, carbon dioxide concentration and desflurane partial pressure to carbon monoxide concentration during low fresh gas flows to a circle anaesthetic breathing system. *Eur. J. Anaesthesiol.* 25 (8): 620–626.

28. Holak, E.J., Mei, D.A., Dunning, M.B., 3rd, Gundamraj, R., Noseir, R., Zhang, L., and Woehlck, H.J. (2003). Carbon monoxide production from sevoflurane breakdown: modeling of exposures under clinical conditions. *Anesth. Analg.* 96 (3): 757–764. table of contents.

29. Dunning, M.B., 3rd, Bretscher, L.E., Arain, S.R., Symkowski, Y., and Woehlck, H.J. (2007). Sevoflurane breakdown produces flammable concentrations of hydrogen. *Anesthesiology* 106 (1): 144–148.

30. Strum, D.P., Johnson, B.H., and Eger, E.I., 2nd (1987). Stability of sevoflurane in soda lime. *Anesthesiology* 67 (5): 779–781.

31. Strum, D.P. and Eger, E.I., 2nd (1994). The degradation, absorption, and solubility of volatile anesthetics in soda lime depend on water content. *Anesth. Analg.* 78 (2): 340–348.

32. Neumann, M.A., Laster, M.J., Weiskopf, R.B., Gong, D.H., Dudziak, R., Förster, H., and Eger, E.I., 2nd (1999). The elimination of sodium and potassium hydroxides from desiccated soda lime diminishes degradation of desflurane to carbon monoxide and sevoflurane to compound A but does not compromise carbon dioxide absorption. *Anesth. Analg.* 89 (3): 768–773.

33. Wohlfeil, E.R., Woehlck, H.J., Gottschall, J.L., and Poole, W. (2001). Increased carboxyhemoglobin from hemolysis mistaken as intraoperative desflurane breakdown. *Anesth. Analg.* 92 (6): 1609–1610.

34. Hayashi, M., Takahashi, T., Morimatsu, H., Fujii, H., Taga, N., Mizobuchi, S., Matsumi, M., Katayama, H., Yokoyama, M., Taniguchi, M., and Morita, K. (2004). Increased carbon monoxide concentration in exhaled air after surgery and anesthesia. *Anesth. Analg.* 99 (2): 444–448. table of contents.

35. Tang, C.S., Fan, S.Z., and Chan, C.C. (2001). Smoking status and body size increase carbon monoxide concentrations in the breathing circuit during low-flow anesthesia. *Anesth. Analg.* 92 (2): 542–547.

36. Kaplan, M., Vreman, H.J., Hammerman, C., Leiter, C., Rudensky, B., MacDonald, M.G., and Stevenson, D.K. (1998). Combination of ABO blood group incompatibility and glucose-6-phosphate dehydrogenase deficiency: effect on hemolysis and neonatal hyperbilirubinemia. *Acta Paediatr.* 87 (4): 455–457.

37. Moncure, M., Brathwaite, C.E., Samaha, E., Marburger, R., and Ross, S.E. (1999). Carboxyhemoglobin elevation in trauma victims. *J. Trauma* 46 (3): 424–427.

38. Sylvester, K.P., Patey, R.A., Rafferty, G.F., Rees, D., Thein, S.L., and Greenough, A. (2005). Exhaled carbon monoxide levels in children with sickle cell disease. *Eur. J. Pediatr.* 164 (3): 162–165.

39. Levy, R.J. (2015). Carbon monoxide pollution and neurodevelopment: a public health concern. *Neurotoxicol. Teratol.* 49: 31–40.

40. Ehlers, M., Labaze, G., Hanakova, M., McCloskey, D., and Wilner, G. (2009). Alarming levels of carboxyhemoglobin in banked blood. *J. Cardiothorac. Vasc. Anesth.* 23 (3): 336–338.

41. Adams, K.F., Koch, G., Chatterjee, B., Goldstein, G.M., O'Neil, J.J., Bromberg, P.A., and Sheps, D.S. (1988). Acute elevation of blood carboxyhemoglobin to 6% impairs exercise performance and aggravates symptoms in patients with ischemic heart disease. *J. Am. Coll. Cardiol.* 12 (4): 900–909.

42. Allred, E.N., Bleecker, E.R., Chaitman, B.R., Dahms, T.E., Gottlieb, S.O., Hackney, J.D., Pagano, M., Selvester, R.H., Walden, S.M., and Warren, J. (1989). Short-term effects of carbon monoxide exposure on the exercise performance of subjects with coronary artery disease. *N. Engl. J. Med.* 321 (21): 1426–1432.

43. Allred, E.N., Bleecker, E.R., Chaitman, B.R., Dahms, T.E., Gottlieb, S.O., Hackney, J.D., Pagano, M., Selvester, R.H., Walden, S.M., and Warren, J. (1991). Effects of carbon monoxide on myocardial ischemia. *Environ. Health Perspect.* 91: 89–132.

44. Anderson, E.W., Andelman, R.J., Strauch, J.M., Fortuin, N.J., and Knelson, J.H. (1973). Effect of low-level carbon monoxide exposure on onset and duration of angina pectoris. A study in ten patients with ischemic heart disease. *Ann. Intern. Med.* 79 (1): 46–50.

45. Aronow, W.S. and Isbell, M.W. (1973). Carbon monoxide effect on exercise-induced angina pectoris. *Ann. Intern. Med.* 79 (3): 392–395.

46. Kleinman, M.T., Davidson, D.M., Vandagriff, R.B., Caiozzo, V.J., and Whittenberger, J.L. (1989). Effects

of short-term exposure to carbon monoxide in subjects with coronary artery disease. *Arch. Environ. Health* 44 (6): 361–369.

47. Kleinman, M.T., Leaf, D.A., Kelly, E., Caiozzo, V., Osann, K., and O'Niell, T. (1998). Urban angina in the mountains: effects of carbon monoxide and mild hypoxemia on subjects with chronic stable angina. *Arch. Environ. Health* 53 (6): 388–397.
48. Sheps, D.S., Adams, K.F., Jr., Bromberg, P.A., Goldstein, G.M., O'Neil, J.J., Horstman, D., and Koch, G. (1987). Lack of effect of low levels of carboxyhemoglobin on cardiovascular function in patients with ischemic heart disease. *Arch. Environ. Health* 42 (2): 108–116.
49. Woehlck, H.J., Connolly, L.A., Cinquegrani, M.P., Dunning, M.B., 3rd, and Hoffmann, R.G. (1999). Acute smoking increases ST depression in humans during general anesthesia. *Anesth. Analg.* 89 (4): 856–860.
50. Kao, L.W. and Nañagas, K.A. (2005). Carbon monoxide poisoning. *Med. Clin. North Am.* 89 (6): 1161–1194.
51. Winter, P.M. and Miller, J.N. (1976). Carbon monoxide poisoning. *Jama* 236 (13): 1502.
52. Baker, M.D., Henretig, F.M., and Ludwig, S. (1988). Carboxyhemoglobin levels in children with nonspecific flu-like symptoms. *J. Pediatr.* 113 (3): 501–504.
53. Foster, M., Goodwin, S.R., Williams, C., and Loeffler, J. (1999). Recurrent acute life-threatening events and lactic acidosis caused by chronic carbon monoxide poisoning in an infant. *Pediatrics* 104 (3): e34.
54. Raub, J.A. and Benignus, V.A. (2002). Carbon monoxide and the nervous system. *Neurosci. Biobehav. Rev.* 26 (8): 925–940.
55. Tomaszewski, C. (2002). Carbon Monoxide. Goldfrank's Toxicologic Emergencies, 7e. 1478–1497. New York: McGraw-Hill.
56. Woehlck, H.J., Dunning, M., 3rd, Raza, T., Ruiz, F., Bolla, B., and Zink, W. (2001). Physical factors affecting the production of carbon monoxide from anesthetic breakdown. *Anesthesiology* 94 (3): 453–456.
57. Bauer, I. and Pannen, B.H. (2009). Bench-to-bedside review: carbon monoxide–from mitochondrial poisoning to therapeutic use. *Crit. Care* 13 (4): 220.
58. Bilgi, M., Goksu, S., Mizrak, A., Cevik, C., Gul, R., Koruk, S., and Sahin, L. (2011). Comparison of the effects of low-flow and high-flow inhalational anaesthesia with nitrous oxide and desflurane on mucociliary activity and pulmonary function tests. *Eur. J. Anaesthesiol.* 28 (4): 279–283.
59. Ryter, S.W., Kim, H.P., Nakahira, K., Zuckerbraun, B.S., Morse, D., and Choi, A.M. (2007). Protective functions of heme oxygenase-1 and carbon monoxide in the respiratory system. *Antioxid. Redox Signal.* 9 (12): 2157–2173.
60. Von Bormann, B., Suksompong, S., Weiler, J., and Zander, R. (2014). Pure oxygen ventilation during general anaesthesia does not result in increased postoperative respiratory morbidity but decreases surgical site infection. An observational clinical study. *PeerJ.* 2: e613.
61. Binici, O., Kati, I., Goktas, U., Soyaral, L., and Aytekin, O.C. (2015). Comparing effects of low and high-flow anesthesia on hemorheology and coagulation factors. *Pak. J. Med. Sci.* 31 (3): 683–687.
62. Pirbudak Cocelli, L., Ugur, M.G., and Karadasli, H. (2012). Comparison of effects of low-flow sevoflurane and desflurane anesthesia on neutrophil and T-cell populations. *Curr. Ther. Res. Clin. Exp.* 73 (1–2): 41–51.
63. Mackern-Oberti, J.P., Llanos, C., Carreño, L.J., Riquelme, S.A., Jacobelli, S.H., Anegon, I., and Kalergis, A.M. (2013). Carbon monoxide exposure improves immune function in lupus-prone mice. *Immunology* 140 (1): 123–132.
64. Mackern-Oberti, J.P., Obreque, J., Méndez, G.P., Llanos, C., and Kalergis, A.M. (2015). Carbon monoxide inhibits T cell activation in target organs during systemic lupus erythematosus. *Clin. Exp. Immunol.* 182 (1): 1–13.
65. Simon, T., Pogu, S., Tardif, V., Rigaud, K., Rémy, S., Piaggio, E., Bach, J.M., Anegon, I., and Blancou, P. (2013). Carbon monoxide-treated dendritic cells decrease β1-integrin induction on $CD8^+$ T cells and protect from type 1 diabetes. *Eur. J. Immunol.* 43 (1): 209–218.
66. Kalayci, D., Dikmen, B., Kaçmaz, M., Taspinar, V., Örnek, D., and Turan, Ö. (2014). Plasma levels of interleukin-10 and nitric oxide in response to two different desflurane anesthesia flow rates. *Rev. Bras. Anestesiol.* 64 (4): 292–298.
67. Thom, S.R., Bhopale, V.M., Fisher, D., Zhang, J., and Gimotty, P. (2004). Delayed neuropathology after carbon monoxide poisoning is immune-mediated. *Proc. Natl. Acad. Sci. U. S. A.* 101 (37): 13660–13665.
68. Thom, S.R., Fisher, D., Xu, Y.A., Garner, S., and Ischiropoulos, H. (1999). Role of nitric oxide-derived oxidants in vascular injury from carbon monoxide in the rat. *Am. J. Physiol.* 276 (3): H984–92.
69. Thom, S.R., Fisher, D., Zhang, J., Bhopale, V.M., Cameron, B., and Buerk, D.G. (2004). Neuronal nitric oxide synthase and N-methyl-D-aspartate neurons in experimental carbon monoxide poisoning. *Toxicol. Appl. Pharmacol.* 194 (3): 280–295.
70. Thom, S.R., Ohnishi, S.T., and Ischiropoulos, H. (1994). Nitric oxide released by platelets inhibits

neutrophil B2 integrin function following acute carbon monoxide poisoning. *Toxicol. Appl. Pharmacol.* 128 (1): 105–110.

71. Thom, S.R., Xu, Y.A., and Ischiropoulos, H. (1997). Vascular endothelial cells generate peroxynitrite in response to carbon monoxide exposure. *Chem. Res. Toxicol.* 10 (9): 1023–1031.

72. Truss, N.J. and Warner, T.D. (2011). Gasotransmitters and platelets. *Pharmacol. Ther.* 132 (2): 196–203.

73. Wigmore, T.J., Mohammed, K., and Jhanji, S. (2016). Long-term survival for patients undergoing volatile versus IV anesthesia for cancer surgery: A retrospective analysis. *Anesthesiology* 124 (1): 69–79.

74. Otterbein, L.E., Bach, F.H., Alam, J., Soares, M., Tao Lu, H., Wysk, M., Davis, R.J., Flavell, R.A., and Choi, A.M. (2000). Carbon monoxide has anti-inflammatory effects involving the mitogen-activated protein kinase pathway. *Nat. Med.* 6 (4): 422–428.

75. Otterbein, L.E., Mantell, L.L., and Choi, A.M. (1999). Carbon monoxide provides protection against hyperoxic lung injury. *Am. J. Physiol.* 276 (4): L688–94.

76. Soares, M.P., Lin, Y., Anrather, J., Csizmadia, E., Takigami, K., Sato, K., Grey, S.T., Colvin, R.B., Choi, A.M., Poss, K.D., and Bach, F.H. (1998). Expression of heme oxygenase-1 can determine cardiac xenograft survival. *Nat. Med.* 4 (9): 1073–1077.

77. Dolinay, T., Szilasi, M., Liu, M., and Choi, A.M. (2004). Inhaled carbon monoxide confers antiinflammatory effects against ventilator-induced lung injury. *Am. J. Respir. Crit. Care Med.* 170 (6): 613–620.

78. Hoetzel, A., Dolinay, T., Vallbracht, S., Zhang, Y., Kim, H.P., Ifedigbo, E., Alber, S., Kaynar, A.M., Schmidt, R., Ryter, S.W., and Choi, A.M. (2008). Carbon monoxide protects against ventilator-induced lung injury via PPAR-gamma and inhibition of Egr-1. *Am. J. Respir. Crit. Care Med.* 177 (11): 1223–1232.

79. Hoetzel, A., Schmidt, R., Vallbracht, S., Goebel, U., Dolinay, T., Kim, H.P., Ifedigbo, E., Ryter, S.W., and Choi, A.M. (2009). Carbon monoxide prevents ventilator-induced lung injury via caveolin-1. *Crit. Care Med.* 37 (5): 1708–1715.

80. Fredenburgh, L.E., Kraft, B.D., Hess, D.R., Harris, R.S., Wolf, M.A., Suliman, H.B., Roggli, V.L., Davies, J.D., Winkler, T., Stenzler, A., Baron, R.M., Thompson, B.T., Choi, A.M., Welty-Wolf, K.E., and Piantadosi, C.A. (2015). Effects of inhaled CO administration on acute lung injury in baboons with pneumococcal pneumonia. *Am. J. Physiol. Lung Cell. Mol. Physiol.* 309 (8): L834–46.

81. Nemzek, J.A., Fry, C., and Abatan, O. (2008). Low-dose carbon monoxide treatment attenuates early pulmonary neutrophil recruitment after acid aspiration. *Am. J. Physiol. Lung Cell. Mol. Physiol.* 294 (4): L644–53.

82. Ameredes, B.T., Otterbein, L.E., Kohut, L.K., Gligonic, A.L., Calhoun, W.J., and Choi, A.M. (2003). Low-dose carbon monoxide reduces airway hyperresponsiveness in mice. *Am. J. Physiol. Lung Cell. Mol. Physiol.* 285 (6): L1270–6.

83. Chapman, J.T., Otterbein, L.E., Elias, J.A., and Choi, A.M. (2001). Carbon monoxide attenuates aeroallergen-induced inflammation in mice. *Am. J. Physiol. Lung Cell. Mol. Physiol.* 281 (1): L209–16.

84. Zhang, X., Shan, P., Alam, J., Fu, X.Y., and Lee, P.J. (2005). Carbon monoxide differentially modulates STAT1 and STAT3 and inhibits apoptosis via a phosphatidylinositol 3-kinase/Akt and p38 kinase-dependent STAT3 pathway during anoxia-reoxygenation injury. *J. Biol. Chem.* 280 (10): 8714–8721.

85. Mishra, S., Fujita, T., Lama, V.N., Nam, D., Liao, H., Okada, M., Minamoto, K., Yoshikawa, Y., Harada, H., and Pinsky, D.J. (2006). Carbon monoxide rescues ischemic lungs by interrupting MAPK-driven expression of early growth response 1 gene and its downstream target genes. *Proc. Natl. Acad. Sci. U. S. A.* 103 (13): 5191–5196.

86. Song, R., Kubo, M., Morse, D., Zhou, Z., Zhang, X., Dauber, J.H., Fabisiak, J., Alber, S.M., Watkins, S.C., Zuckerbraun, B.S., Otterbein, L.E., Ning, W., Oury, T.D., Lee, P.J., McCurry, K.R., and Choi, A.M. (2003). Carbon monoxide induces cytoprotection in rat orthotopic lung transplantation via anti-inflammatory and anti-apoptotic effects. *Am. J. Pathol.* 163 (1): 231–242.

87. Zhang, X., Shan, P., Alam, J., Davis, R.J., Flavell, R.A., and Lee, P.J. (2003). Carbon monoxide modulates Fas/Fas ligand, caspases, and Bcl-2 family proteins via the p38alpha mitogen-activated protein kinase pathway during ischemia-reperfusion lung injury. *J. Biol. Chem.* 278 (24): 22061–22070.

88. Fujita, T., Toda, K., Karimova, A., Yan, S.F., Naka, Y., Yet, S.F., and Pinsky, D.J. (2001). Paradoxical rescue from ischemic lung injury by inhaled carbon monoxide driven by derepression of fibrinolysis. *Nat. Med.* 7 (5): 598–604.

89. Kohmoto, J., Nakao, A., Stolz, D.B., Kaizu, T., Tsung, A., Ikeda, A., Shimizu, H., Takahashi, T., Tomiyama, K., Sugimoto, R., Choi, A.M., Billiar, T.R., Murase, N., and McCurry, K.R. (2007). Carbon monoxide protects rat lung transplants from ischemia-reperfusion injury via a mechanism involving p38 MAPK pathway. *Am. J. Transplant.* 7 (10): 2279–2290.

90. Boutros, C., Zegdi, R., Lila, N., Cambillau, M., Fornes, P., Carpentier, A., and Fabini, J.N. (2007). Carbon

monoxide can prevent acute lung injury observed after ischemia reperfusion of the lower extremities. *J. Surg. Res.* 143 (2): 437–442.

91. Goebel, U., Siepe, M., Mecklenburg, A., Stein, P., Roesslein, M., Schwer, C.I., Schmidt, R., Doenst, T., Geiger, K.K., Pahl, H.L., Schlensak, C., and Loop, T. (2008). Carbon monoxide inhalation reduces pulmonary inflammatory response during cardiopulmonary bypass in pigs. *Anesthesiology* 108 (6): 1025–1036.

92. Bathoorn, E., Slebos, D.J., Postma, D.S., Koeter, G.H., Van Oosterhout, A.J., Van Der Toorn, M., Boezen, H.M., and Kerstjens, H.A. (2007). Anti-inflammatory effects of inhaled carbon monoxide in patients with COPD: a pilot study. *Eur. Respir. J.* 30 (6): 1131–1137.

93. Fredenburgh, L.E., Perrella, M.A., Barragan-Bradford, D., Hess, D.R., Peters, E., Welty-Wolf, K.E., Kraft, B.D., Harris, R.S., Maurer, R., Nakahira, K., Oromendia, C., Davies, J.D., Higuera, A., Schiffer, K.T., Englert, J.A., Dieffenbach, P.B., Berlin, D.A., Lagambina, S., Bouthot, M., Sullivan, A.I., Nuccio, P.F., Kone, M.T., Malik, M.J., Porras, M.A.P., Finkelsztein, E., Winkler, T., Hurwitz, S., Serhan, C.N., Piantadosi, C.A., Baron, R.M., Thompson, B.T., and Choi, A.M. (2018). A phase I trial of low-dose inhaled carbon monoxide in sepsis-induced ARDS. *JCI Insight* 3: 23.

94. Fujimoto, H., Ohno, M., Ayabe, S., Kobayashi, H., Ishizaka, N., Kimura, H., Yoshida, K., and Nagai, R. (2004). Carbon monoxide protects against cardiac ischemia–reperfusion injury in vivo via MAPK and Akt–eNOS pathways. *Arterioscler. Thromb. Vasc. Biol.* 24 (10): 1848–1853.

95. Lavitrano, M., Smolenski, R.T., Musumeci, A., Maccherini, M., Slominska, E., Di Florio, E., Bracco, A., Mancini, A., Stassi, G., Patti, M., Giovannoni, R., Froio, A., Simeone, F., Forni, M., Bacci, M.L., D'Alise, G., Cozzi, E., Otterbein, L.E., Yacoub, M.H., Bach, F.H., and Calise, F. (2004). Carbon monoxide improves cardiac energetics and safeguards the heart during reperfusion after cardiopulmonary bypass in pigs. *FASEB J.* 18 (10): 1093–1095.

96. Sato, K., Balla, J., Otterbein, L., Smith, R.N., Brouard, S., Lin, Y., Csizmadia, E., Sevigny, J., Robson, S.C., Vercellotti, G., Choi, A.M., Bach, F.H., and Soares, M.P. (2001). Carbon monoxide generated by heme oxygenase-1 suppresses the rejection of mouse-to-rat cardiac transplants. *J. Immunol.* 166 (6): 4185–4194.

97. Akamatsu, Y., Haga, M., Tyagi, S., Yamashita, K., Graça-Souza, A.V., Ollinger, R., Czismadia, E., May, G.A., Ifedigbo, E., Otterbein, L.E., Bach, F.H., and Soares, M.P. (2004). Heme oxygenase-1-derived carbon monoxide protects hearts from transplant associated ischemia reperfusion injury. *FASEB J.* 18 (6): 771–772.

98. Nakao, A., Toyokawa, H., Abe, M., Kiyomoto, T., Nakahira, K., Choi, A.M., Nalesnik, M.A., Thomson, A.W., and Murase, N. (2006). Heart allograft protection with low-dose carbon monoxide inhalation: effects on inflammatory mediators and alloreactive T-cell responses. *Transplantation* 81 (2): 220–230.

99. McLaughlin, V.V., Shah, S.J., Souza, R., and Humbert, M. (2015). Management of pulmonary arterial hypertension. *J. Am. Coll. Cardiol.* 65 (18): 1976–1997.

100. Dubuis, E., Potier, M., Wang, R., and Vandier, C. (2005). Continuous inhalation of carbon monoxide attenuates hypoxic pulmonary hypertension development presumably through activation of BKCa channels. *Cardiovasc. Res.* 65 (3): 751–761.

101. Zuckerbraun, B.S., Chin, B.Y., Wegiel, B., Billiar, T.R., Czsimadia, E., Rao, J., Shimoda, L., Ifedigbo, E., Kanno, S., and Otterbein, L.E. (2006). Carbon monoxide reverses established pulmonary hypertension. *J. Exp. Med.* 203 (9): 2109–2119.

102. Vieira, H.L., Queiroga, C.S., and Alves, P.M. (2008). Pre-conditioning induced by carbon monoxide provides neuronal protection against apoptosis. *J. Neurochem.* 107 (2): 375–384.

103. Mahan, V.L., Zurakowski, D., Otterbein, L.E., and Pigula, F.A. (2012). Inhaled carbon monoxide provides cerebral cytoprotection in pigs. *PLoS One* 7 (8): e41982.

104. Queiroga, C.S., Tomasi, S., Widerøe, M., Alves, P.M., Vercelli, A., and Vieira, H.L. (2012). Preconditioning triggered by carbon monoxide (CO) provides neuronal protection following perinatal hypoxia-ischemia. *PLoS One* 7 (8): e42632.

105. Wang, B., Cao, W., Biswal, S., and Doré, S. (2011). Carbon monoxide-activated Nrf2 pathway leads to protection against permanent focal cerebral ischemia. *Stroke* 42 (9): 2605–2610.

106. Benagiano, V., Lorusso, L., Coluccia, A., Tarullo, A., Flace, P., Girolamo, F., Bosco, L., Cagiano, R., and Ambrosi, G. (2005). Glutamic acid decarboxylase and GABA immunoreactivities in the cerebellar cortex of adult rat after prenatal exposure to a low concentration of carbon monoxide. *Neuroscience* 135 (3): 897–905.

107. Carratù, M.R., Cagiano, R., Desantis, S., Labate, M., Tattoli, M., Trabace, L., and Cuomo, V. (2000). Prenatal exposure to low levels of carbon monoxide alters sciatic nerve myelination in rat offspring. *Life Sci.* 67 (14): 1759–1772.

108. Cheng, Y., Thomas, A., Mardini, F., Bianchi, S.L., Tang, J.X., Peng, J., Wei, H., Eckenhoff, M.F., Eckenhoff, R.G., and Levy, R.J. (2012). Neurodevelopmental consequences of sub-clinical

108. carbon monoxide exposure in newborn mice. *PLoS One* 7 (2): e32029.
109. Fechter, L.D. (1987). Neurotoxicity of prenatal carbon monoxide exposure. *Res. Rep. Health Eff. Inst.* (12): 3–22.
110. Fechter, L.D., Karpa, M.D., Proctor, B., Lee, A.G., and Storm, J.E. (1987). Disruption of neostriatal development in rats following perinatal exposure to mild, but chronic carbon monoxide. *Neurotoxicol. Teratol.* 9 (4): 277–281.
111. Chan, W.Y., Lorke, D.E., Tiu, S.C., and Yew, D.T. (2002). Proliferation and apoptosis in the developing human neocortex. *Anat. Rec.* 267 (4): 261–276.
112. De Salvia, M.A., Cagiano, R., Carratù, M.R., Di Giovanni, V., Trabace, L., and Cuomo, V. (1995). Irreversible impairment of active avoidance behavior in rats prenatally exposed to mild concentrations of carbon monoxide. *Psychopharmacology (Berl)* 122 (1): 66–71.
113. Fechter, L.D. and Annau, Z. (1980). Prenatal carbon monoxide exposure alters behavioral development. *Neurobehav. Toxicol.* 2 (1): 7–11.
114. Giustino, A., Cagiano, R., Carratù, M.R., Cassano, T., Tattoli, M., and Cuomo, V. (1999). Prenatal exposure to low concentrations of carbon monoxide alters habituation and non-spatial working memory in rat offspring. *Brain Res.* 844 (1–2): 201–205.
115. Brambrink, A.M., Evers, A.S., Avidan, M.S., Farber, N.B., Smith, D.J., Zhang, X., Dissen, G.A., Creeley, C.E., and Olney, J.W. (2010). Isoflurane-induced neuroapoptosis in the neonatal rhesus macaque brain. *Anesthesiology* 112 (4): 834–841.
116. Istaphanous, G.K., Howard, J., Nan, X., Hughes, E.A., McCann, J.C., McAuliffe, J.J., Danzer, S.C., and Loepke, A.W. (2011). Comparison of the neuroapoptotic properties of equipotent anesthetic concentrations of desflurane, isoflurane, or sevoflurane in neonatal mice. *Anesthesiology* 114 (3): 578–587.
117. Istaphanous, G.K. and Loepke, A.W. (2009). General anesthetics and the developing brain. *Curr. Opin. Anaesthesiol.* 22 (3): 368–373.
118. Jevtovic-Todorovic, V., Hartman, R.E., Izumi, Y., Benshoff, N.D., Dikranian, K., Zorumski, C.F., Olney, J.W., and Wozniak, D.F. (2003). Early exposure to common anesthetic agents causes widespread neurodegeneration in the developing rat brain and persistent learning deficits. *J. Neurosci.* 23 (3): 876–882.
119. Stefovska, V.G., Uckermann, O., Czuczwar, M., Smitka, M., Czuczwar, P., Kis, J., Kaindl, A.M., Turski, L., Turski, W.A., and Ikonomidou, C. (2008). Sedative and anticonvulsant drugs suppress postnatal neurogenesis. *Ann. Neurol.* 64 (4): 434–445.
120. Olney, J.W., Young, C., Wozniak, D.F., Ikonomidou, C., and Jevtovic-Todorovic, V. (2004). Anesthesia-induced developmental neuroapoptosis. Does it happen in humans? *Anesthesiology* 101 (2): 273–275.
121. Rizzi, S., Ori, C., and Jevtovic-Todorovic, V. (2010). Timing versus duration: determinants of anesthesia-induced developmental apoptosis in the young mammalian brain. *Ann. N. Y. Acad. Sci.* 1199: 43–51.
122. Bai, X., Yan, Y., Canfield, S., Muravyeva, M.Y., Kikuchi, C., Zaja, I., Corbett, J.A., and Bosnjak, Z.J. (2013). Ketamine enhances human neural stem cell proliferation and induces neuronal apoptosis via reactive oxygen species-mediated mitochondrial pathway. *Anesth. Analg.* 116 (4): 869–880.
123. Boscolo, A., Milanovic, D., Starr, J.A., Sanchez, V., Oklopcic, A., Moy, L., Ori, C.C., Erisir, A., and Jevtovic-Todorovic, V. (2013). Early exposure to general anesthesia disturbs mitochondrial fission and fusion in the developing rat brain. *Anesthesiology* 118 (5): 1086–1097.
124. Zhang, Y., Dong, Y., Wu, X., Lu, Y., Xu, Z., Knapp, A., Yue, Y., Xu, T., and Xie, Z. (2010). The mitochondrial pathway of anesthetic isoflurane-induced apoptosis. *J. Biol. Chem.* 285 (6): 4025–4037.
125. Cheng, Y. and Levy, R.J. (2014). Subclinical carbon monoxide limits apoptosis in the developing brain after isoflurane exposure. *Anesth. Analg.* 118 (6): 1284–1292.
126. Cheng, Y., Mitchell-Flack, M.J., Wang, A., and Levy, R.J. (2015). Carbon monoxide modulates cytochrome oxidase activity and oxidative stress in the developing murine brain during isoflurane exposure. *Free Radic. Biol. Med.* 86: 191–199.

17

Natural Products that Generate Carbon Monoxide: Chemistry and Nutritional Implications

Ladie Kimberly De La Cruz and Binghe Wang

Department of Chemistry, Center for Diagnostics and Therapeutics, Georgia State University, Atlanta, GA, USA

Introduction

Augmenting the nutrients derived from biosynthesis by the human body, the diet provides a myriad of compounds as sources or precursors of essential nutrients, including gaseous signaling molecules. For NO and H_2S, two well-known gaseous signaling molecules, their dietary, endogenous, and nonenzymatic sources are well documented [1–5]. For carbon monoxide (CO), the picture is limited and incomplete. Two most often cited sources of CO are inhalation from environment and through endogenous oxidative catabolism of heme. It is widely recognized that oxidative catabolism of heme by heme oxygenase (HMOX) is the major source of endogenous CO production accounting for around 88% [6]. Based on the quantification of bilirubin production, around 80% of heme-derived CO arises from red blood cell turnover, while the remainder is generated from catabolism of other heme-containing proteins [7,8]. Nonheme processes such as lipid peroxidation [9], photooxidation [10], xenobiotic degradation [11], and bacterial metabolism [12] are believed to account for the remaining percentage of endogenous CO production. Chapter 1 of this book describes in detail this subject.

Although considered as minor contributions, non-heme-derived CO production in the human body may potentially be important under conditions where O_2 and heme substrates are both limiting. As a biological mediator with roles in the regulation of tissue homeostasis, alternative local CO production aside from oxygen-dependent catabolism of heme is plausible when the HMOX activity is constrained by the limited availability of either O_2 or the heme substrate.

In this chapter, we discuss additional CO sources coming from (i) HO-1 induction of natural products, (ii) nonenzymatic decarbonylation reactions of both diet-derived and endogenously produced organic compounds, and (iii) microbiome-mediated enzymatic CO production from dietary or endogenous substrates.

CO from HMOX1 Induction of Natural Products

HMOX1 is one of the widely recruited cellular responses to perturbations of homeostasis [13]. In addition to heme, HMOX1's natural substrate, various cellular stimuli such as UV radiation [14], heat shock [15], hypoxia [16], and various chemical agents such as hydrogen peroxide [17], NO [18], heavy metals [19], and bacterial endotoxins [20], among many others, lead to increased expression of HMOX1. The induction of HMOX1 is a protective mechanism employed by cells in response to oxidative stress. Paradoxically, it is now increasingly recognized that HMOX1's cytoprotective role is in part mediated by its end products, CO and bilirubin (Chapter 10), previously regarded as merely toxic by-products [21].

Different inducers of HMOX1 regulate its expression mostly at the transcription level through a variety of signal transduction pathways and transcription factors via several response elements (Figure 17.1) [22,23]. The major upstream signaling pathways invoked by HO-1 inducers include mitogen-activated protein kinases (MAPKs) [24], protein kinase C (PKC) [25], 5'-AMP-activated protein kinase [26], and phosphoinositide 3-kinase/RAC-alpha serine/threonine protein kinase (PI3K/Akt) [27]. Signal is transduced through coordinated interaction between several transcription factors such as NF-E2-related factor-2

Carbon Monoxide in Drug Discovery: Basics, Pharmacology, and Therapeutic Potential, First Edition. Edited by Binghe Wang and Leo E. Otterbein.
© 2022 John Wiley & Sons, Inc. Published 2022 by John Wiley & Sons, Inc.

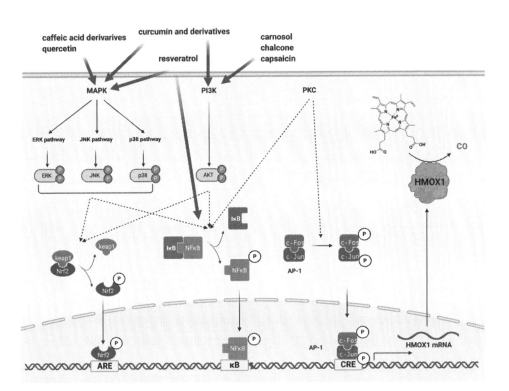

Figure 17.1 Natural product-based HMOX1 inducers set in motion protein phosphorylation-dependent signaling pathways that in turn activate a variety of transcription factors, ultimately leading to enhanced expression of HMOX1.

Table 17.1 Some examples of natural product-based HO-1 inducers that are present in food.

Natural product	Cell line/model	Concentration	Signaling pathways involved	Reference
Curcumin	Human nasal fibroblasts	5 μM	ERK MAPK	[35]
	H9c2 cells	15 μM	PI3K/Akt	[36]
	Lung mesenchymal stem cells	10 μM	PI3K/Akt	[37]
Bisdemethoxycurcumin	Primary mouse cardiomyocytes	100 μM	PI3K/Akt	[38]
Carnosol	PC12 cells	10 μM	PI3K/Akt	[39]
Resveratrol	Human aortic smooth muscle cells	1–10 μM	NF-κB	[40]
	PC12 neurons	15 μM	ERK and Akt	[41]
Piceatannol	Endothelial cells	10–50 μM	PKC and tyrosine kinase	[42]
Caffeic acid phenethyl ester	Murine macrophages	10 μM	P38 MAPK	[43]
Quercetin	Human hepatocytes	10–200 μM	P38 and ERK MAPK	[44]
	Rat aortic smooth muscle cells, human hepatocyte	50 μM	P38 MAPK	[45]
2′-Hydroxychalcone	RAW 264.7 macrophages	15 μM	PI3K/Akt	[46]
Capsaicin	HepG2 cells	200 μM	PI3K/Akt	[47]

(Nrf2) [24–26], activator protein-1 [28-30], nuclear factor-κB (NF-κB) [30], signal transducer and activator of transcription 3 [31], Yin Yang 1 [32], and hypoxia-inducible factor-1α [33]. Another route by which HO-1 can be induced is through the suppression of the activity of the HMOX1 transcriptional repressor protein, Bach1 [34].

Numerous natural products that are commonly found in the human diet have been shown to be inducers of HMOX1 (Table 17.1). By transitivity, these natural products may be regarded as sources of CO based on heme degradation. However, most of these studies do not really measure HMOX1 activity but instead quantify mRNA or HMOX1 abundance.

Curcumin, the main polyphenol present in the rhizome of *Curcuma* species, is reported to possess multiple health benefits and thus is widely incorporated in diets from different cultures, and even cosmetics [48]. Previously, curcumin has been shown to induce both HMOX1 expression and activity in a dose- and time-dependent manner through the Nrf2/ARE pathway in renal epithelial cells, hepatocytes, and smooth muscles [49,50]. More recently, curcumin has been shown to reduce ROS production from human nasal fibroblasts upon exposure to urban particulate matters. This was attributed to HO induction via the NRf2/ARE pathway [35]. In H_2O_2-induced apoptosis in H9c2 cells, the antiapoptotic activity of curcumin was demonstrated via induction of HO-1 through the PI3K/Akt signaling pathway [36]. A similar finding was presented in H_2O_2-mediated damage of lung mesenchymal stem cells [37]. Curcuminoid derivatives have also been shown to induce HO-1 expression through Nrf2 translocation via the PI3K/Akt pathway [38].

Resveratrol, a nonflavonoid polyphenolic natural product that is present in commonly consumed foods such as grapes, apples, blueberries, plums, and peanuts, has also been shown to induce HMOX1 expression [50]. At a concentration less than 15 μM, resveratrol was shown to induce HMOX1 expression in human aortic smooth muscle cells and PC12 cells [40,41]. There are a number of other natural products that have been shown to induce the expression of HMOX1. These were expertly and adequately reviewed by Ferrándiz and Devesa [50] and Kalergis and coworkers [51] and are not duplicated here.

While HMOX1 induction may be a viable way to elicit the beneficial effects of the enzyme in certain instances, it should be noted that this approach is severely limited by the local pool of heme substrate.

Therefore, there is an advantage to harness the cytoprotective effects of HMOX1 through its end product CO. For example, it has been shown that the anti-inflammatory effect of interleukin-10 is dependent on the induction of HMOX1 [13,52] and that the HMOX1 effect can be fulfilled by externally administered CO [52,53].

Sources of CO Through Nonenzymatic Pathways

In this section, we explore nonenzymatic sources of CO from natural products commonly found in the human diet.

Vitamin C

Ascorbic acid is a ketolactone. It can function as a reducing agent by undergoing two consecutive, one-electron oxidations, and the rate of autooxidation is affected by pH and presence of transition metals such as iron [54,55]. In the presence of molecular oxygen and ferric or cupric ions, ascorbic acid induces a "metal redox cycle" cascade producing superoxide anion and hydroxy radicals [56]. The generated superoxide anions interact with ferrous iron to produce hydrogen peroxide. It is through the further reaction of ascorbic acid with these generated highly reactive radicals that its autooxidation takes place. Several autooxidation products have been identified including 2,3-diketogulonic acid, L-threonolactone, threonic acid, and oxalic acid, among many others (Figure 17.2A), upon reaction of ascorbic acid with singlet oxygen [57,58]. The monomethoxy analogue of ascorbic acid was reported to undergo superoxide-mediated conversion to an oxyester with concomitant production of CO and carbon dioxide under base catalysis (Figure 17.2B) [59]. This was proposed to undergo conversion

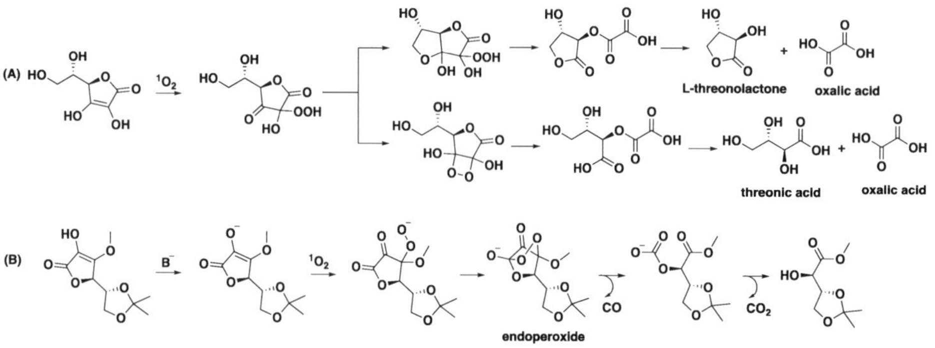

Figure 17.2 (A) Some autooxidation products of ascorbic acid in the presence of singlet oxygen [58]. (B) CO production from base-catalyzed oxidative degradation of monomethoxy ascorbic acid derivative [59].

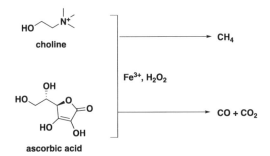

Figure 17.3 CO generation from ascorbic acid in the presence of choline, ferric ions, and H_2O_2.

through an endoperoxide intermediate that undergoes CO release followed by a decarboxylation.

In addition to these well-characterized oxidation products, there are indications that autooxidation of ascorbic acid also produces CO. In 1952, Sjöstrand reported the formation of CO from myoglobin solutions upon addition of ascorbic acid [60]. This was described as analogous to the decomposition of hemoglobin even though the data suggest no clear relationship between bile pigment formation and CO formation. In 2003, CO together with carbon dioxide and methane was detected from ascorbic acid under *in vitro* conditions in the presence of choline, hydrogen peroxide, and catalytic iron (Figure 17.3) [61]. Ascorbate (0.5 M) incubated along with $FeCl_3$ (10 mM) and increasing concentrations of H_2O_2 (0–2 M) produced a maximum of 100 µmol CO/total gas volume after 24 h of incubation. The formation of both CO and CO_2 was attributed to the autooxidation of ascorbic acid. No mechanism was provided, but it might be from the coupled decarboxylation and decarbonylation of a ketoacid intermediate or through an endoperoxide intermediate (see Figure 17.2). As an indirect indication of CO generation from ascorbate, analysis of the spectroscopic changes of myoglobin associated with the addition of sodium ascorbate as a preservative on tuna meat fish revealed very close similarity to that of CO-treated tuna myoglobin [62]. Similar to CO treatment, tuna meat color is preserved as fresh-looking upon treatment with ascorbate.

Phenols and Flavonoids

In 1970, Miyahara and Takahashi studied the autooxidation-induced evolution of CO from alkaline solutions of simple phenolic compounds such as pyrogallol, L-DOPA, hydroquinone, catechol, gallic acid, aminophenols, and resorcinol incubated at room temperature for 20 h [63]. Among the phenols tested, pyrogallol gave the highest volume of CO evolved (around 6% CO yield). L-DOPA, an intermediate in the melanogenesis pathway, gave 0.8% CO yield under *in vitro* conditions. In the presence of tyrosinase, an enzyme in the melanogenesis pathway that converts L-DOPA to dopaquinone, enzymatic oxidation of L-DOPA resulted in 0.4% of CO yield, a value around 10-fold higher compared to the heat-inactivated tyrosinase control. In all instances where CO was generated, an accompanying consumption of O_2 was also observed. In another study, CO production from L-DOPA decreased in the presence of antioxidants and radical scavengers such as superoxide dismutase (36 U/mL, −75%), ascorbic acid (10 µM, −87%), and α-tocopherol (1 mM, −55%) and increased in the presence of radical initiators such as *tert*-butyl hydroperoxide (10 mM, +77%) [64].

Flavonoid compounds, specially flavonols and their glycosides, are natural products that are ubiquitous in the human diet. These compounds exhibit a wide range of biological activities, including antioxidant, anti-inflammatory, antibacterial, and anticancer effects [65]. These compounds may be potential sources of CO through its oxidative nonenzymatic degradation. CO production is initiated upon base-catalyzed oxygenation of the enol form of quercetin forming a peroxo intermediate that cyclizes to form an oxoperoxide that eventually fragments to form the depside and CO (Figure 17.4) using DMF as solvent [66]. CO may also form via a radical mechanism involving a cyclic peroxo intermediate [67]. However, under milder aqueous conditions at pH ≈ 7–8, quercetin undergoes oxidative degradation preferentially via oxidative decarboxylation as opposed to the previously reported oxidative decarbonylation [68]. The mechanism proposed specifies two equivalents of oxygen to account for the formation of one equivalent of CO_2 and the same depside product. A key step in their proposed mechanism is the addition of oxygen across the α-diketone C–C bond to generate a dicarboxylic acid intermediate. This was rationalized by citing two literature precedents: (i) base-catalyzed cleavage of C–C bond in α-diketone via a ketyl anion radical intermediate [69,70] and (ii) superoxide anion radical-mediated cleavage of the α-diketone moiety of ascorbic acid [59].

α-Ketoacids

α-Ketoacids, specifically the ones derived from the deamination of naturally occurring α-amino acids, are ubiquitous and fulfill a variety of roles in

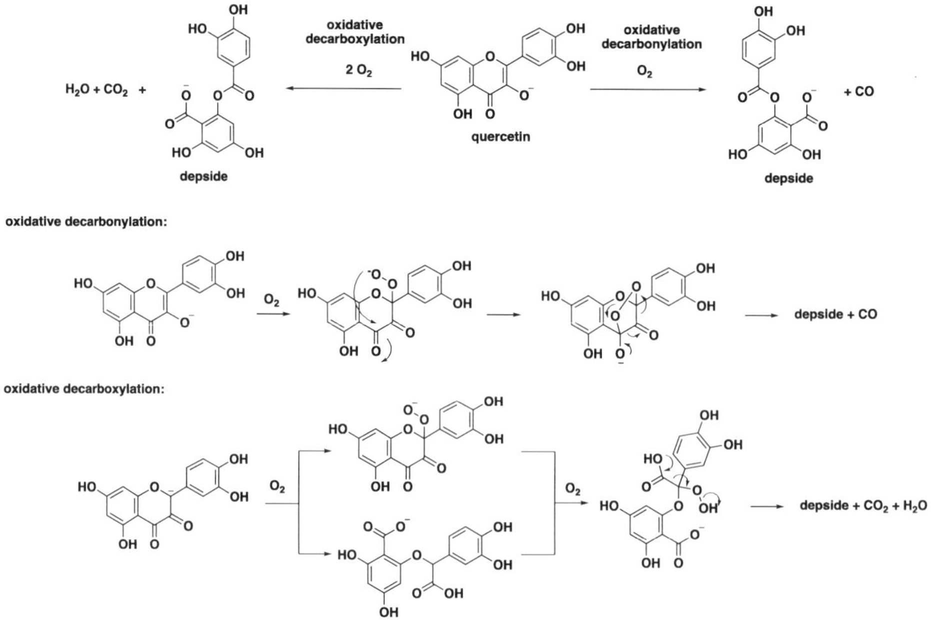

Figure 17.4 Proposed mechanisms for the oxidative degradation of quercetin. (A) base-catalyzed oxidation of quercetin to form depside and CO. [66]. (B) Oxidative decarboxylation to form depside, carbon dioxide, and water [68].

metabolism [71]. Chemically, this group of compounds is well known to undergo decarboxylation reactions. However, a number of reports have shown that aromatic α-ketoacids with β-hydrogens may also undergo decarbonylation reactions. The thermal decomposition of phenylpyruvic acid, an aromatic α-ketoacid, at temperatures between 250 and 500 °C yields 66% CO in addition to phenylacetic acid [72]. 4-Hydroxyphenylpyruvic acid, another aromatic α-ketoacid, is oxygenated in the enol form to yield 71% CO when boiled with the radical initiator azobisisobutyronitrile in acetonitrile at 78 °C for 6 h. When irradiated with visible light for 6 h in acetonitrile in the presence of the dye sensitizer, riboflavin, 74% yield of CO was obtained [73]. Though the CO production yields were clearly impressive, these conditions are not physiologically relevant. In 1987, Hino and Tauchi reported that several strains of bacteria produced additional amounts of CO that cannot be attributed to heme degradation by the HMOX system [74]. The additional amount of CO was determined to come from aromatic pyruvic acids, metabolites of aromatic amino acids added to the culture medium of *Morganella morganii*. Remarkably, aromatic pyruvic acids produced significant amounts of CO even in the absence of bacteria but only when hemin was present. Specifically, 53%, 38%, 16%, and 19% of CO were detected from phenylpyruvic acid, 4-hydroxyphenylpyruvic acid, indole pyruvic acid, and imidazole pyruvic acid, respectively, after 4-h incubation with 2.5 mol% of hemin at 30 °C in phosphate buffer.

There are several possible mechanism(s) of CO generation from arylpyruvic acids [73,75]. Arylpyruvic acids with a methylene carbon adjacent to the ketoacid moiety exist largely as the enol tautomer. This enol tautomer undergoes autoxidation to yield the unstable hydroperoxyl ketoacid intermediate (Figure 17.5). The hydroperoxyl ketoacid may decompose to yield CO, CO_2, and aromatic aldehyde either through linear fragmentation or via the cyclic intermediate, peroxylactone. A non-CO-producing pathway is also possible through the dioxetane route wherein the products are oxalic acid and aromatic aldehyde.

Dihydroxyacetone

1,3-Dihydroxyacetone (DHA), the simplest ketose, is an important glycolytic intermediate. DHA was reported to release CO *in vitro* in the presence of catalytic amounts of the Fenton reagent. Theoretical mechanistic studies using B3LYP/6-311++G(d,p), M06-2X/6-311++G(d,p), and CCSD(T)//M06-2X/6-311++G(d,p) level of theories in water with dielectric constant of 78.39 (using the polarized continuum model) suggested that DHA decomposes completely to form two equivalents of CO, one equivalent of CO_2, and six equivalents of water (Figure 17.6) [76]. Further analysis of probable reaction mechanisms revealed a fast pre-equilibrium between DHA and hydroxyl radical formed from the Fenton reaction followed by the abstraction of the two alcoholic protons of DHA by the hydroxy radical. The formed

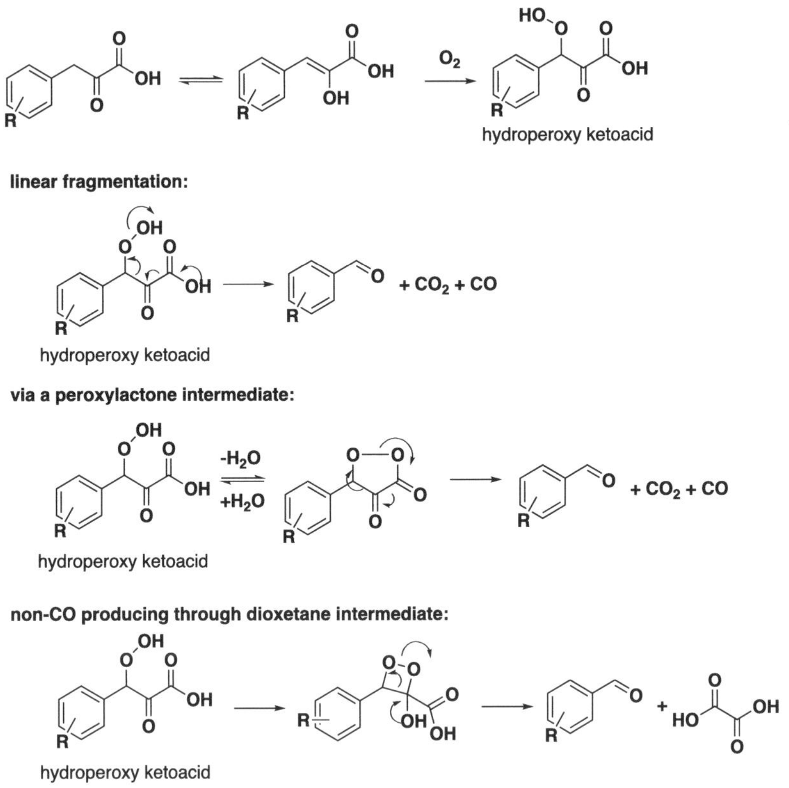

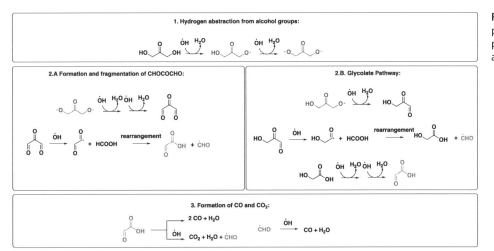

Figure 17.5 Postulated mechanisms of decomposition of hydroperoxyl ketoacid leading to CO or oxalic acid release.

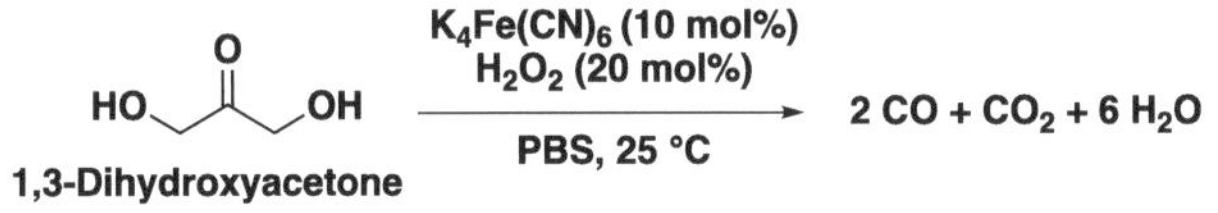

Figure 17.6 The Fenton reagent may catalyze the decomposition of 1,3-dihydroxyacetone to produce CO, CO_2, and water [76].

Figure 17.7 Proposed probable reaction pathways between DHA and hydroxyl radical [76].

dihydroxy DHA diradical undergoes a rotation and successive dehydration reactions to form the triketo intermediate (Figure 17.7). This intermediate then undergoes fragmentation to give one equivalent of COCHO radical and formic acid, followed by immediate rearrangement to yield the CHO radical and glyoxylic acid.

An alternative mechanism through the glycolate pathway is also postulated wherein dihydroxy DHA radical undergoes elimination of alpha hydrogen followed by a fragmentation and rearrangement to give glycolate and CHO radical. Then, two hydroxy radicals abstract two protons to form glyoxylic acid. The glyoxylic acid can then dissociate directly to CO and H_2O. Alternatively, the hydroxy radical can facilitate the dissociation of glyoxylic acid and the conversion of the unstable CHO radical to CO and H_2O.

The proposed mechanisms based on quantum mechanical calculations are partially supported by experimental results from measuring *in vitro* CO production using the myoglobin assay. However, the stoichiometry of the reaction was not established. Although this report requires further validation, its potential physiological impact is something to consider since both DHA and the Fenton reaction [77] are ubiquitous components in the biological milieu.

Sources of CO Through Microbiome Action Independent of Heme Degradation

Human colonic microbiota harbors more than 10^{11} organisms per gram of content [78]. Microbial HMOX or other heme-degrading enzymes are expressed by the microbiota for several reasons [79]. Therefore, in the presence of heme substrate, it is conceivable to have microbiome-derived CO production in the human gut. Aside from heme-degrading enzymes, there are a number of microbial enzymes that produce CO from the catabolism of ingested natural products.

Bacterial Metabolism of Flavonoids

Flavonoid pigments may be sources of CO. In 1959, Simpson et al. showed that CO is formed stoichiometrically from the enzymatic degradation of rutin by the mold species: *Pullularia fermentas, Aspergillus flavus*, and *A. niger* [80]. The enzymatic transformation required stoichiometric amounts of O_2, and forms stoichiometric amounts of CO arising from C-3 of rutin. Specifically, it was shown that rutin is degraded by *A. niger* to rutinose, phloroglucinol carboxylic acid, protocatechuic acid, and CO (Figure 17.8).

Pirins, a group of nonheme metalloproteins, are involved in many biological processes and are highly conserved in all taxa, including prokaryotic microorganisms, fungi, plants, and mammals [81]. In humans, pirin proteins are expressed in all tissues and act as regulators of transcription factors [82,83] such as in NF-κB wherein the oxidative state of the iron cofactor of hPirin (human Pirin) acts as a redox sensor [84]. YhhW, a Pirin homologue isolated from *Escherichia coli* [85], and hPirin were shown to possess quercetinase-like function [86]. These two pirin orthologues are capable of degrading a broad range of flavonol analogues under oxidative conditions to release CO. Strikingly, CO release was observed with the incubation of homogenized onion juice with pirin orthologues at both room temperature and 37 °C. The broad substrate specificity of pirin including flavonols that are commonly found in the human diet along with the ubiquity of the enzyme in human tissues and the microbiome suggests that this could be a significant source of endogenous CO, independent of heme catabolism.

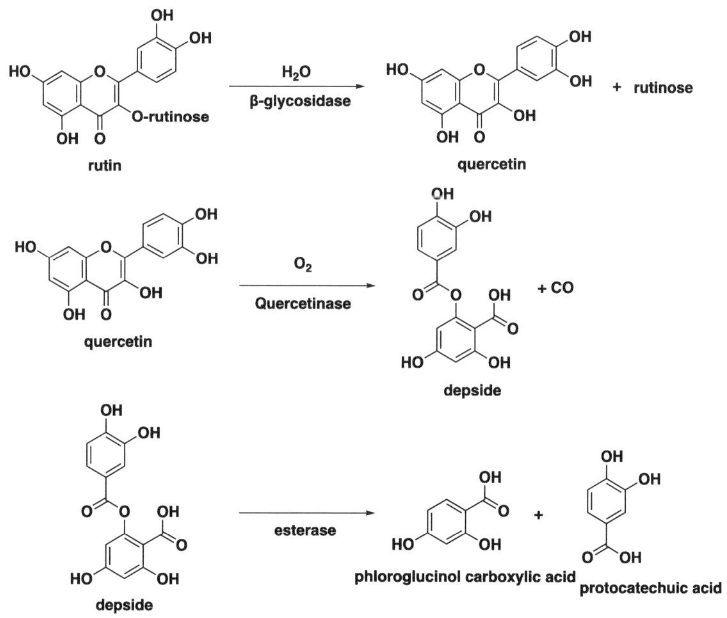

Figure 17.8 Degradation of rutin by A. niger to produce CO, protocatechuic acid, and phloroglucinol carboxylic acid.

It is interesting to note that pirin is a nuclear protein [87] and its substrate flavonols such as quercetin are also known to accumulate in the nucleus [88]. Such observations promote the notion that quercetin-associated biological effects may be exerted through CO production. As mentioned earlier, quercetin can also induce HO-1 to produce CO. However, under limiting heme pools, CO production through HO-1 induction may be compromised. Pirin-mediated degradation of quercetin to produce CO may be an alternative adaptive response to stress. However, there are cases wherein pirin-mediated degradation of quercetin blocks its biological effects such as antiviral activity. For example, quercetin derivatives have been shown to inhibit poliovirus RNA synthesis through inhibition of the PI3K-dependent Akt activation pathway. In the presence of a high intracellular concentration of pirin, quercetin's antiviral activity is modulated. The poliovirus replication in HeLa cells, characterized by higher expression levels of pirin, exhibited higher resistance to quercetin as opposed to normal kidney epithelial cells, wherein pirin expression level is lower [89].

Bacterial Metabolism of Acireductone from the Methionine Salvage Pathway

Acireductone dioxygenases (ARDs) belong to a structural class known as cupins and are recognized as "moonlighting" enzymes with functions involving regulation and participation in several processes such as carcinogenesis, tumor metastasis, hepatitis C infection, and Down's syndrome-associated congenital heart defects [90]. Initially discovered in *Klebsiella oxytoca*, two ARD isoforms were shown to have different chromatographic properties as well as enzymatic activities despite having identical polypeptide sequence only differing in the bound active site divalent transition metal ion [91,92]. ARD is the only known enzyme with regioselective oxidation based on the identity of the bound metal ion. Fe^{2+}-ARD in the presence of oxygen converts acireductone to the α-ketoacid precursor of methionine and formate. This conversion facilitates the recycling of methionine in the salvage pathway. On the other hand, Ni^{2+}-ARD incorporates oxygen and fragments acireductone to methylthiopropionate, formate, and CO, which funnels acireductone away from the methionine salvage pathway (Figure 17.9). A model reaction monitored by ^{18}O and ^{14}C tracer experiments revealed how these two isoforms differentially incorporate O_2 to fragment the acireductone based on the identity of the metal ion (Figure 17.9B). The binding of these different metals on the same protein ligands with the same pseudo-octahedral geometry propagates structural protein changes such as repacking of hydrophobic cores, instigating displacement of helices leading to differential active site accessibility. Either a five- or a four-membered cyclic peroxide intermediate is formed when bound to Ni^{2+} and Fe^{2+}, respectively, leading to different oxidative cleavage pathways for each.

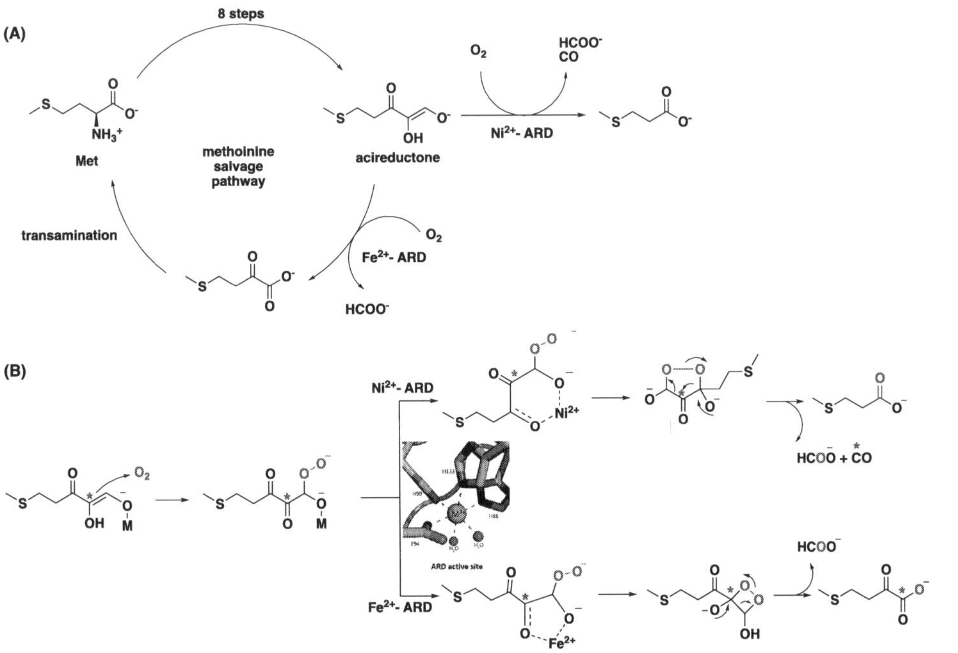

Figure 17.9 (A) CO is formed from acireductone, an intermediate in the methionine salvage pathway, through the action of Ni^{2+}-ARD. ARD bound to Fe^{2+} does not produce CO. (B) A postulated mechanism for oxygen incorporation and subsequent fragmentation based on ^{18}O and ^{14}C tracer experiments. Inset: ARD active site Reprinted with permission from Ref. [90]. Copyright (2017) American Chemical Society.

This interesting metal-dependent enzymatic activity was subsequently demonstrated *in vitro* in recombinant *Mus musculus* ARD [93]. Shortly after, a recombinant human equivalent was also identified, human ARD (HsARD), wherein Fe^{2+}-bound HsARD converts acireductone to the ketoacid intermediate for the methionine salvage pathway, while the Ni^{2+}, Co^{2+}, and Mn^{2+} forms catalyze the CO-producing pathway [94]. While endogenous CO production arising from ARD action has not been reported in normal mammalian tissues, significant CO production from human microbiota expressing ARDs might be a possibility. The CO-producing pathway of ARD is becoming of interest as it has been postulated to be cytoprotective based on the methionine dependence of many carcinogenic processes as well as CO's ability to modulate inflammation and apoptosis [95]. The ARD gene is either upregulated or downregulated in several carcinoma, and linked to the regulation of the activity of membrane-type 1 matrix metalloproteinase, an enzyme associated with tumor metastasis [96]. The ability to produce CO after simple metal exchanges with Ni, Co, and Mn has been postulated to protect tumor cells from apoptosis [95].

Bacterial Metabolism of Amino Acids Such as Tyrosine

Around 71% of the 341 gastrointestinal microbial species listed in the Human Microbiome Project Gastrointestinal Tract reference genome database encode for hydrogenases wherein 60% specifically encode [FeFe]-hydrogenases [78]. Hydrogenases allow microbes to either utilize molecular hydrogen as an energy source or convert excess reducing power as molecular hydrogen [97]. [FeFe]-hydrogenases contain a catalytic cofactor (H-cluster) that in turn is comprised of a 2Fe subcluster wherein the two Fe atoms are coordinated by the nonprotein ligands CO and cyanide (CN^-) for stabilization [98]. Microbes have evolved to biosynthesize these diatomic ligands. HydG, one of the radical *S*-adenosylmethionine (AdoMet) enzymes among the three accessory proteins necessary for the synthesis of the 2Fe subcluster, was reported to catalyze the formation of CO from tyrosine (Figure 17.10) [99]. CO has been detected using deoxyhemoglobin as a reporter. A radical AdoMet-mediated reaction mechanism was postulated through a dehydroglycine intermediate based on a prior report of its decarbonylation [100]. Another mechanism was put forward wherein the formation of CO and

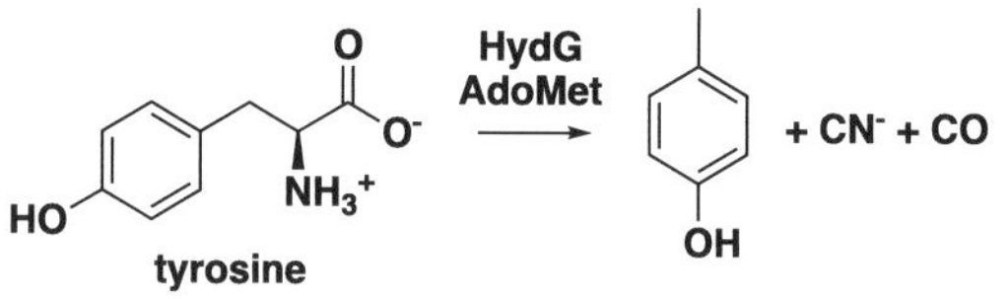

Figure 17.10 Formation of CO from tyrosine from the action of HydG, a radical AdoMet enzyme.

CN^- precursor was observed in the absence of the C-terminal [4Fe–4S] cluster indicating that the conversion does not require a metal activator [101]. It was hypothesized that the key intermediate is a glycyl radical generated from an energetically favorable homolytic cleavage of the tyrosine radical. Glycyl radical decarboxylation produces the reactive CO_2^- radical that has to be reduced by the [4Fe–4S] cluster to CO.

A detailed reaction mechanism was proposed wherein a radical SAM reaction at the N-terminal of the [4Fe–4S] cluster generates a tyrosine radical bound at the C-terminal end by the abstraction of the phenolic proton [102]. The tyrosine radical then undergoes heterolysis at the bond tethering the side chain to the alpha carbon to form a transient 4-oxidobenzyl radical and a dehydroglycine bound to the C-terminal [4Fe–4S] cluster (Figure 17.11). After electron and proton transfer, 4-oxidobenzyl radical is converted to *p*-cresol, while the dehydroglycine falls apart to Fe-bound CO and CN^-. It should be noted that the formed CO and CN^- are then utilized to generate the $Fe(CO)_2CN$ complex, which is subsequently shuttled into the hydrogenase maturation sequence [103]. The CO from tyrosine cleavage by HydG is immediately iron bound and not in the free form.

In the active site of another hydrogenase, [NiFe] hydrogenase, one CO ligand together with two CN^- ligands is needed. HypX (hydrogenase pleiotropic maturation X), a maturation protein for [NiFe] hydrogenase, is required for the synthesis of CO under aerobic conditions. CO is generated from N^{10}-formyltetrahydrofolate as the substrate wherein CO is formed from the decarbonylation of a formyl CoA intermediate (Figure 17.12) [104,105]. HypX catalysis is proposed to occur in two steps by two HypX modules. First, the N-terminal module transfers the formyl group from the substrate N^{10}-formyltetrahydrofolate to CoA producing formyl CoA. The C-terminal module then converts formyl CoA to CO with the regeneration of CoA. HypX is the first identified CO-producing enzyme that does not require a metal cofactor.

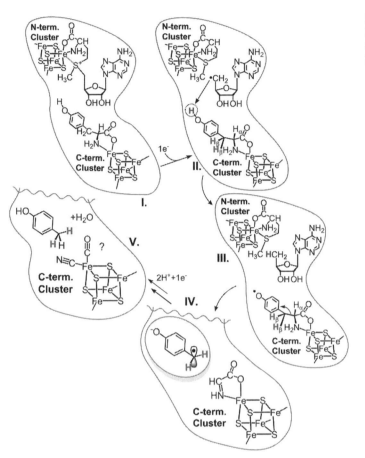

Figure 17.11 Proposed mechanism for the HydG-catalyzed production of Fe-bound CO and CN⁻ ligands from free tyrosine.
Figure reproduced with permission from Ref. [102].

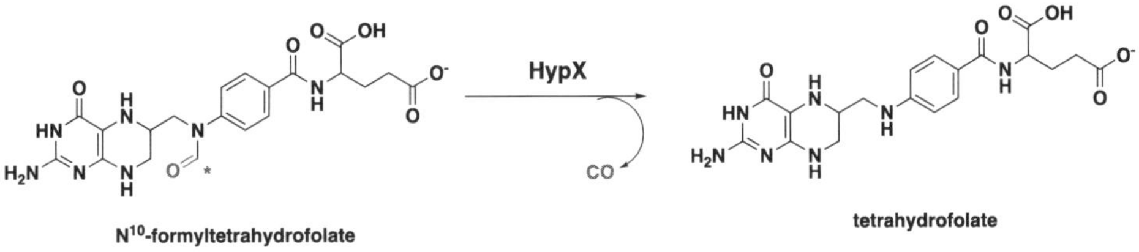

Figure 17.12 HypX catalyze the decarbonylation of N^{10}-formyltetrahydrofolate.

Bacterial Metabolism of Intermediates in Nucleotide Biosynthesis

Thiamin-pyrimidine (HMP-P) synthase is one of the enzymes in the *de novo* synthesis of thiamine, an essential cofactor for many enzymes such as pyruvate dehydrogenase [106]. It catalyzes the rearrangement of 5-aminoimidazole ribotide (AIR) to hydroxymethyl pyrimidine phosphate (HMP-P). This enzymatic reaction was demonstrated to produce formate and CO from C1′ and C3′, respectively, along with HMP-P and 5′-deoxyadenosine (Figure 17.13) [107]. HMP-P synthase, a [4Fe–4S] cluster protein and a radical SAM enzyme, upon treatment with SAM and under reducing conditions, generates an organic radical that is postulated to react with AIR as substrate through a series of iterative hydrogen atom abstractions.

Bacterial Metabolism of Glucose Metabolites

In *Desulfovibrio vulgaris*, an alternative route to pyruvate fermentation with concomitant generation of hydrogen or CO was reported (Figure 17.14) [108]. Using lactate (38 mM) as an electron donor and sulfate as an electron sink, a CO burst of 6000 ppm (0.24 mM) was observed in hyd mutants where genes for Fe-only hydrogenases were deleted. Similarly, using pyruvate (30 mM) as substrate, the hyd mutant produced 4000 ppm of CO (0.16 mM). However, it should be noted that

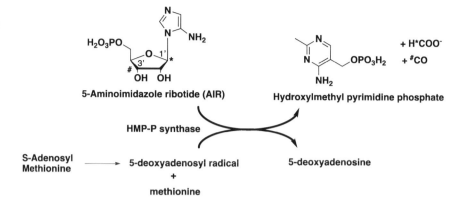

Figure 17.13 AIR undergoes a complex rearrangement reaction catalyzed by thiamin-pyrimidine (HMP-P) synthase to yield CO, formate, and hydroxymethyl pyrimidine phosphate.

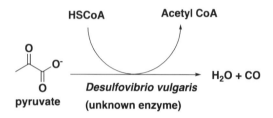

Figure 17.14 Pyruvate fermentation in *Desulfovibrio vulgaris* may lead to formation of CO through the action of an unknown enzyme.

despite the significant production of CO, its proportion relative to the starting concentration of the substrate and the amount of acetate formed is very small. In wild-type *D. vulgaris*, the steady-state concentration of CO is at around 4 μM due to its conversion by the cytoplasmic CODH and CO-dependent hydrogenase to CO_2 and H_2. Aside from hydrogen cycling, the significance of CO production during growth on lactate or pyruvate in *D. vulgaris* is not clear. Interestingly, Fe-only hydrogenase is inhibited by 0.1 μM CO [109]. On the other hand, CO is also required as a nonprotein ligand in the active site of hydrogenases.

Other Microbial CO-Generating Enzymes

There are other microbial CO-generating enzymes such bacterial 2,4-dioxygenases [110–112], methane monooxygenase [113], and CO dehydrogenase [114], but are not discussed here since their substrates are not commonly ingested in the diet or formed endogenously in the human body.

Conclusions

The enzyme HMOX is recognized as the major source of CO in the human body. As is becoming evident, CO has pathophysiological roles encompassing many biological processes [115]. However, CO production from HMOX action is contingent on the availability of heme as a substrate. Therefore, the question of alternative biological sources of CO arises, even in the context of its pathophysiological significance.

Despite the low CO yields from nonenzymatic pathways, compounds such as vitamin C and flavonoids from the diet, amino acid derivatives, and glycolytic intermediates are ubiquitous. Therefore, it might be worth re-examining their secondary pathophysiological effects in the context of CO production.

The other often overlooked CO source is the microbiome specifically in the gut. CO is not among the major gases detected in the gut. However, there are indications that CO may be a messenger molecule utilized among the microbiome as well as within the human host [79]. The gut microbiome may be a significant but underexplored source of CO. Interestingly, survey of the literature reveals that some routes to produce CO from natural products and biologically relevant substrates require the activation of molecular oxygen in the presence of transition metals either in the free form or bound in metalloenzymes. Therefore, the availability of oxygen in the gut is a consideration when exploring these possibilities.

Acknowledgments

CO-related work in the Wang lab was partially supported by a grant from the National Institutes of Health (R01DK119202). We also acknowledge the support of the Georgia Research Alliance for an Eminent Scholar endowment, the Center for Diagnostics and Therapeutics for a graduate fellowship to L.K.D.L.C., and internal financial sources at Georgia State University. Figure 17.1 was created using Biorender.com.

References

1. Kobayashi, J., Ohtake, K., and Uchida, H. (2015). NO-rich diet for lifestyle-related diseases. *Nutrients* 7: 4911–4937.
2. Kurtz, T.W., DiCarlo, S.E., Pravenec, M., and Morris, R.C. (2018). Functional foods for augmenting nitric oxide activity and reducing the risk for salt-induced hypertension and cardiovascular disease in Japan. *J. Cardiol.* 72: 42–49.
3. Zand, J., Lanza, F., Garg, H.K., and Bryan, N.S. (2011). All-natural nitrite and nitrate containing dietary supplement promotes nitric oxide production and reduces triglycerides in humans. *Nutr. Res.* 31: 262–269.
4. Carbonero, F., Benefiel, A., Alizadeh-Ghamsari, A., and Gaskins, H.R. (2012). Microbial pathways in colonic sulfur metabolism and links with health and disease. *Front. Physiol.* 3.
5. Wang, R. (2012). Physiological implications of hydrogen sulfide: A whiff exploration that blossomed. *Physiol. Rev.* 92: 791–896.
6. Berk, P.D., Rodkey, F.L., Blaschke, T.F., Collison, H.A., and Waggoner, J.G. (1974). Comparison of plasma bilirubin turnover and carbon monoxide production in man. *J. Lab. Clin. Med.* 83: 29–37.
7. Berk, P.D., Blaschke, T.F., Scharschmidt, B.F., Waggoner, J.G., and Berlin, N.I. (1976). A new approach to quantitation of the various sources of bilirubin in man. *J. Lab. Clin. Med.* 87: 767–780.
8. Vreman, H.J., Wong, R.J., Kadotani, T., and Stevenson, D.K. (2005). Determination of carbon monoxide (CO) in rodent tissue: effect of heme administration and environmental CO exposure. *Anal. Biochem.* 341: 280–289.
9. Vreman, H.J., Wong, R.J., Sanesi, C.A., Dennery, P.A., and Stevenson, D.K. (1998). Simultaneous production of carbon monoxide and thiobarbituric acid reactive substances in rat tissue preparations by an iron-ascorbate system. *Can. J. Physiol. Pharmacol.* 76: 1057–1065.
10. Vremana, H.J., Gillman, M.J., Downum, K.R., and Stevenson, D.K. (1990). In vitro generation of carbon monoxide from organic molecules and synthetic metalloporphyrins mediated by light. *Dev. Pharmacol. Ther.* 15: 112–124.
11. Kurppa, K., Kivistö, H., and Vainio, H. (1981). Dichloromethane and carbon monoxide inhalation: Carboxyhemoglobin addition, and drug metabolizing enzymes in rat. *Int. Arch. Occup. Environ. Health* 49: 83–87.
12. Engel, R.R., Matsen, J.M., Chapman, S.S., and Schwartz, S. (1972). Carbon monoxide production from heme compounds by bacteria. *J. Bacteriol.* 112: 1310–1315.
13. Soares, M.P. and Bach, F.H. (2009). Heme oxygenase-1: From biology to therapeutic potential. *Trends. Mol. Med.* 15: 50–58.
14. Nisar, M.F., Parsons, K.S.G., Bian, C.X., and Zhong, J.L. (2015). UVA irradiation induced heme oxygenase-1: A novel phototherapy for morphea. *Photochem. Photobiol.* 91: 210–220.
15. Ewing, J.F. and Maines, M.D. (1991). Rapid induction of heme oxygenase 1 mRNA and protein by hyperthermia in rat brain: heme oxygenase 2 is not a heat shock protein. *Proc. Natl. Acad. Sci. U. S. A.* 88: 5364–5368.
16. Panchenko, M.V., Farber, H.W., and Korn, J.H. (2000). Induction of heme oxygenase-1 by hypoxia and free radicals in human dermal fibroblasts. *Am. J. Physiol. Cell Physiol.* 278: C92–c101.
17. Cisowski, J., Loboda, A., Józkowicz, A., Chen, S., Agarwal, A., and Dulak, J. (2005). Role of heme oxygenase-1 in hydrogen peroxide-induced VEGF synthesis: effect of HO-1 knockout. *Biochem. Biophys. Res. Commun.* 326: 670–676.
18. Durante, W., Kroll, M.H., Christodoulides, N., Peyton, K.J., and Schafer, A.I. (1997). Nitric oxide induces heme oxygenase-1 gene expression and carbon monoxide production in vascular smooth muscle cells. *Circ. Res.* 80: 557–564.
19. Suzuki, H., Tashiro, S., Sun, J., Doi, H., Satomi, S., and Igarashi, K. (2003). Cadmium induces nuclear export of bach1, a transcriptional repressor of heme oxygenase-1 gene. *J. Biol. Chem.* 278: 49246–49253.
20. Carraway, M.S., Ghio, A.J., Taylor, J.L., and Piantadosi, C.A. (1998). Induction of ferritin and heme oxygenase-1 by endotoxin in the lung. *Am. J. Physiol.* 275: L583–92.
21. Ryter, S.W., Otterbein, L.E., Morse, D., and Choi, A.M.K. (2002). Heme oxygenase/carbon monoxide signaling pathways: Regulation and functional significance. *Mol. Cell Biochem.* 234–235: 249–263.
22. Gozzelino, R., Jeney, V., and Soares, M.P. (2010). Mechanisms of cell protection by heme oxygenase-1. *Annu. Rev. Pharmacol. Toxicol.* 50: 323–354.
23. Ayer, A., Zarjou, A., Agarwal, A., and Stocker, R. (2016). Heme oxygenases in cardiovascular health and disease. *Physiol. Rev.* 96: 1449–1508.
24. Naidu, S., Vijayan, V., Santoso, S., Kietzmann, T., and Immenschuh, S. (2009). Inhibition and genetic deficiency of p38 MAPK up-regulates heme oxygenase-1 gene expression via Nrf2. *J. Immunol.* 182: 7048–7057.
25. Rushworth, S.A., Chen, X.-L., Mackman, N., Ogborne, R.M., and O'Connell, M.A. (2005). Lipopolysaccharide-induced heme oxygenase-1

expression in human monocytic cells is mediated via Nrf2 and protein kinase C. *J. Immunol.* 175: 4408–4415.

26. Liu, X.-M., Peyton, K.J., Shebib, A.R., Wang, H., Korthuis, R.J., and Durante, W. (2011). Activation of AMPK stimulates heme oxygenase-1 gene expression and human endothelial cell survival. *Am. J. Physiol. Heart Circ. Physiol.* 300: H84–H93.

27. Shi, J., Yu, J., Zhang, Y., Wu, L., Dong, S., Wu, L., Wu, L., Du, S., Zhang, Y., and Ma, D. (2019). PI3K/Akt pathway-mediated HO-1 induction regulates mitochondrial quality control and attenuates endotoxin-induced acute lung injury. *Lab. Investig.* 99: 1795–1809.

28. Inamdar, N.M., Ahn, Y.I., and Alam, J. (1996). The heme-responsive element of the mouse heme oxygenase-1 gene is an extended AP-1 binding site that resembles the recognition sequences for MAF and NF-E2 transcription factors. *Biochem. Biophys. Res. Commun.* 221: 570–576.

29. Camhi, S.L., Alam, J., Otterbein, L., Sylvester, S.L., and Choi, A.M. (1995). Induction of heme oxygenase-1 gene expression by lipopolysaccharide is mediated by AP-1 activation. *Am. J. Respir. Cell Mol. Biol.* 13: 387–398.

30. Lavrovsky, Y., Schwartzman, M.L., Levere, R.D., Kappas, A., and Abraham, N.G. (1994). Identification of binding sites for transcription factors NF-kappa B and AP-2 in the promoter region of the human heme oxygenase 1 gene. *Proc. Natl. Acad. Sci. U.S.A.* 91: 5987–5991.

31. Tron, K., Samoylenko, A., Musikowski, G., Kobe, F., Immenschuh, S., Schaper, F., Ramadori, G., and Kietzmann, T. (2006). Regulation of rat heme oxygenase-1 expression by interleukin-6 via the Jak/STAT pathway in hepatocytes. *J. Hepatol.* 45: 72–80.

32. Beck, K., Wu, B.J., Ni, J., Santiago, F.S., Malabanan, K.P., Li, C., Wang, Y., Khachigian, L.M., and Stocker, R. (2010). Interplay between heme oxygenase-1 and the multifunctional transcription factor Yin Yang 1 in the inhibition of intimal hyperplasia. *Circ. Res.* 107: 1490–1497.

33. Dawn, B. and Bolli, R. (2005). HO-1 induction by HIF-1: a new mechanism for delayed cardioprotection? *Am. J. Physiol. Heart Circ. Physiol.* 289: H522–4.

34. Sun, J., Hoshino, H., Takaku, K., Nakajima, O., Muto, A., Suzuki, H., Tashiro, S., Takahashi, S., Shibahara, S., Alam, J., Taketo, M.M., Yamamoto, M., and Igarashi, K. (2002). Hemoprotein Bach1 regulates enhancer availability of heme oxygenase-1 gene. *EMBO J.* 21: 5216–5224.

35. Kim, J.-S., Oh, J.-M., Choi, H., Kim, S.W., Kim, S.W., Kim, B.G., Cho, J.H., Lee, J., and Lee, D.C. (2020). Activation of the Nrf2/HO-1 pathway by curcumin inhibits oxidative stress in human nasal fibroblasts exposed to urban particulate matter. *BMC Complement. Med. Ther.* 20: 101–101.

36. Yang, X., Jiang, H., and Shi, Y. (2017). Upregulation of heme oxygenase-1 expression by curcumin conferring protection from hydrogen peroxide-induced apoptosis in H9c2 cardiomyoblasts. *Cell Biosci.* 7: 20–20.

37. Ke, S., Zhang, Y., Lan, Z., Li, S., Zhu, W., and Liu, L. (2020). Curcumin protects murine lung mesenchymal stem cells from H_2O_2 by modulating the Akt/Nrf2/HO-1 pathway. *J. Int. Med. Res.* 48: 300060520910665–300060520910665.

38. Li, X., Huo, C., Xiao, Y., Xu, R., Liu, Y., Jia, X., and Wang, X. (2019). Bisdemethoxycurcumin protection of cardiomyocyte mainly depends on Nrf2/HO-1 activation mediated by the PI3K/AKT pathway. *Chem. Res. Toxicol.* 32: 1871–1879.

39. Martin, D., Rojo, A.I., Salinas, M., Diaz, R., Gallardo, G., Alam, J., De Galarreta, C.M., and Cuadrado, A. (2004). Regulation of heme oxygenase-1 expression through the phosphatidylinositol 3-kinase/Akt pathway and the Nrf2 transcription factor in response to the antioxidant phytochemical carnosol. *J. Biol. Chem.* 279: 8919–8929.

40. Juan, S.-H., Cheng, T.-H., Lin, H.-C., Chu, Y.-L., and Lee, W.-S. (2005). Mechanism of concentration-dependent induction of heme oxygenase-1 by resveratrol in human aortic smooth muscle cells. *Biochem. Pharmacol.* 69: 41–48.

41. Chen, C.Y., Jang, J.H., Li, M.H., and Surh, Y.J. (2005). Resveratrol upregulates heme oxygenase-1 expression via activation of NF-E2-related factor 2 in PC12 cells. *Biochem. Biophys. Res. Commun.* 331: 993–1000.

42. Wung, B.S., Hsu, M.C., Wu, C.C., and Hsieh, C.W. (2006). Piceatannol upregulates endothelial heme oxygenase-1 expression via novel protein kinase C and tyrosine kinase pathways. *Pharmacol. Res.* 53: 113–122.

43. Stähli, A., Maheen, C.U., Strauss, F.J., Eick, S., Sculean, A., and Gruber, R. (2019). Caffeic acid phenethyl ester protects against oxidative stress and dampens inflammation via heme oxygenase 1. *Int. J. Oral Sci.* 11: 6–6.

44. Yao, P., Nussler, A., Liu, L., Hao, L., Song, F., Schirmeier, A., and Nussler, N. (2007). Quercetin protects human hepatocytes from ethanol-derived oxidative stress by inducing heme oxygenase-1 via the MAPK/Nrf2 pathways. *J. Hepatol.* 47: 253–261.

45. Lin, H.C., Cheng, T.H., Chen, Y.C., and Juan, S.H. (2004). Mechanism of heme oxygenase-1 gene induction by quercetin in rat aortic smooth muscle cells. *Pharmacology* 71: 107–112.

46. Abuarqoub, H., Foresti, R., Green, C.J., and Motterlini, R. (2006). Heme oxygenase-1 mediates the anti-inflammatory actions of 2′-hydroxychalcone in RAW 264.7 murine macrophages. *Am. J. Physiol. Cell Physiol.* 290: C1092–C1099.
47. Joung, E.J., Li, M.H., Lee, H.G., Somparn, N., Jung, Y.S., Na, H.K., Kim, S.H., Cha, Y.N., and Surh, Y.J. (2007). Capsaicin induces heme oxygenase-1 expression in HepG2 cells via activation of PI3K-Nrf2 signaling: NAD(P)H: quinoneoxidoreductase as a potential target. *Antioxid. Redox Signal.* 9: 2087–2098.
48. Hewlings, S.J. and Kalman, D.S. (2017). Curcumin: A review of its effects on human health. *Foods* 6: 92.
49. Balogun, E., Hoque, M., Gong, P., Killeen, E., Green, C.J., Foresti, R., Alam, J., and Motterlini, R. (2003). Curcumin activates the haem oxygenase-1 gene via regulation of Nrf2 and the antioxidant-responsive element. *Biochem. J.* 371: 887–895.
50. Ferrándiz, M.L. and Devesa, I. (2008). Inducers of heme oxygenase-1. *Curr. Pharm. Des.* 14: 473–486.
51. Funes, S.C., Rios, M., Fernández-Fierro, A., Covián, C., Bueno, S.M., Riedel, C.A., Mackern-Oberti, J.P., and Kalergis, A.M. (2020). Naturally derived heme-oxygenase 1 inducers and their therapeutic application to immune-mediated diseases. *Front. Immunol.* 11.
52. Lee, T.S. and Chau, L.Y. (2002). Heme oxygenase-1 mediates the anti-inflammatory effect of interleukin-10 in mice. *Nat. Med.* 8: 240–246.
53. Otterbein, L.E., Bach, F.H., Alam, J., Soares, M., Tao Lu, H., Wysk, M., Davis, R.J., Flavell, R.A., and Choi, A.M.K. (2000). Carbon monoxide has anti-inflammatory effects involving the mitogen-activated protein kinase pathway. *Nat. Med.* 6: 422–428.
54. Buettner, G.R. and Jurkiewicz, B.A. (1996). Catalytic metals, ascorbate and free radicals: combinations to avoid. *Radiat. Res.* 145: 532–541.
55. Du, J., Cullen, J.J., and Buettner, G.R. (2012). Ascorbic acid: chemistry, biology and the treatment of cancer. *Biochim. Biophys. Acta* 1826: 443–457.
56. Harel, S. (1994). Oxidation of ascorbic acid and metal ions as affected by NaCl. *J. Agric. Food Chem.* 42 (11): 2402–2406.
57. Tikekar, R.V., Anantheswaran, R.C., Elias, R.J., and LaBorde, L.F. (2011). Ultraviolet-induced oxidation of ascorbic acid in a model juice system: Identification of degradation products. *J. Agric. Food Chem.* 59: 8244–8248.
58. Miyake, N., Otsuka, Y., and Kurata, T. (1997). Autoxidation reaction mechanism for l-ascorbic acid in methanol without metal ion catalysis. *Biosci. Biotechnol. Biochem.* 61: 2069–2075.
59. Frimer, A.A. and Gilinsky-Sharon, P. (1995). Reaction of superoxide with ascorbic acid derivatives: Insight into the superoxide-mediated oxidation of dehydroascorbic acid. *J. Org. Chem.* 60: 2796–2801.
60. Sjöstrand, T. (1952). Formation of carbon monoxide by coupled oxidation of myoglobin with ascorbic acid. *Acta Physiol. Scan.* 26: 334–337.
61. Ghyczy, M., Torday, C., and Boros, M. (2003). Simultaneous generation of methane, carbon dioxide, and carbon monoxide from choline and ascorbic acid - a defensive mechanism against reductive stress? *FASEB J.* 17: 1124–1126.
62. Howes, B.D., Milazzo, L., Droghetti, E., Nocentini, M., and Smulevich, G. (2019). Addition of sodium ascorbate to extend the shelf-life of tuna meat fish: a risk or a benefit for consumers? *J. Inorg. Biochem.* 200: 110813.
63. Miyahara, S. and Takahasi, H. (1971). Biological CO evolution: Carbon monoxide evolution during auto- and enzymatic oxidation of phenols. *J. Biochem.* 69: 231–233.
64. Cooper, M.J. and Engel, R.R. (1991). Carbon monoxide production from L-3,4-dihydroxyphenylalanine; a method for assessing the oxidant/antioxidant properties of drugs. *Clin. Chim. Acta* 202: 105–109.
65. Jucá, M.M., Cysne Filho, F.M.S., De Almeida, J.C., Mesquita, D.D.S., Barriga, J.R.M., Dias, K.C.F., Barbosa, T.M., Vasconcelos, L.C., Leal, L., Ribeiro, J.E., and Vasconcelos, S.M.M. (2020). Flavonoids: biological activities and therapeutic potential. *Nat. Prod. Res.* 34: 692–705.
66. Brown, S.B., Rajananda, V., Holroyd, J.A., and Evans, E.G. (1982). A study of the mechanism of quercetin oxygenation by 18O labelling. A comparison of the mechanism with that of haem degradation. *Biochem. J.* 205: 239–244.
67. Krishnamachari, V., Levine, L.H., Zhou, C., and Paré, P.W. (2004). In vitro flavon-3-ol oxidation mediated by a B ring hydroxylation pattern. *Chem. Res. Toxicol.* 17: 795–804.
68. Zenkevich, I.G., Eshchenko, A.Y., Makarova, S.V., Vitenberg, A.G., Dobryakov, Y.G., and Utsal, V.A. (2007). Identification of the products of oxidation of quercetin by air oxygen at ambient temperature. *Molecules (Basel, Switzerland)* 12: 654–672.
69. Korobitsina, I., Gurevich, L., and Rodina, L. (1969). Chemistry of ketones of furanidine series. IX. alkaline cleavage of 2, 2, 5, 5-Tetraalkyl-3, 4-furanidindiones. *Russ. J. Org. Chem* 5: 567–570.
70. Kutnevich, A., Rudenko, A., Rodina, L., and Pragst, F. (1978). Anion-radicals from cyclic diketones with alkyl substituents. *Russ. J. Org. Chem.* 14: 1343–1343.

71. Cooper, A.J.L., Ginos, J.Z., and Meister, A. (1983). Synthesis and properties of the .alpha.-keto acids. *Chem. Rev.* 83: 321–358.

72. Hurd, C.D. and Raterink, H.R. (1934). The decomposition of alpha keto acids. *J. Am. Chem. Soc.* 56: 1348–1350.

73. Jefford, C.W., Knoepfel, W., and Cadby, P.A. (1978). Oxygenation of 3-aryl-2-hydroxyacrylic acids. The question of linear fragmentation vs. cyclization and cleavage of intermediates. *J. Am. Chem. Soc.* 100: 6432–6436.

74. Hino, S. and Tauchi, H. (1987). Production of carbon monoxide from aromatic amino acids by Morganella morganii. *Arch. Microbiol.* 148: 167–171.

75. Zinner, K., Vidigal-Martinelli, C., Durán, N., Marsaioli, A.J., and Cilento, G. (1980). A new source of carbon oxides in biochemical systems. Implications regarding dioxetane intermediates. *Biochem. Biophys. Res. Commun.* 92: 32–37.

76. Sadhukhan, T., Das, D., Kalekar, P., Avasare, V., and Pal, S. (2017). Fenton's reagent catalyzed release of carbon monooxide from 1,3-Dihydroxy acetone. *J. Phys. Chem. A* 121: 4569–4577.

77. Enami, S., Sakamoto, Y., and Colussi, A.J. (2014). Fenton chemistry at aqueous interfaces. *Proc. Natl. Acad. Sci. U.S.A.* 111: 623–628.

78. Wolf, P.G., Biswas, A., Morales, S.E., Greening, C., and Gaskins, H.R. (2016). H2 metabolism is widespread and diverse among human colonic microbes. *Gut. Microbes* 7: 235–245.

79. Hopper, C.P., De La Cruz, L.K., Lyles, K.V., Wareham, L.K., Gilbert, J.A., Eichenbaum, Z., Magierowski, M., Poole, R.K., Wollborn, J., and Wang, B. (2020). Role of carbon monoxide in host–gut microbiome communication. *Chem. Rev.* 120 (24): 13273–13311.

80. Simpson, F.J., Talbot, G., and Westlake, D.W.S. (1960). Production of carbon monoxide in the enzymatic degradation of rutin. *Biochem. Biophys. Res. Commun.* 2: 15–18.

81. Soo, P.C., Horng, Y.T., Lai, M.J., Wei, J.R., Hsieh, S.C., Chang, Y.L., Tsai, Y.H., and Lai, H.C. (2007). Pirin regulates pyruvate catabolism by interacting with the pyruvate dehydrogenase E1 subunit and modulating pyruvate dehydrogenase activity. *J. Bacteriol.* 189: 109–118.

82. Pang, H., Bartlam, M., Zeng, Q., Miyatake, H., Hisano, T., Miki, K., Wong, -L.-L., Gao, G.F., and Rao, Z. (2004). Crystal structure of human pirin: An iron-binding nuclear protein and transcription cofactor. *J. Biol. Chem.* 279: 1491–1498.

83. Dechend, R., Hirano, F., Lehmann, K., Heissmeyer, V., Ansieau, S., Wulczyn, F.G., Scheidereit, C., and Leutz, A. (1999). The Bcl-3 oncoprotein acts as a bridging factor between NF-κB/Rel and nuclear co-regulators. *Oncogene* 18: 3316–3323.

84. Liu, F., Rehmani, I., Esaki, S., Fu, R., Chen, L., De Serrano, V., and Liu, A. (2013). Pirin is an iron-dependent redox regulator of NF-κB. *Proc. Natl. Acad. Sci. U.S.A.* 110: 9722–9727.

85. Adams, M. and Jia, Z. (2005). Structural and biochemical analysis reveal pirins to possess quercetinase activity. *J. Biol. Chem.* 280: 28675–28682.

86. Guo, B., Zhang, Y., Hicks, G., Huang, X., Li, R., Roy, N., and Jia, Z. (2019). Structure-dependent modulation of substrate binding and biodegradation activity of pirin proteins toward plant flavonols. *ACS Chem. Biol.* 14: 2629–2640.

87. Wendler, W.M.F., Kremmer, E., Förster, R., and Winnacker, E.-L. (1997). Identification of pirin, a novel highly conserved nuclear protein. *J. Biol. Chem.* 272: 8482–8489.

88. Notas, G., Nifli, A.-P., Kampa, M., Pelekanou, V., Alexaki, V.-I., Theodoropoulos, P., Vercauteren, J., and Castanas, E. (2012). Quercetin accumulates in nuclear structures and triggers specific gene expression in epithelial cells. *J. Nutr. Biochem.* 23: 656–666.

89. Neznanov, N., Kondratova, A., Chumakov, K., Neznanova, L., Kondratov, R., Banerjee, A., and Gudkov, A. (2008). Quercetinase pirin makes poliovirus replication resistant to flavonoid quercetin. *DNA Cell. Biol.* 27: 191–198.

90. Deshpande, A.R., Pochapsky, T.C., and Ringe, D. (2017). The metal drives the chemistry: Dual functions of acireductone dioxygenase. *Chem. Rev.* 117: 10474–10501.

91. Dai, Y., Pochapsky, T.C., and Abeles, R.H. (2001). Mechanistic studies of two dioxygenases in the methionine salvage pathway of Klebsiella pneumoniae. *Biochemistry* 40: 6379–6387.

92. Wray, J.W. and Abeles, R.H. (1993). A bacterial enzyme that catalyzes formation of carbon monoxide. *J. Biol. Chem.* 268: 21466–21469.

93. Deshpande, A.R., Wagenpfeil, K., Pochapsky, T.C., Petsko, G.A., and Ringe, D. (2016). Metal-dependent function of a mammalian acireductone dioxygenase. *Biochemistry* 55: 1398–1407.

94. Deshpande, A.R., Pochapsky, T.C., Petsko, G.A., and Ringe, D. (2017). Dual chemistry catalyzed by human acireductone dioxygenase. *Protein Eng. Des. Sel.* 30: 197–204.

95. Liu, X. and Pochapsky, T.C. (2019). Human acireductone dioxygenase (HsARD), cancer and human health: Black hat, white hat or gray? *Inorganics* 7: 101.

96. Uekita, T., Gotoh, I., Kinoshita, T., Itoh, Y., Sato, H., Shiomi, T., Okada, Y., and Seiki, M. (2004). Membrane-type 1 matrix metalloproteinase cytoplasmic tail-binding protein-1 is a new member of the cupin superfamily: A possible multifunctional protein acting as an invasion suppressor down-regulated in tumors. *J. Biol. Chem.* 279: 12734–12743.

97. Vignais, P.M. and Billoud, B. (2007). Occurrence, classification, and biological function of hydrogenases: An overview. *Chem. Rev.* 107: 4206–4272.
98. Lampret, O., Esselborn, J., Haas, R., Rutz, A., Booth, R.L., Kertess, L., Wittkamp, F., Megarity, C.F., Armstrong, F.A., Winkler, M., and Happe, T. (2019). The final steps of [FeFe]-hydrogenase maturation. *Proc. Natl. Acad. Sci. U.S.A.* 116: 15802–15810.
99. Shepard, E.M., Duffus, B.R., George, S.J., McGlynn, S.E., Challand, M.R., Swanson, K.D., Roach, P.L., Cramer, S.P., Peters, J.W., and Broderick, J.B. (2010). [FeFe]-hydrogenase maturation: HydG-catalyzed synthesis of carbon monoxide. *J. Am. Chem. Soc.* 132: 9247–9249.
100. Dean, R.T., Padgett, H.C., and Rapoport, H. (1976). A high yield regiospecific preparation of iminium salts. *J. Am. Chem. Soc.* 98: 7448–7449.
101. Nicolet, Y., Martin, L., Tron, C., and Fontecilla-Camps, J.C. (2010). A glycyl free radical as the precursor in the synthesis of carbon monoxide and cyanide by the [FeFe]-hydrogenase maturase HydG. *FEBS Lett.* 584: 4197–4202.
102. Kuchenreuther, J.M., Myers, W.K., Stich, T.A., George, S.J., NejatyJahromy, Y., Swartz, J.R., Britt, R.D., and Radical, A. (2013). Intermediate in tyrosine scission to the CO and CN⁻ ligands of FeFe hydrogenase. *Science* 342: 472–475.
103. Rao, G., Tao, L., Suess, D.L.M., and Britt, R.D. (2018). A [4Fe–4S]-Fe(CO)(CN)-l-cysteine intermediate is the first organometallic precursor in [FeFe] hydrogenase H-cluster bioassembly. *Nat. Chem.* 10: 555–560.
104. Schulz, A.-C., Frielingsdorf, S., Pommerening, P., Lauterbach, L., Bistoni, G., Neese, F., Oestreich, M., and Lenz, O. (2020). Formyltetrahydrofolate decarbonylase synthesizes the active site CO Ligand of O2-Tolerant [NiFe] hydrogenase. *J. Am. Chem. Soc.* 142: 1457–1464.
105. Bürstel, I., Siebert, E., Frielingsdorf, S., Zebger, I., Friedrich, B., and Lenz, O. (2016). CO synthesized from the central one-carbon pool as source for the iron carbonyl in O_2-tolerant [NiFe]-hydrogenase. *Proc. Natl. Acad. Sci. U.S.A.* 113: 14722–14726.
106. Guan, J.-C., Hasnain, G., Garrett, T.J., Chase, C.D., Gregory, J., Hanson, A.D., and McCarty, D.R. (2014). Divisions of labor in the thiamin biosynthetic pathway among organs of maize. *Front. Plant Sci.* 5.
107. Chatterjee, A., Hazra, A.B., Abdelwahed, S., Hilmey, D.G., and Begley, T.P. (2010). A "radical dance" in thiamin biosynthesis: mechanistic analysis of the bacterial hydroxymethylpyrimidine phosphate synthase. *Angew. Chem. Int. Ed. Engl.* 49: 8653–8656.
108. Voordouw, G. (2002). Carbon monoxide cycling by Desulfovibrio vulgaris Hildenborough. *J. Bacteriol.* 184: 5903–5911.
109. Fauque, G., Peck, H.D., Jr., Moura, J.J., Huynh, B.H., Berlier, Y., DerVartanian, D.V., Teixeira, M., Przybyla, A.E., Lespinat, P.A., Moura, I. et al. (1988). The three classes of hydrogenases from sulfate-reducing bacteria of the genus Desulfovibrio. *FEMS Microbiol. Rev.* 4: 299–344.
110. Fischer, F., Künne, S., and Fetzner, S. (1999). Bacterial 2,4-dioxygenases: New members of the α/β hydrolase-fold superfamily of enzymes functionally related to serine hydrolases. *J. Bacteriol.* 181: 5725–5733.
111. Bauer, I., De Beyer, A., Tshisuaka, B., Fetzner, S., and Lingens, F. (1994). A novel type of oxygenolytic ring cleavage: 2,4-oxygenation and decarbonylation of 1H-3-hydroxy-4-oxoquinaldine and 1H-3-hydroxy-4-oxoquinoline. *FEMS Microbiol. Lett.* 117: 299–304.
112. Bauer, I., Max, N., Fetzner, S., and Lingens, F. (1996). 2,4-dioxygenases catalyzing N-heterocyclic-ring cleavage and formation of carbon monoxide. Purification and some properties of 1H-3-hydroxy-4-oxoquinaldine 2,4-dioxygenase from Arthrobacter sp. Rü61a and comparison with 1H-3-hydroxy-4-oxoquinoline 2,4-dioxygenase from Pseudomonas putida 33/1. *Eur. J. Biochem.* 240: 576–583.
113. Nicolaidis, A.A. and Sargent, A.W. (1987). Isolation of methane monooxygenase-deficient mutants from *Methylosinus trichosporium* OB3b using dichloromethane. *FEMS Microbiol. Lett.* 41: 47–52.
114. Can, M., Armstrong, F.A., and Ragsdale, S.W. (2014). Structure, function, and mechanism of the nickel metalloenzymes, CO dehydrogenase, and acetyl-CoA synthase. *Chem. Rev.* 114: 4149–4174.
115. Yang, X., Lu, W., Hopper, C.P., Ke, B., and Wang, B. (2020). Nature's marvels endowed in gaseous molecules I: Carbon monoxide and its physiological and therapeutic roles. *Acta Pharm. Sin. B* https://doi.org/10.1016/j.apsb.2020.10.010.

Section III.

Carbon Monoxide Sensing and Scavenging

18

Fluorescent Probes for Intracellular Carbon Monoxide Detection

Ryan R. Walvoord[1], Morgan R. Schneider[2], and Brian W. Michel[2]

[1]Department of Chemistry, Ursinus College, Collegeville, PA 19426, USA
[2]Department of Chemistry and Biochemistry, University of Denver, Denver, CO 80210, USA

Introduction

Over the past two decades, carbon monoxide (CO) has been revealing itself not just as a tragic and toxic molecule, but also as an endogenous signaling molecule and increasingly as a potential therapeutic [1–5]. The development of CO probes has been critical to the understanding of endogenous production and for the monitoring of CO-releasing molecules (CORMs) [6–10]. While early molecular approaches for chromogenic detection of CO used various metal complexes to show a visual response to CO in an atmosphere [11–15], seminal reports of fluorogenic detection in 2012 opened the field of cellular CO detection [16,17]. Challenges associated with designing CO probes have limited the number of mechanistic approaches toward detection; however, this has not limited creative design and improvements.

Most small molecule probes for biologically relevant analytes fall into two categories: (i) coordination-based probes or (ii) reaction-based probes [18–24]. The reported coordination-based probes are for metal ions such as Ca, Zn, Cu, Hg, etc. On the other hand, reaction-based probes, also known as activity-based sensors (ABS), are generally designed around the inherent reactivity of species such as H_2O_2 and other reactive oxygen, sulfur, and nitrogen species [25]. It should be noted that there are exceptions to both of these categories with ABS probes for metal ions (Fe/Cu) and coordination-based probes for reactive small molecules (H_2S). Since CO is a small organic molecule, it would seem logical to investigate an ABS approach toward its detection. However, the main challenge in designing probes for the intracellular detection of CO arises from the fact that CO is generally not reactive with main group elements. It is not electrophilic/Lewis acidic and is a relatively poor Lewis base. As a result, prior to 2012 detection of CO from biological systems utilized the hemoglobin assay [6]. This assay takes advantage of a colorimetric readout upon coordination of CO to hemoglobin, which points to a critical aspect of CO reactivity with transition metals. Indeed, the high affinity of CO for hemoglobin is one of its best-known roles and is associated with its toxic effects. CO has a very high affinity for hemoglobin, binding with 210–250-fold higher affinity as compared to O_2 [26]. While a small amount of this interaction arises from the aforementioned Lewis basicity of the lone pair on C of CO, it is the presence of π-backbonding from the partially filled d-orbitals present in transition metals that leads to this strong binding interaction. This model of bonding is described by a combination of orbital interactions that influence the strength of bonding and reactivity of CO (Figure 18.1). Indeed, transition metals play a fundamental role in the approaches described below for the detection of CO.

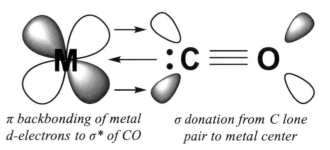

Figure 18.1 Model of bonding between CO and a metal with d-electrons.

π backbonding of metal d-electrons to σ* of CO

σ donation from C lone pair to metal center

There are many important areas of chemistry that take advantage of the interaction of CO with transition metals. These range from industrially important transformations such as hydroformylation [27] to a physical organic tool for reading out the electron richness of transition metals via the modulation of the CO infrared stretch [28]. As a result, it is not surprising that probes capable of detecting CO draw heavily from known synthetic systems for CO reactivity. There have been a number of advances in CO probes in less than a decade of research, with a number of questions remaining. Efforts over the past decade have results in numerous advances in CO probe capabilities, including sensitivity, subcellular targeting, and longer excitation/emission wavelengths. However, several important questions remain, including the following: Are transition metals necessary for CO detection? What metals are best from the standpoint of reactivity and potential interactions within cellular environments? Further, we must take caution in evaluating these probes, since metal carbonyls are frequently used as CO donors, and their reactivity cannot be ignored. A number of recent reviews have highlighted various aspects of CO probe development [9,29–40]. This chapter will look specifically at the mechanistic approaches toward CO detection, key advances of specific probe classes that allow for lower limits of detection (LODs), and other important characteristics such as shifted emission wavelengths. The method for determining LOD is included for each reported value where feasible, with the $3\sigma/k$ (σ = standard deviation of blank and k = slope of calibration curve) and S/N = 3 (concentration of analyte that provides a signal three times the standard deviation of noise) being the most common. We urge future investigations to report thorough experimental information when determining LOD values in order to provide informative and reproducible metrics [41].

Types of Probes

The successful approaches toward the detection of intracellular CO all utilize a metal to facilitate the interaction of the probe with CO. We broadly categorize these efforts into three main categories: (i) fluorescent protein; (ii) metal displacement; and (iii) metal reduction. These are conceptually depicted in Figure 18.2. Key contributions to these areas will be highlighted in the sections to follow. A number of similar and creatively distinct approaches have been taken for the detection of CO in the air with colorimetric or fluorescent responses. Additionally, alternative imaging modalities derived from the strategies described herein have been developed; however, an in-depth discussion of these systems will not be included in this chapter. Finally, some recent reports have claimed "metal-free" CO detection. We will address these probes and note caution to the field. Owing to its unique place in the field as the first reported approach to fluorescent CO detection in a cellular environment and the only fluorescent protein, we will begin our discussion with the fluorescent protein CO sensor (COSer).

Fluorescent Protein COSer

In 2012, the He group reported the first approach for detection of CO in a cellular environment that elicited a fluorescent response [16]. They took advantage of a known heme protein, CooA, from *Rhodospirillum rubrum*, a photosynthetic bacterium, which had previously been reported to selectively sense and oxidize CO [42]. The He group modified the CooA to utilize the native conformational change as a fluorescent CO sensor by appending a circularly permutated variant of the yellow fluorescent protein, which is sensitive to conformational changes. The resulting profluorescent CO sensor (COSer)

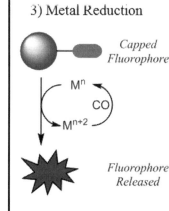

Figure 18.2 Categories of probes for intracellular CO detection.

demonstrated a modest turn-on of twofold and good selectivity for CO over other potential heme ligands. Remarkably, similar linear heme ligands such as NO and cyanide did not produce a significant increase in fluorescence as compared to CO. COSer was transfected into HeLa and HEK293T cells. In both cases, a significant increase in fluorescence was observed both with the application of CO saturated water and by the use of the CO-releasing molecule CORM-2. Finally, they were able to show real-time detection of CO within just a few minutes by treating transfected cells with CORM-2.

It is interesting to note that COSer is the sole example of a fluorescent protein for the detection of CO and that this system has not been used extensively or improved upon. It is possible that this is related to the stability of this sensor, as air-free handling was required, or the ease of use of small molecule fluorescent probes has limited the need for a fluorescent protein. It should be noted that there is one related hemoprotein model complex for extracellular quantification and removal of CO in cells [43]. Additionally, an interesting aspect of COSer is that it is the only probe reported to date with a strong chance of acting as a reversible chemosensor, which would be capable of showing temporal fluctuations in CO levels. While the reversibility of CO detection was not discussed by the authors, this is certainly a missing tool in the field of intracellular CO detection.

Metal Displacement

A relatively recently developed strategy in fluorescent probe design has been to displace a fluorophore bonded to a metal. A number of probes have been developed that take advantage of this chemistry, including those beyond the scope of CO detection. For CO, these probes have primarily used palladium, owing to the wide application of synthetic carbonylation chemistry affected by palladium. However, there are some recent reports using other metals, such as ruthenium, for CO detection. Probes where a metal–fluorophore bond is displaced in response to CO are discussed below.

COP-1

Contemporaneous to the development of the genetically encoded COSer, the Chang group reported the first small molecule fluorescent probe capable of detecting CO in live cells [17]. The authors recognized that the adoption of well-established palladium-mediated carbonylation chemistry was a viable approach to detecting CO [44–46]. At that time, reports had come out suggesting that a number of palladium-mediated reactions, including Tsuji–Trost-type chemistry [47,48] and Suzuki cross-couplings [48], were viable in cellular environments. As a result, a CO probe (COP-1) was designed, based on known cyclopalladated complexes with established CO reactivity (Figure 18.3). The palladacycle was appended with a BODIPY fluorophore to provide a fluorescent readout. In the presence of palladium, the BODIPY fluorescence was largely quenched ($\Phi = 0.01$); however, upon reaction with CO, an ~10-fold increase in fluorescence was observed within 60 min. It is notable that a twofold response is observed in 5 min and a fourfold response in 15 min. Exposure of the probe to CO in the presence of water with an organic cosolvent yielded a carboxylic acid product in 96% yield, which was identified by ^{1}H, ^{13}C nuclear magnetic resonance (NMR), mass spectrometry (MS), and later X-ray diffraction [49]. Further MS analysis of *in vitro* experiments with CO gas confirmed the carboxylic acid product. It is worthwhile

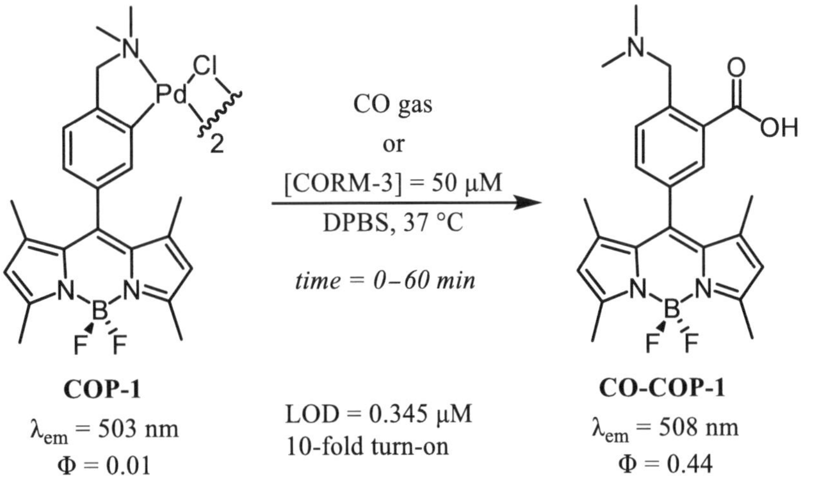

Figure 18.3 COP-1 reacts with CO gas or a CORM.

COP-1
$\lambda_{em} = 503$ nm
$\Phi = 0.01$

CO gas
or
[CORM-3] = 50 μM
DPBS, 37 °C
time = 0–60 min

LOD = 0.345 μM
10-fold turn-on

CO-COP-1
$\lambda_{em} = 508$ nm
$\Phi = 0.44$

to note here that both COSer and COP-1 were evaluated with CO gas in addition to CORMs. As we will discuss later, this is an important validation for all new CO probes.

Further, COP-1 demonstrated good selectivity over other reactive oxygen, nitrogen, and sulfur species with a notable slight response to H_2S. While the LOD was not explicitly determined in this initial report, it was later determined to be 0.345 μM ($3\sigma/k$) [50]. The authors demonstrated application in live HEK293T cells with response to exogenously added CO with ~1.8-fold turn-on in fluorescence response to 5 μM CORM-3. No toxicity was observed in response to the probe or the CO–COP-1 adduct. This cyclopalladated probe for CO laid the basis for other CO probes to be discussed in this section. The structure of COP-1 as well as other cyclopalladated probes has been both empirically and systematically evaluated. These structure–activity relationships (SARs) will be discussed at the end of this section.

Owing to the ease of use and ready availability of COP-1, it has been the most widely used fluorescent probe for the detection of CO. Applications have primarily been in the evaluation of new CORMs [51–64]. Additionally, COP-1 and other probes have been used to help elucidate some aspects of CO biology [50]. Improved CO probes are key to our further understanding of basal CO biology and potential therapeutic applications.

Variations on Cyclopalladated Probes

Following the initial report of COP-1, a number of variations on cyclopalladated probes have been reported. Important contributions will be highlighted in this section bringing attention to probes that have significantly shifted emission, high turn-on, or particularly good LODs. Examples with evidence to support endogenous CO detection will also be highlighted.

In 2014, Lin and coworkers reported a two-photon probe based on a carbazole–coumarin fluorophore [65]. This probe used the same dimethylamino chelating ligand as COP-1 (Figure 18.4). In contrast to COP-1, protodemetallation was reported as the adduct of the probe reacting with CO. A slightly improved 11-fold turn-on was observed with a blue-shifted emission at 477 nm using one-photon excitation at 378 nm. Additionally, this probe (CC-CO) could be excited with two photons at 740 nm providing a two-photon cross section of 50.1 GM. Live cells were imaged with single-photon excitation and liver tissue was used for two-photon excitation with a 180 μm penetration depth.

Tang and coworkers reported two probes bearing an azobenzene–cyclopalladium structure (ACP-1 and ACP-2, Figure 18.5) [66]. This was the first reported example of an sp^2 hybridized chelating ligand for a CO probe. Again, the product of CO and the probe was identified as protodemetallation, although previous reports of similar compounds had shown carbonylation [67]. ACP-2 showed a 10-fold turn-on similar to previously reported probes. It was noted that this higher turn-on for ACP-2 as compared to ACP-1 was the result of greater fluorescent quenching of the cyclopalladated probe. The authors were able to show good performance in live HEPG2 cells in response to exogenous CORM-2 application. It is particularly noteworthy that the authors observed an increase in signal following hypoxic treatment of cells, suggesting that endogenous CO was observed. This was further supported by applying the heme oxygenase-1 (HO-1) inhibitor zinc protoporphyrin (ZnPP). As with many examples in this section, the detection of endogenous CO would be strengthened by supplemental evidence such as Western blot analysis showing upregulation of HO-1 or another validation method [41].

Lin and coworkers reported a unique fluorescent probe based on the Nile red dye [68]. The key feature of this probe is that the design actually uses a Lewis basic heteroatom of the dye to chelate to palladium (Figure 18.6). That is, the dye structure actually

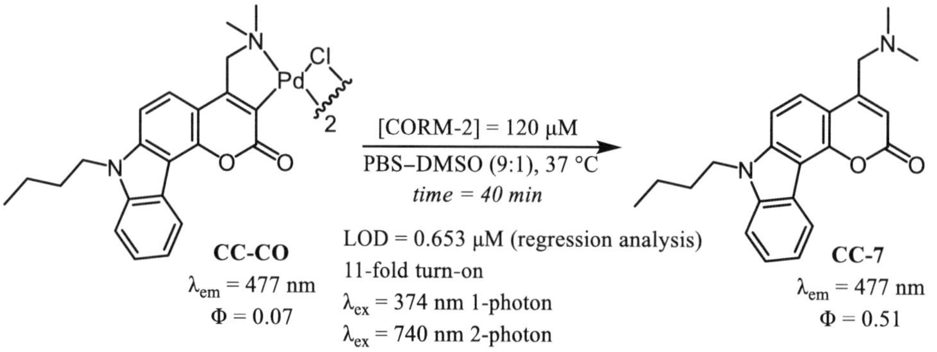

Figure 18.4 CC-CO as a two-photon CO probe reported by Lin and coworkers.

CC-CO
λ_{em} = 477 nm
Φ = 0.07

[CORM-2] = 120 μM
PBS–DMSO (9:1), 37 °C
time = 40 min

LOD = 0.653 μM (regression analysis)
11-fold turn-on
λ_{ex} = 374 nm 1-photon
λ_{ex} = 740 nm 2-photon

CC-7
λ_{em} = 477 nm
Φ = 0.51

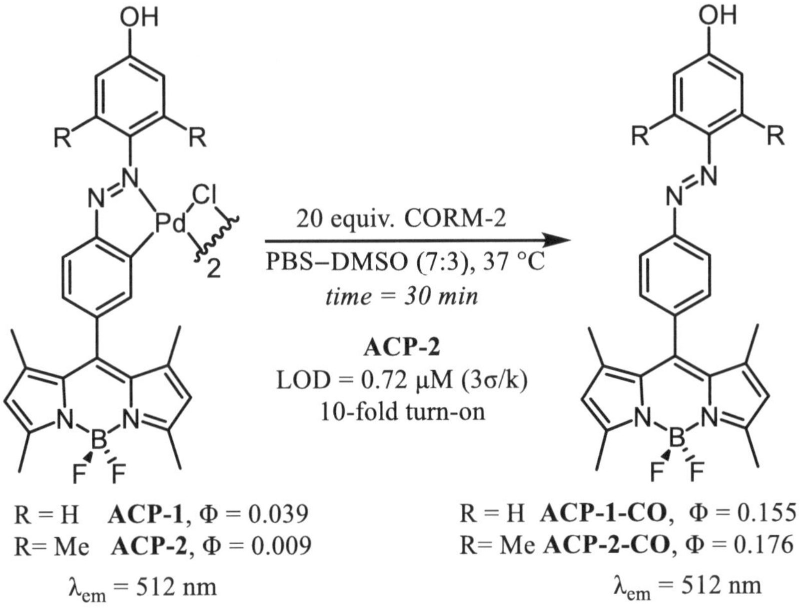

Figure 18.5 ACP-1 and ACP-2 reported by Tang and coworkers show differences in the quenching efficiency depending on the identity of R (H versus Me).

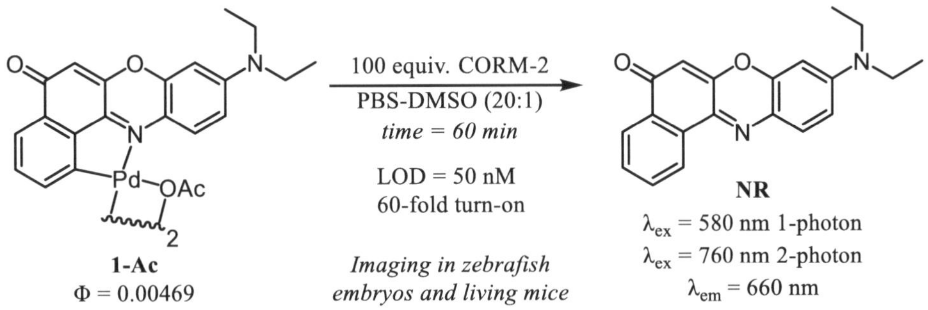

R = H **ACP-1**, Φ = 0.039
R = Me **ACP-2**, Φ = 0.009
λ_{em} = 512 nm

20 equiv. CORM-2
PBS–DMSO (7:3), 37 °C
time = 30 min
ACP-2
LOD = 0.72 μM (3σ/k)
10-fold turn-on

R = H **ACP-1-CO**, Φ = 0.155
R = Me **ACP-2-CO**, Φ = 0.176
λ_{em} = 512 nm

1-Ac
Φ = 0.00469

100 equiv. CORM-2
PBS-DMSO (20:1)
time = 60 min
LOD = 50 nM
60-fold turn-on

Imaging in zebrafish embryos and living mice

NR
λ_{ex} = 580 nm 1-photon
λ_{ex} = 760 nm 2-photon
λ_{em} = 660 nm

Figure 18.6 A Nile red-based probe for detecting CO with direct coordination of palladium to a Lewis basic nitrogen of the fluorophore.

provides the bidentate binding to palladium without further functionalization. Possibly, as a result of a shorter distance between the palladium and the fluorophore, a very low quantum yield of the probe is observed (Φ_f = 0.00469). As a result, upon reaction with CO (CORM-2), a 60-fold turn-on in emission intensity is observed. The consequence of this large turn-on is a low 50 nM LOD (3σ/k). Further, this probe provides a redshifted emission of 660 nm in the near-infrared (NIR) region with excitation at 580 nm for *in vitro* experiments and 561 nm during microscopy experiments.

A slight drawback to this probe is the extended reaction time of 4 h to reach full turn-on. However, the magnitude of turn-on still provides good response over shorter experimental courses. For example, cells treated with probe for 30 min showed a significant turn-on in both one- and two-photon modes in response to hypoxic conditions. The authors were also able to show application of their probe in larval zebrafish and in mouse liver tissue slices with penetration up to 130 μm. Live mice were treated with lipopolysaccharide (LPS), hemin, or ZnPP for 4 days followed by animal sacrifice and organ harvesting. Increased fluorescence detection in the presence of LPS and hemin along with inhibited fluorescence from ZnPP is strong evidence for detection of endogenous CO. Further, the red NIR emission of this platform allowed for imaging in live mice with responses as quick as 2 min. The preponderance of evidence again suggests that endogenous CO is being observed in these experiments, but supplemental validation would be beneficial to demonstrate the power of these probes.

Zhang and coworkers reported another variation on the Nile red-based CO probe [69]. They used a positively charged tetraalkylammonium to block cell penetration and anchor to the cellular membrane (Figure 18.7). They reported an *in vitro* LOD of 0.23 μM (3σ/k). Further, the authors looked at endogenous CO production from different cell lines and tissues. Live tissue imaging of multiple organs showed

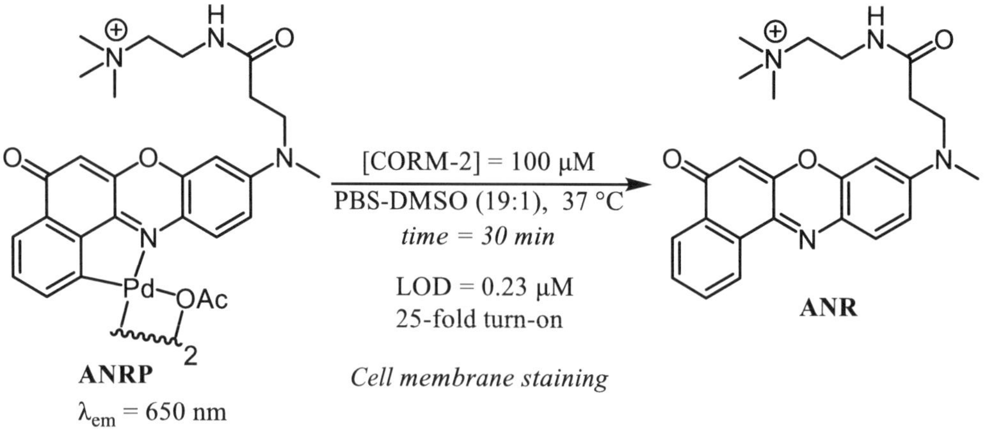

Figure 18.7 A cell membrane-anchored probe for detecting CO reported by Zhang and coworkers.

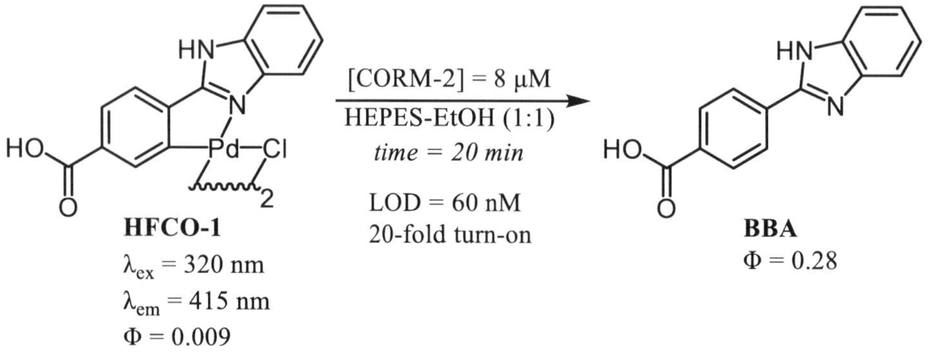

Figure 18.8 A benzimidazole-based fluorescent probe for CO reported by Huang and coworkers.

much brighter fluorescence from liver tissue. This supports detection of CO and that the liver is the major organ for CO production. It is known that HO levels are significantly higher in certain tissues, including liver [70].

Further Nile red SAR and mechanism studies were performed by Klán and coworkers [71]. This investigation will be discussed in the section below on ligand tuning.

Huang and coworkers reported another cyclopalladated CO probe with the fluorophore structure again acting as the ligand on palladium (Figure 18.8) [72]. In this work, a benzimidazole moiety is chelated to palladium. A 20-fold turn-on was reported in response to CO and the LOD was determined to be 0.06 μM ($3\sigma/k$). However, this is a blueshifted probe with emission maximum at 415 nm and excitation at 320 nm, which limits the applicability of this probe.

Ligand Tuning

An intriguing aspect of probes based on the reactivity of transition metals is that the reactive handle can be significantly modulated based on the identity of the ligands coordinated to the metal. The tuning of transition metal reactivity is a field unto itself in the realms of catalysis and synthetic inorganic and organometallic chemistry. Through the design of new probes with varying fluorophores, it is possible to empirically evaluate different ligands on palladium. For example, since the report of COP-1, the majority of new probes contain an sp^2 hybridized nitrogen ligand chelating to palladium. However, in the absence of purposeful comparative studies, it is difficult to make significant conclusions from these reports. A couple of reports have looked to explicitly optimize the structure of the Nile red-based and COP-1-based probes, with improved probes resulting from both studies.

Klán and coworkers set out to evaluate structural variations and the mechanism within the Nile red platform [71]. A number of variations were synthesized focusing on the aryl ring directly bonded to palladium, water-solubilizing groups off the aniline nitrogen, and the nature of the bridging carboxylate ligand (Figure 18.9). In modifying the aryl ring bonded to palladium, the authors observed some slight variations in quantum yields and reactivity. Specifically, the 2-hydroxy variant **2** showed a greater quenching for the palladacycle probe, which results in a greater probe turn-on for CO detection. Additionally, the presence of the electron-donating hydroxyl group accelerated the rate of CO detection as compared to the unsubstituted or 3-fluoro analogues (**1-Ac** and **3**). Other variations made by the

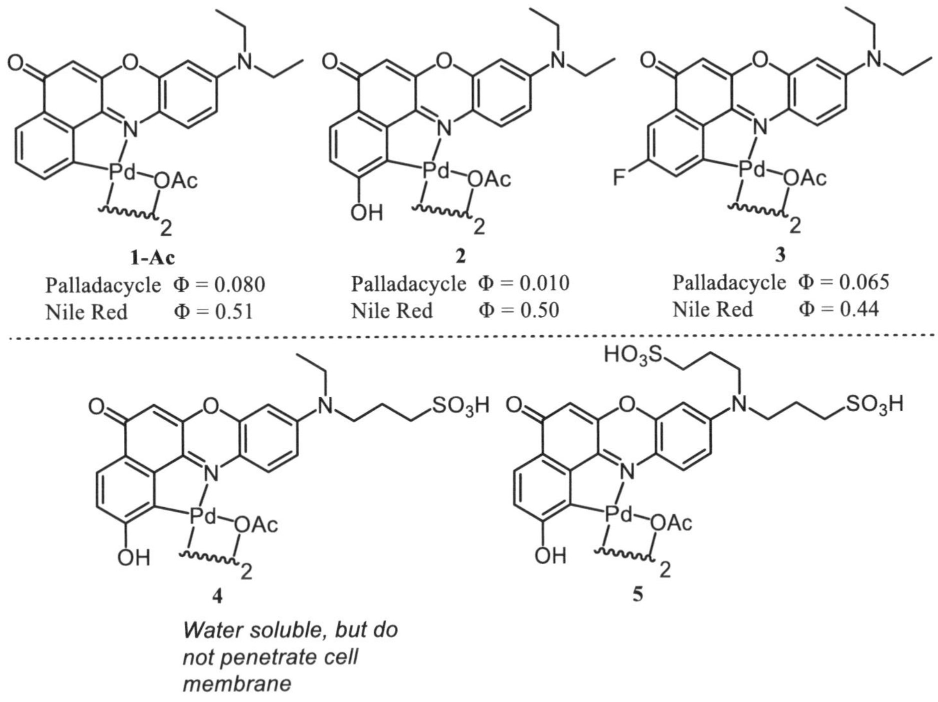

Figure 18.9 Systematically modified Nile red-based CO probes reported by Klán and coworkers.

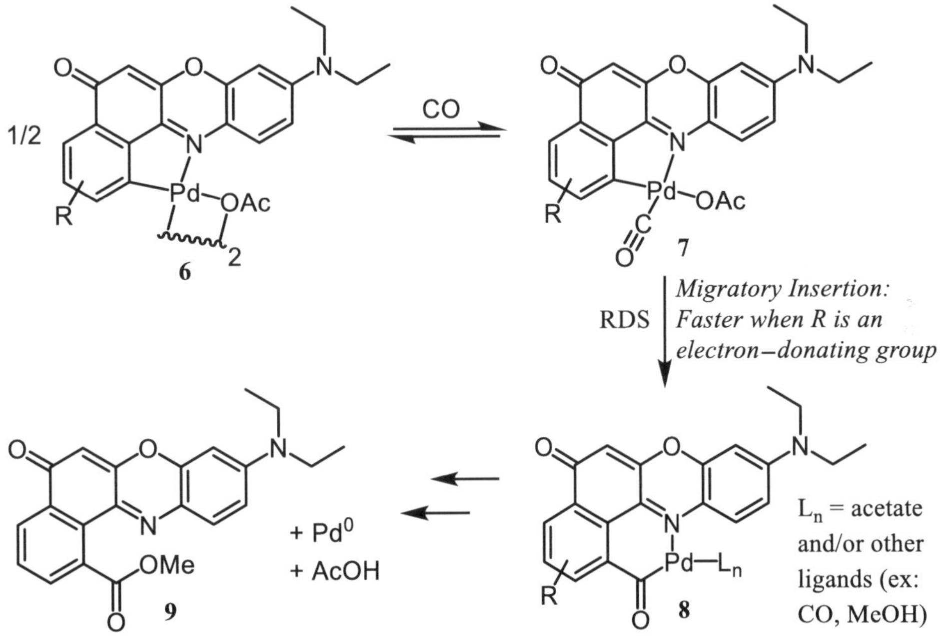

1-Ac
Palladacycle Φ = 0.080
Nile Red Φ = 0.51

2
Palladacycle Φ = 0.010
Nile Red Φ = 0.50

3
Palladacycle Φ = 0.065
Nile Red Φ = 0.44

4

5

Water soluble, but do not penetrate cell membrane

Figure 18.10 Proposed mechanism based on studies by Klán and coworkers.

RDS | *Migratory Insertion: Faster when R is an electron–donating group*

L_n = acetate and/or other ligands (ex: CO, MeOH)

+ Pd⁰
+ AcOH

authors, while insightful, did not provide probes with improved characteristics for intracellular CO detection. For example, the addition of water-solubilizing groups in **4** and **5** provided less aggregation and increased aqueous quantum yields *in vitro*; however, the probes were not able to penetrate cell membranes. Interestingly, this suggests that esterase-sensitive groups that increase water solubility could be an attractive area for probe improvements [73].

Furthermore, mechanistic studies by Klán and coworkers suggested the rapid formation of a probe–CO adduct **7** that was reversible by sparging the solution with argon (Figure 18.10). It is suggested that the migratory insertion process into the palladacycle is rate determining, which is also supported by the rate of detection being sensitive to substitution on the aryl ring. Interestingly, the authors noted different products of probes reacting with CO based on the

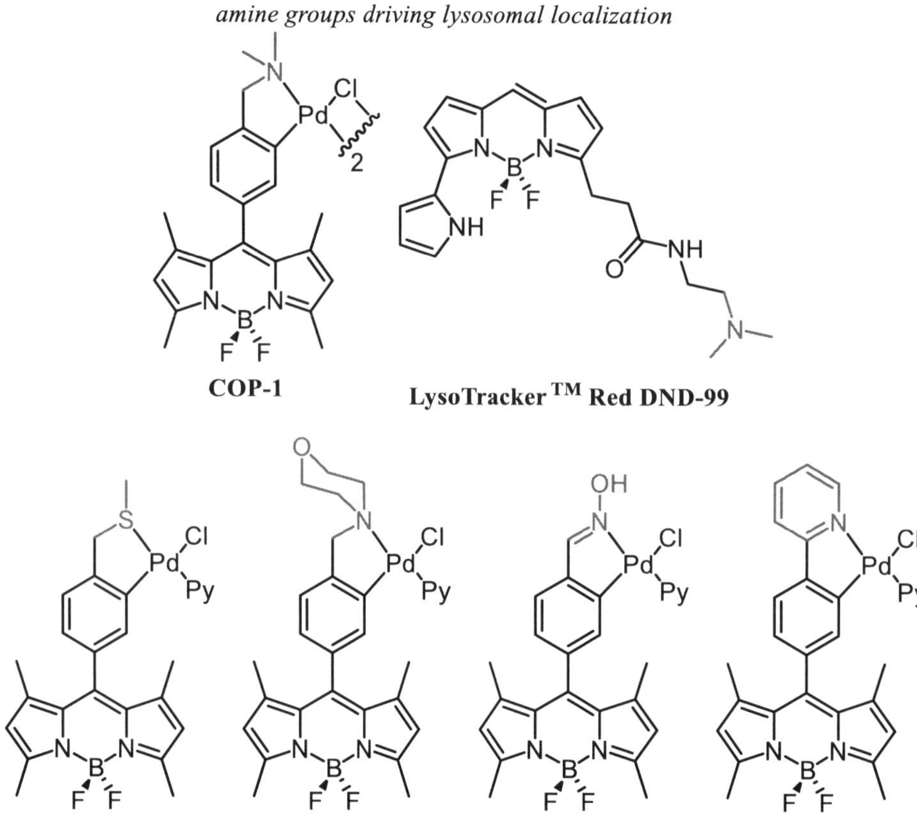

Figure 18.11 Comparison of the amine groups of COP-1 and commercially available lysosome stain. Probes with varied chelating groups.

substituents on the probe and the conditions of the reaction. The evidence provided by the authors and the observations from other probes in this class make it apparent that both the carbonylation product and reductive protodemetallation are feasible products. An insightful experiment would be to observe these various products directly from a cellular environment by mass spectrometry. However, the precise product may often be less important than the probe providing a sensitive response to CO.

Morstein, Michel, Chang, and coworkers recently reported the systematic evaluation of the influence of ligand environment on reactivity derived from the COP-1 platform [49]. Driven by improving staining pattern and applications in live cells, the authors determined that the dimethylamine group of COP-1 likely imparted lysosomal localization (Figure 18.11). Bridge splitting of the μ-chloro dimer with pyridine did not provide any improvement to the probes, but did facilitate improved characterization of a monomeric species. Most cyclopalladated probes are isolated as dimeric species, although this is likely disrupted either in preparation for cellular applications or upon interacting with the cellular milieu.

The authors then sought to evaluate the influence of the chelating ligand. A range of disparate chelating ligands were evaluated and a morpholino amine COP-3-Py was found to retain the reactivity of COP-1 with an improved diffuse staining pattern in live cells.

Further addition of ester groups on the BODIPY core in COP-3E-Py provided a probe with improved penetration as well as improved aqueous solubility and cellular retention after presumed esterase-mediated cleavage of the ester group (Figure 18.12). Through these modifications, the authors were able to shift the subcellular localization from the lysosome to the endoplasmic reticulum (ER) membrane. ER localization is desirable for CO probes as HO-1 is also found there [74]. These modifications should allow for improved detection of CO fluxes, which was demonstrated through detection of endogenous CO release in fly brain models.

The systematic modifications described in this section highlight an intriguing aspect of probes containing a discrete ligated metal complex. Ligand tuning can improve turn-on response, reactivity, and cellular staining. Combining chemical insight from traditional

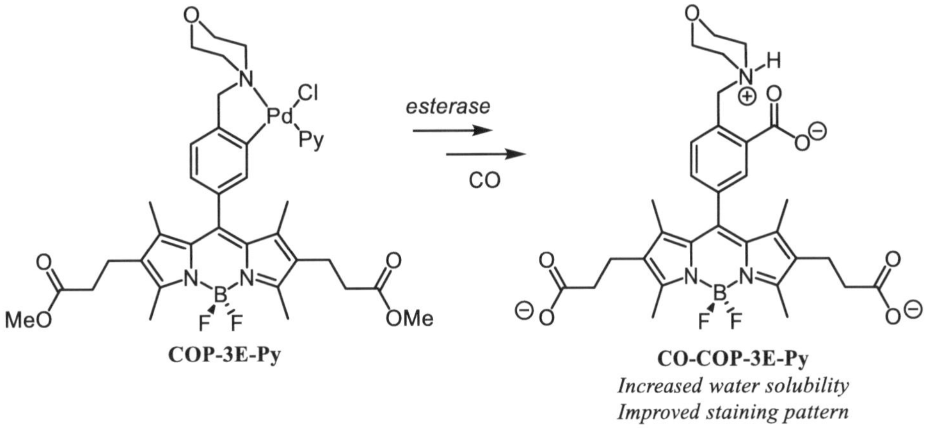

Figure 18.12 Improved probe COP-3E-Py, with morpholino chelating group and esterase cleavable groups to increase solubility and staining pattern.

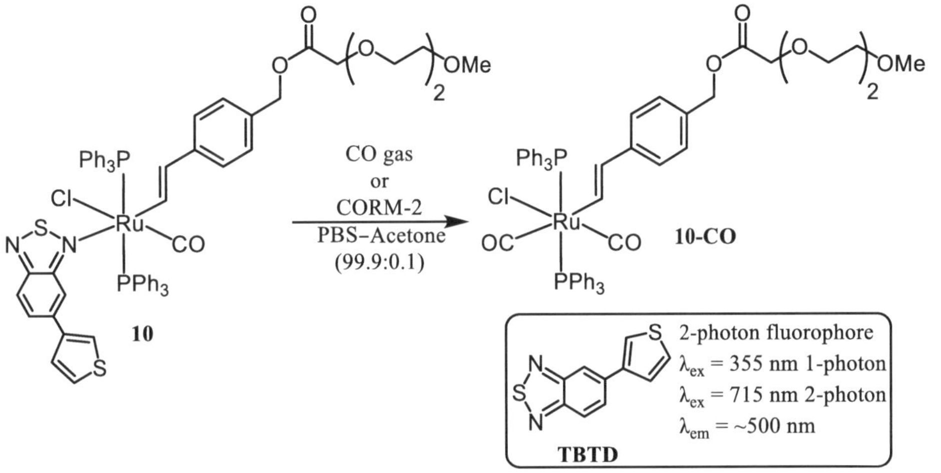

Figure 18.13 Vinyl–Ru(II) complex for CO detection via coordination and ligand displacement.

probe work, such as appending groups to improve solubility and/or cellular retention, along with structural variations of ligands will continue to have a significant impact on the fields of CO detection and other analytes that use metal-based reaction triggers.

Ruthenium-Based Probes

Collaborative work from Martínez-Máñez, Wilton-Ely, and coworkers has resulted in a number of Rh-, Ru-, and Os-based molecules for the detection of CO [37], the earliest of which predates the seminal work for cellular CO detection from the He and Chang groups [14]. While most of this work has focused primarily on detection of atmospheric CO, recent reports have demonstrated applications in cellular environments. In this approach, it is the coordination of CO to the metal center that displaces another ligand leading to a fluorogenic response. As such, these examples fall into the category of metal displacement for the purposes of our categorization, although it

should be noted that these operate quite distinctly from the cyclopalladated probes discussed above.

In 2017, Wilton-Ely, Martínez-Máñez, and coworkers reported a ruthenium(II) complex **10** capable of *ex vivo* detection of endogenous CO (Figure 18.13) [75]. A two-photon fluorophore, 5-(3-thienyl)-2,1,3-benzothiadiazole (TBTD), was coordinated directly to Ru through a nitrogen, which was confirmed by X-ray diffraction of single crystals. CO gas or a CORM could be used to provide a CO ligand that would displace the TBTD to give **10-CO** and provide a fluorescent readout. This probe could also be excited in two-photon mode, which was utilized for microscopy experiments detecting CO in RAW 264.7 macrophages treated with CORM-3 or hemin.

Another variation of this strategy was recently reported by Kuimova, Wilton-Ely, and coworkers, which provided not just the detection of CO, but also a measure of viscosity changes in cells [76]. Here, a pendant BODIPY fluorophore was quenched not by

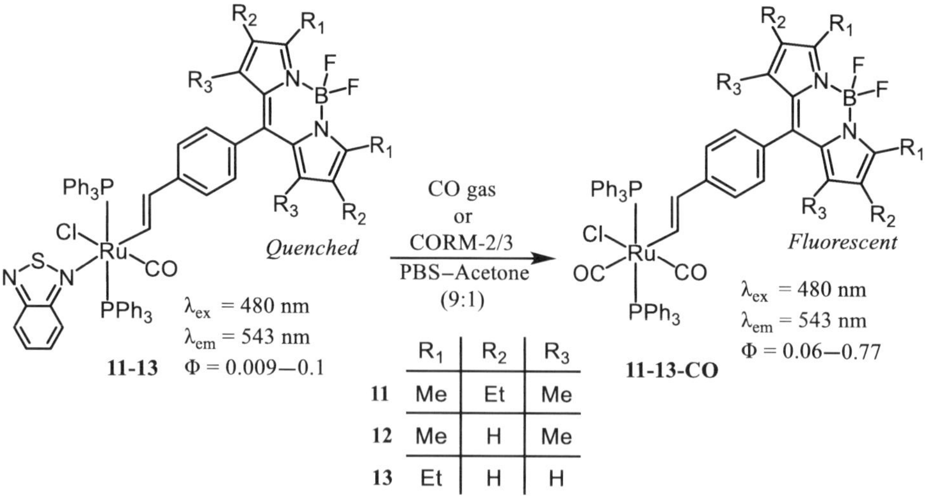

Figure 18.14 Vinyl–Ru(II) complexes for CO and viscosity detection with a BODIPY fluorophore.

the metal, but due to 2,1,3-benzothiadiazole ligand on the metal (Figure 18.14). When this ligand was replaced by CO, a significant enhancement in the BODIPY fluorescence was observed. Using CORM-2, a 0.53 µM LOD (method not specified) was determined. Further, using time-resolved microscopy, the authors were able to correlate the cellular environment of the probe to the relative viscosity, demonstrating a unique aspect of this approach.

While these Ru-based probes that rely on CO coordination have not found widespread use yet, it is important for the field to explore a variety of mechanisms for detecting CO. Remarkably, these purely coordination-based probes demonstrate good selectivity for CO as the ligand exchange appears to be quite selective. While the reported selectivity suggests this approach may find important applications, additional information about CO binding, such as K_d values, would be instructive to the field.

Overall, the advantages of metal displacement approaches to CO detection are that they are easy to use, only require the addition of one component (as compared to the deallylative examples below), and allow for ligand tuning of reactivity and other parameters. Furthermore, this class of probes is well established and has been widely used by many different labs, which has validated them as useful tools for detecting CO. One disadvantage of metal displacement probes for CO detection is that they may not be trivial to prepare for labs not familiar with organometallic synthesis. However, many labs will readily share prepared probes with other academic researchers, so this can alleviate that concern. Additionally, while these probes are used in very low concentrations, they do release a stoichiometric amount of metal upon CO detection. In the case of palladium, this is reduced Pd^0, which is less likely to interfere with cellular processes as compared to Pd in the 2+ oxidation state. The Ru complexes are still largely coordinated by strong ligands. Overall, authors have reported that toxicity is not a significant issue associated with metal displacement probes in cellular environments.

Metal Reduction by CO

General Strategy and Mechanism

The most widely reported fluorescence detection strategy for CO functions via a palladium-mediated deallylation sequence invokes a ternary system of a Pd^{2+} salt in addition to an allylated fluorophore. Design of these probes typically encompasses converting a bright hydroxy or amino fluorophore to the corresponding allyl ether, carbonate, or carbamate. The allyl moiety serves both as a reactive trigger and to inhibit fluorescence by attenuating internal charge transfer (ICT) processes or disrupting conjugation (e.g. via spirocyclization). Rather than directly interacting with the sensing molecule itself, CO undergoes oxidation with the Pd^{2+} to produce Pd^0. This process occurs via initial coordination of CO, followed by either insertion or direct acyl attack by water to form the unstable metal carboxylate (Figure 18.15) [77,78]. Subsequent decomposition produces carbon dioxide and the reduced palladium species. The more electron-rich Pd^0 is capable of reacting with the probe via oxidative addition to the allyl moiety, a process commonly encountered in Tsuji–Trost reactions. Dissociation of the "free" deallylated probe yields a fluorogenic response typically due to restored ICT processes and/or conjugation.

Figure 18.15 General chemical signaling mechanism for CO-responsive fluorescent probes: (A) reduction of Pd^{2+} with CO and (B) subsequent deallylation of allyl ethers or carbonates.

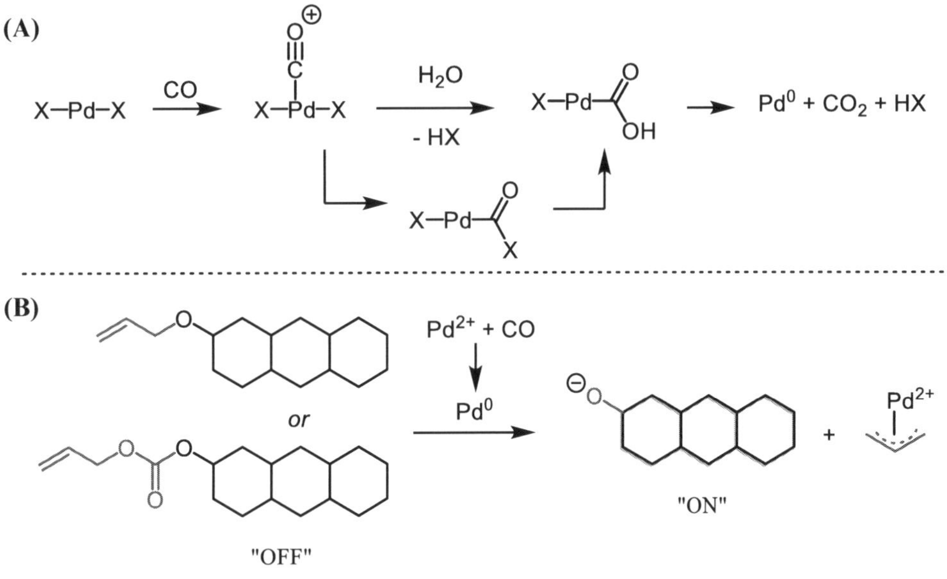

Figure 18.16 First deallylation probe PCO-1 and turn-on mechanism.

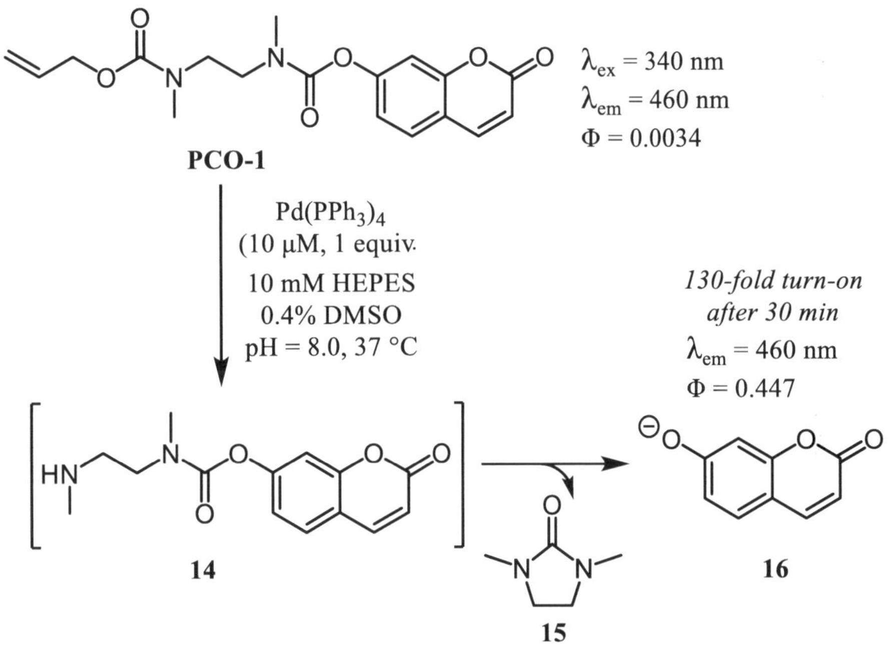

$\lambda_{ex} = 340$ nm
$\lambda_{em} = 460$ nm
$\Phi = 0.0034$

130-fold turn-on after 30 min
$\lambda_{em} = 460$ nm
$\Phi = 0.447$

Although metal-mediated deallylation sequences had previously been reported for Pd detection [79], Dhara and coworkers reported the initial probe, PCO-1, employing this strategy for fluorescent detection of CO in 2015 [80]. This hydroxycoumarin is functionalized with an allyl carbamate linker, which, upon exposure to $PdCl_2$ and CORM-3, undergoes a deallylation–cyclization–elimination sequence to release the brightly fluorescent coumarin phenolate **16** (Figure 18.16). *In vitro* studies revealed high selectivity for CORM-3 compared to common reactive oxygen and sulfur species in the presence of $PdCl_2$, and the probe was capable of up to a 150-fold turn-on response with a very low reported LOD of 8.49 nM ($3\sigma/k$). The probe was also capable of detecting exogenous CO in cells incubated with $PdCl_2$ and CORM-3. Despite the drawbacks of rather short excitation and emission wavelengths, this first example illustrated the high sensitivity possible from this mechanism.

Following the report of PCO-1, the deallylation trigger strategy has been applied to a wide array of fluorescent scaffolds, and a comprehensive review of each probe is beyond the scope of this chapter.

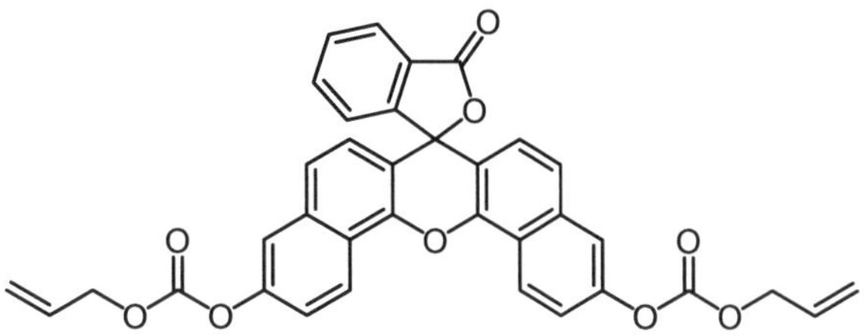

Figure 18.17 Fluorescein-based deallylation probes with product excitation/emission wavelengths and in vitro turn-on response conditions.

NF-APC

Product:
λ_{ex} = 620 nm
λ_{em} = 670 nm

35-fold turn-on response (45 min):
[NF-APC] = 10 μM, [Na$_2$PdCl$_4$] = 60 μM, [CORM-3] = 100 μM in PBS with 40% DMSO, pH 7.4, 25 °C

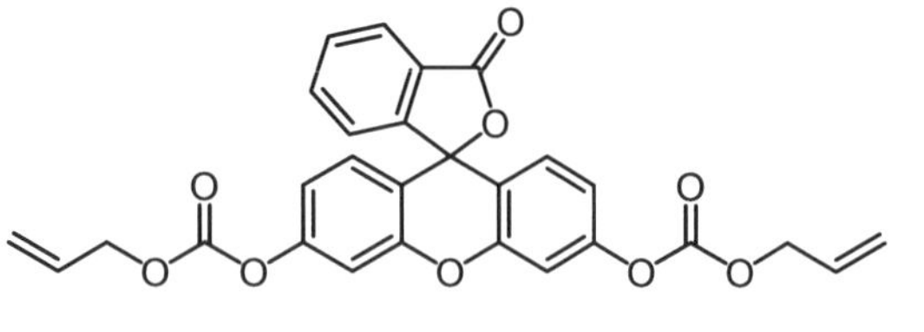

FL-CO-1

Product:
λ_{ex} = 490 nm
λ_{em} = 520 nm

>100-fold turn-on response (15 min):
[FL-CO-1] = [PdCl$_2$] = 5 mM, [CORM-3] = 50 μM in 10 mM PBS with 0.5% DMSO, pH 7.4, 37 °C

Rh-NIR-CO

Product:
λ_{ex} = 541 nm
λ_{em} = 676 nm

28-fold turn-on response (5 min):
[Rh-NIR-CO] = [PdCl$_2$] = 10 μM, [CORM-3] = 100 μM in 10 mM PBS with 1% DMSO, pH 7.4, 37 °C

Representative examples exhibiting notable photophysical properties or application in novel imaging and biological experiments are included herein.

Overview of Deallylative Fluorescent Probes

Several probes have been developed that employ fluorescein-based scaffolds. Alkylation or acylation of these dyes traps them into a near-nonemissive spirocyclic form, which reopens upon deallylation. Zhang and coworkers first reported naphthofluorescein probe NF-APC, which contains two allyl carbonate groups as the reactive triggers (Figure 18.17) [81]. Despite the sensitivity of this moiety, the probe displayed good selectivity to a combination of Pd^{2+} with CO in the form of CORM-3 with only mild interference from H$_2$S. The extended conjugation of this xanthene yielded an early example of a CO probe with emission in the NIR (λ_{em} = 670 nm). In vitro studies indicated a 35-fold response to excess Na$_2$PdCl$_4$ and CORM-3 after 45 min and a 127 nM LOD (3σ/k). The use of 40% DMSO in PBS as solvent suggests limited aqueous solubility of the probe. However, confocal imaging studies with living HeLa cells demonstrated cell permeability and feasibility for live cell imaging using exogenous CO.

Feng next reported the bis-allyl carbonate fluorescein probe FL-CO-1, which functions in a similar deallylation ring-opening process [82]. This probe exhibits notable sensitivity, yielding a greater than 100-fold emission increase at 570 nm after 15 min of treatment with PdCl$_2$ (1 equiv.) and CORM-3 (10 equiv.). A 37 nM LOD was calculated (S/N = 3). Cell imaging studies exhibited a dose-dependent fluorescent response to exogenous CO in the form of CORM-3 or dissolved aqueous CO. Based on reports that heme may stimulate heme oxygenase levels and corresponding endogenous CO levels, A549 cells were preincubated with 100 μM heme prior to incubation with probe (1 μM) and PdCl$_2$ (1 μM). Noticeably increased fluorescent response was observed in cells 4 and 10 h after heme treatment compared to cells with only probe system. This is the first example of fluorescent imaging of CO produced via heme stimulation in cells. A follow-up report described a bis-allyl ether variant using fluorescein or 2,7-dichlorofluorescein [83]. These probes showed improved stability in solution, particularly at higher pH, and high sensitivity, while retaining high sensitivity (~150-fold turn-on after 20 min with 10 equiv. CORM-3) and efficacy in similar exogenous and endogenous cell imaging studies. Two recent reports have installed allyl carbonate triggers onto modified rhodol scaffolds, which provide further redshifted emissions [84,85]. For example, probe Rh-NIR-CO exhibits emission in the NIR (676 nm) with a notably large 135 nm Stokes shift, which is uncommon for most fluorescein derivatives. Treatment with PdCl$_2$ and CORM-3 produced a complete turn-on response (28-fold) in just 5 min. This tool was successfully applied in fluorescence imaging of living cells, zebrafish, and living mice for detecting exogenous CO.

The 1,8-naphthalimide fluorophore provides a useful scaffold for developing probes capable of ratiometric sensing. Installation of allyl carbamate or allyl ether moieties at the 4-position has thus yielded useful tools for CO detection (Figure 18.18A) [86]. For example, allyl ether probe Ratio-CO is emissive at 455 nm [87]. Upon reaction with Pd^{2+} and CO, the deallylated probe is formed with a shifted emission peak at 545 nm. The ratiometric signal was determined to provide an *in vitro* LOD of 17.9 nM ($3\sigma/k$) and was applied for exogenous CO imaging in live cells. Recently, Zhang and coworkers reported the related allyl ether probe, Mito-Ratio-CO, which exhibits similar photophysical properties to its parent but includes a methyl pyridinium moiety for mitochondrial targeting [88]. This probe was applied in cell imaging experiments for detecting endogenous CO via heme stimulation and oxidative stress.

Preincubation of HepG2 cells with either LPS or phorbol 12-myristate 13-acetate followed by probe system yielded 13.6- and 22.4-fold increases in fluorescence ratio, respectively, supporting an increase in mitochondrial CO under oxidative stress.

An alternate strategy for reaction-based ratiometric imaging is displayed by probe BTCV-CO, which is red-emissive (Figure 18.18B) [89]. Reaction with Pd^0 (via PdCl$_2$ reduction by CO) induces a deallylation–cyclization sequence to form a green emissive 2-benzothiazolyl coumarin. The ratiometric signal ($I_{545\ nm}/I_{710\ nm}$) displayed a maximum 39-fold increase to CORM-2 and PdCl$_2$ and a 21.6 nM LOD ($3\sigma/k$). This probe was capable of fluorescence imaging of exogenous CO in living cells and live mice. Additionally, the product molecule displayed enhanced emission in high concentrations of toluene, indicating the potential for aggregation-induced emission in hydrophobic environments.

Fluorogenic probes with far-red or NIR emission benefit from greater tissue penetration, photostability, and reduced background from autofluorescence, among other advantages. The past few years have

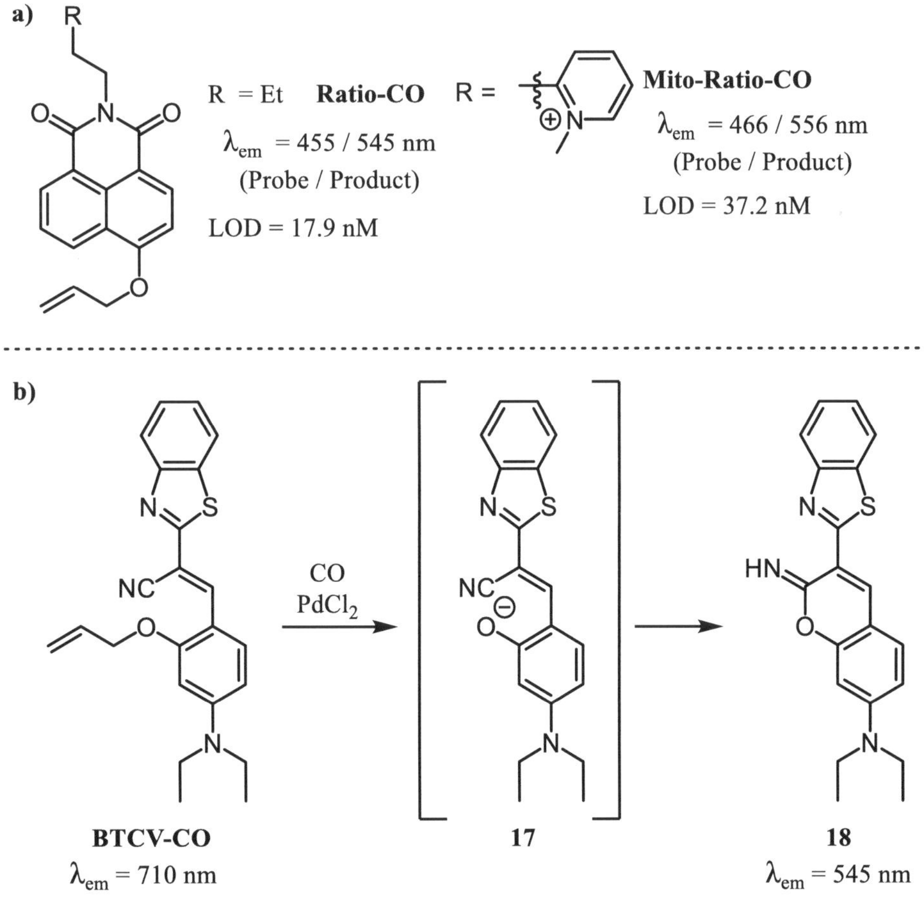

Figure 18.18 Examples of ratiometric CO deallylation probes: (A) naphthalimide-based probes and (B) deallylation–cyclization sequence and change in emission wavelength for BTCV-CO.

yielded numerous examples of incorporating heteroallyl moieties into cyanine, hemicyanine, 2-dicyanomethylene-3-cyano-4,5,5-trimethyl-2,5-dihydrofuran (TCF), dicyanoisophorone, and other extended conjugated scaffolds to produce far-emitting CO fluorescent probes [90–98]. Feng and coworkers have reported several of these probes, including hemicyanine dye **19** (Figure 18.19) [99]. This initially nonemissive allyl carbonate probe undergoes rapid deallylation to produce an NIR emission at 714 nm (λ_{ex} = 678 nm). Notably, solutions of **19** with PdCl$_2$ provided an ~400-fold increase in emission intensity after 3 min of exposure to CORM-3 (10 equiv.), and a 3.2 nM LOD (S/N = 3) was reported. **19** is capable of detecting exogenous and endogenous CO in living cells and was also the first NIR probe for imaging CO in living mice (via CORM-3 injection). As a notable example of efforts to push this emission even further into the NIR, Li and coworkers recently reported FDX-CO, a dicyanoisophorone-based probe featuring an allyl carbonate trigger [100]. While less sensitive to CO than **19**, FDX-CO displays notable photophysical properties, including a large Stokes shift (190 nm) and long emission wavelength (770 nm). This probe proved viable for endogenous CO detection in cell studies using heme or LPS stimulation as well as exogenous CO in living mice. Structurally similar probe NIR-Ratio-CO employs an allyl carbamate trigger and provides a ratiometric signal with long-wavelength emissions and large Stokes shift (λ_{ex} = 440 nm, λ_{em} = 592/655 nm) [101]. The ratiometric signal yielded an LOD of 61 nM ($3\sigma/k$) and was applied for cellular imaging of exogenous and endogenous CO via heme stimulation.

Detection of nonmetal species by fluorescent dyes largely operates via reaction-based strategies, which consume analyte in the sensing process. Growing interest has developed in replacement probes that are able to sense analyte without disrupting cellular homeostasis. In this context, Berreau and coworkers have recently reported a novel "sense and release" strategy based on an allylated flavonol **22** for detecting and replenishing CO (Figure 18.20) [102]. Upon reaction with Pd^{2+} and CO, the probe can deallylate to release the flavonol **21** with concomitant shift in emission from ~465 to ~600 nm. A controllable subsequent reaction with oxygen and light yields a molecule of CO and a nonemissive organic product **22**. *In vitro* studies with this probe were hindered by metal coordination of Pd or Rh (from CORM

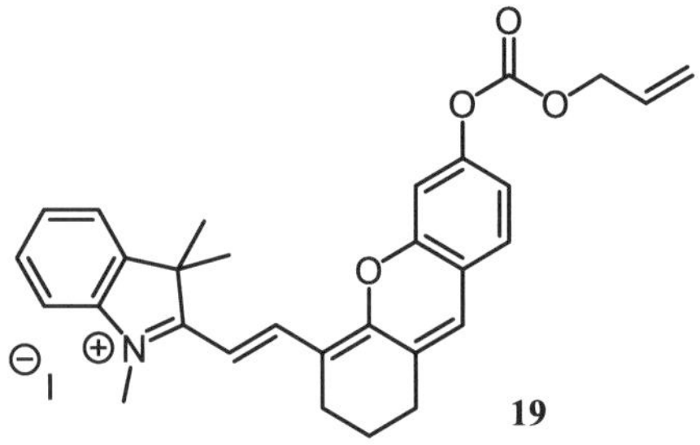

Figure 18.19 Select examples of long-wavelength emission deallylation probes for CO detection.

Product:
λ_{ex} = 678 nm
λ_{em} = 714 nm

~400-fold turn-on response (3 min):
[probe] = [PdCl$_2$] = 5 μM, [CORM-3] = 50 μM in 10 mM PBS with 0.5% DMSO, pH 7.4, 37 °C

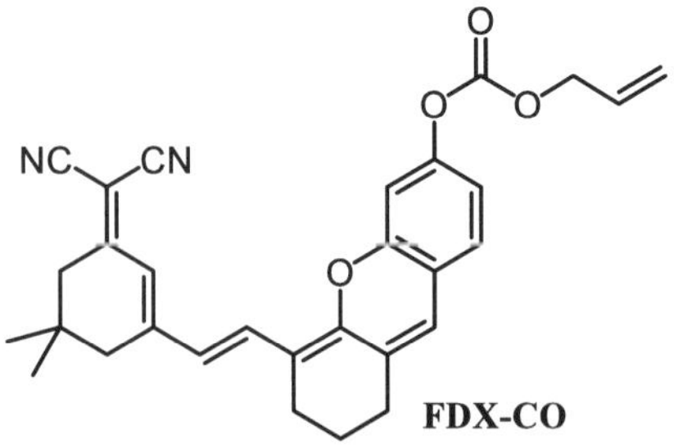

19

Product:
λ_{ex} = 580 nm
λ_{em} = 770 nm

~4-fold turn-on response (8 min):
[FDX-CO] = [PdCl$_2$] = 10 μM, [CORM-3] = 30 μM in 100 mM PBS with 10% DMF, pH 7.4

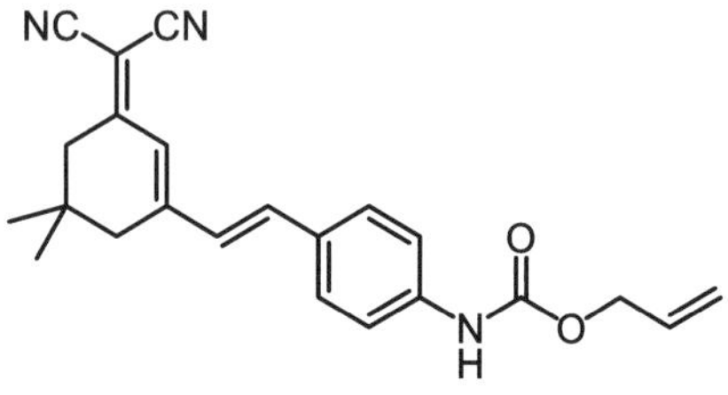

FDX-CO

λ_{ex} = 440 nm
λ_{em} = 592 / 655 nm
(*probe / product*)

~7-fold ratiometric fluorescence response (15 min):
[NIR-Ratio-CO] = 5 μM, [PdCl$_2$] = 10 μM, [CORM-3] = 100 μM in 10 mM PBS with 20% DMSO, pH 7.4, 37 °C

NIR-Ratio-CO

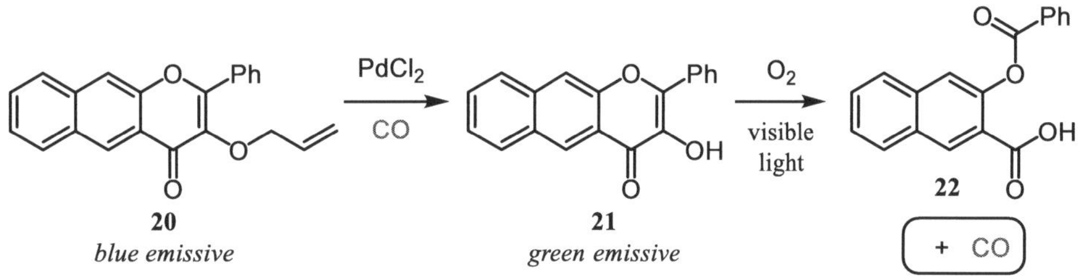

Figure 18.20 Analyte replacement strategy employing a flavonol-derived deallylative probe.

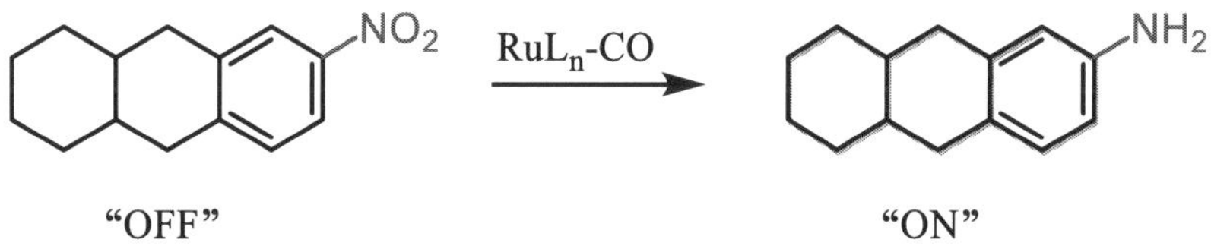

Figure 18.21 General strategy for nitroarene reduction probes for CORM detection.

fragments) to the flavonol, which altered fluorescent signal and inhibited the ensuing CO release. However, both CO detection and subsequent release by the probe were demonstrated in living cells, ostensibly owing to species in the cellular environment able to outcompete this metal binding.

The deallylation strategy for CO detection is synthetically straightforward and amenable to a large range of sensing molecules, providing opportunities for tuning excitation and emission wavelengths, subcellular targeting, and other properties. These probes generally provide rapid response and high sensitivity. Allyl carbonates have provided exceptional responses, but several reports indicate potentially problematic background hydrolysis and/or interference from nucleophilic analytes. Allyl ethers and carbamates provide improved stability. The deallylation approach is inherently limited by the ternary nature of the probe "system," which requires a Pd^{2+} source in addition to CO for response. Accordingly, the signal results from interaction with Pd^0 rather than CO directly, thus limiting the fidelity of spatial imaging conclusions. A final consideration is that the reducing cellular environment or other chemical reductants may lead to additional complications in producing accurate signal from CO itself.

Application in Other Imaging Modalities

Beyond fluorescent probes, the sequential Pd^{2+} reduction–deallylation strategy has been applied for bioluminescence imaging. An O-allylated luciferin probe has been reported that is unreactive toward luciferase but undergoes selective deallylation with Pd^{2+} and CO to release free D-luciferin [103]. This probe and $PdCl_2$-containing liposomes were utilized in bioluminescent imaging of exogenous CO (via CORM-3) and endogenous CO (via heme stimulation) in luciferase-transfected cells, although cell toxicity is a concern at the indicated probe and $PdCl_2$ concentrations (100 μM each). Modest luminescence response was also observed toward exogenous (10–15 mM CORM-3) and endogenous CO (20–40 mM heme preincubation) studies in live mice.

Nitroarene Reduction

General Mechanism and Reactivity

A series of fluorescent probes featuring the reduction of an aryl nitro group to aryl amine has been reported for detecting CO in the form of CORMs. Mechanistically, these probes are proposed to proceed via sequential six-electron reduction with 3 equiv. of metal carbonyl via nitroso and hydroxyl intermediates [104]. In general, these compounds take advantage of the accompanying change in internal charge transfer from this transformation to effect a significant fluorogenic response (Figure 18.21). The largely biorthogonal reactivity of the nitro group imparts high selectivity toward common biological reactive species to these probes, although potential interference from nitroreductase activity has not been evaluated.

A thorough investigation recently established that the reduction of the nitroaryl unit in this class of probes requires a ruthenium carbonyl moiety [105]. Specifically, independent synthesis and evaluation of several probes, including LysoFP-NO$_2$ and NIR-CO, reproduced their strong fluorogenic responses to CORM-3 or CORM-2, both of which are ruthenium-based CORMs. However, these compounds failed to respond to pure CO gas. Similar results were established using p-nitrobenzamide as a model substrate, firmly establishing the differential ability of the Ru-based CORMs in reducing the nitroarene

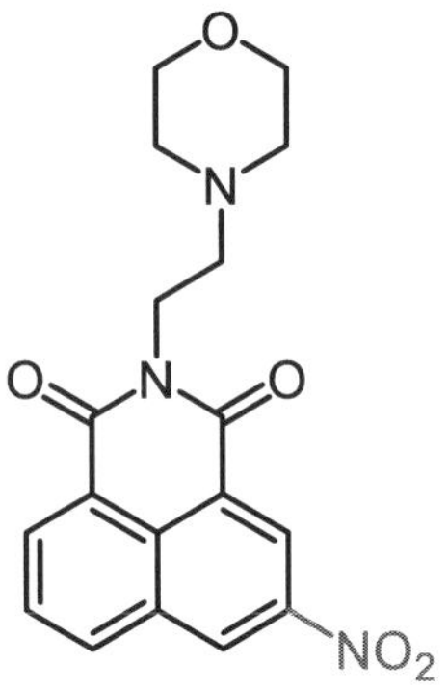

Figure 18.22 Examples of nitroarene reduction probes for CORM detection.

LysoFP-NO$_2$

$\lambda_{ex/em}$ = 440 / 558 nm
$\Phi_{(probe)}$ = 0.0016
$\Phi_{(product)}$ = 0.1025
LOD = 0.60 μM

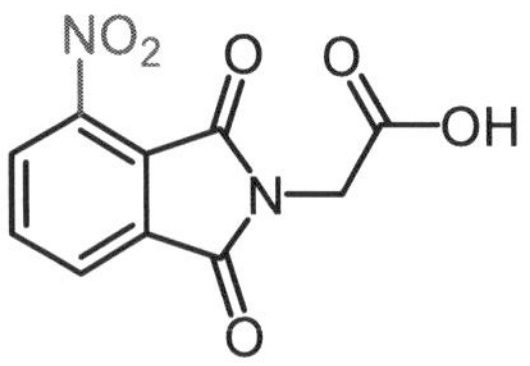

CORM-green

$\lambda_{ex/em}$ = 420 / 503 nm
$\Phi_{(probe)}$ < 0.001
$\Phi_{(product)}$ = 0.318
LOD = 16 nM

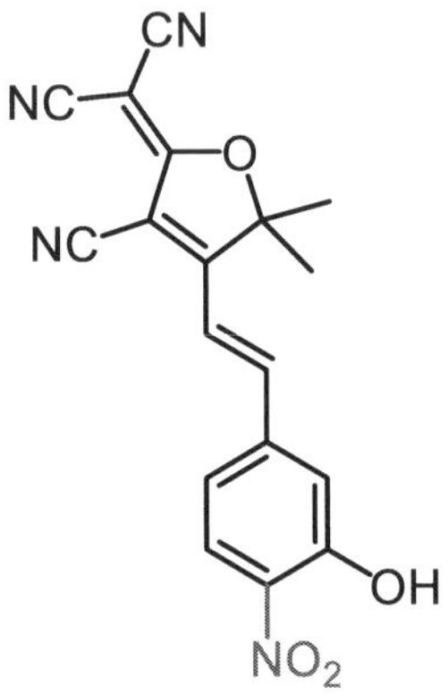

NIR-CO

$\lambda_{ex/em}$ = 580 / 665 nm
LOD = 6.1 nM

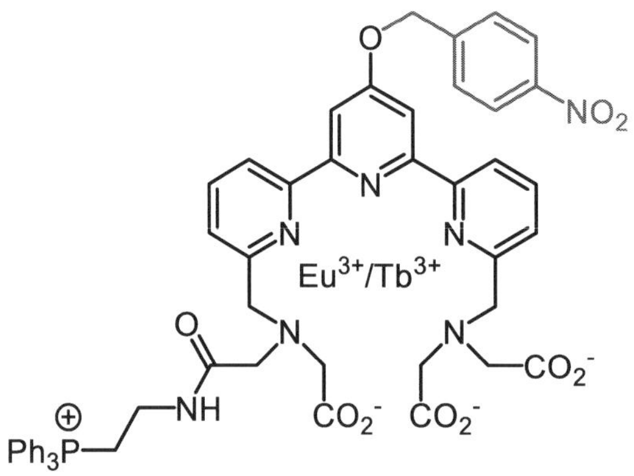

Mito-NBTTA-Tb^{3+}/Eu^{3+}

48-fold ratiometric response
λ_{ex} = 330 nm
λ_{em} = 610 / 540 nm (*probe / product*)
LOD = 0.44 μM

functionality to the corresponding amine. In the same study, palladacycle probe COP-1 produced fluorescent turn-on responses to a metal-free CO prodrug as well as pure CO gas. Thus, probes functioning via nitroaryl reduction are more accurately able to sense Ru-based CORMs but not free CO, thereby obscuring biological conclusions regarding endogenous CO. One critical lesson from this study is the need to use CO from different sources, including gaseous CO, in validating the performance of a CO probe.

Nitroaryl Reduction Fluorescent Probes

Dhara and coworkers reported the initial entry in this family, LysoFP-NO$_2$, in 2018 (Figure 18.22) [106]. This probe is based on a 3-nitro-1,8-naphthalimide core and includes an *N*-ethyl morpholino linker. *In vitro* studies using 10 μM probe and 100 μM CORM-3 revealed a greater than 75-fold increase in fluorescence after 45 min (λ_{ex} = 440 nm, λ_{ex} = 528 nm). This response was rationalized via strengthened ICT and/or suppression of photoinduced electron transfer

(PET) to the nitro group via reduction to the corresponding amine. LysoFP-NO$_2$ displayed notable response to CORM-3 in living MCF7 cells, establishing the viability of this mechanistic modality for intracellular detection applications. As predicted by the morpholino moiety, lysosomal targeting was supported via colocalization imaging with LysoTracker Blue. This group has since reported two additional probes based on the 3-nitro-naphthalimide motif compatible with live cell imaging [107,108]. The most recent example is notable for including a pyridinium moiety for localization in cell nuclei.

Several additional tools employing the nitroarene reduction strategy for CORM detection have been reported using alternative fluorescent scaffolds. Feng and coworkers reported CORM-green, a 3-nitrophthalimide-based dye that is nearly nonemissive [109]. Reduction with CORM-3 to the amino product produces a strong fluorescent response (λ_{ex} = 365 nm, λ_{em} = 503 nm) due to an excited-state intramolecular proton process. This tool was shown to exhibit high selectivity to a wide range of reactive analytes and an LOD of ~16 nM (S/N = 3). CORM-green produced clear responses to CORM-3 incubation in fluorescent imaging studies in living cells, living zebrafish, and living mice.

Sheng and coworkers recently reported NIR-CO, which is based on the TCF fluorescent scaffold [110]. Reduction of the nitro group to amine upon exposure to CORM-2 restores the distinct donor–acceptor structure and ICT fluorescence with emission into the NIR region (λ_{em} = 665 nm). The LOD was calculated to be 6.1 nM (S/N = 3) based on *in vitro* studies with CORM-2. The use of 30% DMSO for *in vitro* studies likely indicates limited aqueous solubility of this probe. Despite this limitation, NIR-CO was used in fluorescence imaging in live cells with exogenous CORM-2, as well as in glucose deprivation studies in both live cells and zebrafish. Greater fluorescent response was observed in both cells and zebrafish incubated with lower levels of glucose, and incubation with ZnPP, a HO-1 inhibitor, attenuated fluorescence in samples with comparable glucose. These observations are consistent with the postulated upregulation of HO-1 via transient glucose inhibition. However, caution is warranted in inferring endogenous CO levels from these biological experiments due to the nonreactivity of NIR-CO to free CO.

A unique lanthanide-based time-gated luminescent probe for ratiometric CORM detection has been reported by Song and coworkers [111]. This probe system is comprised of a terpyridyl polyacid core coordinated to Eu^{3+} or Tb^{3+} along with a phosphonium tag. A *p*-nitrobenzyloxy group provides the mechanistic turn-on handle, reacting with CORM-3 to eliminate the corresponding phenolate after reduction. The concomitant change in emission wavelength and intensity from the different Eu and Tb complexes provides a ratiometric signal ($I_{540\ nm}$/$I_{610\ nm}$) with a 48-fold response and a 0.44 µM LOD ($3\sigma/k$). Imaging studies revealed that incubated probe (100 µM) in living HeLa cells yielded ratiometric response to CORM-3 and confirmed mitochondrial localization. Increased signal was observed in cells preincubated with heme or LPS, consistent with increased HO-1 expression. The probe also displayed turn-on response to CORM-3 in mouse liver tissue, *Daphnia magna*, and living mice.

Fluorescent probes based on this nitroarene reduction strategy commonly offer simple preparation, high selectivity to other biological analytes, and can produce excellent sensitivity toward Ru-based CORMs. As researchers continue to explore new reaction-based approaches to CO detection, we caution against sole evaluation with CORMs and encourage validation with CO gas.

Miscellaneous Response Mechanisms

The overwhelming majority of CO-sensing fluorescent probes operate via the previously described strategies. However, a few novel approaches have been reported and may spur future study. Zhou and coworkers have described RCO, a rhodamine-based probe functionalized with a pyridylhydrazone moiety (Figure 18.23) [112]. *In vitro* treatment with CORM-3 produced a 46-fold increase in emission at 578 nm after 30 min, and an approximate 10 nM LOD was determined (method unspecified). The fluorogenic response to CORM-3 is attributed to ring opening of the spirolactam and hydrolysis to release the bright parent rhodamine B dye. Using density functional theory calculations, the authors propose reaction of CO with the hydrazone to form a hydrolytically unstable diazetidone. However, a greatly attenuated response is observed to gaseous CO, which may indicate that this interesting trigger operates through metal coordination with CORMs. Cell imaging studies indicated mitochondrial targeting and turn-on response to CORM-3 as well as increased signal in cells preincubated with heme.

A unique response mechanism has been reported using iodocarbazole platform MPVC-I. This probe undergoes an azidocarbonylation reaction when reacted with CO in the presence of azide ions and Pd^{2+} to produce a fluorescent product **24**. This cross-coupling-based platform is compatible with two-photon excitation and suitable for CO analysis of gaseous

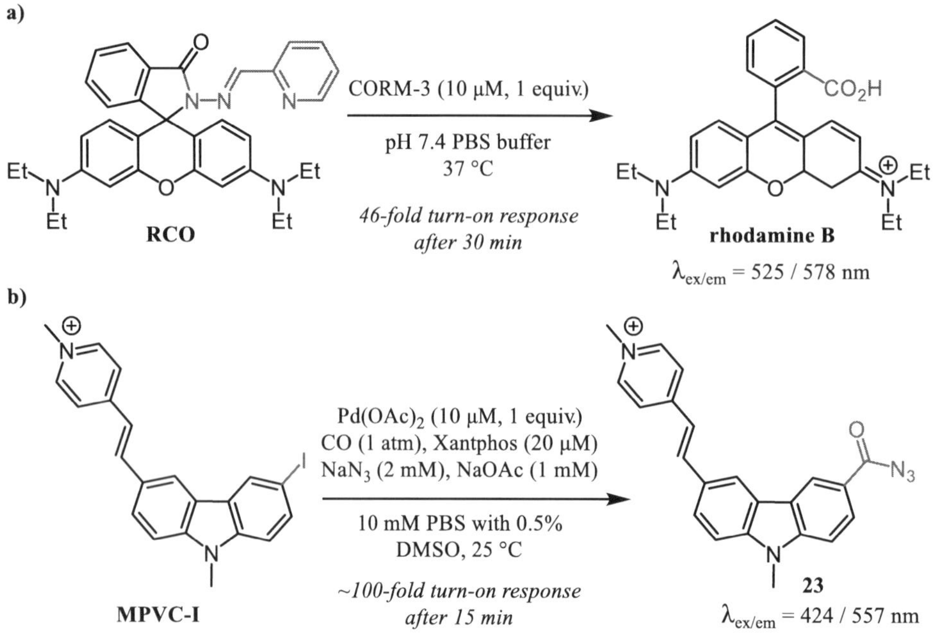

Figure 18.23 Fluorogenic CO probes with alternate turn-on mechanisms: (A) ring opening and hydrolysis of a pyridyl hydrazone and (B) palladium-catalyzed azidocarbonylation.

and blood samples. Interference with thiols and toxicity with azide limit the viability of this probe system in living systems.

Conclusions

In the first decade of development, metal displacement and deallylation approaches have been the most prevalent approaches to CO detection, in terms of both development and application. However, new strategies are still needed. For example, there are currently no probes for CO detection that operate via reversible binding or sensing. Such chemosensors would be valuable for temporal monitoring of CO levels in biological applications. Another key question is whether metals will always be required for CO detection. Limited reactivity of CO in the absence of a transition metal suggests that metals are required. Again, we note caution in analyte selection/CO source and sensing conclusions during the development of new probe structures and triggers. It is critical to evaluate a new system with CO gas. Organic CO prodrugs can aid in this evaluation as well. In the coming years, researchers will continue to use CO probes for answering questions about the endogenous roles of CO and the continued development of CO-related therapeutics. Sensitive, selective, and tunable fluorescent CO probes are critical to these advancements.

Acknowledgments

We gratefully acknowledge the partial financial support by the National Science Foundation (CHE-1900482) and the National Institutes of Health (R21GM135824) to B.W.M. and M.R.S. for our work on fluorescent probe design. R.R.W acknowledges acquisition of an NMR spectrometer through the National Science Foundation (CHE-1726836) in partial support of fluorescent probe development.

References

1. Ryter, S.W., Alam, J., and Choi, A.M.K. (2006). Heme oxygenase-1/carbon monoxide: From basic science to therapeutic applications. *Physiol. Rev.* 86: 583–650.
2. Motterlini, R. and Otterbein, L.E. (2010). The therapeutic potential of carbon monoxide. *Nat. Rev. Drug Discovery* 9: 728–743.
3. Yang, X., De Caestecker, M., Otterbein, L.E., and Wang, B. (2020). Carbon monoxide: An emerging therapy for acute kidney injury. *Med. Res. Rev.* 40: 1147–1177.
4. Hopper, C.P., De La Cruz, L.K., Lyles, K.V., Wareham, L.K., Gilbert, J.A., Eichenbaum, Z., Magierowski, M., Poole, R.K., Wollborn, J., and Wang, B. (2020). Role of carbon monoxide in host–gut microbiome communication. *Chem. Rev.* 120: 13273–13311.

5. Wu, L. and Wang, R. (2005). Carbon monoxide: Endogenous production, physiological functions, and pharmacological applications. *Pharmacol. Rev.* 57: 585.

6. Motterlini, R., Clark James, E., Foresti, R., Sarathchandra, P., Mann Brian, E., and Green Colin, J. (2002). Carbon monoxide-releasing molecules. *Circ. Res.* 90: e17–e24.

7. Alberto, R. and Motterlini, R. (2007). Chemistry and biological activities of CO-releasing molecules (CORMs) and transition metal complexes. *Dalton Trans.* 36: 1651–1660.

8. Roberto, M., Brian, E.M., Tony, R.J., James, E.C., Roberta, F., and Colin, J.G. (2003). Bioactivity and pharmacological actions of carbon monoxide-releasing molecules. *Curr. Pharm. Des.* 9: 2525–2539.

9. Heinemann, S.H., Hoshi, T., Westerhausen, M., and Schiller, A. (2014). Carbon monoxide – physiology, detection and controlled release. *Chem. Commun.* 50: 3644–3660.

10. Romão, C.C., Blättler, W.A., Seixas, J.D., and Bernardes, G.J.L. (2012). Developing drug molecules for therapy with carbon monoxide. *Chem. Soc. Rev.* 41: 3571–3583.

11. Itou, M., Araki, Y., Ito, O., and Kido, H. (2006). Carbon monoxide ligand-exchange reaction of triruthenium cluster complexes induced by photosensitized electron transfer: A new type of photoactive CO color sensor. *Inorg. Chem.* 45: 6114–6116.

12. Paul, S., Amalraj, F., and Radhakrishnan, S. (2009). CO sensor based on polypyrrole functionalized with iron porphyrin. *Synth. Met.* 159: 1019–1023.

13. Benito-Garagorri, D., Puchberger, M., Mereiter, K., and Kirchner, K. (2008). Stereospecific and reversible CO binding at iron pincer complexes. *Angew. Chem. Int. Ed.* 47: 9142–9145.

14. Moragues, M.E., Esteban, J., Ros-Lis, J.V., Martínez-Máñez, R., Marcos, M.D., Martínez, M., Soto, J., and Sancenón, F. (2011). Sensitive and selective chromogenic sensing of carbon monoxide via reversible axial CO coordination in binuclear rhodium complexes. *J. Am. Chem. Soc.* 133: 15762–15772.

15. Gulino, A., Gupta, T., Altman, M., Lo Schiavo, S., Mineo, P.G., Fragalà, I.L., Evmenenko, G., Dutta, P., and Van Der Boom, M.E. (2008). Selective monitoring of parts per million levels of CO by covalently immobilized metal complexes on glass. *Chem. Commun.* 2900–2902.

16. Wang, J., Karpus, J., Zhao, B.S., Luo, Z., Chen, P.R., and He, C. (2012). A selective fluorescent probe for carbon monoxide imaging in living cells. *Angew. Chem. Int. Ed.* 51: 9652–9656.

17. Michel, B.W., Lippert, A.R., and Chang, C.J. (2012). A reaction-based fluorescent probe for selective imaging of carbon monoxide in living cells using a palladium-mediated carbonylation. *J. Am. Chem. Soc.* 134: 15668–15671.

18. Aron, A.T., Ramos-Torres, K.M., Cotruvo, J.A., and Chang, C.J. (2015). Recognition- and reactivity-based fluorescent probes for studying transition metal signaling in living systems. *Acc. Chem. Res.* 48: 2434–2442.

19. Chan, J., Dodani, S.C., and Chang, C.J. (2012). Reaction-based small-molecule fluorescent probes for chemoselective bioimaging. *Nat. Chem.* 4: 973–984.

20. Yang, Y., Zhao, Q., Feng, W., and Li, F. (2013). Luminescent chemodosimeters for bioimaging. *Chem. Rev.* 113: 192–270.

21. Ashton, T.D., Jolliffe, K.A., and Pfeffer, F.M. (2015). Luminescent probes for the bioimaging of small anionic species in vitro and in vivo. *Chem. Soc. Rev.* 44: 4547–4595.

22. Lou, Z., Li, P., and Han, K. (2015). Redox-responsive fluorescent probes with different design strategies. *Acc. Chem. Res.* 48: 1358–1368.

23. New, E.J. (2016). Harnessing the potential of small molecule intracellular fluorescent sensors. *ACS Sens.* 1: 328–333.

24. Lavis, L.D. (2017). Chemistry is dead. Long live chemistry! *Biochemistry* 56: 5165–5170.

25. Bruemmer, K.J., Crossley, S.W.M., and Chang, C.J. (2020). Activity-based sensing: a synthetic methods approach for selective molecular imaging and beyond. *Angew. Chem. Int. Ed.* 59: 13734–13762.

26. Douglas, C.G., Haldane, J.S., and Haldane, J.B.S. (1912). The laws of combination of hæmoglobin with carbon monoxide and oxygen. *J. Physiol.* 44: 275–304.

27. Franke, R., Selent, D., and Börner, A. (2012). Applied hydroformylation. *Chem. Rev.* 112: 5675–5732.

28. Tolman, C.A. (1977). Steric effects of phosphorus ligands in organometallic chemistry and homogeneous catalysis. *Chem. Rev.* 77: 313–348.

29. Yang, M., Fan, J., Du, J., and Peng, X. (2020). Small-molecule fluorescent probes for imaging gaseous signaling molecules: current progress and future implications. *Chem. Sci.* 11: 5127–5141.

30. Mukhopadhyay, S., Sarkar, A., Chattopadhyay, P., and Dhara, K. (2020). Recent advances in fluorescence light-up endogenous and exogenous carbon monoxide detection in biology. *Chem. – Asian J.* 15: 3162–3179.

31. Liu, X., Li, N., Li, M., Chen, H., Zhang, N., Wang, Y., and Zheng, K. (2020). Recent progress in fluorescent probes for detection of carbonyl species: formaldehyde, carbon monoxide and phosgene. *Coord. Chem. Rev.* 404: 213109.

32. Alday, J., Mazzeo, A., and Suarez, S. (2020). Selective detection of gasotransmitters using fluorescent probes based on transition metal complexes. *Inorg. Chim. Acta* 510: 119696.
33. Tang, Y., Ma, Y., Yin, J., and Lin, W. (2019). Strategies for designing organic fluorescent probes for biological imaging of reactive carbonyl species. *Chem. Soc. Rev.* 48: 4036–4048.
34. Ohata, J., Bruemmer, K.J., and Chang, C.J. (2019). Activity-based sensing methods for monitoring the reactive carbon species carbon monoxide and formaldehyde in living systems. *Acc. Chem. Res.* 52: 2841–2848.
35. Iovan, D.A., Jia, S., and Chang, C.J. (2019). Inorganic chemistry approaches to activity-based sensing: From metal sensors to bioorthogonal metal chemistry. *Inorg. Chem.* 58: 13546–13560.
36. Strianese, M. and Pellecchia, C. (2016). Metal complexes as fluorescent probes for sensing biologically relevant gas molecules. *Coord. Chem. Rev.* 318: 16–28.
37. Marín-Hernández, C., Toscani, A., Sancenón, F., Wilton-Ely, J.D.E.T., and Martínez-Máñez, R. (2016). Chromo-fluorogenic probes for carbon monoxide detection. *Chem. Commun.* 52: 5902–5911.
38. Marhenke, J., Trevino, K., and Works, C. (2016). The chemistry, biology and design of photochemical CO releasing molecules and the efforts to detect CO for biological applications. *Coord. Chem. Rev.* 306: 533–543.
39. Zhou, X., Lee, S., Xu, Z., and Yoon, J. (2015). Recent progress on the development of chemosensors for gases. *Chem. Rev.* 115: 7944–8000.
40. Yuan, L., Lin, W., Tan, L., Zheng, K., and Huang, W. (2013). Lighting up carbon monoxide: Fluorescent probes for monitoring CO in living cells. *Angew. Chem. Int. Ed.* 52: 1628–1630.
41. Bezner, B.J., Ryan, L.S., and Lippert, A.R. (2020). Reaction-based luminescent probes for reactive sulfur, oxygen, and nitrogen species: Analytical techniques and recent progress. *Anal. Chem.* 92: 309–326.
42. Roberts, G.P., Kerby, R.L., Youn, H., and Conrad, M. (2008). *The Smallest Biomolecules: Diatomics and Their Interactions with Heme Proteins* (ed. A. Ghosh), 498–523. Amsterdam: Elsevier.
43. Minegishi, S., Yumura, A., Miyoshi, H., Negi, S., Taketani, S., Motterlini, R., Foresti, R., Kano, K., and Kitagishi, H. (2017). Detection and removal of endogenous carbon monoxide by selective and cell-permeable hemoprotein model complexes. *J. Am. Chem. Soc.* 139: 5984–5991.
44. Dupont, J., Consorti, C.S., and Spencer, J. (2005). The potential of palladacycles: More than just precatalysts. *Chem. Rev.* 105: 2527–2572.
45. Dupont, J., Pfeffer, M., Daran, J.C., and Jeannin, Y. (1987). Reactivity of cyclopalladated compounds. Part 17. Influence of the donor atom in metallacyclic rings on the insertion of tert-butyl isocyanide and carbon monoxide into their palladium-carbon bonds. X-ray molecular structure of cyclo-[Pd(.eta.-(CN)-.mu.-C(C6H4CH2SMe): nBu-tert)Br]2. *Organometallics* 6: 899–901.
46. Li, H., Cai, G.-X., and Shi, Z.-J. (2010). LiCl-Promoted Pd(ii)-catalyzed ortho carbonylation of N,N-dimethylbenzylamines. *Dalton Trans.* 39: 10442–10446.
47. Santra, M., Ko, S.-K., Shin, I., and Ahn, K.H. (2010). Fluorescent detection of palladium species with an O-propargylated fluorescein. *Chem. Commun.* 46: 3964–3966.
48. Yusop, R.M., Unciti-Broceta, A., Johansson, E.M.V., Sánchez-Martín, R.M., and Bradley, M. (2011). Palladium-mediated intracellular chemistry. *Nat. Chem.* 3: 239–243.
49. Morstein, J., Höfler, D., Ueno, K., Jurss, J.W., Walvoord, R.R., Bruemmer, K.J., Rezgui, S.P., Brewer, T.F., Saitoe, M., Michel, B.W., and Chang, C.J. (2020). Ligand-directed approach to activity-based sensing: Developing palladacycle fluorescent probes that enable endogenous carbon monoxide detection. *J. Am. Chem. Soc.* 142: 15917–15930.
50. Ueno, K., Morstein, J., Ofusa, K., Naganos, S., Suzuki-Sawano, E., Minegishi, S., Rezgui, S.P., Kitagishi, H., Michel, B.W., Chang, C.J., Horiuchi, J., and Saitoe, M. (2020). Carbon monoxide, a retrograde messenger generated in postsynaptic mushroom body neurons, evokes noncanonical dopamine release. *J. Neurosci.* 40: 3533.
51. Kueh, J.T.B., Stanley, N.J., Hewitt, R.J., Woods, L.M., Larsen, L., Harrison, J.C., Rennison, D., Brimble, M.A., Sammut, I.A., and Larsen, D.S. (2017). Norborn-2-en-7-ones as physiologically-triggered carbon monoxide-releasing prodrugs. *Chem. Sci.* 8: 5454–5459.
52. Jin, Z., Wen, Y., Xiong, L., Yang, T., Zhao, P., Tan, L., Wang, T., Qian, Z., Su, B.-L., and He, Q. (2017). Intratumoral H2O2-triggered release of CO from a metal carbonyl-based nanomedicine for efficient CO therapy. *Chem. Commun.* 53: 5557–5560.
53. Ji, X., Ji, K., Chittavong, V., Yu, B., Pan, Z., and Wang, B. (2017). An esterase-activated click and release approach to metal-free CO-prodrugs. *Chem. Commun.* 53: 8296–8299.
54. Ji, X., De La Cruz, L.K.C., Pan, Z., Chittavong, V., and Wang, B. (2017). pH-Sensitive metal-free carbon monoxide prodrugs with tunable and predictable release rates. *Chem. Commun.* 53: 9628–9631.

55. Diring, S., Carné-Sánchez, A., Zhang, J., Ikemura, S., Kim, C., Inaba, H., Kitagawa, S., and Furukawa, S. (2017). Light responsive metal–organic frameworks as controllable CO-releasing cell culture substrates. *Chem. Sci.* 8: 2381–2386.

56. Ji, X., Zhou, C., Ji, K., Aghoghovbia, R.E., Pan, Z., Chittavong, V., Ke, B., and Wang, B. (2016). Click and release: a chemical strategy toward developing gasotransmitter prodrugs by using an intramolecular Diels–Alder reaction. *Angew. Chem. Int. Ed.* 55: 15846–15851.

57. Fujita, K., Tanaka, Y., Abe, S., and Ueno, T. (2016). A photoactive carbon-monoxide-releasing protein cage for dose-regulated delivery in living cells. *Angew. Chem. Int. Ed.* 55: 1056–1060.

58. Fayad-Kobeissi, S., Ratovonantenaina, J., Dabiré, H., Wilson, J.L., Rodriguez, A.M., Berdeaux, A., Dubois-Randé, J.-L., Mann, B.E., Motterlini, R., and Foresti, R. (2016). Vascular and angiogenic activities of CORM-401, an oxidant-sensitive CO-releasing molecule. *Biochem. Pharmacol.* 102: 64–77.

59. Albuquerque, I.S., Jeremias, H.F., Chaves-Ferreira, M., Matak-Vinkovic, D., Boutureira, O., Romão, C.C., and Bernardes, G.J.L. (2015). An artificial CO-releasing metalloprotein built by histidine-selective metallation. *Chem. Commun.* 51: 3993–3996.

60. Pai, S., Hafftlang, M., Atongo, G., Nagel, C., Niesel, J., Botov, S., Schmalz, H.-G., Yard, B., and Schatzschneider, U. (2014). New modular manganese(i) tricarbonyl complexes as PhotoCORMs: in vitro detection of photoinduced carbon monoxide release using COP-1 as a fluorogenic switch-on probe. *Dalton Trans.* 43: 8664–8678.

61. Ji, X., Pan, Z., Li, C., Kang, T., De La Cruz, L.K.C., Yang, L., Yuan, Z., Ke, B., and Wang, B. (2019). Esterase-sensitive and pH-controlled carbon monoxide prodrugs for treating systemic inflammation. *J. Med. Chem.* 62: 3163–3168.

62. Amorim, A.L., Peterle, M.M., Guerreiro, A., Coimbra, D.F., Heying, R.S., Caramori, G.F., Braga, A.L., Bortoluzzi, A.J., Neves, A., Bernardes, G.J.L., and Peralta, R.A. (2019). Synthesis, characterization and biological evaluation of new manganese metal carbonyl compounds that contain sulfur and selenium ligands as a promising new class of CORMs. *Dalton Trans.* 48: 5574–5584.

63. Gao, C., Liang, X., Guo, Z., Jiang, B.-P., Liu, X., and Shen, X.-C. (2018). Diiron hexacarbonyl complex induces site-specific release of carbon monoxide in cancer cells triggered by endogenous glutathione. *ACS Omega* 3: 2683–2689.

64. De La Cruz, L.K.C., Benoit, S.L., Pan, Z., Yu, B., Maier, R.J., Ji, X., and Wang, B. (2018). Click, release, and fluoresce: a chemical strategy for a cascade prodrug system for codelivery of carbon monoxide, a drug payload, and a fluorescent reporter. *Org. Lett.* 20: 897–900.

65. Zheng, K., Lin, W., Tan, L., Chen, H., and Cui, H. (2014). A unique carbazole–coumarin fused two-photon platform: development of a robust two-photon fluorescent probe for imaging carbon monoxide in living tissues. *Chem. Sci.* 5: 3439–3448.

66. Li, Y., Wang, X., Yang, J., Xie, X., Li, M., Niu, J., Tong, L., and Tang, B. (2016). Fluorescent probe based on azobenzene-cyclopalladium for the selective imaging of endogenous carbon monoxide under hypoxia conditions. *Anal. Chem.* 88: 11154–11159.

67. Takahashi, H. and Tsuji, J. (1967). Organic syntheses by means of noble metal compounds: XXXIII. Carbonylation of azobenzene-palladium chloride complexes. *J. Organomet. Chem.* 10: 511–517.

68. Liu, K., Kong, X., Ma, Y., and Lin, W. (2017). Rational design of a robust fluorescent probe for the detection of endogenous carbon monoxide in living zebrafish embryos and mouse tissue. *Angew. Chem. Int. Ed.* 56: 13489–13492.

69. Xu, S., Liu, H.-W., Yin, X., Yuan, L., Huan, S.-Y., and Zhang, X.-B. (2019). A cell membrane-anchored fluorescent probe for monitoring carbon monoxide release from living cells. *Chem. Sci.* 10: 320–325.

70. Tenhunen, R., Marver, H.S., and Schmid, R. (1970). The enzymatic catabolism of hemoglobin: stimulation of microsomal heme oxygenase by hemin. *J. Lab. Clin. Med.* 75: 410–421.

71. Madea, D., Martínek, M., Muchová, L., Váňa, J., Vítek, L., and Klán, P. (2020). Structural modifications of nile red carbon monoxide fluorescent probe: Sensing mechanism and applications. *J. Org. Chem.* 85: 3473–3489.

72. Sun, M., Yu, H., Zhang, K., Wang, S., Hayat, T., Alsaedi, A., and Huang, D. (2018). Palladacycle based fluorescence turn-on probe for sensitive detection of carbon monoxide. *ACS Sens.* 3: 285–289.

73. Li, X., Gao, X., Shi, W., and Ma, H. (2014). Design strategies for water-soluble small molecular chromogenic and fluorogenic probes. *Chem. Rev.* 114: 590–659.

74. Kim, H.P., Pae, H.-O., Back, S.H., Chung, S.W., Woo, J.M., Son, Y., and Chung, H.-T. (2011). Heme oxygenase-1 comes back to endoplasmic reticulum. *Biochem. Biophys. Res. Commun.* 404: 1–5.

75. De La Torre, C., Toscani, A., Marín-Hernández, C., Robson, J.A., Terencio, M.C., White, A.J.P., Alcaraz,

M.J., Wilton-Ely, J.D.E.T., Martínez-Máñez, R., and Sancenón, F. (2017). Ex vivo tracking of endogenous CO with a Ruthenium(II) complex. *J. Am. Chem. Soc.* 139: 18484–18487.

76. Robson, J.A., Kubánková, M., Bond, T., Hendley, R.A., White, A.J.P., Kuimova, M.K., and Wilton-Ely, J.D.E.T. (2020). Simultaneous detection of carbon monoxide and viscosity changes in cells. *Angew. Chem. Int. Ed.* 59: *n/a*. 21431–21435.

77. Garrou, P.E. and Heck, R.F. (1976). The mechanism of carbonylation of halo(bis ligand)organoplatinum(II), -palladium(II), and -nickel(II) complexes. *J. Am. Chem. Soc.* 98: 4115–4127.

78. Van Leeuwen, P.W.N.M., Zuideveld, M.A., Swennenhuis, B.H.G., Freixa, Z., Kamer, P.C.J., Goubitz, K., Fraanje, J., Lutz, M., and Spek, A.L. (2003). Alcoholysis of Acylpalladium(II) complexes relevant to the alternating copolymerization of ethene and carbon monoxide and the alkoxycarbonylation of alkenes: The importance of Cis-coordinating phosphines. *J. Am. Chem. Soc.* 125: 5523–5539.

79. Song, F., Garner, A.L., and Koide, K. (2007). A highly sensitive fluorescent sensor for palladium based on the allylic oxidative insertion mechanism. *J. Am. Chem. Soc.* 129: 12354–12355.

80. Pal, S., Mukherjee, M., Sen, B., Mandal, S.K., Lohar, S., Chattopadhyay, P., and Dhara, K. (2015). A new fluorogenic probe for the selective detection of carbon monoxide in aqueous medium based on Pd(0) mediated reaction. *Chem. Commun.* 51: 4410–4413.

81. Yan, J.-W., Zhu, J.-Y., Tan, Q.-F., Zhou, L.-F., Yao, P.-F., Lu, Y.-T., Tan, J.-H., and Zhang, L. (2016). Development of a colorimetric and NIR fluorescent dual probe for carbon monoxide. *RSC Adv.* 6: 65373–65376.

82. Feng, W., Liu, D., Feng, S., and Feng, G. (2016). Readily available fluorescent probe for carbon monoxide imaging in living cells. *Anal. Chem.* 88: 10648–10653.

83. Feng, S., Liu, D., Feng, W., and Feng, G. (2017). Allyl fluorescein ethers as promising fluorescent probes for carbon monoxide imaging in living cells. *Anal. Chem.* 89: 3754–3760.

84. Wang, Z., Zhao, Z., Wang, R., Yuan, R., Liu, C., Duan, Q., Zhu, W., Li, X., and Zhu, B. (2019). A mitochondria-targetable colorimetric and far-red fluorescent probe for the sensitive detection of carbon monoxide in living cells. *Anal. Methods* 11: 288–295.

85. Hong, J., Xia, Q., Zhou, E., and Feng, G. (2020). NIR fluorescent probe based on a modified rhodol-dye with good water solubility and large Stokes shift for monitoring CO in living systems. *Talanta* 215: 120914.

86. Feng, W., Hong, J., and Feng, G. (2017). Colorimetric and ratiometric fluorescent detection of carbon monoxide in air, aqueous solution, and living cells by a naphthalimide-based probe. *Sens. Actuators, B* 251: 389–395.

87. Wang, Z., Geng, Z., Zhao, Z., Sheng, W., Liu, C., Lv, X., He, Q., and Zhu, B. (2018). A highly specific and sensitive ratiometric fluorescent probe for carbon monoxide and its bioimaging applications. *New J. Chem.* 42: 14417–14423.

88. Zang, S., Shu, W., Shen, T., Gao, C., Tian, Y., Jing, J., and Zhang, X. (2020). Palladium-triggered ratiometric probe reveals CO's cytoprotective effects in mitochondria. *Dyes Pigm.* 173: 107861.

89. Wang, J., Li, C., Chen, Q., Li, H., Zhou, L., Jiang, X., Shi, M., Zhang, P., Jiang, G., and Tang, B.Z. (2019). An easily available ratiometric reaction-based AIE probe for carbon monoxide light-up imaging. *Anal. Chem.* 91: 9388–9392.

90. Yan, L., Nan, D., Lin, C., Wan, Y., Pan, Q., and Qi, Z. (2018). A near-infrared fluorescent probe for rapid detection of carbon monoxide in living cells. *Spectrochim. Acta, Part A* 202: 284–289.

91. Li, S.-J., Zhou, D.-Y., Li, Y.-F., Yang, B., Ou-Yang, J., Jie, J., Liu, J., and Li, C.-Y. (2018). Mitochondria-targeted near-infrared fluorescent probe for the detection of carbon monoxide in vivo. *Talanta* 188: 691–700.

92. Zhang, W., Wang, Y., Dong, J., Zhang, Y., Zhu, J., and Gao, J. (2019). Rational design of stable near-infrared cyanine-based probe with remarkable large stokes shift for monitoring carbon monoxide in living cells and in vivo. *Dyes Pigm.* 171: 107753.

93. Gong, S., Hong, J., Zhou, E., and Feng, G. (2019). A near-infrared fluorescent probe for imaging endogenous carbon monoxide in living systems with a large Stokes shift. *Talanta* 201: 40–45.

94. Wang, Z., Zhao, Z., Liu, C., Geng, Z., Duan, Q., Jia, P., Li, Z., Zhu, H., Zhu, B., and Sheng, W. (2019). A long-wavelength ultrasensitive colorimetric fluorescent probe for carbon monoxide detection in living cells. *Photochem. Photobiol. Sci.* 18: 1851–1857.

95. Deng, Y., Hong, J., Zhou, E., and Feng, G. (2019). Near-infrared fluorescent probe with a super large Stokes shift for tracking CO in living systems based on a novel coumarin-dicyanoisophorone hybrid. *Dyes Pigm.* 170: 107634.

96. Zhou, E., Gong, S., and Feng, G. (2019). Rapid detection of CO in vitro and in vivo with a ratiometric probe showing near-infrared turn-on fluorescence, large Stokes shift, and high signal-to-noise ratio. *Sens. Actuators, B* 301: 127075.

97. Liu, Y., Wang, W.-X., Tian, Y., Tan, M., Du, Y., Jie, J., and Li, C.-Y. (2020). A near-infrared fluorescence probe with a large Stokes shift for detecting carbon monoxide in living cells and mice. *Dyes Pigm.* 180: 108517.

98. Zhang, Y., Kong, X., Tang, Y., Li, M., Yin, Y., and Lin, W. (2020). The development of a hemicyanine-based ratiometric CO fluorescent probe with a long emission wavelength and its applications for imaging CO in vitro and in vivo. *New J. Chem.* 44: 12107–12112.
99. Feng, W. and Feng, G. (2018). A readily available colorimetric and near-infrared fluorescent turn-on probe for detection of carbon monoxide in living cells and animals. *Sens. Actuators, B* 255: 2314–2320.
100. Tian, Y., Jiang, W.-L., Wang, W.-X., Peng, J., Li, X.-M., Li, Y., and Li, C.-Y. (2021). The construction of a near-infrared fluorescent probe with dual advantages for imaging carbon monoxide in cells and in vivo. *Analyst.* 146: Advance Article. 118–123.
101. Zhou, E., Gong, S., Hong, J., and Feng, G. (2020). Development of a new ratiometric probe with near-infrared fluorescence and a large Stokes shift for detection of gasotransmitter CO in living cells. *Spectrochim. Acta, Part A* 227: 117657.
102. Popova, M., Lazarus, L.S., Benninghoff, A.D., and Berreau, L.M. (2020). CO sense and release flavonols: progress toward the development of an analyte replacement photoCORM for use in living cells. *ACS Omega* 5: 10021–10033.
103. Tian, X., Liu, X., Wang, A., Lau, C., and Lu, J. (2018). Bioluminescence imaging of carbon monoxide in living cells and nude mice based on Pd0-mediated Tsuji–Trost reaction. *Anal. Chem.* 90: 5951–5958.
104. Tafesh, A.M. and Weiguny, J. (1996). A review of the selective catalytic reduction of aromatic nitro compounds into aromatic amines, isocyanates, carbamates, and ureas using CO. *Chem. Rev.* 96: 2035–2052.
105. Yuan, Z., Yang, X., De La Cruz, L.K., and Wang, B. (2020). Nitro reduction-based fluorescent probes for carbon monoxide require reactivity involving a ruthenium carbonyl moiety. *Chem. Commun.* 56: 2190–2193.
106. Dhara, K., Lohar, S., Patra, A., Roy, P., Saha, S.K., Sadhukhan, G.C., and Chattopadhyay, P. (2018). A new lysosome-targetable turn-on fluorogenic probe for carbon monoxide imaging in living cells. *Anal. Chem.* 90: 2933–2938.
107. Das, B., Lohar, S., Patra, A., Ahmmed, E., Mandal, S.K., Bhakta, J.N., Dhara, K., and Chattopadhyay, P. (2018). A naphthalimide-based fluorescence "turn-on" chemosensor for highly selective detection of carbon monoxide: imaging applications in living cells. *New J. Chem.* 42: 13497–13502.
108. Sarkar, A., Fouzder, C., Chakraborty, S., Ahmmed, E., Kundu, R., Dam, S., Chattopadhyay, P., and Dhara, K. (2020). A nuclear-localized naphthalimide-based fluorescent light-up probe for selective detection of carbon monoxide in living cells. *Chem. Res. Toxicol.* 33: 651–656.
109. Feng, W., Feng, S., and Feng, G. (2019). A fluorescent ESIPT probe for imaging CO-releasing molecule-3 in living systems. *Anal. Chem.* 91: 8602–8606.
110. Wang, Z., Liu, C., Wang, X., Duan, Q., Jia, P., Zhu, H., Li, Z., Zhang, X., Ren, X., Zhu, B., and Sheng, W. (2019). A metal-free near-infrared fluorescent probe for tracking the glucose-induced fluctuations of carbon monoxide in living cells and zebrafish. *Sens. Actuators, B* 291: 329–336.
111. Tang, Z., Song, B., Ma, H., Luo, T., Guo, L., and Yuan, J. (2019). Mitochondria-targetable ratiometric time-gated luminescence probe for carbon monoxide based on lanthanide complexes. *Anal. Chem.* 91: 2939–2946.
112. Zhang, C., Xie, H., Zhan, T., Zhang, J., Chen, B., Qian, Z., Zhang, G., Zhang, W., and Zhou, J. (2019). A new mitochondrion targetable fluorescent probe for carbon monoxide-specific detection and live cell imaging. *Chem. Commun.* 55: 9444–9447.

Section IV.

Therapeutic Applications

19

CO in Solid Organ Transplantation

Roberta Foresti[1], Roberto Motterlini[1], and Stephan Immenschuh[2]

[1]*IMRB, INSERM, University Paris-Est Créteil, F-94010 Créteil, France*
[2]*Institute of Transfusion Medicine and Transplant Engineering, Hannover Medical School, Hannover, Germany*

Introduction

During transplantation procedures, organs from donors are exposed to a series of stressful conditions that might impair their function once transplanted into recipients. In fact, harvested organs are first subjected to ischemia during storage in preservation solutions and then undergo reperfusion during transplantation that will inevitably cause a degree of injury, thus affecting organ function and viability. In addition, the immune response to the newly transplanted graft is also responsible for exacerbating tissue damage and contributing to organ rejection. Carbon monoxide (CO) generated by heme oxygenases (HO-1 and HO-2) is now recognized as an essential signaling mediator endowed with a variety of physiological effects that may be useful in the context of organ transplantation. Since scientists have developed different modes of CO delivery to biological systems, it became evident that organ transplantation is an ideal setting for testing the hypothesis that CO-based treatments could be useful for preventing tissue injury during grafting. This chapter will update the current knowledge on the outcome of transplantation procedures in different organs where CO was used as an adjuvant in the management of graft dysfunction and rejection.

Modes of CO Delivery in Organ Transplantation

The increasing number of publications pointing to a protective role of endogenously derived CO in a variety of pathological disorders instilled the notion that this gaseous signaling molecule could be utilized for therapeutic applications. The first study to address this possibility was published in 1999 by Otterbein and coworkers who administered CO gas by inhalation to mice subjected to hyperoxia, a condition known to result in severe lung injury [1]. Even though high concentrations of CO can impair O_2 transport by red blood cells and ultimately O_2 delivery to organs and tissues [2], it was shown that the amount of CO gas administered and the time of CO exposure (250 ppm for 72 h) were such that the transient increase in CO-hemoglobin (15%) was well tolerated by the animals. The data indicated that mice exposed to CO gas exhibited a higher resistance to hyperoxia and displayed a significant decrease in lung tissue damage, pulmonary edema, and airway inflammation [1]. A few years later, the same laboratory utilized CO gas in a mouse-to-rat transplantation model and demonstrated that rats exposed to CO gas (400 ppm) for 2 weeks after transplantation resulted in suppression of graft rejection and restoration of long-term graft survival (see below) [3]. The increased survival rate was associated with inhibition of platelet aggregation and reductions of thrombosis and myocardial infarction. In parallel with the development of approaches to study the effect of CO gas in experimental models, the use of "transition metal carbonyls" as a new strategy to deliver CO for therapeutic purposes was proposed in 2002 [4]. These compounds, termed CO-releasing molecules (CORMs), have the ability to release CO in a controlled fashion *in vitro* and *in vivo* and simulate the action of endogenously produced CO (see Chapter 12 for more details on CORMs). The pharmacological effects of one of these compounds, CORM-3, were investigated in a model of cardiac allograft rejection in mice. It was found that CORM-3 administered to mice recipients significantly protected against myocardial ischemia–reperfusion injury (IRI) and prolonged heart graft survival,

Carbon Monoxide in Drug Discovery: Basics, Pharmacology, and Therapeutic Potential, First Edition. Edited by Binghe Wang and Leo E. Otterbein.
© 2022 John Wiley & Sons, Inc. Published 2022 by John Wiley & Sons, Inc.

Table 19.1 Organ transplants worldwide in 2018.[a]

Organ	Kidney	Liver	Heart	Lung	Pancreas	Small bowel
Number	95 479	34 074	8311	6475	2338	163

[a] A total of 146 840 organ transplants were performed in 2018 according to Global Observatory on Donation and Transplantation home page (Data from WHO-ON).

while an inactive CORM-3 incapable of releasing CO was without effect [5]. These reports indicated that both CO gas and CORMs are pharmacologically active in transplantation procedures and remain to date the mostly used strategies to deliver CO for prolonging graft survival and the function of transplanted organs (see below).

CO Ameliorates Organ and Tissue Function After Transplantation

The clinical success of transplantation is affected by various donor- and host-dependent factors, including donor age, human leukocyte antigen matching, and duration of warm and cold graft ischemia times. Notably, the severity of tissue injury caused by ischemia–reperfusion in transplanted organs is of major importance for early graft dysfunction as well as long-term transplant rejection and directly correlates with the lengths of ischemia times that activate inflammatory responses [6,7]. While intracellular inflammatory responses in isolated individual cell types have been extensively studied and are fairly well understood, pathophysiological aspects of intercellular cooperation and cross-talk between various cell types in distinct organs are less clear. Notably, organ-specific cellular interactions appear to be important for preservation procedures of various grafts and how they can be improved by targeted interventions such as *ex vivo* machine perfusion [8,9]. As presented in the sections below, CO can principally interfere with various transplantation-associated pathophysiological events, including IRI, endothelial cell (EC) death and proliferation, or cellular inflammatory responses, all of which are critical for acute and chronic graft rejection [10]. Applications of CO and CORMs as a potential therapeutic strategy in the setting of various organ transplantations are discussed and pertinent findings from preclinical animal models and clinical trials are presented. This chapter covers two major aspects of using CO in organ transplantation: CO administered to recipients of transplantation and CO used in *ex vivo* organ preservation. Because there are various chapters in the book that focus on organ injury including those related to IRI, this chapter provides an overview of the various organ transplant literature and the impact of CO.

CO in the Kidney

The kidney was the first solid organ transplanted in 1954 [11], and currently, transplantations of kidneys largely outnumber those of other organs (95 479 kidneys out of 146 840 total organs transplanted worldwide in 2018; see Table 19.1). Hence, extensive clinical experience on the various immunological and nonimmunological mechanisms of graft rejection in kidney transplantation is available [12]. Due to the prolonged duration of cold ischemia time in deceased donor kidneys, administration of CO gas by inhalation or CORMs appears to hold major promise for potential targeted interventions in this setting [13].

In a preclinical model of combined rat renal and cardiac transplantation, it has been demonstrated that treatment with the HO products, CO and biliverdin, provides protection against IRI-mediated tissue damage in both organs [14]. In a follow-up study, the specific beneficial effects of gaseous CO were confirmed in a rat model, in which CO protected against chronic allograft nephropathy via blocking chronic fibroinflammation [15]. In parallel, application of CORM-2, a ruthenium-based metal carbonyl complex known to release CO, was also shown to counteract renal graft rejection via its anti-inflammatory effects in an allogenic rat model [16]. Moreover, the anti-inflammatory effects of low-dose gaseous CO have been demonstrated to be salutary in chronic allograft nephropathy [17]. Notably, intraoperative administration of gaseous CO during the surgical kidney transplantation procedure has been shown to protect against delayed graft function in a pig model and to restore renal function more rapidly than in controls [18].

CO in the Heart

Heart transplantation has emerged as an efficient therapeutic strategy in the treatment of patients with end-stage heart failure [19]. In contrast to kidney transplantation, in which living kidney donors are now a feasible clinical option, in heart transplantation grafts are only available from brain-dead donors [20].

CO generated by HO-1 has been demonstrated to alleviate transplant rejection via inhibiting platelet aggregation in a mouse-to-rat heart transplantation model [3]. Subsequently, CO was shown to be beneficial via antiapoptotic effects that ameliorated IRI-associated damage after transplantation [21]. Similarly, gaseous CO specifically protected against chronic graft rejection by preventing the formation of arteriosclerotic lesions in a rodent model of aorta transplantation [22]. Independently, other authors corroborated that CO elicited beneficial effects against IRI in a model of combined heterotopic heart and orthotopic kidney transplantation [14]. These findings have been confirmed in a model of fully allogeneic rat heterotopic heart transplantation for low doses of gaseous CO over a time period of up to 100 days [23]. As an extension of these studies, a combination of CO with hydrogen was shown to have salutary effects in a syngeneic heterotopic heart transplantation model [24]. Finally, application of CORM-3 has also been reported using a mouse model of cardiac allograft rejection. In this model, recipient mice that were administered with CORM-3 showed a prolonged survival rate with 60% of transplanted hearts still beating after 25 days, while all mice treated with the inactive counterpart showed heart rejection within 9 days from transplantation [5]. Based on published findings, it appears that reduction of inflammation and oxidative stress are important mechanisms mediating the protective effect of CO in organ transplantation [25].

CO in the Lung

The clinical outcome in transplantation of lungs is markedly worse when compared with that of other solid organs due to the high incidence of acute graft rejection [26]. Moreover, because the low availability of suitable donor organs is a main obstacle in lung transplantation, major efforts aim at increasing the donor pool and the quality of lung grafts [27], which appear to make CO an ideal candidate for potential treatments in transplantation settings of this organ. In a rat model of orthotopic lung transplantation, administration of high concentrations of CO gas had salutary effects via upregulating a number of anti-inflammatory and antiapoptotic genes [28]. Independently, others have shown in a rat model of isogeneic lung transplantation that ischemic lungs were protected by exposure to gaseous CO via inhibition of the nuclear factor Egr-1, a nuclear activator of proinflammatory and prothrombotic gene expression [29]. Similarly, in a preclinical model of lung transplantation, low levels of CO were able to counteract IRI-mediated damage in grafted organs [30]. Also of interest is a recent report on tracheal grafts transplanted in mice in which recipients were previously treated with CORM-2. Allografts treated with CORM-2 displayed a striking reduction of thickening in epithelial and subepithelial airway layers and a reduction of luminal obliteration. These positive effects in the transplanted grafts were associated with a reduction in the number of $CD3^+$ lymphocytes and macrophages as well as a marked decrease in the expression of proinflammatory mediators such as interferon-γ, interleukin (IL)-2, and IL-17A [31].

CO in the Liver

The number of liver transplantations has continuously increased in recent years and has been ranked second among worldwide transplanted organs (34 074 organs in 2018; see Table 19.1). Hepatic grafts are less prone to graft rejection after transplantation if compared with other organs, because the liver is more tolerant due to its high exposure to various antigens from the gut [32].

In preclinical studies, inhalation of CO gas has been shown to be beneficial in liver transplantation, because it provides protection against IRI-induced damage as demonstrated in a rat model of orthotopic transplantation [33]. Others have shown that the salutary effects of CO are mediated via downregulating the number of Kupffer cells during the early phase after reperfusion of liver during transplantation [34]. Interestingly, it has been demonstrated in a rat hepatic IRI model that the protective effects exerted by CORM-2 were mediated via the inflammation signaling molecule high-mobility group box 1 [35]. Furthermore, in a recent investigation, CO (750 ppm) was included among a group of anti-inflammatory strategies during liver perfusion prior to transplantation in pigs [36]. The authors reasoned that a combination of strategies could improve the protective effects of *ex vivo* organ perfusion and demonstrated that livers perfused with CO, N-acetylcysteine, sevoflurane, and alprostadil at 33 °C exhibited reduced IL-6, tumor necrosis factor (TNF)-α, and other damage markers during perfusion. After transplantation, livers in the treatment group displayed lower serum activity of aspartate amino transferase (a marker of liver damage) and bilirubin, suggesting a positive effect of this approach. Thus, the possibility of adding CO together with a series of compounds that could prevent an

CO in Intestine and Pancreas

The number of transplanted non-liver gastrointestinal organs such as small intestine and pancreatic islet cells is relatively low compared to other organs. Consequently, much less is known about the clinical success rate in these organs (Table 19.1).

In a rat model of syngeneic small intestine transplantation, it has been shown that exposure to gaseous CO protected against IRI via antiapoptotic and anti-inflammatory effects [37,38]. Similarly, exposure to CO provided protection in a syngeneic murine marginal mass islet transplantation model and subsequent improvement of islet cell functions via antiapoptotic effects [39]. These findings were confirmed in a follow-up study with an allogeneic mouse islet cell transplantation model [40]. It has been shown that CO treatment of the donor, the islet transplants, or the recipient can improve the outcome of allogeneic islet transplantation in mice [41]. Finally, a recent clinical trial revealed the possibility of exploiting CO as a therapeutic adjuvant to improve the outcome of islet transplantation [42]. Harvested islets from patients undergoing pancreatectomy and islet autotransplantation were placed in control or CO-saturated solutions prior to grafting. No adverse events directly related to the use of CO were observed at 6 months post-transplantation. Moreover, subjects receiving CO-treated islets had less β-cell death, decreased CCL23, and increased CXCL12 levels at 1 or 3 days post-transplantation compared with controls. This pilot trial showed for the first time that harvesting human islets in CO-saturated solutions is a safe strategy in the management of pancreatectomy and islet autotransplantation patients.

CO Gas and CORMs in Organ Preservation

The growing evidence that CO gas and CORMs could be utilized in a controlled fashion to treat inflammation, vascular dysfunction, and ischemic events in the context of organ transplantation [43] prompted scientists to explore the possibility that CO could also function as an adjuvant to improve preservation of organs for transplantation. This strategy would have the obvious advantage of avoiding the systemic administration of CO while targeting the specific organs and tissues *ex vivo* by simply supplementing preservation solutions with either CO gas or CORMs. In fact, placing organs harvested from donors in cold storage solutions is the most common procedure in clinical use and a simple and effective way to preserve and transport organs prior to transplantation [44]. The composition of these hypothermic storage solutions varies and, although Celsior, University Wisconsin (UW), St Thomas, and Euro-Collins solutions are the most commonly used, different formulations have been studied and optimized over the years in order to increase the preservation time and organ function once the graft is transplanted [45]. It is only at the start of this century that CO gas and CORMs have been suggested as additives of cold storage solutions for organ preservation.

CO in Kidney Storage

The first report exploring how CO affects the function of an organ stored in a cold preservation solution was published by Sandouka and colleagues in 2005 [46]. In this study, rabbit kidneys were harvested and immediately flushed with cold Celsior solution alone (control) or supplemented with either CORM-3 or CORM-A1 (50 μM) prior to storage for 24 h at 4 °C. Renal hemodynamic parameters were then assessed on the isolated kidneys using a perfusion apparatus at 37 °C. The rate and amount of CO released from CORMs in the preservation solutions at 4 °C were determined over the 24-h storage period revealing that CORM-3 liberated CO more rapidly ($t_{1/2} \approx 1$ h) than CORM-A1 ($t_{1/2} \approx 18$ h) confirming previous studies on the different kinetics of CO release between these two compounds [5,47,48]. Nevertheless, the total amount of CO accumulated in the cold solution after 24 h was in the same order of magnitude (CORM-3 ≈ 50 μM CO; CORM-A1 ≈ 35 μM CO). Most importantly, it was found that kidneys stored in the presence of CORMs displayed significantly improved renal hemodynamics compared to untreated kidneys once perfused on the organ isolated apparatus. This was revealed by a marked increase in perfusion flow rate, urine flow, and glomerular filtration rate, all key parameters that reflect the function of the kidney. It was also confirmed that the renal protective effects mediated by CORMs were due to CO liberated by the compounds since kidneys stored with inactive CORM-A1 or CORM-3 that were depleted of CO (iCORMs) displayed renal functions similar to control untreated organs. Moreover, this study showed that mitochondria of kidneys stored in either CORM-3 or CORM-A1 solutions were much better protected as indicated by the preservation of mitochondrial respiration, which was significantly impaired in kidneys stored either with cold solution alone or in the presence of iCORMs. Mechanistically, the work of Sandouka and collaborators also revealed that the effects of CO liberated by CORMs were mediated by soluble guanylate cyclase (sGC) via the production of the second messenger cyclic guanosine phosphate (cGMP) since the increase in perfusion flow rate was abolished in kidneys stored with CORM-A1 in the presence of ODQ, an inhibitor of guanylate cyclase. Altogether, these data confirmed for the first time the feasibility of using CORMs and thus CO as effective adjuvants for organ preservation solutions to ameliorate renal function in the context of kidney transplantation.

In line with the experimental approach and findings described by Sandouka et al., subsequent studies also reported the use of organ preservation solutions supplemented with CO gas. Nakao and coworkers investigated whether *ex vivo* delivery of CO gas to the kidney would ameliorate renal injury after cold storage and transplantation. In this study, kidney transplantation was performed in rats following 24 h of cold preservation in UW solutions equilibrated with or without CO gas (final concentration = 40 μM) [49]. It was found that while untreated kidneys during cold storage resulted in progressive deterioration of renal function and increased inflammation following transplantation, kidney grafts preserved with CO gas had significantly less tissue injury. This was associated with increased survival of transplanted animals compared to the control group. It was also found that renal injury in the control group showed considerable degradation of cytochrome P450 enzymes and increased intracellular free heme levels, whereas CO gas-treated kidney grafts maintained their cytochrome P450 protein levels and displayed a decrease in heme-mediated oxidative stress. In a similar study published by the same group, kidneys harvested from pigs were initially stored for 48 h in cold UW solutions or UW supplemented with CO gas and then autotransplanted in a 14-day follow-up study [50]. The results showed that animal survival after transplantation of untreated kidneys was 80% and this was accompanied by increased fibrosis, renal tubular damage, production of inflammatory markers, and a significant increase in serum creatinine and urea levels, specific markers of renal dysfunction. In contrast, CO gas-treated kidneys displayed a better profile with significantly less severe histopathological damage, reduced inflammation, decreased serum creatinine and urea, and leading to a 100% survival of transplanted animals. More recently, a manganese-containing CORM (CORM-401) that has been developed to deliver CO *in vivo* with high efficiency [51,52] has been reported to reduce IRI associated with prolonged cold storage of renal allografts obtained from donation after circulatory death in a porcine model of transplantation. In this model, treatment with CORM-401 reduced urinary protein excretion, attenuated renal damage, and prevented intrarenal hemorrhage and vascular clotting during reperfusion. In addition, CORM-401 appeared to exert anti-inflammatory actions by suppressing the activation of Toll-like receptors [53].

Thus, these findings strongly indicate that *ex vivo* treatment of kidney grafts with CORMs or CO gas during cold storage may be an effective and safe strategy to reduce tissue injury during transplantation procedures.

CO in Liver Storage

Liver has also been investigated for its propensity to be stored successfully in cold solutions supplemented with either CO or CORMs prior to transplantation. Ikeda and colleagues reported that syngeneic transplantation (i.e., between animals genetically identical) of rat liver grafts preserved for 24 h in cold UW solutions containing 5% CO gas resulted in a much higher survival rate (80%) compared to control and untreated transplanted livers (32%) [54]. The CO-treated livers showed less damaged sinusoidal ECs, which are critical for the proper function of hepatocytes, and this was associated with a marked reduction in hepatic necrosis, neutrophil extravasation, expression of adhesion molecules, and decreased levels of serum alanine aminotransferase. Of interest, it has to be noted that liver grafts in UW solutions bubbled with 5% CO gas were kept in tightly sealed container for the whole cold preservation period. Soluble CO levels in these solutions were maintained in the order of 40 μmol/l and, after 18 h of cold preservation, CO levels were 18 pmol/mg of tissue compared to 0.84 pmol/mg in untreated livers.

The benefit of CO gas applied by continuous persufflation during liver preservation was also investigated on graft recovery in an isolated rat liver model. Livers were subjected to 18 h of cold storage with or without persufflation with CO (dissolved in nitrogen) at a concentration of 50 or 250 ppm and graft viability was assessed thereafter upon warm reperfusion *ex vivo* [55]. CO treatment significantly reduced the release of cellular enzymes such as lactate dehydrogenase and increased bile production and the energetic status of livers upon reperfusion by about 50%. These effects were associated with a reduction of free radical-induced lipid peroxidation and vascular perfusion resistance as well as improved mitochondrial ultrastructure. Similarly, Pizarro and colleagues reported a significant improvement in hepatic functions after storage of rat livers for 48 h in UW solutions supplemented with CORM-3 (50 µM) but not with its inactive counterpart (iCORM-3) [56]. The effect of CORM-3 was evident by a marked increase in perfusion flow and decrease in intrahepatic resistance observed when the organs were reperfused on an isolated system. CORM-3-treated livers also displayed a better metabolic capacity as indicated by augmented hepatic oxygen consumption and glycogen content, a reduced release of lactate dehydrogenase, and a better assessment of histological hepatocyte structure. These findings altogether reveal that CO gas and CORMs limit the injury sustained by the liver during cold storage confirming the previous results obtained on the kidney.

CO in the Storage of Other Organs

Following the reports that preservation of livers and kidneys can benefit from the addition of CO to the cold storage solutions, other organs have been investigated with similar results. For instance, bubbling CO gas (5%) into cold UW storage solutions containing lung, intestine, or vein grafts provided increased tissue protection and function when the viability of these organs was evaluated either *ex vivo* or after transplantation *in vivo*. The major beneficial effects mediated by CO were once again manifested by decreased production of inflammatory markers, reduced cellular infiltrate, and preserved vascular and tissue integrity [57–59]. In the case of vascular graft implantation, data showed that CO upregulated the expression of vascular endothelial growth factor and protected ECs against apoptosis leading to inhibition of smooth muscle cell (SMC) proliferation and thus reduced neointimal hyperplasia [59]. These data have important implications in the clinical setting of myocardial diseases because excessive proliferation of SMCs and vascular remodeling typify the major obstacles in coronary heart bypass surgeries and heart transplantation outcomes. Finally, the importance of CO as signaling and effector mediator for the viability of myocardial function *ex vivo* has been confirmed with the use of CORMs. Musameh and coworkers studied isolated rat hearts undergoing cold ischemic storage for 6 h using St Thomas Hospital solution that was supplemented with either CORM-3 or iCORM-3. They found that addition of CORM-3 to the preservation solution resulted in a significant improvement in systolic and diastolic function, a higher coronary flow, and a lower release of cardiac enzymes creatine kinase and lactate dehydrogenase when compared with hearts treated with iCORM-3 [60].

Cellular and Molecular Mechanisms Mediating the Protective Effects of CO in Organ Transplantation

CO can affect various functions in distinct cell populations that are critically involved in graft rejection after solid organ transplantation. In particular, CO can attenuate proinflammatory responses in cells of the vascular system that are critical for tolerance of grafted organs [7,61]. In the following, the role of CO in various cell types that are involved in the pathophysiology of organ transplantation will be discussed.

Interaction of CO with ECs

In solid organ transplantation, the endothelium is the main barrier of the grafted organ to the host immune system, because immune cells of the recipient make their first contact with the grafted organ via the endothelium [61]. Moreover, pathologies related to endothelial dysfunction caused by events either before or after transplantation (e.g., ischemia-dependent tissue damage or immunological graft–host interactions) may compromise graft ECs and their interplay with other cell types of the transplanted organ. Notably, the endothelium is easily accessible by gaseous transmitters via the blood circulation [13].

In ECs, CO has been shown to modulate gene expression of mitogen-activated protein kinases (MAPKs) via modulating the cellular cGMP content. This effect has been proposed to occur via a paracrine mechanism because CO produced by HO activity in SMCs appears to be responsible for this regulation [62]. Moreover, others have shown that HO-1-derived CO can prevent cell death of ECs via activation of p38 MAPK and activation of antiapoptotic pathways [63,64]. Notably, the beneficial effects of low CO concentrations have recently been demonstrated to be mediated via a shift of the metabolic balance in ECs from glycolysis to oxidative phosphorylation caused by mitochondrial uncoupling [65].

Interaction of CO with SMCs

SMCs play a critical role in transplantation, because excessive proliferation of these cells in grafted solid organs leads to transplant vasculopathy, a major cause of chronic rejection in heart and kidney transplantation [61]. CO gas modulates various complex physiological and pathophysiological regulatory pathways in SMCs. Not only SMCs produce CO via HO enzyme activity, but, in addition, CO is a critical modulator of SMC functions in graft rejection. Independently, CO has been shown to be generated via HO-1 induction in SMCs, which, in turn, activates sGC to produce the second messenger cGMP [66,67]. Moreover, CO has also been shown to be generated by other exogenous stimuli in SMCs, including cAMP-dependent activation of protein kinase A [68]. Notably, a number of key regulatory events are controlled by CO in SMCs, such as cell proliferation, suggesting that this gaseous molecule can play an autocrine regulatory role. For example, it has been shown that CO can specifically control proliferation of vascular SMCs in hypoxia via an sGC/cGMP-dependent pathway [69]. Similarly, MAPKs, including p38 MAPK, have been demonstrated to mediate inhibition of proliferation and migration in SMCs [70,71]. Moreover, a CO-activated cGMP-dependent pathway was able to provide protection against ischemic lung injury by attenuating the inflammatory response [72]. In addition to cGMP-dependent pathways, regulation by CO may also be mediated without sGC [70,73–75] and the migration of vascular SMCs is inhibited in this manner [76]. Finally, the CO donor CORM-3 provided protection in an *in vivo* mouse model of pulmonary hypertension via inhibiting the cell cycle inhibitor p21 in SMCs [77].

Interaction of CO with Macrophages

The innate immune system plays a critical role in solid organ transplantation [7,78]. In particular, mechanisms of innate immunity associated with IRI such as cell death in response to damage-associated molecular patterns (DAMPs) or signaling via extracellular nucleotides (eg. ATP) cause transplant rejection [6,79]. Macrophages are key regulators of innate immune responses and have attracted major attention as therapeutic targets in organ transplantation [80].

In macrophages, CO has potent anti-inflammatory effects in various *in vitro* and *in vivo* settings. Specifically, CO can counteract the upregulation of the proinflammatory cytokines TNF-α and IL-6, while simultaneously upregulating the expression of the anti-inflammatory cytokine IL-10 in a p38-dependent manner [71]. Moreover, it has been shown that anti-inflammatory effects of CO in macrophages are mediated via distinct signaling cascades, including c-Jun N-terminal kinase, *in vitro* and *in vivo* [81,82]. As described above for SMCs, anti-inflammatory effects of CO are mediated via an sGC-dependent pathway that counteracts platelet aggregation and thrombosis in IRI [22,72,83]. Notably, anti-inflammatory effects of CO have also been shown to inhibit the formation of arteriosclerotic lesions associated with chronic graft rejection in a mouse heart transplantation model [22]. Interestingly, removal of CO from lipopolysaccharide (LPS)-stimulated macrophages in cell culture by a potent CO scavenger led to an increase in the production of TNF-α and reactive oxygen species (ROS) suggesting that endogenous CO is obligatory to maintain both redox and inflammatory homeostasis [84]. These findings are supported by *in vivo* studies showing that hybrid compounds capable of inducing HO-1 and simultaneously releasing CO reduce macrophage activation and inflammation in mice challenged with LPS [85].

Interaction of CO with T-Cells

T-cells are key regulators of the adaptive immune system and are primary targets of pharmacological interventions in solid organ transplantation by various immunosuppressants [7]. In these cells, CO has independently been shown to block the production of IL-2, the key regulatory cytokine of T-cell proliferation [86,87]. A principal immunomodulatory role of HO-1-derived CO via activation-induced cell death has been demonstrated in a murine model of the T-cell alloantigen response [88]. Moreover, a regulatory role of CO in modulating regulatory T-cells (FoxP3 + T cells), a subset of T-cells that is crucially involved in controlling immunological tolerance, has been shown in a setting of islet transplantation [40] (for a review, see [89]). CO-dependent immunomodulation via T-cells may also be mediated via dendritic cells, a cell population that links innate and adaptive immunity. Although not directly demonstrated for gaseous CO or CO-RMs, targeting of HO-1-dependent dendritic cell functions regulated specific alloreactive T-cell proliferation in humans and rats [90]. More recently, the immunomodulatory protective effects of CO via a T-cell-mediated mechanism have also been shown in a mouse model of cerebral malaria [91] and treatment of T-cells with the CO donor CORM-2 inhibited ROS-dependent T-cell activation *in vitro* and *in vivo* [92].

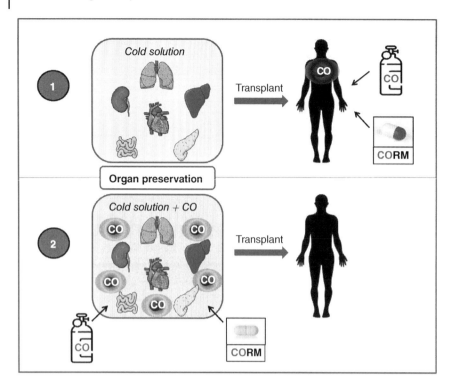

Figure 19.1 Strategies to improve the outcome of organ transplantation by CO. (1) Recipients are treated with either CO gas or CORM prior to transplantation to increase their resistance to graft rejection. (2) Organs are stored in cold preservation solutions saturated with CO gas or supplemented with CORMs before grafts are transplanted into the recipients.

To summarize, the fact that CO beneficially affects the function of different cell types that have critical roles in organ transplantation suggests a multilayered and coordinated action of CO in improving of graft outcome and survival.

Translation of CO in the Clinical Settings of Organ Transplantation

Due to its known toxicity when applied at high doses, a critical question is whether and how CO may become applicable in its gaseous or solid form (CORMs or other formulations) in the clinic. A recent study suggested the application of CO (molybdenum-based CORM) via a membrane-controlled extracorporeal CO release system in clinical settings. Here, the delivery of CO is exclusive to the extracorporeal blood and was shown to be precisely monitored via measurement of systemic CO–hemoglobin [93]. Questions remain whether CO–hemoglobin will then donate CO to the tissue of interest or whether CO–hemoglobin per se could be protective in a systemic manner. It has also been recently shown by the same authors in a pig model of extracorporeal resuscitation that treatment with CO (via CORM) was associated with improvement of cardiac micro- and macrocirculation due to reduced DAMP signaling [94]. Notably, in a recent phase I trial low-dose inhaled CO gas has been shown to be a feasible, well-tolerated, and safe treatment modality in patients with sepsis-mediated acute respiratory distress syndrome [95]. These findings suggest that comparable modes of CO administration may also be applicable in settings of solid organ transplantation.

Conclusions

This chapter highlighted the usefulness of CO in reducing or preventing organ dysfunction associated with transplantation based on convincing evidence emerging from several preclinical and human studies. From a therapeutic perspective, we can delineate two feasible strategies for the exploitation of CO in this context (see Figure 19.1). On one side, patients receiving organs from donors could be treated with CO prior to, during, and/or after the transplant procedure. This would have a dual effect on the immune response of the host as well as on the graft itself. One could envision that this strategy will require more stringent regulations and extensive clinical trials before approval. On the other side, the applicability of CO for organ preservation purposes appears to be easier. For example, CO gas or CORMs could be supplemented as adjuvants to preservation solutions before transplantation and during the use of machine perfusion. This strategy would target only the organ, thus sparing the patient from any potential side effects of CO. Even in this case it would be important to determine the maximum duration period of cold ischemia and the potential storability of various organs exposed to CO, taking into consideration that organ-specific properties could influence the protocols for the use of CO. Thus, a major focus of future work should be on validating the specific protective potential of CO and CORMs in the setting of various solid organ transplantations.

References

1. Otterbein, L.E., Mantell, L.L., and Choi, A.M.K. (1999). Carbon monoxide provides protection against hyperoxic lung injury. *Am. J. Physiol.* 276 (4 Pt 1): L688–L694.
2. Piantadosi, C.A. (2002). Biological chemistry of carbon monoxide. *Antioxid. Redox Signal.* 4 (2): 259–270.
3. Sato, K., Balla, J., Otterbein, L., Smith, R.N., Brouard, S., Lin, Y., Csizmadia, E., Sevigny, J., Robson, S.C., Vercellotti, G., Choi, A.M., Bach, F.H., and Soares, M.P. (2001). Carbon monoxide generated by heme oxygenase-1 suppresses the rejection of mouse-to-rat cardiac transplants. *J. Immunol.* 166 (6): 4185–4194.
4. Motterlini, R., Clark, J.E., Foresti, R., Sarathchandra, P., Mann, B.E., and Green, C.J. (2002). Carbon monoxide-releasing molecules: characterization of biochemical and vascular activities. *Circ. Res.* 90 (2): E17–24.
5. Clark, J.E., Naughton, P., Shurey, S., Green, C.J., Johnson, T.R., Mann, B.E., Foresti, R., and Motterlini, R. (2003). Cardioprotective actions by a water-soluble carbon monoxide-releasing molecule. *Circ. Res.* 93 (2): e2–8.
6. Land, W.G., Agostinis, P., Gasser, S., Garg, A.D., and Linkermann, A. (2016). Transplantation and damage-associated molecular patterns (DAMPs). *Am. J. Transplant.* 16 (12): 3338–3361.
7. Wood, K.J. and Goto, R. (2012). Mechanisms of rejection: current perspectives. *Transplantation* 93 (1): 1–10.
8. Xu, J., Buchwald, J.E., and Martins, P.N. (2020). Review of current machine perfusion therapeutics for organ preservation. *Transplantation* 104 (9): 1792–1803.
9. Pober, J.S., Jane-wit, D., Qin, L., and Tellides, G. (2014). Interacting mechanisms in the pathogenesis of cardiac allograft vasculopathy. *Arterioscler. Thromb. Vasc. Biol.* 34 (8): 1609–1614.
10. Motterlini, R. and Otterbein, L.E. (2010). The therapeutic potential of carbon monoxide. *Nat. Rev. Drug Discov.* 9 (9): 728–743.
11. Sayegh, M.H. and Carpenter, C.B. (2004). Transplantation 50 years later–progress, challenges, and promises. *N. Engl. J. Med.* 351 (26): 2761–2766.
12. Nankivell, B.J. and Alexander, S.I. (2010). Rejection of the kidney allograft. *N. Engl. J. Med.* 363: 1451–1462.
13. Snijder, P.M., Van Den Berg, E., Whiteman, M., Bakker, S.J., Leuvenink, H.G., and Van Goor, H. (2013). Emerging role of gasotransmitters in renal transplantation. *Am. J. Transplant.* 13 (12): 3067–3075.
14. Nakao, A., Neto, J.S., Kanno, S., Stolz, D.B., Kimizuka, K., Liu, F., Bach, F.H., Billiar, T.R., Choi, A.M., Otterbein, L.E., and Murase, N. (2005). Protection against ischemia/reperfusion injury in cardiac and renal transplantation with carbon monoxide, biliverdin and both. *Am. J. Transplant.* 5 (2): 282–291.
15. Neto, J.S., Nakao, A., Toyokawa, H., Nalesnik, M.A., Romanosky, A.J., Kimizuka, K., Kaizu, T., Hashimoto, N., Azhipa, O., Stolz, D.B., Choi, A.M., and Murase, N. (2006). Low-dose carbon monoxide inhalation prevents development of chronic allograft nephropathy. *Am. J. Physiol. Renal Physiol.* 290 (2): F324–334.
16. Caumartin, Y., Stephen, J., Deng, J.P., Lian, D., Lan, Z., Liu, W., Garcia, B., Jevnikar, A.M., Wang, H., Cepinskas, G., and Luke, P.P. (2011). Carbon monoxide-releasing molecules protect against ischemia-reperfusion injury during kidney transplantation. *Kidney Int.* 79 (10): 1080–1089.
17. Nakao, A., Faleo, G., Nalesnik, M.A., Seda-Neto, J., Kohmoto, J., and Murase, N. (2009). Low-dose carbon monoxide inhibits progressive chronic allograft nephropathy and restores renal allograft function. *Am. J. Physiol. Renal Physiol.* 297 (1): F19–26.
18. Hanto, D.W., Maki, T., Yoon, M.H., Csizmadia, E., Chin, B.Y., Gallo, D., Konduru, B., Kuramitsu, K., Smith, N.R., Berssenbrugge, A., Attanasio, C., Thomas, M., Wegiel, B., and Otterbein, L.E. (2010). Intraoperative administration of inhaled carbon monoxide reduces delayed graft function in kidney allografts in Swine. *Am. J. Transplant.* 10 (11): 2421–2430.
19. Taylor, D.O., Edwards, L.B., Boucek, M.M., Trulock, E.P., Waltz, D.A., Keck, B.M., and Hertz, M.I. International Society for Heart and Lung Transplantation (2006). Registry of the International Society for Heart and Lung Transplantation: twenty-third official adult heart transplantation report–2006. *J. Heart Lung Transplant.* 25 (8): 869–879.
20. See Hoe, L.E., Wells, M.A., Bartnikowski, N., Obonyo, N.G., Millar, J.E., Khoo, A., Ki, K.K., Shuker, T., Ferraioli, A., Colombo, S.M., Chan, W., McGiffin, D.C., Suen, J.Y., and Fraser, J.F. (2020). Heart transplantation from brain dead donors: A systematic review of animal models. *Transplantation* 104 (11): 2272–2289.

21. Akamatsu, Y., Haga, M., Tyagi, S., Yamashita, K., Graca-Souza, A.V., Ollinger, R., Czismadia, E., May, G.A., Ifedigbo, E., Otterbein, L.E., Bach, F.H., and Soares, M.P. (2004). Heme oxygenase-1-derived carbon monoxide protects hearts from transplant associated ischemia reperfusion injury. *Faseb J.* 18 (6): 771–772.

22. Otterbein, L.E., Zuckerbraun, B.S., Haga, M., Liu, F., Song, R., Usheva, A., Stachulak, C., Bodyak, N., Smith, R.N., Csizmadia, E., Tyagi, S., Akamatsu, Y., Flavell, R.J., Billiar, T.R., Tzeng, E., Bach, F.H., Choi, A.M., and Soares, M.P. (2003). Carbon monoxide suppresses arteriosclerotic lesions associated with chronic graft rejection and with balloon injury. *Nat. Med.* 9 (2): 183–190.

23. Nakao, A., Toyokawa, H., Abe, M., Kiyomoto, T., Nakahira, K., Choi, A.M., Nalesnik, M.A., Thomson, A.W., and Murase, N. (2006). Heart allograft protection with low-dose carbon monoxide inhalation: effects on inflammatory mediators and alloreactive T-cell responses. *Transplantation* 81 (2): 220–230.

24. Nakao, A., Kaczorowski, D.J., Wang, Y., Cardinal, J.S., Buchholz, B.M., Sugimoto, R., Tobita, K., Lee, S., Toyoda, Y., Billiar, T.R., and McCurry, K.R. (2010). Amelioration of rat cardiac cold ischemia/reperfusion injury with inhaled hydrogen or carbon monoxide, or both. *J. Heart Lung Transplant.* 29 (5): 544–553.

25. Nakao, A. and Toyoda, Y. (2012). Application of carbon monoxide for transplantation. *Curr. Pharm. Biotechnol.* 13 (6): 827–836.

26. Hsiao, H.M., Scozzi, D., Gauthier, J.M., and Kreisel, D. (2017). Mechanisms of graft rejection after lung transplantation. *Curr. Opin. Organ Transplant.* 22 (1): 29–35.

27. Young, K.A. and Dilling, D.F. (2019). The future of lung transplantation. *Chest* 155 (3): 465–473.

28. Song, R., Kubo, M., Morse, D., Zhou, Z., Zhang, X., Dauber, J.H., Fabisiak, J., Alber, S.M., Watkins, S.C., Zuckerbraun, B.S., Otterbein, L.E., Ning, W., Oury, T.D., Lee, P.J., McCurry, K.R., and Choi, A.M. (2003). Carbon monoxide induces cytoprotection in rat orthotopic lung transplantation via anti-inflammatory and anti-apoptotic effects. *Am. J. Pathol.* 163 (1): 231–242.

29. Mishra, S., Fujita, T., Lama, V.N., Nam, D., Liao, H., Okada, M., Minamoto, K., Yoshikawa, Y., Harada, H., and Pinsky, D.J. (2006). Carbon monoxide rescues ischemic lungs by interrupting MAPK-driven expression of early growth response 1 gene and its downstream target genes. *Proc. Natl. Acad. Sci. U. S. A.* 103 (13): 5191–5196.

30. Kohmoto, J., Nakao, A., Kaizu, T., Tsung, A., Ikeda, A., Tomiyama, K., Billiar, T.R., Choi, A.M., Murase, N., and McCurry, K.R. (2006). Low-dose carbon monoxide inhalation prevents ischemia/reperfusion injury of transplanted rat lung grafts. *Surgery* 140 (2): 179–285.

31. Ohtsuka, T., Kaseda, K., Shigenobu, T., Hato, T., Kamiyama, I., Goto, T., Kohno, M., and Shimoda, M. (2014). Carbon monoxide-releasing molecule attenuates allograft airway rejection. *Transpl. Int.* 27 (7): 741–747.

32. Thorgersen, E.B., Barratt-Due, A., Haugaa, H., Harboe, M., Pischke, S.E., Nilsson, P.H., and Mollnes, T.E. (2019). The role of complement in liver injury, regeneration, and transplantation. *Hepatology* 70 (2): 725–736.

33. Kaizu, T., Nakao, A., Tsung, A., Toyokawa, H., Sahai, R., Geller, D.A., and Murase, N. (2005). Carbon monoxide inhalation ameliorates cold ischemia/reperfusion injury after rat liver transplantation. *Surgery* 138 (2): 229–235.

34. Tomiyama, K., Ikeda, A., Ueki, S., Nakao, A., Stolz, D.B., Koike, Y., Afrazi, A., Gandhi, C., Tokita, D., Geller, D.A., and Murase, N. (2008). Inhibition of Kupffer cell-mediated early proinflammatory response with carbon monoxide in transplant-induced hepatic ischemia/reperfusion injury in rats. *Hepatology* 48 (5): 1608–1620.

35. Sun, J., Guo, E., Yang, J., Yang, Y., Liu, S., Hu, J., Jiang, X., Dirsch, O., Dahmen, U., Dong, W., and Liu, A. (2017). Carbon monoxide ameliorates hepatic ischemia/reperfusion injury via sirtuin 1-mediated deacetylation of high-mobility group box 1 in rats. *Liver Transpl.* 23 (4): 510–526.

36. Goldaracena, N., Echeverri, J., Spetzler, V.N., Kaths, J.M., Barbas, A.S., Louis, K.S., Adeyi, O.A., Grant, D.R., Selzner, N., and Selzner, M. (2016). Anti-inflammatory signaling during ex vivo liver perfusion improves the preservation of pig liver grafts before transplantation. *Liver Transpl.* 22 (11): 1573–1583.

37. Nakao, A., Kimizuka, K., Stolz, D.B., Neto, J.S., Kaizu, T., Choi, A.M., Uchiyama, T., Zuckerbraun, B.S., Nalesnik, M.A., Otterbein, L.E., and Murase, N. (2003). Carbon monoxide inhalation protects rat intestinal grafts from ischemia/reperfusion injury. *Am. J. Pathol.* 163 (4): 1587–1598.

38. Nakao, A., Kimizuka, K., Stolz, D.B., Seda Neto, J., Kaizu, T., Choi, A.M., Uchiyama, T., Zuckerbraun, B.S., Bauer, A.J., Nalesnik, M.A., Otterbein, L.E., Geller, D.A., and Murase, N. (2003). Protective effect of carbon monoxide inhalation for cold-preserved small intestinal grafts. *Surgery* 134 (2): 285–292.

39. Gunther, L., Berberat, P.O., Haga, M., Brouard, S., Smith, R.N., Soares, M.P., Bach, F.H., and Tobiasch, E. (2002). Carbon monoxide protects pancreatic beta-cells from apoptosis and improves islet function/

survival after transplantation. *Diabetes* 51 (4): 994–999.

40. Lee, S.S., Gao, W., Mazzola, S., Thomas, M.N., Csizmadia, E., Otterbein, L.E., Bach, F.H., and Wang, H. (2007). Heme oxygenase-1, carbon monoxide, and bilirubin induce tolerance in recipients toward islet allografts by modulating T regulatory cells. *FASEB J.* 21 (13): 3450–3457.

41. Wang, H., Lee, S.S., Gao, W., Czismadia, E., McDaid, J., Ollinger, R., Soares, M.P., Yamashita, K., and Bach, F.H. (2005). Donor treatment with carbon monoxide can yield islet allograft survival and tolerance. *Diabetes* 54 (5): 1400–1406.

42. Wang, H., Gou, W., Strange, C., Wang, J., Nietert, P.J., Cloud, C., Owzarski, S., Shuford, B., Duke, T., Luttrell, L., Lesher, A., Papas, K.K., Herold, K.C., Clark, P., Usmani-Brown, S., Kitzmann, J., Crosson, C., Adams, D.B., and Morgan, K.A. (2019). Islet harvest in carbon monoxide-saturated medium for chronic pancreatitis patients undergoing islet autotransplantation. *Cell Transplant.* 28 (1_suppl): 25S–36S.

43. Motterlini, R. and Foresti, R. (2014). Heme oxygenase-1 as a target for drug discovery. *Antioxid. Redox Signal.* 20 (11): 1810–1826.

44. Brook, N.R., Waller, J.R., and Nicholson, M.L. (2003). Nonheart-beating kidney donation: current practice and future developments. *Kidney Int.* 63 (4): 1516–1529.

45. Fuller, B., Froghi, F., and Davidson, B. (2018). Organ preservation solutions: linking pharmacology to survival for the donor organ pathway. *Curr. Opin. Organ Transplant.* 23 (3): 361–368.

46. Sandouka, A., Fuller, B.J., Mann, B.E., Green, C.J., Foresti, R., and Motterlini, R. (2006). Treatment with carbon monoxide-releasing molecules (CO-RMs) during cold storage improves renal function at reperfusion. *Kidney Int.* 69 (2): 239–247.

47. Motterlini, R., Sawle, P., Bains, S., Hammad, J., Alberto, R., Foresti, R., and Green, C.J. (2005). CORM-A1: a new pharmacologically active carbon monoxide-releasing molecule. *FASEB J.* 19 (2): 284–286.

48. Motterlini, R., Mann, B.E., and Foresti, R. (2005). Therapeutic applications of carbon monoxide-releasing molecules (CO-RMs). *Expert Opin. Investig. Drugs* 14 (11): 1305–1318.

49. Nakao, A., Faleo, G., Shimizu, H., Nakahira, K., Kohmoto, J., Sugimoto, R., Choi, A.M., McCurry, K.R., Takahashi, T., and Murase, N. (2008). Ex vivo carbon monoxide prevents cytochrome P450 degradation and ischemia/reperfusion injury of kidney grafts. *Kidney Int.* 74: 1009–1016.

50. Yoshida, J., Ozaki, K.S., Nalesnik, M.A., Ueki, S., Castillo-Rama, M., Faleo, G., Ezzelarab, M., Nakao, A., Ekser, B., Echeverri, G.J., Ross, M.A., Stolz, D.B., and Murase, N. (2010). Ex vivo application of carbon monoxide in UW solution prevents transplant-induced renal ischemia/reperfusion injury in pigs. *Am. J. Transplant.* 10: 763–772.

51. Fayad-Kobeissi, S., Ratovonantenaina, J., Dabire, H., Wilson, J.L., Rodriguez, A.M., Berdeaux, A., Dubois-Rande, J.L., Mann, B.E., Motterlini, R., and Foresti, R. (2016). Vascular and angiogenic activities of CORM-401, an oxidant-sensitive CO-releasing molecule. *Biochem. Pharmacol.* 102: 64–77.

52. Braud, L., Pini, M., Muchova, L., Manin, S., Kitagishi, H., Sawaki, D., Czibik, G., Ternacle, J., Derumeaux, G., Foresti, R., and Motterlini, R. (2018). Carbon monoxide-induced metabolic switch in adipocytes improves insulin resistance in obese mice. *JCI Insight* 3 (22).

53. Bhattacharjee, R.N., Richard-Mohamed, M., Sun, Q., Haig, A., Aboalsamh, G., Barrett, P., Mayer, R., Alhasan, I., Pineda-Solis, K., Jiang, L., Alharbi, H., Saha, M., Patterson, E., Sener, A., Cepinskas, G., Jevnikar, A.M., and Luke, P.P.W. (2018). CORM-401 reduces ischemia reperfusion injury in an ex vivo renal porcine model of the donation after circulatory death. *Transplantation* 102 (7): 1066–1074.

54. Ikeda, A., Ueki, S., Nakao, A., Tomiyama, K., Ross, M.A., Stolz, D.B., Geller, D.A., and Murase, N. (2009). Liver graft exposure to carbon monoxide during cold storage protects sinusoidal endothelial cells and ameliorates reperfusion injury in rats. *Liver Transpl.* 15 (11): 1458–1468.

55. Koetting, M., Leuvenink, H., Dombrowski, F., and Minor, T. (2010). Gaseous persufflation with carbon monoxide during ischemia protects the isolated liver and enhances energetic recovery. *Cryobiology* 61 (1): 33–37.

56. Pizarro, M.D., Rodriguez, J.V., Mamprin, M.E., Fuller, B.J., Mann, B.E., Motterlini, R., and Guibert, E.E. (2009). Protective effects of a carbon monoxide-releasing molecule (CORM-3) during hepatic cold preservation. *Cryobiology* 58 (3): 248–255.

57. Kohmoto, J., Nakao, A., Sugimoto, R., Wang, Y., Zhan, J., Ueda, H., and McCurry, K.R. (2008). Carbon monoxide-saturated preservation solution protects lung grafts from ischemia-reperfusion injury. *J. Thorac. Cardiovasc. Surg.* 136 (4): 1067–1075.

58. Nakao, A., Toyokawa, H., Tsung, A., Nalesnik, M.A., Stolz, D.B., Kohmoto, J., Ikeda, A., Tomiyama, K., Harada, T., Takahashi, T., Yang, R., Fink, M.P., Morita, K., Choi, A.M., and Murase, N. (2006). Ex vivo application of carbon monoxide in university of wisconsin solution to prevent intestinal cold ischemia/reperfusion injury. *Am. J. Transplant.* 6: 2243–2255.

59. Nakao, A., Huang, C.S., Stolz, D.B., Wang, Y., Franks, J.M., Tochigi, N., Billiar, T.R., Toyoda, Y., Tzeng, E., and McCurry, K.R. (2011). Ex vivo carbon monoxide delivery inhibits intimal hyperplasia in arterialized vein grafts. *Cardiovasc. Res.* 89 (2): 457–463.

60. Musameh, M.D., Green, C.J., Mann, B.E., Fuller, B.J., and Motterlini, R. (2007). Improved myocardial function after cold storage with preservation solution supplemented with a carbon monoxide-releasing molecule (CORM-3). *J. Heart Lung Transplant.* 26 (11): 1192–1198.

61. Abrahimi, P., Liu, R., and Pober, J.S. (2015). Blood vessels in allotransplantation. *Am. J. Transplant.* 15 (7): 1748–1754.

62. Morita, T. and Kourembanas, S. (1995). Endothelial cell expression of vasoconstrictors and growth factors is regulated by smooth muscle cell-derived carbon monoxide. *J. Clin. Invest.* 96: 2676–2682.

63. Brouard, S., Otterbein, L.E., Anrather, J., Tobiasch, E., Bach, F.H., Choi, A.M., and Soares, M.P. (2000). Carbon monoxide generated by heme oxygenase 1 suppresses endothelial cell apoptosis. *J. Exp. Med.* 192 (7): 1015–1026.

64. Brouard, S., Berberat, P.O., Tobiasch, E., Seldon, M.P., Bach, F.H., and Soares, M.P. (2002). Heme oxygenase-1-derived carbon monoxide requires the activation of transcription factor NF-kappa B to protect endothelial cells from tumor necrosis factor-alpha-mediated apoptosis. *J. Biol. Chem.* 277 (20): 17950–17961.

65. Kaczara, P., Motterlini, R., Rosen, G.M., Augustynek, B., Bednarczyk, P., Szewczyk, A., Foresti, R., and Chlopicki, S. (2015). Carbon monoxide released by CORM-401 uncouples mitochondrial respiration and inhibits glycolysis in endothelial cells: a role for mitoBKCa channels. *Biochim. Biophys. Acta* 1847 (10): 1297–1309.

66. Morita, T., Perrella, M.A., Lee, M.-E., and Kourembanas, S. (1995). Smooth muscle cell-derived carbon monoxide is a regulator of vascular cGMP. *Proc. Natl. Acad. Sci. U. S. A.* 92: 1475–1479.

67. Christodoulides, N., Durante, W., Kroll, M.H., and Schafer, A.I. (1995). Vascular smooth muscle cell heme oxygenases generate guanylyl cyclase-stimulatory carbon monoxide. *Circulation* 91: 2306–2309.

68. Durante, W., Christodoulides, N., Cheng, K., Peyton, K.J., Sunahara, R.K., and Schafer, A.I. (1997). cAMP induces heme oxygenase-1 gene expression and carbon monoxide production in vascular smooth muscle. *Am. J. Physiol.* 273 (1 Pt 2): H317–323.

69. Morita, T., Mitsialis, S.A., Koike, H., Liu, Y., and Kourembanas, S. (1997). Carbon monoxide controls the proliferation of hypoxic vascular smooth muscle cells. *J. Biol. Chem.* 272 (52): 32804–32809.

70. Kim, H.P., Wang, X., Nakao, A., Kim, S.I., Murase, N., Choi, M.E., Ryter, S.W., and Choi, A.M. (2005). Caveolin-1 expression by means of p38beta mitogen-activated protein kinase mediates the antiproliferative effect of carbon monoxide. *Proc. Natl. Acad. Sci. U. S. A.* 102 (32): 11319–11324.

71. Otterbein, L.E., Bach, F.H., Alam, J., Soares, M., Tao Lu, H., Wysk, M., Davis, R.J., Flavell, R.A., and Choi, A.M. (2000). Carbon monoxide has anti-inflammatory effects involving the mitogen-activated protein kinase pathway. *Nat. Med.* 6 (4): 422–428.

72. Fujita, T., Toda, K., Karimova, A., Yan, S.F., Naka, Y., Yet, S.F., and Pinsky, D.J. (2001). Paradoxical rescue from ischemic lung injury by inhaled carbon monoxide driven by derepression of fibrinolysis. *Nat. Med.* 7 (5): 598–604.

73. Taille, C., El-Benna, J., Lanone, S., Boczkowski, J., and Motterlini, R. (2005). Mitochondrial respiratory chain and NAD(P)H oxidase are targets for the antiproliferative effect of carbon monoxide in human airway smooth muscle. *J. Biol. Chem.* 280 (27): 25350–25360.

74. Kim, H.P., Ryter, S.W., and Choi, A.M. (2006). CO as a cellular signaling molecule. *Annu. Rev. Pharmacol. Toxicol.* 46: 411–449.

75. Song, R., Mahidhara, R.S., Liu, F., Ning, W., Otterbein, L.E., and Choi, A.M. (2002). Carbon monoxide inhibits human airway smooth muscle cell proliferation via mitogen-activated protein kinase pathway. *Am. J. Respir. Cell Mol. Biol.* 27 (5): 603–610.

76. Rodriguez, A.I., Gangopadhyay, A., Kelley, E.E., Pagano, P.J., Zuckerbraun, B.S., and Bauer, P.M. (2010). HO-1 and CO decrease platelet-derived growth factor-induced vascular smooth muscle cell migration via inhibition of Nox1. *Arterioscler. Thromb. Vasc. Biol.* 30 (1): 98–104.

77. Abid, S., Houssaini, A., Mouraret, N., Marcos, E., Amsellem, V., Wan, F., Dubois-Rande, J.L., Derumeaux, G., Boczkowski, J., Motterlini, R., and Adnot, S. (2014). P21-dependent protective effects of a carbon monoxide-releasing molecule-3 in pulmonary hypertension. *Arterioscler. Thromb. Vasc. Biol.* 34 (2): 304–312.

78. Jane-Wit, D., Fang, C., and Goldstein, D.R. (2016). Innate immune mechanisms in transplant allograft vasculopathy. *Curr. Opin. Organ Transplant.* 21 (3): 253–257.

79. Yeudall, S., Leitinger, N., and Laubach, V.E. (2020). Extracellular nucleotide signaling in solid organ transplantation. *Am. J. Transplant.* 20 (3): 633–640.

80. Salehi, S. and Reed, E.F. (2015). The divergent roles of macrophages in solid organ transplantation. *Curr. Opin. Organ Transplant.* 20 (4): 446–453.

81. Morse, D., Pischke, S.E., Zhou, Z., Davis, R.J., Flavell, R.A., Loop, T., Otterbein, S.L., Otterbein, L.E., and Choi, A.M. (2003). Suppression of inflammatory cytokine production by carbon monoxide involves the JNK pathway and AP-1. *J. Biol. Chem.* 278 (39): 36993–36998.
82. Kim, H.P., Wang, X., Zhang, J., Suh, G.Y., Benjamin, I.J., Ryter, S.W., and Choi, A.M. (2005). Heat shock protein-70 mediates the cytoprotective effect of carbon monoxide: involvement of p38 beta MAPK and heat shock factor-1. *J. Immunol.* 175 (4): 2622–2629.
83. Brune, B. and Ullrich, V. (1987). Inhibition of platelet aggregation by carbon monoxide is mediated by activation of guanylate cyclase. *Mol. Pharmacol.* 32 (4): 497–504.
84. Minegishi, S., Yumura, A., Miyoshi, H., Negi, S., Taketani, S., Motterlini, R., Foresti, R., Kano, K., and Kitagishi, H. (2017). Detection and removal of endogenous carbon monoxide by selective and cell-permeable hemoprotein model complexes. *J. Am. Chem. Soc.* 139 (16): 5984–5991.
85. Motterlini, R., Nikam, A., Manin, S., Ollivier, A., Wilson, J.L., Djouadi, S., Muchova, L., Martens, T., Rivard, M., and Foresti, R. (2019). HYCO-3, a dual CO-releaser/Nrf2 activator, reduces tissue inflammation in mice challenged with lipopolysaccharide. *Redox Biol.* 20: 334–348.
86. Song, R., Mahidhara, R.S., Zhou, Z., Hoffman, R.A., Seol, D.W., Flavell, R.A., Billiar, T.R., Otterbein, L.E., and Choi, A.M. (2004). Carbon monoxide inhibits T lymphocyte proliferation via caspase-dependent pathway. *J. Immunol.* 172 (2): 1220–1226.
87. Pae, H.O., Oh, G.S., Choi, B.M., Chae, S.C., Kim, Y.M., Chung, K.R., and Chung, H.T. (2004). Carbon monoxide produced by heme oxygenase-1 suppresses T cell proliferation via inhibition of IL-2 production. *J. Immunol.* 172 (8): 4744–4751.
88. McDaid, J., Yamashita, K., Chora, A., Ollinger, R., Strom, T.B., Li, X.C., Bach, F.H., and Soares, M.P. (2005). Heme oxygenase-1 modulates the alloimmune response by promoting activation-induced cell death of T cells. *FASEB J.* 19 (3): 458–460.
89. Brusko, T.M., Wasserfall, C.H., Agarwal, A., Kapturczak, M.H., and Atkinson, M.A. (2005). An integral role for heme oxygenase-1 and carbon monoxide in maintaining peripheral tolerance by CD4+CD25+ regulatory T cells. *J. Immunol.* 174 (9): 5181–5186.
90. Chauveau, C., Remy, S., Royer, P.J., Hill, M., Tanguy-Royer, S., Hubert, F.X., Tesson, L., Brion, R., Beriou, G., Gregoire, M., Josien, R., Cuturi, M.C., and Anegon, I. (2005). Heme oxygenase-1 expression inhibits dendritic cell maturation and proinflammatory function but conserves IL-10 expression. *Blood* 106 (5): 1694–1702.
91. Jeney, V., Ramos, S., Bergman, M.L., Bechmann, I., Tischer, J., Ferreira, A., Oliveira-Marques, V., Janse, C.J., Rebelo, S., Cardoso, S., and Soares, M.P. (2014). Control of disease tolerance to malaria by nitric oxide and carbon monoxide. *Cell Rep.* 8 (1): 126–136.
92. Yan, Y., Wang, L., Chen, S., Zhao, G., Fu, C., Xu, B., Tan, X., Xiang, Y., and Chen, G. (2020). Carbon monoxide inhibits T cell proliferation by suppressing reactive oxygen species signaling. *Antioxid. Redox Signal.* 32 (7): 429–446.
93. Wollborn, J., Hermann, C., Goebel, U., Merget, B., Wunder, C., Maier, S., Schafer, T., Heuler, D., Muller-Buschbaum, K., Buerkle, H., Meinel, L., Schick, M.A., and Steiger, C. (2018). Overcoming safety challenges in CO therapy – Extracorporeal CO delivery under precise feedback control of systemic carboxyhemoglobin levels. *J. Control. Release* 279: 336–344.
94. Wollborn, J., Steiger, C., Ruetten, E., Benk, C., Kari, F.A., Wunder, C., Meinel, L., Buerkle, H., Schick, M.A., and Goebel, U. (2020). Carbon monoxide improves haemodynamics during extracorporeal resuscitation in pigs. *Cardiovasc. Res.* 116 (1): 158–170.
95. Fredenburgh, L.E., Perrella, M.A., Barragan-Bradford, D., Hess, D.R., Peters, E., Welty-Wolf, K.E., Kraft, B.D., Harris, R.S., Maurer, R., Nakahira, K., Oromendia, C., Davies, J.D., Higuera, A., Schiffer, K.T., Englert, J.A., Dieffenbach, P.B., Berlin, D.A., Lagambina, S., Bouthot, M., Sullivan, A.I., Nuccio, P.F., Kone, M.T., Malik, M.J., Porras, M.A.P., Finkelsztein, E., Winkler, T., Hurwitz, S., Serhan, C.N., Piantadosi, C.A., Baron, R.M., Thompson, B.T., and Choi, A.M. (2018). A phase I trial of low-dose inhaled carbon monoxide in sepsis-induced ARDS. *JCI Insight* 3 (23).

20

Carbon Monoxide in Lung Injury and Disease

Stefan W. Ryter[1,2]

[1]*Weill Cornell Medicine, New York, NY, USA*
[2]*Proterris, Inc., Boston, MA, USA*

Introduction

Carbon monoxide (CO) is a low molecular weight (molecular weight: 28.01) diatomic gas that can paradoxically interact with biological systems as both an endogenous physiological mediator and an external effector with potential harmful sequelae [1]. Unlike other endogenously produced small gaseous mediators such as nitric oxide (NO) and H_2S (Chapters 8 and 9), CO has limited reactivity in biological systems. The primary reactivity of CO is restricted to iron centers, such as those found in hemoglobin, mitochondrial cytochromes, and other heme-containing enzymes [2]. Since the middle of the twentieth century, the scientific community has recognized that CO can evolve endogenously in humans as the by-product of endogenous metabolic activity [3–5], and that this endogenously derived CO can appear as a fraction of human exhaled breath [6]. The endogenous production of CO is largely attributed to the degradation of heme, with the majority (~88%) originating from erythrocyte-dependent hemoglobin turnover [3–5], and with a minor component of heme-derived CO arising from the turnover of cellular hemoproteins such as mitochondrial cytochromes [2,7]. The heme oxygenase enzyme system (EC 1:14:14:18) [8,9], which includes a stress-inducible form, HO-1 [10,11], and a constitutive form, HO-2 [2], provides the heme catalytic activity responsible for endogenous CO production (Chapter 1).

CO gas is known as a common contaminant of indoor and outdoor air, which originates primarily in the combustion of organic matter, including fossil fuels [12]. Since the primary mode of entry of CO is via inhalation, the lung and lung vasculature represent natural primary targets for both the therapeutic and toxicological effects of CO. There are other minor sources of endogenous CO production, including cytochrome P450-dependent drug metabolism, exposure to CO precursors such as carbon tetrachloride [12], and possibly the gut microbiota (Chapters 1, 3, and 17).

Inhaled CO (iCO) diffuses rapidly across alveolar and capillary membranes, with the majority forming a tight complex with the oxygen carrier protein hemoglobin to form carboxyhemoglobin (COHb), with an affinity for hemoglobin that is approximately 200–250 times that of oxygen [12,13]. Partial occupation by CO at the O_2 binding sites of hemoglobin hinders the release of O_2 from the remaining heme groups. These effects of CO inhibit the capacity of the blood to deliver O_2 to peripheral tissue, thus promoting tissue hypoxia [12]. The formation of COHb is reversible by removal of the CO source in favor of O_2 inspiration. Thus, oxygen therapy is a common antidote for CO poisoning [13]. The basal COHb level in humans is ~0.1–1% in the absence of environmental contamination or smoking. Habitation of heavily populated urban areas with high levels of ambient CO, such as those originating from automobile exhaust, or exposure to cigarette or wood smoke, may increase this background [12,14,15].

The therapeutic potential of CO versus its lethal potential, as with any other pharmacologically applied substance or "drug," is highly dose dependent. Since the toxic potential of this gas has been well publicized with accounts of CO lethality and death from accidental exposure, there has been some resistance among the medical community and public at large in accepting a therapeutic window for CO and its potential as a therapy. Signs of physiological and cognitive toxicity begin at CO exposure doses eliciting 20–50% COHb levels, while lethal doses exceed this range [13,16]. In contrast, application of iCO in recently completed clinical safety trials has targeted 6–8% COHb and not to exceed 10% [17]. These concentrations were found safe for human application and caused no adverse effects [17].

As an alternative to CO gas, transitional metal carbonyl-based CO-releasing molecules (CORMs), as well as newer organic CO prodrugs, have been developed for preclinical experimentation and for potential clinical application (Chapters 12–15). These compounds have the advantage of CO delivery to tissues without excessive COHb buildup [18]. Prototype compounds such as the ruthenium-based CORM-2 and CORM-3 have been used for several studies described in this chapter, including preclinical studies in acute lung injury (ALI) and sepsis [18].

A number of biochemical and cellular targets of CO have been proposed. CO can regulate cellular processes (including anti-inflammatory and tissue protective effects) via modulation of cellular signal transduction pathways [19,20]. Since CO binds primarily to iron-containing prosthetic groups, mainly represented by heme, and has little biological reactivity outside this sphere, primary CO-sensing targets are likely represented by cytochromes and heme-containing proteins. This is exemplified by soluble guanylate cyclase (sGC), which generates the second messenger cyclic 3′5′ guanosine monophosphate (cGMP) upon gaseous ligand binding to its intrinsic heme moiety. The heme of sGC has a far greater affinity for NO, the classical ligand, than for CO. Thus, the competition of CO for sGC is of unclear physiological significance but may be relevant under conditions of reduced NO bioavailability or therapeutic application of CO to the lung [21]. Other heme-containing enzymes that serve as targets of CO based on model studies include inducible NO synthase (iNOS), cytochrome c oxidase, heme-containing transcription factors (e.g., Rev-Erb-α/β), and NADPH oxidases (NOXs) that generate superoxide anion (O_2^-) for innate immunity. The characterization and relevance of these targets vary in a cell type-specific manner [22].

Mitochondria, key organelles responsible for cellular energy production, have been implicated as one of the major proximal targets for CO-dependent regulation of cellular physiology and signaling (Figure 20.1). Mitochondria are complex organelles that are subjected to genetically regulated programs for their biosynthesis (mitochondrial biogenesis), turnover by autophagy (mitophagy), and dynamic regulation of morphology (fusion and fission). The complex relationship between CO and mitochondrial function has been described elsewhere (Chapter 6), and in recent reviews [22–24]. In brief, these include modulation of mitochondrial respiration and energy production, modulation of mitochondrial reactive oxygen species (mtROS) production for signaling, induction of mitochondrial biogenesis in various cell types, and potential modulation of mitochondrial dynamics. A number of secondary targets and effector molecules activated by CO, presumably downstream of primary targets including mitochondrial perturbation, have emerged from model studies, which include mitogen-activated protein kinases (MAPKs) and nuclear factor kappa-B (NF-κB) [1]. These effects can vary in a cell type-specific fashion, and thus will be covered in the following sections.

Lung Cell Type-Specific Effects

The lung is a complex organ consisting of multiple cell types and exerts a primary function of gas exchange. The principal cell types of the lung include epithelial cells of the airway, bronchi, and alveoli; interstitial fibroblasts and endothelial cells of the pulmonary vasculature; smooth muscle cells (SMCs) of the airway and pulmonary vasculature; and alveolar and systemic macrophages [25]. This section will focus on cellular data relevant to pulmonary system. Effects of CO on biochemical phenomenon and cellular signaling will be discussed for each major cell type, as results vary on a cell type-specific basis (Figure 20.2).

Macrophages

Macrophages are a vital component of lung physiology that provide a "first-line" defense against invading pathogens. Macrophages exert a variety of specialized functions, including maintenance of pulmonary homeostasis, immune surveillance, bacterial clearance, elimination of cell debris, responses to infection, and resolution of inflammation [26–28]. Two distinct macrophage populations are present in the lung, including alveolar macrophages, which contact with type I and II epithelial cells of the alveolus, and interstitial macrophages residing in the lung parenchyma [26–28]. These are further divided in subpopulations (M1, M2) whose distributions change during exposure to environmental or inflammatory stimuli. M1 macrophages (classically activated macrophages) respond to proinflammatory cytokines such as interferon-γ and tumor necrosis factor-α (TNF-α), produce proinflammatory cytokines, eliminate intracellular pathogens via phagocytosis, and promote a local Th1 environment. M2 macrophages (alternatively activated macrophages) represent a

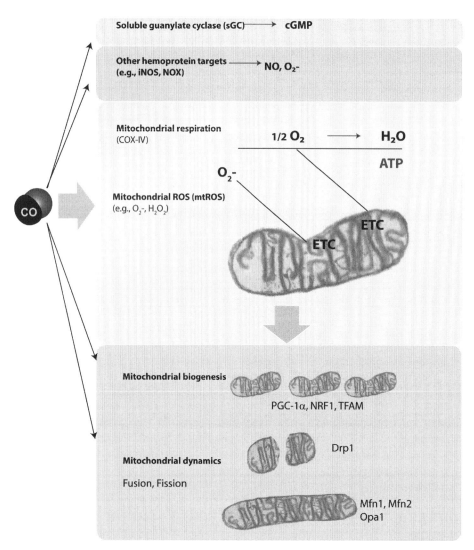

Figure 20.1 Hypothetical proximal targets of CO as relevant to pulmonary cells. CO is known to interact with several biochemical targets in the various cell types of the lung. The classical target is sGC that, upon gaseous ligand binding, can increase the production of cGMP. CO can potentially target other cellular hemoproteins, including iNOS and NOX isoforms, potentially leading to modulation of NO and superoxide anion (O_2^-) production, respectively. Mitochondria have emerged as a primary target for CO action, based on numerous *in vitro* studies. CO can inhibit cellular respiration at high concentration and modulate mtROS production via interactions with cytochrome c oxidase and other potential sites of the electron transport chain. CO can also impact the mitochondrial biogenesis program through modulation of its various regulatory factors, including PGC-1α, NRF1, and TFAM. Modulation of mitochondrial dynamics (fusion and fission) is an area of potential novel investigations.

more diverse phenotype that orchestrates type 2 immune responses, and whose delayed accumulation may contribute to tissue fibrosis or repair [27]. Due to difficulties in harvesting and culturing primary alveolar macrophages, many studies on CO regulation have used alternate macrophage cell lines, such as murine RAW 264.7 cells, and alternate primary cell cultures, including peritoneal macrophages and bone marrow-derived macrophages, mainly for ease of collection for model studies.

TLR-Dependent Responses

The anti-inflammatory effects of CO were first described in cultured RAW 264.7 murine macrophages challenged with bacterial lipopolysaccharide (LPS) [29]. Exogenous CO application (250 ppm, 2 h pretreatment) or HO-1 expression inhibited the LPS-induced production of proinflammatory cytokines such as TNF-α, interleukin-1β (IL-1β), and macrophage inflammatory protein-1β (MIP-1β) in cultured macrophages, whereas increased the production of

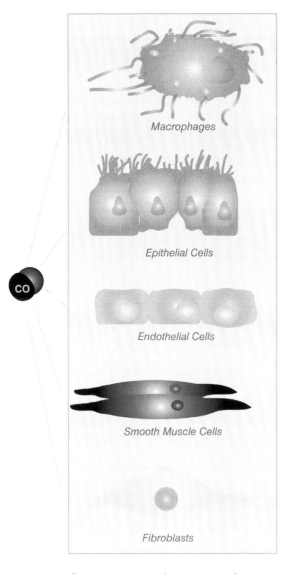

Figure 20.2 Low-concentration CO elicits cell type-specific protective effects in *in vitro* injury models. The lung has a number of distinct cell types that contribute to function, including fibroblasts, alveolar and systemic macrophages, endothelial cells and SMCs (of the pulmonary vasculature), airway SMCs, and alveolar and bronchial epithelial cells. CO has been shown to exert anti-inflammatory effects in macrophages and epithelial cells, which include demonstration of inflammasome or autophagy modulation, respectively. Additionally, antiapoptotic effects have been demonstrated primarily in pulmonary endothelial cells. CO can inhibit the proliferation of pulmonary SMCs (vascular and airway) and fibroblasts. Finally, CO has been shown to exert antifibrotic effects, including inhibition of extracellular matrix (ECM) production in fibroblasts.

the anti-inflammatory cytokine IL-10 during LPS challenge. The original characterization of these effects excluded cGMP signaling and implicated differential activation of mitogen-activated protein kinases including p38 MAPK and c-Jun NH$_2$-terminal kinase in CO-treated cells [29,30].

Further studies have identified a novel mechanism by which CO may exert anti-inflammatory effects involving the downregulation of Toll-like receptor (TLR) trafficking to lipid rafts [31]. The effects of CO on cytokine production were investigated in RAW 264.7 cells induced with various TLR ligands. CO (250 ppm, 2 h pretreatment) inhibited TLR4 (e.g., LPS) and TLR2, TLR5, and TLR9 ligand-induced TNF-α production, but did not affect TLR3 ligand (e.g., poly[I:C])-induced signaling or TNF-α production. The trafficking of TLRs to plasma membrane lipid raft domains may constitute an early upstream event in the activation of TLR-dependent signaling pathways associated with proinflammatory stimuli. Trafficking of TLR4 to lipid rafts in response to LPS was ROS dependent, since it was abrogated by chemical inhibitors of phagocyte NOX and absent in macrophages deficient in the gp91^{phox} component of the oxidase. CO selectively inhibited LPS-induced recruitment of TLR4 to lipid rafts, which was associated with the inhibition of NOX activity and LPS-inducible ROS production in macrophages. CO also inhibited the translocation of TLR adaptors to the lipid raft, including MyD88 and TRIF, as well as their association with TLR4. In summary, CO differentially regulated TLR signaling pathways by inhibiting translocation of TLR4 to lipid rafts through inhibition of NOX-dependent ROS generation [31].

In macrophages, upregulation of mtROS by CO application at low concentration (50–500 ppm, 1 h) was linked to inhibition of cytochrome *c* oxidase [32]. In turn, elevated mtROS for signaling evoked anti-inflammatory effects through p38 MAPK activation. Furthermore, the effects of CO were associated with maintenance of cellular ATP and increase of mitochondrial membrane potential in the presence of CO at the concentrations used [32]. Furthermore, upregulation of mtROS in cultured macrophages following CO treatment (250 ppm, 1 h) was attributed to adaptive cellular signaling through upregulation of PPAR-γ, and concomitant downregulation of the proinflammatory mediator Egr-1 [33]. Further studies revealed that regulation of PPAR-γ by CO was dependent on post-translational SUMOylation of this protein. CO modulation of EGR-1 expression required mtROS, under regulation by mitochondrial uncoupling protein-2, and was also p38 MAPK dependent [34]. Additionally, CO treatment (250 ppm, 0.5–4 h) stabilized hypoxia-inducible factor 1-alpha via mtROS signaling in macrophages, which was associated with anti-inflammatory effects, in part by upregulation of transforming growth factor beta 1 (TGF-β1) gene expression [35].

Taken together, these studies reveal a complex signaling network, likely mediated by mitochondrial perturbation, by which CO can modulate the inflammatory responses of cultured macrophages.

Inflammasome-Dependent Responses

Inflammasomes are cytosolic protein complexes present in immune cells and other cell types that promote the cleavage of caspase-1, which leads to the maturation and secretion of proinflammatory cytokines, including IL-1β and IL-18. Among the known inflammasomes, the NOD-, leucine-rich region-, and pyrin domain-containing 3 (NLRP3)-dependent inflammasome is crucially involved in the pathogenesis of various acute or chronic inflammatory diseases [36,37].

CO has been shown to both positively and negatively modulate NLRP3 activation, depending on experimental context. Exposure to CO gas (250 ppm, either as 2 h pretreatment or when initiated up to 3 h after LPS challenge) was observed to inhibit caspase-1 activation and secretion of IL-1β and IL-18 in bone marrow-derived macrophages in response to exogenous LPS and ATP treatment, an *in vitro* model of NLRP3 inflammasome activation [38]. CO also inhibited IL-18 secretion in response to LPS and nigericin, an alternate NLRP3 inflammasome activator. LPS and ATP stimulation induced the formation of complexes between NLRP3 and ASC, or NLRP3 and caspase-1. CO treatment (250 ppm) inhibited these molecular interactions induced by LPS and ATP. In addition, CO inhibited mitochondrial depolarization induced by LPS and ATP in macrophages. These results suggest that CO negatively regulates NLRP3 inflammasome activation in part by preventing mitochondrial dysfunction [38]. Corresponding *in vivo* studies demonstrated that CO (250 ppm) reduced IL-1β production in response to LPS treatment alone [38].

Application of the CO donor compound CORM-2 (20–80 μM) downregulated caspase-1 activation and IL-1β maturation and secretion in the context of ER stress-induced inflammation [39]. In contrast to these observations, another report has demonstrated that CO (250 ppm) augmented macrophage NLRP3 inflammasome activation in the presence of live bacteria, when initiated 1–6 h after bacterial challenge. These studies found that CO can promote ATP secretion from bacteria, which in turn can upregulate NLRP3 inflammasome activation by stimulating the purinergic receptor (P2X7R) in bacteria-treated macrophages [40]. Consistent with *in vitro* observations, CO treatment (250 ppm) elicited a two- to threefold increase in ATP in peritoneal lavage fluid of bacterially infected mice in association with enhanced bacterial clearance [40].

The reasons for contrasting findings between studies remain unclear but may be related to differences between live bacterial sepsis [40] and inflammation induced by cell free LPS [38], as well as timing of CO administration. In mice, CO likely exerted an anti-inflammatory effect in the context of challenge with cell free LPS, by antagonizing the TLR4-mediated pathway (first activating signal). In contrast, CO demonstrated a marked antibacterial innate immune response in a live bacteria model, resulting in increased mobilization bacterial ATP (second signal) for inflammasome activation as a component of the host defense response. Further studies will be needed to elucidate the precise mechanisms by which CO shows potential as novel regulator of the NLRP3 inflammasome and secretion of IL-1β and IL-18 in the context of both inflammation and infection.

Resolution of Inflammation

Therapeutic CO has been shown to modulate the resolution of inflammation. Lipid mediators (LMs), which include eicosanoids omega-3-derived specialized pro-resolution mediators (SPMs), are key signaling molecules in resolution of inflammation [41]. CO was shown to modulate resolution active programs during self-limited inflammation in mice [42]. iCO gas (250 ppm) significantly inhibited polymorphonuclear neutrophil (PMN) infiltration into peritoneum and shortened the resolution interval. LM profiling revealed that CO exposure reduced leukotriene B4 levels, and elevated SPMs such as resolvin

D1 (RvD1) and maresin-1. In human macrophages, SPMs increase HO-1 expression and accumulation of RvD1 and RvD5, which were reversed by inhibition of 15-lipoxygenase type-1. CO increased phagocytosis of opsonized zymosan by human macrophages, which was augmented by SPM. Further, CO-dependent phagocytosis could be reversed by 15-lipoxygenase inhibition, and SPM-stimulated phagocytosis was diminished by HO-1 inhibition. In a murine peritonitis model, iCO increased the number of macrophages carrying ingested apoptotic PMN in exudates and enhanced PMN apoptosis. Taken together, CO was found to accelerate the resolution of acute inflammation, shorten resolution intervals, enhance macrophage efferocytosis, and transiently regulate SPM levels. These results support suggest pro-resolving mechanisms for CO [42].

Pulmonary Epithelial Cells

Epithelial cells maintain the integrity of the alveolar–capillary barrier and defend against oxidative injury. Compromised epithelial cell function may permit fluid and macromolecules to leak into the airspace, resulting in clinical respiratory failure and death [43]. Apoptosis signaling pathways can have an important role in hyperoxia-induced lung cell death, regardless of whether the final phenotypic outcome resembles apoptosis or necrosis.

Protective Effects of CO on Epithelial Cell Death

Overexpression of HO-1 or exogenous CO (250 ppm, 3 h pretreatment) protected A549 alveolar epithelial cells against cell death from exposure to hyperoxia [44,45]. Treatment with a p38 MAPK inhibitor or transient transfection with dominant negative mutants of p38β or mitogen-activated protein kinase kinase-3 (MKK3) abolished the cytoprotective effect of CO against hyperoxia-mediated stress [45]. Further mechanistic studies have also implicated the signal transducer and activator of transcription protein-3 (STAT3) as a mediator of CO-dependent antiapoptotic protection in hyperoxic lung cell injury [46].

CO-Dependent Regulation of Epithelial Cell Autophagy

In addition to modulation of apoptosis programs, CO has been shown to modulate cellular autophagy as a component of its cytoprotective properties. Autophagy, a cellular autodigestive program, plays an important role in maintaining cellular homeostasis during environmental stress [47]. In this dynamic process, cellular organelles and protein are sequestered into double membrane-bound autophagosomes, which are subsequently transported to the lysosomes where the contents are degraded by proteases. Small molecules regenerated from degraded macromolecules are salvaged for anabolic pathways as a form of metabolic component recycling. Additionally, autophagy plays an integrated role in host defense by functioning as a bacterial and viral clearance mechanism. The mechanisms underlying CO-dependent lung cell protection and the role of autophagy in this process remain unclear. Continuous CO exposure (250 ppm) time-dependently increased the expression and activation of the autophagic protein, microtubule-associated protein-1 light chain-3B (LC3B), in mouse lung (24–72 h), and in cultured human alveolar (A549) or human bronchial epithelial cells (0–24 h) [48]. Furthermore, CO increased autophagosome formation in epithelial cells. ROS are believed to play an important role in the activation of cellular autophagy. CO exposure upregulated mtROS in epithelial cells in conjunction with activation of autophagy markers. Furthermore, CO-dependent induction of LC3B expression was inhibited by the antioxidant N-acetyl-L-cysteine and the mitochondria-targeting antioxidant, Mito-TEMPO. These data suggest that CO promotes autophagy through mtROS generation and signaling. The relationships between autophagy proteins and CO-dependent cytoprotection were investigated using the hyperoxia model. CO protected against hyperoxia-induced cell death and inhibited excess hyperoxia-elicited ROS production. The ability of CO to protect against hyperoxia-induced cell death and caspase-3 activation was abrogated in epithelial cells infected with LC3B-targeting small interfering RNA, indicating a contributory role for this key autophagosome-associated protein. These studies uncover a new mechanism for the protective action of CO, in support of potential therapeutic application of this gas [48].

Pulmonary Endothelial Cells

Endothelial cells are fundamental for providing a basal lining for the pulmonary and systemic vasculature, which contributes to the control of vascular permeability. Endothelial cells are a primary target for toxicity by free heme and other pro-oxidants. CO has been shown to confer endothelial cell protection in part by modulating apoptotic signaling pathways in model studies [49–52]. When applied concentrations of 250–10,000 ppm, CO inhibited cell death caused by proapoptotic agents such as TNF-α in endothelial cells, which required the p38 MAPK pathway, and modulation of NF-κB signaling [49,50]. Low-dose CO application (15 ppm) was found to inhibit anoxia–reoxygenation (A/R)-induced lung endothelial cell apoptosis and caspase-3 and -8 activation. Exposure of primary rat pulmonary artery endothelial cells to CO inhibited apoptosis and enhanced p38 MAPK activation in A/R. Transfection of p38α dominant negative mutant or chemical inhibition of p38 MAPK activity abrogated the antiapoptotic effects of CO in the A/R model [51,52]. Additional targets for CO-dependent regulation of cellular apoptosis in the A/R include the STAT3 and PI3K/Akt pathways [53].

In cultured mouse lung endothelial cells, application of CO (250 ppm) prevented hyperoxia-induced apoptosis by inhibition of both extrinsic and intrinsic apoptotic pathways, via downregulation of hyperoxia-induced ROS production. CO was shown to inhibit caspase-8 activation, Bax and Bid activation, cytochrome c release from the mitochondria, and downstream caspase-3 and -9 activation. Additionally, CO inhibited plasma membrane DISC formation by attenuating ERK1/2-dependent NOX activation and ROS production, and by preventing DISC trafficking to the plasma membrane. In this cell type, a novel interaction between HO-1 and Bax was found to be enhanced by CO treatment [54]. CO (250 ppm) was also shown to inhibit Fas-mediated apoptosis induced by FAS activating antibody in mouse lung endothelial cells, in part by recruiting the antiapoptotic molecule FLIP to the activated death-inducing signal complex [55].

Pulmonary Fibroblasts

Pulmonary fibroblasts are the most abundant cell type of the lung interstitium and can serve an important role in the repair and remodeling processes following injury. The controlled accumulation of fibroblasts to sites of inflammation is crucial to effective tissue repair after injury.

CO exposure was shown to have antifibrotic effects during experimental pulmonary fibrosis in animal model studies [56]. Early studies in this domain focused on determining the effects of low-dose CO on fibroblast proliferation and production of ECM proteins. In human fetal lung fibroblasts (MRC-5), exposure to CO (250 ppm, 7 days) reduced fibroblast proliferation in a manner dependent on sGC, but independent of MAPK pathways. CO exposure led to increased cellular levels of p21$^{Waf1/Cip1}$ and decreased levels of cyclins A/D. CO suppressed collagen-1 and fibronectin gene expression in pulmonary fibroblasts in a manner that was highly dependent on the basic helix–loop–helix transcription factor ID-1 (inhibitor of differentiation/DNA synthesis), a regulator of the myofibroblast phenotype. In contrast, these effects were found to be independent of sGC and MAPK signaling [56]. Gene expression profiling of fibroblasts treated with TGF-β1 and CO (250 ppm) revealed several differentially expressed gene families including those related to muscular system development and the small proline-rich family of proteins. In fibroblasts, CO (250 ppm) inhibited α-smooth muscle actin expression and enhanced small proline-rich protein 1a expression. These effects were shown to depend on modulation of the ERK1/2 MAPK pathway [57]. Additionally, endogenous HO-1-derived CO was implicated in the antifibrotic effects of the natural antioxidant compound, quercetin [58]. Taken together, these data suggest that CO exerts an antifibrotic effect in the lung, and this effect may be due to suppression of fibroblast proliferation and/or suppression of matrix deposition by fibroblasts via novel transcriptional targets.

Human Airway SMCs

SMCs make up a critical component of the pulmonary airways and represent a primary target for inflammatory responses triggered by inhaled irritants and allergens. CO exerts antiproliferative effects when applied to various types of SMCs, and thus has potential application in preventing tissue remodeling [59–62]. CO (250 ppm) inhibits smooth muscle proliferation by activation of sGC and p38β MAPK, and increased expression of the cyclin-dependent kinase inhibitor p21$^{Waf1/Cip1}$ [59,60,63].

Continuous CO exposure (250 ppm) was also found to inhibit the proliferation of human airway epithelial cells via modulation of the ERK1/2 pathway [61]. Further experiments in this cell type revealed that the antiproliferative effects of CO derived from CORM-2 (10 μM) involved the increased production of mtROS, and/or the downregulation of cytosolic ROS as a consequence of NOX inhibition, leading to ERK1/2 downregulation and decreased cyclin D expression [62]. Caveolin-1, a candidate tumor suppressor protein, and the major structural component of plasma membrane caveolae, may serve as an accessory in the antiproliferative effects of CO treatments (250 ppm) [63]. Additional mechanisms for antiproliferative effects of CO include the downregulation of endothelial-derived smooth muscle mitogenic factors [64].

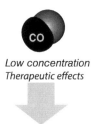
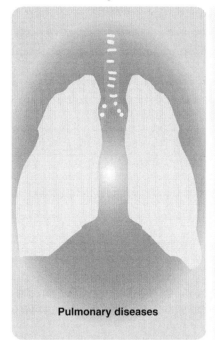

Figure 20.3 Therapeutic potential of iCO in preclinical models of pulmonary disease. CO, when inhaled at low concentration (e.g., 250 ppm), or applied as CORMs, has been shown to confer tissue protection in preclinical models of pulmonary disease. These include various forms of ALI induced by oxygen, pro-inflammatory stress, ischemia/reperfusion, and sepsis. Further, CO has been shown to confer graft protection in models of lung transplant. CO can confer these effects mainly by antiapoptotic and anti-inflammatory effects, and/or antimicrobial effects (as in sepsis). Finally, antiproliferative effects of CO (on vascular smooth muscle and fibroblasts) have been implicated in pulmonary hypertension and pulmonary fibrosis, respectively. Additionally, CO can confer antifibrotic effects in models of pulmonary fibrosis.

CO as an Experimental Protective Mediator in Pulmonary Disease and Critical Illness

The therapeutic potential of CO has been tested in numerous models of lung injury (Figure 20.3). Observations made in *in vivo* models are categorized below and recapitulate many of the mechanistic observations made in *in vitro* models as described earlier. Recent progress in the clinical development of iCO has been discussed in Chapter 30.

Acute Lung Injury

Experimental ALI

Low doses of CO have been shown to provide anti-inflammatory protection against ALI in rodent models. The continuous administration of CO (250 ppm, with 1 h pretreatment) during hyperoxia exposure prolonged the survival of rats and mice subjected to a lethal dose of hyperoxia, and dramatically reduced histological indices of lung injury, including airway neutrophil infiltration, fibrin deposition, alveolar proteinosis, pulmonary edema, and apoptosis, relative to animals exposed to hyperoxia without CO [45,65]. In mice, hyperoxia was shown to induce the expression of proinflammatory cytokines (i.e., TNF-α, IL-1β, and IL-6) and activate major MAPK pathways in lung tissue. The protection afforded by CO treatment (250 ppm) against the lethal effects of hyperoxia correlated with the inhibited release of proinflammatory cytokines in the bronchoalveolar lavage fluid (BALF). The protective effects of CO (250 ppm) in this model were found to depend on the MKK3/p38β MAPK pathway [45]. More recent studies report protective effects of HO-1 or CO in a model of hyperoxia-induced bronchopulmonary dysplasia in neonatal mice [66]. Lung-specific transgenic overexpression of HO-1 alleviated hyperoxia-induced lung inflammation, edema, arterial remodeling, and right ventricular hypertrophy. Similar protective responses were elicited by intermittent CO inhalation (250 ppm, 1 h/day for 14 days) in this model. However, neither CO nor HO-1 expression prevented alveolar simplification in this model [66].

LPS-Induced Injury

Systemic inflammation can cause injury to various organs, including the lung. Similar to effects with HO-1 gene transfer, anti-inflammatory effects of low-dose iCO (i.e., 250 ppm) were demonstrated *in vivo* using a mouse model of endotoxemia [29]. CO preconditioning (250 ppm, 1 h pretreatment) reduced the production of serum TNF and IL-1β and increased the production of IL-10, reduced organ injury, and prolonged survival following systemic LPS challenge [29]. In further model studies, the anti-inflammatory protection afforded by CO (250 ppm, 1 h pretreatment) against LPS-induced organ injury was associated with inhibition of iNOS expression and activity in the lung [67].

Anti-inflammatory effects of CO have also been demonstrated in preclinical models of inflammatory diseases in higher mammals. CO exposure (250 ppm, 1 h pretreatment) ameliorated the development of disseminated intravascular coagulation in swine and inhibited serum levels of the proinflammatory cytokine IL-1β in response to LPS challenge while upregulated the anti-inflammatory cytokine IL-10 levels [68].

A less robust anti-inflammatory effect was observed in a nonhuman primate (NHP) model of cynomolgus macaques subjected to LPS challenge [69]. CO exposure (500 ppm, 6 h) following LPS inhalation decreased TNF-α release in BALF but had no apparent effect on IL-6 and IL-8 release, in addition to reducing pulmonary neutrophilia at the highest concentration (500 ppm). This reduction of pulmonary neutrophilia was comparable to pretreatment with a standard inhaled corticosteroid budesonide. However, the therapeutic efficacy of CO in this model required higher doses that resulted in relatively high COHb levels (>30%). This work highlights the complexity of interspecies variation in lung responses to CO, and of dose–response relationships of CO to COHb levels and anti-inflammatory effects [69]. This study was the first to examine the therapeutic index and dose–response relationships of CO therapy in NHPs using an acute inflammation model.

Hemorrhagic Shock

Hemorrhagic shock and resuscitation (HSR) can elicit pulmonary inflammation leading to the propagation of ALI. A series of studies have shown the potential for protective anti-inflammatory effects of iCO or CORM treatment in protection against ALI and other injuries secondary to HSR. In the rat model, HSR was shown to cause ALI and pulmonary edema, cellular infiltration and hemorrhage, upregulation of inflammatory gene expression, and increased lung cell apoptosis. iCO administration (250 ppm, 1 h pretreatment and 3 h post-resuscitation) significantly prevented ALI, and reduced markers of inflammation and apoptosis, without affecting hemodynamic status or tissue oxygenation [70,71]. In animal models of HSR, therapeutic application of CORM-3 (4 mg/kg, i.v.) also alleviated HSR-induced ALI and pulmonary edema. CORM-3 also inhibited HSR-dependent upregulation of pro-inflammatory mediator genes (i.e., TNF, iNOS, and IL-1β) and expression of IL-1β and MIP-2. The expression of anti-inflammatory IL-10 was induced by CORM-3, which also inhibited lung cell apoptosis. The protective effects of CORM-3 (4 mg/kg, i.v.) against HSR-induced ALI were associated with upregulation PPAR-γ. The study concluded that CORM-3 ameliorated HSR-induced lung injury through anti-inflammatory and antiapoptotic effects, without detrimental effects on oxygenation and hemodynamics [72].

Ventilator-Induced Lung Injury

The therapeutic potential of CO has been shown in a specialized clinically relevant model of ventilator-induced lung injury (VILI) in rodents [73–76]. Rats ventilated with an injurious (high tidal volume) ventilator setting in combination with LPS injection exhibited lung injury. The inclusion of CO (250 ppm) during mechanical ventilation reduced the inflammatory cell infiltration in BALF. In the absence of cardiovascular effects, CO dose-dependently decreased TNF-α and increased IL-10 in the BALF [73]. CO application has also been reported to confer tissue protection in a mouse model of VILI, at moderate tidal volume ventilation [74–76]. In this model, mechanical ventilation caused significant lung injury reflected by increases in protein concentration, and total cell and neutrophil counts in BALF. CO (250 ppm, 8 h) reduced mechanical ventilation-induced cytokine and chemokine production and prevented lung injury during ventilation, involving the inhibition of mechanical ventilation-induced increases in BALF protein concentration and cell count, lung neutrophil recruitment, and pulmonary edema [74,75]. To date, these effects of CO were associated with the activation of caveolin-1 [74], activation of PPAR-γ, and the inhibition of Egr-1 signaling [75]. These studies, taken together, suggest that mechanical ventilation in the presence of CO may provide protection in animal models of VILI. However, more studies are required to determine the exact mechanisms underlying the therapeutic potential of CO in VILI models.

Sepsis- and Pneumonia-Associated Lung Injury

Sepsis is characterized by aberrant inflammatory responses to systemic infection. CORM-2 treatment (30 mg/kg, i.p.) was found to protect endotoxemic mice or mice subjected to cecal ligation and puncture (CLP)-induced polymicrobial sepsis [77]. These effects were accomplished in part by reduction of high-mobility group box 1 and proinflammatory cytokine production [77]. CO has been shown to promote bacterial clearance in sepsis models through incompletely understood mechanisms [40,78]. The pharmacological application of CORM-2 by injection enhanced bacterial phagocytosis *in vivo* and rescued heme oxygenase-deficient mice from sepsis-induced mortality [78]. In an *Escherichia coli* infection model, endogenous HO-derived CO was implicated as a mediator of macrophage phagocytosis [40]. CO inhalation (250 ppm) when applied either as pretreatment or post-treatment improved mouse survival in the CLP model of polymicrobial sepsis. The protective effects of CO in CLP were related to the reduction of inflammation and enhanced bacterial clearance from lungs, blood, and other organs [79].

Low-dose iCO was found to accelerate resolution of ALI in a clinically relevant baboon model of pneumococcal pneumonia [80]. Administration of iCO (200 ppm for 60 min) at 48 h following *Streptococcus pneumoniae* inoculation significantly attenuated histological signs of ALI at 8 days, with evidence of enhanced mitochondrial biogenesis in alveolar type-2 epithelial cells and macrophages, and systemic antioxidant effects with augmentation of SOD2 expression in the kidney [80]. Application of iCO also reduced proinflammatory urinary cysteinyl leukotrienes and improved levels of circulating SPMs in baboons [81]. Plasma obtained from *S. pneumoniae*-infected baboons displayed significantly reduced levels of eicosapentaenoic acid-derived E-series resolvins (RvE) and lipoxins. These observations suggested that pneumococcal pneumonia can deregulate pro-resolution programs in baboons. In these animals, iCO exposure led to increased plasma RvE and lipoxin levels. These studies concluded that altered SPM profiles during pneumonia can be partially restored with low-dose iCO [81].

Human ARDS

Acute respiratory distress syndrome ("ARDS") is a type of respiratory failure characterized by widespread inflammation in the lungs that has a high (~40%) mortality rate. The high morbidity and mortality of ARDS reflect the inefficacy of currently available diagnostic markers and therapeutic modalities [82]. To date, limited studies have explored the potential role of CO in clinical ARDS, based on preclinical studies in baboon pneumonia [80,81]. Clinical studies in iCO therapy for ARDS have largely been directed at establishing dosimetry and confirming safety and feasibility. However, limited biomarker data from these trial efforts have been published. In one study, iCO (200 ppm) versus placebo displayed a reduction of plasma mtDNA [17], a biomarker shown to be highly predictive of ICU mortality [83]. Thus, encouraging data in this trial have provided a basis for dosimetry and efficacy measurements in the ongoing phase 2 ARDS study (NCT03799874), and for future planned phase 3 efficacy trials [17].

Malaria-Induced ALI/ARDS

Malaria is a serious parasitic disease with a high incidence of mortality. Malaria-associated (MA) ALI/ARDS is one of the major clinical complications of severe malaria, which is characterized by a high mortality rate and can occur even after therapeutic interventions. DBA/2 mice infected with *Plasmodium berghei ANKA* (PbA) develop symptoms of ALI/ARDS that mimic the human syndrome, including pulmonary edema, hemorrhage, pleural effusion, and hypoxemia. Elevated HO-1 expression in inflammatory cells has been observed in severe human malaria [84] and has been described as a potential therapeutic target in animal models of malaria in which animals are infected with *Plasmodium* [84]. Protective roles of HO-1 were demonstrated experimentally in mouse models of MA-ALI/ARDS [85] and experimental cerebral malaria (CM) [86]. These protective effects were attributed to either the HO-1-dependent degradation of free heme, a putative proinflammatory agent in malaria [85], or the endogenous production of HO-derived CO [86], which can confer anti-inflammatory effects. Increased lung endothelial permeability and upregulation of vascular endothelial growth factor (VEGF) and other proinflammatory cytokines were associated with MA-ALI/ARDS and were inhibited in experimental models by HO-1 induction using heme [87].

Mice were found to be protected against experimental malaria after treatment with either CORMs or iCO [84,86,88,89]. Application of CO (250 ppm) for 72 h, starting at day 2 after infection, prevented death of *P. berghei* ANKA-infected DBA/2 mice by ALI without alterations in the parasitemia, but associated with inhibition of circulating VEGF levels. CO administration also reduced hemorrhages and pulmonary edema in this model [88]. Exogenous iCO also conferred protection in a model of experimental CM, following PbA infection by preventing blood–brain barrier disruption, brain microvasculature congestion, and neuroinflammation, including CD8$^+$ T-cell brain sequestration. CO also reduced oxidation of free hemoglobin, and prevented heme release, which were associated with its antimalarial effects [86]. ALF-492, a novel CORM compound (36.7 mg/kg), protected against PbA-induced CM, via release of CO and secondary induction of HO-1 [89]. When used in combination with the antimalarial drug artesunate, ALF-492 was an effective adjunctive and adjuvant treatment for CM, conferring protection after onset of severe disease [89]. These investigations, taken together, suggest that modulation iCO and/or pharmacological application of CORMs may represent a prospective therapy for MA-ALI and CM associated with *Plasmodium* infection, and potentially enhance the efficacy of other antimalarial drugs when used in combination therapies [84].

Pulmonary Hypertension

Pulmonary arterial hypertension (PAH) is a terminal disease characterized by a progressive increase in pulmonary vascular resistance leading to right ventricular failure. Administration of CO was shown to provide protection in rodent models of monocrotaline-induced and hypoxia-induced PAH. Exposure to CO (250 ppm, 1 h/day) reversed established PAH and right ventricular hypertrophy, and restored right ventricular and pulmonary arterial pressures, as well as pulmonary vascular morphology, to that of controls [90]. The ability of CO to reverse PAH was dependent on endothelial NOS3 and NO generation, since CO failed to reverse chronic hypoxia-induced PAH in mice genetically deficient for eNOS ($nos3^{-/-}$) [90]. The protective effect of CO in this model was endothelial cell dependent, and associated with increased apoptosis and decreased cellular proliferation of vascular SMCs [90]. Additional studies have shown that continuous CO exposure (50 ppm, 21 days) decreased pulmonary artery vascular resistance and inhibited hypoxic vasoconstriction, through activation of the sGC/cGMP, and the hyperpolarization of potassium channels [91]. A clinical trial for iCO in severe PAH was previously registered (NCT01523548) but withdrawn.

Pulmonary Fibrosis

Idiopathic pulmonary fibrosis (IPF) is a chronic, progressive lung disease characterized by lung scarring or thickening of lung tissues associated with fibroblast hyperproliferation and extracellular matrix (ECM) remodeling with no known etiology or effective treatment [92]. The incidence of IPF is age dependent, and has a high rate of mortality, with a median survival of 3 years. The pathogenesis of IPF involves dysregulated wound healing with continuous damage to lung epithelium, fibroblast/myofibroblast activation, and excessive ECM production, leading to aberrant lung remodeling [2]. IPF affects primarily the lower respiratory tract resulting in compromised efficiency of alveolar gas exchange [92]. Bleomycin (BLM), a redox cycling compound that generates O_2^- and H_2O_2, and causes lesions in mouse lung after intratracheal administration, is used to model IPF in animals. Exogenous CO treatment can provide protection against BLM-induced fibrotic lung injury in mice [56]. In mice treated with BLM intratracheally and then exposed to CO (250 ppm, 3 h/day for 14 days) or ambient air, the lungs from CO-treated animals displayed reduced lung hydroxyproline, collagen, and fibronectin levels relative to air-treated BLM-injured controls. The protective effect of CO (250 ppm) in this model was associated with an antiproliferative effect of CO on fibroblast proliferation associated with the increased expression of p21$^{\text{Waf1/Cip1}}$ and inhibition of cyclin A/D expression [56].

The feasibility of iCO to human therapy for IPF was recently evaluated in a multicenter phase IIa, double-blinded, sham-controlled, clinical trial, in which IPF patients were randomized to receive iCO (100–200 ppm, 2 h/day, twice a week) or air for 12 weeks. The study found no differences in physiological measurements, incidence of acute exacerbations, hospitalization, death, patient-reported outcomes, or secondary endpoints, such as matrix metalloproteinase-7. Furthermore, no differences were reported in the distribution of adverse events between iCO and placebo-treated patients. Thus, the study concluded that iCO can be safely administered to IPF patients.

Lung Ischemia/Reperfusion Injury and Primary Graft Dysfunction (Lung Transplant)

In an *in vivo* lung ischemia/reperfusion (I/R) model and corresponding *in vitro* model using pulmonary endothelial cells, I/R was shown to cause lung cell injury consistent with activation of Fas and caspase-3-dependent apoptosis. In mice, CO was applied at 500 ppm with 1 h pretreatment and continuous exposure during I/R protocols in mice, which resulted in protection from lung I/R injury [51].

Corresponding *in vitro* experiments used 15 ppm CO applied to rat pulmonary artery endothelial cells. CO was shown to confer antiapoptotic protection via activating the p38α MAPK isoform and the upstream MAPK kinase MKK3 to inhibit Fas/Fas ligand expression, caspase-3, -8, and -9, mitochondrial cytochrome *c* release, Bid/Bax activation, and poly(ADP-ribose) polymerase cleavage, while upregulating antiapoptotic Bcl-2 family proteins [51,52]. CO (500 ppm) failed to confer protection from I/R injury in MKK3-deficient mice [52]. Taken together, these studies demonstrated that CO protects the lung against I/R-induced endothelial cell apoptosis *in vivo* via the MKK3/p38α MAPK pathway [52].

In organ transplantation, I/R injury subsequent to transplantation may represent a major causative factor in graft dysfunction, which increases the probability of eventual graft failure. CO has been intensively studied as an anti-inflammatory therapeutic in experimental organ transplantation (see Chapter 19). CO has a demonstrated potential for reducing transplant associated I/R injury in the lung and also reducing the incidence of primary graft dysfunction when added to organ preservation fluid or when applied to donors and/or recipients in gaseous form at low concentration [93–96]. The application of CO can confer protection during transplantation of multiple organs. In a model of orthotopic lung transplantation in rats, exogenous application of CO (500 ppm) to recipients significantly protected the graft, and reduced hemorrhage, fibrosis, and thrombosis after transplantation [93]. Furthermore, CO inhibited lung cell apoptosis and inhibited lung and systemic proinflammatory cytokine production [93,94]. Similar observations were made using the vascular transplantation model; when transplant recipients of aortic grafts were exposed to CO (250 ppm), these animals displayed reduced intimal hyperplasia, and reduced leukocyte, macrophage, and T-cell infiltration in the graft [60]. The modulation of responses, and the inhibition of apoptosis and inflammation associated with I/R injury, likely represents the main mechanisms by which CO promotes the survival and function of transplanted organs, though effects on circulation and cell proliferation may also contribute [60].

Conclusions

CO has emerged as an endogenous gaseous mediator with pleiotropic roles in the regulation of cell physiology and function via its regulation of cellular signal transduction pathways. Although the effects of CO may vary in a cell type-specific manner, the basic mechanisms of action involve anti-inflammatory, antiapoptotic, and antiproliferative effects, which act alone or in concert to mitigate injury. These effects are likely to be relevant in diseases that have a strong component of inflammation and/or cell death in the pathology or in hyperproliferative disorders. Among the putative biochemical targets of CO, emerging evidence suggests that mitochondria serve as a proximal target for CO, such that modulation of the functioning of this vital organelle triggers downstream cellular responses. Nevertheless, other non-mitochondrial targets of CO have been proposed, including enzymes such as sGC, iNOS, and NOXs, and gas-binding transcription factors.

Gaseous molecules continue to show future promise as they join the ranks of experimental and clinical therapeutics. Among the known medical gases, CO inhalation has been demonstrated to have potential applications in pulmonary diseases and other inflammatory diseases. To date, salutary effects of CO have been demonstrated in a number rodent model studies, though recent studies have attempted to recapitulate findings in larger animals such as swine, monkeys, and baboons [68,69,80,81]. Thus, differences in lung physiological responses to CO between rodents, large animals (e.g., NHPs), and humans require further investigation.

Further research will determine the translatability of iCO, CORMs, or other forms of CO donors as therapeutics in human pulmonary diseases. As discussed elsewhere (Chapter 30), clinical trials have been conducted, which demonstrate safety of iCO application in sepsis-induced ARDS and other chronic lung diseases, such as chronic obstructive pulmonary disease and IPF [17,97,98]. Additional studies will be required to confirm the safety and efficacy of iCO as a treatment for inflammatory lung diseases. The demonstration of therapeutic efficacy of iCO remains unrealized and will be the subject of forthcoming clinical trials. Pharmacological application of CO using other donors provides attractive alternatives to inhalation gas [18,99–101]. However, further understanding of the pharmacokinetics and toxicological responses of CORMs, including hemodynamic effects and metabolism of the backbone and/or transition metal components, must be achieved before employing CO donors as clinical therapy. The effectiveness of CO donors as therapeutics in human diseases, including sepsis, renal transplantation, pulmonary fibrosis, and pulmonary hypertension, awaits the outcome of additional planned preclinical testing, whereas clinical testing lags behind that of iCO. Forthcoming efforts will provide a resolution on the feasibility and efficacy of exploiting the therapeutic index of CO and CO donors in human diseases.

Conflict of interest

Stefan Ryter is an employee of Proterris, Inc., a company involved in development of CO-based therapeutics.

References

1. Ryter, S.W. and Otterbein, L.E. (2004). Carbon monoxide in biology and medicine. *Bioessays* 26 (3): 270–280.
2. Maines, M.D. (1997). The heme oxygenase system: a regulator of second messenger gases. *Annu. Rev. Pharmacol. Toxicol.* 37: 517–554.
3. Coburn, R.F., Blakemore, W.S., and Forster, R.E. (1963). Endogenous carbon monoxide production in man. *J. Clin. Invest.* 42: 1172–1178.
4. Sjostrand, T. (1949). Endogenous production of carbon monoxide in man under normal and pathophysiological conditions. *Scand. J. Clin. Lab Invest.* 1: 201–214.
5. Sjostrand, T. (1952). The formation of carbon monoxide by the decomposition of hemoglobin in vivo. *Acta Physiol. Scand.* 26: 338–344.
6. Ryter, S.W. and Choi, A.M. (2013). Carbon monoxide in exhaled breath testing and therapeutics. *J. Breath Res.* 7(1): 017111.
7. Kutty, R.K., Daniel, R.F., Ryan, D.E., Levin, W., and Maines, M.D. (1988). Rat liver cytochrome P-450b, P-420b, and P-420c are degraded to biliverdin by heme oxygenase. *Arch. Biochem. Biophys.* 260: 638–644.
8. Tenhunen, R., Marver, H.S., and Schmid, R. (1969). Microsomal heme oxygenase. Characterization of the enzyme. *J. Biol. Chem.* 244: 6388–6394.
9. Tenhunen, R., Ross, M.E., Marver, H.S., and Schmid, R. (1970). Reduced nicotinamide-adenine dinucleotide phosphate dependent biliverdin reductase: partial purification and characterization. *Biochemistry* 9: 298–303.
10. Keyse, S.M. and Tyrrell, R.M. (1989). Heme oxygenase is the major 32-kDa stress protein induced in human skin fibroblasts by UVA radiation, hydrogen peroxide, and sodium arsenite. *Proc. Natl. Acad. Sci. U.S.A.* 86: 99–103.
11. Ryter, S.W., Alam, J., and Choi, A.M. (2006). Heme oxygenase-1/carbon monoxide: from basic science to therapeutic applications. *Physiol. Rev.* 86: 583–650.
12. Von Burg, R. (1999). Carbon monoxide. *J. Appl. Toxicol.* 19: 379–386.
13. Gorman, D., Drewry, A., Huang, Y.L., and Sames, C. (2003). The clinical toxicology of carbon monoxide. *Toxicology* 187: 25–38.
14. Rudra, C.B., Williams, M.A., Sheppard, L., Koenig, J.Q., Schiff, M.A., Frederick, I.O., and Dills, R. (2010). Relation of whole blood carboxyhemoglobin concentration to ambient carbon monoxide exposure estimated using regression. *Am. J. Epidemiol.* 171 (8): 942–951.
15. Russell, M.A. (1973). Blood carboxyhaemoglobin changes during tobacco smoking. *Postgrad. Med. J.* 49: 684–687.
16. Piantadosi, C.A. (1999). Diagnosis and treatment of carbon monoxide poisoning. *Respir. Care Clin. N. Am.* 5: 183–202.
17. Fredenburgh, L.E., Perrella, M.A., Barragan-Bradford, D., Hess, D.R., Peters, E., Welty-Wolf, K.E., Kraft, B.D., Harris, R.S., Maurer, R., Nakahira, K., Oromendia, C., Davies, J.D., Higuera, A., Schiffer, K.T., Englert, J.A., Dieffenbach, P.B., Berlin, D.A., Lagambina, S., Bouthot, M., Sullivan, A.I., Nuccio, P.F., Kone, M.T., Malik, M.J., Porras, M.A.P., Finkelsztein, E., Winkler, T., Hurwitz, S., Serhan, C.N., Piantadosi, C.A., Baron, R.M., Thompson, B.T., and Choi, A.M. (2018). A phase I trial of low-dose inhaled carbon monoxide in sepsis-induced ARDS. *JCI Insight* 3 (23).
18. Motterlini, R. and Otterbein, L.E. (2010). The therapeutic potential of carbon monoxide. *Nat. Rev. Drug Discov.* 9: 728–743.
19. Ryter, S.W., Ma, K.C., and Choi, A.M.K. (2018). Carbon monoxide in lung cell physiology and disease. *Am. J. Physiol. Cell Physiol.* 314: C211–C227.
20. Kim, H.P., Ryter, S.W., and Choi, A.M. (2006). CO as a cellular signaling molecule. *Annu. Rev. Pharmacol. Toxicol.* 46: 411–449.
21. Tsai, A.L., Berka, V., Martin, E., and Olson, J.S. (2012). A "sliding scale rule" for selectivity among NO, CO, and O_2 by heme protein sensors. *Biochemistry* 51: 172–186.
22. Ryter, S.W. (2019). Heme oxygenase-1/carbon monoxide as modulators of autophagy and inflammation. *Arch. Biochem. Biophys.* 678: 108186.
23. Ryter, S.W. (2020). Therapeutic potential of heme oxygenase-1 and carbon monoxide in acute organ injury, critical illness, and inflammatory disorders. *Antioxidants (Basel)* 9 (11): 1153.
24. Figueiredo-Pereira, C., Dias-Pedroso, D., Soares, N.L., and Vieira, H.L.A. (2020). CO-mediated cytoprotection is dependent on cell metabolism modulation. *Redox Biol.* 32: 101470.
25. Menzel, D.B. and Amdur, M.O. (1986). *Toxic Response of the Respiratory System. In Casarett and Doull's*

Toxicology, the Basic Science of Poisons, 3e (Ed. K. Klaassen, M.O. Amdur and J. Doull), 330–358. New York, NY: MacMillan Publishing Company.

26. Hu, G. and Christman, J.W. (2019). Editorial: Alveolar macrophages in lung inflammation and resolution. *Front. Immunol.* 10: 2275.
27. Byrne, A.J., Mathie, S.A., Gregory, L.G., and Lloyd, C.M. (2015). Pulmonary macrophages: key players in the innate defence of the airways. *Thorax* 70: 1189–1196.
28. Misharin, A.V., Budinger, S.G.R., and Perlman, H. (2011). The lung macrophage: A Jack of all trades. *Am. J. Respir. Crit. Care Med.* 184 (5): 497–498.
29. Otterbein, L.E., Bach, F.H., Alam, J., Soares, M., Tao Lu, H., Wysk, M., Davis, R.J., Flavell, R.A., and Choi, A.M. (2000). Carbon monoxide has anti-inflammatory effects involving the mitogen-activated protein kinase pathway. *Nat. Med.* 6: 422–428.
30. Morse, D., Pischke, S.E., Zhoum, Z., Davism, R.J., Flavell, R.A., Loop, T., Otterbein, S.L., Otterbein, L.E., and Choi, A.M. (2003). Suppression of inflammatory cytokine production by carbon monoxide involves the JNK pathway and AP-1. *J. Biol. Chem.* 278: 36993–36998.
31. Nakahira, K., Kim, H.P., Geng, X.H., Nakao, A., Wang, X., Murase, N., Drain, P.F., Wang, X., Sasidhar, M., Nabel, E.G., Takahashi, T., Lukacs, N.W., Ryter, S.W., Morita, K., and Choi, A.M. (2006). Carbon monoxide differentially inhibits TLR signaling pathways by regulating ROS-induced trafficking of TLRs to lipid rafts. *J. Exp. Med.* 203: 2377–2389.
32. Zuckerbraun, B.S., Chin, B.Y., Bilban, M., d'Avila, J.C., Raom, J., Billiar, T.R., and Otterbein, L.E. (2007). Carbon monoxide signals via inhibition of cytochrome c oxidase and generation of mitochondrial reactive oxygen species. *FASEB J.* 21: 1099–1106.
33. Bilban, M., Bach, F.H., Otterbein, S.L., Ifedigbo, E., d'Avila, J.C., Esterbauer, H., Chin, B.Y., Usheva, A., Robson, S.C., Wagner, O., and Otterbein, L.E. (2006). Carbon monoxide orchestrates a protective response through PPARgamma. *Immunity* 24 (5): 601–610.
34. Haschemi, A., Chin, B.Y., Jeitler, M., Esterbauer, H., Wagner, O., Bilban, M., and Otterbein, L.E. (2011). Carbon monoxide induced PPARγ SUMOylation and UCP2 block inflammatory gene expression in macrophages. *PLoS One* 6 (10): e26376.
35. Chin, B.Y., Jiang, G., Wegiel, B., Wang, H.J., Macdonald, T., Zhang, X.C., Gallo, D., Cszimadia, E., Bach, F.H., Lee, P.J., and Otterbein, L.E. (2007). Hypoxia-inducible factor 1alpha stabilization by carbon monoxide results in cytoprotective preconditioning. *Proc. Natl. Acad. Sci U. S. A.* 104 (12): 5109–5114.
36. Schroder, K. and Tschopp, J. (2010). The inflammasomes. *Cell* 140: 821–832.
37. Davis, B.K., Wen, H., and Ting, J.P. (2011). The inflammasome NLRs in immunity, inflammation, and associated diseases. *Annual. Rev. Imunol.* 29: 707–735.
38. Jung, S.S., Moon, J.S., Xu, J., Ifedigbo, E., Ryter, S.W., Choi, A.M., and Nakahira, K. (2015). Carbon monoxide negatively regulates NLRP3 inflammasome activation in macrophages. *Am. J. Physiol. Lung Cell Mol. Physiol.* 308: L1058–L1067.
39. Kim, S., Joe, Y., Jeong, S.O., Zheng, M., Back, S.H., Park, S.W., Ryter, S.W., and Chung, H.T. (2014). Endoplasmic reticulum stress is sufficient for the induction of IL-1β production via activation of the NF-κB and inflammasome pathways. *Innate Immun.* 20: 799–815.
40. Wegiel, B., Larsen, R., Gallo, D., Chin, B.Y., Harris, C., Mannam, P., Kaczmarek, E., Lee, P.J., Zuckerbraun, B.S., Flavell, R., Soares, M.P., and Otterbein, L.E. (2014). Macrophages sense and kill bacteria through carbon monoxide-dependent inflammasome activation. *J. Clin. Invest.* 124: 4926–4940.
41. Buckley, C.D., Gilroy, D.W., and Serhan, C.N. (2014). Proresolving lipid mediators and mechanisms in the resolution of acute inflammation. *Immunity* 40: 315–327.
42. Chiang, N., Shinohara, M., Dalli, J., Mirakaj, V., Kibi, M., Choi, A.M., and Serhan, C.N. (2013). Inhaled carbon monoxide accelerates resolution of inflammation via unique proresolving mediator-heme oxygenase-1 circuits. *J. Immunol.* 190 (12): 6378–6388.
43. Mantell, L.L. and Lee, P.J. (2000). Signal transduction pathways in hyperoxia-induced lung cell death. *Mol. Genet. Metab.* 71 (1–2): 359–370.
44. Lee, P.J., Alam, J., Wiegand, G.W., and Choi, A.M. (1996). Overexpression of heme oxygenase-1 in human pulmonary epithelial cells results in cell growth arrest and increased resistance to hyperoxia. *Proc. Natl. Acad. Sci. U S A.* 93: 10393–10398.
45. Otterbein, L.E., Otterbein, S.L., Ifedigbo, E., Liu, F., Morse, D.E., Fearns, C., Ulevitch, R.J., Knicklebein, R., Flavell, R.A., and Choi, A.M. (2003). MKK3 mitogen activated protein kinase pathway mediates carbon monoxide-induced protection against oxidant induced lung injury. *Am. J. Pathol.* 163: 2555–2563.
46. Zhang, X., Shan, P., Jiang, G., Zhang, S.S., Otterbein, L.E., Fu, X.Y., and Lee, P.J. (2006). Endothelial STAT3 is essential for the protective effects of HO-1 in oxidant-induced lung injury. *FASEB J* 20: 2156–2158.
47. Choi, A.M., Ryter, S.W., and Levine, B. (2013). Autophagy in human health and disease. *N. Engl. J. Med.* 368 (7): 651–662.

48. Lee, S.J., Ryter, S.W., Xu, J.F., Nakahira, K., Kim, H.P., Choi, A.M., and Kim, Y.S. (2011). Carbon monoxide activates autophagy via mitochondrial reactive oxygen species formation. *Am. J. Respir. Cell Mol. Biol.* 45: 867–873.

49. Brouard, S., Otterbein, L.E., Anrather, J., Tobiasch, E., Bach, F.H., Choi, A.M., and Soares, M.P. (2000). Carbon monoxide generated by heme oxygenase-1 suppresses endothelial cell apoptosis. *J. Exp. Med.* 192: 1015–1026.

50. Brouard, S., Berberat, P.O., Tobiasch, E., Seldon, M.P., Bach, F.H., and Soares, M.P. (2002). Heme oxygenase-1-derived carbon monoxide requires the activation of transcription factor NF-kappa B to protect endothelial cells from tumor necrosis factor-alpha-mediated apoptosis. *J. Biol. Chem.* 277: 17950–17961.

51. Zhang, X., Shan, P., Otterbein, L.E., Alam, J., Flavell, R.A., Davis, R.J., Choi, A.M., and Lee, P.J. (2003). Carbon monoxide inhibition of apoptosis during ischemia-reperfusion lung injury is dependent on the p38 mitogen-activated protein kinase pathway and involves caspase 3. *J. Biol. Chem.* 278 (2): 1248–1258.

52. Zhang, X., Shan, P., Alam, J., Davis, R.J., Flavell, R.A., and Lee, P.J. (2003). Carbon monoxide modulates Fas/Fas ligand, caspases, and Bcl-2 family proteins via the p38alpha mitogen-activated protein kinase pathway during ischemia-reperfusion lung injury. *J. Biol. Chem.* 278 (24): 22061–22070.

53. Zhang, X., Shan, P., Alam, J., Fu, X.Y., and Lee, P.J. (2005). Carbon monoxide differentially modulates STAT1 and STAT3 and inhibits apoptosis via a phosphatidylinositol 3-kinase/Akt and p38 kinase-dependent STAT3 pathway during anoxia-reoxygenation injury. *J. Biol. Chem.* 280: 8714–8721.

54. Wang, X., Wang, Y., Lee, S.J., Kim, H.P., Choi, A.M., and Ryter, S.W. (2011). Carbon monoxide inhibits Fas activating antibody-induced apoptosis in endothelial cells. *Med. Gas Res.* 1 (1): 8.

55. Wang, X., Wang, Y., Kim, H.P., Nakahira, K., Ryter, S.W., and Choi, A.M. (2007). Carbon monoxide protects against hyperoxia-induced endothelial cell apoptosis by inhibiting reactive oxygen species formation. *J. Biol. Chem.* 282 (3): 1718–1726.

56. Zhou, Z., Song, R., Fattman, C.L., Greenhill, S., Alber, S., Oury, T.D., Choi, A.M., and Morse, D. (2005). Carbon monoxide suppresses bleomycin-induced lung fibrosis. *Am. J. Pathol.* 166 (1): 27–37.

57. Zheng, L., Zhou, Z., Lin, L., Alber, S., Watkins, S., Kaminski, N., Choi, A.M., and Morse, D. (2009). Carbon monoxide modulates alpha-smooth muscle actin and small proline rich-1a expression in fibrosis. *Am. J. Respir. Cell Mol. Biol.* 41 (1): 85–92.

58. Nakamura, T., Matsushima, M., Hayashi, Y., Shibasaki, M., Imaizumi, K., Hashimoto, N., Shimokata, K., Hasegawa, Y., and Kawabe, T. (2011). Attenuation of transforming growth factor-β-stimulated collagen production in fibroblasts by quercetin-induced heme oxygenase-1. *Am. J. Respir. Cell Mol. Biol.* 44 (5): 614–620.

59. Morita, T., Mitsialis, S.A., Koike, H., Liu, Y., and Kourembanas, S. (1997). Carbon monoxide controls the proliferation of hypoxic vascular smooth muscle cells. *J. Biol. Chem.* 272: 32804–32809.

60. Otterbein, L.E., Zuckerbraun, B.S., Haga, M., Liu, F., Song, R., Usheva, A., Stachulak, C., Bodyak, N., Smith, R.N., Csizmadia, E., Tyagi, S., Akamatsu, Y., Flavell, R.J., Billiar, T.R., Tzeng, E., Bach, F.H., Choi, A.M., and Soares, M.P. (2003). Carbon monoxide suppresses arteriosclerotic lesions associated with chronic graft rejection and with balloon injury. *Nat. Med.* 9: 183–190.

61. Song, R., Mahidhara, R.S., Liu, F., Ning, W., Otterbein, L.E., and Choi, A.M. (2002). Carbon monoxide inhibits human airway smooth muscle cell proliferation via mitogen-activated protein kinase pathway. *Am. J. Respir. Cell Mol. Biol.* 27: 603–606.

62. Taille, C., El-Benna, J., Lanone, S., Boczkowski, J., and Motterlini, R. (2005). Mitochondrial respiratory chain and NAD(P)H oxidase are targets for the antiproliferative effect of carbon monoxide in human airway smooth muscle. *J. Biol. Chem.* 280: 25350–25360.

63. Kim, H.P., Wang, X., Nakao, A., Kim, S.I., Murase, N., Choi, M.E., Ryter, S.W., and Choi, A.M.K. (2005). Caveolin-1 expression by means of p38β mitogen activated protein kinase mediates the antiproliferative effect of carbon monoxide. *Proc. Natl. Acad. Sci.* 102: 11319–11324.

64. Morita, T. and Kourembanas, S. (1995). Endothelial cell expression of vasoconstrictors and growth factors is regulated by smooth muscle cell-derived carbon monoxide. *J. Clin. Invest.* 96: 2676–2682.

65. Otterbein, L.E., Mantell, L.L., and Choi, A.M. (1999). Carbon monoxide provides protection against hyperoxic lung injury. *Am. J. Physiol.* 276: L688–L694.

66. Fernandez-Gonzalez, A., Alex Mitsialis, S., Liu, X., and Kourembanas, S. (2012). Vasculoprotective effects of heme oxygenase-1 in a murine model of hyperoxia-induced bronchopulmonary dysplasia. *Am. J. Physiol. Lung Cell Mol. Physiol.* 302: L775–L784.

67. Sarady, J.K., Zuckerbraun, B.S., Bilban, M., Wagner, O., Usheva, A., Liu, F., Ifedigbo, E., Zamora, R., Choi, A.M., and Otterbein, L.E. (2004). Carbon monoxide protection against endotoxic shock involves reciprocal effects on iNOS in the lung and liver. *FASEB J* 18: 854–856.

68. Mazzola, S., Forni, M., Albertini, M., Bacci, M.L., Zannoni, A., Gentilini, F., Lavitrano, M., Bach, F.H.,

Otterbein, L.E., and Clement, M.G. (2005). Carbon monoxide pretreatment prevents respiratory derangement and ameliorates hyperacute endotoxic shock in pigs. *FASEB J* 19: 2045–2047.

69. Mitchell, L.A., Channell, M.M., Royer, C.M., Ryter, S.W., Choi, A.M., and McDonald, J.D. (2010). Evaluation of inhaled carbon monoxide as an anti-inflammatory therapy in a nonhuman primate model of lung inflammation. *Am. J. Physiol. Lung Cell Mol. Physiol.* 299: L891–L897.

70. Kawanishi, S., Takahashi, T., Morimatsu, H., Shimizu, H., Omori, E., Sato, K., Matsumi, M., Maeda, S., Nakao, A., and Morita, K. (2013). Inhalation of carbon monoxide following resuscitation ameliorates hemorrhagic shock-induced lung injury. *Mol. Med. Rep.* 7: 3–10.

71. Kanagawa, F., Takahashi, T., Inoue, K., Shimizu, H., Omori, E., Morimatsu, H., Maeda, S., Katayama, H., Nakao, A., and Morita, K. (2010). Protective effect of carbon monoxide inhalation on lung injury after hemorrhagic shock/resuscitation in rats. *J. Trauma* 69: 185–194.

72. Kumada, Y., Takahashi, T., Shimizu, H., Nakamura, R., Omori, E., Inoue, K., and Morimatsu, H. (2019). Therapeutic effect of carbon monoxide-releasing molecule-3 on acute lung injury after hemorrhagic shock and resuscitation. *Exp. Ther. Med.* 17: 3429–3440.

73. Dolinay, T., Szilasi, M., Liu, M., and Choi, A.M. (2004). Inhaled carbon monoxide confers anti-inflammatory effects against ventilator-induced lung injury. *Am. J. Respir. Crit. Care Med.* 170: 613–620.

74. Hoetzel, A., Schmidt, R., Vallbracht, S., Goebel, U., Dolinay, T., Kim, H.P., Ifedigbo, E., Ryter, S.W., and Choi, A.M. (2009). Carbon monoxide prevents ventilator-induced lung injury via caveolin-1. *Crit. Care Med.* 37: 1708–1715.

75. Hoetzel, A., Dolinay, T., Vallbracht, S., Zhang, Y., Kim, H.P., Ifedigbo, E., Alber, S., Kaynar, A.M., Schmidt, R., Ryter, S.W., and Choi, A.M. (2008). Carbon monoxide protects against ventilator-induced lung injury via PPAR-gamma and inhibition of Egr-1. *Am. J. Respir. Crit. Care Med.* 177: 1223–1232.

76. Faller, S., Foeckler, M., Strosing, K.M., Spassov, S., Ryter, S.W., Buerkle, H., Loop, T., Schmidt, R., and Hoetzel, A. (2012). Kinetic effects of carbon monoxide inhalation on tissue protection in ventilator-induced lung injury. *Lab. Invest.* 92: 999–1012.

77. Tsoyi, K., Lee, T.Y., Lee, Y.S., Kim, H.J., Seo, H.G., Lee, J.H., and Chang, K.C. (2009). Heme-oxygenase-1 induction and carbon monoxide-releasing molecule inhibit lipopolysaccharide (LPS)-induced high-mobility group box 1 release in vitro and improve survival of mice in LPS- and cecal ligation and puncture-induced sepsis model in vivo. *Mol. Pharmacol.* 76: 173–182.

78. Chung, S.W., Liu, X., Macias, A.A., Baron, R.M., and Perrella, M.A. (2008). Heme oxygenase-1-derived carbon monoxide enhances the host defense response to microbial sepsis in mice. *J. Clin. Invest.* 118: 239–247.

79. Lee, S., Lee, S.J., Coronata, A.A., Fredenburgh, L.E., Chung, S.W., Perrella, M.A., Nakahira, K., Ryter, S.W., and Choi, A.M. (2014). Carbon monoxide confers protection in sepsis by enhancing beclin 1-dependent autophagy and phagocytosis. *Antioxid. Redox Signal.* 20 (3): 432–442.

80. Fredenburgh, L.E., Kraft, B.D., Hess, D.R., Harris, R.S., Wolf, M.A., Suliman, H.B., Roggli, V.L., Davies, J.D., Winkler, T., Stenzler, A., Baron, R.M., Thompson, B.T., Choi, A.M., Welty-Wolf, K.E., and Piantadosi, C.A. (2015). Effects of inhaled CO administration on acute lung injury in baboons with pneumococcal pneumonia. *Am. J. Physiol. Lung Cell Mol. Physiol.* 309: L834–L846.

81. Dalli, J., Kraft, B.D., Colas, R.A., Shinohara, M., Fredenburgh, L.E., Hess, D.R., Chiang, N., Welty-Wolf, K., Choi, A.M., Piantadosi, C.A., and Serhan, C.N. (2015). The regulation of proresolving lipid mediator profiles in baboon pneumonia by inhaled carbon monoxide. *Am. J. Respir. Cell Mol. Biol.* 53 (3): 314–325.

82. Bellani, G., Laffey, J.G., Pham, T., Fan, E., Brochard, L., Esteban, A., Gattinoni, L., Van Haren, F., Larsson, A., McAuley, D.F., Ranieri, M., Rubenfeld, G., Thompson, B.T., Wrigge, H., Slutsky, A.S., and Pesenti, A. (2016). LUNG SAFE Investigators; ESICM Trials Group. Epidemiology, patterns of care, and mortality for patients with acute respiratory distress syndrome in intensive care units in 50 countries. *JAMA* 315: 788–800.

83. Nakahira, K., Kyung, S.Y., Rogers, A.J., Gazourian, L., Youn, S., Massaro, A.F., Quintana, C., Osorio, J.C., Wang, Z., Zhao, Y., Lawler, L.A., Christie, J.D., Meyer, N.J., McCausland, F.R., Waikar, S.S., Waxman, A.B., Chung, R.T., Bueno, R., Rosas, I.O., Fredenburgh, L.E., Baron, R.M., Christiani, D.C., Hunninghake, G.M., and Choi, A.M. (2013). Circulating mitochondrial DNA in patients in the ICU as a marker of mortality: derivation and validation. *PLoS Med.* 10 (12): e1001577. discussion e1001577.

84. Pereira, M.L.M., Marinho, C.R.F., and Epiphanio, S. (2018). Could heme oxygenase-1 be a new target for therapeutic intervention in malaria-associated acute lung injury/acute respiratory distress syndrome? *Front. Cell Infect. Microbiol.* 8: 161.

85. Seixas, E., Gozzelino, R., Chora, A., Ferreira, A., Silva, G., Larsen, R., Rebelo, S., Penido, C., Smith, N.R.,

Coutinho, A., and Soares, M.P. (2009). Heme oxygenase-1 affords protection against noncerebral forms of severe malaria. *Proc. Natl. Acad. Sci. U.S.A.* 106: 15837–15842.

86. Pamplona, A., Ferreira, A., Balla, J., Jeney, V., Balla, G., Epiphanio, S., Chora, A., Rodrigues, C.D., Gregoire, I.P., Cunha-Rodrigues, M., Portugal, S., Soares, M.P., and Mota, M.M. (2007). Heme oxygenase-1 and carbon monoxide suppress the pathogenesis of experimental cerebral malaria. *Nat. Med.* 13: 703–710.

87. Pereira, M.L., Ortolan, L.S., Sercundes, M.K., Debone, D., Murillo, O., Lima, F.A., Marinho, C.R., and Epiphanio, S. (2016). Association of heme oxygenase 1 with lung protection in malaria-associated ALI/ARDS. *Mediators Inflamm.* 2016: 4158698.

88. Epiphanio, S., Campos, M.G., Pamplona, A., Carapau, D., Pena, A.C., Ataíde, R., Monteiro, C.A., Félix, N., Costa-Silva, A., Marinho, C.R., Dias, S., and Mota, M.M. (2010). VEGF promotes malaria-associated acute lung injury in mice. *PLoS Pathog.* 6: e1000916.

89. Pena, A.C., Penacho, N., Mancio-Silva, L., Neres, R., Seixas, J.D., Fernandes, A.C., Romão, C.C., Mota, M.M., Bernardes, G.J., and Pamplona, A. (2012). A novel carbon monoxide-releasing molecule fully protects mice from severe malaria. *Antimicrob. Agents Chemother.* 56: 1281–1290.

90. Zuckerbraun, B.S., Chin, B.Y., Wegiel, B., Billiar, T.R., Czsimadia, E., Rao, J., Shimoda, L., Ifedigbo, E., Kanno, S., and Otterbein, L.E. (2006). Carbon monoxide reverses established pulmonary hypertension. *J. Exp. Med.* 203: 2109–2119.

91. Dubuis, E., Potier, M., Wang, R., and Vandier, C. (2005). Continuous inhalation of carbon monoxide attenuates hypoxic pulmonary hypertension development presumably through activation of BKCa channels. *Cardiovasc. Res.* 65: 751–761.

92. Martinez, F.J., Collard, H.R., Pardo, A., Raghu, G., Richeldi, L., Selman, M., Swigris, J.J., Taniguchi, H., and Wells, A.U. (2017). Idiopathic pulmonary fibrosis. *Nat. Rev. Dis. Primers.* 3: 17074.

93. Song, R., Kubo, M., Morse, D., Zhou, Z., Zhang, X., Dauber, J.H., Fabisiak, J., Alber, S.M., Watkins, S.C., Zuckerbraun, B.S., Otterbein, L.E., Ning, W., Oury, T.D., Lee, P.J., McCurry, K.R., and Choi, A.M. (2003). Carbon monoxide induces cytoprotection in rat orthotopic lung transplantation via anti-inflammatory and anti-apoptotic effects. *Am. J. Pathol.* 163: 231–242.

94. Kohmoto, J., Nakao, A., Kaizu, T., Tsung, A., Ikeda, A., Tomiyama, K., Billiar, T.R., Choi, A.M., Murase, N., and McCurry, K.R. (2006). Low-dose carbon monoxide inhalation prevents ischemia/reperfusion injury of transplanted rat lung grafts. *Surgery* 140 (2): 179–185.

95. Kohmoto, J., Stolz, D.B., Kaizu, T., Tsung, A., Ikeda, A., Shimizu, H., Takahashi, T., Tomiyama, K., Sugimoto, R., Choi, A.M., Billiar, T.R., Murase, N., and McCurry, K.R. (2007). Carbon monoxide protects rat lung transplants from ischemia-reperfusion injury via a mechanism involving p38 MAPK pathway. *Am. J. Transplant.* 7 (10): 2279–2290.

96. Kohmoto, J., Nakao, A., Sugimoto, R., Wang, Y., Zhan, J., Ueda, H., and McCurry, K.R. (2008). Carbon monoxide saturated preservation solution protects lung grafts from ischemia reperfusion injury. *J. Thorac. Cardiovasc. Surg.* 136: 1067–1075.

97. Bathoorn, E., Slebos, D.J., Postma, D.S., Koeter, G.H., Van Oosterhout, A.J., Van Der Toorn, M., Boezen, H.M., and Kerstjens, H.A. (2007). Anti-inflammatory effects of inhaled carbon monoxide in patients with COPD: a pilot study. *Eur. Resp. J.* 30: 1131–1137.

98. Rosas, I.O., Goldberg, H.J., Collard, H.R., El-Chemaly, S., Flaherty, K., Hunninghake, G.M., Lasky, J.A., King, T.E., Jr., Lederer, D.J., Machado, R., Martinez, F.J., Maurer, R., Teller, D., Noth, I., Peters, E., Raghu, G., Garcia, J.G.N., and Choi, A.M.K. (2017). A Phase II clinical trial of low dose inhaled carbon monoxide in idiopathic pulmonary fibrosis. *Chest* 153 (1): 94–104.

99. Ling, K., Men, F., Wang, W.C., Zhou, Y.Q., Zhang, H.W., and Ye, D.W. (2018). Carbon monoxide and its controlled release: therapeutic application, detection, and development of carbon monoxide releasing molecules (CORMs). *J. Med. Chem.* 61: 2611–2635.

100. Romão, C.C., Blättler, W.A., Seixas, J.D., and Bernardes, G.J. (2012). Developing drug molecules for therapy with carbon monoxide. *Chem. Soc. Rev.* 41: 3571–3583.

101. Yang, X., Ke, B.W., Lu, W., and Wang, B.H. (2020). CO as a therapeutic agent: discovery and delivery forms. *Chin. J. Nat. Med.* 18 (4): 284–295.

21

Carbon Monoxide in Acute Brain Injury and Brain Protection

Alexandra Mazur[1], Madison Fangman[1], Rani Ashouri[1], Hannah Pamplin[1], Shruti Patel[1], and Sylvain Doré[1,2]

[1]*Department of Anesthesiology, Center for Translational Research in Neurodegenerative Disease and McKnight Brain Institute, University of Florida College of Medicine, 1275 Center Drive, Biomed Sci J493, Gainesville, FL 32610, USA*
[2]*Departments of Neurology, Psychiatry, Pharmaceutics, and Neuroscience, University of Florida College of Medicine, Gainesville, FL, USA*

Introduction

Over the past several decades, carbon monoxide (CO) has been studied as a potential cytoprotective treatment for neurological diseases. At normal physiological concentrations, CO functions as a signaling molecule that is released during heme degradation, and it is known to activate antiapoptotic, vasodilatory, antithrombotic, proangiogenic, and anti-inflammatory pathways [1]. The relatively high affinity that CO exhibits for hemoglobin is a primary reason for its reported toxicity, which can result in acute or chronic neurological conditions. Although CO poisoning is widely studied, the use of low-dose exogenous CO and CO-releasing molecules (CORMs) to treat neurological diseases has shown promising results in preclinical studies. One potential theory for the perceived beneficial results is the ability of CO to acutely upregulate downstream antioxidant and anti-inflammatory pathways, which are discussed later in this chapter. CO gas and CORMs are being tested in clinical trials as putative therapeutic agents for various conditions, including chronic headaches, acute respiratory distress syndrome, vascular disorders, pulmonary fibrosis, chronic obstructive pulmonary disease, pulmonary hypertension, and traumatic brain injury (TBI). Although CO exposure can be toxic at high doses, the use of a lower therapeutic dose of CO gas is subhypoxic and has not been shown to cause CO poisoning in preclinical models. Additionally, administration of low-dose CO gas did not cause neurological symptoms or carboxyhemoglobin (COHb) poisoning in healthy human volunteers [2].

One method to administer CO in a localized and controlled fashion is the use of CORMs. CORMs are undergoing development to better control the amount of CO released and the specific delivery mechanisms throughout the body to potentially target regions of interest. To date, the overwhelming majority of studied CORMs are metal carbonyl complexes, although boranocarbonates and other organic prodrugs have also shown promise as viable CORMs [3–5]. Preclinical models have revealed the effectiveness of CORMs in alleviating the detrimental side effects of various central nervous system diseases. Many of the specific mechanisms of release used by CORMs remain unknown because CORMs (e.g., CORM-2 and CORM-3) cannot be tracked via fluorescence. Another reason is their molecule complexity, which can interfere with desired reactions and produce side products. According to Lazarus, Benninghoff, and Berreau [3], CORMs that are simple metal carbonyl complexes have been shown to spontaneously release CO in aqueous environments. Targeted delivery of CO remains unprogressive when using metal-free CORMs. There are several classes of organic CO prodrugs that have tunable release rates and triggered release in pH-, reactive oxygen species-, esterase-, and concentration-dependent fashions [5,6]. Chapters 11–17 in Section II of this book discuss in detail the various delivery forms and their applications. This chapter will mainly discuss the mechanisms by which CO is neuroprotective after acute brain injury through Nrf2 and heme oxygenase 1 (HO-1) upregulation. It will provide a synopsis of preclinical studies using CO gas or CORMs to treat acute brain injuries, including ischemic stroke, neonatal hypoxic–ischemic encephalopathy (HIE), hemorrhagic stroke, and TBI. Finally, clinical trials (at www.clinicaltrials.gov) that use CO gas and CORMs to treat disease in humans will be discussed, as will the risks and benefits of beginning clinical trials using CO gas or CORMs to treat acute brain injuries in humans. Chapter 30 also covers various clinical trials involving CO gas and CO delivery forms.

Potential Role of CO in the Injured Brain

Vasodilatory Effect of CO

The neuroprotective effects of CO after acute brain injury may be due in part to its vasodilatory activity. CO, which has a much longer half-life than nitric oxide, modulates soluble guanylyl cyclase and induces smooth muscle vasorelaxation. Furthermore, the binding of CO to heme causes cerebral arteriolar dilation through increases of Ca^{2+} sparks, the opening of Ca^{2+}-activated K^+ channels, and hyperpolarization of smooth muscle cells [7]. Thus, CO, through its vasodilatory activity, could help restore blood flow to the brain after ischemic stroke (Figure 21.1).

Anti-inflammatory Effect of CO in the Brain

CO exhibits neuroprotective effects partly through regulation of the Nrf2 pathway. In response to stress caused by various neurological conditions, such as stroke or brain trauma, CO administered in low doses has been found to have an anti-inflammatory effect through the regulation of various kinases. CO acts as a second messenger, subsequently increasing Nrf2 translocation to the nucleus [1,8]. Nrf2 increases the expression of many downstream anti-inflammatory proteins, including HO-1, glutathione reductase, and NADPH quinone dehydrogenase 1 [9]. Wang et al. [8] highlighted the anti-inflammatory and neuroprotective effects of CO-regulated Nrf2 expression to treat ischemic stroke. This experiment revealed worsened behavioral tests and neurological outcomes in Nrf2 knockout mice, indicating that there is evidence of CO and Nrf2 anti-inflammatory neuroprotection. Additionally, a study conducted by Otterbein [10] reviewed the effects of CO in numerous inflammatory disorders and revealed its ability to modulate cytokine production, protecting the lungs and hearts of rodents from endotoxins. Although this study did not review inflammatory conditions specifically within the brain, it adds data supporting the anti-inflammatory effects of CO and lays the groundwork for future research in this area.

Anti-Cell Death Effect of CO

Low-dose CO has also been shown to have anti-cell death (or so-called antiapoptotic) effects that are seen in the reduction of organ rejection after transplant (Chapter 19). Treatment with low-dose CO or overexpression of HO-1 prevented TNF-induced apoptosis in mouse fibroblast cells. Interestingly, CO exposure was found to protect against lung ischemia–reperfusion injury by reducing cellular apoptosis-like death [11]. Chapter 20 discusses CO and lung injury in detail.

CO has also been shown to have anti-cell death effects after peroxynitrite formation. Li et al. [12] demonstrated CO-mediated protection against peroxynitrite in PC12 cells, which are widely used to model neural differentiation in the brain [13]. Although peroxynitrite damage has been assessed in

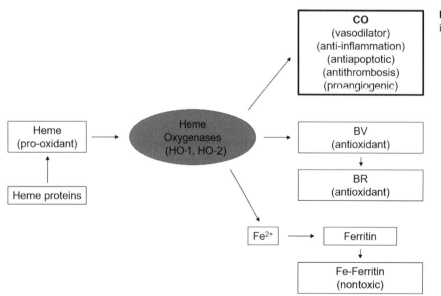

Figure 21.1 Potential roles of CO in the injured brain.

multiple brain models, such as TBI and Huntington disease, there has yet to be research that directly examines CO as a mediator for peroxynitrite damage in the brain [14,15]. Moreover, another study by Cheng and Levy [16] revealed the anti-cell death potential of CO in developing brains. By looking specifically at isoflurane exposure in newborn CD-1 mice, the study showed that CO could limit cell death by inhibiting cytochrome c peroxidase activation. CO therapy has also been evaluated after subarachnoid hemorrhage (SAH) injury, and studies found a significantly increased lumen area and wall thickness in animals treated with CO [17]. These studies provide a foundation for CO-mediated anti-cell death in a wide range of neurological events, although continued research in this area is needed.

Antithrombotic Effect of CO

CO has been reported to have early thrombolytic effects after ischemia, improving endothelial cell survival after arterial thrombosis [18]. Researchers used C57BL/6 × 129 ApoE$^{-/-}$ mice to model carotid thrombosis. This model works by introducing a modified 29-gauge needle roughed and coated with an α-cyanoacrylate bead to the common carotid artery via the internal carotid artery. The needle was used to abrade the artery four times. Either a viral vector or control saline was injected into the arterial lumen. The researchers found that treating arterial thrombosis with 250 ppm of CO for 2 h led to earlier thrombolysis and restoration of blood flow than control air. It follows from these reported outcomes that if CO is administered soon after the onset of stroke symptoms, it could aid in dissolving cerebral arterial blockages and prevent additional brain tissue damage.

Proangiogenic Effect of CO

Li Volti et al. [19] found that HO-1-deficient endothelial cells showed increased angiogenesis after CORM-1 and CORM-3 exposure. HO-1 antisense genes were developed, retrovirally transduced, and used to impair HO-1 expression. CORM-3 was prepared by one of the researchers and shown to be an especially potent enhancer of angiogenesis. The capillary formation was assessed by using a growth factor-induced Matrigel matrix. The researchers also quantified vascular endothelial growth factor (VEGF) expression via ELISA. Notably, HO-1 upregulation resulted in increased VEGF expression. Furthermore, endothelial cells transduced with the antisense HO-1 gene, which resulted in impaired HO-1 expression, showed less VEGF production. However, the production of VEGF was rescued via exposure to CO but not to bilirubin. The results of this experiment suggest that HO-1-mediated angiogenesis is dependent on endogenous HO-mediated CO production. Extrapolation of these results indicates that CO administered exogenously hours to days after the onset of stroke symptoms could enhance cerebral angiogenesis to restore normal oxygenation to the brain and prevent further tissue damage.

Examples of Neuroprotective CO Mechanisms

CO and Heme Oxygenase in the Brain

During acute brain injuries, such as stroke or TBI, excess amounts of heme are released from brain tissue or erythrocytes [20]. HO1, the inducible isoform of heme oxygenase, plays an important role in acute brain injury by degrading the pro-oxidative and proinflammatory heme molecules and reducing tissue damage after stroke. Multiple studies have been conducted on the effects of HO enzymes on acute brain injuries. A study by Goto et al. [21] examined the expression of HO-2 in wild-type (WT) and HO-2 knockout mice that were subjected to focal brain ischemia. They found that CO derived from HO-2 may play a neuroprotective role by limiting unnecessary acceleration of glucose oxidation, improving metabolic tolerance against ischemia. Fu et al. [22] examined HO1 expression after middle cerebral artery occlusion (MCAO) in rats. This study found a swift increase in HO1 from 1 to 6 h after MCAO, which indicates that the cerebral protection provided by HO1 occurs rapidly. This study also suggests that glial cells' ability to resist damage from ischemia is stronger than neuronal resistance and that HO-1-positive glial cells may help reduce ischemic damage to neurons. In another study, transgenic mice that overexpressed HO-1 in neurons were subjected to MCAO and developed smaller brain lesions, both 6 and 24 h after MCAO [23]. HO1 activity is upregulated after brain injury, potentially causing endogenous CO production to increase significantly, although the heme substrate under normal conditions may be sparse. Similarly, the immense diversity of molecules, such as endotoxins, prostaglandins, metals, and cytokines, that induce HO-1 expression hints at the crucial role that HO-1 plays in maintaining cellular homeostasis [24]. Upregulation of HO-1 and possible increased endogenous CO levels are associated with cerebral protection and improved functional neurological outcomes; therefore, it is possible that HO-1 and/or CO play an important role in cytoprotection and cellular signaling after brain injury, including ischemic events, as well as other pathologies involving hemolysis.

CO and Nrf2

Because Nrf2, an anti-inflammatory and antioxidant transcription factor, upregulates HO-1 transcription after brain damage by binding to the antioxidant response element (ARE) region of the HO-1 promoter, researchers wanted to know whether exogenous CO upregulates Nrf2 activity by increasing Nrf2 translocation into the nucleus. Wang et al. [8] subjected Nrf2 knockout mice to 250 ppm of CO for 18 h immediately after permanent MCAO. CO exposure in WT mice led to a significant reduction in infarct size, decreased brain edema, and decreased neurological deficits compared to an air-controlled group with identical flow rate. Remarkably, this CO treatment increased dissociation of Nrf2 from Keap1, Nrf2 nuclear translocation in the brain, and binding of Nrf2 to ARE in the HO-1 promoter 6 h after CO exposure, as seen in the electrophoresis mobility shift assay. It also increased HO-1 protein expression even hours after CO exposure was terminated. Essentially, no benefits were observed in the Nrf2 knockout mice. This study suggests that treatment with exogenous CO can provide a pleiotropic neuroprotective effect after stroke by activating the anti-inflammatory/antioxidant Nrf2 transcription factor, which increases the expression of several anti-inflammatory proteins, including HO-1. This study measured mouse COHb levels and found that CO levels returned to normal physiological levels relatively quickly after exposure.

Furthermore, to build on the mechanism shown in Figure 21.2, the proposed method of CO activation of Nrf2 is via protein kinase R-like endoplasmic reticulum kinase (PERK) and a downstream eukaryotic initiation factor 2-α (eIF2α) pathway. Kim et al. [25] are responsible for the elucidation of this pathway via observation of phosphorylated, and thus activated, PERK, eIF2α, and Nrf2 when complementary siRNA was added to each component of the pathway.

Therapeutic Role of CO in Acute Brain Injury

Ischemic Stroke

Ischemic stroke is caused by a cerebral arterial blockage, which leads to an inadequate supply of oxygen and blood to the brain and results in the death of brain tissue. Notably, ischemic stroke is the most common form of stroke worldwide [26]. Symptoms include vision problems, slurred speech, weakness, loss of coordination, seizures, confusion, numbness, incontinence, and paralysis [27]. It is critical to treat ischemic stroke immediately after the onset of symptoms to reduce brain tissue death as much as possible.

Currently, the only acute treatments for ischemic stroke are tissue plasminogen activator (tPA) and thrombectomy [28]. tPA is a thrombolytic agent that can be administered intravenously and dissolves cerebral arterial blood clots [29]; however, it must be administered within 4 h of the onset of stroke symptoms. An alternative acute treatment for ischemic stroke is endovascular surgery or mechanical thrombectomy, in which the cerebral arterial blockage is

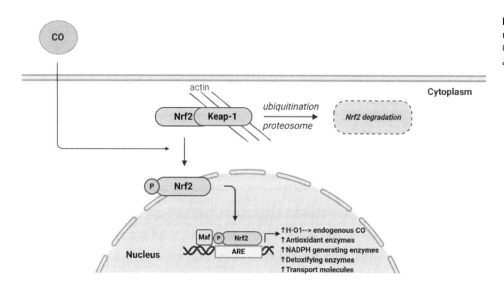

Figure 21.2 Postulated neuroprotective role of CO through the Nrf2 and HO-1 pathway.

surgically removed [30]. Importantly, mechanical thrombectomy also has a small window of time in which it can be useful in breaking down arterial blockages after the initial onset of stroke symptoms. Stroke outcomes are dependent on characteristics such as age, gender, and severity of stroke [26]. Even after acute treatment, roughly half of patients with stroke still suffer from physical or cognitive impairment [31]. The window to administer tPA or perform a thrombectomy for patients with stroke is brief; this fact, coupled with the limited efficacy of these treatments, demonstrates the need for new therapies that effectively reduce brain death after ischemic stroke and that can also be administered hours to days after the stroke has occurred. A summary of preclinical studies using inhaled CO gas or CORMs to treat ischemic stroke is provided below.

Several studies have been conducted on the use of CORMs to treat ischemic stroke in mice. A study on mice with transient MCAO (tMCAO) found that CORM-3 improved the outcome of stroke injury by reducing neuroinflammation when administered 1 h after the onset of stroke [32]. Wang et al. [33] measured the effect of CORM-2 on rats with global cerebral ischemia induced by cardiac arrest. CORM-2 improved mitochondrial biogenesis, which led to improved neurological outcomes when administered 30 min after stroke. Neurological deficit scores were measured in both studies, and neither showed that CO had any significant adverse effects on the animals. Additionally, the tMCAO mice that were administered CORM had lower neurological deficit scores than the tMCAO mice treated with saline.

Other studies have investigated the effect of CO inhalation on ischemic stroke outcomes in mice. In a study by Zeynalov and Doré [1], mice were exposed to either 125 or 250 ppm of CO at the onset of reperfusion (cohort 1) or 250 ppm of CO at 1 and 3 h after reperfusion (cohort 2) after ischemia. All mice were subjected to CO therapy for an 18-h period. This study found that the mice subjected to 125 and 250 ppm of CO developed significantly smaller lesions and less edema than those exposed to regular air. In addition, it was found that 250 ppm of CO offered the best cytoprotection when it was administered less than 1 h after reperfusion. This study measured COHb levels and found that CO levels returned to physiological levels within 2 h of the CO exposure.

Neonatal Hypoxic–Ischemic Encephalopathy

Neonatal HIE is brain damage that occurs in newborns and is caused by oxygen deprivation and reduced blood flow to the brain during or after childbirth. It is associated with conditions such as cerebral palsy, seizure disorders, and intellectual and developmental disabilities [34]. HIE is characterized by breathing problems, difficulty feeding, reduced muscle tone, abnormal responses to light, seizures, and abnormal consciousness [35]. It is diagnosed by using the Apgar scale, brain imaging, umbilical cord blood gas tests, and electroencephalography. Currently, the only established treatment used to minimize permanent brain damage from HIE is hypothermia therapy [34]. This involves cooling the newborn to a body temperature below homeostasis to allow for recovery. Hypothermia can be induced through "selective brain cooling" by using a cooling cap or whole-body cooling [35]. To be effective, hypothermia therapy must be administered shortly after the onset of ischemia.

Two studies examined the effect of inhaled CO on newborn HIE. Douglas-Escobar et al. [36] subjected neonatal mouse pups to a hypoxia chamber for 20 min. One hour after hypoxia exposure, the mice were given 250 ppm of CO for 1h/day over the course of 3 days. This study was the first to demonstrate that CO preserved cortical tissue volumes after HIE and COHb concentrations were found to be 22% for 200 ppm of CO and 27% for 250 ppm of CO, both of which are below levels that could induce neurological abnormalities [36]. A study by Queiroga et al. [37] exposed rat pups to 250 ppm of CO 1 h before HIE was induced. This study found a neuroprotective role for CO, as indicated by an increase in Bcl-2 expression, a protein known to regulate cellular apoptosis. A separate study looked at the effects of HO-1, HIF-1, and CO content in an HIE model in newborn rat pups [38]. In this study, electrical acupuncture was used to increase the HO-1 expression level, which led to an increase in CO content in cortex cells. The electrical acupuncture therapy showed a decrease in cortex damage and downregulation of cystathionine beta-synthase expression compared to the control groups. These preclinical studies indicate that inhaled CO may be an effective treatment for newborns with HIE.

Subarachnoid Hemorrhage and Intracerebral Hemorrhage

SAH is a type of hemorrhagic stroke triggered by bleeding in the space between the brain and the skull. It is typically rooted in a ruptured aneurysm or major head injury. About 51% of patients diagnosed with SAH die, and another third are permanently disabled with brain damage and require lifelong care. Current treatments for SAH include surgery, such as ventriculostomy or craniotomy, or catheter-based therapy [39].

Intracerebral hemorrhage (ICH) is another type of hemorrhagic stroke caused by bleeding inside the brain. It is caused predominantly by high blood pressure and trauma, requiring emergency treatment, close monitoring, and possibly surgery [40]. CO is a potential therapeutic agent that has been associated with neuroprotection in both SAH and ICH preclinical models.

SAH often results in cerebral vasospasms, delayed cerebral ischemia, and loss in behavioral function; interestingly, in preclinical models, exogenous CO has been shown to reduce vasospasm as well as alleviate behavioral issues induced by SAH, including preservation of locomotor and motor function skills [17]. When exposed to 250 ppm of CO at 1 L/min flow, male C57BL/6 mice in the CO-treated group exhibited a higher lumen area to wall thickness ratio than the air-treated group did, showing that controlled CO inhalation decreased cerebral vasospasm [17]. CO has also been shown to use a circadian rhythm pathway to alleviate the effects of SAH. Male C57BL/6 mice in an SAH stroke model involving exposure to 250 ppm of CO for 1 h every 7 days showed considerably less neuronal injury, regulated by CO through circadian gene expression [41]. Increased neuronal apoptosis-like death was reversed in CO-treated groups, whereas such cell death remained increased in control groups [41]. Various CO pathways can reduce the harmful effects of SAH, including CO regulation of microglial erythrophagocytosis via CD36 expression. The increased CD36 receptor activity on macrophages in the brain after SAH allows macrophages to increase erythrophagocytosis and protects against oxidative damage to neurons caused by erythrocytes, heme, and excess free iron in the brain after SAH. Because of the lack of CD36 that leads to significant levels of brain injury after hemorrhage, CO has been found to use CD36 to improve microglial uptake of erythrocytes to elicit neuroprotection [42]. Because HO enzymes, mostly HO-1, generate CO in microglia, CO aids in RBC clearance through CD36 microglial receptor regulation; thus, it alleviates SAH-induced neuronal injury and improves neuronal health through increased phagocytic ability [43]. In one experiment looking at CORM-3 neuroprotection against ICH in Sprague–Dawley rats, 4 or 8 mg/kg of CORM-3 was administered intravenously, and TNF-α was used as a marker of inflammation for all treatment groups. CORM-3 administration resulted in protective measures against ICH, reducing inflammation when administered 5 min before ICH or 3 days after surgery but increasing inflammation when administered 3 h after surgery [44]. These results indicate that CORM-3 administration is a potential therapeutic method for functional recovery after a hemorrhagic stroke, depending on administration time.

Traumatic Brain Injury

Along with hemorrhagic stroke, CO has been tested in TBI models and shown cytoprotective effects. TBI, which is typically caused by an external mechanical insult to the brain, often results in brain tissue damage. The resulting brain dysfunction can be temporary, lasting days to years, or permanent, with symptoms potentially worsening over time. These effects include disabilities in thinking, memory, movement, sensation, and emotional function. TBI is a major cause of disability and death; current medications to manage symptoms include diuretics, and decompressive craniectomy surgeries can be performed in severe cases [45].

Inhaled CO at low concentrations can be beneficial in protecting against ischemic injury [1]. When 4 mg/kg of CORM-3 was injected via the femoral vein, it reduced pyroptosis and apoptosis-like cell death and improved blood flow in the amygdala, thereby reducing depression and anxiety-like behaviors in rats in a TBI and hemorrhagic shock and resuscitation model [46]. Times of immobility, center entries, rearing/learning scores, and grooming also improved on behavior marker tests, indicating that CORM-3 can improve neurological alterations after TBI and neuronal degeneration in the amygdala [46]. Whereas TBI results in pericyte cell death, CO exposure resulted in pericyte rescue that assisted in functional recovery and some degree of neurogenesis [47]. Other carriers of CO, such as polynitroxylated PEGylated hemoglobin, have also been studied as a resuscitation fluid and provide evidence of the neuroprotective abilities of CO against secondary neurodegeneration in a model of hemorrhagic shock and TBI [48].

Considerations, Limitations, and Future Directions

Given the prevalence of acute brain injuries and predominantly poor outcomes in these patients, there is a need to develop more effective treatments. A combination of acute therapies using a multipronged approach needs to be investigated to reduce brain injury, neuronal death, and neurological deficits;

therefore, research is focused on the use of CO and CORMs as an additional treatment option in combination with the current standard of care. The clinical applications of CO in the various delivery forms previously discussed in this book are limited and focused on the brain. However, as preclinical experiments continue to show more promise, specifically in the realm of TBI, we expect that neurologically focused clinical trials will be conducted in the near future. Furthermore, as current trials make headway in determining safe standards of administering CO, more clinical trials on the use of CO to treat brain-related pathologies will likely begin. In general, CO and CORMs did not reveal unexpected toxicity at therapeutic doses in human clinical trials for other diseases, indicating that these doses are safe to administer in humans. Multiple preclinical studies have been conducted using both CO gas and CORMs to treat ischemic stroke, neonatal HIE, hemorrhagic stroke, and TBI and have reproducible results. In most preclinical experiments, CO reduced brain damage and neurological deficits in these disease models significantly more than the placebo control treatment. Most recently, a clinical trial has begun for Sanguinate, a form of bovine COHb, also called PP-07, for intravenous use in acute ischemic stroke (NCT04677777). There are currently no published results proving bovine COHb efficacy in increasing COHb; however, these studies appear promising. Based on the results of future clinical trials, CO and CORMs could provide an additional beneficial acute treatment for various forms of brain injury and significantly reduce death and disability.

Acknowledgments

This work was supported in part by grants from the Brain Aneurysm Foundation, the American Heart Association, the National Institutes of Health (R21NS095166, NS103036, NS110008, and 1R56NS116076), and the Department of Defense (AZ180127).

References

1. Zeynalov, E. and Doré, S. (2009). *Neurotox. Res.* 15: 133–137.
2. Mahan, V.L. (2012). *Med. Gas Res.* 2: 32.
3. Lazarus, L.S., Benninghoff, A.D., and Berreau, L.M. (2020). *Acc. Chem. Res.* 53: 2273–2285.
4. Schatzschneider, U. (2015). *Br. J. Pharmacol.* 172: 1638–1650.
5. Ji, X. and Wang, B. (2018). *Acc. Chem. Res.* 51: 1377–1385.
6. Yang, X.-X., Ke, B.-W., Lu, W., and Wang, B.-H. (2020). *Chin. J. Nat. Med.* 18: 284–295.
7. Leffler, C.W., Parfenova, H., Jaggar, J.H., and Wang, R. (2006). *J. Appl. Physiol.* 100: 1065–1076.
8. Wang, B., Cao, W., Biswal, S., and Doré, S. (2011). *Stroke* 42: 2605–2610.
9. Ahmed, S.M.U., Luo, L., Namani, A., Wang, X.J., and Tang, X. (2017). *Biochim. Biophys. Acta Mol. Basis Dis.* 1863: 585–597.
10. Otterbein, L.E. (2002). *Antioxid. Redox Signal.* 4: 309–319.
11. Ryter, S.W. and Otterbein, L.E. (2004). *Bioessays* 26: 270–280.
12. Li, M.-H., Cha, Y.-N., and Surh, Y.-J. (2006). *Biochem. Biophys. Res. Commun.* 342: 984–990.
13. Westerink, R.H.S. and Ewing, A.G. (2008). *Acta Physiol. (Oxf)* 192: 273–285.
14. Singh, I.N., Sullivan, P.G., and Hall, E.D. (2007). *J. Neurosci. Res.* 85: 2216–2223.
15. Pérez-De La Cruz, V., González-Cortés, C., Galván-Arzate, S., Medina-Campos, O.N., Pérez-Severiano, F., Ali, S.F., Pedraza-Chaverrí, J., and Santamaría, A. (2005). *Neuroscience* 135: 463–474.
16. Cheng, Y. and Levy, R.J. (2014). *Anesth. Analg.* 118: 1284–1292.
17. Kamat, P.K., Ahmad, A.S., and Doré, S. (2019). *Arch. Biochem. Biophys.* 676: 108117.
18. Chen, Y.H., Tsai, H.L., Chiang, M.T., and Chau, L.Y. (2006). *J. Biomed. Sci.* 13: 721–730.
19. Li Volti, G., Sacerdoti, D., Sangras, B., Vanella, A., Mezentsev, A., Scapagnini, G., Falck, J.R., and Abraham, N.G. (2005). *Antioxid. Redox Signal.* 7: 704–710.
20. Li, R., Saleem, S., Zhen, G., Cao, W., Zhuang, H., Lee, J., Smith, A., Altruda, F., Tolosano, E., and Doré, S. (2009). *J. Cereb. Blood Flow Metab.* 29: 953–964.
21. Goto, S., Morikawa, T., Kubo, A., Takubo, K., Fukuda, K., Kajimura, M., and Suematsu, M. (2018). *J. Clin. Biochem. Nutr.* 63: 70–79.
22. Fu, R., Zhao, Z.Q., Zhao, H.Y., Zhao, J.S., and Zhu, X.L. (2006). *Neurol. Res.* 28: 38–45.
23. Panahian, N., Yoshiura, M., and Maines, M.D. (1999). *J. Neurochem.* 72: 1187–1203.
24. Choi, A.M. and Alam, J. (1996). *Am. J. Respir. Cell Mol. Biol.* 15: 9–19.
25. Kim, K.M., Pae, H.-O., Zheng, M., Park, R., Kim, Y.-M., and Chung, H.-T. (2007). *Circ. Res.* 101: 919–927.
26. Mozaffarian, D., Benjamin, E.J., Go, A.S., Arnett, D.K., Blaha, M.J., Cushman, M., Das, S.R., De Ferranti, S., Després, J.P., Fullerton, H.J., Howard, V.J., Huffman, M.D., Isasi, C.R., Jiménez, M.C., Judd, S.E.,

Kissela, B.M., Lichtman, J.H., Lisabeth, L.D., Liu, S., Mackey, R.H., Magid, D.J., McGuire, D.K., Mohler, E.R., Moy, C.S., Muntner, P., Mussolino, M.E., Nasir, K., Neumar, R.W., Nichol, G., Palaniappan, L., Pandey, D.K., Reeves, M.J., Rodriguez, C.J., Rosamond, W., Sorlie, P.D., Stein, J., Towfighi, A., Turan, T.N., Virani, S.S., Woo, D., Yeh, R.W., and Turner, M.B. (2016). American heart association statistics committee; Stroke statistics subcommittee; Writing group members. *Circulation* 133: e38–360.

27. Prabhakaran, S., Ruff, I., and Bernstein, R.A. (2015). *JAMA* 313: 1451–1462.

28. Saver, J.L., Goyal, M., Bonafe, A., Diener, H.-C., Levy, E.I., Pereira, V.M., Albers, G.W., Cognard, C., Cohen, D.J., Hacke, W., Jansen, O., Jovin, T.G., Mattle, H.P., Nogueira, R.G., Siddiqui, A.H., Yavagal, D.R., Baxter, B.W., Devlin, T.G., Lopes, D.K., Reddy, V.K., Du Mesnil De Rochemont, R., Singer, O.C., and Jahan, R. (2015). Swift prime investigators. *N. Engl. J. Med.* 372: 2285–2295.

29. National Institute of Neurological Disorders and Stroke rt-PA Stroke Study Group (1995). *N. Engl. J. Med.* 333: 1581–1587.

30. Ciccone, A., Valvassori, L., Nichelatti, M., Sgoifo, A., Ponzio, M., Sterzi, R., and Boccardi, E. (2013). Synthesis expansion investigators. *N. Engl. J. Med.* 368: 904–913.

31. Wardlaw, J.M., Murray, V., Berge, E., Del Zoppo, G., Sandercock, P., Lindley, R.L., and Cohen, G. (2012). *Lancet* 379: 2364–2372.

32. Wang, J., Zhang, D., Fu, X., Yu, L., Lu, Z., Gao, Y., Liu, X., Man, J., Li, S., Li, N., Chen, X., Hong, M., Yang, Q., and Wang, J. (2018). *J. Neuroinflamm.* 15: 188.

33. Wang, P., Yao, L., Zhou, -L.-L., Liu, Y.-S., Chen, M., Wu, H.-D., Chang, R.-M., Li, Y., Zhou, M.-G., Fang, X.-S., Yu, T., Jiang, L.-Y., and Huang, Z.-T. (2016). *Int. J. Biol. Sci.* 12: 1000–1009.

34. Glass, H.C. (2018). *Continuum (Minneap., Minn.)* 24: 57–71.

35. Douglas-Escobar, M. and Weiss, M.D. (2015). *JAMA Pediatr.* 169: 397–403.

36. Douglas-Escobar, M., Mendes, M., Rossignol, C., Bliznyuk, N., Faraji, A., Ahmad, A.S., Doré, S., and Weiss, M.D. (2018). *Front. Pediatr.* 6: 120.

37. Queiroga, C.S.F., Tomasi, S., Widerøe, M., Alves, P.M., Vercelli, A., and Vieira, H.L.A. (2012). *PLoS One* 7: e42632.

38. Liu, Y., Li, Z., Shi, X., Liu, Y., Li, W., Duan, G., Li, H., Yang, X., Zhang, C., and Zou, L. (2014). *Neurochem. Res.* 39: 1724–1732.

39. Suarez, J.I., Tarr, R.W., and Selman, W.R. (2006). *N. Engl. J. Med.* 354: 387–396.

40. Qureshi, A.I., Mendelow, A.D., and Hanley, D.F. (2009). *Lancet* 373: 1632–1644.

41. Schallner, N., Lieberum, J.-L., Gallo, D., LeBlanc, R.H., Fuller, P.M., Hanafy, K.A., and Otterbein, L.E. (2017). *Stroke* 48: 2565–2573.

42. Kaiser, S., Selzner, L., Weber, J., and Schallner, N. (2020). *Glia* 68: 2427–2445.

43. Schallner, N., Pandit, R., LeBlanc, R., Thomas, A.J., Ogilvy, C.S., Zuckerbraun, B.S., Gallo, D., Otterbein, L.E., and Hanafy, K.A. (2015). *J. Clin. Invest.* 125: 2609–2625.

44. Yabluchanskiy, A. The effect of CORM-3 on the inflammatory nature of haemorrhagic stroke. Doctoral dissertation, 2011.

45. Risdall, J.E. and Menon, D.K. (2011). *Philos. Trans. R. Soc. Lond. B, Biol. Sci.* 366: 241–250.

46. Li, Y., Zhang, L.-M., Zhang, D.-X., Zheng, W.-C., Bai, Y., Bai, J., Fu, L., and Wang, X.-P. (2020). *Neurochem. Int.* 140: 104842.

47. Choi, Y.K., Maki, T., Mandeville, E.T., Koh, S.-H., Hayakawa, K., Arai, K., Kim, Y.-M., Whalen, M.J., Xing, C., Wang, X., Kim, K.-W., and Lo, E.H. (2016). *Nat. Med.* 22: 1335–1341.

48. Seno, S., Wang, J., Cao, S., Saraswati, M., Park, S., Simoni, J., Ma, L., Soltys, B., Hsia, C.J.C., Koehler, R.C., and Robertson, C.L. (2020). *BMC Neurosci.* 21: 22.

22

CO as a Protective Mediator of Liver Injury: The Role of PERK in HO-1/CO-Mediated Maintenance of Cellular Homeostasis in the Liver

Yeonsoo Joe[1,2], Jeongmin Park[1], Mihyang Do[1], Stefan W. Ryter[3], Young-Joon Surh[4], Uh-Hyun Kim[5], and Hun Taeg Chung[1,2]

[1]Department of Biological Sciences, University of Ulsan, Ulsan 44610, Republic of Korea
[2]Mycos Therapeutics Inc., Ulsan 44610, Republic of Korea
[3]Joan and Sanford I. Weill Department of Medicine, and Division of Pulmonary and Critical Care Medicine, Weill Cornell Medical Center, New York, NY 10065, USA
[4]Tumor Microenvironment Global Core Research Center and Research Institute of Pharmaceutical Sciences, College of Pharmacy, Seoul National University, Seoul 08733, Republic of Korea
[5]National Creative Research Laboratory for Ca^{2+} Signaling Network, Chonbuk National University Medical School, Jeonju 54907, Republic of Korea

Introduction

Carbon monoxide (CO) was first described in 1949 as an endogenously produced substance found in human exhaled breath [1]. Two decades later, Tenhunen and colleagues discovered that CO originates from the oxidative catabolism of heme catalyzed by heme oxygenases (HOs) [2].

HO catalyzes the first and rate-limiting step of heme degradation by opening the Fe–protoporphyrin IX ring to release CO, free Fe^{2+}, and biliverdin IXα, the latter of which is converted to bilirubin IXα by NAD(P)H biliverdin reductase (Chapter 10). The HO-catalyzed reaction requires molecular oxygen (O_2) and NADPH cytochrome P450 reductase as the electron donor [3]. Thirty years after the discovery of HO, Poss and Tonegawa, using HO-1 knockout mice (*Hmox1$^{-/-}$*), described HO-1 as a protective protein, which provides strong antioxidant defense *in vivo* and *in vitro* by scavenging pro-oxidant heme and producing cytoprotective products [4]. Among the HO-1 reaction products, CO has recently emerged as a signaling molecule with various cytoprotective functions [5–8], including antiapoptotic, anti-inflammatory, and antiproliferative effects, modulation of cell differentiation, regulation of cell metabolism, maintenance of tissue homeostasis, and promotion of tolerance states. This chapter will describe the potential of CO as a therapeutic molecule in various hepatic diseases, including acute liver injury and hepatitis, and nonalcoholic steatohepatitis (NASH). Furthermore, this chapter will explore the mechanisms by which low-dose CO can regulate various metabolic pathways to maintain tissue homeostasis under stressful environmental conditions, with an emphasis on metabolic regulation of endoplasmic reticulum (ER) stress pathways.

CO as a Candidate Therapeutic in Liver Disease

Acute Liver Injury

Early studies have established CO as a potential regulator of hepatic vascular tone. Suematsu et al., in experiments using isolated perfused rat liver, estimated CO flux in the venous effluent at 0.7 nmol/min per gram of liver. These investigators found that administration of an HO inhibitor zinc protoporphyrin IX (1 µM) abolished baseline CO generation, and increased vascular resistance and hepatic sinusoidal constriction. These effects were reversed by including CO (1 µM) in the perfusate. These studies suggested that CO can function as an endogenous modulator of hepatic sinusoidal perfusion [9].

These investigators also observed that endogenously HO-derived CO was associated with protection against hepatobiliary dysfunction in an endotoxin shock model in rats. Under nitric oxide (NO)-depleting conditions, supplementation with CO (4–8 µmol/l) improved sinusoidal perfusion and restored bile formation without causing an elevation of tissue cGMP content in the liver of endotoxin-treated rats [10]. Further proposed mechanisms underlying protection of hepatobiliary function by the HO/CO system include HO-2-derived CO-dependent upregulation of soluble guanylate cyclase (sGC) in the hepatic stellate cells of the parenchyma, which improves circulation, as well as upregulation of HO-1 and increased HO-1-derived CO production in hepatic Kupffer cells [11]. In an experimental endotoxemia model, Sarady et al. observed that CO gas inhalation (250 ppm) protected against the lethal effects of LPS challenge in mice associated with multiorgan failure. The hepatoprotective effect of CO in this model was evidenced by reduced expression of serum alanine aminotransferase (ALT), a marker of liver injury, and reduced proinflammatory cytokine production. In this model, hepatoprotection by inhaled CO was associated with elevated expression of inducible NO synthase (iNOS) and NO production in the liver, with reciprocal effects observed in the lung [12].

In a model of hemorrhagic shock and resuscitation (HSR) in mice, CO inhalation (250 ppm, continuous) conferred protection against systemic inflammation and end-organ damage [13]. CO inhalation during HSR decreased serum levels of proinflammatory cytokines and increased serum levels of the anti-inflammatory cytokine IL-10. Additionally, CO reduced hemorrhage-induced hepatic cellular hypoxia. *In vitro* experiments also suggested that CO (250 ppm) protected mouse hepatocytes from hypoxia-induced death while preserving ATP levels. Taken together, the study concluded that CO protects against systemic effects of HSR [13]. Furthermore, CO inhalation (250 ppm) was shown to promote liver regeneration in a mouse model of partial hepatectomy. Exposure to inhaled CO (250 ppm, 1 h) prior to 70% hepatectomy improved animal survival. CO-treated mice displayed enhanced hepatic expression of hepatocyte growth factor and its receptor c-Met in the liver in association with enhancement of hepatocyte proliferation. CO also enhanced hepatocyte proliferation *in vitro*. The authors suggested that CO may offer a viable therapeutic option to promote liver regeneration after hepatectomy [14].

The prototype CO-releasing molecule (CORM) tricarbonyldichlororuthenium(II) dimer (CORM-2) conferred protection against the lethal effects of cecal ligation and puncture (CLP)-induced sepsis [15]. CLP promoted neutrophil accumulation, intercellular adhesion molecule 1 (ICAM-1) expression, and activation of NF-κB in the liver. These effects were significantly attenuated by systemic administration of CORM-2 (8 mg/kg). These results suggest that pharmacological application of CO can attenuate sepsis-induced acute liver injury through modulation of the inflammatory response and promotion of the endothelial pro-adhesion phenotype [15].

Recent studies have also demonstrated protection against LPS-induced organ injury using HYCO-3, a dual-specificity hybrid CORM (CORM-401) conjugated to the NF-E2-related factor 2 (Nrf2) activating moiety dimethyl fumarate. In mice subjected to lipopolysaccharide (LPS) challenge, oral HYCO-3 (40 mg/kg) inhibited the expression of proinflammatory cytokines (i.e., tumor necrosis factor [TNF], IL-1β, and IL-6) while increasing the expression of anti-inflammatory cytokines (i.e., IL-10) and arginase-1 in the liver and other organs. The anti-inflammatory effects of HYCO-3 were conserved in the liver of Nrf2 knockout mice, indicating that they were conferred mainly by the CO-releasing moiety of HYCO-3 [16].

In a new approach to drug delivery, CO was targeted to mitochondria. This was achieved through enrichment-triggered release by conjugating a bimolecular CO prodrug pair with the triphenylphosphonium moiety, which is known to accumulate in the mitochondrion. This approach led to a remarkable level of potency with EC_{50} of approximately 0.4 mg/kg as measured by serum aspartate aminotransferase (AST) and ALT levels, when tested in a chemically induced liver injury model [17].

Fulminant Hepatitis

Fulminant hepatitis is associated with hepatocyte apoptosis, hemorrhagic necrosis, and liver inflammation.

Treatment of experimental animals with D-galactosamine (D-GalN) in combination with LPS or TNF, a model of acute liver failure, causes lethal liver injury associated with elevated ROS production from inflammatory cells, and increased hepatocyte apoptosis. CO inhalation (250 ppm, 1 h pretreatment) prevented TNF/D-GalN-induced liver injury and hepatocyte apoptosis. These effects were found to require NF-κB activation and downstream activation of iNOS. Furthermore, CO-mediated protection in this model required NO production and NO-dependent HO-1 activation [18]. CO inhalation (250 ppm, 2 h/day) was also found to protect against LPS/D-GalN-induced injury [19]. This protective effect was associated with reduction of proinflammatory cytokine (i.e., IL-1β) production and cytosolic mtDNA levels, a marker of mitochondrial dysfunction, and with activation of mitochondrial homeostatic mechanisms, including upregulation of mitophagy and mitochondrial biogenesis (see the "Nuclear TFEB Enhances Both Autophagy and Mitophagy" section) [19]. Additionally, CO was found to improve hepatocyte mitochondrial bioenergetics and ATP production via an sGC-dependent mechanism [20]. Similar anti-inflammatory effects in this mouse model were observed with injection of CO-saturated solution (8–15 ml/kg), including inhibition of hepatic NO production, hepatocyte apoptosis, and a reduction of proinflammatory cytokine levels [21].

Autoimmune Hepatitis

In mice, acute hepatitis can be induced by concanavalin A (ConA) treatment, which causes rapid activation of CD1d-positive natural killer T-cells. These activated cells produce excessive proinflammatory cytokines that provoke inflammatory liver injury [22]. CO inhalation (500 ppm, 1 h duration, 4 h after injury) improved mouse survival in a model of ConA-induced autoimmune hepatitis [20]. In related *in vitro* models, 1% CO exposure (10 000 ppm) was shown to improved energy production (ATP generation) of hepatocytes [20]. Similar protective effects of CO (500 ppm for 1 h, beginning at 4 h postinjury) were observed in anti-Fas monoclonal antibody-induced hepatitis [20]. More recent studies have shown that CO released from CORM-A1 (2 mg/kg, i.p.) can confer protection in ConA-induced autoimmune hepatitis through the modulation of Nrf2-regulated gene expression [23].

Acetaminophen-Induced Liver Toxicity

Acetaminophen (APAP) overdose causes hepatic injury after metabolic transformation by the cytochrome P450 system. In a mouse model of APAP-induced liver injury, CO gas inhalation (250 ppm for 2 h after 6 and 18 h administration of APAP) alleviated APAP-induced liver damage *in vivo*. This protection was associated with reduced serum ALT and AST levels as well as reduced proinflammatory cytokine production. CO reduced the expression of pro-apoptotic molecule CHOP in liver tissues of APAP challenged mice, while dramatically increasing the expression of hepatic HO-1 and Parkin, a mitophagy regulator protein [24]. Similar results were obtained with CORM-A1-mediated delivery of CO in this model. CORM-A1 (20 mg/kg) treatment of APAP challenged mice reduced proinflammatory gene expression (i.e., TNF, IL-1β, and IL-6) and led to restoration of hepatic reduced glutathione (GSH) levels and activation of the Nrf2-mediated antioxidant response [25].

Hepatic Ischemia/Reperfusion Injury

Ischemia/reperfusion (I/R), the stoppage and subsequent restoration of blood flow, causes hepatic injury associated with aggravated inflammatory responses. Hepatic I/R injury can arise as a complication of liver surgery and transplantation. CO inhalation (250 ppm for 12 h) was found to protect against warm hepatic I/R injury in male C57BL/6 mice. Several underlying mechanisms were proposed, including inhibition of glycogen synthase kinase-3-beta [26]. Additional studies implicated Sirt1 modulation in the hepatoprotective effect of CO in the I/R injury model. CO enhanced Sirt1 expression via inhibition of a regulatory microRNA (miR-34a), leading to protection against liver injury through deacetylation of NF-κB p65/p53. These events were found to mitigate inflammatory responses and inhibit hepatocellular apoptosis [27].

In a rat model of hepatic I/R injury, a systemic infusion with CORM-2 protected the liver from I/R injury and reduced serum AST/ALT levels. Treatment with CORM-2 also reduced hepatocyte apoptosis after I/R injury. Furthermore, treatment with CORM-2 inhibited hepatic NF-κB activity, reduced pro-inflammatory cytokines TNF and IL-6, and downregulated adhesion molecule ICAM-1 expression in hepatic endothelial cells. CORM-2 treatment also reduced the accumulation of neutrophils in the liver upon I/R injury. These studies implicated CORM-2 as a potential therapeutic in hepatic I/R injury [28].

In a rat model of orthotopic liver transplantation, rats were exposed to air or CO (100 ppm) either as a 1-h pretreatment or 24 h after liver transplantation [29]. CO inhalation significantly decreased serum AST and ALT levels and suppressed hepatic necrosis formation and neutrophil influx 6–48 h after liver transplantation, compared with air-treated controls. The expression of TNF, ICAM-1, and iNOS in the liver graft was significantly inhibited in the CO-treated group at 1 h post-reperfusion [29]. Taken together, these studies demonstrate that CO can confer multimodal protection in hepatic I/R injury models.

Nonalcoholic Steatohepatitis

NASH is a condition characterized by liver inflammation and injury caused by a buildup of fat in the liver. We have observed that inhaled CO inhalation (250 ppm, 2 h/day for 8 weeks) can provide protection in models of ER stress-induced and high-fat diet-induced hepatic steatosis [30]. These protective effects were dependent on upregulation of fibroblast growth factor 21 (FGF21) expression and secretion via PERK pathway activation as described below.

In mice fed a high-fat and high-fructose (HFHF) diet for 16 weeks, CO derived from CORM-A1 (2 mg/kg/day i.p., from the 9th to 16th week) protected against hepatic steatosis in this model [31]. The protective effects of CORM-A1 in HFHF-fed mice were associated with improved lipid homeostasis and activation of the Nrf2-dependent antioxidant response. Furthermore, the protection afforded by CORM-A1 was associated with improvement of mitochondrial function and enhanced ATP production.

Mitochondrial-Dependent Mechanisms Underlying CO Protection

To fully realize the therapeutic potential of CO in the liver, it will be essential to determine the mechanisms and targets by which CO alters hepatocyte function. Despite numerous reports describing the various functions of CO in biological systems, only a few have identified detailed molecular mechanisms. The following sections will describe targeted molecular mechanisms for CO-dependent cytoprotection based on observations of modulation of mitochondrial function and ER stress pathways in the liver.

CO Elicits ROS Production from Mitochondria

CO is an inherently stable molecule with little biological reactivity with respect to non-iron compounds [32]. CO can bind to heme groups at the iron center and form complexes with hemoproteins, including cytochrome c oxidase [33–37].

Mitochondria contain several heme-containing proteins needed for their bioenergetic functions, including respiration and subsequent ATP production; these proteins can be targeted by CO. One of these, cytochrome c oxidase, was first described for its CO binding properties [38,39].

CO elicits mitochondrial reactive oxygen species (mtROS) generation by binding and inhibiting cytochrome c oxidase, resulting in the accumulation of electrons at the site of the mitochondrial electron transport chain [35]. Impaired electron flow increases the chance of oxygen reduction into superoxide, which in turn can be converted to hydrogen peroxide (H_2O_2) by superoxide dismutase (SOD) [40–42]. The diffusible H_2O_2 can act as a paracrine and endocrine signaling molecule. The signaling effects of H_2O_2 are detailed in the next section.

Mitochondrial ROS Activate PERK

Mitochondria are known as a major source of ROS production. Oxidative phosphorylation in the electron transport chain converts O_2 into H_2O by four-electron reduction, but a small percentage of O_2 is converted into superoxide anion (O_2^-) by one-electron reduction. Superoxide produced in the mitochondrial matrix is readily reduced by SOD to hydrogen peroxide (H_2O_2), which is then reduced by glutathione peroxidase or peroxiredoxin to water. Although ROS are produced as by-products of cellular respiration in the mitochondria, cells can actively generate ROS for cellular signaling via NADPH oxidase (NOX) enzymes. When mitochondria are under stressed conditions, including any impairment in the components of the respiratory chain, increased amounts of ROS are produced that modulate cellular signaling. Several previous reports described CO as a mtROS inducer via inhibiting cytochrome c oxidase [42,43] (Figure 22.1).

Over a decade ago, we sought to investigate the effects of CO on the fate of cells under ER stress induced by thapsigargin or tunicamycin. To our surprise, we found that cells treated with low concentrations of exogenous CO (250 ppm by gas inhalation for 1–2 h) in the absence of other stimuli displayed phosphorylation of PERK and survived ER stress-mediated cell death. We found that CO protected the cells from ER stress-induced apoptosis via the p38 mitogen-activated protein kinase-dependent pathway [44]. However, it has remained unclear why only PERK, among the three branches of ER sensors, was phosphorylated in CO-treated cells. More recently, we reported that CO-induced mtROS are responsible for the selective activation of PERK [30].

Accumulating studies have demonstrated that the signals derived from a subtle mitochondrial functional disturbance are transferred to the nucleus, leading to the altered expression of nuclear-encoded genes, which includes mitochondrial proteins (i.e., mitochondrial retrograde signaling [MRS]). MRS alleviates mitochondrial perturbation (i.e., mitohormesis) and is considered a homeostatic stress response against intrinsic (e.g., aging) and extrinsic (e.g., chemical or pathogenic) stimuli. There are increasing observations showing the importance of the integrated stress response (ISR)–ATF4 pathway in MRS [45]. Moreover, Nrf2 cooperates with ATF4 to increase the expression of cell survival genes while decreasing apoptotic genes, including CHOP [44,46].

After revealing the selective activation of PERK by CO, we demonstrated that CO could form stress granules (SGs) via eukaryotic initiation factor 2α (eIF2α) phosphorylation and ATF4 expression via inducing the ISR [47]. Thus, CO could induce the expression of Nrf2 and ATF4, two strong cytoprotective transcription factors that can function individually or cooperatively to enhance the survival of target cells, tissues, and organisms.

PERK Phosphorylates Three Interactomes (eIF2α, Nrf2, and Filamin A)

One of the most consistent observations in our studies describing the effect of CO on the ER is the selective activation of PERK by phosphorylation [44]. PERK as an ER stress sensor is largely recognized as a

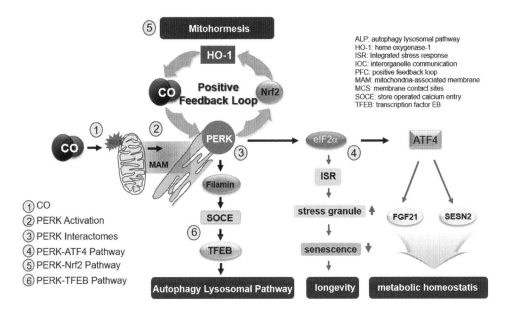

Figure 22.1 A PERK-centric view of CO functions. This scheme shows signaling steps from the exposure of cells or tissues to endogenous CO or low levels of exogenous CO to the final effects of CO.

1. CO inactivates COX4 of the mitochondrial ETC, resulting in the generation of ROS.
2. mtROS activates PERKs that are in the ER or mitochondria-associated membrane (MAM). However, the exact mechanism of CO-induced PERK activation remains unresolved.
3. PERK interacts with three molecules.
   a. The classical substrate of PERK, eIF2α, is phosphorylated by PERK resulting in ISR that permits selective expression of ATF4 during the global suppression of protein synthesis.
   b. PERK phosphorylates Nrf2, which translates to the nucleus for the expression of many cytoprotective genes, including HO-1.
   c. The most recently discovered PERK interactor shown using a biotin labeling method is filamin, which is responsible for the nuclear translocation of TFEB and activation of the autophagy–lysosomal pathway (ALP).
4. The PERK–eIF2α–ATF4 pathway induces many cytoprotective genes, including FGF21 and SESN2.
5. The PERK–Nrf2–antioxidant gene pathway confers mitohormesis through a positive feedforward loop that may be attenuated by CO scavenging.
6. The PERK–filamin–SOCE–TFEB pathway increases ALP that promotes mitophagy and mitochondrial biogenesis with the participation of Nrf2 and ATF4.

key mediator of the unfolded protein response (UPR). However, growing evidence suggests that PERK may govern signaling pathways through UPR-independent functions [48–51]. In this chapter, we discuss both the canonically adaptive signaling cascade of PERK involving eIF2α phosphorylation (ISR) and the emerging noncanonical pathways with particular relevance to the induction of interorganellar communications.

CO-induced, mtROS-dependent activation of PERK mediates phosphorylation of its main cytosolic substrate, Ser51 of eIF2α. PERK-mediated eIF2α phosphorylation causes a temporal translational block, which largely prevents protein translation, while causing the activation of transcription factor ATF4, which is the main mediator of the transcriptional program governed by PERK [46,52–54].

The second substrate of PERK is Nrf2, whose phosphorylation enables its dissociation from Kelch-like ECH-associated protein 1, nuclear translocation, and activation of the Nrf2-driven transcriptional program [55].

Most recently, PERK was found to interact with Filamin A (FLNA) by using proximity-dependent biotin identification methods [56] (Figure 22.1).

The PERK–eIF2α–ATF Pathway

CO Induces SG Assembly

SGs are cytoplasmic RNA–protein complexes found frequently in cultured eukaryotic cells. In the presence of CORM-2 (40 µM) for 6 h [47]SGs are caused by the assembly of the eIF2/GTP/tRNAi^{Met} ternary complex that delivers initiator tRNAi^{Met} to the 40S ribosomal subunit, which is another checkpoint of translation initiation [57]. The ternary complex and the mRNA/eIF4F complex combine with the 40S ribosomal subunit before the recruitment of the 60S ribosomal subunit to initiate protein synthesis [58].

In the presence of CO, eIF2 is a substrate for PERK that phosphorylates serine residue 51 of eIF2α [44]. Phosphorylated eIF2α (p-eIF2α) inhibits GDP–GTP exchange, leading to a decrease in ternary complexes, and inhibits the initiation of translation [59]. p-eIF2α leads to polysome disassembly and accumulates noncanonical 48S preinitiation complexes that lack charged tRNAi^{Met} and early initiation factors eIF2 and eIF5. The influx of stalled 48S complexes leads to SG assembly. The majority of SG proteins contain either a low-complexity (prion-like) domain, which facilitates phase transition, or RNA binding domains, which allows interaction with core SG components, or both. The functions of SGs include protection of cellular RNAs from degradation, mRNA triage during stress, participation in stress signaling, and protection from cell death.

For most stresses including low-dose exogenous CO, SG assembly is initiated by increased p-eIF2α formation via the phosphorylation of eIF2α by CO-induced mtROS-mediated PERK activation [47] (Figure 22.1).

CO-Induced ATF4 Maintains Hepatic Homeostasis

ATF4 is a master transcription factor that binds to the cAMP response element or amino acid response element (AARE) in the promoter regions of its target genes involved in resistance to cellular stress [60,61]. ATF4 is activated by ER stress as well as by the mtROS-induced PERK pathway in MAM. ATF4 is required for the induction of genes involved in the oxidative stress response in the liver. We reported that CO-generated mtROS induces ATF4 via the PERK pathway in MAM, and that ATF4 in turn triggers the increased transcription of target genes that can protect cells and sustain liver homeostasis [19,30,46].

Fibroblast Growth Factor 21

FGF21, a member of the FGF family, plays a key role in various metabolic functions, including insulin sensitivity, glucose tolerance, and lipid metabolism [26,62,63]. FGF21 is produced predominantly in the liver and, to a lesser extent, in extrahepatic tissues, such as skeletal muscle, white adipose tissue, and brown adipose tissue [64–67]. The FGF21-induced intracellular signaling cascades are initiated by binding to the FGF receptor and cofactor β-Klotho on the cell surface [68].

It has been reported that FGF21 is regulated by various molecules [69–74]. During fasting, FGF21 expression is increased by the recruitment of the peroxisome proliferator-activated receptor-α (PPARα), a nuclear hormone receptor, to the FGF21 promoter region. Other molecules, including thyroid hormone receptor-α, retinoic acid receptor-α, and retinoic acid receptor-related orphan receptor-α, are associated with stimulation of FGF21 expression [75–78]. Furthermore, PERK, one of the ER stress transducers, is closely related to FGF21 expression. ATF4 activated by PERK signaling binds to two AARE sequences on FGF21, which results in enhancing of FGF21 expression [72,73].

Several studies have shown that CO inhalation (250 ppm in C57BL/6 or cell lines) or CORMs (20 mg/kg in C57BL/6 or 20 µM in cell lines) can selectively induce PERK activation through production of mild mtROS [24,30,44,46,67]. PERK activation induces activation of downstream molecules, eIF2α and ATF4 [79]. CO-induced ATF4 increased FGF21 expression, which then ameliorated high-fat diet- and tunicamycin-induced obesity, insulin resistance, and hepatic steatosis [30]. However, inhibition of intracellular ROS or mtROS using N-acetyl-L-cysteine or mito-TEMPO, respectively, blocked the increase of FGF21 expression by CO. Moreover, in $Fgf21^{-/-}$ mice CO was unable to protect against changes in obesity, insulin resistance, and hepatic steatosis [30] (Figure 22.2). Such results indicate that CO initiates the mtROS–PERK–eIF2α–ATF4 pathway to induce the expression of FGF21 (Figure 22.2).

Sestrin-2

Sestrin-2 (Sesn2) is a member of the highly conserved stress-inducible antioxidant proteins, which induces AMP-activated protein kinase (AMPK) activation and inhibition of rapamycin complex 1 (mTORC1) and intracellular ROS levels [80,81]. Sesn2 plays an important role in autophagy induction and reduction of metabolic disease, including nonalcoholic fatty liver disease and type 2 diabetes [75,76,80].

While Sesn2 possesses antioxidant functions, the production of ROS can stimulate Sesn2 expression through antioxidant response elements (AREs) [77,78]. Nrf2 is a master regulator of various antioxidant genes via binding to AREs [82]. Oxidative stress can activate Nrf2 through nuclear translocation, which then binds to AREs and stimulates Sesn2 expression [77]. Similar to FGF21 regulation, Sesn2 is

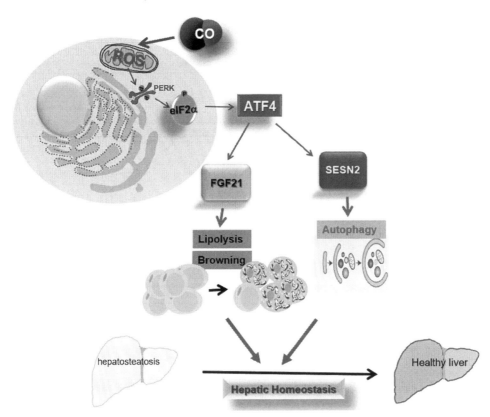

Figure 22.2 CO-induced ATF4 maintains hepatic homeostasis via inducing FGF21 and SESN2. ROS production by CO activates the PERK–eIF2α–ATF4 pathway. Enhancement of ATF4 expression can induce FGF21 and SESN2 levels. The increase in FGF21 expression by the CO–ATF pathway promotes lipolysis and browning to ameliorate the hepatosteatosis. CO-induced SESN2 expression induces activation of autophagy resulting in hepatic homeostasis.

also related to PERK signaling [83]. ATF4 activation by PERK can induce Sesn2 expression [84]. ER stress-induced PERK activation can in turn activate ATF4 and Nrf2, which directly binds to the Sesn2 promoter [83,84]. Consistent with FGF21, CO has been reported to induce Sesn2 expression, which is mediated by PERK-activated ATF4 [46]. CO or CORMs increased mtROS production, and CO-mediated mtROS were found to increase Sesn2 expression via the PERK–eIF2α–ATF4 pathway, but not through IRE1a and ATF6, in hepatocytes [46]. CO-induced Sesn2 expression activated AMPK and inhibited mTORC1, which then promoted autophagy [46]. Finally, CO-activated autophagy via Sesn2 was shown to ameliorate methionine- and choline-deficient (MCD) diet-induced hepatitis [46] (Figure 22.2).

CO–PERK–Nrf2–HO-1 Axis: A Positive Feedforward Loop

A majority of the biological activities of PERK, including inhibition of translation and cell cycle arrest, have been attributed to its function as an eIF2α kinase [85]. However, PERK functions are known to be essential for cell survival following ER stress. The mechanism of how PERK signaling promotes cell survival is unclear. Thus, the possibility that PERK targeted additional downstream substrates that function as cell protecting effectors was addressed. In 2003, Cullinan et al. performed a yeast two-hybrid screen with the PERK catalytic domain to identify novel PERK substrates [55]. They described the identification of the Cap'n'collar transcription factor Nrf2 as a PERK substrate. In 2007, we revealed that CO activates PERK through the formation of mtROS and that the phosphorylated form of PERK protects cells against ER stress via Nrf2 activation and expression of its downstream target genes, including HO-1. Thus, a low concentration of exogenous CO induces further endogenous CO production from HO-1 through a positive feedforward loop [44].

Taken together, the available data suggest a model wherein under conditions of low concentration of exogenous CO, mtROS are formed and activation of PERK occurs near MAM or on the ER surface. The activated PERK phosphorylates both Nrf2 and eIF2α. While protein translation is globally inhibited as a result, translation increases for selected proteins, including ATF4. On the other hand, Nrf2 phosphorylation facilitates Nrf2 nuclear import. Once in the nucleus, Nrf2 heteromerizes with small Maf proteins or ATF4 to increase the expression of target genes whose actions maintain cellular redox homeostasis [86,87] (Figure 22.1).

PERK–Nrf2 Pathway Induces Mitohormesis

Mild physiological or pharmacological stresses, such as exercise and exposure to low concentration of exogenous CO, elicit beneficial effects through a process known as "mitohormesis." Exercise is known to activate Nrf2 through the induction of NOX4 expression and to increase endurance capacity [88]. A low level of CO exposure elicits the PERK–Nrf2 pathway to contribute to the homeostasis of mitochondrial function.

Usually, hormesis is known as a process in which previous exposure to a low level of stress, including CO, promotes adaptive changes to the cell that enable it to tolerate subsequent stress. When this concept applies to the mitochondria, it is called mitohormesis, suggesting that in response to perturbation (i.e., via COX inhibition) mitochondria can initiate retrograde signals to the nucleus that coordinate transcriptional responses of ATF4 (PERK–eIF2α pathway) and Nrf2 (PERK–Nrf2 pathway). In the case of CO-induced mitohormesis, CO can perturb mitochondrial function. The perturbation is relayed by ROS to PERK in the cytosol, and subsequent activation of PERK by phosphorylation induces the expression of several transcription factors in the nucleus. These retrograde signaling pathways and subsequent nuclear transcriptional changes including ATF4 and Nrf2 induce various cytoprotective pathways that augment stress resistance for protection from a wide array of subsequent stresses (Figure 22.3).

Crosstalk of ATF4 and Nrf2 Maintains Redox Homeostasis

Two of the most important molecules for the maintenance of homeostasis and redox balance are GSH and NADPH. Reduced GSH is the major antioxidant component of the cell and its recycling depends on NADPH as a source of electrons; NADPH, on the other hand, is known to be produced primarily by the pentose phosphate pathway (PPP).

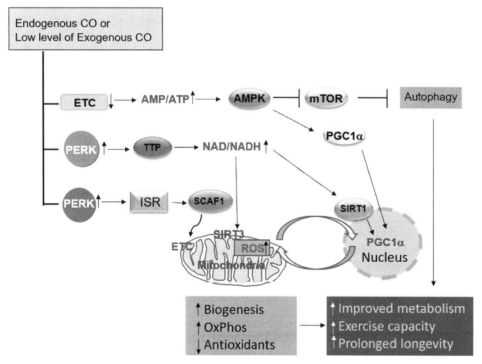

Figure 22.3 Exposure to CO elicits a strong antioxidant defense response through mitohormesis. Mitohormesis caused by mild ROS from endogenous or exogenous CO elicits a strong antioxidant defense response. Exposure to endogenous or a small amount of exogenous CO causes an initial decline in ATP, an increase in the NAD$^+$/NADH ratio, and an upregulation of respiratory supercomplex assembly factor 1 (SCAF1). Initial changes caused by CO exposure, such as small decrease in ATP levels and mtROS production, are sensed by downstream pathways that increase PERK phosphorylation, resulting in the promotion of mitochondrial biogenesis and increased ATP synthesis. An increase in AMP/ATP activates AMPK, which further activates PGC-1α for mitochondrial biogenesis and inhibits mTORC1 for autophagy and mitophagy. An increase in NAD$^+$/NADH activates SIRT1 and SIRT3, which are also responsible for ATP synthesis and mitochondrial biogenesis, respectively. Finally, the PERK–ISR pathway drives the transcription of the respiratory chain SCAF1. The upregulation of SCAF1 promotes respiratory supercomplex assembly, higher respiratory capacity, and improved bioenergetics.

According to Almeida et al., CORM-A1 (2 mg/kg, i.p.) stimulates mRNA expression of 6-phosphogluconate dehydrogenase (PHGDH) and transketolase, which are enzymes for the oxidative and nonoxidative phases of the PPP, respectively, and the protein expression and activity of glucose 6-phosphate dehydrogenase, the rate-limiting enzyme of the PPP. These investigators found that CORM-A1 increased neuronal differentiation yield from retinoic acid-treated postmitotic SH-SY5Y cells [89].

In contrast, DeNicola et al. reported that Nrf2 controls the expression of the key serine/glycine biosynthesis enzyme genes such as PHGDH, phosphoserine aminotransferase 1, phosphoserine phosphatase, and serine hydroxymethyltransferase 2 via ATF4 to support GSH and NADPH production [90] (Figure 22.4).

CO–PERK–TFEB Pathway

Transcription factor EB (TFEB) was the first member of the MiTF/TFE family identified as a master regulator of lysosomal biogenesis, which is activated in response to multiple stimuli, including ER stress, mitochondrial stress, and pathogen exposure [91–94]. TFEB directly binds to the coordinated lysosomal expression and regulation elements in the promoters of many genes implicated in lysosome-related processes, such as autophagy and endocytosis [92,95,96]. TFEB activity is regulated by phosphorylation, which is mediated mainly by mTORC1, a major kinase complex that controls cell growth and negatively regulates autophagy [97]. Dephosphorylated TFEB translocates to the nucleus to activate transcriptional target genes under inactivation of mTORC1 or activation of the Ca^{2+}-dependent phosphatase calcineurin [98]. Thus, PERK activation with CO contributes to transport of TFEB into the nucleus via calcineurin activation, thereby regulating autophagy/mitophagy and mitochondrial biogenesis [19]. Therefore, CO ameliorates acute hepatitis-induced liver injury, in part by preserving mitochondrial homeostasis via the PERK–TFEB pathway.

The CO–PERK Pathway Induces Nuclear Translocation of TFEB

Under starvation, inactivation of mTORC1 allows nuclear translocation of TFEB, resulting in adaptation to stress by enhancing lysosomal biogenesis and autophagy [95,97,99,100]. ER stress causes the translocation of TFEB into the nucleus in a process that is dependent on PERK but not on mTORC1 [92]. Furthermore, calmodulin-regulated protein phosphatase calcineurin promotes transport of TFEB to the nucleus by dephosphorylation [98]. Moreover, PERK, as a tethering molecule between the ER and cytoplasm, plays an important role in coordinating Ca^{2+} dynamics [56,101]. Agostinis and coworkers reported that PERK regulates intracellular Ca^{2+} fluxes and store-operated Ca^{2+} entry (SOCE), the main process used to replenish ER Ca^{2+} levels upon ER Ca^{2+} release. This group sought to find novel PERK interactors by using proximity-dependent biotin identification, and thereby identified FLNA as a PERK interacting molecule. FLNA regulates the filamentous actin (F-actin) cytoskeleton, and forms cross-linked,

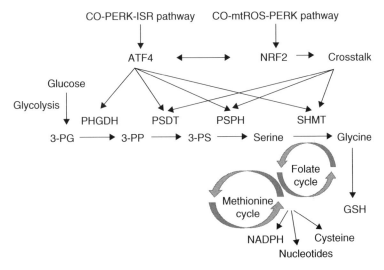

Figure 22.4 CO-induced transcription factors, ATF4 and Nrf2, crosstalk to maintain redox homeostasis. Endogenous or low-concentration exogenous CO activates ATF4 and Nrf2, which generate reduced GSH and NADPH through serine/glycine biosynthesis. Mechanistically, CO-induced PERK activation increases transcriptional activities of ATF4 and Nrf2, which increase the expression of serine/glycine biosynthetic enzymes through the crosstalk of the two transcription factors [44]. These enzymes produce serine and glycine from the glycolytic intermediate 3-PG and funnel the carbon into GSH and NADPH via the folate and transsulfuration cycles. 3-PG, 3-phosphoglycerate; 3-PP, 3-phosphohydroxypyruvate; 3-PS, 3-phosphoserine.

orthogonal networks of F-actin fibers. The functional consequences of the interaction between PERK and FLNA are that formation of the ER–PM (plasma membrane) contact site occurs through an ER Stim1 to PM Orai-1 interaction. The increase in Ca^{2+} by interaction of PERK and FLNA activates the Ca^{2+}–calcineurin signaling cascade. On the other hand, calcineurin is a downstream molecule of PERK signaling by directly interacting with PERK [102]. Thus, PERK/calcineurin signaling-regulated RyR2 opening is a possible mechanism of $[Ca^{2+}]_i$ elevation in myocytes [103]. In addition, CO-induced Ca^{2+} is triggered by calcineurin-mediated disassociation of the FKBP from the RyR. Therefore, treatment with GSK2606414, a PERK inhibitor, or Ryanodine as an inhibitor of RyR channel activity, inhibited CO-induced Ca^{2+} levels in hepatocytes. Further, CO-induced increases in Ca^{2+} levels promote TFEB nuclear translocation through calcineurin activity. Finally, TFEB nuclear translocation with CO may be increased by a PERK–Ca^{2+}–calcineurin pathway [19] (Figure 22.5).

Nuclear TFEB Enhances Both Autophagy and Mitophagy

Autophagy, a lysosome-dependent catabolic process, is induced by starvation, and the resulting breakdown products are subsequently used for new cellular components and energy. The starvation-activated lysosomal–autophagic pathway plays an important role in both cellular clearance and lipid catabolism [94]. Moreover, the increase in nuclear TFEB is induced by starvation through an autoregulatory feedback loop and exerts global transcriptional control on lipid catabolism via PGC-1α and PPARα [104]. Enhancement of TFEB activity has emerged as a potential therapeutic approach for various lysosomal and protein aggregation disorders [105]. TFEB positively regulates the expression of genes encoding molecules involved in phagocytosis, lipid catabolism, and overall lysosomal function [106]. CO has been shown to induce autophagy through augmented mtROS production. CO-activated autophagy requires TFEB. Lysosomal biogenesis can be triggered by TFEB, which activates lysosomal and autophagy-related genes, thereby increasing the number of lysosomes and promoting autophagic degradation. Inhibition of mtROS generation with mitoTEMPO decreased CO-induced autophagy and lysosomal gene expression. The increase of mtROS at low levels may regulate mitochondrial quality by balancing mitochondrial autophagy (mitophagy) with mitochondrial biogenesis. Mitophagy is the process by which damaged mitochondria are eliminated via autophagy [19]. Under conditions of loss of mitochondrial membrane potential, PINK1 kinase induces

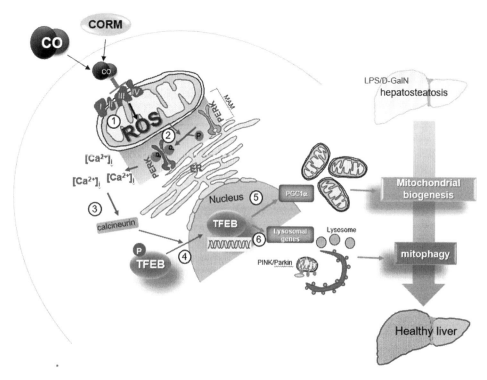

Figure 22.5 CO-induced TFEB nuclear translocation increases mitochondrial biogenesis and mitophagy to alleviate liver damage. PERK as a member of MAMs is activated by CO-induced mtROS (1), which induces PERK phosphorylation. (2) Activated PERK increases intracellular Ca^{2+} and consequently activates calcineurin. (3) The dephosphorylation of TFEB by calcineurin results in TFEB translocation into the nucleus. (4) Activated TFEB induces autophagy/mitophagy activation (5) and lysosome biogenesis (6) to ameliorate liver damage.

recruitment of the cytosolic E3 ligase Parkin to the outer mitochondrial membrane. Parkin-mediated ubiquitination of select outer mitochondrial membrane proteins, such as mitofusins, Miro1, and Tom70, recruits key regulators of autophagosome formation, resulting in the elimination of impaired mitochondria [107]. The positive transcriptional feedback loop between PGC-1α and TFEB is critical for modulating mitochondrial quality and function in different tissues. PGC-1α is a master regulator of mitochondrial biogenesis by regulating TFEB expression [108]. Likewise, TFEB promotes mitochondrial degradation and biogenesis by inducing the expression of PGC-1α and lysosomal genes [100]. Accordingly, CO-induced TFEB activation attenuated liver damage induced by LPS/D-GalN. Primary hepatocytes lacking TFEB inhibited CO-induced mitochondrial biogenesis and mitophagy [19]. Therefore, CO-induced TFEB nuclear translocation regulates in part the autophagic turnover of damaged mitochondria from cells affected by liver tissue injury (Figure 22.5).

Conclusions

CO, in the form of gaseous inhalation therapy or administration of CO donors, has emerged as a potential therapeutic molecule in liver injury and disease. In particular, as summarized in this chapter, these include acute liver injuries associated with proinflammatory conditions, I/R, HSR, sepsis, and hepatotoxins. In this chapter, we explored multiple mechanisms by which CO, a candidate hepatoprotective molecule, can protect against liver injury via hepatocyte cytoprotection and restoration of organ homeostasis. Collectively, the mechanisms associated with CO-mediated protection in these models include vascular effects, anti-inflammatory effects, and antiapoptotic effects, mitigation of ER stress, and potential pro-proliferative effects in hepatocytes. We emphasize that mitochondria represent a primary cellular target of CO, leading to the generation of mtROS that transduces the effects of CO on mitochondrial respiratory activity to downstream signaling pathways associated with cytoprotection and liver homeostasis, including regulation of apoptosis and adaptive gene expression.

ER stress represents a component of hepatic injury that can cause cell death and activate adaptive responses. We have discovered that, among the branches of the UPR associated with ER stress, CO exerts a selective preference for PERK activation via phosphorylation. PERK phosphorylates three interactomes (i.e., eIF2α, Nrf2, and FLNA). The PERK–eIF2α–ATF axis is associated with the ISR and responsible for SG formation and inhibition of cellular senescence. ATF4 promotes the activation of cytoprotective genes (i.e., FGF21 and SESN2) that contribute to metabolic homeostasis. The PERK–Nrf2 axis activates a global antioxidant response, which includes expression of HO-1 in a feedforward loop, and the expression of other cytoprotective or detoxification-associated proteins. Finally, the FLNA–PERK interaction is responsible for nuclear translocation of TFEB to activate mitophagy and *de novo* synthesis of mitochondria. We conclude that targeting the ER stress pathway, and PERK in particular, using either CO, CO donors, and/or related pharmacotherapeutics, may provide a strategy for therapeutic development for hepatic diseases such as hepatitis and hepatosteatosis. Although the safety of inhaled CO has been tested in recent clinical trials involving acute and chronic lung diseases [32], to date no clinical studies have been initiated in metabolic or hepatic diseases. Nevertheless, inhaled CO and CO donors continue to show promise as experimental therapeutics for liver disease.

Acknowledgment

This work was supported by the Priority Research Centers Program through the National Research Foundation of Korea funded by the Ministry of Education (2014R1A6A1030318).

References

1. Sjostrand, T. (1949). Endogenous formation of carbon monoxide in man. *Nature* 164: 580.
2. Tenhunen, R., Marver, H.S., and Schmid, R. (1968). The enzymatic conversion of heme to bilirubin by microsomal heme oxygenase. *Proc. Natl. Acad. Sci. U. S. A.* 61: 748–755.
3. Tenhunen, R., Marver, H.S., and Schmid, R. (1969). Microsomal heme oxygenase. Characterization of the enzyme. *J. Biol. Chem.* 244: 6388–6394.
4. Poss, K.D. and Tonegawa, S. (1997). Reduced stress defense in heme oxygenase 1-deficient cells. *Proc. Natl. Acad. Sci. U. S. A.* 94: 10925–10930.
5. Motterlini, R. and Foresti, R. (2017). Biological signaling by carbon monoxide and carbon monoxide-releasing molecules. *Am. J. Physiol. Cell Physiol.* 312: C302–c313.
6. Motterlini, R. and Otterbein, L.E. (2010). The therapeutic potential of carbon monoxide. *Nat. Rev. Drug Discov.* 9: 728–743.

7. Oliveira, S.R., Queiroga, C.S., and Vieira, H.L. (2016). Mitochondria and carbon monoxide: cytoprotection and control of cell metabolism - a role for Ca(2+)? *J. Physiol.* 594: 4131–4138.
8. Queiroga, C.S., Vercelli, A., and Vieira, H.L. (2015). Carbon monoxide and the CNS: challenges and achievements. *Br. J. Pharmacol.* 172: 1533–1545.
9. Suematsu, M., Goda, N., Sano, T., Kashiwagi, S., Egawa, T., Shinoda, Y., and Ishimura, Y. (1995). Carbon monoxide: an endogenous modulator of sinusoidal tone in the perfused rat liver. *J. Clin. Invest.* 96: 2431–2437.
10. Kyokane, T., Norimizu, S., Taniai, H., Yamaguchi, T., Takeoka, S., Tsuchida, E., Naito, M., Nimura, Y., Ishimura, Y., and Suematsu, M. (2001). Carbon monoxide from heme catabolism protects against hepatobiliary dysfunction in endotoxin-treated rat liver. *Gastroenterol* 120: 1227–1240.
11. Suematsu, M., Tsukada, K., Tajima, T., Yamamoto, T., Ochiai, D., Watanabe, H., Yoshimura, Y., and Goda, N. (2005). Carbon monoxide as a guardian against hepatobiliary dysfunction. *Alcohol Clin. Exp. Res.* 29: 134S–139S.
12. Sarady, J.K., Zuckerbraun, B.S., Bilban, M., Wagner, O., Usheva, A., Liu, F., Ifedigbo, E., Zamora, R., Choi, A.M., and Otterbein, L.E. (2004). Carbon monoxide protection against endotoxic shock involves reciprocal effects on iNOS in the lung and liver. *FASEB J.* 18: 854–856.
13. Zuckerbraun, B.S., McCloskey, C.A., Gallo, D., Liu, F., Ifedigbo, E., Otterbein, L.E., and Billiar, T.R. (2005). Carbon monoxide prevents multiple organ injury in a model of hemorrhagic shock and resuscitation. *Shock* 23: 527–532.
14. Kuramitsu, K., Gallo, D., Yoon, M., Chin, B.Y., Csizmadia, E., Hanto, D.W., and Otterbein, L.E. (2011). Carbon monoxide enhances early liver regeneration in mice after hepatectomy. *Hepatology* 53: 2016–2026.
15. Cepinskas, G., Katada, K., Bihari, A., and Potter, R.F. (2008). Carbon monoxide liberated from carbon monoxide-releasing molecule CORM-2 attenuates inflammation in the liver of septic mice. *Am. J. Physiol. Gastrointest. Liver Physiol.* 294: G184–91.
16. Motterlini, R., Nikam, A., Manin, S., Ollivier, A., Wilson, J.L., Djouadi, S., Muchova, L., Martens, T., Rivard, M., and Foresti, R. (2019). HYCO-3, a dual CO-releaser/Nrf2 activator, reduces tissue inflammation in mice challenged with lipopolysaccharide. *Redox Biol.* 20: 334–348.
17. Zheng, Y., Ji, X., Yu, B., Ji, K., Gallo, D., Csizmadia, E., Zhu, M., Choudhury, M.R., De La Cruz, L.K.C., Chittavong, V., Pan, Z., Yuan, Z., Otterbein, L.E., and Wang, B. (2018). Enrichment-triggered prodrug activation demonstrated through mitochondria-targeted delivery of doxorubicin and carbon monoxide. *Nat. Chem.* 10: 787–794.
18. Zuckerbraun, B.S., Billiar, T.R., Otterbein, S.L., Kim, P.K., Liu, F., Choi, A.M., Bach, F.H., and Otterbein, L.E. (2003). Carbon monoxide protects against liver failure through nitric oxide-induced heme oxygenase 1. *J. Exp. Med.* 198: 1707–1716.
19. Kim, H.J., Joe, Y., Rah, S.Y., Kim, S.K., Park, S.U., Park, J., Kim, J., Ryu, J., Cho, G.J., Surh, Y.J., Ryter, S.W., Kim, U.H., and Chung, H.T. (2018). Carbon monoxide-induced TFEB nuclear translocation enhances mitophagy/mitochondrial biogenesis in hepatocytes and ameliorates inflammatory liver injury. *Cell Death Dis.* 9: 1060.
20. Tsui, T.Y., Obed, A., Siu, Y.T., Yet, S.F., Prantl, L., Schlitt, H.J., and Fan, S.T. (2007). Carbon monoxide inhalation rescues mice from fulminant hepatitis through improving hepatic energy metabolism. *Shock* 27: 165–171.
21. Wen, Z., Liu, Y., Li, F., and Wen, T. (2013). Low dose of carbon monoxide intraperitoneal injection provides potent protection against GalN/LPS-induced acute liver injury in mice. *J. Appl. Toxicol.* 33: 1424–1432.
22. Fang, X., Wang, R., Ma, J., Ding, Y., Shang, W., and Sun, Z. (2012). Ameliorated ConA-induced hepatitis in the absence of PKC-theta. *PLoS One* 7: e31174.
23. Mangano, K., Cavalli, E., Mammana, S., Basile, M.S., Caltabiano, R., Pesce, A., Puleo, S., Atanasov, A.G., Magro, G., Nicoletti, F., and Fagone, P. (2018). Involvement of the Nrf2/HO-1/CO axis and therapeutic intervention with the CO-releasing molecule CORM-A1, in a murine model of autoimmune hepatitis. *J. Cell Physiol.* 233: 4156–4165.
24. Chen, Y., Park, H.J., Park, J., Song, H.C., Ryter, S.W., Surh, Y.J., Kim, U.H., Joe, Y., and Chung, H.T. (2019). Carbon monoxide ameliorates acetaminophen-induced liver injury by increasing hepatic HO-1 and Parkin expression. *FASEB J.* 33: 13905–13919.
25. Upadhyay, K.K., Jadeja, R.N., Thadani, J.M., Joshi, A., Vohra, A., Mevada, V., Patel, R., Khurana, S., and Devkar, R.V. (2018). Carbon monoxide releasing molecule A-1 attenuates acetaminophen-mediated hepatotoxicity and improves survival of mice by induction of Nrf2 and related genes. *Toxicol. Appl. Pharmacol.* 360: 99–108.
26. Kim, H.J., Joe, Y., Kong, J.S., Jeong, S.O., Cho, G.J., Ryter, S.W., and Chung, H.T. (2013). Carbon monoxide protects against hepatic ischemia/reperfusion injury via ROS-dependent Akt signaling and inhibition of glycogen synthase kinase 3beta. *Oxid. Med. Cell Longev.* 2013: 306421.

27. Kim, H.J., Joe, Y., Yu, J.K., Chen, Y., Jeong, S.O., Mani, N., Cho, G.J., Pae, H.O., Ryter, S.W., and Chung, H.T. (2015). Carbon monoxide protects against hepatic ischemia/reperfusion injury by modulating the miR-34a/SIRT1 pathway. *Biochim. Biophys. Acta* 1852: 1550–1559.

28. Wei, Y., Chen, P., De Bruyn, M., Zhang, W., Bremer, E., and Helfrich, W. (2010). Carbon monoxide-releasing molecule-2 (CORM-2) attenuates acute hepatic ischemia reperfusion injury in rats. *BMC Gastroenterol.* 10: 42.

29. Kaizu, T., Nakao, A., Tsung, A., Toyokawa, H., Sahai, R., Geller, D.A., and Murase, N. (2005). Carbon monoxide inhalation ameliorates cold ischemia/reperfusion injury after rat liver transplantation. *Surgery* 138: 229–235.

30. Joe, Y., Kim, S., Kim, H.J., Park, J., Chen, Y., Park, H.J., Jekal, S.J., Ryter, S.W., Kim, U.H., and Chung, H.T. (2018). FGF21 induced by carbon monoxide mediates metabolic homeostasis via the PERK/ATF4 pathway. *FASEB J.* 32: 2630–2643.

31. Upadhyay, K.K., Jadeja, R.N., Vyas, H.S., Pandya, B., Joshi, A., Vohra, A., Thounaojam, M.C., Martin, P.M., Bartoli, M., and Devkar, R.V. (2020). Carbon monoxide releasing molecule-A1 improves nonalcoholic steatohepatitis via Nrf2 activation mediated improvement in oxidative stress and mitochondrial function. *Redox Biol.* 28: 101314.

32. Ryter, S.W., Ma, K.C., and Choi, A.M.K. (2018). Carbon monoxide in lung cell physiology and disease. *Am. J. Physiol. Cell Physiol.* 314: C211–c227.

33. Boczkowski, J., Poderoso, J.J., and Motterlini, R. (2006). CO-metal interaction: Vital signaling from a lethal gas. *Trends Biochem. Sci.* 31: 614–621.

34. Brown, S.D. and Piantadosi, C.A. (1989). Reversal of carbon monoxide-cytochrome c oxidase binding by hyperbaric oxygen in vivo. *Adv. Exp. Med. Biol.* 248: 747–754.

35. Chance, B., Erecinska, M., and Wagner, M. (1970). Mitochondrial responses to carbon monoxide toxicity. *Ann. N. Y. Acad. Sci.* 174: 193–204.

36. Volpe, J.A., O'Toole, M.C., and Caughey, W.S. (1975). Quantitative infrared spectroscopy of CO complexes of cytochrome c oxidase, hemoglobin and myoglobin: evidence for one CO per heme. *Biochem. Biophys. Res. Commun.* 62: 48–53.

37. Romão, C.C. and Vieira, H.L. (2013). Metal Carbonyls for CO-based Therapies: challenges and successes. In: *Advances in Organometallic Chemistry and Catalysis ICOMC Silver/Gold Jubilee Book.* Armando J. L. Pombeiro (Editor) Wiley ISBN: 978-1-118-51014-8 545–561.

38. Ishigami, I., Zatsepin, N.A., Hikita, M., Conrad, C.E., Nelson, G., Coe, J.D., Basu, S., Grant, T.D., Seaberg, M.H., Sierra, R.G., Hunter, M.S., Fromme, P., Fromme, R., Yeh, S.R., and Rousseau, D.L. (2017). Crystal structure of CO-bound cytochrome c oxidase determined by serial femtosecond X-ray crystallography at room temperature. *Proc. Natl. Acad. Sci. U. S. A.* 114: 8011–8016.

39. Piantadosi, C.A., Sylvia, A.L., Saltzman, H.A., and Jöbsis-Vandervliet, F.F. (1985). Carbon monoxide-cytochrome interactions in the brain of the fluorocarbon-perfused rat. *J. Appl. Physiol.* 58: 665–672.

40. Bilban, M., Haschemi, A., Wegiel, B., Chin, B.Y., Wagner, O., and Otterbein, L.E. (2008). Heme oxygenase and carbon monoxide initiate homeostatic signaling. *J. Mol. Med.* 86: 267–279.

41. Fridovich, I. (1997). Superoxide anion radical (O_2^-.), superoxide dismutases, and related matters. *J. Biol. Chem.* 272: 18515–18517.

42. Zuckerbraun, B.S., Chin, B.Y., Bilban, M., d'Avila, J.D.C., Rao, J., Billiar, T.R., and Otterbein, L.E. (2007). Carbon monoxide signals via inhibition of cytochrome c oxidase and generation of mitochondrial reactive oxygen species. *FASEB J.* 21: 1099–1106.

43. Otterbein, L.E., Foresti, R., and Motterlini, R. (2016). Heme oxygenase-1 and carbon monoxide in the heart: the balancing act between danger signaling and pro-survival. *Circ. Res.* 118: 1940–1959.

44. Kim, K.M., Pae, H.O., Zheng, M., Park, R., Kim, Y.M., and Chung, H.T. (2007). Carbon monoxide induces heme oxygenase-1 via activation of protein kinase R-like endoplasmic reticulum kinase and inhibits endothelial cell apoptosis triggered by endoplasmic reticulum stress. *Circ. Res.* 101: 919–927.

45. Kasai, S., Yamazaki, H., Tanji, K., Engler, M.J., Matsumiya, T., and Itoh, K. (2019). Role of the ISR-ATF4 pathway and its cross talk with Nrf2 in mitochondrial quality control. *J. Clin. Biochem. Nutr.* 64: 1–12.

46. Kim, H.J., Joe, Y., Kim, S.K., Park, S.U., Park, J., Chen, Y., Kim, J., Ryu, J., Cho, G.J., Surh, Y.J., Ryter, S.W., Kim, U.H., and Chung, H.T. (2017). Carbon monoxide protects against hepatic steatosis in mice by inducing sestrin-2 via the PERK-eIF2α-ATF4 pathway. *Free Radic. Biol. Med.* 110: 81–91.

47. Chen, Y., Joe, Y., Park, J., Song, H.-C., Kim, U.-H., and Chung, H.T. (2019). Carbon monoxide induces the assembly of stress granule through the integrated stress response. *Biochem. Biophys. Res. Commun.* 512: 289–294.

48. Del Vecchio, C.A., Feng, Y., Sokol, E.S., Tillman, E.J., Sanduja, S., Reinhardt, F., and Gupta, P.B. (2014). De-differentiation confers multidrug resistance via noncanonical PERK-Nrf2 signaling. *PLoS Biol.* 12: e1001945.

49. McQuiston, A. and Diehl, J.A. (2017). Recent insights into PERK-dependent signaling from the stressed endoplasmic reticulum. *F1000Res.* 6.
50. Peñaranda-Fajardo, N.M., Meijer, C., Liang, Y., Dijkstra, B.M., Aguirre-Gamboa, R., Den Dunnen, W.F., and Kruyt, F.A. (2019). ER stress and UPR activation in glioblastoma: identification of a noncanonical PERK mechanism regulating GBM stem cells through SOX2 modulation. *Cell Death Dis.* 10: 1–16.
51. Van Vliet, A.R., Garg, A.D., and Agostinis, P. (2016). Coordination of stress, Ca2+, and immunogenic signaling pathways by PERK at the endoplasmic reticulum. *Biol. Chem.* 397: 649–656.
52. Harding, H.P., Zhang, Y., and Ron, D. (1999). Protein translation and folding are coupled by an endoplasmic-reticulum-resident kinase. *Nature* 397: 271–274.
53. Shi, Y., Vattem, K.M., Sood, R., An, J., Liang, J., Stramm, L., and Wek, R.C. (1998). Identification and characterization of pancreatic eukaryotic initiation factor 2 α-subunit kinase, PEK, involved in translational control. *Mol. Cell Biol.* 18: 7499–7509.
54. Sood, R., Porter, A.C., Ma, K., Quilliam, L.A., and Wek, R.C. (2000). Pancreatic eukaryotic initiation factor-2α kinase (PEK) homologues in humans, Drosophila melanogaster and Caenorhabditis elegans that mediate translational control in response to endoplasmic reticulum stress. *Biochem. J.* 346: 281–293.
55. Cullinan, S.B., Zhang, D., Hannink, M., Arvisais, E., Kaufman, R.J., and Diehl, J.A. (2003). Nrf2 is a direct PERK substrate and effector of PERK-dependent cell survival. *Mol. Cell Biol.* 23: 7198–7209.
56. Van Vliet, A.R., Giordano, F., Gerlo, S., Segura, I., Van Eygen, S., Molenberghs, G., Rocha, S., Houcine, A., Derua, R., Verfaillie, T., Vangindertael, J., De Keersmaecker, H., Waelkens, E., Tavernier, J., Hofkens, J., Annaert, W., Carmeliet, P., Samali, A., Mizuno, H., and Agostinis, P. (2017). The ER stress sensor PERK coordinates ER-plasma membrane contact site formation through interaction with filamin-A and F-actin remodeling. *Mol. Cell* 65: 885–899.e6.
57. Kedersha, N., Chen, S., Gilks, N., Li, W., Miller, I.J., Stahl, J., and Anderson, P. (2002). Evidence that ternary complex (eIF2-GTP-tRNAi Met)–deficient preinitiation complexes are core constituents of mammalian stress granules. *Mol. Biol. Cell* 13: 195–210.
58. Jackson, R.J., Hellen, C.U., and Pestova, T.V. (2010). The mechanism of eukaryotic translation initiation and principles of its regulation. *Nat. Rev. Mol. Cell Biol.* 11: 113–127.
59. Krishnamoorthy, T., Pavitt, G.D., Zhang, F., Dever, T.E., and Hinnebusch, A.G. (2001). Tight binding of the phosphorylated α subunit of initiation factor 2 (eIF2α) to the regulatory subunits of guanine nucleotide exchange factor eIF2B is required for inhibition of translation initiation. *Mol. Cell Biol.* 21: 5018–5030.
60. Ameri, K. and Harris, A.L. (2008). Activating transcription factor 4. *Int. J. Biochem. Cell Biol.* 40: 14–21.
61. Ben-Porath, I. and Weinberg, R.A. (2005). The signals and pathways activating cellular senescence. *Int. J. Biochem. Cell Biol.* 37: 961–976.
62. De Sousa-Coelho, A.L., Relat, J., Hondares, E., Pérez-Martí, A., Ribas, F., Villarroya, F., Marrero, P.F., and Haro, D. (2013). FGF21 mediates the lipid metabolism response to amino acid starvation. *J. Lipid Res.* 54: 1786–1797.
63. Kharitonenkov, A., Wroblewski, V.J., Koester, A., Chen, Y.-F., Clutinger, C.K., Tigno, X.T., Hansen, B.C., Shanafelt, A.B., and Etgen, G.J. (2007). The metabolic state of diabetic monkeys is regulated by fibroblast growth factor-21. *Endocrinol* 148: 774–781.
64. Hondares, E., Iglesias, R., Giralt, A., Gonzalez, F.J., Giralt, M., Mampel, T., and Villarroya, F. (2011). Thermogenic activation induces FGF21 expression and release in brown adipose tissue. *J. Biol. Chem.* 286: 12983–12990.
65. Hotta, Y., Nakamura, H., Konishi, M., Murata, Y., Takagi, H., Matsumura, S., Inoue, K., Fushiki, T., and Itoh, N. (2009). Fibroblast growth factor 21 regulates lipolysis in white adipose tissue but is not required for ketogenesis and triglyceride clearance in liver. *Endocrinol* 150: 4625–4633.
66. Izumiya, Y., Bina, H.A., Ouchi, N., Akasaki, Y., Kharitonenkov, A., and Walsh, K. (2008). FGF21 is an Akt-regulated myokine. *FEBS Lett.* 582: 3805–3810.
67. Nishimura, T., Nakatake, Y., Konishi, M., and Itoh, N. (2000). Identification of a novel FGF, FGF-21, preferentially expressed in the liver. *Biochim. Biophys. Acta* 1492: 203–206.
68. Kharitonenkov, A., Dunbar, J.D., Bina, H.A., Bright, S., Moyers, J.S., Zhang, C., Ding, L., Micanovic, R., Mehrbod, S.F., and Knierman, M.D. (2008). FGF-21/FGF-21 receptor interaction and activation is determined by βKlotho. *J. Cell Physiol.* 215: 1–7.
69. Adams, A.C., Astapova, I., Badman, M.K., Kurgansky, K.E., Flier, J.S., Hollenberg, A.N., and Maratos-Flier, E. (2010). Thyroid hormone regulates hepatic expression of fibroblast growth factor 21 in a PPARα-dependent manner. *J. Biol. Chem.* 285: 14078–14082.
70. Li, Y., Wong, K., Walsh, K., Gao, B., and Zang, M. (2013). Retinoic acid receptor β stimulates hepatic induction of fibroblast growth factor 21 to promote fatty acid oxidation and control whole-body energy homeostasis in mice. *J. Biol. Chem.* 288: 10490–10504.

71. Lundåsen, T., Hunt, M.C., Nilsson, L.-M., Sanyal, S., Angelin, B., Alexson, S.E., and Rudling, M. (2007). PPARα is a key regulator of hepatic FGF21. *Biochem. Biophys. Res. Commun.* 360: 437–440.

72. Maruyama, R., Shimizu, M., Li, J., Inoue, J., and Sato, R. (2016). Fibroblast growth factor 21 induction by activating transcription factor 4 is regulated through three amino acid response elements in its promoter region. *Biosci., Biotechnol. Biochem.* 80: 929–934.

73. Wan, X.-S., Lu, X.-H., Xiao, Y.-C., Lin, Y., Zhu, H., Ding, T., Yang, Y., Huang, Y., Zhang, Y., and Liu, Y.-L. (2014). ATF4-and CHOP-dependent induction of FGF21 through endoplasmic reticulum stress. *BioMed Res. Int.* 2014.

74. Wang, Y., Solt, L.A., and Burris, T.P. (2010). Regulation of FGF21 expression and secretion by retinoic acid receptor-related orphan receptor α. *J. Biol. Chem.* 285: 15668–15673.

75. Lee, J.H., Budanov, A.V., Talukdar, S., Park, E.J., Park, H.L., Park, H.-W., Bandyopadhyay, G., Li, N., Aghajan, M., and Jang, I. (2012). Maintenance of metabolic homeostasis by Sestrin2 and Sestrin3. *Cell Metab.* 16: 311–321.

76. Park, H.-W., Park, H., Ro, S.-H., Jang, I., Semple, I.A., Kim, D.N., Kim, M., Nam, M., Zhang, D., and Yin, L. (2014). Hepatoprotective role of Sestrin2 against chronic ER stress. *Nat. Commun.* 5: 1–11.

77. Seo, K., Seo, S., Ki, S.H., and Shin, S.M. (2016). Compound C increases sestrin2 expression via mitochondria-dependent ROS production. *Biol. Pharmaceut. Bull.* 39: 799–806.

78. Shin, B.Y., Jin, S.H., Cho, I.J., and Ki, S.H. (2012). Nrf2-ARE pathway regulates induction of Sestrin-2 expression. *Free Radic. Biol. Med.* 53: 834–841.

79. Lu, P.D., Harding, H.P., and Ron, D. (2004). Translation reinitiation at alternative open reading frames regulates gene expression in an integrated stress response. *J. Cell Biol.* 167: 27–33.

80. Bae, S.H., Sung, S.H., Oh, S.Y., Lim, J.M., Lee, S.K., Park, Y.N., Lee, H.E., Kang, D., and Rhee, S.G. (2013). Sestrins activate Nrf2 by promoting p62-dependent autophagic degradation of Keap1 and prevent oxidative liver damage. *Cell Metab.* 17: 73–84.

81. Budanov, A.V. and Karin, M. (2008). p53 target genes sestrin1 and sestrin2 connect genotoxic stress and mTOR signaling. *Cell* 134: 451–460.

82. Venugopal, R. and Jaiswal, A.K. (1996). Nrf1 and Nrf2 positively and c-Fos and Fra1 negatively regulate the human antioxidant response element-mediated expression of NAD (P) H: quinone oxidoreductase1 gene. *Proc. Natl. Acad. Sci. U.S.A.* 93: 14960–14965.

83. Ding, B., Parmigiani, A., Divakaruni, A.S., Archer, K., Murphy, A.N., and Budanov, A.V. (2016). Sestrin2 is induced by glucose starvation via the unfolded protein response and protects cells from non-canonical necroptotic cell death. *Sci. Rep.* 6: 22538.

84. Brüning, A., Rahmeh, M., and Friese, K. (2013). Nelfinavir and bortezomib inhibit mTOR activity via ATF4-mediated sestrin-2 regulation. *Mol. Oncol.* 7: 1012–1018.

85. Hamanaka, R.B., Bennett, B.S., Cullinan, S.B., and Diehl, J.A. (2005). PERK and GCN2 contribute to eIF2α phosphorylation and cell cycle arrest after activation of the unfolded protein response pathway. *Mol. Biol. Cell* 16: 5493–5501.

86. Bellezza, I., Giambanco, I., Minelli, A., and Donato, R. (2018). Nrf2-Keap1 signaling in oxidative and reductive stress. *Biochim. Biophys. Acta, Mol. Cell Res.* 1865: 721–733.

87. Itoh, K., Chiba, T., Takahashi, S., Ishii, T., Igarashi, K., Katoh, Y., Oyake, T., Hayashi, N., Satoh, K., and Hatayama, I. (1997). An Nrf2/small Maf heterodimer mediates the induction of phase II detoxifying enzyme genes through antioxidant response elements. *Biochem. Biophys. Res. Comm.* 236: 313–322.

88. Ma, Q. (2013). Role of nrf2 in oxidative stress and toxicity. *Annu. Rev. Pharmacol. Toxicol.* 53: 401–426.

89. Almeida, A.S., Soares, N.L., Sequeira, C.O., Pereira, S.A., Sonnewald, U., and Vieira, H.L.A. (2018). Improvement of neuronal differentiation by carbon monoxide: Role of pentose phosphate pathway. *Redox Biol.* 17: 338–347.

90. DeNicola, G.M., Chen, P.H., Mullarky, E., Sudderth, J.A., Hu, Z., Wu, D., Tang, H., Xie, Y., Asara, J.M., Huffman, K.E., Wistuba, I.I., Minna, J.D., DeBerardinis, R.J., and Cantley, L.C. (2015). NRF2 regulates serine biosynthesis in non-small cell lung cancer. *Nat. Genet.* 47: 1475–1481.

91. Ma, X., Liu, H., Murphy, J.T., Foyil, S.R., Godar, R.J., Abuirqeba, H., Weinheimer, C.J., Barger, P.M., and Diwan, A. (2015). Regulation of the transcription factor EB-PGC1α axis by beclin-1 controls mitochondrial quality and cardiomyocyte death under stress. *Mol. Cell. Biol.* 35: 956–976.

92. Martina, J.A., Diab, H.I., Brady, O.A., and Puertollano, R. (2016). TFEB and TFE3 are novel components of the integrated stress response. *EMBO J.* 35: 479–495.

93. Pastore, N., Brady, O.A., Diab, H.I., Martina, J.A., Sun, L., Huynh, T., Lim, J.A., Zare, H., Raben, N., Ballabio, A., and Puertollano, R. (2016). TFEB and TFE3 cooperate in the regulation of the innate immune response in activated macrophages. *Autophagy* 12: 1240–1258.

94. Raben, N. and Puertollano, R. (2016). TFEB and TFE3: Linking lysosomes to cellular adaptation to stress. *Annu. Rev. Cell Dev. Biol.* 32: 255–278.

95. Martina, J.A., Diab, H.I., Lishu, L., Jeong-A, L., Patange, S., Raben, N., and Puertollano, R. (2014). The nutrient-responsive transcription factor TFE3

promotes autophagy, lysosomal biogenesis, and clearance of cellular debris. *Sci. Signal.* 7: ra9–ra9.

96. Palmieri, M., Impey, S., Kang, H., Di Ronza, A., Pelz, C., Sardiello, M., and Ballabio, A. (2011). Characterization of the CLEAR network reveals an integrated control of cellular clearance pathways. *Human Mol. Genet.* 20: 3852–3866.

97. Martina, J.A., Chen, Y., Gucek, M., and Puertollano, R. (2012). MTORC1 functions as a transcriptional regulator of autophagy by preventing nuclear transport of TFEB. *Autophagy* 8: 903–914.

98. Medina, D.L., Di Paola, S., Peluso, I., Armani, A., De Stefani, D., Venditti, R., Montefusco, S., Scotto-Rosato, A., Prezioso, C., Forrester, A., Settembre, C., Wang, W., Gao, Q., Xu, H., Sandri, M., Rizzuto, R., De Matteis, M.A., and Ballabio, A. (2015). Lysosomal calcium signalling regulates autophagy through calcineurin and TFEB. *Nat. Cell Biol.* 17: 288–299.

99. Roczniak-Ferguson, A., Petit, C.S., Froehlich, F., Qian, S., Ky, J., Angarola, B., Walther, T.C., and Ferguson, S.M. (2012). The transcription factor TFEB links mTORC1 signaling to transcriptional control of lysosome homeostasis. *Sci. Signal.* 5: ra42–ra42.

100. Settembre, C., Zoncu, R., Medina, D.L., Vetrini, F., Erdin, S., Erdin, S., Huynh, T., Ferron, M., Karsenty, G., and Vellard, M.C. (2012). A lysosome-to-nucleus signalling mechanism senses and regulates the lysosome via mTOR and TFEB. *EMBO J.* 31: 1095–1108.

101. Verfaillie, T., Rubio, N., Garg, A.D., Bultynck, G., Rizzuto, R., Decuypere, J.P., Piette, J., Linehan, C., Gupta, S., Samali, A., and Agostinis, P. (2012). PERK is required at the ER-mitochondrial contact sites to convey apoptosis after ROS-based ER stress. *Cell Death Differ.* 19: 1880–1891.

102. Wang, R., McGrath, B.C., Kopp, R.F., Roe, M.W., Tang, X., Chen, G., and Cavener, D.R. (2013). Insulin secretion and Ca^{2+} dynamics in β-cells are regulated by PERK (EIF2AK3) in concert with calcineurin. *J. Biol. Chem.* 288: 33824–33836.

103. Liu, Z., Cai, H., Zhu, H., Toque, H., Zhao, N., Qiu, C., Guan, G., Dang, Y., and Wang, J. (2014). Protein kinase RNA-like endoplasmic reticulum kinase (PERK)/calcineurin signaling is a novel pathway regulating intracellular calcium accumulation which might be involved in ventricular arrhythmias in diabetic cardiomyopathy. *Cell Signal.* 26: 2591–2600.

104. Settembre, C., De Cegli, R., Mansueto, G., Saha, P.K., Vetrini, F., Visvikis, O., Huynh, T., Carissimo, A., Palmer, D., Klisch, T.J., Wollenberg, A.C., Di Bernardo, D., Chan, L., Irazoqui, J.E., and Ballabio, A. (2013). TFEB controls cellular lipid metabolism through a starvation-induced autoregulatory loop. *Nat. Cell Biol.* 15: 647–658.

105. Zhang, W., Li, X., Wang, S., Chen, Y., and Liu, H. (2020). Regulation of TFEB activity and its potential as a therapeutic target against kidney diseases. *Cell Death Disc* 6: 1–10.

106. Samie, M. and Cresswell, P. (2015). The transcription factor TFEB acts as a molecular switch that regulates exogenous antigen-presentation pathways. *Nat. Immunol.* 16: 729–736.

107. Narendra, D., Walker, J.E., and Youle, R. (2012). Mitochondrial quality control mediated by PINK1 and Parkin: links to parkinsonism. *Cold Spring Harb. Perspect. Biol.* 4: a011338.

108. Tsunemi, T. and La Spada, A.R. (2012). PGC-1α at the intersection of bioenergetics regulation and neuron function: from Huntington's disease to Parkinson's disease and beyond. *Prog. Neurobiol.* 97: 142–151.

23

CO and Cancer

James N. Arnold and Joanne E. Anstee

Faculty of Life Sciences and Medicine, School of Cancer and Pharmaceutical Sciences, King's College London, Guy's Hospital, London SE1 1UL, UK

Introduction

Heme oxygenase (HO) activity was first associated with cancer in a study of patients with chronic leukemia in 1976, where a change in HO activity in the bone marrow was detected [1]. Subsequently, tumors were identified to be a site of increased HO activity, relative to the healthy tissue, in models of liver cancer in mice [2]. However, it was not until the late 1980s when HOs were more formally characterized [3,4], and HO-1 became associated with the stress and oxidative damage response [5], that the implications of HO activity in cancer progression became apparent. In the proceeding years, our knowledge of the role of HOs and carbon monoxide (CO) in inflammation and cancer greatly expanded, with now close to 2000 articles recorded on PubMed.gov in the field.

HO-1 is expressed in a variety of different tumor types, including bladder [6], breast [7], colorectal [8], glioblastoma [9], head and neck [10], leukemia [11], lung [12], melanoma [13], neuroblastoma [14], prostate [15], and renal [16]. Expression of HO-1 is also associated with poor prognosis in patients with cancer [7,14,17]. Although all the metabolites of heme degradation have been demonstrated to play a role in tumor progression [17,18], CO has emerged as an important molecule, facilitating several vital protumoral processes, including cytoprotection [19,20], angiogenesis [21], metastasis [22], and immune suppression [23]. However, the role of HO-1, and CO in particular, remains controversial in relation to cancer with antitumoral properties having also been described. Most pertinent is the observation that exposing tumor-bearing mice to low-dose CO (250 ppm) can result in tumor control and increased sensitivity to chemotherapy [15]. This observation highlights the importance of understanding the fundamental biological contexts that account for the divergent roles of CO in cancer. The situation is further complicated by the observation that some tumor cells can generate a proteolytically cleaved catalytically inactive form of HO-1 that can translocate to the nucleus and regulate gene transcription [24]. The truncated form of HO-1 (which is missing the C-terminal 23 amino acids) has been associated with a tumor cell response to environments of low oxygen tension [24,25], commonly referred to as "hypoxia," which is a common feature of solid cancers. The presence of nuclear HO-1 has now been observed in a range of cancers, but is most well characterized in prostate cancer where it has been associated with disease progression [10,26]. The biological importance of CO in cancer progression, and its influence on different cell types, requires careful consideration to fully understand how to most effectively capitalize on our current knowledge for translation into the clinic. This chapter will consider our current knowledge of CO in relation to cancer and will discuss its well-established functions and the controversies that still remain in the field.

CO Production in the Tumor: The Where and Why

In cancers, HO-1 can be expressed by both the malignant tumor cells and also the nonmalignant cells that surround the cancer, collectively referred to as the stroma [7,18,22]. The stroma represents a complex crosstalk of both immune and mesenchymal cell populations that provide a tissue protective microenvironment to allow the tumor cells to evade antitumoral immune responses [27–29] and suppress the effects of anticancer therapeutic agents [30]. These

Carbon Monoxide in Drug Discovery: Basics, Pharmacology, and Therapeutic Potential, First Edition. Edited by Binghe Wang and Leo E. Otterbein.
© 2022 John Wiley & Sons, Inc. Published 2022 by John Wiley & Sons, Inc.

biological responses are analogous to the stromal reactions found in a healing wound [22], a concept originally outlined by Harold Dvorak in his seminal article in 1986 [31], which has since underpinned many of the contextual advances in cancer stromal biology. As such, it is not surprising that HO-1 expression is also upregulated in the healing wound [22], and CO plays a role in wound repair [32–34]. Tumor cells actively exploit the wound healing response to facilitate tumor progression through influencing the phenotype and response of the stroma [22], which in cancer is catastrophically misguided and promotes disease progression and eventual metastasis, rather than favoring eradication of the abnormal malignant cells.

Cellular Sources of CO in the Tumor Microenvironment

There are numerous reports of both malignant tumor cells and stromal cell populations expressing HO-1 across different cancers [18,22,35]. Although the expression of HO-1 by tumor cells can be influenced by the tumor microenvironment (TME; discussed in the "TME Cues for HO-1 Induction" section), there are instances where HO-1 can be constitutively expressed due to oncogenes driving the *HMOX1* promoter [11] or genetic mutations within the promoter itself [36]. The human *HMOX1* gene promoter has a 5′-flanking region containing differing numbers of GT microsatellite repeats. The number of GT dinucleotide repeats (which range between 11 and 40) can influence basal *HMOX1* transcription and inducibility of the gene in response to stimuli [37,38]. Interestingly, polymorphisms in the number of GT repeats have been associated with cancer predisposition. Individuals with longer GT repeats in the *HMOX1* promoter (that express lower HO-1) have a higher prevalence of gastric, lung, and oral squamous cancer compared to individuals with shorter GT repeats (that display higher HO-1 expression) [39]. There is also evidence of post-transcriptional regulation of *HMOX1* by microRNA-378, a small noncoding RNA that specifically targets and destabilizes *HMOX1* mRNA, preventing translation [40].

Several stromal cell populations have been described to express HO-1 [18]. Tumor-associated macrophages (TAMs), which are a prevalent stromal cell type in many cancers, have emerged as a major expressor of the enzyme across a variety of tumors in both murine models [7,22,28] and the human disease [7,22]. TAMs predominantly derive from peripheral circulating CCR2$^+$ monocytic precursors that are recruited into the TME [41–43]. However, some TAMs also derive from tissue-resident sources that have an ability to self-renew [44–46]. Macrophages are a highly plastic cell type and are capable of responding to the TME to adopt a variety of distinct phenotypes, which differ in their gene expression profiles, creating a "spectrum" of possible pro- and anti-inflammatory polarized states [47]. The extremes of the macrophage polarization spectrum are often referred to as M1 (proinflammatory) and M2 (anti-inflammatory) phenotypes [48] (Figure 23.1). M1 macrophages are associated with interferon-gamma (IFN-γ) signaling, which promotes microbial killing, Th1 responses, antigen presentation, and proinflammatory cytokine and enzyme release [49]. Conversely, M2 macrophages are poor antigen presenters, immune suppressive, and facilitate tissue remodeling/angiogenesis [49] (Figure 23.1). HO-1 has been associated with an M2 macrophage phenotype and may even directly contribute to skewing the macrophage program [50] (Figure 23.1). TAMs are often compared to an M2 phenotype, and the classification edges have been broadened to encompass "M2-like" phenotypes, which share some, but not all, of the M2 signature.

The ability of macrophages to degrade heme was first acknowledged in 1971 by Pimstone and colleagues [51]. However, it has become clear that HO-1 is not a pan-TAM marker but instead restricted to a refined phenotypic subset of these cells [7,13,22,28,52–54]. A subset of high HO-1-expressing TAMs are marked by the surface expression of fibroblast activation protein-alpha (FAP), a surface protein often associated with fibroblasts [22,27,28]. In a subcutaneous Lewis lung carcinoma (LL2) tumor, FAP$^+$ TAMs accounted for approximately 10% of the total TAM population and were the major tumoral expressor of HO-1 [28]. The FAP$^+$ HO-1$^+$ TAM subset can be derived from a macrophage polarization response to IL-6 in a collagen-rich TME, analogous to the TME found in a cutaneous healing wound [22].

Dendritic cells (DCs), a professional antigen-presenting cell type, have also been demonstrated to express HO-1 in the TME. DC expression of HO-1 compromises their immune function [23,55], which can be attributed directly to the effects of CO on these cells [23]. There are also reports from several groups that an immunomodulatory CD4$^+$ CD25$^+$ Foxp3$^+$ Treg population plays an important role in immune suppression in the TME and can constitutively express HO-1 [56,57]. Foxp3 is a transcription factor that is fundamental to the immunomodulatory program of the CD4$^+$ CD25$^+$ Treg subset [56], and HO-1 may be directly linked to the Foxp3-regulated gene expression program [58]. This observation is

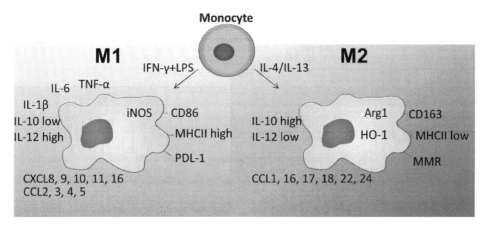

Figure 23.1 The M1/M2 polarization extremes of macrophages. An overview of characteristic gene expression associated with M1 and M2 macrophage polarization. Arg1, arginase-1; iNOS, inducible nitric oxide synthase; LPS, lipopolysaccharide; MMR, macrophage mannose receptor (CD206); TNF-α, tumor necrosis factor-alpha.

human Treg specific as naturally activated murine CD4$^+$ CD25$^+$ Tregs isolated from the spleen and lymph nodes do not express higher HO-1 levels than CD4$^+$ CD25$^-$ T-cells [59].

The diversity of cell types reported to express HO-1, and their prevalence in the TME, are dependent on a number of TME-specific variables, some of which will be discussed further in the "TME Cues for HO-1 Induction" section. However, as CO can diffuse outside of a cell to influence non-HO-1-expressing cells [60], CO's influence on the TME is not necessarily restricted to any given subset and can be further reaching. The widespread expression of HO-1 in cancer also highlights the fundamental relationship between the enzyme and the disease.

TME Cues for HO-1 Induction

As highlighted earlier, there are instances where tumor cells [11,36] or Tregs [56,57] can constitutively express HO-1; however, generally HO-1 is expressed at a low level in most mammalian cells and highly upregulated in response to signals received within the TME. The promoter of the *HMOX1* gene is sensitive to a variety of signaling pathways associated with stress stimuli, which are present in the TME, including heme [61], reactive oxygen species (ROS) [62], hypoxia [63], inflammatory cytokines [22], and prostaglandins [64] (Figure 23.2). As such, it is not surprising that HO-1 is expressed in such a wide variety of cancers.

The promoter region of the *HMOX1* gene contains several transcription factor binding elements that allow *HMOX1* to be transcribed in response to a range of biological pathways and external stimuli. Of these, Nrf2 has been demonstrated to be a major transcription factor for the *HMOX1* gene binding the *ARE* site in the promoter region. Nrf2 is a vital transcription factor controlling over 200 genes associated with the antioxidant response, for which HO-1 plays a significant role [65]. Nrf2 has been considered as a therapeutic target in cancer, and has even been proposed as an oncogene [65]. Nrf2-regulated *HMOX1* gene transcription is, however, tightly regulated. Under basal redox conditions, a repressor protein Kelch-like ECH-associated protein 1 (Keap1) binds to Nrf2 preventing its nuclear translocation and activity, while promoting the recruitment of Cullin-3-dependent E3 ubiquitin ligase, which promotes its ubiquitination and proteasomal degradation [66,67]. However, where there is oxidative stress and high ROS, as is found in the TME, Keap1 undergoes oxidation of its sulfhydryl groups, leading to the release, stabilization, and nuclear translocation of Nrf2 [68]. In the nucleus, Nrf2 heterodimerizes with small Maf (sMaf) proteins to mediate *HMOX1* transcription through binding the *ARE* site in the promoter (Figure 23.2). However, Nrf2-directed gene transcription can also be negatively regulated by Bach1, a transcriptional repressor and heme-binding protein, which also heterodimerizes with sMaf proteins and competes with Nrf2 for binding the *ARE*, restricting access to the site [69] (Figure 23.2). The TME is a site of considerable cell death, with tumor cell growth outstripping nutrient and oxygen supply from inefficient vasculature, which results in the release of heme-containing proteins. Heme is capable of binding Bach1 causing its release from the *ARE* site [70], which in turn then favors Nrf2 access to drive the expression of *HMOX1* [71] (Figure 23.2). This regulation allows for an exquisite level of control of *HMOX1* gene expression in the TME in response to

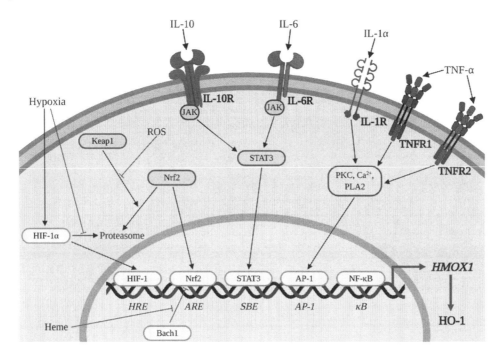

Figure 23.2 The diverse range of signals that induce HMOX1 expression. HMOX1 mRNA expression is induced by a range of molecular and physical signals. In normal conditions, Nrf2 is inhibited by Keap1 that prevents nuclear translocation and promotes proteasomal degradation of Nrf2. ROS generated by oxidative stress inhibits the interaction between Keap1 and Nrf2 in the cytoplasm, allowing nuclear factor erythroid 2-related factor 2 (Nrf2) to translocate to the nucleus and bind to the antioxidant response element (ARE) site of the HMOX1 promoter. In the presence of heme, Bach1 is prevented from inhibiting Nrf2 binding the ARE, allowing HMOX1 expression. Under conditions of hypoxia, HIF-1α is stabilized and escapes proteasomal degradation and translocates into the nucleus where it forms a complex with HIF-1β and binds to the hypoxia-responsive element (HRE) in the HMOX1 promoter to drive gene expression. IL-10 and IL-6 can induce expression via the JAK/STAT3 pathway. IL-1α- and TNF-α-induced HMOX1 expression requires protein kinase C (PKC), Ca^{2+}, and phospholipase A2 (PLA2) signaling via the transcription factor activator protein 1 (AP-1) binding the AP-1 site in the HMOX1 promoter. The transcription factor NF-κB (nuclear factor kappa B) also has a binding element in the HMOX1 promoter, although the key signals under which NF-κB can drive HMOX1 expression in the TME are less well known.
Image was created using the BioRender software.

heme released through necrotic events. However, controlled cell death such as is observed through apoptosis can also induce HO-1 via Nrf2 through the release of sphingosine-1-phosphate (S1P), which binds sphingosine-1-phosphate receptor, a G protein-coupled receptor [50].

Tumors are sites of chronic inflammation with a complex cytokine milieu of both pro- and anti-inflammatory cytokines. Although the cellular response to cytokines relies on the expression of the relevant receptors, a variety of cytokines have been demonstrated to regulate *HMOX1* gene expression. Stimulating murine bone marrow (BM)-derived macrophages with a variety of cytokines commonly found in the TME demonstrates the breadth of cytokine-mediated expression of the *Hmox1* gene (Figure 23.3). IL-6 [22,72] and IL-10 [73] can induce *HMOX1* gene expression via the JAK/STAT3 pathway through the STAT binding element located in the promoter region of the *HMOX1* gene [72,74] (Figures 23.2 and 23.3). It is also likely that IL-4 can regulate HO-1 as part of the "M2" alternative activation program via STAT3 signaling [75] (Figure 23.3).

HO-1 expression in murine FAP⁺ TAMs was demonstrated to be also primarily mediated by IL-6, which directly regulates *Hmox1* expression through STAT3 (Figure 23.2) and indirectly controls FAP expression through an interaction of the macrophage with type I collagen [22]. *HMOX1* expression can also be regulated by proinflammatory cytokines, allowing HO-1 to play a role in the resolution phase of the inflammatory response, which is perpetuated in chronic inflammatory diseases such as cancer. Proinflammatory IL-1α and TNF-α induce *HMOX1* expression through activating PKC, calcium (Ca^{2+}) signaling, and PLA2 [76], via the transcription factor AP-1 binding the *AP-1* binding site in the *HMOX1* promoter [77] (Figure 23.2). NF-κB signaling is also

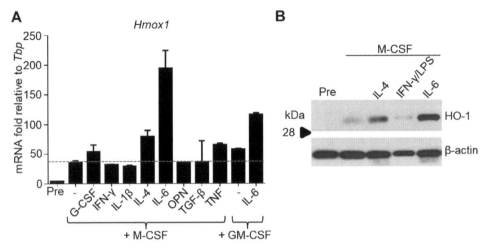

Figure 23.3 TME cytokine regulation of HO-1 in macrophages. (A) qRT-PCR analysis of Hmox1 mRNA expression relative to the housekeeping gene Tbp in murine bone marrow-derived (BM) cells before (Pre) and after macrophage differentiation in the presence of indicated cytokines at 50 ng/mL for 72 h (n = 2 wells). (B) Western blot for HO-1 and the loading control β-actin in murine BM cells before (Pre) and after 72h culture in the presence of 50 ng/ml M-CSF with/without 50 ng/mL IL-4 (M2), IFN-γ + LPS (M1), or IL-6. Bar charts are presented as mean + SEM. G-CSF, granulocyte-colony stimulating factor; GM-CSF, granulocyte-macrophage CSF; M-CSF, macrophage CSF; OPN, osteopontin; TGF-β, transforming growth factor-beta. Data are representative of multiple experiments. Panel (B) is adapted from ref. [22].
Figure 04, p 07 / Springer Nature / CC BY 4.0.

an important pathway in tumor progression [78] and the *HMOX1* promoter contains an NF-κB binding element [79]. However, IFN-γ and LPS both signal through NF-κB and do not regulate HO-1 expression, at least in murine macrophages (Figure 23.3). This may indicate a cell type-, species- or context-dependent utility of the NF-κB binding element in the *HMOX1* promoter.

Hypoxia is also an important physiological regulator of HO-1, and the *HMOX1* promoter contains an *HRE* to enable this. Hypoxia is a common characteristic of most solid TMEs and, as such, could represent a major driving signal for HO-1 expression. Hypoxia stabilizes the transcription factor HIF-1α in the cell cytoplasm allowing it to escape proteasomal degradation and permits its translocation to the nucleus where it complexes with HIF-1β to initiate expression from the *HMOX1* gene [63]. Hypoxia can also influence cellular relocation of HO-1, increasing its mitochondrial localization that results in mitochondrial dysfunction with inhibition of the heme-containing terminal oxidase cytochrome *c* oxidase, increased ROS, and promotion of mitochondrial autophagy [80]. Furthermore, hypoxia acts as a signal for proteolytic cleavage of HO-1 to its truncated catalytically inactive form and subsequent chromosomal maintenance-1-mediated nuclear–cytoplasmic shuttling [24]. Nuclear HO-1 has transcriptional activity and has been demonstrated to downregulate NF-κB and SP-1 DNA binding activity, and upregulate oxidant-responsive transcription factors, such as AP-1, AP-2, Brn-3, and core-binding factor [24]. Overexpression of nuclear HO-1 was also demonstrated to increase the promoter activity of the *HMOX1* gene itself, creating a positive feedback loop of HO-1 expression [24]. However, truncation and nuclear localization of HO-1 is specifically associated with malignant tumor cells rather than healthy stromal cells in the TME [15]. Although the observation has now been validated in multiple studies [64], the *in vivo* biological importance of nuclear HO-1 remains an important question to be fully addressed.

Despite the broad range of signals that can induce *HMOX1* gene expression *in vitro*, it is surprising, and somewhat unexpected, that HO-1 expression is restricted to only a small subset of cells when analyzed within the TME (Figure 23.4). As HO-1 has the potential to be expressed in a variety of cells exposed to the appropriate stimuli, it is important to elucidate the key signals within the TME, which control the expression of the enzyme.

CO and Cell Fate in the TME

HO-1 is a well-established cytoprotective protein; however, the role of CO in dictating cell fate provides an interesting dichotomy in the context of the TME. CO modulates the activity of a variety of kinases in the cell, in particular p38 MAPK, which can prevent apoptosis in some stromal cell types such as

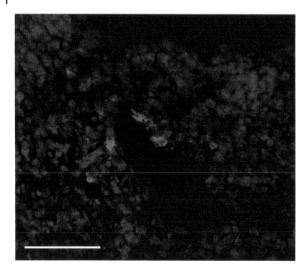

Figure 23.4 HO-1 expression in the TME. Representative image of a frozen section of a murine orthotopic 4T1 tumor grown in the mammary gland of syngeneic BALB/c mice stained with DAPI (nuclei; blue) and an antibody for HO-1 (R&D Systems) that was detected using donkey anti-goat Alexa Fluor 568 (red) (Thermo Fisher Scientific). The scale bar represents 50 μm. Unpublished data.

endothelial cells [60], but conversely inhibit proliferation and enhance cell death in malignant tumor cells [15,81]. These divergent outcomes in response to CO are most likely to arise from differences in the underlying metabolism of the respective cell types [82]. Treatment of tumor cells with CO results in their metabolic exhaustion and apoptosis as a result of an increase in oxygen consumption, which is associated with the generation of mitochondrial ROS [15]. This is not a tumor cell-specific response per se as the same effect can be observed in healthy macrophages, but this conversely results in their survival [83]. CO can protect healthy cells from cell death through promoting DNA repair processes [60,84]. These divergent responses from CO exposure suggest that healthy cells may be more resilient than tumor cells to changes in the metabolic demands placed on them. Exposing mice bearing prostate or lung cancers daily to 1 h of CO (250 ppm) resulted in a control of tumor growth [15]. As such, therapeutically supplying exogenous CO to the tumor provides an opportunity to protect healthy tissues while targeting the malignant cells.

CO and Suppression of the Antitumor Immune Response

The dawn of "immunotherapy," the concept of harnessing the patients' own immune system to target and attack malignant tumor cells, has provided an exciting new era for cancer medicine and transformed the way we look to approach the treatment of the disease. However, it is apparent that the TME is not conducive to permitting efficient immune reactions, with an overtly immune-suppressive and tissue repair-like stromal reaction [31], which protects the tumor cells from immune eradication. A family of regulatory proteins that act to suppress immune responses, and T-cell activation/activity, are collectively referred to as "immune checkpoints." These "checkpoints" are exploited in cancer to facilitate immune evasion and have provided therapeutic opportunities for intervention as a means to "release the breaks" and boost the antitumor immune response [85]. Therapeutically blocking the signaling of immune checkpoint molecules PD-1 and CTLA-4 has delivered unprecedented clinical responses in cancer patients [85]. Combining these therapies also improves the overall clinical response compared to the monotherapies [86]. The success of combination therapies highlights the redundancy within this family of receptors. However, a significant proportion of patients still do not respond to the current therapies and other targets may still exist [86,87]. Evidence suggests that HO-1 may in itself represent an innate immune checkpoint in the TME [7]. In a side-by-side comparison using the cytotoxic chemotherapeutic (CCT) 5-fluorouracil (5-FU) to elicit an antitumor immune response (discussed in the "Cytotoxic Chemotherapies" section), pharmacological inhibition of HO activity using tin mesoporphyrin (SnMP) displayed superior antitumor efficacy compared to neutralizing anti-PD-1 antibodies in a spontaneous murine model of breast cancer (MMTV-PyMT) [7]. This observation highlighted that HO-1, in some TMEs, could be more hierarchically important as an immune checkpoint than the clinically targeted PD-1.

CO and the T-Cell Response in Cancer

HO-1 activity has been demonstrated to suppress T-cell effector function [7,28] (Figure 23.5). In mice bearing subcutaneous LL2 tumors that were

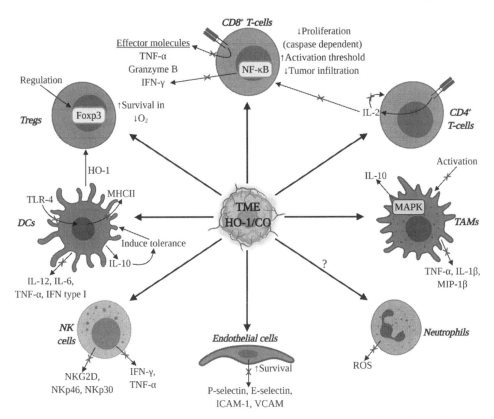

Figure 23.5 Effects of CO on the stroma. HO-1 and HO-1-derived CO in the TME can affect a wide range of immune cell types to suppress the antitumor immune response. CO is able to suppress NF-κB signaling within CD8$^+$ T-cells, which is vital for IFN-γ secretion. CO can also suppress CD8$^+$ T-cell effector functions mediated via TCR engagement. HO-1 has also been implicated in suppressing CD8$^+$ T-cell proliferation, tumor infiltration, and increasing the activation threshold. CO can also block TCR-dependent IL-2 production in CD4$^+$ T-cells, influencing the proliferation of both CD8$^+$ and CD4$^+$ T-cells. Additionally, HO-1 has been implicated as part of the Foxp3 regulatory program in Tregs and CO itself has been demonstrated to improve Treg survival in conditions of hypoxia. CO allows DCs to secrete IL-10, which further helps maintain these cells in a tolerogenic state. CO can also inhibit DCs from producing ROS and several proinflammatory cytokines and can suppress Toll-like receptor-4 (TLR-4) signaling and MHCII expression. HO-1 itself can be directly expressed by DCs and this can influence the regulatory activity of Tregs. In TAMs, CO inhibits activation as well as suppresses the release of proinflammatory cytokines while eliciting an increase in IL-10 production. CO also influences endothelial cells and can promote survival and suppress the expression of adhesion molecules. Additionally, HO-1 has the potential to inhibit ROS expression from heme-activated neutrophils and decrease the cytotoxicity of natural killer (NK) cells through suppressing their ability to secrete IFN-γ and TNF-α, and express activatory cell receptors.
Image was created using the BioRender software.

rendered immunogenic through the expression of ovalbumin protein (LL2/OVA), pharmacologically targeting HO activity using SnMP resulted in immunological control of tumor growth [27].

However, importantly, SnMP did not affect tumor growth in the absence of a robust antitumor immune response in LL2 tumors [28], as well as other tumors [7,22]. Such results highlight the need to consider the antigenicity of the tumor, and quality of the antitumor immune response, in pharmacologically targeting HO-1 as a therapeutic strategy. In MMTV-PyMT tumors, pharmacological inhibition of HO activity using SnMP, or genetic inactivation of the *Hmox1* gene in the myeloid compartment (the major tumoral source of HO-1), improved the proportion of cytotoxic T-cells capable of producing the effector molecules granzyme B, IFN-γ, and TNF-α [7] (Figure 23.5). In agreement with this, an intradermally injected OVA-expressing murine thymic lymphoma (EG7-OVA) model reaffirmed the role of HO-1-expressing TAMs in the maintenance of the immunesuppressive TME [52], where myeloid-specific genetic inactivation of the *Hmox1* gene resulted in an increased proliferation, tumor infiltration, and cytotoxic activity of CD8$^+$ T-cells [52]. CO represents an important metabolite of HO activity in modulating the antitumor immune response in the TME and can suppress the ability for CD8$^+$ T-cells to elicit effector functions upon T-cell receptor (TCR) engagement [28]. CO's role in modulating inflammatory pathways such as downregulating STAT1 and upregulating STAT3 [88] and suppressing NF-κB signaling [89,90] can directly compromise the antitumor immune response. NF-κB in particular is vital for CD8⁺ T-cell effector function [91] and exposing activated CD8⁺ T-cells to CO suppresses NF-κB signaling within these cells (Figure 23.6).

The immune-suppressive actions of the immunomodulatory CD4⁺ CD25⁺ Foxp3⁺ Treg subset have also been associated with the HO-1/CO axis (Figure 23.5). CO has been demonstrated to improve Treg survival in hypoxic regions [92] and loss of tumoral HO-1 activity is associated with a loss of Tregs in the TME [35]. Interestingly, HO-1 has been implicated as part of the regulatory program of the transcription factor Foxp3 and is constitutively expressed in human Tregs [58]. In human Tregs, HO-1 may directly contribute to the immunomodulatory capabilities of these cells [58]. However, the importance of HO-1/CO for Treg function remains controversial and other studies have concluded that HO-1 activity is insufficient to account for the immunomodulatory capabilities of Tregs [59,93].

A role of CO in preventing proliferation of T-cell populations is now well established by several groups [93–95]. CO can block CD4⁺ T-cell production of IL-2, a principal cytokine required for T-cell proliferation and cell cycle entry; the mechanism of IL-2 is postulated to be via inhibiting TCR-dependent ERK phosphorylation [95] (Figure 23.5). CO can also inhibit CD8⁺ T-cell proliferation by suppressing the expression and activity of caspase-3 and -8. Such inhibition is partially mediated through upregulation of a cyclin-dependent kinase inhibitor, $p21^{cip}$ [1], a potent inhibitor against cell cycle progression [94].

The role of CO in suppression of T-cell proliferation is not restricted to effector cells and can also influence the expansion of Tregs [93]. Interestingly, CO's ability to suppress T-cell proliferation is lost once TCR signaling has been received and the cell cycle has been initiated [95]. Such results suggest that CO plays a role in increasing the threshold for activation, which may contribute to maintaining immune tolerance. The immune-suppressive effects of CO on T-cell proliferation and function highlight the importance of HO-1 in supporting immune evasion and preventing immunological eradication by the tumor.

CO and Its Modulation of the Broader TME

The wider TME represents a collection of almost all cells of the immune system, exploited to aid tumor progression. CO has several functions, outside of directly influencing T-cell responses, which contribute to the overall immune-suppressive nature of the TME. DCs, a professional antigen-presenting cell, are vital to processing tumor antigens and orchestrating T-cell antitumor immune responses [96]. As described earlier, HO-1 expression by DCs has been demonstrated to directly influence their immune function [23,55], which can be attributed directly to the effect of CO on these cells [23]. CO can directly influence the antigen-presenting capabilities of DCs through downregulating their MHCII expression [23,97,98] (Figure 23.5). HO-1-induced IL-10 expression in the TME also plays a role in blocking DC maturation and maintaining these cells in a tolerogenic state [14]. CO also influences the effector functions of

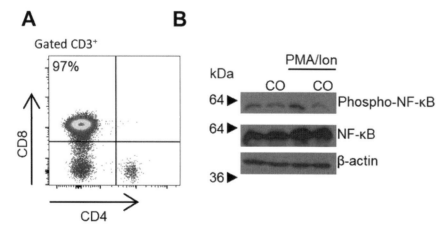

Figure 23.6 CO modulates NF-κB signaling in CD8⁺ T-cells. (A) Representative purity of MACS bead-purified spleen-derived effector CD8⁺ T-cells cultured for 72 h on anti-CD3/CD28-coated plates with 10 ng/mL IL-2, stained for the markers indicated, and assessed using flow cytometry. Dead cells were excluded as negative staining for 7-aminoactinomycin D (7AAD) (not shown). (B) Western blot for total and phosphorylated NF-κB in lysates from effector CD8⁺ T-cells in panel (A), pretreated for 3 h with 250 ppm CO in an enclosed chamber or ambient conditions, with or without PMA/ionomycin stimulation as indicated. CO was maintained in the culture during the stimulation through the inclusion of 100 μM of CORM-2 in the CO group. Each lane represents CD8⁺ T-cell preparations pooled from four mice. Unpublished data.

DCs through modulating their cytokine secretion and inhibiting the production of the pro-inflammatory cytokines IL-12, IL-6, TNF-α, and IFN type I, while concurrently maintaining their ability to secrete the immune-suppressive cytokine IL-10 [23,55]. TLR-4 signaling is an important maturation signal for DCs. TLR-4 is activated in the TME through endogenous TLR-4 ligands associated with stressed and dying cells, known as damage-associated molecular patterns, such as high-mobility group box 1 (HMGB1) protein and heat shock protein 90 [99]. CO can act to suppress TLR-4 signaling through regulating the interaction of the receptor with caveolin-1, the principal structural protein of the plasmalemmal caveolae, which regulates inflammatory signals from the cell membrane [100,101]. The role of CO in suppressing DC maturation and activity has the potential to compromise the ability of these cells to support the generation of antitumor immune responses.

TAM polarization can also be directly modulated by CO, leading to a skewed anti-inflammatory phenotype in these cells [102] (Figure 23.5). CO inhibition of TLR signaling (described earlier), iNOS expression, and the release of HMGB1 may also lead to inhibition of macrophage activation [103,104]. *In vitro* studies using the macrophage cell line RAW 264.7 demonstrated that CO could suppress macrophage expression of proinflammatory TNF-α, IL-1β, and macrophage inflammatory protein 1-β, while increasing IL-10 production that was dependent on MAPK signaling [105] (Figure 23.5). Interestingly, the relationship between TAMs and HO-1 does not always promote tumor progression. In an A549 lung carcinoma model, external CO exposure was capable of downregulating HO-1 protein expression and modulated the infiltration and polarization of macrophages to an inflammatory antitumor phenotype, which resulted in the control of tumor growth [82]. The presence of inflammatory CD86⁺ myeloid cells was required for tumor control, and when anti-CD86 antibodies were administered (to deplete the cells), the inhibitory effects of CO were reversed [82].

CO has also been demonstrated to suppress endothelial cell activation, which may modulate the ability of immune cells to attach and extravasate from the blood to the TME. *In vitro* exposure of endothelial cells to CO, using a CO-releasing molecule (CORM), suppressed the proinflammatory cytokine-mediated expression of the adhesion molecules P- and E-selectin, ICAM-1, and VCAM-1, which reduced the ability of neutrophils to attach to the endothelial cells [106] (Figure 23.5). More broadly, it is possible that HO-1 activity may also inhibit heme-mediated neutrophil activation, which is associated with their ability to generate ROS [107,108]. HO-1 can also decrease the cytotoxicity of NK cells through suppressing their ability to secrete IFN-γ and TNF-α, and express activatory NK cell receptors such as NKG2D, NKp46, and NKp30 [109] (Figure 23.5). Also, HO-1 activity protects vemurafenib-treated melanoma cells from NK cell-mediated killing [110]. However, the role of CO in neutrophils and NK cells is relatively underdeveloped and the extent of the influence of CO on their protumoral function remains to be fully explored.

The diverse roles of CO in modulating T-cell responses and in maintaining an immune-suppressive TME underpin the innate role of HO-1 in tissue protection and tolerance, which are exploited by cancer to facilitate disease progression.

CO and Resistance to Anticancer Therapeutics

Cytotoxic Chemotherapies

CCTs have been utilized for the treatment of cancer for nearly 70 years. These drugs were first administered for their ability to target various aspects of the cell cycle, including DNA replication and the inhibition of the dynamic processes of mitosis. However, these drugs rarely represent cures for late-stage cancers as tumor cells adapt to circumvent the cytotoxic nature of these compounds [111]. HO-1 has also been demonstrated to be induced in response to exposure to CCTs [112,113] and HO-1 expression can provide a resistance mechanism to a variety of CCTs, including etoposide, doxorubicin, gemcitabine, and cisplatin [18,113–115]. The role of HO-1 in drug resistance is not just restricted to CCTs and may include other drugs such as sapatinib, a pan-HER family kinase inhibitor [116].

As our knowledge of the interplay between the immune system and cancer has deepened, it has also highlighted an unexpected, but important, role for CCTs in stimulating antitumor immune responses. The ability for CCTs to modulate antitumor responses can be through the release of tumor antigens, infiltration of immune cells, and the preferential targeting of immune-suppressive stromal populations [99], which may underpin a significant proportion of the clinical efficacy of these compounds [99,117–120]. As such, the immune-suppressive effects of CO (Section "CO and Suppression of the Antitumor Immune Response") have the ability to compromise the overall efficacy of CCTs from an immunological perspective. In MMTV-PyMT tumors, pharmacological targeting of HO activity using SnMP, or genetic inactivation of the *Hmox1* gene, potentiated the immune-stimulating effects of 5-FU, permitting a CCT-elicited CD8$^+$ T-cell response to infiltrate and immunologically control tumor growth [7]. Inhibition of HO-1 using SnMP did not affect the number of tumor infiltrating T-cells, but did improve the proportion of cytotoxic T-cells capable of producing effector molecules such as granzyme B, IFN-γ, and TNF-α [7]. However, our ability to predict these responses is poorly understood and likely resides in the intrinsic response of the tumor cells to the CCT [99], and the broader TME, for which CO could contribute. Short-term starvation can sensitize cancer cells to CCTs while concomitantly protecting normal cells [121,122], through a mechanism involving insulin-like growth factor-1 signaling [121]. Tumor-bearing mice on a fasting-mimicking diet resulted in a lower tumor cell expression of HO-1, which rendered the cells more sensitive to doxorubicin and cyclophosphamide in the 4T1 model of breast cancer, and to doxorubicin in the B16 melanoma model. Such sensitization resulted in CD8$^+$ T-cell-mediated tumor control that was associated with a loss of Tregs [35].

CO can also directly influence the intrinsic tumor cell sensitivity to CCTs [15]. CCTs can exacerbate the effect of CO targeting mitochondrial function in tumor cells resulting in cell death (discussed in the "CO and Cell Fate in the TME" section). Exposure of a prostate cancer cell line (PC3) to a low concentration of CO rendered the cells significantly more sensitive to DNA-damaging CCTs camptothecin and doxorubicin, but not non-DNA-damaging agents TNF-α or cycloheximide [15]. Tumor control observed in mice bearing subcutaneous PC3 tumors exposed to low-dose CO (250 ppm, 1 h/day) was further augmented by the coadministration of doxorubicin. Such effects were partially attributed to a mitochondrion-dependent increase in ROS [15]. Interestingly, CO conversely protects healthy stromal cells from the effects of CCTs through reducing their oxygen consumption [15] and upregulating their DNA repair pathways [84].

Other Treatment Modalities

More than 50% of cancer patients receive radiotherapy at some point within their treatment, and there is a significant research interest in resolving the biological resistance mechanisms, which suppress the effects of this treatment. Radioresistance has broadly been associated with increased cellular functions associated with the stress response and DNA repair capabilities [123]. The generation of ROS is a key event that underpins the anticancer effects of ionizing radiation [124]. Radiotherapy can induce HO-1 expression, and activity of the enzyme can act as a resistance pathway for this therapy [125]. HO-1 inhibition using zinc protoporphyrin (ZnPP) enhanced the radiosensitivity of human non-small cell cancer cell line A529 [125,126]. CO could also reduce radiation-induced bystander effects (RIBEs), which can lead to secondary cancers. In Chinese hamster ovary cells, CO appeared to suppress RIBE-associated proliferation and chromosomal abnormalities [127]. This effect could be attributed to CO's ability to induce DNA damage repair pathways facilitating repair of double-stranded DNA breaks [84].

The therapeutic effects of photodynamic therapy (PDT) can also be influenced by HO-1/CO. PDT utilizes an activating laser light source, a photosensitizing agent, and molecular oxygen to generate ROS, which results in tumor destruction. Using 5-aminolevulinic acid-based PDT, it was demonstrated that pharmacological inhibition of HO activity using tin protoporphyrin (SnPP) resulted in a significant increase in the sensitivity of melanoma cells (WM541) to the therapy [128]. ZnPP also increased the sensitivity of colon adenocarcinoma cells (C26) and ovarian carcinoma (MDAH 2774) cells to PDT using photosensitizer verteporfin [129]. Though the specific contribution of CO to these observations requires further investigation, these studies provide compelling evidence for the possibility of targeting HO activity to improve the efficacy of radiotherapy and PDT.

CO in Vascular Function and Metastasis

CO has a well-established role in influencing vascular function and maintaining homeostasis [130]. The wide-ranging effects of CO on the vasculature also mean that CO has the capability to influence the TME and disease progression in dramatic ways. Importantly, CO has been demonstrated to influence angiogenesis and metastasis, both crucial traits that determine the ultimate success of cancer progression [18] (Figure 23.7).

CO in Vascular Function

CO has an important role in vascular function through exerting its effects on both vascular smooth muscle cells (VSMCs) and endothelial cells [131]. CO has also been demonstrated to influence vascular tone through modulating vasodilation and

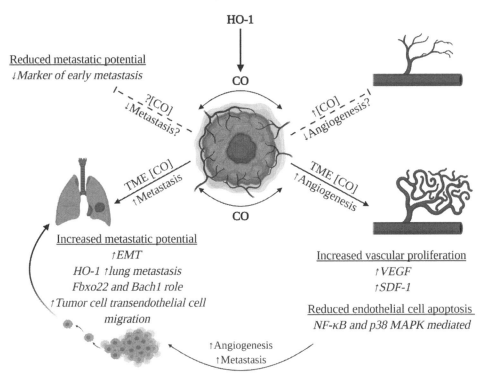

Figure 23.7 The effects of CO on angiogenesis and metastasis within the TME. CO is highly influential in both angiogenesis and metastasis in the TME. CO has been widely reported to promote angiogenesis. One mechanism for this is through increasing vascular proliferation by inducing the expression of vascular endothelial growth factor (VEGF) and stromal cell-derived factor-1 (SDF-1) by the endothelium. Additionally, CO can reduce endothelial cell apoptosis via NF-κB- and p38 MAPK-dependent mechanisms. Conversely, some in vitro studies have demonstrated high concentrations of CO to suppress VEGF-induced endothelial cell proliferation in the TME and this could potentially result in reduced angiogenesis. The success of metastasis is directly related to the success of angiogenesis and therefore the role of CO in angiogenesis indirectly influences this phenotypic gain. Alongside this, CO also influences metastasis directly. HO-1 has been demonstrated to be critical for the epithelial–mesenchymal transition (EMT), and CO specifically has also been demonstrated to facilitate transendothelial migration and can be required for successful metastasis to the lung, a trait that may be Fbxo22 and Bach1 dependent. However, in another experimental system CO appeared to suppress a marker of early metastasis, further highlighting the diverse roles of CO in the TME. Image was created using the BioRender software.

vasoconstriction and thus also the blood pressure of the system, the regulation of which is essential to maintaining homeostasis [131]. Experiments using CORM-2 and CORM-A1 have validated a role for CO in vasodilation in a model utilizing aortic rings [132,133]. In this model, the vasodilatory properties of CO have been attributed to the activation of soluble guanylyl cyclase (sGC) and the subsequent generation of cyclic guanosine monophosphate (cGMP) in VSMCs [134]. This finding of CO-dependent cGMP elevation in the vasculature has also been seen in models examining rat aortic and pulmonary artery smooth muscle cells [135]. Alongside the CO-dependent cGMP-mediated vasodilatory effects, aortic SMC-generated CO can also cause cGMP-dependent downregulation of the potent vasoconstrictor endothelin 1, as well as downregulation of the mitogen platelet-derived growth factor B in endothelial cells, thereby indirectly inhibiting SMC proliferation [135]. In regard to the vasodilatory findings, some studies have suggested this observation may be due to CO's effect on potassium (K^+) channel activity. A study examining porcine cerebral arteries, for example, demonstrated that CO-dependent dilation was abolished by K^+ channel blockers, demonstrating a dependence on this mechanism [136]. Interestingly, one experiment using blockers of ATP-dependent K^+ channels and sGC activity demonstrated that the vasoactive properties of CO are likely partly mediated by both mechanisms [137].

Conversely, there is evidence that CO can also facilitate vasoconstriction through inhibiting the endothelium-dependent generation of NO [138]. CO may promote vasodilation by acting on VSMCs while simultaneously exerting a vasoconstrictive effect by inhibiting NO production [138]. The effect of CO on influencing vascular tone, and therefore function, all comes together to demonstrate the important role of CO in maintaining homeostasis. In the TME where homeostasis is no longer maintained, and neoangiogenesis is active, CO is able to affect the vasculature in different ways, and in particular can play a key role in influencing the success of angiogenesis.

CO in Angiogenesis

Angiogenesis describes the process of new blood vessel formation from pre-existing vasculature. It is an extremely critical process in tumor growth and many known external factors, including CO, have been demonstrated to influence this process [131]. Many studies have demonstrated CO to be proangiogenic [64,131,139,140] (Figure 23.7). An essential criterion for this phenotype is the ability to enhance the proliferation of endothelial cells, and a critical regulator of this process is VEGF. Inhibition of HO activity using SnPP in VSMCs prevented VEGF production, and overexpression of HO-1 resulted in an induction of VEGF protein expression [141]. CO has been demonstrated directly to increase VEGF expression in both VSMCs and endothelial cells [64,141,142]. More recent *in vitro* studies have alluded to this mechanism further and indicated that VEGF upregulation via CO is HIF-1α dependent [143,144]. Additionally, this effect may be further enhanced via an endogenous positive feedback loop whereby VEGF itself can induce HO-1 in endothelial cells and therefore result in subsequent CO production [145]. CO can also influence SDF-1-dependent angiogenesis, whereby SDF-1 is highly influential in the migration, recruitment, and retention of endothelial progenitor cells [140] (Figure 23.7). Capillary sprouting in response to SDF-1 has been demonstrated to require the presence of HO-1/CO [140]. Alongside these pro-proliferative roles of CO on vascular forming cells, CO has also been demonstrated to prevent endothelial cell apoptosis and therefore indirectly influence angiogenesis, which is reversed by SnMP [60,105]. This effect on endothelial cells has been demonstrated to be both NF-κB and p38 MAPK dependent, further demonstrating the far-ranging actions of CO in the TME [60,146]. The above findings support experiments examining models of pancreatic cancer, urothelial cancer, and melanoma, which demonstrate HO-1 to be critical for tumor-promoting angiogenesis [147–149].

Many studies outline the proangiogenic role of CO. However, some studies have suggested CO to have an inhibitory role in such a process. A study examining the prostate cancer cell line PC3 demonstrated that HO-1 overexpression downregulated the expression of proangiogenic factors such as VEGFA and HIF-1α [150], which was demonstrated to be CO mediated [81]. Also, therapeutically targeting HO-1 activity using SnMP, as a single agent, has shown to have no effect on tumor growth in a variety of tumor models [7,22,28]. Furthermore, SnMP did not control tumor growth [7] nor display an observable effect on angiogenesis in MMTV-PyMT mice (Figure 23.8).

This suggests that an effect of CO on angiogenesis may be TME dependent. Although SnMP did not

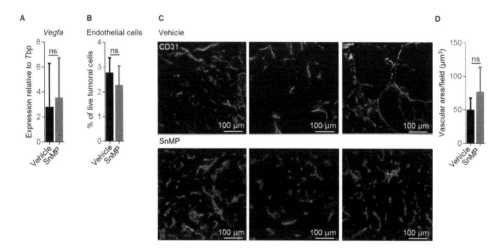

Figure 23.8 HO inhibition and angiogenesis in the TME. (A) qRT-PCR analysis of mRNA expression of *Vegfa* relative to the housekeeping gene *Tbp* in terminal size MMTV-PyMT tumors treated with vehicle (black) or SnMP (red, 25 μmol/kg/day) ($n = 3$). (B) Abundance of live (7AAD$^-$) CD45$^-$ CD31$^+$ endothelial cells quantitated using flow cytometry of enzyme-dispersed 1500 mm^3 tumors from MMTV-PyMT mice treated with vehicle or SnMP ($n = 3$). (C) Representative confocal microscopy images of frozen tumor sections taken from MMTV-PyMT mice treated with vehicle or SnMP (25 μmol/kg/day), stained with antibodies against CD31. (D) Quantification of CD31 staining in panel (C) ($n = 3$). Data are represented as mean + SD. ns, $P > 0.05$. Unpublished data generated by Dr Tamara Muliaditan.

affect tumor growth in a subcutaneous LL2 model, it did permit immunological control of tumor growth in the immunogenic LL2/OVA tumors through a mechanism, which was IFN-γ and TNF-α dependent [28]. T-cell cytokines IFN-γ and TNF-α act synergistically to activate endothelial cells [151] and CO can suppress the endothelial cell response to TNF-α [60]. It is possible that activation of the endothelium contributed to the tumor control observed through promoting coagulation within the tumor vasculature. In agreement with this, SnMP treatment increased the expression of tissue factor, a protein that initiates blood coagulation, on the endothelial cells in the tumor [28]. Such results suggest that HO-1 activity in the TME provides a tissue-protective role for the endothelium, suppressing activation.

Select *in vitro* experiments using human umbilical vein endothelial cells demonstrated that CO could also have an antiangiogenic role and actually suppress VEGF-induced endothelial cell proliferation [152]. However, other studies using the same cells and source of CO have demonstrated opposing findings [142]. A notable difference between these studies was the concentration of CO used, with the lower concentrations resulting in the more widely reported proangiogenic effect. Therefore, this may indicate that the action of CO is highly concentration dependent in regard to angiogenesis. Consequently, in the physiological context of the TME, lower concentrations of CO may act to limit cellular damage, and therefore would confer an angiogenic advantage. All studies together indicate a complex and diverse role for CO in relation to angiogenesis, with variations in the effects seen between different systems and conditions.

CO in Metastasis and in the Metastatic Niche

Metastasis describes the ability of tumor cells to colonize sites distal to that of the primary tumor, and the metastatic niche describes the environment in a secondary organ where tumor cells are able to colonize [153]. The gain of a metastatic phenotype is a crucial step in cancer progression and tumor invasiveness. The success of metastasis is strictly related to the stimulation of angiogenesis and many studies outline the importance CO in this phenotypic gain [18]. In a number of cancer models, overexpression of HO-1 has correlated with an increase in metastatic potential [149,154,155]. SnMP administration *in vivo* also suppressed pulmonary metastasis of orthotopic 4T1 mammary tumors despite not controlling growth of the primary tumor [28]. Interestingly, SnMP did not affect the seeding of intravenously injected 4T1 tumor cells in the lung, suggesting that the mechanism of HO activity promoting metastasis may be associated with promoting tumor cell intravasation into the blood at the primary tumor site. *In vitro* tumor cell transwell transmigration experiments across an endothelial monolayer (as is required for extravasation into the blood) confirmed a role for CO in promoting transendothelial migration of tumor cells, which was independent of a role of CO on endothelial permeability [22]. HO-1 has been demonstrated to play a role in EMT, which is a key step in the metastatic process, where cells enter a more motile phenotype (Figure 23.7). Inhibiting HO-1 activity using ZnPP has been demonstrated to reverse EMT [156]. As such, CO could promote the ability of tumor cells to undergo transendothelial migration through an EMT-related mechanism.

Several studies also highlight a role for CO in the metastatic niche in the lung. Myeloid HO-1 expression has been demonstrated to promote lung metastatic colonization *in vivo*. Through performing subcutaneous or intravenous injection of tumor cells, it was demonstrated that a reduction in HO-1 significantly reduced lung metastases in both cases, thereby supporting the involvement of hematopoietic HO-1 facilitating tumor colonization at this metastatic niche [157]. Additionally, when HO-1 expression was reduced, treatment with CORM-2 could re-permit this effect, thus demonstrating the importance of CO in this system [157]. The critical role of CO was also demonstrated in a model of human glioma where CORM treatment led to a CO-dependent increase in cell proliferation and migration [158]. Additionally, studies using the B16 melanoma cell line injected intravenously demonstrated a similar propensity for lung colonization when HO-1 was overexpressed [149,159]. Furthermore, in a model of lung cancer, it has been proposed that HO-1 enhances metastatic potential in a Fbxo22- and Bach1-dependent manner [160] (Figure 23.7).

As with angiogenesis, the role of CO in metastasis appears complex and somewhat context and TME dependent. In two models of non-small cell lung cancer, overexpression of HO-1 appeared to reduce distal metastasis to the brain [40] and lungs [161]. Additionally, in an orthotopic prostate cancer model, CO appeared to suppress urokinase-type plasminogen activator, a marker of early metastasis and a key marker of invasion [15]. Therefore, although most studies indicate CO as having a prometastatic role, there are clearly examples where this is not the case. Further exploration is needed to understand these contrasting findings. Overall, these data demonstrate the complexity and context-dependent role of CO in the TME.

Conclusions

As discussed in this chapter, CO influences many aspects of the TME. Many of these pathways act to promote progression of the disease, but there are pertinent examples where CO has antitumoral activities. These differences suggest that biological context and the landscape of the TME may play a major role in dictating the overall outcome of CO exposure. The outcome can be dependent on variables such as the immunological backdrop of the tumor, whether T-cells have infiltrated and the quality of the antitumor immune response [7,28], the cell type exposed to CO (with differential responses between malignant cells and healthy stroma) [15], and the local concentration of CO [143,153]. These findings pose important questions for understanding how to most effectively translate these observations into the clinic and to understand how to harness the HO-1/CO axis for antitumor activity. The preclinical evidence demonstrating low-dose CO exposure can deliver antitumor control provides a compelling translational avenue [15]. In the context of boosting antitumor immunity, as an immunotherapy strategy, HO inhibition may be equally attractive [7]. Both approaches would exploit different fundamental properties of CO within the TME. As HO inhibition does not control tumor growth in many preclinical models, translation to the clinic will likely require its use within combination therapies. Optimal combinations may involve CCTs [7] to promote the antitumor immune response [99] or as a sensitizer for radiotherapy [125,126] or PDT [128,129]. Although the focus on the TME has been overwhelmingly on HO-1, the biological role of HO-2 in cancer and the TME requires further investigation. As our understanding of the intricate crosstalk between the tumor cells and the stroma deepens, no doubt further important roles for CO in the context of cancer will emerge in the years ahead.

References

1. Mahonen, Y., Anttinen, M., Vuopio, P., and Tenhunen, R. (1976). Bone marrow: its contribution to heme catabolism. *Ann. Clin. Res.* 8 (Suppl 17): 35–38.
2. Stout, D.L. and Becker, F.F. (1986). Heme enzyme patterns in genetically and chemically induced mouse liver tumors. *Cancer Res.* 46: 2756–2759.
3. Maines, M.D., Trakshel, G.M., and Kutty, R.K. (1986). Characterization of two constitutive forms of rat liver microsomal heme oxygenase. Only one molecular species of the enzyme is inducible. *J. Biol. Chem.* 261: 411–419.
4. Shibahara, S., Muller, R., Taguchi, H., and Yoshida, T. (1985). Cloning and expression of cDNA for rat heme oxygenase. *Proc. Natl. Acad. Sci. U. S. A.* 82: 7865–7869.
5. Keyse, S.M. and Tyrrell, R.M. (1989). Heme oxygenase is the major 32-kDa stress protein induced in human skin fibroblasts by UVA radiation, hydrogen peroxide, and sodium arsenite. *Proc. Natl. Acad. Sci. U. S. A.* 86: 99–103.
6. Miyata, Y., Kanda, S., Mitsunari, K., Asai, A., and Sakai, H. (2014). Heme oxygenase-1 expression is associated with tumor aggressiveness and outcomes in patients with bladder cancer: a correlation with smoking intensity. *Transl. Res.* 164: 468–476.
7. Muliaditan, T., Opzoomer, J.W., Caron, J., Okesola, M., Kosti, P., Lall, S., Van Hemelrijck, M., Dazzi, F., Tutt, A., Grigoriadis, A., Gillett, C.E., Madden, S.F., Burchell, J.M., Kordasti, S., Diebold, S.S., Spicer, J.F., and Arnold, J.N. (2018). Repurposing Tin mesoporphyrin as an immune checkpoint inhibitor shows therapeutic efficacy in preclinical models of cancer. *Clin. Cancer Res.* 24: 1617–1628.
8. Yin, H., Fang, J., Liao, L., Maeda, H., and Su, Q. (2014). Upregulation of heme oxygenase-1 in colorectal cancer patients with increased circulation carbon monoxide levels, potentially affects chemotherapeutic sensitivity. *BMC Cancer* 14: 436.
9. Deininger, M.H., Meyermann, R., Trautmann, K., Duffner, F., Grote, E.H., Wickboldt, J., and Schluesener, H.J. (2000). Heme oxygenase (HO)-1 expressing macrophages/microglial cells accumulate during oligodendroglioma progression. *Brain Res.* 882: 1–8.
10. Gandini, N.A., Fermento, M.E., Salomon, D.G., Blasco, J., Patel, V., Gutkind, J.S., Molinolo, A.A., Facchinetti, M.M., and Curino, A.C. (2012). Nuclear localization of heme oxygenase-1 is associated with tumor progression of head and neck squamous cell carcinomas. *Exp. Mol. Pathol.* 93: 237–245.
11. Mayerhofer, M., Florian, S., Krauth, M.T., Aichberger, K.J., Bilban, M., Marculescu, R., Printz, D., Fritsch, G., Wagner, O., Selzer, E., Sperr, W.R., Valent, P., and Sillaber, C. (2004). Identification of heme oxygenase-1 as a novel BCR/ABL-dependent survival factor in chronic myeloid leukemia. *Cancer Res.* 64: 3148–3154.
12. Degese, M.S., Mendizabal, J.E., Gandini, N.A., Gutkind, J.S., Molinolo, A., Hewitt, S.M., Curino, A.C., Coso, O.A., and Facchinetti, M.M. (2012).

Expression of heme oxygenase-1 in non-small cell lung cancer (NSCLC) and its correlation with clinical data. *Lung Cancer* 77: 168–175.

13. Torisu-Itakura, H., Furue, M., Kuwano, M., and Ono, M. (2000). Co-expression of thymidine phosphorylase and heme oxygenase-1 in macrophages in human malignant vertical growth melanomas. *Jpn. J. Cancer Res.* 91: 906–910.

14. Fest, S., Soldati, R., Christiansen, N.M., Zenclussen, M.L., Kilz, J., Berger, E., Starke, S., Lode, H.N., Engel, C., Zenclussen, A.C., and Christiansen, H. (2016). Targeting of heme oxygenase-1 as a novel immune regulator of neuroblastoma. *Int. J. Cancer* 138: 2030–2042.

15. Wegiel, B., Gallo, D., Csizmadia, E., Harris, C., Belcher, J., Vercellotti, G.M., Penacho, N., Seth, P., Sukhatme, V., Ahmed, A., Pandolfi, P.P., Helczynski, L., Bjartell, A., Persson, J.L., and Otterbein, L.E. (2013). Carbon monoxide expedites metabolic exhaustion to inhibit tumor growth. *Cancer Res.* 73: 7009–7021.

16. Goodman, A.I., Choudhury, M., Da Silva, J.L., Schwartzman, M.L., and Abraham, N.G. (1997). Overexpression of the heme oxygenase gene in renal cell carcinoma. *Proc. Soc. Exp. Biol. Med.* 214: 54–61.

17. Jozkowicz, A., Was, H., and Dulak, J. (2007). Heme oxygenase-1 in tumors: is it a false friend? *Antioxid. Redox Signal.* 9: 2099–2117.

18. Nitti, M., Piras, S., Marinari, U.M., Moretta, L., Pronzato, M.A., and Furfaro, A.L. (2017). HO-1 Induction in cancer progression: A matter of cell adaptation. *Antioxidants* 6: 29.

19. Lin, H.H., Lai, S.C., and Chau, L.Y. (2011). Heme oxygenase-1/carbon monoxide induces vascular endothelial growth factor expression via p38 kinase-dependent activation of Sp1. *J. Biol. Chem.* 286: 3829–3838.

20. Zhang, X., Shan, P., Otterbein, L.E., Alam, J., Flavell, R.A., Davis, R.J., Choi, A.M., and Lee, P.J. (2003). Carbon monoxide inhibition of apoptosis during ischemia-reperfusion lung injury is dependent on the p38 mitogen-activated protein kinase pathway and involves caspase 3. *J. Biol. Chem.* 278: 1248–1258.

21. Li Volti, G., Sacerdoti, D., Sangras, B., Vanella, A., Mezentsev, A., Scapagnini, G., Falck, J.R., and Abraham, N.G. (2005). Carbon monoxide signaling in promoting angiogenesis in human microvessel endothelial cells. *Antioxid. Redox Signal.* 7: 704–710.

22. Muliaditan, T., Caron, J., Okesola, M., Opzoomer, J.W., Kosti, P., Georgouli, M., Gordon, P., Lall, S., Kuzeva, D.M., Pedro, L., Shields, J.D., Gillett, C.E., Diebold, S.S., Sanz-Moreno, V., Ng, T., Hoste, E., and Arnold, J.N. (2018). Macrophages are exploited from an innate wound healing response to facilitate cancer metastasis. *Nat. Commun.* 9: 2951.

23. Remy, S., Blancou, P., Tesson, L., Tardif, V., Brion, R., Royer, P.J., Motterlini, R., Foresti, R., Painchaut, M., Pogu, S., Gregoire, M., Bach, J.M., Anegon, I., and Chauveau, C. (2009). Carbon monoxide inhibits TLR-induced dendritic cell immunogenicity. *J. Immunol.* 182 (4): 1877–1884.

24. Lin, Q., Weis, S., Yang, G., Weng, Y.H., Helston, R., Rish, K., Smith, A., Bordner, J., Polte, T., Gaunitz, F., and Dennery, P.A. (2007). Heme oxygenase-1 protein localizes to the nucleus and activates transcription factors important in oxidative stress. *J. Biol. Chem.* 282: 20621–20633.

25. Schaefer, B. and Behrends, S. (2017). Translocation of heme oxygenase-1 contributes to imatinib resistance in chronic myelogenous leukemia. *Oncotarget* 8: 67406–67421.

26. Sacca, P., Meiss, R., Casas, G., Mazza, O., Calvo, J.C., Navone, N., and Vazquez, E. (2007). Nuclear translocation of haeme oxygenase-1 is associated to prostate cancer. *Br. J. Cancer* 97: 1683–1689.

27. Kraman, M., Bambrough, P.J., Arnold, J.N., Roberts, E.W., Magiera, L., Jones, J.O., Gopinathan, A., Tuveson, D.A., and Fearon, D.T. (2010). Suppression of antitumor immunity by stromal cells expressing fibroblast activation protein-alpha. *Science* 330: 827–830.

28. Arnold, J.N., Magiera, L., Kraman, M., and Fearon, D.T. (2014). Tumoral immune suppression by macrophages expressing fibroblast activation protein-alpha and heme oxygenase-1. *Cancer Immunol. Res.* 2: 121–126.

29. Gonzalez, H., Hagerling, C., and Werb, Z. (2018). Roles of the immune system in cancer: from tumor initiation to metastatic progression. *Genes Dev.* 32: 1267–1284.

30. DeNardo, D.G., Brennan, D.J., Rexhepaj, E., Ruffell, B., Shiao, S.L., Madden, S.F., Gallagher, W.M., Wadhwani, N., Keil, S.D., Junaid, S.A., Rugo, H.S., Hwang, E.S., Jirstrom, K., West, B.L., and Coussens, L.M. (2011). Leukocyte complexity predicts breast cancer survival and functionally regulates response to chemotherapy. *Cancer Discov.* 1: 54–67.

31. Dvorak, H.F. (1986). Tumors: wounds that do not heal. Similarities between tumor stroma generation and wound healing. *N. Engl. J. Med.* 315: 1650–1659.

32. Grochot-Przeczek, A., Lach, R., Mis, J., Skrzypek, K., Gozdecka, M., Sroczynska, P., Dubiel, M., Rutkowski, A., Kozakowska, M., Zagorska, A., Walczynski, J., Was, H., Kotlinowski, J., Drukala, J., Kurowski, K., Kieda, C., Herault, Y., Dulak, J., and Jozkowicz, A. (2009). Heme oxygenase-1 accelerates cutaneous wound healing in mice. *PloS One* 4: e5803.

33. Takagi, T., Naito, Y., Uchiyama, K., Mizuhima, K., Suzuki, T., Horie, R., Hirata, I., Tsuboi, H., and Yoshikawa, T. (2016). Carbon monoxide promotes gastric wound healing in mice via the protein kinase C pathway. *Free Radic. Res.* 50: 1098–1105.

34. Ahanger, A.A., Prawez, S., Kumar, D., Prasad, R., Amarpal, Tandan, S.K., and Kumar, D. (2011). Wound healing activity of carbon monoxide liberated from CO-releasing molecule (CO-RM). *Naunyn Schmiede Arch. Pharmacol.* 384: 93–102.

35. Di Biase, S., Lee, C., Brandhorst, S., Manes, B., Buono, R., Cheng, C.W., Cacciottolo, M., Martin-Montalvo, A., De Cabo, R., Wei, M., Morgan, T.E., and Longo, V.D. (2016). Fasting-mimicking diet reduces HO-1 to Promote T cell-mediated tumor cytotoxicity. *Cancer Cell* 30: 136–146.

36. Vashist, Y.K., Uzungolu, G., Kutup, A., Gebauer, F., Koenig, A., Deutsch, L., Zehler, O., Busch, P., Kalinin, V., Izbicki, J.R., and Yekebas, E.F. (2011). Heme oxygenase-1 germ line GTn promoter polymorphism is an independent prognosticator of tumor recurrence and survival in pancreatic cancer. *J. Surg. Oncol.* 104: 305–311.

37. Exner, M., Minar, E., Wagner, O., and Schillinger, M. (2004). The role of heme oxygenase-1 promoter polymorphisms in human disease. *Free Radic. Biol. Med.* 37: 1097–1104.

38. Taha, H., Skrzypek, K., Guevara, I., Nigisch, A., Mustafa, S., Grochot-Przeczek, A., Ferdek, P., Was, H., Kotlinowski, J., Kozakowska, M., Balcerczyk, A., Muchova, L., Vitek, L., Weigel, G., Dulak, J., and Jozkowicz, A. (2010). Role of heme oxygenase-1 in human endothelial cells: lesson from the promoter allelic variants. *Arterioscler. Thromb. Vasc. Biol.* 30: 1634–1641.

39. Lo, S.S., Lin, S.C., Wu, C.W., Chen, J.H., Yeh, W.I., Chung, M.Y., and Lui, W.Y. (2007). Heme oxygenase-1 gene promoter polymorphism is associated with risk of gastric adenocarcinoma and lymphovascular tumor invasion. *Ann. Surg. Oncol.* 14: 2250–2256.

40. Skrzypek, K., Tertil, M., Golda, S., Ciesla, M., Weglarczyk, K., Collet, G., Guichard, A., Kozakowska, M., Boczkowski, J., Was, H., Gil, T., Kuzdzal, J., Muchova, L., Vitek, L., Loboda, A., Jozkowicz, A., Kieda, C., and Dulak, J. (2013). Interplay between heme oxygenase-1 and miR-378 affects non-small cell lung carcinoma growth, vascularization, and metastasis. *Antioxid. Redox Signal.* 19: 644–660.

41. Stephens, T.C., Currie, G.A., and Peacock, J.H. (1978). Repopulation of gamma-irradiated Lewis lung carcinoma by malignant cells and host macrophage progenitors. *Br. J. Cancer* 38: 573–582.

42. Franklin, R.A., Liao, W., Sarkar, A., Kim, M.V., Bivona, M.R., Liu, K., Pamer, E.G., and Li, M.O. (2014). The cellular and molecular origin of tumor-associated macrophages. *Science* 344: 921–925.

43. Movahedi, K., Laoui, D., Gysemans, C., Baeten, M., Stange, G., Van Den Bossche, J., Mack, M., Pipeleers, D., In't Veld, P., De Baetselier, P., and Van Ginderachter, J.A. (2010). Different tumor microenvironments contain functionally distinct subsets of macrophages derived from Ly6C(high) monocytes. *Cancer Res.* 70: 5728–5739.

44. Stewart, C.C. and Beetham, K.L. (1978). Cytocidal activity and proliferative ability of macrophages infiltrating the EMT6 tumor. *Int. J. Cancer* 22: 152–159.

45. Mahoney, K.H. and Heppner, G.H. (1987). Facs analysis of tumor-associated macrophage replication – differences between metastatic and nonmetastatic murine mammary-tumors. *J. Leukoc. Biol.* 41: 205–211.

46. Bottazzi, B., Erba, E., Nobili, N., Fazioli, F., Rambaldi, A., and Mantovani, A. (1990). A paracrine circuit in the regulation of the proliferation of macrophages infiltrating murine sarcomas. *J. Immunol.* 144: 2409–2412.

47. Mantovani, A., Sozzani, S., Locati, M., Allavena, P., and Sica, A. (2002). Macrophage polarization: tumor-associated macrophages as a paradigm for polarized M2 mononuclear phagocytes. *Trends Immunol.* 23: 549–555.

48. Stein, M., Keshav, S., Harris, N., and Gordon, S. (1992). Interleukin 4 potently enhances murine macrophage mannose receptor activity: a marker of alternative immunologic macrophage activation. *J. Exp. Med.* 176: 287–292.

49. Mills, C.D., Kincaid, K., Alt, J.M., Heilman, M.J., and Hill, A.M. (2000). M-1/M-2 macrophages and the Th1/Th2 paradigm. *J. Immunol.* 164: 6166–6173.

50. Weis, N., Weigert, A., Von Knethen, A., and Brune, B. (2009). Heme oxygenase-1 contributes to an alternative macrophage activation profile induced by apoptotic cell supernatants. *Mol. Biol. Cell* 20: 1280–1288.

51. Pimstone, N.R., Tenhunen, R., Seitz, P.T., Marver, H.S., and Schmid, R. (1971). The enzymatic degradation of hemoglobin to bile pigments by macrophages. *J. Exp. Med.* 133 (6): 1264–1281.

52. Alaluf, E., Vokaer, B., Detavernier, A., Azouz, A., Splittgerber, M., Carrette, A., Boon, L., Libert, F., Soares, M., Le Moine, A., and Goriely, S. (2020). Heme oxygenase-1 orchestrates the immunosuppressive program of tumor-associated macrophages. *JCI Insight* 5: e133929.

53. Nemeth, Z., Li, M., Csizmadia, E., Dome, B., Johansson, M., Persson, J.L., Seth, P., Otterbein, L., and Wegiel, B. (2015). Heme oxygenase-1 in

macrophages controls prostate cancer progression. *Oncotarget* 6: 33675–33688.

54. Sierra-Filardi, E., Vega, M.A., Sanchez-Mateos, P., Corbi, A.L., and Puig-Kroger, A. (2010). Heme Oxygenase-1 expression in M-CSF-polarized M2 macrophages contributes to LPS-induced IL-10 release. *Immunobiology* 215: 788–795.

55. Chauveau, C., Remy, S., Royer, P.J., Hill, M., Tanguy-Royer, S., Hubert, F.X., Tesson, L., Brion, R., Beriou, G., Gregoire, M., Josien, R., Cuturi, M.C., and Anegon, I. (2005). Heme oxygenase-1 expression inhibits dendritic cell maturation and proinflammatory function but conserves IL-10 expression. *Blood* 106: 1694–1702.

56. Hori, S., Nomura, T., and Sakaguchi, S. (2003). Control of regulatory T cell development by the transcription factor Foxp3. *Science* 299 (5609): 1057–1061.

57. Pae, H.O., Oh, G.S., Choi, B.M., Chae, S.C., and Chung, H.T. (2003). Differential expressions of heme oxygenase-1 gene in CD25- and CD25+ subsets of human CD4+ T cells. *Biochem. Biophys. Res. Commun.* 306: 701–705.

58. Choi, B.M., Pae, H.O., Jeong, Y.R., Kim, Y.M., and Chung, H.T. (2005). Critical role of heme oxygenase-1 in Foxp3-mediated immune suppression. *Biochem. Biophys. Res. Commun.* 327: 1066–1071.

59. Zelenay, S., Chora, A., Soares, M.P., and Demengeot, J. (2007). Heme oxygenase-1 is not required for mouse regulatory T cell development and function. *Int. Immunol.* 19: 11–18.

60. Brouard, S., Otterbein, L.E., Anrather, J., Tobiasch, E., Bach, F.H., Choi, A.M., and Soares, M.P. (2000). Carbon monoxide generated by heme oxygenase 1 suppresses endothelial cell apoptosis. *J. Exp. Med.* 192: 1015–1026.

61. Alam, J., Shibahara, S., and Smith, A. (1989). Transcriptional activation of the heme oxygenase gene by heme and cadmium in mouse hepatoma cells. *J. Biol. Chem.* 264: 6371–6375.

62. Ryter, S.W. and Choi, A.M. (2005). Heme oxygenase-1: redox regulation of a stress protein in lung and cell culture models. *Antioxid. Redox Signal.* 7: 80–91.

63. Lee, P.J., Jiang, B.H., Chin, B.Y., Iyer, N.V., Alam, J., Semenza, G.L., and Choi, A.M. (1997). Hypoxia-inducible factor-1 mediates transcriptional activation of the heme oxygenase-1 gene in response to hypoxia. *J. Biol. Chem.* 272: 5375–5381.

64. Jozkowicz, A., Huk, I., Nigisch, A., Weigel, G., Weidinger, F., and Dulak, J. (2002). Effect of prostaglandin-J(2) on VEGF synthesis depends on the induction of heme oxygenase-1. *Antioxid. Redox Signal.* 4: 577–585.

65. Rojo De La Vega, M., Chapman, E., and Zhang, D.D. (2018). NRF2 and the hallmarks of cancer. *Cancer Cell* 34: 21–43.

66. Itoh, K., Wakabayashi, N., Katoh, Y., Ishii, T., Igarashi, K., Engel, J.D., and Yamamoto, M. (1999). Keap1 represses nuclear activation of antioxidant responsive elements by Nrf2 through binding to the amino-terminal Neh2 domain. *Genes Dev.* 13: 76–86.

67. Katoh, Y., Iida, K., Kang, M.I., Kobayashi, A., Mizukami, M., Tong, K.I., McMahon, M., Hayes, J.D., Itoh, K., and Yamamoto, M. (2005). Evolutionary conserved N-terminal domain of Nrf2 is essential for the Keap1-mediated degradation of the protein by proteasome. *Arch. Biochem. Biophys.* 433: 342–350.

68. Dinkova-Kostova, A.T., Holtzclaw, W.D., Cole, R.N., Itoh, K., Wakabayashi, N., Katoh, Y., Yamamoto, M., and Talalay, P. (2002). Direct evidence that sulfhydryl groups of Keap1 are the sensors regulating induction of phase 2 enzymes that protect against carcinogens and oxidants. *Proc. Natl. Acad. Sci. U. S. A.* 99: 11908–11913.

69. Sun, J., Hoshino, H., Takaku, K., Nakajima, O., Muto, A., Suzuki, H., Tashiro, S., Takahashi, S., Shibahara, S., Alam, J., Taketo, M.M., Yamamoto, M., and Igarashi, K. (2002). Hemoprotein Bach1 regulates enhancer availability of heme oxygenase-1 gene. *EMBO J.* 21 (19): 5216–5224.

70. Ogawa, K., Sun, J., Taketani, S., Nakajima, O., Nishitani, C., Sassa, S., Hayashi, N., Yamamoto, M., Shibahara, S., Fujita, H., and Igarashi, K. (2001). Heme mediates derepression of Maf recognition element through direct binding to transcription repressor Bach1. *EMBO J.* 20: 2835–2843.

71. Dhakshinamoorthy, S., Jain, A.K., Bloom, D.A., and Jaiswal, A.K. (2005). Bach1 competes with Nrf2 leading to negative regulation of the antioxidant response element (ARE)-mediated NAD(P)H:quinoneoxidoreductase 1 gene expression and induction in response to antioxidants. *J. Biol. Chem.* 280: 16891–16900.

72. Tron, K., Samoylenko, A., Musikowski, G., Kobe, F., Immenschuh, S., Schaper, F., Ramadori, G., and Kietzmann, T. (2006). Regulation of rat heme oxygenase-1 expression by interleukin-6 via the Jak/STAT pathway in hepatocytes. *J. Hepatol.* 45: 72–80.

73. Lee, T.S. and Chau, L.Y. (2002). Heme oxygenase-1 mediates the anti-inflammatory effect of interleukin-10 in mice. *Nat. Med.* 8: 240–246.

74. Lai, C.F., Ripperger, J., Morella, K.K., Jurlander, J., Hawley, T.S., Carson, W.E., Kordula, T., Caligiuri, M.A., Hawley, R.G., Fey, G.H., and Baumann, H. (1996). Receptors for interleukin (IL)-10 and IL-6-type cytokines use similar signaling mechanisms for

inducing transcription through IL-6 response elements. *J. Biol. Chem.* 271: 13968–13975.

75. Bhattacharjee, A., Shukla, M., Yakubenko, V.P., Mulya, A., Kundu, S., and Cathcart, M.K. (2013). IL-4 and IL-13 employ discrete signaling pathways for target gene expression in alternatively activated monocytes/macrophages. *Free Radic. Biol. Med.* 54: 1–16.

76. Terry, C.M., Clikeman, J.A., Hoidal, J.R., and Callahan, K.S. (1999). TNF-alpha and IL-1alpha induce heme oxygenase-1 via protein kinase C, Ca2+, and phospholipase A2 in endothelial cells. *Am. J. Physiol.* 276: H1493–501.

77. Terry, C.M., Clikeman, J.A., Hoidal, J.R., and Callahan, K.S. (1998). Effect of tumor necrosis factor-alpha and interleukin-1 alpha on heme oxygenase-1 expression in human endothelial cells. *Am. J. Physiol.* 274: H883–91.

78. Xia, Y., Shen, S., and Verma, I.M. (2014). NF-kappaB, an active player in human cancers. *Cancer Immunol. Res.* 2: 823–830.

79. Lavrovsky, Y., Schwartzman, M.L., Levere, R.D., Kappas, A., and Abraham, N.G. (1994). Identification of binding sites for transcription factors NF-kappa B and AP-2 in the promoter region of the human heme oxygenase 1 gene. *Proc. Natl. Acad. Sci. U. S. A.* 91: 5987–5991.

80. Bansal, S., Biswas, G., and Avadhani, N.G. (2014). Mitochondria-targeted heme oxygenase-1 induces oxidative stress and mitochondrial dysfunction in macrophages, kidney fibroblasts and in chronic alcohol hepatotoxicity. *Redox Biol.* 2: 273–283.

81. Vitek, L., Gbelcova, H., Muchova, L., Vanova, K., Zelenka, J., Konickova, R., Suk, J., Zadinova, M., Knejzlik, Z., Ahmad, S., Fujisawa, T., Ahmed, A., and Ruml, T. (2014). Antiproliferative effects of carbon monoxide on pancreatic cancer. *Dig. Liver Dis.* 46: 369–375.

82. Nemeth, Z., Csizmadia, E., Vikstrom, L., Li, M., Bisht, K., Feizi, A., Otterbein, S., Zuckerbraun, B., Costa, D.B., Pandolfi, P.P., Fillinger, J., Dome, B., Otterbein, L.E., and Wegiel, B. (2016). Alterations of tumor microenvironment by carbon monoxide impedes lung cancer growth. *Oncotarget* 7: 23919–23932.

83. Chin, B.Y., Jiang, G., Wegiel, B., Wang, H.J., Macdonald, T., Zhang, X.C., Gallo, D., Cszimadia, E., Bach, F.H., Lee, P.J., and Otterbein, L.E. (2007). Hypoxia-inducible factor 1alpha stabilization by carbon monoxide results in cytoprotective preconditioning. *Proc. Natl. Acad. Sci. U. S. A.* 104: 5109–5114.

84. Otterbein, L.E., Hedblom, A., Harris, C., Csizmadia, E., Gallo, D., and Wegiel, B. (2011). Heme oxygenase-1 and carbon monoxide modulate DNA repair through ataxia-telangiectasia mutated (ATM) protein. *Proc. Natl. Acad. Sci. U. S. A.* 108: 14491–14496.

85. Sharma, P. and Allison, J.P. (2015). The future of immune checkpoint therapy. *Science* 348: 56–61.

86. Wolchok, J.D., Kluger, H., Callahan, M.K., Postow, M.A., Rizvi, N.A., Lesokhin, A.M., Segal, N.H., Ariyan, C.E., Gordon, R.A., Reed, K., Burke, M.M., Caldwell, A., Kronenberg, S.A., Agunwamba, B.U., Zhang, X., Lowy, I., Inzunza, H.D., Feely, W., Horak, C.E., Hong, Q., Korman, A.J., Wigginton, J.M., Gupta, A., and Sznol, M. (2013). Nivolumab plus ipilimumab in advanced melanoma. *N. Engl. J. Med.* 369: 122–133.

87. Kim, J.E., Patel, M.A., Mangraviti, A., Kim, E.S., Theodros, D., Velarde, E., Liu, A., Sankey, E.W., Tam, A., Xu, H., Mathios, D., Jackson, C.M., Harris-Bookman, S., Garzon-Muvdi, T., Sheu, M., Martin, A.M., Tyler, B.M., Tran, P.T., Ye, X., Olivi, A., Taube, J.M., Burger, P.C., Drake, C.G., Brem, H., Pardoll, D.M., and Lim, M. (2017). Combination therapy with anti-PD-1, Anti-TIM-3, and focal radiation results in regression of murine gliomas. *Clin. Cancer Res.* 23: 124–136.

88. Zhang, X., Shan, P., Alam, J., Fu, X.Y., and Lee, P.J. (2005). Carbon monoxide differentially modulates STAT1 and STAT3 and inhibits apoptosis via a phosphatidylinositol 3-kinase/Akt and p38 kinase-dependent STAT3 pathway during anoxia-reoxygenation injury. *J. Biol. Chem.* 280: 8714–8721.

89. Cepinskas, G., Katada, K., Bihari, A., and Potter, R.F. (2008). Carbon monoxide liberated from carbon monoxide-releasing molecule CORM-2 attenuates inflammation in the liver of septic mice. *Am. J. Physiol. Gastrointest. Liver Physiol.* 294: G184–91.

90. Megias, J., Busserolles, J., and Alcaraz, M.J. (2007). The carbon monoxide-releasing molecule CORM-2 inhibits the inflammatory response induced by cytokines in Caco-2 cells. *Br. J. Pharmacol.* 150: 977–986.

91. Clavijo, P.E. and Frauwirth, K.A. (2012). Anergic CD8+ T lymphocytes have impaired NF-kappaB activation with defects in p65 phosphorylation and acetylation. *J. Immunol.* 188: 1213–1221.

92. Dey, M., Chang, A.L., Wainwright, D.A., Ahmed, A.U., Han, Y., Balyasnikova, I.V., and Lesniak, M.S. (2014). Heme oxygenase-1 protects regulatory T cells from hypoxia-induced cellular stress in an experimental mouse brain tumor model. *J. Neuroimmunol.* 266: 33–42.

93. Biburger, M., Theiner, G., Schadle, M., Schuler, G., and Tiegs, G. (2010). Pivotal Advance: heme oxygenase 1 expression by human CD4+ T cells is not sufficient for their development of immunoregulatory capacity. *J. Leukoc. Biol.* 87: 193–202.

94. Song, R., Mahidhara, R.S., Zhou, Z., Hoffman, R.A., Seol, D.W., Flavell, R.A., Billiar, T.R., Otterbein, L.E., and Choi, A.M. (2004). Carbon monoxide inhibits T lymphocyte proliferation via caspase-dependent pathway. *J. Immunol.* 172: 1220–1226.

95. Pae, H.O., Oh, G.S., Choi, B.M., Chae, S.C., Kim, Y.M., Chung, K.R., and Chung, H.T. (2004). Carbon monoxide produced by heme oxygenase-1 suppresses T cell proliferation via inhibition of IL-2 production. *J. Immunol.* 172: 4744–4751.

96. Roberts, E.W., Broz, M.L., Binnewies, M., Headley, M.B., Nelson, A.E., Wolf, D.M., Kaisho, T., Bogunovic, D., Bhardwaj, N., and Krummel, M.F. (2016). Critical role for CD103(+)/CD141(+) dendritic cells bearing CCR7 for tumor antigen trafficking and priming of T cell immunity in melanoma. *Cancer Cell* 30: 324–336.

97. Chora, A.A., Fontoura, P., Cunha, A., Pais, T.F., Cardoso, S., Ho, P.P., Lee, L.Y., Sobel, R.A., Steinman, L., and Soares, M.P. (2007). Heme oxygenase-1 and carbon monoxide suppress autoimmune neuroinflammation. *J. Clin. Invest.* 117: 438–447.

98. Liu, Y., Li, P., Lu, J., Xiong, W., Oger, J., Tetzlaff, W., and Cynader, M. (2008). Bilirubin possesses powerful immunomodulatory activity and suppresses experimental autoimmune encephalomyelitis. *J. Immunol.* 181: 1887–1897.

99. Opzoomer, J.W., Sosnowska, D., Anstee, J.E., Spicer, J.F., and Arnold, J.N. (2019). Cytotoxic chemotherapy as an immune stimulus: A molecular perspective on turning up the immunological heat on cancer. *Front. Immunol.* 10: 1654.

100. Ding, J.L., Li, Y., Zhou, X.Y., Wang, L., Zhou, B., Wang, R., Liu, H.X., and Zhou, Z.G. (2009). Potential role of the TLR4/IRAK-4 signaling pathway in the pathophysiology of acute pancreatitis in mice. *Inflamm. Res.* 58: 783–790.

101. Botto, S., Gustin, J.K., and Moses, A.V. (2017). The heme metabolite carbon monoxide facilitates KSHV infection by inhibiting TLR4 signaling in endothelial cells. *Front. Microbiol.* 8: 568.

102. Lee, T.S., Tsai, H.L., and Chau, L.Y. (2003). Induction of heme oxygenase-1 expression in murine macrophages is essential for the anti-inflammatory effect of low dose 15-deoxy-Delta 12,14-prostaglandin J2. *J. Biol. Chem.* 278: 19325–19330.

103. Nakahira, K., Kim, H.P., Geng, X.H., Nakao, A., Wang, X., Murase, N., Drain, P.F., Wang, X., Sasidhar, M., Nabel, E.G., Takahashi, T., Lukacs, N.W., Ryter, S.W., Morita, K., and Choi, A.M. (2006). Carbon monoxide differentially inhibits TLR signaling pathways by regulating ROS-induced trafficking of TLRs to lipid rafts. *J. Exp. Med.* 203: 2377–2389.

104. Tsoyi, K., Nizamutdinova, I.T., Jang, H.J., Mun, L., Kim, H.J., Seo, H.G., Lee, J.H., and Chang, K.C. (2010). Carbon monoxide from CORM-2 reduces HMGB1 release through regulation of IFN-beta/JAK2/STAT-1/INOS/NO signaling but not COX-2 in TLR-activated macrophages. *Shock* 34: 608–614.

105. Otterbein, L.E., Bach, F.H., Alam, J., Soares, M., Tao Lu, H., Wysk, M., Davis, R.J., Flavell, R.A., and Choi, A.M. (2000). Carbon monoxide has anti-inflammatory effects involving the mitogen-activated protein kinase pathway. *Nat. Med.* 6: 422–428.

106. Nassour, I., Kautza, B., Rubin, M., Escobar, D., Luciano, J., Loughran, P., Gomez, H., Scott, J., Gallo, D., Brumfield, J., Otterbein, L.E., and Zuckerbraun, B.S. (2015). Carbon monoxide protects against hemorrhagic shock and resuscitation-induced microcirculatory injury and tissue injury. *Shock* 43: 166–171.

107. Graca-Souza, A.V., Arruda, M.A., De Freitas, M.S., Barja-Fidalgo, C., and Oliveira, P.L. (2002). Neutrophil activation by heme: implications for inflammatory processes. *Blood* 99: 4160–4165.

108. Porto, B.N., Alves, L.S., Fernandez, P.L., Dutra, T.P., Figueiredo, R.T., Graca-Souza, A.V., and Bozza, M.T. (2007). Heme induces neutrophil migration and reactive oxygen species generation through signaling pathways characteristic of chemotactic receptors. *J. Biol. Chem.* 282 (33): 24430–24436.

109. Gómez-Lomelí, P., Bravo-Cuellar, A., Hernández-Flores, G., Jave-Suárez, L.F., Aguilar-Lemarroy, A., Lerma-Díaz, J.M., Domínguez-Rodríguez, J.R., Sánchez-Reyes, K., and Ortiz-Lazareno, P.C. (2014). Increase of IFN-gamma and TNF-alpha production in CD107a + NK-92 cells co-cultured with cervical cancer cell lines pre-treated with the HO-1 inhibitor. *Cancer Cell Int.* 14: 100.

110. Furfaro, A.L., Ottonello, S., Loi, G., Cossu, I., Piras, S., Spagnolo, F., Queirolo, P., Marinari, U.M., Moretta, L., Pronzato, M.A., Mingari, M.C., Pietra, G., and Nitti, M. (2020). HO-1 downregulation favors BRAF(V600) melanoma cell death induced by Vemurafenib/PLX4032 and increases NK recognition. *Int. J. Cancer* 146: 1950–1962.

111. Raguz, S. and Yague, E. (2008). Resistance to chemotherapy: new treatments and novel insights into an old problem. *Br. J. Cancer* 99: 387–391.

112. Berberat, P.O., Dambrauskas, Z., Gulbinas, A., Giese, T., Giese, N., Kunzli, B., Autschbach, F., Meuer, S., Buchler, M.W., and Friess, H. (2005). Inhibition of heme oxygenase-1 increases responsiveness of pancreatic cancer cells to

anticancer treatment. *Clin. Cancer Res.* 11: 3790–3798.

113. Tan, Q., Wang, H., Hu, Y., Hu, M., Li, X., Aodengqimuge, Ma, Y., Wei, C., and Song, L. (2015). Src/STAT3-dependent heme oxygenase-1 induction mediates chemoresistance of breast cancer cells to doxorubicin by promoting autophagy. *Cancer Sci.* 106: 1023–1032.

114. Tang, Q.F., Sun, J., Yu, H., Shi, X.J., Lv, R., Wei, H.C., and Yin, P.H. (2016). The Zuo Jin Wan formula induces mitochondrial apoptosis of cisplatin-resistant gastric cancer cells via cofilin-1. *Evid. Based Complement Alternat. Med.* 2016: 8203789.

115. Miyake, M., Fujimoto, K., Anai, S., Ohnishi, S., Nakai, Y., Inoue, T., Matsumura, Y., Tomioka, A., Ikeda, T., Okajima, E., Tanaka, N., and Hirao, Y. (2010). Inhibition of heme oxygenase-1 enhances the cytotoxic effect of gemcitabine in urothelial cancer cells. *Anticancer Res.* 30: 2145–2152.

116. Tracey, N., Creedon, H., Kemp, A.J., Culley, J., Muir, M., Klinowska, T., and Brunton, V.G. (2020). HO-1 drives autophagy as a mechanism of resistance against HER2-targeted therapies. *Breast Cancer Res. Treat.* 179: 543–555.

117. Casares, N., Pequignot, M.O., Tesniere, A., Ghiringhelli, F., Roux, S., Chaput, N., Schmitt, E., Hamai, A., Hervas-Stubbs, S., Obeid, M., Coutant, F., Metivier, D., Pichard, E., Aucouturier, P., Pierron, G., Garrido, C., Zitvogel, L., and Kroemer, G. (2005). Caspase-dependent immunogenicity of doxorubicin-induced tumor cell death. *J. Exp. Med.* 202: 1691–1701.

118. Geary, S.M., Lemke, C.D., Lubaroff, D.M., and Salem, A.K. (2013). The combination of a low-dose chemotherapeutic agent, 5-fluorouracil, and an adenoviral tumor vaccine has a synergistic benefit on survival in a tumor model system. *PloS One* 8: e67904.

119. Bracci, L., Schiavoni, G., Sistigu, A., and Belardelli, F. (2014). Immune-based mechanisms of cytotoxic chemotherapy: implications for the design of novel and rationale-based combined treatments against cancer. *Cell Death Differ.* 21: 15–25.

120. Pfirschke, C., Engblom, C., Rickelt, S., Cortez-Retamozo, V., Garris, C., Pucci, F., Yamazaki, T., Poirier-Colame, V., Newton, A., Redouane, Y., Lin, Y.J., Wojtkiewicz, G., Iwamoto, Y., Mino-Kenudson, M., Huynh, T.G., Hynes, R.O., Freeman, G.J., Kroemer, G., Zitvogel, L., Weissleder, R., and Pittet, M.J. (2016). Immunogenic chemotherapy sensitizes tumors to checkpoint blockade therapy. *Immunity* 44 (2): 343–354.

121. Lee, C., Raffaghello, L., Brandhorst, S., Safdie, F.M., Bianchi, G., Martin-Montalvo, A., Pistoia, V., Wei, M., Hwang, S., Merlino, A., Emionite, L., De Cabo, R., and Longo, V.D. (2012). Fasting cycles retard growth of tumors and sensitize a range of cancer cell types to chemotherapy. *Sci. Transl. Med.* 4: 124ra27.

122. Raffaghello, L., Lee, C., Safdie, F.M., Wei, M., Madia, F., Bianchi, G., and Longo, V.D. (2008). Starvation-dependent differential stress resistance protects normal but not cancer cells against high-dose chemotherapy. *Proc. Natl. Acad. Sci. U. S. A.* 105: 8215–8220.

123. Tapio, S. and Jacob, V. (2007). Radioadaptive response revisited. *Radiat. Environ. Biophys.* 46: 1–12.

124. Caputo, F., Vegliante, R., and Ghibelli, L. (2012). Redox modulation of the DNA damage response. *Biochem. Pharmacol.* 84: 1292–1306.

125. Chen, N., Wu, L., Yuan, H., and Wang, J. (2015). ROS/Autophagy/Nrf2 pathway mediated low-dose radiation induced radio-resistance in human lung adenocarcinoma A549 cell. *Int. J. Biol. Sci.* 11: 833–844.

126. Zhang, W., Qiao, T., and Zha, L. (2011). Inhibition of heme oxygenase-1 enhances the radiosensitivity in human nonsmall cell lung cancer a549 cells. *Cancer Biother. Radiopharm.* 26: 639–645.

127. Tong, L., Yu, K.N., Bao, L., Wu, W., Wang, H., and Han, W. (2014). Low concentration of exogenous carbon monoxide protects mammalian cells against proliferation induced by radiation-induced bystander effect. *Mutat. Res.* 759: 9–15.

128. Frank, J., Lornejad-Schafer, M.R., Schoffl, H., Flaccus, A., Lambert, C., and Biesalski, H.K. (2007). Inhibition of heme oxygenase-1 increases responsiveness of melanoma cells to ALA-based photodynamic therapy. *Int. J. Oncol.* 31: 1539–1545.

129. Nowis, D., Legat, M., Grzela, T., Niderla, J., Wilczek, E., Wilczynski, G.M., Glodkowska, E., Mrowka, P., Issat, T., Dulak, J., Jozkowicz, A., Was, H., Adamek, M., Wrzosek, A., Nazarewski, S., Makowski, M., Stoklosa, T., Jakobisiak, M., and Golab, J. (2006). Heme oxygenase-1 protects tumor cells against photodynamic therapy-mediated cytotoxicity. *Oncogene* 25: 3365–3374.

130. Durante, W., Johnson, F.K., and Johnson, R.A. (2006). Role of carbon monoxide in cardiovascular function. *J. Cell. Mol. Med.* 10: 672–686.

131. Loboda, A., Jazwa, A., Grochot-Przeczek, A., Rutkowski, A.J., Cisowski, J., Agarwal, A., Jozkowicz, A., and Dulak, J. (2008). Heme oxygenase-1 and the vascular bed: from molecular mechanisms to therapeutic opportunities. *Antioxid. Redox Signal.* 10: 1767–1812.

132. Motterlini, R., Clark, J.E., Foresti, R., Sarathchandra, P., Mann, B.E., and Green, C.J. (2002). Carbon

monoxide-releasing molecules. *Circ. Res.* 90: e17–e24.

133. Motterlini, R., Sawle, P., Hammad, J., Bains, S., Alberto, R., Foresti, R., and Green, C.J. (2005). CORM-A1: a new pharmacologically active carbon monoxide-releasing molecule. *FASEB J.* 19: 284–286.

134. Hussain, A.S., Marks, G.S., Brien, J.F., and Nakatsu, K. (1997). The soluble guanylyl cyclase inhibitor 1H-[1,2,4]oxadiazolo[4,3-alpha]quinoxalin-1-one (ODQ) inhibits relaxation of rabbit aortic rings induced by carbon monoxide, nitric oxide, and glyceryl trinitrate. *Can. J. Physiol. Pharmacol.* 75: 1034–1037.

135. Morita, T., Perrella, M.A., Lee, M.E., and Kourembanas, S. (1995). Smooth muscle cell-derived carbon monoxide is a regulator of vascular cGMP. *Proc. Natl. Acad. Sci. U. S. A.* 92: 1475–1479.

136. Leffler, C.W., Nasjletti, A., Yu, C., Johnson, R.A., Fedinec, A.L., and Walker, N. (1999). Carbon monoxide and cerebral microvascular tone in newborn pigs. *Am. J. Physiol.-Heart Circ. Physiol.* 276: H1641–H1646.

137. Foresti, R., Hammad, J., Clark, J.E., Johnson, T.R., Mann, B.E., Friebe, A., Green, C.J., and Motterlini, R. (2004). Vasoactive properties of CORM-3, a novel water-soluble carbon monoxide-releasing molecule. *Brit. J. Pharmacol.* 142: 453–460.

138. Johnson, F.K. and Johnson, R.A. (2003). Carbon monoxide promotes endothelium-dependent constriction of isolated gracilis muscle arterioles. *Am. J. Physiol. Regul. Integr. Comp. Physiol.* 285: R536–41.

139. Dulak, J., Jozkowicz, A., Foresti, R., Kasza, A., Frick, M., Huk, I., Green, C.J., Pachinger, O., Weidinger, F., and Motterlini, R. (2002). Heme oxygenase activity modulates vascular endothelial growth factor synthesis in vascular smooth muscle cells. *Antioxid. Redox Signal.* 4: 229–240.

140. Deshane, J., Chen, S., Caballero, S., Grochot-Przeczek, A., Was, H., Li Calzi, S., Lach, R., Hock, T.D., Chen, B., Hill-Kapturczak, N., Siegal, G.P., Dulak, J., Jozkowicz, A., Grant, M.B., and Agarwal, A. (2007). Stromal cell-derived factor 1 promotes angiogenesis via a heme oxygenase 1-dependent mechanism. *J. Exp. Med.* 204: 605–618.

141. Dulak, J., Józkowicz, A., Foresti, R., Kasza, A., Frick, M., Huk, I., Green, C.J., Pachinger, O., Weidinger, F., and Motterlini, R. (2002). Heme oxygenase activity modulates vascular endothelial growth factor synthesis in vascular smooth muscle cells. *Antioxid. Redox Signal.* 4: 229–240.

142. Józkowicz, A., Huk, I., Nigisch, A., Weigel, G., Dietrich, W., Motterlini, R., and Dulak, J. (2003). Heme oxygenase and angiogenic activity of endothelial cells: stimulation by carbon monoxide and inhibition by tin protoporphyrin-IX. *Antioxid. Redox Signal.* 5: 155–162.

143. Choi, Y.K., Kim, C.K., Lee, H., Jeoung, D., Ha, K.S., Kwon, Y.G., Kim, K.W., and Kim, Y.M. (2010). Carbon monoxide promotes VEGF expression by increasing HIF-1alpha protein level via two distinct mechanisms, translational activation and stabilization of HIF-1alpha protein. *J. Biol. Chem.* 285: 32116–32125.

144. Cheng, -C.-C., Guan, -S.-S., Yang, H.-J., Chang, -C.-C., Luo, T.-Y., Chang, J., and Ho, A.-S. (2016). Blocking heme oxygenase-1 by zinc protoporphyrin reduces tumor hypoxia-mediated VEGF release and inhibits tumor angiogenesis as a potential therapeutic agent against colorectal cancer. *J. Biomed. Sci.* 23: 18.

145. Bussolati, B., Ahmed, A., Pemberton, H., Landis, R.C., Di Carlo, F., Haskard, D.O., and Mason, J.C. (2004). Bifunctional role for VEGF-induced heme oxygenase-1 in vivo: induction of angiogenesis and inhibition of leukocytic infiltration. *Blood* 103: 761–766.

146. Brouard, S., Berberat, P.O., Tobiasch, E., Seldon, M.P., Bach, F.H., and Soares, M.P. (2002). Heme oxygenase-1-derived carbon monoxide requires the activation of transcription factor NF-kappa B to protect endothelial cells from tumor necrosis factor-alpha-mediated apoptosis. *J. Biol. Chem.* 277: 17950–17961.

147. Sunamura, M., Duda, D.G., Ghattas, M.H., Lozonschi, L., Motoi, F., Yamauchi, J.-I., Matsuno, S., Shibahara, S., and Abraham, N.G. (2003). Heme oxygenase-1 accelerates tumor angiogenesis of human pancreatic cancer. *Angiogenesis* 6: 15–24.

148. Miyake, M., Fujimoto, K., Anai, S., Ohnishi, S., Kuwada, M., Nakai, Y., Inoue, T., Matsumura, Y., Tomioka, A., Ikeda, T., Tanaka, N., and Hirao, Y. (2011). Heme oxygenase-1 promotes angiogenesis in urothelial carcinoma of the urinary bladder. *Oncol. Rep.* 25: 653–660.

149. Was, H., Cichon, T., Smolarczyk, R., Rudnicka, D., Stopa, M., Chevalier, C., Leger, J.J., Lackowska, B., Grochot, A., Bojkowska, K., Ratajska, A., Kieda, C., Szala, S., Dulak, J., and Jozkowicz, A. (2006). Overexpression of heme oxygenase-1 in murine melanoma: increased proliferation and viability of tumor cells, decreased survival of mice. *Am. J. Pathol.* 169: 2181–2198.

150. Ferrando, M., Gueron, G., Elguero, B., Giudice, J., Salles, A., Leskow, F.C., Jares-Erijman, E.A., Colombo, L., Meiss, R., Navone, N., De Siervi, A., and Vazquez, E. (2011). Heme oxygenase 1 (HO-1)

challenges the angiogenic switch in prostate cancer. *Angiogenesis* 14: 467–479.

151. Pober, J.S., Doukas, J., Hughes, C.C., Savage, C.O., Munro, J.M., and Cotran, R.S. (1990). The potential roles of vascular endothelium in immune reactions. *Hum. Immunol.* 28 (2): 258–262.

152. Ahmad, S., Hewett, P.W., Fujisawa, T., Sissaoui, S., Cai, M., Gueron, G., Al-Ani, B., Cudmore, M., Ahmed, S.F., Wong, M.K., Wegiel, B., Otterbein, L.E., Vitek, L., Ramma, W., Wang, K., and Ahmed, A. (2015). Carbon monoxide inhibits sprouting angiogenesis and vascular endothelial growth factor receptor-2 phosphorylation. *Thromb. Haemost.* 113: 329–337.

153. Nguyen, D.X., Bos, P.D., and Massagué, J. (2009). Metastasis: from dissemination to organ-specific colonization. *Nat. Rev. Cancer* 9: 274–284.

154. Tertil, M., Skrzypek, K., Florczyk, U., Weglarczyk, K., Was, H., Collet, G., Guichard, A., Gil, T., Kuzdzal, J., Jozkowicz, A., Kieda, C., Pichon, C., and Dulak, J. (2014). Regulation and novel action of thymidine phosphorylase in non-small cell lung cancer: crosstalk with Nrf2 and HO-1. *PloS One* 9: e97070–e97070.

155. Huang, S.-M., Lin, C., Lin, H.-Y., Chiu, C.-M., Fang, C.-W., Liao, K.-F., Chen, D.-R., and Yeh, W.-L. (2015). Brain-derived neurotrophic factor regulates cell motility in human colon cancer. *Endocr. Relat. Cancer* 22: 455–464.

156. Zhao, Z., Zhao, J., Xue, J., Zhao, X., and Liu, P. (2016). Autophagy inhibition promotes epithelial-mesenchymal transition through ROS/HO-1 pathway in ovarian cancer cells. *Am. J. Can. Res.* 6: 2162–2177.

157. Lin, H.H., Chiang, M.T., Chang, P.C., and Chau, L.Y. (2015). Myeloid heme oxygenase-1 promotes metastatic tumor colonization in mice. *Cancer Sci.* 106: 299–306.

158. Castruccio Castracani, C., Longhitano, L., Distefano, A., Di Rosa, M., Pittalà, V., Lupo, G., Caruso, M., Corona, D., Tibullo, D., and Li Volti, G. (2020). Heme oxygenase-1 and carbon monoxide regulate growth and progression in glioblastoma cells. *Mol. Neurobiol.* 57: 2436–2446.

159. Liu, Q., Wang, B., Yin, Y., Chen, G., Wang, W., Gao, X., Wang, P., and Zhou, H. (2013). Overexpressions of HO-1/HO-1G143H in C57/B6J mice affect melanoma B16F10 lung metastases rather than change the survival rate of mice-bearing tumours. *Exp. Biol. Med. (Maywood)* 238: 696–704.

160. Lignitto, L., LeBoeuf, S.E., Homer, H., Jiang, S., Askenazi, M., Karakousi, T.R., Pass, H.I., Bhutkar, A.J., Tsirigos, A., Ueberheide, B., Sayin, V.I., Papagiannakopoulos, T., and Pagano, M. (2019). Nrf2 activation promotes lung cancer metastasis by inhibiting the degradation of bach1. *Cell* 178: 316–329.e18.

161. Tertil, M., Golda, S., Skrzypek, K., Florczyk, U., Weglarczyk, K., Kotlinowski, J., Maleszewska, M., Czauderna, S., Pichon, C., Kieda, C., Jozkowicz, A., and Dulak, J. (2015). Nrf2-heme oxygenase-1 axis in mucoepidermoid carcinoma of the lung: Antitumoral effects associated with down-regulation of matrix metalloproteinases. *Free Radic. Biol. Med.* 89: 147–157.

24

CO and Diabetes

Rebecca P. Chow[1] and Hongjun Wang[1,2]

[1]*Department of Surgery, Medical University of South Carolina, 173 Ashley Avenue, Charleston, SC 29425, USA*
[2]*Ralph Johnson Veteran Medical Center, Charleston, SC, USA*

Introduction

Diabetes is a metabolic disorder in which the body does not produce insulin or use insulin efficiently, resulting in elevated glucose levels in the blood and urine. It is one of the major causes of blindness, kidney failure, vascular diseases, heart attacks, strokes, and other complications [1,2]. According to the World Health Organization, there were 422 million people worldwide who were diagnosed with diabetes in 2014. The global prevalence of diabetes among adults rose from 4.7% in 1980 to 8.5% in 2014, and it is estimated to rise to 10.9% (700 million) by 2045 [2]. By 2030, diabetes is expected to be the seventh leading cause of death globally. The majority of patients have type 2 diabetes, while only 5–10% have type 1 diabetes. Type 1 diabetes, also known as juvenile diabetes or insulin-dependent diabetes, is an autoimmune disease caused by the immune destruction of insulin-secreting pancreatic β-cells, which leads to an absolute deficiency of insulin in the body. Type 2 diabetes, also called adult-onset diabetes or insulin-independent diabetes, is often caused by insulin resistance, when the peripheral tissues (the muscle, liver, and adipose) no longer absorb and use blood glucose for energy in response to insulin. β-Cell dysfunction also occurs in patients with type 2 diabetes. β-cell dysfunction and insulin resistance are often interrelated in triggering the pathogenesis of type 2 diabetes [3].

As of now, there is currently no cure for type 1 or type 2 diabetes. The major goal of therapy is to achieve good glycemic control, minimize diabetes complications, and improve the quality of life of patients [4]. Multiple pathophysiological mechanisms contribute to the development of type 2 diabetes; therefore, drugs are designed to target those pathophysiological pathways [5]. Major classes of oral antidiabetic medications include sulfonylureas, biguanides, meglitinide, thiazolidinedione, dipeptidyl peptidase 4 inhibitors, sodium–glucose cotransporter inhibitors, and α-glucosidase inhibitors. Although some drugs have excellent safety records, long-term use may still cause side effects [6–8]. Insulin and insulin analogues are the safest drugs for controlling hyperglycemia and are used in type 1 and some type 2 diabetics [9]. However, insulin therapy is a lifelong treatment, which leads to treatment burden and affects the patient's quality of life. In recent years, significant progress has been made in the artificial pancreas technology field, but more time is needed to achieve precise glucose control [10]. Therefore, for patients with type 1 diabetes, the most promising approach to mimic the physiological functions of a healthy pancreas is β-cell replacement therapy, either through whole pancreas transplantation or through islet cell transplantation. Pancreas transplantation restores or at least improves glycemic control, and the patient often attains insulin independence [11]. Islet transplantation focuses on replacing only the islet cells and is associated with fewer microvascular complications, but is less effective in achieving long-term insulin independence compared to pancreas transplantation [12,13]. Nevertheless, major advances have been made in improving the efficacy of islet transplantation, with a focus on improving immunesuppressive regimes, islet quality, and graft survival after transplantation [14].

Carbon monoxide (CO) is produced endogenously in the body as a by-product of heme degradation catalyzed by the action of heme oxygenases (HOs) [15]. CO prevents normal oxidation in the tissues and has long been considered a catabolic waste product that causes poisoning in humans [16]. CO poisoning (COP) is a risk factor for diabetes. It causes hypoxia and

Carbon Monoxide in Drug Discovery: Basics, Pharmacology, and Therapeutic Potential, First Edition. Edited by Binghe Wang and Leo E. Otterbein.
© 2022 John Wiley & Sons, Inc. Published 2022 by John Wiley & Sons, Inc.

inflammation, and induces immunological reactions in organs, including the pancreas [17]. In 1993, Verma and colleagues showed that CO acted as a neurotransmitter [18]. Since then, accumulating studies demonstrated that low concentrations of CO confer cytoprotective functions, including, but not limited to, antioxidant, vasomodulation, immune modulation, and regulation of cell death [19]. Not only have the beneficial effects of low-dose CO inhalation been explored, but exhaled CO levels also showed potential as a new tool for diabetes monitoring [20]. The dichotomy of CO creates research opportunities in studying the pathophysiology of diabetes and the therapeutic implications for CO use in clinical settings. This review highlights the role of CO and CO-releasing molecules (CORMs) in obesity, insulin resistance, β-cell dysfunction, and other symptoms related to type 2 diabetes and its application to islet transplantation for the treatment of type 1 diabetes and chronic pancreatitis in both preclinical and clinical settings.

CO in Obesity, Insulin Resistance, and Type 2 Diabetes

CO in Obesity

Obesity is one of the major risk factors for metabolic dysfunction and type 2 diabetes [21]. It is characterized by excessive visceral fat accumulation along with chronic low-grade inflammation, which contributes to health deterioration, metabolic dysfunction, and type 2 diabetes. CO has potential as a therapeutic agent for the treatment of obesity. For example, it has been found that aortic tissue CO levels were significantly lower in adult Zucker diabetic fatty (ZDF) rats, a widely used model for obesity and diabetes, compared to lean control rats [22]. In contrast, ZDF rats treated with cobalt protoporphyrin (CoPP), an HO-1 inducer and thus CO inducer, showed increased HO-1 activity and CO levels in kidney and aorta, accompanied by reduced adipose tissue volumes and remodeling compared to nontreated ZDF rats [22]. These results support the potential role of CO as a therapeutic agent for obesity and its associated health risks.

In a model of diet-induced obesity, mice receiving chronic CO inhalation at 28 or 200 ppm showed loss of body fat and whole body weight until 10 weeks after treatment where the effect was diminished after week 12. Nonetheless, inhalation of CO at 28 ppm was able to prevent the development of obesity in mice fed a high-fat diet, indicating that time to treatment is a determining factor for a successful outcome [23]. Metal-based CORMs have also been tested as therapeutic tools to reverse diet-induced obesity. The water-soluble CORM-A1 is a second-generation CORM with a longer half-life compared to first-generation molecules [24,25]. CORM-A1 treatment not only prevented the development of diet-induced obesity, hyperglycemia, and insulin resistance in high-fat diet-fed mice, but also reversed these phenotypes [26,27]. In addition, CORM-A1 treatment reduced production of proinflammatory cytokines in white adipose tissue of diet-induced obese mice [27]. In concurrence with the results observed with CORM-A1, CORM-401, another CORM, also led to body weight reduction with improved insulin sensitivity when given to diet-induced obese mice [28]. The therapeutic mechanism of CORM is hypothesized to be a CO-mediated alternation in adipose tissue metabolism toward glycolysis and hence leading to glucose tolerance [28]. Taken together, both CO gas and CORM show protective effects in diet-induced obesity in mice.

CO in Insulin Resistance

Insulin resistance is a metabolic condition in which cells in the peripheral tissues, including liver, muscle, and adipose tissue, become resistant to insulin. Insulin resistance increases the risks for diabetes and cardiovascular diseases. Several studies have demonstrated that CO promotes insulin sensitivity in animal models. For example, treatment of the ob/ob (B6v-Lep ob/J) mice with the HO-1 inducer CoPP improved insulin sensitivity by increasing serum adiponectin and reducing tumor necrosis factor alpha (TNF-α) and reactive oxygen species (ROS) levels, factors that contribute to insulin resistance in adipose tissue [29,30]. The therapeutic role of CO in insulin sensitivity is further supported by the use of CORMs that diminished insulin resistance in mouse models of diet-induced obesity [28]. It is hypothesized that CO directly modulates Akt phosphorylation, a major pathway involved in insulin signaling, and thus enhancing insulin sensitivity [28]. In addition, CO has been shown to abrogate insulin resistance through modulating adipose tissue macrophages. Accumulation of adipose tissue macrophages, predominantly activated M1 macrophages, contributes to the development of insulin resistance [31]. Activated M1 macrophages express CD11c, and depletion of CD11c$^+$ macrophages reduces local and systemic inflammatory markers linked to the development of obesity-associated insulin resistance, including TNF-α and IL-6, in diet-induced obesity in mice [32,33]. Similarly, during human menopause, loss of ovarian function leads to the infiltration of CD11c cells into adipose tissue, contributing to insulin resistance [34,35]. Choi et al. demonstrated that CO attenuated

CD11c+ cell infiltration into adipose tissue in ovariectomized mice and increased insulin sensitivity [35]. CO exerted such protection by activating soluble guanylate cyclase to generate cyclic GMP (cGMP), and consequently induced HO-1 expression and reduced ROS generation, leading to decreased CD11c expression in macrophages.

CO in Other Symptoms of Type 2 Diabetes

CO acts like a double-edged sword in type 2 diabetes. Low doses of CO have protective effects against type 2 diabetes. For example, a prominent event in the pathogenesis of type 2 diabetes is impaired glucose-stimulated insulin secretion (GSIS) [36]. Evidence showed that CO stimulates GSIS in rodents, thus revealing its antidiabetic properties [37]. Hyperglycemia, another symptom of type 2 diabetes, contributes to endothelial dysfunction by inducing oxidative stress and upregulating intracellular adhesion molecule-1 (ICAM-1) [38,39]. Studies showed that CORMs diminished ICAM-1 expression under high-glucose conditions through regulating the AMPK and peroxisome proliferator activated receptor γ (PPARγ) pathways [40]. In addition, the induction of CO-producing enzyme, HO-1, improves vascular dysfunction in diabetic rats, likely via the inhibition of ROS production and suppression of diabetes-induced cyclooxygenase-2 expression [41].

Although appropriate levels of CO can ameliorate type 2 diabetes, overexposure to CO increases the risk of developing diabetes. COP may induce the production of ROS and trigger immunological and inflammatory reactions in all organs, resulting in impaired suppression of glucagon secretion and reduced insulin secretion [42,43]. Indeed, patients with COP showed a higher risk of diabetes than the non-COP group [17]. Moreover, long-term exposure to CO at 1025 ppb and PM2.5 levels increased proteinuria in type 2 diabetes [44]. Enhanced levels of exhaled CO have been observed in patients with diabetes, and CO levels correlated with glucose concentration in the blood [20]. The measurement of exhaled CO may therefore prove to be a noninvasive way to monitor disease severity and therapeutic efficacy [45,46].

CO in Type 1 Diabetes

CO in the Treatment of Type 1 Diabetes

Type 1 diabetes is an autoimmune disease caused by the autoimmune response targeting and destroying insulin-secreting pancreatic β-cells. Daily insulin injection is the most common therapy for type 1 diabetes. As the disease progresses, patients often develop hypoglycemia unawareness and defects in their counterregulatory defense [47]. Intensive insulin therapy therefore may cause severe hypoglycemia, and insulin therapy cannot prevent diabetes complications such as cardiovascular diseases. As a result, safe and effective approaches to treat type 1 diabetes are needed. The therapeutic effects of CO have been demonstrated in a type 1 diabetes mouse model by suppressing activation of dendritic cells, which can prime $CD8^+$ T-cells that contribute to the onset of type 1 diabetes in nonobese diabetic (NOD) mice [48]. In this study, the authors showed that injection of *ex vivo* primed dendritic cells with gaseous CO inhibited β1-integrin expressing $CD8^+$ T-cell infiltration and consequently protected β-cells from death, therefore contributing to reduced incidence of type 1 diabetes. These data suggested that CO might inhibit islet destruction via downregulation of β1-integrin expression in CD8 T-cells. Hu and colleagues also confirmed that CO inhalation at 250 ppm for 2 h twice per week over a 15-week period protected against type 1 diabetes in the NOD mice [49]. However, excessive exposure to CO levels in ambient air was associated with a high risk of type 1 diabetes, highlighting the importance of determining precise quantities of CO in order to fully benefit from its therapeutic effects [50].

CO in Regulation of Islet Insulin Release

β-cells secrete insulin via at least three interconnected intracellular signaling pathways, including the ATP-dependent K^+ (KATP) channel, and the metabolic and neurohormonal amplifying pathways where intracellular Ca^{2+} concentrations play a central role [51]. A critical potentiator of GSIS is a gut peptide hormone, glucagon-like peptide-1 [52], which increases cyclic AMP concentrations and in turn triggers a cascade of signals including Ca^{2+} signaling, leading to insulin secretion through the sequential production of two Ca^{2+} secondary messengers, nicotinic acid adenine dinucleotide phosphate and cyclic ADP-ribose (cADPR) [53]. Nitric oxide (NO), which is produced by islets, activates guanylyl cyclase to generate cGMP and ultimately induces cADPR formation [54,55]. Like NO, CO is also secreted by islets and able to activate cGMP production. Exogenous CO has been shown to amplify cGMP formation, Ca^{2+} signal strength, and insulin secretion [56]. CO coordinates the secretory activity of β-cells by promoting the generation of cytoplasmic Ca^{2+} concentration transients [57]. Interestingly, in mouse islets, high concentrations of CO from CORM inhibited cGMP formation (200–400 μM), whereas lower concentrations (10–100 μM) potentiated cGMP formation in a dose-dependent manner [56]. However, mouse islets that were incubated with 1 mM saturated CO released more insulin than those treated with 100 μM CO, suggesting that the potential mechanistic differences exist between pharmacological probes (CORMs) and nonpharmacological agents (CO) on how they affect the GSIS pathway [58]. In nonobese type 2 diabetic Goto–Kakizaki (GK) rats, a marked impairment of insulin response to glucose was at least in part caused by impairment of the glucose–HO–CO signaling pathway, in which decrease in glucose stimulation of islet CO production and a reduced expression of HO-2 were observed in the GK islets [59].

CO in Islet Survival and Transplantation

Allogeneic islet transplantation is a promising approach to treat patients with type 1 diabetes, and autologous islet transplantation is performed in chronic pancreatitis patients after total pancreatectomy to avoid surgical diabetes [60]. However, the efficacy of islet transplantation is not ideal. Type 1 diabetes patients often require islets isolated from two to three cadaveric pancreases to achieve insulin independence and patients need to take lifelong immunosuppressants, while only one-third of nonprediabetic chronic pancreatitis patients remained insulin independent after autologous islet transplantation [61–63]. Although significant improvements have been made on immunosuppression and other aspects of the transplant procedures, transplanted β-cells frequently undergo apoptosis prior to full engraftment, and islet function is compromised. It is estimated that about 70% of transplanted islet mass is lost after 30 days of transplantation even in syngeneic islet transplantation settings [64]. Therefore, one of the critical tasks in the islet transplantation field is to improve islet graft survival, especially focusing on graft revascularization after transplantation into the liver, a new environment for islets [65].

Multiple approaches incorporating CO into islet transplantation have been investigated in order to preserve islet survival and function during and after transplantation. These include the use of exogenous CO, pharmacological probes, and genetic tools. Diabetic mice that received CO-treated murine islets demonstrated faster return to normoglycemia when islets were pre-exposed to CO for 2 h [66]. In a syngeneic islet transplantation model, transplantation of 250 marginal mass islets resulted in a significant delay of normoglycemia after transplantation compared to transplantation of 500 islets; i.e., it took 12–18 days for recipient mice to reach normoglycemia. In contrast, in mice receiving islet cultured in solution presaturated with CO, normoglycemia was reached in about 7 days. CO exerted such protective effects by suppression of nonspecific inflammation-induced islet cell apoptosis [66]. In addition, CO treatment to only the islet donor led to long-term tolerance in recipients in which the islets survived for more than 3 months in an allogeneic islet transplantation setting [67]. The results were confirmed using human islets: treatment with CO reduced inflammation and prevented apoptosis, and enhanced long-term graft survival when transplanted into diabetic NOD-SCID mice [68].

In an experiment comparing the effect of CoPP pre- and post-treatment during islet transplantation, diabetic mice receiving islets harvested from CoPP-treated donors and/or a 9-day CoPP injection regimen post-transplantation had an 80% chance of achieving normoglycemia after 56 days, while mice that received nontreated islets failed to achieve normoglycemia [69]. Furthermore, mice that received CoPP-treated islets followed by 9 days of CoPP injections post-transplantation showed greater improvement of blood glucose levels and body weight compared to those that only received CoPP-treated islets [69]. This suggests that production of CO in donors and recipients could impact overall outcome.

Alternatively, diabetic mice receiving CO-treated islets induced via transduction with HO-1 adenoviral vector showed a lower degree of lymphocytic infiltration and protection against TNF and cycloheximide-mediated cytotoxicity compared to those receiving mock transduced islets [70]. Likewise, CO generated from chemical donors, such as CORMs, or HO-1 induction (CoPP injection) also reduces hyperglycemia and preserves islet function [69,71,72]. Islets treated with CORMs before transplantation demonstrated better resistance to exogenous inflammatory cytokines, thus increasing islet survivability [71]. Taken together, treatment of islet donors and recipients with CO improves outcomes after islet transplantation.

Mechanisms of CO in Promoting Islet Survival

The mechanisms by which CO protects islet graft survival during islet transplantation have been explored through *in vivo* and *in vitro* models. Post-transplantation stress responses contribute to >50% islet graft loss 2–3 days post-transplantation due to stress-induced β-cell death [73]. Hypoxia, inflammation, hyperglycemia, nutrient deprivation, and lipotoxicity have been identified as key mechanisms contributing to β-cell apoptosis [74–76].

Administration of CO to donors suppresses the proinflammatory response, which typically occurs in transplanted islets in the recipient [67]. The expression of proinflammatory genes in the transplanted islets, including TNF-α, inducible NO synthase, and Fas ligand, was notably lower in CO-treated animals compared to the nontreated group 10 days after transplantation. In contrast, expression of the antiapoptotic gene bcl-2 was upregulated in the CO-treated group 10 days post-transplantation. In the CO-treated group, long-term tolerance of islet grafts and less macrophage infiltration were also observed in the transplanted islets, and the observation was confirmed by reduced gene expression of monocyte chemoattractant protein-1, a chemokine that regulates macrophage migration to an inflammatory site. Taken together, these results outline a possible mechanism by which CO decreases inflammation in islets after transplantation, leading to a decrease in rejection; the upregulation of protective genes such as bcl-2 boosts transplant tolerance by shifting the dynamic from transplant rejection to tolerance [67].

CO enhances islet survival by regulating autophagy in β-cells. Autophagy is a cellular self-degradation process that involves delivery of cytoplasmic materials to the lysosome. This self-degradation provides nutrients to generate energy during starvation, and replaces damaged or other unnecessary organelles with new ones [77]. Autophagy plays a vital role in preventing high-fat diet-induced β-cell dysfunction and protecting human islets from amyloid polypeptide-induced toxicity [78,79]. When islets were exposed to hypoxia, CO treatment enhanced gene expression of the autophagy markers LC-3II and bcl-2, and reduced gene expression of proapoptotic and anti-inflammatory proteins, contributing to improved engraftment and islet graft survival in streptozotocin-induced diabetic mice [68]. This CO-induced autophagy is mediated by ATG16L1, a regulator of autophagy in hypoxia-induced osteoclast differentiation [68,80]. Deficiency of ATG16L1 in islets abolished autophagy and antiapoptotic effects of CO under hypoxic conditions, suggesting that ATG16L1 is necessary for the protective and prosurvival effects of CO. Furthermore, exposure to CO suppressed endoplasmic reticulum (ER) stress-regulated gene expression, including glucose-regulated protein 78 (grp78), grp94, and CCAAT-enhancer-binding protein homologous protein (CHOP), suggesting that suppression of ER stress by CO may be partially responsible for the antiapoptotic effect observed in CO-treated islets (Figure 24.1) [68].

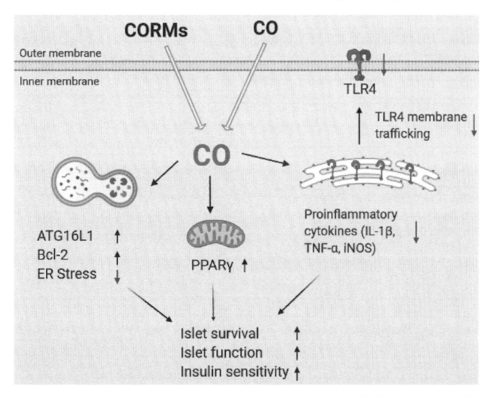

Figure 24.1 Nonheme-dependent mechanisms involved in CO-mediated islet function. Gaseous CO or CORMs promote islet function via multiple pathways. These include (i) ATG16L1 and blc-2 expression and reduction of ER stress; (ii) the activation of PPARγ; and (iii) the inhibition of membrane TLR4 expression. CO modulates these pathways in islet cells, contributing to reduced inflammation and thus prolonged islet survival.

CO may also function via suppressing Toll-like receptors (TLRs) when given to islet donor. TLRs are pattern recognition receptors that play a critical role in sensing pathogen invasion [81]. Upon activation, they trigger an inflammatory response that is mediated by innate immune cells, including macrophages and monocytes [82,83]. Among the TLR family, TLR4 is one of the most well-studied TLRs. TLR4 recognizes lipopolysaccharide (LPS), a constituent of the outer cell layer of Gram-negative bacteria [84]. In addition to LPS, a group of molecules derived from the host during stress or cell injury, also known as endogenous TLR ligands, initiates immune responses via the activation of TLR4 [85]. Studies have shown that TLR4 activation is directly involved in transplant rejection [86–88]. TLR4 expression is upregulated during the islet isolation process [89], and TLR4 serves as a key mediator in islet damage and rejection. Deletion of TLR4 ($Tlr4^{-/-}$) in islets protected those islets from immune rejection when transplanted into allogeneic recipients. This was confirmed when TLR4 was deleted in islets using a TLR4 dominant negative adenovirus. In addition, CO treatment of the donor suppressed TLR4 expression in isolated islets and islet grafts, which suggest that CO likely contributes to islet survival in part by inhibiting TLR4 expression in β-cells (Figure 24.1) [89].

Another potential mechanism of CO protection is the upregulation of PPARγ in islets [90]. PPARγ is a nuclear receptor that modulates inflammatory response [91]. Deficient expression of PPARγ is associated with diabetes, cardiovascular disease, and obesity [92,93]. CO upregulates expression of PPARγ via the mitochondria in macrophages [94]. In islets, PPARγ activation via treatment with the PPARγ agonist 15d-PGJ$_2$ suppressed inflammation in islet grafts and contributed to long-term islet survival in streptozotocin-induced diabetic mice [90]. Upon exposing islet donors to CO, PPARγ expression was remarkably increased in isolated islets compared to islets from nondonors. The beneficial effects of CO on islet survival were abolished by the PPARγ antagonist, GW9662, further confirming that PPARγ activity is needed for CO to impart benefit in the islet allogeneic transplantation setting.

Taken together, it seems that CO exerts protection in islets in a nonheme-dependent fashion, via regulating autophagy, suppressing membrane expression of pattern recognition receptors, upregulating PPARγ gene expression (likely via the mitochondria), and others (Figure 24.1). It may be worth finding out whether these effects are correlated or independent of each other.

CO in Diabetes Complications

Diabetes is a well-established risk factor for cardiovascular diseases, and people with diabetes have a two- to fourfold increased risk of developing vascular diseases [95]. These vascular complications are separated into macrovascular (i.e., stroke, coronary artery, and peripheral arterial diseases) and microvascular (i.e., retinopathy, neuropathy, and nephropathy) complications. The underlying mechanism by which diabetes leads to vascular complications is thought to be associated with endothelial dysfunction and ROS-induced injury and inflammation [96,97]. Multiple studies have shown that vasorelaxant effects of both exogenously applied and endogenously generated CO occur via the cGMP signaling pathway and calcium-activated K channels [98–100]. CO has been shown to prevent endothelial cell death by rendering endothelial cells resistant to oxidative stress-mediated injury, which contributes to the antiapoptotic and anti-inflammatory effects of CO in prophylactic vascular protection in type 1 diabetes [101]. CORM treatment demonstrated cardioprotection in cardiac cells and isolated mouse hearts. Moreover, CORM reverses cardiac muscle damage after ischemia and hence could potentially be used to ameliorate cardiac dysfunction [25]. Diabetes-induced neuronal dysfunction caused by oxidative stress, inflammation, and apoptosis is also heavily mediated by HO-1 [102]. In a mouse model of diabetic neuropathy, administration of 10 mg/kg of CORM-2 inhibited allodynia and hyperalgesia and reduced the expression of neuropathic pain markers *Cybb* and *Bdkrb1* [103]. However, CORM-2 increased plasma levels of cholesterol, non-high-density lipoprotein (HDL) and HDL cholesterol. This suggests that CORM-2 produces antinociceptive effects in diabetic mice but without favorable effects on dyslipidemia. The antinociceptive effects of CORM-2 in neuropathic pain were found to be mediated via the reduction in oxidative stress and inflammation as CORM-2 treatment restricted microglial activation and NO synthesis following nerve injury [104].

CO in Clinical Islet Transplantation

The beneficial effect of harvesting islet in CO-saturated medium in animal models was expanded to chronic pancreatitis patients who undergo total pancreatectomy and islet autotransplantation in a blinded, controlled phase 1 clinical trial (clinicaltrials.gov, NCT02567240) [19]. The goal of the trial was to test whether stress-induced apoptosis of transplanted islets can be minimized by harvesting islets in a CO-rich environment leading to increased islet survival and function. In this trial, islets of participants assigned to the CO group were isolated in a CO-saturated medium. Islets of control patients were isolated using the normal medium. The study enrolled 16 patients, where 10 subjects received CO-treated islets and 6 subjects received nontreated control islets. CO-treated islets showed significantly higher viability/quality before transplantation as measured by oxygen consumption rate. Patients receiving CO-treated islets showed a trend toward reduced β-cell death as measured by reduced serum concentrations of unmethylated insulin DNA, a β-cell death marker. In addition, CO patients showed decreased C–C motif chemokine ligand 23 (CCL23) and increased C–X–C motif chemokine ligand 12 (CXCL12) levels in serum at 1 and 3 days post-transplantation compared to controls. CCL23 and CXCL12 are chemokines related to inflammation. CCL23 attracts inflammatory cells and upregulates inflammatory cytokines such as TNF [105]. In contrast, CXCL12 protects β-cells from inflammatory and oxidative stress-induced apoptosis [106,107], and increases cell survival in allogeneic and xenogeneic islet transplantation [108]. The altered CCL23 and CXCL12 levels suggest less inflammation in CO-treated islets post-transplantation. Reduced inflammation might have contributed to the outcome that resulted in CO patients having a higher chance of achieving insulin independence compared to non-treated control patients. Most importantly, no adverse events directly related to the infusion of CO islets were observed. Taken together, this trial showed that harvesting human islets in CO-saturated medium is safe for autologous islet transplantation patients.

Conclusions

The elaborate research on CO and CORMs over the past two decades has clearly illustrated that CO levels within a specific dose range manifest protective effects and can be used for therapeutic purposes in various pathological conditions, including diabetes and diabetes complications. CO notably showed beneficial effects on islet function, survival, and insulin resistance and is making its own space in research and clinical trials (Figure 24.1). The success of CO therapy could be an effective strategy to minimize donor tissue damage by maintaining long-term islet survival, and hence more patients who suffer from diabetes may benefit from islet transplantation. In addition, CO and CORMs showed protective effects in reducing obesity, insulin resistance, hyperglycemia, and diabetes complications. It is encouraging that the therapeutic benefits of CO treatment for those diagnosed with diabetes have been tested in clinical trials. More bench to bedside clinical trials are needed to enable the translation of this potent molecule into therapy.

Acknowledgments

H. Wang was supported by the National Institutes of Health (1R01DK105183, R01DK120394, R01DK126454 and R01DK118529) and the Department of Veterans Affairs (VA-ORD BLR&D Merit I01BX004536). We thank Mr Michael Lee and Ms Lindsay Swaby for language editing.

References

1. Hippisley-Cox, J. and Coupland, C. (2016). Diabetes treatments and risk of amputation, blindness, severe kidney failure, hyperglycaemia, and hypoglycaemia: open cohort study in primary care. *BMJ* 352: i1450.
2. Sarwar, N., Gao, P., Kondapally Seshasai, S.R. et al. (2010). Diabetes mellitus, fasting blood glucose concentration, and risk of vascular disease: a collaborative meta-analysis of 102 prospective studies. *Lancet*.
3. Cerf, M.E. (2013). Beta cell dysfunction and insulin resistance. *Front. Endocrinol. (Lausanne)* 4: 37.
4. Lorenzo-Vizcaya, A. and Isenberg, D.A. (2020). The use of anti-TNF-alpha therapies for patients with systemic lupus erythematosus. Where are we now? *Expert Opin. Biol. Ther.*
5. Defronzo, R.A. (2009). Banting lecture. From the triumvirate to the ominous octet: a new paradigm for the treatment of type 2 diabetes mellitus. *Diabetes* 58: 773–795.
6. Asche, C., LaFleur, J., and Conner, C. (2011). A review of diabetes treatment adherence and the association with clinical and economic outcomes. *Clin. Ther* 33: 74–109.
7. Leroux, C., Brazeau, A.S., Gingras, V. et al. (2014). Lifestyle and cardiometabolic risk in adults with type 1 diabetes: a review. *Can. J. Diabetes* 38: 62–69.
8. Chaudhury, A., Duvoor, C., Reddy Dendi, V.S. et al. (2017). Clinical review of antidiabetic drugs: implications for type 2 diabetes mellitus management. *Front. Endocrinol. (Lausanne)* 8: 6.
9. Rys, P., Pankiewicz, O., Łach, K. et al. (2011). Efficacy and safety comparison of rapid-acting insulin aspart and regular human insulin in the treatment of type 1

and type 2 diabetes mellitus: a systematic review. *Diabetes Metab.* 37: 190–200.

10. Nijhoff, M.F. and De Koning, E.J.P. (2018). Artificial pancreas or novel beta-cell replacement therapies: A race for optimal glycemic control? *Curr. Diabetes Rep.* 18: 110–110.

11. Niclauss, N., Morel, P., and Berney, T. (2014). Has the gap between pancreas and islet transplantation closed? *Transplantation* 98: 593–599.

12. Ludwig, B., Ludwig, S., Steffen, A. et al. (2010). Islet versus pancreas transplantation in type 1 diabetes: competitive or complementary? *Curr. Diab. Rep.* 10: 506–511.

13. Vardanyan, M., Parkin, E., Gruessner, C. et al. (2010). Pancreas vs. islet transplantation: a call on the future. *Curr. Opin. Organ. Transplant.* 15: 124–130.

14. Roep, B.O. (2020). Improving clinical islet transplantation outcomes. *Diabetes Care* 43 (698): LP–700.

15. Ryter, S.W. and Choi, A.M. (2009). Heme oxygenase-1/carbon monoxide: from metabolism to molecular therapy. *Am. J. Respir. Cell Mol. Biol.* 41: 251–260.

16. Nasmith, G.G. and Graham, D.A.L. (1906). The hæmatology of carbon-monoxide poisoning. *J. Physiol.*

17. Huang, C.C., Ho, C.H., Chen, Y.C. et al. (2017). Increased risk for diabetes mellitus in patients with carbon monoxide poisoning. *Oncotarget.*

18. Verma, A., Hirsch, D.J., Glatt, C.E. et al. (1993). Carbon monoxide: a putative neural messenger. *Science.*

19. Wang, H., Gou, W., Strange, C. et al. (2019). Islet harvest in carbon monoxide-saturated medium for chronic pancreatitis patients undergoing islet autotransplantation. *Cell Transplant.*

20. Paredi, P., Biernacki, W., Invernizzi, G. et al. (1999). Exhaled carbon monoxide levels elevated in diabetes and correlated with glucose concentration in blood: a new test for monitoring the disease? *Chest*

21. Stumvoll, M., Goldstein, B.J., and Van Haeften, T.W. (2005). Type 2 diabetes: principles of pathogenesis and therapy. *Lancet (London, England)* 365: 1333–1346.

22. Nicolai, A., Li, M., Kim, D.H. et al. (2009). Heme oxygenase-1 induction remodels adipose tissue and improves insulin sensitivity in obesity-induced diabetic rats. *Hypertension.*

23. Hosick, P.A., Ahmed, E.K., Gousset, M.U. et al. (2014). Inhalation of carbon monoxide is ineffective as a long-term therapy to reduce obesity in mice fed a high fat diet. *BMC Obes.*

24. Motterlini, R., Clark, J.E., Foresti, R. et al. (2002). Carbon monoxide-releasing molecules: characterization of biochemical and vascular activities. *Circ. Res.*

25. Clark, J.E., Naughton, P., Shurey, S. et al. (2003). Cardioprotective actions by a water-soluble carbon monoxide-releasing molecule. *Circ. Res.*

26. Hosick, P.A., AlAmodi, A.A., Storm, M.V. et al. (2014). Chronic carbon monoxide treatment attenuates development of obesity and remodels adipocytes in mice fed a high-fat diet. *Int. J. Obes. (Lond)* 38: 132–139.

27. Hosick, P.A., AlAmodi, A.A., Hankins, M.W. et al. (2016). Chronic treatment with a carbon monoxide releasing molecule reverses dietary induced obesity in mice. *Adipocyte.*

28. Braud, L., Pini, M., Muchova, L. et al. (2018). Carbon monoxide-induced metabolic switch in adipocytes improves insulin resistance in obese mice. *JCI Insight* 3.

29. Li, M., Kim, D.H., Tsenovoy, P.L. et al. (2008). Treatment of obese diabetic mice with a heme oxygenase inducer reduces visceral and subcutaneous adiposity, increases adiponectin levels, and improves insulin sensitivity and glucose tolerance. *Diabetes*

30. Lin, Y., Berg, A.H., Iyengar, P. et al. (2005). The hyperglycemia-induced inflammatory response in adipocytes: the role of reactive oxygen species. *J. Biol. Chem.*

31. Lumeng, C.N., Deyoung, S.M., Bodzin, J.L. et al. (2007). Increased inflammatory properties of adipose tissue macrophages recruited during diet-induced obesity. *Diabetes* 56: 16–23.

32. Li, P., Lu, M., Nguyen, M.T. et al. (2010). Functional heterogeneity of CD11c-positive adipose tissue macrophages in diet-induced obese mice. *J. Biol. Chem.* 285: 15333–15345.

33. Castoldi, A., Naffah De Souza, C., Camara, N.O. et al. (2015). The macrophage switch in obesity development. *Front. Immunol.* 6: 637.

34. Vieira Potter, V.J., Strissel, K.J., Xie, C. et al. (2012). Adipose tissue inflammation and reduced insulin sensitivity in ovariectomized mice occurs in the absence of increased adiposity. *Endocrinology* 153: 4266–4277.

35. Choi, E.-K., Park, H.-J., Sul, O.-J. et al. (2015). Carbon monoxide reverses adipose tissue inflammation and insulin resistance upon loss of ovarian function. *Am. J. Physiol. Endocrinol. Metab.* 308: E621–E630.

36. Meda, P. and Schuit, F. (2013). Glucose-stimulated insulin secretion: The hierarchy of its multiple cellular and subcellular mechanisms. *Diabetologia.*

37. Mohammed Al-Amily, I., Lundquist, I., and Salehi, A. (2019). Expression levels of enzymes generating NO and CO in islets of murine and human diabetes. *Biochem. Biophys. Res. Commun.*

38. Szekanecz, Z., Shah, M.R., Pearce, W.H. et al. (1994). Intercellular adhesion molecule-1 (ICAM-1) expression and soluble ICAM-1 (sICAM-1) production by cytokine-activated human aortic endothelial cells: a possible role for ICAM-1 and sICAM-1 in atherosclerotic aortic aneurysms. *Clin. Exp. Immunol.* 98: 337–343.

39. Sattar, N., Murray, H.M., Welsh, P. et al. (2009). Are elevated circulating intercellular adhesion molecule 1 levels more strongly predictive of diabetes than vascular risk? Outcome of a prospective study in the elderly. *Diabetologia* 52: 235–239.

40. Nizamutdinova, I.T., Kim, Y.M., Kim, H.J. et al. (2009). Carbon monoxide (from CORM-2) inhibits high glucose-induced ICAM-1 expression via AMP-activated protein kinase and PPAR-γ activations in endothelial cells. *Atherosclerosis*.

41. Wang, Y., Ying, L., Chen, Y. et al. (2014). Induction of heme oxygenase-1 ameliorates vascular dysfunction in streptozotocin-induced type 2 diabetic rats. *Vasc. Pharmacol.*

42. Weaver, L.K. (2020). Carbon monoxide poisoning. *Undersea Hyperb. Med.*

43. Hampson, N.B., Rudd, R.A., and Hauff, N.M. (2009). Increased long-term mortality among survivors of acute carbon monoxide poisoning. *Crit. Care Med.*

44. Chin, W.S., Chang, Y.K., Huang, L.F. et al. (2018). Effects of long-term exposure to CO and PM2.5 on microalbuminuria in type 2 diabetes. *Int. J. Hyg. Environ. Health*

45. Martinez, S.A., Quaife, S.L., Hasan, A. et al. (2020). Contingency management for smoking cessation among individuals with type 2 diabetes: protocol for a multi-center randomized controlled feasibility trial. *Pilot Feasibility Stud.* 6: 82–82.

46. Ismail, A.A., Gill, G.V., Lawton, K. et al. (2000). Comparison of questionnaire, breath carbon monoxide and urine cotinine in assessing the smoking habits of Type 2 diabetic patients. *Diabet. Med.*

47. McCall, A.L. and Farhy, L.S. (2013). Treating type 1 diabetes: from strategies for insulin delivery to dual hormonal control. *Minerva Endocrinol.* 38: 145–163.

48. Simon, T., Pogu, S., Tardif, V. et al. (2013). Carbon monoxide-treated dendritic cells decrease β1-integrin induction on CD8+ T cells and protect from type 1 diabetes. *Eur. J. Immunol.*

49. Hu, C.M., Lin, H.H., Chiang, M.T. et al. (2007). Systemic expression of heme oxygenase-1 ameliorates type 1 diabetes in NOD mice. *Diabetes*.

50. Michalska, M., Zorena, K., Wąż, P. et al. (2020). Gaseous pollutants and particulate matter (PM) in ambient air and the number of new cases of type 1 diabetes in children and adolescents in the pomeranian voivodeship, Poland. *Biomed Res. Int.*

51. Skelin Klemen, M., Dolenšek, J., Slak Rupnik, M. et al. (2017). The triggering pathway to insulin secretion: functional similarities and differences between the human and the mouse β cells and their translational relevance. *Islets* 9: 109–139.

52. Kieffer, T.J. and Habener, J.F. (1999). The glucagon-like peptides.

53. Kim, B.J., Park, K.H., Yim, C.Y. et al. (2008). Generation of nicotinic acid adenine dinucleotide phosphate and cyclic ADP-ribose by glucagon-like peptide-1 evokes Ca2+ signal that is essential for insulin secretion in mouse pancreatic islets. *Diabetes*.

54. Kots, A.Y., Martin, E., Sharina, I.G., et al. (2009). A short history of cGMP, guanylyl cyclases, and cGMP-dependent protein kinases.

55. Southern, C., Schulster, D., and Green, I.C. (1990). Inhibition of insulin secretion by interleukin-1β and tumour necrosis factor-α via an L-arginine-dependent nitric oxide generating mechanism. *FEBS Lett.*

56. Rahman, F.U., Park, D.R., Joe, Y. et al. (2019). Critical roles of carbon monoxide and nitric oxide in Ca 2+ signaling for insulin secretion in pancreatic islets. *Antioxid. Redox Signal.*

57. Lundquist, I., Alm, P., Salehi, A. et al. (2003). Carbon monoxide stimulates insulin release and propagates Ca2+ signals between pancreatic β-cells. *Am. J. Physiol. Endocrinol. Metab.* 285: 1055–1063.

58. Mosén, H., Salehi, A., Henningsson, R. et al. (2006). Nitric oxide inhibits, and carbon monoxide activates, islets acid α-glucoside hydrolase activities in parallel with glucose-stimulated insulin secretion. *J. Endocrinol.*

59. Mosén, H., Salehi, A., Alm, P. et al. (2005). Defective glucose-stimulated insulin release in the diabetic Goto-Kakizaki (GK) rat coincides with reduced activity of the islet carbon monoxide signaling pathway. *Endocrinology*.

60. Morgan, K.A., Lancaster, W.P., Owczarski, S.M. et al. (2018). Patient selection for total pancreatectomy with islet autotransplantation in the surgical management of chronic pancreatitis. *J. Am. Coll. Surg.* 226: 446–451.

61. Ryan, E.A., Lakey, J.R.T., Paty, B.W. et al. (2002). Successful islet transplantation: Continued insulin reserve provides long-term glycemic control. *Diabetes*.

62. Shapiro, A.M.J., Ryan, E.A., and Lakey, J.R.T. (2001). Islet cell transplantation. *Lancet* 358: S21–S21.

63. Wang, H., Desai, K.D., Dong, H. et al. (2013). Prior surgery determines islet yield and insulin requirement in patients with chronic pancreatitis. *Transplantation* 95: 1051–1057.

64. Davalli, A.M., Ogawa, Y., Ricordi, C. et al. (1995). A selective decrease in the beta cell mass of human

islets transplanted into diabetic nude mice. *Transplantation* 59: 817–820.

65. Plesner, A. and Verchere, C.B. (2011). Advances and challenges in islet transplantation: Islet procurement rates and lessons learned from suboptimal islet transplantation. *J. Transplant.* 2011: 979527.

66. Günther, L., Berberat, P.O., Haga, M. et al. (2002). Carbon monoxide protects pancreatic β-cells from apoptosis and improves islet function/survival after transplantation. *Diabetes.*

67. Wang, H., Lee, S.S., Gao, W. et al. (2005). Donor treatment with carbon monoxide can yield islet allograft survival and tolerance. *Diabetes.*

68. Kim, D.S., Song, L., Wang, J. et al. (2018). Carbon monoxide inhibits islet apoptosis via induction of autophagy. *Antioxid. Redox Signal.*

69. Fu, S.H., Hsu, B.R.S., Juang, J.H. et al. Cobalt-protoporphyrin treatment enhances murine isoislets engraftment.

70. Li, Y., Li, G., Dong, W. et al. (2006). Transplantation of rat islets transduced with human heme oxygenase-1 gene using adenovirus vector. *Pancreas.*

71. Cai, X.-H., Wang, G.-Q., Liang, R. et al. (2020). CORM-2 pretreatment attenuates inflammation-mediated islet dysfunction. *Cell Transplant.* 29: 0963689720903691–0963689720903691.

72. Pileggi, A., Damaris Molano, R., Berney, T. et al. (2001). Heme oxygenase-1 induction in islet cells results in protection from apoptosis and improved in vivo function after transplantation. *Diabetes.*

73. Biarnés, M., Montolio, M., Nacher, V. et al. (2002). Beta-cell death and mass in syngeneically transplanted islets exposed to short- and long-term hyperglycemia. *Diabetes* 51: 66–72.

74. Lee, Y., Ravazzola, M., Park, B.H. et al. (2007). Metabolic mechanisms of failure of intraportally transplanted pancreatic β-cells in rats: role of lipotoxicity and prevention by leptin. *Diabetes.*

75. McCall, M. and James Shapiro, A.M. Update on islet transplantation. *Cold Spring Harb. Perspect. Med.* 2012.

76. Wang, X., Meloche, M., Verchere, C.B. et al. (2011). Improving islet engraftment by gene therapy. *J. Transplant.*

77. Glick, D., Barth, S., and Macleod, K.F. (2010). Autophagy: cellular and molecular mechanisms. *J. Pathol.* 221: 3–12.

78. Ebato, C., Uchida, T., Arakawa, M. et al. (2008). Autophagy is important in islet homeostasis and compensatory increase of beta cell mass in response to high-fat diet. *Cell Metab.*

79. Rivera, J.F., Costes, S., Gurlo, T. et al. (2014). Autophagy defends pancreatic β cells from Human islet amyloid polypeptide-induced toxicity. *J. Clin. Investig.*

80. Sun, K.T., Chen, M.Y.C., Tu, M.G. et al. (2015). MicroRNA-20a regulates autophagy related protein-ATG16L1 in hypoxia-induced osteoclast differentiation. *Bone.*

81. Qureshi, S.T., Gros, P., and Malo, D. (1999). Host resistance to infection genetic control of lipopolysaccharide responsiveness by TOLL-like receptor genes.

82. Lester, S.N. and Li, K. (2014). Toll-like receptors in antiviral innate immunity. *J. Mol. Biol.* 426: 1246–1264.

83. O'Brien, G.C., Wang, J.H., and Redmond, H.P. (2005). Bacterial lipoprotein induces resistance to gram-negative sepsis in TLR4-deficient mice via enhanced bacterial clearance. *J. Immunol.*

84. Poltorak, A., He, X., Smirnova, I. et al. (1998). Defective LPS signaling in C3H/HeJ and C57BL/10ScCr mice: Mutations in Tlr4 gene. *Science.*

85. Yu, L., Wang, L., and Chen, S. (2010). Endogenous toll-like receptor ligands and their biological significance. *J. Cell. Mol. Med.*

86. Methe, H., Zimmer, E., Grimm, C. et al. (2004). Evidence for a role of toll-like receptor 4 in development of chronic allograft rejection after cardiac transplantation. *Transplantation.*

87. Palmer, S.M., Burch, L.H., Mir, S. et al. (2006). Donor polymorphisms in Toll-like receptor-4 influence the development of rejection after renal transplantation. *Clin. Transplant.*

88. Thornley, T.B., Brehm, M.A., Markees, T.G. et al. (2006). TLR agonists abrogate costimulation blockade-induced prolongation of skin allografts. *J. Immunol.*

89. Goldberg, A., Parolini, M., Chin, B.Y. et al. (2007). Toll-like receptor 4 suppression leads to islet allograft survival. *FASEB J.*

90. Wang, H., Wu, H., Rocuts, F. et al. (2012). Activation of peroxisome proliferator-activated receptor γ prolongs islet allograft survival. *Cell Transplant.*

91. Miles, P.D.G., Barak, Y., He, W. et al. (2000). Improved insulin-sensitivity in mice heterozygous for PPAR-γ deficiency. *J. Clin. Investig.*

92. Halabi, C.M., Beyer, A.M., De Lange, W.J. et al. (2008). Interference with PPARγ function in smooth muscle causes vascular dysfunction and hypertension. *Cell Metab.*

93. Lehrke, M. and Lazar, M.A. (2005). The many faces of PPARγ. *Cell* 123: 993–999.

94. Bilban, M., Bach, F.H., Otterbein, S.L. et al. (2006). Carbon monoxide orchestrates a protective response through PPARgamma. *Immunity* 24: 601–610.

95. Fox, C.S., Coady, S., Sorlie, P.D. et al. (2004). Trends in cardiovascular complications of diabetes. *J. Am. Med. Assoc.*
96. Sena, C.M., Pereira, A.M., and Seiça, R. (2013). Endothelial dysfunction - A major mediator of diabetic vascular disease.
97. Giacco, F. and Brownlee, M. (2010). Oxidative stress and diabetic complications.
98. Wang, R. (1998). Resurgence of carbon monoxide: an endogenous gaseous vasorelaxing factor. *Can. J. Physiol. Pharmacol.* 76: 1–15.
99. Wang, R. and Wu, L. (1997). The chemical modification of KCa channels by carbon monoxide in vascular smooth muscle cells. *J. Biol. Chem.* 272: 8222–8226.
100. Wang, R., Wang, Z., and Wu, L. (1997). Carbon monoxide-induced vasorelaxation and the underlying mechanisms. *Br. J. Pharmacol.* 121: 927–934.
101. Rodella, L., Lamon, B.D., Rezzani, R. et al. (2006). Carbon monoxide and biliverdin prevent endothelial cell sloughing in rats with type I diabetes. *Free Radic. Biol. Med.*
102. Negi, G., Nakkina, V., Kamble, P., et al. (2015). Heme oxygenase-1, a novel target for the treatment of diabetic complications: Focus on diabetic peripheral neuropathy.
103. Méndez-Lara, K.A., Santos, D., Farré, N. et al. (2018). Administration of CORM-2 inhibits diabetic neuropathy but does not reduce dyslipidemia in diabetic mice. *PLoS ONE*.
104. Hervera, A., Leánez, S., Negrete, R. et al. (2012). Carbon monoxide reduces neuropathic pain and spinal microglial activation by inhibiting nitric oxide synthesis in mice. *PLoS ONE*.
105. Singh, U.P., Singh, N.P., Murphy, E.A. et al. (2016). Chemokine and cytokine levels in inflammatory bowel disease patients. *Cytokine*.
106. Cowley, M.J., Weinberg, A., Zammit, N.W. et al. (2012). Human islets express a marked proinflammatory molecular signature prior to transplantation. *Cell Transplant.*
107. Dinić, S., Grdović, N., Usković, A. et al. (2016). CXCL12 protects pancreatic β-cells from oxidative stress by a Nrf2-induced increase in catalase expression and activity. *Proc. Jpn. Acad., Ser. B.*
108. Chen, T., Yuan, J., Duncanson, S. et al. (2015). Alginate encapsulant incorporating CXCL12 supports long-term allo- and xenoislet transplantation without systemic immune suppression. *Am. J. Transplant.*

25

Carbon Monoxide and Acute Kidney Injury

Mark de Caestecker

Division of Nephrology, Department of Medicine, Vanderbilt University Medical Center, Nashville, TN, USA

Introduction

Acute Kidney Injury: Definition, Causes, and the Clinical Problem

Acute kidney injury (AKI) complicates ~20% of hospitalizations, is an independent predictor of mortality, and often results in incomplete recovery, giving rise to chronic kidney disease (CKD) and, in some cases, end-stage renal disease (ESRD) [1–3]. Simply defined in clinical practice as a rapid decline in kidney function that occurs over a 7-day period or less [1], AKI is in fact a highly heterogeneous clinical syndrome that is precipitated by different injurious events, often in combination. These include the effects of reduced renal blood flow due to dehydration, blood loss, and heart failure treatment; sepsis; major surgery (including cardiac bypass surgery); cold and warm ischemia–reperfusion injury (IRI) as a result of cold storage and subsequent warm renal transplant surgery; crush injury resulting in extensive muscle damage causing rhabdomyolysis; and drug toxicities [1–3]. AKI also often occurs in association with pre-existing comorbidities, including CKD, old age, and diabetes, all of which increase the risk of AKI and are also likely to influence the underlying pathobiology and response to treatment [2–4]. In addition, there are marked differences in individual responses to the same injury, with some otherwise indistinguishable patients showing complete functional recovery after AKI, while others have delayed or incomplete recovery [4].

Mechanisms of AKI

Several common cellular and molecular responses to AKI have been identified from different rodent models. These findings have to be tempered because concerns about the risks of renal biopsies in patients with AKI have limited our ability to examine renal tissues in order to determine whether patients share common cellular and molecular responses to those seen in rodent models of AKI [5]. Having said that, limited postmortem studies indicate that there are shared elements in rodent and human AKI. These include tubular injury in the corticomedullary region, including occasional necrosis and apoptosis of tubular epithelium, extensive epithelial sloughing, increased tubular proliferation, and peritubular inflammatory infiltrates [6–8]. Furthermore, data from several large-scale single-cell transcriptomic studies of kidneys from mice and humans with IRI-induced AKI (IRI-AKI) indicate that at least one of the more commonly used experimental models of AKI, IRI-AKI [9], shares common epithelial, inflammatory, and metabolic responses to those seen in human AKI [10–13].

Experimental Models of AKI

There are several frequently used rodent models of AKI. The first is IRI-AKI, induced by surgical clamping of the renal pedicle. This model provides a defined and controllable injury, in which there is dominant injury to one of the most metabolically active tubular segments of the nephron, the S3 segment of proximal tubular epithelial cells (PTECs) [14]. While the severity of cellular damage is rarely seen in human AKI biopsies [15], this model has been widely used to model effects of reduced renal blood flow associated with cardiac surgery, renal transplantation, nephron sparing surgery for renal tumors (which requires clamping of the renal pedicle), hypotension, and shock [9], and is the predominant model that has been used to study the mechanisms of maladaptive repair and CKD progression after AKI [16]. The second type of models is chemically induced, including drug- or toxin-induced AKI (including the effects of endogenously produced toxins). A variety of exogenous and endogenous toxins have been used in experimental models to recapitulate the effects of nephrotoxic agents in AKI [17]. Commonly used models include (i) cisplatin-induced AKI, both short-term high-dose and long-term, repeat low-dose models [18,19], modeling the effects of kidney injury resulting from the use of cisplatin chemotherapy for patients with a variety of different solid organ malignancies [20]; (ii) rhabdomyolysis-induced AKI through intramuscular glycerol injection leading to muscle damage and subsequent systemic release of myoglobin and heme-associated oxidative damage, modeling the effects of trauma and crush injury [21]; and (iii) heme-associated kidney injury resulting from intravascular hemolysis in patients with malaria, sepsis-associated AKI (SA-AKI), and cardiac surgery [22–24]. The primary site of injury induced by these toxins depends largely on cellular uptake. For example, cisplatin is taken up and concentrated in PTECs following uptake from the circulation by the basolateral organic cation transporters, while myoglobin (and free hemoglobin) is filtered by the glomerulus and concentrated in PTECs by the apical membrane megalin/LRP2 endocytic receptor system [24,25]. The third type of models is SA-AKI. A variety of models are used to recapitulate the mechanisms of SA-AKI [9]. A number of features of SA-AKI in humans, including an early hyperdynamic circulatory state associated the systemic inflammatory response syndrome, followed by hypotension and septic shock, are recapitulated in a model of polymicrobial sepsis induced by surgical induction of cecal ligation and puncture (CLP), but it has been a challenge for many laboratories to induce survivable injury that is associated with reduced renal function [26]. However, a feature of this model that has garnered much attention is the observation that like humans with severe SA-AKI [6], rodents with even marked loss of renal function usually show relatively mild cellular damage in the kidney. Studies in sepsis models have also provided important insights into how inflammation, and microcirculatory and mitochondrial dysfunction contribute to the pathophysiology of AKI even in the absence of widespread cell death [27].

Cellular Responses in AKI

Irrespective of the initial causes of injury, the cellular responses to AKI can be divided into primary responses that are the direct result of the injurious event, for example, PTEC injury after IRI and toxin-induced AKI, and secondary responses that extend the injury [28]. The latter include the effects of sterile inflammatory responses to cellular damage, which propagate tissue injury through release of inflammatory proteases and reactive oxygen species. In addition, there is often renal vasoconstriction and thrombotic occlusion of the renal microvasculature, which gives rise to secondary ischemic tissue injury. These early events are usually followed by a phase of resolution in which localized renal blood flow is restored, dead cells are removed, and cellular repair is initiated [29]. It is in this last phase that surviving damaged cells undergo proliferative repair and redifferentiation into functional epithelium and vasculature, or if there is a failure of repair, it results in the formation of dedifferentiated, proinflammatory inflammatory niduses of senescent cells that promote fibrosis and are thought to underly CKD progression [16,30].

Primary cellular responses to the initiating injury vary depending on the nature of the injury, but include (i) apoptosis and necrosis precipitated by ATP depletion in IRI-AKI [14]; (ii) apoptosis caused by heme-induced oxidative stress and lipid peroxidation in rhabdomyolysis, and mitochondrial damage caused by low-dose cisplatin injury [18,25]; and (iii) in SA-AKI, mitochondrial damage and renal tubular dysfunction often without cell death [31]. After the initiating injury, release of damage-associated molecular patterns (DAMPs) by dead and dying cells, upregulation of adhesion molecules by activated endothelium, and secretion of chemokines by damaged epithelial cells promote a robust inflammatory response, with early infiltration by neutrophils, monocytes/macrophages, and natural killer cells, leading to further cellular damage by local release of reactive oxygen species and proteases, as well as activation of damaging adaptive immune responses involving both T and B lymphocytes [32]. Local vasoconstriction also plays an important role in amplifying injury after IRI-AKI, toxin-induced AKI, and SA-AKI. In part, this results from activation of a maladaptive physiological feedback mechanism, called tubuloglomerular feedback (TGF). In TGF, proximal tubular dysfunction induced by injury increases delivery of chloride ions to more distal tubules, triggering vasoconstriction of the afferent glomerular arteriole, which in turn reduces glomerular blood flow (and thereby reducing glomerular filtration and decreasing chloride delivery to the distal tubules), but also reduces postglomerular capillary blood flow, which is required for oxygenation of the remaining renal tubular structures [33]. Other mechanisms include vasoconstriction due to endothelial dysfunction and microvessel occlusion resulting from inflammatory cell adhesion and activation of procoagulant pathways by injured endothelium [33,34].

Survival of injured epithelium after AKI is dependent on restoration of normal homeostasis during the secondary phase of injury. Since mitochondrial dysfunction associated with depolarization, fragmentation, and release of damaging reactive oxygen species plays a major role in most experimental forms of AKI [18,31,35], restoration of mitochondrial homeostasis understandably plays a critical role in determining long-term outcomes after AKI. Mechanisms promoting mitochondrial homeostasis that have been shown to ameliorate AKI include activation of mitochondrial fusion and biogenesis by Pgc1α and mitophagy to remove damaged mitochondria [35,36].

Cellular repair is initiated once the primary and secondary cellular responses to injury start to resolve. Since all compartments in the kidney may be affected, this includes restoration of functioning renal tubular epithelium by proliferative repair and redifferentiation of tubular epithelium, resolution and removal of inflammatory cells, and angiogenesis to repair and/or replace severely damaged peritubular capillaries. In addition, there is remodeling and resolution of increased extracellular matrix deposition and fibroblast proliferation that occurs in the interstitial space between the renal tubules and contributes to post-AKI fibrosis [16,37]. Failed repair to any of these compartments contributes to CKD progression after AKI. Failed tubular repair results in the persistence of epithelium that has molecular features of senescent cells and secretes proinflammatory and profibrotic factors that promote local inflammation and fibrosis [16,30]. Incomplete repair of damaged endothelium results in peritubular capillary rarefaction, which reduces oxygen and nutrient delivery to affected areas of the kidney promoting fibrosis and is closely associated with CKD progression in different models of AKI [29,38]. Finally, accumulation of inflammatory foci promotes further injury and results in local tissue destruction [37].

Therapeutic Opportunities for AKI

Despite the clinical importance of AKI, no therapeutic interventions have been shown to definitively prevent AKI, improve recovery, or improve long-term outcomes after AKI, including CKD, ESRD, or death [39–42]. For this reason, there is a broad consensus among clinical and basic AKI research communities that development of the first effective disease-modifying treatments for AKI will address an important, unmet medical need [43]. There are a variety of reasons for these therapeutic failures. In part, this stems from an inherent weakness in clinical study design for patients with AKI due to the lack of reliable clinical and/or biological predictors to determine who will develop and recover from AKI, limiting our ability to recruit homogeneous populations of patients with similar disease pathophysiologies [44–46]. However, there is also concern that preclinical models used to establish the focus of therapeutic interventions in humans have failed to model the major pathophysiologies and comorbidities associated with human AKI, and have not been used to identify biomarkers that could be used to predict drug dosing required in clinical trials [9,43]. Both of these limitations represent challenges but also opportunities for therapeutic development. First, it is anticipated that clinical, genetic, and transcriptomic data from research on kidney biopsies obtained from patients with AKI through the NIDDK-funded Kidney Precision Medicine Project will allow us to more accurately phenotype patients with AKI for inclusion in clinical trials [46,47]. These studies will also allow us to better model the molecular mechanisms of AKI using new experimental model systems (including the use of humanized mouse models, established models of AKI in the context of common comorbidities, notably diabetes and CKD, and using humanized organs on a chip and human kidney organoids to study AKI responses) [9,48]. However, pending these developments, there are opportunities for therapeutic development using agents that target a broad array of cellular responses that contribute to pathogenesis of AKI. This obviates the need to identify a single molecular target that can be expected to ameliorate injury and/or enhance repair after AKI in a heterogeneous population of patients with AKI. Two main approaches to achieve this include the use of phenotypic, rather than target-based, drug discovery platforms that aim to target common cellular responses rather than isolated molecular targets [49], and the exploitation of therapeutics that are known to target a broad array of cellular responses that are common to diverse forms of AKI. As we will argue in the ensuing sections, the highly pleotropic effects of CO-based therapeutics offer this opportunity for AKI [50].

Heme Oxygenases and CO in AKI

Protective Role of Heme Oxygenases in AKI

Protective effects of CO in the kidney were first suggested from studies on the role of heme oxygenase (HO) in rhabdomyolysis-induced AKI [51]. Of the HOs, HO-1 is a ubiquitous, inducible stress response gene that is upregulated in all cells in response to cellular stress, while HO-2 is constitutively expressed, largely expressed in neurons and the vasculature. Both play essential roles in cellular detoxification of heme proteins. HOs catalyze oxidative degradation of heme, generating CO, ferrous iron, and biliverdin, which is reduced to bilirubin, as obligate by-products of heme metabolism [52]. Heme is a prosthetic group that is found not only in hemoglobin and myoglobin, but also in hundreds of other cellular proteins, including mitochondrial respiratory cytochromes and cytochrome P450s, regulating cellular energetics and metabolism; nitric oxide (NO) synthase, guanylate cyclase, and prostaglandin synthase, which regulate vascular tone; antioxidant enzymes catalases and peroxidases; and oxidant-generating NADPH oxidase [53]. While the major fraction of heme is bound to hemoproteins, low levels of free intracellular and extracellular heme are required to facilitate dynamic exchange of heme between hemoproteins depending on the cellular needs. This is facilitated by a large family of intracellular heme transporters and trafficking proteins, while heme buffering proteins such as hemopexin and haptoglobin bind to and mitigate heme toxicity [53], as discussed earlier. Homeostatic control of heme is critical to maintain low levels of unbound heme, and results from a balance in synthesis versus HO-dependent catabolism. Low micromolar quantities of free heme induce p21-dependent cell cycle arrest and apoptosis, while high levels of heme promote oxidative stress, with damaging lipid peroxidation and activation of inflammatory responses [54]. High levels of extracellular and circulating heme in patients with rhabdomyolysis and intravascular hemolysis [24,25] act as DAMPs, which can activate TLR4-dependent inflammatory responses [55].

In patients with rhabdomyolysis, myoglobin, the second most abundant hemoprotein in the body, is released from damaged skeletal muscle into the circulation, filtered into the glomerular filtrate, and concentrated in PTECs by megalin/LRP2-dependent endocytosis, increasing intracellular free heme, which promotes oxidative damage to the kidney leading to AKI [25]. Given the role of the HO system in detoxifying unbound heme, Nath et al. studied the role of HO-1 in mitigating injury in a rat model of rhabdomyolysis-induced AKI [51]. In an elegant series of experiments conducted nearly 30 years ago, they showed that HO-1 was rapidly upregulated in rat kidneys after induction of rhabdomyolysis by intramuscular glycerol injection. They demonstrated that further stimulating HO-1 expression by intravenous injection of hemoglobin prior to injury (preconditioning) and inhibition of HO-1 activity using the competitive inhibitor of HO, tin protoporphyrin, attenuated and exacerbated AKI, respectively. At the time, the protective effect of HO-1 induction was thought to be largely mediated by detoxification of free heme released from excess myoglobin in the kidney by HO-1. However, since that time, a number of other pharmacological and more definitive genetic loss- and gain-of-function studies have shown that HO-1 is protective in a number of different models of AKI, including SA-AKI [56], cisplatin-induced AKI [57–60], and IRI-AKI [61–63]. While there is evidence that increased circulating levels of cell-free hemoglobin contributes to SA-AKI [22], and that heme released from cytochrome P450 enzymes after IRI also contributes to tissue injury after IRI-AKI [64], these data suggest that other, noncanonical effects of HO-1 might also mediate the salutary effects of HO-1 in AKI.

HO-1 in Human AKI

The clinical importance of the HO-1 pathway is underscored by the enhanced oxidative stress, hyperinflammatory state, and endothelial dysfunction described in rare cases of HO-1 deficiency in humans [65]. Promotor mapping indicates that while many of the same stress responses regulate HO-1 expression in humans and mice, there are differences that need to be considered when comparing the role of HO-1 in human and experimental models of AKI [66]. Despite this, HO-1 is strongly induced and is protective in rhabdomyolysis and high-dose cisplatin-induced AKI in a humanized mouse model in which HO-1 expression was reconstituted in *HO-1*-deficient mice using a human BAC transgene containing the human HO-1 gene locus (also known as *HMOX1*) [67]. These data indicate that despite differences in gene regulation, common mechanisms regulate renal HO-1 expression in mice and humans with AKI. In line with this, there is increased urinary and plasma HO-1 in mice after IRI-AKI, high-dose cisplatin-induced AKI, and rhabdomyolysis-induced AKI [68], and increased urinary and/or plasma HO-1 has been identified in ICU patients with AKI, and in patients with AKI after cardiac surgery and SA-AKI [68–70]. These findings are reinforced by evidence that there are common sequence length polymorphisms in the regulatory region of the *HMOX1* gene locus that are risk factors for the development of AKI after cardiac surgery and sepsis [70,71], and are associated with reduced plasma HO-1 levels in septic patients with AKI [70]. These data provide strong evidence that HO-1 and its downstream effectors, including CO, are likely to be valid therapeutic targets for human AKI.

Therapeutic Manipulation HO-1 Expression in AKI

One approach that has been used to modulate HO-1 activity after AKI has been to increase HO-1 expression therapeutically. Unlike CO (or bilirubin) therapeutics, this has the advantage of not only increasing downstream effectors of HO-1 activity, but also decreasing toxic levels of unbound heme that accumulates in the kidney in a variety of different forms of AKI [22–25,64]. Low doses of heme analogues and iron-containing porphyrins (including hemin), which are insufficient to promote tissue injury, are potent inducers of HO-1 expression in the kidney, and have been used as preconditioning agents to reduce the severity in cisplatin-induced AKI, IRI-AKI, and rhabdomyolysis-induced AKI [51,72–75]. It is also notable that HO-1 induction by heme and iron-containing porphyrins decreases insulin resistance, reduces systemic inflammation, and reduces blood pressure in obese, leptin-deficient mice [76]; reduces proteinuria and renal interstitial inflammation in obese Zucker rats [77]; and ameliorates the severity of AKI in aged mice that otherwise show a marked reduction of renal HO-1 induction after IRI-AKI [78]. These data suggest that HO-1 induction may have additional protective effects in AKI associated with diabetes, CKD, and aging. However, because high levels of unbound heme promote oxidative stress, lipid peroxidation, and activation of inflammatory responses in the kidney [54,55], dosing and timing of treatment are critical since delayed treatment and/or increased heme or porphyrin doses delivered systemically also exacerbate IRI-AKI and rhabdomyolysis-induced AKI [79,80]. Genetic approaches have also been exploited to enhance HO-1 expression for renal xenotransplantation, with the generation of transgenic minipigs that ubiquitously overexpress HO-1 [81–83], and intravenously infused bone marrow-derived macrophages that have been transduced *in vitro* by an adenoviral HO-1 expression vector improve functional recovery after IRI-AKI [84]. However, despite their promise, likely because of their potential toxicities and concerns about dosing and timing of therapy, none of these approaches has yet been translated into humans.

Other approaches to enhance HO-1 expression have utilized a variety of nonspecific approaches to activate NRF2-dependent induction of HO-1 [66]. These have included the use of resveratrol and bardoxolone, both of which activate NRF2, increase HO-1 expression, and ameliorate IRI-AKI, toxin-induced AKI, and SA-AKI [85–87]. However, it is unclear whether these effects are mediated through HO-1, or other NRF2-dependent antioxidant responses. Interestingly, CO itself is a potent inducer of HO-1 expression through activation of an NRF2-dependent positive feedback loop [88]. Thus, CO therapy has the potential not only to activate CO-dependent protective responses, but also to activate potent NRF2-dependent antioxidant defenses, including free heme detoxification by HO-1.

CO as a Mediator of HO-1 Activity in AKI

HO-1 induction has several effects that mitigate injury in AKI. These include (i) anti-inflammatory, antioxidant, cytoprotective, and antiapoptotic effects; (ii) increased injury-associated autophagy and mitochondrial quality control; and (iii) vasodilatory, antithrombotic, and proangiogenic effects in the vasculature [52,66,89–91]. These effects may in part be mediated by HO-1-dependent degradation of free heme, but there is increasing evidence that many of these salutary effects of HO-1 in diverse tissue injuries may be mediated through noncanonical effects of the by-products of heme metabolism, biliverdin/bilirubin, ferrous iron, and CO, all of which have been shown to have salutary effects in alleviating tissue injury [88]. Biliverdin and bilirubin are potent free radical scavengers that inhibit damaging lipid peroxidation (the reader is referred to Chapter 10 on the protective effects of bile pigments). They have anti-inflammatory effects on leucocyte attachment and migration to activated endothelial surfaces, and bilirubin ameliorates injury in IRI-AKI in rats [91]. Though ferrous iron released from degraded heme is also toxic due to its ability to induce lipid peroxidation and iron-dependent oxidative programmed cell death, or ferroptosis [92], it also induces expression of the iron storage protein, ferritin, in PTECs and macrophages after AKI, which itself has antioxidant and anti-inflammatory effects, and is protective in mice with rhabdomyolysis- and cisplatin-induced AKI [91,93]. However, there is also a substantial body of data to indicate that CO released by heme degradation locally accounts for the many of the salutary effects of HO-1 [88]. While definitive evidence showing that CO treatment rescues HO-1 knockout mice from increased susceptibility to AKI is lacking, extensive work indicates CO therapies have many of the same effects as HO-1 in AKI. Moreover, CO reverses the increased mortality of HO-1 knockout mice with sepsis, and the cytoprotective effects of HO-1 in endothelial cells require CO [94,95]. These data suggest that salutatory effects of CO would compensate for loss of HO-1 in AKI.

CO is a freely diffusible, stable gas, most of which binds to hemoglobin to form carboxyhemoglobin (COHb) in the lungs where it is eliminated in the expired air [50]. Most CO production is an integral part of red blood cell heme turnover by the reticuloendothelial system in the spleen and liver. During periods of stress, for example, associated with infections and hemolytic sickle cell crises, increased CO production in the body can be detected in the exhaled breath that is solely due to HO-1 activity. Much of our understanding of the role of CO in mediating the protective effects of HO-1 is based on the therapeutic use of CO in which nontoxic doses of CO have been shown to have similar effects to HO-1 in experimental models of tissue injury. This has been shown in acute lung injury [96–99], sepsis [94,100], cardiac bypass [101], and cardiovascular disease [102], in addition to AKI [50]. These effects are very similar to the effects of HO-1 induction after tissue injuries and include (i) cytoprotective effects through activation of antioxidant, mitochondrial biogenesis, and mitophagy pathways; (ii) potent anti-inflammatory effects; and (iii) vasoprotective and vasodilatory effects [103–106]. These effects would be expected to ameliorate initial injury, and decrease secondary injury resulting from inflammation and reduced blood flow. CO also protects against ER stress-induced endothelial apoptosis and promotes reparative angiogenesis [88]. In addition, CO has been shown to enhance DNA damage response pathways after injury [107], which would reduce maladaptive repair associated with epithelial senescence after AKI [16]. Taken together, these data indicate that therapeutic use of CO has many of the same salutary effects on tissue injury and repair as HO-1 induction, and suggest that bypassing HO-1 induction with pharmacological delivery of CO would provide similar therapeutic benefits in AKI.

CO Therapeutics for AKI

A variety of delivery systems have been used to achieve therapeutic levels of CO in experimental and human tissue injuries. We will focus the rest of our discussion on those CO therapies that have been evaluated for the prevention and/or treatment of AKI. The reader is referred to other chapters for more comprehensive discussions of other CO therapeutics and CO delivery systems that have yet to be evaluated for therapeutic efficacy in AKI.

Inhaled CO Gas

The first demonstration that CO could be used therapeutically to ameliorate organ injury used inhaled CO (iCO) to mitigate lung injury in rats exposed to 100% oxygen. In their landmark 1999 paper, Otterbein et al. showed that 250 ppm of CO gas reduced mortality in rats exposed to hyperoxia from 100% to 0% at 72 h [99]. Since that time, a number of studies have used iCO in a variety of experimental models of organ injury, with evidence that even short-term exposure to iCO (1 h) has salutary effects in rodent and large animal models [108]. iCO studies in AKI have been largely restricted to models of IRI. One study, which was principally focused on evaluating the protective effects of iCO on short-term (4-h) lipopolysaccharide (LPS)-induced acute lung injury, also showed that pretreatment with 250 ppm of iCO for 1 h improved renal function in this model [109]. However, this study also showed that iCO pretreatment improved cardiac function in this model, so it is unclear whether beneficial effects of iCO pretreatment in a hyperacute model of severe sepsis reflect direct renal protection or improved renal function resulting from increased cardiac output and increased renal perfusion pressures. Direct evidence for renal protection by iCO therapy has been provided in a series of studies modeling IRI associated with renal transplantation. In the first study, Neto et al. demonstrated that treatment with 250 ppm of iCO 1 h before and 24 h after orthotopic, syngeneic transplantation in rats using kidneys maintained for 24 h in cold storage solution at 4°C showed improved survival and graft function [110]. This was associated with peak COHb levels of 25% (increasing from baseline of 2.3%). In subsequent studies, the same group demonstrated that 20 ppm of iCO with maximal COHb levels of 6.1% was enough to improve renal recovery at 24 h using the same orthotopic renal transplant model [111]. These findings have since been recapitulated in a 6-day pig study of allogeneic renal transplantation in which animals were placed on immunosuppressive therapy (FK506) at the time of transplantation [112]. In these studies, animals were treated with iCO for only 1 h, starting at the time of surgery. Untreated animals had impaired functional recovery over the 6-day course of these studies. This models delayed graft function seen in clinical practice, which commonly occurs in patients transplanted with kidneys that have undergone prolonged cold storage, and is a harbinger of bad short- and long-term renal outcomes after kidney transplantation [113,114]. While iCO had no effect on renal function for the first 2 days after transplantation, iCO treatment markedly accelerated functional recovery over the next 4 days. These findings suggested that short-term treatment with iCO at the time of renal transplant surgery might provide a practical approach to deliver iCO in a controlled perioperative environment, and that this might have long-term beneficial effects on renal transplantation outcomes. Moreover, by protecting kidneys from long-term effects of prolonged cold ischemia, iCO therapy might expand patient access to donor kidneys that might otherwise have been discarded because they had undergone more prolonged periods of cold ischemia. This led to the first clinical trial of perioperative iCO to prevent delayed graft function in renal transplant recipients (NCT00531856) [101]. This phase 1/2 study was sponsored by Ikaria who hold the patent on COVOX, an iCO delivery device for ventilated patients that allows precise CO dosing adjusted for patient weight. The study was recruiting patients between 2007 and 2011 when the study was withdrawn for business reasons. No adverse events were reported within the allowable limit of COHb of 14%, but results of this study have not been made public [50,115].

Aside from renal transplantation, perioperative iCO at 250 ppm has also been shown to improve renal function and reduce renal injury in short-term (2-h) pig studies of cardiopulmonary bypass surgery (CBP), where maximal COHb levels of 9.1% were sufficient to ameliorate CBP-induced AKI [116], and in a mouse model of IRI-AKI, pretreatment with 250 ppm of iCO improved functional recovery 24 h after inducing bilateral IRI surgically [117]. Taken together, these data indicate that iCO is effective in IRI-associated AKI when given prior to or at the time of the initial injury, and suggest iCO as a promising perioperative treatment for human transplantation. However, this promise has yet to be proven in a properly powered clinical trial and key questions remain, including whether iCO is effective in other forms of AKI, whether treatment must be initiated before the injury occurs, or whether the treatment can be delayed in models of AKI that have different kinetics and mechanisms of injury. This has important implications, since aside from surgery-associated AKI, or therapy-induced AKI initiated in a hospital setting, most patients presenting with AKI already have had their renal insult before they are first evaluated by a physician [43]. In addition to these questions, there are intrinsic portability, dosing, and environmental safety concerns outside of the controlled and closely monitored surgical and ICU environments of ventilated patients that present significant challenges for widespread implementation of iCO as a therapeutic for many different forms of AKI.

CO-Enriched Solutions and Hemoglobin Products

Because of the practical challenges in implementing iCO therapy for all but the most intensively monitored patients in perioperative and ICU environments, there has also been considerable interest in developing alternative delivery systems for CO. For example, one group has shown that CO-enriched cold storage solution (generated by bubbling UW cold storage solution with 5% CO gas for 10 min) improves renal function and survival in a syngeneic rat transplant model of prolonged cold storage (24 h) [118]. The same group has shown that CO-enriched UW solution improves renal functional recovery over a 14-day period in a 48-h cold storage renal autotransplantation pig model, and that this is associated with an approximately 10-fold increase in tissue CO levels (with no change in recipient COHb levels) at the time of transplantation [119]. This indicates that despite the relatively low aqueous solubility of CO, enough is dissolved in the storage solution to increase tissue CO levels. As an alternative delivery vehicle, Hillhurst Pharmaceuticals developed a patented orally active liquid CO formulation, HBI-002, that has been primarily developed as a therapy for sickle cell disease (by reducing vascular occlusion and inflammation), and reproducibly increases COHb to 4–6% shortly after gavage in mice [120]. Pretreatment with HBI-002 also improves renal function after bilateral IRI-AKI in mice [117]. This is also notable since there is an ongoing, dose escalation phase 1 safety trial of HBI-002 in normal volunteers (NCT03926819).

CO-enriched hemoglobin products have also been shown to ameliorate AKI. For example, transfusion with CO-enriched red blood cells, generated by bubbling CO gas through red cell preparations *ex vivo* [121], has been shown to improve survival, renal function, and kidney injury in rats 24 h after rhabdomyolysis-induced AKI resulting from crush injury or glycerol-induced muscle damage, with or without associated hemorrhagic shock [122]. As an alternative more stable product, Sanguinate, which is CO-enriched pegylated bovine hemoglobin, acts as an oxygen carrier and blood substitute that releases CO rapidly after infusion and is safe for use in humans [123]. This has been shown to improve renal function and kidney injury in a short-term (1-h) model of severe hemorrhagic shock in rats when compared with resuscitation using blood [124], and to improve renal cortical microcirculatory blood flow and oxygenation (but not renal function) in a short-term model of LPS-induced shock in rats [125]. It is notable that unlike other forms of acellular hemoglobin [126], there is no evidence that Sanguinate sequesters NO and thereby reduces vasodilatory effects of NO [124]. In fact, Sanguinate has been shown to act a vasodilator in different vascular beds. Further studies are needed to determine whether protective effects of Sanguinate are mediated by CO release, but this has promise as a safe alternative to iCO for renal protection from AKI in ER or ICU settings. Of note, a phase 2 study using Sanguinate to prevent cadaveric donor delayed graft function in renal transplant recipients was completed in 2016, but the study results have not been published (NCT024490202).

Metal-Based CO-Releasing Molecules

A number of approaches have been used to develop small molecules capable of releasing CO under physiological conditions [115]. To this end, Motterlini and others began to study metal-based CO-releasing molecules (CORMs), which vary in the metal used (Mn^{2+}, Fe^{2+}, Ru^{2+}, etc.), and properties modulating CO release rates and triggers for CO release (water, light, and enzymes) [127,128]. Much of this discussion is restricted to two of the most extensively studied CORMs, CORM-2 and CORM-3, which are Ru(II) based and capable of spontaneously transferring CO to a CO acceptor, such as myoglobin, in aqueous solution [129,130]. CORM-2 improves renal function and tissue injury in short-term (24-h) LPS-induced AKI in mice [131,132] and in a longer-term model of polymicrobial sepsis induced by surgical CLP in rats [133]. CORM-2 also improves cardiac function in the CLP sepsis model [127], so these protective effects may result from improved renal perfusion. However, pretreatment of donor rats with CORM-2 or CORM-3 before kidney harvesting increases survival and improves renal function in recipients after allogenic transplantation [134], and addition of CORM-3 to cold perfusates improves renal function and blood flow in *ex vivo* pig kidneys harvested after 30 min of warm ischemia and prolonged cold storage [135]. This indicates that CORM-2 and CORM-3 have direct renoprotective effects. Further, CORM-3 reduces injury and improves renal functional recovery after bilateral IRI-AKI in mice [136], and in rats with moderately severe cisplatin-induced AKI [137]. It is notable, however, that with the exception of the CLP-AKI study in which CORM-2 was given 3 h after surgery [133], in all other published studies, CORMs were given before the initiating injury. Also, CORM-3 does not improve renal functional recovery after bilateral IRI-AKI if treatment is delayed even 1 h after surgery [136]. Consistent with iCO IRI and renal transplantation studies discussed earlier, these data suggest that at least in models of IRI-AKI, CORM-2 and CORM-3 have to be given in advance of the initiating event. While there are no published data on their use in models of AKI, other CORMs have been developed with distinct CO release characteristics that protect against prolonged cold ischemic injury when mixed with kidney preservation solutions in rabbit and pig isolated perfused kidney preparations. These include CORM-A1 [138], which contains the metal-like element, boron, as the CO acceptor and releases CO more reproducibly than either CORM-2 or CORM-3 *in vivo* [139,140], and manganese-based CORM-401 [141], which releases three molecules of CO per CORM [142], and is more potent than CORM-A1 [143]. In this context, it is notable that unlike iCO, doses of CORM-3 that are effective in ameliorating injury in different mouse models have no detectable effect on COHb levels [128,144], while CORM-A1 and CORM-401 treatments induce a clear dose-dependent increase in COHb levels in mice over time, each with different decay kinetics [145,146]. Further, while there is no increase in COHb levels, there is a two- to threefold increase in CO tissue levels in kidney, liver, and spleen in mice within 10 min of an i.v. injection with CORM-3. This suggests that differences in CO release characteristics of different CORMs are likely to influence tissue distribution and kinetics, and potentially timing of therapeutic interventions in different models of AKI.

CORMs also have nonrenal effects that may favorably influence AKI outcomes in the context of common comorbidities associated with AKI. For example, in addition to CORM-2 improving cardiac function in the CLP model of polymicrobial sepsis [127], CORM-3 reduces cardiac damage and improves functional recovery in rodent models of acute myocardial infarction [128,147], and CORM-401 improves insulin sensitivity in obese, diabetic mice on high-fat diets [145]. These findings raise the possibility that CORMs may have further protective effects in AKI associated with cardiac disease and diabetes, and so also need to be evaluated in models of AKI with cardiorenal syndrome and diabetes [9,148]. With that said, there are some intrinsic issues with the more commonly used metal-based CORMs that have raised concerns about further development for the treatment of AKI in humans [115]. First, the most widely used CORMs have been shown to have CO-independent activities [144,149–163], which may confound analysis and dose scheduling based on COHb and/or tissue CO levels in AKI. Second, residual transition metal products after release of CO, so-called iCORMs, can have direct cytotoxic and proinflammatory effects *in vitro* [164], and with the need for reoccurring administration of transition metal complexes and known toxicity of transition metals at nonphysiological concentrations [165,166], and chemical reactivity of CORM functional groups, including cysteine thiols [164], there is concern that iCORMs may accumulate and have long-term toxicities. For this reason, perhaps, there are currently no published or posted studies using CORMs in humans (ClinicalTrials.gov).

Organic CO Prodrugs

To address toxicity concerns and CO-independent effects of metal-based CORMs, the laboratory of Binghe Wang first described the chemistry for organic CO prodrugs with tunable CO release properties in 2014 [167]. A large series of these has since been developed with different CO release rates [168], triggered by exposure to aqueous solution, changes in pH [168,169], esterases [169,170], and reactive oxygen species [171]. More recently, organic CO prodrugs have been described that selectively target mitochondria [172]. These prodrugs have also been shown to ameliorate tissue injury in animal models of liver injury [169,172], chemical gastritis [173], systemic inflammation [169], and colitis [174], but there are only limited data describing their use in AKI. The only published work using organic CO prodrugs in AKI showed that pretreatment with BW-101, one of the earliest prototypes developed [168], was as effective as iCO in preserving renal function after bilateral IRI-AKI in mice [117]. Since that time, advances in CO prodrug chemistry have optimized prodrug stability and CO release kinetics, as described in two recent reviews [175,176]. Furthermore, detailed pharmacokinetics of a series of orally, intraperitoneally, and intravenously administered CO prodrugs have identified prodrugs and delivery routes with different *in vivo* half-lives and COHb kinetics [177], but it remains to be established whether these are also effective in different models of AKI.

Mechanisms of Action

One of the challenges for therapeutic development of CO is defining its mechanism of action in complex pathophysiological systems. This is important for a variety of reasons, but perhaps most importantly, this knowledge will allow us to establish whether the molecular pathways targeted by CO in experimental models of AKI are conserved in humans. In the case of CO, the single biggest challenge in establishing its mechanism of action is because its salutary effects are mediated not by binding to a single molecular target, but through its ability to bind to and modify activities of a large family of heme proteins [103,104]. This is compounded by the fact that CO binding to heme proteins in physiological systems is not a static process determined exclusively by their *in vitro* binding affinities. Since CO does not form covalent bonds, is highly diffusible, and is able to cross virtually any cellular compartment, CO binding profiles are dependent on heme protein environments that exist in and around a cell at a given time [102]. Furthermore, because CO competes with oxygen and other gasotransmitters, such as NO and H_2S, CO effects vary depending on other factors in the environment at that particular instant in time and are not consistent even in the same cell or tissue type. Our understanding of this process is further constrained since it has thus far proven impossible to define which proteins are targeted by CO in tissues because CO binding is reversible and tends to dissociate after tissue contents are disrupted. In addition, unlike NO and H_2S, CO is not a reactive gas, so does not generate reactive intermediates with target proteins that can easily be measured [103]. Because of these challenges, we currently lack the tools to identify the dominant molecular targets that are engaged by CO, so that much of our understanding of the salutary effects of CO in tissue injury is based on analyses of downstream responses noted earlier that occur at a cellular or physiological level. These responses are often indirect, mediated by molecular events that may occur many steps downstream of the heme protein targets. For example, CO has been shown to upregulate circadian rhythm protein PER2, as well as the adenosine receptors, Adora2a/2b, after AKI, and genetic studies indicate that these pathways are required to mediate the salutary effects of CO after IRI-AKI [117]. However, both PER2 and purinergic signaling suppress inflammatory responses [178,179], so it is possible that by enhancing renal inflammation after AKI through different mechanisms in PER2 and purinergic signaling mouse mutants, CO is now ineffective because inflammatory responses in the kidney are now uncontrolled. In other words, these studies do not necessarily define the proximal mechanism of action by which CO ameliorates AKI, as this may be upstream of the inflammatory responses in AKI. In the absence of a defined molecular target, this process will be difficult to untangle. Further studies may be able to identify proximal effectors of CO action, such as mitochondria [105], which may drive many of its downstream actions in AKI.

Future Directions

Because of its pleotropic effects that mitigate diverse mechanisms of injury, its relatively wide safety margins and proven efficacy in different experimental models of AKI, and because the HO-1/CO pathway is activated in human AKI, CO therapeutics have considerable potential to be an effective new therapy for patients with AKI. However, in addition to establishing its dominant mechanisms of action in different forms of AKI, there are several areas where further work is needed before the therapeutic CO should be rigorously studied in human AKI.

Timing CO Therapies

There is a need to establish optimal dosing regimens for CO therapeutics in different forms of AKI. This is of importance given differences in the kinetics of injury in models of AKI, and even greater heterogeneity of renal injuries in patients with AKI [28]. Limited data indicate that CO therapies are ineffective if given even 1 h after IRI-AKI in mice [117,136], and perhaps because of these findings, with few exceptions [133], studies in models of IRI-AKI, sepsis-induced AKI, and toxin-induced AKI have included pretreatment with CO as part of the treatment regimens. In one study, a single dose of CORM-2 given 3 h after surgery was shown to improve survival and renal function in a rat with CLP-induced polymicrobial sepsis [133], while a number of other studies, in addition to pretreatment, include repeat dosing with CO therapies at various intervals after the initiation of IRI [110,111] and cisplatin-induced injuries [137]. Further work is needed to establish whether these delayed dosing regimens without pretreatment using CO therapeutics with different CO release rates are effective in models of AKI with distinct kinetics of tissue injury. These studies will have important implications for therapeutic intervention in humans where the initiating events causing AKI often occur hours or even days before the patient is seen or noted to have AKI [9].

Measuring Tissue CO and Pharmacodynamic Biomarker Discovery

One of the reasons clinical trials have failed to show benefits of promising AKI therapies has been the failure to develop biomarkers that indicate whether the treatment has engaged its molecular targets with dosing regimens used in clinical trials [44,180]. As a result, it is often impossible to establish optimal dosing for patients, so that negative results in a number of therapeutic trials for AKI may have occurred because the drugs were being given at suboptimal doses. In the case of CO therapeutics, COHb levels provide an accurate measure of systemic CO exposure, but this does not necessarily reflect all important changes in tissue CO. This is particularly important for determining effective dosing with CORMs and CO prodrugs, where differences in compound stability and CO release rates *in vivo* may result in increased CO delivery to tissues without necessarily increasing COHb levels [144]. However, measuring CO tissue levels is much more challenging, requiring sensitive GC with special mercury reduction detectors, which are not readily available in research labs. Palladium-based fluorescent probes have been developed that allow for the detection of CO in cultured cells [181,182], but these lack sensitivity for quantitative detection of CO in harvested tissues. Furthermore, while the development of better tools to detect tissue CO will boost preclinical evaluation of CO therapeutics for AKI, concerns about risks of renal biopsies in patients with AKI will always limit our ability to examine renal tissues to determine tissue CO levels in humans [5]. For this reason, there is also a need to develop urinary or serum biomarkers that reflect CO target engagement in the kidney. Two complementary approaches could be used for this: (i) unbiased proteomic and/or metabolomic approaches to predict CO biomarker signatures that are indicative of effective target engagement in the kidney; however, these studies may not translate to clinical practice; and (ii) a more methodical approach, based on improved understanding of the dominant molecular pathways that mediate the salutary effect of CO in different models of AKI. These may provide insights that might that lead to the identification of pathway-specific biomarkers of CO target engagement, provided these are shown to be regulated in human AKI. Further work in this space will be critical for successful clinical development of CO therapeutics for AKI.

Pharmacological Properties to Optimize CO Delivery to the Kidney

The development of CO prodrugs selectively targeting the kidney may also be used to preserve efficiency while reducing systemic CO exposures. Strategies for this might include the development of CORM or CO prodrugs with folate conjugates, which have been used to target high-affinity folate receptors in PTECs [183], or using peptide conjugates to more selectively target CO prodrugs to renal tubules and/or inflammatory cells [184]. While previous studies have developed approaches to enrich CO delivery to tumors (e.g., using a folate-conjugated protein nanoemulsion loaded with CORM-2 to target lymphomas *in vivo*) [185,186], to enable photoactivation-dependent release of CO for temporal and spatial control of CO release [108], and for targeting CO prodrug release to mitochondria [187], there are no published instances targeting CO delivery to the kidney.

Conclusions

Since the discovery, nearly 30 years ago, that HO-1 protects the kidney from heme-induced injury resulting from muscle damage [51], it has become increasingly apparent that manipulating the pleotropic vasoprotective, anti-inflammatory, and cytoprotective effects of CO could present a solution to the clinical challenges of developing disease-modifying therapeutics for the heterogeneous clinical syndrome of AKI. As discussed in this chapter, as well as other medicinal chemistry chapters in this book, important questions remain about optimal CO delivery and pharmacodynamic monitoring of CO therapeutics for AKI. However, the fact that different CO delivery systems are effective in diverse animal models and species with AKI, and the fact that CO therapeutics are being advanced in a variety of different clinical scenarios, suggests that CO therapeutics may end up being one of the first to be shown to improve clinically significant short- and long-term outcomes in patients with AKI. This would address one of the most important unmet medical needs in nephrology practice worldwide.

References

1. Rewa, O. and Bagshaw, S.M. (2014). Acute kidney injury-epidemiology, outcomes and economics. *Nat. Rev. Nephrol.* 10: 193–207.
2. Hoste, E.A.J., Kellum, J.A., Selby, N.M., Zarbock, A., Palevsky, P.M., Bagshaw, S.M., and Goldstein, S.L. (2018). Cerda J and Chawla LS. Global epidemiology and outcomes of acute kidney injury. *Nat. Rev. Nephrol.* 14: 607–625.
3. Bellomo, R. (2012). Kellum JA and Ronco C. Acute kidney injury. *Lancet (London, England)* 380: 756–766.
4. Chawla, L.S., Bellomo, R., Bihorac, A., Goldstein, S.L., Siew, E.D., Bagshaw, S.M., Bittleman, D., Cruz, D., Endre, Z., Fitzgerald, R.L., Forni, L., Kane-Gill, S.L., Hoste, E., Koyner, J., Liu, K.D., Macedo, E., Mehta, R., Murray, P., Nadim, M., Ostermann, M., Palevsky, P.M., Pannu, N., Rosner, M., Wald, R., Zarbock, A., Ronco, C., and Kellum, J.A. (2017). Acute disease quality initiative W. Acute kidney disease and renal recovery: consensus report of the acute disease quality initiative (ADQI) 16 workgroup. *Nat. Rev. Nephrol.* 13: 241–257.
5. Fiorentino, M., Bolignano, D., Tesar, V., Pisano, A., Van Biesen, W., D'Arrigo, G., Tripepi, G., and Gesualdo, L. Group E-EIW (2016). Renal biopsy in 2015–from epidemiology to evidence-based indications. *Am. J. Nephrol.* 43: 1–19.
6. Takasu, O., Gaut, J.P., Watanabe, E., To, K., Fagley, R.E., Sato, B., Jarman, S., Efimov, I.R., Janks, D.L., Srivastava, A., Bhayani, S.B., and Drewry, A. (2013). Swanson PE and hotchkiss RS. Mechanisms of cardiac and renal dysfunction in patients dying of sepsis. *Am. J. Respir. Crit. Care Med.* 187: 509–517.
7. Solez, K. (1979). Morel-Maroger L and Sraer JD. The morphology of "acute tubular necrosis" in man: analysis of 57 renal biopsies and a comparison with the glycerol model. *Medicine* 58: 362–376.
8. Solez, K., Racusen, L.C., Marcussen, N., Slatnik, I., and Keown, P. (1993). Burdick JF and Olsen S. Morphology of ischemic acute renal failure, normal function, and cyclosporine toxicity in cyclosporine-treated renal allograft recipients. *Kidney Int.* 43: 1058–1067.
9. Skrypnyk, N.I. and Siskind, L.J. (2016). Faubel S and de Caestecker MP. Bridging translation for acute kidney injury with better preclinical modeling of human disease. *Am. J. Physiol. Renal. Physiol.* 310: F972–F984.
10. Legouis, D., Ricksten, S.-E., Faivre, A., Verissimo, T., Gariani, K., Verney, C., Galichon, P., Berchtold, L., Feraille, E., Fernandez, M., Placier, S., Koppitch, K., Hertig, A., Martin, P.Y., Naesens, M., Pugin, J., and McMahon, A.P. (2020). Cippa PE and de Seigneux S. Altered proximal tubular cell glucose metabolism during acute kidney injury is associated with mortality. *Nat. Metab.* 2: 732–743.
11. Cippa, P.E., Liu, J., Sun, B., and Kumar, S. (2019). Naesens M and McMahon AP. A late B lymphocyte action in dysfunctional tissue repair following kidney injury and transplantation. *Nat. Commun.* 10: 1157.
12. Cippa, P.E., Sun, B., Liu, J., and Chen, L. (2018). Naesens M and McMahon AP. Transcriptional trajectories of human kidney injury progression. *JCI Insight* 3.
13. Kirita, Y., Wu, H., and Uchimura, K. (2020). Wilson PC and Humphreys BD. Cell profiling of mouse acute

kidney injury reveals conserved cellular responses to injury. *Proc. Natl. Acad. Sci. U. S. A.* 117: 15874–15883.

14. Sharfuddin, A.A. and Molitoris, B.A. (2011). Pathophysiology of ischemic acute kidney injury. *Nat. Rev. Nephrol.* 7: 189–200.

15. Heyman, S.N. (2009). Rosen S and Rosenberger C. Animal models of renal dysfunction: acute kidney injury. *Expert. Opin. Drug Discov.* 4: 629–641.

16. Ferenbach, D.A. and Bonventre, J.V. (2015). Mechanisms of maladaptive repair after AKI leading to accelerated kidney ageing and CKD. *Nat. Rev. Nephrol.* 11: 264–276.

17. Singh, A.P., Junemann, A., Muthuraman, A., Jaggi, A.S., and Singh, N. (2012). Grover K and Dhawan R. Animal models of acute renal failure. *Pharmacol. Rep.* 64: 31–44.

18. Pabla, N. and Dong, Z. (2008). Cisplatin nephrotoxicity: mechanisms and renoprotective strategies. *Kidney Int.* 73: 994–1007.

19. Sharp, C.N. and Siskind, L.J. (2017). Developing better mouse models to study cisplatin-induced kidney injury. *Am. J. Physiol. Renal. Physiol.* 313: F835–F841.

20. Ghosh, S. (2019). Cisplatin: the first metal based anticancer drug. *Bioorg. Chem.* 88: 102925.

21. Bosch, X. (2009). Poch E and Grau JM. Rhabdomyolysis and acute kidney injury. *N. Engl. J. Med.* 361: 62–72.

22. Kerchberger, V.E. and Ware, L.B. (2020). The role of circulating cell-free hemoglobin in sepsis-associated acute kidney injury. *Semin. Nephrol.* 40: 148–159.

23. O'Neal, J.B., Shaw, A.D., and Billings, F. (2016). Acute kidney injury following cardiac surgery: current understanding and future directions. *Crit. Care* 20: 187.

24. Van Avondt, K. (2019). Nur E and Zeerleder S. Mechanisms of haemolysis-induced kidney injury. *Nat. Rev. Nephrol.* 15: 671–692.

25. Panizo, N., Rubio-Navarro, A., Amaro-Villalobos, J.M., Egido, J., and Moreno, J.A. (2015). Molecular mechanisms and novel therapeutic approaches to rhabdomyolysis-induced acute kidney injury. *Kidney Blood Press. Res.* 40: 520–532.

26. Doi, K., Leelahavanichkul, A., Yuen, P.S., and Star, R.A. (2009). Animal models of sepsis and sepsis-induced kidney injury. *J. Clin. Invest.* 119: 2868–2878.

27. Peerapornratana, S. and Manrique-Caballero, C.L. (2019). Gomez H and Kellum JA. Acute kidney injury from sepsis: current concepts, epidemiology, pathophysiology, prevention and treatment. *Kidney Int.* 96: 1083–1099.

28. Desanti De Oliveira, B., Xu, K., Shen, T.H., Callahan, M., Kiryluk, K., D'Agati, V.D., Tatonetti, N.P., Barasch, J., and Devarajan, P. (2019). Molecular nephrology: types of acute tubular injury. *Nat. Rev. Nephrol.* 15: 599–612.

29. Molitoris, B.A. (2014). Therapeutic translation in acute kidney injury: the epithelial/endothelial axis. *J. Clin. Invest.* 124: 2355–2363.

30. Docherty, M.H. and O'Sullivan, E.D. (2019). Bonventre JV and Ferenbach DA. Cellular senescence in the kidney. *J. Am. Soc. Nephrol.* 30: 726–736.

31. Parikh, S.M., Yang, Y., He, L., and Tang, C. (2015). Zhan M and Dong Z. Mitochondrial function and disturbances in the septic kidney. *Semin. Nephrol.* 35: 108–119.

32. Jang, H.R. and Rabb, H. (2015). Immune cells in experimental acute kidney injury. *Nat. Rev. Nephrol.* 11: 88–101.

33. Ferenbach, D.A. and Bonventre, J.V. (2016). Kidney tubules: intertubular, vascular, and glomerular cross-talk. *Curr. Opin. Nephrol. Hypertens.* 25: 194–202.

34. Ince, C., Mayeux, P.R., Nguyen, T., Gomez, H., Kellum, J.A., Ospina-Tascon, G.A., Hernandez, G., and Murray, P. (2016). De Backer D and Workgroup AX. The endothelium in sepsis. *Shock* 45: 259–270.

35. Emma, F. and Montini, G. (2016). Parikh SM and Salviati L. Mitochondrial dysfunction in inherited renal disease and acute kidney injury. *Nat. Rev. Nephrol.* 12: 267–280.

36. Wang, Y., Cai, J., Tang, C., and Dong, Z. (2020). Mitophagy in acute kidney injury and kidney repair. *Cells* 9.

37. Sato, Y. (2020). Takahashi M and Yanagita M. Pathophysiology of AKI to CKD progression. *Semin. Nephrol.* 40: 206–215.

38. Menshikh, A., Scarfe, L., Delgado, R., Finney, C., and Zhu, Y. (2019). Yang H and de Caestecker MP. Capillary rarefaction is more closely associated with CKD progression after cisplatin, rhabdomyolysis, and ischemia-reperfusion-induced AKI than renal fibrosis. *Am. J. Physiol. Renal. Physiol.* 317: F1383–F1397.

39. Billings, F. and Shaw, A.D. (2014). Clinical trial endpoints in acute kidney injury. *Nephron Clin. Pract.* 127: 89–93.

40. Okusa, M.D., Molitoris, B.A., Palevsky, P.M., Chinchilli, V.M., Liu, K.D., Cheung, A.K., Weisbord, S.D., Faubel, S., Kellum, J.A., Wald, R., Chertow, G.M., Levin, A., Waikar, S.S., Murray, P.T., Parikh, C.R., Shaw, A.D., Go, A.S., Chawla, L.S., Kaufman, J.S., Devarajan, P., Toto, R.M., Hsu, C.Y., Greene, T.H., Mehta, R.L., Stokes, J.B., Thompson, A.M., Thompson, B.T., Westenfelder, C.S.,

Tumlin, J.A., Warnock, D.G., Shah, S.V., Xie, Y., and Duggan, E.G. (2012). Kimmel PL and Star RA. Design of clinical trials in acute kidney injury: a report from an NIDDK workshop–prevention trials. *Clin. J. Am. Soc. Nephrol.* 7: 851–855.

41. Weisbord, S.D. and Palevsky, P.M. (2016). Design of clinical trials in acute kidney injury: Lessons from the past and future directions. *Semin. Nephrol.* 36: 42–52.

42. Molitoris, B.A., Okusa, M.D., Palevsky, P.M., Chawla, L.S., Kaufman, J.S., Devarajan, P., Toto, R.M., Hsu, C.Y., Greene, T.H., Faubel, S.G., Kellum, J.A., Wald, R., Chertow, G.M., Levin, A., Waikar, S.S., Murray, P.T., Parikh, C.R., Shaw, A.D., Go, A.S., Chinchilli, V.M., Liu, K.D., Cheung, A.K., Weisbord, S.D., Mehta, R.L., Stokes, J.B., Thompson, A.M., Thompson, B.T., Westenfelder, C.S., Tumlin, J.A., Warnock, D.G., Shah, S.V., Xie, Y., and Duggan, E.G. (2012). Kimmel PL and Star RA. Design of clinical trials in AKI: a report from an NIDDK workshop. Trials of patients with sepsis and in selected hospital settings. *Clin. J. Am. Soc. Nephrol.* 7: 856–860.

43. Zuk, A., Palevsky, P.M., Fried, L., Harrell, F.E., Jr., Khan, S., McKay, D.B., Devey, L., Chawla, L., De Caestecker, M., Kaufman, J.S., Thompson, B.T., Agarwal, A., Greene, T., Okusa, M.D., Bonventre, J.V., Dember, L.M., Liu, K.D., Humphreys, B.D., Gossett, D., Xie, Y., and Norton, J.M. (2018). Kimmel PL and Star RA. overcoming translational barriers in acute kidney injury: A report from an NIDDK workshop. *Clin. J. Am. Soc. Nephrol.* 13: 1113–1123.

44. De Caestecker, M. and Harris, R. (2018). Translating knowledge into therapy for acute kidney injury. *Semin. Nephrol.* 38: 88–97.

45. Zuk, A., Palevsky, P.M., Fried, L., Harrell, F.E., Jr., Khan, S., McKay, D.B., Devey, L., Chawla, L., De Caestecker, M., Kaufman, J.S., Thompson, B.T., Agarwal, A., Greene, T., Okusa, M.D., Bonventre, J.V., Dember, L.M., Liu, K.D., Humphreys, B.D., Gossett, D., Xie, Y., and Norton, J.M. (2018). Kimmel PL and Star RA. Overcoming translational barriers in acute kidney injury: A report from an NIDDK workshop. *Clin. J. Am. Soc. Nephrol.* 13: 1113–1123.

46. Desanti De Oliveira, B., Xu, K., Shen, T.H., Callahan, M., Kiryluk, K., D'Agati, V.D., Tatonetti, N.P., Barasch, J., and Devarajan, P. (2019). Molecular nephrology: types of acute tubular injury. *Nat. Rev. Nephrol.* 15: 599–612.

47. Lake, B.B., Chen, S., Hoshi, M., Plongthongkum, N., Salamon, D., Knoten, A., Vijayan, A., Venkatesh, R., Kim, E.H., Gao, D., and Gaut, J. (2019). Zhang K and Jain S. A single-nucleus RNA-sequencing pipeline to decipher the molecular anatomy and pathophysiology of human kidneys. *Nat. Commun.* 10: 2832.

48. Przepiorski, A., Crunk, A.E., and Espiritu, E.B. (2020). Hukriede NA and Davidson AJ. The utility of human kidney organoids in modeling kidney disease. *Semin. Nephrol.* 40: 188–198.

49. Hukriede, N. (2017). Vogt A and de Caestecker M. Drug discovery to halt the progression of acute kidney injury to chronic kidney disease: a case for phenotypic drug discovery in acute kidney injury. *Nephron* 137: 268–272.

50. Yang, X., De Caestecker, M., Otterbein, L.E., and Wang, B. (2020). Carbon monoxide: An emerging therapy for acute kidney injury. *Med. Res. Rev.* 40: 1147–1177.

51. Nath, K.A., Balla, G., Vercellotti, G.M., Balla, J., and Jacob, H.S. (1992). Levitt MD and Rosenberg ME. Induction of heme oxygenase is a rapid, protective response in rhabdomyolysis in the rat. *J. Clin. Invest.* 90: 267–270.

52. Ryter, S.W. (2019). Heme oxygenase-1/carbon monoxide as modulators of autophagy and inflammation. *Arch. Biochem. Biophys.* 678: 108186.

53. Donegan, R.K. and Moore, C.M. (2019). Hanna DA and Reddi AR. Handling heme: the mechanisms underlying the movement of heme within and between cells. *Free Radic. Biol. Med.* 133: 88–100.

54. Tracz, M.J. (2007). Alam J and Nath KA. Physiology and pathophysiology of heme: implications for kidney disease. *J. Am. Soc. Nephrol.* 18: 414–420.

55. Janciauskiene, S. (2020). Vijayan V and Immenschuh S. TLR4 Signaling by Heme and the role of heme-binding blood proteins. *Front. Immunol.* 11: 1964.

56. Kang, K., Nan, C., Fei, D., Meng, X., Liu, W., Zhang, W., Jiang, L., and Zhao, M. (2013). Pan S and Zhao M. Heme oxygenase 1 modulates thrombomodulin and endothelial protein C receptor levels to attenuate septic kidney injury. *Shock* 40: 136–143.

57. Bolisetty, S., Traylor, A.M., Kim, J., Joseph, R., and Ricart, K. (2010). Landar A and Agarwal A. Heme oxygenase-1 inhibits renal tubular macroautophagy in acute kidney injury. *J. Am. Soc. Nephrol.* 21: 1702 1712.

58. Bolisetty, S., Traylor, A., and Joseph, R. (2016). Zarjou A and Agarwal A. Proximal tubule-targeted heme oxygenase-1 in cisplatin-induced acute kidney injury. *Am. J. Physiol. Renal. Physiol.* 310: F385–94.

59. Kim, J., Zarjou, A., Traylor, A.M., Bolisetty, S., Jaimes, E.A., Hull, T.D., George, J.F., Mikhail, F.M., and Agarwal, A. (2012). In vivo regulation of the heme oxygenase-1 gene in humanized transgenic mice. *Kidney Int.* 82: 278–291.

60. Agarwal, A., Balla, J., and Alam, J. (1995). Croatt AJ and Nath KA. Induction of heme oxygenase in toxic renal injury: a protective role in cisplatin nephrotoxicity in the rat. *Kidney Int.* 48: 1298–1307.

61. Blydt-Hansen, T.D., Katori, M., Lassman, C., Ke, B., Coito, A.J., Iyer, S., Buelow, R., Ettenger, R., Busuttil, R.W., and Kupiec-Weglinski, J.W. (2003). Gene transfer-induced local heme oxygenase-1 overexpression protects rat kidney transplants from ischemia/reperfusion injury. *J. Am. Soc. Nephrol.* 14: 745–754.
62. Pittock, S.T., Norby, S.M., Grande, J.P., Croatt, A.J., Bren, G.D., Badley, A.D., Caplice, N.M., Griffin, M.D., and Nath, K.A. (2005). MCP-1 is up-regulated in unstressed and stressed HO-1 knockout mice: Pathophysiologic correlates. *Kidney Int.* 68: 611–622.
63. Hull, T.D., Kamal, A.I., Boddu, R., Bolisetty, S., Guo, L., Tisher, C.C., Rangarajan, S., Chen, B., Curtis, L.M., George, J.F., and Agarwal, A. (2015). Heme oxygenase-1 regulates myeloid cell trafficking in AKI. *J. Am. Soc. Nephrol.* 26: 2139–2151.
64. Paller, M.S. and Jacob, H.S. (1994). Cytochrome P-450 mediates tissue-damaging hydroxyl radical formation during reoxygenation of the kidney. *Proc. Natl. Acad. Sci. U. S. A.* 91: 7002–7006.
65. Yachie, A., Niida, Y., Wada, T., Igarashi, N., Kaneda, H., Toma, T., and Ohta, K. (1999). Kasahara Y and Koizumi S. Oxidative stress causes enhanced endothelial cell injury in human heme oxygenase-1 deficiency. *J. Clin. Invest.* 103: 129–135.
66. Lever, J.M., Boddu, R., George, J.F., and Agarwal, A. (2016). Heme oxygenase-1 in kidney health and disease. *Antioxid. Redox Signal.* 25: 165–183.
67. Kim, J., Zarjou, A., Traylor, A.M., Bolisetty, S., Jaimes, E.A., Hull, T.D., George, J.F., Mikhail, F.M., and Agarwal, A. (2012). In vivo regulation of the heme oxygenase-1 gene in humanized transgenic mice. *Kidney Int.* 82: 278–291.
68. Zager, R.A. (2012). Johnson AC and Becker K. Plasma and urinary heme oxygenase-1 in AKI. *J. Am. Soc. Nephrol.* 23: 1048–1057.
69. Billings, F., Yu, C., Byrne, J.G., Petracek, M.R., and Pretorius, M. (2014). Heme oxygenase-1 and acute kidney injury following cardiac surgery. *Cardiorenal Med.* 4: 12–21.
70. Vilander, L.M., Vaara, S.T., Donner, K.M., Lakkisto, P., and Kaunisto, M.A. (2019). Pettila V and Group FS. Heme oxygenase-1 repeat polymorphism in septic acute kidney injury. *PLoS One* 14: e0217291.
71. Leaf, D.E., Body, S.C., Muehlschlegel, J.D., McMahon, G.M., Lichtner, P., Collard, C.D., and Shernan, S.K. (2016). Fox AA and Waikar SS. Length polymorphisms in heme oxygenase-1 and AKI after cardiac surgery. *J. Am. Soc. Nephrol.* 27: 3291–3297.
72. Al-Kahtani, M.A., Abdel-Moneim, A.M., Elmenshawy, O.M., and El-Kersh, M.A. (2014). Hemin attenuates cisplatin-induced acute renal injury in male rats. *Oxid. Med. Cell. Longev.* 2014: 476430.
73. Rossi, M., Delbauve, S., Roumeguere, T., Wespes, E., Leo, O., Flamand, V., Le Moine, A., and Hougardy, J.M. (2019). HO-1 mitigates acute kidney injury and subsequent kidney-lung cross-talk. *Free Radic. Res.* 53: 1035–1043.
74. Pan, H., Shen, K., Wang, X., and Meng, H. (2014). Wang C and Jin B. Protective effect of metalloporphyrins against cisplatin-induced kidney injury in mice. *PLoS One* 9: e86057.
75. Zager, R.A. (2016). Johnson AC and Frostad KB. Combined iron sucrose and protoporphyrin treatment protects against ischemic and toxin-mediated acute renal failure. *Kidney Int.* 90: 67–76.
76. Burgess, A., Li, M., Vanella, L., Kim, D.H., Rezzani, R., Rodella, L., Sodhi, K., Canestraro, M., Martasek, P., and Peterson, S.J. (2010). Kappas A and Abraham NG. Adipocyte heme oxygenase-1 induction attenuates metabolic syndrome in both male and female obese mice. *Hypertension* 56: 1124–1130.
77. Ndisang, J.F. and Tiwari, S. (2014). Mechanisms by which heme oxygenase rescue renal dysfunction in obesity. *Redox Biol.* 2: 1029–1037.
78. Ferenbach, D.A., Nkejabega, N.C., McKay, J., Choudhary, A.K., Vernon, M.A., Beesley, M.F., Clay, S., Conway, B.C., and Marson, L.P. (2011). Kluth DC and Hughes J. The induction of macrophage hemeoxygenase-1 is protective during acute kidney injury in aging mice. *Kidney Int.* 79: 966–976.
79. Rossi, M., Delbauve, S., Wespes, E., Roumeguere, T., Leo, O., and Flamand, V. (2018). Le Moine A and Hougardy JM. Dual effect of hemin on renal ischemia-reperfusion injury. *Biochem. Biophys. Res. Commun.* 503: 2820–2825.
80. Nath, K.A., Belcher, J.D., Nath, M.C., Grande, J.P., Croatt, A.J., and Ackerman, A.W. (2018). Katusic ZS and Vercellotti GM. Role of TLR4 signaling in the nephrotoxicity of heme and heme proteins. *Am. J. Physiol. Renal. Physiol.* 314: F906–F914.
81. Park, S.J., Cho, B., Koo, O.J., Kim, H., Kang, J.T., Hurh, S., Kim, S.J., Yeom, H.J., Moon, J., Lee, E.M., Choi, J.Y., Hong, J.H., Jang, G., Hwang, J.I., and Yang, J. (2014). Lee BC and Ahn C. Production and characterization of soluble human TNFRI-Fc and human HO-1(HMOX1) transgenic pigs by using the F2A peptide. *Transgenic Res.* 23: 407–419.
82. Yeom, H.J., Koo, O.J., Yang, J., Cho, B., Hwang, J.I., Park, S.J., Hurh, S., Kim, H., Lee, E.M., Ro, H., Kang, J.T., Kim, S.J., Won, J.K., O'Connell, P.J., Kim, H., and Surh, C.D. (2012). Lee BC and Ahn C. Generation and characterization of human heme oxygenase-1 transgenic pigs. *PLoS One* 7: e46646.

83. Petersen, B., Ramackers, W., Lucas-Hahn, A., Lemme, E., Hassel, P., Queisser, A.L., Herrmann, D., Barg-Kues, B., Carnwath, J.W., Klose, J., Tiede, A., Friedrich, L., Baars, W., and Schwinzer, R. (2011). Winkler M and Niemann H. Transgenic expression of human heme oxygenase-1 in pigs confers resistance against xenograft rejection during ex vivo perfusion of porcine kidneys. *Xenotransplantation* 18: 355–368.

84. Ferenbach, D.A., Ramdas, V., Spencer, N., Marson, L., and Anegon, I. (2010). Hughes J and Kluth DC. Macrophages expressing heme oxygenase-1 improve renal function in ischemia/reperfusion injury. *Mol. Ther.* 18: 1706–1713.

85. Wu, J., Liu, X., Fan, J., Chen, W., Wang, J., Zeng, Y., and Feng, X. (2014). Yu X and Yang X. Bardoxolone methyl (BARD) ameliorates aristolochic acid (AA)-induced acute kidney injury through Nrf2 pathway. *Toxicology* 318: 22–31.

86. Wu, Q.Q., Wang, Y., Senitko, M., Meyer, C., Wigley, W.C., Ferguson, D.A., Grossman, E., Chen, J., Zhou, X.J., Hartono, J., Winterberg, P., and Chen, B. (2011). Agarwal A and Lu CY. Bardoxolone methyl (BARD) ameliorates ischemic AKI and increases expression of protective genes Nrf2, PPARgamma, and HO-1. *Am. J. Physiol. Renal. Physiol.* 300: F1180–92.

87. Wang, Y., Feng, F., and Liu, M. (2018). Xue J and Huang H. Resveratrol ameliorates sepsis-induced acute kidney injury in a pediatric rat model via Nrf2 signaling pathway. *Exp. Ther. Med.* 16: 3233–3240.

88. Grochot-Przeczek, A. (2012). Dulak J and Jozkowicz A. Haem oxygenase-1: non-canonical roles in physiology and pathology. *Clin. Sci. (Lond)* 122: 93–103.

89. Dulak, J. (2008). Loboda A and Jozkowicz A. Effect of heme oxygenase-1 on vascular function and disease. *Curr. Opin. Lipidol.* 19: 505–512.

90. Nath, K.A. (2014). Heme oxygenase-1 and acute kidney injury. *Curr. Opin. Nephrol. Hypertens.* 23: 17–24.

91. Bolisetty, S., Zarjou, A., and Agarwal, A. (2017). Heme oxygenase 1 as a therapeutic target in acute kidney injury. *Am. J. Kidney Dis.* 69: 531–545.

92. Sharma, S. and Leaf, D.E. (2019). Iron chelation as a potential therapeutic strategy for AKI prevention. *J. Am. Soc. Nephrol.* 30: 2060–2071.

93. Zarjou, A., Bolisetty, S., Joseph, R., Traylor, A., Apostolov, E.O., Arosio, P., Balla, J., Verlander, J., and Darshan, D. (2013). Kuhn LC and Agarwal A. Proximal tubule H-ferritin mediates iron trafficking in acute kidney injury. *J. Clin. Invest.* 123: 4423–4434.

94. Chung, S.W., Liu, X., and Macias, A.A. (2008). Baron RM and Perrella MA. Heme oxygenase-1-derived carbon monoxide enhances the host defense response to microbial sepsis in mice. *J. Clin. Invest.* 118: 239–247.

95. Brouard, S., Otterbein, L.E., Anrather, J., Tobiasch, E., and Bach, F.H. (2000). Choi AM and Soares MP. Carbon monoxide generated by heme oxygenase 1 suppresses endothelial cell apoptosis. *J. Exp. Med.* 192: 1015–1026.

96. Bilban, M., Bach, F.H., Otterbein, S.L., and Ifedigbo, E. (2006). d'Avila JC, Esterbauer H, Chin BY, Usheva A, Robson SC, Wagner O and Otterbein LE. Carbon monoxide orchestrates a protective response through PPARgamma. *Immunity* 24: 601–610.

97. Cheng, Y. and Rong, J. (2017). Therapeutic potential of heme oxygenase-1/carbon monoxide system against ischemia-reperfusion injury. *Curr. Pharm. Des.* 23: 3884–3898.

98. Fujita, T., Toda, K., Karimova, A., Yan, S.F., and Naka, Y. (2001). Yet SF and Pinsky DJ. Paradoxical rescue from ischemic lung injury by inhaled carbon monoxide driven by derepression of fibrinolysis. *Nat. Med.* 7: 598–604.

99. Otterbein, L.E. (1999). Mantell LL and Choi AM. Carbon monoxide provides protection against hyperoxic lung injury. *Am. J. Physiol.* 276: L688–94.

100. Lancel, S., Hassoun, S.M., Favory, R., and Decoster, B. (2009). Motterlini R and Neviere R. Carbon monoxide rescues mice from lethal sepsis by supporting mitochondrial energetic metabolism and activating mitochondrial biogenesis. *J. Pharmacol. Exp. Ther.* 329: 641–648.

101. Goebel, U. and Wollborn, J. (2020). Carbon monoxide in intensive care medicine-time to start the therapeutic application?! *Intensive Care Med. Exp.* 8: 2.

102. Otterbein, L.E., Foresti, R., and Motterlini, R. (2016). Heme oxygenase-1 and carbon monoxide in the heart: The balancing act between danger signaling and pro-survival. *Circ. Res.* 118: 1940–1959.

103. Motterlini, R. and Foresti, R. (2017). Biological signaling by carbon monoxide and carbon monoxide-releasing molecules. *Am. J. Physiol. Cell Physiol.* 312: C302–c313.

104. Motterlini, R. and Otterbein, L.E. (2010). The therapeutic potential of carbon monoxide. *Nat. Rev. Drug Discov.* 9: 728–743.

105. Schallner, N. and Otterbein, L.E. (2015). Friend or foe? Carbon monoxide and the mitochondria *Front. Physiol.* 6: 17.

106. Stucki, D. and Stahl, W. (2020). Carbon monoxide – beyond toxicity? *Toxicol. Lett.* 333: 251–260.

107. Otterbein, L.E., Hedblom, A., Harris, C., and Csizmadia, E. (2011). Gallo D and Wegiel B. Heme oxygenase-1 and carbon monoxide modulate DNA repair through ataxia-telangiectasia mutated (ATM)

protein. *Proc. Natl. Acad. Sci. U. S. A.* 108: 14491–14496.
108. Ji, X., Damera, K., Zheng, Y., Yu, B., Otterbein, L.E., and Wang, B. (2016). Toward carbon monoxide-based therapeutics: Critical drug delivery and developability issues. *J. Pharm. Sci.* 105: 406–416. and references cited therein.
109. Mazzola, S., Forni, M., Albertini, M., Bacci, M.L., Zannoni, A., Gentilini, F., Lavitrano, M., and Bach, F.H. (2005). Otterbein LE and Clement MG. Carbon monoxide pretreatment prevents respiratory derangement and ameliorates hyperacute endotoxic shock in pigs. *FASEB J.* 19: 2045–2047.
110. Neto, J.S., Nakao, A., Kimizuka, K., Romanosky, A.J., Stolz, D.B., Uchiyama, T., and Nalesnik, M.A. (2004). Otterbein LE and Murase N. Protection of transplant-induced renal ischemia-reperfusion injury with carbon monoxide. *Am. J. Physiol. Renal. Physiol.* 287: F979–89.
111. Nakao, A., Neto, J.S., Kanno, S., Stolz, D.B., Kimizuka, K., Liu, F., Bach, F.H., Billiar, T.R., and Choi, A.M. (2005). Otterbein LE and Murase N. Protection against ischemia/reperfusion injury in cardiac and renal transplantation with carbon monoxide, biliverdin and both. *Am. J. Transplant.* 5: 282–291.
112. Hanto, D.W., Maki, T., Yoon, M.H., Csizmadia, E., Chin, B.Y., Gallo, D., Konduru, B., Kuramitsu, K., Smith, N.R., Berssenbrugge, A., Attanasio, C., and Thomas, M. (2010). Wegiel B and Otterbein LE. Intraoperative administration of inhaled carbon monoxide reduces delayed graft function in kidney allografts in Swine. *Am. J. Transplant.* 10: 2421–2430.
113. Mannon, R.B. (2018). Delayed graft function: The AKI of kidney transplantation. *Nephron* 140: 94–98.
114. Siedlecki, A. (2011). Irish W and Brennan DC. Delayed graft function in the kidney transplant. *Am. J. Transplant.* 11: 2279–2296.
115. Katsnelson, A. (2019). The good side of carbon monoxide. *Acs Central Sci.* 5: 1632–1635.
116. Goebel, U., Siepe, M., Schwer, C.I., Schibilsky, D., Foerster, K., Neumann, J., Wiech, T., and Priebe, H.J. (2010). Schlensak C and Loop T. Inhaled carbon monoxide prevents acute kidney injury in pigs after cardiopulmonary bypass by inducing a heat shock response. *Anesth. Analg.* 111: 29–37.
117. Correa-Costa, M., Gallo, D., Csizmadia, E., Gomperts, E., Lieberum, J.L., Hauser, C.J., Ji, X., Wang, B., and Camara, N.O.S. (2018). Robson SC and Otterbein LE. Carbon monoxide protects the kidney through the central circadian clock and CD39. *Proc. Natl. Acad. Sci. U. S. A.* 115: E2302–E2310.
118. Ozaki, K.S., Yoshida, J., Ueki, S., Pettigrew, G.L., Ghonem, N., Sico, R.M., Lee, L.Y., Shapiro, R., and Lakkis, F.G. (2012). Pacheco-Silva A and Murase N. Carbon monoxide inhibits apoptosis during cold storage and protects kidney grafts donated after cardiac death. *Transpl. Int.* 25: 107–117.
119. Yoshida, J., Ozaki, K.S., Nalesnik, M.A., Ueki, S., Castillo-Rama, M., Faleo, G., Ezzelarab, M., Nakao, A., Ekser, B., Echeverri, G.J., and Ross, M.A. (2010). Stolz DB and Murase N. Ex vivo application of carbon monoxide in UW solution prevents transplant-induced renal ischemia/reperfusion injury in pigs. *Am. J. Transplant.* 10: 763–772.
120. Belcher, J.D., Gomperts, E., Nguyen, J., Chen, C., Abdulla, F., Kiser, Z.M., Gallo, D., and Levy, H. (2018). Otterbein LE and Vercellotti GM. Oral carbon monoxide therapy in murine sickle cell disease: Beneficial effects on vaso-occlusion, inflammation and anemia. *PLoS One* 13: e0205194.
121. Cabrales, P. (2007). Tsai AG and Intaglietta M. Hemorrhagic shock resuscitation with carbon monoxide saturated blood. *Resuscitation* 72: 306–318.
122. Taguchi, K., Ogaki, S., Nagasaki, T., Yanagisawa, H., Nishida, K., Maeda, H., Enoki, Y., Matsumoto, K., Sekijima, H., Ooi, K., Ishima, Y., Watanabe, H., and Fukagawa, M. (2020). Otagiri M and Maruyama T. Carbon monoxide rescues the developmental lethality of experimental rat models of rhabdomyolysis-induced acute kidney injury. *J. Pharmacol. Exp. Ther.* 372: 355–365.
123. Abuchowski, A. (2016). PEGylated bovine carboxyhemoglobin (SANGUINATE): Results of clinical safety testing and use in patients. *Adv. Exp. Med. Biol.* 876: 461–467.
124. Guerci, P., Ergin, B., Kapucu, A., Hilty, M.P., and Jubin, R. (2019). Bakker J and Ince C. Effect of polyethylene-glycolated carboxyhemoglobin on renal microcirculation in a rat model of hemorrhagic shock. *Anesthesiology* 131: 1110–1124.
125. Guerci, P., Ergin, B., Kandil, A., Ince, Y., Heeman, P., and Hilty, M.P. (2020). Bakker J and Ince C. Resuscitation with PEGylated carboxyhemoglobin preserves renal cortical oxygenation and improves skeletal muscle microcirculatory flow during endotoxemia. *Am. J. Physiol. Renal. Physiol.* 318: F1271–F1283.
126. Bachert, S.E. (2020). Dogra P and Boral LI. Alternatives to Transfusion. *Am. J. Clin. Pathol.* 153: 287–293.

127. Wang, X., Qin, W., Qiu, X., and Cao, J. (2014). Liu D and Sun B. A novel role of exogenous carbon monoxide on protecting cardiac function and improving survival against sepsis via mitochondrial energetic metabolism pathway. *Int. J. Biol. Sci.* 10: 777–788.

128. Guo, Y., Stein, A.B., Wu, W.J., Tan, W., Zhu, X., Li, Q.H., and Dawn, B. (2004). Motterlini R and Bolli R. Administration of a CO-releasing molecule at the time of reperfusion reduces infarct size in vivo. *Am. J. Physiol. Heart Circ. Physiol.* 286: H1649–53.

129. Motterlini, R., Clark, J.E., Foresti, R., and Sarathchandra, P. (2002). Mann BE and Green CJ. Carbon monoxide-releasing molecules: characterization of biochemical and vascular activities. *Circ. Res.* 90: E17–24.

130. Motterlini, R., Mann, B.E., Johnson, T.R., and Clark, J.E. (2003). Foresti R and Green CJ. Bioactivity and pharmacological actions of carbon monoxide-releasing molecules. *Curr. Pharm. Des.* 9: 2525–2539.

131. Uddin, M.J. (2018). Pak ES and Ha H. Carbon monoxide releasing molecule-2 protects mice against acute kidney injury through inhibition of ER stress. *Korean J. Physiol. Pharmacol.* 22: 567–575.

132. Shiohira, S., Yoshida, T., and Shirota, S. (2007). Tsuchiya K and Nitta K. Protective effect of carbon monoxide donor compounds in endotoxin-induced acute renal failure. *Am. J. Nephrol.* 27: 441–446.

133. Wang, P., Huang, J., Li, Y., Chang, R., and Wu, H. (2015). Lin J and Huang Z. Exogenous carbon monoxide decreases sepsis-induced acute kidney injury and inhibits NLRP3 inflammasome activation in rats. *Int. J. Mol. Sci.* 16: 20595–20608.

134. Caumartin, Y., Stephen, J., Deng, J.P., Lian, D., Lan, Z., Liu, W., Garcia, B., Jevnikar, A.M., and Wang, H. (2011). Cepinskas G and Luke PP. Carbon monoxide-releasing molecules protect against ischemia-reperfusion injury during kidney transplantation. *Kidney Int.* 79: 1080–1089.

135. Bagul, A. and Hosgood, S.A. (2008). Kaushik M and Nicholson ML. Carbon monoxide protects against ischemia-reperfusion injury in an experimental model of controlled nonheartbeating donor kidney. *Transplantation* 85: 576–581.

136. Vera, T., Henegar, J.R., and Drummond, H.A. (2005). Rimoldi JM and Stec DE. Protective effect of carbon monoxide-releasing compounds in ischemia-induced acute renal failure. *J. Am. Soc. Nephrol.* 16: 950–958.

137. Tayem, Y., Johnson, T.R., and Mann, B.E. (2006). Green CJ and Motterlini R. Protection against cisplatin-induced nephrotoxicity by a carbon monoxide-releasing molecule. *Am. J. Physiol. Renal. Physiol.* 290: F789–94.

138. Sandouka, A., Fuller, B.J., Mann, B.E., and Green, C.J. (2006). Foresti R and Motterlini R. Treatment with CO-RMs during cold storage improves renal function at reperfusion. *Kidney Int.* 69: 239–247.

139. Motterlini, R., Sawle, P., Hammad, J., Bains, S., Alberto, R., Foresti, R., and Green, C.J. (2005). CORM-A1: a new pharmacologically active carbon monoxide-releasing molecule. *FASEB J.* 19: 284–286.

140. Klein, M. and Neugebauer, U. (2016). Schmitt M and Popp J. Elucidation of the CO-Release Kinetics of CORM-A1 by means of vibrational spectroscopy. *Chemphyschem* 17: 985–993.

141. Bhattacharjee, R.N., Richard-Mohamed, M., Sun, Q., Haig, A., Aboalsamh, G., Barrett, P., Mayer, R., Alhasan, I., Pineda-Solis, K., Jiang, L., Alharbi, H., Saha, M., Patterson, E., Sener, A., Cepinskas, G., Jevnikar, A.M., and Luke, P.P.W. (2018). CORM-401 reduces ischemia reperfusion injury in an Ex Vivo renal porcine model of the donation after circulatory death. *Transplantation* 102: 1066–1074.

142. Crook, S.H., Mann, B.E., Meijer, A.J., Adams, H., and Sawle, P. (2011). Scapens D and Motterlini R. [Mn(CO)4{S2CNMe(CH2CO2H)}], a new water-soluble CO-releasing molecule. *Dalton Trans.* 40: 4230–4235.

143. Fayad-Kobeissi, S., Ratovonantenaina, J., Dabire, H., Wilson, J.L., Rodriguez, A.M., Berdeaux, A., Dubois-Rande, J.L., and Mann, B.E. (2016). Motterlini R and Foresti R. Vascular and angiogenic activities of CORM-401, an oxidant-sensitive CO-releasing molecule. *Biochem. Pharmacol.* 102: 64–77.

144. Seixas, J.D., Santos, M.F., Mukhopadhyay, A., Coelho, A.C., Reis, P.M., Veiros, L.F., Marques, A.R., Penacho, N., Goncalves, A.M., Romao, M.J., and Bernardes, G.J. (2015). Santos-Silva T and Romao CC. A contribution to the rational design of Ru(CO)3Cl2L complexes for in vivo delivery of CO. *Dalton Trans.* 44: 5058–5075.

145. Braud, L., Pini, M., Muchova, L., Manin, S., Kitagishi, H., Sawaki, D., Czibik, G., Ternacle, J., Derumeaux, G., Foresti, R., and Motterlini, R. (2018). Carbon monoxide-induced metabolic switch in adipocytes improves insulin resistance in obese mice. *JCI Insight* 3.

146. Fagone, P., Mangano, K., Quattrocchi, C., Motterlini, R., Di Marco, R., Magro, G., and Penacho, N. (2011). Romao CC and Nicoletti F. Prevention of clinical and histological signs of proteolipid protein (PLP)-induced experimental allergic

encephalomyelitis (EAE) in mice by the water-soluble carbon monoxide-releasing molecule (CORM)-A1. *Clin. Exp. Immunol.* 163: 368–374.

147. Segersvard, H., Lakkisto, P., Hanninen, M., Forsten, H., Siren, J., Immonen, K., Kosonen, R., and Sarparanta, M. (2018). Laine M and Tikkanen I. Carbon monoxide releasing molecule improves structural and functional cardiac recovery after myocardial injury. *Eur. J. Pharmacol.* 818: 57–66.

148. Funahashi, Y., Chowdhury, S., Eiwaz, M.B. and Hutchens, M.P. (2020). Acute cardiorenal syndrome: models and heart-kidney connectors. *Nephron* 144: 629–633.

149. Santos-Silva, T., Mukhopadhyay, A., Seixas, J.D., and Bernardes, G.J. (2011). Romao CC and Romao MJ. Towards improved therapeutic CORMs: understanding the reactivity of CORM-3 with proteins. *Curr. Med. Chem.* 18: 3361–3366.

150. Southam, H.M., Smith, T.W., Lyon, R.L., Liao, C., Trevitt, C.R., Middlemiss, L.A., Cox, F.L., Chapman, J.A., El-Khamisy, S.F., Hippler, M., and Williamson, M.P. (2018). Henderson PJF and Poole RK. A thiol-reactive Ru(II) ion, not CO release, underlies the potent antimicrobial and cytotoxic properties of CO-releasing molecule-3. *Redox Biol.* 18: 114–123.

151. Juszczak, M. and Kluska, M. (2020). Wysokiński D and Woźniak K. DNA damage and antioxidant properties of CORM-2 in normal and cancer cells. *Sci. Rep.* 10: 12200.

152. Nielsen, V.G. (2020). The anticoagulant effect of Apis mellifera phospholipase A(2) is inhibited by CORM-2 via a carbon monoxide-independent mechanism. *J. Thromb. Thrombolysis* 49: 100–107.

153. Nielsen, V.G. (2020). Wagner MT and Frank N. Mechanisms responsible for the anticoagulant properties of neurotoxic dendroaspis venoms: A viscoelastic analysis. *Int. J. Mol. Sci.* 21: 2082.

154. Nielsen, V.G. (2020). Ruthenium, not carbon monoxide, inhibits the procoagulant activity of atheris, echis, and pseudonaja venoms. *Int. J. Mol. Sci.* 21: 2970.

155. Stucki, D., Krahl, H., Walter, M., Steinhausen, J., and Hommel, K. (2020). Brenneisen P and Stahl W. Effects of frequently applied carbon monoxide releasing molecules (CORMs) in typical CO-sensitive model systems – A comparative in vitro study. *Arch. Biochem. Biophys.* 687: 108383.

156. Rossier, J., Delasoie, J., Haeni, L., and Hauser, D. (2020). Rothen-Rutishauser B and Zobi F. Cytotoxicity of Mn-based photoCORMs of ethynyl-α-diimine ligands against different cancer cell lines: The key role of CO-depleted metal fragments. *J. Inorg. Biochem.* 209: 111122.

157. Santos-Silva, T., Mukhopadhyay, A., Seixas, J.D., Bernardes, G.J., Romão, C.C., and Romão, M.J. (2011). CORM-3 reactivity toward proteins: the crystal structure of a Ru(II) dicarbonyl-lysozyme complex. *J. Am. Chem. Soc.* 133: 1192–1195.

158. Nobre, L.S. and Jeremias, H. (2016). Romao CC and Saraiva LM. Examining the antimicrobial activity and toxicity to animal cells of different types of CO-releasing molecules. *Dalton Trans.* 45: 1455–1466.

159. Yuan, Z. and Yang, X. (2020). De La Cruz LKC and Wang B. Nitro reduction-based fluorescent probes for carbon monoxide require reactivity involving a ruthenium carbonyl moiety. *Chem. Commun.* 56: 2190–2193.

160. Wareham, L.K., Poole, R.K., and Tinajero-Trejo, M. (2015). CO-releasing metal carbonyl compounds as antimicrobial agents in the post-antibiotic era. *J. Biol. Chem.* 290: 18999–19007.

161. Gessner, G., Sahoo, N., Swain, S.M., Hirth, G., Schönherr, R., Mede, R., Westerhausen, M., Brewitz, H.H., Heimer, P., and Imhof, D. (2017). Hoshi T and Heinemann SH. CO-independent modification of K(+) channels by tricarbonyldichlororuthenium(II) dimer (CORM-2). *Eur. J. Pharmacol.* 815: 33–41.

162. Dong, D.L., Chen, C., Huang, W., Chen, Y., Zhang, X.L., and Li, Z. (2008). Li Y and Yang BF. Tricarbonyldichlororuthenium (II) dimer (CORM2) activates non-selective cation current in human endothelial cells independently of carbon monoxide releasing. *Eur. J. Pharmacol.* 590: 99–104.

163. Nielsen, V.G. and Garza, J.I. (2014). Comparison of the effects of CORM-2, CORM-3 and CORM-A1 on coagulation in human plasma. *Blood Coagul. Fibrinolysis* 25: 801–805.

164. Faizan, M., Muhammad, N., Niazi, K.U.K., Hu, Y., Wang, Y., Wu, Y., Sun, H., Liu, R., Dong, W., Zhang, W., and Gao, Z. (2019). CO-releasing materials: An emphasis on therapeutic implications, as release and subsequent cytotoxicity are the part of therapy. *Materials (Basel)* 12.

165. Jomova, K. and Vondrakova, D. (2010). Lawson M and Valko M. Metals, oxidative stress and neurodegenerative disorders. *Mol. Cell. Biochem.* 345: 91–104.

166. Aisen, P. and Cohen, G. (2013). Transition metal toxicity. *Int. Rev. Exp. Pathol.* 31.

167. Wang, D., Viennois, E., Ji, K., Damera, K., Draganov, A., Zheng, Y., Dai, C., Merlin, D., and Wang, B. (2014). A click-and-release approach to CO prodrugs. *Chem. Commun.* 50: 15890–15893.

168. Pan, Z., Ji, X., Chittavong, V., Li, W., Ji, K., Zhu, M., Zhang, J., and Wang, B. (2017). Organic

CO-prodrugs: structure CO-release rate relationship Studies. *Chem. Eur. J.* 23: 9838–9845.

169. Ji, X., Pan, Z., Li, C., Kang, T., De La Cruz, L.K.C., Yang, L., Yuan, Z., Ke, B., and Wang, B. (2019). Esterase-sensitive and pH-controlled carbon monoxide prodrugs for treating systemic inflammation. *J. Med. Chem.* 62: 3163–3168.

170. Ji, X., Ji, K., Chittavong, V., Aghoghovbia, R.E., Zhu, M., and Wang, B. (2017). Click and fluoresce: A bioorthogonally activated smart probe for wash-free fluorescent labeling of biomolecules. *J. Org. Chem.* 82: 1471–1476.

171. Pan, Z., Zhang, J., Chittavong, V., Ji, X., and Wang, B. (2018). Organic CO prodrugs activated by endogenous ROS. *Org. Lett.* 20: 8–11.

172. Zheng, Y., Ji, X., Yu, B., Ji, K., Gallo, D., Csizmadia, E., Zhu, M., De La Cruz, L.K.C., Choudhary, M.R., Chittavong, V., Pan, Z., Yuan, Z., Otterbein, L.E., and Wang, B. (2018). Enrichment-triggered prodrug activation demonstrated through mitochondria-targeted delivery of doxorubicin and carbon monoxide. *Nat. Chem.* 10: 787–794.

173. Bakalarz, D., Surmiak, M., Wójcik, D., Yang, X., Edyta, K., Pan, Z., Śliwowski, Z., De La Cruz, L.K., Ginter, G., Buszewicz, G., Brzozowski, T., Cieszkowski, J., Głowacka, U., Magierowska, K., Wang, B., and Magicrowski, M. (2020). Organic carbon monoxide prodrug, BW-CO-111, in protection against chemically-induced gastric mucosal damage. *Acta Pharm. Sin. B.* in press. 10.1016/j.apsb.2020.08.005.

174. Ji, X., Zhou, C., Ji, K., Aghoghovbia, R., Pan, Z., Chittavong, V., Ke, B., and Wang, B. (2016). Click and release: A chemical strategy toward developing gasotransmitter prodrugs by using an intramolecular diels-alder reaction. *Angew. Chem. Int. Ed. Engl.* 55: 15846–15851.

175. Ji, X. and Wang, B. (2018). Strategies toward organic carbon monoxide prodrugs. *Acc. Chem. Res.* 51: 1377–1385. and references cited therein.

176. Yang, X.X. and Ke, B.W. (2020). Lu W and Wang BH. CO as a therapeutic agent: discovery and delivery forms. *Chin. J. Nat. Med.* 18: 284–295.

177. Wang, M., Yang, X., Pan, Z., Wang, Y., De La Cruz, L.K., Wang, B., and Tan, C. (2020). Towards "CO in a pill": pharmacokinetic studies of carbon monoxide prodrugs in mice. *J. Control. Release* 327: 174–185.

178. Lee, G.R., Shaefi, S., and Otterbein, L.E. (2019). HO-1 and CD39: it Takes Two to Protect the Realm. *Front. Immunol.* 10: 1765.

179. Arjona, A. and Silver, A.C. (2012). Walker WE and Fikrig E. Immunity's fourth dimension: approaching the circadian-immune connection. *Trends Immunol.* 33: 607–612.

180. De Caestecker, M., Humphreys, B.D., Liu, K.D., Fissell, W.H., Cerda, J., Nolin, T.D., Askenazi, D., Mour, G., Harrell, F.E., Jr., and Pullen, N. (2015). Okusa MD and Faubel S. Bridging translation by improving preclinical study design in AKI. *J. Am. Soc. Nephrol.* 26: 2905–2916.

181. Michel, B.W. (2012). Lippert AR and Chang CJ. A reaction-based fluorescent probe for selective imaging of carbon monoxide in living cells using a palladium-mediated carbonylation. *J. Am. Chem. Soc.* 134: 15668–15671.

182. Ohata, J. (2019). Bruemmer KJ and Chang CJ. Activity-based sensing methods for monitoring the reactive carbon species carbon monoxide and formaldehyde in living systems. *Acc. Chem. Res.* 52: 2841–2848.

183. Hilgenbrink, A.R. and Low, P.S. (2005). Folate receptor-mediated drug targeting: from therapeutics to diagnostics. *J. Pharm. Sci.* 94: 2135–2146.

184. Wang, J. (2017). Masehi-Lano JJ and Chung EJ. Peptide and antibody ligands for renal targeting: nanomedicine strategies for kidney disease. *Biomater. Sci.* 5: 1450–1459.

185. Garcia-Gallego, S. and Bernardes, G.J. (2014). Carbon-monoxide-releasing molecules for the delivery of therapeutic CO in vivo. *Angew. Chem. Int. Ed. Engl.* 53: 9712–9721.

186. Loureiro, A., Bernardes, G.J., Shimanovich, U., Sarria, M.P., Nogueira, E., Preto, A., Gomes, A.C., and Cavaco-Paulo, A. (2015). Folic acid-tagged protein nanoemulsions loaded with CORM-2 enhance the survival of mice bearing subcutaneous A20 lymphoma tumors. *Nanomedicine* 11: 1077–1083.

187. Zheng, Y., Ji, X., Yu, B., Ji, K., Gallo, D., Csizmadia, E., Zhu, M., Choudhury, M.R., De La Cruz, L.K.C., Chittavong, V., Pan, Z., and Yuan, Z. (2018). Otterbein LE and Wang B. Enrichment-triggered prodrug activation demonstrated through mitochondria-targeted delivery of doxorubicin and carbon monoxide. *Nat. Chem.* 10: 787–794.

26

CO as an Antiplatelet Agent: An Energy Metabolism Perspective

Patrycja Kaczara[1], Kamil Przyborowski[1], Roberto Motterlini[2], and Stefan Chlopicki[1]

[1]*Jagiellonian Centre for Experimental Therapeutics (JCET), Jagiellonian University, Krakow, Poland*
[2]*INSERM Unit 955, Faculty of Health, University Paris-Est, Creteil, France*

Introduction

Antiplatelet therapy represents a major pharmacological strategy aimed at reducing the risk of atherothrombosis-associated cardiovascular events. A number of antiplatelet drugs have been developed and are currently used clinically, and a series of novel antiplatelet agents is on its way to clinical practice. Yet, the majority of available antiplatelet agents have been developed with the specific aim of targeting platelet surface receptors, or signaling intraplatelet mechanisms, but not to modify or alter intracellular metabolism. In this chapter, we will review the evidence supporting the importance of heme oxygenase (HO)-derived carbon monoxide (CO) in the regulation of thromboresistance *in vivo*, in particular exemplified by CO-dependent inhibition of platelet function. We will also review emerging evidence showing that CO affords antiplatelet activities that could be mechanistically explained by the effects of CO in modulating intraplatelet energy metabolism resulting in NAD^+ and ATP deficiency. These very recent findings might open a new avenue for antiplatelet strategies based on inhibition of platelet energy metabolism.

Endogenously Generated CO from HO Enzymes Regulates Thromboresistance *In Vivo*

The first attempts to assess the effects of exogenous CO on platelet functions were undertaken in the 1970s to 1980s, still prior to the discovery of intracellular CO-generating enzymes, namely HO-1 and HO-2 [1]. The aim of these studies was to characterize the detrimental effects of cigarette smoke constituents on the cardiovascular system. In several studies *in vivo*, inhalation of CO gas or cigarette smoking resulted in platelet hyperactivation, as measured using *ex vivo* approaches in animals or humans [2–4]. For example, in the paper by Birnstingl and coauthors [2], platelets obtained from rabbits exposed to gaseous CO for 6–14 h displayed an increase in stickiness immediately after CO exposure, while on the next day they had displayed reduced activation below the level observed prior to CO exposure. At the time, some authors hypothesized that CO inhalation could cause morphological changes to the vascular wall associated with damage of endothelial cells (ECs), thereby leading to platelet activation *in vivo* [3]. Consistent with these reports, smoking-induced detrimental effects on platelets were ascribed to some activities by CO, and not to the action of nicotine, another well-known component of cigarette smoke [4]. Indeed, in humans, smoking of two cigarettes over a 15-min period increased platelet aggregation in whole blood *ex vivo* as observed after completion of smoking and then the platelet function returned to basal pre-experimental levels by 90 min after smoking. Interestingly, opposite effects were observed under *in vitro* conditions in studies reported by Mansouri and Perry [5], in which platelet aggregation measured after incubation of platelets with cigarette smoke or CO gas in a test tube was reduced. These results would suggest that a lower nontoxic concentration of CO, applied *in vivo* as a gas, afforded antiplatelet effects in accordance with reports of the inhaled CO *in vivo* that inhibited platelet–leukocyte interaction [6].

The convincing evidence demonstrating that CO exerts antithrombotic actions was provided when genetically engineered mouse models with deletion of the HO-1 enzyme ($Hmox1^{-/-}$) were available.

Carbon Monoxide in Drug Discovery: Basics, Pharmacology, and Therapeutic Potential, First Edition. Edited by Binghe Wang and Leo E. Otterbein.
© 2022 John Wiley & Sons, Inc. Published 2022 by John Wiley & Sons, Inc.

Indeed, mice lacking HO-1 are characterized by accelerated arterial thrombus formation induced by photochemical vessel injury; however, most interestingly, these mice display preserved overall hemostatic functions as evidenced by measurement of bleeding time, prothrombin time, and platelet count, which were all unchanged in HO-1-deficient mice [7]. In contrast, following injury of arterial vessels, $Hmox1^{-/-}$ mice exhibited increased endothelial apoptosis in association with denudation of the endothelium. This damage to the endothelium was accompanied by (i) enhanced expression of tissue factor (TF), the most important initiator of intravascular coagulation; and (ii) elevated plasma levels of von Willebrand factor, which plays a key role in platelet adhesion to the wound sites. Arteries from $Hmox1^{-/-}$ mice subjected to injury were also characterized by overproduction of reactive oxygen species [7]. Altogether, these results infer that lack of HO-1, and consequently diminished production of vasoprotective CO, severely altered the intrinsic antithrombotic properties of the vessel wall in response to photochemical injury, leading to the development of a prothrombotic phenotype. Importantly, the increased thrombosis in $Hmox1^{-/-}$ mice was rescued by inhalation of gaseous CO for 24 h prior to vessel injury. These experiments support a potential role of endogenously produced CO from heme degradation by HO-1 in maintenance of vascular thromboresistance.

The importance of the HO-1/CO pathway in the suppression of thrombosis through vascular mechanisms was also supported by other studies. In a model of arterial thrombosis using $ApoE^{-/-}$ mice following mechanical injury of the carotid artery, delivery of the HO-1 gene to injured arteries using an HO-1 adenovirus initiated thrombolysis and blood flow restoration as compared with untreated arteries. A similar thrombolytic effect in mice with established thrombosis was observed following treatment for 2 h with CO gas inhalation. The enhanced thrombolytic activity found in the above-mentioned studies was attributed to downregulation of arterial expression of plasminogen activator inhibitor-1 (PAI-1), an inhibitor of plasmin generation [8]. In addition, local overexpression of HO-1 in injured arteries protected vessels against neointimal hyperplasia and an inflammatory response [8]. Although venous thrombosis differs in its pathogenesis from arterial thrombosis, it is important to emphasize that HO-1 activity also contributes to thromboresistance in veins. HO-1-deficient mice were reported to develop greater thrombi in response to blood stasis induced by ligation of the infrarenal inferior vena cava (IVC) due to enhanced thrombus generation and impaired thrombolysis [9]. Most likely, this effect resulted from the fact that HO-1 deficiency contributed to a wide range of thrombosis-induced inflammatory responses, including, among others, the reprogramming of the endothelium from an anticoagulant to a procoagulant phenotype. Indeed, increased expression of TF was observed in thrombus-containing ligated IVC in $Hmox1^{-/-}$ mice as compared with wild-type counterparts [9].

Thus, a number of studies suggest that the HO-1/CO pathway elicits an antithrombotic effect in arteries and veins due to multiple protective mechanisms localized to the vessel wall. Nevertheless, based on the results of these studies, it cannot be excluded a priori that increased thrombus formation in $Hmox1^{-/-}$ mice in the setting of vascular injury is linked to the lack of antiplatelet effects of CO generated by HO-1 activity. This interesting aspect of the HO-1/CO pathway has not been investigated in the elegant studies described above [7–9].

Endogenously Generated CO from HO Enzymes Displays Antiplatelet Effects

The evidence that endogenous HO-1/CO safeguards against thromboresistance *in vivo* by limiting platelet adhesion to the endothelium came from the experimental model of hepatic ischemia–reperfusion injury in rats using intravital microscopy. The induction of HO-1 *in vivo* by cobalt protoporphyrin inhibited platelet adhesion to sinusoidal ECs in response to reperfusion [10]. Further convincing evidence showing that CO suppresses thrombosis by effects on platelets was provided in studies related to experimental organ transplantation where HO-1 activity protected against the rejection of mouse-to-rat cardiac transplants [11]. Under conditions of HO-1 inhibition, an animal treated with gaseous CO suppressed graft rejection and restored long-term graft survival. This effect was associated with inhibition of thrombosis in coronary arterioles and protection against myocardial infarction [11]. In an elegant complementary series of experiments by the same group [11], ECs *in vitro* were treated with either an inducer (CoPPIX) or inhibitor (SnPPIX) of HO-1 activity, followed by coincubation with platelets. It was shown that aggregation of platelets exposed to ECs was inhibited when platelets were incubated with ECs overexpressing HO-1, but platelet aggregation was enhanced after coincubation of ECs in the presence of HO-1 inhibition. These experiments indicate that CO at the levels produced endogenously by HO-1 might affect platelet activity under *in vitro* conditions; in particular, increased HO-1 activity in ECs has the ability to regulate platelet function. Therefore, the antiplatelet activity of CO, similar to its anti-inflammatory, antiapoptotic, and antiproliferative effects [12], contributes to protection against transplant rejections.

The role of HO-1-derived CO as an antiplatelet mediator *in vivo* has been further confirmed in a report showing that treatment of $Hmox1^{-/-}$ aortic graft recipients with CO delivered by systemic administration of a CO-releasing molecule (CORM), CORM-2 ($[Ru(CO)_3Cl_2]_2$), prevented thrombosis within aortic grafts and prolonged the otherwise compromised survival rate observed in $Hmox1^{-/-}$ untreated mice [13]. Furthermore, $Hmox1^{-/-}$ recipients receiving HO-1-expressing wild-type platelets showed an improved hindlimb function and prolonged post-transplant survival, supporting CO-mediated protection also provided by platelets generating CO [13]. Furthermore, the antithrombotic effect of CO endogenously generated by HO-1, observed in a model of $FeCl_3$-induced arterial thrombosis *in vivo*, could have been mediated mainly by platelet inhibition [14]. Indeed, experimental upregulation of HO-1 activity in mice significantly prolonged thrombus formation not only *in vivo* but also in *ex vivo* thrombus formation assay when blood was perfused over a collagen-coated capillary tube [14].

To summarize, endogenously produced HO-1-derived CO appears to regulate vessel thromboresistance and the effects observed could be linked to an effect on either the coagulation and fibrinolysis systems or platelet function, or both (Figure 26.1). Needless to say, it is rather difficult to distinguish the effects of CO on these two systems (coagulation/fibrinolysis versus platelets) under *in vivo* conditions as they are tightly interlinked. Activated platelets are essential in thrombosis development, as they contribute to platelet plug formation, but are also involved in the regulation of coagulation and fibrinolysis. Consequently, a crosstalk between platelets, coagulation, and fibrinolysis safeguards optimal and balanced thrombus growth, in which a number of mechanisms are involved. For example, activated platelets express on their surface phosphatidylserine that confers a procoagulant surface by providing binding sites for assembly of procoagulant factors to generate thrombin responsible for fibrin formation [15]. Furthermore, platelets undergo degranulation and release a wide variety of factors, including, for example, procoagulant factor V [16] and antifibrinolytic PAI-1 [17]. On the other hand, the inhibition of platelets can diminish thrombin generation taking place on the surface of activated platelets as

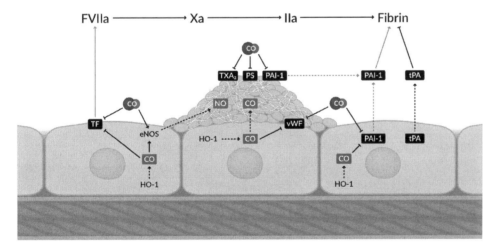

Figure 26.1 Major mechanisms underlying the antithrombotic activity of CO under *in vivo* conditions. CO has beneficial effect on vessel wall function, thereby it ensures protection against transformation of endothelial phenotype from antithrombotic to prothrombotic. CO generated endogenously by HO-1 or CO delivered exogenously (e.g., using CORMs) can stimulate endothelial production of nitric oxide (NO), a well-known inhibitor of platelet aggregation [21,22], and downregulate inflammation-induced vascular responses, including expression of PAI-1 [8], TF [7], and release of von Willebrand factor (vWF) [7]. Therefore, CO, by acting on the vascular endothelium, can suppress platelet adhesion [10], aggregation [21], and coagulation as well as enhance fibrinolysis [9]. However, in this chapter we provide evidence for the major role of direct effects of CO on platelets in the antithrombotic actions of CO, based on impairment of intraplatelet energy metabolism (see details below). This mechanism contributes to inhibitory effects of CO on platelet aggregation [23], and may contribute to subsequent platelet-dependent anticoagulant and profibrinolytic effects. CO inhibits various aspects of platelet function, including thromboxane A_2 (TXA_2) production [24], release of platelet-derived PAI-1 [21], and exposure of procoagulant phosphatidylserine (PS) on the platelet surface [20].

well as accelerate fibrinolysis through the reduction of platelet PAI-1 release.

Therefore, it might well be that the antithrombotic activity of CO *in vivo* involves indirect anticoagulant and profibrinolytic effects resulting from direct action of CO on platelets. Still, the evidence supporting the notion that exogenous and endogenous CO as a regulator of platelet function has not been appreciated. Here, we review the evidence that HO-1-derived CO affords important antiplatelet effects and describe mechanisms of antiplatelet activity of exogenous CO with a particular attention to the effects of CO on energy metabolism. We do not address here CO-related effects on coagulation and fibrinolysis, which is outside the scope of this chapter and was studied and reviewed previously [18–20].

Exogenous CO Inhibits Activation of Platelets

Convincing evidence on the effects of CO on platelets *in vivo* was provided by Kramkowski and colleagues in 2012 [20]. In their experiments, the authors utilized two CORMs developed by Motterlini's laboratory, CORM-3 ($Ru(CO)_3Cl(glycinate)$) and CORM-A1 ($Na_2H_3BCO_2$), that differed in their chemical structure, reactivity, and kinetics of CO release in biological systems. Specifically, CORM-3 is a water-soluble ruthenium-based carbonyl that liberates CO with a "fast kinetics" ($t_{1/2}$ = 1 min) in biological media [25], cells, and tissues, whereas CORM-A1 is a boron-containing carboxylic acid that generates and releases CO with a "slow rate" at physiological pH ($t_{1/2}$ = 21 min) [26]. Not only these compounds provided an excellent tool to deliver CO *in vivo*, but their comparative effect was also instrumental to better characterize the effects of CO on platelet biology and to explore the potential of using CO as an antiplatelet therapy. Although it was known from previous studies that CORMs exert an antiaggregatory activity on human platelets [27,28], Kramkowski's study was the first *in vivo* report on the antithrombotic effects of two water-soluble CORMs having different rates of CO release [20]. It was found that both CORM-A1 and CORM-3 administered intravenously into rats inhibited thrombus formation in electrically injured carotid arteries. However, the "slow CO releaser" CORM-A1 significantly reduced thrombus weight, while the "fast CO releaser" CORM-3 displayed only a weak effect. Most importantly, in an intravital laser-induced venous thrombosis, both CORM-A1 and CORM-3 demonstrated significant antithrombotic activities, but only CORM-A1 showed strong antiplatelet effects as evidenced by the expression levels of platelet surface phosphatidylserine in the growing thrombus. Furthermore, CORM-3 but not CORM-A1, at doses used in the experiments, was hypotensive. These findings suggest that although both CORM-3 and CORM-A1 inhibited thrombosis *in vivo*, CORM-A1 displayed a relatively weak hypotensive effect and induced a stronger inhibition of platelet aggregation accompanied by a decrease in concentration of active PAI-1. In contrast, CORM-3 displayed a stronger hypotensive effect and decreased fibrin generation, but without a direct influence on platelet aggregation and fibrinolysis. Indeed, CORM-3 did not affect fibrinolysis and did not modify plasma PAI-1 concentration, while CORM-A1 caused a decrease in plasma level of active PAI-1. The difference in the effects on PAI-1 by CORM-3 and CORM-A1 might be due to a difference in the degree of antiplatelet effects achieved by these compounds, as higher doses of CORM-3 significantly inhibited arterial thrombus formation in rats with $FeCl_3$-induced thrombosis, and this effect was accompanied by a fall in plasma level of active PAI-1 [21]. Interestingly, in the report by Soni and colleagues, CORM-3-induced effects on platelet aggregation, time to occlusion, change in thrombus weight, and plasma PAI-1 were significantly blocked by ODQ (inhibitor of soluble guanylate cyclase [sGC]), and by L-NAME (inhibitor of NO synthase), leading to the conclusion that the antithrombotic activity of CORM-3 *in vivo* is mediated by NO and sGC [21]. Interestingly, under *in vitro* conditions, the effects of CO delivered by CORM-A1 could not be mimicked by CO derived from CORM-3. In preparations of human platelet-rich plasma, washed platelets, and the whole blood, CORM-A1 was more potent than CORM-3 in inhibiting platelets, suggesting that indeed CORM-A1 is the prototypic compound for CO delivery to achieve therapeutically relevant amounts of CO that would mediate an antiplatelet effect.

Thus, the antithrombotic effect of CORM-A1 *in vivo* is mediated by an effect of CO on platelet aggregation, and the kinetics of CO release appeared to play a key role in achieving antiplatelet effects of CORMs *in vivo* without causing undesired hypotensive effects [20]. Similarly, under *in vitro* conditions, CORM-A1 appears to be a stronger antiplatelet agent than CORM-3 [28]. Accordingly, these two comparative reports provide important information on the biological effects of CO on platelets; that is, (i) the type of CORMs for therapeutic applications needs to be selected for a given pathological indication based on various aspects of the chemical reactivity of the chosen CORM that may lead to different mechanisms of CO release within a given biological environment; and (ii) CORM-A1, a nonmetal organic CO generator that releases CO at a slow rate, appears to be an excellent prototypic compound for CO delivery to achieve therapeutically relevant antiplatelet effects *in vivo*. Although additional studies are required to substantiate the use of CORMs to target platelet function, these findings clearly indicate the importance of fully characterizing the chemical reactivity and biological effects of different classes of CORMs for given pathological indications. In the case of the effect of CO on platelet aggregation and function, it is crucial to compare the biological action of different CORMs particularly *in vivo* since the metabolism of metal-based CORMs (including CORM-3 and CORM-2) and organic CO-releasing agents (including CORM-A1) is practically unknown. That is, we still do not know how CO is distributed to the various tissues and cells once it is administered *in vivo*. It should be noted that CO distribution in an organism will also depend on the different routes of CORM administration (intravenously versus intraperitoneally versus orally). Therefore, a thorough and robust investigation on the efficacy of these compounds in animal models of disease is warranted. Chapter 4 of this book specifically addresses pharmacokinetic issues of CO.

The effectiveness of CORMs slowly liberating CO as antiplatelet agents was further confirmed by another type of metal-based CORMs, *cis*-rhenium(II)-dicarbonyl-vitamin B12 complexes (B12-ReCORMs) with a tunable CO release rate [29]. Quite paradoxically, although this study was intended as a proof of concept to show that alterations in the electronic properties at the Co^{III} center of the B12 biovectors translate directly into variations of the CO release kinetics at the Re^{II} metal ion, the differences in CO-releasing kinetics among the different complexes were still relatively small with the release rate quite close to that of CORM-A1. Most of the B12-ReCORMs displayed pronounced antiplatelet activities similar to or even slightly higher than those of CORM-A1, confirming the notion that slow CO releasers display desirable pharmacological profiles as antiplatelet agents.

The mechanism of action of CO was in various systems compared to the mechanism known for NO [30]. In this context, it must be emphasized that potent antiplatelet effects of compounds with slow kinetics of CO delivery, as compared with fast CO releasers, were in striking contrast to the antiplatelet effects of NO donors with various kinetics of NO release. Inhibition of platelet aggregation by NO donors was more potent with a fast NO releaser (DEA-NO, $t_{1/2}$ = 2 min) than slow NO releasers, such as PAPA-NO ($t_{1/2}$ = 15 min) or other slow NO donors [28]. Furthermore, the inhibitory effect of NO on platelet aggregation is well known to be mediated by sGC, and is reversed by ODQ and potentiated by phosphodiesterase-5 inhibitor, sildenafil. In contrast, inhibition of platelet aggregation by CORM-A1 was not significantly affected by either ODQ or sildenafil [28].

Of note, Brüne and Ullrich already demonstrated in 1987 that platelets bubbled for 15–30 s with pure CO gas failed to aggregate, and it was demonstrated that this effect is mediated by the activation of sGC [24]. Bubbling with gaseous CO enables relatively rapid delivery of high concentrations of CO. The reversibility of the observed effect by visible light enabled scientists to conclude that CO binds to the heme regulatory subunit of sGC in its ferrous state. Then, Friebe and coauthors demonstrated that inhibition of platelet aggregation in response to CO was potentiated by YC-1 (an activator of sGC), and a combination of gaseous CO and YC-1 inhibited aggregation and increased cGMP production in platelets [31]. These results seem to be in contrast with the mechanisms of action of CORM-A1, shown to be sGC independent [28]. It is possible that the divergent mechanisms of action mediated by CO gas and CORM-A1 might be due to the differences in concentration of the two systems, distinct chemical reactivity toward biological and intracellular components, distinct kinetics of CO delivery, or combination of these factors. Furthermore, comparative studies of these important aspects that determine biological activity of exogenous CO, in relation to endogenous HO-1-derived CO, are needed.

Altogether, the results suggest that there are divergent mechanisms of action mediated by CO gas and CORM-A1 that seem to refute the generally accepted view that both ways of CO delivery promote biological effects through the same mechanism. In the case of CO-mediated effects on platelets, this generally accepted notion in the field of CO research cannot be substantiated. Indeed, the mechanism of action for the antiplatelet effects of CO was shown to be sGC dependent when delivered in the gaseous form [24] and sGC independent when delivered slowly by CORM-A1 [28]. Recently, we directly compared effects on platelet aggregation and energy metabolism of CO delivered rapidly by gaseous CO-saturated buffer (CO_G) or slowly by CORM-A1, at the concentrations leading to similar inhibition of platelet aggregation, and demonstrated distinct mechanisms of CO action – CO_G activated sGC, but did not affect energy metabolism of platelets, whereas CORM-A1 impaired energy metabolism of platelets without effects on sGC. These experiments clearly demonstrated that the source and the kinetics of CO delivery determine the mechanism of antiplatelet action of CO, and that different systems of CO delivery should not be applied interchangeably in studies on platelets [32].

It remains to be established whether endogenous HO-1-derived CO affords antiplatelet effects by sGC-dependent or -independent way. Interestingly, hemin-induced cGMP production in platelets was not blunted in HO-1-deficient mice in comparison with wild-type animals suggesting that HO-1/CO action *in vivo* was not mediated by the activation of sGC [14]. It is possible that both HO-1-derived CO and CO slowly liberated by CORM-A1 induce effects, which are mechanistically similar, because CORM-A1 mimics the kinetics of endogenous CO gradually produced over time during the degradation of heme. Although endogenously produced CO has not been measured yet, we hypothesize that HO-1-derived CO has a distinct kinetics of release as compared with CO gas delivery or CO released from CORM-3.

Taken together, the slow CO releasers, CORM-A1 and B12-ReCORMs, have been reported to act as superior antiplatelet agents as compared with CORM-3, which releases CO instantly. The antiplatelet action of CORM-A1 does not involve sGC activation, as also suggested earlier for other CORMs [27]. Accordingly, CORM-A1 was identified as the first among slow releasing compounds to display a promising pharmacological profile in inhibiting platelet aggregation that seems to be clearly distinct from CO gas, and more potent compared with the faster releasing CORMs. Still, the mechanism of action of CORM-A1 in platelets remains undefined. A number of possible targets for CO known to be operative in other cells such as ATP-activated potassium channels and p38MAPK could not explain the antiplatelet effects of CO [21]. Although it has been reported that CO delivered by CORM-2 suppressed lipopolysaccharide (LPS)-induced overactivation of platelets by interfering with glycoprotein-mediated HS1 phosphorylation [33], whether this mechanism of CO action contributes to inhibition of platelets exposed to classical agonists, such as collagen or thrombin, was not examined. The same group proposed another mechanism of antiplatelet action of CO mediated by the PI3K–Akt–GSK3β pathway [34]. Our recent work published in *Arteriosclerosis, Thrombosis, and Vascular Biology* and highlighted by Editorial comment was the first to point out a novel action of CORM-A1 in modulating energy metabolism in platelets that could explain the antiplatelet effects of CO (see summary in Figure 26.2) [23,35].

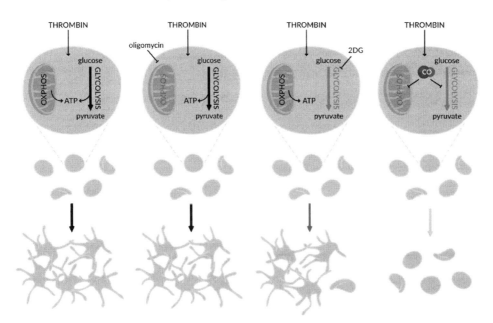

Figure 26.2 Simultaneous inhibition of mitochondrial respiration and glycolysis by CO overcomes metabolic plasticity of platelets and blocks platelet aggregation. Platelet aggregation is an energetically demanding process, which requires activation of mitochondrial respiration or glycolysis to deliver ATP. Inhibition of mitochondrial respiration does not inhibit platelet aggregation, leading to a compensatory mechanism that accelerates glycolysis. Inhibition of glycolysis slightly reduces platelet aggregation, despite accelerated mitochondrial respiration, but still platelets are ready to aggregate. Simultaneous inhibition of mitochondrial respiration and glycolysis by CO or combined treatment with inhibitors for these processes (e.g., oligomycin – inhibitor of ATP synthase in mitochondria; and 2-deoxy-D-glucose (2DG) – inhibitor of glycolysis) efficiently inhibits platelet aggregation. Adapted from [23].

Effects of Exogenous CO on Energy Metabolism in Platelets

Resting platelets are discoid-shape enucleated blood cells with homogeneously distributed granules. However, upon activation, the platelet cytoskeleton undergoes rearrangement resulting in a change of shape, formation of plasma membrane pseudopodia, and release of platelet-derived extracellular vesicles. Generally, early receptor signaling of this complex process of platelet activation can be induced by soluble platelet agonists (such as thrombin, ADP, and TXA_2), adhesion receptor ligands (such as collagen), or inflammatory stimuli (such as LPS) [36]. Platelet aggregation is further amplified by platelet secretion products (such as ADP or TXA_2) and local prothrombotic factors contributing to thrombin generation (such as TF) [37]. All these highly complex signals converge in platelet activation and induce platelet adhesion, changes in platelet shape, and their aggregation, with a final upregulation of platelet surface receptors. ATP-derived energy is required for several of these processes, among them are impressive and instant cytoskeleton rearrangements contributing to the change of platelet shape and to aggregation itself, secretion of dense, alpha and acid hydrolase-containing granules, protein phosphorylation, activation of phospholipase C, and other signaling cascades [38–41].

Thus, platelet function and their ability to regulate hemostasis, which could be life-saving upon bleeding to prevent loss of the blood, heavily depend on intercellular ATP availability provided by mitochondrial respiration and glycolysis. Naturally, mitochondria are the organelles considered as the major source for ATP production, forming a milieu for tricarboxylic acid cycle, β-oxidation, and oxidative phosphorylation. Of note, mitochondria not only play a role in energy metabolism, but also participate in generation of reactive oxygen species, calcium homeostasis, regulation of apoptosis, and ER stress response mechanisms [42]. Interestingly, despite preserved mitochondrial function, activated platelets rely on ATP derived primarily from glycolysis rather than from mitochondrial respiration [38,43]. To sustain energy metabolism, platelets can utilize diverse metabolic fuels that supply glycolysis or oxidative phosphorylation, mainly glucose, fatty acids, glutamine, glycogen, and citrate, but also albumin and acetate [38,43–47]. A number of reports confirm a high substrate plasticity of energy metabolism in platelets [43–45,48,49], which enables platelet aggregation even after inhibition of a single metabolic pathway (Figure 26.2) [23,43,45,50]. This extreme metabolic plasticity of platelets is important to maintain platelet function even under limited access to glucose or oxygen and, in terms of the role of platelets, it seems to be highly beneficial. However, in the case of chronically elevated blood glucose or fatty acids, the metabolic plasticity may lead to hyperreactivity of platelets [51–53].

Importantly, altered energy metabolism was described in platelets derived from patients with diabetes mellitus [54], but also other diseases such as asthma [55], sickle cell disease [56], Parkinson's disease [57], or sepsis [58]. Furthermore, certain physiological processes, like aging, are also accompanied by development of platelet hyperreactivity [59]. Thus, two conclusions emerge: (i) systemic metabolic disorders may contribute to changes in platelet activity and function; and (ii) targeting cellular metabolism – oxidative phosphorylation in mitochondria and glycolysis – may be a strategy to regulate the function of platelets.

In this context, given that CO can modulate energy metabolism in various cell types [60–68], and that platelet aggregation is an energy-demanding process, we started to reason on the intriguing possibility that the antiplatelet effects observed with CO delivered by CORMs, being independent of sGC activation [27,28], could result from modulation of platelet energy metabolism. In our recent work, indeed we found that CO liberated from CORM-A1 inhibited both mitochondrial respiration and glycolysis (Figure 26.2). An inhibition of mitochondrial respiration was expected, as CO efficiently inhibits cytochrome c oxidase [69,70]. Interestingly, in platelets treated with CORM-A1, fumarate and malate concentrations decreased, indicating that succinate dehydrogenase (SDH) could be another possible target for CO in mitochondria [23]. Nevertheless, application of dimethyl malonate, a cell-permeable precursor of malonate and thus an inhibitor of SDH, did not affect platelet function, suggesting rather a minor role for SDH in modulation of platelet mitochondrial respiration and aggregation.

Oxidative phosphorylation was expected to be inhibited by CO; however, inhibition of glycolysis was not so obvious. Even though it was demonstrated that CO released from CORM-401 inhibited glycolysis in various cell types [65–68,71], CO released from CORM-A1 was shown either to increase glycolysis in microglia cells [65] or to be without an effect on glycolysis in neuroblastoma cells [72]. Interestingly, the changes in platelet glycolysis under CORM-A1 treatment (monitored by the Seahorse XF technique as changes of ECAR) were biphasic – slightly increased with lower concentrations and decreased with higher concentrations of CORM-A1 [23]. Mass spectrometry-based metabolomic analysis revealed that CO induced an inhibition of proximal glycolysis and shunted glucose through the pentose phosphate pathway (Figure 26.3). Our results might suggest that glycolysis could be inhibited at the level of GAPDH, as GAPDH contains a heme moiety and theoretically can be targeted by CO. Surprisingly, GAPDH activity was unaffected by CORM-A1. Importantly, in the presence of exogenous pyruvate, the antiplatelet effect of CORM-A1 was reversed and glycolysis was accelerated, even in platelets treated with CORM-A1. Exogenous pyruvate resulted in regeneration of NAD^+ from NADH and an increase in its availability for GAPDH. Measurements of NAD^+ and NADH demonstrated that CORM-A1-activated process(es) consuming NAD^+, however, these processes still need to be identified. The study revealed for the first time that antiplatelet actions of CO result from NAD^+ and ATP depletion, thus effectively overcoming the plasticity of energy metabolism in platelets (Figure 26.3).

Accordingly, NAD^+ seems to be a key regulator of platelet function, but still the knowledge about NAD-dependent processes in platelets is insufficient, and further studies are needed to identify major pathways of nonredox NAD^+ consumption in platelets. NAD is required for approximately 500 different enzymatic reactions and undergoes constant synthesis, degradation, and recycling processes [73]. Proteomic analysis identified in platelets several proteins consuming NAD, among them are poly(ADP-ribose) polymerases (1, 2, 6, 9, and 14), sirtuin 3, and CD157 [74]. It was shown that regulation of protein acetylation by sirtuin 2 plays a central role in platelet function [75], whereas the inhibition of sirtuins induces apoptosis-like changes in platelets [76], suggesting an important role for sirtuins in determination of platelet function and aging. Another NAD-consuming protein – CD38 – was shown to play an important role in platelet function via Ca^{2+} signaling, which was mediated by the products of CD38 activity [77]. Still, the mechanism regulating the NAD metabolome and homeostasis in platelets remains controversial. Delabie and coauthors demonstrated that platelets contain enzymatic machineries enabling transformation of nicotinamide riboside (NR) to NAD at rates comparable to other cells [78]. In contrast, Lee and coauthors did not identify the presence of NR kinase in platelets [74].

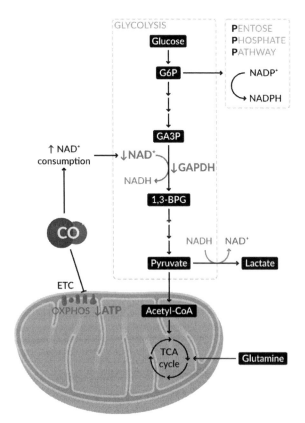

Figure 26.3 CO inhibits mitochondrial respiration and glycolysis, shunting glucose through the pentose phosphate pathway. CO inhibits cytochrome *c* oxidase in mitochondrial electron transport chain (ETC; leading to inhibition of oxidative phosphorylation), and activates nonredox processes consuming NAD^+ (leading to NAD depletion). Depletion of NAD^+ – a coenzyme necessary for glyceraldehyde 3-phosphate dehydrogenase (GAPDH), which is a rate-limiting enzyme in glycolysis – results in inhibition of glycolysis. Glucose 6-phosphate (G6P) is directed to the pentose phosphate pathway. Adapted from [23].

Perspectives

The identification of the mechanisms by which CO inhibits platelet aggregation by modulating energy metabolism enabled us to better understand the difference between CO and another endothelium-derived mediator, NO, which induces antiplatelet effects through activation of sGC [24,27,28,79,80]. Although it has been demonstrated that NO, similarly to CO, inhibits cytochrome c oxidase in platelet mitochondria, at the same time NO activates glycolysis in a broad concentration range [80]. However, unlike CO, NO inhibits platelet aggregation through activation of sGC and independently of energy metabolism inhibition. Similarly, other vasoprotective mediators produced by the endothelium, such as prostacyclin and CD39-derived adenosine, which is involved with antiplatelet action, induce their effects via cAMP, and not by modulation of energy metabolism. The data so far suggest a uniqueness of CO in the mechanisms of antiplatelet action, as compared with other endothelium-derived antiplatelet mediators. Currently used antiplatelet drugs employ different mechanisms of action as compared with the mechanisms of antiplatelet action of CO. For example, ADP antagonists, such as clopidogrel or prasugrel, act through P2Y12 receptor inhibition; aspirin (acetylsalicylic acid) inhibits TXA_2 synthesis through cyclooxygenase-1 blockade; vorapaxar inhibits PAR1 receptor; and eptifibatide inhibits receptor $\alpha_{IIb}\beta_3$. A number of novel antiplatelet drugs have been developed, including GP (glycoprotein) Ib-IX-V or GPVI antagonists [81]. It appears that the majority of antiplatelet strategies are currently targeted at platelet surface receptors, though with the notable example of thymidine phosphorylase and recently discovered antiplatelet effects of tipiracil hydrochloride (TPI) [82]. In fact, even aspirin through the inhibition of TXA_2 generation blocks subsequent TP receptor-dependent amplification of platelet activation. In this context, CORMs that possess a slow kinetic of CO liberation are unique as they seem to form a novel class of antiplatelet agents, which act via NAD^+ depletion through intracellular energy metabolism and via ATP deficiency. These findings might open a new avenue for antiplatelet strategies based on the selective inhibition of pathways involved in energy metabolism in platelets. Their application might not be limited to atherothrombosis-associated diseases to prevent thrombotic events and cardiovascular risk, but also may be useful in various other diseases associated with platelets' activation [83,84].

References

1. Maines, M.D., Trakshel, G.M., and Kutty, R.K. (1986). Characterization of two constitutive forms of rat liver microsomal heme oxygenase. Only one molecular species of the enzyme is inducible. *J. Biol. Chem.* 261: 411–419.
2. Birnstingl, M.A., Brinson, K., and Chakrabarti, B.K. (1971). The effect of short-term exposure to carbon monoxide on platelet stickiness. *Br. J. Surg.* 58: 837–839.
3. Marshall, M. and Hess, H. (1981). Akute Wirkungen niedriger Kohlenmonoxidkonzentrationen auf Blutrheologie, Thrombozytenfunktion und Arterienwand beim Miniaturschwein. *Res. Exp. Med.* 178: 201–210.
4. Bierenbaum, M.L., Fleischman, A.I., Stier, A., Somol, S.H., and Watson, P.B. (1978). Effect of cigarette smoking upon in vivo platelet function in man. *Thromb. Res.* 12: 1051–1057.
5. Mansouri, A. and Perry, C.A. (1982). Alteration of platelet aggregation by cigarette smoke and carbon monoxide. *Thromb. Haemost.* 48: 286–288.
6. Shinohara, M., Kibi, M., Riley, I.R., Chiang, N., Dalli, J., Kraft, B.D., Piantadosi, C.A., Choi, A.M.K., and Serhan, C.N. (2014). Cell-Cell interactions and bronchoconstrictor eicosanoid reduction with inhaled carbon monoxide and resolvin D1. *Am. J. Physiol. - Lung Cell. Mol. Physiol.* 307: L746–L757.
7. True, A.L., Olive, M., Boehm, M., San, H., Westrick, R.J., Raghavachari, N., Xu, X., Lynn, E.G., Sack, M.N., Munson, P.J., Gladwin, M.T., and Nabel, E.G. (2007). Heme oxygenase-1 deficiency accelerates formation of arterial thrombosis through oxidative damage to the endothelium, which is rescued by inhaled carbon monoxide. *Circ. Res.* 101: 893–901.
8. Chen, Y.H., Tsai, H.L., Chiang, M.T., and Chau, L.Y. (2006). Carbon monoxide-induced early thrombolysis contributes to heme oxygenase-1-mediated inhibition of neointimal growth after vascular injury in hypercholesterolemic mice. *J. Biomed. Sci.* 13: 721–730.
9. Tracz, M.J., Juncos, J.P., Grande, J.P., Croatt, A.J., Ackerman, A.W., Katusic, Z.S., and Nath, K.A. (2008). Induction of heme oxygenase-1 is a beneficial response in a murine model of venous thrombosis. *Am. J. Pathol.* 173: 1882–1890.
10. Tamura, T., Kondo, T., Ogawa, K., Fukunaga, K., and Ohkohchi, N. (2013). Protective effect of heme oxygenase-1 on hepatic ischemia-reperfusion injury through inhibition of platelet adhesion to the sinusoids. *Journal of Gastroenterology and Hepatology* 28: 700–706.

11. Sato, K., Balla, J., Otterbein, L., Smith, R.N., Brouard, S., Lin, Y., Csizmadia, E., Sevigny, J., Robson, S.C., Vercellotti, G., Choi, A.M., Bach, F.H., and Soares, M.P. (2001). Carbon monoxide generated by heme oxygenase-1 suppresses the rejection of mouse-to-rat cardiac transplants. *J. Immunol.* 166: 4185–4194.
12. Nakao, A., Choi, A.M.K., and Murase, N. (2006). Protective effect of carbon monoxide in transplantation. *J. Cell. Mol. Med.* 10: 650–671.
13. Chen, B., Guo, L., Fan, C., Bolisetty, S., Joseph, R., Wright, M.M., Agarwal, A., and George, J.F. (2009). Carbon monoxide rescues heme oxygenase-1-deficient mice from arterial thrombosis in allogeneic aortic transplantation. *Am. J. Pathol.* 175: 422–429.
14. Peng, L., Mundada, L., Stomel, J.M., Liu, J.J., Sun, J., Yet, S.F., and Fay, W.P. (2004). Induction of heme oxygenase-1 expression inhibits platelet-dependent thrombosis. *Antioxid. Redox Signal.* 6: 729–735.
15. Reddy, E.C. and Rand, M.L. (2020). Procoagulant phosphatidylserine-exposing platelets in vitro and in vivo. *Front. Cardiovasc. Med.* 7: 15.
16. Alberio, L., Safa, O., Clemetson, K.J., Esmon, C.T., and Dale, G.L. (2000). Surface expression and functional characterization of α-granule factor V in human platelets: Effects of ionophore A23187, thrombin, collagen, and convulxin. *Blood* 95: 1694–1702.
17. Morrow, G.B., Whyte, C.S., and Mutch, N.J. (2019). Functional plasminogen activator inhibitor 1 is retained on the activated platelet membrane following platelet activation. *Haematologica* 105: 2824–2833.
18. Nielsen, V.G. and Pretorius, E. (2014). Carbon monoxide: Anticoagulant or procoagulant? *Thromb. Res.* 133: 315–321.
19. Donaghy, D., Yoo, S., Johnson, T., Nielsen, V., and Olver, C. (2018). Carbon monoxide-releasing molecule enhances coagulation and decreases fibrinolysis in normal canine plasma. *Basic Clin. Pharmacol. Toxicol.* 123: 257–262.
20. Kramkowski, K., Leszczynska, A., Mogielnicki, A., Chlopicki, S., Grochal, E., Mann, B., Brzoska, T., Urano, T., Motterlini, R., and Buczko, W. (2012). Antithrombotic properties of water-soluble carbon monoxide-releasing molecules. *Arteriosclerosis, Thrombosis, and Vascular Biology* 32: 2149–2157.
21. Soni, H., Jain, M., and Mehta, A.A. (2011). Investigation into the mechanism(s) of antithrombotic effects of carbon monoxide releasing molecule-3 (CORM-3). *Thromb. Res.* 127: 551–559.
22. Iwata, M., Inoue, T., Asai, Y., Hori, K., Fujiwara, M., Matsuo, S., Tsuchida, W., and Suzuki, S. (2020). The protective role of localized nitric oxide production during inflammation may be mediated by the heme oxygenase-1/carbon monoxide pathway. *Biochem. Biophys. Reports.* 23: 100790.
23. Kaczara, P., Sitek, B., Przyborowski, K., Kurpinska, A., Kus, K., Stojak, M., and Chlopicki, S. (2020). Antiplatelet effect of carbon monoxide is mediated by NAD+ and ATP depletion. *Arteriosclerosis, Thrombosis, and Vascular Biology* 40: 2376–2390.
24. Brüne, B. and Ullrich, V. (1987). Inhibition of platelet aggregation by carbon monoxide is mediated by activation of guanylate cyclase. *Mol. Pharmacol.* 32: 497–504.
25. Clark, J.E., Naughton, P., Shurey, S., Green, C.J., Johnson, T.R., Mann, B.E., Foresti, R., and Motterlini, R. (2003). Cardioprotective actions by a water-soluble carbon monoxide-releasing molecule. *Circ. Res.* 93: e2–8.
26. Motterlini, R., Sawle, P., Hammad, J., Bains, S., Alberto, R., Foresti, R., and Green, C.J. (2005). CORM-A1: a new pharmacologically active carbon monoxide-releasing molecule. *FASEB J.* 19: 284–286.
27. Chlopicki, S., Olszanecki, R., Marcinkiewicz, E., Lomnicka, M., and Motterlini, R. (2006). Carbon monoxide released by CORM-3 inhibits human platelets by a mechanism independent of soluble guanylate cyclase. *Cardiovasc. Res.* 71: 393–401.
28. Chlopicki, S., Łomnicka, M., Fedorowicz, A., Grochal, E., Kramkowski, K., Mogielnicki, A., Buczko, W., and Motterlini, R. (2012). Inhibition of platelet aggregation by carbon monoxide-releasing molecules (CO-RMs): comparison with NO donors. *Naunyn Schmiedebergs Arch. Pharmacol.* 385: 641–650.
29. Prieto, L., Rossier, J., Dersznak, K., Dybas, J., Oetterli, R.M., Kottelat, E., Chlopicki, S., Zelder, F., and Zobi, F. (2017). Modified biovectors for the tuneable activation of anti-platelet carbon monoxide release. *Chem. Commun. (Camb)* 53: 6840–6843.
30. Li, L., Hsu, A., and Moore, P.K. (2009). Actions and interactions of nitric oxide, carbon monoxide and hydrogen sulphide in the cardiovascular system and in inflammation - a tale of three gases! *Pharmacol. Ther.* 123: 386–400.
31. Friebe, A., Müllershausen, F., Smolenski, A., Walter, U., Schultz, G., and Koesling, D. (1998). YC-1 potentiates nitric oxide- and carbon monoxide-induced cyclic GMP effects in human platelets. *Mol. Pharmacol.* 54: 962–967.
32. Kaczara, P., Przyborowski, K., Mohaissen, T., and Chlopicki, S. (2021). Distinct pharmacological properties of gaseous co and co-releasing molecule in human platelets. *Int. J. Mol. Sci.* 22: 3584.
33. Liu, D., Liang, F., Wang, X., Cao, J., Qin, W., and Sun, B. (2013). Suppressive effect of CORM-2 on LPS-induced platelet activation by glycoprotein mediated

HS1 phosphorylation interference. *PLoS One* 8: e83112.
34. Liu, D., Wang, X., Qin, W., Chen, J., Wang, Y., Zhuang, M., and Sun, B. (2016). Suppressive effect of exogenous carbon monoxide on endotoxin-stimulated platelet over-activation via the glycoprotein-mediated PI3K-Akt-GSK3β pathway. *Sci. Rep.* 6: 23653.
35. Jain, K., Tyagi, T., and Hwa, J. (2020). "CO"ping with a sticky situation. *Arteriosclerosis, Thrombosis, and Vascular Biology* 40: 2344–2345.
36. Estevez, B. and Du, X. (2017). New concepts and mechanisms of platelet activation signaling. *Physiology* 32: 162–177.
37. Yun, S.H., Sim, E.H., Goh, R.Y., Park, J.I., and Han, J.Y. (2016). Platelet activation: the mechanisms and potential biomarkers. *Biomed. Res. Int.* 2016: 9060143.
38. Akkerman, J.W. and Holmsen, H. (1981). Interrelationships among platelet responses: studies on the burst in proton liberation, lactate production, and oxygen uptake during platelet aggregation and Ca^{2+} secretion. *Blood* 57: 956–966.
39. Holmsen, H., Kaplan, K.L., and Dangelmaier, C.A. (1982). Differential energy requirements for platelet responses. A simultaneous study of aggregation, three secretory processes, arachidonate liberation, phosphatidylinositol breakdown and phosphatidate production. *Biochem. J.* 208: 9–18.
40. Verhoeven, A.J.M., Gorter, G., Mommersteeg, M.E., and Akkerman, J.W.N. (1985). The energetics of early platelet responses. Energy consumption during shape change and aggregation with special reference to protein phosphorylation and the polyphosphoinositide cycle. *Biochem. J.* 228: 451–462.
41. Mant, M.J. (1980). Platelet adherence to collagen: Metabolic energy requirements. *Thromb. Res.* 17: 729–736.
42. Melchinger, H., Jain, K., Tyagi, T., and Hwa, J. (2019). Role of platelet mitochondria: Life in a nucleus-free zone. *Front. Cardiovasc. Med.* 6: 153.
43. Aibibula, M., Naseem, K.M., and Sturmey, R.G. (2018). Glucose metabolism and metabolic flexibility in blood platelets. *J. Thromb. Haemost.* 16: 2300–2314.
44. Karpatkin, S. (1967). Studies on human platelet glycolysis. Effect of glucose, cyanide, insulin, citrate, and agglutination and contraction on platelet glycolysis. *J. Clin. Invest.* 46: 409–417.
45. Ravi, S., Chacko, B., Sawada, H., Kramer, P.A., Johnson, M.S., Benavides, G.A., O'Donnell, V., Marques, M.B., and Darley-Usmar, V.M. (2015). Metabolic plasticity in resting and thrombin activated platelets. *PLoS One* 10: e0123597.
46. Doery, J.C., Hirsh, J., and Cooper, I. (1970). Energy metabolism in human platelets: interrelationship between glycolysis and oxidative metabolism. *Blood* 36: 159–168.
47. Guppy, M., Abas, L., Neylon, C., Whisson, M.E., Whitham, S., Pethick, D., and Niu, X. (1997). Fuel choices by human platelets in human plasma. *Eur. J. Biochem.* 244: 161–167.
48. Cohen, P. and Wittels, B. (1970). Energy substrate metabolism in fresh and stored human platelets. *J. Clin. Invest.* 49: 119–127.
49. Fink, B.D., Herlein, J.A., O'Malley, Y., and Sivitz, W.I. (2012). Endothelial cell and platelet bioenergetics: Effect of glucose and nutrient composition. *PLoS One* 7: e39430.
50. Corona De La Peña, N., Gutiérrez-Aguilar, M., Hernández-Reséndiz, I., Marín-Hernández, Á., and Rodríguez-Enríquez, S. (2017). Glycoprotein Ib activation by thrombin stimulates the energy metabolism in human platelets. *PLoS One* 12: e0182374.
51. Tang, W.H., Stitham, J., Gleim, S., Di Febbo, C., Porreca, E., Fava, C., Tacconelli, S., Capone, M., Evangelista, V., Levantesi, G., Wen, L., Martin, K., Minuz, P., Rade, J., Patrignani, P., and Hwa, J. (2011). Glucose and collagen regulate human platelet activity through aldose reductase induction of thromboxane. *J. Clin. Invest.* 121: 4462–4476.
52. Sudic, D., Razmara, M., Forslund, M., Ji, Q., Hjemdahl, P., and Li, N. (2006). High glucose levels enhance platelet activation: involvement of multiple mechanisms. *Br. J. Haematol.* 133: 315–322.
53. Przyborowski, K., Kassassir, H., Wojewoda, M., Kmiecik, K., Sitek, B., Siewiera, K., Zakrzewska, A., Rudolf, A.M., Kostogrys, R., Watala, C., Zoladz, J.A., and Chlopicki, S. (2017). Effects of a single bout of strenuous exercise on platelet activation in female ApoE/LDLR$^{-/-}$ mice. *Platelets* 28: 657–667.
54. Ran, J., Guo, X., Li, Q., Mei, G., and Lao, G. (2009). Platelets of type 2 diabetic patients are characterized by high ATP content and low mitochondrial membrane potential. *Platelets* 20: 588–593.
55. Xu, W., Cardenes, N., Corey, C., Erzurum, S.C., and Shiva, S. (2015). Platelets from asthmatic individuals show less reliance on glycolysis. *PLoS One* 10: e0132007.
56. Cardenes, N., Corey, C., Geary, L., Jain, S., Zharikov, S., Barge, S., Novelli, E.M., and Shiva, S. (2014). Platelet bioenergetic screen in sickle cell patients reveals mitochondrial complex V inhibition, which contributes to platelet activation. *Blood* 123: 2864–2872.
57. Schapira, A.H., Gu, M., Taanman, J.W., Tabrizi, S.J., Seaton, T., Cleeter, M., and Cooper, J.M. (1998). Mitochondria in the etiology and pathogenesis of Parkinson's disease. *Ann. Neurol.* 44: S89–S98.

58. Sjövall, F., Morota, S., Hansson, M.J., Friberg, H., Gnaiger, E., and Elmér, E. (2010). Temporal increase of platelet mitochondrial respiration is negatively associated with clinical outcome in patients with sepsis. *Crit. Care.* 14: R214.
59. Davizon-Castillo, P., McMahon, B., Aguila, S., Bark, D., Ashworth, K., Allawzi, A., Campbell, R.A., Montenont, E., Nemkov, T., D'Alessandro, A., Clendenen, N., Shih, L., Sanders, N.A., Higa, K., Cox, A., Padilla-Romo, Z., Hernandez, G., Wartchow, E., Trahan, G.D., Nozik-Grayck, E., Jones, K., Pietras, E.M., DeGregori, J., Rondina, M.T., and Di Paola, J. (2019). TNF-a–driven inflammation and mitochondrial dysfunction define the platelet hyperreactivity of aging. *Blood* 134: 727–740.
60. Lo Iacono, L., Boczkowski, J., Zini, R., Salouage, I., Berdeaux, A., Motterlini, R., and Morin, D. (2011). A carbon monoxide-releasing molecule (CORM-3) uncouples mitochondrial respiration and modulates the production of reactive oxygen species. *Free Radic. Biol. Med.* 50: 1556–1564.
61. Long, R., Salouage, I., Berdeaux, A., Motterlini, R., and Morin, D. (2014). CORM-3, a water soluble CO-releasing molecule, uncouples mitochondrial respiration via interaction with the phosphate carrier. *Biochim. Biophys. Acta.* 1837: 201–209.
62. Queiroga, C.S.F., Almeida, A.S., Alves, P.M., Brenner, C., and Vieira, H.L.A. (2011). Carbon monoxide prevents hepatic mitochondrial membrane permeabilization. *BMC Cell Biol.* 12: 10.
63. Reiter, C.E.N. and Alayash, A.I. (2012). Effects of carbon monoxide (CO) delivery by a CO donor or hemoglobin on vascular hypoxia inducible factor 1α and mitochondrial respiration. *FEBS Open Bio* 2: 113–118.
64. Wegiel, B., Gallo, D., Csizmadia, E., Harris, C., Belcher, J., Vercellotti, G.M., Penacho, N., Seth, P., Sukhatme, V., Ahmed, A., Pandolfi, P.P., Helczynski, L., Bjartell, A., Persson, J.L., and Otterbein, L.E. (2013). Carbon monoxide expedites metabolic exhaustion to inhibit tumor growth. *Cancer Res.* 73: 7009–7021.
65. Wilson, J.L., Bouillaud, F., Almeida, A.S., Vieira, H.L., Ouidja, M.O., Dubois-Randé, J.-L., Foresti, R., and Motterlini, R. (2017). Carbon monoxide reverses the metabolic adaptation of microglia cells to an inflammatory stimulus. *Free Radic. Biol. Med.* 104: 311–323.
66. Kaczara, P., Motterlini, R., Rosen, G.M., Augustynek, B., Bednarczyk, P., Szewczyk, A., Foresti, R., and Chlopicki, S. (2015). Carbon monoxide released by CORM-401 uncouples mitochondrial respiration and inhibits glycolysis in endothelial cells: a role for mitoBKCa channels. *Biochim. Biophys. Acta* 1847: 1297–1309.
67. Kaczara, P., Motterlini, R., Zakrzewska, A., Kus, K., Abramov, A.Y., and Chlopicki, S. (2016). Carbon monoxide shifts energetic metabolism from glycolysis to oxidative phosphorylation in endothelial cells. *FEBS Lett.* 590: 3469–3480.
68. Kaczara, P., Proniewski, B., Lovejoy, C., Kus, K., Motterlini, R., Abramov, A.Y., and Chlopicki, S. (2018). CORM-401 induces calcium signalling, NO increase and activation of pentose phosphate pathway in endothelial cells. *FEBS J.* 285: 1346–1358.
69. Cooper, C.E. and Brown, G.C. (2008). The inhibition of mitochondrial cytochrome oxidase by the gasescarbon monoxide, nitric oxide, hydrogen cyanide and hydrogensulfide: chemical mechanism and physiological significance. *J. Bioenerg. Biomembr.* 40: 533–539.
70. Ishigami, I., Zatsepin, N.A., Hikita, M., Conrad, C.E., Nelson, G., Coe, J.D., Basu, S., Grant, T.D., Seaberg, M.H., Sierra, R.G., Hunter, M.S., Fromme, P., Fromme, R., Yeh, S.R., and Rousseau, D.L. (2017). Crystal structure of CO-bound cytochrome c oxidase determined by serial femtosecond X-ray crystallography at room temperature. *Proc. Natl. Acad. Sci. U. S. A.* 114: 8011–8016.
71. Stojak, M., Kaczara, P., Motterlini, R., and Chlopicki, S. (2018). Modulation of cellular bioenergetics by CO-releasing molecules and NO-donors inhibits the interaction of cancer cells with human lung microvascular endothelial cells. *Pharmacol. Res.* 136: 160–171.
72. Almeida, A.S., Figueiredo-Pereira, C., and Vieira, H.L.A. (2015). Carbon monoxide and mitochondria—modulation of cell metabolism, redox response and cell death. *Front. Physiol.* 6: 33.
73. Rajman, L., Chwalek, K., and Sinclair, D.A. (2018). Therapeutic potential of NAD-boosting molecules: The in vivo evidence. *Cell Metab.* 27: 529–547.
74. Lee, H., Chae, S., Park, J., Bae, J., Go, E.B., Kim, S.J., Kim, H., Hwang, D., Lee, S.W., and Lee, S.Y. (2016). Comprehensive proteome profiling of platelet identified a protein profile predictive of responses to an antiplatelet agent sarpogrelate. *Mol. Cell Proteomics* 15: 3461–3472.
75. Moscardó, A., Vallés, J., Latorre, A., Jover, R., and Santos, M.T. (2015). The histone deacetylase sirtuin 2 is a new player in the regulation of platelet function. *J. Thromb. Haemost.* 13: 1335–1344.
76. Kumari, S., Chaurasia, S.N., Nayak, M.K., Mallick, R.L., and Dash, D. (2015). Sirtuin inhibition induces apoptosis-like changes in platelets and thrombocytopenia. *J. Biol. Chem.* 290: 12290–12299.

77. Mushtaq, M., Nam, T.S., and Kim, U.H. (2011). Critical role for CD38-mediated Ca^{2+} signaling in thrombin-induced procoagulant activity of mouse platelets and hemostasis. *J. Biol. Chem.* 286: 12952–12958.
78. Delabie, W., Maes, W., Devloo, R., Van Den Hauwe, M.R., Vanhoorelbeke, K., Compernolle, V., and Feys, H.B. (2020). The senotherapeutic nicotinamide riboside raises platelet nicotinamide adenine dinucleotide levels but cannot prevent storage lesion. *Transfusion* 60: 165–174.
79. Brüne, B., Schmidt, K.U., and Ullrich, V. (1990). Activation of soluble guanylate cyclase by carbon monoxide and inhibition by superoxide anion. *Eur. J. Biochem.* 192: 683–688.
80. Tomasiak, M., Stelmach, H., Rusak, T., and Wysocka, J. (2004). Nitric oxide and platelet energy metabolism. *Acta Biochim. Pol.* 51: 789–803.
81. Bultas, J. (2013). Antiplatelet therapy-a pharmacologist's perspective. *Cor. Vasa.* 55: e86–e94.
82. Belcher, A., Zulfiker, A.H.M., Li, O.Q., Yue, H., Gupta, A.S., and Li, W. (2021). Targeting thymidine phosphorylase with tipiracil hydrochloride attenuates thrombosis without increasing risk of bleeding in mice. *Arteriosclerosis, Thrombosis, and Vascular Biology* 41: 668–682.
83. Laurence, J., Elhadad, S., Robison, T., Terry, H., Varshney, R., Woolington, S., Ghafoory, S., Choi, M.E., and Ahamed, J. (2017). HIV protease inhibitor-induced cardiac dysfunction and fibrosis is mediated by platelet-derived TGF-β1 and can be suppressed by exogenous carbon monoxide. *PLoS One* 12: 0187185.
84. Smeda, M., Przyborowski, K., Stojak, M., and Chlopicki, S. (2020). The endothelial barrier and cancer metastasis: does the protective facet of platelet function matter? *Biochem. Pharmacol.* 176: 113886.

27

CO in Gastrointestinal Physiology and Protection

Katarzyna Magierowska and Marcin Magierowski

Department of Physiology, Jagiellonian University Medical College, Krakow, Poland

Introduction

The gastrointestinal (GI) tract is constantly exposed to external stimuli and thus one of the most dysfunction-susceptible systems. However, the GI mucosal barrier successfully separates internal milieu from luminal environment [1]. The GI barrier provides the first defense against a broad range of antigens, toxins, and pathogens. The GI wall is formed by a single-cell layer of surface epithelium, underlying lamina propria, muscularis mucosae, and serosa. Gastric columnar cells secrete alkaline mucus, which forms a physical barrier against hydrochloric acid and other acids, while the intestinal epithelium with tight junctions is a selectively permeable barrier responsible for the absorption of water, electrolytes, and dietary nutrients [2]. The mucus of the small intestine is composed of a highly organized network of gel-forming glycoproteins produced by goblet cells and not only is a passive barrier but also seems to be involved in interaction with gut microbiota [3].

Undisturbed gastric blood flow (GBF) within the submucosal layer is crucial for the GI barrier homeostasis. Additionally, the main physiological protective factors involved in the maintenance of gastric mucosal integrity are prostaglandins (PGs), newly discovered food intake controlling peptides (such as ghrelin, leptin, and orexin-1), and endogenous gaseous mediators, such as nitric oxide (NO), carbon monoxide (CO), and hydrogen sulfide (H_2S) [4–8]. On the other hand, different types of substances such as bile acid, lactic acid, bacteria, glutamine, and epidermal growth factor have recently been demonstrated to exert a protective effect on the intestinal barrier [9–12].

There are several external factors affecting GI barrier functions: ethanol, drug ingestion (in particular, nonsteroidal anti-inflammatory drugs [NSAIDs]), pathogens, exposure to chronic stress, and ischemia followed by reperfusion.

Chronic alcohol abuse leads to increased intestinal permeability and endotoxin translocation, among others, with increased levels of acetaldehyde as an important contributing factor [13]. Ethanol-induced gastric damage has been shown to be associated with necrosis following exfoliation of surface epithelium. Local contact with ethanol results in histamine-induced venule constrictions following dilation of arteries. Increased blood flow and capillary pressure accompanying membrane permeability lead to sub-epithelial hemorrhage and edema [14].

GI mucosal damage may also be mediated by bacterial factors [15]. *Helicobacter pylori* (Hp) is a Gram-negative microaerophilic bacterial species, which colonizes gastric epithelial surface and evokes antrum gastritis. Hp-induced dysregulated gastrin release and gastric acid hypersecretion lead to an elevated acid load into the duodenal bulb and subsequent appearance of metaplastic gastric epithelium. Spots of gastric metaplasia in the duodenal bulb are a prerequisite for Hp colonization [16]. Hp produces virulence determinants, which facilitate survival in highly acidic conditions. Flagella, the vacuolating cytotoxin, oncogenic cytotoxin-associated antigen (CagA), and urease are involved in colonization, immune escape, and disease development [17]. Urease catalyzes the hydrolysis of urea to carbon dioxide and ammonia. The latter buffers the acidity in the vicinity of the bacteria and is essential for their survival [15,18].

NSAID-induced GI damage is due to their topical erosive effects, and systemic events through inhibition of cyclooxygenase (COX)-1 and COX-2 with a subsequent depletion of PGs. COX exists in two isoforms: COX-1 is constitutively expressed in various tissues and COX-2 is induced in response to

---

Carbon Monoxide in Drug Discovery: Basics, Pharmacology, and Therapeutic Potential, First Edition. Edited by Binghe Wang and Leo E. Otterbein.
© 2022 John Wiley & Sons, Inc. Published 2022 by John Wiley & Sons, Inc.

inflammation [19]. Initially, the undesired adverse reactions and ulcerogenic effect of NSAIDs in the GI tract were considered to be mostly due to inhibition of COX-1, but not COX-2 [20]. However, more recent studies showed that the ulcerogenic properties of NSAIDs occur by inhibition of not only COX-1, but also COX-2 [21,22].

Stress is also strongly involved in the pathophysiology of GI mucosal injury exerting short- and long-term effects [23]. Chronic stress has been implicated in peptic ulcer and irritable bowel syndrome pathogenesis [24,25]. Stress leads to dysregulation of the brain–gut axis and causes changes to the gut microbiome [26]. Moreover, stress-induced ulceration has been linked to changes in gastric acid secretion leading to hyperacidity, hypermotility, and increased permeability of the gastric mucosa to H^+ ions, which in turn affects the microcirculation [26,27].

Impairment of GI barrier functions may also occur as a consequence of ischemia/reperfusion (I/R) episodes [28]. The I/R injury is mainly characterized by the release of free radicals during the initial phase of reperfusion, which is further aggravated by increased tissue lipid peroxidation due to activated neutrophils and overproduction of proinflammatory cytokines [29–31]. Acute I/R-induced gastric mucosal hemorrhagic lesions can progress to deeper chronic ulcerations under prolonged periods of reperfusion [32]. Intestinal I/R results in mucosal barrier function disruption, facilitating bacterial translocation into the circulation and in turn systemic inflammatory responses [33,34].

Taken together, maintaining the balance between protective and damaging factors is crucial to maintaining GI functionality and homeostasis.

Endogenous CO and the Heme Oxygenase System in the GI Tract

Endogenous production of CO in mammalian tissues was first identified in early 1950 by Torgny Sjöstrand and heme was suggested as the source of CO production [35,36]. Later, the identification of hepatic microsomal heme oxygenase (HO) by Tenhunen and colleagues closed the circle in establishing the molecular link between enzymatic degradation of heme and CO production [37,38]. It is estimated that this heme-dependent pathway provides about 88% of endogenous CO, while the remaining part of CO derives from heme-independent sources such as xenobiotic metabolism, lipid peroxidation, photooxidation, and autooxidation [39,40]. HO degrades pro-oxidant heme to generate equivalent amounts of biliverdin IXα, CO, and ferrous iron (Fe^{2+}) [37]. Two isoforms of HO have been identified: an inducible form, HO-1 (33 kDa), and a constitutive HO-2 (36 kDa), which is less regulated [41]. Chapters 1 and 3 cover in depth the subject of HO and CO production.

Importantly, mRNA and/or protein expression for HO has been reported within the GI tract mainly under pathological conditions and localized within the esophageal, gastric, and colonic mucosa [42–44].

Chemical CO Donors and Cellular Targets of CO in GI Tract

There are a few ways of delivering exogenous CO to tissues, including CO inhalation, administration of HO-1 inducers, or application of prodrugs (CO-releasing molecules [CORMs]) liberating CO and involving hepatic metabolism [45]. A large number of CO delivery approaches have been developed to provide preclinical potential of CO as covered in detail in Chapters 11–17. Briefly, these delivery methods include metal-based CORMs [46,47] and their derivatives for controlled delivery [48], and metal-free organic CO prodrugs (Chapters 14–17) [49,50]. Steiger et al. developed an oral CO release system (OCORS) in form of coated tablets using cellulose acetate, which allows for release of precisely controlled amounts of CO in the GI tract [51].

Recent advances in the field of experimental gastroenterology have demonstrated that the HO/CO axis is an important physiological regulator of GI functions and seems to act as an essential component of the complex mechanism of mucosal defense. To better understand the mechanism of CO's contribution in gastric protection and resistance to injury, we need to know the underlying cellular mechanisms responsible for CO-induced beneficial effects.

The action of CO depends on its ability to bind hemoproteins and to alter their biochemical functions. The most recognized and dangerous effects of CO are linked to its competitive inhibition of oxygen binding to hemoglobin and inhibition of mitochondrial cytochrome c oxidase, leading to decreased oxygen delivery to tissues and impediment of the electron transport chain [52]. However, CO binding to the heme domain of soluble guanylyl cyclase (sGC) results in about a four fold increase of this enzyme activity and this pathway has been proposed to be involved in smooth muscle relaxation and vasodilatation also within the GI tract [53]. The interactions of CO with other heme-containing proteins such as cytochrome P450 and iNOS also lead to protein inhibition [54,55].

CO has been reported to contribute significantly to vasorelaxation through the regulation of vascular smooth muscle tone [56]. Vascular smooth muscle cells express numerous types of potassium channels, which participate in muscle contraction [57]. CO-induced vasodilation has been attributed to its stimulation of the large-conductance, Ca^{2+}-activated K^+ channels [58,59]. Interestingly, the hypotensive effect of CO in the GI tract seems to be independent of or only partly dependent on the co-activity of vasodilatory afferent sensory neurons possibly via vanilloid subtype of transient receptor potential channels (TRPV)-1 and calcitonin gene-related peptide (CGRP) [27,42,60].

Over the last few years, the involvement of p38 mitogen-activated protein kinase (p38 MAPK) signaling pathway in anti-inflammatory, antiproliferative, and antiapoptotic effects of CO has been reported [61–63]. This effect might be dose dependent since higher doses of CORM-2 even enhanced ethanol-induced gastric damage formation [64]. Additionally, genes involved in antiapoptotic activities within the cell are activated through CO in a nuclear factor kappa-light-chain-enhancer of activated B cells (NF-κB)-dependent manner [65,66].

CO facilitates dissociation of NF-E2-related factor 2 (Nrf2) from Keap1 and therefore stimulates the translocation and nuclear accumulation of Nrf2. Nrf2 binding to antioxidant response element sequence in the gene promoter is the mechanism responsible for transcription of cytoprotective genes [67]. Contribution of Nrf2 into CO-mediated GI protection has also been reported previously [68–70]. Undoubtedly, CO regulates many physiological processes at a cellular level. Chapter 2 covers in depth the various molecular targets of CO.

CO Donors and HO-1 in GI Protection

CO Donors Against Drug-Induced GI Toxicity

In terms of physicochemical properties, NSAIDs such as aspirin (ASA) or naproxen are weakly acidic and remain nonionized in the acidic environment of the gastric lumen. Lipophilic NSAIDs can diffuse through the phospholipid membrane to enter the cells and shift to a mostly ionized form, leading to intracellular enrichment. This effect known as "ion trapping" is responsible for their observed topical gastrotoxicity. Direct effect of epithelial injury by NSAIDs has been attributed to uncoupling mitochondrial oxidative phosphorylation and reduction in mucus–bicarbonate secretion following a decrease in hydrophobicity of the gastric mucus [71]. Under these circumstances, hydrochloric acid and pepsin attack the gastric surface epithelium, which further amplifies local gastric toxicity.

The aforementioned "trapping" theory does not apply to small intestine, where the surface pH of the intestinal mucosa remains neutral [72]. Hemorrhagic lesions in jejunum and ileum of rats occur within 24 h of administration of conventional NSAIDs [73]. Selective COX-1 or COX-2 inhibitors, SC-560 and rofecoxib, respectively, did not cause any damage to the small intestine by themselves, while coadministration of those two together induced intestinal hemorrhagic lesions [74]. NSAID-induced enteropathy, in addition to PG depletion following COX inhibition, has been linked to intestinal hypermotility [75]. An increase in intestinal motility occurs through changes in amplitude and frequency of contractions and may be precipitated by NSAID-induced NOS inhibition with subsequent endogenous NO deficiency [76]. The effect of increased intestinal motility is exacerbated when NSAIDs are coadministered with antisecretory drugs [77,78].

NSAIDs increase intestinal permeability as the result of inhibition of oxidative phosphorylation; however, this process is not fully understood [79,80]. Increased permeability results in constant exposure of enterocytes to bile acids, pancreatic enzymes, and enterobacteria, which all contribute to inflammation [72]. Human gut dysbiosis has also been reported to be a contributing factor in the pathogenesis of NSAID-induced enteropathy [81]. Interestingly, NSAIDs have no effect on gastric mucosa permeability in humans [82].

Gastroprotection of CO donors, CORM-2 and a CO prodrug (BW-CO-111), against acute ASA-induced gastric damage has recently been demonstrated [43,60,83]. These CO donors decrease hemorrhagic gastric lesion formation, concomitantly increasing GBF. CO-induced vasorelaxation and blood flow occur in an sGC/cGMP-dependent manner. The protective effect of CO toward gastric mucosa has been assigned to its anti-inflammatory activity. CO decreased the level of mRNA expression for the proinflammatory genes COX-2, iNOS, and IL-1β upregulated by administration of high doses of ASA. According to this study, CO-induced gastroprotection involves modulation of the Nrf2 pathway. ASA treatment also results in GBF impairment, local hypoxia, and increased expression in oxygen-sensitive factors. Vasoactive properties of CO are responsible for the observed improvement of microcirculation within gastric mucosa manifested by a decrease in the expression of hypoxia-inducible factor 1α (HIF-1α) [43]. Moreover, CO maintains the antioxidant status of gastric mucosa, reflected in a decrease in lipid peroxidation products, mainly through enhancement of glutathione peroxidase-1 expression [60]. The protective activity of CO against ASA-induced gastropathy seems not to be mediated by neuropeptides such as CGRP released from sensory afferent nerves [60]. At this point, it is important to note the agreement of results from CORM-2 and CO-111. This aspect is essential in light of the newly reported CO-independent activities of CORM-2 and other ruthenium-based CORMs [84–87]. The consistent results from two different types of donors, CORM-2 and CO-111, reinforce the validity of CO being the active agent in these experiments.

On the other hand, the cytoprotective effects of HO-1 may be linked to its mitochondrial translocation observed in experimental models of indomethacin-induced gastric injury [88]. A putative mode of action of HO-1 has been attributed to limiting free heme accumulation as a consequence of NSAID administration [88].

The severity of indomethacin-induced small intestinal damage has been observed to be reduced by proton pump inhibitor (PPI), lansoprazole [89]. It is interesting to note that the protective effect of lansoprazole was abolished by prior administration of the HO-1 inhibitor, tin–protoporphyrin IX. These data suggest that the beneficial impact of PPI on intestinal mucosa may involve induction of the HO-1/CO system [89]. Interestingly, it has been shown that upregulation of HO-1 protein expression in the intestinal mucosa was not observed for the other widely used PPI drug, omeprazole [80]. Other research findings have suggested that the protective effect of PPI against NSAID-induced gastric damage is the result of pantoprazole-induced HO-1 expression via Nrf2 followed by attenuation of inflammatory mediators [90].

Bisphosphonates (BPs) are the most widely used and most effective drugs in osteoporosis therapy. This is also a group of drugs with adverse effects in the upper GI tract [91,92]. It appears that BPs and NSAIDs might synergistically increase the risk and severity of ulceration [93,94]. BPs as topical irritants directly damage the epithelium of the esophagus and stomach, with no impact on the microvasculature [95,96]. CORM-2 has been shown to exert protection against alendronate-induced gastric lesions in rats pre-exposed to mild stress [92]. The underlying mechanism is based on improvement of GBF and hypoxia limitation, as shown by a decrease in HIF-1α expression. Concomitant downregulation of the NF-κB pathway explains the antihypoxic effect of CO in this experimental model of injury [92,97]. Moreover, CO involves the activation of the sGC/cGMP signaling pathway in alendronate-induced gastric damage, similarly to its gastroprotection against NSAID-induced injuries [98]. CO released from its donors exerts anti-inflammatory and anti-oxidative activity, as shown by decreased TNF and IL-1β levels and limited lipid peroxidation process at damaged mucosal sites [92,98]. Overview of the role of CO prodrugs in the maintenance of GI mucosal defense is presented in Figure 27.1.

CO Donors Against Acute Esophageal and Gastric Mucosal Injuries

Published data on the gastroprotective properties of CO seem to suggest its effect within esophageal mucosa as well. However, few studies were conducted in this field until very recently. When the lower esophageal sphincter is weakened by orally administered drugs such as NSAIDs, it may lead to the influx of bile

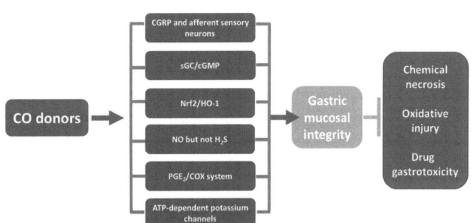

Figure 27.1 Overview of the role of CO prodrugs in maintenance of GI mucosal defense. Abbreviations: CGRP, calcitonin gene-related peptide; sGC/cGMP, soluble guanylyl cyclase/cyclic guanosine monophosphate; Nrf-2/HO-1, nuclear factor erythroid 2-related factor 2; NO, nitric oxide; H₂S, hydrogen sulfide; PGE₂/COX, prostaglandin E₂/cyclooxygenase.

and acidic stomach content. Short-term exposure of the esophagus to low pH results in acute cytokine-mediated reflux esophagitis [99,100]. The first evidence for CO-mediated esophageal protection demonstrated that CO-mediated vasoactivity partially involves afferent sensory neurons and TRPV-1 activity and is independent of the PGE_2/COX system [42]. The NF-κB signaling pathway and NF-κB targeted genes such as IL-1β, IL-8, IL-6, and TNF are involved in proinflammatory changes within the esophageal mucosa [101]. According to recent studies, pretreatment with the CO donor, CORM-2, regulates inflammatory responses through a decrease in esophageal mRNA expression for NF-κB, IL-1β, and TNF [42].

The association between HO-1 polymorphisms and the risk of human cancer has been demonstrated. It has been proven that shorter (GT) repeats are associated with higher transcriptional activity of HO-1 gene expression in humans and these subjects show lower risk of developing esophageal squamous cell carcinoma [102].

Gastroprotective properties of CO have been shown with some efforts toward explaining its pleiotropic effects [7]. CO derived from donors is effective against Hp infection [103,104]. CO inhibits bacterial growth and cellular oxygen consumption due to binding to a metal-containing enzyme terminal oxidase in the Hp respiratory chain [103]. Bacterial urease, another enzyme with metal in its structure and essential for Hp survival in a low-pH environment, is also affected by CO. In contrast to antibiotics, CORM-2 does not induce the formation of coccoid, which is thought to be responsible for Hp antibiotic resistance development [105]. Furthermore, it has also been observed that HO-1 upregulation is associated with inhibition of phosphorylation of the Hp virulence factor CagA [106].

The GI system is remarkably susceptible to stress, which activates the hypothalamic–pituitary–adrenal axis and leads to vasoconstriction within gastric mucosa due to increased catecholamine secretion. Stress-induced gastritis has been characterized not only by microcirculation disturbance but also by increased gastric secretion followed by elevated mucosal permeability to H^+ ions, inflammation, mucosal edema, and free radical formation [26,107]. A widely accepted clinically relevant animal model for investigating stress-induced ulcerogenic changes in gastric mucosa is the model of water immersion and restraint stress (WRS) [23]. Very recent studies revealed that CORM-2-derived CO induced protection against WRS-induced gastric damage [108]. Beneficial effects of CORM-2 within gastric mucosa compromised with WRS have been attributed to increased HO-1 expression, which in turn results in elevated CO content in the gastric mucosa and in the form of carboxyhemoglobin (COHb) in the blood. The hyperemic effect of this gaseous molecule is due to involvement of the endogenous arachidonate metabolites, PGs and sGC/cGMP, but is independent of NO. In agreement with previously described reports, pretreatment with a CO donor reduces inflammation as evidenced by decreased mRNA expression of the proinflammatory isoforms COX-2 and iNOS [108].

It has been generally acknowledged that tissue exposure to I/R episodes leads to xanthine oxidase-derived generation of radicals with subsequent lipid peroxidation. Free radicals play important roles in the pathogenesis of I/R injury through neutrophil infiltration to the site of inflammation. Activated neutrophils cause increases in microvascular permeability and overall endothelial dysfunction, resulting in mucosal hemorrhagic lesions [109]. Interestingly, I/R- induced lesions have been demonstrated for stomach, small intestine, and colon. Among these three, the small intestine exhibits the highest degree of vulnerability to reperfusion injury [110].

Recently published data demonstrate that CORM-2 protects gastric mucosa against I/R-induced gastropathy [31]. In this experimental model, K^+-ATP channels and sGC/cGMP and NO/NOS systems are proposed as the molecular targets for CO-induced gastroprotection. Upregulation of HO-1 mRNA has already been observed as a characteristic of oxidative injury [7]. In gastric mucosa pretreated with CORM-2 and subsequently exposed to I/R, HO-1 expression is even higher compared to I/R alone. In this condition, self-defense gastric mucosal response is undoubtedly enhanced and presumably occurs as the consequence of the dual action of HO-1 through increased removal of free heme and/or increased CO generation. Moreover, CORM-2 prevents I/R-induced decrease in mucosal PGE_2 and presents strong anti-inflammatory and antioxidative properties [31].

In parallel to the above-mentioned observations, CO plays an important role in gastric mucosa exposed to chemical stimuli such as ethanol, which acts as irritating and necrotizing agent [64,111]. It has been shown that models of ethanol-induced gastric mucosal injury are associated with HO-1 mRNA upregulation and pretreatment with CORM-2 enhanced this effect [64]. HO-1 induction by hemin reduced gastric damage mainly through oxidative stress suppression and improvement of antioxidant status of gastric mucosa [111]. Moreover, GBF increase mediated by the activation of sGC/cGMP, PG/COX, and NO/NOS systems has also been shown in CO-induced gastroprotection against ethanol injury [64]. It is also important to note that similar results were observed with an organic CO prodrug, CO-111 [83]. Such results are important mutual validation of the two different types of CO donors.

CO Donors Against Lower GI Tract Injuries

Recent advances toward explaining the role of CO against intestinal injuries have been made. The CO/HO-1 system has been reported to be involved in inflammatory bowel disease (IBD), including ulcerative colitis and Crohn's disease [112–114]. 2,4,6-Trinitrobenzine sulfonic acid (TNBS)-induced colitis is a frequently used animal model sharing pathological features with human Crohn's disease and thereby remains to be useful tool to study a multifactorial pathogenesis of IBD [115]. Formation of reactive oxygen species has been recognized as a crucial mechanism in colonic inflammation [116,117]. It has been revealed that CO inhalation leads to inhibition of TNBS-induced colonic damage in mice as reflected in the decrease in lipid peroxidation. Further, an organic CO prodrug was also shown to be efficacious in treating TNBS-induced colitis [118]. Dextran sodium sulfate (DSS)-induced inflammatory colitis is another relevant model, which clinically and histologically reflects features observed in human [119]. Micelles encapsulating CORM-2 (SMA/CORM-2) have been shown to have antioxidative and anti-inflammatory effects in the DDS model of intestinal injury [120]. Thus, the above examples show the cross-validation of CO's beneficial effects in animal models of colitis by using inhaled CO, metal-based CORM (CORM-2), and an organic CO prodrug. There are other examples that help to provide additional validation [121–124].

The enteric nervous system (ENS) as local network of different cell types is located within the wall of the GI tract forming myenteric and submucosal plexuses [125]. CO is thought to act as neurotransmitter in the ENS affecting intestinal smooth muscle relaxation through cGMP-dependent mechanisms. Additionally, CO is involved in inhibitory nonadrenergic noncholinergic transmission since HO-2-deficient jejunal circular smooth muscle cells expressed reduced inhibitory transmission, while CO inhalations restore this effect [126,127].

Moreover, the involvement of CO in the murine model of postoperative ileus (POI) has also been investigated [128,129]. CO gas inhalation attenuates inflammatory response in the course of POI and this effect was enhanced by HO-1 induction following oxidative stress reduction and endogenous CO production at the sites of inflammation [129]. However, local CO treatment with OCORS tablets appears to be insufficient [128]. Indeed, more research is required to provide a more definite conclusion. On the other hand, small intestinal transplantation is strongly related to inflammatory response and motoric dysfunction of the graft muscularis connected mainly with suppression of GI motility and jejunal muscle contractility [130]. CO has been shown to improve circular muscle function and motility through inhibition of NO and prostanoid production by iNOS and COX-2, respectively [131–133]. Additionally, CO attenuates immune reaction via attenuation of IL-6 and IL-1β pathways [132,133].

HO-1 expression is increased in inflamed colonic mucosa compared to that in the healthy controls [44]. Regulation of chronic intestinal inflammation may also occur through pharmacological inhibition of HO-1, which leads to aggravation of TNBS- or DSS-induced acute colitis in rodents [114,134]. CO administration and increase in endogenous CO production via HO-1 activity reverse Th1-mediated chronic intestinal mucosa inflammatory process in IL-10-deficient mice through interferon-γ signaling pathway inhibition [113]. Upregulation of iNOS followed by NO release has been observed in IBD and is closely associated with oxidative stress and intestinal inflammation [135,136]. HO-1 induction leads to inhibition of iNOS expression in colonic tissues and remains one of the antioxidant responses in chronic and recurrent disorders of the intestinal tract [134]. Increased HO-1 expression in patients with colorectal cancer has been demonstrated and this effect was more evident in well-differentiated cancers. Moreover, this observation was consistent with elevated COHb level in blood [137].

Recent studies have highlighted the relationship between gut dysbiosis and development of colitis [138]. The CO/HO-1 pathway has recently been proposed as one of the crucial factors for understanding the complexity of microbial alterations within the gut [139]. Intestinal infections are known to be an important factor in the development of IBD [140]. It appears that intestinal HO-1 is induced by the enteric microbiota [141]. Moreover, CO exerts antimicrobial activity and promotes bacterial clearance and therefore can modulate intestinal inflammation [141,142].

CO Donors and Gastric Ulcer Healing (Therapeutic Potential of CO in GI Tract)

An experimental chronic acetic acid ulcer model has been established by Takagi et al. and Okabe et al. This model relies on topical serosal application or submucosal injection of acetic acid [143,144]. The procedure leads to the formation of deep ulcers, which share histological and immunological features with those observed in human [145]. The margins and the base of ulcer form during the initial development phase, which last up to 3 days after initial injury. The margin arises from epithelial cells due to angiogenesis, while the base represents granulation tissue, which is characterized mainly by the presence of fibroblasts, macrophages, and proliferating endothelial cells [146]. The ulcer healing phase starts after 3–10 days of ulcer induction and relies on migration of epithelial cells to the ulcer margin, formation of new vessels, and ulcer base contraction. Between 20 and 40 days after initial injury is the reconstruction phase, in which remodeling of glands, muscularis mucosae, and muscularis propria takes place. In the maturation phase, differentiation of specialized types of cells occurs [147].

It turns out that the CO/HO-1 system is extensively involved in the course of gastric ulcer healing, promoting the resolution of inflammation. HO-1 protein expression increases 6 h after ulcer induction, reaching the highest expression between days 3 and 5 and remains elevated until day 15 [148,149]. Several target genes and signaling pathways involved in the multifactorial process of ulcer healing have been elucidated so far. Oral administration of CO-saturated saline solution promotes re-epithelialization in the ulcerative area [150]. Underlying CO-mediated processes include activation of protein C kinase (PKC), but not the MAPK pathway. PKC is a family of serine/threonine kinases that mediates an intracellular signaling cascade, leading to cell proliferation, differentiation, and apoptosis [151]. Moreover, it has been observed that CORM-2 promotes healing of gastric ulcers in animal models [152]. Regulation of the microcirculation through an increase in the activity of ATP-dependent potassium channels, sGC, and NO biosynthesis has been attributed to the beneficial ulcer-healing properties of CO [152]. The above-mentioned observations have been accompanied by increased expression of epidermal growth factor [153]. The wide spectrum of CO-mediated pleiotropic effects within gastric ulcer healing processes may also involve upregulation of hepatocyte growth factor receptor, insulin-like growth receptor, or vascular endothelial growth factor receptor 2, which is also consistent with the role of CO in the maintenance of gastric mucosal integrity [153]. Overview of the role of CO prodrugs in gastric ulcer healing is presented in Figure 27.2.

Crosstalk of CO with H$_2$S and NO in GI Tract Pathophysiology and Pharmacology

It has been recognized that CO could interact with endogenous H$_2$S and NO biosynthesis and signaling pathways and vice versa, given the heme moieties in the enzymes that generate these gases. These subjects are covered in depth in Chapters 8 and 9. Possible crosstalks have been reported in GI pathologies and related treatment [154,155]. CO released

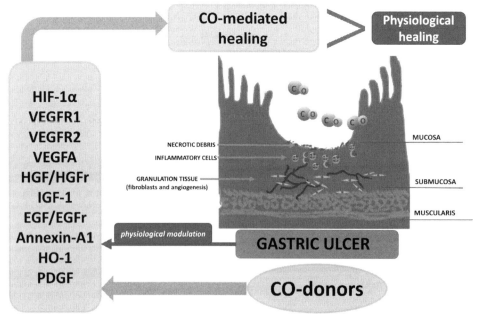

Figure 27.2 An overview of the role of CO donors in the gastric ulcer healing process. Abbreviations: HIF-1α, hypoxia-inducible factor 1-α; VEGFR1, vascular endothelial growth factor receptor 1; VEGFR2, vascular endothelial growth factor receptor 1; VEGFA, vascular endothelial growth factor A; HGF/HGFr, hepatocyte growth factor/hepatocyte growth factor receptor; IGF-1, insulin-like growth factor 1; EGF/EGFr, epidermal growth factor receptor/epidermal growth factor receptor; HO-1, heme oxygenase 1; PDGF, platelet-derived growth factor.

from CORM-2 was shown to prevent gastric mucosa from damages induced by ASA independently of endogenous H_2S biosynthesis. However, H_2S requires CO-producing HO-1 activity to maintain its gastroprotective effects [43]. Additionally, overexpression of HO-1 has been reported to be involved in the protective GI effects of H_2S-releasing derivative of naproxen, ATB-346 [156]. Interestingly, H_2S but not CO-mediated gastric mucosal protection was shown to be dependent on endogenous NO biosynthesis [60]. Moreover, pretreatment with CORM-2 has been reported to decrease stress-induced gastric damage and this effect was accompanied by the decreased NO content in gastric mucosa [108]. On the other hand, CO has been shown to exert therapeutic activity in the course of peptic ulcer in experimental models [152]. CORM-2 applied i.g. for 9 days accelerated gastric ulcer healing independently of endogenous H_2S biosynthesis [157]. However, pharmacological inhibition of HO diminished the ulcer-healing effects of the H_2S donor, NaHS [157]. Such results confirm that the gastroprotective and therapeutic activity of CO is independent of mucosal H_2S and NO biosynthesis, while CO contributes to endogenous H_2S-mediated GI barrier maintenance.

Conclusions and Future Directions in the Development of CO-Based GI Pharmacology

CO produced endogenously by HO-1 or released from pharmacological donors offers protective effects of the physiology of the GI barrier. CO has been shown to modulate molecular signaling pathways and impart antioxidative and anti-inflammatory effects in the GI tract. CO protects esophageal and GI mucosa against damage induced by various noxious stimuli, such as stress, oxidative damage, or drugs. CO also has therapeutic potential since CO donors in various forms have been reported to accelerate gastric ulcer healing and resolution of chemically induced colitis. Further, possible involvements of gut microbiota in various chronic pathologies of the GI tract may be CO mediated and should be further extensively investigated. Moreover, the possible contribution of HO in the progression of chronic GI diseases, such esophageal cancer and Barrett's esophagus, still remains unknown. Thus, further development of novel CO donors could be an important aspect of providing appropriate pharmacological tools to shed light on missing links and unexplained molecular mechanisms of CO activity in the course of treatments for GI disorders.

Acknowledgments

Research activity in the M.M. laboratory on the roles of CO in the GI tract is currently supported by a grant from National Science Centre (Poland), no. UMO-2019/33/B/NZ4/00616.

References

1. Bhatia, V. and Tandon, R.K. (2005). Stress and the gastrointestinal tract. *J. Gastroenterol. Hepatol.* https://doi.org/10.1111/j.1440-1746.2004.03508.x.
2. Groschwitz, K.R. and Hogan, S.P. (2009). Intestinal barrier function: Molecular regulation and disease pathogenesis. *J. Allergy Clin. Immunol.* https://doi.org/10.1016/j.jaci.2009.05.038.
3. Schroeder, B.O. (2019). Fight them or feed them: How the intestinal mucus layer manages the Gut microbiota. *Gastroenterol. Rep.* https://doi.org/10.1093/gastro/goy052.
4. Laine, L., Takeuchi, K., and Tarnawski, A. (2008). Gastric mucosal defense and cytoprotection: Bench to bedside. *Gastroenterology*. https://doi.org/10.1053/j.gastro.2008.05.030.
5. Bülbül, M., Tan, R., Gemici, B., Öngüt, G., and Izgüt-Uysal, V.N. (2008). Effect of orexin-a on ischemia-reperfusion-induced gastric damage in rats. *J. Gastroenterol.* 43 (3): 202–207. https://doi.org/10.1007/s00535-007-2148-3.
6. Magierowski, M., Magierowska, K., Kwiecien, S., and Brzozowski, T. (2015). Gaseous mediators nitric oxide and hydrogen sulfide in the mechanism of gastrointestinal integrity, protection and ulcer healing. *Molecules*. https://doi.org/10.3390/molecules20059099.
7. Magierowska, K., Brzozowski, T., and Magierowski, M. (2018). Emerging role of carbon monoxide in regulation of cellular pathways and in the maintenance of gastric mucosal integrity. *Pharmacol. Res.* https://doi.org/10.1016/j.phrs.2018.01.008.
8. Motawi, T.K., Abd Elgawad, H.M., and Shahin, N.N. (2008). Gastroprotective effect of leptin in indomethacin-induced gastric injury. *J. Biomed. Sci.* 15 (3): 405–412. https://doi.org/10.1007/s11373-007-9227-6.
9. De Diego-cabero, N., Mereu, A., Menoyo, D., Holst, J.J., and Ipharraguerre, I.R. (2015). Bile acid mediated effects on gut integrity and performance of early-weaned piglets. *BMC Vet. Res.* 11 (1). https://doi.org/10.1186/s12917-015-0425-6.
10. Wang, H., Zhang, C., Wu, G., Sun, Y., Wang, B., He, B., Dai, Z., and Wu, Z. (2015). Glutamine enhances tight junction protein expression and modulates

corticotropin-releasing factor signaling in the jejunum of piglets. *J. Nutr.* 145 (1): 25–31. https://doi.org/10.3945/jn.114.202515.

11. Yang, F., Hou, C., Zeng, X., and Qiao, S. (2015). The use of lactic acid bacteria as a probiotic in swine diets. *Pathogens.* https://doi.org/10.3390/pathogens4010034.

12. Tang, X., Liu, H., Yang, S., Li, Z., Zhong, J., and Fang, R. (2016). Epidermal growth factor and intestinal barrier function. *Mediators Inflamm.* https://doi.org/10.1155/2016/1927348.

13. Ferrier, L., Bérard, F., Debrauwer, L., Chabo, C., Langella, P., Buéno, L., and Fioramonti, J. (2006). Impairment of the intestinal barrier by ethanol involves enteric microflora and mast cell activation in rodents. *Am. J. Pathol.* 168 (4): 1148–1154. https://doi.org/10.2353/ajpath.2006.050617.

14. Oates, P.J. and Hakkinen, J.P. (1988). Studies on the mechanism of ethanol-induced gastric damage in rats. *Gastroenterology* 94 (1): 10–21. https://doi.org/10.1016/0016-5085(88)90604-X.

15. Ansari, S. and Yamaoka, Y. (2019). Helicobacter pylori virulence factors exploiting gastric colonization and its pathogenicity. *Toxins.* https://doi.org/10.3390/toxins11110677.

16. Olbe, L., Fandriks, L., Hamlet, A., Svennerholm, A.M., and Thoreson, A.C. (2000). Mechanisms involved in helicobacter pylori induced duodenal ulcer disease: An overview. *World J. Gastroenterol.* https://doi.org/10.3748/wjg.v6.i5.619.

17. Sukri, A., Hanafiah, A., Mohamad Zin, N., and Kosai, N.R. (2020). Epidemiology and role of helicobacter pylori virulence factors in gastric cancer carcinogenesis. *APMIS.* https://doi.org/10.1111/apm.13034.

18. Waldum, H.L., Kleveland, P.M., and Sørdal, Ø.F. (2016). Helicobacter pylori and gastric acid: An intimate and reciprocal relationship. *Therap. Adv. Gastroenterol.* https://doi.org/10.1177/1756283X16663395.

19. Kargman, S., Charleson, S., Cartwright, M., Frank, J., Riendeau, D., Mancini, J., Evans, J., and O'Neill, G. (1996). Characterization of prostaglandin G/H synthase 1 and 2 in rat, dog, monkey, and human gastrointestinal tracts. *Gastroenterology* 111 (2): 445–454. https://doi.org/10.1053/gast.1996.v111.pm8690211.

20. Chan, C.C., Boyce, S., Brideau, C., Ford-Hutchinson, A.W., Gordon, R., Guay, D., Hill, R.G., Li, C.S., Mancini, J., Penneton, M., Prasit, P., Rasori, R., Riendeau, D., Roy, P., Tagari, P., Vickers, P., Wong, E., and Rodger, I.W. (1995). Pharmacology of a selective cyclooxygenase-2 inhibitor, L-745,337: A novel nonsteroidal anti-inflammatory agent with an ulcerogenic sparing effect in rat and nonhuman primate stomach. *J. Pharmacol. Exp. Ther.* 274 (3): 1531–1537.

21. Tanaka, A., Araki, H., Komoike, Y., Hase, S., and Takeuchi, K. (2001). Inhibition of Both COX-1 and COX-2 is required for development of gastric damage in response to nonsteroidal antiinflammatory drugs. *J. Physiol. Paris* 95: 21–27. https://doi.org/10.1016/S0928-4257(01)00005-5.

22. Takeuchi, K., Tanaka, A., Kato, S., Amagase, K., and Satoh, H. (2010). Roles of COX Inhibition in Pathogenesis of NSAID-induced small intestinal damage. *Clin. Chim. Acta.* https://doi.org/10.1016/j.cca.2009.12.026.

23. Inoue, Y., Fujino, Y., Onodera, M., Kikuchi, S., Sato, M., Sato, H., Noda, H., Kkojika, M., Suzuki, Y., and Endo, S. (2015). A case of multiple hemorrhagic gastric ulcers developed via a mechanism similar to water-immersion restraint stress. *Open J. Clin. Diagnostics* 05 (04): 136–140. https://doi.org/10.4236/ojcd.2015.54022.

24. Moody, F.G. and Cheung, L.Y. (1976). Stress ulcers: Their pathogenesis, diagnosis and treatment. *Surg. Clin. North Am.* 56 (6): 1469–1478. https://doi.org/10.1016/S0039-6109(16)41099-6.

25. Öhman, L. and Simrén, M. (2007). New insights into the pathogenesis and pathophysiology of irritable bowel syndrome. *Dig. Liver Dis.* https://doi.org/10.1016/j.dld.2006.10.014.

26. Konturek, P.C., Brzozowski, T., and Konturek, S.J. (2011). Stress and the Gut: Pathophysiology, clinical consequences, diagnostic approach and treatment options. *J. Physiol. Pharmacol.* 62 (6): 591–599.

27. Kwiecien, S., Jasnos, K., Magierowski, M., Sliwowski, Z., Pajdo, R., Brzozowski, B., Mach, T., Wojcik, D., and Brzozowski, T. (2014). Lipid peroxidation, reactive oxygen species and antioxidative factors in the pathogenesis of gastric mucosal lesions and mechanism of protection against oxidative stress – induced gastric injury. *J. Physiol. Pharmacol.* 65 (6): 613–622.

28. Kim, Y.J., Kim, E.H., and Hahm, K.B. (2012). Oxidative stress in inflammation-based gastrointestinal tract diseases: Challenges and opportunities. *J. Gastroenterol. Hepatol.* 27 (6): 1004–1010. https://doi.org/10.1111/j.1440-1746.2012.07108.x.

29. Yoshikawa, T., Naito, Y., Ueda, S., Ichikawa, H., Takahashi, S., Yasuda, M., and Kondo, M. (1992). Ischemia-reperfusion injury and free radical involvement in gastric mucosal disorders. *Adv. Exp. Med. Biol.* 316: 231–238. https://doi.org/10.1007/978-1-4615-3404-4_27.

30. Nadatani, Y., Watanabe, T., Shimada, S., Otani, K., Tanigawa, T., and Fujiwara, Y. (2018). Microbiome and intestinal ischemia/reperfusion injury. *J. Clin. Biochem. Nutr.* https://doi.org/10.3164/jcbn.17-137.

31. Magierowska, K., Korbut, E., Hubalewska-Mazgaj, M., Surmiak, M., Chmura, A., Bakalarz, D., Buszewicz, G., Wójcik, D., Śliwowski, Z., Ginter, G., Gromowski, T., Kwiecień, S., Brzozowski, T., and Magierowski, M. (2019). Oxidative gastric mucosal damage induced by ischemia/reperfusion and the mechanisms of its prevention by carbon monoxide-releasing tricarbonyldichlororuthenium (II) dimer. *Free Radic. Biol. Med.* 145: 198–208. https://doi.org/10.1016/j.freeradbiomed.2019.09.032.

32. Brzozowski, T., Konturek, P.C., Konturek, S.J., Drozdowicz, D., Kwiecień, S., Pajdo, R., Bielanski, W., and Hahn, E.G. (2000). Role of gastric acid secretion in progression of acute gastric erosions induced by ischemia-reperfusion into gastric ulcers. *Eur. J. Pharmacol.* 398 (1): 147–158. https://doi.org/10.1016/S0014-2999(00)00287-9.

33. Collard, C.D. and Gelman, S. (2001). Pathophysiology, clinical manifestations, and prevention of ischemia-reperfusion injury. *Anesthesiology*. https://doi.org/10.1097/00000542-200106000-00030.

34. Grootjans, J., Lenaerts, K., Derikx, J.P.M., Matthijsen, R.A., De Bruïne, A.P., Van Bijnen, A.A., Van Dam, R.M., Dejong, C.H.C., and Buurman, W.A. (2010). Human intestinal ischemia-reperfusion-induced inflammation characterized: Experiences from a new translational model. *Am. J. Pathol.* 176 (5): 2283–2291. https://doi.org/10.2353/ajpath.2010.091069.

35. Sjöstrand, T. (1951). The in vitro formation and disposal of carbon monoxide in blood. *Nature* 168 (4278): 729–730. https://doi.org/10.1038/168729a0.

36. Sjöstrand, T. (1951). A preliminary report on the in vitro formation of carbon monoxide in blood. *Acta Physiol. Scand.* 22 (2–3): 142–143. https://doi.org/10.1111/j.1748-1716.1951.tb00763.x.

37. Tenhunen, R., Marver, H.S., and Schmid, R. (1969). Microsomal heme oxygenase. characterization of the enzyme. *J. Biol. Chem.* 244 (23): 6388–6394. https://doi.org/10.1016/S0021-9258(18)63477-5.

38. Tenhunen, R., Marver, H.S., and Schmid, R. (1968). The enzymatic conversion of heme to bilirubin by microsomal heme oxygenase. *Proc. Natl. Acad. Sci. U. S. A.* 61 (2): 748–755. https://doi.org/10.1073/pnas.61.2.748.

39. Ryter, S.W., Alam, J., and Choi, A.M.K. (2006). Heme oxygenase-1/carbon monoxide: From basic science to therapeutic applications. *Physiol. Rev.* https://doi.org/10.1152/physrev.00011.2005.

40. Ryter, S.W. and Choi, A.M.K. (2009). Heme oxygenase-1/carbon monoxide: From metabolism to molecular therapy. *Am. J. Respir. Cell Mol. Biol.* https://doi.org/10.1165/rcmb.2009-0170TR.

41. Shibahara, S., Yoshizawa, M., Suzuki, H., Takeda, K., Meguro, K., and Endo, K. (1993). Functional analysis of CDNAs for two types of human heme oxygenase and evidence for their separate regulation. *J. Biochem.* 113 (2): 214–218. https://doi.org/10.1093/oxfordjournals.jbchem.a124028.

42. Magierowska, K., Bakalarz, D., Wójcik, D., Korbut, E., Danielak, A., Głowacka, U., Pajdo, R., Buszewicz, G., Ginter, G., Surmiak, M., Kwiecień, S., Chmura, A., Magierowski, M., and Brzozowski, T. (2020). Evidence for cytoprotective effect of carbon monoxide donor in the development of acute esophagitis leading to acute esophageal epithelium lesions. *Cells* 9 (5). https://doi.org/10.3390/cells9051203.

43. Magierowski, M., Magierowska, K., Hubalewska-Mazgaj, M., Adamski, J., Bakalarz, D., Sliwowski, Z., Pajdo, R., Kwiecien, S., and Brzozowski, T. (2016). Interaction between endogenous carbon monoxide and hydrogen sulfide in the mechanism of gastroprotection against acute aspirin-induced gastric damage. *Pharmacol. Res.* 114: 235–250. https://doi.org/10.1016/j.phrs.2016.11.001.

44. Takagi, T., Naito, Y., Mizushima, K., Nukigi, Y., Okada, H., Suzuki, T., Hirata, I., Omatsu, T., Okayama, T., Handa, O., Kokura, S., Ichikawa, H., and Yoshikawa, T. (2008). Increased intestinal expression of heme oxygenase-1 and its localization in patients with ulcerative colitis. *J. Gastroenterol. Hepatol.* 23. https://doi.org/10.1111/j.1440-1746.2008.05443.x.

45. Chauveau, C., Bouchet, D., Roussel, J.C., Mathieu, P., Braudeau, C., Renaudin, K., Tesson, L., Soulillou, J.P., Iyer, S., Buelow, R., and Anegon, I. (2002). Gene transfer of heme oxygenase-1 and carbon monoxide delivery inhibit chronic rejection. *Am. J. Transplant.* 2 (7): 581–592. https://doi.org/10.1034/j.1600-6143.2002.20702.x.

46. Motterlini, R., Clark, J.E., Foresti, R., Sarathchandra, P., Mann, B.E., and Green, C.J. (2002). Carbon monoxide-releasing molecules: Characterization of biochemical and vascular activities. *Circ. Res.* 90 (2). https://doi.org/10.1161/hh0202.104530.

47. Motterlini, R., Mann, B., Johnson, T., Clark, J., Foresti, R., and Green, C. (2005). Bioactivity and pharmacological actions of carbon monoxide-releasing molecules. *Curr. Pharm. Des.* 9 (30): 2525–2539. https://doi.org/10.2174/1381612033453785.

48. Kautz, A.C., Kunz, P.C., and Janiak, C. (2016). CO-releasing molecule (CORM) conjugate systems. *Dalton Trans.* 45 (45): 18045–18063. https://doi.org/10.1039/C6DT03515A.

49. Ji, X. and Wang, B. (2018). Strategies toward organic carbon monoxide prodrugs. *Acc. Chem. Res.* 51 (6): 1377–1385. https://doi.org/10.1021/acs.accounts.8b00019.

50. Yang, X., Lu, W., Hopper, C.P., Ke, B., and Wang, B. (2020). Nature's marvels endowed in gaseous

molecules i: Carbon monoxide and its physiological and therapeutic roles. *Acta Pharm. Sin. B* xxx. https://doi.org/10.1016/j.apsb.2020.10.010.

51. Steiger, C., Lühmann, T., and Meinel, L. (2014). Oral drug delivery of therapeutic gases – carbon monoxide release for gastrointestinal diseases. *J. Control. Release* 189: 46–53. https://doi.org/10.1016/j.jconrel.2014.06.025.

52. Wu, L. and Wang, R. (2005). Carbon monoxide: Endogenous production, physiological functions, and pharmacological applications. *Pharmacol. Rev.* https://doi.org/10.1124/pr.57.4.3.

53. Furchgott, R.F. and Jothianahdan, D. (1991). Endothelium-dependent and -independent vasodilation involving cyclic GMP: Relaxation induced by nitric oxide, carbon monoxide and light. *Blood Vessels* 28: 52–61. https://doi.org/10.1159/000158843.

54. Omura, T. and Sato, R. (1964). The carbon monoxide-binding pigment of liver microsomes. *J. Biol. Chem.* 239: 2370–2378.

55. Srisook, K., Han, S.S., Choi, H.S., Li, M.H., Ueda, H., Kim, C., and Cha, Y.N. (2006). CO from enhanced HO activity or from CORM-2 INHIBITS Both O 2- and NO production and downregulates HO-1 expression in LPS-stimulated macrophages. *Biochem. Pharmacol.* 71 (3): 307–318. https://doi.org/10.1016/j.bcp.2005.10.042.

56. Wang, R., Wang, Z., and Wu, L. (1997). Carbon monoxide-induced vasorelaxation and the underlying mechanisms. *Br. J. Pharmacol.* 121 (5): 927–934. https://doi.org/10.1038/sj.bjp.0701222.

57. Jackson, W.F. (2017). Potassium channels in regulation of vascular smooth muscle contraction and growth. *Adv. Pharmacol.* 78: 89–144. https://doi.org/10.1016/bs.apha.2016.07.001.

58. Jaggar, J.H., Li, A., Parfenova, H., Liu, J., Umstot, E.S., Dopico, A.M., and Leffler, C.W. (2005). Heme is a carbon monoxide receptor for large-conductance Ca 2+-activated K+ channels. *Circ. Res.* 97 (8): 805–812. https://doi.org/10.1161/01.RES.0000186180.47148.7b.

59. Bolognesi, M., Sacerdoti, D., Piva, A., Di Pascoli, M., Zampieri, F., Quarta, S., Motterlini, R., Angeli, P., Merkel, C., and Gatta, A. (2007). Carbon monoxide-mediated activation of large-conductance calcium-activated potassium channels contributes to mesenteric vasodilatation in cirrhotic rats. *J. Pharmacol. Exp. Ther.* 321 (1): 187–194. https://doi.org/10.1124/jpet.106.116665.

60. Magierowski, M., Hubalewska-Mazgaj, M., Magierowska, K., Wojcik, D., Sliwowski, Z., Kwiecien, S., and Brzozowski, T. (2018). Nitric oxide, afferent sensory nerves, and antioxidative enzymes in the mechanism of protection mediated by tricarbonyldichlororuthenium(II) dimer and sodium hydrosulfide against aspirin-induced gastric damage. *J. Gastroenterol.* 53 (1): 52–63. https://doi.org/10.1007/s00535-017-1323-4.

61. Brouard, S., Otterbein, L.E., Anrather, J., Tobiasch, E., Bach, F.H., Choi, A.M.K., and Soares, M.P. (2000). Carbon monoxide generated by heme oxygenase 1 suppresses endothelial cell apoptosis. *J. Exp. Med.* 192 (7): 1015–1025. https://doi.org/10.1084/jem.192.7.1015.

62. Otterbein, L.E., Bach, F.H., Alam, J., Soares, M., Lu, H.T., Wysk, M., Davis, R.J., Flavell, R.A., and Choi, A.M.K. (2000). Carbon monoxide has anti-inflammatory effects involving the mitogen- activated protein kinase pathway. *Nat. Med.* 6 (4): 422–428. https://doi.org/10.1038/74680.

63. Amersi, F., Shen, X.D., Anselmo, D., Melinek, J., Iyer, S., Southard, D.J., Katori, M., Volk, H.D., Busuttil, R.W., Buelow, R., and Kupiec-Weglinski, J.W. (2002). Ex vivo exposure to carbon monoxide prevents hepatic ischemia/reperfusion injury through P38 MAP kinase pathway. *Hepatology* 35 (4): 815–823. https://doi.org/10.1053/jhep.2002.32467.

64. Magierowska, K., Magierowski, M., Hubalewska-Mazgaj, M., Adamski, J., Surmiak, M., Sliwowski, Z., Kwiecien, S., and Brzozowski, T. (2015). Carbon monoxide (CO) released from tricarbonyldichlororuthenium (II) dimer (CORM-2) in gastroprotection against experimental ethanol-induced gastric damage. *PLoS One* 10 (10). https://doi.org/10.1371/journal.pone.0140493.

65. Brouard, S., Berberat, P.O., Tobiasch, E., Seldon, M.P., Bach, F.H., and Soares, M.P. (2002). Heme oxygenase-1-derived carbon monoxide requires the activation of transcription factor NF-κB to protect endothelial cells from tumor necrosis factor-α-mediated apoptosis. *J. Biol. Chem.* 277 (20): 17950–17961. https://doi.org/10.1074/jbc.M108317200.

66. Kim, H.S., Loughran, P.A., Rao, J., Billiar, T.R., and Zuckerbraun, B.S. (2008). Carbon monoxide activates NF-κB via ROS generation and Akt pathways to protect against cell death of hepatocytes. *Am. J. Physiol. – Gastrointest. Liver Physiol.* 295 (1). https://doi.org/10.1152/ajpgi.00105.2007.

67. Wang, B., Cao, W., Biswal, S., and Doré, S. (2011). Carbon monoxide-activated Nrf2 pathway leads to protection against permanent focal cerebral ischemia. *Stroke* 42 (9): 2605–2610. https://doi.org/10.1161/STROKEAHA.110.607101.

68. Su, D., Wang, X., Ma, Y., Hao, J., Wang, J., Lu, Y., Liu, Y., Wang, X., and Zhang, L. (2021). Nrf2-Induced MiR-23a-27a-24-2 cluster modulates damage repair of intestinal mucosa by targeting the bach1/HO-1 axis in inflammatory bowel diseases. *Free Radic. Biol.*

Med. 163: 1–9. https://doi.org/10.1016/j.freeradbiomed.2020.11.006.

69. Harada, S., Nakagawa, T., Yokoe, S., Edogawa, S., Takeuchi, T., Inoue, T., Higuchi, K., and Asahi, M. (2015). Autophagy deficiency diminishes indomethacin-induced intestinal epithelial cell damage through activation of the ERK/Nrf2/HO-1 pathway. *J. Pharmacol. Exp. Ther.* 355 (3). https://doi.org/10.1124/jpet.115.226431.

70. Xu, L., He, S.S., Yin, P., Li, D.Y., Mei, C., Yu, X.H., Shi, Y.R., Jiang, L.S., and Liu, F.H. (2016). Punicalagin induces Nrf2 translocation and HO-1 expression via PI3K/Akt, protecting rat intestinal epithelial cells from oxidative stress. *Int. J. Hyperth.* 32 (5). https://doi.org/10.3109/02656736.2016.1155762.

71. Lim, Y.J., Lee, J.S., Ku, Y.S., and Hahm, K.B. (2009). Rescue strategies against non-steroidal anti-inflammatory drug-induced gastroduodenal damage. *J. Gastroenterol. Hepatol.* https://doi.org/10.1111/j.1440-1746.2009.05929.x.

72. Matsui, H., Shimokawa, O., Kaneko, T., Nagano, Y., Rai, K., and Hyodo, I. (2011). The pathophysiology of non-steroidal anti-inflammatory drug (NSAID)-induced mucosal injuries in stomach and small intestine. *J. Clin. Biochem. Nutr.* https://doi.org/10.3164/jcbn.10-79.

73. Tanaka, A., Kunikata, T., Mizoguchi, H., Kato, S., and Takeuchi, K. (1999). Dual action of nitric oxide in pathogenesis of indomethacin-induced small intestinal ulceration in rats. *J. Physiol. Pharmacol.* 50 (3): 405–417.

74. Tanaka, A., Hase, S., Miyazawa, T., and Takeuchi, K. (2002). Up-regulation of cyclooxygenase-2 by inhibition of cyclooxygenase-1: A key to nonsteroidal anti-inflammatory drug-induced intestinal damage. *J. Pharmacol. Exp. Ther.* 300 (3): 754–761. https://doi.org/10.1124/jpet.300.3.754.

75. Takeuchi, K., Miyazawa, T., Tanaka, A., Kato, S., and Kunikata, T. (2002). Pathogenic importance of intestinal hypermotility in NSAID-induced small intestinal damage in rats. *Digestion* 66 (1): 30–41. https://doi.org/10.1159/000064419.

76. Ohno, R., Yokota, A., Tanaka, A., and Takeuchi, K. (2004). Induction of small intestinal damage in rats following combined treatment with cyclooxygenase-2 and nitric-oxide synthase inhibitors. *J. Pharmacol. Exp. Ther.* 310 (2): 821–827. https://doi.org/10.1124/jpet.104.065961.

77. Satoh, H., Amagase, K., and Takeuchi, K. (2014). Mucosal protective agents prevent exacerbation of NSAID-induced small intestinal lesions caused by antisecretory drugs in rats. *J. Pharmacol. Exp. Ther.* 348 (2): 227–235. https://doi.org/10.1124/jpet.113.208991.

78. Takeuchi, K. and Satoh, H. (2015). NSAID-induced small intestinal damage – roles of various pathogenic factors. *Digestion*. https://doi.org/10.1159/000374106.

79. Bjarnason, I., Macpherson, A., and Hollander, D. (1995). Intestinal permeability: An overview. *Gastroenterology* 108 (5): 1566–1581. https://doi.org/10.1016/0016-5085(95)90708-4.

80. Higuchi, K., Umegaki, E., Watanabe, T., Yoda, Y., Morita, E., Murano, M., Tokioka, S., and Arakawa, T. (2009). Present status and strategy of NSAIDs-induced small bowel injury. *J. Gastroenterol.* https://doi.org/10.1007/s00535-009-0102-2.

81. Le Bastard, Q., Al-Ghalith, G.A., Grégoire, M., Chapelet, G., Javaudin, F., Dailly, E., Batard, E., Knights, D., and Montassier, E. (2018). Systematic review: Human gut dysbiosis induced by non-antibiotic prescription medications. *Aliment. Pharmacol. Ther.* https://doi.org/10.1111/apt.14451.

82. Aabakken, L., Larsen, S., and Osnes, M. (1989). Sucralfate for prevention of naproxen-induced mucosal lesions in the proximal and distal gastrointestinal tract. *Scand. J. Rheumatol.* 18 (6): 361–368. https://doi.org/10.3109/03009748909102097.

83. Bakalarz, D., Surmiak, M., Yang, X., Wójcik, D., Korbut, E., Śliwowski, Z., Ginter, G., Buszewicz, G., Brzozowski, T., Cieszkowski, J., Głowacka, U., Magierowska, K., Pan, Z., Wang, B., and Magierowski, M. (2020). Organic carbon monoxide prodrug, BW-CO-111, in protection against chemically-induced gastric mucosal damage. *Acta Pharm. Sin. B.* https://doi.org/10.1016/j.apsb.2020.08.005.

84. Southam, H.M., Smith, T.W., Lyon, R.L., Liao, C., Trevitt, C.R., Middlemiss, L.A., Cox, F.L., Chapman, J.A., El-Khamisy, S.F., Hippler, M., Williamson, M.P., Henderson, P.J.F., and Poole, R.K. (2018). A thiol-reactive Ru(II) ion, not CO release, underlies the potent antimicrobial and cytotoxic properties of CO-releasing molecule-3. *Redox Biol.* 18: 114–123. https://doi.org/10.1016/j.redox.2018.06.008.

85. Yuan, Z., Yang, X., De La Cruz, L.K., and Wang, B. (2020). Nitro reduction-based fluorescent probes for carbon monoxide require reactivity involving a ruthenium carbonyl moiety. *Chem. Commun.* 56 (14): 2190–2193. https://doi.org/10.1039/c9cc08296d.

86. Nielsen, V.G. (2020). Ruthenium, not carbon monoxide, inhibits the procoagulant activity of atheris, echis, and pseudonaja venoms. *Int. J. Mol. Sci.* 21 (8). https://doi.org/10.3390/ijms21082970.

87. Nielsen, V.G. (2020). The anticoagulant effect of apis mellifera phospholipase A2 is inhibited by CORM-2 via a carbon monoxide-independent mechanism. *J. Thromb. Thrombolysis* 49 (1): 100–107. https://doi.org/10.1007/s11239-019-01980-0.

88. Bindu, S., Pal, C., Dey, S., Goyal, M., Alam, A., Iqbal, M.S., Dutta, S., Sarkar, S., Kumar, R., Maity, P., and Bandyopadhyay, U. (2011). Translocation of heme oxygenase-1 to mitochondria is a novel cytoprotective mechanism against non-steroidal anti-inflammatory drug-induced mitochondrial oxidative stress, apoptosis, and gastric mucosal injury. *J. Biol. Chem.* 286 (45): 39387–39402. https://doi.org/10.1074/jbc.M111.279893.

89. Yoda, Y., Amagase, K., Kato, S., Tokioka, S., Murano, M., Kakimoto, K., Nishio, H., Umegaki, E., Takeuchi, K., and Higuchi, K. (2010). Prevention by lansoprazole, a proton pump inhibitor, of indomethacin-induced small intestinal ulceration in rats through induction of heme oxygenase-1. *J. Physiol. Pharmacol.* 61 (3): 287–294.

90. Lee, H.J., Han, Y.M., Kim, E.H., Kim, Y.J., Hahm, K.B., and Possible, A. (2012). Involvement of Nrf2-mediated heme oxygenase-1 up-regulation in protective effect of the proton pump inhibitor pantoprazole against indomethacin-induced gastric damage in rats. *BMC Gastroenterol.* 12. https://doi.org/10.1186/1471-230X-12-143.

91. Lanza, F.L. (2002). Gastrointestinal adverse effects of bisphosphonates: Etiology, incidence and prevention. *Treat. Endocrinol.* https://doi.org/10.2165/00024677-200201010-00004.

92. Magierowski, M., Magierowska, K., Szmyd, J., Surmiak, M., Sliwowski, Z., Kwiecien, S., and Brzozowski, T. (2016). Hydrogen sulfide and carbon monoxide protect gastric mucosa compromised by mild stress against alendronate injury. *Dig. Dis. Sci.* 61 (11): 3176–3189. https://doi.org/10.1007/s10620-016-4280-5.

93. Graham, D.Y. and Malaty, H.M. (2001). Alendronate and naproxen are synergistic for development of gastric ulcers. *Arch. Intern. Med.* 161 (1): 107–110. https://doi.org/10.1001/archinte.161.1.107.

94. Cryer, B. and Bauer, D.C. (2002). Oral bisphosphonates and upper gastrointestinal tract problems: what is the evidence? *Mayo Clin. Proc.* 77 (10): 1031–1043. https://doi.org/10.4065/77.10.1031.

95. Jeal, W., Barradell, L.B., and McTavish, D. (1997). Alendronate. A review of its pharmacological properties and therapeutic efficacy in postmenopausal osteoporosis. *Drugs* 53 (3): 415–434. https://doi.org/10.2165/00003495-199753030-00006.

96. Wallace, J.L., Dicay, M., McKnight, W., Bastaki, S., and Blank, M.A. (1999). N-bisphosphonates cause gastric epithelial injury independent of effect on the microcirculation. *Aliment. Pharmacol. Ther.* 13 (12): 1675–1682. https://doi.org/10.1046/j.1365-2036.1999.00658.x.

97. Van Uden, P., Kenneth, N.S., and Rocha, S. (2008). Regulation of hypoxia-inducible factor-1alpha by NF-KappaB. *Biochem. J.* 412: 1470–8728. (Electronic).

98. Costa, N.R.D., Silva, R.O., Nicolau, L.A.D., Lucetti, L.T., Santana, A.P.M., Aragão, K.S., Soares, P.M.G., Ribeiro, R.A., Souza, M.H.L.P., Barbosa, A.L.R., and Medeiros, J.V.R. (2013). Role of soluble guanylate cyclase activation in the gastroprotective effect of the HO-1/CO pathway against alendronate-induced gastric damage in rats. *Eur. J. Pharmacol.* 700 (1–3): 51–59. https://doi.org/10.1016/j.ejphar.2012.12.007.

99. Souza, R.F. (2017). Reflux esophagitis and its role in the pathogenesis of barrett's msetaplasia. *J. Gastroenterol.* https://doi.org/10.1007/s00535-017-1342-1.

100. Grossi, L., Ciccaglione, A.F., and Marzio, L. (2017). Esophagitis and its causes: who is 'guilty' when acid is found 'not guilty'? *World J. Gastroenterol.* https://doi.org/10.3748/wjg.v23.i17.3011.

101. Fang, Y., Chen, H., Hu, Y., Djukic, Z., Tevebaugh, W., Shaheen, N.J., Orlando, R.C., Hu, J., and Chen, X. (2013). Gastroesophageal reflux activates the NF-KB pathway and impairs esophageal barrier function in mice. *Am. J. Physiol. – Gastrointest. Liver Physiol.* 305 (1). https://doi.org/10.1152/ajpgi.00438.2012.

102. Hu, J.L., Li, Z.Y., Liu, W., Zhang, R.G., Li, G.L., Wang, T., Ren, J.H., and Wu, G. (2010). Polymorphism in heme oxygenase-1 (HO-1) promoter and alcohol are related to the risk of esophageal squamous cell carcinoma on chinese males. *Neoplasma* 57 (1): 86–92. https://doi.org/10.4149/neo_2010_01_086.

103. Tavares, A.F., Parente, M.R., Justino, M.C., Oleastro, M., Nobre, L.S., and Saraiva, L.M. (2013). The bactericidal activity of carbon monoxide-releasing molecules against helicobacter pylori. *PLoS One* 8 (12). https://doi.org/10.1371/journal.pone.0083157.

104. De La Cruz, L.K.C., Benoit, S.L., Pan, Z., Yu, B., Maier, R.J., Ji, X., and Wang, B. (2018). Click, release, and fluoresce: a chemical strategy for a cascade prodrug system for codelivery of carbon monoxide, a drug payload, and a fluorescent reporter. *Org. Lett.* 20 (4): 897–900. https://doi.org/10.1021/acs.orglett.7b03348.

105. Chu, Y.T., Wang, Y.H., Wu, J.J., and Lei, H.Y. (2010). Invasion and multiplication of helicobacter pylori in gastric epithelial cells and implications for antibiotic resistance. *Infect. Immun.* 78 (10): 4157–4165. https://doi.org/10.1128/IAI.00524-10.

106. Gobert, A.P., Verriere, T., De Sablet, T., Peek, R.M., Chaturvedi, R., and Wilson, K.T. (2013). Haem oxygenase-1 inhibits phosphorylation of the helicobacter pylori oncoprotein caga in gastric epithelial cells. *Cell. Microbiol.* 15 (1): 145–156. https://doi.org/10.1111/cmi.12039.

107. Yisireyili, M., Alimujiang, A., Aili, A., Li, Y., Yisireyili, S., and Abudureyimu, K. (2020). Chronic restraint stress induces gastric mucosal inflammation with enhanced oxidative stress in a murine model. *Psychol. Res. Behav. Manag.* 13: 383–393. https://doi.org/10.2147/PRBM.S250945.

108. Magierowska, K., Magierowski, M., Surmiak, M., Adamski, J., Mazur-Bialy, A.I., Pajdo, R., Sliwowski, Z., Kwiecien, S., and Brzozowski, T. (2016). The protective role of carbon monoxide (CO) produced by heme oxygenases and derived from the CO-releasing molecule CORM-2 in the pathogenesis of stress-induced gastric lesions: evidence for non-involvement of nitric oxide (NO). *Int. J. Mol. Sci.* 17 (4). https://doi.org/10.3390/ijms17040442.

109. Smith, G.S., Mercer, D.W., Cross, J.M., Barreto, J.C., and Miller, T.A. (1996). Gastric injury induced by ethanol and ischemia-reperfusion in the rat: Differing roles for lipid peroxidation and oxygen radicals. *Dig. Dis. Sci.* 41 (6): 1157–1164. https://doi.org/10.1007/BF02088232.

110. Fukuyama, K., Iwakiri, R., Noda, T., Kojima, M., Utsumi, H., Tsunada, S., Sakata, H., Ootani, A., and Fujimoto, K. (2001). Apoptosis induced by ischemia-reperfusion and fasting in gastric mucosa compared to small intestinal mucosa in rats. *Dig. Dis. Sci.* 46 (3): 545–549. https://doi.org/10.1023/A:1005695031233.

111. Gomes, A.S., Gadelha, G.G., Lima, S.J., Garcia, J.A., Medeiros, J.V.R., Havt, A., Lima, A.A., Ribeiro, R.A., Brito, G.A.C., Cunha, F.Q., and Souza, M.H.L.P. (2010). Gastroprotective effect of heme-oxygenase 1/Biliverdin/CO Pathway in ethanol-induced gastric damage in mice. *Eur. J. Pharmacol.* 642 (1–3): 140–145. https://doi.org/10.1016/j.ejphar.2010.05.023.

112. Naito, Y., Takagi, T., and Yoshikawa, T. (2004). Heme oxygenase-1: A new therapeutic target for inflammatory bowel disease. *Aliment. Pharmacol. Ther., Supplement* 20: 177–184. https://doi.org/10.1111/j.1365-2036.2004.01992.x.

113. Hegazi, R.A.F., Rao, K.N., Mayle, A., Sepulveda, A.R., Otterbein, L.E., and Plevy, S.E. (2005). Carbon monoxide ameliorates chronic murine colitis through a heme oxygenase 1-dependent pathway. *J. Exp. Med.* 202 (12): 1703–1713. https://doi.org/10.1084/jem.20051047.

114. Takagi, T., Naito, Y., Mizushima, K., Hirai, Y., Harusato, A., Okayama, T., Katada, K., Kamada, K., Uchiyama, K., Handa, O., Ishikawa, T., and Itoh, Y. (2018). Herne oxygenase-1 prevents murine intestinal inflammation. *J. Clin. Biochem. Nutr.* 63 (3): 169–174. https://doi.org/10.3164/jcbn.17-133.

115. Antoniou, E., Margonis, G.A., Angelou, A., Pikouli, A., Argiri, P., Karavokyros, I., Papalois, A., and Pikoulis, E. (2016). The TNBS-induced colitis animal model: An overview. *Ann. Med. Surg.* https://doi.org/10.1016/j.amsu.2016.07.019.

116. Barros, S.É.D.L., Dias, T.M.D.S., Moura, M.S.B.D., Soares, N.R.M., Pierote, N.R.A., Araújo, C.O.D.D., Maia, C.S.C., Henriques, G.S., Barros, V.C., Moita Neto, J.M., Parente, J.M.L., Marreiro, D.D.N., and Nogueira, N.D.N. (2020). Relationship between selenium status and biomarkers of oxidative stress in crohn disease. *Nutrition* 74. https://doi.org/10.1016/j.nut.2020.110762.

117. Luceri, C., Bigagli, E., Agostiniani, S., Giudici, F., Zambonin, D., Scaringi, S., Ficari, F., Lodovici, M., and Malentacchi, C. (2019). Analysis of oxidative stress-related markers in crohn's disease patients at surgery and correlations with clinical findings. *Antioxidants* 8 (9). https://doi.org/10.3390/antiox8090378.

118. Ji, X., Zhou, C., Ji, K., Aghoghovbia, R.E., Pan, Z., Chittavong, V., Ke, B., and Wang, B. (2016). Click and release: A chemical strategy toward developing gasotransmitter prodrugs by using an intramolecular Diels–Alder reaction. *Angew. Chemie – Int. Ed.* 55 (51): 15846–15851. https://doi.org/10.1002/anie.201608732.

119. Perše, M. and Cerar, A. (2012). Dextran sodium sulphate colitis mouse model: Traps and tricks. *J. Biomed. Biotechnol.* https://doi.org/10.1155/2012/718617.

120. Yin, H., Fang, J., Liao, L., Nakamura, H., and Maeda, H. (2014). Styrene-maleic acid copolymer-encapsulated CORM2, a water-soluble carbon monoxide (CO) donor with a constant CO-releasing property, exhibits therapeutic potential for inflammatory bowel disease. *J. Control. Release* 187: 14–21. https://doi.org/10.1016/j.jconrel.2014.05.018.

121. Lv, C., Su, Q., Fang, J., and Yin, H. (2019). Styrene-maleic acid copolymer-encapsulated carbon monoxide releasing molecule-2 (SMA/CORM-2) suppresses proliferation, migration and invasion of colorectal cancer cells in vitro and in vivo. *Biochem. Biophys. Res. Commun.* 520 (2): 320–326. https://doi.org/10.1016/j.bbrc.2019.09.112.

122. Takagi, T., Naito, Y., Uchiyama, K., Suzuki, T., Hirata, I., Mizushima, K., Tsuboi, H., Hayashi, N., Handa, O., Ishikawa, T., Yagi, N., Kokura, S., Ichikawa, H., and Yoshikawa, T. (2011). Carbon monoxide liberated from carbon monoxide-releasing molecule exerts an anti-inflammatory effect on dextran sulfate sodium-induced colitis in mice. *Dig. Dis. Sci.* 56 (6): 1663–1671. https://doi.org/10.1007/s10620-010-1484-y.

123. Takagi, T., Naito, Y., Tanaka, M., Mizushima, K., Ushiroda, C., Toyokawa, Y., Uchiyama, K., Hamaguchi, M., Handa, O., and Itoh, Y. (2018).

Carbon monoxide ameliorates murine T-cell-dependent colitis through the inhibition of Th17 differentiation. *Free Radic. Res.* 52 (11–12): 1328–1335. https://doi.org/10.1080/10715762.2018.1470327.

124. Fukuda, W., Takagi, T., Katada, K., Mizushima, K., Okayama, T., Yoshida, N., Kamada, K., Uchiyama, K., Ishikawa, T., Handa, O., Konishi, H., Yagi, N., Ichikawa, H., Yoshikawa, T., Cepinskas, G., Naito, Y., and Itoh, Y. (2014). Anti-inflammatory effects of carbon monoxide-releasing molecule on trinitrobenzene sulfonic acid-induced colitis in mice. *Dig. Dis. Sci.* 59 (6): 1142–1151. https://doi.org/10.1007/s10620-013-3014-1.

125. Knauf, C., Abot, A., Wemelle, E., and Cani, P.D. (2020). Targeting the enteric nervous system to treat metabolic disorders? "enterosynes" as therapeutic gut factors. *Neuroendocrinology*. https://doi.org/10.1159/000500602.

126. Zakhary, R., Poss, K.D., Jaffrey, S.R., Ferris, C.D., Tonegawa, S., and Snyder, S.H. (1997). Targeted gene deletion of heme oxygenase 2 reveals neural role for carbon monoxide. *Proc. Natl. Acad. Sci. U. S. A.* 94 (26). https://doi.org/10.1073/pnas.94.26.14848.

127. Xue, L., Farrugia, G., Miller, S.M., Ferris, C.D., Snyder, S.H., and Szurszewski, J.H. (2000). Carbon monoxide and nitric oxide as coneurotransmitters in the enteric nervous system: Evidence from genomic deletion of biosynthetic enzymes. *Proc. Natl. Acad. Sci. U. S. A.* 97 (4). https://doi.org/10.1073/pnas.97.4.1851.

128. Van Dingenen, J., Steiger, C., Zehe, M., Meinel, L., and Lefebvre, R.A. (2018). Investigation of orally delivered carbon monoxide for postoperative ileus. *Eur. J. Pharm. Biopharm.* 130. https://doi.org/10.1016/j.ejpb.2018.07.009.

129. Moore, B.A., Otterbein, L.E., Türler, A., Choi, A.M.K., and Bauer, A.J. (2003). Inhaled carbon monoxide suppresses the development of postoperative ileus in the murine small intestine. *Gastroenterology* 124 (2). https://doi.org/10.1053/gast.2003.50060.

130. Türler, A., Kalff, J.C., Heeckt, P., AbuElmagd, K.M., Schraut, W.H., Bond, G.J., Moore, B.A., Brünagel, G., and Bauer, A.J. (2002). Molecular and functional observations on the donor intestinal muscularis during human small bowel transplantation. *Gastroenterology* 122 (7). https://doi.org/10.1053/gast.2002.33628.

131. Eskandari, M.K., Kalff, J.G., Billiar, T.R., Lee, K.K.W., and Bauer, A.J. (1999). LPS-induced muscularis macrophage nitric oxide suppresses rat jejunal circular muscle activity. *Am. J. Physiol. – Gastrointest. Liver Physiol.* 277 (2): 40–42. https://doi.org/10.1152/ajpgi.1999.277.2.g478.

132. Nakao, A., Moore, B.A., Murase, N., Liu, F., Zuckerbraun, B.S., Bach, F.H., Choi, A.M.K., Nalesnik, M.A., Otterbein, L.E., and Bauer, A.J. (2003). Immunomodulatory effects of inhaled carbon monoxide on rat syngeneic small bowel graft motility. *Gut* 52 (9). https://doi.org/10.1136/gut.52.9.1278.

133. Nakao, A., Kimizuka, K., Stolz, D.B., Neto, J.S., Kaizu, T., Choi, A.M.K., Uchiyama, T., Zuckerbraun, B.S., Nalesnik, M.A., Otterbein, L.E., and Murase, N. (2003). Carbon monoxide inhalation protects rat intestinal grafts from ischemia/reperfusion injury. *Am. J. Pathol.* 163 (4). https://doi.org/10.1016/S0002-9440(10)63515-8.

134. Wang, W.P., Guo, X., Koo, M.W.L., Wong, B.C.Y., Lam, S.K., Ye, Y.N., and Cho, C.H. (2001). Protective role of heme oxygenase-1 on trinitrobenzene sulfonic acid-induced colitis in rats. *Am. J. Physiol. – Gastrointest. Liver Physiol.* 281 (2): 44–2. https://doi.org/10.1152/ajpgi.2001.281.2.g586.

135. Cross, R.K. and Wilson, K.T. (2003). Nitric oxide in inflammatory bowel disease. *Inflamm. Bowel Dis.* https://doi.org/10.1097/00054725-200305000-00006.

136. Kolios, G., Valatas, V., and Ward, S.G. (2004). Nitric oxide in inflammatory bowel disease: a universal messenger in an unsolved puzzle. *Immunology*. https://doi.org/10.1111/j.1365-2567.2004.01984.x.

137. Yin, H., Fang, J., Liao, L., Maeda, H., and Su, Q. (2014). Upregulation of heme oxygenase-1 in colorectal cancer patients with increased circulation carbon monoxide levels, potentially affects chemotherapeutic sensitivity. *BMC Cancer* 14 (1). https://doi.org/10.1186/1471-2407-14-436.

138. Gao, X., Cao, Q., Cheng, Y., Zhao, D., Wang, Z., Yang, H., Wu, Q., You, L., Wang, Y., Lin, Y., Li, X., Wang, Y., Bian, J.S., Sun, D., Kong, L., Birnbaumer, L., and Yang, Y. (2018). Chronic stress promotes colitis by disturbing the Gut microbiota and triggering immune system response. *Proc. Natl. Acad. Sci. U. S. A.* 115 (13): E2960–E2969. https://doi.org/10.1073/pnas.1720696115.

139. Hopper, C.P., De La Cruz, L.K., Lyles, K.V., Wareham, L.K., Gilbert, J.A., Eichenbaum, Z., Magierowski, M., Poole, R.K., Wollborn, J., and Wang, B. (2020). Role of carbon monoxide in host–gut microbiome communication. *Chem. Rev.* https://doi.org/10.1021/acs.chemrev.0c00586.

140. Sartor, R.B. (2008). Microbial influences in inflammatory bowel diseases. *Gastroenterology* 134 (2): 577–594. https://doi.org/10.1053/j.gastro.2007.11.059.

141. Onyiah, J.C., Sheikh, S.Z., Maharshak, N., Steinbach, E.C., Russo, S.M., Kobayashi, T., Mackey, L.C., Hansen, J.J., Moeser, A.J., Rawls, J.F., Borst, L.B., Otterbein, L.E., and Plevy, S.E. (2013). Carbon monoxide and heme oxygenase-1 prevent intestinal inflammation in mice by promoting bacterial clearance. *Gastroenterology* 144 (4): 789–798. https://doi.org/10.1053/j.gastro.2012.12.025.

142. Su, W.C., Liu, X., Macias, A.A., Baron, R.M., and Perrella, M.A. (2008). Heme oxygenase-1-derived carbon monoxide enhances the host defense response to microbial sepsis in mice. *J. Clin. Invest.* 118 (1): 239–247. https://doi.org/10.1172/JCI32730.

143. Takagi, K., Okabe, S., and Saziki, R. (1969). A new method for the production of chronic gastric ulcer in rats and the effect of several drugs on its healing. *Jpn. J. Pharmacol.* 19 (3): 418–426. https://doi.org/10.1254/jjp.19.418.

144. Okabe, S., Roth, J.L.A., and Pfeiffer, C.J. (1971). A method for experimental, penetrating gastric and duodenal ulcers in rats – observations on normal healing. *Am. J. Dig. Dis.* 16 (3): 277–284. https://doi.org/10.1007/BF02235252.

145. Okabe, S. and Amagase, K. (2005). An overview of acetic acid ulcer models: the history and state of the art of peptic ulcer research. *Biol. Pharm. Bull.* https://doi.org/10.1248/bpb.28.1321.

146. Tarnawski, A.S. (2005). Cellular and molecular mechanisms of gastrointestinal ulcer healing. *Dig. Dis. Sci.* https://doi.org/10.1007/s10620-005-2803-6.

147. Schmassmann, A. (1998). Mechanisms of ulcer healing and effects of nonsteroidal anti-inflammatory drugs. *Am. J. Med.* 104: 00211–8. https://doi.org/10.1016/S0002-9343(97).

148. Barton, S.G.R.G., Rampton, D.S., Winrow, V.R., Domizio, P., and Feakins, R.M. (2003). Expression of heat shock protein 32 (Hemoxygenase-1) in the normal and inflamed human stomach and colon: An immunohistochemical study. *Cell Stress Chaperones* 8 (4): 329–334. https://doi.org/10.1379/1466-1268(2003)008<0329:EOHSPH>2.0.CO;2.

149. Guo, J.S., Cho, C.H., Wang, W.P., Shen, X.Z., Cheng, C.L., and Koo, M.W.L. (2003). Expression and activities of three inducible enzymes in the healing of gastric ulcers in rats. *World J. Gastroenterol.* 9 (8): 1767–1771. https://doi.org/10.3748/wjg.v9.i8.1767.

150. Takagi, T., Naito, Y., Uchiyama, K., Mizuhima, K., Suzuki, T., Horie, R., Hirata, I., Tsuboi, H., and Yoshikawa, T. (2016). Carbon monoxide promotes gastric wound healing in mice via the protein kinase C pathway. *Free Radic. Res.* 50 (10): 1098–1105. https://doi.org/10.1080/10715762.2016.1189546.

151. Lim, P.S., Sutton, C.R., and Rao, S. (2015). Protein kinase C in the immune system: From signalling to chromatin regulation. *Immunology* 146 (4): 508–522. https://doi.org/10.1111/imm.12510.

152. Magierowski, M., Magierowska, K., Hubalewska-Mazgaj, M., Sliwowski, Z., Ginter, G., Pajdo, R., Chmura, A., Kwiecien, S., and Brzozowski, T. (2017). Carbon monoxide released from its pharmacological donor, tricarbonyldichlororuthenium (II) dimer, accelerates the healing of pre-existing gastric ulcers. *Br. J. Pharmacol.* 174 (20): 3654–3668. https://doi.org/10.1111/bph.13968.

153. Magierowska, K., Bakalarz, D., Wójcik, D., Chmura, A., Hubalewska-Mazgaj, M., Licholai, S., Korbut, E., Kwiecien, S., Sliwowski, Z., Ginter, G., Brzozowski, T., and Magierowski, M. (2019). Time-dependent course of gastric ulcer healing and molecular markers profile modulated by increased gastric mucosal content of carbon monoxide released from its pharmacological donor. *Biochem. Pharmacol.* 163: 71–83. https://doi.org/10.1016/j.bcp.2019.02.011.

154. Głowacka, U., Brzozowski, T., and Magierowski, M. (2020). Synergisms, discrepancies and interactions between hydrogen sulfide and carbon monoxide in the gastrointestinal and digestive system physiology, pathophysiology and pharmacology. *Biomolecules.* https://doi.org/10.3390/biom10030445.

155. De Araújo, S., Oliveira, A.P., Sousa, F.B.M., Souza, L.K.M., Pacheco, G., Filgueiras, M.C., Nicolau, L.A.D., Brito, G.A.C., Cerqueira, G.S., Silva, R.O., Souza, M.H.L.P., and Medeiros, J.V.R. (2018). AMPK activation promotes gastroprotection through mutual interaction with the gaseous mediators H_2S, NO, and CO. *Nitric Oxide – Biol. Chem.* 78: 60–71. https://doi.org/10.1016/j.niox.2018.05.008.

156. Magierowski, M., Magierowska, K., Surmiak, M., Hubalewska-Mazgaj, M., Kwiecien, S., Wallace, J.L., and Brzozowski, T. (2017). The effect of hydrogen sulfide-releasing naproxen (ATB-346) versus naproxen on formation of stress-induced gastric lesions, the regulation of systemic inflammation, hypoxia and alterations in gastric microcirculation. *J. Physiol. Pharmacol.* 68 (5): 749–756.

157. Magierowski, M., Magierowska, K., Hubalewska-Mazgaj, M., Surmiak, M., Sliwowski, Z., Wierdak, M., Kwiecien, S., Chmura, A., and Brzozowski, T. (2018). Cross-talk between hydrogen sulfide and carbon monoxide in the mechanism of experimental gastric ulcers healing, regulation of gastric blood flow and accompanying inflammation. *Biochem. Pharmacol.* 149: 131–142. https://doi.org/10.1016/j.bcp.2017.11.020.

28

Carbon Monoxide and Sickle Cell Disease

Edward Gomperts[1,2,3], John Belcher[3], Howard Levy[2], and Greg Vercellotti[3]

[1]*Children's Hospital Los Angeles, Los Angeles, CA 90027, USA*
[2]*Hillhurst Biopharmaceuticals, Inc., Montrose, CA 91020, USA*
[3]*Division of Hematology, Oncology and Transplantation, Vascular Biology Center, Department of Medicine, University of Minnesota, Minneapolis, MN 55455, USA*

Introduction

Sickle cell disease (SCD) is an inherited recessive hemoglobinopathy that leads to hemolytic anemia, resulting from a point mutation in the beta-globin chain of hemoglobin (Hb) (glutamic acid 6-valine) with recurrent painful episodes, significantly shortened life expectancy, and poor quality of life. SCD manifests clinically through a variety of severe morbidities resulting from vaso-occlusive crises (VOCs) and chronic ischemia–reperfusion injury (IRI). There are approximately 140 000 SCD patients in the United States, the European Union, and the United Kingdom and millions of affected individuals worldwide. In the United States, the annual cost of treating these patients is ~$1.1 billion. There are limited options for the medical management of SCD and prevention of VOCs, with four licensed therapeutics that are approved to modulate the impact of the disease, with all four having variable efficacy. These four therapeutics target different mechanisms underlying the pathogenesis of SCD. While three of these agents are recent additions to the therapeutic armamentarium for SCD, the short- and long-term serious consequences of the sickle mutation are modulated but not prevented by these agents. To this end, investigators have targeted a cure through bone marrow transplantation and more recently genetically modified cellular therapy. Both of these targets have considerable potential to achieve the intended benefits of preventing VOCs, but both approaches bring their own significant challenges as well as costs and questions related to widespread availability. Consequently, at this time, both approaches have limited applicability to the majority of patients with this serious disease over the short and long term.

As compared with the currently licensed four therapeutics, the gasotransmitter carbon monoxide (CO) imparts significant protective effects via multiple physiological mechanisms demonstrated in four mouse models of SCD and also in two published clinical studies. CO at low, nontoxic doses has been shown to reduce vascular stasis, which is a marker of VOCs, limit inflammation, and modify apoptosis, as well as reduce red blood cell (RBC) sickling and increase RBC survival in two human subjects [1,2]. The substantial mechanistic preclinical data and the proof-of-concept clinical data demonstrate a reasonable likelihood of clinical benefit of CO.

Sickle Cell Disease

SCD is a painful, lifelong hemoglobinopathy with substantial morbidities and premature mortality. It is inherited as a missense point mutation in the Hb beta-globin gene where glutamic acid at position 6 is substituted by valine. This mutation is thought to have taken place in equatorial Africa and in heterozygous individuals endows resistance to falciparum malaria and thus enhanced survival. The prevalence of the mutation is the highest in individuals of equatorial African origin with the highest incidence in equatorial west and central Africa [3]. SCD is also prevalent in individuals of the Middle East and India, among other areas [3,4]. It is estimated that SCD affects approximately 100 000 individuals in the United States, as well as approximately 37 000 patients in the European Union and the United Kingdom, and millions of patients globally [5,6]. The prevalence of the disease is increasing as infant survival rises worldwide [7]. The number of children born with SCD is expected to exceed 14 million worldwide in the next 40 years, with about 79% of these children being born in sub-Saharan Africa where the mortality is the highest under 5 years of age [8].

SCD is characterized by anemia and painful VOCs. The change in molecular structure of homozygous sickle Hb allows the deoxygenated state to form polymers that promote red cell rigidity, hemolysis, Hb autoxidation, RBC membrane damage, decreased RBC deformability, increased RBC adhesion, intravascular and extravascular hemolysis, inflammation, vaso-occlusion, and ultimately multiple organ injury from IRI or infarction. Hemolysis results in the release of heme into the vascular space, which promotes enhanced adhesion of sickle RBCs, leukocytes (white blood cells [WBCs]), and platelets to the endothelium and the formation of multicellular aggregates resulting in vaso-occlusion and IRI pathophysiology [3].

SCD manifests clinically through a variety of severe morbidities: pain crisis, stroke, autosplenectomy, acute chest syndrome, and, over time, life-threatening chronic cardiac, pulmonary, liver, central nervous system and renal injury. A substantial medical need exists to reduce VOCs and consequent morbidities and mortality. There are ~6 million patients worldwide [5]. According to the Centers for Disease Control, the median survival among adults in North America is approximately 42 years [8], although a higher median survival rate was reported in an adult-only study in patients with SCD in the United Kingdom [8,9]. In addition to the reduction of survival, the ravages of this microvascular occlusive disease severely and progressively reduce the quality of life from infancy through adulthood.

Pathophysiology of SCD

Normal deoxygenated hemoglobin (HbA) does not polymerize. However, HbS under conditions of deoxygenation polymerizes to form sickle Hb aggregates and these aggregates result in the morphological shape changes of sickle RBCs (sRBCs). The pathophysiology of SCD is complex, but ultimately takes its origin from the formation of these sickle Hb aggregates as a consequence of the single point mutation in the Hb molecule. The aggregation process under physiological deoxygenation circumstance is highest in the postcapillary venules when the flexible red cell is transformed by the sickling process into rigid bodies [10,11]. Vaso-occlusion, which is the major pathological and clinical consequence of the sickling process, will occur when the transit time of sRBCs into larger diameter vessels in the microvasculature is longer than the time taken for the sickling process to complete.

Both intravascular hemolysis and extravascular hemolysis and thus the release of free Hb into plasma are a consequence of the sickling process. In normal individuals, free Hb is rapidly bound to circulating haptoglobin with high affinity ($K_d \sim 10^{-12}$ M) and carried to CD163 receptors on macrophages for endocytosis and degradation [12–15]. However, in SCD patients, the release of Hb from hemolyzed sRBCs depletes the plasma haptoglobin resulting in free Hb circulating in plasma [15, 16, 17]. Free Hb in the vasculature rapidly quenches the vasodilator nitric oxide (NO) resulting in vasospasm [18,19]. In the pro-oxidative SCD vasculature, the reaction of NO and other oxidants with ferrous Hb oxidizes the heme iron from ferrous Hb to ferric Hb (metHb). MetHb is unstable and releases hemin into the vascular space, and hemin release amplifies the pathophysiological cascade ultimately resulting in the vasculopathy underpinning the widespread tissue and organ damage that is the hallmark of this devastating disease [20–22].

In normal individuals, any released hemin is rapidly bound to circulating hemopexin with high affinity ($K_d < 10^{-13}$ M) and carried to CD91 (LRP1) receptors on hepatocytes for endocytosis and degradation [23–25]. However, in SCD patients the continuous formation of metHb in plasma and consequent continuing release of hemin depletes plasma hemopexin levels in a similar manner to Hb depleting haptoglobin [16,17]. This results in free hemin bound to lower affinity RBC-derived microparticles, albumin, and lipoproteins [26–29]. Hemin is released continuously from these carriers to cell membranes of circulating immune cells, platelets, and endothelial cells in the vessel wall [30]. Hemin is damaging to the cell membrane, is cytotoxic, promotes oxidative stress, and activates innate immune toll-like receptor 4 (TLR4) that signals the activation of monocytes and the endothelium [20,31–35]. The resulting inflammatory cascades include degranulation of endothelial Weibel–Palade bodies, enzymatically derived oxidant production, NF-κB-driven inflammasome activation, proinflammatory cytokine production, and adhesion molecule expression. Hemin, cytokines, lipopolysaccharide, infections, and the microbiome can all promote proinflammatory responses leading to adhesion of sRBC, leukocytes, and platelets to the endothelium and the formation of multicellular aggregates leading to vaso-occlusion [36–40]. The subsequent reopening of occluded vessels promotes ischemia–reperfusion pathophysiology, inflammatory pain, and tissue injury that amplify the innate immune inflammatory responses [41–43].

A substantial medical need exists to reduce VOCs, also termed sickle cell crises, and the consequent morbidities and mortality. There are currently four approved therapeutics for VOC reduction that cover a variety of pathophysiological mechanisms underlying SCD, each targeting a single mechanism to prevent VOCs and each with its own unique risks and benefits. However, none of these four agents eliminates the VOCs or the morbidities and mortality of SCD [44]. The cytotoxic and myelosuppressive agent hydroxyurea (HU) has been shown to reduce, but not eliminate, VOCs by increasing fetal Hb, and thus reducing HbS polymerization. However, HU has variable efficacy and its significant potential toxicity requires frequent blood monitoring [45]. It is the single licensed agent that is currently indicated for use in children with SCD. Recently, L-glutamine has been approved to prevent VOCs. Its mode of action is uncertain, but may promote NAD redox improvement, and it has a moderate level of efficacy [46]. A Hb-modifying agent, voxelotor, has recently been licensed for administration in adults and adolescents on the basis of an increase in Hb and other red cell parameters, and thus improvement in the anemia of SCD. Voxelotor forms a reversible covalent bond with the N-terminal valine of the alpha chain of Hb resulting in an allosteric modification of Hb that increases oxygen affinity. With regard to mechanism, voxelotor is thought to decrease the concentration of deoxygenated HbS and increase Hb concentration ~1 g/dl. However, an improvement in the VOC incidence among those subjects treated was not demonstrated in the pivotal clinical trial [47]. Further research with voxelotor is ongoing to further understand this key issue. A fourth recently licensed agent is the anti-P-selectin antibody crizanlizumab, which is a humanized monoclonal antibody that binds to P-selectin and blocks its interaction with the P-selectin glycoprotein ligand. Studies with this agent demonstrated a decrease in the incidence of VOCs by 45.3% in those subjects receiving the agent versus nontreated subjects [48]. In addition to the above four licensed therapeutics, a number of drugs are at various phases in development to manage or prevent VOCs [44]. Importantly, as with the four licensed agents, the majority of these agents target single rather than multiple mechanisms in the pathogenesis of SCD.

CO Binding to Hb

CO is a tasteless, odorless, noncorrosive gas that exists as a stable diatomic molecule consisting of one carbon and one oxygen atom. It is physiologically present in all species that utilize heme protein, being an endogenous human metabolite produced in the breakdown of heme via the heme oxygenase (HO) enzymes HO-1 and HO-2. CO binds to Hb with an affinity 200–250 times that of oxygen and a small component of CO also binds to other heme proteins, including myoglobin, cytochrome oxidase, cytochrome P450, and the hydroperoxidases [49]. In humans, normal levels of carboxyhemoglobin (COHb) are 0.5–1.5% of total Hb [49]. Blood levels of COHb are routinely determined by point-of-care devices or clinical laboratories by co-oximetry; CO saturation is reported as a percentage of total Hb.

CO as an inhaled gas rapidly diffuses across the pulmonary alveolar membrane binding to Hb to form COHb and is subsequently eliminated unchanged almost entirely through the lungs with the percent of oxygen in the inspired air and rate of ventilation being the determinants for CO elimination. It is rapidly distributed by the bloodstream essentially unchanged except for a very low percentage that is converted to CO_2 [49]. In the tissues, a portion of CO offloads from Hb, much like O_2, where it can impact cellular function. Thus, alveolar ventilation is the major determinant of COHb levels with a biological half-life in healthy resting adults at sea level being 4–6 h [49]. The binding of CO to Hb results in a reduction in the oxygen-carrying capacity of Hb because of competitive binding.

Toxicity of CO is first observed with a peak COHb saturation of approximately 20% and fatality occurs at levels above 40% with duration of exposure being a key determinant of the extent of toxicity and mortality. Although conflicting data exist, this information is based on the Agency for Toxic Substances and Disease Registry that incorporates the extensive epidemiological and clinical data, including long-term toxicology studies in rodents and other animals [50].

Environmental sources of CO are myriad via combustion reactions among other sources, and especially relevant in industrialized economies. Tobacco smoking is associated with COHb levels peaking as high as 20% among chain smokers. Industrial sources of CO include motor vehicle internal combustion engine exhaust gases as well as other fossil fuel energy sources. Solid waste disposal also accounts for a substantial component of total CO emissions. In the home environment, improperly vented equipment utilizing oil, gas, wood, or coal sourced fuels are major causes of inhaled atmospheric CO and thus mortality [49,50].

In contrast to the known paradigm of CO toxicity, CO is now accepted as a signaling molecule capable of regulating a host of physiological and therapeutic processes [51]. In recent years, it has been shown that CO at low concentrations exerts key physiological functions in various models of tissue inflammation and injury, providing potent cytoprotection in models of inflammation, including SCD [52–55], organ transplantation [56], and acute lung injury [57], among others [58,59].

HO-1, Hemolysis, and CO

CO is generated endogenously by all cells when heme is degraded by the HO enzymes, generating the majority of endogenously produced CO as well as equimolar quantities of biliverdin (BV) and iron [60]. Interest in HO has grown beyond heme metabolism, as it has been shown to be a key cytoprotective enzyme [60].

There are two HO enzymes, HO-1 and HO-2. These isoforms differ in their structure, location, and biological role. HO-1 is highly inducible and found in all cells and is upregulated by a variety of agents, including heme, pathogens, IRI, oxidants, and, in certain instances, CO itself [61–68]. HO-2 is constitutively expressed in the brain, vasculature, kidneys, and testes where it is involved in neurotransmission, vasomotor tone, protection from heme toxicity, and spermatogenesis [63, 64]. Of note, heme nephrotoxicity increases in older mice as HO-2 mRNA levels decline relative to younger mice [64]. Although heme is a major inducer of HO-1, a variety of nonheme-containing agents, including CO itself, heavy metals, oxidative stress, and heat shock, are also strong inducers of HO-1 expression [65–68]. This diversity of HO-1 inducers has provided further support that HO-1 can serve as a key biological effector in the adaptation and/or defense against oxidative stresses. This was validated with the development of $Hmox-1^{-/-}$ mice, strengthening the paradigm that HO-1 is indeed an important molecule in the host's defense against oxidant stress, as $Hmox1^{-/-}$ mice exhibit increased susceptibility to inflammation and IRI [69]. In contrast, induction of endogenous HO-1 provides potent protection in various *in vivo* and *in vitro* models [69–72], including SCD [50,73]. Interestingly, the first human case of HO-1 deficiency showed nearly identical phenotypic changes as those observed in $Hmox1^{-/-}$ mice [74]. These important reports provide insightful clues to HO's anti-inflammatory effects that likely involve removal of pro-oxidant heme and production of the bioactive products of HO-1, including BV and CO. Exogenously administered CO can rescue the proinflammatory phenotype of $Hmox1^{-/-}$ mice and closely mimic the protection observed with HO-1 induction [75,76].

Sickle Cell Acute Clinical Events

The formation of irreversibly sickled cells that are rigid rather than flexible occurs largely in vascular sites where deoxygenation is extensive. Thus, postcapillary venules are key sites for SCD-based microvascular obstruction, which is a key initiating mechanism of SCD pathology. Hemolysis will also occur with the formation of the rigid sRBCs, with the consequences of heme release compounding the consequences of the vaso-occlusion [77]. When such obstruction reaches a critical level, clinical VOC becomes apparent. The frequency and extent of these clinically overt VOC events vary considerably from patient to patient with a smaller number of patients having VOC as frequently as every 4–6 weeks, while others as infrequent as once in 2–3 years [78]. While a few genetic modifiers, such as hemoglobin F, hemoglobin C, and beta-thalassemia when co-inherited, are understood to impact the frequency of VOCs, there are likely other factors, probably genetic as well as acquired, that are poorly understood [79].

The acute onset of a VOC can vary in severity depending on the extent of the ischemic lesion. The presentation of the VOC will also depend on the anatomical site of the VOC and its consequences will depend on the extent of tissue necrosis and the impact of the necrosis on the functionality of the surviving surrounding tissue. A VOC can present with the sudden onset of acute pain at any peripheral site, including the chest wall, spinal column, distal leg, and head of the femur, as well as with priapism. The extent, duration, and severity of pain will also be impacted by the frequency and recurrence of painful episodes [79].

The acute onset of a VOC in critical anatomical sites can become life threatening. Acute splenic sequestration is a serious VOC-associated morbidity or mortality in small children. In these events, the microvasculature within the spleen becomes widely obstructed due to VOCs, resulting in a massive sequestration of RBCs within the spleen vasculature, with a consequent fall in Hb levels and RBC count to life-threatening levels. Functional asplenism is almost an invariable consequence of the ongoing microvascular occlusion events taking place in the spleen, with attendant risks of overwhelming sepsis associated with disseminated encapsulated bacterial septicemia, such as pneumococcus [79].

Equally threatening are acute pulmonary failure syndromes, one of which is acute chest syndrome. In acute chest syndrome, widespread pulmonary microvascular obstruction occurs due to VOCs resulting in anoxia and hypercapnia. This life-threatening complication is more frequent in children than adults and is a leading cause of death in children with SCD. Other causes of acute pulmonary insufficiency include acute severe bacterial infection amplified in severity by the absence of a functioning spleen manifesting as an acute respiratory distress syndrome. In addition, a massive long bone infarction can result in extensive microvascular fat pulmonary emboli [79].

In addition to the above, there are major concerns in SCD around cerebral ischemia, as cerebrovascular occlusive episodes in patients with SCD are common and are major clinical events that uniformly determine adverse outcomes. These events can manifest across the age spectrum as overt strokes [80], but cerebral infarcts can also be silent [81]. Silent cerebral infarcts (SCIs) documented by magnetic resonance imaging changes have been reported in 10–30% of patients with SCD, and a prevalence of SCIs of 53% has been documented in adults with SCD [82,83]. The prevalence of stroke in SCD is reported to be as high as 24%, and a crippling cerebrovascular occlusive event can occur as early as 2 years of age, with the highest incidence being within the first decade of life. The incidence of stroke has been demonstrated to be 10% per 36 months in children with SCD who show abnormal transcranial Doppler ultrasound readings [84]. The impact of SCD on the brain in these individuals is a major cause of suboptimal cognitive performance in SCD adults [85].

Sickle Cell Long-Term Consequences

The spectrum of presentation of SCD-associated VOCs is well documented with improved management dependent on early diagnosis and therapeutic intervention. However, ongoing persistent microvascular occlusive events in individuals with SCD may also occur at the subclinical level with long-term major life-impacting as well as life-threatening consequences. The clinical outcomes of these ongoing microvascular occlusive events are as threatening over the long term as the acute VOC events. Thus, major organ dysfunction syndromes are well characterized in SCD and are primarily responsible for the documented shortened life expectancy in this population. These include the consequences of SCIs and pulmonary hypertension syndromes with accompanying cardiac dysfunction, myocardial ischemic fibrosis, and renal failure are major causes of early adult mortality [79]. Also, not unexpectedly, SCD is also associated with long-term microvascular consequences affecting the retina, with consequent vision impairment.

Mechanism of Action of CO in SCD

There has been a great deal of insight into how CO modulates SCD. There are six mechanisms that have been identified, all with substantial data to support the hypotheses (Figure 28.1 and Table 28.1). There are several key mechanisms of action (Table 28.1) supported with substantial data.

CO Inhibits HbS Polymerization [1,86–88]

CO binds avidly to HbS, preventing polymerization. Also, CO can bind to formed HbS polymers and enhance their melting. This reduces the persistence

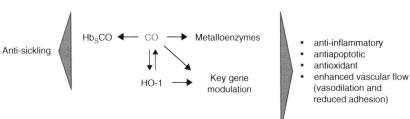

Figure 28.1 Mechanisms of action of CO in SCD.

Table 28.1 Mechanisms of action of CO in SCD.

Potential mechanism	Publications	Potential effect
Antipolymerization	Aroutiounian SK, et al. Biophys Chem 2001;91(2):167–181. Chung LL, et al. Arch Biochem Biophys 1978; 189:535–539. Sirs JA. Lancet1963;1:971–972. Higgins JM, et al. Proc Natl Acad Sci USA 2007;104:20496–20500.	CO binds to HbS, modulating HbS polymerization
Sickle RBC hydration	Vandorpe DH, et al. PLoS One 2010;5:e8732.	Inhibition of Ca^{2+}-permeable cation conductance

(Continued)

Table 28.1 (Continued)

Potential mechanism	Publications	Potential effect
Antioxidative	Meng F, et al. Anal Biochem 2017;521:11–19. Kassa T, et al. FEBS Open Bio 2016;6:876–884. Nakahira K, et al. J Exp Med 2006;203:2377–2389. Chin BY, et al. Proc Natl Acad Sci USA 2007;104:5109–5114. Bilban M, et al. Immunity 2006;24:601–610. Bilban M, et al. J Mol Med 2008;86:267–279. Haschemi A, et al. PLoS One 2011;6:e26376. Bories GFP, et al. Blood 2020;136:1535–1548.	Decreased oxidative stress
Antiinflammatory	Belcher JD, et al. Blood 2013;122:2757–2764. Otterbein LE, et al. Nat Med 2000;4:422–428. Chi PL, et al. Mol Neurobiol 2015;52(1):277–292. Shin DY, et al. Blood 2012;119:2523–2532. Morse D, et al. Free Radic Biol Med 2009;47(1):1–12. Chung J, et al. Mol Immunol 2011;48:1793–1799. Sawle P, et al. Br J Pharmacol 2005;14:800–810. Zhang X, et al. J Biol Chem 2005;280:8714–8721.	Modulation of pro- and anti-inflammatory mediators
Vasodilation	Stone JR, et al. Biochemistry 1994;33:5636–5640. Wegiel B, et al. Circulation 2010;121:537–548. Zuckerbraun BS, et al. J Exp Med 2006;203:2109–2119.	Increased cGMP
Antiapoptosis	Liu XM, et al. Cardiovasc Res 2002;55:396–405. Li MH, et al. Biochem Biophys Res Commun 2006;342:984–990. Kim HS, et al. Biochem Biophys Res Commun 2006;344:1172–1178. Wang X, et al. J Biol Chem 2007;282:1718–1726. Vieira HL, et al. J Neurochem 2008;107:375–384. Queiroga CS, et al. Br J Pharmacol 2014. Mar;172(6):1533–45. Almeida AS, et al. J Biol Chem 2012;287:10762–10770.	Decreased mitochondrial damage and cell death

of HbS polymers in RBCs, consequently reducing the ability of larger HbS polymers to rapidly form during transit in hypoxic tissues. Sirs showed a reduction of sRBCs from 10.2% to 3.9% with a peak COHb saturation of approximately 4% in blood from a single patient, and there have been *in vitro* reports of CO melting sickle polymers and CO preventing occlusion in a microfluidic device at concentrations as low as 0.1% COHb [87].

Anti-inflammatory Effects [89–103]

CO has potent anti-inflammatory activity. This functions through the activation of Nrf2-responsive genes leading to HO-1 upregulation and NF-κB downregulation. A wide spectrum of genes is modulated by CO, including Nrf2, NF-κB, STAT3, CREBH, and various MAP kinases. Clues to the protective effect of CO are delineated in the work of Ferreira et al. showing that mice with HbS are resistant to malaria infection, due in part to the induction of HO-1 and its products, including CO [96]. HO-1 is highly inducible by CO and both HO-1 and CO modulate inflammatory responses. HO-1- and CO-mediated inhibition of NF-κB activation reduces adhesion molecule and proinflammatory cytokine expression, which inhibits vaso-occlusion in SCD mouse models. Other anti-inflammatory effects include increases in phospho-p38 MAPK and phospho-Akt as well as decreases in innate immune TLR4 signaling and TLR4 trafficking to the cell membrane surface. CO also inhibits NLRP3 inflammasome activation.

Antioxidative Effects [104–106]

CO has been documented to provide antioxidative effects via a spectrum of processes. In addition to the potent anti-inflammatory effects of CO on Nrf2 and NF-κB, the activation of these genes also results in antioxidative properties. CO's antioxidative properties also arise by the inhibition by CO of the oxidation of ferrous HbS to ferric HbS and by the stabilization of the HbS molecule by CO binding to heme, thereby limiting heme release from ferric HbS. CO may also inhibit superoxide production by NADPH oxidase 2 through CO binding to heme groups on the gp91phox subunit. In addition, CO activates Nrf2-responsive antioxidant genes, including enzymes involved in glutathione production and regeneration, enzymes involved in NADPH production [107], and enzymes involved in heme and iron metabolism, such as haptoglobin, hemopexin, HO-1, and ferritin [107,108]. These antioxidative effects are enhanced via the CO-induced shift to the pentose phosphate shunt adaptation of macrophages to enhanced heme detoxification [107].

Impact on P-Sickle [109]

CO inhibits the Ca^{2+}-permeable cation conductance channel or P-sickle, thereby inhibiting sickle RBC dehydration. Presumably, an increase in RBC hydration would decrease the propensity for HbS polymer formation.

Vasodilation via Increased cGMP Production [110, 111]

CO binds to the heme protein guanylate cyclase and increases cGMP production leading to vasodilation of blood vessels.

Reduction of Apoptosis [112–115]

CO has antiapoptotic activity by preventing mitochondrial depolarization, inhibiting cytochrome c release, suppressing p53 expression, and inhibiting caspase activation.

CO in SCD

In Vitro/Ex Vivo Studies

The initial studies of CO in SCD took place decades ago. Given the high affinity of CO for Hb, it was hypothesized that CO would reduce HbS polymerization, and thus limit sickle-associated clinical effects. However, the COHb saturations required to produce a clinical benefit in SCD through this mechanism were not known.

The potential for CO to directly limit the sickling phenomenon though binding to HbS has been shown via a number of *in vitro* studies focused on HbS polymerization. It has been documented that the equilibrium solubility of HbS with CO is linearly proportional to the amount of bound CO. As early as 1963, Sirs demonstrated that the presence of CO reduces the proportion of sickled cells *in vitro* [1]. A more recent study utilizing a microfluidic device showed concentrations of CO as low as 0.1% preventing occlusion when perfused with blood from SCD individuals. In addition, CO was shown to directly melt HbS polymers *in vitro*; the melting of polymers may be a key event limiting HbS polymers from acting as a seed for rapid polymerization in hypoxic tissue [86]. Given these *in vitro* experiments, it is a logical supposition that the impact of CO on polymerization could limit sickle-associated clinical effects.

CO in Animal Models of SCD

A number of studies in four different transgenic SCD mouse models demonstrate that modulation of the HO-1/CO pathway is effective in impacting SCD pathophysiology, and suggest low-dose CO as a novel approach to treatment (Table 28.2) [52–55]. The first

Table 28.2 Studies with CO administered via three different modalities carried out in four different mouse SCD genetic constructs.

Study	SCD models	Modality	Peak COHb	Duration;exposure	Key outcomes
Belcher et al. 2006 [52]	• S + S-Antilles • Berkeley	iCO	11%	3 days; 1×/day	Vascular stasis; NF-κB
Beckman et al. 2009 [53]	• S + S-Antilles	iCO	2%	Approximately 10 weeks; 3×/week	WBCs; liver parenchymal necrosis
Belcher et al. 2013 [54]	• NY1DD • Townes	CO prodrug PEG-COHb (MP4CO)	6.5%	1 day; 1×/day	Mortality; vascular stasis; NF-κB; WBCs; P-selectin; vWF
Belcher et al. 2018 [55]	• NY1DD • Townes	Oral CO liquid (HBI-002)	5%	1 day; 1×/day × 5 and 10 days	Vascular stasis; NF-κB; VCAM-1; HO-1; Nrf2; WBCs; red cell parameters

reported study (Belcher et al. 2006) [52] utilized a sickle mouse model of VOCs that employs a skin window and intravital microscopy after 1 h of hypoxia (7% O_2) followed by 1 h of reoxygenation in room air. This study demonstrated upregulation of HO-1 by inhaled CO (iCO), which was associated with modulation of the pathobiology of SCD in S + S-Antilles and Berkeley (Berk) SCD mice. In addition to significantly reducing hypoxia/reoxygenation (H/R)-induced vascular stasis, this protective effect was associated with inflammation mitigation that included reductions in WBC–endothelium interactions and the inhibition of NF-κB, VCAM-1, and ICAM-1 expression. An additional study [73] showed that livers of sickle mice overexpressing a wt-HO-1 transgene in the liver, but not an enzymatically inactive ns-HO-1 transgene, had marked activation of the phospho-p38 MAPK and phospho-Akt cell signaling pathways, reduced levels of NF-κB in liver, and decreased sVCAM-1 in serum. Importantly, H/R-induced vascular stasis was inhibited in SCD mice overexpressing the enzymatically active wt-HO-1, but not the enzymatically inactive ns-HO-1.

Subsequent research (Beckman et al. 2009) [53] extended the investigation of the anti-inflammatory effects of iCO in S + S-Antilles SCD mice in a longer-term treatment setting. Treatment of mice with 25 or 250 ppm of iCO for 1 h/day, 3 days/week for 8–10 weeks with peak COHb saturation of 2% and 22%, respectively, was shown to decrease vascular inflammation and organ pathology. CO treatment in this model significantly reduced elevated total WBC, neutrophil, and lymphocyte counts concurrent with reduced staining for myeloid and lymphoid markers in the bone marrow with the bone marrow exhibiting significant decreases in colony-forming-unit granulocyte macrophages during colony-forming cell assays. Prior observations of the increase in anti-inflammatory signaling pathways phospho-Akt and phospho-p38 MAPK in the livers of CO-treated mice were confirmed. Treated SCD mice had a significant reduction in liver parenchymal necrosis, reflecting the anti-inflammatory and anti-VO benefits of CO administration.

A third study (Belcher et al. 2013) [54] demonstrated protective effects of CO in two additional SCD mouse models (NY1DD and Townes) using a pegylated (PEG) COHb drug product (MP4CO) with intravenous administration. In this study, the administration of CO in NY1DD SCD mice markedly induced HO-1 activity and inhibited NF-κB activation in the vessel wall and H/R-induced microvascular stasis as compared to controls. Importantly, CO administration also induced nuclear Nrf2, an important transcriptional regulator of HO-1 and other cytoprotective antioxidant genes, and improved vascular perfusion. Additionally, in heterozygous HbAS-Townes SCD mice, the administration of CO significantly reduced mortality in a model of induced acute lung injury and cardiac dysfunction. CO resulted in a reduction in NF-κB activation and decreased P-selectin and von Willebrand factor (vWF) on blood vessels, indicating anti-inflammatory and antiadhesion effects. This study demonstrated efficacy at approximately 6.5% peak COHb saturation.

A fourth study carried out (Belcher et al. 2018) [55] in the NY1DD and Townes genetic mouse constructs utilized an oral liquid CO drug product (HBI-002). This research extended the above observations with HBI-002 (10 ml/kg) administered once daily by

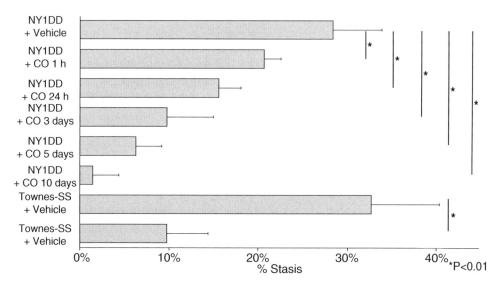

Figure 28.2 HBI-002 provides protection against vascular stasis; degree of protection increases with days of dosing. Bar values represent mean ± SD. *$P < 0.01$, HBI-002 versus vehicle. Full spectrum of results as presented is unpublished.

gavage for 1, 3, 5, or 10 days (Figure 28.2). This dosing of HBI-002 was compared with vehicle control in a duration of dosing-dependent manner. Mice were assessed for efficacy markers (microvascular stasis, markers of inflammation, markers of RBC protection) and mechanistic gene regulation (Nrf2, HO-1, NF-κB, and VCAM-1 expression). In NY1DD mice, H/R-induced stasis was 27% in vehicle-treated mice, 21% when HBI-002 was administered 1 h before H/R, and 16% when administered as a single dose 24 h before H/R. With repeated administration, stasis was 10% after 3 days of once-daily HBI-002 treatments before H/R, 6% after 5 days of once-daily treatments before H/R, and 1% after 10 days of once-daily treatments before H/R. In Townes-SS mice, H/R-induced stasis was 33% in vehicle-treated mice, and 10% after 5 days of once-daily HBI-002 treatments ($P < 0.01$). Importantly, these effects were observed at peak COHb saturations of 4–6%. Nrf2 and HO-1 protein was measured by immunoblot in livers from Townes-SS mice treated once daily for 10 days with HBI-002 or vehicle (10 ml/kg). Hepatic Nrf2 and HO-1 were markedly increased in Townes-SS mice treated with HBI-002 compared with vehicle. Nrf2 and HO-1 were increased 4.9-fold ($P < 0.01$) and 12.2-fold ($P < 0.01$), respectively, consistent with earlier studies showing activation of Nrf2 signaling and HO-1 induction in response to administration of exogenous CO. In addition, Townes-SS mice treated with HBI-002 had markedly decreased NF-κB and VCAM-1 expression compared to vehicle-treated mice.

CO in Human Subjects with SCD

The clinical safety and tolerability of CO at levels up to 13.9% COHb have been demonstrated in 23 clinical trials using iCO or PEG-COHb (Chapter 30) [116]. CO treatment in healthy adult volunteers showed no significant adverse events (unpublished) with rigorous assessment of vital signs (blood pressure, heart, and respiratory rate), electrocardiogram, clinical chemistry, urinalysis, and neurocognitive testing.

Clinical studies with iCO and PEG-COHb have focused on acute treatment of VOC rather than chronic use for the prevention of VOCs due to the limitations of these agents. Sirs [1] first demonstrated that iCO reduced the proportion of sickled RBCs in an SCD patient. Beutler [2] extended this work, demonstrating that the administration of iCO increased RBC survival in two young adult SCD patients. In addition, in a retrospective epidemiological study, a significantly lower rate of hospital admission for VOC was found on days with higher atmospheric levels of CO and not found for the other five atmospheric pollutants investigated. These data prompted two companies (Sangart and Prolong) to initiate clinical programs with PEG-COHb to treat VOCs. The results of phase 1b clinical studies demonstrated safety and tolerability in SCD patients [116–118]. However, the announced intent of these companies as well as Hb toxicity suggests that these products will be limited to acute use for treatment of VOCs rather than chronic use for the prevention of VOCs [119,120]. Importantly, there has been no failure in a clinical CO development program in SCD for safety reasons.

Methods of CO Administration

The majority of reports of CO administration in both animal and human studies used either iCO or drug products that contain a non-CO active pharmaceutical ingredient (API) that releases CO, specifically intravenous PEG-COHb, as the mode of treatment. However, the recent report of studies in two sickle mouse models documents positive outcomes utilizing the administration of CO via the oral liquid drug product (HBI-002) that contains CO as the API [55]. All three modes of administration of CO, iCO, CO prodrug, and oral liquid, showed efficacy of CO and this efficacy has been demonstrated in both chronic and acute settings in four transgenic SCD mouse models, including models of both moderate and severe diseases [52–55]. Importantly, rigorous safety data performed in human volunteers up to a COHb of ~14% showed no serious adverse event. However, each mode of CO administration brings specific requirements, issues, or constraints.

The administration of iCO brings the challenges of breathing CO as an inhaled gas that would likely limit its acceptance and utilization for chronic dosing and/or use in a home setting. These difficulties are associated with the clinical hurdles associated with using a gaseous therapeutic that include issues regarding patient education and compliance with overdosing being a major concern. Further, there are environmental safety concerns associated with the presence of large cylinders of compressed CO because of the possibility of escape of CO into the immediate environment due to mishandling of the cylinder and its controls. This would likely preclude the home use of iCO and also create substantial reluctance from clinical centers in the management of acute VOC episodes in a hospital or emergency room setting. Equally challenging is dosing of the iCO as clinical administration of CO gas requires precise ventilatory and blood gas level control. While accurate dosing can be reliably administered and monitored by frequent blood gas measurement in a controlled ICU or operating room environment, CO gas clinical administration is limited outside of these settings. Breathing the incorrect dose due to variable administration, incorrectly applying a breathing mask, or changes in lung function over time are major hurdles.

Intravenous PEG-COHb solutions bring a different set of challenges. The well-documented toxicity of cell-free Hb likely precludes chronic use of PEG-COHb in SCD [119,120] in the prevention of VOCs, and the requirement for intravenous infusion is a barrier to home use. However, once efficacy and safety are documented, this approach may become useful for the treatment of VOCs in hospital or emergency room environments.

The oral administration of a liquid CO drug product offers a potential solution to overcome the challenges of chronic CO administration, in both the home and hospital settings. The SCD mouse studies showed its feasibility and potential efficacy with the attainment of COHb levels of less than 10%, which are well documented to be associated with efficacy and very unlikely to be associated with adverse outcomes. This mode of administration brings the added advantage that the risk of overdosage can be obviated by requiring a large enough liquid dosage volume to effectively preclude overdose due to maximum stomach volume.

Importantly, during the study of CO across its three different modes of administration in numerous sickle mouse and human studies, there has been no indication of CO-related toxicity at the targeted COHb levels of <10% COHb saturation. This is as expected based on the extensive CO toxicology literature and knowledge of oxygen transport.

Although the efficacy of CO in SCD may rely upon multiple mechanisms of action, the exact dosing regimen to maximize efficacy and ensure safety depends on the interplay of these mechanisms. Preclinical data in SCD mice and the limited clinical data to date indicate that smaller, daily doses with peak COHb levels at 2–6.5% are effective in downregulating pathophysiological mechanisms of SCD [52–55,116–118]. In addition, the experiment with daily oral dosing of HBI-002 over 5 and 10 days suggests an enhanced effect with extended durations of dosing.

Conclusions

There are substantial data that have been generated over many years of research that support the hypothesis that CO at low nontoxic levels has the potential to improve the clinical course of SCD. This potential of CO is enhanced via the broad array of biological processes effected through multiple key genes and biological modifiers. This encouraging body of information bodes well for the urgent search to limit and possibly prevent the ravages of this aggressive, painful, and life-threatening inherited disorder. This promise is strengthened by the well-documented data indicating safety as well as potential efficacy at COHb saturation below 10%. However, this potential will need to be evaluated for efficacy under formal clinical study to assess the potential for CO to prevent acute episodes, treat acute episodes, and limit the long-term consequences of SCD-related organ damage.

Nevertheless, the potential for the successful application of this highly encouraging therapy with strong current indicators for efficacy and safety is unfortunately guarded. This is primarily due to the substantial challenges of administration of CO to patients across the age spectrum. Fortunately, innovative approaches in the clinical delivery of this gas, such as the HBI-002 oral drug product, that enable the chronic, self-administration of CO in the home care setting are under development and provide a potential solution to this dosing challenge. Section II of this book (Chapters 11–17) also describes various ways of delivering CO in different applications.

References

1. Sirs, J.A. (1963). The use of carbon monoxide to prevent sickle-cell formation. *Lancet* 1: 971–972.
2. Beutler, E. (1975). The effect of carbon monoxide on red cell life span in sickle cell disease. *Blood* 46: 253–259.
3. Bunn, H.F. (1997). Pathogenesis and treatment of sickle cell disease. *New Eng. J. Med.* 337: 762–768.
4. National Heart, Lung, and Blood Institute. Disease and conditions index. Sickle cell anemia: who is at risk? Bethesda, MD: US Department of Health and Human Services, National Institutes of Health, National Heart, Lung, and Blood Institute. Accessed February 17, 2016. Available from: https://www.nhlbi.nih.gov/health-topics/Sickle-cell-disease.
5. Hassell, K.L. (2010). Population estimates of sickle cell disease in the US. *Amer. J. Prev. Med.* 38: 512–521.
6. Haemoglobinopathies on the Move. Is Europe ready? Health and Migration Policy Perspectives. Accessed November 29, 2021. Available from: https://thalassaemia.org.cy/wp-content/uploads/2018/07/Publication-3-Is-Europe-Ready.pdf.
7. Piel, F.B., Hay, S.I., Gupta, S., Weatherall, D.J., and Williams, T.N. (2013). Global burden of sickle cell

anaemia in children under five, 2010–2050: modelling based on demographics, excess mortality, and interventions. *PLoS Med.* 10 (7): e1001484.
8. Platt, O.S., Brambilla, D.J., Rosse, W.F., Milner, P.F., Castro, O., Steinberg, M.H. et al. (1994). Mortality in sickle cell disease. Life expectancy and risk factors for early death. *N. Engl. J. Med.* 330: 1639–1644.
9. Gardner, K., Douiri, A., Drasar, E., Allman, M., Mwirigi, A., Awogbade, M., and Thein, S.L. (2016). Survival in adults with sickle cell disease in a high-income setting. *Blood* 128 (10): 1436–1438. Sep 8.
10. Kassim, A.A. and DeBaum, M.R. (2013). Sickle cell disease, vasculopathy and therapeutics. *Annu. Rev. Med.* 64: 451–466.
11. Hofrichter, J., Ross, P.D., and Eaton, W.A. (1974). Kinetics and mechanism of deoxyhemoglobin S gelation: a new approach to understanding sickle cell disease. *Proc. Natl. Acad. Sci. U.S.A.* 71: 4864–4868.
12. Cooper, C.E., Schaer, D.J., Buehler, P.W., Wilson, M.T., Reeder, B.J., Silkstone, G., Svistunenko, D.A., Bulow, L., and Alayash, A.I. (2013). Haptoglobin binding stabilizes hemoglobin ferryl iron and the globin radical on tyrosine beta145. *Antioxid. Redox Signal.* 18: 2264–2273.
13. Moestrup, S.K. and Moller, H.J. (2004). CD163: a regulated hemoglobin scavenger receptor with a role in the anti-inflammatory response. *Ann. Med.* 36: 347–354.
14. Nielsen, M.J., Moller, H.J., and Moestrup, S.K. (2010). Hemoglobin and heme scavenger receptors. *Antioxid. Redox Signal.* 12: 261–273.
15. Muller-Eberhard, U., Javid, J., Liem, H.H., Hanstein, A., and Hanna, M. (1968). Plasma concentrations of hemopexin, haptoglobin and heme in patients with various hemolytic diseases. *Blood* 32: 811–815.
16. Santiago, R.P., Guarda, C.C., Figueiredo, C.V.B., Fiuza, L.M., Aleluia, M.M., Adanho, C.S.A., Carvalho, M.O.S., Pitanga, T.N., Zanette, D.L., Lyra, I.M., Nascimento, V.M.L., Vercellotti, G.M., Belcher, J.D., and Goncalves, M.S. (2018). Serum haptoglobin and hemopexin levels are depleted in pediatric sickle cell disease patients. *Blood. Cells. Mol. Dis.* 72: 34–36.
17. Schaer, D.J., Buehler, P.W., Alayash, A.I., Belcher, J.D., and Vercellotti, G.M. (2013) Hemolysis and free hemoglobin revisited: exploring hemoglobin and hemin scavengers as a novel class of therapeutic proteins. *Blood* 121 (8): 1276–1284.
18. Reiter, C.D., Wang, X., Tanus-Santos, J.E., Hogg, N., Cannon, R.O., 3rd, Schechter, A.N., and Gladwin, M.T. (2002). Cell-free hemoglobin limits nitric oxide bioavailability in sickle-cell disease. *Nat. Med.* 8: 1383–1389.
19. Morris, C.R., Gladwin, M.T., and Kato, G.J. (2008). Nitric oxide and arginine dysregulation: a novel pathway to pulmonary hypertension in hemolytic disorders. *Curr. Mol. Med.* 8: 620–632.
20. Belcher, J.D., Chen, C., Nguyen, J., Milbauer, L., Abdulla, F., Alayash, A.I., Smith, A., Nath, K.A., Hebbel, R.P., and Vercellotti, G.M. (2014). Heme triggers TLR4 signaling leading to endothelial cell activation and vaso-occlusion in murine sickle cell disease. *Blood* 123: 377–390.
21. Bunn, H.F. and Jandl, J.H. (1968). Exchange of heme among hemoglobins and between hemoglobin and albumin. *J. Biol. Chem.* 243: 465–475.
22. Umbreit, J. (2007). Methemoglobin–it's not just blue: a concise review. *Am. J. Hematol.* 82: 134–144.
23. Hrkal, Z., Vodrazka, Z., and Kalousek, I. (1974). Transfer of heme from ferrihemoglobin and ferrihemoglobin isolated chains to hemopexin. *Eur. J. Biochem.* 43: 73–78.
24. Hvidberg, V., Maniecki, M.B., Jacobsen, C., Hojrup, P., Moller, H.J., and Moestrup, S.K. (2005). Identification of the receptor scavenging hemopexin-heme complexes. *Blood* 106: 2572–2579.
25. Smith, A. and Morgan, W.T. (1979). Haem transport to the liver by haemopexin. Receptor-mediated uptake with recycling of the protein. *Biochem. J.* 182: 47–54.
26. Balla, J., Jacob, H.S., Balla, G., Nath, K., Eaton, J.W., and Vercellotti, G.M. (1993). Endothelial-cell heme uptake from heme proteins: induction of sensitization and desensitization to oxidant damage. *Proc. Natl. Acad. Sci. U.S.A.* 90: 9285–9289.
27. Belcher, J.D., Marker, P.H., Geiger, P., Girotti, A.W., Steinberg, M.H., Hebbel, R.P., and Vercellotti, G.M. (1999). Low-density lipoprotein susceptibility to oxidation and cytotoxicity to endothelium in sickle cell anemia. *J. Lab. Clin. Med.* 133: 605–612.
28. Camus, S.M., De Moraes, J.A., Bonnin, P., Abbyad, P., Le Jeune, S., Lionnet, F., Loufrani, L., Grimaud, L., Lambry, J.C., Charue, D., Kiger, L., Renard, J.M., Larroque, C., Le Clesiau, H., Tedgui, A., Bruneval, P., Barja-Fidalgo, C., Alexandrou, A., Tharaux, P.L., Boulanger, C.M., and Blanc-Brude, O.P. (2015). Circulating cell membrane microparticles transfer heme to endothelial cells and trigger vasoocclusions in sickle cell disease. *Blood* 125: 3805–3814.
29. Hanson, M.S., Piknova, B., Keszler, A., Diers, A.R., Wang, X., Gladwin, M.T., Hillery, C.A., and Hogg, N. (2011). Methaemalbumin formation in sickle cell disease: effect on oxidative protein modification and HO-1 induction. *Br. J. Haematol.* 154: 502–511.
30. Balla, G., Jacob, H.S., Eaton, J.W., Belcher, J.D., and Vercellotti, G.M. (1991). Hemin: a possible physiological mediator of low density lipoprotein oxidation and endothelial injury. *Arterioscler. Thromb.* 11: 1700–1711.
31. Dutra, F.F., Alves, L.S., Rodrigues, D., Fernandez, P.L., De Oliveira, R.B., Golenbock, D.T., Zamboni, D.S., and Bozza, M.T. (2014). Hemolysis-induced lethality involves inflammasome activation by heme. *Proc. Natl. Acad. Sci. U.S.A.* 111: E4110–4118.

32. Figueiredo, R.T., Fernandez, P.L., Mourao-Sa, D.S., Porto, B.N., Dutra, F.F., Alves, L.S., Oliveira, M.F., Oliveira, P.L., Graca-Souza, A.V., and Bozza, M.T. (2007). Characterization of heme as activator of Toll-like receptor 4. *J. Biol. Chem.* 282: 20221–20229.

33. Jeney, V., Balla, J., Yachie, A., Varga, Z., Vercellotti, G.M., Eaton, J.W., and Balla, G. (2002). Pro-oxidant and cytotoxic effects of circulating heme. *Blood* 100: 879–887.

34. Porto, B.N., Alves, L.S., Fernandez, P.L., Dutra, T.P., Figueiredo, R.T., Graca-Souza, A.V., and Bozza, M.T. (2007). Heme induces neutrophil migration and reactive oxygen species generation through signaling pathways characteristic of chemotactic receptors. *J. Biol. Chem.* 282: 24430–24436.

35. Ghosh, S., Ihunnah, C.A., Hazra, R. et al. (Apr 7 2016). Nonhaemopoietic Nrf2 dominantly impedes adult progression of sickle cell anemia in mice. *JCI Insight* 2016;1(4):e81090.

36. Hidalgo, A., Chang, J., Jang, J.E., Peired, A.J., Chiang, E.Y., and Frenette, P.S. (2009). Heterotypic interactions enabled by polarized neutrophil microdomains mediate thromboinflammatory injury. *Nat. Med.* 15: 384–391.

37. Li, J., Kim, K., Hahm, E., Molokie, R., Hay, N., Gordeuk, V.R., Du, X., and Cho, J. (2014). Neutrophil AKT2 regulates heterotypic neutrophil-platelet interactions during vascular inflammation. *J. Clin. Investig.* 124: 1483–1496.

38. Manwani, D. and Frenette, P.S. (2013). Vaso-occlusion in sickle cell disease: pathophysiology and novel targeted therapies. *Blood* 122: 3892–3898.

39. Turhan, A., Weiss, L.A., Mohandas, N., Coller, B.S., and Frenette, P.S. (2002). Primary role for adherent leukocytes in sickle cell vascular occlusion: a new paradigm. *Proc. Natl. Acad. Sci. U.S.A.* 99: 3047–3051.

40. Zhang, D., Chen, G., Manwani, D., Mortha, A., Xu, C., Faith, J.J., Burk, R.D., Kunisaki, Y., Jang, J.E., Scheiermann, C., Merad, M., and Frenette, P.S. (2015). Neutrophil ageing is regulated by the microbiome. *Nature* 525: 528–532.

41. Osarogiagbon, R.U., Choong, S., Belcher, J.D., and Vercellotti, G.M. (Jul 1 2000). Paller MS and Hebbel RP. Reperfusion injury pathophysiology in sickle transgenic mice. *Blood* 96 (1): 314–320.

42. Hebbel, R.P., Osarogiagbon, R., and Kaul, D. (2004). The endothelial biology of sickle cell disease: inflammation and a chronic vasculopathy. *Microcirculation* 11: 129–151.

43. Rees, D.C., Williams, T.N., and Gladwin, M.T. (2010). Sickle cell disease. *Lancet* 76: 2018–2031.

44. Telen, M.J. (2016). Beyond hydroxyurea: new and old drugs in the pipeline for sickle cell disease. *Blood* 127: 810–819.

45. Smith-Whitley, K. (2014). Reproductive issues in sickle cell disease. *Hematology: Am Soc Hematol Educ Program* 2014(1):418–424.

46. Niihara, Y., Miller, S., Kanter, J., Lanzkron, S., Smith, W., Hsu, L., Gordeuk, V., Viswanathan, K., Sarnaik, S., Osunkwo, I., Gillaume, E., Sadanandan, S., Siegr, L., Lsky, J., Panosyan, E., Blake, O., New, T., Bellwvue, R., Tran, L., Razon, R., Stark, C., Neumayr, L., and Vichinsky, E. (2018). A phase 3 trail of l-glutamine in sickle cell diseae. *N. Eng. J. Med.* 379: 226–235.

47. Vichinsky, E., Hoppe, C.C., Ataga, K.I., Ware, R., Nduba, V., Amal-Beshlawy, Hassab, H., Achebe, M.M., Alkindi, S., Brown, C.R., Diuguid, D.L., Telfer, P., Tsitsikas, D.A., Elghandour, A., Gordeuk, V.R., Kanter, J., Abboud, M.R., Leher-Graiwer, J., Tonda, M., Intondi, A., Tong, B., and Howard, J. (2019). A phase 3 randomized trial of voxelotor in sickle cell disease. *N. Eng. J. Med.* 381: 509–519.

48. Ataga, K.I., Kutlar, A., Kanter, J., Liles, D., Cancado, R., Friedrisch, J., Guthrie, T.H., Knight-Madden, J., Alvarez, O.A., Gordeuk, V.R., Gualandro, S., Colella, M.P., Smith, W.R., Rollins, S.A., Stocker, J.W., and Rother, R.P. (Feb 2 2017). Crizanlizumab for the prevention of pain crises in sickle cell disease. *N. Engl. J. Med.* 376 (5): 429–439.

49. Stewart, R.D. (1975). The effect of carbon monoxide on humans. *Annu. Rev. Pharmacol.* 15: 409–423.

50. Wilbur, S., Williams, M., Williams, R. et al. (Jun 2012). *Toxicological Profile for Carbon Monoxide*. Atlanta (GA): Agency for Toxic Substances and Disease Registry (US). Accessed November 29, 2021. Available from: https://www.atsdr.cdc.gov/toxprofiles/tp201.pdf.

51. Motterlini, R. and Otterbein, L.E. (2010). The therapeutic potential of carbon monoxide. *Nat. Rev. Drug Discov.* 9: 728–743.

52. Belcher, J.D., Mahaseth, H., Welch, T.E., Otterbein, L.E., Hebbel, R.P., and Vercellotti, G.M.J. (2006). Heme oxygenase-1 is a modulator of inflammation and vaso-occlusion in transgenic sickle mice. *Clin. Invest.* 116: 808–816.

53. Beckman, J., Belcher, J.D., Vineyard, J.V., Chen, C., Nguyen, J., Nwaneri, M.D., O'Sullivan, M.G., Gulbahee, E., Hebbel, R.P., and Vercellotti, G.M. (2009). Inhaled carbon monoxide reduces leukocytosis in a murine model of sickle cell disease. *Am. J. Physiol. Heart Circ. Physiol.* 297: H1243–H1253.

54. Belcher, J.D., Young, M., Chen, C., Nguyen, J., Burhop, K., Tran, P., and Vercellotti, G.M. (2013). MP4CO, a pegylated hemoglobin saturated with carbon monoxide, is a modulator of HO-1, inflammation and vaso-occlusion in transgenic sickle cell mice. *Blood* 122: 2757–2764.

55. Belcher, J.D., Gomperts, E., Nguyen, J., Chen, C., Abdulla, F., Kiser, Z.M., Gallo, D., Levy, H., Otterbein,

L.E., and Vercellotti, G.M. (Oct 11 2018). Oral carbon monoxide therapy in murine sickle cell disease: beneficial effects on vaso-occlusion, inflammation and anemia. *PLoS One* 13 (10): e0205194.

56. Sato, K., Balla, J., Otterbein, L., Smith, R.N., Brouard, S., Lin, Y., Csizmadia, E., Sevigny, J., Robson, S.C., Vercellotti, G., Choi, A.M., Bach, F.H., and Soares, M.P. (2001). Carbon monoxide generated by heme oxygenase-1 suppresses the rejection of mouse-to-rat cardiac transplants. *J. Immunol.* 166: 4185–4194.

57. Otterbein, L.E., Mantell, L.L., and Choi, A.M. (1999). Carbon monoxide provides protection against hyperoxic lung injury. *Am. J. Physiol.* 276: L688–L694.

58. Otterbein, L.E., Haga, M., Zuckerbraun, B.S., Bach, F.H., Billiar, T.S., Choi, A.M., and Soares, M.P. (2003). Carbon monoxide inhibits transplant and balloon angioplasty-associated intimal hyperplasia. *Nat. Med.* 9: 183–190. 2002.

59. Kuramitsu, K., Gallo, D., Yoon, M., Chin, B.Y., Csizmadia, E., Hanto, D.W., and Otterbein, L.E. (Jun 2011). Carbon monoxide enhances early liver regeneration in mice after hepatectomy. *Hepatology* 53 (6): 2016–2026.

60. Otterbein, L.E., Soares, M.P., Yamashita, K., and Bach, F.H. (2003). Heme oxygenase-1: unleashing the protective properties of heme. *Trends Immunol.* 24 (8): 449–455.

61. Abraham, N.G., Lin, J.H.-C., Schwartzman, M.L., Levere, R.D., and Shibahara, S. (1988). The physiological significance of heme oxygenase. *Int. J. Biochem.* 20: 543–558.

62. Camhi, S.L., Alam, J., Otterbein, L.E., Sylvester, S.L., and Choi, A.M.K. (1995). Induction of heme oxygenase-1 gene expression by lipopolysaccharide is mediated by AP-1 activation. *Am. J. Resp. Cell Mol. Biol.* 13: 387–398.

63. Maines, M.D. (1997). The heme oxygenase system: a regulator of second messenger gases. *Annu. Rev. Pharmacol. Toxicol.* 37: 517–554.

64. Nath, K.A., Grande, J.P., Farrugia, G., Croatt, A.J., Belcher, J.D., Hebbel, R.P., Vercellotti, G.M., Katusic, Z.S. (2013) Age sensitizes the kidney to heme protein-induced acute kidney injury. *Am J Physiol Renal Physiol.* 304(3): F317–F325.

65. Applegate, L.A., Luscher, P., and Tyrell, R.M. (1991). Induction of heme oxygenase: a general response to oxidant stress in cultured mammalian cells. *Cancer Res.* 51: 974–978.

66. Keyse, S.M. and Tyrell, R.M. (1989). Heme oxygenase is the major 32-kDa stress protein induced in human skin fibroblasts by UVA radiation, hydrogen peroxide, and sodium arsenite. *Proc. Natl., Acad. Sci. U.S.A.* 86: 99–103.

67. Lee, P.J., Alam, J., Sylvester, S.L., Imamdar, N., Otterbein, L.E., and Choi, A.M.K. (1996). Regulation of heme oxygenase-1 expression in vivo and in vitro in hyperoxic lung injury. *Am. J. Respir. Cell Mol. Biol.* 14: 556–468.

68. Agarwal, A. and Nick, H.S. (2000). Renal response to tissue injury: lessons from heme oxygenase-1 GeneAblation and expression. *J. Am. Soc. Nephrol.* 11: 965–973.

69. Kapturczak, M.H., Wasserfall, C., Brusko, T., Campbell-Thompson, M., Ellis, T.M., Atkinson, M.A., and Agarwal, A. (2004). Heme oxygenase-1 modulates early inflammatory responses: evidence from the heme oxygenase-1-deficient mouse. *Am. J. Pathol.* 165: 1045–1053.

70. Abraham, N.G., Lavrovsky, Y., Schwartzman, M.L., Stoltz, R.A., Levere, R.D., Gerritsen, M.E., Shibarhara, S., and Kappas, A. (1995). Transfection of the human heme oxygenase gene into rabbit coronary microvessel endothelial cells: protective effects against heme and hemoglobin toxicity. *Proc. Natl. Acad. Sci. U.S.A.* 92: 6798–6802.

71. Nath, K.A., Balla, G., Vercelotti, G.M., Balla, J., Jacob, H.S., Levitt, M.D., and Rosenberg, M.E. (1992). Induction of heme oxygenase is a rapid, protective response in rhabdomyolysis in the rat. *J. Clin. Invest.* 90: 267–270.

72. Vile, G.F., Basu-Modak, S., Waltner, C., and Tyrell, R.M. (1994). Heme oxygenase-1 mediates an adaptive response to oxidative stress in human skin fibroblasts. *PNAS* 91: 2607–2610.

73. Belcher, J.D., Vineyard, J.V., Bruzzone, C.M. et al. (2010). Heme oxygenase-1 gene delivery by sleeping beauty inhibits vascular stasis in a murine model of sickle cell disease. *J. Mol. Med. (Berl)* 88 (7): 665–675.

74. Yachie, A. et al. (1999). Oxidative stress causes enhanced endothelial cell injury in human heme oxygenase- 1 deficiency. *J. Clin. Invest.* 103: 129–135.

75. Pamplona, A., Ferreira, A., Balla, J., Jeney, V., Balla, G., Epiphanio, S., Chora, A., Rodrigues, C.D., Gregoire, I.P., Cunha-Rodrigues, M., Portugal, S., Soares, M.P., and Mota, M.M. (Jun 2007). Heme oxygenase-1 and carbon monoxide suppress the pathogenesis of experimental cerebral malaria. *Nat. Med.* 13 (6): 703–710.

76. Akamatsu, Y., Haga, M., Tyagi, S., Yamashita, K., Graça-Souza, A.V., Ollinger, R., Csizmadia, E., May, G.A., Ifedigbo, E., Otterbein, L.E., Bach, F.H., and Soares, M.P. (Apr 2004). Heme oxygenase-1-derived carbon monoxide protects hearts from transplant associated ischemia reperfusion injury. *FASEB J.* 18 (6): 771–772.

77. Veluswamy, S., Shah, P., Khaleel, M., Thuptimdang, W., Chalacheva, P., Sunwoo, J., Denton, C., Kato, R.,

Detterich, J., Wood, J., Sposto, R., Khoo, M., Zeltzer, L., and Coates, T. (2020). Progressive vasoconstriction with sequential thermal stimulation indicates vascular dysautonomia in sickle cell disease. *Blood* 136: 1191–1200.

78. Platt, O.S. et al (1991). Pain in sickle cell disease: rate and risk factors. *N. Eng. J. Med.* 325: 11–19.

79. Dover, G.J. and Platt, O.S. (1998). Sickle Cell Disease in Nathan and Oski Hematology in Infancy and Childhood, 5e. 762–809. WB Saunders and Co. Chapter 20.

80. DeBaum, M.R. and Kirkham, F.J. (2016). Central nervous system complications and management in sickle cell disease. *Blood* 127: 829–838.

81. Dowling, M.M., Quinn, C.T., Plumb, P. et al (2012). Acute silent cerebral ischemia and infarction during acute anemia in children with and without sickle cell disease. *Blood* 120: 3891–3897.

82. Kwiatkowski, J.L., Zimmerman, R.A., Pollock, A.N. et al (2009). Silent infarcts in young children with sickle cell disease. *Br. J. Haematol.* 146: 300–305.

83. Ohene-Frempong, K., Weiner, S.J., Sleeper, L.A. et al (1998). Cerebrovascular accidents in sickle cell disease: rates and risk factors. *Blood* 91: 288–294.

84. Bernaudin, F., Verlhac, S., Arnaud, C. et al (2011). Impact of early transcranial doppler screening and intensive therapy on cerebral vasculopathy outcome in a new born sickle cell anemia cohort. *Blood* 117: 1130–1140.

85. Schatz, J., Brown, R.T., Pascual, J.M., Hsu, L., and DeBaum, M.R. (2001). Poor school and cognitive functioning with silent cerebral infarcts and sickle cell disease. *Neurology* 56: 1109–1111.

86. Higgins, J.M., Eddington, B.T., Bhatia, S.N., and Mahadevan, L. (2007). Sickle cell vaso-occlusion and rescue in a micro-fluidic device. *Proc. Natl. Acad. Sci. U.S.A.* 104: 20496–20500.

87. Aroutiounian, S.K., Louderback, J.G., Ballas, S.K., and Kim-Shapiro, D.B. (2001). Evidence for carbon monoxide binding to sickle cell polymers during melting. *Biophys. Chem.* 91 (2): P167–181.

88. Chung, L.L. and Magdorff-Fairchild, B. (1978). Extent of polymerization in partially liganded sickle hemoglobin. *Arch. Biochem. Biophys.* 189 (2): 535–539.

89. Chi, P.L., Lin, C.C., Chen, Y.W., Hsiao, L.D., and Yang, C.M. (2015). CO induces Nrf2-dependent heme oxygenase-1 transcription by cooperating with Sp1 and c-Jun in rat brain astrocytes. *Mol. Neurobiol.* 52 (1): 277–292.

90. Shin, D.Y., Chung, J., Joe, Y., Pae, H.O., Chang, K.C., Cho, G.J., Ryter, S.W., and Chung, H.T. (Mar 15 2012). Pretreatment with CO-releasing molecules suppresses hepcidin expression during inflammation and endoplasmic reticulum stress through inhibition of the STAT3 and CREBH pathways. *Blood* 119 (11): 2523–2532.

91. Morse, D., Lin, L., Choi, A.M., and Ryter, S.W. (Jul 1 2009). Heme oxygenase-1, a critical arbitrator of cell death pathways in lung injury and disease. *Free Radic. Biol. Med.* 47 (1): 1–12.

92. Chung, J., Shin, D.Y., Zheng, M., Joe, Y., Pae, H.O., Ryter, S.W., and Chung, H.T. (Sep 2011). Carbon monoxide, a reaction product of heme oxygenase-1, suppresses the expression of C-reactive protein by endoplasmic reticulum stress through modulation of the unfolded protein response. *Mol. Immunol.* 48 (15–16): 1793–1799.

93. Otterbein, L.E. et al. (2000). Carbon monoxide has anti-inflammatory effects involving the mitogen-activated protein kinase pathway. *Nat. Med.* 6 (4): 422–428.

94. Sawle, P., Foresti, R., Mann, B.E., Johnson, T.R., Green, C.J., and Motterlini, R. (Jul 2005). Carbon monoxide-releasing molecules (CO-RMs) attenuate the inflammatory response elicited by lipopolysaccharide in RAW264.7 murine macrophages. *Br. J. Pharmacol.* 145 (6): 800–810.

95. Zhang, X., Shan, P., Alam, J., Fu, X.Y., and Lee, P.J. (Mar 11 2005). Carbon monoxide differentially modulates STAT1 and STAT3 and inhibits apoptosis via a phosphatidylinositol 3-kinase/Akt and p38 kinase-dependent STAT3 pathway during anoxia-reoxygenation injury. *J. Biol. Chem.* 280 (10): 8714–8721.

96. Ferreira, A., Marguti, I., Bechmann, I., Jeney, V., Chora, A., Palha, N.R., Rebelo, S., Henri, A., Beuzard, Y., and Soares, M.P. (2011). Sickle hemoglobin confers tolerance to Plasmodium infection. *Cell* 145: 398–409.

97. Wang, X.M., Kim, H.P., Nakahira, K., Ryter, S.W., and Choi, A.M. (Mar 15 2009). The heme oxygenase-1/carbon monoxide pathway suppresses TLR4 signaling by regulating the interaction of TLR4 with caveolin-1. *J. Immunol.* 182 (6): 3809–3818.

98. Rocuts, F., Ma, Y., Zhang, X., Gao, W., Yue, Y., Vartanian, T., and Wang, H. (Aug 15 2010). Carbon monoxide suppresses membrane expression of TLR4 via myeloid differentiation factor-2 in betaTC3 cells. *J. Immunol.* 185 (4): 2134–2139.

99. Riquelme, S.A. 1., Bueno, S.M., and Kalergis, A.M. (Feb 2015). Carbon monoxide down-modulates Toll-like receptor 4/MD2 expression on innate immune cells and reduces endotoxic shock susceptibility. *Immunology* 144 (2): 321–332.

100. Jung, S.S., Moon, J.S., Xu, J.F., Ifedigbo, E., Ryter, S.W., Choi, A.M., and Nakahira, K. (May 15 2015). Carbon monoxide negatively regulates NLRP3 inflammasome activation in macrophages. *Am. J. Physiol. Lung Cell. Mol. Physiol.* 308 (10): L1058–67.

101. Wang, P., Huang, J., Li, Y., Chang, R., Wu, H., Lin, J., and Huang, Z. (Aug 31 2015). Exogenous carbon monoxide decreases sepsis-induced acute kidney injury and inhibits NLRP3 inflammasome activation in rats. *Int.J. Mol. Sci.* 16 (9): 20595–20608.
102. Jiang, L., Fei, D., Gong, R., Yang, W., Yu, W., Pan, S., Zhao, M., and Zhao, M. (Nov 2016). CORM-2 inhibits TXNIP/NLRP3 inflammasome pathway in LPS-induced acute lung injury. *Inflamm. Res.* 65 (11): 905–915.
103. Zhang, W., Tao, A., Lan, T., Cepinskas, G., Kao, R., Martin, C.M., and Rui, T. (Mar 2017). Carbon monoxide releasing molecule-3 improves myocardial function in mice with sepsis by inhibiting NLRP3 inflammasome activation in cardiac fibroblasts. *Basic Res. Cardiol.* 112 (2): 16.
104. Meng, F. and Alayash, A.I. (Mar 15 2017). Determination of extinction coefficients of human hemoglobin in various redox states. *Anal Biochem.* 521: 11–19. Epub 2017 Jan 6. doi: 10.1016/j.ab.2017.01.002.
105. Kassa, T., Jana, S., Meng, F., and Alayash, A.I. (Aug 8 2016). Differential heme release from various hemoglobin redox states and the upregulation of cellular heme oxygenase-1. *FEBS Open Bio.* 6 (9): 876–884.
106. Nakahira, K., Kim, H.P., Geng, X.H., Nakao, A., Wang, X., Murase, N., Drain, P.F., Wang, X., Sasidhar, M., Nabel, E.G., Takahashi, T., Lukacs, N.W., Ryter, S.W., Morita, K., and Choi, A.M. (Oct 2 2006). Carbon monoxide differentially inhibits TLR signaling pathways by regulating ROS-induced trafficking of TLRs to lipid rafts. *J. Exp. Med.* 203 (10): 2377–2389.
107. Bories, G.F.P., Yeudall, S., Serbulea, V., Fox, T.E., Isakson, B.E., and Leitinger, N. (2020). Macrophage metabolic adaptation to heme detoxification involves CO-dependent activation of the pentose phosphate pathway. *Blood* 136: 1535–1548.
108. Metere, A., Iorio, E., Scorza, G., Camerini, S., Casella, M., Crescenzi, M., Minetti, M., and Pietraforte, D. (2014). Carbon monoxide signaling in human red blood cells: Evidence for pentose phosphate pathway activation and protein deglutathionylation. *Antioxid. Redox Signal.* 20 (3).
109. Vandorpe, D.H. 1., Xu, C., Shmukler, B.E., Otterbein, L.E., Trudel, M., Sachs, F., Gottlieb, P.A., Brugnara, C., and Alper, S.L. (Jan 15 2010). Hypoxia activates a Ca2+-permeable cation conductance channel sensitive to CO and to GsMTx-4 in human and mouse sickle erythrocytes. *PLoS One* 5 (1): e8732.
110. Wegiel, B., Gallo, D.J., Raman, K.G., Karlsson, J.M., Ozanich, B., Chin, B.Y., Tzeng, E., Ahmad, S., Ahmed, A., Baty, C.J., and Otterbein, L.E. (Feb 2 2010). Nitric oxide-dependent bone marrow progenitor mobilization by carbon monoxide enhances endothelial repair after vascular injury. *Circulation* 121 (4): 537–548.
111. Zuckerbraun, B.S., Chin, B.Y., Wegiel, B., Billiar, T.R., Czsimadia, E., Rao, J., Shimoda, L., Ifedigbo, E., Kanno, S., and Otterbein, L.E. (Sep 4 2006). Carbon monoxide reverses established pulmonary hypertension. *J. Exp. Med.* 203 (9): 2109–2119.
112. Liu, X.M., Chapman, G.B., Peyton, K.J., Schafer, A.I., and Durante, W. (Aug 1 2002). Carbon monoxide inhibits apoptosis in vascular smooth muscle cells. *Cardiovasc. Res.* 55 (2): 396–405.
113. Li, M.H., Cha, Y.N., and Surh, Y.J. (Apr 14 2006). Carbon monoxide protects PC12 cells from peroxynitrite-induced apoptotic death by preventing the depolarization of mitochondrial transmembrane potential. *Biochem. Biophys. Res. Commun.* 342 (3): 984–990.
114. Kim, H.S., Loughran, P.A., Kim, P.K., Billiar, T.R., and Zuckerbraun, B.S. (Jun 16 2006). Carbon monoxide protects hepatocytes from TNF-alpha/Actinomycin D by inhibition of the caspase-8-mediated apoptotic pathway. *Biochem. Biophys. Res. Commun.* 344 (4): 1172–1178.
115. Wang, X., Wang, Y., Kim, H.P., Nakahira, K., Ryter, S.W., and Choi, A.M. (Jan 19 2007). Carbon monoxide protects against hyperoxia-induced endothelial cell apoptosis by inhibiting reactive oxygen species formation. *J. Biol. Chem.* 282 (3): 1718–1726.
116. US National Library of Medicine. September 2020. Retrieved from https://clinicaltrials.gov.
117. Howard, J., Thein, S.L., Galactéros, F., Inati, A., Reid, M.E., Keipert, P.E., Small, T., and Booth, F. (2013). Safety and tolerability Of MP4CO: A dose escalation study in stable patients with sickle cell disease. *Blood* 122 (21): 2205.
118. Buontempo, P., Jubin, R., Buontempo, C., Real, R., Kazo, F., and O'Brien, S. (2017). Adeel F and Abuchowski A. Pegylated carboxyhemoglobin bovine (SANGUINATE®) Restores RBCs roundness and reduces pain during a sickle cell vaso-occlusive crisis. *Blood* 130: 969.
119. Natanson, C., Kern, S.J., Lurie, P., Banks, S.M., and Wolfe, S.M. (2008). Cell-free hemoglobin-based blood substitutes and risk of myocardial infarction and death: a meta-analysis. *JAMA* 299 (19): 2304–2312. May 21.
120. Alayash, A.I. (Jan 4 2017). Hemoglobin-based blood substitutes and the treatment of sickle cell disease: More harm than help? *Biomolecules* 7 (1).

29

CO and Pain Management

Olga Pol[1,2]

[1]*Grup de Neurofarmacologia Molecular, Institut d'Investigació Biomèdica Sant Pau, Hospital de la Santa Creu i Sant Pau, 08041 Barcelona, Spain*
[2]*Grup de Neurofarmacologia Molecular, Institut de Neurociències, Universitat Autònoma de Barcelona, 08193 Barcelona, Spain*

Introduction

Pain is a multifaceted and complex pathology that is a worldwide health problem not only from the medical perspective but also from social and economic perspectives [1]. Acute pain is an essential survival mechanism since it alerts to potential dangers, providing chances to avoid and to prevent damage [2]. However, when pain becomes persistent, it causes great suffering in patients and presents an important clinical problem that imposes enormous economic and social burdens [3]. Chronic pain is defined as a syndrome that is characterized by persistent suffering for long periods of time [4,5]. Chronic pain can originate and manifest in different forms, including diabetic neuropathy, osteoarthritic pain, inflammatory pain, or neuropathic pain, and affects a high percentage of people [6].

Chronic pain is also commonly accompanied by several emotional disorders, such as depression, anxiety, and cognitive deficits, which exert a negative impact on the perception of pain and lead to a significant deterioration in the quality of life, which can even induce suicidal behaviors [7–9]. The management of chronic pain is an unmet medical need and a major challenge in pain research because the present therapies have limited effectiveness and important side effects [5,10–13].

Numerous studies have evaluated the anti-inflammatory and antinociceptive effects of carbon monoxide (CO) as well as those produced by nuclear factor-2 erythroid factor-2 (Nrf2) transcription factor or inducible heme oxygenase (HO-1) activators in several animal pain models [14–24]. In this chapter, we summarize the data regarding the antinociceptive actions of CO and of the Nrf2 or HO-1 activators and the mechanisms implicated in their effects during chronic pain. The interaction of CO, Nrf2, and HO-1 activators with opioids, cannabinoids, and other painkilling drugs during chronic pain has also been reviewed.

CO and Pain Treatment

Previous studies demonstrated that CO gas inhalation reduces acute pain and inflammatory pain caused by the subplantar injection of prostaglandin E_2 (PGE_2) [25] or formalin [26–28]. Arthritic and neuropathic pain can also be inhibited after treatment with unspecific HO inducers, such as hemin [29–31]. Similarly, central, spinal, or local injection of heme lysinate also diminishes the acute inflammatory pain provoked by PGE_2 [25], interleukin (IL)-1β [25], or formalin [26,27] and acute thermal pain [32]. These studies demonstrate that the stimulation of the HO/CO pathway can protect against acute and chronic pain.

Considering the great promise of CO as a therapeutic agent, its use requires a pharmaceutical formulation for delivery in a safe and controllable manner to avoid the direct inhalation of CO and to control the dose of gas administered [33]. Some types of CO donors are available and classified into different categories, such as metal-based CO-releasing molecules (CO-RMs), organic CO prodrugs, CO in solution, and CO immobilized in modified hemoglobin [34–36]. CO-RMs are an effective alternative to inhaled gaseous CO since they can be administered orally or systematically and allow for controlled release of CO [33], and consequently, numerous types of CO-RMs have been developed, such as tricarbonyldichlororuthenium(II) dimer (CORM-2),

Carbon Monoxide in Drug Discovery: Basics, Pharmacology, and Therapeutic Potential, First Edition. Edited by Binghe Wang and Leo E. Otterbein.
© 2022 John Wiley & Sons, Inc. Published 2022 by John Wiley & Sons, Inc.

tricarbonylchloro(glycinato)ruthenium(II) (CORM-3), $Mn_2(CO)_{10}$ (CORM-1), $Na_2(H_3BCO_2)$ (CORM-A1), ALF492, ALF186, or CORM-401 [37,38]. In pain studies, the most utilized CO-RM is CORM-2, which has been tested in different preclinical pain models, including neuropathic [16,18], inflammatory [17], and osteoarthritic [39] pain and in diabetic neuropathy [40].

Recent studies have identified new compounds that release CO in an even more controlled manner than CORM-2, such as the new series of small-molecule hybrids that possess a carbonic anhydrase inhibitor (CAI) warhead linked to a CO-RM tail (CAI-CORMs) [41]; CORM-2 loaded in solid lipid nanoparticles (SLNs; CORM-2-SLNs) that causes a slow release of CO with a half-life 50 times longer than that of CORM-2 [42]; CO-RMs that liberate CO under irradiation, including unsaturated cyclic diketones, xanthene carboxylic acids, *meso*-carboxy BODIPYs, and hydroxyflavone, which produce potent anti-inflammatory effects [43]; and HYCOs, which are hybrid compounds that consist of a CO-RM conjugated to an Nrf2/HO-1 activator and generate potent anti-inflammatory effects in skin wounds, psoriasis, and multiple sclerosis [44–48].

Another class of CO donors includes metal-free organic CO prodrugs in which carbonyls in the form of amides, ketones, carboxylic acids, oxalyl groups, and esters, among others, are the precursors for CO [35]. Organic prodrugs allow for a diverse set of chemistry for triggered release, cell/tissue targeting, tunable pharmacokinetic properties, and addressing toxicity issues by the design of "carrier" molecules. The activation of CO release is regulated by certain physiological conditions, such as water, enzymes, reduction/oxidation, or enrichment in specific cells and organelles, and photoinduced conditions [34,35].

Effects of CO in Inflammatory and Arthritic Pain

Inflammatory pain is commonly associated with tissue damage and the resulting inflammatory process characterized by the presence of allodynia (pain response to normally innocuous stimuli), hyperalgesia (increased response to painful stimuli), and spontaneous pain [49,50]. Inflammatory pain results from the production of several proinflammatory mediators, such as IL-1, IL-6, tumor necrosis factor alpha (TNF-α), inducible nitric oxide synthase (NOS2), PGE_2, and bradykinin, which sensitize nociceptors and induce allodynia and the hypersensitivity state that accompanies inflammation [51]. Osteoarthritis is a chronic degenerative joint disorder characterized by the destruction of articular cartilage, causing subchondral bone alterations, inflammation, and intense pain [52]. Chronic osteoarthritis pain is characterized by persistent pain with inflammatory and neuropathic components that cause physical inability to perform daily tasks, including difficulty walking, among others [53]. Chronic osteoarthritis pain is also accompanied by affective disorders, such as anxiety and depression [54], and therapies to treat it are limited, with low effectiveness and important adverse effects.

The intraperitoneal, intrathecal, and/or oral administration of CO-RMs, such as CORM-2 and/or CORM-3, inhibits acute thermal pain [55] and inflammatory pain provoked by the administration of formalin or complete Freund's adjuvant (CFA) (Figure 29.1) [17,39]. These treatments also inhibit inflammatory and arthritic pain [14,17,19,39,56,57]. Interestingly, CAI-CORMs, which release CO in a more controlled form, induce greater pain-relieving effects in rheumatoid arthritis, enhancing the pain-relieving effects of CO-RMs and demonstrating

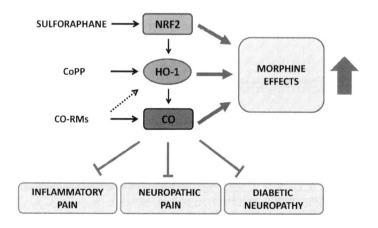

Figure 29.1 The antinociceptive actions of CO and of the Nrf2 or HO-1 activators and their interaction with μ-opioid receptor (MOR) agonists during chronic pain. Activation of the Nrf2 transcription factor by sulforaphane, HO-1 by cobalt protoporphyrin IX (CoPP), or CO by CO-RMs inhibits inflammatory and neuropathic pain induced by nerve injury or associated with diabetes and further potentiates the local analgesic effects of the MOR agonist, morphine.

that the slow release of CO produced by CAI-CORMs is a very good option to relieve chronic arthritic pain [41].

Effects of CO in Neuropathic Pain

Neuropathic pain originating from nerve injury or associated with several pathologies, such as diabetes, cancer, and multiple sclerosis, provokes peripheral and central sensitization [58,59]. Neuropathic pain is also characterized by the presence of allodynia, hyperalgesia, and spontaneous pain [60]. In addition to sensory dysfunction, chronic nerve injury also evokes behavioral disabilities, such as depressive-like and anxiety-like behaviors, disrupted social interactions, and sleep disturbances [61–63]. This type of pain is difficult to treat with the most potent analgesic compounds, such as opioids, especially MOR agonists, which have reduced efficacy in this type of pain [64], or other common analgesics, including antiepileptics (e.g., gabapentin or pregabalin) and antidepressants (e.g., amitriptyline or duloxetine), which are relatively ineffective in neuropathic pain and have important side effects [65]. Thus, an important challenge in the current research of pain is the investigation of optimal treatments for this type of pain.

Several studies have demonstrated that acute intrathecal and intraperitoneal administration of CORM-2 alleviates the neuropathic pain symptoms induced by chronic constriction of the sciatic nerve (CCI) [66], partial sciatic nerve ligation (PSNL) [39], and spared nerve injury (SNI) [67]. Importantly, complete blockade of neuropathic pain was obtained with the repetitive administration of CORM-2 in animals with CCI-induced neuropathic pain [18,31,59,68] or SNL [20,69] as well as CORM-3, which also blocked the allodynia and hyperalgesia caused by CCI (Figure 29.1) [18].

The administration of CORM-2-SLNs capable of slow release of CO further potentiates the antiallodynic and antihyperalgesic effects of CORM-2 in CCI-induced neuropathic pain [42], supporting the enhanced effects induced by CAI-CORMs in arthritic pain [41] and confirming the idea that the administration of slow-releasing CO compounds is a good strategy for chronic pain treatment. However, on several occasions, it could be difficult to distinguish whether CO-RM effects are produced by the release of CO gas or any radical intermediate formed during CO release from ruthenium (Ru)-based CORM-2 or CORM-3, as occurs in the inhibition of diverse snake and lizard venoms induced by CORM-2 [70–72] and with the antimicrobial and cytotoxic properties of CORM-3 [73]. CORM-2 also activates nonselective cation currents in endothelial cells independent of CO release [74]. The lack of analgesic effects induced by inactive CO-RMs (iCORM-2 and iCORM-3) in mice with neuropathic pain suggested that the antinociceptive effects produced by CORM-2 and CORM-3 in these animals are mainly mediated by CO released by these compounds [16].

Effects of CO in Diabetic Neuropathy

Painful diabetic neuropathy is described as a tingling or burning sensation that is accompanied by allodynia and hyperalgesia [75] and is one of the most common complications of diabetes in humans that greatly diminishes quality of life [76]. However, despite being an important problem for diabetic patients, current treatments are not fully effective [77]. The search for new alternatives for improving painful neuropathy represents a major challenge in pain research. Only a few studies have evaluated the effects of CO on diabetic neuropathy. These studies demonstrated that while the acute administration of CORM-2 only mitigates diabetic neuropathy [40,78], repeated administration of this compound completely inhibited the allodynia and hyperalgesia manifested in type 1 diabetic mice (Figure 29.1) [79]. Thus, similar to chronic inflammatory and neuropathic pain induced by nerve injury, repetitive treatment with CORM-2 also inhibits neuropathic pain associated with type 1 diabetes. More experiments are required to evaluate the effects of this treatment in different preclinical models of type 2 diabetes.

The Activation of Nrf2/HO-1 Signaling Pathway in Pain Modulation

Nrf2 Activators in Chronic Pain

The activation of the Nrf2 transcription factor is a defense mechanism against oxidative stress by activating the transcription of several genes involved in redox balance, detoxification, and inflammation, such as HO-1, NAD(P)H:quinone oxidoreductase 1 (NQO1), superoxide dismutase (SOD), glutathione peroxidase, glutathione reductase, glutathione S-transferase (GST), glutamate cysteine ligase, and glucose 6-phosphate dehydrogenase [80]. Oxidative stress is a key mechanism implicated in the development of chronic pain. Reactive oxygen species (ROS) are elevated in the central and peripheral nervous systems of animals with chronic inflammatory or neuropathic pain [81] and ROS scavengers alleviate hyperalgesia [82]. Moreover, the potential role of antioxidants in pain alleviation has been demonstrated with treatment with Nrf2 activators [21,23].

The Nrf2 activator sulforaphane inhibits nitroglycerin-induced hyperalgesia and acute and chronic inflammatory pain provoked by formalin and CFA, respectively [15,21,83]. Treatment with other Nrf2 activators, such as oltipraz, plumbagin, and dimethyl fumarate, also inhibits neuropathic pain caused by nerve injury [84–87], chemotherapeutic medications [88], or accompanying diabetes in rodents (Figure 29.1) [22,23]. These studies demonstrate that Nrf2 activation modulates the proinflammatory and pronociceptive responses generated by peripheral inflammation, nerve injury, chemotherapeutic agents, or hyperglycemia and is a good target for pain treatment.

HO-1 Activators in Chronic Pain

The acute administration of the HO-1 inducer CoPP only alleviated the allodynia and/or hyperalgesia incited by CCI [89], PSNL [39], or SNI [67], while chronic treatment with CoPP completely blocked the development of neuropathic pain originating from CCI or SNL (Figure 29.1) [18,20,59]. Interestingly, the intraspinal injection of a recombinant lentivirus overexpressing HO-1 also suppressed neuropathic pain induced by spinal nerve ligation [20] and alleviated vincristine-induced neuropathic pain in rodents [90]. Similar to inflammatory and neuropathic pain induced by nerve injury, painful diabetic neuropathy associated with type 2 or type 1 diabetes [91,92] was also inhibited by CoPP treatment. These results demonstrate that the activation of the Nrf2/HO-1/CO pathway is an important therapeutic target against different types of pain.

Mechanisms Implicated in the Antinociceptive Effects of CO

CO-Releasing Compounds

The mechanisms implicated in the analgesic actions of CO during chronic pain have been principally studied by assessing the effects of CO-RMs, specifically CORM-2. Intraperitoneal administration of CORM-2 activates the expression of HO-1 in paws [17] and loci coerulei of animals with peripheral inflammatory pain [19] as well as in the dorsal root ganglia of animals with CCI-induced neuropathic pain [16,18]. This treatment also modulates the oxidative stress induced by nerve injury in the spinal cord and specific brain zones, for example, the amygdala, prefrontal cortex, hippocampus, and/or the hypothalamus, as demonstrated by the normalization of the decreased expression of HO-1 manifested in these specific areas [59]. CORM-2 treatment also increases the protein levels of HO-1 in the spinal cords, dorsal root ganglia, and sciatic nerves of type 1 diabetic mice [40,92]. Thus, the powerful antioxidant and defensive role that CORM-2 plays in chronic pain conditions is mainly mediated by enhancing the synthesis of the antioxidant enzyme HO-1. Although more studies are required, an increase in the nuclear levels of Nrf2 and/or in the transfer of Nrf2 from the cytoplasm to the nucleus is a probable mechanism to explain the increased expression of HO-1 produced by CORM-2 in animals with neuropathic pain [59], as happens with CORM-A1 in hepatic cells [93].

The inhibitory actions of CORM-2 during neuropathic pain were highly potentiated with CORM-2-SLNs [42] whose improved actions were in part mediated by the greater increased expression of HO-1 and the full blockade of HO-2 produced by CORM-2-SLNs in the spinal cords of animals with CCI-induced neuropathic pain [42]. The development and maintenance of inflammatory or neuropathic pain triggered by nerve injury or linked with diabetes also involve the activation of glial cells as well as the induction of several plasticity changes in different nociceptive pathways [85,94–96]. Thus, the analgesic effects induced by CORM-2 in inflammatory pain are also mediated by the inhibition of activated astroglia caused by peripheral inflammation in the locus coeruleus [19]. In neuropathic pain, CORM-2 and/or CORM-2-SLNs block the microglial and/or astroglial activation triggered by CCI in the spinal cords, amygdalas, and/or hippocampi of these animals [18,42,59,66,68,84], thus contributing to the analgesic action produced by CO-RMs.

Nitric oxide also mediates numerous inflammatory and neuropathic pain symptoms [97–99]. Peripheral inflammation increases the expression of neuronal (NOS1) and/or inducible (NOS2) nitric oxide synthases in the dorsal root ganglia, spinal cord, and locus coeruleus [19,100–103], and CORM-2 normalizes the high levels of NOS1 in the dorsal root ganglia and loci coerulei of these animals [17,19]. In the same way, CORM-2 inhibited the increased expression of

NOS1 and NOS2 induced by nerve injury [16,18] or was linked with type 1 diabetes [40,92,104] in the spinal cord and sciatic nerves. Interestingly, CORM-2-SLN is more potent than CORM-2 in inhibiting the NOS1 and NOS2 upregulation provoked by CCI in the spinal cord and dorsal root ganglia [42], whose effects also contribute to the enhancement of the antiallodynic and antihyperalgesic actions of CORM-2-SLN as compared with CORM-2 in neuropathic pain.

Mitogen-activated protein kinases (MAPKs) are another pathway implicated in the nociceptive effects caused by inflammation and nerve injury. Accordingly, peripheral inflammation provokes the activation of Jun N-terminal kinase (JNK), extracellular signal-regulated kinase 1/2 (ERK 1/2), and/or P38 in the spinal cord and locus coeruleus [19,21,105–107]. Nerve injury also induces the upregulation of p-ERK 1/2 in the amygdala, prefrontal cortex, hippocampus, and spinal cord [59,66,84,85,108]. CORM-2 is capable of inhibiting the overexpression of p-ERK 1/2 in the amygdala and prefrontal cortex but not in the hippocampi, spinal cords, and/or dorsal root ganglia of rodents with neuropathic pain [59,66]. ERK 1/2 phosphorylation induced by peripheral inflammation in the locus coeruleus was not reversed by CORM-2 treatment [19]. The enhanced levels of p-JNK caused by sciatic nerve injury in the spinal cord, amygdala, and hippocampus [59,84,85,105] were further inhibited by CORM-2 in the spinal cord [90] and amygdala but not in the hippocampus [59]. In the same way, CORM-2 also diminished the upregulation of p-P38 in the dorsal root ganglia, prefrontal cortices, and hippocampi of sciatic nerve-injured animals but not in spinal cords [59,66,90]. Thus, the phosphorylation of ERK, JNK, or P38 induced by nerve injury in different tissues is not always modulated by CORM-2, revealing that the effects of this treatment vary according to the tissue [59].

Exogenous administration of CO further prevents local cytokine production triggered by inflammation [109,110], and the upregulation of IL-1β, IL-18, IL-6, and TNF-α induced by nerve injury in the spinal cord [66]. CORM-2-SLNs, but not CORM-2, potentiate the synthesis of anti-inflammatory IL-10 [42,66] and produced a higher inhibition of the spinal cord levels of TNF-α than those produced by CORM-2 in sciatic nerve-injured animals [42].

The activated phosphoinositide 3-kinase (PI3K)/phosphoprotein kinase B (p-Akt) signaling pathway is also involved in neuropathic pain and is further blocked by CORM-2 in the spinal cord and dorsal root ganglia [66]. CORM-2 treatment is also effective in inhibiting the upregulation of purinergic ionotropic subtype 4 (P2X4) receptors in microglia of the dorsal horn, whose action also contributes to the antinociceptive properties of CORM-2 in neuropathic pain [66,111,112].

Nrf2 Activators

The principal mechanisms implicated in the inhibition of nociception induced by Nrf2 transcription factor activation in animals with acute and chronic pain are the augmentation and/or regularization of the expression of HO-1, NQO1, GST, or SOD, together with the inhibition of NOS2 overexpression, microglial activation, and MAPK phosphorylation in the spinal cord, dorsal root ganglia, and/or paw tissues [15,21,83,86]. That is, Nrf2 activators modulate inflammatory responses by reducing the synthesis of NOS2, cyclooxygenase-2, and several cytokines, such as TNF-α, IL-1β, and/or IL-6 [21,24,86,113,114], as well as the oxidative stress provoked by nerve injury by normalizing the decreased levels of HO-1 in the spinal cords and several brain areas of animals with neuropathic pain [84,85]. The induction of Nrf2 also potentiates the synthesis of HO-1, restores the decreased levels of NQO1, and normalizes the NOS2 and/or COX-2 overexpression provoked by hyperglycemia in the sciatic nerves of diabetic mice [22,23]. These data demonstrate that the inhibition of the synthesis of several inflammatory mediators and the activation of the antioxidant enzymes HO-1 and/or NQO1 are two important mechanisms used by Nrf2 activators to inhibit chronic inflammatory and neuropathic pain [21,85] as well as diabetic neuropathy [22,23].

The microglial activation provoked by nerve injury in the spinal cord and/or hippocampus is also blocked by sulforaphane and/or oltipraz [84,85]. Both treatments further inhibit the phosphorylation of JNK, ERK 1/2, and/or P38 in the spinal cord of animals with inflammatory or neuropathic pain [21,23,84] as well as the activation of PI3K/p-Akt and p-IKBα in the spinal cord of sciatic nerve-injured mice [84]. Dimethyl fumarate, another Nrf2 activator, upregulates the expression of hydroxyl carboxylic acid receptor type 2 in neuropathic pain [115] and enhances the expression of Nrf2, HO-1, and Mn-SOD in a model of nitroglycerin-induced migraine in mice, demonstrating the important role of Nrf2 in the reduction of oxidative stress related to migraine [116].

In summary, the protective role played by the Nrf2 transcription factor during chronic pain is mediated by inhibiting the inflammation, oxidative stress, and plasticity changes induced by injury.

HO-1 Activators

Similar to Nrf2 activators, treatment with CoPP also avoids the decreased expression of HO-1 in the spinal cord and several brain areas implicated in the modulation of neuropathic pain [59]. CoPP further potentiates the expression of HO-1 in the peripheral nervous system of animals with CCI-induced neuropathic pain [16,18] or type 1 diabetes [40,92]. CoPP further augments the expression of HO-1 in the dorsal root ganglia and loci coerulei of animals with CFA-induced inflammatory pain [19,56] and HO-1 released by a lentivirus further enhances the expression of HO-1 in the spinal cords of rodents with vincristine-induced neuropathic pain [90].

NQO1 is one of the most important proteins that protects cells from oxidative stress induced by diabetes, as verified by the increased hyperglycemia and higher ROS levels observed in NQO1-KO diabetic animals [117]. The reduced expression of this coenzyme in the sciatic nerves of type 2 diabetic animals was completely reversed by CoPP, showing the involvement of NQO1 in the analgesic activities induced by CoPP in diabetic animals [91].

The contribution of microglia to the development of diabetic neuropathy has been demonstrated [118], with the inhibition of neuropathic pain associated with diabetes induced by minocycline, an inhibitor of microglia [119,120]. CoPP also blocks the microglial activation induced by diabetes [92], and the enhanced expression of astrocytes and microglia provoked by chemotherapeutic agents [90], as well as the activated microglia caused by CCI in the spinal cord [16] and the astroglial activation caused by CFA in the locus coeruleus [19].

The systemic administration of CoPP also normalizes the upregulation of NOS1 induced by peripheral inflammation in the locus coeruleus [19] and those of NOS1 and NOS2 provoked by nerve injury or associated with type 1 diabetes in the spinal cord and sciatic nerves [16,18,92,104]. This HO-1 inducer also normalized ERK 1/2 phosphorylation in the loci coerulei of animals with peripheral inflammation [19] and in the amygdalas and prefrontal cortices but not in the hippocampi, spinal cords, and/or dorsal root ganglia of animals with CCI-induced neuropathic pain [59]. CoPP is further capable of preventing ERK 1/2 phosphorylation in the spinal cords of animals with chemotherapy-induced neuropathic pain [90] showing the different effects induced by this drug in accordance with the type of pain.

Regarding JNK, CoPP inhibits its phosphorylation in the spinal cord [90] and/or the amygdala [59], but not in the hippocampus of animals with neuropathic pain [59]. CoPP further decreases the upregulation of p-P38 in the dorsal root ganglia, prefrontal cortices, hippocampi, and spinal cords of sciatic nerve-injured animals [59,66,90]. These data revealed the different effects induced by CoPP in the regulation of MAPK during inflammatory and neuropathic pain.

The Role of Ion Channels in HO-1/CO-Induced Antinociception

In recent years, ion channels have been recognized as important effectors in the actions of CO under normal and pathological circumstances [121–123]. The ability of CO to regulate different classes of ion channels, including K^+ activated by calcium [124,125], voltage-activated K^+ (Kv) channels [126,127], L-type Ca^{2+} channels [128–130], tandem P domain K^+ channels (TREK1) [131], or the epithelial Na^+ channels [132], is crucial for several functions exerted by CO, such as the regulation of vasodilatation, neuronal excitability, contractility, apoptosis, etc. [122].

The contribution of some Kv channels and sodium channels (NaV) in chronic pain disorders is well demonstrated. Treatment with NaV modulators [133,134] and that with Kv activators [135–137] are two strategies used for chronic pain management. Considering that CORM-3 inhibits NaV channels in alveolar epithelial cells [132] and activates Kv channels in the endothelium of cirrhotic animals [138], the possible contribution of NaV and/or Kv channels to the inhibition of chronic pain elicited by CO-RMs is a possible mechanism of action of these compounds to investigate further.

Recent studies have also demonstrated that TREK1 is involved in many physiological and pathological processes, such as neuroprotection and pain [139]. Pain is also modulated by CO-RMs or HO-1 inducers [140], but the possible involvement of TREK1 in their analgesic effects remains unexplored.

The participation of ATP-sensitive K^+ channels in HO/CO pathway-induced antinociception during inflammatory and arthritic pain has also been demonstrated with the reversion of the analgesic activity of hemin [141] and dimethyl decarbonate with the administration of glibenclamide (an ATP-sensitive K^+ channel inhibitor) [30]. No evidence of the contribution of ATP-sensitive K^+ channels to the antinociceptive effects induced by specific HO-1 activators or CO-RMs during neuropathic pain has been demonstrated.

Interaction Between CO and Analgesics in Pain

Interaction Between CO Donors and Nrf2–HO-1 Activators with Opioids

The administration of MOR agonists, such as morphine and fentanyl, produces potent analgesic effects during inflammatory but not in neuropathic pain caused by nerve injury [142]. Moreover, the administration of high doses or the repetitive administration of MOR agonists is accompanied by abundant undesirable effects, such as constipation, tolerance, and respiratory depression [143–146]. Thus, the search for new options to potentiate their effectiveness in neuropathic pain is crucial. Several studies have examined the role played by the HO-1/CO signaling pathway in the analgesic effects of MOR agonists in chronic pain, revealing that treatment with CoPP, CORM-2, and/or CORM-3 improved the local pain-relieving effects of morphine in acute pain induced by a thermal stimulus [55], chronic CFA-induced inflammatory pain [56] (Figure 29.2), and neuropathic pain provoked by CCI [16] or PSNL (Figure 29.3) [39]. CORM-2 also potentiates the antinociceptive actions produced by the systemic injection of morphine in animals with CCI-induced neuropathic pain [68] and CoPP improves the inhibition of the allodynia and hyperalgesia induced by morphine in type 1 diabetic mice [92].

In addition, δ-opioid receptors (DORs) are another type of opioid receptor highly implicated in the analgesic properties of opioids [147–150]. DOR agonists are less effective in inhibiting inflammatory pain than MOR agonists, but they relieve both inflammatory and neuropathic pain with comparable efficacy and few adverse effects [56,142,145,151–153]. The administration of CoPP and/or CORM-2 enhanced the antiallodynic and antihyperalgesic effects of [D-Pen2,5]-enkephalin hydrate (DPDPE), a DOR agonist, in inflammatory pain [56] and in type 1 diabetes [78]. In animals with neuropathic pain induced by

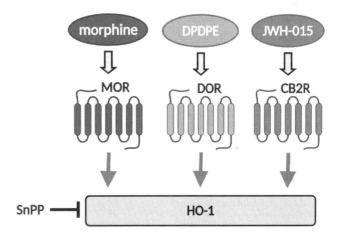

Figure 29.2 The role of HO-1 in the analgesic effects of MOR, DOR, and CB2R agonists in inflammatory pain and diabetic neuropathy. In inflammatory pain or diabetic neuropathy, the local administration of morphine (a MOR agonist), DPDPE (a DOR agonist), or JWH-015 (a CB2R agonist) might induce their analgesic effects by activating HO-1 enzyme. The local pain-relieving effects of MOR, DOR, and CB2R agonists were reversed with the HO-1 inhibitor, SnPP.

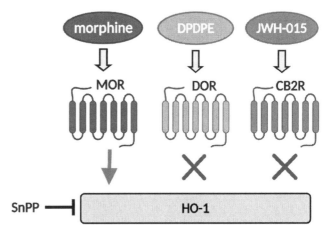

Figure 29.3 The role of HO-1 in the analgesic effects of MOR, DOR, and CB2R agonists in sciatic nerve injury-induced neuropathic pain. In sciatic nerve injury-induced neuropathic pain, the local administration of morphine (a MOR agonist), but not DPDPE (a DOR agonist) or JWH-015 (a CB2R agonist), might induce its analgesic effects by activating the HO-1 enzyme. The local pain-relieving effects of MOR, but not DOR or CB2R agonists, were reversed with the HO-1 inhibitor, SnPP.

CCI, the administration of CO delivered exogenously (CORM-2) or endogenously produced by HO-1 (CoPP) decreases the local antinociceptive effects of DPDPE [16]. Moreover, whereas the local pain-relieving effects of DOR agonists in inflammatory pain or diabetic neuropathy were reversed with the HO-1 inhibitor tin protoporphyrin IX (SnPP) (Figure 29.2) [56,78], this did not take place in CCI-induced neuropathic pain (Figure 29.3) [16]. These studies showed the different mechanisms of action implicated in the analgesic effects of DOR agonists in inflammatory, diabetic, and nerve injury-induced neuropathic pain. In this way, in inflammatory pain and diabetic neuropathy, both MOR and DOR agonists might activate the HO-1 pathway to induce their effects, similar to that which happens with MOR in CCI-induced neuropathic pain but not with DOR, revealing that HO-1 does not seem to play a significant role in the analgesic actions of DOR in animals with nerve injury-induced neuropathic pain [16]. CORM-2 and/or CoPP treatment also increases and/or normalizes the expression of MOR in the dorsal root ganglia of animals with inflammatory pain [56], CCI-induced neuropathic pain [16], or diabetic neuropathy [92], as well as the expression of DOR in inflammatory pain [56]. In contrast, these treatments did not change the protein levels of DOR in mice with nerve injury-induced neuropathic pain [16], thus supporting the idea that other mechanisms of action than HO-1 activation are implicated in the analgesic actions of DOR agonists under nerve injury-induced neuropathic pain.

The power of the Nrf2 transcription factor to potentiate the antinociceptive effects of opioids during inflammatory and neuropathic pain has also been shown [21,23,85]. Sulforaphane improves the local antinociceptive effects of morphine in preclinical models of inflammatory [21] and neuropathic pain [85] as well as those of DOR agonists such as H-Dmt-Tic-NH-CH(CH$_2$-COOH)-Bid in inflammatory pain [105] and those of DPDPE and (+)-4-[(αR) α ((2S-,5R)-4-allyl-2,5-dimethyl-1-piperazinyl)-3-methoxybenzyl]-N,N-diethylbenzamide in diabetic neuropathy [23]. The enhanced effects of morphine induced by sulforaphane during inflammatory pain might be explained by the upregulation of MOR induced by this Nrf2 activator in paw tissues [21], while in CCI-induced neuropathic pain, sulforaphane normalized the downregulation of MOR in the spinal cords and hippocampi of these animals [85]. In type 2 diabetic mice, sulforaphane also prevented the downregulation of DOR in the sciatic nerve [23].

In view of the copious adverse effects caused by the chronic administration of MOR agonists [142,154,155], the potentiation of their analgesic effects through their coadministration with CORM-2 and HO-1 or Nrf2 activators represents a new approach for the management of inflammatory pain, nerve injury-induced neuropathic pain, and diabetic neuropathy. In addition, considering the diminished side effects induced by DOR agonists, treatment with HO-1 and Nrf2 activators combined with DOR agonists is another interesting approach for improving chronic inflammatory pain and diabetic neuropathy [140].

Interaction Between CO Releasers and Nrf2–HO-1 Activators with Cannabinoids

The endogenous cannabinoid system is highly involved in the modulation of acute and chronic pain, and treatment with cannabinoid receptor 1 (CB1R) or 2 (CB2R) agonists inhibits different types of pain [40,103,156–158]. Since the activation of CB2R inhibits nociception with fewer adverse reactions than CB1R [156–160], several works assessed the analgesic properties of specific CB2R agonists, such as 2-methyl-1-propyl-1H-indol-3-yl)-1-naphthalenylmethanone (JWH-015) and (6aR,10aR)-3-(1,1-dimethylbutyl)-6a,7,10,10a-tetrahydro-6,6,9-trimethyl-6H-dibenzo[b,d]pyran (JWH-133), in animals with inflammatory or neuropathic pain [40,91,158,161–163].

The interaction of CO, HO-1, and Nrf2 with CB2R in chronic pain has also been demonstrated. First, CO delivered exogenously or endogenously by HO-1 increased the local antiallodynic and antihyperalgesic effects of JWH-015 in inflammatory pain [56] or neuropathic pain accompanying type 1 diabetes [40] and those produced by JWH-015 and JWH-133 in type 2 diabetic mice [91]. In contrast, CORM-2 and CoPP treatments decreased the effects of JWH-015 in animals with CCI-induced neuropathic pain [16], revealing the differential effects induced by CO and/or HO-1 in the improvement of pain induced by CB2R agonists according to the type of pain. In accordance, the improvements in the analgesic effects of JWH-015 induced by CoPP during inflammatory pain are mainly mediated by activating the synthesis of endogenous opioid peptides [56]. CORM-2 and CoPP did not augment the expression of CB2R in sciatic nerve-injured mice [16], but their expression increased in the sciatic nerves of type 1 and type 2 diabetic mice treated with CORM-2 and/or CoPP [40,91]. These effects also correlated with the fact that SnPP, an HO-1 inhibitor, blocked only the effects of JWH-015 and/or JWH-133 in inflammatory and diabetic pain conditions (Figures 29.2 and 29.3), suggesting, as occurs with DOR, that the enzyme HO-1 only participates in the effects of CB2R alleviating pain caused by peripheral inflammation or associated with diabetes. The increased expression of CB2R triggered by CORM-2 and/or CoPP in type 1 and type 2 diabetic mice might explain the enhanced antinociceptive effects of JWH-015 [40].

Interaction Between CO/HO-1 and Other Drugs

CoPP, CORM-2, or CORM-3 also enhances the analgesic actions of other drugs. Thus, CoPP and CORM-2 treatments improve the effects of pregabalin and gabapentin in animals with neuropathic pain by increasing the expression of HO-1 and Ca^{2+} channel α2δ1 subunit gene and inhibiting microglial activation and the upregulation of TNF-α in the spinal cord [67]. The analgesic actions of buprenorphine during neuropathic pain were also intensified by CORM-2 treatment, and the effects were mainly due to the inhibition of glial activation via P2X4 receptors [68]. Moreover, treatment with CORM-3 further increases the antiallodynic and antihyperalgesic effect of *Harpagophytum procumbens* during inflammatory pain [57].

In addition to these studies, it is well known that several drugs inhibit pain by activating the Nrf2, HO-1, and/or NQO1 signaling pathway. For example, rosiglitazone, a selective agonist of PPARγ [164], and 5-fluoro-2-oxindole, an oxindole [165], inhibit neuropathic pain; 7-deacetylgedunin, a limonoid [166], some sulfated polysaccharides [167], flavanones [168], oxindoles [169], and phenolic compounds [170] alleviate inflammatory pain; and allylpyrocatechol, a phytoconstituent of *Piper betle* leaves [171], and quercetin, a flavonoid [172], attenuate arthritis, among others.

More studies are required to demonstrate the potential improvement in the analgesic effects of several drugs when combined with CO and HO-1 or Nrf2 activators.

Conclusions and Perspectives

The potent analgesic effects induced by the activation of the Nrf2/HO-1/CO signaling pathway in different types of chronic pain, such as inflammatory and neuropathic pain provoked by nerve injury or associated with diabetes, have been established. Interestingly, new compounds that release CO in a more controlled manner (CAI-CORMs) produce stronger improvements in pain than classic CO-RMs in arthritis- or nerve injury-induced neuropathic pain. Nevertheless, more research is needed to validate the potential improvement made by CAI-CORMs and/or other new HO-1 or Nrf2 activators in other preclinical pain models. Novel formulations that permit the release of CO in combination with other specific drugs at specific sites must also be developed.

The possible mechanisms implicated in the analgesic activities of CO-RMs and HO-1 or Nrf2 activators during chronic pain have been described. In inflammatory pain, these compounds mediate their analgesic effects mainly by activating the synthesis of antioxidant enzymes and inhibiting the inflammatory and plasticity changes induced by inflammation in the peripheral and/or central nervous system. The inhibition of MAPK, NF-κB, and PI3K/Akt phosphorylation and glial cell activation induced by nerve injury, chemotherapeutic agents, or hyperglycemia and the activation of the endogenous antioxidant system are the principal factors responsible for the analgesic effects of CO-RMs and HO-1 or Nrf2 activators during neuropathic pain.

New strategies to potentiate the efficacy of morphine under neuropathic pain conditions with its coadministration with CO-RMs, HO-1, or Nrf2 activators have been demonstrated. The endogenous activation of HO-1 or Nrf2 and the exogenous induction of CO also enhance the analgesic properties of MOR, DOR, and CB2R agonists during inflammatory pain and diabetic neuropathy, representing significant progress in the use of opioids and/or cannabinoids for their management. Nonetheless, additional studies evaluating the effects of more potent HO-1 inducers or Nrf2 activators, or of new compounds that combine CO-RMs with these activators, are fundamental to obtain new treatments capable of alleviating chronic pain with few side effects.

Acknowledgments

This work was supported by Ministerio de Economía y Competitividad, Instituto de Salud Carlos III and Fondo Europeo de Desarrollo Regional, Unión Europea (grants PS0900968 and PI1400927), Instituto de Salud Carlos III and Comunitat Autònoma de Catalunya (grant 983161), Fundació La Marató de TV3, Barcelona, Spain (grant 070810), and Ministerio de Ciencia, Innovación y Universidades, Instituto de Salud Carlos III and Fondo Europeo de Desarrollo Regional, Unión Europea (grant PI1800645).

References

1. Monti, S. and Caporali, R. (2015). *Reumatismo* 67: 35–44.

2. Bell, A. (2018). *Vet. J.* 237: 55–62.
3. Goldberg, D.S. and McGee, S.J. (2011). *BMC Public Health* 11: 770.
4. Dale, R. and Stacey, B. (2016). *Med. Clin. North Am.* 100: 55–64.
5. Hylands-White, N., Duarte, R.V., and Raphael, J.H. (2017). *Rheumatol. Int.* 37: 29–42.
6. Dydyk, A.M. and Grandhe, S. (2020). StatPearls. Treasure Island (FL): StatPearls Publishing.
7. Maletic, V. and Raison, C.L. (2009). *Front. Biosci.* 14: 5291–5338.
8. Attal, N., Lanteri-Minet, M., Laurent, B., Fermanian, J., and Bouhassira, D. (2011). *Pain* 152: 2836–2843.
9. Kawai, K., Kawai, A.T., Wollan, P., and Yawn, B.P. (2017). *Fam. Pract.* 34: 656–661.
10. Angst, M.S. and Clark, J.D. (2006). *Anesthesiology* 104: 570–587.
11. Atkinson, T.J. and Fudin, J. (2020). *Phys. Med. Rehabil. Clin. N. Am.* 31: 219–231.
12. Uddin, M.S., Mamun, A.A., Rahman, M.A., Kabir, M.T., Alkahtani, S., Alanazi, I.S., Perveen, A., Ashraf, G.M., Bin-Jumah, M.N., and Abdel-Daim, M.M. (2020). *Front. Neurosci.* 14: 478.
13. Beal, B.R. and Wallac, E.M.S. (2016). *Med. Clin. North Am.* 100: 65–79.
14. Ferrándiz, M.L., Maicas, N., Garcia-Arnandis, I., Terencio, M.C., Motterlini, R., Devesa, I., Joosten, L.A., Van Den Berg, W.B., and Alcaraz, M.J. (2008). *Ann. Rheum. Dis.* 67: 1211–1217.
15. Rosa, A.O., Egea, J., Lorrio, S., Rojo, A.I., Cuadrado, A., and López, M.G. (2008). *Pain* 137: 332–339.
16. Hervera, A., Leánez, S., Motterlini, R., and Pol, O. (2013). *Anesthesiology* 118: 1180–1197.
17. Negrete, R., Hervera, A., Leánez, S., and Pol, O. (2014). *Psychopharmacology* 231: 853–861.
18. Hervera, A., Leánez, S., Negrete, R., Motterlini, R., and Pol, O. (2012). *PLoS One* 7: e43693.
19. Moreno, P., Cazuza, R.A., Mendes-Gomes, J., Díaz, A.F., Polo, S., Leánez, S., Leite-Panissi, C., and Pol, O. (2019). *Int. J. Mol. Sci.* 20: 2211.
20. Liu, X., Zhang, Z., Cheng, Z., Zhang, J., Xu, S., Liu, H., Jia, H., and Jin, Y. (2016). *Pain Med.* 17: 220–229.
21. Redondo, A., Chamorro, P.A.F., Riego, G., Leánez, S., and Pol, O. (2017). *J. Pharmacol. Exp. Ther.* 363: 293–302.
22. Negi, G., Kumar, A., and Sharma, S.S. (2011). *Curr. Neurovasc. Res.* 8: 294–304.
23. McDonnell, C., Leánez, S., and Pol, O. (2017). *PLoS One* 12: e0180998.
24. Wang, C. and Wang, C. (2017). *Inflammopharmacology* 25: 99–106.
25. Steiner, A.A., Branco, L.G., Cunha, F.Q., and Ferreira, S.H. (2000). *Br. J. Pharmacol.* 132: 1673–1682.
26. Nascimento, C.G. and Branco, L.G. (2007). *Eur. J. Pharmacol.* 556: 55–61.
27. Nascimento, C.G. and Branco, L.G. (2008). *Eur. J. Pharmacol.* 581: 71–76.
28. Nascimento, C.G. and Branco, L.G. (2009). *Braz. J. Med. Biol. Res.* 42: 141–147.
29. Kaur, S., Bijjem, K.R., and Sharma, P.L. (2011). *Inflammopharmacology* 19: 265–272.
30. Chaves, H.V., Do Val, D.R., Ribeiro, K.A., Lemos, J.C., Souza, R.B., Gomes, F., Da Cunha, R., De Paulo Teixeira Pinto, V., Filho, G.C., De Souza, M., Bezerra, M.M., and De Castro Brito, G.A. (2018). *Inflamm. Res.* 67: 407–422.
31. Bijjem, K.R., Padi, S.S., and Lal Sharma, P. (2013). *Naunyn Schmiedebergs Arch. Pharmacol.* 386: 79–90.
32. Carvalho, P.G., Branco, L.G., and Panissi, C.R. (2011). *Brain Res.* 1385: 107–113.
33. Motterlini, R. and Foresti, R. (2017). *Am. J. Physiol. Cell Physiol.* 312: C302–C313.
34. Ji, X., Damera, K., Zheng, Y., Yu, B., Otterbein, L.E., and Wang, B. (2016). *J. Pharm. Sci.* 105: 406–416.
35. Yang, X.X., Ke, B.W., Lu, W., and Wang, B.H. (2020). *Chin. J. Nat. Med.* 18: 284–295.
36. Motterlini, R., Mann, B.E., and Foresti, R. (2005). *Expert Opin. Investig. Drugs* 14: 1305–1318.
37. Pena, A.C., Penacho, N., Mancio-Silva, L., Neres, R., Seixas, J.D., Fernandes, A.C., Romão, C.C., Mota, M.M., Bernardes, G.J., and Pamplona, A. (2012). *Antimicrob. Agents Chemother.* 56: 1281–1290.
38. Fayad-Kobeissi, S., Ratovonantenaina, J., Dabiré, H., Wilson, J.L., Rodriguez, A.M., Berdeaux, A., Dubois-Randé, J.L., Mann, B.E., Motterlini, R., and Foresti, R. (2016). *Biochem. Pharmacol.* 102: 64–77.
39. Hervera, A., Gou, G., Leánez, S., and Pol, O. (2013). *Psychopharmacology* 228: 463–477.
40. Castany, S., Carcolé, M., Leánez, S., and Pol, O. (2016). *Psychopharmacology* 233: 2209–2219.
41. Berrino, E., Milazzo, L., Micheli, L., Vullo, D., Angeli, A., Bozdag, M., Nocentini, A., Menicatti, M., Bartolucci, G., Di Cesare Mannelli, L., Ghelardini, C., Supuran, C.T., and Carta, F.J. (2019). *Med. Chem.* 62: 7233–7249.
42. Joshi, H.P., Kim, S.B., Kim, S., Kumar, H., Jo, M.J., Choi, H., Kim, J., Kyung, J.W., Sohn, S., Kim, K.T., Kim, J.K., and Han, I.B. (2019). *Mol. Neurobiol.* 56: 5539–5554.
43. Abeyrathna, N., Washington, K., Bashur, C., and Liao, Y. (2017). *Org. Biomol. Chem.* 15: 8692–8699.
44. Wilson, J.L., Fayad Kobeissi, S., Oudir, S., Haas, B., Michel, B., Dubois Randé, J.L., Ollivier, A., Martens, T., Rivard, M., Motterlini, R., and Foresti, R. (2014). *Chemistry* 20: 14698–14704.
45. Nikam, A., Ollivier, A., Rivard, M., Wilson, J.L., Mebarki, K., Martens, T., Dubois-Randé, J.L.,

Motterlini, R., and Foresti, R. (2016). *J. Med. Chem.* 59: 756–762.

46. Motterlini, R., Nikam, A., Manin, S., Ollivier, A., Wilson, J.L., Djouadi, S., Muchova, L., Martens, T., Rivard, M., and Foresti, R. (2019). *Redox Biol.* 20: 334–348.

47. Ollivier, A., Foresti, R., El Ali, Z., Martens, T., Kitagishi, H., Motterlini, R., and Rivard, M. (2019). *ChemMedChem* 14: 1684–1691.

48. El Ali, Z., Ollivier, A., Manin, S., Rivard, M., Motterlini, R., and Foresti, R. (2020). *Redox Biol.* 34: 101521.

49. Marchand, F., Perretti, M., and McMahon, S.B. (2005). Role of the immune system in chronic pain. *Nat. Rev. Neurosci.* 6: 521–532.

50. Xu, Q. and Yaksh, T.L. (2011). *Curr. Opin. Anaesthesiol.* 24: 400–407.

51. Kidd, B.L. and Urban, L.A. (2001). *Br. J. Anaesth.* 87: 3–11.

52. Alcaraz, M.J., Megías, J., García-Arnandis, I., Clérigues, V., and Guillén, M.I. (2010). *Biochem. Pharmacol.* 80: 13–21.

53. Eitner, A., Hofmann, G.O., and Schaible, H.G. (2017). *Front. Mol. Neurosci.* 10: 349.

54. Sharma, A., Kudesia, P., Shi, Q., and Gandhi, R. (2016). *Open Access Rheumatol.* 8: 103–113.

55. Gou, G., Leánez, S., and Pol, O. (2014). *Eur. J. Pharmacol.* 737: 41–46.

56. Carcolé, M., Castany, S., Leánez, S., and Pol, O. (2014). *J. Pharmacol. Exp. Ther.* 351: 224–232.

57. Parenti, C., Aricò, G., Chiechio, S., Di Benedetto, G., Parenti, R., and Scoto, G.M. (2015). *Molecules* 20: 16758–16769.

58. Ji, R.R. and Suter, M.R. (2007). *Mol. Pain.* 3: 33.

59. Riego, G., Redondo, A., Leánez, S., and Pol, O. (2018). *Biochem. Pharmacol.* 148: 52–63.

60. Austin, P.J., Wu, A., and Moalem-Taylor, G. (2012). *J. Vis. Exp.* 61: 3393.

61. Monassi, C.R., Bandler, R., and Keay, K.A. (2003). *Eur. J. Neurosci.* 17: 1907–1920.

62. Roeska, K., Doods, H., Arndt, K., Treede, R.D., and Ceci, A. (2008). *Pain* 139: 349–357.

63. Colloca, L., Ludman, T., Bouhassira, D., Baron, R., Dickenson, A.H., Yarnitsky, D., Freeman, R., Truini, A., Attal, N., Finnerup, N.B., Eccleston, C., Kalso, E., Bennett, D.L., Dworkin, R.H., and Raja, S.N. (2017). *Nat. Rev. Dis. Primers.* 3: 17002.

64. Pilat, D., Piotrowska, A., Rojewska, E., Jurga, A., Ślusarczyk, J., Makuch, W., Basta-Kaim, A., Przewlocka, B., and Mika, J. (2016). *Mol. Cell. Neurosci.* 71: 114–124.

65. Selvy, M., Cuménal, M., Kerckhove, N., Courteix, C., Busserolles, J., and Balayssac, D. (2020). *Expert Opin. Drug Saf.* 19: 707–733.

66. Jurga, A.M., Piotrowska, A., Makuch, W., Przewlocka, B., and Mika, J. (2017). *Front. Pharmacol.* 8: 48.

67. Godai, K. and Kanmura, Y. (2018). *Pain Rep.* 3: e677.

68. Jurga, A.M., Piotrowska, A., Starnowska, J., Rojewska, E., Makuch, W., and Mika, J. (2016). *Pharmacol. Rep.* 68: 206–213.

69. Wang, H. and Sun, X. (2017). *J. Mol. Neurosci.* 63: 58–69.

70. Gessner, G., Sahoo, N., Swain, S.M., Hirth, G., Schönherr, R., Mede, R., Westerhausen, M., Brewitz, H.H., Heimer, P., Imhof, D., Hoshi, T., and Heinemann, S.H. (2017). *Eur. J. Pharmacol.* 815: 33–41.

71. Nielsen, V.G. (2020). *J. Thromb. Thrombolysis* 49: 100–107.

72. Nielsen, V.G. (2020). *Int. J. Mol. Sci.* 21: 2970.

73. Southam, H.M., Smith, T.W., Lyon, R.L., Liao, C., Trevitt, C.R., Middlemiss, L.A., Cox, F.L., Chapman, J.A., El-Khamisy, S.F., Hippler, M., Williamson, M.P., Henderson, P., and Poole, R.K. (2018). *Redox Biol.* 18: 114–123.

74. Dong, D.L., Chen, C., Huang, W., Chen, Y., Zhang, X.L., Li, Z., Li, Y., and Yang, B.F. (2008). *Eur. J. Pharmacol.* 590: 99–104.

75. Schreiber, A.K., Nones, C.F., Reis, R.C., Chichorro, J.G., and Cunha, J.M. (2015). *World J. Diabetes* 6: 432–444.

76. Vincent, A.M., Callaghan, B.C., Smith, A.L., and Feldman, E.L. (2011). *Nat. Rev. Neurol.* 7: 573–583.

77. Peltier, A., Goutman, S.A., and Callaghan, B.C. (2014). *BMJ* 348: g1799.

78. Castany, S., Carcolé, M., Leánez, S., and Pol, O. (2016). *Neurosci. Lett.* 614: 49–54.

79. Méndez-Lara, K.A., Santos, D., Farré, N., Ruiz-Nogales, S., Leánez, S., Sánchez-Quesada, J.L., Zapico, E., Lerma, E., Escolà-Gil, J.C., Blanco-Vaca, F., Martín-Campos, J.M., Julve, J., and Pol, O. (2018). *PLoS One* 13: e0204841.

80. Cuadrado, A., Manda, G., Hassan, A., Alcaraz, M.J., Barbas, C., Daiber, A., Ghezzi, P., León, R., López, M.G., Oliva, B., Pajares, M., Rojo, A.I., Robledinos-Antón, N., Valverde, A.M., Guney, E., and Schmidt, H.H.H.W. (2018). *Pharmacol. Rev.* 70: 348–383.

81. Nishio, N., Taniguchi, W., Sugimura, Y.K., Takiguchi, N., Yamanaka, M., Kiyoyuki, Y., Yamada, H., Miyazaki, N., Yoshida, M., and Nakatsuka, T. (2013). *Neuroscience* 247: 201–212.

82. Kim, H.K., Park, S.K., Zhou, J.L., Taglialatela, G., Chung, K., Coggeshall, R.E., and Chung, J.M. (2004). *Pain* 111: 116–124.

83. Di, W., Shi, X., Lv, H., Liu, J., Zhang, H., Li, Z., and Fang, Y.J. (2016). *Headache Pain* 17: 99.

84. Díaz, A.F., Polo, S., Gallardo, N., Leánez, S., and Pol, O. (2019). *J. Clin. Med.* 8: 890.

85. Ferreira-Chamorro, P., Redondo, A., Riego, G., Leánez, S., and Pol, O. (2018). *Front. Pharmacol.* 9: 1332.

86. Li, J., Ma, J., Lacagnina, M.J., Lorca, S., Odem, M.A., Walters, E.T., Kavelaars, A., and Grace, P.M. (2020). *Anesthesiology* 132: 343–356.

87. Arruri, V., Komirishetty, P., Areti, A., Dungavath, S.K.N., and Kumar, A. (2017). *Pharmacol. Rep.* 69: 625–632.

88. Singh, J., Saha, L., Singh, N., Kumari, P., Bhatia, A., and Chakrabarti, A. (2019). *J. Pharm. Pharmacol.* 71: 797–805.

89. Chen, Y., Chen, H., Xie, K., Liu, L., Li, Y., Yu, Y., and Wang, G. (2015). *Inflammation* 38: 1835–1846.

90. Shen, Y., Zhang, Z.J., Zhu, M.D., Jiang, B.C., Yang, T., and Gao, Y. (2015). *J. Neurobiol. Dis.* 79: 100–110.

91. McDonnell, C., Leánez, S., and Pol, O. (2017). *Int. J. Mol. Sci.* 18: 2268.

92. Castany, S., Carcolé, M., Leánez, S., and Pol, O. (2016). *PLoS One* 11: e0146427.

93. Upadhyay, K.K., Jadeja, R.N., Vyas, H.S., Pandya, B., Joshi, A., Vohra, A., Thounaojam, M.C., Martin, P.M., Bartoli, M., and Devkar, R.V. (2020). *Redox Biol.* 28: 101314.

94. Raghavendra, V., Tanga, F.Y., and DeLeo, J.A. (2004). *Eur. J. Neurosci.* 20: 467–473.

95. Mika, J., Osikowicz, M., Rojewska, E., Korostynski, M., Wawrzczak-Bargiela, A., Przewlocki, R., and Przewlocka, B. (2009). *Eur. J. Pharmacol.* 623: 65–72.

96. Ji, R.R., Nackley, A., Huh, Y., Terrando, N., and Maixner, W. (2018). *Anesthesiology* 129: 343–366.

97. De Alba, J., Clayton, N.M., Collins, S.D., Colthup, P., Chessell, I., and Knowles, R.G. (2006). *Pain* 120: 170–181.

98. La Buda, C.J., Koblish, M., Tuthill, P., Dolle, R.E., and Little, P.J. (2006). *Eur. J. Pain* 10: 505–512.

99. Schmidtko, A., Gao, W., König, P., Heine, S., Motterlini, R., Ruth, P., Schlossmann, J., Koesling, D., Niederberger, E., Tegeder, I., Friebe, A., and Geisslinger, G. (2008). *J. Neurosci.* 28: 8568–8576.

100. Chu, Y.C., Guan, Y., Skinner, J., Raja, S.N., Johns, R.A., and Tao, Y.X. (2005). *Pain* 119: 113–123.

101. Boettger, M.K., Uceyler, N., Zelenka, M., Schmitt, A., Reif, A., Chen, Y., and Sommer, C. (2007). *Eur. J. Pain* 11: 810–818.

102. Chen, Y., Boettger, M.K., Reif, A., Schmitt, A., Uçeyler, N., and Sommer, C. (2010). *Mol. Pain* 6: 13.

103. Negrete, R., Hervera, A., Leánez, S., Martín-Campos, J.M., and Pol, O. (2011). *PLoS One* 6: e26688.

104. Negi, G., Nakkina, V., Kamble, P., and Sharma, S.S. (2015). *Pharmacol. Res.* 102: 158–167.

105. Polo, S., Díaz, A.F., Gallardo, N., Leánez, S., Balboni, G., and Pol, O. (2019). *Front. Pharmacol.* 10: 283.

106. Gao, Y.J. and Ji, R.R. (2008). *Neurosci. Lett.* 437: 180–183.

107. Ji, R.R., Gereau, R.W.I.V., Malcangio, M., and Strichartz, G.R. (2009). *Brain Res. Rev.* 60: 135–148.

108. Rojewska, E., Popiolek-Barczyk, K., Kolosowska, N., Piotrowska, A., Zychowska, M., Makuch, W., Przewlocka, B., and Mika, J. (2015). *PLoS One* 10: e0138583.

109. Guillén, M.I., Megías, J., Clérigues, V., Gomar, F., and Alcaraz, M.J. (2008). *Rheumatology* 47: 1323–1328.

110. Motterlini, R., Haas, B., and Foresti, R. (2012). *Med. Gas Res.* 2: 28.

111. Wilkinson, W.J. and Kemp, P.J. (2011). *Purinergic Signal.* 7: 57–64.

112. Tsuda, M., Masuda, T., Tozaki-Saitoh, H., and Inoue, K. (2013). *Front. Cell Neurosci.* 7: 191.

113. Kim, H.A., Yeo, Y., Kim, W.U., and Kim, S. (2009). *Rheumatology* 48: 932–938.

114. Davidson, R.K., Jupp, O., De Ferrars, R., Kay, C.D., Culley, K.L., Norton, R., Driscoll, C., Vincent, T.L., Donell, S.T., Bao, Y., and Clark, I.M. (2013). *Arthritis Rheum.* 65: 3130–3140.

115. Boccella, S., Guida, F., De Logu, F., De Gregorio, D., Mazzitelli, M., Belardo, C., Iannotta, M., Serra, N., Nassini, R., De Novellis, V., Geppetti, P., Maione, S., and Luongo, L. (2019). *FASEB J.* 33: 1062–1073.

116. Casili, G., Lanza, M., Filippone, A., Campolo, M., Paterniti, I., Cuzzocrea, S., and Esposito, E. (2020). *J. Neuroinflammation* 17: 59.

117. Yeo, S.H., Noh, J.R., Kim, Y.H., Gang, G.T., Kim, S.W., Kim, K.S., Hwang, J.H., Shong, M., and Lee, C.H. (2013). *Toxicol. Lett.* 219: 35–41.

118. Wang, D., Couture, R., and Hong, Y. (2014). *Eur. J. Pharmacol.* 728: 59–66.

119. Ismail, C.A.N., Suppian, R., Aziz, C.B.A., and Long, I. (2019). *J. Diabetes Metab. Disord.* 18: 181–190.

120. Pabreja, K., Dua, K., Sharma, S., Padi, S.S., and Kulkarni, S.K. (2011). *Eur. J. Pharmacol.* 661: 15–21.

121. Peers, C., Dallas, M.L., and Scragg, J.L. (2009). *Commun. Integr. Biol.* 2: 241–242.

122. Peers, C. (2011). *Exp. Physiol.* 96: 836–839.

123. Wilkinson, W.J. and Kemp, P.J. (2011). *J. Physiol.* 589: 3055–3062.

124. Jaggar, J.H., Li, A.L., Parfenova, H., Liu, J.X., Umstot, E.S., Dopico, A.M., and Leffler, C.W. (2005). *Circ. Res.* 97: 805–812.

125. Telezhkin, V., Brazier, S.P., Mears, R., Müller, C.T., Riccardi, D., and Kemp, P.J. (2011). *Pflugers Arch.* 461: 665–675.

126. Dallas, M.L., Boyle, J.P., Milligan, C.J., Sayer, R., Kerrigan, T.L., McKinstry, C., Lu, P., Mankouri, J.,

Harris, M., Scragg, J.L., Pearson, H.A., and Peers, C. (2011). *FASEB J.* 25: 1519–1530.

127. Al-Owais, M.M., Hettiarachchi, N.T., Boyle, J.P., Scragg, J.L., Elies, J., Dallas, M.L., Lippiat, J.D., Steele, D.S., and Peers, C. (2017). *Cell Death Dis.* 8: e3163.

128. Lim, I., Gibbons, S.J., Lyford, G.L., Miller, S.M., Strege, P.R., Sarr, M.G., Chatterjee, S., Szurszewski, J.H., Shah, V.H., and Farrugia, G. (2005). *Am. J. Physiol. Gastrointest. Liver Physiol.* 288: G7–14.

129. Scragg, J.L., Dallas, M.L., Wilkinson, J.A., Varadi, G., and Peers, C. (2008). *J. Biol. Chem.* 283: 24412–24419.

130. Duckles, H., Al-Owais, M.M., Elies, J., Johnson, E., Boycott, H.E., Dallas, M.L., Porter, K.E., Boyle, J.P., Scragg, J.L., and Peers, C. (2015). *Adv. Exp. Med. Biol.* 860: 291–300.

131. Dallas, M.L., Scragg, J.L., and Peers, C. (2008). *Neuroreport* 19;: 345–348.

132. Althaus, M., Fronius, M., Buchäckert, Y., Vadász, I., Clauss, W.G., Seeger, W., Motterlini, R., and Morty, R.E. (2009). *Am. J. Respir. Cell Mol. Biol.* 41: 639–650.

133. Cardoso, F.C. and Lewis, R.J. (2018). *Br. J. Pharmacol.* 175: 2138–2157.

134. Dib-Hajj, S.D. and Waxman, S.G. (2019). *Annu. Rev. Neurosci.* 42: 87–106.

135. Busserolles, J., Tsantoulas, C., Eschalier, A., and López García, J.A. (2016). *Pain* 157: S7–S14.

136. Abd-Elsayed, A., Jackson, M., Gu, S.L., Fiala, K., and Gu, J. (2019). *Mol. Pain* 15: 1744806919864256.

137. Zhang, J., Rong, L., Shao, J., Zhang, Y., Liu, Y., Zhao, S., Li, L., Yu, W., Zhang, M., Ren, X., Zhao, Q., Zhu, C., Luo, H., Zang, W., and Cao, J. (2020). *J. Neurochem.* 10.1111/jnc.15117.

138. Bolognesi, M., Sacerdoti, D., Piva, A., Di Pascoli, M., Zampieri, F., Quarta, S., Motterlini, R., Angeli, P., Merkel, C., and Gatta, A. (2007). *J. Pharmacol. Exp. Ther.* 321: 187–194.

139. Djillani, A., Mazella, J., Heurteaux, C., and Borsotto, M. (2019). *Front. Pharmacol.* 10: 379.

140. Pol, O. (2021). *Med. Res. Rev.* 41: 136–155.

141. Pereira De Ávila, M.A., Giusti-Paiva, A., and Giovani De Oliveira Nascimento, C. (2014). *Eur. J. Pharmacol.* 726: 41–48.

142. Obara, I., Parkitna, J.R., Korostynski, M., Makuch, W., Kaminska, D., Przewlocka, B., and Przewlocki, R. (2009). *Pain* 141: 283–291.

143. Benyamin, R., Trescot, A.M., Datta, S., Buenaventura, R., Adlaka, R., Sehgal, N., Glaser, S.E., and Vallejo, R. (2008). *Pain Physician* 11: S105–S120.

144. Law, P.Y., Reggio, P.H., and Loh, H.H. (2013). *Trends Biochem. Sci.* 38: 275–282.

145. Hervera, A., Leánez, S., and Pol, O. (2012). *Eur. J. Pharmacol.* 685: 42–51.

146. Fernández-Dueñas, V., Pol, O., García-Nogales, P., Hernández, L., Planas, E., and Puig, M.M. (2007). *J. Pharmacol. Exp. Ther.* 322: 360–368.

147. Pol, O. and Puig, M.M. (2004). *Curr. Top. Med. Chem.* 4: 51–61.

148. Pol, O. (2007). *Curr. Med. Chem.* 14: 1945–1955.

149. Stein, C. (2018). *Expert Opin. Investig. Drugs* 27: 765–775.

150. Bodnar, R.J. (2019). *Peptides* 124: 170223.

151. Conibear, A.E., Asghar, J., Hill, R., Henderson, G., Borbely, E., Tekus, V., Helyes, Z., Palandri, J., Bailey, C., Starke, I., Von Mentzer, B., Kendall, D., and Kelly, E. (2020). *J. Pharmacol. Exp. Ther.* 372: 224–236.

152. Aguila, B., Coulbault, L., Boulouard, M., Léveillé, F., Davis, A., Tóth, G., Borsodi, A., Balboni, G., Salvadori, S., Jauzac, P., and Allouche, S. (2007). *Br. J. Pharmacol.* 152: 1312–1324.

153. Martínez-Navarro, M., Cabañero, D., Wawrzczak-Bargiela, A., Robe, A., Gavériaux-Ruff, C., Kieffer, B.L., Przewlocki, R., Baños, J.E., and Maldonado, R. (2020). *Br. J. Pharmacol.* 177: 1187–1205.

154. Zychowska, M., Rojewska, E., Kreiner, G., Nalepa, I., Przewlocka, B., and Mika, J. (2013). *J. Neuroimmunol.* 262: 35–45.

155. Narita, M., Imai, S., Nakamura, A., Ozeki, A., Asato, M., Rahmadi, M., Sudo, Y., Hojo, M., Uezono, Y., Devi, L.A., Kuzumaki, N., and Suzuki, T. (2013). *Addict. Biol.* 18: 614–622.

156. Hervera, A., Negrete, R., Leánez, S., Martín-Campos, J., and Pol, O. (2010). *J. Pharmacol. Exp. Ther.* 334: 887–896.

157. Hama, A. and Sagen, J. (2011). *Brain Res.* 1412: 44–54.

158. Ikeda, H., Ikegami, M., Kai, M., Ohsawa, M., and Kamei, J. (2013). *Neuroscience* 250: 446–454.

159. Fox, A. and Bevan, S. (2005). *Expert Opin. Investig. Drugs* 14: 695–703.

160. Hsieh, G.C., Pai, M., Chandran, P., Hooker, B.A., Zhu, C.Z., Salyers, A.K., Wensink, E.J., Zhan, C., Carroll, W.A., Dart, M.J., Yao, B.B., Honore, P., and Meyer, M.D. (2011). *Br. J. Pharmacol.* 162: 428–440.

161. Vincenzi, F., Targa, M., Corciulo, C., Tabrizi, M.A., Merighi, S., Gessi, S., Saponaro, G., Baraldi, P.G., Borea, P.A., and Varani, K. (2013). *Pain* 154: 864–873.

162. Aly, E., Khajah, M.A., and Masocha, W. (2019). *Molecules* 25: 106.

163. Kumawat, V.S. and Kaur, G. (2019). *Eur. J. Pharmacol.* 862: 172628.

164. Zhou, Y.Q., Liu, D.Q., Chen, S.P., Chen, N., Sun, J., Wang, X.M., Li, D.Y., Tian, Y.K., and Ye, D.W. (2020). *Biomed. Pharmacother.* 129: 110356.

165. Ferreira-Chamorro, P., Redondo, A., Riego, G., and Pol, O. (2021). *Cell. Mol. Neurobiol.* 41: 995–1008.
166. Chen, J.Y., Zhu, G.Y., Su, X.H., Wang, R., Liu, J., Liao, K., Ren, R., Li, T., and Liu, L. (2017). *Oncotarget* 8: 55051–55063.
167. Ribeiro, N.A., Chaves, H.V., Da Conceição Rivanor, R.L., Do Val, D.R., De Assis, E.L., Silveira, F.D., Gomes, F., Freitas, H.C., Vieira, L.V., Da Silva Costa, D.V., De Castro Brito, G.A., Bezerra, M.M., and Benevides, N. (2020). *Int. J. Biol. Macromol.* 150: 253–260.
168. Afridi, R., Khan, A.U., Khalid, S., Shal, B., Rasheed, H., Ullah, M.Z., Shehzad, O., Kim, Y.S., and Khan, S. (2019). *BMC Pharmacol. Toxicol.* 20: 57.
169. Redondo, A., Riego, G., and Pol, O. (2020). *Antioxidants* 9: 1249.
170. Khalid, S., Ullah, M.Z., Khan, A.U., Afridi, R., Rasheed, H., Khan, A., Ali, H., Kim, Y.S., and Khan, S. (2018). *Front. Pharmacol.* 9: 140.
171. De, S., Manna, A., Kundu, S., De Sarkar, S., Chatterjee, U., Sen, T., Chattopadhyay, S., and Chatterjee, M. (2017). *J. Pharmacol. Exp. Ther.* 360: 249–259.
172. Borghi, S.M., Mizokami, S.S., Pinho-Ribeiro, F.A., Fattori, V., Crespigio, J., Clemente-napimoga, J.T., Napimoga, M.H., Pitol, D.L., Issa, J., Fukada, S.Y., Casagrande, R., and Verri, W.A., Jr. (2018). *J. Nutr. Biochem.* 53: 81–95.

30

Clinical Trials of Low-Dose Carbon Monoxide

Edward Gomperts[1], Andrew Gomperts[2], and Howard Levy[2]

[1]*Children's Hospital Los Angeles, Los Angeles, CA 90027, USA*
[2]*Hillhurst Biopharmaceuticals, Inc., 2029 Verdugo Blvd, Montrose, CA 91020, USA*

Introduction

The assessment of low-dose carbon monoxide (CO) as a therapeutic in the clinic is unique because of the diversity of data of the effects of low-dose CO in humans. These data include a wide range of exposure to CO, from the use of commercially available medical grade CO gas in therapeutic studies to environmental exposure to CO as a by-product of combustion (whether due to industrial activity, home heating, motor vehicle exhaust, fire, or smoking) with exposures ranging from acute, extremely high levels, as in a fire, to chronic, low levels, as from regularly smoking cigarettes. These reports have resulted in a substantial literature indicating the effects of CO in humans. This includes a large clinical and nonclinical safety literature, numerous clinical experience studies, many epidemiological studies, and even formal prospective clinical studies. Preclinical data in animals as well as a number of therapeutic studies in humans suggest protective effects of CO in a number of disease models. Moreover, the safety literature indicates that peak carboxyhemoglobin (COHb) saturation of up to ~20% is safe and tolerable [1], though there are conflicting data and many studies do not take into account critical pharmacokinetic aspects of toxicology, such as area under the curve/duration of exposure [1]. Thus, perhaps paradoxically in view of the historic perception of CO as always being toxic, there are persistent efforts to demonstrate clinically the potential therapeutic benefits of low-dose CO.

Most reported clinical therapeutic studies have been performed with commercially available CO gas, attempting to recapitulate the purported benefits of CO observed in animal studies. However, the purity of the CO gas used in some studies was not verified and some studies used CO sources such as hydrocarbon combustion, both of which confound the results due to the presence of non-CO substances/contaminants. Moreover, epidemiological data exist in various patient populations examining either smoking or environmental CO exposure. Although these epidemiological data provide some preliminary guidance as to patient populations that may benefit from low-dose CO, these data fall short of providing definitive evidence of efficacy and safety due to the known limitations of epidemiological studies. Further, while such epidemiological studies are focused on CO, there are typically other substances in breathed air or CO sources such as smoking, including other gases, chemicals, and particles, that can confound outcomes in such studies. Such substances other than CO include sulfur oxides, nitrogen oxides, volatile organic compounds, polycyclic aromatic hydrocarbons, simple hydrocarbons, trace metals, ozone, particulates, and allergens that vary in content depending on the source of CO and the environmental conditions [2].

Although the safety literature, clinical experience studies, and epidemiological studies are helpful in guiding CO drug development, the rigor of these studies and thus the reliability of the results are highly variable [1]. This leads to conflicting outcomes and difficulty with interpretation. For this reason, the focus in this chapter is on formal clinical studies with appropriately tested low-dose CO that have been conducted on a prospective basis. These include clinical studies of low-dose CO that were industry sponsored and/or listed in the ClinicalTrials.gov database with a ClinicalTrials.gov identifier (NCT number). All available industry-sponsored studies (whether or not completed and whether or not listed in the ClinicalTrials.gov database) are included in this review, and non-industry-sponsored studies that were completed or are ongoing and are included in the ClinicalTrials.gov database are also included. Table ClinicalTrials.gov30.1 includes a list of the 23 clinical studies discussed.

These clinical studies have focused on safety, pharmacokinetics, and proof-of-concept efficacy in a variety of indications and patient populations. The 23 clinical studies provide data on approximately 570 subjects, include 13 phase 1 studies, 8 phase 2 and phase 2/3 clinical studies, and 2 ongoing/planned clinical studies. To date, no pivotal phase 3 clinical study has been reported with low-dose CO. Phase 1 safety and tolerability studies have included both normal healthy volunteers (HVs) and patients with a variety of diseases. Methods of administration of low-dose CO include both (i) low-dose CO drug products with CO as the active pharmaceutical ingredient (API) and (ii) low-dose CO drug products that do not have CO itself as the API but that provide exogenous CO upon administration (designated in this chapter a "non-CO API"). The low-dose CO drug products with a non-CO API consist of a variety of classes of molecules with varying and overlapping nomenclature. These include small inorganic molecules, small organic molecules, and large molecules. These molecules have been termed CO-releasing molecules (CORMs), organic CO prodrugs, and pegylated COHb (PEG-COHb), among other designations.

Both inhaled CO gas (iCO) and an oral liquid CO drug product, designated HBI-002, are NCT-listed drug products with CO as the API. With regard to drug products with non-CO API, only intravenous drug products consisting of dilute parental PEG-COHb solutions (designated MP4CO and Sanguinate*) have been studied in the clinic. Targets of the various clinical studies include both short-term dosing for acute disease management and longer-term dosing for chronic disease management with varying safety and efficacy outcomes. Clinical results for the different API types are discussed separately because, as expected, CO API and non-CO API drug products have proven to be substantially different in absorption, distribution, metabolism, and toxicology/safety as well as achievable CO level. Importantly, CO itself is largely not metabolized, whereas non-CO APIs consist of chemical agents and/or excipients that alter and complicate the toxicity/safety profile as compared with CO itself and potentially also alter the overall efficacy profile.

For these reasons, the clinical studies described here have been reported separately with low-dose CO drug products, where

(i) CO is the API and
(ii) CO itself is not the API (a non-CO API).

Overall, with CO as the API, these clinical studies demonstrate safety and tolerability up to 13.9% peak COHb saturation, with approximately 250 subjects in total treated. Importantly, no imbalance in adverse events (AEs) as compared with untreated control subjects and no serious AEs (SAEs) were seen up to this 13.9% COHb peak saturation. Studies up to 22% peak COHb with CO as the API have also been conducted without an imbalance in AEs as compared to control, but, as expected in the study design and per the literature and as described below, headache (a study outcome) was often seen in study subjects at this higher peak COHb.

The non-CO APIs studied thus far in the clinic are PEG-COHb molecules, and these molecules have been dosed to achieve up to approximately 4.5% COHb saturation (8 ml/kg Sanguinate). Studies with these molecules, which have included a total of 320 subjects, have reported a toxicology profile similar to those reported for other cell-free hemoglobin (Hb) products (also termed Hb-based oxygen carriers). Reported AEs include cardiovascular issues (hypertension, troponin I increase, and a few myocardial infarction occurrences) as well as hematuria. These AEs are attributed to the effects of circulating free Hb, heme, and ferric iron and/or scavenging of nitric oxide (NO) by the heme moiety [3,4]. It is also important to note that these PEG-COHb molecules also provide increased oxygen delivery capacity after CO release from the PEG-Hb, which may provide additional benefit.

Regardless, the benefit–risk profile of all of these CO-based therapeutic products needs to be assessed individually for each specific drug product and in each studied patient population and therapeutic indication.

Clinical Studies with CO as the API

As summarized in Table 30.1, a substantial number of phase 1 and phase 2 clinical trials studying the safety, tolerability, dosing, and efficacy of low-dose CO in both HVs (six studies) and patients (five studies) have been reported with CO as the API (entries 1–11, Table 30.1). These clinical trials with CO as the API are described in further detail below and summarized in Table 30.2.

Table 30.1 Clinical trials with low-dose CO.

Entry	API type	Drug product	Administration	Phase	Subjects	Sponsor	Description/indication	Status	NCT/publication
1	CO	iCO	Inhaled	1	HVs	INO	Single dose safety	Completed	Mahan 2012 [5]
2	CO	iCO	Inhaled	1	HVs	INO	Repeat dose safety	Completed	Mahan 2012 [5]
3	CO	iCO	Inhaled	1	HVs	Academic	Headache/migraine	Completed	NCT02066558 (Arngrim 2017 [6])
4	CO	iCO	Inhaled	1	HVs	Academic	Headache/migraine	Completed	NCT03385174
5	CO	iCO	Inhaled	1	HVs	Academic	Lung inflammation prevention	Completed	NCT00094406 (Narute 2014 [7])
6	CO	iCO	Inhaled	1	HVs	Academic	Pulmonary blood vessel function	Completed	NCT03067701
7	CO	iCO	Inhaled	1	Patients	Academic	Migraine	Completed	NCT03075020 (Ghanizada 2018 [8])
8	CO	iCO	Inhaled	1	Patients	Academic	ARDS	Completed	NCT02425579 (Fredenbergh 2018 [9])
9	CO	iCO	Inhaled	2	Patients	INO	Kidney transplant – DGF	Withdrawn	NCT00531856
10	CO	iCO	Inhaled	2	Patients	Academic	COPD	Completed	NCT00122694 (Bathoorn 2007 [10])
11	CO	iCO	Inhaled	2	Patients	Academic	IPF	Completed	NCT01214187 (Rosas 2017 [11])
12	CO	iCO	Inhaled	2	Patients	Academic	ARDS	Ongoing	NCT03799874
13	CO	HBI-002	Oral	1	HVs	Hillhurst	Single and repeat dose safety	Planned	NCT03926819
14	Non-CO	PEG-COHb	IV	1	HVs	Prolong	Single dose safety	Completed	Misra 2014 [12]
15	Non-CO	PEG-COHb	IV	1	Patients	Prolong	End-stage renal disease	Completed	NCT02437422 (Abu Jawdeh 2017 [13])
16	Non-CO	PEG-COHb	IV	1	Patients	Prolong	Severe anemia	Completed	NCT02754999
17	Non-CO	PEG-COHb	IV	1	Patients	Prolong	SCD	Completed	NCT01848925 (Misra 2016 [14])
18	Non-CO	PEG-COHb	IV	1	Patients	Sangart	SCD	Completed	NCT01356485 (Howard 2013 [15])
19	Non-CO	PEG-COHb	IV	2	Patients	Prolong	SCD leg ulcer	Completed	NCT02600390 (Valentino 2017 [16])
20	Non-CO	PEG-COHb	IV	2	Patients	Prolong	SCD VOC (ex-United States)	Completed	NCT02672540
21	Non-CO	PEG-COHb	IV	2	Patients	Prolong	SCD VOC (United States)	Completed	NCT02411708
22	Non-CO	PEG-COHb	IV	2	Patients	Prolong	Subarachnoid hemorrhage	Completed	NCT02323685 (Dhar 2017 [17])
23	Non-CO	PEG-COHb	IV	2/3	Patients	Prolong	Kidney transplant – DGF	Completed	NCT02490202

Table 30.2 Clinical trials with CO as API.

Entry	Phase	Subjects	Drug product	Administration	Sponsor	Description/indication	No. of subjects	Dosing	Outcomes	NCT/publication
1	1	HVs	iCO	Inhaled	Ikaria/INO	Single dose safety	32	Increasing dose to 8.8% peak COHb	All doses well tolerated; no significant difference in AEs	Mahan 2012 [5]
2	1	HVs	iCO	Inhaled	Ikaria/INO	Repeat dose safety	12	Increasing dose to 13.9% peak COHb daily for 10 days	All doses well tolerated; no significant difference in AEs	Mahan 2012 [5]
3	1	HVs	iCO	Inhaled	Academic	Headache/migraine	12	Single dose resulting in 22% COHb peak	Mild headache more frequent in test subjects; no significant difference in AEs	NCT02066558 (Arngrim 2017 [6])
4	1	HVs	iCO	Inhaled	Academic	Headache/migraine	15	NR	NR	NCT03385174
5	1	HVs	iCO	Inhaled	Academic	Lung inflammation prevention	36	100 ppm CO gas for 6 h	Altered miRNA expression after LPS-induced lung inflammation	NCT00094406 (Narute 2014 [7])
6	1	HVs	iCO	Inhaled	Academic	Pulmonary blood vessel function	9	1000 ppm CO gas for 30–40 min	NR	NCT03067701
7	1	Patients	iCO	Inhaled	Academic	Migraine	12	Single dose resulting in 22% COHb peak	No difference in migraine occurrence; mild headache more frequent in test subjects; no significant difference in AEs	NCT03075020 (Ghanizada 2018 [8])
8	1	Patients	iCO	Inhaled	Academic	ARDS	12	100 or 200 ppm CO for 90 min (reaching 3.5% and 4.9% COHb peak, respectively) for 1–5 days	No imbalance in AEs or SAEs; no related SAE. Significant reduction in circulating mitochondrial DNA levels	NCT02425579 (Fredenbergh 2018 [9])

Entry	Phase	Subjects	Drug product	Administration	Sponsor	Description/indication	No. of subjects	Dosing	Outcomes	NCT/publication
9	2	Patients	iCO	Inhaled	INO	Kidney transplant – DGF	32	Four dose groups, single dose; two postoperative (0.7 or 2.0 mg/kg/h) and two intraoperative (2.0 or 3.0 mg/kg/h) reaching 7.6% and 11.9% COHb peak, respectively)	No imbalance in AEs or SAEs; no related SAE. No difference for ECG or neurocognitive testing. Trend toward improved renal function in highest dose group intraoperatively. Study withdrawn	NCT00531856
10	2	Patients	iCO	Inhaled	Academic	COPD	20	CO (100 or 125 ppm) or air (crossover) for 2 h daily for 4 days resulting in 2.6% and 3.1% COHb peak	Well tolerated. No significantly related AE imbalance and no effects on hemodynamics. Difference in exacerbation of COPD deemed unrelated	NCT00122694 (Bathoorn 2007 [10])
11	2	Patients	iCO	Inhaled	Academic	IPF	58	CO (100 or 200 ppm) or air for 2 h 2×/week for 12 weeks (reaching up to 3.7% COHb peak with 200 ppm)	No statistically significant differences in AEs or SAEs. No related SAE. No statistically significant differences in efficacy biomarker (MMP7), pulmonary function testing, functional assessments, or patient-reported outcomes	NCT01214187 (Rosas 2017 [11])

Safety and Tolerability Studies (Phase 1 Studies)

In the eight phase 1 clinical studies of CO as the API (entries 1–8, Table 30.1), tolerability was demonstrated and there was no apparent safety signal or difference in AEs between test and control groups with COHb peak up to 13.9% COHb (entry 2, Table 30.1). In clinical studies where a 22% COHb peak was reached, and which should not be characterized as low-dose CO, there was no apparent safety signal or significant difference in AEs, though a higher incidence of prolonged mild headache was reported, as expected in the study (headache was an outcome) and from the literature [1], indicating a potential lack of safety and tolerability at a 22% COHb saturation peak with CO as an API (entry 3, Table 30.1) .

In a phase 1 study in acute respiratory distress syndrome (ARDS) patients in intensive care, a mean peak COHb of 4.9% and a maximum peak COHb of 6.87% were reached; there was no imbalance in AEs or SAEs between treated and control subjects; and there was no related SAE (entry 8, Table 30.1). Of the eight phase 1 safety and tolerability studies conducted with CO as the API, all used iCO as the drug product. Of these, two were industry sponsored and six non-industry sponsored.

Industry Sponsored

INO Therapeutics LLC sponsored two phase 1 trials of iCO, consisting of (i) a single ascending dose (SAD) clinical study and (ii) a multiple ascending dose (MAD) clinical study, both of which demonstrated safety and tolerability (entries 1 and 2, Table 30.1) [5]. These studies were randomized, single-blind, and placebo-controlled studies. Administration included a single 1 h of iCO gas or air, with varying CO gas concentrations. The SAD study consisted of 32 subjects in four groups of eight subjects randomized 3:1 iCO:placebo and dosed at 0.2, 0.75, 2.0, and 2.3 mg/kg/h, resulting in mean peak COHb levels of 2.0%, 3.4%, 7.7%, and 8.8%, respectively. All doses were well tolerated, and there was no imbalance in AE and there was no SAE. No clinically important abnormalities, effects, changes over time, or difference between dose groups could be found for laboratory safety assessments, vital signs, or electrocardiogram (ECG) results. In addition, neurocognitive testing was performed and no difference was observed.

The MAD study consisted of four panels of 12 HVs, where 10 subjects received iCO and 2 subjects received air placebo. In panel 1, the subjects were given repeated doses of 2.3 mg CO/kg or placebo for 10 consecutive days. In panel 2, the 12 subjects received a single dose of 3.0 mg CO/kg or placebo.

Table 30.3 Summary results of phase 2 kidney transplant study with iCO.

	iCO	Placebo
Total number of subjects	23	8
Pharmacokinetic testing		
Individual subject peak COHb	14.8%	1.3%
Dosing group mean COHb saturation (range)	3.0 mg/kg/h: 11.9% (9.8–14.8%)[a] 2.0 mg/kg/h: 7.6% (7.2–8.3%)[a] 0.7 mg/kg/h: NR	1.0% (0.6–1.3%)[a]
Neurocognitive testing		
MMSE (mean)	29.2 ± 0.9	29.3 ± 1.2
Spatial memory accuracy score (mean)	89.5 ± 13.2	94 ± 7.9
Impairment in power of attention (number of subjects)	1/22	1/7
Impairment in continuity of attention (number of subjects)	0/22	1/7
Impairment in picture recognition (number of subjects)	1/22	0/7

[a]For COHb data only, number of subjects = 15; placebo group, $n = 4$; medium-CO group, $n = 6$; high-CO group, $n = 5$.

Table 30.4 Serum creatinine results of phase 2 kidney transplant study with iCO.

		iCO2.0 mg/kg/h intraoperatively	iCO3.0 mg/kg/h intraoperatively	Placebo
Baseline value (mg/dl)	N	6	5	4
	Mean (SD)	5.04 (2.35)	5.96 (4.59)	6.71 (0.76)
Percent change from baseline to day 3	N	6	5	4
	Mean (SD)	−60.84 (23.99)	−68.52 (12.42)	−54.82 (56.40)
Percent change from baseline to day 4	N	6	4	4
	Mean (SD)	−66.57 (17.53)	−62.48 (17.81)	−51.20 (64.13)

Note: Most subjects discharged at postoperative days 3 and 4.

In panel 3, the 12 subjects received repeated doses of 3.0 mg CO/kg or placebo for 10 consecutive days. In panel 4, the subjects received a 3.0 mg CO/kg single dose sourced from two CO gas sources with different concentrations. The highest level of COHb measured was 13.9% in the 3.0 mg/kg/h dose. All doses were well tolerated. Further details on study design and the results of this study have not been reported.

Non-industry Sponsored

Six non-industry-sponsored phase 1 clinical studies have been carried out with iCO, including four studies in HVs and two in patients. The patient studies were in a lower morbidity patient group (migraine) and a higher morbidity patient group (ARDS patients in ICU and on a ventilator). These non-industry-sponsored phase 1 clinical studies with iCO vary substantially in terms of dosing, measured outcomes, and availability of results.

A headache/migraine-oriented study in 12 HVs (NCT02066558) [6] studied iCO in a double blind, placebo-controlled, crossover clinical trial (entry 3, Table 30.1). Subjects were given a single dose of either iCO or room air through a mask, and crossed over with a minimum of a week between doses. The target peak COHb was 22%, and this level was expected to cause headache. CO was administered in three inhalation periods of approximately 3 min each, with each inhalation period approximately 10 min apart. For the first inhalation period, CO gas (99.997% purity) was administered with room air. The volume of pure CO gas in the subsequent administrations was titrated to reach the target peak COHb saturation. A number of parameters, including mean blood velocity of the middle cerebral artery (VMCA), the diameter of the superficial temporal artery and radial artery, headache intensity, blood flow in the facial microvasculature, and vital signs (e.g., ECG, mean arterial blood pressure, heart rate, respiration frequency, and end-tidal CO_2), were measured along with blood chemistry. Hemodynamic assessments other than facial blood flow were taken at baseline (pre-dose) and every 10 min from 30 to 120 min after the first dose. Vital sign assessments, headache intensity, and facial blood flow were assessed at baseline and every 10 min through 120 min after the first dose. Blood samples were taken at baseline, every 10 min through 30 min after the first dose, and then every 30 min from 30 to 120 min after the first dose. A headache diary was kept for the subsequent 12 h. iCO at peak COHb of ~22% was associated with prolonged mild headache, as well as increased mean VMCA and facial skin blood flow, whereas no changes were observed in the diameter of the superficial temporal artery and the radial artery. Despite the relatively high peak COHb, there were no differences in the incidence of AEs between CO and placebo.

A second headache/migraine study in HVs by the same research group (NCT03385174) further investigated the cause of headaches from higher dose CO administration (entry 4, Table 30.1). In this study, 15 HVs were dosed with CO, and in addition to assessment of headache, magnetic resonance imaging was used to assess changes in the circumference of intra- and extracerebral blood vessels. Further details on study design and the results of this study have not been reported.

A third clinical study examined lung inflammation prevention in 36 HVs (NCT00094406) (entry 5, Table 30.1) [7]. In this study, *Escherichia coli* endotoxin (LPS) was instilled in a lung segment and saline in a contralateral segment followed by CO at 100 ppm, NO, or room air for 6 h. Bronchoalveolar lavage was assessed for changes in miRNA. Both CO and NO altered the miRNA expression in the lungs of both LPS-exposed and saline-exposed lung segments. Thematic miRNA analysis predicted that major biological themes represented by these signatures include MAP kinases, STAT3, GSK3, and p53 pathways. Further details on study design and the results of this study have not been reported.

The fourth non-industry-sponsored phase 1 clinical trial with CO as an API studied pulmonary blood vessel function in an open-label manner in HVs (NCT03067701) (entry 6, Table 30.1). Nine subjects intermittently (once per minute) inhaled 1000 ppm CO gas for 30–40 min in order to mimic tobacco smoking from a hookah (water pipe). Endothelial function was measured by brachial artery flow-mediated dilation. Further details on study design and the results of this study have not been reported.

The fifth study was conducted in migraine patients by the same research group that conducted the previously discussed headache/migraine studies in HVs (NCT03075020) (entry 7, Table 30.1) [8]. In this randomized, double-blind, placebo-controlled crossover study, 12 migraine patients were treated in a similar manner as described earlier, with a single dose resulting in a target 22% COHb peak. iCO did not provoke more migraine attacks compared to placebo, though, as found in HVs and as reported in the literature for COHb at greater than 20% COHb saturation [1], CO induced headache more often than placebo. Also, as shown before, iCO increased facial skin blood flow, no SAE was reported, and there was no significant difference in vital signs or AEs between treated and control patients.

The sixth phase 1 study with CO as an API assessed low-dose iCO in mechanically ventilated sepsis-induced ARDS patients in intensive care (NCT02425579) (entry 8, Table 30.1) [9]. This blinded, randomized, placebo-controlled, dose escalation study included 12 patients. Patients were dosed with 100 ppm CO, 200 ppm CO, or room air (control) for 90 min once daily for at least 1 day and up to 5 consecutive days. In the 100 ppm CO group, the mean peak COHb saturation was 3.48%, and the maximum COHb was 4.4%. In the 200 ppm CO group, the mean peak COHb was 4.9%, and the maximum COHb peak was 6.87%. There was no imbalance in AEs or SAEs between treated and control subjects, and there was no related SAE. CO-treated patients had significant reduction in plasma mtDNA levels on day 2 compared with pretreatment baseline levels versus the respective mtDNA changes in placebo-treated subjects, and there was a trend toward improved respiratory and systemic secondary endpoints in CO-treated patients, particularly the lung injury score and sequential organ failure assessment scores on day 7.

Phase 2 Clinical Studies with CO as the API

As indicated in Table 30.2, three phase 2 studies have been reported with CO as the API, all using iCO as the drug product (entries 9–11, Table 30.1). Of these, one was industry sponsored and two were non-industry sponsored. In these studies, tolerability was demonstrated and there was no apparent safety signal or difference in AEs between test and control groups with COHb peak up to 11.9% COHb. Although these studies were not statistically powered for efficacy, the industry study in delayed graft function (DGF) in kidney transplant showed a strong trend to efficacy at the highest intraoperative dose (discussed later).

Industry Sponsored

INO Therapeutics LLC sponsored a phase 2 trial of iCO in kidney transplant recipients assessing DGF (NCT00531856) (entry 9, Table 30.1). This randomized, blinded, placebo-controlled study consisted of dosing iCO for 1 h to 31 kidney transplant recipients in four groups of approximately eight, each randomized 3:1. In total, 23 subjects received iCO, and 8 subjects received air placebo. Twenty-nine of the 31 subjects were living donor kidney transplant recipients. Groups A and B (n = 8 per group) were dosed postoperatively 12–48 h after transplant with placebo or at 0.7 or 2.0 mg CO/kg/h, respectively. Groups C (n = 8) and D (n = 7) were dosed intraoperatively during kidney transplant with placebo or at 2.0 or 3.0 mg CO/kg/h, respectively. The 3.0 mg/kg/h group resulted in an average peak COHb of 11.9% (individual maximum of 14.8%) and the 2.0 mg/kg/h groups resulted in an average peak COHb of 7.6% (individual maximum of 8.3%).

Safety and tolerability were assessed, as were pharmacokinetics (PK) and renal function related to DGF, including changes in glomerular filtration rate (GFR) as measured by creatinine clearance. With regard to safety, there were no clinically significant impairments in cardiovascular and neurocognitive function for CO as compared with placebo-dosed subjects.

The overall incidence of treatment-emergent AEs was comparable between CO and placebo groups, and there was no study drug-related SAE. Twelve-lead ECG was conducted at screening, baseline, and 8 h after dosing. No difference in treatment groups was observed for ECG. Neurocognitive testing included spatial working memory task and the Mini-Mental State Examination (MMSE) administered at baseline and at week 16 and 24 follow-up visits assessed by blinded review by an independent neurologist (Table 30.3). Results in the 29 subjects demonstrated (i) no difference in treatment groups for spatial working memory task and the MMSE administered at weeks 16 and 24; and (ii) no signs/symptoms of extrapyramidal impairment on neurological examination in any subject at 16/24-week follow-up visit. In addition, a trend in improved kidney function versus placebo controls as measured by serum creatinine and GFR was seen in the intraoperative cohorts (Table 30.4). Further details on study design and the results of this study have not been reported.

Non-industry Sponsored

A phase 2 clinical study in chronic obstructive pulmonary disorder (COPD) patients (NCT00122694) [10] studied iCO in 20 patients in a blinded, randomized, placebo-controlled crossover study (entry 10, Table 30.1). CO gas (100 or 125 ppm) or room air was administered for 2 h daily for 4 consecutive days. Patients were crossed over after a minimum 1-week washout period. The first nine patients were treated with 100 ppm CO, resulting in mean peak COHb of 2.6%, with a maximum peak COHb of 3.5%. The last 10 patients were treated with 125 ppm CO, resulting in a median peak COHb of 3.1% and a maximum peak COHb of 4.5%. With regard to safety, there were two exacerbations of COPD, though these were thought to be unrelated to CO. One exacerbation occurred 2 weeks after administration, and the second occurred on the third day of CO administration. There were no differences in other AEs between CO and placebo. There were no significant differences in vital signs, change of cardiac frequency, and change in blood pressure between CO and placebo groups. With regard to inflammation and lung function, there was no difference in sputum neutrophil percentage or in blood leukocytes or C-reactive protein, but there was a trend toward reduction in sputum eosinophil percentage. There was no effect of CO on lung function or health measures, though CO inhalation resulted in a trend in improvement of responsiveness to methacholine.

An additional phase 2 study with iCO was conducted in idiopathic pulmonary fibrosis (IPF) patients (NCT01214187) (entry 11, Table 30.1) [11]. Fifty-eight patients were randomized to iCO or air (1:1). iCO patients were titrated to the highest possible dose of the two CO gas doses (100 and 200 ppm) with the requirement that their peak COHb remain below 8%. Patients were treated under close supervision in an ambulatory setting for 2 h twice a week for 12 weeks. Patients receiving the 200 ppm dose reached a peak COHb range of 2.4–3.7%. With regard to safety, there were no statistically significant differences in AEs or SAEs, and there was no related SAE. Importantly, there was no increase in neurological or cardiovascular events in CO-treated patients as compared with placebo. No differences were observed in physiological measures, incidence of acute exacerbations, hospitalization, death, or patient-reported outcomes. CO was well tolerated. With regard to efficacy, there were no statistically significant differences in the efficacy biomarker serum matrix metalloproteinase-7 (MMP7), pulmonary function testing, functional assessments, or patient-reported outcomes.

Clinical Studies with Low-Dose CO Drug Products with Non-CO API

As summarized in Table 30.1, a substantial number of phase 1 and phase 2 clinical trials studying the safety, tolerability, dosing, and efficacy of low-dose CO in both HVs and patients have been conducted/reported with drug products with non-CO APIs (PEG-COHb) that provide exogenous CO (entries 14–23, Table 30.1). These clinical trials with non-CO APIs are described in further detail later and in Table 30.5.

In the clinical studies conducted thus far with non-CO APIs, including PEG-COHb with human Hb (MP4CO) and PEG-COHb with bovine Hb (Sanguinate), the AE profiles generally reflected the widely reported AE profiles with cell-free Hb solutions, and included cardiovascular AEs such as hypertension, troponin I elevation, hematuria, and myocardial infarction [3,4]. The peak COHb achieved was approximately 4.5%. Given the AE profile at this relatively low peak COHb saturation, the potential exists for dose-limiting toxicity of PEG-COHb drug products due to toxicity associated with the PEG-Hb itself and non-CO breakdown products such as Hb and PEG, and their respective breakdown products. Importantly, the risk of toxicity of the non-CO components of the API exists with all low-dose CO drug products that utilize non-CO APIs.

Table 30.5 Clinical trials with low-dose CO drug products with non-CO API.

Entry	Phase	Subjects	Drug product	Administration	Sponsor	Description/ indication	No. of subjects	Dosing	Outcomes	NCT/publication
14	1	HVs	PEG-COHb	IV	Prolong	Single dose safety	24	Ascending dose at 2, 3, or 4 ml/kg; COHb peak NR	All doses well tolerated; Hb-related AEs seen, including mild to moderate hypertension and hematuria	Misra 2014 [12]
15	1	Patients	PEG-COHb	IV	Prolong	End-stage renal disease	5	Weekly infusion for up to 3 weeks at 8 ml/kg. COHb peak NR	Hb-related AEs seen, including hypertension, troponin I, and non-ST elevation myocardial infarction (SAE). Study terminated early due to SAE	NCT02437422 (Abu Jawdeh 2017 [13])
16	1	Patients	PEG-COHb	IV	Prolong	Severe anemia	103	As needed (PRN) infusions of 500 ml of Sanguinate	NR	NCT02754999
17	1	Patients	PEG-COHb	IV	Prolong	SCD	24	4 or 8 ml/kg single dose	Hb-related AEs/SAEs seen, including transient troponin I elevation, hypertension, and hematuria as well as one SAE (pulmonary hypertension). Arthralgia significantly increased in PEG-COHb subjects	NCT01848925 (Misra 2016 [14])
18	1	Patients	PEG-COHb	IV	Sangart	SCD	24	Single ascending dose to 8 ml/kg; COHb peak 3.7%	No treatment-emergent abnormalities	NCT01356485 (Howard 2013 [15])

Entry	Phase	Subjects	Drug product	Administration	Sponsor	Description/indication	No. of subjects	Dosing	Outcomes	NCT/publication
19	2	Patients	PEG-COHb	IV	Prolong	SCD leg ulcer	10	Sanguinate (8 ml/kg) weekly for either 4 or 6 weeks. Open label	Some treatment-emergent AEs, including hypertension and ECG interval changes, and one SAE. No significant changes in wound healing	NCT02600390 (Valentino 2017 [16])
20	2	Patients	PEG-COHb	IV	Prolong	SCD VOC (ex-United States)	34	Sanguinate or saline infusion of 8 ml/kg daily for 2 days after presentation for VOC	NA	NCT02672540
21	2	Patients	PEG-COHb	IV	Prolong	SCD VOC (United States)	24	One Sanguinate or saline infusion of 8 ml/kg after presentation for VOC	NA	NCT02411708
22	2	Patients	PEG-COHb	IV	Prolong	Subarachnoid hemorrhage	12	Open label. Single dose of 4, 6, or 8 ml/kg, resulting in approximate peak COHb of 3.5%, 4%, and 4.5%	Significant increase in global cerebral blood flow in high-dose group, and significant increase in regional cerebral blood flow to vulnerable regions. Some treatment-emergent AEs, including hypertension	NCT02323685 (Dhar 2017 [17])
23	2/3	Patients	PEG-COHb	IV	Prolong	Kidney transplant – DGF	60	Sanguinate or saline infusion of 8 ml/kg on day of surgery and 24 h after surgery	NA	NCT02490202

Phase 1 Clinical Studies with Low-Dose CO Drug Products with Non-CO API

As indicated in Table 30.5, five phase 1 safety and tolerability studies have been conducted with a non-CO API, all using an intravenous PEG-COHb, and all were conducted by industry (entries 14–18, Table 30.1). Two companies sponsored these clinical trials, including four with Prolong Pharmaceuticals, LLC's Sanguinate and one with Sangart, Inc.'s MP4CO. These drug products were both designed to provide both CO as a therapeutic and also oxygen-carrying capacity by the PEG-Hb after the CO is released. Also, as discussed earlier and per the extensive safety literature with cell-free Hb drug products, there are significant safety concerns as well as concerns with regard to dose-limiting toxicity with cell-free Hb drug products [3,4].

One study has been reported with a low-dose CO drug product with a non-CO API in HVs (entry 14, Table 30.1). This blinded, placebo-controlled, SAD study with the drug product Sanguinate, a PEG-COHb drug product, included 24 subjects [12]. Three dose groups of 2, 3, or 4 ml/kg (80, 120, or 160 mg/kg) included eight subjects each, randomized 3:1 to drug product or saline control. Although pharmacokinetic assessments of the PEG-Hb molecule were conducted, no COHb data were reported. Drug product was well tolerated, and there were no SAEs in this study. No clinically significant changes in ECGs, echocardiograms, or blood biochemistry and hematology parameters were observed. However, some subjects receiving Sanguinate showed mild to moderate hypertension and hematuria, both of which are AEs associated with the administration of cell-free Hb [3,4].

Four phase 1 clinical studies have been conducted with non-CO API drug products in patients (entries 15–18, Table 30.1). A phase 1 clinical trial studied Sanguinate in stable end-stage renal disease patients on chronic dialysis (NCT02437422) (entry 15, Table 30.1) [13]. In this open-label study, 5 patients were given weekly infusions at 8 ml/kg for up to 3 weeks, though 10 patients were initially targeted. The study was terminated due to an SAE (myocardial infarction). Two subjects experienced troponin I elevation, one of which was associated with nonspecific ECG changes and chest pain leading to the diagnosis of a non-ST elevation myocardial infarction. Worsening hypertension was also observed in one subject. These AEs and SAEs have been associated with the administration of cell-free Hb [3,4]. Although pharmacokinetic assessments of the PEG-Hb molecule were conducted, no COHb data were reported.

The second clinical study was conducted with Sanguinate in severe anemia patients who were unwilling or unable to receive red blood cell transfusion (NCT02754999) (entry 16, Table 30.1). It is expected that these patients would be under hospital care within a critical care facility. This study comprised 103 patients, who were administered 500 ml of Sanguinate as needed. Further details on study design and the results of this study are not reported.

A phase 1 clinical study in SCD (sickle cell disease) patients was also conducted with Sanguinate (NCT01848925) (entry 17, Table 30.1) [14]. In this open-label, randomized, single-dose study with two dose levels, 24 patients in two groups of 12 randomized 2:1 were administered either Sanguinate (at 4 or 8 ml/kg, respectively) or hydroxyurea (15 mg/kg). Although pharmacokinetic assessments of the PEG-Hb molecule were conducted, no COHb data were reported. More AEs were reported in the Sanguinate groups than reported in the hydroxyurea groups. More Sanguinate subjects showed mild to moderate hypertension, troponin I elevation, and hematuria, and there was one SAE in the Sanguinate group due to moderate pulmonary hypertension (increase in tricuspid regurgitant velocity without clinical signs), all of which have been associated with the administration of cell-free Hb [3,4].

Sangart, Inc. also conducted a clinical trial in SCD patients with its MP4CO drug product. This blinded, placebo-controlled, SAD study included 24 subjects (NCT01356485) (entry 18, Table 30.1) [15]. There were six dose groups of four subjects each. These groups consisted of the following dose levels (in ml MP4CO/kg): 0.35, 1, 2, 4, 4 (split into 2 + 2), and 8 (split into 4 + 4), with the last two dose groups comprised of split doses administered 24 h apart. Each group was randomized 3:1 MP4CO:placebo control. In the highest single-dose group (4 ml/kg), COHb increased by 2% to a peak saturation of 3.7%. There was no SAE, no evidence of hypertension or troponin I increase, and AE rates were similar between groups.

Phase 2 Clinical Studies with Low-Dose CO Drug Products with Non-CO API

As indicated in Table 30.5, five phase 2 studies have been conducted with a low-dose CO drug product with a non-CO API, all using the Sanguinate intravenous PEG-COHb drug product made with bovine Hb (entries 19–23, Table 30.1). In these clinical studies, the AE profile reflected the widely reported AEs with cell-free Hb solutions, and included cardiovascular AEs such as hypertension and ECG interval changes [3,4]. The highest peak COHb saturation achieved was approximately 4.5%. The observation of AEs at this and lower COHb saturations, as compared with the lack of comparable AEs at these and substantially higher peak COHb saturations with CO as the API, indicates the potential for dose-limiting toxicity of the non-CO component and its breakdown products. This drug product was designed to provide both CO as a therapeutic and also oxygen-carrying capacity by the PEG-Hb after the CO is released. It is important to note that these phase 2 studies were generally low in study subject number and not powered to show efficacy, and a full publication of results is available for only one of these studies.

An open-label phase 2 clinical study in SCD patients assessing leg ulcer healing with Sanguinate was also conducted by Prolong Pharmaceuticals, LLC (NCT02600390) (entry 19, Table 30.1) [16]. Ten SCD patients suffering from chronic SCD-associated leg ulceration were infused with Sanguinate (8 ml/kg) weekly in two cohorts of 4 and 6 weeks of dosing, with five patients per cohort. With regard to safety, treatment-emergent AEs related to study drug were reported in 2 of 10 patients. Hypertension and changes in ECG interval were also reported, though the ECG interval changes were not considered clinically meaningful. One SAE was reported in the 6-week dose group. With regard to efficacy, no significant changes in leg ulcer pain or wound surface area were reported. Further details on study design and results are not available.

Two randomized, placebo-controlled, blinded phase 2 studies in vaso-occlusive crisis (VOC) in SCD patients were also conducted with Sanguinate in the United States and ex-United States (NCT02411708 and NCT02672540, respectively), dosing after presentation for VOC (entries 20 and 21, Table 30.1). The US study included 24 patients infused one time with drug product or saline control infusion at 8 ml/kg after presentation for VOC. The ex-US study included 34 SCD patients admitted to hospital for VOC. These patients were infused with drug product or saline infusion at 8 ml/kg on the day of admission to hospital for VOC and on the day after admission for VOC. Further details on study design and results are not available.

The final completed phase 2 clinical trial with the Sanguinate was conducted in subarachnoid hemorrhage (SAH) patients in an ascending dose, open-label manner (NCT02323685) (entry 22, Table 30.1) [17]. Twelve patients were included. These patients were admitted to the Neurology/Neurosurgery Intensive Care Unit with aneurysmal SAH, had the ruptured aneurysm secured by surgical or endovascular means, and were at risk of delayed cerebral ischemia. These patients were dosed in three ascending dose groups of four patients each with a single dose of drug product at 4, 6, or 8 ml/kg, resulting in approximate mean COHb peaks of 3.5%, 4%, and 4.5%. With regard to safety, no subjects had deterioration in neurological status and no safety concerns were identified, though transient hypertension was observed. With regard to efficacy, a significant increase in global cerebral blood flow as evaluated by positron emission tomography (^{15}O-PET) in the high-dose group was observed, as was a significant increase in regional cerebral blood flow to vulnerable regions.

Prolong Pharmaceuticals, LLC sponsored a randomized, placebo-controlled, blinded phase 2/3 clinical trial of Sanguinate, a PEG-COHb drug product, in kidney transplant patients assessing DGF (NCT02490202) (entry 23, Table 30.1). Sixty kidney transplant patient recipients of a donation after brain death donor kidney were given infusions of either Sanguinate or saline (randomized 1:1) of 8 ml/kg on day of surgery and 8 ml/kg 24 h after surgery. The phase 2 and phase 3 parts of the trial utilized the same protocol, but the phase 2 patients were followed up for 30 days, while the phase 3 patients were followed up over 1 year. Further details on study design and the results of this study are not available.

Ongoing and Planned Clinical Studies

As indicated in Table 30.1, two clinical studies are currently ongoing or planned, both with CO as the API. The first is a planned industry-sponsored phase 1 study with the oral CO drug product HBI-002, and the second is an ongoing non-industry-sponsored phase 2 study with iCO in ARDS.

A randomized open-label phase 1 study in HVs with the oral liquid low-dose CO drug product

HBI-002 is planned by Hillhurst Biopharmaceuticals, Inc. (NCT03926819). This study plans to include 20 subjects in a SAD/multiple dose design. HBI-002 was designed to enable the reliable and precise administration of low-dose CO to realize the substantial therapeutic potential of CO, and HBI-002 was designed to overcome the considerable barriers to use with iCO and with low-dose CO drug products with a non-CO API (see later and Table 30.6).

A non-industry-sponsored randomized, placebo-controlled, blinded phase 2 clinical study with iCO in mechanically ventilated patients with ARDS is currently recruiting subjects (NCT03799874). Patients will be given CO gas at 200 ppm or air for 90 min daily for 5 days. This study plans to include 100 patients.

Discussion

Overall, 23 clinical studies, as defined earlier, including a total of approximately 570 subjects have been conducted with low-dose CO, either with CO as the API or with a drug product with a non-CO API (all PEG-COHb drug products). The safety profiles and peak COHb saturations achieved with these two types of CO drug products (CO as API versus non-CO API) in these clinical trials differ substantially.

With low-dose CO drug products with CO as the API, a total of 11 studies have been completed, all with iCO, and 2 additional studies are ongoing. The completed studies include a total of approximately 250 subjects, including HVs, stable patients, and seriously ill patients. With regard to peak COHb saturations in HVs, safety and tolerability were demonstrated up to a peak 13.9% COHb saturation, and in kidney transplant patients safety and tolerability were demonstrated up to a peak 14.8% COHb saturation. With regard to duration of dosing, in HVs, 10 days of dosing were achieved (up to a daily 13.9% COHb peak), and in highly morbid IPF patients a 12 week exposure two times per week (up to 3.7% peak COHb) was achieved. In all of these 11 studies, no imbalances in AEs between CO and placebo control were observed, and no treatment-related significant AEs were reported. The safety and tolerability data with CO as API reported in these clinical trials provide support for further development of a drug product with CO as the API. However, no formal phase 3 studies have been performed and benefit has yet to be demonstrated. As with all drug products, the benefit as compared to the risk for drug products with CO as an API needs to be demonstrated in controlled clinical studies with sufficient numbers of study subjects.

A number of concerns and challenges exist with the use of an iCO drug product outside of surgery. First, although inhaled drugs are accepted intraoperatively, there is substantial patient and healthcare provider resistance to inhaled drugs in a nonoperative setting, especially where the administration takes place over a longer period of time as with the administration in the clinical trials described herein. Patient compliance is thus a barrier for iCO drug products, except when used intraoperatively. A second important issue with iCO is the need for compressed gas cylinders of CO. These cylinders present a safety risk due to accidental inhalation if the cylinders are not fully closed or damaged, as CO is a colorless, odorless gas that is toxic at high levels. In the hospital setting, this risk is to both the patients themselves and hospital staff. In the home, this risk is to patients, family members, and home caregivers. Finally, although intraoperative dosing of gases can be well controlled due to the constant monitoring of blood gases and other measures, nonoperative therapeutic gas dosing is often less accurate. There is a substantial risk of a patient not breathing the correct dose, whether due to incorrectly used administration equipment (e.g., incorrectly applied breathing mask), variable inhalation (differences in breathing volume and rate during administration), poor compliance with duration of inhalation, or changes in lung function (e.g., decreased due to infection) (Table 30.6).

New drug products utilizing innovative technology for the administration of CO, such as HBI-002 (Tables 30.1 and 30.4) and others, for the delivery of CO as an API are being developed to address these issues. Although the API of these drug products is CO, because different route(s) of administration are used, bridging safety studies will be necessary to confirm that the substantial safety data with iCO apply. In addition, as with iCO, the benefit as compared to the risk for CO as an API needs to be demonstrated in much larger clinical studies.

With low-dose CO drug products where CO is not the API, a total of 10 studies have been conducted, including a total of approximately 320 subjects, including HVs, stable patients, and seriously ill patients, all with PEG-COHb products. Results of only 6 of the 10 total studies with a non-CO API have been reported. In the reported studies, peak COHb saturation was not reported in HV, though the highest dose in HVs (4 ml/kg) was lower than the highest dose in patients (8 ml/kg). The highest peak COHb saturation reported with a non-CO API drug product was 4.5% at a dose of 8 ml/kg, in SAH patients. With regard to duration of dosing, a 6-week exposure, once per week (at 8 ml/kg Sanguinate),

Table 30.6 Low-dose CO drug product comparison.

API type	Drug product	Method of delivery	Advantages	Barriers to commercial use
CO	iCO	Inhaled	• Safety dataset for API and method of delivery • Demonstration of appropriate clinical PK • Appropriate delivery method for use during surgery	• Safety risk of accidental inhalation due to presence of CO canisters • Healthcare provider resistance due to risk of accidental inhalation • Dosing inaccuracy (except during surgery) • Patient resistance to wearing inhalation mask/device • Development and approval of medical device to administer CO gas need to be completed • Clinical testing of specific CO gas product with device needs to be completed
CO	HBI-002	Oral	• Safety dataset for API • Demonstration of appropriate preclinical PK • Optimal delivery method for almost all use settings	• Appropriate clinical PK needs to be demonstrated • Clinical testing needs to be started/completed
Non-CO	PEG-COHb	IV	• Numerous completed clinical trials	• Demonstrated carrier molecule toxicity • PK may not be appropriate (likely due to dose-limiting toxicity) • Patient resistance to intravenous dosing for certain indications

was reported in stable SCD leg ulcer patients. In these studies, AEs typically seen with cell-free Hb products were reported, including cardiovascular events (hypertension, troponin I increase, ECG interval changes, and, in one instance, myocardial infarction) as well as hematuria [3,4]. It is important to note that not all of these AEs were seen in all the studies, and one of the studies reported no treatment-emergent abnormalities. One of the studies, in end-stage renal disease patients, was terminated early due to a patient experiencing myocardial infarction. Thus far, the safety profile expected of a cell-free Hb solution has been documented with PEG-COHb [3,4]. However, as with all drug products, the benefit as compared to the risk for a PEG-COHb has not been fully evaluated. It is important to note that the PEG-Hb carrier provides oxygen-carrying capacity after releasing CO, which could provide benefit in certain clinical indications.

As discussed earlier, one of the concerns with low-dose CO drug products with non-CO APIs as compared with CO as an API is the potential for toxicity/safety concerns due to the non-CO breakdown products of the non-CO APIs. In the case of PEG-COHb molecules, the clinical trials discussed herein clearly indicate a negative or adverse safety profile associated with the non-CO breakdown product cell-free Hb. As expected, per the extensive literature [3,4], cell-free Hb causes toxicity even when delivered at doses that result in relatively low peak COHb saturations, suggesting the possibility of dose-limiting toxicity for these drug products.

Various types of other low-dose CO drug products with non-CO APIs are in preclinical stages of development. These agents include small inorganic molecules (also termed CORMs), small organic molecules (also termed organic CO prodrugs or organic CORMs), and hybrid molecules consisting of a CORM combined with another active molecule (also termed HYCOs) such as hybrids of a CORM and a nuclear factor erythroid 2-related factor 2 activator. However, there is no information that any of these agents have entered the clinic, and no clinical results have been reported with any of these classes of molecules. Issues of stability, release kinetics, and potency have previously been reported with these molecules. In addition, for these molecules it will be necessary to consider the toxicity/safety and thus benefit–risk ratio separately from drug products with CO itself as the API, as indicated by the reportedly problematic toxicity of a number of the CORMs being developed, especially in chronic use settings [18–20]. Please see

Chapters 12–17 for further details on small-molecule drug products with non-CO APIs.

Conclusions

In conclusion, a substantial literature including general nonclinical and clinical safety studies, clinical experience studies, epidemiological studies, and formal prospective clinical studies provides a large database with which to evaluate CO as a therapeutic. The 23 prospective clinical studies carried out and documented in the literature and/or the ClinicalTrials. gov database have included a total of 570 subjects and provide important information to help inform the benefit/risk of low-dose CO, either with CO as the API or with a drug product with a non-CO API.

However, the use of CO as an API with inhaled administration and the use of low-dose CO drug products with non-CO API, regardless of administration method, present substantial barriers to patient use. For iCO, these barriers include accidental inhalation exposure of patients and healthcare workers and the need for compressed CO gas cylinders, imprecise dosing, and patient resistance to impractical chronic inhaled dosing. For non-CO API drug products, these barriers include toxicity concerns with carrier molecules and more complex metabolism, as well as potential stability CO release kinetics, and potency issues. Delivering CO as an API through non-inhaled means, as with the oral low-dose CO drug product HBI-002, could potentially build on the substantial safety literature with CO as an API while avoiding the respective barriers associated with iCO and non-CO APIs.

Nevertheless, the development of CO as a therapeutic agent is gaining momentum given the highly encouraging clinical safety data generated to date and the large body of preclinical information indicating therapeutic benefit across a broad spectrum of serious diseases with current unmet clinical needs.

References

1. Wilbur, S., Williams, M., Williams, R. et al. (Jun 2012). *Toxicological Profile for Carbon Monoxide*. Atlanta (GA): Agency for Toxic Substances and Disease Registry (US). Available from: https://www.ncbi.nlm.nih.gov/books/NBK153693.
2. Manisalidis, I., Stavropoulou, E., Stavropoulos, A., and Bezirtzoglou, E. (Feb 20 2020). Environmental and health impacts of air pollution: A review. *Front. Public Health* 8: 14. PMID: 32154200; PMCID: PMC7044178. doi: 10.3389/fpubh.2020.00014.
3. Natanson, C., Kern, S.J., Lurie, P., Banks, S.M., and Wolfe, S.M. (May 21 2008). Cell-free hemoglobin-based blood substitutes and risk of myocardial infarction and death: a meta-analysis. *JAMA* 299 (19): 2304–2312.
4. Alayash, A.I. (Jan 4 2017). Hemoglobin-based blood substitutes and the treatment of sickle cell disease: more harm than help? *Biomolecules* 7 (1).
5. Mahan, V.L. (Dec 27 2012). Neuroprotective, neurotherapeutic, and neurometabolic effects of carbon monoxide. *Med. Gas Res.* 2 (1): 32. PMID: 23270619; PMCID: PMC3599315. doi: 10.1186/2045-9912-2-32.
6. Arngrim, N., Schytz, H.W., Britze, J., Vestergaard, M.B., Sander, M., Olsen, K.S., Olesen, J., and Ashina, M. (Apr 2018). Carbon monoxide inhalation induces headache in a human headache model. *Cephalalgia* 38 (4): 697–706. Epub 2017 May 5. PMID: 28474984. doi: 10.1177/0333102417708768.
7. Effects Of Inhaled Nitric Oxide Or Carbon Monoxide On MicroRNAs In Bronchoalveolar Lavage Of Healthy Humans After Segmental Challenge With Saline Or Endotoxin. Purushottam S. Narute, Roberto F. Machado, Rongman Cai, Junfeng Sun, Carolea Logun, James H. Shelhamer, Anthony F. Suffredini. D36. SEPSIS, ACUTE RESPIRATORY DISTRESS SYNDROME, AND ACUTE LUNG INJURY. May 1, 2014, A5768-A5768.
8. Ghanizada, H., Arngrim, N., Schytz, H.W., Olesen, J., and Ashina, M. (Nov 2018). Carbon monoxide inhalation induces headache but no migraine in patients with migraine without aura. *Cephalalgia* 38 (13): 1940–1949. Epub 2018 Mar 14. PMID: 29540069. doi: 10.1177/0333102418765771.
9. Fredenburgh, L.E., Perrella, M.A., Barragan-Bradford, D., Hess, D.R., Peters, E., Welty-Wolf, K.E., Kraft, B.D., Harris, R.S., Maurer, R., Nakahira, K., Oromendia, C., Davies, J.D., Higuera, A., Schiffer, K.T., Englert, J.A., Dieffenbach, P.B., Berlin, D.A., Lagambina, S., Bouthot, M., Sullivan, A.I., Nuccio, P.F., Kone, M.T., Malik, M.J., Porras, M.A.P., Finkelsztein, E., Winkler, T., Hurwitz, S., Serhan, C.N., Piantadosi, C.A., Baron, R.M., Thompson, B.T., and Choi, A.M. (Dec 6 2018). A phase I trial of low-dose inhaled carbon monoxide in sepsis-induced ARDS. *JCI Insight* 3 (23): e124039. PMID: 30518685; PMCID: PMC6328240. doi: 10.1172/jci.insight.124039.
10. Bathoorn, E., Slebos, D.J., Postma, D.S. et al. (2007). Anti-inflammatory effects of inhaled carbon monoxide in patients with COPD: a pilot study. *Eur. Respir. J.* 30: 1131–1137.

11. Rosas, I.O., Goldberg, H.J., Collard, H.R., El-Chemaly, S., Flaherty, K., Hunninghake, G.M., Lasky, J.A., Lederer, D.J., Machado, R., Martinez, F.J., Maurer, R., Teller, D., Noth, I., Peters, E., Raghu, G., Garcia, J.G.N., and Choi, A.M.K. (Jan 2018). A phase II clinical trial of low-dose inhaled carbon monoxide in idiopathic pulmonary fibrosis. *Chest* 153 (1): 94–104.
12. Misra, H., Lickliter, J., Kazo, F., Abuchowski, A. (Aug 2014). PEGylated carboxyhemoglobin bovine (SANGUINATE): results of a phase I clinical trial. *Artif Organs.* 38 (8): 702–707. doi: 10.1111/aor.12341. Epub 2014 Aug 12. PMID: 25113835.
13. Abu Jawdeh, B.G., Woodle, E.S., Leino, A.D. et al. (2018). A phase Ib, open-label, single arm study to assess the safety, pharmacokinetics, and impact on humoral sensitization of SANGUINATE infusion in patients with end-stage renal disease. *Clin. Transplant.* 32: e13155.
14. Misra, H., Bainbridge, J., Berryman, J., Abuchowski, A., Galvez, K.M., Uribe, L.F., Hernandez, A.L., and Sosa, N.R. (Jan – Mar 2017). A phase Ib open label, randomized, safety study of SANGUINATE™ in patients with sickle cell anemia. *Rev. Bras. Hematol. Hemoter.* 39 (1): 20–27.
15. Howard, J., Thein, S.L., Galactéros, F., Inati, A., Reid, M.E., Keipert, P.E., Small, T., and Booth, F. (2013). Safety and tolerability of MP4CO: A dose escalation study in stable patients with sickle cell disease. *Blood* 122 (21): 2205.
16. Valentino, J., Misra, H., Lopez, L., and Paulino, G.M. (2017). Use of pegylated-carboxyhemoglobin bovine for the treatment of sickle cell disease associated leg ulcers: results from a phase 2 safety study. *Haematologica* 102: 243–244.
17. Dhar, R., Misra, H., and Diringer, M.N. (2017). SANGUINATE™ (PEGylated carboxyhemoglobin bovine) improves cerebral blood flow to vulnerable brain regions at risk of delayed cerebral ischemia after subarachnoid hemorrhage. *Neurocrit. Care* 27: 341.
18. Winburn, I.C., Gunatunga, K., McKernan, R.D., Walker, R.J., Sammut, I.A., and Harrison, J.C. (Jul 2012). Cell damage following carbon monoxide releasing molecule exposure: implications for therapeutic applications. *Basic Clin. Pharmacol. Toxicol.* 111 (1): 31–41.
19. Marques, A.R., Kromer, L., Gallo, D.J., Penacho, N., Rodrigues, S.S., Seixas, J.D., Bernardes, G.J., Reis, P.M., Otterbein, S.L., Ruggieri, R.A., Goncalves, A.S.G., Goncalves, A.M.L., De Matos, M.N., Bento, I., Otterbein, L.E., Blattler, W.A., and Romao, C.C. (2012). *Organometallics* 31: 5810–5822.
20. Wang, P., Liu, H., Zhao, Q., Chen, Y., Liu, B., Zhang, B., and Zheng, Q. (2014). *Eur. J. Med. Chem.* 74: 199–215.

Index

Note: Page numbers followed by "*f*" denotes figure and "*t*" denotes table respectively.

a

AARE. *See* amino acid response element (AARE)
Abbott Laboratories 288–289
ABS. *See* activity-based sensors (ABS)
acetaminophen (APAP) 387
acetaminophen-induced liver toxicity 387
acireductone dioxygenases (ARDs) 309, 309*f*, 310
  bacterial metabolism of 309–310
active pharmaceutical ingredient (APIs) 524, 525
activity-based sensors (ABS) 321
acute brain injury and brain protection 377–378
  CO:
    and heme oxygenase 379–380
    and Nrf2 380
  ischemic stroke 380–381
  neonatal hypoxic–ischemic encephalopathy 381
  potential role of 378, 378*f*
    anti-cell death effect of 378–379
    anti-inflammatory effect in brain 378
    antithrombotic effect of 379
    proangiogenic effect of 379
    vasodilatory effect of 378
  prevalence of 382–383
  subarachnoid hemorrhage and intracerebral hemorrhage 381–382
  traumatic brain injury (TBI) 382
acute inflammatory pain 497
acute kidney injury (AKI) 442
  cellular and molecular responses 434–436
  CO prodrugs in 442
  CO therapeutics for 439
    enriched solutions and hemoglobin products 440–441
    inhaled CO gas 439–440
    mechanisms of action 442
    metal-based molecules 441–442
    organic prodrugs 442
    timing therapies 443
  definition, causes and clinical problem 434
  diverse forms of 436
  downstream actions in 442
  experimental forms of 436
  experimental models of 434–435
  heme oxygenases and, protective role of 436–437
  HO-1:
    activity in 438–439
    expression in 438
    in human 437
  human and experimental models of 437
  inflammatory responses in 442
  injured epithelium after 436
  mechanisms of 434
  models of 443
  nephrotoxic agents in 435
  optimize CO delivery to kidney 443–444
  pathophysiology of 435
  and pharmacodynamic biomarker discovery 443
  rodent models of 434
  therapeutic efficacy in 439
  therapeutic opportunities for 436
  tissue CO levels in 441–442
acute liver failure, CO prodrugs protect against 250, 250*f*
acute liver injury 385–386
  acetaminophen-induced liver toxicity 387
  autoimmune hepatitis 386–387
  fulminant hepatitis 386
  hepatic ischemia/reperfusion injury 387
  nonalcoholic steatohepatitis (NASH) 387
acute lung injury:
  experimental 367
  hemorrhagic shock 368
  human ARDS 369
  LPS-induced injury 368
  malaria-induced ALI/ARDS 369–370
  sepsis and pneumonia-associated lung injury 369
  ventilator-induced lung injury (VILI) 368
acute respiratory distress syndrome (ARDS) 369, 516
acute VOC events 486
Adair constant 53

adenine nucleotide translocator
    protein 31
adenosine triphosphate (ATP) 88
    cell energy metabolism
        and 88–89
    cellular levels of 88
    CO enhancement of 88
    deficiency 461
    generation and bioenergy
        production 90
    mitochondria-derived 118
    mitochondrial production
        of 166
    and phosphocreatine levels 89
    primary source of 89
    production 88, 90
adult-onset diabetes 423
adverse events (AEs) 512
    treatment-emergent 519
AEs. See adverse events (AEs)
Agency for Toxic Substances and
    Disease Registry 484
AhR. See aryl hydrocarbon
    receptor (AhR)
air–blood interface 49
airway neuroepithelial bodies 120
AKI. See acute kidney injury (AKI)
Akt1 112, 113f
Akt activation pathway 31, 309
ALA. See aminolevulinic acid
    (ALA)
alanine aminotransferase
    (ALT) 248, 250, 386, 387
albumin-based delivery of
    CORM 273
alkali hydroxide composition 287
alkaline mucus 466
allodynia 499, 500
Allogeneic islet transplantation 426
α-ketoacids 305–306
ALT. See alanine aminotransferase
    (ALT)
5-aminoimidazole ribotide
    (AIR) 311, 312f
amino acid response element
    (AARE) 390
amino acids, bacterial metabolism
    of 310–312
aminolevulinic acid (ALA) 97–98
amygdala 500–502
    neuronal degeneration in 382
analbuminemia 178
analgesic effects of drugs 505
Anesthesia Patient Safety
    Foundation (APSF) 289

anesthesia-related carbon
    monoxide exposure
    286–288, 287f
    biological effects:
        of LFA 292–293
        overt toxicity vs. low-dose
            exposure 291–292
        policy 291
        potential therapeutic role of
            perioperative CO
            exposure 293–295
    breathing systems 287
    general endotracheal
        anesthesia 288–289
    endogenous sources
        290–291
    exogenous sources 289–290,
        289f
    machine 286, 287f
anesthetics, degradation of 290
angiogenesis 412–413, 472
    and metastasis 410, 411f
    stimulation of 413
antiapoptotic genes 349
antibacterial studies 250–251
anti-cell death effect of
    CO 378–379
anti-inflammatory applications,
    nanogenerator for
    274–275, 275f
anti-inflammatory effects:
    of BW-CO-103 247, 248f
    of CO 378
anti-inflammatory genes 349
antinociception 502
antioxidant response element
    (ARE) region 380
antioxidants 424
    defense 91–92
    responsive genes 93–94
antiplatelet agent 453
    endogenously generated CO
        from HO enzymes
        454–456
    exogenous CO inhibits activation
        of platelets 456–458
    exogenous CO on energy
        metabolism in
        platelets 459–461
    thromboresistance in
        vivo 453–454
antithrombotic actions 453–454
antithrombotic activity of CO
    455, 455f
antithrombotic effect in CO 379

APAP. See acetaminophen (APAP)
APAP-induced liver injury
    250, 250f
APIs. See active pharmaceutical
    ingredient (APIs)
APSF. See Anesthesia Patient Safety
    Foundation (APSF)
ARDS. See acute respiratory
    distress syndrome (ARDS)
ARDs. See acireductone
    dioxygenases (ARDs)
ARE region. See antioxidant
    response element (ARE)
    region
aromatic pyruvic acids 306
arthritic pain 497, 498–499
aryl hydrocarbon receptor
    (AhR) 178
arylpyruvic acids 306
ascorbic acid 304
    in presence of choline
        304–305, 305f
    in presence of singlet
        oxygen 304, 304f
aspartate transaminase
    (AST) 248
aspirin 468
AST. See aspartate transaminase
    (AST)
astrocytes, cultures of 93
ATF4:
    activation:
        by PERK 391
        of transcription factor 389
    crosstalk of 392
    hepatic homeostasis 390–391
    pathway in MRS 388, 391,
        393, 393f
ATP. See adenosine triphosphate
    (ATP)
attenuate arthritis 505
autoimmune hepatitis 386–387
autophagy 361, 427
Avogadro constant 225
azobenzene–cyclopalladium
    structure 324

b

baboon pneumonia 369
bacterial metabolism:
    of acireductone from methionine
        salvage pathway 309–310
    of amino acids such as
        tyrosine 310–312
    of flavonoids 308–309

bacterial phagocytosis 369
BCN 234–235, 235f
benzimidazole-based fluorescent probe 325, 326f
Berkeley (Berk) SCD mice 489
β-cell replacement therapy 423
β-elimination strategy 242–243, 243f
  esterase-triggered release system 244–245, 245f
  ROS-triggered elimination 243–244
β-elimination-triggered CO prodrugs 241, 242
biguanides 423
bile pigments 179
  direct administration of 181
biliverdin/bilirubin 438, 467, 485
  biological activities of 180
  biological functions of 178
  cell targets of 181t
  direct administration of 181
  evolutionary aspects of 174–176
  and immune system 178–179
  metabolic functions 179–180
  metabolism 176–178
  metabolites and derivatives 180–181
  protective actions and effects of 181
  signaling 180
  therapeutic potential of modulation 181–182
  unconjugated 178
biliverdin reductase (BLVR) 176–177
bimolecular DAR 234–235, 235f
bimolecular prodrug system for delivering CO and floxuridine 241
bioenergy 88
biofilm metabolic activity 168
biological activity 210–211
biological clock, TTFL in 98f
biotinylated photoCORM 274, 274f
bisphosphonates (BPs) 469
bleomycin (BLM) 370
BLM. See bleomycin (BLM)
blood oxygen saturation 118
BLVR. See biliverdin reductase (BLVR)
BLVRA 180
BMAL1/CLOCK, bHLH-PAS domains of 100f

B16 melanoma cell line 413
BODIPY:
  carboxylic acids 228, 229
  fluorescence 330
  fluorophore 323
bone marrow (BM) 491. See bone marrow (BM)
Bowman's space 8
BPs. See bisphosphonates (BPs)
"braking" mechanism 4
breast cancer cells 90
brown adipose tissue 390
Brown–Norway (BN) rats 165
BW-CO-103 238, 245, 247, 249f

C
CAI. See carbonic anhydrase inhibitor (CAI)
calcitonin gene-related peptide (CGRP) 468
camptothecin 410
cancer cells 89
  delivery to 270f, 274
cancer, CO and 401
  angiogenesis 412–413
  and cell fate in TME 405–406
  cytotoxic chemotherapies 409–410
  metastasis and metastatic niche 413–414
  modulation of broader TME 408–409
  production in tumor 401–402
    cellular sources of 402–403
    TME cues for HO-1 induction 403–405
  and suppression of antitumor immune response 406
  T-cell response 406–408
  treatment modalities 410
  vascular function and metastasis 410–414
Candida rugosa 263
cannabinoid receptor 1 (CB1R) agonists 504
cannabinoid receptor 2 (CB2R) agonists 504
carbon dioxide absorption 290
carbonic acid 287
carbonic anhydrase inhibitor (CAI) 498
carbon monoxide (CO):
  ameliorates organ and tissue function after transplantation 348

anti-inflammatory and antinociceptive effects of 497
antinociceptive actions of 498, 498f
chemical biology of 137
chemical properties 136–137
covalent bonding of 137
effects of altitude on endogenous 124
environmental sources of 484
gastroprotective properties of 470
host-derived 169
hypoxia 125
independent effects of 127–128
inhibition by 142
interactions of 30
interplay in pathophysiology 144–147
marine mammals 123–124
mechanism of action of 457
nanogenerator, components of 274–275, 275f
in nature, roles of 122–123
and oxygen:
  in mitochondria 127–128
  in vasculature 126–127
prodrugs, representative structures of 239
in pulmonary hypertension 126, 126f
regulation of 143
signal transduction 144
therapeutic levels of 198
vasoactive properties of 468–469
carbon monoxide and energy metabolism:
  and ATP 88–89
  diabetes and obesity 93–94
  and lipids 92–93
  mitochondrial metabolism 89–90
  modulation of OXPHOS 89–90
  nonoxidative metabolism, glycolysis and 90–91, 91f
  pentose phosphate pathway and antioxidant defense 91–92
carbon monoxide-releasing molecules (CO-RMs) 197, 203, 223, 347, 348, 361, 377, 386, 409, 441, 455, 456–457, 497, 525
  administration 457
  albumin-based delivery of 273

chemical reactivity and biological activity 210–211
classes of 457
COHb 209–210
CORM-401 207–209, 207f, 208f
mechanism(s) and efficiency of co delivery 207–208, 207f
models of metabolic dysfunction and inflammation 208–209, 208f
effectiveness of 457
enzyme-triggered 263
with high efficiency in vitro and in vivo 206–207
hybrid CO-releasing molecules 214
dual activity molecules 214
Mn- and Ru-based 215–216, 215–216f
inactive 212–214
low toxicity 229
metal-based 211, 246, 424
CORM-3 211–212
transition metals 211
metallic 224
monitoring of 321
into nanomaterials 274
nonmetallic 224
in obesity 424
structures of 72, 73f
for therapeutic applications 456
transition metal carbonyls as pharmacologically active 203, 203f
CORM-3 205–206
CORM-A1 206
CORM-1 and CORM-2 205–205
types of 497–498
use of 199–200
carbonmonoxy myoglobin (CO) 203
levels 55–56
carbonylation 324, 328
carboxyhemoglobin (COHb) 44, 53, 110, 120, 198, 209–210, 439, 470, 484, 511
association of levels of 10–11t
blood levels 54
CFK model for predicting 64, 64f
concentrations 45, 52, 56
data of oral administration of CORMs 73, 74t

direct and rapid measurements of 46
elevated levels of 12
elevation in mice 246f
endogenously generated 64
in femoral artery blood 62
high level of 69
kinetics 442
levels 27, 53, 57, 60, 69–70, 72, 73f, 288
in preeclamptic women 8
maximal level 60
pharmacokinetics of 46
poisoning 377
saturations 52–53, 484, 490, 512, 524
steady state of 60
cardiac allograft rejection 349
cardiac myocytes 29
cardiomyocytes 125
cardiopulmonary bypass (CPB) 293
cardiovascular pathophysiology, interactions in 145
CBS proteins. See cystathionine β-synthase (CBS) proteins
CCT. See cytotoxic chemotherapeutic (CCT)
CD11c cells 424–425
CD163 receptors 483
CD8$^+$ T-cells 370, 407f, 408f, 410, 425
CD8 T-cells 425
cecal ligation and puncture (CLP) 369, 435
cell cultures, CO in 46
cell cycle progression 408
cell metabolism 88
CO modulation of 90–91, 91f
cell transplantation model 350
cell types, diversity of 403f
cellular energy homeostasis 113
cellular repair 436
cellular signaling 361
cellular stress 437
cerebral arterial blood clots 380
cerebral artery occlusion (MCAO) in rats 379
cerebral ischemia 486
cerebral oxygen delivery 57
cerebrospinal fluid (CSF) 48
CFA. See complete Freund's adjuvant (CFA)

CFK model. See Coburn–Forster–Kane (CFK) model
cGMP. See cyclic guanosine monophosphate (cGMP)
CGRP. See calcitonin gene-related peptide (CGRP)
cheletropic reactions 233
chemical reactivity 210–211
chemical strategies for delivering CO 233, 234
chemokine ligand 12 (CXCL12) levels 429
chlorophyll degradation products 175
cholesterol biosynthesis 93
chronic alcohol abuse 466
chronic fibroinflammation 348
chronic hypoxia, cytoprotection during 123
chronic kidney disease (CKD) 165, 434
progression 436
chronic obstructive pulmonary disorder (COPD) patients 519
chronic osteoarthritis pain 498
chronic pain:
definition of 497
HO-1 activators in 500
Kv channels and sodium channels (NaV) in 502
MOR agonists in 503
Nrf2 activators in 499–500
cigarette smoking 12–13, 453
circadian clock, role of CO in 97
biological rhythms:
circadian system 102
HemoCD1, 102–104, 104f
with circadian regulatory factors 97–99, 98f
CLOCK and NPAS2 99–101, 100f, 101f
REV-ERB 101–102, 101f, 103f
circadian rhythmicity 30
circadian system 102
CKD. See chronic kidney disease (CKD)
C-labeled heme 7
clinical islet transplantation 428–429
ClinicalTrials.gov database 511–512
CLOCK 99–101, 100f, 101f

clock proteins 97
CLP. *See* cecal ligation and puncture (CLP)
CO. *See* carbon monoxide (CO); carbonmonoxy myoglobin (CO)
coagulation 456
cobalt protoporphyrin (CoPP) 504
  administration of 502, 503
  treatment with 499, 502
  upregulation of p-P 38, 502
Coburn–Forster–Kane (CFK) model 63–65
  "ad hoc" modifications 65
  calculations based on 64, 64*t*
  model for predicting 64, 64*f*
CO-COX:
  inhibition by 68
  saturation level 67, 68*f*
CO-enriched solutions 201
COHb. *See* carboxyhemoglobin (COHb)
CO/HO-1 pathway 471
CO/HO-1 system 472
colon adenocarcinoma cells 410
colonic microbiota biofilms 168
complete Freund's adjuvant (CFA) 498
concanavalin A (ConA) treatment 386–387
continuous prenatal exposure 294
COP. *See* CO poisoning (COP)
COP-1 323–324
CO poisoning (COP) 423–424
CoPP. *See* cobalt protoporphyrin (CoPP)
CO-releasing molecule-2 (CORM-2) 197, 197*f*
CO-responsive fluorescent probes 331*f*
CORM-1 205–205
  chemical structures and properties of 204*f*
  homogeneous distribution of 200
  presence of 204
CORM-2 205–205, 349, 413, 441, 472–473
  administration of 204, 503
  bactericidal action of 204–205
  chemical structures and properties of 204*f*
  coadministration with 504
  effects of 205

folic acid-tagged protein nanoemulsions containing 273
  inhibitory actions of 500
  intraperitoneal administration of 499
CORM-3 205–206, 211–212, 441
  chemical structures and properties of 204*f*
  design and synthesis of 205
CORM-401:
  mechanism(s) and efficiency of co delivery 207–208, 207*f*
  models of metabolic dysfunction and inflammation 208–209, 208*f*
CORM-A1 206, 459
  chemical structures and properties of 204*f*
  in vivo 456
CO-RMs. *See* carbon monoxide-releasing molecules (CO-RMs)
CORM-2-SLNs 499
CO sensor (COSer) 322, 323
counteract renal graft rejection 348
CPB. *See* cardiopulmonary bypass (CPB)
crizanlizumab 484
CSF. *See* cerebrospinal fluid (CSF)
curcumin 304
CXCL12 levels. *See* chemokine ligand 12 (CXCL12) levels
cyclic ADP-ribose (cADPR) 425
cyclic guanosine monophosphate (cGMP) 411, 425
  production 457
  signaling 363, 428
cyclooxygenase (COX)-1 466–467
cyclooxygenase (COX)-2 466–467
cyclopalladated probes, variations on 324–326
cystathionine β-synthase (CBS) proteins 165–166
cytochrome *c* oxidase (COX) 58–59, 59*f*, 89, 110, 271
cytochromes 97, 108, 109
  P450 enzymes 351
cytoprotective effects of CO 201
cytoprotective genes 468
cytosolic CO release, mitochondria-targeted *vs.* 272, 272*f*
cytotoxic chemotherapeutic (CCT) 409

ability for 409
  efficacy of 409–410
  response to exposure to 409
cytotoxic chemotherapies 409–410
cytotoxicity 200

d

damage-associated molecular patterns (DAMPs) 435
  signaling 354
DAMPs. *See* damage-associated molecular patterns (DAMPs)
DAR. *See* Diels–Alder reaction (DAR)
DCs. *See* dendritic cells (DCs)
dead cells 408*f*, 409
deallylation trigger strategy 331
decarboxylation–decarbonylation sequence of reactions 233
decarboxylation–decarbonylation strategy 245, 246*f*
d-electrons 321, 321*f*
delivery systems and noncarrier formulations 197
  advantages to 197
  CO-enriched solutions 201
  electrospun formulations of CO 199–200, 200*f*
  extracorporeal delivery of CO 199, 199*f*
  oral CO release systems (OCORS) 197–198, 197*f*
  therapeutic gas releasing systems 198–199, 198*f*
  transition metals 197
dendritic cells (DCs) 402
Desulfovibrio vulgaris 311
  pyruvate fermentation in 311–312, 312*f*
dextran sodium sulfate (DSS) 471
DHA diradical 306–307, 307*f*
diabetes, CO and 179, 423–424
  in clinical islet transplantation 428–429
  complications 428
  global prevalence of 423
  in insulin resistance 424–425
  and obesity 93–94, 424
  pathogenesis of 147
  in type 1 diabetes 425–428
    promoting islet survival 426–428

in regulation of islet insulin release 425–426
   treatment of 425
   in type 2 diabetes, symptoms of 425
diabetic neuropathy 499, 502
   DOR agonists in 504
diabetic pain conditions 504–505
Diels–Alder reaction (DAR):
   hydrophobicity-facilitated 237
   into single molecule 237
dienophile (alkyne) 241
diet-induced obesity 424
dihalomethane exposures 70–72
diketone 228
dimerization, regulation of 142
dimethyl fumarate 500, 501
dimethyl sulfoxide (DMSO) 205
dinitrobenzene sulfonic acid (DNBS) 168
direct assessment, distribution in tissues 60
DMSO. See dimethyl sulfoxide (DMSO)
DNA:
   repair pathways 410
   repair processes 406
   replication 409
dorsal root ganglia 501
dose dependence 33
Down's syndrome 309
doxorubicin 410
DPDPE, antinociceptive effects of 504
drinks/homogeneous formulations 70
drug molecules, targeted delivery of 235–236
duodenal bulb 466

e
ECCORS. See extracorporeal CO-releasing system (ECCORS)
Egr-1 signaling 368
eIF2/GTP/tRNAiMet ternary complex 390
electrochemical sensors 47
elimination-based water-soluble CO prodrugs 243
ELISA 379
endogenous cannabinoid system 504
endoplasmic reticulum (ER) 427
   membrane 328

stress pathways 385
endothelial cells 409, 410–411, 472
endothelial dysfunction 435–436, 470
endothelial NOS (eNOS) 29
endothelium 454
end-stage renal disease (ESRD) 434
energy metabolism 458
   in platelets 459–460
enhanced permeation and retention (EPR) 274
eNOS. See endothelial NOS (eNOS)
enrichment-triggered strategy 236
ENS. See enteric nervous system (ENS)
enteric nervous system (ENS) 471
environmental exposure 292
enzyme HO 27–28
enzyme-triggered photoCORMS 263, 264f, 268–269, 269f
epithelial cell autophagy, CO-dependent regulation of 365
epithelial cell death, protective effects of CO on 365
EPR. See enhanced permeation and retention (EPR)
ER. See endoplasmic reticulum (ER)
erythrocytes 3
erythropoietin, production of 121
Escherichia coli 167, 168
   infection model 369
ESRD. See end-stage renal disease (ESRD)
esterase-sensitive CO prodrugs 241, 242, 245
esterase sensitivity 241
ethanol 466
eukaryotic initiation factor 2-α (eIF2α) pathway 380
exercise, CO production in 8–9
exogenous administration of CO 501
external stimuli 263
extracellular CO release 272, 272f
extracorporeal CO-releasing system (ECCORS) 199
extracorporeal membrane oxygenation 27

extracorporeal resuscitation 354
ex vivo machine perfusion 348
ex vivo organ perfusion 349

f
F-actin fibers 394
fentanyl 503
Fenton reaction 308
Fenton reagent 307f
ferroptosis, iron prodrug for 276
FGF. See fresh gas flow (FGF)
fibrinolysis 455, 456
fibroblast growth factor 21 (FGF-21) 93, 390, 391f
   expression 388
Fick's first law 49
FID. See flame ionization detector (FID)
Filamin A (FLNA)
   Ca^{2+} by interaction of 394
flagella 466
flame ionization detector (FID) 46
flavonoids 305, 505
   bacterial metabolism of 308–309
   compounds 305
flavonol-derived deallylative probe 334, 335f
fluorescein-based deallylation probes 331, 332f
fluorescent detection of CO 331
fluorescent probes for intracellular carbon monoxide detection 47, 321–322
   categories of 322, 323f
   cyclopalladated probes, variations on 324–326
   ligand tuning 326–329
   metal reduction by CO:
      deallylative fluorescent probes, overview of 332–335
      general strategy and mechanism 330–332
   nitroarene reduction:
      general mechanism and reactivity 335–336, 336f
      miscellaneous response mechanisms 337–338, 338f
      nitroaryl reduction fluorescent probes 336–337
      proposed mechanism 336, 337f
   ruthenium-based probes 329–330
   types of probes 322

COP-1   323–324
fluorescent protein COSer   322–323
metal displacement   323
fluorescent protein COSer   322–323
fluorogenic CO probes   337, 338*f*
folic acid-tagged protein nanoemulsions containing CORM-2   273
formate   309
free energy   88
free heme   92
fresh gas flow (FGF)   288
fulminant hepatitis   386
fusion/fission processes   33

g

gasotransmitters   167
gastric blood flow (GBF)   466
gastric columnar cells   466
gastric mucosa   470
gastric mucosal integrity   472
gastric ulcer healing   471–472
gastrointestinal physiology and protection   466–467
  CO:
    donors and cellular targets of   467–468
    donors and HO-1 in   468–473
    and heme oxygenase system   467
    with H_2S and NO in   472–473
    prodrugs in maintenance of   469*f*
  donors:
    against acute esophageal and gastric mucosal injuries   469–471
    against drug-induced   468–469
    and gastric ulcer healing   471–472
    against lower GI tract injuries   471
gastrointestinal (GI) tract:
  barrier functions:
    external factors affecting   466
    impairment of   467
  mucosal damage   466
  NSAID-induced damage   466
GBF. *See* gastric blood flow (GBF)

general endotracheal anesthesia   288–289
  endogenous sources   290–291
  exogenous sources   289–290, 289*f*
GFR. *See* glomerular filtration rate (GFR)
Gilbert's syndrome   6, 182
GI tract. *See* gastrointestinal (GI) tract
glibenclamide   502
glomerular filtration rate (GFR)   518, 519
gluconeogenesis   99
glucose metabolism   99
glucose-stimulated insulin secretion (GSIS)   425
glucose tolerance   424
glutathione (GSH)   91, 262–263, 263*f*
glycolysis   90–91, 91*f*, 460*f*
  by CO   458, 458*f*
  rate-limiting enzyme of   92
  reinforcement of   93
Goto–Kakizaki (GK) rats   426
GSH. *See* glutathione (GSH)
GSIS. *See* glucose-stimulated insulin secretion (GSIS)
gut microbiome   312

h

Haldane coefficient   53
Haldane constant   57
haptoglobin   437
Harpagophytum procumbens   505
Hb. *See* hemoglobin (Hb)
HBI-002   489–490, 489*f*
HBO. *See* hyperbaric oxygen (HBO)
HbS polymerization   486–487
headaches   517
healthy volunteers (HVs):
  headache/migraine study in   517
  open-label phase 1 study in   523–524
heart transplantation   348–349
heat shock protein 90 (Hsp90)   141
HEK293T cells   323
HeLa cells   275–276, 309, 323
helices, displacement of   309
Helicobacter pylori   241, 246, 466
Heliopora coerulea   176

hematological diseases   9
heme   437
  ability of macrophages   402
  α-methene of   3
  arginate administration models   60
  availability of   123, 163–164
  basal level regulation of   98–99
  binding   97
  biosynthesis of   97–98, 98, 98*f*, 164
  chemical structures of   163–164, 163*f*
  degradation of   5, 458
  enzymatic degradation of   6
  homeostatic control of   437
  structural and functional roles of   163–164
  successive oxygenation steps of   4, 5*f*
heme and iron metabolism   162–164
  heme synthesis   164–165
heme biosynthesis   30
  physiological process of   164
heme catabolism   290–291
heme degradation   3, 385
heme metabolism   438
heme oxygenase-1(HO-1):
  activators   498, 498*f*
  in chronic pain   500
  coadministration with   504
  cytoprotective effects of   469
  deficiency   454
  deficient mice   454
  diversity of   485
  downstream effectors of   438
  enzymes   453–454
  expression of   7, 401, 406*f*, 410, 413, 438, 471
  genes in adaptive responses   35*f*
  inactive form of   401
  induction   403–405
  inhibition   454–455
  in mammalian tissue   123
  mRNA   470
  nuclear   405
  overexpression   412
  pathway   438
  pharmacological inhibition of   471
  polymorphisms   470
  protective effects of   439
  protective roles of   369

protein expression 469
truncated form of 401
upregulation of 9, 166, 379
heme oxygenase (HO) 118, 143, 160, 176, 347, 423–424, 436–437, 453
   activity 401
   competitive inhibitor of 438
   cytoprotective function of 4
   enzymatic function of 3–4
   enzymatic reactions of 4
   enzymes 13, 27, 28–29, 177, 259–260, 260f, 360, 484, 485
   heme degradation via 125
   pathway 177
   pharmacological inhibition of 406
Heme Protein Database 108
heme proteins (Hp) 108, 109t, 442
   cellular detoxification of 437
heme synthesis 31, 164–165
   regulation of 164
heme targets 29–31, 30f
hemin treatment 33
HemoCD1, 102–104, 104f
hemoglobin (Hb) 482, 483
   concentrations of 123
   heme units in 53
   oxygen binding sites of 28
hemolysis 483, 485
hemopexin 437
hemoproteins 28
   function of 29
   ligands for 69
hemorrhagic necrosis 386
hemorrhagic shock and resuscitation (HSR) 368, 386, 440
hemorrhagic stroke 382
Henry's law 49–50
hepatic ischemia/reperfusion injury 387
hepatocyte apoptosis 386
HEPG2 cells 324
high-content prodrugs 240–241, 241f
high-fat and high-fructose (HFHF) diet 388
high-performance liquid chromatography (HPLC) method 238
Hillhurst Pharmaceuticals 440
HMOX1, 177
   expression 404
   inducers of 302–303

   induction of natural products 182, 302–304, 303f, 303t
   inducers 302
HMOX1 gene 402, 403–404, 404, 405, 410, 437, 491
HMP-P. See hydroxymethyl pyrimidine phosphate (HMP-P)
HO. See heme oxygenase (HO)
HO-1. See heme oxygenase-1(HO-1)
Ho-1 gene 34, 98–99
HOMO–LUMO gap 234
hormesis 392
host–virus interaction 170–171
Hp. See heme proteins (Hp)
HSR. See hemorrhagic shock and resuscitation (HSR)
H_2S Sensing 268
human airway SMCs 366, 367f
human ARDS 369
human colonic microbiota 308
human gut dysbiosis 308, 468
humans, CO production in:
   during exercise 8–9
   in healthy adults 6–7
   and levels in diseases 8–9
      from induction of P450, 13
   overview of pathological conditions 9–13
   in neonates and infants 7–8
   during pregnancy 8
   through heme degradation 3–6, 5f
human umbilical vein endothelial cells (HUVECs) 263, 413
HUVECs. See human umbilical vein endothelial cells (HUVECs)
HVs. See healthy volunteers (HVs)
hybrid CO-releasing molecules 214
   dual activity molecules 214
   Mn- and Ru-based 215–216, 215–216f
hydroformylation 322
hydrogenase 310
hydrogen sulfide (H_2S) and CO 126, 160
   bactericidal effects of 167–169
   biosynthesis and transformation of 161, 161f
   endogenous production of 162
   against heavy metal toxicity 162

   heme and iron metabolism 162–164
      heme synthesis 164–165
   host–virus interaction 170–171
   in human plasma 161
   ischemia/reperfusion damage 162
   microbiota 167
      in host cells 167
      self-protection 169–170
   mitochondrial production of 166
   molecular mechanisms for signaling processes 160
   oxygen sensing 165–167
   productions of 162
      in mammalian cells 160–162, 161f
      reciprocal effects 162
   vasoactive effects of 166
hydroperoxyl ketoacid, decomposition of 307f
hydroxymethyl pyrimidine phosphate (HMP-P) 311
hydroxy radicals 308
hyperalgesia 499, 500
hyperbaric oxygen (HBO) 48
   therapy 62–63
hyperbilirubinemia 178, 182
hyperglycemia 423, 425, 426–427
hypernatremia 138
hyperoxia 347, 439
   persistent exposure to 120
hypoxia 111, 120–121, 128, 165, 405, 423–424, 426–427
   benefits of intermittent 121
   cellular physiology under 122, 122f
   and HIF-1α 122, 122f
   intermittent 121
hypoxia-induced vasodilation 127
hypoxia-inducible factor-1α (HIF-1α) 120–121
hypoxia-treated macrophages 128
hypoxic pulmonary hypertension hypoxia 123

i

IBD. See inflammatory bowel disease (IBD)
iCO. See inhaled CO (iCO)
iCORMs 441–442
idiopathic pulmonary fibrosis (IPF) patients 519

IH. See intermittent hypoxia (IH)
IMM. See inner mitochondrial membrane (IMM)
immune eradication 406
immune modulation 424
immunosuppressive therapy (FK506) 439
immunotherapy 406
inducible NO synthase (iNOS) 361, 386
  uncoupling of 142
inferior vena cava (IVC) 454
inflammasome-dependent responses 364
inflammation, resolution of 364–365
inflammatory airway disorders 9
inflammatory bowel disease (IBD) 471
inflammatory cell infiltration 201
inflammatory cellular response 89
inflammatory pain 498–499
  DOR agonists in 504
inhaled anesthetic agents 289, 289f
inhaled CO (iCO) 360, 439, 440, 512
  clinical studies with 490
  drug product 524
  to human therapy for IPF 370
  low-dose 369
  phase 2 kidney transplant study with 516–517t
  therapeutic potential of 367f
  therapy, renal protection by 439
innate immunity 146–147
inner mitochondrial membrane (IMM) 114
iNOS. See inducible NO synthase (iNOS)
insulin 423
  dependent diabetes 423
  independent diabetes 423
  resistance 424–425
  secretion 147
  sensitivity 424
interleukin (IL)-10, 293
intermittent hypoxia (IH) 121, 165
internal alkynes 237–238
internal charge transfer (ICT) processes 330
internal stimuli-sensitive donors 260–262

intestine 350
intracellular heme proteins, destabilization of 163
intracellular signaling 28
intracerebral hemorrhage 381–382
intramolecular DARs 237–240, 238–239f, 241–242, 242f
ion trapping 468
IPF patients. See idiopathic pulmonary fibrosis (IPF) patients
IRI. See ischemia–reperfusion injury (IRI)
iron 485
  metabolism 162–164
  prodrug for ferroptosis 276
ischemia–reperfusion injury (IRI) 121, 251, 347–348, 434
ischemic stroke 380–381
islet autotransplantation 428–429
islet transplantation 426
  benefit from 429
isotope-labeling techniques 46
IVC. See inferior vena cava (IVC)

j

JAK/STAT3 pathway 404f
juvenile diabetes 423

k

KCa channels 35
Keilin's discovery of cytochrome 110
ketolactone 304
kidney:
  effective target engagement in 443
  ischemia–reperfusion injury 251
  storage 350–351
  transplantation 439–440
  transplant patients 523
Klan's group 227
Klebsiella oxytoca 309
Kupffer cells 385

l

lactate ringers (LRs) 201
Lewis lung carcinoma (LL2) tumor 402, 407
LFA. See low-flow anesthesia (LFA)
Liao's group 226–227
ligand tuning 328–329
linear tetrapyrroles 175, 176t

lipid mediators (LMs) 364
lipids 92–93
  metabolism 93
  oxidation 3
lipophilic NSAIDs 468
lipopolysaccharide (LPS) 138, 167, 428, 458
  induced injury 368
  stimulated systemic inflammation 247–249, 248f
lipotoxicity 426–427
liver:
  inflammation 386
  storage 351–352
  transplantations 349–350
liver injury, protective mediator of 385
  acute liver injury 385–386
    acetaminophen-induced liver toxicity 387
    autoimmune hepatitis 386–387
    fulminant hepatitis 386
    hepatic ischemia/reperfusion injury 387
    nonalcoholic steatohepatitis (NASH) 387
  CO–PERK–Nrf2–HO-1 axis 391
    ATF4 and Nrf2 redox homeostasis 392–393
    mitohormesis 392
  CO–PERK–TFEB pathway 393
    nuclear translocation of TFEB 393–394
    TFEB, autophagy and mitophagy 394–395
  mitochondrial ROS activate PERK 388
  PERK–eIF2α–ATF pathway:
    ATF4 maintains hepatic homeostasis 390
    fibroblast growth factor 21 390
    sestrin-2 (Sesn2) 390–391
    SG assembly 389–390
  PERK phosphorylates 388–389
  ROS production from mitochondria 388
LLC's Sanguinate 522
LL2 tumor. See Lewis lung carcinoma (LL2) tumor
LMs. See lipid mediators (LMs)
locus coeruleus 501

long-wavelength emission deallylation probes 334, 334f
low-dose carbon monoxide (CO) 511–512
  assessment of 511
  clinical studies with 522, 524–525
    industry sponsored 518–519
    non-industry sponsored 519
  clinical trials with 512, 513t
  CO as API 512–516, 518
    clinical trials with 514–515t
    industry sponsored 516–517, 518–519
    non-industry sponsored 517–518, 519
    safety and tolerability studies 516
  drug development 511
  non-CO API 522–523
    drug products with 512, 519–521, 520–521t
  ongoing and planned clinical studies 523–524
  preclinical data in animals 511
  product comparison 524–525, 525t
lower limits of detection (LODs) 322
low-flow anesthesia (LFA) 287, 288, 291, 292–293
  biological effects of 292–293
  context of 291
LPS. See lipopolysaccharide (LPS)
LRs. See lactate ringers (LRs)
luminal obliteration 349
lung:
  endothelial cells 90–91
  inflammation prevention 517–518
  ischemia/reperfusion injury 370–371
  transplantation of 349
lung injury and disease 360–361
  acute lung injury:
    experimental 367
    hemorrhagic shock 368
    human ARDS 369
    LPS-induced injury 368
    malaria-induced ALI/ARDS 369–370
    sepsis and pneumonia-associated lung injury 369
    ventilator-induced lung injury (VILI) 368
  cell type-specific protective effects 362, 363f
  human airway SMCs 366, 367f
  lung ischemia/reperfusion injury and primary graft dysfunction (lung transplant) 370–371
  pulmonary arterial hypertension (PAH) 370
  pulmonary endothelial cells 365–366
  pulmonary fibroblasts 366
  pulmonary fibrosis 370
  specific effects 361
    CO-dependent regulation of epithelial cell autophagy 365
    inflammasome-dependent responses 364
    macrophages 361–362
    protective effects of CO on epithelial cell death 365
    pulmonary epithelial cells 365
    resolution of inflammation 364–365
    TLR-dependent responses 362–364
  therapeutic potential of CO 360–361

m
macrophages 29, 91, 167, 177, 361–362
  accumulation of 121
  efferocytosis 365
  HO-1 in 405f
  hypoxia-treated 128
  M1/M2 polarization extremes of 403f
  phagocytosis 369
Maf recognition element (MARE) 34
magnetic heating-promoted CO release 270, 270f
malaria-induced ALI/ARDS 369–370
MAPKs. See mitogen-activated protein kinases (MAPKs)
MARE. See Maf recognition element (MARE)
master clock 97
maximal COHb level 60, 60f
Mb. See myoglobin (Mb)
MbCO:
  levels 65
  saturation 52, 54f, 55
meglitinide 423
membrane-based approach 199, 199f
mesoporous silica nanoparticles 275
mesoporphyrin (SnMP) 406
  administration in $vivo$ 413
  treatment 413
metabolic syndrome 179
metal carbonyls 205, 260
metal displacement 323
metal displacement approaches 330
metal-free CO detection 322
metal-free donors 260
metallic photoCORMs 223, 223f
metal redox cycle cascade 304
metal reduction by CO:
  deallylative fluorescent probes, overview of 332–335
  general strategy and mechanism 330–332
metastatic niche 413–414
methacholine 519
MetHb. See methemoglobin (MetHb)
methemoglobin (MetHb) 46
methionine 309
methylene chloride (MC) 224
metronidazole 250
Michael acceptor 247
Michaelis–Menten constant 110
microbial CO-generating enzymes 312
microbial HMOX 308
microbiota 167
  in host cells 167
  metabolism of 167
  self-protection 169–170
microenvironment 119–120, 120t
microglia 89
  contribution of 502
microglial erythrophagocytosis 382
microscale OCORS (M-OCORS) 198
middle cerebral artery occlusion (MCAO) 380
Mini-Mental State Examination (MMSE) 519
minocycline 502
miscellaneous response mechanisms 337–338, 338f
mitoBKCa channel 114

mitochondria, carbon monoxide
and   31, 108, 119, 260, 361,
386
cytochrome *c* oxidase and
oxygen
availability   110–111
effects   113–114
electron transport chain
108–110, 109*f*, 109*t*
metabolization in   93
oxidant production   111
quality control   111–113, 113*f*
for targeting   89, 235–236,
236–237*f*
mitochondrial autophagy   33
mitochondrial biogenesis   31–32,
32*f*
basic redox pathway for   113*f*
mitochondrial
depolarization   90–91
mitochondrial electron transport
chain (ETC)   118
mitochondrial membrane   395
mitochondrial metabolism   89–90
mitochondrial oxidant
production   111
mitochondrial oxidases   118
mitochondrial perturbation   364
mitochondrial pro-oxidants   112
mitochondrial quality
control   32–33
mitochondrial reactive oxygen
species (mtROS):
generation   388
production   361
mitochondrial respiration   460*f*
by CO   458, 458*f*
simultaneous inhibition of   458,
458*f*
mitochondrial uncoupling   352
mitochondria-targeted CO
release   271, 271*f*
vs. cytosolic CO release   272, 272*f*
mitochondrion targeting   224
mitogen-activated protein kinases
(MAPKs)   302, 352, 361, 501
signaling   171, 179, 353, 409
mitohormesis   392, 392*f*
MMSE. *See* Mini-Mental State
Examination (MMSE)
molecular effector systems   29, 31
heme targets   29–31, 30*f*
and mitochondrial
biogenesis   31–32, 32*f*
and mitochondrial quality
control   32–33

and NADPH oxidase   31
nonheme targets   33–35, 35*f*
molecular mechanisms of actions
and CO:
interaction with NO and
H$_2$S   28–29, 29*f*
introduction and therapeutic
perspectives   27
molecular effector systems. *see*
molecular effector systems
physiological responses
to   27–28
"moonlighting" enzymes   309
MOR agonists   503
chronic administration of   504
Morganella morganii   306
morphine   503
morpholino   328–329, 329*f*
Mtb respiration and
bioenergetics   168–169
mTORC1, inactivation of   393
mtROS. *See* mitochondrial reactive
oxygen species (mtROS)
mucosal defense   467
multiorgan failure   385–386
multiple ascending dose (MAD)
clinical study   516
multistage strategy assembly/
disassembly   275–276, 276*f*
murine, domain structures
of   100*f*
Mus musculus ARD   310
Mycobacterium tuberculosis
(Mtb)   168
myeloperoxidase (MPO)   198, 249
myocardial
bioenergetics   293–294
myocardial function, viability
of   352
myocardial infarction   347
myocytes   181
myoglobin (Mb)   97, 437
binding affinity of oxygen to   55
compartment   52, 52*f*
concentrations of   123
high abundance   54

n
NADPH:
cytochrome P450, 385
oxidase   31
nanomaterials   274
NASH. *See* nonalcoholic
steatohepatitis (NASH)
National Ambient Air Quality
Standard for CO   291

natural apoptosis   295
natural products, CO   302
from HMOX1 induction
of   302–304, 303*f*, 303*t*
microbiome action independent
of heme
degradation   308–312
through nonenzymatic
pathways   304
1,3-Dihydroxyacetone
(DHA)   306–308
α-ketoacids   305–306
phenols and flavonoids   305
vitamin C   304–305
Nb. *See* neuroglobin (Nb)
NbCO saturation   57, 58*f*
neonatal hypoxic–ischemic
encephalopathy   381
neonatal jaundice   179
neonates, red blood cells in   7
nerve injury, microglial
activation   501
neurocognitive testing   519
neuroglobin (Nb)   56–58
distribution   57
Neurology/Neurosurgery Intensive
Care Unit   523
neuropathic pain   497, 499
CCI-induced   504
development of   499
inhibitory actions of CORM-2
500
symptoms   500
neurotoxins   295
nicotinamide riboside (NR),
transformation of   460
NIDDK-funded Kidney Precision
Medicine Project   436
Nile red-based CO probes   327*f*
Nile red dye   324
nitric oxide (NO) in human
physiology   136, 385, 483,
500
activation of sGC by   144
activity and regulation of nitric
oxide synthases   138
Ca^{2+}/CaM signaling   140–141
chemical properties   136–137
chemical reactivity of   28, 29*f*
covalent bonding of   137
endogenous production and
regulation of   138
free radical species   136–137
heat shock protein 90
(Hsp90)   141
heme oxygenase by   143

historical depiction of 136
hydrophobic cell
    membrane 136
induced autoinactivation 142
insulin secretion and
    pathogenesis of
    diabetes 147
interactions of 30, 145
interplay in pathophysiology
    144–145
NOS1 138
NOS2 139
NOS3 139–140
platelet aggregation and
    thrombosis 145–146
post-translational regulation
    of 143
production, steps of 140
reactions with free
    radicals 137–138
reactivity of 137
redox chemistry of 137
regulation of dimerization 142
roles in innate immunity and
    septic shock 146–147
serine/threonine
    phosphorylation 141
shared affinity of 137
signal transduction 144
steps of production 140
transcriptional regulation of
    HO-1 by 143
transition metals 137
vascular tone 145
VSMC:
    apoptosis 146
    proliferation and
        migration 146
nitric oxide synthases (NOSs) 138
    cytochrome 467
    enzymes 28, 30
    functional "uncoupling" of 142
    post-translational regulation
        of 140
    structure 140
    superoxide production
        by 142–143
nitroarene reduction:
    general mechanism and
        reactivity 335–336, 336f
    miscellaneous response
        mechanisms 337–338,
        338f
    nitroaryl reduction fluorescent
        probes 336–337

probes 334, 335f
nitroaryl reduction fluorescent
    probes 336–337
NK cells, cytotoxicity of 409
nonalcoholic steatohepatitis
    (NASH) 179, 385, 387–388
nonheme-dependent
    mechanisms 427, 427f
nonheme intracellular
    pathways 33, 34t
nonheme processes 302
nonheme targets 33–35, 35f
nonhuman primate (NHP) model
    of cynomolgus macaques
    368
nonmetallic CORMs 224
nonoxidative metabolism 90–91,
    91f
nonprotein-bound (free)
    heme 163
nonsteroidal anti-inflammatory
    drugs (NSAIDs)
    epithelial injury by 468
    lipophilic 468
nonsynaptic isolated
    mitochondria 89
nonsynaptic mitochondria 31
norbornadienones:
    cheletropic extrusion
        from 234–235
    spontaneous nature for 233
normobaric hypoxia
    treatment 121
normoxia, cellular physiology
    under 122, 122f
NOS1 138
NOS2 139
    expression and iNOS
        activity 139
    mechanism for synergistic
        induction of 139
NOS3 139–140
NOSs. See nitric oxide synthases
    (NOSs)
NPAS2 99–101, 100f, 101f
NPAS2 PASA heme sensor
    domain 101f
NQO1 502
Nrf2 162, 393, 393f, 403, 404
    activation of 162
    activators 498, 498f, 499–500
    nuclear translocation 380
    postulated neuroprotective role
        of CO through 380, 380f
    transcription factor 380, 504

NSAIDs. See nonsteroidal anti-
    inflammatory drugs
    (NSAIDs)
nuclear magnetic resonance
    (NMR) 323
nucleophilic cysteine 161
nutrient deprivation 426–427

O
obesity 179, 424
    diabetes and 93–94
OCORS. See oral CO release
    systems (OCORS)
1,3-Dihydroxyacetone
    (DHA) 306–308
optic neuritis 58
optic neuropathy 57–58
oral CO release systems
    (OCORS) 197–198,
    197f, 467
    ruthenium-based 198
organic carbon monoxide
    prodrugs 232–233
    cheletropic extrusion from
        norbornadienones:
        β-elimination strategy
        starting from norbornene-
        7-ones 242–245, 243f,
        245f
    bimolecular DAR 234–235,
        235f
    CO for targeting
        mitochondria 235–236,
        236–237f
    high-content prodrugs
        240–241, 241f
    intramolecular DAR
        237–240, 238–239f
    from prodrugs based on
        intramolecular DARs
        241–242, 242f
    chemistry concepts and design
        principles used in
        developing 233, 234f
    decarboxylation–
        decarbonylation
        strategy 245, 246f
    in various disease
        models 246–247
    antibacterial
        studies 250–251
    APAP-induced liver
        injury 250, 250f
    chemically induced gastric
        injury 252

kidney ischemia–reperfusion injury   251
LPS-stimulated systemic inflammation   247–249, 248f
rhabdomyolysis-induced acute kidney injury   252, 253f
TNBS-induced colitis   249
vasorelaxation effects   251
organic CO donors   223–224, 224f
organic CO prodrugs in disease models   247, 248f
organic CORMs   224
organic fluorinated compounds   286
organic photoCORMs   223
current   226
development of   224–226, 226f
discovery of   226–227
photochemical properties   227–229, 228f
photoreactions of   227–229, 228, 228f
types of   226–227
organic prodrugs   73–74, 498
organ transplantation, outcome of   354f
orthotopic liver transplantation   387
osteoarthritis   498
ovarian carcinoma   410
overt toxicity vs. low-dose exposure   291–292
oxidation of carbon fuel   88
oxidation-sensitive CO donors   261–262, 262f
oxidative derivatives   175, 176t
oxidative mitochondrial metabolism   92
oxidative phosphorylation (OXPHOS)   120, 127–128, 163, 459, 468
and mitochondrial metabolism   90
modulation of   89–90
oxidative stress   164, 500
oxindole   505
oxoperoxide   305
OXPHOS. See oxidative phosphorylation (OXPHOS)
oxygen   118–119
cell survival and death   120
delivery to tissues and impediment   467
dissociation curves   51, 51f

hypoxia   120–121
benefits of intermittent   121
and HIF-1α   122, 122f
independent effects of   127–128
paradox   119–120, 119t
sensing   165–167
oxygen–hemoglobin dissociation curve   122
oxyhemoglobin   46

p

PAH. See pulmonary arterial hypertension (PAH)
pain management   497
antinociceptive effects of CO:
HO-1 activators   502
in HO-1/CO-induced antinociception   502
Nrf2 activators   501
releasing compounds   500–501
CO   497–498
in diabetic neuropathy   499
donors and Nrf2–HO-1 activators with opioids   503–504, 503f
in inflammatory and arthritic pain   498–499
in neuropathic pain   499
releasers and Nrf2–HO-1 activators with cannabinoids   504–505
CO/HO-1 other drugs   505
description of   497
Nrf2 activators in chronic pain   499–500
HO-1 activators in chronic pain   500
palladacycle   327
probe   326
palladium species   330
pancreas   350
parenchyma, hepatic stellate cells of   385
partial sciatic nerve ligation (PSNL)   499
pattern recognition receptors   428
PBMCs. See peripheral blood mononuclear cells (PBMCs)
PC3. See prostate cancer cell line (PC3)
PCO-1   331f
P450 cytochrome   467
Pd detection   331
PDT. See photodynamic therapy (PDT)

PEG-COHb. See pegylated COHb (PEG-COHb)
pegylated COHb (PEG-COHb):
clinical studies with   490
products   524
solutions   491
pentose phosphate pathway (PPP)   91–92, 392
CO modulation of   92
pericyclic reactions   233
peripheral blood mononuclear cells (PBMCs)   68
PERK:
biological activities of   391
Ca2$^+$ by interaction of   394
centric view of CO functions   388–389, 389f
phosphorylates   388–389
signaling   391
peroxisome proliferator-activated receptors (PPAR)   178
PGs. See prostaglandins (PGs)
pharmaceutical development, features for   232
pharmacodynamic biomarker discovery   443
pharmacokinetic characteristics of carbon monoxide   44–45
after inhalation   47–48
absorption   49–50
distribution   50–56
PK profiles   48–49
animal models and postmortem human subjects   61t
dihalomethane exposures   70–72
direct assessment, distribution in tissues   60
distribution in tissues:
with cytochrome c oxidase   58–59
with neuroglobin   56–58
drinks/homogeneous formulations   70
elimination   60–63, 62f
organic prodrugs   73–74
PK   63
CFK model   63–65
characteristics of endogenously generated   66
other models   65–66
properties from various donors   69–70
quantification, in biological samples   45–46

electrochemical detection 47
fluorescent probes 47
gas chromatography 46–47
spectrophotometry/
  CO-oximetry 46
releasing materials 74–75
releasing molecules 72–73
target engagements and
  therapeutic implications
  66–69
PHDs. See prolyl hydroxylases
  (PHDs)
phenols 305
phenylpyruvic acid, thermal
  decomposition of 306
phloroglucinol carboxylic
  acid 308, 308f
phosphoinositide 3-kinase
  (PI3K) 501
photoCORMs 263–267
  aromatic carboxylic 227
  boron-containing 224, 224f
  CO content of 225
  CO delivery using 265
  concentration of 225
  criteria for 227
  development of 226
  enzyme-triggered 268–269, 269f
  ideal 226
  intracellular vs. extracellular
    delivery using 272–273
  luminescent 265
  metal carbonyl-based 264,
    265f, 266
  metal-free 266, 266f
  peptide conjugation to 273
  small molecule CO donors
    and 271–272, 271–272f
  structures of organic 226, 226f
  toxicity of 225
  types of 227, 228f
  vitamin B_{12}-appended
    manganese 274
photodynamic therapy
  (PDT) 225, 410
photoexcitation 228
phycocyanobilin 175
phytochromobilin 175
PI3K. See phosphoinositide
  3-kinase (PI3K)
P450 inhibitors 13
Piper betle leaves 505
pirin 309
PKC. See protein C kinase (PKC)

PLA fiber. See polylactic acid (PLA)
  fiber
plasmalemmal caveolae 409
Plasmodium infection 370
platelet aggregation 145–146,
  456, 461
  by CORM-A1 457
  inhibition of 457
platelets:
  cGMP production in 458
  energy metabolism in 459–460
  hyperactivation 453
  hyperreactivity 459
  NR kinase in 460
  without effects on sGC 458
pluripotent cells 90
p38 MAPK 293, 405–406
p38 mitogen-activated protein
  kinase (p38 MAPK) 468
pneumococcal pneumonia 369
pneumonia 120, 369
POI. See postoperative ileus (POI)
polylactic acid (PLA) fiber 200
polymicrobial sepsis 369
porcine liver esterase (PLE) 263
postoperative ileus (POI) 198, 471
PPARγ in islets 428
PPI. See proton pump inhibitor (PPI)
PPP. See pentose phosphate
  pathway (PPP)
preclinical models 382
preclinical pain models 498
preeclampsia 8, 166
pregnancy, CO production
  during 8
primary graft dysfunction (lung
  transplant) 370–371
proangiogenic effect of CO 379
prodrugs 232
  BW-CO-48 249f
  for delivering CO and
    metronidazole 240
  general structures of 73–74, 74f
Prolong Pharmaceuticals 522, 523
prolyl hydroxylases (PHDs) 122
prostaglandins (PGs) 466
prostate cancer cell line
  (PC3) 410
protein C kinase (PKC) 472
proteinuria 8
protocatechuic acid 308, 308f
proton pump inhibitor (PPI) 469
proximal tubular epithelial cells
  (PTECs) 434

pseudohypoxia 118
  cellular physiology under
    122, 122f
Pseudomonas aeruginosa
  168, 204
P-sickle, impact on 488
PTECs. See proximal tubular
  epithelial cells (PTECs)
pulmonary arterial hypertension
  (PAH) 126, 294, 370
pulmonary cells 361–362, 362f
pulmonary endothelial
  cells 365–366
pulmonary epithelial cells 365
pulmonary fibroblasts 366
pulmonary fibrosis 370
pulmonary hypertension 353
pulmonary neutrophilia 368
purinergic receptor (P2X7R) 364

q

quercetin 305, 505
  oxidative degradation of 306f
  pirin-mediated degradation
    of 309

r

ratiometric CO deallylation
  probes 333, 333f
RBC. See red blood cell (RBC)
reactive nitrogen oxide species
  (RNOS) 138
reactive nitrogen species
  (RNS) 33–34
reactive oxygen species (ROS) 29,
  118–119, 138, 424, 500
  alleviation and CO release 276
  elevated levels of 262
  generation 90, 125
  mitochondrion-dependent
    increase in 410
  production 125
    from mitochondria 388
    by specific hemoproteins 31
    and suppression 425
    and tissue damage 120
  sensing 267
  sensitive CO prodrugs 243
  signaling 90
receptor potential channels
  (TRPV)-1 468
red blood cell (RBC) 482
  membrane damage 483
  sequestration of 485

reduction gas detector (RGD)   60
regression formula   64
renal functional recovery   440, 441–442
renal perfusion   441
renal transplantation   440
renal tubular structures   435–436
renal tumors, nephron sparing surgery for   434
reparative angiogenesis   439
resveratrol   304
REV-ERB   101–102, 101f, 103f
RGD. See reduction gas detector (RGD)
rhabdomyolysis   252, 253f, 437
Rhodospirillum rubrum   322
RNA synthesis   309
RNOS. See reactive nitrogen oxide species (RNOS)
Root effect   52
ROS. See reactive oxygen species (ROS)
Ru-based probes   330

S

SA-AKI. See sepsis-associated AKI (SA-AKI)
SAD/multiple dose design   524
SAEs. See serious AEs (SAEs)
SAH. See subarachnoid hemorrhage (SAH)
Salmonella typhimurium   167
SAM enzyme   311
Sangart, Inc.   522
Sanguinate   441, 522–523
SCD. See sickle cell disease (SCD)
SCI. See spinal cord injury (SCI)
sciatic nerve   499
sciatic nerve-injured mice   501
SCIs. See silent cerebral infarcts (SCIs)
SCN. See suprachiasmatic nucleus (SCN)
sepsis   369
  bacterial model of   128
  models, studies in   435
sepsis-associated AKI (SA-AKI)   435, 437
sepsis-mediated acute respiratory distress syndrome   354
septic shock   146–147

serine/threonine phosphorylation   141
serious AEs (SAEs)   512
serum creatinine   519
Sesn2. See sestrin-2 (Sesn2)
sestrin-2 (Sesn2)   390–391, 391f
Sevoflurane   290
sGC. See soluble guanylate cyclase (sGC)
SHRs. See spontaneously hypertensive rats (SHRs)
SH-SY5Y cell model   90
sickle cell disease (SCD)   482–483
  acute clinical events   485–486
  anemia of   484
  animal models of   488–490
  cause of death in children with   485
  CO binding to Hb   484
  dysfunction syndromes   486
  HO-1, hemolysis, and CO   485
  human subjects with   490
  in vitro/ex vivo studies   488
  leg ulcer patients   525
  long-term consequences   486
  mechanism of action of CO in   486–487t, 486f
    anti-inflammatory effects   487
    antioxidative effects   487–488
    cGMP production   488
    HbS polymerization   486–487
    impact on P-sickle   488
    reduction of apoptosis   488
  methods of administration   490–491
  morbidities and mortality of   484
  mouse models of   482, 489
  pathogenesis of   482
  pathology   485
  pathophysiology of   483–484
  preclinical data in   491
  protection against vascular stasis   489f
  severe morbidities   483
  vaso-occlusive crisis (VOC) in   523
sickle RBCs (sRBCs)   483

signaling molecules   28
signal transduction   144
silent cerebral infarcts (SCIs)   486
single active principal (CO)   235
single ascending dose (SAD) clinical study   516
sinusoidal perfusion   385
skeletal muscle   32, 32f, 390
sliding scale rule   28
small interfering RNA (siRNA) transfection   166
small-molecule hybrids   498
SMCs. See smooth muscle cells (SMCs)
smooth muscle cells (SMCs)   125, 127, 361
  antiproliferative effects on   128
  human airway   366, 367f
  interaction of CO with   353
  proliferation   352
SNI. See spared nerve injury (SNI)
soda lime   287
solid organ transplantation, CO   347
  ameliorates organ and tissue function after transplantation   348
    heart   348–349
    in intestine and pancreas   350
    kidney   348
    liver   349–350
    lung   349
  cellular and molecular mechanisms mediating   352
    in clinical settings of organ transplantation   354
    with ECs   352
    with macrophages   353
    with SMCs   353
    with T-cells   353–354, 354f
  gas and CORMs in organ preservation   350
    in kidney storage   350–351
    in liver storage   351–352
    in storage of organs   352
  modes of delivery in   347–348, 348t
  worldwide   348t
solid waste disposal   484
solubility ratio   50

soluble guanylate cyclase
(sGC) 28, 351, 361, 385, 411, 467
  activation of 30, 30f
  heme of 361
  interactions of 30
spared nerve injury (SNI) 499
specialized pro-resolution mediators (SPMs) 364
spinal cord injury (SCI) 121
spinal cord, phosphorylation in 502
SPMs. See specialized pro-resolution mediators (SPMs)
spontaneously hypertensive rats (SHRs) 165
spontaneously releasing CO donors 260–261, 261f
spontaneous pain 499
Sprague–Dawley (SD) rats 165
sRBCs. See sickle RBCs (sRBCs)
Staphylococcus aureus 31, 200, 269
stereotypes 118
sterile inflammatory responses 435
Strelitzia reginae 176
Streptococcus aureus 168
Streptococcus pneumoniae 369
StREs. See stress response elements (StREs)
stress 467
  and inflammation 120
stress response elements (StREs) 32
stroma 401–402
  phenotype and response of 402
subarachnoid hemorrhage (SAH) 381–382
  injury 379
  patients 523
sulfonylureas 423
suprachiasmatic nucleus (SCN) 97
surface epithelium 466
synergistic anti-inflammation 276, 277f

t

TAMs. See tumor-associated macrophages (TAMs)
targeted delivery of carbon monoxide 259
albumin-based delivery of CORM to cancer cells and tumors 273
appended flavonols 267, 268f
delivery to cancer cells 274
donors and photoCORMs 271–272, 271–272f
endogenous production 259–260
folic acid-tagged protein nanoemulsions containing CORM-2, 273
highly controlled, environment-sensing CORMs 267
enzyme-triggered photoCORMS for 268–269, 269f
H_2S sensing 268
magnetic heating-promoted CO release 270, 270f
ROS sensing 267
thiol sensing 267, 268f
tumor microenvironment 269–270, 270f
ultrasound-mediated release 270–271
intracellular vs. extracellular delivery using photoCORMs 272–273
iron prodrug for ferroptosis 276
metal carbonyl and metal-free donors 260
multistage strategy assembly/disassembly strategy 275–276, 276f
nanogenerator for tumor therapy and anti-inflammatory applications 274–275, 275f
nanomaterials and 274
NIR-triggered nanoparticles for targeted CO delivery 277
peptide conjugation to photoCORMs 273
ROS alleviation and 276
triggered donors 260
  enzyme-triggered CORMs 263
  external stimuli 263
  glutathione (GSH) 262–263, 263f
internal stimuli-sensitive donors 260–262
internal stimuli to enhance control of release 263
photoCORMs 263–267
visible light-triggered tandem deprotection 266, 267f
vitamin B_{12}-appended manganese photoCORM 274
TATA box 178
TBCO. See total blood CO (TBCO)
TBI. See traumatic brain injury (TBI)
TCA cycle. See tricarboxylic acid (TCA) cycle
T-cell receptor (TCR) 407
T-cells 410
  alloantigen response 353
  cytokines 413
  differentiation of 179
  effector function 406–407, 407f
  populations 408
  proliferation 408
TCR. See T-cell receptor (TCR)
tetrapyrrolic structure 181
TFEB. See transcription factor EB (TFEB)
TGF. See tubuloglomerular feedback (TGF)
TGRS. See therapeutic gas delivery systems (TGRS)
therapeutic delivery of CO 203
therapeutic gas delivery systems (TGRS) 198–199, 198f
therapeutic gas releasing systems 198–199, 198f
thiamin-pyrimidine (HMP-P) synthase 311
thiazolidinedione 423
thiol sensing 267, 268f
thrombin generation 459
thrombosis 145–146, 347
  suppression of 454
tipiracil hydrochloride (TPI), antiplatelet effects of 461
tissue factor (TF), enhanced expression of 454
tissue plasminogen activator (tPA) 380
tissues:
  CO distribution in 60
  fibrosis 362

homeostasis 302
TLRs. *See* toll-like receptors (TLRs)
TLR-4 signaling. *See* toll-like receptor-4 (TLR-4) signaling
tMCAO. *See* transient MCAO (tMCAO)
TME. *See* tumor microenvironment (TME)
TNBS-induced colitis 249
tobacco smoke 12–13
toll-like receptor 4 (TLR4) 409, 483
toll-like receptors (TLRs) 428
  dependent responses 362–364
  signaling 363, 409
  trafficking to lipid rafts 363
toll-like receptor-4 (TLR-4) signaling 407*f*
total blood CO (TBCO) 48
total pancreatectomy 428–429
tPA. *See* tissue plasminogen activator (tPA)
TPCPD 234–235, 235*f*
TPP. *See* triphenylphosphonium (TPP)
transcription factor EB (TFEB) 32, 393
  autophagy and mitophagy 394–395
  nuclear translocation of 393–394, 394*f*
transcription–translation feedback loop (TTFL) 97
  in biological clock 98*f*
transient MCAO (tMCAO) 381
transition metal carbonyls 203, 203*f*
  CORM-3, 205–206
  CORM-A1 206
  CORM-1 and CORM-2 205–205
transition metals 197, 211
  known toxicity of 442
  reactivity 326
trapping theory 468
traumatic brain injury (TBI) 382
tricarboxylic acid (TCA) cycle 89, 90
triggered donors 260
  enzyme-triggered CORMs 263
  external stimuli 263

glutathione (GSH) 262–263, 263*f*
internal stimuli-sensitive donors 260–262
internal stimuli to enhance control of release 263
photoCORMs 263–267
triphenylphosphonium (TPP) 236
tripyrrolic biopyrrins 181
Tsuji–Trost reactions 330
TTFL. *See* transcription–translation feedback loop (TTFL)
tubular epithelium 436
tubuloglomerular feedback (TGF) 435
tumor:
  antigens 409
  models, variety of 412
  therapy, nanogenerator for 274–275, 275*f*
  transendothelial migration of 413
  treatment of 406
tumor-associated macrophages (TAMs) 402, 409
tumor microenvironment (TME) 269–270, 270*f*, 402, 410, 412
  CO on angiogenesis and metastasis within 410, 411*f*
  cytokine regulation 405*f*
  HMOX1 gene expression in 403–404
  HO inhibition and angiogenesis in 412, 412*f*
  immune-suppressive 409
  immune-suppressive nature of 408
type 1 diabetes:
  in islet survival and transplantation 426
  promoting islet survival 426–428
  prophylactic vascular protection in 428
  in regulation of islet insulin release 425–426
  treatment of 425*f*
type 2 diabetes 423
tyrosine, bacterial metabolism of 310–312

u

UGT1A1 expression 178
  activities of 181–182
ulcerative colitis 249
ulcer, healing process 472*f*
ultrasound-mediated release 270–271
unfolded protein response (UPR) 389
unimolecular CO prodrugs 238
United States Environmental Protection Agency 291
UPR. *See* unfolded protein response (UPR)
urobilinoids 178

v

vascular endothelial growth factor (VEGF) 121, 379, 410, 411*f*, 412
  in macrophages 128
  production of 121
vascular endothelium 121
vascular smooth muscle cells (VSMCs) 162, 410–411
  apoptosis 146
  proliferation and migration 146
vascular tone 145
vasodilatory effect of CO 378
vasomodulation 424
vaso-occlusion 483
vaso-occlusive crises (VOCs) 482, 483–484
  acute onset of 485
  anemia and painful 483
  reduction 484
  treatment of 404
vasorelaxation effects 251
vasospasm 483
VEGF. *See* vascular endothelial growth factor (VEGF)
velocity of the middle cerebral artery (VMCA) 517
ventilator-induced lung injury (VILI) 368
verdoheme 6
verteporfin 410
VILI. *See* ventilator-induced lung injury (VILI)
Vinyl–Ru(II) complex 329*f*, 330*f*
vitamin C 304–305
VMCA. *See* velocity of the middle cerebral artery (VMCA)

VOCs. *See* vaso-occlusive crises (VOCs)
von Willebrand factor 182
VSMCs. *See* vascular smooth muscle cells (VSMCs)

w

water immersion and restraint stress (WRS) 470
Weddell seals 123
Weibel–Palade bodies 483
white adipose tissue 390
WRS. *See* water immersion and restraint stress (WRS)

x

xanthene 228
xenobiotics 181–182
    detoxification function 163
Xenopus laevis 175–176

y

Yamada's group 228

z

zinc protoporphyrin (ZnPP) 324, 410
Zucker diabetic fatty (ZDF) rats 424